DES ⟡ **W9-DBB-445**

DISCRETE-TIME SIGNAL PROCESSING Third Edition

On-line Study Guide and MyPearsoneBook Access Card

Thank you for purchasing a new copy of *Discrete-Time Signal Processing,* Third Edition, by Alan V. Oppenheim and Ronald W. Schafer. The information below provides instruction on how to access the **On-line Study Guide** and **MyPearsoneBook**, Pearson's new visual tools for teaching and reinforcing key concepts and techniques.

The On-line Study Guide (created by Mark Yoder and Wayne Padgett) allows for self-paced exploration of topics including **Live Figures** where you can interact directly with waveforms, **Build-a-Figure** exercises which walk you through the creation of figures from the text, **Matlab Homework**, **Matlab Projects**, and **Demos** of real world examples.

MyPearsoneBook is Pearson's new eBook platform that gives you the same high-quality experience as the printed book, only electronically. While viewing the text at full fidelity, this new platform allows you to interact and annotate the text. MyPearsoneBook also gives you a quick and easy way to navigate through content, either by chapter and topic, or by search terms.

To access On-line Study Guide and/or MyPearsoneBook for Oppenheim & Schafer, 3e:

1. Go to www.pearsonhighered.com/oppenheim
2. Click on the title *Discrete-Time Signal Processing*, Third Edition.
3. Click on the link to Register. There you can register as a First-Time User and Returning User.
4. Use a coin to scratch off the metallic coating below and reveal your student access code.

   **Do not use a knife or other sharp object as it may damage the code.

5. On the registration page, enter your student access code. Do not type the dashes. You can use lower or uppercase letters.
6. Follow the on-screen instructions. If you need help during the online registration process, simply click on **Need Help?**
7. Once your personal Login Name and Password are confirmed, you can begin viewing your On-line Study Guide. You need only register for one product to gain access to both – your same username and password will work in each case.

To login to On-line Study Guide or MyPearsoneBook for the first time after you've registered:
Follow steps 1 and 2 to return to the **On-line Study Guide** or **MyPearsoneBook** link. Then, follow the link for "Returning Users" to enter your Login Name and Password.

Note to Instructors: For access to the Instructor Resource Center, contact your Pearson Representative.

IMPORTANT: The access code on this page can only be used once to establish a subscription to the Oppenheim/Schafer, *Discrete-Time Signal Processing*, Third Edition On-line Study Guide and MyPearsoneBook. If this access code has already been scratched off, it may no longer be valid. If this is the case, you can purchase a subscription by going to the *www.pearsonhighered.com/oppenheim* website and selecting "Get Access."

www.pearsonhighered.com

For help with registration or technical support, visit http://247pearsoned.custhelp.com

THIRD EDITION

DISCRETE-TIME SIGNAL PROCESSING

ALAN V. OPPENHEIM
MASSACHUSETTS INSTITUTE OF TECHNOLOGY

RONALD W. SCHAFER
HEWLETT-PACKARD LABORATORIES

PEARSON

Upper Saddle River · Boston · Columbus · San Francisco · New York
Indianapolis · London · Toronto · Sydney · Singapore · Tokyo · Montreal
Dubai · Madrid · Hong Kong · Mexico City · Munich · Paris · Amsterdam · Cape Town

Vice President and Editorial Director, ECS: *Marcia J. Horton*
Acquisition Editor: *Andrew Gilfillan*
Editorial Assistant: *William Opaluch*
Director of Team-Based Project Management: *Vince O'Brien*
Senior Marketing Manager: *Tim Galligan*
Marketing Assistant: *Mack Patterson*
Senior Managing Editor: *Scott Disanno*
Production Project Manager: *Clare Romeo*
Senior Operations Specialist: *Alan Fischer*
Operations Specialist: *Lisa McDowell*
Art Director: *Kristine Carney*
Cover Designer: *Kristine Carney*
Cover Photo: *Librado Romero/New York Times—Maps and Graphics*
Manager, Cover Photo Permissions: *Karen Sanatar*
Composition: *PreTeX Inc.: Paul Mailhot*
Printer/Binder: *Courier Westford*
Typeface: 10/12 TimesTen

Pearson Education Ltd., London
Pearson Education Singapore, Pte. Ltd.
Pearson Education Canada, Inc., Toronto
Pearson Education–Japan, Tokyo
Pearson Education Australia Pty. Ltd., Sydney
Pearson Education North Asia Ltd., Hong Kong
Pearson Education de Mexico, S.A. de C.V.
Pearson Education Malaysia, Pte. Ltd.
Pearson Education, Inc., Upper Saddle River, New Jersey

Prentice Hall
is an imprint of

www.pearsonhighered.com

10 9 8 7 6 5 4 3 2 1

ISBN-13: 978-0-13-198842-2
ISBN-10:   0-13-198842-5

To Phyllis, Justine, and Jason

*To Dorothy, Bill, Tricia, Ken, and Kate
and in memory of John*

CONTENTS

PREFACE

This third edition of *Discrete-Time Signal Processing* is a descendent of our original textbook *Digital Signal Processing* published in 1975. That very successful text appeared at a time when the field was young and just beginning to develop rapidly. At that time the topic was taught only at the graduate level and at only a very few schools. Our 1975 text was designed for such courses. It is still in print and is still used successfully at a number of schools in the United States and internationally.

By the 1980's, the pace of signal processing research, applications and implementation technology made it clear that digital signal processing (DSP) would realize and exceed the potential that had been evident in the 1970's. The burgeoning importance of DSP clearly justified a revision and updating of the original text. In organizing that revision, it was clear that so many changes had occurred both in the field and in the level and style with which the topic was taught, that it was most appropriate to develop a new textbook, strongly based on our original text, while keeping the original text in print. We titled the new book, published in 1989, *Discrete-Time Signal Processing* to emphasize that most of the DSP theory and design techniques discussed in the text apply to discrete-time systems in general, whether analog or digital.

In developing *Discrete-Time Signal Processing* we recognized that the basic principles of DSP were being commonly taught at the undergraduate level, sometimes even as part of a first course on discrete-time linear systems, but more often, at a more advanced level in third-year, fourth-year, or beginning graduate subjects. Therefore, it was appropriate to expand considerably the treatment of such topics as linear systems, sampling, multirate signal processing, applications, and spectral analysis. In addition, more examples were included to emphasize and illustrate important concepts. Consistent with the importance that we placed on well constructed examples and homework problems, that new text contained more than 400 problems.

While the field continued to advance in both theory and applications, the underlying fundamentals remained largely the same, albeit with a refinement of emphasis, understanding and pedagogy. Consequently, the Second Edition of *Discrete-Time Signal Processing* was published in 1999. That new edition was a major revision, with the intent of making the subject of discrete-time signal processing even more accessible to students and practicing engineers, without compromising on the coverage of what we considered to be the essential concepts that define the field.

This third edition of *Discrete-Time Signal Processing* is a major revision of our Second Edition. The new edition is in response to changes in the way the subject is taught and to changes in scope of typical courses at the undergraduate and first-year graduate level. It continues the tradition of emphasizing the accessibility of the topics to students and practicing engineers and focusing on the fundamental principles with broad applicability. A major feature of the new edition is the incorporation and expansion of some of the more advanced topics, the understanding of which are now essential in order to work effectively in the field. Every chapter of the second edition has undergone significant review and changes, one entirely new chapter has been added, and one chapter has been restored and significantly up-dated from the first edition. With this third edition, a closely integrated and highly interactive companion web site has been developed by Professors Mark Yoder and Wayne Padgett of Rose-Hulman Institute of Technology. A more complete discussion of the website is given in the website overview section following this Preface.

As we have continued to teach the subject over the ten years since the second edition, we have routinely created new problems for homework assignments and exams. Consistent with the importance that we have always placed on well constructed examples and homework problems, we have selected over 130 of the best of these to be included in the third edition, which now contains a total of more than 700 homework problems overall. The homework problems from the second edition that do not appear in this new edition are available on the companion web site.

As in the earlier generations of this text, it is assumed that the reader has a background of advanced calculus, along with a good understanding of the elements of complex numbers and complex variables. A background in linear system theory for continuous-time signals, including Laplace and Fourier transforms, as taught in most undergraduate electrical and mechanical engineering curricula, remains a basic prerequisite. It is also now common in most undergraduate curricula to include an early exposure to discrete-time signals and systems, discrete-time Fourier transforms and discrete-time processing of continuous-time signals.

Our experience in teaching discrete-time signal processing at the advanced undergraduate level and the graduate level confirms that it is essential to begin with a careful review of these topics so that students move on to the more advanced topics from a solid base of understanding and a familiarity with a consistent notational framework that is used throughout the course and the accompanying textbook. Most typically in a first exposure to discrete-time signal processing in early undergraduate courses, students learn to carry out many of the mathematical manipulations, but it is in revisiting the topics that they learn to reason more deeply with the underlying concepts. Therefore in this edition we retain the coverage of these fundamentals in the first five chapters, enhanced with new examples and expanded discussion. In later sections of some chapters, some topics such as quantization noise are included that assume a basic background in random signals. A brief review of the essential background for these sections is included in Chapter 2 and in Appendix A.

An important major change in DSP education that has occurred in the past decade or so is the widespread use of sophisticated software packages such as MATLAB, Lab-VIEW, and Mathematica to provide an interactive, "hands-on" experience for students. The accessibility and ease of use of these software packages provide the opportunity to connect the concepts and mathematics that are the basis for discrete-time signal processing, to applications involving real signals and real-time systems. These software packages are well documented, have excellent technical support, and have excellent user interfaces. These make them easily accessible to students without becoming a distraction from the goal of developing insight into and intuition about the fundamentals. It is now common in many signal processing courses to include projects and exercises to be done using one or several of the available software packages. Of course, this needs to be done carefully in order to maximize the benefit to student learning by emphasizing experimentation with the concepts, parameters, and so on, rather than simple cookbook exercises. It is particularly exciting that with one of these powerful software packages installed, every student's laptop computer becomes a state-of-the-art laboratory for experimenting with discrete-time signal processing concepts and systems.

As teachers, we have consistently looked for the best way to use computer resources to improve the learning environment for our students. We continue to believe in textbooks as the best way to encapsulate knowledge in the most convenient and stable form. Textbooks necessarily evolve on a relatively slow time scale. This ensures a certain stability and provides the time to sort through developments in the field and to test ways of presenting new ideas to students. On the other hand, changes in computer software and hardware technology are on a much faster time scale. Software revisions often occur semi-annually, and hardware speeds continue to increase yearly. This, together with the availability of the world-wide-web, provides the opportunity to more frequently update the interactive and experimental components of the learning environment. For these reasons, providing separate environments for the basic mathematics and basic concepts in the form of the textbook and the hands-on interactive experience primarily through the world-wide-web seems to be a natural path.

With these thoughts in mind, we have created this third edition of *Discrete-Time Signal Processing*, incorporating what we believe to be the fundamental mathematics and concepts of discrete-time signal processing and with tight coupling to a companion website created by our colleagues Mark Yoder and Wayne Padgett of Rose-Hulman Institute of Technology. The website contains a variety of interactive and software resources for learning that both reinforce and expand the impact of the text. This website is described in more detail in the introductory section following this Preface. It is designed to be dynamic and continually changing to rapidly incorporate new resources developed by the authors of the text and by the website authors. The website will be sensitive to the continually changing hardware and software environments that serve as the platform for visualization of abstract concepts and experimentation with real signal processing problems. We are excited by the virtually limitless potential for this companion website environment to significantly improve our ability to teach and our students' ability to learn the subject of discrete-time signal processing.

The material in this book is organized in a way that provides considerable flexibility in its use at both the undergraduate and graduate level. A typical one-semester undergraduate elective might cover in depth Chapter 2, Sections 2.0–2.9; Chapter 3; Chapter 4, Sections 4.0–4.6; Chapter 5, Sections 5.0–5.3; Chapter 6, Sections 6.0–6.5; and Chapter 7, Sections 7.0–7.3 and a brief overview of Sections 7.4–7.6. If students have studied discrete-time signals and systems in a previous signals and systems course, it would be possible to move more quickly through the material of Chapters 2, 3, and 4, thus freeing time for covering Chapter 8. A first-year graduate course or senior elective could augment the above topics with the remaining topics in Chapter 5, a discussion of multirate signal processing (Section 4.7), an exposure to some of the quantization issues introduced in Section 4.8, and perhaps an introduction to noise shaping in A/D and D/A converters as discussed in Section 4.9. A first-year graduate course should also include exposure to some of the quantization issues addressed in Sections 6.6–6.9, a discussion of optimal FIR filters as incorporated in Sections 7.7–7.9, and a thorough treatment of the discrete Fourier transform (Chapter 8) and its computation using the FFT (Chapter 9). The discussion of the DFT can be effectively augmented with many of the examples in Chapter 10. In a two-semester graduate course, the entire text including the new chapters on parametric signal modeling (Chapter 11) and the cepstrum (Chapter 13) can be covered along with a number of additional advanced topics. In all cases, the homework problems at the end of each chapter can be worked with or without the aid of a computer, and problems and projects from the website can be assigned to strengthen the connection between theory and computer implementation of signal processing systems.

We conclude this Preface with a summary of chapter contents highlighting the significant changes in the third edition.

In Chapter 2, we introduce the basic class of discrete-time signals and systems and define basic system properties such as linearity, time invariance, stability, and causality. The primary focus of the book is on linear time-invariant systems because of the rich set of tools available for designing and analyzing this class of systems. In particular, in Chapter 2 we develop the time-domain representation of linear time-invariant systems through the convolution sum and discuss the class of linear time-invariant systems represented by linear constant-coefficient difference equations. In Chapter 6, we develop this class of systems in considerably more detail. Also in Chapter 2 we discuss the frequency-domain representation of discrete-time signals and systems through the discrete-time Fourier transform. The primary focus in Chapter 2 is on the representation of sequences in terms of the discrete-time Fourier transform, i.e., as a linear combination of complex exponentials, and the development of the basic properties of the discrete-time Fourier transform.

In Chapter 3, we develop the z-transform as a generalization of the Fourier transform. This chapter focuses on developing the basic theorems and properties of the z-transform and the development of the partial fraction expansion method for the inverse transform operation. A new section on the unilateral z-transform has been added in this edition. In Chapter 5, the results developed in Chapters 2 and 3 are used extensively in a detailed discussion of the representation and analysis of linear time-invariant systems. While the material in Chapters 2 and 3 might be review for many students, most introductory signals and systems courses will not contain either the depth or breadth of coverage of these chapters. Furthermore, these chapters establish notation that will be used throughout the text. Thus, we recommend that Chapters 2 and 3 be studied as carefully as is necessary for students to feel confident of their grasp of the fundamentals of discrete-time signals and systems.

Chapter 4 is a detailed discussion of the relationship between continuous-time and discrete-time signals when the discrete-time signals are obtained through periodic sampling of continuous-time signals. This includes a development of the Nyquist sampling theorem. In addition, we discuss upsampling and downsampling of discrete-time signals, as used, for example, in multirate signal processing systems and for sampling rate conversion. The chapter concludes with a discussion of some of the practical issues encountered in conversion from continuous time to discrete time including prefiltering to avoid aliasing, modeling the effects of amplitude quantization when the discrete-time signals are represented digitally, and the use of oversampling in simplifying the A/D and D/A conversion processes. This third edition includes new examples of quantization noise simulations, a new discussion of interpolation filters derived from splines, and new discussions of multi-stage interpolation and two-channel multi-rate filter banks.

In Chapter 5 we apply the concepts developed in the previous chapters to a detailed study of the properties of linear time-invariant systems. We define the class of ideal, frequency-selective filters and develop the system function and pole-zero representation for systems described by linear constant-coefficient difference equations, a class of systems whose implementation is considered in detail in Chapter 6. Also in Chapter 5, we define and discuss group delay, phase response and phase distortion, and the relationships between the magnitude response and the phase response of systems, including a discussion of minimum-phase, allpass, and generalized linear phase systems. Third edition changes include a new example of the effects of group delay and attenuation, which is also available on the website for interactive experimentation.

In Chapter 6, we focus specifically on systems described by linear constant-coefficient difference equations and develop their representation in terms of block diagrams and linear signal flow graphs. Much of this chapter is concerned with developing a variety of the important system structures and comparing some of their properties. The importance of this discussion and the variety of filter structures relate to the fact that in a practical implementation of a discrete-time system, the effects of coefficient inaccuracies and arithmetic error can be very dependent on the specific structure used. While these basic issues are similar for digital and discrete-time analog implementations, we illustrate them in this chapter in the context of a digital implementation through a discussion of the effects of coefficient quantization and arithmetic roundoff noise for digital filters. A new section provides a detailed discussion of FIR and IIR lattice filters for implementing linear constant-coefficient difference equations. As discussed in Chapter 6 and later in Chapter 11, this class of filter structures has become extremely important in many applications because of their desirable properties. It is common in discussions of lattice filters in many texts and papers to tie their importance intimately to linear prediction analysis and signal modeling. However the importance of using lattice implementations of FIR and IIR filters is independent of how the difference equation to be implemented is obtained. For example the difference equation might have resulted from the use of filter design techniques as discussed in Chapter 7, the use of parametric signal modeling as discussed in Chapter 11 or any of a variety of other ways in which a difference equation to be implemented arises.

While Chapter 6 is concerned with the representation and implementation of linear constant-coefficient difference equations, Chapter 7 is a discussion of procedures for obtaining the coefficients of this class of difference equations to approximate a desired system response. The design techniques separate into those used for infinite impulse response (IIR) filters and those used for finite impulse response (FIR) filters. New examples of IIR filter design provide added insight into the properties of the different approximation methods. A new example on filter design for interpolation provides a framework for comparing IIR and FIR filters in a practical setting.

In continuous-time linear system theory, the Fourier transform is primarily an analytical tool for representing signals and systems. In contrast, in the discrete-time case, many signal processing systems and algorithms involve the explicit computation of the Fourier transform. While the Fourier transform itself cannot be computed, a sampled version of it, the discrete Fourier transform (DFT), can be computed, and for finite-length signals the DFT is a complete Fourier representation of the signal. In Chapter 8, the DFT is introduced and its properties and relationship to the discrete-time Fourier transform (DTFT) are developed in detail. In this chapter we also provide an introduction to the discrete cosine transform (DCT) which plays a very important role in applications such as audio and video compression.

In Chapter 9, the rich and important variety of algorithms for computing or generating the DFT is introduced and discussed, including the Goertzel algorithm, the fast Fourier transform (FFT) algorithms, and the chirp transform. In this third edition, the basic upsampling and downsampling operations discussed in Chapter 4 are used to provide additional insight into the derivation of FFT algorithms. As also discussed in this chapter, the evolution of technology has considerably altered the important metrics in evaluating the efficiency of signal processing algorithms. At the time of our first book in the 1970's both memory and arithmetic computation (multiplications and also floating point additions) were costly and the efficiency of algorithms was typically judged by how much of these resources were required. Currently it is commonplace to use additional memory to increase speed and to reduce the power requirements in the implementation of signal processing algorithms. In a similar sense, multi-core platforms have in some contexts resulted in favoring parallel implementation of algorithms even at the cost of increased computation. Often the number of cycles of data exchange, communication on a chip, and power requirements are now key metrics in choosing the structure for implementing an algorithm. As discussed in chapter 9, while the FFT is more efficient in terms of the required multiplications than the Goertzel algorithm or the direct computation of the DFT, it is less efficient than either if the dominant metric is communication cycles since direct computation or the Goertzel algorithm can be much more highly parallelized than the FFT.

With the background developed in the earlier chapters and particularly Chapters 2, 3, 5, and 8, we focus in Chapter 10 on Fourier analysis of signals using the DFT. Without a careful understanding of the issues involved and the relationship between the continuous-time Fourier transform, the DTFT, and the DFT, using the DFT for practical signal analysis can often lead to confusion and misinterpretation. We address a number of these issues in Chapter 10. We also consider in some detail the Fourier analysis of signals with time-varying characteristics by means of the time-dependent Fourier transform. New in the third edition is a more detailed discussion of filter bank analysis including an illustration of the MPEG filter bank, new examples of time-dependent Fourier analysis of chirp signals illustrating the effect of window length, and more detailed simulations of quantization noise analysis.

Chapter 11 is an entirely new chapter on the subject of parametric signal modeling. Starting with the basic concept of representing a signal as the output of an LTI system, Chapter 11 shows how the parameters of the signal model can be found by solution of a set of linear equations. Details of the computations involved in setting up and solving the equations are discussed and illustrated by examples. Particular emphasis is on the Levinson–Durbin solution algorithm and the many properties of the solution that are easily derived from the details of the algorithm such as the lattice filter interpretation.

Chapter 12 is concerned with the discrete Hilbert transform. This transform arises in a variety of practical applications, including inverse filtering, complex representations for real bandpass signals, single-sideband modulation techniques, and many others. With the advent of increasingly sophisticated communications systems and the growing richness of methods for efficiently sampling wide-band and multi-band continuous-time signals, a basic understanding of Hilbert transforms is becoming increasingly important. The Hilbert transform also plays an important role in the discussion of the cepstrum in Chapter 13.

Our first book in 1975 and the first edition of this book in 1989 included a detailed treatment of the class of nonlinear techniques referred to as cepstral analysis and homomorphic deconvolution. These techniques have become increasingly important and now have widespread use in applications such as speech coding, speech and speaker recognition, analysis of geophysical and medical imaging data, and in many other applications in which deconvolution is an important theme. Consequently with this edition we reintroduce those topics with expanded discussion and examples. The chapter contains a detailed discussion of the definition and properties of the cepstrum and the variety of ways of computing it including new results on the use of polynomial rooting as a basis for computing the cepstrum. An exposure to the material in Chapter 13 also offers the reader the opportunity to develop new insights into the fundamentals presented in the early chapters, in the context of a set of nonlinear signal analysis techniques with growing importance and that lend themselves to the same type of rich analysis enjoyed by linear techniques. The chapter also includes new examples illustrating the use of homomorphic filtering in deconvolution.

We look forward to using this new edition in our teaching and hope that our colleagues and students will benefit from the many enhancements from the previous editions. Signal processing in general and discrete-time signal processing in particular have a richness in all their dimensions that promises even more exciting developments ahead.

Alan V. Oppenheim
Ronald W. Schafer

THE COMPANION WEBSITE

A companion website has been developed for this text by Mark A. Yoder and Wayne T. Padgett of Rose-Hulman Institute of Technology and is accessible at www.pearsonhighered.com/oppenheim. This web companion, which will continuously evolve, is designed to reinforce and enhance the material presented in the textbook by providing visualizations of important concepts and a framework for obtaining "hands-on" experience with using the concepts. It contains six primary elements: *Live Figures*, *Build-a-Figures*, *MATLAB-based homework problems*, *MATLAB-based projects*, *Demos*, and additional *Traditional Homework Problems*, each tying into specific sections and pages in the book.

Live Figures

The Live Figures element reinforces concepts in the text by presenting "live" versions of select figures. With these, the reader is able to interactively investigate how parameters and concepts interoperate using graphics and audio. Live Figures were created with NI LabVIEW signal processing tools. The following three examples provide a glimpse of what is available with this element of the website:

Figure 2.10(a)-(c) in Section 2.3 on page 28 shows the graphical method for computing a discrete-convolution with the result shown in Figure 2.10(d). The corresponding Live Figure allows the user to choose the input signals and manually slide the flipped input signal past the impulse response and see the result being calculated and plotted. Users can quickly explore many different configurations and quickly understand how graphical convolution is applied.

Figure 4.73 on page 231 of Section 4.9.2 shows the power spectral density of the quantization noise and signal after noise shaping. The Live Figure shows the spectrum of the noise and signal as a live audio file plays. The reader can see and hear the noise as the noise shaping is enabled or disabled and as a lowpass filter is applied to remove the noise.

Figure 5.5(a) on page 282 of Section 5.1.2 shows three pulses, each of a different frequency, which enter an LTI system. Figure 5.6 on page 283 shows the output of the LTI system. The associated Live Figure allows students to experiment with the location of the poles and zeros in the system as well as the amplitude, frequency, and position of the pulses to see the effect on the output. These are just three examples of the many web-based Live Figures accessible on the companion website.

Build-a-Figure

The Build-a-Figure element extends the concept of the Live Figure element. It guides the student in recreating selected figures from the text using MATLAB to reinforce the understanding of the basic concepts. Build-a-Figures are not simply step-by-step recipes for constructing a figure. Rather, they assume a basic understanding of MATLAB and introduce new MATLAB commands and techniques as they are needed to create the figures. This not only further reinforces signal processing concepts, but also develops MATLAB skills in the context of signal processing. As an example, Figures 2.3 and 2.5 on pages 12 and 16 in Section 2.1 of the text are plots of several sequences. The corresponding Build-a-Figures introduce the MATLAB plot command techniques for labeling plots, incorporating Greek characters, and including a legend. Later Build-a-Figures use this knowledge as needed in creating plots. The Noise Shaping and Group Delay Build-a-Figures (Figure 4.73, page 231 and Figure 5.5, page 282) have instructions for recreating the Live Figures discussed above. Rather than giving step-by-step instructions, they introduce new MATLAB commands and suggest approaches for recreating the figures with considerable latitude for experimentation.

MATLAB Homework Problems

Through the MATLAB Homework Problems element, the companion website provides a primary mechanism for combining MATLAB with homework exercises. One aspect of this is the use of homework to practice using MATLAB somewhat in the style of Build-a-Figures. These exercises are much like non-MATLAB exercises but with MATLAB used to facilitate certain parts, such as in plotting results. The second avenue is the use of MATLAB to explore and solve problems that cannot be done by mathematical analysis. The MATLAB problems are all classroom tested and tend to be short exercises, comparable to the Basic Problems in the textbook, in which the user is asked to complete straightforward signal processing tasks using MATLAB. These problems are modest in scope as would be typical of one of several problems in a weekly homework assignment. Some problems are directly linked to analytic problems in the text, while others will stand on their own. Many of the problems blend analytic solutions with MATLAB, emphasizing the complementary value of each approach.

MATLAB-Based Projects

The MATLAB-based Projects element contains longer and more sophisticated projects or exercises than the homework problems. The projects explore important concepts from the textbook in greater depth and are relatively extensive. Projects are linked to sections of the text and can be used once that section is understood. For example, the first project is somewhat tutorial in nature and can be used at any stage. It introduces MATLAB and shows how it is used to create and manipulate discrete-time signals and systems. It assumes that the students have some programming experience, but not necessarily in MATLAB. Many of the other projects require some filter design techniques and therefore tie in with Chapter 7 (Filter Design Techniques) or later. They explore topics such as FIR and IIR filter design, filter design for sample rate conversion, testing a "Folk Theorem" about humans not being able to hear phase in a signal, enhancing speech by removing noise, hardware considerations for removing noise, spectral estimation and more. All have been classroom tested and some have led to student publications.

Demos

The Demos are interactive demonstrations that relate to specific chapters. Unlike the Live Figures, they do not tie directly to a given figure. Rather, they illustrate a bigger idea that the student can understand after completing the chapter. For example, one demo shows the importance of using a linear-phase filter when it is essential to preserve the shape of a bandlimited pulse.

Additional Traditional Homework Problems

A sixth important component of the website is a collection of problems that were removed from the second edition to make room for new problems. These problems can be used to supplement the problems in the text. Each of these problems is given in `.pdf` and `.tex` form along with any figures needed to create the problem.

In summary, the companion web site is a rich set of resources which are closely tied to the textbook. The resources range from the Live Figures which reinforce new concepts to the MATLAB-based projects which challenge the students to go beyond the textbook to explore new ideas. This website will continuously evolve as new teaching resources are developed by the authors of the text and by the website authors, Mark Yoder and Wayne Padgett.

THE COVER

In this third edition of Discrete-Time Signal Processing we continue the cover theme of "waves" as a symbol of our book and of signal processing. The cover of the first edition was a colorful rendering of a time-varying spectral waterfall plot. For the second edition, the artist Vivian Berman carried the theme forward resulting in a combination of spectral plots and artistic wave patterns. In considering possibilities for the cover, we were particularly drawn to a striking photograph by Librado Romero in a *New York Times* article (May 7, 2009). This article by Holland Cotter entitled "Storm King Wavefield" was about the artist Maya Lin's new work at the Storm King Art Center.[1] With this suggestion, Kristine Carney at Pearson/Prentice-Hall produced the beautiful cover for this edition.

To us, the grass-covered earthen waves in Ms. Lin's sculpture symbolize much about the field of Signal Processing and suggest the perfect evolution of our covers. As the *New York Times* article states,

> "Like any landscape, it is a work in progress. Vegetation is still coming in, drainage issues are in testing mode, and there are unruly variables: woodchucks have begun converting one wave into an apartment complex."

Change a few words here and there, and it provides an intriguing description of the field of Discrete-Time Signal Processing. It has a beautiful solid framework. Furthermore, new ideas, constraints, and opportunities keep the field fluid and dynamically changing, and there will always be a few "unruly variables." As Mr. Cotter also notes, Ms. Lin's work

> "sharpens your eye to existing harmonies and asymmetries otherwise overlooked."

Even after more than 40 years of living and working in the field of signal processing, we are consistently surprised and delighted by the harmonies, symmetries, and asymmetries that continue to be revealed.

[1] Information about the Storm King Art Center can be found at www.stormking.org and about Ms. Lin and her beautiful art at www.mayalin.com.

ACKNOWLEDGMENTS

This third edition of *Discrete-Time Signal Processing* has evolved from the first two editions (1989, 1999) which originated from our first book *Digital Signal Processing* (1975). The influence and impact of the many colleagues, students and friends who have assisted, supported and contributed to those earlier works remain evident in this new edition and we would like to express again our deep appreciation to all whom we have acknowledged more explicitly in those previous editions.

Throughout our careers we both have had the good fortune of having extraordinary mentors. We would each like to acknowledge several who have had such a major impact on our lives and careers.

Al Oppenheim was profoundly guided and influenced as a graduate student and throughout his career by Professor Amar Bose, Professor Thomas Stockham, and Dr. Ben Gold. As a teaching assistant for several years with and as a doctoral student supervised by Professor Bose, Al was significantly influenced by the inspirational teaching, creative research style and extraordinary standards which are characteristic of Professor Bose in everything that he does. Early in his career Al Oppenheim was also extremely fortunate to develop a close collaboration and friendship with both Dr. Ben Gold and Professor Thomas Stockham. The incredible encouragement and role model provided by Ben was significant in shaping Al's style of mentoring and research. Tom Stockham also provided significant mentoring, support and encouragement as well as ongoing friendship and another wonderful role model. The influence of these extraordinary mentors flows throughout this book.

Most notable among the many teachers and mentors who have influenced Ron Schafer are Professor Levi T. Wilson, Professor Thomas Stockham, and Dr. James L. Flanagan. Professor Wilson introduced a naive small town boy to the wonders of mathematics and science in a way that was memorable and life changing. His dedication to teaching was an inspiration too strong to resist. Professor Stockham was a great teacher, a friend at a crucial time, a valued colleague, and a wonderfully creative engineer. Jim Flanagan is a giant in the area of speech science and engineering and an inspiration to all who are so lucky as to have worked with him. Not all great teachers carry the title "Professor". He taught Ron and many others the value of careful thought, the value of dedication to a field of learning, and the value of clear and lucid writing and expression. Ron Schafer freely admits appropriating many habits of thought and expression from these great mentors, and does so with confidence that they don't mind at all.

Throughout our academic careers, MIT and Georgia Tech have provided us with a stimulating environment for research and teaching and have provided both encouragement and support for this evolving project. Since 1977 Al Oppenheim has spent several sabbaticals and almost every summer at the Woods Hole Oceanographic Institution (WHOI) and he is deeply appreciative of this special opportunity and association. It was during those periods and in the wonderful WHOI environment that much of the writing of the various editions of this book were carried out.

AT MIT and at Georgia Tech we have both received generous financial support from a number of sources. Al Oppenheim is extremely grateful for the support from Mr. Ray Stata and Analog Devices, Inc., the Bose Foundation, and the Ford Foundation for the funding of research and teaching at MIT in various forms. Both of us have also enjoyed the support of Texas Instruments, Inc. for our teaching and research activities. In particular, Gene Frantz at TI has been a dedicated supporter of our work and DSP education in general at both academic institutions. Ron Schafer is also grateful for the generous support from the John and Mary Franklin Foundation, which funded the John and Marilu McCarty Chair at Georgia Tech. Demetrius Paris, long time director of the School of ECE at Georgia Tech, and W. Kelly Mosley and Marilu McCarty of the Franklin Foundation, deserve special thanks for their friendship and support for over 30 years. Ron Schafer is appreciative of the opportunity to be a part of the research team at Hewlett-Packard Laboratories, first through research support at Georgia Tech over many years, and since 2004, as an HP Fellow. The third edition could not have been completed without the encouragement and support of HP Labs managers Fred Kitson, Susie Wee, and John Apostolopoulos.

Our association with Prentice Hall Inc. began several decades ago with our first book published in 1975 and has continued through all three editions of this book as well as with other books. We feel extremely fortunate to have worked with Prentice Hall. The encouragement and support provided by Marcia Horton and Tom Robbins through this and many other writing projects and by Michael McDonald, Andrew Gilfillan, Scott Disanno, and Clare Romeo with this edition have significantly enhanced the enjoyment of writing and completing this project.

As with the previous editions, in producing this third edition, we were fortunate to receive the help of many colleagues, students, and friends. We greatly appreciate their generosity in devoting their time to help us with this project. In particular we express our thanks and appreciation to:

Professor John Buck for his significant role in the preparation of the second edition and his continued time and effort during the life of that edition,

Professors Vivek Goyal, Jae Lim, Gregory Wornell, Victor Zue and Drs. Babak Ayazifar, Soosan Beheshti, and Charles Rohrs who have taught at MIT from various editions and have made many helpful comments and suggestions,

Professors Tom Barnwell, Russ Mersereau, and Jim McClellan, long-time friends and colleagues of Ron Schafer, who have taught frequently from various editions and have influenced many aspects of the book,

Professor Bruce Black of Rose-Hulman Institute of Technology for carefully organizing ten years worth of new problems, selecting the best of these, and updating and integrating them into the chapters,

Professor Mark Yoder and Professor Wayne Padgett for the development of an outstanding companion web site for this edition,

Ballard Blair for his assistance in updating the bibliography,

Eric Strattman, Darla Secor, Diane Wheeler, Stacy Schultz, Kay Gilstrap, and Charlotte Doughty for their administrative assistance in the preparation of this revision and continued support of our teaching activities,

Tom Baran for his help with many of the computer issues associated with managing the files for this edition and for his significant help with the examples in a number of the chapters,

Shay Maymon who meticulously read through most of the chapters, reworked many of the problems in the more advanced chapters, and made important corrections and suggestions,

To all who helped in careful reviewing of the manuscript and page proofs: Berkin Bilgic, Albert Chang, Myung Jin Choi, Majid Fozunbal, Reeve Ingle, Jeremy Leow, Ying Liu, Paul Ryu, Sanquan Song, Dennis Wei, and Zahi Karam.

And to the many teaching assistants who have influenced this edition directly or indirectly while working with us in teaching the subject at MIT and at Georgia Tech.

Introduction

The rich history and future promise of signal processing derive from a strong synergy between increasingly sophisticated applications, new theoretical developments and constantly emerging new hardware architectures and platforms. Signal processing applications span an immense set of disciplines that include entertainment, communications, space exploration, medicine, archaeology, geophysics, just to name a few. Signal processing algorithms and hardware are prevalent in a wide range of systems, from highly specialized military systems and industrial applications to low-cost, high-volume consumer electronics. Although we routinely take for granted the extraordinary performance of multimedia systems, such as high definition video, high fidelity audio, and interactive games, these systems have always relied heavily on state-of-the-art signal processing. Sophisticated digital signal processors are at the core of all modern cell phones. MPEG audio and video and JPEG[1] image data compression standards rely heavily on many of the signal processing principles and techniques discussed in this text. High-density data storage devices and new solid-state memories rely increasingly on the use of signal processing to provide consistency and robustness to otherwise fragile technologies. As we look to the future, it is clear that the role of signal processing is expanding, driven in part by the convergence of communications, computers, and signal processing in both the consumer arena and in advanced industrial and government applications.

The growing number of applications and demand for increasingly sophisticated algorithms go hand-in-hand with the rapid development of device technology for implementing signal processing systems. By some estimates, even with impending limitations

---

[1] The acronyms MPEG and JPEG are the terms used in even casual conversation for referring to the standards developed by the "Moving Picture Expert Group (MPEG)" and the "Joint Photographic Expert Group (JPEG)" of the "International Organization for Standardization (ISO)."

on Moore's Law, the processing capability of both special-purpose signal processing microprocessors and personal computers is likely to increase by several orders of magnitude over the next 10 years. Clearly, the importance and role of signal processing will continue to expand at an accelerating rate well into the future.

Signal processing deals with the representation, transformation, and manipulation of signals and the information the signals contain. For example, we may wish to separate two or more signals that have been combined by some operation, such as addition, multiplication, or convolution, or we may want to enhance some signal component or estimate some parameter of a signal model. In communications systems, it is generally necessary to do preprocessing such as modulation, signal conditioning, and compression prior to transmission over a communications channel, and then to carry out postprocessing at the receiver to recover a facsimile of the original signal. Prior to the 1960s, the technology for such signal processing was almost exclusively continuous-time analog technology.[2] A continual and major shift to digital technologies has resulted from the rapid evolution of digital computers and microprocessors and low-cost chips for analog to digital (A/D) and digital to analog (D/A) conversion. These developments in technology have been reinforced by many important theoretical developments, such as the fast Fourier transform (FFT) algorithm, parametric signal modeling, multirate techniques, polyphase filter implementation, and new ways of representing signals, such as with wavelet expansions. As just one example of this shift, analog radio communication systems are evolving into reconfigurable "software radios" that are implemented almost exclusively with digital computation.

Discrete-time signal processing is based on processing of numeric sequences indexed on integer variables rather than functions of a continuous independent variable. In digital signal processing (DSP), signals are represented by sequences of finite-precision numbers, and processing is implemented using digital computation. The more general term *discrete-time signal processing* includes digital signal processing as a special case but also includes the possibility that sequences of samples (sampled data) could be processed with other discrete-time technologies. Often the distinction between the terms discrete-time signal processing and digital signal processing is of minor importance, since both are concerned with discrete-time signals. This is particularly true when high-precision computation is employed. Although there are many examples in which signals to be processed are inherently discrete-time sequences, most applications involve the use of discrete-time technology for processing signals that originate as continuous-time signals. In this case, a continuous-time signal is typically converted into a sequence of samples, i.e., a discrete-time signal. Indeed, one of the most important spurs to widespread application of digital signal processing was the development of low-cost A/D, D/A conversion chips based on differential quantization with noise shaping. After discrete-time processing, the output sequence is converted back to a continuous-time signal. Real-time operation is often required or desirable for such systems. As computer speeds have increased, discrete-time processing of continuous-time signals in real time has become commonplace in communication systems, radar and sonar, speech and video coding and enhancement, biomedical engineering, and many

---

[2]In a general context, we shall refer to the independent variable as "time," even though in specific contexts, the independent variable may take on any of a broad range of possible dimensions. Consequently, continuous time and discrete time should be thought of as generic terms referring to a continuous independent variable and a discrete independent variable, respectively.

other areas of application. Non-real-time applications are also common. The compact disc player and MP3 player are examples of asymmetric systems in which an input signal is processed only once. The initial processing may occur in real time, slower than real time, or even faster than real time. The processed form of the input is stored (on the compact disc or in a solid state memory), and final processing for reconstructing the audio signal is carried out in real time when the output is played back for listening. The compact disc and MP3 recording and playback systems rely on many of the signal processing concepts that we discuss in this book.

Financial Engineering represents another rapidly emerging field which incorporates many signal processing concepts and techniques. Effective modeling, prediction and filtering of economic data can result in significant gains in economic performance and stability. Portfolio investment managers, for example, are relying increasingly on using sophisticated signal processing since even a very small increase in signal predictability or signal-to-noise ratio (SNR) can result in significant gain in performance.

Another important area of signal processing is *signal interpretation*. In such contexts, the objective of the processing is to obtain a characterization of the input signal. For example, in a speech recognition or understanding system, the objective is to interpret the input signal or extract information from it. Typically, such a system will apply digital pre-processing (filtering, parameter estimation, and so on) followed by a pattern recognition system to produce a symbolic representation, such as a phonemic transcription of the speech. This symbolic output can, in turn, be the input to a symbolic processing system, such as a rules-based expert system, to provide the final signal interpretation.

Still another relatively new category of signal processing involves the symbolic manipulation of signal processing expressions. This type of processing is potentially useful in signal processing workstations and for the computer-aided design of signal processing systems. In this class of processing, signals and systems are represented and manipulated as abstract data objects. Object-oriented programming languages provide a convenient environment for manipulating signals, systems, and signal processing expressions without explicitly evaluating the data sequences. The sophistication of systems designed to do signal expression processing is directly influenced by the incorporation of fundamental signal processing concepts, theorems, and properties, such as those that form the basis for this book. For example, a signal processing environment that incorporates the property that convolution in the time domain corresponds to multiplication in the frequency domain can explore a variety of rearrangements of filtering structures, including those involving the direct use of the discrete Fourier transform (DFT) and the FFT algorithm. Similarly, environments that incorporate the relationship between sampling rate and aliasing can make effective use of decimation and interpolation strategies for filter implementation. Similar ideas are currently being explored for implementing signal processing in network environments. In this type of environment, data can potentially be tagged with a high-level description of the processing to be done, and the details of the implementation can be based dynamically on the resources available on the network.

Many of the concepts and design techniques discussed in this text are now incorporated into the structure of sophisticated software systems such as MATLAB, Simulink, Mathematica, and LabVIEW. In many cases where discrete-time signals are acquired and stored in computers, these tools allow extremely sophisticated signal processing operations to be formed from basic functions. In such cases, it is not generally necessary to know the details of the underlying algorithm that implements the computation of an operation like the FFT, but nevertheless it is essential to understand what is computed and how it should be interpreted. In other words, a good understanding of the concepts considered in this text is essential for intelligent use of the signal processing software tools that are now widely available.

Signal processing problems are not confined, of course, to one-dimensional signals. Although there are some fundamental differences in the theories for one-dimensional and multidimensional signal processing, much of the material that we discuss in this text has a direct counterpart in multidimensional systems. The theory of multidimensional digital signal processing is presented in detail in a variety of references including Dudgeon and Mersereau (1984), Lim (1989), and Bracewell (1994).[3] Many image processing applications require the use of two-dimensional signal processing techniques. This is the case in such areas as video coding, medical imaging, enhancement and analysis of aerial photographs, analysis of satellite weather photos, and enhancement of video transmissions from lunar and deep-space probes. Applications of multidimensional digital signal processing to image processing are discussed, for example, in Macovski (1983), Castleman (1996), Jain (1989), Bovic (ed.) (2005), Woods (2006), Gonzalez and Woods (2007), and Pratt (2007). Seismic data analysis as required in oil exploration, earthquake measurement, and nuclear test monitoring also uses multidimensional signal processing techniques. Seismic applications are discussed in, for example, Robinson and Treitel (1980) and Robinson and Durrani (1985).

Multidimensional signal processing is only one of many advanced and specialized topics that build on the fundamentals covered in this text. Spectral analysis based on the use of the DFT and the use of signal modeling is another particularly rich and important aspect of signal processing. We discuss many facets of this topic in Chapters 10 and 11, which focus on the basic concepts and techniques relating to the use of the DFT and parametric signal modeling. In Chapter 11, we also discuss in some detail high resolution spectrum analysis methods, based on representing the signal to be analyzed as the response of a discrete-time linear time-invariant (LTI) filter to either an impulse or to white noise. Spectral analysis is achieved by estimating the parameters (e.g., the difference equation coefficients) of the system and then evaluating the magnitude squared of the frequency response of the model filter. Detailed discussions of spectrum analysis can be found in the texts by Kay (1988), Marple (1987), Therrien (1992), Hayes (1996) and Stoica and Moses (2005).

Signal modeling also plays an important role in data compression and coding, and here again, the fundamentals of difference equations provide the basis for understanding many of these techniques. For example, one class of signal coding techniques, referred to as linear predictive coding (LPC), exploits the notion that if a signal is the response of a certain class of discrete-time filters, the signal value at any time index is a linear function of (and thus linearly predictable from) previous values. Consequently, efficient signal representations can be obtained by estimating these prediction parameters and using them along with the prediction error to represent the signal. The signal can then be regenerated when needed from the model parameters. This class of signal

---

[3]Authors names and dates are used throughout the text to refer to books and papers listed in the Bibliography at the end of the book.

coding techniques has been particularly effective in speech coding and is described in considerable detail in Jayant and Noll (1984), Markel and Gray (1976), Rabiner and Schafer (1978) and Quatieri (2002), and is also discussed in some detail in Chapter 11.

Another advanced topic of considerable importance is adaptive signal processing. Adaptive systems represent a particular class of time-varying and, in some sense, non-linear systems with broad application and with established and effective techniques for their design and analysis. Again, many of these techniques build from the fundamentals of discrete-time signal processing covered in this text. Details of adaptive signal processing are given by Widrow and Stearns (1985), Haykin (2002) and Sayed (2008).

These represent only a few of the many advanced topics that extend from the content covered in this text. Others include advanced and specialized filter design procedures, a variety of specialized algorithms for evaluation of the Fourier transform, specialized filter structures, and various advanced multirate signal processing techniques, including wavelet transforms. (See Burrus, Gopinath, and Guo (1997), Vaidyanathan (1993) and Vetterli and Kovačević (1995) for introductions to these topics.)

It has often been said that the purpose of a fundamental textbook should be to uncover, rather than cover, a subject. In choosing the topics and depth of coverage in this book, we have been guided by this philosophy. The preceding brief discussion and the Bibliography at the end of the book make it abundantly clear that there is a rich variety of both challenging theory and compelling applications to be uncovered by those who diligently prepare themselves with a study of the fundamentals of DSP.

HISTORIC PERSPECTIVE

Discrete-time signal processing has advanced in uneven steps over time. Looking back at the development of the field of discrete-time signal processing provides a valuable perspective on fundamentals that will remain central to the field for a long time to come. Since the invention of calculus in the 17th century, scientists and engineers have developed models to represent physical phenomena in terms of functions of continuous variables and differential equations. However, numeric techniques have been used to solve these equations when analytical solutions are not possible. Indeed, Newton used finite-difference methods that are special cases of some of the discrete-time systems that we present in this text. Mathematicians of the 18th century, such as Euler, Bernoulli, and Lagrange, developed methods for numeric integration and interpolation of functions of a continuous variable. Interesting historic research by Heideman, Johnson, and Burrus (1984) showed that Gauss discovered the fundamental principle of the FFT (discussed in Chapter 9) as early as 1805—even before the publication of Fourier's treatise on harmonic series representation of functions.

Until the early 1950s, signal processing as we have defined it was typically carried out with analog systems implemented with electronic circuits or even with mechanical devices. Even though digital computers were becoming available in business environments and in scientific laboratories, they were expensive and had relatively limited capabilities. About that time, the need for more sophisticated signal processing in some application areas created considerable interest in discrete-time signal processing. One of the first uses of digital computers in DSP was in geophysical exploration, where relatively low frequency seismic signals could be digitized and recorded on magnetic tape for later processing. This type of signal processing could not generally be done in real time; minutes or even hours of computer time were often required to process only seconds of data. Even so, the flexibility of the digital computer and the potential payoffs made this alternative extremely inviting.

Also in the 1950s, the use of digital computers in signal processing arose in a different way. Because of the flexibility of digital computers, it was often useful to simulate a signal processing system on a digital computer before implementing it in analog hardware. In this way, a new signal processing algorithm or system could be studied in a flexible experimental environment before committing economic and engineering resources to constructing it. Typical examples of such simulations were the vocoder simulations carried out at Massachusetts Institute of Technology (MIT) Lincoln Laboratory and Bell Telephone Laboratories. In the implementation of an analog channel vocoder, for example, the filter characteristics affected the perceived quality of the coded speech signal in ways that were difficult to quantify objectively. Through computer simulations, these filter characteristics could be adjusted and the perceived quality of a speech coding system evaluated prior to construction of the analog equipment.

In all of these examples of signal processing using digital computers, the computer offered tremendous advantages in flexibility. However, the processing could not be done in real time. Consequently, the prevalent attitude up to the late 1960s was that the digital computer was being used to *approximate*, or *simulate*, an analog signal processing system. In keeping with that style, early work on digital filtering concentrated on ways in which a filter could be programmed on a digital computer so that with A/D conversion of the signal, followed by digital filtering, followed by D/A conversion, the overall system approximated a good analog filter. The notion that digital systems might, in fact, be practical for the actual real-time implementation of signal processing in speech communication, radar processing, or any of a variety of other applications seemed, even at the most optimistic times, to be highly speculative. Speed, cost, and size were, of course, three of the important factors in favor of the use of analog components.

As signals were being processed on digital computers, researchers had a natural tendency to experiment with increasingly sophisticated signal processing algorithms. Some of these algorithms grew out of the flexibility of the digital computer and had no apparent practical implementation in analog equipment. Thus, many of these algorithms were treated as interesting, but somewhat impractical, ideas. However, the development of such signal processing algorithms made the notion of all-digital implementation of signal processing systems even more tempting. Active work began on the investigation of digital vocoders, digital spectrum analyzers, and other all-digital systems, with the hope that eventually, such systems would become practical.

The evolution of a new point of view toward discrete-time signal processing was further accelerated by the disclosure by Cooley and Tukey (1965) of an efficient class of algorithms for computation of Fourier transforms known collectively as the FFT. The FFT was significant for several reasons. Many signal processing algorithms that had been developed on digital computers required processing times several orders of magnitude greater than real time. Often, this was because spectrum analysis was an important component of the signal processing and no efficient means were available for implementing it. The FFT reduced the computation time of the Fourier transform by orders of magnitude, permitting the implementation of increasingly sophisticated signal processing algorithms with processing times that allowed interactive experimentation with the system. Furthermore, with the realization that the FFT algorithms might, in fact, be implementable with special-purpose digital hardware, many signal processing algorithms that previously had appeared to be impractical began to appear feasible.

Another important implication of the FFT was that it was an inherently discrete-time concept. It was directed toward the computation of the Fourier transform of a discrete-time signal or sequence and involved a set of properties and mathematics that was exact in the discrete-time domain—it was not simply an approximation to a continuous-time Fourier transform. This had the effect of stimulating a reformulation of many signal processing concepts and algorithms in terms of discrete-time mathematics, and these techniques then formed an exact set of relationships in the discrete-time domain. Following this shift away from the notion that signal processing on a digital computer was merely an approximation to analog signal processing techniques, there emerged the current view that discrete-time signal processing is an important field of investigation in its own right.

Another major development in the history of discrete-time signal processing occurred in the field of microelectronics. The invention and subsequent proliferation of the microprocessor paved the way for low-cost implementations of discrete-time signal processing systems. Although the first microprocessors were too slow to implement most discrete-time systems in real time except at very low sampling rates, by the mid-1980s, integrated circuit technology had advanced to a level that permitted the implementation of very fast fixed-point and floating-point microcomputers with architectures specially designed for implementing discrete-time signal processing algorithms. With this technology came, for the first time, the possibility of widespread application of discrete-time signal processing techniques. The rapid pace of development in microelectronics also significantly impacted the development of signal processing algorithms in other ways. For example, in the early days of real-time digital signal processing devices, memory was relatively costly and one of the important metrics in developing signal processing algorithms was the efficient use of memory. Digital memory is now so inexpensive that many algorithms purposely incorporate more memory than is absolutely required so that the power requirements of the processor are reduced. Another area in which technology limitations posed a significant barrier to widespread deployment of DSP was in conversion of signals from analog to discrete-time (digital) form. The first widely available A/D and D/A converters were stand-alone devices costing thousands of dollars. By combining digital signal processing theory with microelectronic technology, over-sampled A/D and D/A converters costing a few dollars or less have enabled a myriad of real-time applications.

In a similar way, minimizing the number of arithmetic operations, such as multiplies or floating point additions, is now less essential, since multicore processors often have several multipliers available and it becomes increasingly important to reduce communication between cores, even if it then requires more multiplications. In a multicore environment, for example, direct computation of the DFT (or the use of the Goertzel algorithm) is more "efficient" than the use of an FFT algorithm since, although many more multiplications are required, communication requirements are significantly reduced because the processing can be more efficiently distributed among multiple processors or cores. More broadly, the restructuring of algorithms and the development of new ones to exploit the opportunity for more parallel and distributed processing is becoming a significant new direction in the development of signal processing algorithms.

FUTURE PROMISE

Microelectronics engineers continue to strive for increased circuit densities and production yields, and as a result, the complexity and sophistication of microelectronic systems continually increase. The complexity, speed, and capability of DSP chips have grown exponentially since the early 1980s and show no sign of slowing down. As wafer-scale integration techniques become highly developed, very complex discrete-time signal processing systems will be implemented with low cost, miniature size, and low power consumption. Furthermore, technologies such as microelectronic mechanical systems (MEMS) promise to produce many types of tiny sensors whose outputs will need to be processed using DSP techniques that operate on distributed arrays of sensor inputs. Consequently, the importance of discrete-time signal processing will continue to increase, and the future development of the field promises to be even more dramatic than the course of development that we have just described.

Discrete-time signal processing techniques have already promoted revolutionary advances in some fields of application. A notable example is in the area of telecommunications, where discrete-time signal processing techniques, microelectronic technology, and fiber optic transmission have combined to change the nature of communication systems in truly revolutionary ways. A similar impact can be expected in many other areas. Indeed, signal processing has always been, and will always be, a field that thrives on new applications. The needs of a new field of application can sometimes be filled by knowledge adapted from other applications, but frequently, new application needs stimulate new algorithms and new hardware systems to implement those algorithms. Early on, applications to seismology, radar, and communication provided the context for developing many of the core signal processing techniques that we discuss in this book. Certainly, signal processing will remain at the heart of applications in national defense, entertainment, communication, and medical care and diagnosis. Recently, we have seen applications of signal processing techniques in new areas as disparate as finance and DNA sequence analysis.

Although it is difficult to predict where other new applications will arise, there is no doubt that they will be obvious to those who are prepared to recognize them. The key to being ready to solve new signal processing problems is, and has always been, a thorough grounding in the fundamental mathematics of signals and systems and in the associated design and processing algorithms. While discrete-time signal processing is a dynamic, steadily growing field, its fundamentals are well formulated, and it is extremely valuable to learn them well. Our goal in this book is to uncover the fundamentals of the field by providing a coherent treatment of the theory of discrete-time linear systems, filtering, sampling, discrete-time Fourier analysis, and signal modeling. This text should provide the reader with the knowledge necessary for an appreciation of the wide scope of applications for discrete-time signal processing and a foundation for contributing to future developments in this exciting field.

2

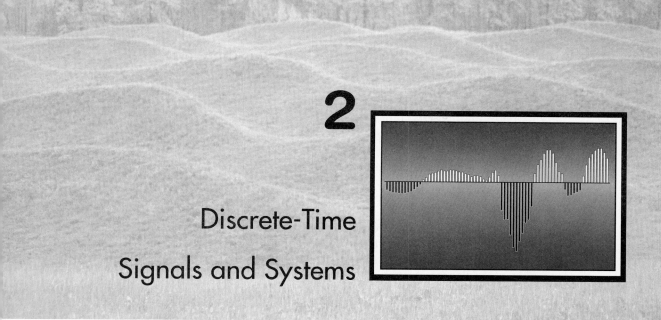

Discrete-Time
Signals and Systems

2.0 INTRODUCTION

The term *signal* is generally applied to something that conveys information. Signals may, for example, convey information about the state or behavior of a physical system. As another class of examples, signals are synthesized for the purpose of communicating information between humans or between humans and machines. Although signals can be represented in many ways, in all cases, the information is contained in some pattern of variations. Signals are represented mathematically as functions of one or more independent variables. For example, a speech signal is represented mathematically as a function of time, and a photographic image is represented as a brightness function of two spatial variables. A common convention—and one that usually will be followed in this book—is to refer to the independent variable of the mathematical representation of a signal as time, although in specific examples, the independent variable may not in fact correspond to time.

The independent variable in the mathematical representation of a signal may be either continuous or discrete. *Continuous-time signals* are defined along a continuum of time and are thus represented by a continuous independent variable. Continuous-time signals are often referred to as *analog signals*. *Discrete-time signals* are defined at discrete times, and thus, the independent variable has discrete values; that is, discrete-time signals are represented as sequences of numbers. Signals such as speech or images may have either a continuous- or a discrete-variable representation, and if certain conditions hold, these representations are entirely equivalent. Besides the independent variables being either continuous or discrete, the signal amplitude may be either continuous or discrete. *Digital signals* are those for which both time and amplitude are discrete.

9

Signal-processing systems may be classified along the same lines as signals. That is, continuous-time systems are systems for which both the input and the output are continuous-time signals, and discrete-time systems are those for which both the input and the output are discrete-time signals. Similarly, a digital system is a system for which both the input and the output are digital signals. Digital signal processing, then, deals with the transformation of signals that are discrete in both amplitude and time. The principal focus of this book is on discrete-time—rather than digital—signals and systems. However, the theory of discrete-time signals and systems is also exceedingly useful for digital signals and systems, particularly if the signal amplitudes are finely quantized. The effects of signal amplitude quantization are considered in Sections 4.8, 6.8–6.10, and 9.7.

In this chapter, we present the basic definitions, establish notation, and develop and review the basic concepts associated with discrete-time signals and systems. The presentation of this material assumes that the reader has had previous exposure to some of this material, perhaps with a different emphasis and notation. Thus, this chapter is primarily intended to provide a common foundation for material covered in later chapters.

In Section 2.1, we discuss the representation of discrete-time signals as sequences and describe the basic sequences such as the unit impulse, the unit step, and complex exponential, which play a central role in characterizing discrete-time systems and form building blocks for more general sequences. In Section 2.2, the representation, basic properties, and simple examples of discrete-time systems are presented. Sections 2.3 and 2.4 focus on the important class of linear time-invariant (LTI) systems and their time-domain representation through the convolution sum, with Section 2.5 considering the specific class of LTI systems represented by linear, constant–coefficient difference equations. Section 2.6 develops the frequency domain representation of discrete-time systems through the concept of complex exponentials as eigenfunctions, and Sections 2.7, 2.8, and 2.9 develop and explore the Fourier transform representation of discrete-time signals as a linear combination of complex exponentials. Section 2.10 provides a brief introduction to discrete-time random signals.

2.1 DISCRETE-TIME SIGNALS

Discrete-time signals are represented mathematically as sequences of numbers. A sequence of numbers x, in which the nth number in the sequence is denoted $x[n]$,[1] is formally written as

$$x = \{x[n]\}, \qquad -\infty < n < \infty, \tag{2.1}$$

where n is an integer. In a practical setting, such sequences can often arise from periodic sampling of an analog (i.e., continuous-time) signal $x_a(t)$. In that case, the numeric value of the nth number in the sequence is equal to the value of the analog signal, $x_a(t)$, at time nT: i.e.,

$$x[n] = x_a(nT), \qquad -\infty < n < \infty. \tag{2.2}$$

The quantity T is the *sampling period,* and its reciprocal is the *sampling frequency.* Although sequences do not always arise from sampling analog waveforms, it is convenient

---

[1]Note that we use [ ] to enclose the independent variable of discrete-variable functions, and we use ( ) to enclose the independent variable of continuous-variable functions.

to refer to $x[n]$ as the "n^{th} sample" of the sequence. Also, although, strictly speaking, $x[n]$ denotes the nth number in the sequence, the notation of Eq. (2.1) is often unnecessarily cumbersome, and it is convenient and unambiguous to refer to "the sequence $x[n]$" when we mean the entire sequence, just as we referred to "the analog signal $x_a(t)$." We depict discrete-time signals (i.e., sequences) graphically, as shown in Figure 2.1. Although the abscissa is drawn as a continuous line, it is important to recognize that $x[n]$ is defined only for integer values of n. It is not correct to think of $x[n]$ as being zero when n is not an integer; $x[n]$ is simply undefined for noninteger values of n.

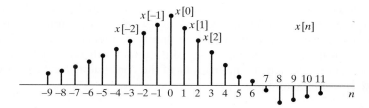

Figure 2.1   Graphic representation of a discrete-time signal.

As an example of a sequence obtained by sampling, Figure 2.2(a) shows a segment of a speech signal corresponding to acoustic pressure variation as a function of time, and Figure 2.2(b) presents a sequence of samples of the speech signal. Although the original speech signal is defined at all values of time t, the sequence contains information about the signal only at discrete instants. The sampling theorem, discussed in Chapter 4,

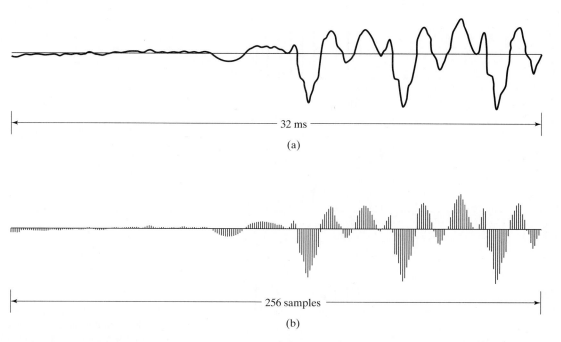

Figure 2.2   (a) Segment of a continuous-time speech signal $x_a(t)$. (b) Sequence of samples $x[n] = x_a(nT)$ obtained from the signal in part (a) with $T = 125\ \mu$s.

guarantees that the original signal can be reconstructed as accurately as desired from a corresponding sequence of samples if the samples are taken frequently enough.

In discussing the theory of discrete-time signals and systems, several basic sequences are of particular importance. These sequences are shown in Figure 2.3 and will be discussed next.

The *unit sample sequence* (Figure 2.3a) is defined as the sequence

$$\delta[n] = \begin{cases} 0, & n \neq 0, \\ 1, & n = 0. \end{cases} \tag{2.3}$$

The unit sample sequence plays the same role for discrete-time signals and systems that the unit impulse function (Dirac delta function) does for continuous-time signals and systems. For convenience, we often refer to the unit sample sequence as a discrete-time impulse or simply as an impulse. It is important to note that a discrete-time impulse does not suffer from the mathematic complications of the continuous-time impulse; its definition in Eq. (2.3) is simple and precise.

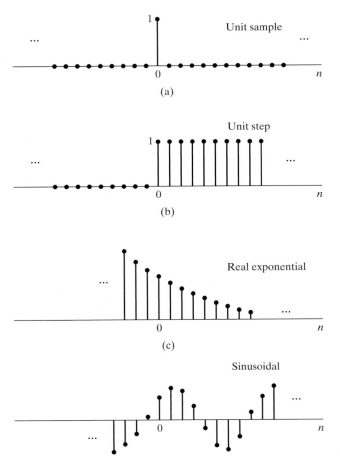

Figure 2.3 Some basic sequences. The sequences shown play important roles in the analysis and representation of discrete-time signals and systems.

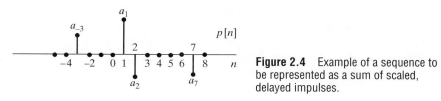

Figure 2.4 Example of a sequence to be represented as a sum of scaled, delayed impulses.

One of the important aspects of the impulse sequence is that an arbitrary sequence can be represented as a sum of scaled, delayed impulses. For example, the sequence $p[n]$ in Figure 2.4 can be expressed as

$$p[n] = a_{-3}\delta[n+3] + a_1\delta[n-1] + a_2\delta[n-2] + a_7\delta[n-7]. \tag{2.4}$$

More generally, any sequence can be expressed as

$$x[n] = \sum_{k=-\infty}^{\infty} x[k]\delta[n-k]. \tag{2.5}$$

We will make specific use of Eq. (2.5) in discussing the representation of discrete-time linear systems.

The unit step sequence (Figure 2.3b) is defined as

$$u[n] = \begin{cases} 1, & n \ge 0, \\ 0, & n < 0. \end{cases} \tag{2.6}$$

The unit step is related to the unit impulse by

$$u[n] = \sum_{k=-\infty}^{n} \delta[k]; \tag{2.7}$$

that is, the value of the unit step sequence at (time) index n is equal to the accumulated sum of the value at index n and all previous values of the impulse sequence. An alternative representation of the unit step in terms of the impulse is obtained by interpreting the unit step in Figure 2.3(b) in terms of a sum of delayed impulses, as in Eq. (2.5). In this case, the nonzero values are all unity, so

$$u[n] = \delta[n] + \delta[n-1] + \delta[n-2] + \cdots \tag{2.8a}$$

or

$$u[n] = \sum_{k=0}^{\infty} \delta[n-k]. \tag{2.8b}$$

As yet another alternative, the impulse sequence can be expressed as the first backward difference of the unit step sequence, i.e.,

$$\delta[n] = u[n] - u[n-1]. \tag{2.9}$$

Exponential sequences are another important class of basic signals. The general form of an exponential sequence is

$$x[n] = A\alpha^n. \tag{2.10}$$

If A and α are real numbers, then the sequence is real. If $0 < \alpha < 1$ and A is positive, then the sequence values are positive and decrease with increasing n, as in Figure 2.3(c).

For $-1 < \alpha < 0$, the sequence values alternate in sign but again decrease in magnitude with increasing n. If $|\alpha| > 1$, then the sequence grows in magnitude as n increases.

The exponential sequence $A\alpha^n$ with α complex has real and imaginary parts that are exponentially weighted sinusoids. Specifically, if $\alpha = |\alpha|e^{j\omega_0}$ and $A = |A|e^{j\phi}$, the sequence $A\alpha^n$ can be expressed in any of the following ways:

$$x[n] = A\alpha^n = |A|e^{j\phi}|\alpha|^n e^{j\omega_0 n}$$

$$= |A||\alpha|^n e^{j(\omega_0 n + \phi)} \tag{2.11}$$

$$= |A||\alpha|^n \cos(\omega_0 n + \phi) + j|A||\alpha|^n \sin(\omega_0 n + \phi).$$

The sequence oscillates with an exponentially growing envelope if $|\alpha| > 1$ or with an exponentially decaying envelope if $|\alpha| < 1$. (As a simple example, consider the case $\omega_0 = \pi$.)

When $|\alpha| = 1$, the sequence has the form

$$x[n] = |A|e^{j(\omega_0 n + \phi)} = |A|\cos(\omega_0 n + \phi) + j|A|\sin(\omega_0 n + \phi); \tag{2.12}$$

that is, the real and imaginary parts of $e^{j\omega_0 n}$ vary sinusoidally with n. By analogy with the continuous-time case, the quantity ω_0 is called the *frequency* of the complex sinusoid or complex exponential, and ϕ is called the *phase*. However, since n is a dimensionless integer, the dimension of ω_0 is radians. If we wish to maintain a closer analogy with the continuous-time case, we can specify the units of ω_0 to be radians per sample and the units of n to be samples.

The fact that n is always an integer in Eq. (2.12) leads to some important differences between the properties of discrete-time and continuous-time complex exponential sequences and sinusoidal sequences. Consider, for example, a frequency $(\omega_0 + 2\pi)$. In this case,

$$x[n] = A e^{j(\omega_0 + 2\pi)n}$$

$$= A e^{j\omega_0 n} e^{j2\pi n} = A e^{j\omega_0 n}. \tag{2.13}$$

Generally, complex exponential sequences with frequencies $(\omega_0 + 2\pi r)$, where r is an integer, are indistinguishable from one another. An identical statement holds for sinusoidal sequences. Specifically, it is easily verified that

$$x[n] = A \cos[(\omega_0 + 2\pi r)n + \phi]$$

$$= A \cos(\omega_0 n + \phi). \tag{2.14}$$

The implications of this property for sequences obtained by sampling sinusoids and other signals will be discussed in Chapter 4. For now, we conclude that, when discussing complex exponential signals of the form $x[n] = A e^{j\omega_0 n}$ or real sinusoidal signals of the form $x[n] = A \cos(\omega_0 n + \phi)$, we need only consider frequencies in an interval of length 2π. Typically, we will choose either $-\pi < \omega_0 \leq \pi$ or $0 \leq \omega_0 < 2\pi$.

Another important difference between continuous-time and discrete-time complex exponentials and sinusoids concerns their periodicity in n. In the continuous-time case, a sinusoidal signal and a complex exponential signal are both periodic in time with the period equal to 2π divided by the frequency. In the discrete-time case, a periodic sequence is a sequence for which

$$x[n] = x[n + N], \qquad \text{for all } n, \tag{2.15}$$

where the period N is necessarily an integer. If this condition for periodicity is tested for the discrete-time sinusoid, then

$$A \cos(\omega_0 n + \phi) = A \cos(\omega_0 n + \omega_0 N + \phi), \qquad (2.16)$$

which requires that

$$\omega_0 N = 2\pi k, \qquad (2.17)$$

where k is an integer. A similar statement holds for the complex exponential sequence $Ce^{j\omega_0 n}$; that is, periodicity with period N requires that

$$e^{j\omega_0 (n+N)} = e^{j\omega_0 n}, \qquad (2.18)$$

which is true only for $\omega_0 N = 2\pi k$, as in Eq. (2.17). Consequently, complex exponential and sinusoidal sequences are not necessarily periodic in n with period $(2\pi/\omega_0)$ and, depending on the value of ω_0, may not be periodic at all.

Example 2.1   Periodic and Aperiodic Discrete-Time Sinusoids

Consider the signal $x_1[n] = \cos(\pi n/4)$. This signal has a period of $N = 8$. To show this, note that $x[n+8] = \cos(\pi(n+8)/4) = \cos(\pi n/4 + 2\pi) = \cos(\pi n/4) = x[n]$, satisfying the definition of a discrete-time periodic signal. Contrary to continuous-time sinusoids, increasing the value of ω_0 for a discrete-time sinusoid does not necessarily decrease the period of the signal. Consider the discrete-time sinusoid $x_2[n] = \cos(3\pi n/8)$, which has a higher frequency than $x_1[n]$. However, $x_2[n]$ is not periodic with period 8, since $x_2[n + 8] = \cos(3\pi(n + 8)/8) = \cos(3\pi n/8 + 3\pi) = -x_2[n]$. Using an argument analogous to the one for $x_1[n]$, we can show that $x_2[n]$ has a period of $N = 16$. Thus, increasing the value of $\omega_0 = 2\pi/8$ to $\omega_0 = 3\pi/8$ also increases the period of the signal. This occurs because discrete-time signals are defined only for integer indices n.

The integer restriction on n results in some sinusoidal signals not being periodic at all. For example, there is no integer N such that the signal $x_3[n] = \cos(n)$ satisfies the condition $x_3[n + N] = x_3[n]$ for all n. These and other properties of discrete-time sinusoids that run counter to their continuous-time counterparts are caused by the limitation of the time index n to integers for discrete-time signals and systems.

When we combine the condition of Eq. (2.17) with our previous observation that ω_0 and $(\omega_0 + 2\pi r)$ are indistinguishable frequencies, it becomes clear that there are N distinguishable frequencies for which the corresponding sequences are periodic with period N. One set of frequencies is $\omega_k = 2\pi k/N$, $k = 0, 1, \ldots, N - 1$. These properties of complex exponential and sinusoidal sequences are basic to both the theory and the design of computational algorithms for discrete-time Fourier analysis, and they will be discussed in more detail in Chapters 8 and 9.

Related to the preceding discussion is the fact that the interpretation of high and low frequencies is somewhat different for continuous-time and discrete-time sinusoidal and complex exponential signals. For a continuous-time sinusoidal signal $x(t) = A \cos(\Omega_0 t + \phi)$, as Ω_0 increases, $x(t)$ oscillates progressively more rapidly. For the discrete-time sinusoidal signal $x[n] = A \cos(\omega_0 n + \phi)$, as ω_0 increases from $\omega_0 = 0$ toward $\omega_0 = \pi$, $x[n]$ oscillates progressively more rapidly. However, as ω_0 increases from $\omega_0 = \pi$ to $\omega_0 = 2\pi$, the oscillations become slower. This is illustrated in Figure 2.5. In

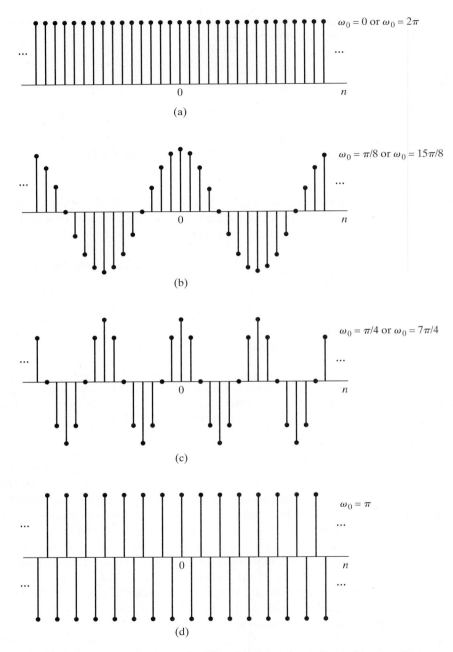

Figure 2.5 $\cos \omega_0 n$ for several different values of ω_0. As ω_0 increases from zero toward π (parts a–d), the sequence oscillates more rapidly. As ω_0 increases from π to 2π (parts d–a), the oscillations become slower.

fact, because of the periodicity in ω_0 of sinusoidal and complex exponential sequences, $\omega_0 = 2\pi$ is indistinguishable from $\omega_0 = 0$, and, more generally, frequencies around $\omega_0 = 2\pi$ are indistinguishable from frequencies around $\omega_0 = 0$. As a consequence, for sinusoidal and complex exponential signals, values of ω_0 in the vicinity of $\omega_0 = 2\pi k$ for any integer value of k are typically referred to as low frequencies (relatively slow oscillations), whereas values of ω_0 in the vicinity of $\omega_0 = (\pi + 2\pi k)$ for any integer value of k are typically referred to as high frequencies (relatively rapid oscillations).

2.2 DISCRETE-TIME SYSTEMS

A discrete-time system is defined mathematically as a transformation or operator that maps an input sequence with values $x[n]$ into an output sequence with values $y[n]$. This can be denoted as

$$y[n] = T\{x[n]\} \tag{2.19}$$

and is indicated pictorially in Figure 2.6. Equation (2.19) represents a rule or formula for computing the output sequence values from the input sequence values. It should be emphasized that the value of the output sequence at each value of the index n may depend on input samples $x[n]$ for all values of n, i.e., y at time n can depend on all or part of the entire sequence x. The following examples illustrate some simple and useful systems.

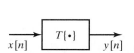

Figure 2.6   Representation of a discrete-time system, i.e., a transformation that maps an input sequence $x[n]$ into a unique output sequence $y[n]$.

Example 2.2   The Ideal Delay System

The ideal delay system is defined by the equation

$$y[n] = x[n - n_d], \qquad -\infty < n < \infty, \tag{2.20}$$

where n_d is a fixed positive integer representing the delay of the system. In other words, the ideal delay system shifts the input sequence to the right by n_d samples to form the output. If, in Eq. (2.20), n_d is a fixed negative integer, then the system would shift the input to the left by $|n_d|$ samples, corresponding to a time advance.

In the system of Example 2.2, only one sample of the input sequence is involved in determining a certain output sample. In the following example, this is not the case.

Example 2.3   Moving Average

The general moving-average system is defined by the equation

$$y[n] = \frac{1}{M_1 + M_2 + 1} \sum_{k=-M_1}^{M_2} x[n-k]$$

$$= \frac{1}{M_1 + M_2 + 1} \big\{ x[n+M_1] + x[n+M_1-1] + \cdots + x[n] \qquad (2.21)$$

$$+ x[n-1] + \cdots + x[n-M_2] \big\} .$$

This system computes the nth sample of the output sequence as the average of $(M_1 + M_2 + 1)$ samples of the input sequence around the nth sample. Figure 2.7 shows an input sequence plotted as a function of a dummy index k and the samples (solid dots) involved in the computation of the output sample $y[n]$ for $n = 7$, $M_1 = 0$, and $M_2 = 5$. The output sample $y[7]$ is equal to one-sixth of the sum of all the samples between the vertical dotted lines. To compute $y[8]$, both dotted lines would move one sample to the right.

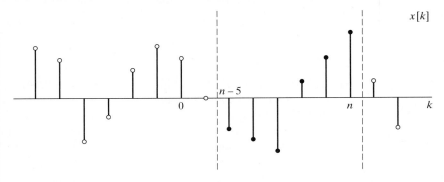

Figure 2.7   Sequence values involved in computing a moving average with $M_1 = 0$ and $M_2 = 5$.

Classes of systems are defined by placing constraints on the properties of the transformation $T\{\cdot\}$. Doing so often leads to very general mathematical representations, as we will see. Of particular importance are the system constraints and properties, discussed in Sections 2.2.1–2.2.5.

2.2.1 Memoryless Systems

A system is referred to as memoryless if the output $y[n]$ at every value of n depends only on the input $x[n]$ at the same value of n.

Example 2.4   A Memoryless System

An example of a memoryless system is a system for which $x[n]$ and $y[n]$ are related by

$$y[n] = (x[n])^2, \qquad \text{for each value of } n. \qquad (2.22)$$

The system in Example 2.2 is not memoryless unless $n_d = 0$; in particular, that system is referred to as having "memory" whether n_d is positive (a time delay) or negative (a time advance). The moving average system in Example 2.3 is not memoryless unless $M_1 = M_2 = 0$.

2.2.2 Linear Systems

The class of linear systems is defined by the principle of superposition. If $y_1[n]$ and $y_2[n]$ are the responses of a system when $x_1[n]$ and $x_2[n]$ are the respective inputs, then the system is linear if and only if

$$T\{x_1[n] + x_2[n]\} = T\{x_1[n]\} + T\{x_2[n]\} = y_1[n] + y_2[n] \tag{2.23a}$$

and

$$T\{ax[n]\} = aT\{x[n]\} = ay[n], \tag{2.23b}$$

where a is an arbitrary constant. The first property is the *additivity property*, and the second the *homogeneity* or *scaling property*. These two properties together comprise the principle of superposition, stated as

$$T\{ax_1[n] + bx_2[n]\} = aT\{x_1[n]\} + bT\{x_2[n]\} \tag{2.24}$$

for arbitrary constants a and b. This equation can be generalized to the superposition of many inputs. Specifically, if

$$x[n] = \sum_k a_k x_k[n], \tag{2.25a}$$

then the output of a linear system will be

$$y[n] = \sum_k a_k y_k[n], \tag{2.25b}$$

where $y_k[n]$ is the system response to the input $x_k[n]$.

By using the definition of the principle of superposition, it is easily shown that the systems of Examples 2.2 and 2.3 are linear systems. (See Problem 2.39.) An example of a nonlinear system is the system in Example 2.4.

Example 2.5   The Accumulator System

The system defined by the input–output equation

$$y[n] = \sum_{k=-\infty}^{n} x[k] \tag{2.26}$$

is called the accumulator system, since the output at time n is the accumulation or sum of the present and all previous input samples. The accumulator system is a linear system. Since this may not be intuitively obvious, it is a useful exercise to go through the steps of more formally showing this. We begin by defining two arbitrary inputs $x_1[n]$ and $x_2[n]$ and their corresponding outputs

$$y_1[n] = \sum_{k=-\infty}^{n} x_1[k], \tag{2.27}$$

$$y_2[n] = \sum_{k=-\infty}^{n} x_2[k]. \tag{2.28}$$

When the input is $x_3[n] = ax_1[n] + bx_2[n]$, the superposition principle requires the output $y_3[n] = ay_1[n] + by_2[n]$ for all possible choices of a and b. We can show this by starting from Eq. (2.26):

$$y_3[n] = \sum_{k=-\infty}^{n} x_3[k], \tag{2.29}$$

$$= \sum_{k=-\infty}^{n} \left(ax_1[k] + bx_2[k]\right), \tag{2.30}$$

$$= a \sum_{k=-\infty}^{n} x_1[k] + b \sum_{k=-\infty}^{n} x_2[k], \tag{2.31}$$

$$= ay_1[n] + by_2[n]. \tag{2.32}$$

Thus, the accumulator system of Eq. (2.26) satisfies the superposition principle for all inputs and is therefore linear.

Example 2.6   A Nonlinear System

Consider the system defined by

$$w[n] = \log_{10}(|x[n]|). \tag{2.33}$$

This system is not linear. To prove this, we only need to find one counterexample— that is, one set of inputs and outputs which demonstrates that the system violates the superposition principle, Eq. (2.24). The inputs $x_1[n] = 1$ and $x_2[n] = 10$ are a counterexample. However, the output for $x_1[n] + x_2[n] = 11$ is

$$\log_{10}(1 + 10) = \log_{10}(11) \neq \log_{10}(1) + \log_{10}(10) = 1.$$

Also, the output for the first signal is $w_1[n] = 0$, whereas for the second, $w_2[n] = 1$. The scaling property of linear systems requires that, since $x_2[n] = 10x_1[n]$, if the system is linear, it must be true that $w_2[n] = 10w_1[n]$. Since this is not so for Eq. (2.33) for this set of inputs and outputs, the system is *not* linear.

2.2.3 Time-Invariant Systems

A time-invariant system (often referred to equivalently as a shift-invariant system) is a system for which a time shift or delay of the input sequence causes a corresponding shift in the output sequence. Specifically, suppose that a system transforms the input sequence with values $x[n]$ into the output sequence with values $y[n]$. Then, the system is said to be time invariant if, for all n_0, the input sequence with values $x_1[n] = x[n - n_0]$ produces the output sequence with values $y_1[n] = y[n - n_0]$.

As in the case of linearity, proving that a system is time invariant requires a general proof making no specific assumptions about the input signals. On the other hand, proving non-time invariance only requires a counter example to time invariance. All of the systems in Examples 2.2–2.6 are time invariant. The style of proof for time invariance is illustrated in Examples 2.7 and 2.8.

Example 2.7   The Accumulator as a Time-Invariant System

Consider the accumulator from Example 2.5. We define $x_1[n] = x[n - n_0]$. To show time invariance, we solve for both $y[n-n_0]$ and $y_1[n]$ and compare them to see whether they are equal. First,

$$y[n - n_0] = \sum_{k=-\infty}^{n-n_0} x[k]. \tag{2.34}$$

Next, we find

$$y_1[n] = \sum_{k=-\infty}^{n} x_1[k] \tag{2.35}$$

$$= \sum_{k=-\infty}^{n} x[k - n_0]. \tag{2.36}$$

Substituting the change of variables $k_1 = k - n_0$ into the summation gives

$$y_1[n] = \sum_{k_1=-\infty}^{n-n_0} x[k_1]. \tag{2.37}$$

Since the index k in Eq. (2.34) and the index k_1 in Eq. (2.37) are dummy indices of summation, and can have any label, Eqs. (2.34) and (2.37) are equal and therefore $y_1[n] = y[n - n_0]$. The accumulator is a time-invariant system.

The following example illustrates a system that is not time invariant.

Example 2.8   The Compressor System

The system defined by the relation

$$y[n] = x[Mn], \qquad -\infty < n < \infty, \tag{2.38}$$

with M a positive integer, is called a compressor. Specifically, it discards $(M - 1)$ samples out of M; i.e., it creates the output sequence by selecting every M^{th} sample. This system is not time invariant. We can show that it is not by considering the response $y_1[n]$ to the input $x_1[n] = x[n - n_0]$. For the system to be time invariant, the output of the system when the input is $x_1[n]$ must be equal to $y[n - n_0]$. The output $y_1[n]$ that results from the input $x_1[n]$ can be directly computed from Eq. (2.38) to be

$$y_1[n] = x_1[Mn] = x[Mn - n_0]. \tag{2.39}$$

Delaying the output $y[n]$ by n_0 samples yields

$$y[n - n_0] = x[M(n - n_0)]. \tag{2.40}$$

Comparing these two outputs, we see that $y[n - n_0]$ is not equal to $y_1[n]$ for all M and n_0, and therefore, the system is not time invariant.

It is also possible to prove that a system is not time invariant by finding a single counterexample that violates the time-invariance property. For instance, a counterexample for the compressor is the case when $M = 2$, $x[n] = \delta[n]$, and $x_1[n] = \delta[n - 1]$. For this choice of inputs and M, $y[n] = \delta[n]$, but $y_1[n] = 0$; thus, it is clear that $y_1[n] \neq y[n - 1]$ for this system.

2.2.4 Causality

A system is causal if, for every choice of n_0, the output sequence value at the index $n = n_0$ depends only on the input sequence values for $n \leq n_0$. This implies that if $x_1[n] = x_2[n]$ for $n \leq n_0$, then $y_1[n] = y_2[n]$ for $n \leq n_0$. That is, the system is *nonanticipative*. The system of Example 2.2 is causal for $n_d \geq 0$ and is noncausal for $n_d < 0$. The system of Example 2.3 is causal if $-M_1 \geq 0$ and $M_2 \geq 0$; otherwise it is noncausal. The system of Example 2.4 is causal, as is the accumulator of Example 2.5 and the nonlinear system in Example 2.6. However, the system of Example 2.8 is noncausal if $M > 1$, since $y[1] = x[M]$. Another noncausal system is given in the following example.

Example 2.9   The Forward and Backward Difference Systems

The system defined by the relationship

$$y[n] = x[n + 1] - x[n] \tag{2.41}$$

is referred to as the *forward difference system*. This system is not causal, since the current value of the output depends on a future value of the input. The violation of causality can be demonstrated by considering the two inputs $x_1[n] = \delta[n - 1]$ and $x_2[n] = 0$ and their corresponding outputs $y_1[n] = \delta[n] - \delta[n - 1]$ and $y_2[n] = 0$ for all n. Note that $x_1[n] = x_2[n]$ for $n \leq 0$, so the definition of causality requires that $y_1[n] = y_2[n]$ for $n \leq 0$, which is clearly not the case for $n = 0$. Thus, by this counterexample, we have shown that the system is not causal.

The *backward difference system,* defined as

$$y[n] = x[n] - x[n - 1], \tag{2.42}$$

has an output that depends only on the present and past values of the input. Because $y[n_0]$ depends only on $x[n_0]$ and $x[n_0 - 1]$, the system is causal by definition.

2.2.5 Stability

A number of somewhat different definitions are commonly used for stability of a system. Throughout this text, we specifically use bounded-input bounded-output stability.

A system is stable in the bounded-input, bounded-output (BIBO) sense if and only if every bounded input sequence produces a bounded output sequence. The input $x[n]$ is bounded if there exists a fixed positive finite value B_x such that

$$|x[n]| \leq B_x < \infty, \qquad \text{for all } n. \tag{2.43}$$

Stability requires that, for every bounded input, there exists a fixed positive finite value B_y such that

$$|y[n]| \leq B_y < \infty, \qquad \text{for all } n. \tag{2.44}$$

It is important to emphasize that the properties we have defined in this section are properties of *systems,* not of the inputs to a system. That is, we may be able to find inputs for which the properties hold, but the existence of the property for some inputs does not mean that the system has the property. For the system to have the property, it must hold for *all* inputs. For example, an unstable system may have some bounded inputs for which the output is bounded, but for the system to have the property of stability, it must be true that for *all* bounded inputs, the output is bounded. If we can find just one input for which the system property does not hold, then we have shown that the system does *not* have that property. The following example illustrates the testing of stability for several of the systems that we have defined.

Example 2.10   Testing for Stability or Instability

The system of Example 2.4 is stable. To see this, assume that the input $x[n]$ is bounded such that $|x[n]| \leq B_x$ for all n. Then $|y[n]| = |x[n]|^2 \leq B_x^2$. Thus, we can choose $B_y = B_x^2$ and prove that $y[n]$ is bounded.

Likewise, we can see that the system defined in Example 2.6 is unstable, since $y[n] = \log_{10}(|x[n]|) = -\infty$ for any values of the time index n at which $x[n] = 0$, even though the output will be bounded for any input samples that are not equal to zero.

The accumulator, as defined in Example 2.5 by Eq. (2.26), is also not stable. For example, consider the case when $x[n] = u[n]$, which is clearly bounded by $B_x = 1$. For this input, the output of the accumulator is

$$y[n] = \sum_{k=-\infty}^{n} u[k] \tag{2.45}$$

$$= \begin{cases} 0, & n < 0, \\ (n+1), & n \geq 0. \end{cases} \tag{2.46}$$

There is no finite choice for B_y such that $(n+1) \leq B_y < \infty$ for all n; thus, the system is unstable.

Using similar arguments, it can be shown that the systems in Examples 2.2, 2.3, 2.8, and 2.9 are all stable.

2.3 LTI SYSTEMS

As in continuous time, a particularly important class of discrete-time systems consists of those that are both linear and time invariant. These two properties in combination lead to especially convenient representations for such systems. Most important, this class of systems has significant signal-processing applications. The class of linear systems is defined by the principle of superposition in Eq. (2.24). If the linearity property is combined with the representation of a general sequence as a linear combination of delayed impulses as in Eq. (2.5), it follows that a linear system can be completely characterized by its impulse response. Specifically, let $h_k[n]$ be the response of the system to the input $\delta[n - k]$, an impulse occurring at $n = k$. Then, using Eq. (2.5) to represent the input, it follows that

$$y[n] = T\left\{ \sum_{k=-\infty}^{\infty} x[k]\delta[n-k] \right\}, \tag{2.47}$$

and the principle of superposition in Eq. (2.24), we can write

$$y[n] = \sum_{k=-\infty}^{\infty} x[k]T\{\delta[n-k]\} = \sum_{k=-\infty}^{\infty} x[k]h_k[n]. \tag{2.48}$$

According to Eq. (2.48), the system response to any input can be expressed in terms of the responses of the system to the sequences $\delta[n - k]$. If only linearity is imposed, then $h_k[n]$ will depend on both n and k, in which case the computational usefulness of Eq. (2.48) is somewhat limited. We obtain a more useful result if we impose the additional constraint of time invariance.

The property of time invariance implies that if $h[n]$ is the response to $\delta[n]$, then the response to $\delta[n - k]$ is $h[n - k]$. With this additional constraint, Eq. (2.48) becomes

$$y[n] = \sum_{k=-\infty}^{\infty} x[k]h[n - k], \qquad \text{for all } n. \tag{2.49}$$

As a consequence of Eq. (2.49), an LTI system is completely characterized by its impulse response $h[n]$ in the sense that, given the sequences $x[n]$ and $h[n]$ for all n, it is possible to use Eq. (2.49) to compute each sample of the output sequence $y[n]$.

Equation (2.49) is referred to as the *convolution sum*, and we represent this by the operator notation

$$y[n] = x[n] * h[n]. \tag{2.50}$$

The operation of discrete-time convolution takes two sequences $x[n]$ and $h[n]$ and produces a third sequence $y[n]$. Equation (2.49) expresses each sample of the output sequence in terms of all of the samples of the input and impulse response sequences.

The notation of Eq. (2.50) for the operation of convolution as shorthand for Eq. (2.49) is convenient and compact but needs to be used with caution. The basic definition of the convolution of two sequences is embodied in Eq. (2.49) and any use of the shorthand form in Eq. (2.50) should always be referred back to Eq. (2.49). For example, consider $y[n - n_0]$. From Eq. (2.49) we see that

$$y[n - n_0] = \sum_{k=-\infty}^{\infty} x[k]h[n - n_0 - k] \tag{2.51}$$

or in short hand notation

$$y[n - n_0] = x[n] * h[n - n_0] \tag{2.52}$$

Substituting $(n - n_0)$ for n in Eq. (2.49) leads to the correct result and conclusion, but blindly trying the same substitution in Eq. (2.50) does not. In fact, $x[n - n_0] * h[n - n_0]$ results in $y[n - 2n_0]$.

The derivation of Eq. (2.49) suggests the interpretation that the input sample at $n = k$, represented as $x[k]\delta[n - k]$, is transformed by the system into an output sequence $x[k]h[n - k]$, for $-\infty < n < \infty$, and that, for each k, these sequences are superimposed (summed) to form the overall output sequence. This interpretation is illustrated in Figure 2.8, which shows an impulse response, a simple input sequence having three nonzero samples, the individual outputs due to each sample, and the composite output due to all

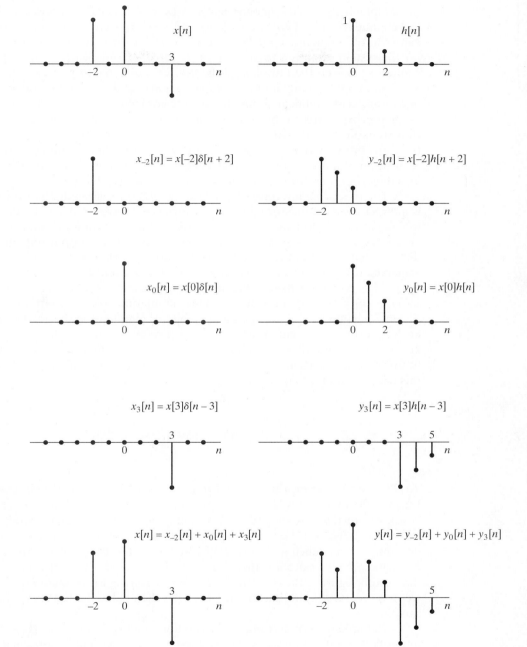

Figure 2.8   Representation of the output of an LTI system as the superposition of responses to individual samples of the input.

the samples in the input sequence. Specifically, $x[n]$ can be decomposed as the sum of the three sequences $x[-2]\delta[n+2]$, $x[0]\delta[n]$, and $x[3]\delta[n-3]$ representing the three nonzero values in the sequence $x[n]$. The sequences $x[-2]h[n+2]$, $x[0]h[n]$, and $x[3]h[n-3]$ are the system responses to $x[-2]\delta[n+2]$, $x[0]\delta[n]$, and $x[3]\delta[n-3]$, respectively. The response to $x[n]$ is then the sum of these three individual responses.

Although the convolution-sum expression is analogous to the convolution integral of continuous-time linear system theory, the convolution sum should not be thought of as an approximation to the convolution integral. The convolution integral is mainly a tool of mathematical analysis in continuous-time linear system theory; we will see that the convolution sum, in addition to its analytical importance, often serves as an explicit realization of a discrete-time linear system. Thus, it is important to gain some insight into the properties of the convolution sum in actual calculations.

The preceding interpretation of Eq. (2.49) emphasizes that the convolution sum is a direct result of linearity and time invariance. However, a slightly different way of looking at Eq. (2.49) leads to a particularly useful computational interpretation. When viewed as a formula for computing a single value of the output sequence, Eq. (2.49) dictates that $y[n]$ (i.e., the nth value of the output) is obtained by multiplying the input sequence (expressed as a function of k) by the sequence whose values are $h[n-k]$, $-\infty < k < \infty$ for any fixed value of n, and then summing all the values of the products $x[k]h[n-k]$, with k a counting index in the summation process. Therefore, the operation of convolving two sequences involves doing the computation specified by Eq. (2.49) for each value of n, thus generating the complete output sequence $y[n]$, $-\infty < n < \infty$. The key to carrying out the computations of Eq. (2.49) to obtain $y[n]$ is understanding how to form the sequence $h[n-k]$, $-\infty < k < \infty$, for all values of n that are of interest. To this end, it is useful to note that

$$h[n-k] = h[-(k-n)]. \tag{2.53}$$

To illustrate the interpretation of Eq. (2.53), suppose $h[k]$ is the sequence shown in Figure 2.9(a) and we wish to find $h[n-k] = h[-(k-n)]$. Define $h_1[k]$ to be $h[-k]$, which is shown in Figure 2.9(b). Next, define $h_2[k]$ to be $h_1[k]$, delayed, by n samples on the k axis, i.e., $h_2[k] = h_1[k-n]$. Figure 2.9(c) shows the sequence that results from delaying the sequence in Figure 2.9(b) by n samples. Using the relationship between $h_1[k]$ and $h[k]$, we can show that $h_2[k] = h_1[k-n] = h[-(k-n)] = h[n-k]$, and thus, the bottom figure is the desired signal. To summarize, to compute $h[n-k]$ from $h[k]$, we first reverse $h[k]$ in time about $k = 0$ and then delay the time-reversed signal by n samples.

To implement discrete-time convolution, the two sequences $x[k]$ and $h[n-k]$ are multiplied together sample by sample for $-\infty < k < \infty$, and the products are summed to compute the output sample $y[n]$. To obtain another output sample, the origin of the sequence $h[-k]$ is shifted to the new sample position, and the process is repeated. This computational procedure applies whether the computations are carried out numerically on sampled data or analytically with sequences for which the sample values have simple formulas. The following example illustrates discrete-time convolution for the latter case.

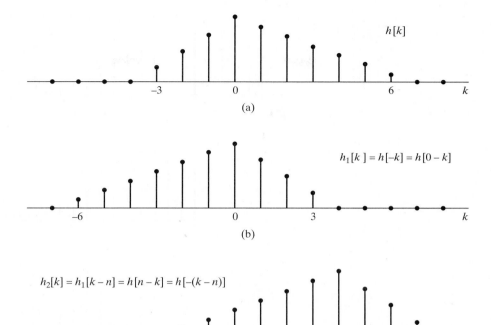

Figure 2.9   Forming the sequence $h[n - k]$. (a) The sequence $h[k]$ as a function of k. (b) The sequence $h[-k]$ as a function of k. (c) The sequence $h[n - k] = h[-(k - n)]$ as a function of k for $n = 4$.

Example 2.11   Analytical Evaluation of the Convolution Sum

Consider a system with impulse response

$$h[n] = u[n] - u[n - N]$$

$$= \begin{cases} 1, & 0 \le n \le N - 1, \\ 0, & \text{otherwise.} \end{cases}$$

The input is

$$x[n] = \begin{cases} a^n, & n \ge 0, \\ 0, & n < 0, \end{cases}$$

or equivalently,

$$x[n] = a^n u[n].$$

To find the output at a particular index n, we must form the sums over all k of the product $x[k]h[n - k]$. In this case, we can find formulas for $y[n]$ for different sets of values of n. To do this, it is helpful to sketch the sequences $x[k]$ and $h[n - k]$ as functions of k for different representative values of n. For example, Figure 2.10(a) shows the sequences $x[k]$ and $h[n - k]$, plotted for n a negative integer. Clearly, all

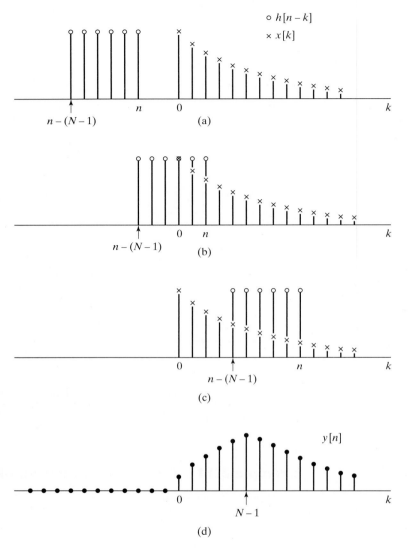

Figure 2.10 Sequence involved in computing a discrete convolution. (a)–(c) The sequences $x[k]$ and $h[n-k]$ as a function of k for different values of n. (Only nonzero samples are shown.) (d) Corresponding output sequence as a function of n.

negative values of n give a similar picture; i.e., the nonzero portions of the sequences $x[k]$ and $h[n-k]$ do not overlap, so

$$y[n] = 0, \qquad n < 0.$$

Figure 2.10(b) illustrates the two sequences when $0 \le n$ and $n - N + 1 \le 0$. These two conditions can be combined into the single condition $0 \le n \le N - 1$. By considering Figure 2.10(b), we see that since

$$x[k]h[n-k] = a^k, \qquad \text{for } 0 \le k \le n$$

when $0 \le n \le N - 1$.

it follows that

$$y[n] = \sum_{k=0}^{n} a^k, \qquad \text{for } 0 \leq n \leq N - 1. \tag{2.54}$$

The limits on the sum can be seen directly from Figure 2.10(b). Equation (2.54) shows that $y[n]$ is the sum of $n + 1$ terms of a geometric series in which the ratio of terms is a. This sum can be expressed in closed form using the general formula

$$\sum_{k=N_1}^{N_2} \alpha^k = \frac{\alpha^{N_1} - \alpha^{N_2+1}}{1 - \alpha}, \qquad N_2 \geq N_1. \tag{2.55}$$

Applying this formula to Eq. (2.54), we obtain

$$y[n] = \frac{1 - a^{n+1}}{1 - a}, \qquad 0 \leq n \leq N - 1. \tag{2.56}$$

Finally, Figure 2.10(c) shows the two sequences when $0 < n - N + 1$ or $N - 1 < n$. As before,

$$x[k]h[n - k] = a^k, \qquad n - N + 1 \leq k \leq n,$$

but now the lower limit on the sum is $n - N + 1$, as seen in Figure 2.10(c). Thus,

$$y[n] = \sum_{k=n-N+1}^{n} a^k, \qquad \text{for } N - 1 < n. \tag{2.57}$$

Using Eq. (2.55), we obtain

$$y[n] = \frac{a^{n-N+1} - a^{n+1}}{1 - a},$$

or

$$y[n] = a^{n-N+1} \left(\frac{1 - a^N}{1 - a} \right). \tag{2.58}$$

Thus, because of the piecewise-exponential nature of both the input and the unit sample response, we have been able to obtain the following closed-form expression for $y[n]$ as a function of the index n:

$$y[n] = \begin{cases} 0, & n < 0, \\ \dfrac{1 - a^{n+1}}{1 - a}, & 0 \leq n \leq N - 1, \\ a^{n-N+1} \left(\dfrac{1 - a^N}{1 - a} \right), & N - 1 < n. \end{cases} \tag{2.59}$$

This sequence is shown in Figure 2.10(d).

Example 2.11 illustrates how the convolution sum can be computed analytically when the input and the impulse response are given by simple formulas. In such cases, the sums may have a compact form that may be derived using the formula for the sum of a geometric series or other "closed-form" formulas.[2] When no simple form is available,

---

[2]Such results are discussed, for example, in Grossman (1992) and Jolley (2004).

the convolution sum can still be evaluated numerically using the technique illustrated in Example 2.11 whenever the sums are finite, which will be the case if either the input sequence or the impulse response is of finite length, i.e., has a finite number of nonzero samples.

2.4 PROPERTIES OF LINEAR TIME-INVARIANT SYSTEMS

Since all LTI systems are described by the convolution sum of Eq. (2.49), the properties of this class of systems are defined by the properties of discrete-time convolution. Therefore, the impulse response is a complete characterization of the properties of a specific LTI system.

Some general properties of the class of LTI systems can be found by considering properties of the convolution operation.[3] For example, the convolution operation is commutative:

$$x[n] * h[n] = h[n] * x[n]. \tag{2.60}$$

This can be shown by applying a substitution of variables to the summation index in Eq. (2.49). Specifically, with $m = n - k$,

$$y[n] = \sum_{m=\infty}^{-\infty} x[n-m]h[m] = \sum_{m=-\infty}^{\infty} h[m]x[n-m] = h[n] * x[n], \tag{2.61}$$

so the roles of $x[n]$ and $h[n]$ in the summation are interchanged. That is, the order of the sequences in a convolution operator is unimportant; hence, the system output is the same if the roles of the input and impulse response are reversed. Accordingly, an LTI system with input $x[n]$ and impulse response $h[n]$ will have the same output as an LTI system with input $h[n]$ and impulse response $x[n]$. The convolution operation also distributes over addition; i.e.,

$$x[n] * (h_1[n] + h_2[n]) = x[n] * h_1[n] + x[n] * h_2[n]. \tag{2.62}$$

This follows in a straightforward way from Eq. (2.49) and is a direct result of the linearity and commutativity of convolution. Equation (2.62) is represented pictorially in Figure 2.11, where Figure 2.11(a) represents the right-hand side of Eq. (2.62) and Figure 2.11(b) the left-hand side.

The convolution operation also satisfies the associative property, i.e.,

$$y[n] = (x[n] * h_1[n]) * h_2[n] = x[n] * (h_1[n] * h_2[n]). \tag{2.63}$$

Also since the convolution operation is commutative, Eq. (2.63) is equivalent to

$$y[n] = x[n] * (h_2[n] * h_1[n]) = (x[n] * h_2[n]) * h_1[n]. \tag{2.64}$$

These equivalences are represented pictorially in Figure 2.12. Also, Eqs. (2.63) and (2.64) clearly imply that if two LTI systems with impulse responses $h_1[n]$ and $h_2[n]$ are cascaded in either order, the equivalent overall impulse response $h[n]$ is

$$h[n] = h_1[n] * h_2[n] = h_2[n] * h_1[n]. \tag{2.65}$$

---

[3]In our discussion below and throughout the text, we use the shorthand notation of Eq. (2.50) for the operation of convolution, but again emphasize that the properties of convolution are derived from the definition of Eq. (2.49).

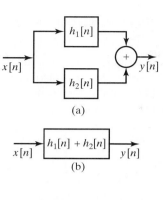

(a)

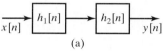

(b)

Figure 2.11   (a) Parallel combination of LTI systems. (b) An equivalent system.

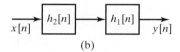

(a)

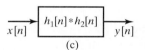

(b)

(c)

Figure 2.12   (a) Cascade combination of two LTI systems. (b) Equivalent cascade. (c) Single equivalent system.

In a parallel combination, the systems have the same input, and their outputs are summed to produce an overall output. It follows from the distributive property of convolution that the connection of two LTI systems in parallel is equivalent to a single system whose impulse response is the sum of the individual impulse responses; i.e.,

$$h[n] = h_1[n] + h_2[n]. \tag{2.66}$$

The constraints of linearity and time invariance define a class of systems with very special properties. Stability and causality represent additional properties, and it is often important to know whether an LTI system is stable and whether it is causal. Recall from Section 2.2.5 that a stable system is a system for which every bounded input produces a bounded output. LTI systems are stable if and only if the impulse response is absolutely summable, i.e., if

$$B_h = \sum_{k=-\infty}^{\infty} |h[k]| < \infty. \tag{2.67}$$

This can be shown as follows. From Eq. (2.61),

$$|y[n]| = \left| \sum_{k=-\infty}^{\infty} h[k]x[n-k] \right| \le \sum_{k=-\infty}^{\infty} |h[k]|\,|x[n-k]|. \tag{2.68}$$

If $x[n]$ is bounded, so that

$$|x[n]| \le B_x,$$

then substituting B_x for $|x[n - k]|$ can only strengthen the inequality. Hence,

$$|y[n]| \leq B_x B_h. \tag{2.69}$$

Thus, $y[n]$ is bounded if Eq. (2.67) holds; in other words, Eq. (2.67) is a sufficient condition for stability. To show that it is also a necessary condition, we must show that if $B_h = \infty$, then a bounded input can be found that will cause an unbounded output. Such an input is the sequence with values

$$x[n] = \begin{cases} \dfrac{h^*[-n]}{|h[-n]|}, & h[n] \neq 0, \\ 0, & h[n] = 0, \end{cases} \tag{2.70}$$

where $h^*[n]$ is the complex conjugate of $h[n]$. The sequence $x[n]$ is clearly bounded by unity. However, the value of the output at $n = 0$ is

$$y[0] = \sum_{k=-\infty}^{\infty} x[-k]h[k] = \sum_{k=-\infty}^{\infty} \frac{|h[k]|^2}{|h[k]|} = B_h. \tag{2.71}$$

Therefore, if $B_h = \infty$, it is possible for a bounded input sequence to produce an unbounded output sequence.

The class of causal systems was defined in Section 2.2.4 as comprising those systems for which the output $y[n_0]$ depends only on the input samples $x[n]$, for $n \leq n_0$. It follows from Eq. (2.49) or Eq. (2.61) that this definition implies the condition

$$h[n] = 0, \qquad n < 0, \tag{2.72}$$

for causality of LTI systems. (See Problem 2.69.) For this reason, it is sometimes convenient to refer to a sequence that is zero for $n < 0$ as a *causal sequence*, meaning that it could be the impulse response of a causal system.

To illustrate how the properties of LTI systems are reflected in the impulse response, let us consider again some of the systems defined in Examples 2.2–2.9. First, note that only the systems of Examples 2.2, 2.3, 2.5, and 2.9 are linear and time invariant. Although the impulse response of nonlinear or time-varying systems can be found by simply using an impulse input, it is generally of limited interest, since the convolution-sum formula and Eqs. (2.67) and (2.72), expressing stability and causality, do not apply to such systems.

First, let us determine the impulse responses of the systems in Examples 2.2, 2.3, 2.5, and 2.9. We can do this by simply computing the response of each system to $\delta[n]$, using the defining relationship for the system. The resulting impulse responses are as follows:

Ideal Delay (Example 2.2)

$$h[n] = \delta[n - n_d], \qquad n_d \text{ a positive fixed integer.} \tag{2.73}$$

Moving Average (Example 2.3)

$$h[n] = \frac{1}{M_1 + M_2 + 1} \sum_{k=-M_1}^{M_2} \delta[n - k]$$

$$= \begin{cases} \dfrac{1}{M_1 + M_2 + 1}, & -M_1 \leq n \leq M_2, \\ 0, & \text{otherwise.} \end{cases} \tag{2.74}$$

Accumulator (Example 2.5)

$$h[n] = \sum_{k=-\infty}^{n} \delta[k] = \begin{cases} 1, & n \geq 0, \\ 0, & n < 0, \end{cases} = u[n]. \tag{2.75}$$

Forward Difference (Example 2.9)

$$h[n] = \delta[n+1] - \delta[n]. \tag{2.76}$$

Backward Difference (Example 2.9)

$$h[n] = \delta[n] - \delta[n-1]. \tag{2.77}$$

Given the impulse responses of these basic systems [Eqs. (2.73)–(2.77)], we can test the stability of each one by computing the sum

$$B_h = \sum_{n=-\infty}^{\infty} |h[n]|.$$

For the ideal delay, moving-average, forward difference, and backward difference examples, it is clear that $B_h < \infty$, since the impulse response has only a finite number of nonzero samples. In general, a system with a finite-duration impulse response (henceforth referred to as an FIR system) will always be stable, as long as each of the impulse response values is finite in magnitude. The accumulator, however, is unstable because

$$B_h = \sum_{n=0}^{\infty} u[n] = \infty.$$

In Section 2.2.5, we also demonstrated the instability of the accumulator by giving an example of a bounded input (the unit step) for which the output is unbounded.

The impulse response of the accumulator has infinite duration. This is an example of the class of systems referred to as *infinite-duration impulse response* (IIR) systems. An example of an IIR system that is stable is a system whose impulse response is $h[n] = a^n u[n]$ with $|a| < 1$. In this case,

$$B_h = \sum_{n=0}^{\infty} |a|^n. \tag{2.78}$$

If $|a| < 1$, the formula for the sum of the terms of an infinite geometric series gives

$$B_h = \frac{1}{1 - |a|} < \infty. \tag{2.79}$$

If, on the other hand, $|a| \geq 1$, then the sum is infinite and the system is unstable.

To test causality of the LTI systems in Examples 2.2, 2.3, 2.5, and 2.9, we can check to see whether $h[n] = 0$ for $n < 0$. As discussed in Section 2.2.4, the ideal delay [$n_d \geq 0$ in Eq. (2.20)] is causal. If $n_d < 0$, then the system is noncausal. For the moving average, causality requires that $-M_1 \geq 0$ and $M_2 \geq 0$. The accumulator and backward difference systems are causal, and the forward difference system is noncausal.

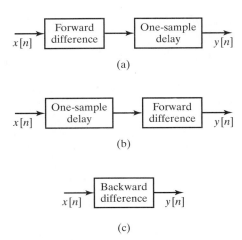

Figure 2.13 Equivalent systems found by using the commutative property of convolution.

The concept of convolution as an operation between two sequences leads to the simplification of many problems involving systems. A particularly useful result can be stated for the ideal delay system. Since the output of the delay system is $y[n] = x[n - n_d]$, and since the delay system has impulse response $h[n] = \delta[n - n_d]$, it follows that

$$x[n] * \delta[n - n_d] = \delta[n - n_d] * x[n] = x[n - n_d]. \tag{2.80}$$

That is, the convolution of a shifted impulse sequence with any signal $x[n]$ is easily evaluated by simply shifting $x[n]$ by the displacement of the impulse.

Since delay is a fundamental operation in the implementation of linear systems, the preceding result is often useful in the analysis and simplification of interconnections of LTI systems. As an example, consider the system of Figure 2.13(a), which consists of a forward difference system cascaded with an ideal delay of one sample. According to the commutative property of convolution, the order in which systems are cascaded does not matter, as long as they are linear and time invariant. Therefore, we obtain the same result when we compute the forward difference of a sequence and delay the result (Figure 2.13a) as when we delay the sequence first and then compute the forward difference (Figure 2.13b). Also, as indicated in Eq. (2.65) and in Figure 2.12, the overall impulse response of each cascade system is the convolution of the individual impulse responses. Consequently,

$$\begin{aligned} h[n] &= (\delta[n + 1] - \delta[n]) * \delta[n - 1] \\ &= \delta[n - 1] * (\delta[n + 1] - \delta[n]) \\ &= \delta[n] - \delta[n - 1]. \end{aligned} \tag{2.81}$$

Thus, $h[n]$ is identical to the impulse response of the backward difference system; that is, the cascaded systems of Figures 2.13(a) and 2.13(b) can be replaced by a backward difference system, as shown in Figure 2.13(c).

Note that the noncausal forward difference systems in Figures 2.13(a) and (b) have been converted to causal systems by cascading them with a delay. In general, any noncausal FIR system can be made causal by cascading it with a sufficiently long delay.

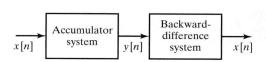

Figure 2.14 An accumulator in cascade with a backward difference. Since the backward difference is the inverse system for the accumulator, the cascade combination is equivalent to the identity system.

Another example of cascaded systems introduces the concept of an *inverse system.* Consider the cascade of systems in Figure 2.14. The impulse response of the cascade system is

$$
\begin{aligned}
h[n] &= u[n] * (\delta[n] - \delta[n-1]) \\
&= u[n] - u[n-1] \\
&= \delta[n].
\end{aligned}
\tag{2.82}
$$

That is, the cascade combination of an accumulator followed by a backward difference (or vice versa) yields a system whose overall impulse response is the impulse. Thus, the output of the cascade combination will always be equal to the input, since $x[n] * \delta[n] = x[n]$. In this case, the backward difference system compensates exactly for (or inverts) the effect of the accumulator; that is, the backward difference system is the *inverse system* for the accumulator. From the commutative property of convolution, the accumulator is likewise the inverse system for the backward difference system. Note that this example provides a system interpretation of Eqs. (2.7) and (2.9). In general, if an LTI system has impulse response $h[n]$, then its inverse system, if it exists, has impulse response $h_i[n]$ defined by the relation

$$
h[n] * h_i[n] = h_i[n] * h[n] = \delta[n].
\tag{2.83}
$$

Inverse systems are useful in many situations where it is necessary to compensate for the effects of a system. In general, it is difficult to solve Eq. (2.83) directly for $h_i[n]$, given $h[n]$. However, in Chapter 3, we will see that the z-transform provides a straightforward method of finding the inverse of an LTI system.

2.5 LINEAR CONSTANT-COEFFICIENT DIFFERENCE EQUATIONS

An important class of LTI systems consists of those systems for which the input $x[n]$ and the output $y[n]$ satisfy an Nth-order linear constant-coefficient difference equation of the form

$$
\sum_{k=0}^{N} a_k y[n-k] = \sum_{m=0}^{M} b_m x[n-m].
\tag{2.84}
$$

The properties discussed in Section 2.4 and some of the analysis techniques introduced there can be used to find difference equation representations for some of the LTI systems that we have defined.

Example 2.12   Difference Equation Representation of the Accumulator

The accumulator system is defined by

$$y[n] = \sum_{k=-\infty}^{n} x[k]. \tag{2.85}$$

To show that the input and output satisfy a difference equation of the form of Eq. (2.84), we rewrite Eq. (2.85) as

$$y[n] = x[n] + \sum_{k=-\infty}^{n-1} x[k] \tag{2.86}$$

Also, from Eq. (2.85)

$$y[n-1] = \sum_{k=-\infty}^{n-1} x[k]. \tag{2.87}$$

Substituting Eq. (2.87) into Eq. (2.86) yields

$$y[n] = x[n] + y[n-1], \tag{2.88}$$

and equivalently,

$$y[n] - y[n-1] = x[n]. \tag{2.89}$$

Thus, in addition to satisfying the defining relationship of Eq. (2.85), the input and output of an accumulator satisfy a linear constant-coefficient difference equation of the form Eq. (2.84), with $N = 1$, $a_0 = 1$, $a_1 = -1$, $M = 0$, and $b_0 = 1$.

The difference equation in the form of Eq. (2.88) suggests a simple implementation of the accumulator system. According to Eq. (2.88), for each value of n, we add the current input value $x[n]$ to the previously accumulated sum $y[n-1]$. This interpretation of the accumulator is represented in block diagram form in Figure 2.15.

Equation (2.88) and the block diagram in Figure 2.15 are referred to as a recursive representation of the system, since each value is computed using previously computed values. This general notion will be explored in more detail later in this section.

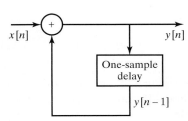

Figure 2.15   Block diagram of a recursive difference equation representing an accumulator.

Example 2.13   Difference Equation Representation of the Moving-Average System

Consider the moving-average system of Example 2.3, with $M_1 = 0$ so that the system is causal. In this case, from Eq. (2.74), the impulse response is

$$h[n] = \frac{1}{(M_2 + 1)}\left(u[n] - u[n - M_2 - 1]\right), \tag{2.90}$$

from which it follows that

$$y[n] = \frac{1}{(M_2 + 1)}\sum_{k=0}^{M_2} x[n - k], \tag{2.91}$$

which is a special case of Eq. (2.84), with $N = 0, a_0 = 1, M = M_2$, and $b_k = 1/(M_2+1)$ for $0 \le k \le M_2$.

Also, the impulse response can be expressed as

$$h[n] = \frac{1}{(M_2 + 1)}\left(\delta[n] - \delta[n - M_2 - 1]\right) * u[n], \tag{2.92}$$

which suggests that the causal moving-average system can be represented as the cascade system of Figure 2.16. We can obtain a difference equation for this block diagram by noting first that

$$x_1[n] = \frac{1}{(M_2 + 1)}\left(x[n] - x[n - M_2 - 1]\right). \tag{2.93}$$

From Eq. (2.89) of Example 2.12, the output of the accumulator satisfies the difference equation

$$y[n] - y[n - 1] = x_1[n],$$

so that

$$y[n] - y[n - 1] = \frac{1}{(M_2 + 1)}(x[n] - x[n - M_2 - 1]). \tag{2.94}$$

Again, we have a difference equation in the form of Eq. (2.84), but this time $N = 1$, $a_0 = 1, a_1 = -1, M = M_2 + 1$ and $b_0 = -b_{M_2+1} = 1/(M_2 + 1)$, and $b_k = 0$ otherwise.

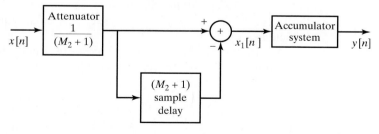

Figure 2.16   Block diagram of the recursive form of a moving-average system.

In Example 2.13, we showed two different difference-equation representations of the moving-average system. In Chapter 6, we will see that many distinct difference equations can be used to represent a given LTI input–output relation.

Just as in the case of linear constant-coefficient differential equations for contin-uous-time systems, without additional constraints or other information, a linear constant-coefficient difference equation for discrete-time systems does not provide a unique specification of the output for a given input. Specifically, suppose that, for a given input $x_p[n]$, we have determined by some means one output sequence $y_p[n]$, so that an equa-tion of the form of Eq. (2.84) is satisfied. Then, the same equation with the same input is satisfied by any output of the form

$$y[n] = y_p[n] + y_h[n], \tag{2.95}$$

where $y_h[n]$ is any solution to Eq. (2.84) with $x[n] = 0$, i.e., a solution to the equation

$$\sum_{k=0}^{N} a_k y_h[n-k] = 0. \tag{2.96}$$

Equation (2.96) is called the *homogeneous difference equation* and $y_h[n]$ the homoge-neous solution. The sequence $y_h[n]$ is in fact a member of a family of solutions of the form

$$y_h[n] = \sum_{m=1}^{N} A_m z_m^n, \tag{2.97}$$

where the coefficients A_m can be chosen to satisfy a set of auxiliary conditions on $y[n]$. Substituting Eq. (2.97) into Eq. (2.96) shows that the complex numbers z_m must be roots of the polynomial

$$A(z) = \sum_{k=0}^{N} a_k z^{-k}. \tag{2.98}$$

i.e., $A(z_m) = 0$ for $m = 1, 2, \ldots, N$. Equation (2.97) assumes that all N roots of the polynomial in Eq. (2.98) are distinct. The form of terms associated with multiple roots is slightly different, but there are always N undetermined coefficients. An example of the homogeneous solution with multiple roots is considered in Problem 2.50.

Since $y_h[n]$ has N undetermined coefficients, a set of N auxiliary conditions is required for the unique specification of $y[n]$ for a given $x[n]$. These auxiliary conditions might consist of specifying fixed values of $y[n]$ at specific values of n, such as $y[-1]$, $y[-2], \ldots, y[-N]$, and then solving a set of N linear equations for the N undetermined coefficients.

Alternatively, if the auxiliary conditions are a set of auxiliary values of $y[n]$, the other values of $y[n]$ can be generated by rewriting Eq. (2.84) as a recurrence formula, i.e., in the form

$$y[n] = -\sum_{k=1}^{N} \frac{a_k}{a_0} y[n-k] + \sum_{k=0}^{M} \frac{b_k}{a_0} x[n-k]. \tag{2.99}$$

If the input $x[n]$ for all n, together with a set of auxiliary values, say, $y[-1], y[-2], \ldots,$ $y[-N]$, is specified, then $y[0]$ can be determined from Eq. (2.99). With $y[0], y[-1], \ldots,$ $y[-N+1]$ now available, $y[1]$ can then be calculated, and so on. When this procedure is used, $y[n]$ is said to be computed *recursively;* i.e., the output computation involves not only the input sequence, but also previous values of the output sequence.

To generate values of $y[n]$ for $n < -N$ (again assuming that the values $y[-1]$, $y[-2], \ldots, y[-N]$ are given as auxiliary conditions), we can rearrange Eq. (2.84) in the form

$$y[n - N] = -\sum_{k=0}^{N-1} \frac{a_k}{a_N} y[n - k] + \sum_{k=0}^{M} \frac{b_k}{a_N} x[n - k], \tag{2.100}$$

from which $y[-N - 1]$, $y[-N - 2]$, $\ldots$ can be computed recursively in the backward direction.

Our principal interest in this text is in systems that are linear and time invariant, in which case the auxiliary conditions must be consistent with these additional requirements. In Chapter 3, when we discuss the solution of difference equations using the z-transform, we implicitly incorporate conditions of linearity and time invariance. As we will see in that discussion, even with the additional constraints of linearity and time invariance, the solution to the difference equation, and therefore the system, is not uniquely specified. In particular, there are, in general, both causal and noncausal LTI systems consistent with a given difference equation.

If a system is characterized by a linear constant-coefficient difference equation and is further specified to be linear, time invariant, and causal, then the solution is unique. In this case, the auxiliary conditions are often stated as *initial-rest conditions*. In other words, the auxiliary information is that if the input $x[n]$ is zero for n less than some time n_0, then the output $y[n]$ is constrained to be zero for n less than n_0. This then provides sufficient initial conditions to obtain $y[n]$ for $n \geq n_0$ recursively using Eq. (2.99).

To summarize, for a system for which the input and output satisfy a linear constant-coefficient difference equation:

- The output for a given input is not uniquely specified. Auxiliary information or conditions are required.
- If the auxiliary information is in the form of N sequential values of the output, later values can be obtained by rearranging the difference equation as a recursive relation running forward in n, and prior values can be obtained by rearranging the difference equation as a recursive relation running backward in n.
- Linearity, time invariance, and causality of the system will depend on the auxiliary conditions. If an additional condition is that the system is initially at rest, then the system will be linear, time invariant, and causal.

The preceding discussion assumed that $N \geq 1$ in Eq. (2.84). If, instead, $N = 0$, no recursion is required to use the difference equation to compute the output, and therefore, no auxiliary conditions are required. That is,

$$y[n] = \sum_{k=0}^{M} \left(\frac{b_k}{a_0} \right) x[n - k]. \tag{2.101}$$

Equation (2.101) is in the form of a convolution, and by setting $x[n] = \delta[n]$, we see that the corresponding impulse response is

$$h[n] = \sum_{k=0}^{M} \left(\frac{b_k}{a_0} \right) \delta[n - k],$$

or

$$h[n] = \begin{cases} \left(\dfrac{b_n}{a_0}\right), & 0 \le n \le M, \\ 0, & \text{otherwise.} \end{cases} \tag{2.102}$$

The impulse response is obviously finite in duration. Indeed, the output of any FIR system can be computed nonrecursively where the coefficients are the values of the impulse response sequence. The moving-average system of Example 2.13 with $M_1 = 0$ is an example of a causal FIR system. An interesting feature of that system was that we also found a recursive equation for the output. In Chapter 6 we will show that there are many possible ways of implementing a desired signal transformation. Advantages of one method over another depend on practical considerations, such as numerical accuracy, data storage, and the number of multiplications and additions required to compute each sample of the output.

2.6 FREQUENCY-DOMAIN REPRESENTATION OF DISCRETE-TIME SIGNALS AND SYSTEMS

In the previous sections, we summarized some of the fundamental concepts of the theory of discrete-time signals and systems. For LTI systems, we saw that a representation of the input sequence as a weighted sum of delayed impulses leads to a representation of the output as a weighted sum of delayed impulse responses. As with continuous-time signals, discrete-time signals may be represented in a number of different ways. For example, sinusoidal and complex exponential sequences play a particularly important role in representing discrete-time signals. This is because complex exponential sequences are eigenfunctions of LTI systems, and the response to a sinusoidal input is sinusoidal with the same frequency as the input and with amplitude and phase determined by the system. These fundamental properties of LTI systems make representations of signals in terms of sinusoids or complex exponentials (i.e., Fourier representations) very useful in linear system theory.

2.6.1 Eigenfunctions for Linear Time-Invariant Systems

The eigenfunction property of complex exponentials for discrete-time systems follows directly from substitution into Eq. (2.61). Specifically, with input $x[n] = e^{j\omega n}$ for $-\infty < n < \infty$, the corresponding output of an LTI system with impulse response $h[n]$ is easily shown to be

$$y[n] = H(e^{j\omega})e^{j\omega n}, \tag{2.103}$$

where

$$H(e^{j\omega}) = \sum_{k=-\infty}^{\infty} h[k]e^{-j\omega k}. \tag{2.104}$$

Consequently, $e^{j\omega n}$ is an eigenfunction of the system, and the associated eigenvalue is $H(e^{j\omega})$. From Eq. (2.103), we see that $H(e^{j\omega})$ describes the change in complex amplitude of a complex exponential input signal as a function of the frequency ω. The eigenvalue $H(e^{j\omega})$ is the *frequency response* of the system. In general, $H(e^{j\omega})$ is complex and can be expressed in terms of its real and imaginary parts as

$$H(e^{j\omega}) = H_R(e^{j\omega}) + jH_I(e^{j\omega}) \qquad (2.105)$$

or in terms of magnitude and phase as

$$H(e^{j\omega}) = |H(e^{j\omega})|e^{j\angle H(e^{j\omega})}. \qquad (2.106)$$

Example 2.14   Frequency Response of the Ideal Delay System

As a simple and important example, consider the ideal delay system defined by

$$y[n] = x[n - n_d], \qquad (2.107)$$

where n_d is a fixed integer. With input $x[n] = e^{j\omega n}$ from Eq. (2.107), we have

$$y[n] = e^{j\omega(n-n_d)} = e^{-j\omega n_d}e^{j\omega n}.$$

The frequency response of the ideal delay is therefore

$$H(e^{j\omega}) = e^{-j\omega n_d}. \qquad (2.108)$$

As an alternative method of obtaining the frequency response, recall that the impulse response for the ideal delay system is $h[n] = \delta[n - n_d]$. Using Eq. (2.104), we obtain

$$H(e^{j\omega}) = \sum_{n=-\infty}^{\infty} \delta[n - n_d]e^{-j\omega n} = e^{-j\omega n_d}.$$

The real and imaginary parts of the frequency response are

$$H_R(e^{j\omega}) = \cos(\omega n_d), \qquad (2.109a)$$

$$H_I(e^{j\omega}) = -\sin(\omega n_d). \qquad (2.109b)$$

The magnitude and phase are

$$|H(e^{j\omega})| = 1, \qquad (2.110a)$$

$$\angle H(e^{j\omega}) = -\omega n_d. \qquad (2.110b)$$

In Section 2.7, we will show that a broad class of signals can be represented as a linear combination of complex exponentials in the form

$$x[n] = \sum_k \alpha_k e^{j\omega_k n}. \qquad (2.111)$$

From the principle of superposition and Eq. (2.103), the corresponding output of an LTI system is

$$y[n] = \sum_k \alpha_k H(e^{j\omega_k})e^{j\omega_k n}. \qquad (2.112)$$

Thus, if we can find a representation of $x[n]$ as a superposition of complex exponential sequences, as in Eq. (2.111), we can then find the output using Eq. (2.112) if we know the frequency response of the system at all frequencies ω_k. The following simple example illustrates this fundamental property of LTI systems.

Example 2.15   Sinusoidal Response of LTI Systems

Let us consider a sinusoidal input

$$x[n] = A\cos(\omega_0 n + \phi) = \frac{A}{2}e^{j\phi}e^{j\omega_0 n} + \frac{A}{2}e^{-j\phi}e^{-j\omega_0 n}. \tag{2.113}$$

From Eq. (2.103), the response to $x_1[n] = (A/2)e^{j\phi}e^{j\omega_0 n}$ is

$$y_1[n] = H(e^{j\omega_0})\frac{A}{2}e^{j\phi}e^{j\omega_0 n}. \tag{2.114a}$$

The response to $x_2[n] = (A/2)e^{-j\phi}e^{-j\omega_0 n}$ is

$$y_2[n] = H(e^{-j\omega_0})\frac{A}{2}e^{-j\phi}e^{-j\omega_0 n}. \tag{2.114b}$$

Thus, the total response is

$$y[n] = \frac{A}{2}[H(e^{j\omega_0})e^{j\phi}e^{j\omega_0 n} + H(e^{-j\omega_0})e^{-j\phi}e^{-j\omega_0 n}]. \tag{2.115}$$

If $h[n]$ is real, it can be shown (see Problem 2.78) that $H(e^{-j\omega_0}) = H^*(e^{j\omega_0})$. Consequently,

$$y[n] = A\,|H(e^{j\omega_0})|\cos(\omega_0 n + \phi + \theta), \tag{2.116}$$

where $\theta = \angle H(e^{j\omega_0})$ is the phase of the system function at frequency ω_0.

For the simple example of the ideal delay, $|H(e^{j\omega_0})| = 1$ and $\theta = -\omega_0 n_d$, as we determined in Example 2.14. Therefore,

$$\begin{aligned} y[n] &= A\cos(\omega_0 n + \phi - \omega_0 n_d) \\ &= A\cos[\omega_0(n - n_d) + \phi], \end{aligned} \tag{2.117}$$

which is identical to what we would obtain directly using the definition of the ideal delay system.

The concept of the frequency response of LTI systems is essentially the same for continuous-time and discrete-time systems. However, an important distinction arises because the frequency response of discrete-time LTI systems is always a periodic function of the frequency variable ω with period 2π. To show this, we substitute $\omega + 2\pi$ into Eq. (2.104) to obtain

$$H(e^{j(\omega+2\pi)}) = \sum_{n=-\infty}^{\infty} h[n]e^{-j(\omega+2\pi)n}. \tag{2.118}$$

Using the fact that $e^{\pm j2\pi n} = 1$ for n an integer, we have

$$e^{-j(\omega+2\pi)n} = e^{-j\omega n}e^{-j2\pi n} = e^{-j\omega n}.$$

Therefore,

$$H(e^{j(\omega+2\pi)}) = H(e^{j\omega}), \qquad \text{for all } \omega, \tag{2.119}$$

and, more generally,

$$H(e^{j(\omega+2\pi r)}) = H(e^{j\omega}), \qquad \text{for } r \text{ an integer.} \tag{2.120}$$

That is, $H(e^{j\omega})$ is periodic with period 2π. Note that this is obviously true for the ideal delay system, since $e^{-j(\omega+2\pi)n_d} = e^{-j\omega n_d}$ when n_d is an integer.

The reason for this periodicity is related directly to our earlier observation that the sequence

$$\{e^{j\omega n}\}, \qquad -\infty < n < \infty,$$

is indistinguishable from the sequence

$$\{e^{j(\omega+2\pi)n}\}, \qquad -\infty < n < \infty.$$

Because these two sequences have identical values for all n, the system must respond identically to both input sequences. This condition requires that Eq. (2.119) hold.

Since $H(e^{j\omega})$ is periodic with period 2π, and since the frequencies ω and $\omega+2\pi$ are indistinguishable, it follows that we need only specify $H(e^{j\omega})$ over an interval of length 2π, e.g., $0 \leq \omega \leq 2\pi$ or $-\pi < \omega \leq \pi$. The inherent periodicity defines the frequency response everywhere outside the chosen interval. For simplicity and for consistency with the continuous-time case, it is generally convenient to specify $H(e^{j\omega})$ over the interval $-\pi < \omega \leq \pi$. With respect to this interval, the "low frequencies" are frequencies close to zero, whereas the "high frequencies" are frequencies close to $\pm\pi$. Recalling that frequencies differing by an integer multiple of 2π are indistinguishable, we might generalize the preceding statement as follows: The "low frequencies" are those that are close to an even multiple of π, while the "high frequencies" are those that are close to an odd multiple of π, consistent with our earlier discussion in Section 2.1.

An important class of LTI systems includes those systems for which the frequency response is unity over a certain range of frequencies and is zero at the remaining frequencies, corresponding to ideal frequency-selective filters. The frequency response of an ideal lowpass filter is shown in Figure 2.17(a). Because of the inherent periodicity of the discrete-time frequency response, it has the appearance of a multiband filter, since frequencies around $\omega = 2\pi$ are indistinguishable from frequencies around $\omega = 0$. In effect, however, the frequency response passes only low frequencies and rejects high frequencies. Since the frequency response is completely specified by its behavior over the interval $-\pi < \omega \leq \pi$, the ideal lowpass filter frequency response is more typically shown only in the interval $-\pi < \omega \leq \pi$, as in Figure 2.17(b). It is understood that the frequency response repeats periodically with period 2π outside the plotted interval. With this implicit assumption, the frequency responses for ideal highpass, bandstop, and bandpass filters are as shown in Figures 2.18(a), (b), and (c), respectively.

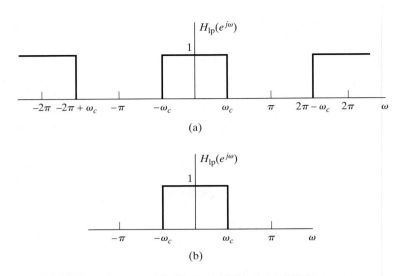

(a)

(b)

Figure 2.17 Ideal lowpass filter showing (a) periodicity of the frequency response and (b) one period of the periodic frequency response.

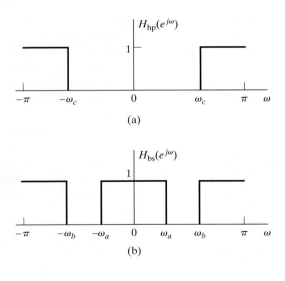

(a)

(b)

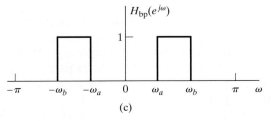

(c)

Figure 2.18 Ideal frequency-selective filters. (a) Highpass filter. (b) Bandstop filter. (c) Bandpass filter. In each case, the frequency response is periodic with period 2π. Only one period is shown.

Example 2.16   Frequency Response of the Moving-Average System

The impulse response of the moving-average system of Example 2.3 is

$$h[n] = \begin{cases} \dfrac{1}{M_1 + M_2 + 1}, & -M_1 \le n \le M_2, \\ 0, & \text{otherwise.} \end{cases}$$

Therefore, the frequency response is

$$H(e^{j\omega}) = \frac{1}{M_1 + M_2 + 1} \sum_{n=-M_1}^{M_2} e^{-j\omega n}. \tag{2.121}$$

For the causal moving average system, $M_1 = 0$ and Eq. (2.121) can be expressed as

$$H(e^{j\omega}) = \frac{1}{M_2 + 1} \sum_{n=0}^{M_2} e^{-j\omega n}. \tag{2.122}$$

Using Eq. (2.55), Eq. (2.122) becomes

$$H(e^{j\omega}) = \frac{1}{M_2 + 1} \left(\frac{1 - e^{-j\omega(M_2+1)}}{1 - e^{-j\omega}} \right)$$

$$= \frac{1}{M_2 + 1} \frac{(e^{j\omega(M_2+1)/2} - e^{-j\omega(M_2+1)/2})e^{-j\omega(M_2+1)/2}}{(e^{j\omega/2} - e^{-j\omega/2})e^{-j\omega/2}}$$

$$= \frac{1}{M_2 + 1} \frac{\sin[\omega(M_2 + 1)/2]}{\sin \omega/2} e^{-j\omega M_2/2}. \tag{2.123}$$

The magnitude and phase of $H(e^{j\omega})$ for this case, with $M_2 = 4$, are shown in Figure 2.19.

If the moving-average filter is symmetric, i.e., if $M_1 = M_2$, then Eq. (2.123) is replaced by

$$H(e^{j\omega}) = \frac{1}{2M_2 + 1} \frac{\sin[\omega(2M_2 + 1)/2]}{\sin(\omega/2)}. \tag{2.124}$$

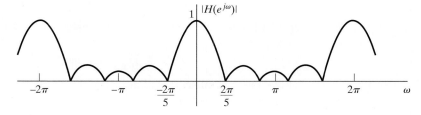

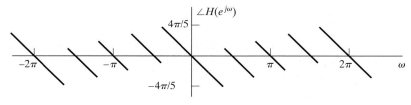

Figure 2.19   (a) Magnitude and (b) phase of the frequency response of the moving-average system for the case $M_1 = 0$ and $M_2 = 4$.

Note that in both cases $H(e^{j\omega})$ is periodic, as is required of the frequency response of a discrete-time system. Note also that $|H(e^{j\omega})|$ falls off at "high frequencies" and $\angle H(e^{j\omega})$, i.e., the phase of $H(e^{j\omega})$, varies linearly with ω. This attenuation of the high frequencies suggests that the system will smooth out rapid variations in the input sequence; in other words, the system is a rough approximation to a lowpass filter. This is consistent with what we would intuitively expect about the behavior of the moving-average system.

2.6.2 Suddenly Applied Complex Exponential Inputs

We have seen that complex exponential inputs of the form $e^{j\omega n}$ for $-\infty < n < \infty$ produce outputs of the form $H(e^{j\omega})e^{j\omega n}$ for LTI systems. Models of this kind are important in the mathematical representation of a wide range of signals, even those that exist only over a finite domain. We can also gain additional insight into LTI systems by considering inputs of the form

$$x[n] = e^{j\omega n}u[n],\qquad(2.125)$$

i.e., complex exponentials that are suddenly applied at an arbitrary time, which for convenience here we choose as $n = 0$. Using the convolution sum in Eq. (2.61), the corresponding output of a causal LTI system with impulse response $h[n]$ is

$$y[n] = \begin{cases} 0, & n < 0, \\ \left(\displaystyle\sum_{k=0}^{n} h[k]e^{-j\omega k}\right)e^{j\omega n}, & n \geq 0. \end{cases}$$

If we consider the output for $n \geq 0$, we can write

$$y[n] = \left(\sum_{k=0}^{\infty} h[k]e^{-j\omega k}\right)e^{j\omega n} - \left(\sum_{k=n+1}^{\infty} h[k]e^{-j\omega k}\right)e^{j\omega n}\qquad(2.126)$$

$$= H(e^{j\omega})e^{j\omega n} - \left(\sum_{k=n+1}^{\infty} h[k]e^{-j\omega k}\right)e^{j\omega n}.\qquad(2.127)$$

From Eq. (2.127), we see that the output consists of the sum of two terms, i.e., $y[n] = y_{\text{ss}}[n] + y_t[n]$. The first term,

$$y_{\text{ss}}[n] = H(e^{j\omega})e^{j\omega n},$$

is the steady-state response. It is identical to the response of the system when the input is $e^{j\omega n}$ for all n. In a sense, the second term,

$$y_t[n] = -\sum_{k=n+1}^{\infty} h[k]e^{-j\omega k}e^{j\omega n},$$

is the amount by which the output differs from the eigenfunction result. This part corresponds to the transient response, because it is clear that in some cases it may approach zero. To see the conditions for which this is true, let us consider the size of the second term. Its magnitude is bounded as follows:

$$|y_t[n]| = \left| \sum_{k=n+1}^{\infty} h[k]e^{-j\omega k}e^{j\omega n} \right| \leq \sum_{k=n+1}^{\infty} |h[k]|.\qquad(2.128)$$

From Eq. (2.128), it should be clear that if the impulse response has finite length, so that $h[n] = 0$ except for $0 \leq n \leq M$, then the term $y_t[n] = 0$ for $n + 1 > M$, or $n > M - 1$. In this case,

$$y[n] = y_{ss}[n] = H(e^{j\omega})e^{j\omega n}, \qquad \text{for } n > M - 1.$$

When the impulse response has infinite duration, the transient response does not disappear abruptly, but if the samples of the impulse response approach zero with increasing n, then $y_t[n]$ will approach zero. Note that Eq. (2.128) can be written

$$|y_t[n]| = \left| \sum_{k=n+1}^{\infty} h[k]e^{-j\omega k}e^{j\omega n} \right| \leq \sum_{k=n+1}^{\infty} |h[k]| \leq \sum_{k=0}^{\infty} |h[k]|. \tag{2.129}$$

That is, the transient response is bounded by the sum of the absolute values of *all* of the impulse response samples. If the right-hand side of Eq. (2.129) is bounded, i.e., if

$$\sum_{k=0}^{\infty} |h[k]| < \infty,$$

then the system is stable. From Eq. (2.129), it follows that, for stable systems, the transient response must become increasingly smaller as $n \to \infty$. Thus, a sufficient condition for the transient response to decay asymptotically is that the system be stable.

Figure 2.20 shows the real part of a complex exponential signal with frequency $\omega = 2\pi/10$. The solid dots indicate the samples $x[k]$ of the suddenly applied complex

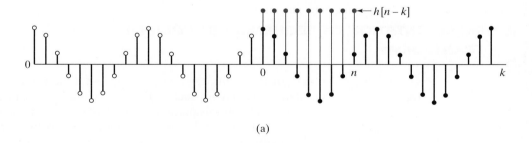

(a)

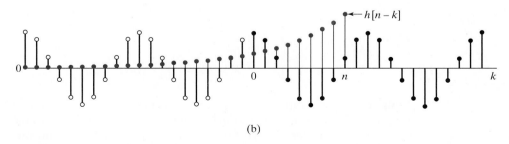

(b)

Figure 2.20   Illustration of a real part of suddenly applied complex exponential input with (a) FIR and (b) IIR.

exponential, while the open circles indicate the samples of the complex exponential that are "missing," i.e., that would be nonzero if the input were of the form $e^{j\omega n}$ for all n. The shaded dots indicate the samples of the impulse response $h[n-k]$ as a function of k for $n=8$. In the finite-length case shown in Figure 2.20(a), it is clear that the output would consist only of the steady-state component for $n \geq 8$, whereas in the infinite-length case, it is clear that the "missing" samples have less and less effect as n increases, owing to the decaying nature of the impulse response.

The condition for stability is also a sufficient condition for the existence of the frequency response function. To see this, note that, in general,

$$|H(e^{j\omega})| = \left| \sum_{k=-\infty}^{\infty} h[k]e^{-j\omega k} \right| \leq \sum_{k=-\infty}^{\infty} |h[k]e^{-j\omega k}| \leq \sum_{k=-\infty}^{\infty} |h[k]|,$$

so the general condition

$$\sum_{k=-\infty}^{\infty} |h[k]| < \infty$$

ensures that $H(e^{j\omega})$ exists. It is no surprise that the condition for existence of the frequency response is the same as the condition for dominance of the steady-state solution. Indeed, a complex exponential that exists for all n can be thought of as one that is applied at $n = -\infty$. The eigenfunction property of complex exponentials depends on stability of the system, since at finite n, the transient response must have become zero, so that we only see the steady-state response $H(e^{j\omega})e^{j\omega n}$ for all finite n.

2.7 REPRESENTATION OF SEQUENCES BY FOURIER TRANSFORMS

One of the advantages of the frequency-response representation of an LTI system is that interpretations of system behavior such as the one we made in Example 2.16 often follow easily. We will elaborate on this point in considerably more detail in Chapter 5. At this point, however, let us return to the question of how we may find representations of the form of Eq. (2.111) for an arbitrary input sequence.

Many sequences can be represented by a Fourier integral of the form

$$x[n] = \frac{1}{2\pi} \int_{-\pi}^{\pi} X(e^{j\omega})e^{j\omega n} d\omega, \tag{2.130}$$

where

$$X(e^{j\omega}) = \sum_{n=-\infty}^{\infty} x[n]e^{-j\omega n}. \tag{2.131}$$

Equations (2.130) and (2.131) together form a Fourier representation for the sequence. Equation (2.130), the *inverse Fourier transform,* is a synthesis formula. That is, it represents $x[n]$ as a superposition of infinitesimally small complex sinusoids of the form

$$\frac{1}{2\pi} X(e^{j\omega})e^{j\omega n} d\omega,$$

with ω ranging over an interval of length 2π and with $X(e^{j\omega})$ determining the relative amount of each complex sinusoidal component. Although, in writing Eq. (2.130), we have chosen the range of values for ω between $-\pi$ and $+\pi$, any interval of length 2π can be used. Equation (2.131), the *Fourier transform*,[4] is an expression for computing $X(e^{j\omega})$ from the sequence $x[n]$, i.e., for analyzing the sequence $x[n]$ to determine how much of each frequency component is required to synthesize $x[n]$ using Eq. (2.130).

In general, the Fourier transform is a complex-valued function of ω. As with the frequency response, we may either express $X(e^{j\omega})$ in rectangular form as

$$X(e^{j\omega}) = X_R(e^{j\omega}) + jX_I(e^{j\omega}) \tag{2.132a}$$

or in polar form as

$$X(e^{j\omega}) = |X(e^{j\omega})|e^{j\angle X(e^{j\omega})}. \tag{2.132b}$$

With $|X(e^{j\omega})|$ representing the magnitude and $\angle X(e^{j\omega})$ the phase.

The phase $\angle X(e^{j\omega})$ is not uniquely specified by Eq. (2.132b), since any integer multiple of 2π may be added to $\angle X(e^{j\omega})$ at any value of ω without affecting the result of the complex exponentiation. When we specifically want to refer to the principal value, i.e., $\angle X(e^{j\omega})$ restricted to the range of values between $-\pi$ and $+\pi$, we denote this as $\mathrm{ARG}[X(e^{j\omega})]$. If we want to refer to a phase function that is a continuous function of ω for $0 < \omega < \pi$, i.e., not evaluated modulo 2π, we use the notation $\arg[X(e^{j\omega})]$.

As is clear from comparing Eqs. (2.104) and (2.131), the frequency response of an LTI system is the Fourier transform of the impulse response. The impulse response can be obtained from the frequency response by applying the inverse Fourier transform integral; i.e.,

$$h[n] = \frac{1}{2\pi} \int_{-\pi}^{\pi} H(e^{j\omega})e^{j\omega n} d\omega. \tag{2.133}$$

As discussed previously, the frequency response is a periodic function of ω. Likewise, the Fourier transform is periodic in ω with period 2π. A Fourier series is commonly used to represent periodic signals, and it is worth noting that indeed, Eq. (2.131) is of the form of a Fourier series for the periodic function $X(e^{j\omega})$. Eq. (2.130), which expresses the sequence values $x[n]$ in terms of the periodic function $X(e^{j\omega})$, is of the form of the integral that would be used to obtain the coefficients in the Fourier series. Our use of Eqs. (2.130) and (2.131) focuses on the representation of the sequence $x[n]$. Nevertheless, it is useful to be aware of the equivalence between the Fourier series representation of continuous-variable periodic functions and the Fourier transform representation of discrete-time signals, since all the familiar properties of Fourier series can be applied, with appropriate interpretation of variables, to the Fourier transform representation of a sequence. (Oppenheim and Willsky (1997), McClellan, Schafer and Yoder (2003).)

Determining the class of signals that can be represented by Eq. (2.130) is equivalent to considering the convergence of the infinite sum in Eq. (2.131). That is, we are concerned with the conditions that must be satisfied by the terms in the sum in Eq. (2.131) such that

$$|X(e^{j\omega})| < \infty \qquad \text{for all } \omega,$$

---

[4]Eq. (2.131) is sometimes more explicitly referred to as the discrete-time Fourier transform, or DTFT, particularly when it is important to distinguish it from the continuous-time Fourier transform.

where $X(e^{j\omega})$ is the limit as $M \to \infty$ of the finite sum

$$X_M(e^{j\omega}) = \sum_{n=-M}^{M} x[n]e^{-j\omega n}. \qquad (2.134)$$

A sufficient condition for convergence can be found as follows:

$$|X(e^{j\omega})| = \left| \sum_{n=-\infty}^{\infty} x[n]e^{-j\omega n} \right|$$

$$\leq \sum_{n=-\infty}^{\infty} |x[n]| \, |e^{-j\omega n}|$$

$$\leq \sum_{n=-\infty}^{\infty} |x[n]| < \infty.$$

Thus, if $x[n]$ is *absolutely summable,* then $X(e^{j\omega})$ exists. Furthermore, in this case, the series can be shown to converge uniformly to a continuous function of ω (Körner (1988), Kammler (2000)). Since a stable sequence is, by definition, absolutely summable, all stable sequences have Fourier transforms. It also follows, then, that any stable *system,* i.e., one having an absolutely summable impulse response, will have a finite and continuous frequency response.

Absolute summability is a sufficient condition for the existence of a Fourier transform representation. In Examples 2.14 and 2.16, we computed the Fourier transforms of the impulse response of the delay system and the moving average system. The impulse responses are absolutely summable, since they are finite in length. Clearly, any finite-length sequence is absolutely summable and thus will have a Fourier transform representation. In the context of LTI systems, any FIR system will be stable and therefore will have a finite, continuous frequency response. However, when a sequence has infinite length, we must be concerned about convergence of the infinite sum. The following example illustrates this case.

Example 2.17   Absolute Summability for a Suddenly-Applied Exponential

Consider $x[n] = a^n u[n]$. The Fourier transform of this sequence is

$$X(e^{j\omega}) = \sum_{n=0}^{\infty} a^n e^{-j\omega n} = \sum_{n=0}^{\infty} (ae^{-j\omega})^n$$

$$= \frac{1}{1 - ae^{-j\omega}} \qquad \text{if } |ae^{-j\omega}| < 1 \quad \text{or} \quad |a| < 1.$$

Clearly, the condition $|a| < 1$ is the condition for the absolute summability of $x[n]$; i.e.,

$$\sum_{n=0}^{\infty} |a|^n = \frac{1}{1 - |a|} < \infty \qquad \text{if } |a| < 1. \qquad (2.135)$$

Absolute summability is a *sufficient* condition for the existence of a Fourier transform representation, and it also guarantees uniform convergence. Some sequences are not absolutely summable, but are square summable, i.e.,

$$\sum_{n=-\infty}^{\infty} |x[n]|^2 < \infty. \tag{2.136}$$

Such sequences can be represented by a Fourier transform if we are willing to relax the condition of uniform convergence of the infinite sum defining $X(e^{j\omega})$. Specifically, in this case, we have mean-square convergence; that is, with

$$X(e^{j\omega}) = \sum_{n=-\infty}^{\infty} x[n]e^{-j\omega n} \tag{2.137a}$$

and

$$X_M(e^{j\omega}) = \sum_{n=-M}^{M} x[n]e^{-j\omega n}, \tag{2.137b}$$

it follows that

$$\lim_{M \to \infty} \int_{-\pi}^{\pi} |X(e^{j\omega}) - X_M(e^{j\omega})|^2 d\omega = 0. \tag{2.138}$$

In other words, the error $|X(e^{j\omega}) - X_M(e^{j\omega})|$ may not approach zero at each value of ω as $M \to \infty$, but the total "energy" in the error does. Example 2.18 illustrates this case.

Example 2.18   Square-Summability for the Ideal Lowpass Filter

In this example we determine the impulse response of the ideal lowpass filter discussed in Section 2.6. The frequency response is

$$H_{\text{lp}}(e^{j\omega}) = \begin{cases} 1, & |\omega| < \omega_c, \\ 0, & \omega_c < |\omega| \le \pi, \end{cases} \tag{2.139}$$

with periodicity 2π also understood. The impulse response $h_{\text{lp}}[n]$ can be found using the Fourier transform synthesis equation (2.130):

$$\begin{aligned}
h_{\text{lp}}[n] &= \frac{1}{2\pi} \int_{-\omega_c}^{\omega_c} e^{j\omega n} d\omega \\
&= \frac{1}{2\pi jn} \left[e^{j\omega n} \right]_{-\omega_c}^{\omega_c} = \frac{1}{2\pi jn} (e^{j\omega_c n} - e^{-j\omega_c n}) \\
&= \frac{\sin \omega_c n}{\pi n}, \qquad -\infty < n < \infty.
\end{aligned} \tag{2.140}$$

We note that, since $h_{\text{lp}}[n]$ is nonzero for $n < 0$, the ideal lowpass filter is noncausal. Also, $h_{\text{lp}}[n]$ is not absolutely summable. The sequence values approach zero as $n \to \infty$, but only as $1/n$. This is because $H_{\text{lp}}(e^{j\omega})$ is discontinuous at $\omega = \omega_c$. Since $h_{\text{lp}}[n]$ is not absolutely summable, the infinite sum

$$\sum_{n=-\infty}^{\infty} \frac{\sin \omega_c n}{\pi n} e^{-j\omega n}$$

does not converge uniformly for all values of ω. To obtain an intuitive feeling for this, let us consider $H_M(e^{j\omega})$ as the sum of a finite number of terms:

$$H_M(e^{j\omega}) = \sum_{n=-M}^{M} \frac{\sin \omega_c n}{\pi n} e^{-j\omega n}. \tag{2.141}$$

The function $H_M(e^{j\omega})$ is evaluated in Figure 2.21 for several values of M. Note that as M increases, the oscillatory behavior at $\omega = \omega_c$ (often referred to as the Gibbs phenomenon) is more rapid, but the size of the ripples does not decrease. In fact, it can be shown that as $M \to \infty$, the maximum amplitude of the oscillations does not approach zero, but the oscillations converge in location toward the points $\omega = \pm\omega_c$. Thus, the infinite sum does not converge uniformly to the discontinuous function $H_{lp}(e^{j\omega})$ of Eq. (2.139). However, $h_{lp}[n]$, as given in Eq. (2.140), is square summable, and correspondingly, $H_M(e^{j\omega})$ converges in the mean-square sense to $H_{lp}(e^{j\omega})$; i.e.,

$$\lim_{M \to \infty} \int_{-\pi}^{\pi} |H_{lp}(e^{j\omega}) - H_M(e^{j\omega})|^2 d\omega = 0.$$

Although the error between $H_M(e^{j\omega})$ and $H_{lp}(e^{j\omega})$ as $M \to \infty$ might seem unimportant because the two functions differ only at $\omega = \omega_c$, we will see in Chapter 7 that the behavior of finite sums such as Eq. (2.141) has important implications in the design of discrete-time systems for filtering.

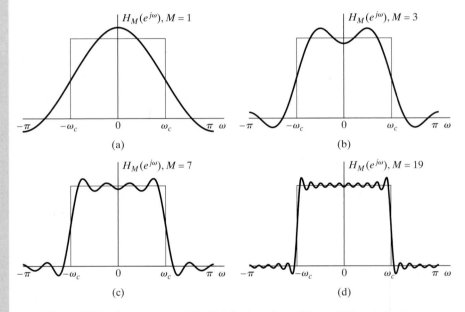

Figure 2.21 Convergence of the Fourier transform. The oscillatory behavior at $\omega = \omega_c$ is often called the Gibbs phenomenon.

It is sometimes useful to have a Fourier transform representation for certain sequences that are neither absolutely summable nor square summable. We illustrate several of these in the following examples.

Example 2.19   Fourier Transform of a Constant

Consider the sequence $x[n] = 1$ for all n. This sequence is neither absolutely summable nor square summable, and Eq. (2.131) does not converge in either the uniform or mean-square sense for this case. However, it is possible and useful to define the Fourier transform of the sequence $x[n]$ to be the periodic impulse train

$$X(e^{j\omega}) = \sum_{r=-\infty}^{\infty} 2\pi\delta(\omega + 2\pi r). \tag{2.142}$$

The impulses in this case are functions of a continuous variable and therefore are of "infinite height, zero width, and unit area," consistent with the fact that Eq. (2.131) does not converge in any regular sense. (See Oppenheim and Willsky (1997) for a discussion of the definition and properties of the impulse function.) The use of Eq. (2.142) as a Fourier representation of the sequence $x[n] = 1$ is justified principally because formal substitution of Eq. (2.142) into Eq. (2.130) leads to the correct result. Example 2.20 represents a generalization of this example.

Example 2.20   Fourier Transform of Complex Exponential Sequences

Consider a sequence $x[n]$ whose Fourier transform is the periodic impulse train

$$X(e^{j\omega}) = \sum_{r=-\infty}^{\infty} 2\pi\delta(\omega - \omega_0 + 2\pi r). \tag{2.143}$$

We show in this example that $x[n]$ is the complex exponential sequence $e^{j\omega_0 n}$, with $-\pi < \omega_0 \leq \pi$.

We can determine $x[n]$ by substituting $X(e^{j\omega})$ into the inverse Fourier transform integral of Eq. (2.130). Because the integration of $X(e^{j\omega})$ extends only over one period, from $-\pi < \omega < \pi$, we need include only the $r = 0$ term from Eq. (2.143). Consequently, we can write

$$x[n] = \frac{1}{2\pi} \int_{-\pi}^{\pi} 2\pi\delta(\omega - \omega_0)e^{j\omega n}d\omega. \tag{2.144}$$

From the definition of the impulse function, it follows that

$$x[n] = e^{j\omega_0 n} \qquad \text{for any } n.$$

For $\omega_0 = 0$, this reduces to the sequence considered in Example 2.19.

Clearly, $x[n]$ in Example 2.20 is not absolutely summable, nor is it square summable, and $|X(e^{j\omega})|$ is not finite for all ω. Thus, the mathematical statement

$$\sum_{n=-\infty}^{\infty} e^{j\omega_0 n}e^{-j\omega n} = \sum_{r=-\infty}^{\infty} 2\pi\delta(\omega - \omega_0 + 2\pi r) \tag{2.145}$$

must be interpreted in the context of generalized functions (Lighthill, 1958). Using that theory, the concept of a Fourier transform representation can be extended to the class of sequences that can be expressed as a sum of discrete frequency components, such as

$$x[n] = \sum_{k} a_k e^{j\omega_k n}, \qquad -\infty < n < \infty. \tag{2.146}$$

From the result of Example 2.20, it follows that

$$X(e^{j\omega}) = \sum_{r=-\infty}^{\infty} \sum_{k} 2\pi a_k \delta(\omega - \omega_k + 2\pi r) \tag{2.147}$$

is a consistent Fourier transform representation of $x[n]$ in Eq. (2.146).

Another sequence that is neither absolutely summable nor square summable is the unit step sequence $u[n]$. Although it is not completely straightforward to show, this sequence can be represented by the following Fourier transform:

$$U(e^{j\omega}) = \frac{1}{1 - e^{-j\omega}} + \sum_{r=-\infty}^{\infty} \pi \delta(\omega + 2\pi r). \tag{2.148}$$

2.8 SYMMETRY PROPERTIES OF THE FOURIER TRANSFORM

In using Fourier transforms, it is useful to have a detailed knowledge of the way that properties of the sequence manifest themselves in the Fourier transform and vice versa. In this section and Section 2.9, we discuss and summarize a number of such properties.

Symmetry properties of the Fourier transform are often very useful for simplifying the solution of problems. The following discussion presents these properties. The proofs are considered in Problems 2.79 and 2.80. Before presenting the properties, however, we begin with some definitions.

A *conjugate-symmetric sequence* $x_e[n]$ is defined as a sequence for which $x_e[n] = x_e^*[-n]$, and a *conjugate-antisymmetric sequence* $x_o[n]$ is defined as a sequence for which $x_o[n] = -x_o^*[-n]$, where * denotes complex conjugation. Any sequence $x[n]$ can be expressed as a sum of a conjugate-symmetric and conjugate-antisymmetric sequence. Specifically,

$$x[n] = x_e[n] + x_o[n], \tag{2.149a}$$

where

$$x_e[n] = \tfrac{1}{2}(x[n] + x^*[-n]) = x_e^*[-n] \tag{2.149b}$$

and

$$x_o[n] = \tfrac{1}{2}(x[n] - x^*[-n]) = -x_o^*[-n]. \tag{2.149c}$$

Adding Eqs. (2.149b) and (2.149c) confirms that Eq. (2.149a) holds. A real sequence that is conjugate symmetric such that $x_e[n] = x_e[-n]$ is referred to as an *even sequence*, and a real sequence that is conjugate antisymmetric such that $x_o[n] = -x_o[-n]$ is referred to as an *odd sequence*.

A Fourier transform $X(e^{j\omega})$ can be decomposed into a sum of conjugate-symmetric and conjugate-antisymmetric functions as

$$X(e^{j\omega}) = X_e(e^{j\omega}) + X_o(e^{j\omega}), \tag{2.150a}$$

where

$$X_e(e^{j\omega}) = \tfrac{1}{2}[X(e^{j\omega}) + X^*(e^{-j\omega})] \tag{2.150b}$$

and

$$X_o(e^{j\omega}) = \tfrac{1}{2}[X(e^{j\omega}) - X^*(e^{-j\omega})]. \tag{2.150c}$$

By substituting $-\omega$ for ω in Eqs. (2.150b) and (2.150c), it follows that $X_e(e^{j\omega})$ is conjugate symmetric and $X_o(e^{j\omega})$ is conjugate antisymmetric; i.e.,

$$X_e(e^{j\omega}) = X_e^*(e^{-j\omega}) \tag{2.151a}$$

and

$$X_o(e^{j\omega}) = -X_o^*(e^{-j\omega}). \tag{2.151b}$$

If a real function of a continuous variable is conjugate symmetric, it is referred to as an *even function*, and a real conjugate-antisymmetric function of a continuous variable is referred to as an *odd function*.

The symmetry properties of the Fourier transform are summarized in Table 2.1. The first six properties apply for a general complex sequence $x[n]$ with Fourier transform $X(e^{j\omega})$. Properties 1 and 2 are considered in Problem 2.79. Property 3 follows from properties 1 and 2, together with the fact that the Fourier transform of the sum of two sequences is the sum of their Fourier transforms. Specifically, the Fourier transform of $\mathcal{R}e\{x[n]\} = \frac{1}{2}(x[n]+x^*[n])$ is the conjugate-symmetric part of $X(e^{j\omega})$, or $X_e(e^{j\omega})$. Similarly, $j\mathcal{I}m\{x[n]\} = \frac{1}{2}(x[n] - x^*[n])$, or equivalently, $j\mathcal{I}m\{x[n]\}$ has a Fourier transform that is the conjugate-antisymmetric component $X_o(e^{j\omega})$ corresponding to property 4. By considering the Fourier transform of $x_e[n]$ and $x_o[n]$, the conjugate-symmetric and conjugate-antisymmetric components, respectively, of $x[n]$, it can be shown that properties 5 and 6 follow.

If $x[n]$ is a real sequence, these symmetry properties become particularly straightforward and useful. Specifically, for a real sequence, the Fourier transform is conjugate symmetric; i.e., $X(e^{j\omega}) = X^*(e^{-j\omega})$ (property 7). Expressing $X(e^{j\omega})$ in terms of its real and imaginary parts as

$$X(e^{j\omega}) = X_R(e^{j\omega}) + jX_I(e^{j\omega}), \tag{2.152}$$

TABLE 2.1   SYMMETRY PROPERTIES OF THE FOURIER TRANSFORM

Sequence $x[n]$	Fourier Transform $X(e^{j\omega})$
1. $x^*[n]$	$X^*(e^{-j\omega})$
2. $x^*[-n]$	$X^*(e^{j\omega})$
3. $\mathcal{R}e\{x[n]\}$	$X_e(e^{j\omega})$ (conjugate-symmetric part of $X(e^{j\omega})$)
4. $j\mathcal{I}m\{x[n]\}$	$X_o(e^{j\omega})$ (conjugate-antisymmetric part of $X(e^{j\omega})$)
5. $x_e[n]$ (conjugate-symmetric part of $x[n]$)	$X_R(e^{j\omega}) = \mathcal{R}e\{X(e^{j\omega})\}$
6. $x_o[n]$ (conjugate-antisymmetric part of $x[n]$)	$jX_I(e^{j\omega}) = j\mathcal{I}m\{X(e^{j\omega})\}$
The following properties apply only when $x[n]$ is real:	
7. Any real $x[n]$	$X(e^{j\omega}) = X^*(e^{-j\omega})$ (Fourier transform is conjugate symmetric)
8. Any real $x[n]$	$X_R(e^{j\omega}) = X_R(e^{-j\omega})$ (real part is even)
9. Any real $x[n]$	$X_I(e^{j\omega}) = -X_I(e^{-j\omega})$ (imaginary part is odd)
10. Any real $x[n]$	$|X(e^{j\omega})| = |X(e^{-j\omega})|$ (magnitude is even)
11. Any real $x[n]$	$\angle X(e^{j\omega}) = -\angle X(e^{-j\omega})$ (phase is odd)
12. $x_e[n]$ (even part of $x[n]$)	$X_R(e^{j\omega})$
13. $x_o[n]$ (odd part of $x[n]$)	$jX_I(e^{j\omega})$

we can derive properties 8 and 9—specifically,

$$X_R(e^{j\omega}) = X_R(e^{-j\omega}) \tag{2.153a}$$

and

$$X_I(e^{j\omega}) = -X_I(e^{-j\omega}). \tag{2.153b}$$

In other words, the real part of the Fourier transform is an even function, and the imaginary part is an odd function, if the sequence is real. In a similar manner, by expressing $X(e^{j\omega})$ in polar form as

$$X(e^{j\omega}) = |X(e^{j\omega})|e^{j\angle X(e^{j\omega})}, \tag{2.154}$$

we can show that, for a real sequence $x[n]$, the magnitude of the Fourier transform, $|X(e^{j\omega})|$, is an even function of ω and the phase, $\angle X(e^{j\omega})$, can be chosen to be an odd function of ω (properties 10 and 11). Also, for a real sequence, the even part of $x[n]$ transforms to $X_R(e^{j\omega})$, and the odd part of $x[n]$ transforms to $jX_I(e^{j\omega})$ (properties 12 and 13).

Example 2.21   Illustration of Symmetry Properties

Let us return to the sequence of Example 2.17, where we showed that the Fourier transform of the real sequence $x[n] = a^n u[n]$ is

$$X(e^{j\omega}) = \frac{1}{1 - ae^{-j\omega}} \qquad \text{if } |a| < 1. \tag{2.155}$$

Then, from the properties of complex numbers, it follows that

$$X(e^{j\omega}) = \frac{1}{1 - ae^{-j\omega}} = X^*(e^{-j\omega}) \qquad \text{(property 7)},$$

$$X_R(e^{j\omega}) = \frac{1 - a\cos\omega}{1 + a^2 - 2a\cos\omega} = X_R(e^{-j\omega}) \qquad \text{(property 8)},$$

$$X_I(e^{j\omega}) = \frac{-a\sin\omega}{1 + a^2 - 2a\cos\omega} = -X_I(e^{-j\omega}) \qquad \text{(property 9)},$$

$$|X(e^{j\omega})| = \frac{1}{(1 + a^2 - 2a\cos\omega)^{1/2}} = |X(e^{-j\omega})| \qquad \text{(property 10)},$$

$$\angle X(e^{j\omega}) = \tan^{-1}\left(\frac{-a\sin\omega}{1 - a\cos\omega}\right) = -\angle X(e^{-j\omega}) \qquad \text{(property 11)}.$$

These functions are plotted in Figure 2.22 for $a > 0$, specifically, $a = 0.75$ (solid curve) and $a = 0.5$ (dashed curve). In Problem 2.32, we consider the corresponding plots for $a < 0$.

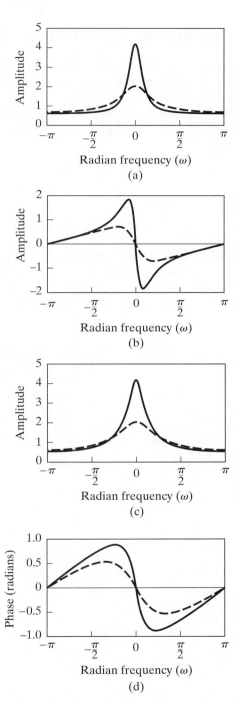

Figure 2.22 Frequency response for a system with impulse response $h[n] = a^n u[n]$. (a) Real part. $a > 0$; $a = 0.75$ (solid curve) and $a = 0.5$ (dashed curve). (b) Imaginary part. (c) Magnitude. $a > 0$; $a = 0.75$ (solid curve) and $a = 0.5$ (dashed curve). (d) Phase.

2.9 FOURIER TRANSFORM THEOREMS

In addition to the symmetry properties, a variety of theorems (presented in Sections 2.9.1–2.9.7) relate operations on the sequence to operations on the Fourier transform. We will see that these theorems are quite similar in most cases to corresponding theorems for continuous-time signals and their Fourier transforms. To facilitate the statement of the theorems, we introduce the following operator notation:

$$X(e^{j\omega}) = \mathcal{F}\{x[n]\},$$

$$x[n] = \mathcal{F}^{-1}\{X(e^{j\omega})\},$$

$$x[n] \xleftrightarrow{\mathcal{F}} X(e^{j\omega}).$$

That is, $\mathcal{F}$ denotes the operation of "taking the Fourier transform of $x[n]$," and $\mathcal{F}^{-1}$ is the inverse of that operation. Most of the theorems will be stated without proof. The proofs, which are left as exercises (Problem 2.81), generally involve only simple manipulations of variables of summation or integration. The theorems in this section are summarized in Table 2.2.

TABLE 2.2   FOURIER TRANSFORM THEOREMS

Sequence	Fourier Transform
$x[n]$	$X(e^{j\omega})$
$y[n]$	$Y(e^{j\omega})$
1. $ax[n] + by[n]$	$aX(e^{j\omega}) + bY(e^{j\omega})$
2. $x[n - n_d]$ (n_d an integer)	$e^{-j\omega n_d} X(e^{j\omega})$
3. $e^{j\omega_0 n} x[n]$	$X(e^{j(\omega - \omega_0)})$
4. $x[-n]$	$X(e^{-j\omega})$ $X^*(e^{j\omega})$ if $x[n]$ real.
5. $nx[n]$	$j\dfrac{dX(e^{j\omega})}{d\omega}$
6. $x[n] * y[n]$	$X(e^{j\omega})Y(e^{j\omega})$
7. $x[n]y[n]$	$\dfrac{1}{2\pi}\displaystyle\int_{-\pi}^{\pi} X(e^{j\theta})Y(e^{j(\omega-\theta)})d\theta$

Parseval's theorem:

8. $\displaystyle\sum_{n=-\infty}^{\infty} |x[n]|^2 = \frac{1}{2\pi}\int_{-\pi}^{\pi} |X(e^{j\omega})|^2 d\omega$

9. $\displaystyle\sum_{n=-\infty}^{\infty} x[n]y^*[n] = \frac{1}{2\pi}\int_{-\pi}^{\pi} X(e^{j\omega})Y^*(e^{j\omega})d\omega$

2.9.1 Linearity of the Fourier Transform

If

$$x_1[n] \xleftrightarrow{\mathcal{F}} X_1(e^{j\omega})$$

and

$$x_2[n] \xleftrightarrow{\mathcal{F}} X_2(e^{j\omega}),$$

then it follows by substitution into the definition of the DTFT that

$$ax_1[n] + bx_2[n] \xleftrightarrow{\mathcal{F}} aX_1(e^{j\omega}) + bX_2(e^{j\omega}). \tag{2.156}$$

2.9.2 Time Shifting and Frequency Shifting Theorem

If

$$x[n] \xleftrightarrow{\mathcal{F}} X(e^{j\omega}),$$

then, for the time-shifted sequence $x[n - n_d]$, a simple transformation of the index of summation in the DTFT yields

$$x[n - n_d] \xleftrightarrow{\mathcal{F}} e^{-j\omega n_d} X(e^{j\omega}). \tag{2.157}$$

Direct substitution proves the following result for the frequency-shifted Fourier transform:

$$e^{j\omega_0 n} x[n] \xleftrightarrow{\mathcal{F}} X(e^{j(\omega-\omega_0)}). \tag{2.158}$$

2.9.3 Time Reversal Theorem

If

$$x[n] \xleftrightarrow{\mathcal{F}} X(e^{j\omega}),$$

then if the sequence is time reversed,

$$x[-n] \xleftrightarrow{\mathcal{F}} X(e^{-j\omega}). \tag{2.159}$$

If $x[n]$ is real, this theorem becomes

$$x[-n] \xleftrightarrow{\mathcal{F}} X^*(e^{j\omega}). \tag{2.160}$$

2.9.4 Differentiation in Frequency Theorem

If

$$x[n] \xleftrightarrow{\mathcal{F}} X(e^{j\omega}),$$

then, by differentiating the DTFT, it is seen that

$$nx[n] \xleftrightarrow{\mathcal{F}} j\frac{dX(e^{j\omega})}{d\omega}. \tag{2.161}$$

2.9.5 Parseval's Theorem

If

$$x[n] \xleftrightarrow{\mathcal{F}} X(e^{j\omega}),$$

then

$$E = \sum_{n=-\infty}^{\infty} |x[n]|^2 = \frac{1}{2\pi} \int_{-\pi}^{\pi} |X(e^{j\omega})|^2 d\omega. \tag{2.162}$$

The function $|X(e^{j\omega})|^2$ is called the *energy density spectrum,* since it determines how the energy is distributed in the frequency domain. Necessarily, the energy density spectrum is defined only for finite-energy signals. A more general form of Parseval's theorem is shown in Problem 2.84.

2.9.6 The Convolution Theorem

If

$$x[n] \xleftrightarrow{\mathcal{F}} X(e^{j\omega})$$

and

$$h[n] \xleftrightarrow{\mathcal{F}} H(e^{j\omega}),$$

and if

$$y[n] = \sum_{k=-\infty}^{\infty} x[k]h[n-k] = x[n] * h[n], \tag{2.163}$$

then

$$Y(e^{j\omega}) = X(e^{j\omega})H(e^{j\omega}). \tag{2.164}$$

Thus, convolution of sequences implies multiplication of the corresponding Fourier transforms. Note that the time-shifting property is a special case of the convolution property, since

$$\delta[n - n_d] \xleftrightarrow{\mathcal{F}} e^{-j\omega n_d} \tag{2.165}$$

and if $h[n] = \delta[n - n_d]$, then $y[n] = x[n] * \delta[n - n_d] = x[n - n_d]$. Therefore,

$$H(e^{j\omega}) = e^{-j\omega n_d} \qquad \text{and} \qquad Y(e^{j\omega}) = e^{-j\omega n_d} X(e^{j\omega}).$$

A formal derivation of the convolution theorem is easily achieved by applying the definition of the Fourier transform to $y[n]$ as expressed in Eq. (2.163). This theorem can also be interpreted as a direct consequence of the eigenfunction property of complex exponentials for LTI systems. Recall that $H(e^{j\omega})$ is the frequency response of the LTI system whose impulse response is $h[n]$. Also, if

$$x[n] = e^{j\omega n},$$

then

$$y[n] = H(e^{j\omega})e^{j\omega n}.$$

That is, complex exponentials are *eigenfunctions* of LTI systems, where $H(e^{j\omega})$, the Fourier transform of $h[n]$, is the eigenvalue. From the definition of integration, the Fourier transform synthesis equation corresponds to the representation of a sequence $x[n]$ as a superposition of complex exponentials of infinitesimal size; that is,

$$x[n] = \frac{1}{2\pi} \int_{-\pi}^{\pi} X(e^{j\omega})e^{j\omega n} d\omega = \lim_{\Delta\omega \to 0} \frac{1}{2\pi} \sum_{k} X(e^{jk\Delta\omega})e^{jk\Delta\omega n} \Delta\omega.$$

By the eigenfunction property of linear systems and by the principle of superposition, the corresponding output will be

$$y[n] = \lim_{\Delta\omega \to 0} \frac{1}{2\pi} \sum_{k} H(e^{jk\Delta\omega})X(e^{jk\Delta\omega})e^{jk\Delta\omega n} \Delta\omega = \frac{1}{2\pi} \int_{-\pi}^{\pi} H(e^{j\omega})X(e^{j\omega})e^{j\omega n} d\omega.$$

Thus, we conclude that

$$Y(e^{j\omega}) = H(e^{j\omega})X(e^{j\omega}),$$

as in Eq. (2.164).

2.9.7 The Modulation or Windowing Theorem

If

$$x[n] \overset{\mathcal{F}}{\longleftrightarrow} X(e^{j\omega})$$

and

$$w[n] \overset{\mathcal{F}}{\longleftrightarrow} W(e^{j\omega}),$$

and if

$$y[n] = x[n]w[n], \tag{2.166}$$

then

$$Y(e^{j\omega}) = \frac{1}{2\pi} \int_{-\pi}^{\pi} X(e^{j\theta})W(e^{j(\omega-\theta)})d\theta. \tag{2.167}$$

Equation (2.167) is a periodic convolution, i.e., a convolution of two periodic functions with the limits of integration extending over only one period. The duality inherent in most Fourier transform theorems is evident when we compare the convolution and modulation theorems. However, in contrast to the continuous-time case, where this duality is complete, in the discrete-time case fundamental differences arise because the Fourier transform is a sum, whereas the inverse transform is an integral with a periodic integrand. Although for continuous time, we can state that convolution in the time domain is represented by multiplication in the frequency domain and vice versa; in discrete time, this statement must be modified somewhat. Specifically, discrete-time convolution of sequences (the convolution sum) is equivalent to multiplication of corresponding periodic Fourier transforms, and multiplication of sequences is equivalent to *periodic* convolution of corresponding Fourier transforms.

　　The theorems of this section and a number of fundamental Fourier transform pairs are summarized in Tables 2.2 and 2.3, respectively. One of the ways that knowledge of Fourier transform theorems and properties is useful is in determining Fourier transforms

TABLE 2.3   FOURIER TRANSFORM PAIRS

Sequence	Fourier Transform
1. $\delta[n]$	1
2. $\delta[n - n_0]$	$e^{-j\omega n_0}$
3. $1 \qquad (-\infty < n < \infty)$	$\displaystyle\sum_{k=-\infty}^{\infty} 2\pi\delta(\omega + 2\pi k)$
4. $a^n u[n] \quad (\lvert a \rvert < 1)$	$\dfrac{1}{1 - ae^{-j\omega}}$
5. $u[n]$	$\dfrac{1}{1 - e^{-j\omega}} + \displaystyle\sum_{k=-\infty}^{\infty} \pi\delta(\omega + 2\pi k)$
6. $(n + 1)a^n u[n] \quad (\lvert a \rvert < 1)$	$\dfrac{1}{(1 - ae^{-j\omega})^2}$
7. $\dfrac{r^n \sin\omega_p(n + 1)}{\sin\omega_p} u[n] \quad (\lvert r \rvert < 1)$	$\dfrac{1}{1 - 2r\cos\omega_p e^{-j\omega} + r^2 e^{-j2\omega}}$
8. $\dfrac{\sin\omega_c n}{\pi n}$	$X(e^{j\omega}) = \begin{cases} 1, & \lvert\omega\rvert < \omega_c, \\ 0, & \omega_c < \lvert\omega\rvert \le \pi \end{cases}$
9. $x[n] = \begin{cases} 1, & 0 \le n \le M \\ 0, & \text{otherwise} \end{cases}$	$\dfrac{\sin[\omega(M + 1)/2]}{\sin(\omega/2)} e^{-j\omega M/2}$
10. $e^{j\omega_0 n}$	$\displaystyle\sum_{k=-\infty}^{\infty} 2\pi\delta(\omega - \omega_0 + 2\pi k)$
11. $\cos(\omega_0 n + \phi)$	$\displaystyle\sum_{k=-\infty}^{\infty} [\pi e^{j\phi}\delta(\omega - \omega_0 + 2\pi k) + \pi e^{-j\phi}\delta(\omega + \omega_0 + 2\pi k)]$

or inverse transforms. Often, by using the theorems and known transform pairs, it is possible to represent a sequence in terms of operations on other sequences for which the transform is known, thereby simplifying an otherwise difficult or tedious problem. Examples 2.22–2.25 illustrate this approach.

Example 2.22   Determining a Fourier Transform Using Tables 2.2 and 2.3

Suppose we wish to find the Fourier transform of the sequence $x[n] = a^n u[n - 5]$. This transform can be computed by exploiting Theorems 1 and 2 of Table 2.2 and transform pair 4 of Table 2.3. Let $x_1[n] = a^n u[n]$. We start with this signal because it is the most similar signal to $x[n]$ in Table 2.3. The table states that

$$X_1(e^{j\omega}) = \frac{1}{1 - ae^{-j\omega}}. \tag{2.168}$$

To obtain $x[n]$ from $x_1[n]$, we first delay $x_1[n]$ by five samples, i.e., $x_2[n] = x_1[n - 5]$. Theorem 2 of Table 2.2 gives the corresponding frequency-domain relationship, $X_2(e^{j\omega}) = e^{-j5\omega}X_1(e^{j\omega})$, so

$$X_2(e^{j\omega}) = \frac{e^{-j5\omega}}{1 - ae^{-j\omega}}. \tag{2.169}$$

To get from $x_2[n]$ to the desired $x[n]$, we need only multiply by the constant a^5, i.e., $x[n] = a^5 x_2[n]$. The linearity property of the Fourier transform, Theorem 1 of Table 2.2, then yields the desired Fourier transform,

$$X(e^{j\omega}) = \frac{a^5 e^{-j5\omega}}{1 - ae^{-j\omega}}. \tag{2.170}$$

Example 2.23   Determining an Inverse Fourier Transform Using Tables 2.2 and 2.3

Suppose that

$$X(e^{j\omega}) = \frac{1}{(1 - ae^{-j\omega})(1 - be^{-j\omega})}. \tag{2.171}$$

Direct substitution of $X(e^{j\omega})$ into Eq. (2.130) leads to an integral that is difficult to evaluate by ordinary real integration techniques. However, using the technique of partial fraction expansion, which we discuss in detail in Chapter 3, we can expand $X(e^{j\omega})$ into the form

$$X(e^{j\omega}) = \frac{a/(a-b)}{1 - ae^{-j\omega}} - \frac{b/(a-b)}{1 - be^{-j\omega}}. \tag{2.172}$$

From Theorem 1 of Table 2.2 and transform pair 4 of Table 2.3, it follows that

$$x[n] = \left(\frac{a}{a-b}\right)a^n u[n] - \left(\frac{b}{a-b}\right)b^n u[n]. \tag{2.173}$$

Example 2.24   Determining the Impulse Response from the Frequency Response

The frequency response of a highpass filter with linear phase is

$$H(e^{j\omega}) = \begin{cases} e^{-j\omega n_d}, & \omega_c < |\omega| < \pi, \\ 0, & |\omega| < \omega_c, \end{cases} \tag{2.174}$$

where a period of 2π is understood. This frequency response can be expressed as

$$H(e^{j\omega}) = e^{-j\omega n_d}(1 - H_{lp}(e^{j\omega})) = e^{-j\omega n_d} - e^{-j\omega n_d}H_{lp}(e^{j\omega}),$$

where $H_{lp}(e^{j\omega})$ is periodic with period 2π and

$$H_{lp}(e^{j\omega}) = \begin{cases} 1, & |\omega| < \omega_c, \\ 0, & \omega_c < |\omega| < \pi. \end{cases}$$

Using the result of Example 2.18 to obtain the inverse transform of $H_{lp}(e^{j\omega})$, together with properties 1 and 2 of Table 2.2, we have

$$h[n] = \delta[n - n_d] - h_{lp}[n - n_d]$$

$$= \delta[n - n_d] - \frac{\sin \omega_c(n - n_d)}{\pi(n - n_d)}.$$

Example 2.25   Determining the Impulse Response for a Difference Equation

In this example, we determine the impulse response for a stable LTI system for which the input $x[n]$ and output $y[n]$ satisfy the linear constant-coefficient difference equation

$$y[n] - \tfrac{1}{2}y[n-1] = x[n] - \tfrac{1}{4}x[n-1]. \tag{2.175}$$

In Chapter 3, we will see that the z-transform is more useful than the Fourier transform for dealing with difference equations. However, this example offers a hint of the utility of transform methods in the analysis of linear systems. To find the impulse response, we set $x[n] = \delta[n]$; with $h[n]$ denoting the impulse response, Eq. (2.175) becomes

$$h[n] - \tfrac{1}{2}h[n-1] = \delta[n] - \tfrac{1}{4}\delta[n-1]. \tag{2.176}$$

Applying the Fourier transform to both sides of Eq. (2.176) and using properties 1 and 2 of Table 2.2, we obtain

$$H(e^{j\omega}) - \tfrac{1}{2}e^{-j\omega}H(e^{j\omega}) = 1 - \tfrac{1}{4}e^{-j\omega}, \tag{2.177}$$

or

$$H(e^{j\omega}) = \frac{1 - \tfrac{1}{4}e^{-j\omega}}{1 - \tfrac{1}{2}e^{-j\omega}}. \tag{2.178}$$

To obtain $h[n]$, we want to determine the inverse Fourier transform of $H(e^{j\omega})$. Toward this end, we rewrite Eq. (2.178) as

$$H(e^{j\omega}) = \frac{1}{1 - \tfrac{1}{2}e^{-j\omega}} - \frac{\tfrac{1}{4}e^{-j\omega}}{1 - \tfrac{1}{2}e^{-j\omega}}. \tag{2.179}$$

From transform 4 of Table 2.3,

$$\left(\tfrac{1}{2}\right)^n u[n] \quad \overset{\mathcal{F}}{\longleftrightarrow} \quad \frac{1}{1 - \tfrac{1}{2}e^{-j\omega}}.$$

Combining this transform with property 2 of Table 2.2, we obtain

$$-\left(\tfrac{1}{4}\right)\left(\tfrac{1}{2}\right)^{n-1} u[n-1] \quad \overset{\mathcal{F}}{\longleftrightarrow} \quad -\frac{\tfrac{1}{4}e^{-j\omega}}{1 - \tfrac{1}{2}e^{-j\omega}}. \tag{2.180}$$

Based on property 1 of Table 2.2, then,

$$h[n] = \left(\tfrac{1}{2}\right)^n u[n] - \left(\tfrac{1}{4}\right)\left(\tfrac{1}{2}\right)^{n-1} u[n-1]. \tag{2.181}$$

2.10 DISCRETE-TIME RANDOM SIGNALS

The preceding sections have focused on mathematical representations of discrete-time signals and systems and the insights that derive from such mathematical representations. Discrete-time signals and systems have both a time-domain and a frequency-domain representation, each with an important place in the theory and design of discrete-time signal-processing systems. Until now, we have assumed that the signals are deterministic, i.e., that each value of a sequence is uniquely determined by a mathematical expression, a table of data, or a rule of some type.

In many situations, the processes that generate signals are so complex as to make precise description of a signal extremely difficult or undesirable, if not impossible. In such cases, modeling the signal as a random process is analytically useful.[5] As an example, we will see in Chapter 6 that many of the effects encountered in implementing digital signal-processing algorithms with finite register length can be represented by additive noise, i.e., a random sequence. Many mechanical systems generate acoustic or vibratory signals that can be processed to diagnose potential failure; again, signals of this type are often best modeled in terms of random signals. Speech signals to be processed for automatic recognition or bandwidth compression and music to be processed for quality enhancement are two more of many examples.

A random signal is considered to be a member of an ensemble of discrete-time signals that is characterized by a set of probability density functions. More specifically, for a particular signal at a particular time, the amplitude of the signal sample at that time is assumed to have been determined by an underlying scheme of probabilities. That is, each individual sample $x[n]$ of a particular signal is assumed to be an outcome of some underlying random variable x_n. The entire signal is represented by a collection of such random variables, one for each sample time, $-\infty < n < \infty$. This collection of random variables is referred to as a *random process,* and we assume that a particular sequence of samples $x[n]$ for $-\infty < n < \infty$ has been generated by the random process that underlies the signal. To completely describe the random process, we need to specify the individual and joint probability distributions of all the random variables.

The key to obtaining useful results from such models of signals lies in their description in terms of averages that can be computed from assumed probability laws or estimated from specific signals. While random signals are not absolutely summable or square summable and, consequently, do not directly have Fourier transforms, many (but not all) of the properties of such signals can be summarized in terms of averages such as the autocorrelation or autocovariance sequence, for which the Fourier transform often exists. As we will discuss in this section, the Fourier transform of the autocorrelation sequence has a useful interpretation in terms of the frequency distribution of the power in the signal. The use of the autocorrelation sequence and its transform has another important advantage: The effect of processing random signals with a discrete-time linear system can be conveniently described in terms of the effect of the system on the autocorrelation sequence.

In the following discussion, we assume that the reader is familiar with the basic concepts of random processes, such as averages, correlation and covariance functions, and the power spectrum. A brief review and summary of notation and concepts is provided in Appendix A. A more detailed presentation of the theory of random signals can be found in a variety of excellent texts, such as Davenport (1970), and Papoulis (2002), Gray and Davidson (2004), Kay (2006) and Bertsekas and Tsitsiklis (2008).

Our primary objective in this section is to present a specific set of results that will be useful in subsequent chapters. Therefore, we focus on wide-sense stationary random signals and their representation in the context of processing with LTI systems.

---

[5]It is common in the signal processing literature to use the terms "random" and "stochastic" interchangeably. In this text, we primarily refer to this class of signals as random signals or random processes.

Although, for simplicity, we assume that $x[n]$ and $h[n]$ are real valued, the results can be generalized to the complex case.

Consider a stable LTI system with real impulse response $h[n]$. Let $x[n]$ be a real-valued sequence that is a sample sequence of a wide-sense stationary discrete-time random process. Then, the output of the linear system is also a sample sequence of a discrete-time random process related to the input process by the linear transformation

$$y[n] = \sum_{k=-\infty}^{\infty} h[n-k]x[k] = \sum_{k=-\infty}^{\infty} h[k]x[n-k].$$

As we have shown, since the system is stable, $y[n]$ will be bounded if $x[n]$ is bounded. We will see shortly that if the input is stationary,[6] then so is the output. The input signal may be characterized by its mean m_x and its autocorrelation function $\phi_{xx}[m]$, or we may also have additional information about 1st- or even 2nd-order probability distributions. In characterizing the output random process $y[n]$ we desire similar information. For many applications, it is sufficient to characterize both the input and output in terms of simple averages, such as the mean, variance, and autocorrelation. Therefore, we will derive input–output relationships for these quantities.

The means of the input and output processes are, respectively,

$$m_{x_n} = \mathcal{E}\{\mathrm{x}_n\}, \qquad m_{y_n} = \mathcal{E}\{\mathrm{y}_n\}, \tag{2.182}$$

where $\mathcal{E}\{\cdot\}$ denotes the expected value of a random variable. In most of our discussion, it will not be necessary to carefully distinguish between the random variables x_n and y_n and their specific values $x[n]$ and $y[n]$. This will simplify the mathematical notation significantly. For example, Eqs. (2.182) will alternatively be written

$$m_x[n] = \mathcal{E}\{x[n]\}, \qquad m_y[n] = \mathcal{E}\{y[n]\}. \tag{2.183}$$

If $x[n]$ is stationary, then $m_x[n]$ is independent of n and will be written as m_x, with similar notation for $m_y[n]$ if $y[n]$ is stationary.

The mean of the output process is

$$m_y[n] = \mathcal{E}\{y[n]\} = \sum_{k=-\infty}^{\infty} h[k]\mathcal{E}\{x[n-k]\},$$

where we have used the fact that the expected value of a sum is the sum of the expected values. Since the input is stationary, $m_x[n-k] = m_x$, and consequently,

$$m_y[n] = m_x \sum_{k=-\infty}^{\infty} h[k]. \tag{2.184}$$

From Eq. (2.184), we see that the mean of the output is also constant. An equivalent expression to Eq. (2.184) in terms of the frequency response is

$$m_y = H(e^{j0})m_x. \tag{2.185}$$

---

[6]In the remainder of the text, we will use the term *stationary* to mean "wide-sense stationary," i.e., that $E\{x[n_1]x[n_2]\}$ for all n_1, n_2 depends only on the difference $(n_1 - n_2)$. Equivalently, the autocorrelation is only a function of the time difference $(n_1 - n_2)$.

Assuming temporarily that the output is nonstationary, the autocorrelation function of the output process for a real input is

$$\phi_{yy}[n, n + m] = \mathcal{E}\{y[n]y[n + m]\}$$

$$= \mathcal{E}\left\{ \sum_{k=-\infty}^{\infty} \sum_{r=-\infty}^{\infty} h[k]h[r]x[n - k]x[n + m - r] \right\}$$

$$= \sum_{k=-\infty}^{\infty} h[k] \sum_{r=-\infty}^{\infty} h[r]\mathcal{E}\{x[n - k]x[n + m - r]\}.$$

Since $x[n]$ is assumed to be stationary, $\mathcal{E}\{x[n-k]x[n+m-r]\}$ depends only on the time difference $m + k - r$. Therefore,

$$\phi_{yy}[n, n + m] = \sum_{k=-\infty}^{\infty} h[k] \sum_{r=-\infty}^{\infty} h[r]\phi_{xx}[m + k - r] = \phi_{yy}[m]. \qquad (2.186)$$

That is, the output autocorrelation sequence also depends only on the time difference m. Thus, for an LTI system having a wide-sense stationary input, the output is also wide-sense stationary.

By making the substitution $\ell = r - k$, we can express Eq. (2.186) as

$$\phi_{yy}[m] = \sum_{\ell=-\infty}^{\infty} \phi_{xx}[m - \ell] \sum_{k=-\infty}^{\infty} h[k]h[\ell + k]$$

$$= \sum_{\ell=-\infty}^{\infty} \phi_{xx}[m - \ell]c_{hh}[\ell], \qquad (2.187)$$

where we have defined

$$c_{hh}[\ell] = \sum_{k=-\infty}^{\infty} h[k]h[\ell + k]. \qquad (2.188)$$

The sequence $c_{hh}[\ell]$ is referred to as the *deterministic autocorrelation sequence* or, simply, the *autocorrelation sequence of h[n]*. It should be emphasized that $c_{hh}[\ell]$ is the autocorrelation of an aperiodic—i.e., finite-energy—sequence and should not be confused with the autocorrelation of an infinite-energy random sequence. Indeed, it can be seen that $c_{hh}[\ell]$ is simply the discrete convolution of $h[n]$ with $h[-n]$. Equation (2.187), then, can be interpreted to mean that the autocorrelation of the output of a linear system is the convolution of the autocorrelation of the input with the aperiodic autocorrelation of the system impulse response.

Equation (2.187) suggests that Fourier transforms may be useful in characterizing the response of an LTI system to a random input. Assume, for convenience, that $m_x = 0$; i.e., the autocorrelation and autocovariance sequences are identical. Then, with $\Phi_{xx}(e^{j\omega})$, $\Phi_{yy}(e^{j\omega})$, and $C_{hh}(e^{j\omega})$ denoting the Fourier transforms of $\phi_{xx}[m]$, $\phi_{yy}[m]$, and $c_{hh}[\ell]$, respectively, from Eq. (2.187),

$$\Phi_{yy}(e^{j\omega}) = C_{hh}(e^{j\omega})\Phi_{xx}(e^{j\omega}). \qquad (2.189)$$

Also, from Eq. (2.188),

$$C_{hh}(e^{j\omega}) = H(e^{j\omega})H^*(e^{j\omega})$$

$$= |H(e^{j\omega})|^2,$$

so

$$\Phi_{yy}(e^{j\omega}) = |H(e^{j\omega})|^2 \Phi_{xx}(e^{j\omega}). \tag{2.190}$$

Equation (2.190) provides the motivation for the term *power density spectrum*. Specifically,

$$\mathcal{E}\{y^2[n]\} = \phi_{yy}[0] = \frac{1}{2\pi} \int_{-\pi}^{\pi} \Phi_{yy}(e^{j\omega})\, d\omega \tag{2.191}$$
$$= \text{total average power in output.}$$

Substituting Eq. (2.190) into Eq. (2.191), we have

$$\mathcal{E}\{y^2[n]\} = \phi_{yy}[0] = \frac{1}{2\pi} \int_{-\pi}^{\pi} |H(e^{j\omega})|^2 \Phi_{xx}(e^{j\omega})\, d\omega. \tag{2.192}$$

Suppose that $H(e^{j\omega})$ is an ideal bandpass filter, as shown in Figure 2.18(c). Since $\phi_{xx}[m]$ is a real, even sequence, its Fourier transform is also real and even, i.e.,

$$\Phi_{xx}(e^{j\omega}) = \Phi_{xx}(e^{-j\omega}).$$

Likewise, $|H(e^{j\omega})|^2$ is an even function of ω. Therefore, we can write

$$\phi_{yy}[0] = \text{average power in output}$$
$$= \frac{1}{2\pi} \int_{\omega_a}^{\omega_b} \Phi_{xx}(e^{j\omega})\, d\omega + \frac{1}{2\pi} \int_{-\omega_b}^{-\omega_a} \Phi_{xx}(e^{j\omega})\, d\omega. \tag{2.193}$$

Thus, the area under $\Phi_{xx}(e^{j\omega})$ for $\omega_a \leq |\omega| \leq \omega_b$ can be taken to represent the mean-square value of the input in that frequency band. We observe that the output power must remain nonnegative, so

$$\lim_{(\omega_b - \omega_a) \to 0} \phi_{yy}[0] \geq 0.$$

This result, together with Eq. (2.193) and the fact that the band $\omega_a \leq \omega \leq \omega_b$ can be arbitrarily small, implies that

$$\Phi_{xx}(e^{j\omega}) \geq 0 \qquad \text{for all } \omega. \tag{2.194}$$

Hence, we note that the power density function of a real signal is real, even, and non-negative.

Example 2.26   White Noise

The concept of white noise is exceedingly useful in a wide variety of contexts in the design and analysis of signal processing and communications systems. A white-noise signal is a signal for which $\phi_{xx}[m] = \sigma_x^2\delta[m]$. We assume in this example that the signal has zero mean. The power spectrum of a white-noise signal is a constant, i.e.,

$$\Phi_{xx}(e^{j\omega}) = \sigma_x^2 \qquad \text{for all } \omega.$$

The average power of a white-noise signal is therefore

$$\phi_{xx}[0] = \frac{1}{2\pi}\int_{-\pi}^{\pi}\Phi_{xx}(e^{j\omega})\,d\omega = \frac{1}{2\pi}\int_{-\pi}^{\pi}\sigma_x^2\,d\omega = \sigma_x^2.$$

The concept of white noise is also useful in the representation of random signals whose power spectra are not constant with frequency. For example, a random signal $y[n]$ with power spectrum $\Phi_{yy}(e^{j\omega})$ can be assumed to be the output of an LTI system with a white-noise input. That is, we use Eq. (2.190) to define a system with frequency response $H(e^{j\omega})$ to satisfy the equation

$$\Phi_{yy}(e^{j\omega}) = |H(e^{j\omega})|^2\sigma_x^2,$$

where σ_x^2 is the average power of the assumed white-noise input signal. We adjust the average power of this input signal to give the correct average power for $y[n]$. For example, suppose that $h[n] = a^n u[n]$. Then,

$$H(e^{j\omega}) = \frac{1}{1 - ae^{-j\omega}},$$

and we can represent all random signals whose power spectra are of the form

$$\Phi_{yy}(e^{j\omega}) = \left|\frac{1}{1 - ae^{-j\omega}}\right|^2\sigma_x^2 = \frac{\sigma_x^2}{1 + a^2 - 2a\cos\omega}.$$

Another important result concerns the cross-correlation between the input and output of an LTI system:

$$\phi_{yx}[m] = \mathcal{E}\{x[n]y[n+m]\}$$

$$= \mathcal{E}\left\{x[n]\sum_{k=-\infty}^{\infty}h[k]x[n+m-k]\right\} \qquad (2.195)$$

$$= \sum_{k=-\infty}^{\infty}h[k]\phi_{xx}[m-k].$$

In this case, we note that the cross-correlation between input and output is the convolution of the impulse response with the input autocorrelation sequence.

The Fourier transform of Eq. (2.195) is

$$\Phi_{yx}(e^{j\omega}) = H(e^{j\omega})\Phi_{xx}(e^{j\omega}). \qquad (2.196)$$

This result has a useful application when the input is white noise, i.e., when $\phi_{xx}[m] = \sigma_x^2\delta[m]$. Substituting into Eq. (2.195), we note that

$$\phi_{yx}[m] = \sigma_x^2 h[m]. \qquad (2.197)$$

That is, for a zero-mean white-noise input, the cross-correlation between input and output of a linear system is proportional to the impulse response of the system. Similarly, the power spectrum of a white-noise input is

$$\Phi_{xx}(e^{j\omega}) = \sigma_x^2, \qquad -\pi \leq \omega \leq \pi. \tag{2.198}$$

Thus, from Eq. (2.196),

$$\Phi_{yx}(e^{j\omega}) = \sigma_x^2 H(e^{j\omega}). \tag{2.199}$$

In other words, the cross power spectrum is in this case proportional to the frequency response of the system. Equations (2.197) and (2.199) may serve as the basis for estimating the impulse response or frequency response of an LTI system if it is possible to observe the output of the system in response to a white-noise input. An example application is in the measurement of the acoustic impulse response of a room or concert hall.

2.11 SUMMARY

In this chapter, we have reviewed and discussed a number of basic definitions relating to discrete-time signals and systems. We considered the definition of a set of basic sequences, the definition and representation of LTI systems in terms of the convolution sum, and some implications of stability and causality. The class of systems for which the input and output satisfy a linear constant-coefficient difference equation with initial rest conditions was shown to be an important subclass of LTI systems. The recursive solution of such difference equations was discussed and the classes of FIR and IIR systems defined.

An important means for the analysis and representation of LTI systems lies in their frequency-domain representation. The response of a system to a complex exponential input was considered, leading to the definition of the frequency response. The relation between impulse response and frequency response was then interpreted as a Fourier transform pair.

We called attention to many properties of Fourier transform representations and discussed a variety of useful Fourier transform pairs. Tables 2.1 and 2.2 summarize the properties and theorems, and Table 2.3 contains some useful Fourier transform pairs.

The chapter concluded with an introduction to discrete-time random signals. These basic ideas and results will be developed further and used in later chapters.

Problems

Basic Problems with Answers

2.1. For each of the following systems, determine whether the system is (1) stable, (2) causal, (3) linear, (4) time invariant, and (5) memoryless:

(a) $T(x[n]) = g[n]x[n]$ with $g[n]$ given

(b) $T(x[n]) = \sum_{k=n_0}^{n} x[k] \qquad n \neq 0$

(c) $T(x[n]) = \sum_{k=n-n_0}^{n+n_0} x[k]$

(d) $T(x[n]) = x[n - n_0]$

(e) $T(x[n]) = e^{x[n]}$
(f) $T(x[n]) = ax[n] + b$
(g) $T(x[n]) = x[-n]$
(h) $T(x[n]) = x[n] + 3u[n+1]$.

2.2. **(a)** The impulse response $h[n]$ of an LTI system is known to be zero, except in the interval $N_0 \leq n \leq N_1$. The input $x[n]$ is known to be zero, except in the interval $N_2 \leq n \leq N_3$. As a result, the output is constrained to be zero, except in some interval $N_4 \leq n \leq N_5$. Determine N_4 and N_5 in terms of N_0, N_1, N_2, and N_3.
(b) If $x[n]$ is zero, except for N consecutive points, and $h[n]$ is zero, except for M consecutive points, what is the maximum number of consecutive points for which $y[n]$ can be nonzero?

2.3. By direct evaluation of the convolution sum, determine the unit step response $(x[n] = u[n])$ of an LTI system whose impulse response is

$$h[n] = a^{-n}u[-n], \qquad 0 < a < 1.$$

2.4. Consider the linear constant-coefficient difference equation

$$y[n] - \tfrac{3}{4}y[n-1] + \tfrac{1}{8}y[n-2] = 2x[n-1].$$

Determine $y[n]$ for $n \geq 0$ when $x[n] = \delta[n]$ and $y[n] = 0, n < 0$.

2.5. A causal LTI system is described by the difference equation

$$y[n] - 5y[n-1] + 6y[n-2] = 2x[n-1].$$

(a) Determine the homogeneous response of the system, i.e., the possible outputs if $x[n] = 0$ for all n.
(b) Determine the impulse response of the system.
(c) Determine the step response of the system.

2.6. **(a)** Determine the frequency response $H(e^{j\omega})$ of the LTI system whose input and output satisfy the difference equation

$$y[n] - \tfrac{1}{2}y[n-1] = x[n] + 2x[n-1] + x[n-2].$$

(b) Write a difference equation that characterizes a system whose frequency response is

$$H(e^{j\omega}) = \frac{1 - \tfrac{1}{2}e^{-j\omega} + e^{-j3\omega}}{1 + \tfrac{1}{2}e^{-j\omega} + \tfrac{3}{4}e^{-j2\omega}}.$$

2.7. Determine whether each of the following signals is periodic. If the signal is periodic, state its period.
(a) $x[n] = e^{j(\pi n/6)}$
(b) $x[n] = e^{j(3\pi n/4)}$
(c) $x[n] = [\sin(\pi n/5)]/(\pi n)$
(d) $x[n] = e^{j\pi n/\sqrt{2}}$.

2.8. An LTI system has impulse response $h[n] = 5(-1/2)^n u[n]$. Use the Fourier transform to find the output of this system when the input is $x[n] = (1/3)^n u[n]$.

2.9. Consider the difference equation

$$y[n] - \frac{5}{6}y[n-1] + \frac{1}{6}y[n-2] = \frac{1}{3}x[n-1].$$

(a) What are the impulse response, frequency response, and step response for the causal LTI system satisfying this difference equation?
(b) What is the general form of the homogeneous solution of the difference equation?
(c) Consider a different system satisfying the difference equation that is neither causal nor LTI, but that has $y[0] = y[1] = 1$. Find the response of this system to $x[n] = \delta[n]$.

2.10. Determine the output of an LTI system if the impulse response $h[n]$ and the input $x[n]$ are as follows:

(a) $x[n] = u[n]$ and $h[n] = a^n u[-n-1]$, with $a > 1$.
(b) $x[n] = u[n-4]$ and $h[n] = 2^n u[-n-1]$.
(c) $x[n] = u[n]$ and $h[n] = (0.5)2^n u[-n]$.
(d) $h[n] = 2^n u[-n-1]$ and $x[n] = u[n] - u[n-10]$.

Use your knowledge of linearity and time invariance to minimize the work in parts (b)–(d).

2.11. Consider an LTI system with frequency response

$$H(e^{j\omega}) = \frac{1 - e^{-j2\omega}}{1 + \frac{1}{2}e^{-j4\omega}}, \qquad -\pi < \omega \le \pi.$$

Determine the output $y[n]$ for all n if the input $x[n]$ for all n is

$$x[n] = \sin\left(\frac{\pi n}{4}\right).$$

2.12. Consider a system with input $x[n]$ and output $y[n]$ that satisfy the difference equation

$$y[n] = ny[n-1] + x[n].$$

The system is causal and satisfies initial-rest conditions; i.e., if $x[n] = 0$ for $n < n_0$, then $y[n] = 0$ for $n < n_0$.

(a) If $x[n] = \delta[n]$, determine $y[n]$ for all n.
(b) Is the system linear? Justify your answer.
(c) Is the system time invariant? Justify your answer.

2.13. Indicate which of the following discrete-time signals are eigenfunctions of stable, LTI discrete-time systems:

(a) $e^{j2\pi n/3}$
(b) 3^n
(c) $2^n u[-n-1]$
(d) $\cos(\omega_0 n)$
(e) $(1/4)^n$
(f) $(1/4)^n u[n] + 4^n u[-n-1]$.

2.14. A single input–output relationship is given for each of the following three systems:

(a) System A: $x[n] = (1/3)^n$, $\quad y[n] = 2(1/3)^n$.
(b) System B: $x[n] = (1/2)^n$, $\quad y[n] = (1/4)^n$.
(c) System C: $x[n] = (2/3)^n u[n]$, $\quad y[n] = 4(2/3)^n u[n] - 3(1/2)^n u[n]$.

Based on this information, pick the strongest possible conclusion that you can make about each system from the following list of statements:

(i) The system cannot possibly be LTI.
(ii) The system must be LTI.

(iii) The system can be LTI, and there is only one LTI system that satisfies this input–output constraint.

(iv) The system can be LTI, but cannot be uniquely determined from the information in this input–output constraint.

If you chose option (iii) from this list, specify either the impulse response $h[n]$ or the frequency response $H(e^{j\omega})$ for the LTI system.

2.15. Consider the system illustrated in Figure P2.15. The output of an LTI system with an impulse response $h[n] = \left(\frac{1}{4}\right)^n u[n+10]$ is multiplied by a unit step function $u[n]$ to yield the output of the overall system. Answer each of the following questions, and briefly justify your answers:

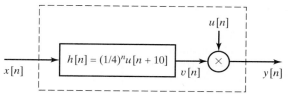

Figure P2.15

(a) Is the overall system LTI?
(b) Is the overall system causal?
(c) Is the overall system stable in the BIBO sense?

2.16. Consider the following difference equation:

$$y[n] - \frac{1}{4}y[n-1] - \frac{1}{8}y[n-2] = 3x[n].$$

(a) Determine the general form of the homogeneous solution to this difference equation.
(b) Both a causal and an anticausal LTI system are characterized by this difference equation. Find the impulse responses of the two systems.
(c) Show that the causal LTI system is stable and the anticausal LTI system is unstable.
(d) Find a particular solution to the difference equation when $x[n] = (1/2)^n u[n]$.

2.17. (a) Determine the Fourier transform of the sequence

$$r[n] = \begin{cases} 1, & 0 \le n \le M, \\ 0, & \text{otherwise.} \end{cases}$$

(b) Consider the sequence

$$w[n] = \begin{cases} \dfrac{1}{2}\left[1 - \cos\left(\dfrac{2\pi n}{M}\right)\right], & 0 \le n \le M, \\ 0, & \text{otherwise.} \end{cases}$$

Sketch $w[n]$ and express $W(e^{j\omega})$, the Fourier transform of $w[n]$, in terms of $R(e^{j\omega})$, the Fourier transform of $r[n]$. (*Hint:* First express $w[n]$ in terms of $r[n]$ and the complex exponentials $e^{j(2\pi n/M)}$ and $e^{-j(2\pi n/M)}$.)
(c) Sketch the magnitude of $R(e^{j\omega})$ and $W(e^{j\omega})$ for the case when $M = 4$.

2.18. For each of the following impulse responses of LTI systems, indicate whether or not the system is causal:

(a) $h[n] = (1/2)^n u[n]$
(b) $h[n] = (1/2)^n u[n-1]$
(c) $h[n] = (1/2)^{|n|}$
(d) $h[n] = u[n+2] - u[n-2]$
(e) $h[n] = (1/3)^n u[n] + 3^n u[-n-1]$.

2.19. For each of the following impulse responses of LTI systems, indicate whether or not the system is stable:

(a) $h[n] = 4^n u[n]$
(b) $h[n] = u[n] - u[n-10]$
(c) $h[n] = 3^n u[-n-1]$
(d) $h[n] = \sin(\pi n/3)u[n]$
(e) $h[n] = (3/4)^{|n|}\cos(\pi n/4 + \pi/4)$
(f) $h[n] = 2u[n+5] - u[n] - u[n-5]$.

2.20. Consider the difference equation representing a causal LTI system

$$y[n] + (1/a)y[n-1] = x[n-1].$$

(a) Find the impulse response of the system, $h[n]$, as a function of the constant a.
(b) For what range of values of a will the system be stable?

Basic Problems

2.21. A discrete-time signal $x[n]$ is shown in Figure P2.21.

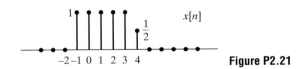

Figure P2.21

Sketch and label carefully each of the following signals:

(a) $x[n-2]$
(b) $x[4-n]$
(c) $x[2n]$
(d) $x[n]u[2-n]$
(e) $x[n-1]\delta[n-3]$.

2.22. Consider a discrete-time LTI system with impulse response $h[n]$. If the input $x[n]$ is a periodic sequence with period N (i.e., if $x[n] = x[n+N]$), show that the output $y[n]$ is also a periodic sequence with period N.

2.23. For each of the following systems, determine whether the system is (1) stable, (2) causal, (3) linear, and (4) time invariant.

(a) $T(x[n]) = (\cos \pi n)x[n]$

(b) $T(x[n]) = x[n^2]$

(c) $T(x[n]) = x[n] \sum_{k=0}^{\infty} \delta[n-k]$

(d) $T(x[n]) = \sum_{k=n-1}^{\infty} x[k].$

2.24. Consider an arbitrary linear system with input $x[n]$ and output $y[n]$. Show that if $x[n] = 0$ for all n, then $y[n]$ must also be zero for all n.

2.25. Consider a system for which the input $x[n]$ and the output $y[n]$ satisfy the following relation.

$$8y[n] + 2y[n-1] - 3y[n-2] = x[n] \qquad \text{(P2.25-1)}$$

(a) For $x[n] = \delta[n]$, show that a *particular* sequence satisfying the difference equation is
$$y_p[n] = \frac{3}{40}\left(-\frac{3}{4}\right)^n u[n] + \frac{1}{20}\left(\frac{1}{2}\right)^n u[n].$$

(b) Determine the homogeneous solution(s) to the difference equation specified in Eq. (P2.25-1).

(c) Determine $y[n]$ for $-2 \le n \le 2$ when $x[n]$ is equal to $\delta[n]$ in Eq. (P2.25-1) and the *initial rest condition* is assumed in solving the difference equation. Note that the initial rest condition implies the system described by Eq. (P2.25-1) is causal.

2.26. For each of the systems in Figure P2.26, pick the strongest valid conclusion that you can make about each system from the following list of statements:

(i) The system must be LTI and is uniquely specified by the information given.

(ii) The system must be LTI, but cannot be uniquely determined from the information given.

(iii) The system could be LTI and if it is, the information given uniquely specifies the system.

(iv) The system could be LTI, but cannot be uniquely determined from the information given.

(v) The system could not possibly be LTI.

For each system for which you choose option (i) or (iii), give the impulse response $h[n]$ for the uniquely specified LTI system. One example of an input and its corresponding output are shown for each system.

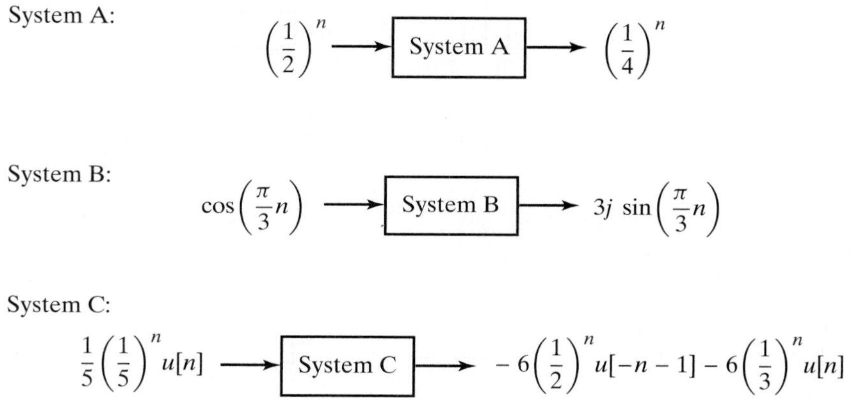

System A:

$$\left(\frac{1}{2}\right)^n \longrightarrow \boxed{\text{System A}} \longrightarrow \left(\frac{1}{4}\right)^n$$

System B:

$$\cos\left(\frac{\pi}{3}n\right) \longrightarrow \boxed{\text{System B}} \longrightarrow 3j \sin\left(\frac{\pi}{3}n\right)$$

System C:

$$\frac{1}{5}\left(\frac{1}{5}\right)^n u[n] \longrightarrow \boxed{\text{System C}} \longrightarrow -6\left(\frac{1}{2}\right)^n u[-n-1] - 6\left(\frac{1}{3}\right)^n u[n]$$

Figure P2.26

2.27. For each of the systems in Figure P2.27, pick the strongest valid conclusion that you can make about each system from the following list of statements:

(i) The system must be LTI and is uniquely specified by the information given.

(ii) The system must be LTI, but cannot be uniquely determined from the information given.

(iii) The system could be LTI, and if it is, the information given uniquely specifies the system.

(iv) The system could be LTI, but cannot be uniquely determined from the information given.

(v) The system could not possibly be LTI.

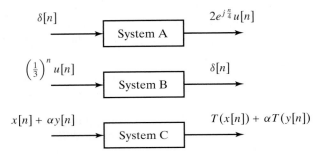

For all choices of $x[n]$, $y[n]$, and the constant α

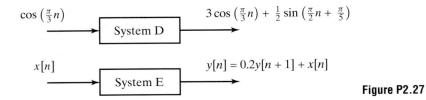

Figure P2.27

2.28. Four input–output pairs of a particular system S are specified in Figure P2.28-1:

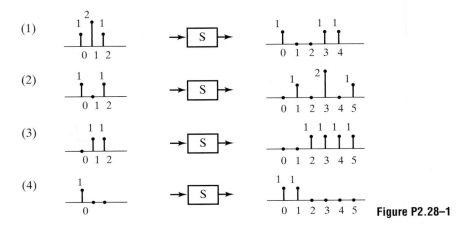

Figure P2.28–1

(a) Can system S be time-invariant? Explain.

(b) Can system S be linear? Explain.

(c) Suppose (2) and (3) are input–output pairs of a particular system S_2, and the system is known to be LTI. What is $h[n]$, the impulse response of the system?

(d) Suppose (1) is the input–output pair of an LTI system S_3. What is the output of this system for the input in Figure P2.28-2:

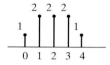

Figure P2.28–2

2.29. An LTI system has impulse response defined by

$$h[n] = \begin{cases} 0 & n < 0 \\ 1 & n = 0, 1, 2, 3 \\ -2 & n = 4, 5 \\ 0 & n > 5 \end{cases}$$

Determine and plot the output $y[n]$ when the input $x[n]$ is:

(a) $u[n]$

(b) $u[n-4]$

(c) $u[n] - u[n-4]$.

2.30. Consider the cascade connection of two LTI systems in Figure P2.30:

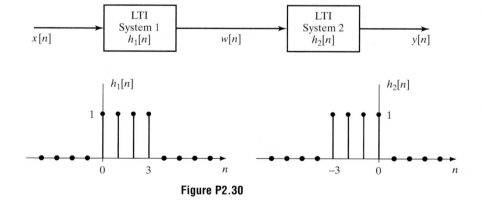

Figure P2.30

(a) Determine and sketch $w[n]$ if $x[n] = (-1)^n u[n]$. Also, determine the overall output $y[n]$.

(b) Determine and sketch the overall impulse response of the cascade system; i.e., plot the output $y[n] = h[n]$ when $x[n] = \delta[n]$.

(c) Now consider the input $x[n] = 2\delta[n] + 4\delta[n-4] - 2\delta[n-12]$. Sketch $w[n]$.

(d) For the input of part (c), write an expression for the output $y[n]$ in terms of the overall impulse response $h[n]$ as defined in part (b). Make a carefully labeled sketch of your answer.

2.31. If the input and output of a causal LTI system satisfy the difference equation

$$y[n] = ay[n-1] + x[n],$$

then the impulse response of the system must be $h[n] = a^n u[n]$.

(a) For what values of a is this system stable?

(b) Consider a causal LTI system for which the input and output are related by the difference equation

$$y[n] = ay[n-1] + x[n] - a^N x[n-N],$$

where N is a positive integer. Determine and sketch the impulse response of this system. *Hint*: Use linearity and time-invariance to simplify the solution.

(c) Is the system in part (b) an FIR or an IIR system? Explain.

(d) For what values of a is the system in part (b) stable? Explain.

2.32. For $X(e^{j\omega}) = 1/(1 - ae^{-j\omega})$, with $-1 < a < 0$, determine and sketch the following as a function of ω:

(a) $\mathcal{R}e\{X(e^{j\omega})\}$

(b) $\mathcal{I}m\{X(e^{j\omega})\}$

(c) $|X(e^{j\omega})|$

(d) $\angle X(e^{j\omega})$.

2.33. Consider an LTI system defined by the difference equation

$$y[n] = -2x[n] + 4x[n-1] - 2x[n-2].$$

(a) Determine the impulse response of this system.

(b) Determine the frequency response of this system. Express your answer in the form

$$H(e^{j\omega}) = A(e^{j\omega})e^{-j\omega n_d},$$

where $A(e^{j\omega})$ is a real function of ω. Explicitly specify $A(e^{j\omega})$ and the delay n_d of this system.

(c) Sketch a plot of the magnitude $|H(e^{j\omega})|$ and a plot of the phase $\angle H(e^{j\omega})$.

(d) Suppose that the input to the system is

$$x_1[n] = 1 + e^{j0.5\pi n} \qquad -\infty < n < \infty.$$

Use the frequency response function to determine the corresponding output $y_1[n]$.

(e) Now suppose that the input to the system is

$$x_2[n] = (1 + e^{j0.5\pi n})u[n] \qquad -\infty < n < \infty.$$

Use the defining difference equation or discrete convolution to determine the corresponding output $y_2[n]$ for $-\infty < n < \infty$. Compare $y_1[n]$ and $y_2[n]$. They should be equal for certain values of n. Over what range of values of n are they equal?

2.34. An LTI system has the frequency response

$$H(e^{j\omega}) = \frac{1 - 1.25e^{-j\omega}}{1 - 0.8e^{-j\omega}} = 1 - \frac{0.45e^{-j\omega}}{1 - 0.8e^{-j\omega}}.$$

(a) Specify the difference equation that is satisfied by the input $x[n]$ and the output $y[n]$.

(b) Use one of the above forms of the frequency response to determine the impulse response $h[n]$.

(c) Show that $|H(e^{j\omega})|^2 = G^2$, where G is a constant. Determine the constant G. (This is an example of an *allpass filter* to be discussed in detail in Chapter 5.)

(d) If the input to the above system is $x[n] = \cos(0.2\pi n)$, the output should be of the form $y[n] = A\cos(0.2\pi n + \theta)$. What are A and θ?

2.35. An LTI system has impulse response given by the following plot:

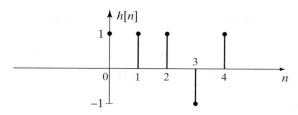

Figure P2.35-1

The input to the system, $x[n]$, is plotted below as a function of n.

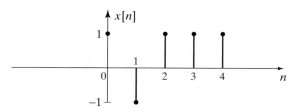

Figure P2.35-2

(a) Use discrete convolution to determine the output of the system $y[n] = x[n] * h[n]$ for the above input. Give your answer as a carefully labeled sketch of $y[n]$ over a range sufficient to define it completely.

(b) The deterministic autocorrelation of a signal $x[n]$ is defined in Eq. (2.188) as $c_{xx}[n] = x[n] * x[-n]$. The system defined by Figure P2.35-1 is a *matched filter* for the input in Figure P2.35-2. Noting that $h[n] = x[-(n-4)]$, express the output in part (a) in terms of $c_{xx}[n]$.

(c) Determine the output of the system whose impulse response is $h[n]$ when the input is $x[n] = u[n+2]$. Sketch your answer.

2.36. An LTI discrete-time system has frequency response given by

$$H(e^{j\omega}) = \frac{(1 - je^{-j\omega})(1 + je^{-j\omega})}{1 - 0.8e^{-j\omega}} = \frac{1 + e^{-j2\omega}}{1 - 0.8e^{-j\omega}} = \frac{1}{1 - 0.8e^{-j\omega}} + \frac{e^{-j2\omega}}{1 - 0.8e^{-j\omega}}.$$

(a) Use one of the above forms of the frequency response to obtain an equation for the impulse response $h[n]$ of the system.

(b) From the frequency response, determine the difference equation that is satisfied by the input $x[n]$ and the output $y[n]$ of the system.

(c) If the input to this system is

$$x[n] = 4 + 2\cos(\omega_0 n) \quad \text{for} - \infty < n < \infty,$$

for what value of ω_0 will the output be of the form

$$y[n] = A = \text{constant}$$

for $-\infty < n < \infty$? What is the constant A?

2.37. Consider the cascade of LTI discrete-time systems shown in Figure P2.37.

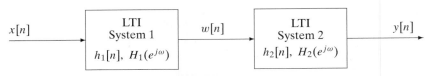

Figure P2.37

The first system is described by the frequency response

$$H_1(e^{j\omega}) = e^{-j\omega} \begin{cases} 0 & |\omega| \le 0.25\pi \\ 1 & 0.25\pi < |\omega| \le \pi \end{cases}$$

and the second system is described by

$$h_2[n] = 2\frac{\sin(0.5\pi n)}{\pi n}$$

(a) Determine an equation that defines the frequency response, $H(e^{j\omega})$, of the overall system over the range $-\pi \le \omega \le \pi$.
(b) Sketch the magnitude, $|H(e^{j\omega})|$, and the phase, $\angle H(e^{j\omega})$, of the overall frequency response over the range $-\pi \le \omega \le \pi$.
(c) Use any convenient means to determine the impulse response $h[n]$ of the overall cascade system.

2.38. Consider the cascade of two LTI systems shown in Figure P2.38.

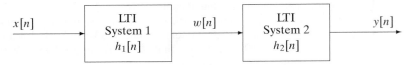

Figure P2.38

The impulse responses of the two systems are:

$$h_1[n] = u[n-5] \quad \text{and} \quad h_2[n] = \begin{cases} 1 & 0 \le n \le 4 \\ 0 & \text{otherwise.} \end{cases}$$

(a) Make a sketch showing both $h_2[k]$ and $h_1[n-k]$ (for some arbitrary $n < 0$) as functions of k.
(b) Determine $h[n] = h_1[n] * h_2[n]$, the impulse response of the overall system. Give your answer as an equation (or set of equations) that define $h[n]$ for $-\infty < n < \infty$ or as a carefully labelled plot of $h[n]$ over a range sufficient to define it completely.

2.39. Using the definition of linearity (Eqs. (2.23a)–(2.23b)), show that the ideal delay system (Example 2.2) and the moving-average system (Example 2.3) are both linear systems.

2.40. Determine which of the following signals is periodic. If a signal is periodic, determine its period.

(a) $x[n] = e^{j(2\pi n/5)}$
(b) $x[n] = \sin(\pi n/19)$
(c) $x[n] = ne^{j\pi n}$
(d) $x[n] = e^{jn}$.

2.41. Consider an LTI system with $|H(e^{j\omega})| = 1$, and let $\arg[H(e^{j\omega})]$ be as shown in Figure P2.41. If the input is

$$x[n] = \cos\left(\frac{3\pi}{2}n + \frac{\pi}{4}\right),$$

determine the output $y[n]$.

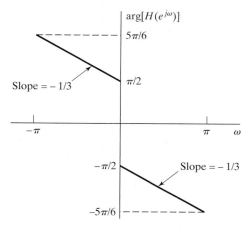

Figure P2.41

2.42. The sequences $s[n]$, $x[n]$, and $w[n]$ are sample sequences of wide-sense stationary random processes where

$$s[n] = x[n]w[n].$$

The sequences $x[n]$ and $w[n]$ are zero-mean and statistically independent. The autocorrelation function of $w[n]$ is

$$E\{w[n]w[n+m]\} = \sigma_w^2 \delta[m],$$

and the variance of $x[n]$ is σ_x^2.

Show that $s[n]$ is white, with variance $\sigma_x^2 \sigma_w^2$.

Advanced Problems

2.43. The operator T represents an LTI system. As shown in the following figures, if the input to the system T is $(\frac{1}{3})^n u[n]$, the output of the system is $g[n]$. If the input is $x[n]$, the output is $y[n]$.

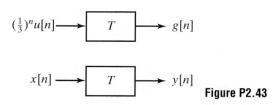

Figure P2.43

Express $y[n]$ in terms of $g[n]$ and $x[n]$.

2.44. $X(e^{j\omega})$ denotes the Fourier transform of the complex-valued signal $x[n]$, where the real and imaginary parts of $x[n]$ are given in Figure P2.44. (*Note*: The sequence is zero outside the interval shown.)

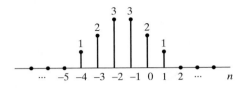

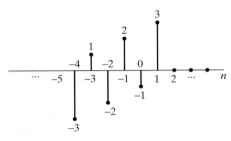

Figure P2.44

Perform the following calculations without explicitly evaluating $X(e^{j\omega})$.

(a) Evaluate $X(e^{j\omega})|_{\omega=0}$.
(b) Evaluate $X(e^{j\omega})|_{\omega=\pi}$.
(c) Evaluate $\int_{-\pi}^{\pi} X(e^{j\omega})\,d\omega$.
(d) Determine and sketch the signal (in the time domain) whose Fourier transform is $X(e^{-j\omega})$.
(e) Determine and sketch the signal (in the time domain) whose Fourier transform is $j\operatorname{Im}\{X(e^{j\omega})\}$.

2.45. Consider the cascade of LTI discrete-time systems shown in Figure P2.45.

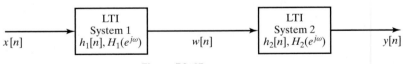

Figure P2.45

System 1 is described by the difference equation

$$w[n] = x[n] - x[n-1],$$

and System 2 is described by

$$h_2[n] = \frac{\sin(0.5\pi n)}{\pi n} \iff H_2(e^{j\omega}) = \begin{cases} 1 & |\omega| < 0.5\pi \\ 0 & 0.5\pi < |\omega| < \pi. \end{cases}$$

The input $x[n]$ is

$$x[n] = \cos(0.4\pi n) + \sin(.6\pi n) + 5\delta[n-2] + 2u[n].$$

Determine the overall output $y[n]$.

(*With careful thought, you will be able to use the properties of LTI systems to write down the answer by inspection.*)

2.46. The DTFT pair

$$a^n u[n] \iff \frac{1}{1 - ae^{-j\omega}} \qquad |a| < 1 \qquad \text{(P2.46-1)}$$

is given.

(a) Using Eq. (P2.46-1), determine the DTFT, $X(e^{j\omega})$, of the sequence

$$x[n] = -b^n u[-n-1] = \begin{cases} -b^n & n \leq -1 \\ 0 & n \geq 0. \end{cases}$$

What restriction on b is necessary for the DTFT of $x[n]$ to exist?

(b) Determine the sequence $y[n]$ whose DTFT is

$$Y(e^{j\omega}) = \frac{2e^{-j\omega}}{1 + 2e^{-j\omega}}.$$

2.47. Consider a "windowed cosine signal"

$$x[n] = w[n] \cos(\omega_0 n).$$

(a) Determine an expression for $X(e^{j\omega})$ in terms of $W(e^{j\omega})$.

(b) Suppose that the sequence $w[n]$ is the finite-length sequence

$$w[n] = \begin{cases} 1 & -L \leq n \leq L \\ 0 & \text{otherwise.} \end{cases}$$

Determine the DTFT $W(e^{j\omega})$. *Hint*: Use Tables 2.2 and 2.3 to obtain a "closed form" solution. You should find that $W(e^{j\omega})$ is a real function of ω.

(c) Sketch the DTFT $X(e^{j\omega})$ for the window in (b). For a given ω_0, how should L be chosen so that your sketch shows two distinct peaks?

2.48. The system T in Figure P2.48 is known to be *time invariant*. When the inputs to the system are $x_1[n]$, $x_2[n]$, and $x_3[n]$, the responses of the system are $y_1[n]$, $y_2[n]$, and $y_3[n]$, as shown.

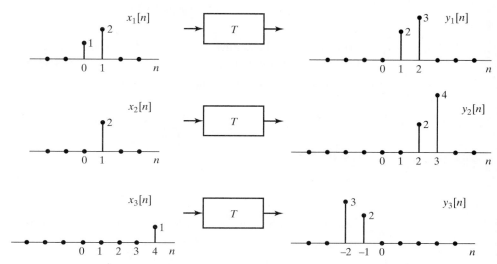

Figure P2.48

(a) Determine whether the system T could be linear.

(b) If the input $x[n]$ to the system T is $\delta[n]$, what is the system response $y[n]$?

(c) What are all possible inputs $x[n]$ for which the response of the system T can be determined from the given information alone?

2.49. The system L in Figure P2.49 is known to be *linear*. Shown are three output signals $y_1[n]$, $y_2[n]$, and $y_3[n]$ in response to the input signals $x_1[n]$, $x_2[n]$, and $x_3[n]$, respectively.

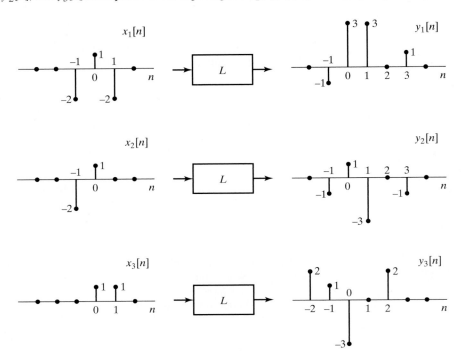

Figure P2.49

(a) Determine whether the system L could be time invariant.

(b) If the input $x[n]$ to the system L is $\delta[n]$, what is the system response $y[n]$?

2.50. In Section 2.5, we stated that the solution to the homogeneous difference equation

$$\sum_{k=0}^{N} a_k y_h[n-k] = 0$$

is of the form

$$y_h[n] = \sum_{m=1}^{N} A_m z_m^n, \qquad (\text{P2.50-1})$$

with the A_m's arbitrary and the z_m's the N roots of the polynomial

$$A(z) = \sum_{k=0}^{N} a_k z^{-k}; \qquad (\text{P2.50-2})$$

i.e.,

$$A(z) = \sum_{k=0}^{N} a_k z^{-k} = \prod_{m=1}^{N} (1 - z_m z^{-1}).$$

(a) Determine the general form of the homogeneous solution to the difference equation

$$y[n] - \tfrac{3}{4} y[n-1] + \tfrac{1}{8} y[n-2] = 2x[n-1].$$

(b) Determine the coefficients A_m in the homogeneous solution if $y[-1] = 1$ and $y[0] = 0$.

(c) Now consider the difference equation

$$y[n] - y[n-1] + \tfrac{1}{4} y[n-2] = 2x[n-1]. \qquad \text{(P2.50-3)}$$

If the homogeneous solution contains only terms of the form of Eq. (P2.50-1), show that the initial conditions $y[-1] = 1$ and $y[0] = 0$ cannot be satisfied.

(d) If Eq. (P2.50-2) has two roots that are identical, then, in place of Eq. (P2.50-1), $y_h[n]$ will take the form

$$y_h[n] = \sum_{m=1}^{N-1} A_m z_m^n + n B_1 z_1^n, \qquad \text{(P2.50-4)}$$

where we have assumed that the double root is z_1. Using Eq. (P2.50-4), determine the general form of $y_h[n]$ for Eq. (P2.50-3). Verify explicitly that your answer satisfies Eq. (P2.50-3) with $x[n] = 0$.

(e) Determine the coefficients A_1 and B_1 in the homogeneous solution obtained in part (d) if $y[-1] = 1$ and $y[0] = 0$.

2.51. Consider a system with input $x[n]$ and output $y[n]$. The input–output relation for the system is defined by the following two properties:

   1. $y[n] - ay[n-1] = x[n]$,
   2. $y[0] = 1$.

(a) Determine whether the system is time invariant.

(b) Determine whether the system is linear.

(c) Assume that the difference equation (property 1) remains the same, but the value $y[0]$ is specified to be zero. Does this change your answer to either part (a) or part (b)?

2.52. Consider the LTI system with impulse response

$$h[n] = \left(\frac{j}{2} \right)^n u[n], \qquad \text{where } j = \sqrt{-1}.$$

Determine the steady-state response, i.e., the response for large n, to the excitation

$$x[n] = \cos(\pi n) u[n].$$

2.53. An LTI system has frequency response

$$H(e^{j\omega}) = \begin{cases} e^{-j\omega 3}, & |\omega| < \dfrac{2\pi}{16} \left(\dfrac{3}{2} \right), \\ 0, & \dfrac{2\pi}{16} \left(\dfrac{3}{2} \right) \le |\omega| \le \pi. \end{cases}$$

The input to the system is a periodic unit-impulse train with period $N = 16$; i.e.,

$$x[n] = \sum_{k=-\infty}^{\infty} \delta[n + 16k].$$

Determine the output of the system.

2.54. Consider the system in Figure P2.54.

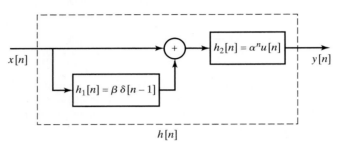

$h[n]$ **Figure P2.54**

    (a) Determine the impulse response $h[n]$ of the overall system.
    (b) Determine the frequency response of the overall system.
    (c) Specify a difference equation that relates the output $y[n]$ to the input $x[n]$.
    (d) Is this system causal? Under what condition would the system be stable?

2.55. Let $X(e^{j\omega})$ denote the Fourier transform of the signal $x[n]$ shown in Figure P2.55. Perform the following calculations without explicitly evaluating $X(e^{j\omega})$:

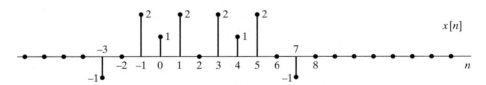

Figure P2.55

    (a) Evaluate $X(e^{j\omega})|_{\omega=0}$.
    (b) Evaluate $X(e^{j\omega})|_{\omega=\pi}$.
    (c) Find $\angle X(e^{j\omega})$.
    (d) Evaluate $\int_{-\pi}^{\pi} X(e^{j\omega})d\omega$.
    (e) Determine and sketch the signal whose Fourier transform is $X(e^{-j\omega})$.
    (f) Determine and sketch the signal whose Fourier transform is $\mathcal{R}e\{X(e^{j\omega})\}$.

2.56. For the system in Figure P2.56, determine the output $y[n]$ when the input $x[n]$ is $\delta[n]$ and $H(e^{j\omega})$ is an ideal lowpass filter as indicated, i.e.,

$$H(e^{j\omega}) = \begin{cases} 1, & |\omega| < \pi/2, \\ 0, & \pi/2 < |\omega| \le \pi. \end{cases}$$

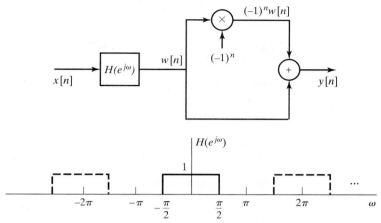

Figure P2.56

2.57. A sequence has the DTFT

$$X(e^{j\omega}) = \frac{1 - a^2}{(1 - ae^{-j\omega})(1 - ae^{j\omega})}, \qquad |a| < 1.$$

(a) Find the sequence $x[n]$.

(b) Calculate $1/2\pi \int_{-\pi}^{\pi} X(e^{j\omega}) \cos(\omega) d\omega$.

2.58. An LTI system is described by the input–output relation

$$y[n] = x[n] + 2x[n-1] + x[n-2].$$

(a) Determine $h[n]$, the impulse response of the system.

(b) Is this a stable system?

(c) Determine $H(e^{j\omega})$, the frequency response of the system. Use trigonometric identities to obtain a simple expression for $H(e^{j\omega})$.

(d) Plot the magnitude and phase of the frequency response.

(e) Now consider a new system whose frequency response is $H_1(e^{j\omega}) = H(e^{j(\omega+\pi)})$. Determine $h_1[n]$, the impulse response of the new system.

2.59. Let the real discrete-time signal $x[n]$ with Fourier transform $X(e^{j\omega})$ be the input to a system with the output defined by

$$y[n] = \begin{cases} x[n], & \text{if } n \text{ is even,} \\ 0, & \text{otherwise.} \end{cases}$$

(a) Sketch the discrete-time signal $s[n] = 1 + \cos(\pi n)$ and its (generalized) Fourier transform $S(e^{j\omega})$.

(b) Express $Y(e^{j\omega})$, the Fourier transform of the output, as a function of $X(e^{j\omega})$ and $S(e^{j\omega})$.

(c) Suppose that it is of interest to approximate $x[n]$ by the interpolated signal $w[n] = y[n] + (1/2)(y[n+1] + y[n-1])$. Determine the Fourier transform $W(e^{j\omega})$ as a function of $Y(e^{j\omega})$.

(d) Sketch $X(e^{j\omega})$, $Y(e^{j\omega})$, and $W(e^{j\omega})$ for the case when $x[n] = \sin(\pi n/a)/(\pi n/a)$ and $a > 1$. Under what conditions is the proposed interpolated signal $w[n]$ a good approximation for the original $x[n]$?

2.60. Consider a discrete-time LTI system with frequency response $H(e^{j\omega})$ and corresponding impulse response $h[n]$.

    (a) We are first given the following three clues about the system:

        (i) The system is causal.

        (ii) $H(e^{j\omega}) = H^*(e^{-j\omega})$.

        (iii) The DTFT of the sequence $h[n + 1]$ is real.

    Using these three clues, show that the system has an impulse response of finite duration.

    (b) In addition to the preceding three clues, we are now given two more clues:

        (iv) $\dfrac{1}{2\pi} \displaystyle\int_{-\pi}^{\pi} H(e^{j\omega}) d\omega = 2.$

        (v) $H(e^{j\pi}) = 0$.

    Is there enough information to identify the system uniquely? If so, determine the impulse response $h[n]$. If not, specify as much as you can about the sequence $h[n]$.

2.61. Consider the three sequences

$$v[n] = u[n] - u[n - 6],$$

$$w[n] = \delta[n] + 2\delta[n - 2] + \delta[n - 4],$$

$$q[n] = v[n] * w[n].$$

    (a) Find and sketch the sequence $q[n]$.

    (b) Find and sketch the sequence $r[n]$ such that $r[n] * v[n] = \displaystyle\sum_{k=-\infty}^{n-1} q[k].$

    (c) Is $q[-n] = v[-n] * w[-n]$? Justify your answer.

2.62. Consider an LTI system with frequency response

$$H(e^{j\omega}) = e^{-j[(\omega/2) + (\pi/4)]}, \qquad -\pi < \omega \le \pi.$$

Determine $y[n]$, the output of this system, if the input is

$$x[n] = \cos\left(\frac{15\pi n}{4} - \frac{\pi}{3}\right)$$

for all n.

2.63. Consider a system S with input $x[n]$ and output $y[n]$ related according to the block diagram in Figure P2.63-1.

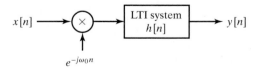

Figure P2.63–1

The input $x[n]$ is multiplied by $e^{-j\omega_0 n}$, and the product is passed through a stable LTI system with impulse response $h[n]$.

    (a) Is the system S linear? Justify your answer.

    (b) Is the system S time invariant? Justify your answer.

    (c) Is the system S stable? Justify your answer.

    (d) Specify a system C such that the block diagram in Figure P2.63-2 represents an alternative way of expressing the input–output relationship of the system S. (*Note:* The system C does not have to be an LTI system.)

Figure P2.63–2

2.64. Consider an ideal lowpass filter with impulse response $h_{lp}[n]$ and frequency response

$$H_{lp}(e^{j\omega}) = \begin{cases} 1, & |\omega| < 0.2\pi, \\ 0, & 0.2\pi \leq |\omega| \leq \pi. \end{cases}$$

(a) A new filter is defined by the equation $h_1[n] = (-1)^n h_{lp}[n] = e^{j\pi n} h_{lp}[n]$. Determine an equation for the frequency response of $H_1(e^{j\omega})$, and plot the equation for $|\omega| < \pi$. What kind of filter is this?

(b) A second filter is defined by the equation $h_2[n] = 2h_{lp}[n]\cos(0.5\pi n)$. Determine the equation for the frequency response $H_2(e^{j\omega})$, and plot the equation for $|\omega| < \pi$. What kind of filter is this?

(c) A third filter is defined by the equation

$$h_3[n] = \frac{\sin(0.1\pi n)}{\pi n} h_{lp}[n].$$

Determine the equation for the frequency response $H_3(e^{j\omega})$, and plot the equation for $|\omega| < \pi$. What kind of filter is this?

2.65. The LTI system

$$H(e^{j\omega}) = \begin{cases} -j, & 0 < \omega < \pi, \\ j, & -\pi < \omega < 0, \end{cases}$$

is referred to as a 90° phase shifter and is used to generate what is referred to as an analytic signal $w[n]$ as shown in Figure P2.65-1. Specifically, the analytic signal $w[n]$ is a complex-valued signal for which

$$Re\{w[n]\} = x[n],$$

$$Im\{w[n]\} = y[n].$$

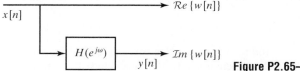

Figure P2.65–1

If $Re\{X(e^{j\omega})\}$ is as shown in Figure P2.65-2 and $Im\{X(e^{j\omega})\} = 0$, determine and sketch $W(e^{j\omega})$, the Fourier transform of the analytic signal $w[n] = x[n] + jy[n]$.

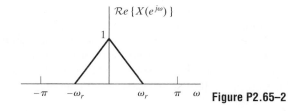

Figure P2.65–2

2.66. The autocorrelation sequence of a signal $x[n]$ is defined as

$$R_x[n] = \sum_{k=-\infty}^{\infty} x^*[k]x[n+k].$$

(a) Show that for an appropriate choice of the signal $g[n]$, $R_x[n] = x[n] * g[n]$, and identify the proper choice for $g[n]$.

(b) Show that the Fourier transform of $R_x[n]$ is equal to $|X(e^{j\omega})|^2$.

2.67. The signals $x[n]$ and $y[n]$ shown in Figure P2.67-1 are the input and corresponding output for an LTI system.

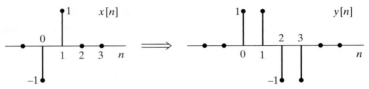

Figure P2.67-1

(a) Find the response of the system to the sequence $x_2[n]$ in Figure P2.67-2.

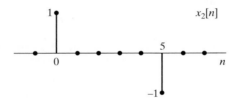

Figure P2.67-2

(b) Find the impulse response $h[n]$ for this LTI system.

2.68. Consider a system for which the input $x[n]$ and output $y[n]$ satisfy the difference equation

$$y[n] - \frac{1}{2}y[n-1] = x[n]$$

and for which $y[-1]$ is constrained to be zero for every input. Determine whether or not the system is stable. If you conclude that the system is stable, show your reasoning. If you conclude that the system is not stable, give an example of a bounded input that results in an unbounded output.

Extension Problems

2.69. The causality of a system was defined in Section 2.2.4. From this definition, show that, for an LTI system, causality implies that the impulse response $h[n]$ is zero for $n < 0$. One approach is to show that if $h[n]$ is not zero for $n < 0$, then the system cannot be causal. Show also that if the impulse response is zero for $n < 0$, then the system will necessarily be causal.

2.70. Consider a discrete-time system with input $x[n]$ and output $y[n]$. When the input is

$$x[n] = \left(\frac{1}{4}\right)^n u[n],$$

the output is

$$y[n] = \left(\frac{1}{2}\right)^n \qquad \text{for all } n.$$

Determine which of the following statements is correct:
- The system must be LTI.
- The system could be LTI.
- The system cannot be LTI.

If your answer is that the system must or could be LTI, give a possible impulse response. If your answer is that the system could not be LTI, explain clearly why not.

2.71. Consider an LTI system whose frequency response is

$$H(e^{j\omega}) = e^{-j\omega/2}, \qquad |\omega| < \pi.$$

Determine whether or not the system is causal. Show your reasoning.

2.72. In Figure P2.72, two sequences $x_1[n]$ and $x_2[n]$ are shown. Both sequences are zero for all n outside the regions shown. The Fourier transforms of these sequences are $X_1(e^{j\omega})$ and $X_2(e^{j\omega})$, which, in general, can be expected to be complex and can be written in the form

$$X_1(e^{j\omega}) = A_1(\omega)e^{j\theta_1(\omega)},$$

$$X_2(e^{j\omega}) = A_2(\omega)e^{j\theta_2(\omega)},$$

where $A_1(\omega), \theta_1(\omega), A_2(\omega)$, and $\theta_2(\omega)$ are all real functions chosen so that both $A_1(\omega)$ and $A_2(\omega)$ are nonnegative at $\omega = 0$, but otherwise can take on both positive and negative values. Determine appropriate choices for $\theta_1(\omega)$ and $\theta_2(\omega)$, and sketch these two phase functions in the range $0 < \omega < 2\pi$.

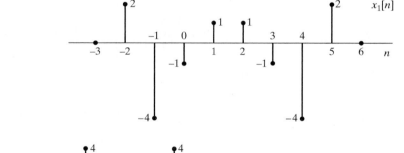

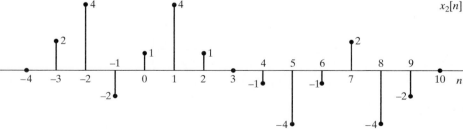

Figure P2.72

2.73. Consider the cascade of discrete-time systems in Figure P2.73. The time-reversal systems are defined by the equations $f[n] = e[-n]$ and $y[n] = g[-n]$. Assume throughout the problem that $x[n]$ and $h_1[n]$ are real sequences.

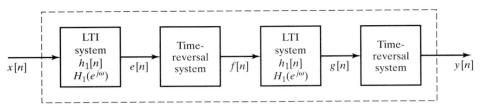

Figure P2.73

(a) Express $E(e^{j\omega})$, $F(e^{j\omega})$, $G(e^{j\omega})$, and $Y(e^{j\omega})$ in terms of $X(e^{j\omega})$ and $H_1(e^{j\omega})$.

(b) The result from part (a) should convince you that the overall system is LTI. Find the frequency response $H(e^{j\omega})$ of the overall system.

(c) Determine an expression for the impulse response $h[n]$ of the overall system in terms of $h_1[n]$.

2.74. The overall system in the dotted box in Figure P2.74 can be shown to be linear and time invariant.

(a) Determine an expression for $H(e^{j\omega})$, the frequency response of the overall system from the input $x[n]$ to the output $y[n]$, in terms of $H_1(e^{j\omega})$, the frequency response of the internal LTI system. Remember that $(-1)^n = e^{j\pi n}$.

(b) Plot $H(e^{j\omega})$ for the case when the frequency response of the internal LTI system is

$$H_1(e^{j\omega}) = \begin{cases} 1, & |\omega| < \omega_c, \\ 0, & \omega_c < |\omega| \leq \pi. \end{cases}$$

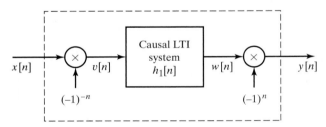

Figure P2.74

2.75. Figure P2.75-1 shows the input–output relationships of Systems A and B, while Figure P2.75-2 contains two possible cascade combinations of these systems.

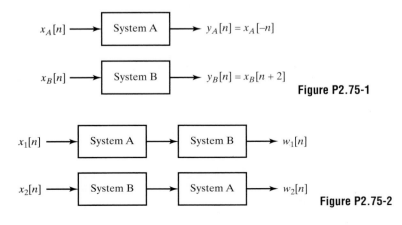

Figure P2.75-1

Figure P2.75-2

If $x_1[n] = x_2[n]$, will $w_1[n]$ and $w_2[n]$ necessarily be equal? If your answer is *yes*, clearly and concisely explain why and demonstrate with an example. If your answer is *not necessarily*, demonstrate with a counterexample.

2.76. Consider the system in Figure P2.76, where the subsystems S_1 and S_2 are LTI.

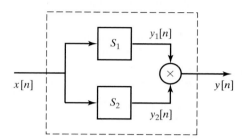

Figure P2.76

(a) Is the overall system enclosed by the dashed box, with input $x[n]$ and output $y[n]$ equal to the product of $y_1[n]$ and $y_2[n]$, guaranteed to be an LTI system? If so, explain your reasoning. If not, provide a counterexample.

(b) Suppose S_1 and S_2 have frequency responses $H_1(e^{j\omega})$ and $H_2(e^{j\omega})$ that are known to be zero over certain regions. Let

$$H_1(e^{j\omega}) = \begin{cases} 0, & |\omega| \leq 0.2\pi, \\ \text{unspecified}, & 0.2\pi < |\omega| \leq \pi, \end{cases}$$

$$H_2(e^{j\omega}) = \begin{cases} \text{unspecified}, & |\omega| \leq 0.4\pi, \\ 0, & 0.4\pi < |\omega| \leq \pi. \end{cases}$$

Suppose also that the input $x[n]$ is known to be bandlimited to 0.3π, i.e.,

$$X(e^{j\omega}) = \begin{cases} \text{unspecified}, & |\omega| < 0.3\pi, \\ 0, & 0.3\pi \leq |\omega| \leq \pi. \end{cases}$$

Over what region of $-\pi \leq \omega < \pi$ is $Y(e^{j\omega})$, the DTFT of $y[n]$, guaranteed to be zero?

2.77. A commonly used numerical operation called the first backward difference is defined as

$$y[n] = \nabla(x[n]) = x[n] - x[n-1],$$

where $x[n]$ is the input and $y[n]$ is the output of the first-backward-difference system.

(a) Show that this system is linear and time invariant.

(b) Find the impulse response of the system.

(c) Find and sketch the frequency response (magnitude and phase).

(d) Show that if

$$x[n] = f[n] * g[n],$$

then

$$\nabla(x[n]) = \nabla(f[n]) * g[n] = f[n] * \nabla(g[n]).$$

(e) Find the impulse response of a system that could be cascaded with the first-difference system to recover the input; i.e., find $h_i[n]$, where

$$h_i[n] * \nabla(x[n]) = x[n].$$

2.78. Let $H(e^{j\omega})$ denote the frequency response of an LTI system with impulse response $h[n]$, where $h[n]$ is, in general, complex.

(a) Using Eq. (2.104), show that $H^*(e^{-j\omega})$ is the frequency response of a system with impulse response $h^*[n]$.

(b) Show that if $h[n]$ is real, the frequency response is conjugate symmetric, i.e., $H(e^{-j\omega}) = H^*(e^{j\omega})$.

2.79. Let $X(e^{j\omega})$ denote the Fourier transform of $x[n]$. Using the Fourier transform synthesis or analysis equations (Eqs. (2.130) and (2.131)), show that

(a) the Fourier transform of $x^*[n]$ is $X^*(e^{-j\omega})$,

(b) the Fourier transform of $x^*[-n]$ is $X^*(e^{j\omega})$.

2.80. Show that for $x[n]$ real, property 7 in Table 2.1 follows from property 1 and that properties 8–11 follow from property 7.

2.81. In Section 2.9, we stated a number of Fourier transform theorems without proof. Using the Fourier synthesis or analysis equations (Eqs. (2.130) and (2.131)), demonstrate the validity of Theorems 1–5 in Table 2.2.

2.82. In Section 2.9.6, it was argued intuitively that

$$Y(e^{j\omega}) = H(e^{j\omega})X(e^{j\omega}), \tag{P2.82-1}$$

when $Y(e^{j\omega})$, $H(e^{j\omega})$, and $X(e^{j\omega})$ are, respectively, the Fourier transforms of the output $y[n]$, impulse response $h[n]$, and input $x[n]$ of an LTI system; i.e.,

$$y[n] = \sum_{k=-\infty}^{\infty} x[k]h[n-k]. \tag{P2.82-2}$$

Verify Eq. (P2.82-1) by applying the Fourier transform to the convolution sum given in Eq. (P2.82-2).

2.83. By applying the Fourier synthesis equation (Eq. (2.130)) to Eq. (2.167) and using Theorem 3 in Table 2.2, demonstrate the validity of the modulation theorem (Theorem 7, Table 2.2).

2.84. Let $x[n]$ and $y[n]$ denote complex sequences and $X(e^{j\omega})$ and $Y(e^{j\omega})$ their respective Fourier transforms.

(a) By using the convolution theorem (Theorem 6 in Table 2.2) and appropriate properties from Table 2.2, determine, in terms of $x[n]$ and $y[n]$, the sequence whose Fourier transform is $X(e^{j\omega})Y^*(e^{j\omega})$.

(b) Using the result in part (a), show that

$$\sum_{n=-\infty}^{\infty} x[n]y^*[n] = \frac{1}{2\pi} \int_{-\pi}^{\pi} X(e^{j\omega})Y^*(e^{j\omega})d\omega. \tag{P2.84-1}$$

Equation (P2.84-1) is a more general form of Parseval's theorem, as given in Section 2.9.5.

(c) Using Eq. (P2.84-1), determine the numerical value of the sum

$$\sum_{n=-\infty}^{\infty} \frac{\sin(\pi n/4)}{2\pi n} \frac{\sin(\pi n/6)}{5\pi n}.$$

2.85. Let $x[n]$ and $X(e^{j\omega})$ represent a sequence and its Fourier transform, respectively. Determine, in terms of $X(e^{j\omega})$, the transforms of $y_s[n]$, $y_d[n]$, and $y_e[n]$ as defined below. In each case, sketch the corresponding output Fourier transform $Y_s(e^{j\omega})$, $Y_d(e^{j\omega})$, and $Y_e(e^{j\omega})$, respectively for $X(e^{j\omega})$ as shown in Figure P2.85.

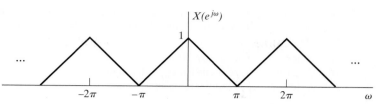

Figure P2.85

(a) Sampler:

$$y_s[n] = \begin{cases} x[n], & n \text{ even,} \\ 0, & n \text{ odd.} \end{cases}$$

Note that $y_s[n] = \frac{1}{2}\{x[n] + (-1)^n x[n]\}$ and $-1 = e^{j\pi}$.

(b) Compressor:

$$y_d[n] = x[2n].$$

(c) Expander:

$$y_e[n] = \begin{cases} x[n/2], & n \text{ even,} \\ 0, & n \text{ odd.} \end{cases}$$

2.86. The two-frequency correlation function $\Phi_x(N, \omega)$ is often used in radar and sonar to evaluate the frequency and travel-time resolution of a signal. For discrete-time signals, we define

$$\Phi_x(N, \omega) = \sum_{n=-\infty}^{\infty} x[n + N]x^*[n - N]e^{-j\omega n}.$$

(a) Show that

$$\Phi_x(-N, -\omega) = \Phi_x^*(N, \omega).$$

(b) If

$$x[n] = A\,a^n u[n], \qquad 0 < a < 1,$$

find $\Phi_x(N, \omega)$. (Assume that $N \geq 0$.)

(c) The function $\Phi_x(N, \omega)$ has a frequency domain dual. Show that

$$\Phi_x(N, \omega) = \frac{1}{2\pi} \int_{-\pi}^{\pi} X\left(e^{j[v+(\omega/2)]}\right) X^*\left(e^{j[v-(\omega/2)]}\right) e^{j2vN}\,dv.$$

2.87. Let $x[n]$ and $y[n]$ be stationary, uncorrelated random signals. Show that if

$$w[n] = x[n] + y[n],$$

then

$$m_w = m_x + m_y \quad \text{and} \quad \sigma_w^2 = \sigma_x^2 + \sigma_y^2.$$

2.88. Let $e[n]$ denote a white-noise sequence, and let $s[n]$ denote a sequence that is uncorrelated with $e[n]$. Show that the sequence

$$y[n] = s[n]e[n]$$

is white, i.e., that

$$E\{y[n]y[n + m]\} = A\,\delta[m],$$

where A is a constant.

2.89. Consider a random signal $x[n] = s[n] + e[n]$, where both $s[n]$ and $e[n]$ are independent zero-mean stationary random signals with autocorrelation functions $\phi_{ss}[m]$ and $\phi_{ee}[m]$, respectively.

    (a) Determine expressions for $\phi_{xx}[m]$ and $\Phi_{xx}(e^{j\omega})$.
    (b) Determine expressions for $\phi_{xe}[m]$ and $\Phi_{xe}(e^{j\omega})$.
    (c) Determine expressions for $\phi_{xs}[m]$ and $\Phi_{xs}(e^{j\omega})$.

2.90. Consider an LTI system with impulse response $h[n] = a^n u[n]$ with $|a| < 1$.

    (a) Compute the deterministic autocorrelation function $\phi_{hh}[m]$ for this impulse response.
    (b) Determine the magnitude-squared function $|H(e^{j\omega})|^2$ for the system.
    (c) Use Parseval's theorem to evaluate the integral

$$\frac{1}{2\pi} \int_{-\pi}^{\pi} |H(e^{j\omega})|^2 d\omega$$

for the system.

2.91. The input to the first-backward-difference system (Example 2.9) is a zero-mean white-noise signal whose autocorrelation function is $\phi_{xx}[m] = \sigma_x^2 \delta[m]$.

    (a) Determine and plot the autocorrelation function and the power spectrum of the corresponding output of the system.
    (b) What is the average power of the output of the system?
    (c) What does this problem tell you about the first backward difference of a noisy signal?

2.92. Let $x[n]$ be a real, stationary, white-noise process, with zero mean and variance σ_x^2. Let $y[n]$ be the corresponding output when $x[n]$ is the input to an LTI system with impulse response $h[n]$. Show that

    (a) $E\{x[n]y[n]\} = h[0]\sigma_x^2$,
    (b) $\sigma_y^2 = \sigma_x^2 \sum_{n=-\infty}^{\infty} h^2[n]$.

2.93. Let $x[n]$ be a real stationary white-noise sequence, with zero mean and variance σ_x^2. Let $x[n]$ be the input to the cascade of two causal LTI discrete-time systems, as shown in Figure P2.93.

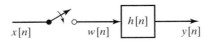

$x[n]$ $h_1[n]$ $y[n]$ $h_2[n]$ $w[n]$ **Figure P2.93**

    (a) Is $\sigma_y^2 = \sigma_x^2 \sum_{k=0}^{\infty} h_1^2[k]$?
    (b) Is $\sigma_w^2 = \sigma_y^2 \sum_{k=0}^{\infty} h_2^2[k]$?
    (c) Let $h_1[n] = a^n u[n]$ and $h_2[n] = b^n u[n]$. Determine the impulse response of the overall system in Figure P2.93, and, from this, determine σ_w^2. Are your answers to parts (b) and (c) consistent?

2.94. Sometimes we are interested in the statistical behavior of an LTI system when the input is a suddenly applied random signal. Such a situation is depicted in Figure P2.94.

$x[n]$ $w[n]$ $h[n]$ $y[n]$

(switch closed at $n = 0$)   **Figure P2.94**

Let $x[n]$ be a stationary white-noise process. The input to the system, $w[n]$, given by

$$w[n] = \begin{cases} x[n], & n \geq 0, \\ 0, & n < 0, \end{cases}$$

is a nonstationary process, as is the output $y[n]$.

(a) Derive an expression for the mean of the output in terms of the mean of the input.
(b) Derive an expression for the autocorrelation sequence $\phi_{yy}[n_1, n_2]$ of the output.
(c) Show that, for large n, the formulas derived in parts (a) and (b) approach the results for stationary inputs.
(d) Assume that $h[n] = a^n u[n]$. Find the mean and mean-square values of the output in terms of the mean and mean-square values of the input. Sketch these parameters as a function of n.

2.95. Let $x[n]$ and $y[n]$ respectively denote the input and output of a system. The input–output relation of a system sometimes used for the purpose of noise reduction in images is given by

$$y[n] = \frac{\sigma_s^2[n]}{\sigma_x^2[n]}(x[n] - m_x[n]) + m_x[n],$$

where

$$\sigma_x^2[n] = \frac{1}{3} \sum_{k=n-1}^{n+1} (x[k] - m_x[n])^2,$$

$$m_x[n] = \frac{1}{3} \sum_{k=n-1}^{n+1} x[k],$$

$$\sigma_s^2[n] = \begin{cases} \sigma_x^2[n] - \sigma_w^2, & \sigma_x^2[n] \geq \sigma_w^2, \\ 0, & \text{otherwise,} \end{cases}$$

and σ_w^2 is a known constant proportional to the noise power.

(a) Is the system linear?
(b) Is the system shift invariant?
(c) Is the system stable?
(d) Is the system causal?
(e) For a fixed $x[n]$, determine $y[n]$ when σ_w^2 is very large (large noise power) and when σ_w^2 is very small (small noise power). Does $y[n]$ make sense for these extreme cases?

2.96. Consider a random process $x[n]$ that is the response of the LTI system shown in Figure P2.96. In the figure, $w[n]$ represents a real zero-mean stationary white-noise process with $E\{w^2[n]\} = \sigma_w^2$.

$$w[n] \longrightarrow \boxed{H(e^{j\omega}) = \frac{1}{1 - 0.5\, e^{-j\omega}}} \longrightarrow x[n]$$ **Figure P2.96**

(a) Express $\mathcal{E}\{x^2[n]\}$ in terms of $\phi_{xx}[n]$ or $\Phi_{xx}(e^{j\omega})$.
(b) Determine $\Phi_{xx}(e^{j\omega})$, the power density spectrum of $x[n]$.
(c) Determine $\phi_{xx}[n]$, the correlation function of $x[n]$.

2.97. Consider an LTI system whose impulse response is real and is given by $h[n]$. Suppose the responses of the system to the two inputs $x[n]$ and $v[n]$ are, respectively, $y[n]$ and $z[n]$, as shown in Figure P2.97.

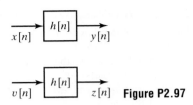

Figure P2.97

The inputs $x[n]$ and $v[n]$ in the figure represent real zero-mean stationary random processes with autocorrelation functions $\phi_{xx}[n]$ and $\phi_{vv}[n]$, cross-correlation function $\phi_{xv}[n]$, power spectra $\Phi_{xx}(e^{j\omega})$ and $\Phi_{vv}(e^{j\omega})$, and cross power spectrum $\Phi_{xv}(e^{j\omega})$.

(a) Given $\phi_{xx}[n]$, $\phi_{vv}[n]$, $\phi_{xv}[n]$, $\Phi_{xx}(e^{j\omega})$, $\Phi_{vv}(e^{j\omega})$, and $\Phi_{xv}(e^{j\omega})$, determine $\Phi_{yz}(e^{j\omega})$, the cross power spectrum of $y[n]$ and $z[n]$, where $\Phi_{yz}(e^{j\omega})$ is defined by

$$\phi_{yz}[n] \overset{\mathcal{F}}{\longleftrightarrow} \Phi_{yz}(e^{j\omega}),$$

with $\phi_{yz}[n] = E\{y[k]z[k-n]\}$.

(b) Is the cross power spectrum $\Phi_{xv}(e^{j\omega})$ always nonnegative; i.e., is $\Phi_{xv}(e^{j\omega}) \geq 0$ for all ω? Justify your answer.

2.98. Consider the LTI system shown in Figure P2.98. The input to this system, $e[n]$, is a stationary zero-mean white-noise signal with average power σ_e^2. The first system is a backward-difference system as defined by $f[n] = e[n] - e[n-1]$. The second system is an ideal lowpass filter with frequency response

$$H_2(e^{j\omega}) = \begin{cases} 1, & |\omega| < \omega_c, \\ 0, & \omega_c < |\omega| \leq \pi. \end{cases}$$

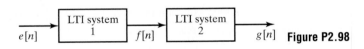

Figure P2.98

(a) Determine an expression for $\Phi_{ff}(e^{j\omega})$, the power spectrum of $f[n]$, and plot this expression for $-2\pi < \omega < 2\pi$.

(b) Determine an expression for $\phi_{ff}[m]$, the autocorrelation function of $f[n]$.

(c) Determine an expression for $\Phi_{gg}(e^{j\omega})$, the power spectrum of $g[n]$, and plot this expression for $-2\pi < \omega < 2\pi$.

(d) Determine an expression for σ_g^2, the average power of the output.

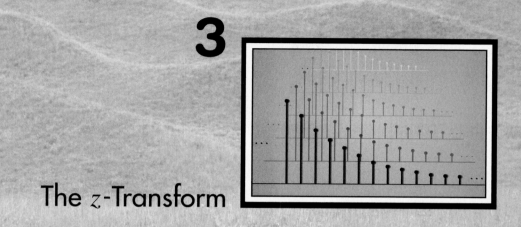

3

The z-Transform

3.0 INTRODUCTION

In this chapter, we develop the z-transform representation of a sequence and study how the properties of a sequence are related to the properties of its z-transform. The z-transform for discrete-time signals is the counterpart of the Laplace transform for continuous-time signals, and they each have a similar relationship to the corresponding Fourier transform. One motivation for introducing this generalization is that the Fourier transform does not converge for all sequences, and it is useful to have a generalization of the Fourier transform that encompasses a broader class of signals. A second advantage is that in analytical problems, the z-transform notation is often more convenient than the Fourier transform notation.

3.1 z-TRANSFORM

The Fourier transform of a sequence $x[n]$ was defined in Chapter 2 as

$$X(e^{j\omega}) = \sum_{n=-\infty}^{\infty} x[n]e^{-j\omega n}. \tag{3.1}$$

The z-transform of a sequence $x[n]$ is defined as

$$X(z) = \sum_{n=-\infty}^{\infty} x[n]z^{-n}. \tag{3.2}$$

This equation is, in general, an infinite sum or infinite power series, with z considered to be a complex variable. Sometimes it is useful to consider Eq. (3.2) as an operator that transforms a sequence into a function. That is, the *z-transform operator* $\mathcal{Z}\{\cdot\}$, defined as

$$\mathcal{Z}\{x[n]\} = \sum_{n=-\infty}^{\infty} x[n]z^{-n} = X(z), \qquad (3.3)$$

transforms the sequence $x[n]$ into the function $X(z)$, where z is a continuous complex variable. The unique correspondence between a sequence and its z-transform will be indicated by the notation

$$x[n] \overset{\mathcal{Z}}{\longleftrightarrow} X(z). \qquad (3.4)$$

The z-transform, as we have defined it in Eq. (3.2), is often referred to as the *two-sided* or *bilateral z-transform*, in contrast to the *one-sided* or *unilateral z-transform*, which is defined as

$$\mathcal{X}(z) = \sum_{n=0}^{\infty} x[n]z^{-n}. \qquad (3.5)$$

Clearly, the bilateral and unilateral transforms are identical if $x[n] = 0$ for $n < 0$, but they differ otherwise. We shall give a brief introduction to the properties of the unilateral z-transform in Section 3.6.

It is evident from a comparison of Eqs. (3.1) and (3.2) that there is a close relationship between the Fourier transform and the z-transform. In particular, if we replace the complex variable z in Eq. (3.2) with the complex quantity $e^{j\omega}$, then the z-transform reduces to the Fourier transform. This is the motivation for the notation $X(e^{j\omega})$ for the Fourier transform. When it exists, the Fourier transform is simply $X(z)$ with $z = e^{j\omega}$. This corresponds to restricting z to have unity magnitude; i.e., for $|z| = 1$, the z-transform corresponds to the Fourier transform. More generally, we can express the complex variable z in polar form as

$$z = re^{j\omega}.$$

With z expressed in this form, Eq. (3.2) becomes

$$X(re^{j\omega}) = \sum_{n=-\infty}^{\infty} x[n](re^{j\omega})^{-n},$$

or

$$X(re^{j\omega}) = \sum_{n=-\infty}^{\infty} (x[n]r^{-n})e^{-j\omega n}. \qquad (3.6)$$

Equation (3.6) can be interpreted as the Fourier transform of the product of the original sequence $x[n]$ and the exponential sequence r^{-n}. For $r = 1$, Eq. (3.6) reduces to the Fourier transform of $x[n]$.

Since the z-transform is a function of a complex variable, it is convenient to describe and interpret it using the complex z-plane. In the z-plane, the contour corresponding to $|z| = 1$ is a circle of unit radius, as illustrated in Figure 3.1. This contour, referred to as the *unit circle*, is the set of points $z = e^{j\omega}$ for $0 \le \omega < 2\pi$. The z-transform evaluated on the unit circle corresponds to the Fourier transform. Note that ω is the angle between the vector from the origin to a point z on the unit circle and the real axis

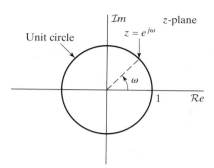

Figure 3.1   The unit circle in the complex z-plane.

of the complex z-plane. If we evaluate $X(z)$ at points on the unit circle in the z-plane beginning at $z = 1$ (i.e., $\omega = 0$) through $z = j$ (i.e., $\omega = \pi/2$) to $z = -1$ (i.e., $\omega = \pi$), we obtain the Fourier transform for $0 \le \omega \le \pi$. Continuing around the unit circle would correspond to examining the Fourier transform from $\omega = \pi$ to $\omega = 2\pi$ or, equivalently, from $\omega = -\pi$ to $\omega = 0$. In Chapter 2, the Fourier transform was displayed on a linear frequency axis. Interpreting the Fourier transform as the z-transform on the unit circle in the z-plane corresponds conceptually to wrapping the linear frequency axis around the unit circle with $\omega = 0$ at $z = 1$ and $\omega = \pi$ at $z = -1$. With this interpretation, the inherent periodicity in frequency of the Fourier transform is captured naturally, since a change of angle of 2π radians in the z-plane corresponds to traversing the unit circle once and returning to exactly the same point.

As we discussed in Chapter 2, the power series representing the Fourier transform does not converge for all sequences; i.e., the infinite sum may not always be finite. Similarly, the z-transform does not converge for all sequences or for all values of z. For any given sequence, the set of values of z for which the z-transform power series converges is called the *region of convergence* (ROC), of the z-transform. As we stated in Section 2.7, if the sequence is absolutely summable, the Fourier transform converges to a continuous function of ω. Applying this criterion to Eq. (3.6) leads to the condition

$$|X(re^{j\omega})| \le \sum_{n=-\infty}^{\infty} |x[n]r^{-n}| < \infty \qquad (3.7)$$

for convergence of the z-transform. From Eq. (3.7) it follows that, because of the multiplication of the sequence by the real exponential r^{-n}, it is possible for the z-transform to converge even if the Fourier transform ($r = 1$) does not. For example, the sequence $x[n] = u[n]$ is not absolutely summable, and therefore, the Fourier transform power series does not converge absolutely. However, $r^{-n}u[n]$ is absolutely summable if $r > 1$. This means that the z-transform for the unit step exists with an ROC $r = |z| > 1$.

Convergence of the power series of Eq. (3.2) for a given sequence depends only on $|z|$, since $|X(z)| < \infty$ if

$$\sum_{n=-\infty}^{\infty} |x[n]||z|^{-n} < \infty, \qquad (3.8)$$

i.e., the ROC of the power series in Eq. (3.2) consists of all values of z such that the inequality in Eq. (3.8) holds. Thus, if some value of z, say, $z = z_1$, is in the ROC,

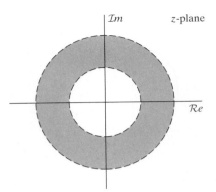

Figure 3.2 The ROC as a ring in the z-plane. For specific cases, the inner boundary can extend inward to the origin, and the ROC becomes a disc. For other cases, the outer boundary can extend outward to infinity.

then all values of z on the circle defined by $|z| = |z_1|$ will also be in the ROC. As one consequence of this, the ROC will consist of a ring in the z-plane centered about the origin. Its outer boundary will be a circle (or the ROC may extend outward to infinity), and its inner boundary will be a circle (or it may extend inward to include the origin). This is illustrated in Figure 3.2. If the ROC includes the unit circle, then this of course implies convergence of the z-transform for $|z| = 1$, or equivalently, the Fourier transform of the sequence converges. Conversely, if the ROC does not include the unit circle, the Fourier transform does not converge absolutely.

A power series of the form of Eq. (3.2) is a Laurent series. Therefore, a number of elegant and powerful theorems from the theory of functions of a complex variable can be employed in the study of the z-transform. (See Brown and Churchill (2007).) For example, a Laurent series, and therefore the z-transform, represents an analytic function at every point inside the ROC; hence, the z-transform and all its derivatives must be continuous functions of z within the ROC. This implies that if the ROC includes the unit circle, then the Fourier transform and all its derivatives with respect to ω must be continuous functions of ω. Also, from the discussion in Section 2.7, the sequence must be absolutely summable, i.e., a stable sequence.

Uniform convergence of the z-transform requires absolute summability of the exponentially weighted sequence, as stated in Eq. (3.7). Neither of the sequences

$$x_1[n] = \frac{\sin \omega_c n}{\pi n}, \qquad -\infty < n < \infty, \tag{3.9}$$

and

$$x_2[n] = \cos \omega_0 n, \qquad -\infty < n < \infty, \tag{3.10}$$

is absolutely summable. Furthermore, neither of these sequences multiplied by r^{-n} would be absolutely summable for any value of r. Thus, neither of these sequences has a z-transform that converges absolutely for any z. However, we showed in Section 2.7 that even though a sequence such as $x_1[n]$ in Eq. (3.9) is not absolutely summable, it does have finite energy (i.e., it is square-summable), and the Fourier transform converges in the mean-square sense to a discontinuous periodic function. Similarly, the sequence $x_2[n]$ in Eq. (3.10) is neither absolutely nor square summable, but a useful Fourier transform for $x_2[n]$ can be defined using impulse functions (i.e., generalized functions or Dirac delta functions). In both cases, the Fourier transforms are not continuous, infinitely differentiable functions, so they cannot result from evaluating a z-transform on the unit circle. Thus, in such cases it is not strictly correct to think of the Fourier transform as being the z-transform evaluated on the unit circle, although we nevertheless continue to use the notation $X(e^{j\omega})$ always to denote the discrete-time Fourier transform.

The z-transform is most useful when the infinite sum can be expressed in closed form, i.e., when it can be "summed" and expressed as a simple mathematical formula. Among the most important and useful z-transforms are those for which $X(z)$ is equal to a rational function inside the ROC, i.e.,

$$X(z) = \frac{P(z)}{Q(z)}, \tag{3.11}$$

where $P(z)$ and $Q(z)$ are polynomials in z. In general, the values of z for which $X(z) = 0$ are the zeros of $X(z)$, and the values of z for which $X(z)$ is infinite are the poles of $X(z)$. In the case of a rational function as in Eq. (3.11), the zeros are the roots of the numerator polynomial and the poles (for finite values of z) are the roots of the denominator polynomial. For rational z-transforms, a number of important relationships exist between the locations of poles of $X(z)$ and the ROC of the z-transform. We discuss these more specifically in Section 3.2. However, we first illustrate the z-transform with several examples.

Example 3.1   Right-Sided Exponential Sequence

Consider the signal $x[n] = a^n u[n]$, where a denotes a real or complex number. Because it is nonzero only for $n \geq 0$, this is an example of the class of *right-sided* sequences, which are sequences that begin at some time N_1 and have nonzero values only for $N_1 \leq n < \infty$; i.e., they occupy the right side of a plot of the sequence. From Eq. (3.2),

$$X(z) = \sum_{n=-\infty}^{\infty} a^n u[n] z^{-n} = \sum_{n=0}^{\infty} (az^{-1})^n.$$

For convergence of $X(z)$, we require that

$$\sum_{n=0}^{\infty} |az^{-1}|^n < \infty.$$

Thus, the ROC is the range of values of z for which $|az^{-1}| < 1$ or, equivalently, $|z| > |a|$. Inside the ROC, the infinite series converges to

$$X(z) = \sum_{n=0}^{\infty} (az^{-1})^n = \frac{1}{1 - az^{-1}} = \frac{z}{z - a}, \qquad |z| > |a|. \tag{3.12}$$

To obtain this closed-form expression, we have used the familiar formula for the sum of terms of a geometric series (see Jolley, 1961). The z-transform of the sequence $x[n] = a^n u[n]$ has an ROC for any finite value of $|a|$. For $a = 1$, $x[n]$ is the unit step sequence with z-transform

$$X(z) = \frac{1}{1 - z^{-1}}, \qquad |z| > 1. \tag{3.13}$$

If $|a| < 1$, the Fourier transform of $x[n] = a^n u[n]$ converges to

$$X(e^{j\omega}) = \frac{1}{1 - ae^{-j\omega}}. \tag{3.14}$$

However, if $a \geq 1$, the Fourier transform of the right-sided exponential sequence does not converge.

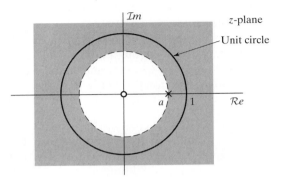

Figure 3.3   Pole–zero plot and ROC for Example 3.1.

In Example 3.1, the infinite sum is equal to a rational function of z inside the ROC. For most purposes, this rational function is a much more convenient representation than the infinite sum. We will see that any sequence that can be represented as a sum of exponentials can equivalently be represented by a rational z-transform. Such a z-transform is determined to within a constant multiplier by its zeros and its poles. For this example, there is one zero, at $z = 0$, and one pole, at $z = a$. The pole–zero plot and the ROC for Example 3.1 are shown in Figure 3.3 where the symbol "$\circ$" denotes the zero and the symbol "$\times$" the pole. For $|a| \geq 1$, the ROC does not include the unit circle, consistent with the fact that, for these values of a, the Fourier transform of the exponentially growing sequence $a^n u[n]$ does not converge.

Example 3.2   Left-Sided Exponential Sequence

Now let

$$x[n] = -a^n u[-n - 1] = \begin{cases} -a^n & n \leq -1 \\ 0 & n > -1. \end{cases}$$

Since the sequence is nonzero only for $n \leq -1$, this is a *left-sided* sequence. The z-transform in this case is

$$X(z) = -\sum_{n=-\infty}^{\infty} a^n u[-n - 1]z^{-n} = -\sum_{n=-\infty}^{-1} a^n z^{-n}$$

$$= -\sum_{n=1}^{\infty} a^{-n} z^n = 1 - \sum_{n=0}^{\infty} (a^{-1} z)^n. \tag{3.15}$$

If $|a^{-1}z| < 1$ or, equivalently, $|z| < |a|$, the last sum in Eq. (3.15) converges, and using again the formula for the sum of terms in a geometric series,

$$X(z) = 1 - \frac{1}{1 - a^{-1}z} = \frac{1}{1 - az^{-1}} = \frac{z}{z - a}, \qquad |z| < |a|. \tag{3.16}$$

The pole–zero plot and ROC for this example are shown in Figure 3.4.

Note that for $|a| < 1$, the sequence $-a^n u[-n-1]$ grows exponentially as $n \to -\infty$, and thus, the Fourier transform does not exist. However, if $|a| > 1$ the Fourier transform is

$$X(e^{j\omega}) = \frac{1}{1 - ae^{-j\omega}}, \tag{3.17}$$

which is identical in form to Eq. (3.14). At first glance, this would appear to violate the uniqueness of the Fourier transform. However, this ambiguity is resolved if we recall that Eq. (3.14) is the Fourier transform of $a^n u[n]$ if $|a| < 1$, while Eq. (3.17) is the Fourier transform of $-a^n u[-n-1]$ when $|a| > 1$.

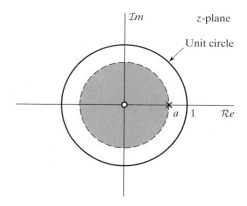

Figure 3.4   Pole–zero plot and ROC for Example 3.2.

Comparing Eqs. (3.12) and (3.16) and Figures 3.3 and 3.4, we see that the sequences and, therefore, the infinite sums are different; however, the algebraic expressions for $X(z)$ and the corresponding pole–zero plots are identical in Examples 3.1 and 3.2. The z-transforms differ only in the ROC. This emphasizes the need for specifying both the algebraic expression and the ROC for the bilateral z-transform of a given sequence. Also, in both examples, the sequences were exponentials and the resulting z-transforms were rational. In fact, as is further suggested by the next example, $X(z)$ will be rational whenever $x[n]$ is a linear combination of real or complex exponentials.

Example 3.3   Sum of Two Exponential Sequences

Consider a signal that is the sum of two real exponentials:

$$x[n] = \left(\frac{1}{2}\right)^n u[n] + \left(-\frac{1}{3}\right)^n u[n]. \tag{3.18}$$

The z-transform is

$$X(z) = \sum_{n=-\infty}^{\infty} \left\{ \left(\frac{1}{2}\right)^n u[n] + \left(-\frac{1}{3}\right)^n u[n] \right\} z^{-n}$$

$$= \sum_{n=-\infty}^{\infty} \left(\frac{1}{2}\right)^n u[n]z^{-n} + \sum_{n=-\infty}^{\infty} \left(-\frac{1}{3}\right)^n u[n]z^{-n} \tag{3.19}$$

$$= \sum_{n=0}^{\infty} \left(\frac{1}{2} z^{-1} \right)^n + \sum_{n=0}^{\infty} \left(-\frac{1}{3} z^{-1} \right)^n$$

$$= \frac{1}{1 - \frac{1}{2} z^{-1}} + \frac{1}{1 + \frac{1}{3} z^{-1}} = \frac{2 \left(1 - \frac{1}{12} z^{-1} \right)}{\left(1 - \frac{1}{2} z^{-1} \right) \left(1 + \frac{1}{3} z^{-1} \right)}$$

$$= \frac{2z \left(z - \frac{1}{12} \right)}{\left(z - \frac{1}{2} \right) \left(z + \frac{1}{3} \right)}. \tag{3.20}$$

For convergence of $X(z)$, both sums in Eq. (3.19) must converge, which requires that both $\left| \frac{1}{2} z^{-1} \right| < 1$ and $\left| \left(-\frac{1}{3} \right) z^{-1} \right| < 1$ or, equivalently, $|z| > \frac{1}{2}$ and $|z| > \frac{1}{3}$. Thus, the ROC is the region of overlap, $|z| > \frac{1}{2}$. The pole–zero plot and ROC for the z-transform of each of the individual terms and for the combined signal are shown in Figure 3.5.

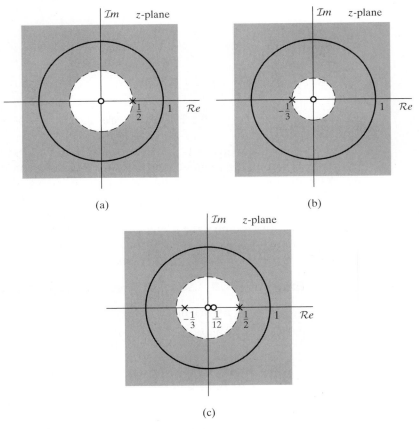

(a)   (b)

(c)

Figure 3.5   Pole–zero plot and ROC for the individual terms and the sum of terms in Examples 3.3 and 3.4. (a) $1/(1 - \frac{1}{2} z^{-1})$, $|z| > \frac{1}{2}$. (b) $1/(1 + \frac{1}{3} z^{-1})$, $|z| > \frac{1}{3}$. (c) $1/(1 - \frac{1}{2} z^{-1}) + 1/(1 + \frac{1}{3} z^{-1})$, $|z| > \frac{1}{2}$.

In each of the preceding examples, we started with the definition of the sequence and manipulated each of the infinite sums into a form whose sum could be recognized. When the sequence is recognized as a sum of exponential sequences of the form of Examples 3.1 and 3.2, the z-transform can be obtained much more simply using the fact that the z-transform operator is linear. Specifically, from the definition of the z-transform in Eq. (3.2), if $x[n]$ is the sum of two terms, then $X(z)$ will be the sum of the corresponding z-transforms of the individual terms. The ROC will be the intersection of the individual ROCs, i.e., the values of z for which both individual sums converge. We have already demonstrated the linearity property in obtaining Eq. (3.19) in Example 3.3. Example 3.4 shows how the z-transform in Example 3.3 can be obtained in a much more straightforward manner by expressing $x[n]$ as the sum of two sequences.

Example 3.4   Sum of Two Exponentials (Again)

Again, let $x[n]$ be given by Eq. (3.18). Then using the general result of Example 3.1 with $a = \frac{1}{2}$ and $a = -\frac{1}{3}$, the z-transforms of the two individual terms are easily seen to be

$$\left(\frac{1}{2}\right)^n u[n] \xleftrightarrow{\;\mathcal{Z}\;} \frac{1}{1 - \frac{1}{2}z^{-1}}, \qquad |z| > \frac{1}{2}, \tag{3.21}$$

$$\left(-\frac{1}{3}\right)^n u[n] \xleftrightarrow{\;\mathcal{Z}\;} \frac{1}{1 + \frac{1}{3}z^{-1}}, \qquad |z| > \frac{1}{3}, \tag{3.22}$$

and, consequently,

$$\left(\frac{1}{2}\right)^n u[n] + \left(-\frac{1}{3}\right)^n u[n] \xleftrightarrow{\;\mathcal{Z}\;} \frac{1}{1 - \frac{1}{2}z^{-1}} + \frac{1}{1 + \frac{1}{3}z^{-1}}, \qquad |z| > \frac{1}{2}, \tag{3.23}$$

as determined in Example 3.3. The pole–zero plot and ROC for the z-transform of each of the individual terms and for the combined signal are shown in Figure 3.5.

All the major points of Examples 3.1–3.4 are summarized in Example 3.5.

Example 3.5   Two-Sided Exponential Sequence

Consider the sequence

$$x[n] = \left(-\frac{1}{3}\right)^n u[n] - \left(\frac{1}{2}\right)^n u[-n - 1]. \tag{3.24}$$

Note that this sequence grows exponentially as $n \to -\infty$. Using the general result of Example 3.1 with $a = -\frac{1}{3}$, we obtain

$$\left(-\frac{1}{3}\right)^n u[n] \xleftrightarrow{\;\mathcal{Z}\;} \frac{1}{1 + \frac{1}{3}z^{-1}}, \qquad |z| > \frac{1}{3},$$

and using the result of Example 3.2 with $a = \frac{1}{2}$ yields

$$-\left(\frac{1}{2}\right)^n u[-n - 1] \xleftrightarrow{\;\mathcal{Z}\;} \frac{1}{1 - \frac{1}{2}z^{-1}}, \qquad |z| < \frac{1}{2}.$$

Thus, by the linearity of the z-transform,

$$X(z) = \frac{1}{1 + \frac{1}{3}z^{-1}} + \frac{1}{1 - \frac{1}{2}z^{-1}}, \quad \frac{1}{3} < |z| \text{ and } |z| < \frac{1}{2},$$

$$= \frac{2\left(1 - \frac{1}{12}z^{-1}\right)}{\left(1 + \frac{1}{3}z^{-1}\right)\left(1 - \frac{1}{2}z^{-1}\right)} = \frac{2z\left(z - \frac{1}{12}\right)}{\left(z + \frac{1}{3}\right)\left(z - \frac{1}{2}\right)}. \tag{3.25}$$

In this case, the ROC is the annular region $\frac{1}{3} < |z| < \frac{1}{2}$. Note that the rational function in this example is identical to the rational function in Example 3.4, but the ROC is different in this case. The pole–zero plot and the ROC for this example are shown in Figure 3.6.

Since the ROC does not contain the unit circle, the sequence in Eq. (3.24) does not have a Fourier transform.

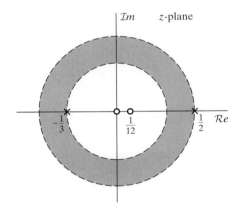

Figure 3.6   Pole–zero plot and ROC for Example 3.5.

In each of the preceding examples, we expressed the z-transform both as a ratio of polynomials in z and as a ratio of polynomials in z^{-1}. From the form of the definition of the z-transform as given in Eq. (3.2), we see that, for sequences that are zero for $n < 0$, $X(z)$ involves only negative powers of z. Thus, for this class of signals, it is particularly convenient for $X(z)$ to be expressed in terms of polynomials in z^{-1} rather than z; however, even when $x[n]$ is nonzero for $n < 0$, $X(z)$ can still be expressed in terms of factors of the form $(1 - az^{-1})$. It should be remembered that such a factor introduces both a pole and a zero, as illustrated by the algebraic expressions in the preceding examples.

These examples show that infinitely long exponential sequences have z-transforms that can be expressed as rational functions of either z or z^{-1}. The case where the sequence has finite length also has a rather simple form. If the sequence is nonzero only in the interval $N_1 \leq n \leq N_2$, the z-transform

$$X(z) = \sum_{n=N_1}^{N_2} x[n]z^{-n} \tag{3.26}$$

has no problems of convergence, as long as each of the terms $|x[n]z^{-n}|$ is finite. In general, it may not be possible to express the sum of a finite set of terms in a closed form, but in such cases it may be unnecessary. For example, if $x[n] = \delta[n] + \delta[n-5]$, then $X(z) = 1 + z^{-5}$, which is finite for $|z| > 0$. An example of a case where a finite number of terms can be summed to produce a more compact representation of the z-transform is given in Example 3.6.

Example 3.6   Finite-Length Truncated Exponential Sequence

Consider the signal

$$x[n] = \begin{cases} a^n, & 0 \le n \le N-1, \\ 0, & \text{otherwise.} \end{cases}$$

Then

$$X(z) = \sum_{n=0}^{N-1} a^n z^{-n} = \sum_{n=0}^{N-1} (az^{-1})^n = \frac{1 - (az^{-1})^N}{1 - az^{-1}} = \frac{1}{z^{N-1}} \frac{z^N - a^N}{z - a}, \tag{3.27}$$

where we have used the general formula in Eq. (2.55) to obtain a closed-form expression for the sum of the finite series. The ROC is determined by the set of values of z for which

$$\sum_{n=0}^{N-1} |az^{-1}|^n < \infty.$$

Since there are only a finite number of nonzero terms, the sum will be finite as long as az^{-1} is finite, which in turn requires only that $|a| < \infty$ and $z \ne 0$. Thus, assuming that $|a|$ is finite, the ROC includes the entire z-plane, with the exception of the origin ($z = 0$). The pole–zero plot for this example, with $N = 16$ and a real and between zero and unity, is shown in Figure 3.7. Specifically, the N roots of the numerator polynomial are at z-plane locations

$$z_k = ae^{j(2\pi k/N)}, \qquad k = 0, 1, \ldots, N-1. \tag{3.28}$$

(Note that these values satisfy the equation $z^N = a^N$, and when $a = 1$, these complex values are the Nth roots of unity.) The zero corresponding to $k = 0$ cancels the pole at $z = a$. Consequently, there are no poles other than the $N - 1$ poles at the origin. The remaining zeros are at z-plane locations

$$z_k = ae^{j(2\pi k/N)}, \qquad k = 1, \ldots, N-1. \tag{3.29}$$

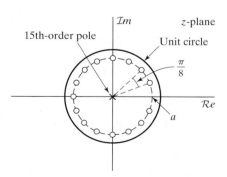

Figure 3.7   Pole–zero plot for Example 3.6 with $N = 16$ and a real such that $0 < a < 1$. The ROC in this example consists of all values of z except $z = 0$.

TABLE 3.1   SOME COMMON z-TRANSFORM PAIRS

Sequence	Transform	ROC
1. $\delta[n]$	1	All z
2. $u[n]$	$\dfrac{1}{1-z^{-1}}$	$|z|>1$
3. $-u[-n-1]$	$\dfrac{1}{1-z^{-1}}$	$|z|<1$
4. $\delta[n-m]$	z^{-m}	All z except 0 (if $m>0$) or ∞ (if $m<0$)
5. $a^n u[n]$	$\dfrac{1}{1-az^{-1}}$	$|z|>|a|$
6. $-a^n u[-n-1]$	$\dfrac{1}{1-az^{-1}}$	$|z|<|a|$
7. $na^n u[n]$	$\dfrac{az^{-1}}{(1-az^{-1})^2}$	$|z|>|a|$
8. $-na^n u[-n-1]$	$\dfrac{az^{-1}}{(1-az^{-1})^2}$	$|z|<|a|$
9. $\cos(\omega_0 n)u[n]$	$\dfrac{1-\cos(\omega_0)z^{-1}}{1-2\cos(\omega_0)z^{-1}+z^{-2}}$	$|z|>1$
10. $\sin(\omega_0 n)u[n]$	$\dfrac{\sin(\omega_0)z^{-1}}{1-2\cos(\omega_0)z^{-1}+z^{-2}}$	$|z|>1$
11. $r^n\cos(\omega_0 n)u[n]$	$\dfrac{1-r\cos(\omega_0)z^{-1}}{1-2r\cos(\omega_0)z^{-1}+r^2z^{-2}}$	$|z|>r$
12. $r^n\sin(\omega_0 n)u[n]$	$\dfrac{r\sin(\omega_0)z^{-1}}{1-2r\cos(\omega_0)z^{-1}+r^2z^{-2}}$	$|z|>r$
13. $\begin{cases}a^n, & 0\le n\le N-1,\\ 0, & \text{otherwise}\end{cases}$	$\dfrac{1-a^N z^{-N}}{1-az^{-1}}$	$|z|>0$

The transform pairs corresponding to some of the preceding examples, as well as a number of other commonly encountered z-transform pairs, are summarized in Table 3.1. We will see that these basic transform pairs are very useful in finding z-transforms given a sequence or, conversely, in finding the sequence corresponding to a given z-transform.

3.2 PROPERTIES OF THE ROC FOR THE z-TRANSFORM

The examples of the previous section suggest that the properties of the ROC depend on the nature of the signal. These properties are summarized in this section with some discussion and intuitive justification. We assume specifically that the algebraic expression for the z-transform is a rational function and that $x[n]$ has finite amplitude, except possibly at $n=\infty$ or $n=-\infty$.

PROPERTY 1: The ROC will either be of the form $0 \le r_R < |z|$, or $|z| < r_L \le \infty$, or, in general the annulus, i.e., $0 \le r_R < |z| < r_L \le \infty$.

PROPERTY 2: The Fourier transform of $x[n]$ converges absolutely if and only if the ROC of the z-transform of $x[n]$ includes the unit circle.

PROPERTY 3: The ROC cannot contain any poles.

PROPERTY 4: If $x[n]$ is a *finite-duration sequence*, i.e., a sequence that is zero except in a finite interval $-\infty < N_1 \le n \le N_2 < \infty$, then the ROC is the entire z-plane, except possibly $z = 0$ or $z = \infty$.

PROPERTY 5: If $x[n]$ is a *right-sided sequence*, i.e., a sequence that is zero for $n < N_1 < \infty$, the ROC extends outward from the *outermost* (i.e., largest magnitude) finite pole in $X(z)$ to (and possibly including) $z = \infty$.

PROPERTY 6: If $x[n]$ is a *left-sided sequence*, i.e., a sequence that is zero for $n > N_2 > -\infty$, the ROC extends inward from the *innermost* (smallest magnitude) nonzero pole in $X(z)$ to (and possibly including) $z = 0$.

PROPERTY 7: A *two-sided sequence* is an infinite-duration sequence that is neither right sided nor left sided. If $x[n]$ is a two-sided sequence, the ROC will consist of a ring in the z-plane, bounded on the interior and exterior by a pole and, consistent with Property 3, not containing any poles.

PROPERTY 8: The ROC must be a connected region.

Property 1 summarizes the general shape of the ROC. As discussed in Section 3.1, it results from the fact that the condition for convergence of Eq. (3.2) is given by Eq. (3.7) repeated here as

$$\sum_{n=-\infty}^{\infty} |x[n]||r^{-n} < \infty \tag{3.30}$$

where $r = |z|$. Equation (3.30) shows that for a given $x[n]$, convergence is dependent only on $r = |z|$ (i.e., not on the angle of z). Note that if the z-transform converges for $|z| = r_0$, then we may decrease r until the z-transform does not converge. This is the value $|z| = r_R$ such that $|x[n]||r^{-n}$ grows too fast (or decays too slowly) as $n \to \infty$, so that the series is not absolutely summable. This defines r_R. The z-transform cannot converge for $r \le r_R$ since r^{-n} will grow even faster. Similarly, the outer boundary r_L can be found by increasing r from r_0 and considering what happens when $n \to -\infty$.

Property 2 is a consequence of the fact that Eq. (3.2) reduces to the Fourier transform when $|z| = 1$. Property 3 follows from the recognition that $X(z)$ is infinite at a pole and therefore, by definition, does not converge.

Property 4 follows from the fact that the z-transform of a finite-length sequence is a finite sum of finite powers of z, i.e.,

$$X(z) = \sum_{n=N_1}^{N_2} x[n]z^{-n}.$$

Therefore, $|X(z)| < \infty$ for all z except $z = 0$ when $N_2 > 0$ and/or $z = \infty$ when $N_1 < 0$.

Properties 5 and 6 are special cases of Property 1. To interpret Property 5 for rational z-transforms, note that a sequence of the form

$$x[n] = \sum_{k=1}^{N} A_k (d_k)^n u[n] \tag{3.31}$$

is an example of a right-sided sequence composed of exponential sequences with amplitudes A_k and exponential factors d_k. While this is not the most general right-sided sequence, it will suffice to illustrate Property 5. More general right-sided sequences can be formed by adding finite-length sequences or shifting the exponential sequences by finite amounts; however, such modifications to Eq. (3.31) would not change our conclusions. Invoking the linearity property, the z-transform of $x[n]$ in Eq. (3.31) is

$$X(z) = \sum_{k=1}^{N} \underbrace{\frac{A_k}{1 - d_k z^{-1}}}_{|z| > |d_k|} . \tag{3.32}$$

Note that for values of z that lie in all of the individual ROCs, $|z| > |d_k|$, the terms can be combined into one rational function with common denominator

$$\prod_{k=1}^{N} (1 - d_k z^{-1});$$

i.e., the poles of $X(z)$ are located at $z = d_1, \ldots, d_N$. Assume for convenience that the poles are ordered so that d_1 has the smallest magnitude, corresponding to the innermost pole, and d_N has the largest magnitude, corresponding to the outermost pole. The least rapidly increasing of these exponentials, as n increases, is the one corresponding to the innermost pole, i.e., d_1, and the most slowly decaying (or most rapidly growing) is the one corresponding to the outermost pole, i.e., d_N. Not surprisingly, d_N determines the inner boundary of the ROC which is the intersection of the regions $|z| > |d_k|$. That is, the ROC of the z-transform of a right-sided sum of exponential sequences is

$$|z| > |d_N| = \max_k |d_k| = r_R, \tag{3.33}$$

i.e., the ROC is outside the outermost pole, extending to infinity. If a right-sided sequence begins at $n = N_1 < 0$, then the ROC will not include $|z| = \infty$.

Another way of arriving at Property 5 is to apply Eq. (3.30) to Eq. (3.31) obtaining

$$\sum_{n=0}^{\infty} \left| \sum_{k=1}^{N} A_k (d_k)^n \right| r^{-n} \leq \sum_{k=1}^{N} |A_k| \left(\sum_{n=0}^{\infty} |d_k/r|^n \right) < \infty, \tag{3.34}$$

which shows that convergence is guaranteed if all the sequences $|d_k/r|^n$ are absolutely summable. Again, since $|d_N|$ is the largest pole magnitude, we choose $|d_N/r| < 1$, or $r > |d_N|$.

For Property 6, which is concerned with left-sided sequences, an exactly parallel argument can be carried out for a sum of left-sided exponential sequences to show that the ROC will be defined by the pole with the smallest magnitude. With the same assumption on the ordering of the poles, the ROC will be

$$|z| < |d_1| = \min_k |d_k| = r_L, \tag{3.35}$$

i.e., the ROC is inside the innermost pole. If the left-sided sequence has nonzero values for positive values of n, then the ROC will not include the origin, $z = 0$. Since $x[n]$ now extends to $-\infty$ along the negative n-axis, r must be restricted so that for each d_k, the exponential sequence $(d_k r^{-1})^n$ decays to zero as n *decreases* toward $-\infty$.

For right-sided sequences, the ROC is dictated by the exponential weighting r^{-n} required to have all exponential terms decay to zero for increasing n; for left-sided sequences, the exponential weighting must be such that all exponential terms decay to zero for decreasing n. Property 7 follows from the fact that for two-sided sequences, the exponential weighting needs to be balanced, since if it decays too fast for increasing n, it may grow too quickly for decreasing n and vice versa. More specifically, for two-sided sequences, some of the poles contribute only for $n > 0$ and the rest only for $n < 0$. The ROC is bounded on the inside by the pole with the largest magnitude that contributes for $n > 0$ and on the outside by the pole with the smallest magnitude that contributes for $n < 0$.

Property 8 is intuitively suggested by our discussion of Properties 4 through 7. Any infinite two-sided sequence can be represented as a sum of a right-sided part (say, for $n \geq 0$) and a left-sided part that includes everything not included in the right-sided part. The right-sided part will have an ROC given by Eq. (3.33), while the ROC of the left-sided part will be given by Eq. (3.35). The ROC of the entire two-sided sequence must be the intersection of these two regions. Thus, if such an intersection exists, it will always be a simply connected annular region of the form

$$r_R < |z| < r_L.$$

There is a possibility of no overlap between the ROCs of the right- and left-sided parts; i.e., $r_L < r_R$. In such cases, the z-transform of the sequence simply does not exist.

Example 3.7   Non-Overlapping Regions of Convergence

An example is the sequence

$$x[n] = \left(\frac{1}{2}\right)^n u[n] - \left(-\frac{1}{3}\right)^n u[-n-1].$$

Applying the corresponding entries from Table 3.1 separately to each part leads to

$$X(z) = \underbrace{\frac{1}{1-\frac{1}{2}z^{-1}}}_{|z| > \frac{1}{2}} + \underbrace{\frac{1}{1+\frac{1}{3}z^{-1}}}_{|z| < \frac{1}{3}} .$$

Since there is no overlap between $|z| > \frac{1}{2}$ and $|z| < \frac{1}{3}$, we conclude that $x[n]$ has no z-transform (nor Fourier transform) representation.

As we indicated in comparing Examples 3.1 and 3.2, the algebraic expression or pole–zero pattern does not completely specify the z-transform of a sequence; i.e., the ROC must also be specified. The properties considered in this section limit the possible ROCs that can be associated with a given pole–zero pattern. To illustrate, consider the pole–zero pattern shown in Figure 3.8(a). From Properties 1, 3, and 8, there are only four possible choices for the ROC. These are indicated in Figures 3.8(b), (c), (d), and (e), each being associated with a different sequence. Specifically, Figure 3.8(b) corresponds

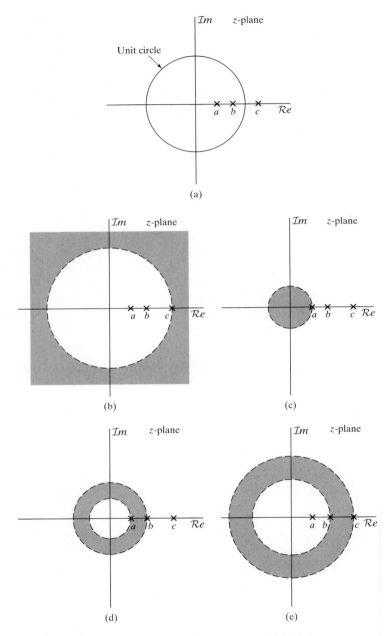

Figure 3.8 Examples of four z-transforms with the same pole–zero locations, illustrating the different possibilities for the ROC, each of which corresponds to a different sequence: (b) to a right-sided sequence, (c) to a left-sided sequence, (d) to a two-sided sequence, and (e) to a two-sided sequence.

to a right-sided sequence, Figure 3.8(c) to a left-sided sequence, and Figures 3.8(d) and 3.8(e) to two different two-sided sequences. If we assume, as indicated in Figure 3.8(a), that the unit circle falls between the pole at $z = b$ and the pole at $z = c$, then the only one of the four cases for which the Fourier transform would converge is that in Figure 3.8(e).

In representing a sequence through its z-transform, it is sometimes convenient to specify the ROC implicitly through an appropriate time-domain property of the sequence. This is illustrated in Example 3.8.

Example 3.8   Stability, Causality, and the ROC

Consider an LTI system with impulse response $h[n]$. As we will discuss in more detail in Section 3.5, the z-transform of $h[n]$ is called the *system function* of the LTI system. Suppose that $H(z)$ has the pole–zero plot shown in Figure 3.9. There are three possible ROCs consistent with Properties 1–8 that can be associated with this pole–zero plot; i.e., $|z| < \frac{1}{2}$, $\frac{1}{2} < |z| < 2$, and $|z| > 2$. However, if we state in addition that the system is stable (or equivalently, that $h[n]$ is absolutely summable and therefore has a Fourier transform), then the ROC must include the unit circle. Thus, stability of the system and Properties 1–8 imply that the ROC is the region $\frac{1}{2} < |z| < 2$. Note that as a consequence, $h[n]$ is two sided; therefore, the system is not causal.

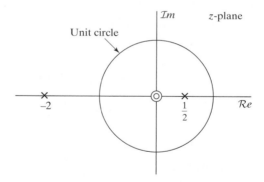

Figure 3.9   Pole–zero plot for the system function in Example 3.8.

If we state instead that the system is causal, and therefore that $h[n]$ is right sided, Property 5 would require that the ROC be the region $|z| > 2$. Under this condition, the system would not be stable; i.e., for this specific pole–zero plot, there is no ROC that would imply that the system is both stable and causal.

3.3 THE INVERSE z-TRANSFORM

In using the z-transform for analysis of discrete-time signals and systems, we must be able to move back and forth between time-domain and z-domain representations. Often, this analysis involves finding the z-transform of sequences and, after some manipulation of the algebraic expressions, finding the inverse z-transform. The inverse z-transform is the following complex contour integral:

$$x[n] = \frac{1}{2\pi j} \oint_C X(z)z^{n-1}dz, \tag{3.36}$$

where C represents a closed contour within the ROC of the z-transform. This integral expression can be derived using the Cauchy integral theorem from the theory of complex variables. (See Brown and Churchill, 2007 for a discussion of the topics of Laurent series and complex integration theorems, all of which are relevant to an in-depth study of fundamental mathematical foundations of the z-transform.) However, for the typical kinds of sequences and z-transforms that we will encounter in the analysis of discrete LTI systems, less formal procedures are sufficient and preferable to techniques based on evaluation of Eq. (3.36). In Sections 3.3.1–3.3.3, we consider some of these procedures, specifically the inspection method, partial fraction expansion, and power series expansion.

3.3.1 Inspection Method

The inspection method consists simply of becoming familiar with, or recognizing "by inspection," certain transform pairs. For example, in Section 3.1, we evaluated the z-transform for sequences of the form $x[n] = a^n u[n]$, where a can be either real or complex. Sequences of this form arise quite frequently, and consequently, it is particularly useful to make direct use of the transform pair

$$a^n u[n] \overset{\mathcal{Z}}{\longleftrightarrow} \frac{1}{1 - az^{-1}}, \qquad |z| > |a|. \tag{3.37}$$

If we need to find the inverse z-transform of

$$X(z) = \left(\frac{1}{1 - \frac{1}{2}z^{-1}}\right), \qquad |z| > \frac{1}{2}, \tag{3.38}$$

and we recall the z-transform pair of Eq. (3.37), we would recognize "by inspection" the associated sequence as $x[n] = \left(\frac{1}{2}\right)^n u[n]$. If the ROC associated with $X(z)$ in Eq. (3.38) had been $|z| < \frac{1}{2}$, we can recall transform pair 6 in Table 3.1 to find by inspection that $x[n] = -\left(\frac{1}{2}\right)^n u[-n - 1]$.

Tables of z-transforms, such as Table 3.1, are invaluable in applying the inspection method. If the table is extensive, it may be possible to express a given z-transform as a sum of terms, each of whose inverse is given in the table. If so, the inverse transform (i.e., the corresponding sequence) can be written from the table.

3.3.2 Partial Fraction Expansion

As already described, inverse z-transforms can be found by inspection if the z-transform expression is recognized or tabulated. Sometimes, $X(z)$ may not be given explicitly in an available table, but it may be possible to obtain an alternative expression for $X(z)$ as a sum of simpler terms, each of which is tabulated. This is the case for any rational function, since we can obtain a partial fraction expansion and easily identify the sequences corresponding to the individual terms.

To see how to obtain a partial fraction expansion, let us assume that $X(z)$ is expressed as a ratio of polynomials in z^{-1}; i.e.,

$$X(z) = \frac{\displaystyle\sum_{k=0}^{M} b_k z^{-k}}{\displaystyle\sum_{k=0}^{N} a_k z^{-k}}. \tag{3.39}$$

Such z-transforms arise frequently in the study of LTI systems. An equivalent expression is

$$X(z) = \frac{z^N \displaystyle\sum_{k=0}^{M} b_k z^{M-k}}{z^M \displaystyle\sum_{k=0}^{N} a_k z^{N-k}}. \tag{3.40}$$

Equation (3.40) explicitly shows that for such functions, there will be M zeros and N poles at nonzero locations in the finite z-plane assuming a_0, b_0, a_N, and b_M are nonzero. In addition, there will be either $M - N$ poles at $z = 0$ if $M > N$ or $N - M$ zeros at $z = 0$ if $N > M$. In other words, z-transforms of the form of Eq. (3.39) always have the same number of poles and zeros in the finite z-plane, and there are no poles or zeros at $z = \infty$. To obtain the partial fraction expansion of $X(z)$ in Eq. (3.39), it is most convenient to note that $X(z)$ could be expressed in the form

$$X(z) = \frac{b_0}{a_0} \frac{\displaystyle\prod_{k=1}^{M}(1 - c_k z^{-1})}{\displaystyle\prod_{k=1}^{N}(1 - d_k z^{-1})}, \tag{3.41}$$

where the c_ks are the nonzero zeros of $X(z)$ and the d_ks are the nonzero poles of $X(z)$. If $M < N$ and the poles are all 1st-order, then $X(z)$ can be expressed as

$$X(z) = \sum_{k=1}^{N} \frac{A_k}{1 - d_k z^{-1}}. \tag{3.42}$$

Obviously, the common denominator of the fractions in Eq. (3.42) is the same as the denominator in Eq. (3.41). Multiplying both sides of Eq. (3.42) by $(1 - d_k z^{-1})$ and evaluating for $z = d_k$ shows that the coefficients, A_k, can be found from

$$A_k = (1 - d_k z^{-1}) X(z)\big|_{z=d_k}. \tag{3.43}$$

Example 3.9   2nd-Order z-Transform

Consider a sequence $x[n]$ with z-transform

$$X(z) = \frac{1}{\left(1 - \frac{1}{4}z^{-1}\right)\left(1 - \frac{1}{2}z^{-1}\right)}, \qquad |z| > \frac{1}{2}. \tag{3.44}$$

The pole–zero plot for $X(z)$ is shown in Figure 3.10. From the ROC and Property 5, Section 3.2, we see that $x[n]$ is a right-sided sequence. Since the poles are both 1st-order, $X(z)$ can be expressed in the form of Eq. (3.42); i.e.,

$$X(z) = \frac{A_1}{\left(1 - \frac{1}{4}z^{-1}\right)} + \frac{A_2}{\left(1 - \frac{1}{2}z^{-1}\right)}.$$

From Eq. (3.43),

$$A_1 = \left(1 - \frac{1}{4}z^{-1}\right)X(z)\bigg|_{z=1/4} = \frac{(1 - \frac{1}{4}z^{-1})}{(1 - \frac{1}{4}z^{-1})(1 - \frac{1}{2}z^{-1})}\bigg|_{z=1/4} = -1,$$

$$A_2 = \left(1 - \frac{1}{2}z^{-1}\right)X(z)\bigg|_{z=1/2} = \frac{(1 - \frac{1}{2}z^{-1})}{(1 - \frac{1}{4}z^{-1})(1 - \frac{1}{2}z^{-1})}\bigg|_{z=1/2} = 2.$$

(Observe that the common factors between the numerator and denominator must be canceled before evaluating the above expressions for A_1 and A_2.) Therefore,

$$X(z) = \frac{-1}{\left(1 - \frac{1}{4}z^{-1}\right)} + \frac{2}{\left(1 - \frac{1}{2}z^{-1}\right)}.$$

Since $x[n]$ is right sided, the ROC for each term extends outward from the outermost pole. From Table 3.1 and the linearity of the z-transform, it then follows that

$$x[n] = 2\left(\frac{1}{2}\right)^n u[n] - \left(\frac{1}{4}\right)^n u[n].$$

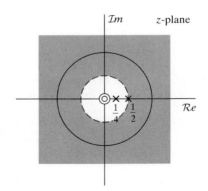

Figure 3.10   Pole–zero plot and ROC for Example 3.9.

Clearly, the numerator that would result from adding the terms in Eq. (3.42) would be at most of degree $(N - 1)$ in the variable z^{-1}. If $M \geq N$, then a polynomial must be added to the right-hand side of Eq. (3.42), the order of which is $(M - N)$. Thus, for $M \geq N$, the complete partial fraction expansion would have the form

$$X(z) = \sum_{r=0}^{M-N} B_r z^{-r} + \sum_{k=1}^{N} \frac{A_k}{1 - d_k z^{-1}}. \tag{3.45}$$

If we are given a rational function of the form of Eq. (3.39), with $M \geq N$, the B_rs can be obtained by long division of the numerator by the denominator, with the division process terminating when the remainder is of lower degree than the denominator. The A_ks can still be obtained with Eq. (3.43).

If $X(z)$ has multiple-order poles and $M \geq N$, Eq. (3.45) must be further modified. In particular, if $X(z)$ has a pole of order s at $z = d_i$ and all the other poles are 1st-order, then Eq. (3.45) becomes

$$X(z) = \sum_{r=0}^{M-N} B_r z^{-r} + \sum_{k=1, k \neq i}^{N} \frac{A_k}{1 - d_k z^{-1}} + \sum_{m=1}^{s} \frac{C_m}{(1 - d_i z^{-1})^m}. \tag{3.46}$$

The coefficients A_k and B_r are obtained as before. The coefficients C_m are obtained from the equation

$$C_m = \frac{1}{(s - m)!(-d_i)^{s-m}} \left\{ \frac{d^{s-m}}{dw^{s-m}} [(1 - d_i w)^s X(w^{-1})] \right\}_{w=d_i^{-1}}. \tag{3.47}$$

Equation (3.46) gives the most general form for the partial fraction expansion of a rational z-transform expressed as a function of z^{-1} for the case $M \geq N$ and for d_i a pole of order s. If there are several multiple-order poles, then there will be a term like the third sum in Eq. (3.46) for each multiple-order pole. If there are no multiple-order poles, Eq. (3.46) reduces to Eq. (3.45). If the order of the numerator is less than the order of the denominator $(M < N)$, then the polynomial term disappears from Eqs. (3.45) and (3.46) leading to Eq. (3.42).

It should be noted that we could have achieved the same results by assuming that the rational z-transform was expressed as a function of z instead of z^{-1}. That is, instead of factors of the form $(1 - az^{-1})$, we could have considered factors of the form $(z - a)$. This would lead to a set of equations similar in form to Eqs. (3.41)–(3.47) that would be convenient for use with a table of z-transforms expressed in terms of z. Since we find it most convenient to express Table 3.1 in terms of z^{-1}, the development we pursued is more useful.

To see how to find the sequence corresponding to a given rational z-transform, let us suppose that $X(z)$ has only 1st-order poles, so that Eq. (3.45) is the most general form of the partial fraction expansion. To find $x[n]$, we first note that the z-transform operation is linear, so that the inverse transform of individual terms can be found and then added together to form $x[n]$.

The terms $B_r z^{-r}$ correspond to shifted and scaled impulse sequences, i.e., terms of the form $B_r \delta[n - r]$. The fractional terms correspond to exponential sequences. To decide whether a term

$$\frac{A_k}{1 - d_k z^{-1}}$$

corresponds to $(d_k)^n u[n]$ or $-(d_k)^n u[-n-1]$, we must use the properties of the ROC that were discussed in Section 3.2. From that discussion, it follows that if $X(z)$ has only simple poles and the ROC is of the form $r_R < |z| < r_L$, then a given pole d_k will correspond to a right-sided exponential $(d_k)^n u[n]$ if $|d_k| \leq r_R$, and it will correspond to a left-sided exponential if $|d_k| \geq r_L$. Thus, the ROC can be used to sort the poles, with all poles inside the inner boundary r_R corresponding to right-sided sequences and all the poles outside the outer boundary corresponding to left-sided sequences. Multiple-order poles also are divided into left-sided and right-sided contributions in the same way. The use of the ROC in finding inverse z-transforms from the partial fraction expansion is illustrated by the following examples.

Example 3.10   Inverse by Partial Fractions

To illustrate the case in which the partial fraction expansion has the form of Eq. (3.45), consider a sequence $x[n]$ with z-transform

$$X(z) = \frac{1 + 2z^{-1} + z^{-2}}{1 - \frac{3}{2}z^{-1} + \frac{1}{2}z^{-2}} = \frac{(1 + z^{-1})^2}{\left(1 - \frac{1}{2}z^{-1}\right)(1 - z^{-1})}, \qquad |z| > 1. \qquad (3.48)$$

The pole–zero plot for $X(z)$ is shown in Figure 3.11. From the ROC and Property 5, Section 3.2, it is clear that $x[n]$ is a right-sided sequence. Since $M = N = 2$ and the poles are all 1st-order, $X(z)$ can be represented as

$$X(z) = B_0 + \frac{A_1}{1 - \frac{1}{2}z^{-1}} + \frac{A_2}{1 - z^{-1}}.$$

The constant B_0 can be found by long division:

$$\frac{1}{2}z^{-2} - \frac{3}{2}z^{-1} + 1 \overline{\left)\begin{array}{l} \phantom{z^{-2} + 2z^{-1} +}2 \\[2pt] z^{-2} + 2z^{-1} + 1 \\[2pt] \underline{z^{-2} - 3z^{-1} + 2} \\[2pt] \phantom{z^{-2} +}5z^{-1} - 1 \end{array}\right.}$$

Since the remainder after one step of long division is of degree 1 in the variable z^{-1}, it is not necessary to continue to divide. Thus, $X(z)$ can be expressed as

$$X(z) = 2 + \frac{-1 + 5z^{-1}}{\left(1 - \frac{1}{2}z^{-1}\right)(1 - z^{-1})}. \qquad (3.49)$$

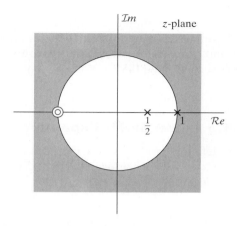

Figure 3.11   Pole–zero plot for the z-transform in Example 3.10.

Now the coefficients A_1 and A_2 can be found by applying Eq. (3.43) to Eq. (3.48) or, equivalently, Eq. (3.49). Using Eq. (3.49), we obtain

$$A_1 = \left[\left(2 + \frac{-1 + 5z^{-1}}{\left(1 - \frac{1}{2}z^{-1}\right)(1 - z^{-1})} \right) \left(1 - \frac{1}{2}z^{-1}\right) \right]_{z=1/2} = -9,$$

$$A_2 = \left[\left(2 + \frac{-1 + 5z^{-1}}{\left(1 - \frac{1}{2}z^{-1}\right)(1 - z^{-1})} \right) (1 - z^{-1}) \right]_{z=1} = 8.$$

Therefore,

$$X(z) = 2 - \frac{9}{1 - \frac{1}{2}z^{-1}} + \frac{8}{1 - z^{-1}}. \tag{3.50}$$

From Table 3.1, we see that since the ROC is $|z| > 1$,

$$2 \overset{\mathcal{Z}}{\longleftrightarrow} 2\delta[n],$$

$$\frac{1}{1 - \frac{1}{2}z^{-1}} \overset{\mathcal{Z}}{\longleftrightarrow} \left(\frac{1}{2}\right)^n u[n],$$

$$\frac{1}{1 - z^{-1}} \overset{\mathcal{Z}}{\longleftrightarrow} u[n].$$

Thus, from the linearity of the z-transform,

$$x[n] = 2\delta[n] - 9\left(\frac{1}{2}\right)^n u[n] + 8u[n].$$

In Section 3.4, we will discuss and illustrate a number of properties of the z-transform that, in combination with the partial fraction expansion, provide a means for determining the inverse z-transform from a given rational algebraic expression and associated ROC, even when $X(z)$ is not exactly in the form of Eq. (3.41). The examples of this section were simple enough so that the computation of the partial fraction expansion was not difficult. However, when $X(z)$ is a rational function with high-degree polynomials in numerator and denominator, the computations to factor the denominator and compute the coefficients become much more difficult. In such cases, software tools such as MATLAB can implement the computations with ease.

3.3.3 Power Series Expansion

The defining expression for the z-transform is a Laurent series where the sequence values $x[n]$ are the coefficients of z^{-n}. Thus, if the z-transform is given as a power series in the form

$$X(z) = \sum_{n=-\infty}^{\infty} x[n]z^{-n}$$

$$= \cdots + x[-2]z^2 + x[-1]z + x[0] + x[1]z^{-1} + x[2]z^{-2} + \cdots, \tag{3.51}$$

we can determine any particular value of the sequence by finding the coefficient of the appropriate power of z^{-1}. We have already used this approach in finding the inverse transform of the polynomial part of the partial fraction expansion when $M \geq N$. This approach is also very useful for finite-length sequences where $X(z)$ may have no simpler form than a polynomial in z^{-1}.

Example 3.11   Finite-Length Sequence

Suppose $X(z)$ is given in the form

$$X(z) = z^2 \left(1 - \frac{1}{2}z^{-1}\right)(1 + z^{-1})(1 - z^{-1}). \tag{3.52}$$

Although $X(z)$ is obviously a rational function of z, it is really not a rational function in the form of Eq. (3.39). Its only poles are at $z = 0$, so a partial fraction expansion according to the technique of Section 3.3.2 is not appropriate. However, by multiplying the factors of Eq. (3.52), we can express $X(z)$ as

$$X(z) = z^2 - \frac{1}{2}z - 1 + \frac{1}{2}z^{-1}.$$

Therefore, by inspection, $x[n]$ is seen to be

$$x[n] = \begin{cases} 1, & n = -2, \\ -\frac{1}{2}, & n = -1, \\ -1, & n = 0, \\ \frac{1}{2}, & n = 1, \\ 0, & \text{otherwise.} \end{cases}$$

Equivalently,

$$x[n] = \delta[n+2] - \frac{1}{2}\delta[n+1] - \delta[n] + \frac{1}{2}\delta[n-1].$$

In finding z-transforms of a sequence, we generally seek to sum the power series of Eq. (3.51) to obtain a simpler mathematical expression, e.g., a rational function. If we wish to use the power series to find the sequence corresponding to a given $X(z)$ expressed in closed form, we must expand $X(z)$ back into a power series. Many power series have been tabulated for transcendental functions such as log, sin, sinh, etc. In some cases, such power series can have a useful interpretation as z-transforms, as we illustrate in Example 3.12. For rational z-transforms, a power series expansion can be obtained by long division, as illustrated in Example 3.13.

Example 3.12   Inverse Transform by Power Series Expansion

Consider the z-transform

$$X(z) = \log(1 + az^{-1}), \qquad |z| > |a|. \tag{3.53}$$

Using the Taylor series expansion for $\log(1 + x)$ with $|x| < 1$, we obtain

$$X(z) = \sum_{n=1}^{\infty} \frac{(-1)^{n+1} a^n z^{-n}}{n}.$$

Therefore,

$$x[n] = \begin{cases} (-1)^{n+1} \dfrac{a^n}{n}, & n \geq 1, \\ 0, & n \leq 0. \end{cases} \tag{3.54}$$

When $X(z)$ is the ratio of polynomials, it is sometimes useful to obtain a power series by long division of the polynomials.

Example 3.13   Power Series Expansion by Long Division

Consider the z-transform

$$X(z) = \frac{1}{1 - az^{-1}}, \qquad |z| > |a|. \tag{3.55}$$

Since the ROC is the exterior of a circle, the sequence is a right-sided one. Furthermore, since $X(z)$ approaches a finite constant as z approaches infinity, the sequence is causal. Thus, we divide, so as to obtain a series in powers of z^{-1}. Carrying out the long division, we obtain

$$
\begin{array}{r}
1 + az^{-1} + a^2 z^{-2} + \cdots \\
1 - az^{-1} \overline{\smash{)}\, 1 } \\
\underline{1 - az^{-1}} \\
az^{-1} \\
\underline{az^{-1} - a^2 z^{-2}} \\
a^2 z^{-2} \ \cdots
\end{array}
$$

or

$$\frac{1}{1 - az^{-1}} = 1 + az^{-1} + a^2 z^{-2} + \cdots.$$

Hence, $x[n] = a^n u[n]$.

By dividing the highest power of z^{-1} in the denominator into the highest power of the numerator in Example 3.13, we obtained a series in z^{-1}. An alternative is to express the rational function as a ratio of polynomials in z and then divide. This leads to a power series in z from which the corresponding left-sided sequence can be determined.

3.4 z-TRANSFORM PROPERTIES

Many of the mathematical properties of the z-transform are particularly useful in studying discrete-time signals and systems. For example, these properties are often used in conjunction with the inverse z-transform techniques discussed in Section 3.3 to obtain the inverse z-transform of more complicated expressions. In Section 3.5 and Chapter 5 we will see that the properties also form the basis for transforming linear constant-coefficient difference equations to algebraic equations in terms of the transform variable z, the solution to which can then be obtained using the inverse z-transform. In this section, we consider some of the most frequently used properties. In the following discussion, $X(z)$ denotes the z-transform of $x[n]$, and the ROC of $X(z)$ is indicated by R_x; i.e.,

$$x[n] \xleftrightarrow{\ \mathcal{Z}\ } X(z), \qquad \mathrm{ROC} = R_x.$$

As we have seen, R_x represents a set of values of z such that $r_R < |z| < r_L$. For properties that involve two sequences and associated z-transforms, the transform pairs will be denoted as

$$x_1[n] \xleftrightarrow{\ \mathcal{Z}\ } X_1(z), \qquad \mathrm{ROC} = R_{x_1},$$

$$x_2[n] \xleftrightarrow{\ \mathcal{Z}\ } X_2(z), \qquad \mathrm{ROC} = R_{x_2}.$$

3.4.1 Linearity

The linearity property states that

$$ax_1[n] + bx_2[n] \xleftrightarrow{\ \mathcal{Z}\ } aX_1(z) + bX_2(z), \qquad \mathrm{ROC\ contains\ } R_{x_1} \cap R_{x_2},$$

and follows directly from the z-transform definition, Eq. (3.2); i.e.,

$$\sum_{n=-\infty}^{\infty} (ax_1[n] + bx_2[n])z^{-n} = a \underbrace{\sum_{n=-\infty}^{\infty} x_1[n]z^{-n}}_{|z| \in R_{x_1}} + b \underbrace{\sum_{n=-\infty}^{\infty} x_2[n]z^{-n}}_{|z| \in R_{x_2}}.$$

As indicated, to split the z-transform of a sum into the sum of corresponding z-transforms, z must be in both ROCs. Therefore, the ROC is at least the intersection of the individual ROCs. For sequences with rational z-transforms, if the poles of $aX_1(z) + bX_2(z)$ consist of all the poles of $X_1(z)$ and $X_2(z)$ (i.e., if there is no pole–zero cancellation), then the ROC will be exactly equal to the overlap of the individual ROCs. If the linear combination is such that some zeros are introduced that cancel poles, then the ROC may be larger. A simple example of this occurs when $x_1[n]$ and $x_2[n]$ are of infinite duration, but the linear combination is of finite duration. In this case the ROC of the linear combination is the entire z-plane, with the possible exception of $z = 0$ or $z = \infty$. An example was given in Example 3.6, where $x[n]$ can be expressed as

$$x[n] = a^n \left(u[n] - u[n - N] \right) = a^n u[n] - a^n u[n - N].$$

Both $a^n u[n]$ and $a^n u[n - N]$ are infinite-extent right-sided sequences, and their z-transforms have a pole at $z = a$. Therefore, their individual ROCs would both be $|z| > |a|$. However, as shown in Example 3.6, the pole at $z = a$ is canceled by a zero at $z = a$, and therefore, the ROC extends to the entire z-plane, with the exception of $z = 0$.

We have already exploited the linearity property in our previous discussion of the use of the partial fraction expansion for evaluating the inverse z-transform. With that procedure, $X(z)$ is expanded into a sum of simpler terms, and through linearity, the inverse z-transform is the sum of the inverse transforms of each of these terms.

3.4.2 Time Shifting

The time-shifting property is,

$$x[n - n_0] \overset{z}{\longleftrightarrow} z^{-n_0} X(z), \qquad \text{ROC} = R_x (\text{except for the possible addition or deletion of } z = 0 \text{ or } z = \infty).$$

The quantity n_0 is an integer. If n_0 is positive, the original sequence $x[n]$ is shifted right, and if n_0 is negative, $x[n]$ is shifted left. As in the case of linearity, the ROC can be changed, since the factor z^{-n_0} can alter the number of poles at $z = 0$ or $z = \infty$.

The derivation of this property follows directly from the z-transform expression in Eq. (3.2). Specifically, if $y[n] = x[n - n_0]$, the corresponding z-transform is

$$Y(z) = \sum_{n=-\infty}^{\infty} x[n - n_0] z^{-n}.$$

With the substitution of variables $m = n - n_0$,

$$Y(z) = \sum_{m=-\infty}^{\infty} x[m] z^{-(m+n_0)}$$

$$= z^{-n_0} \sum_{m=-\infty}^{\infty} x[m] z^{-m},$$

or

$$Y(z) = z^{-n_0} X(z).$$

The time-shifting property is often useful, in conjunction with other properties and procedures, for obtaining the inverse z-transform. We illustrate with an example.

Example 3.14   Shifted Exponential Sequence

Consider the z-transform

$$X(z) = \frac{1}{z - \frac{1}{4}}, \qquad |z| > \frac{1}{4}.$$

From the ROC, we identify this as corresponding to a right-sided sequence. We can first rewrite $X(z)$ in the form

$$X(z) = \frac{z^{-1}}{1 - \frac{1}{4}z^{-1}}, \qquad |z| > \frac{1}{4}. \tag{3.56}$$

This z-transform is of the form of Eq. (3.41) with $M = N = 1$, and its expansion in the form of Eq. (3.45) is

$$X(z) = -4 + \frac{4}{1 - \frac{1}{4}z^{-1}}. \tag{3.57}$$

From Eq. (3.57), it follows that $x[n]$ can be expressed as

$$x[n] = -4\delta[n] + 4\left(\frac{1}{4}\right)^n u[n]. \tag{3.58}$$

An expression for $x[n]$ can be obtained more directly by applying the time-shifting property. First, $X(z)$ can be written as

$$X(z) = z^{-1}\left(\frac{1}{1 - \frac{1}{4}z^{-1}}\right), \qquad |z| > \frac{1}{4}. \tag{3.59}$$

From the time-shifting property, we recognize the factor z^{-1} in Eq. (3.59) as being associated with a time shift of one sample to the right of the sequence $\left(\frac{1}{4}\right)^n u[n]$; i.e.,

$$x[n] = \left(\frac{1}{4}\right)^{n-1} u[n-1]. \tag{3.60}$$

It is easily verified that Eqs. (3.58) and (3.60) are the same for all values of n; i.e., they represent the same sequence.

3.4.3 Multiplication by an Exponential Sequence

The exponential multiplication property is

$$z_0^n x[n] \overset{\mathcal{Z}}{\longleftrightarrow} X(z/z_0), \qquad \text{ROC} = |z_0|R_x.$$

The notation $\text{ROC} = |z_0|R_x$ signifies that the ROC is R_x scaled by the number $|z_0|$; i.e., if R_x is the set of values of z such that $r_R < |z| < r_L$, then $|z_0|R_x$ is the set of values of z such that $|z_0|r_R < |z| < |z_0|r_L$.

This property is easily shown simply by substituting $z_0^n x[n]$ into Eq. (3.2). As a consequence of the exponential multiplication property, all the pole–zero locations are scaled by a factor z_0, since, if $X(z)$ has a pole (or zero) at $z = z_1$, then $X(z/z_0)$ will have a pole (or zero) at $z = z_0 z_1$. If z_0 is a positive real number, the scaling can be interpreted as a shrinking or expanding of the z-plane; i.e., the pole and zero locations change along radial lines in the z-plane. If z_0 is complex with unity magnitude, so that $z_0 = e^{j\omega_0}$, the scaling corresponds to a rotation in the z-plane by an angle of ω_0; i.e., the pole and zero locations change in position along circles centered at the origin. This in turn can be interpreted as a frequency shift or translation of the discrete-time Fourier transform, which is associated with the modulation in the time domain by the complex exponential sequence $e^{j\omega_0 n}$. That is, if the Fourier transform exists, this property has the form

$$e^{j\omega_0 n} x[n] \xleftrightarrow{\mathcal{F}} X(e^{j(\omega-\omega_0)}).$$

Example 3.15   Exponential Multiplication

Starting with the transform pair

$$u[n] \xleftrightarrow{\mathcal{Z}} \frac{1}{1 - z^{-1}}, \qquad |z| > 1, \tag{3.61}$$

we can use the exponential multiplication property to determine the z-transform of

$$x[n] = r^n \cos(\omega_0 n) u[n], \qquad r > 0. \tag{3.62}$$

First, $x[n]$ is expressed as

$$x[n] = \frac{1}{2}(re^{j\omega_0})^n u[n] + \frac{1}{2}(re^{-j\omega_0})^n u[n].$$

Then, using Eq. (3.61) and the exponential multiplication property, we see that

$$\frac{1}{2}(re^{j\omega_0})^n u[n] \xleftrightarrow{\mathcal{Z}} \frac{\frac{1}{2}}{1 - re^{j\omega_0}z^{-1}}, \qquad |z| > r,$$

$$\frac{1}{2}(re^{-j\omega_0})^n u[n] \xleftrightarrow{\mathcal{Z}} \frac{\frac{1}{2}}{1 - re^{-j\omega_0}z^{-1}}, \qquad |z| > r.$$

From the linearity property, it follows that

$$\begin{aligned}
X(z) &= \frac{\frac{1}{2}}{1 - re^{j\omega_0}z^{-1}} + \frac{\frac{1}{2}}{1 - re^{-j\omega_0}z^{-1}}, \qquad |z| > r \\[2mm]
&= \frac{1 - r\cos(\omega_0)z^{-1}}{1 - 2r\cos(\omega_0)z^{-1} + r^2 z^{-2}}, \qquad |z| > r.
\end{aligned} \tag{3.63}$$

3.4.4 Differentiation of *X*(*z*)

The differentiation property states that

$$nx[n] \xleftrightarrow{\mathcal{Z}} -z\frac{dX(z)}{dz}, \qquad \text{ROC} = R_x.$$

This property is verified by differentiating the z-transform expression of Eq. (3.2); i.e., for

$$X(z) = \sum_{n=-\infty}^{\infty} x[n]z^{-n},$$

we obtain

$$-z\frac{dX(z)}{dz} = -z\sum_{n=-\infty}^{\infty} (-n)x[n]z^{-n-1}$$

$$= \sum_{n=-\infty}^{\infty} nx[n]z^{-n} = \mathcal{Z}\{nx[n]\}.$$

We illustrate the use of the differentiation property with two examples.

Example 3.16   Inverse of Non-Rational z-Transform

In this example, we use the differentiation property together with the time-shifting property to determine the inverse z-transform considered in Example 3.12. With

$$X(z) = \log(1 + az^{-1}), \qquad |z| > |a|,$$

we first differentiate to obtain a rational expression:

$$\frac{dX(z)}{dz} = \frac{-az^{-2}}{1 + az^{-1}}.$$

From the differentiation property,

$$nx[n] \overset{\mathcal{Z}}{\longleftrightarrow} -z\frac{dX(z)}{dz} = \frac{az^{-1}}{1 + az^{-1}}, \qquad |z| > |a|. \tag{3.64}$$

The inverse transform of Eq. (3.64) can be obtained by the combined use of the z-transform pair of Example 3.1, the linearity property, and the time-shifting property. Specifically, we can express $nx[n]$ as

$$nx[n] = a(-a)^{n-1}u[n-1].$$

Therefore,

$$x[n] = (-1)^{n+1}\frac{a^n}{n}u[n-1] \overset{\mathcal{Z}}{\longleftrightarrow} \log(1 + az^{-1}), \qquad |z| > |a|.$$

The result of Example 3.16 will be useful in our discussion of the cepstrum in Chapter 13.

Example 3.17   2nd-Order Pole

As another example of the use of the differentiation property, let us determine the z-transform of the sequence
$$x[n] = na^n u[n] = n(a^n u[n]).$$
From the z-transform pair of Example 3.1 and the differentiation property, it follows that

$$X(z) = -z\frac{d}{dz}\left(\frac{1}{1 - az^{-1}}\right), \qquad |z| > |a|$$

$$= \frac{az^{-1}}{(1 - az^{-1})^2}, \qquad |z| > |a|.$$

Therefore,

$$na^n u[n] \overset{\mathcal{Z}}{\longleftrightarrow} \frac{az^{-1}}{(1 - az^{-1})^2}, \qquad |z| > |a|.$$

3.4.5 Conjugation of a Complex Sequence

The conjugation property is expressed as

$$x^*[n] \overset{\mathcal{Z}}{\longleftrightarrow} X^*(z^*), \qquad \text{ROC} = R_x.$$

This property follows in a straightforward manner from the definition of the z-transform, the details of which are left as an exercise (Problem 3.54).

3.4.6 Time Reversal

The time-reversal property is given by

$$x^*[-n] \overset{\mathcal{Z}}{\longleftrightarrow} X^*(1/z^*), \qquad \text{ROC} = \frac{1}{R_x}.$$

The notation ROC=$1/R_x$ implies that R_x is inverted; i.e., if R_x is the set of values of z such that $r_R < |z| < r_L$, then the ROC for $X^*(1/z^*)$ is the set of values of z such that $1/r_L < |z| < 1/r_R$. Thus, if z_0 is in the ROC for $x[n]$, then $1/z_0^*$ is in the ROC for the z-transform of $x^*[-n]$. If the sequence $x[n]$ is real or we do not conjugate a complex sequence, the result becomes

$$x[-n] \overset{\mathcal{Z}}{\longleftrightarrow} X(1/z), \qquad \text{ROC} = \frac{1}{R_x}.$$

As with the conjugation property, the time-reversal property follows easily from the definition of the z-transform, and the details are left as an exercise (Problem 3.54).

Note that if z_0 is a pole (or zero) of $X(z)$, then $1/z_0$ will be a pole (or zero) of $X(1/z)$. The magnitude of $1/z_0$ is simply the reciprocal of the magnitude of z_0. However, the angle of $1/z_0$ is the negative of the angle of z_0. When the poles and zeros of $X(z)$ are all real or in complex conjugate pairs, as they must be when $x[n]$ is real, this complex conjugate pairing is maintained.

Example 3.18   Time-Reversed Exponential Sequence

As an example of the use of the property of time reversal, consider the sequence

$$x[n] = a^{-n}u[-n],$$

which is a time-reversed version of $a^n u[n]$. From the time-reversal property, it follows that

$$X(z) = \frac{1}{1 - az} = \frac{-a^{-1}z^{-1}}{1 - a^{-1}z^{-1}}, \qquad |z| < |a^{-1}|.$$

Note that the z-transform of $a^n u[n]$ has a pole at $z = a$, while $X(z)$ has a pole at $1/a$.

3.4.7 Convolution of Sequences

According to the convolution property,

$$x_1[n] * x_2[n] \xleftrightarrow{\mathcal{Z}} X_1(z)X_2(z), \qquad \text{ROC contains } R_{x_1} \cap R_{x_2}.$$

To derive this property formally, we consider

$$y[n] = \sum_{k=-\infty}^{\infty} x_1[k]x_2[n-k],$$

so that

$$Y(z) = \sum_{n=-\infty}^{\infty} y[n]z^{-n}$$

$$= \sum_{n=-\infty}^{\infty} \left\{ \sum_{k=-\infty}^{\infty} x_1[k]x_2[n-k] \right\} z^{-n}.$$

If we interchange the order of summation (which is allowed for z in the ROC),

$$Y(z) = \sum_{k=-\infty}^{\infty} x_1[k] \sum_{n=-\infty}^{\infty} x_2[n-k]z^{-n}.$$

Changing the index of summation in the second sum from n to $m = n - k$, we obtain

$$Y(z) = \sum_{k=-\infty}^{\infty} x_1[k] \left\{ \sum_{m=-\infty}^{\infty} x_2[m]z^{-m} \right\} z^{-k}$$

$$= \sum_{k=-\infty}^{\infty} x_1[k] \underbrace{X_2(z)}_{|z|\in R_{x_2}} z^{-k} = \left(\sum_{k=-\infty}^{\infty} x_1[k]z^{-k} \right) X_2(z)$$

Thus, for values of z inside the ROCs of both $X_1(z)$ and $X_2(z)$, we can write

$$Y(z) = X_1(z)X_2(z),$$

where the ROC includes the intersection of the ROCs of $X_1(z)$ and $X_2(z)$. If a pole that borders on the ROC of one of the z-transforms is canceled by a zero of the other, then the ROC of $Y(z)$ may be larger.

The use of the z-transform for evaluating convolutions is illustrated by the following example.

Example 3.19   Convolution of Finite-Length Sequences

Suppose that

$$x_1[n] = \delta[n] + 2\delta[n-1] + \delta[n-2]$$

is a finite-length sequence to be convolved with the sequence $x_2[n] = \delta[n] - \delta[n-1]$. The corresponding *z*-transforms are

$$X_1(z) = 1 + 2z^{-1} + z^{-2}$$

and $X_2(z) = 1 - z^{-1}$. The convolution $y[n] = x_1[n] * x_2[n]$ has *z*-transform

$$Y(z) = X_1(z)X_2(z) = (1 + 2z^{-1} + z^{-2})(1 - z^{-1})$$

$$= 1 + z^{-1} - z^{-2} - z^{-3}.$$

Since the sequences are both of finite length, the ROCs are both $|z| > 0$ and therefore so is the ROC of $Y(z)$. From $Y(z)$, we conclude by inspection of the coefficients of the polynomial that

$$y[n] = \delta[n] + \delta[n-1] - \delta[n-2] - \delta[n-3].$$

The important point of this example is that convolution of finite-length sequences is equivalent to polynomial multiplication. Conversely, the coefficients of the product of two polynomials are obtained by discrete convolution of the polynomial coefficients.

The convolution property plays a particularly important role in the analysis of LTI systems as we will discuss in more detail in Section 3.5 and Chapter 5. An example of the use of the *z*-transform for computing the convolution of two infinite-duration sequences is given in Section 3.5.

3.4.8 Summary of Some *z*-Transform Properties

We have presented and discussed a number of the theorems and properties of *z*-transforms, many of which are useful in manipulating *z*-transforms in the analysis of discrete-time systems. These properties and a number of others are summarized for convenient reference in Table 3.2.

3.5 *z*-TRANSFORMS AND LTI SYSTEMS

The properties discussed in Section 3.4 make the *z*-transform a very useful tool for discrete-time system analysis. Since we shall rely on the *z*-transform extensively in Chapter 5 and later chapters, it is worthwhile now to illustrate how the *z*-transform can be used in the representation and analysis of LTI systems.

Recall from Section 2.3 that an LTI system can be represented as the convolution $y[n] = x[n] * h[n]$ of the input $x[n]$ with $h[n]$, where $h[n]$ is the response of the system to the unit impulse sequence $\delta[n]$. From the convolution property of Section 3.4.7, it follows that the *z*-transform of $y[n]$ is

$$Y(z) = H(z)X(z) \tag{3.65}$$

TABLE 3.2   SOME z-TRANSFORM PROPERTIES

Property Number	Section Reference	Sequence	Transform	ROC
		$x[n]$	$X(z)$	R_x
		$x_1[n]$	$X_1(z)$	R_{x_1}
		$x_2[n]$	$X_2(z)$	R_{x_2}
1	3.4.1	$ax_1[n] + bx_2[n]$	$aX_1(z) + bX_2(z)$	Contains $R_{x_1} \cap R_{x_2}$
2	3.4.2	$x[n - n_0]$	$z^{-n_0}X(z)$	R_x, except for the possible addition or deletion of the origin or ∞
3	3.4.3	$z_0^n x[n]$	$X(z/z_0)$	$|z_0|R_x$
4	3.4.4	$nx[n]$	$-z\dfrac{dX(z)}{dz}$	R_x
5	3.4.5	$x^*[n]$	$X^*(z^*)$	R_x
6		$\mathcal{R}e\{x[n]\}$	$\dfrac{1}{2}[X(z) + X^*(z^*)]$	Contains R_x
7		$\mathcal{I}m\{x[n]\}$	$\dfrac{1}{2j}[X(z) - X^*(z^*)]$	Contains R_x
8	3.4.6	$x^*[-n]$	$X^*(1/z^*)$	$1/R_x$
9	3.4.7	$x_1[n] * x_2[n]$	$X_1(z)X_2(z)$	Contains $R_{x_1} \cap R_{x_2}$

where $H(z)$ and $X(z)$ are the z-transforms of $h[n]$ and $x[n]$ respectively. In this context, the z-transform $H(z)$ is called the *system function* of the LTI system whose impulse response is $h[n]$.

The computation of the output of an LTI system using the z-transform is illustrated by the following example.

Example 3.20   Convolution of Infinite-Length Sequences

Let $h[n] = a^n u[n]$ and $x[n] = Au[n]$. To use the z-transform to evaluate the convolution $y[n] = x[n] * h[n]$, we begin by finding the corresponding z-transforms as

$$H(z) = \sum_{n=0}^{\infty} a^n z^{-n} = \frac{1}{1 - az^{-1}}, \qquad |z| > |a|,$$

and

$$X(z) = \sum_{n=0}^{\infty} Az^{-n} = \frac{A}{1 - z^{-1}}, \qquad |z| > 1.$$

The z-transform of the convolution $y[n] = x[n] * h[n]$ is therefore

$$Y(z) = \frac{A}{(1 - az^{-1})(1 - z^{-1})} = \frac{Az^2}{(z - a)(z - 1)}, \qquad |z| > 1,$$

where we assume that $|a| < 1$ so that the overlap of the ROCs is $|z| > 1$.

The poles and zeros of $Y(z)$ are plotted in Figure 3.12, and the ROC is seen to be the overlap region. The sequence $y[n]$ can be obtained by determining the inverse *z*-transform. The partial fraction expansion of $Y(z)$ is

$$Y(z) = \frac{A}{1-a}\left(\frac{1}{1-z^{-1}} - \frac{a}{1-az^{-1}}\right) \qquad |z| > 1.$$

Therefore, taking the inverse *z*-transform of each term yields

$$y[n] = \frac{A}{1-a}(1-a^{n+1})u[n].$$

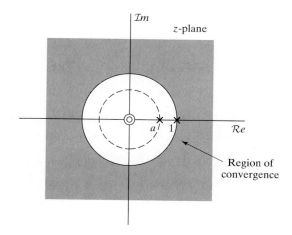

Figure 3.12   Pole–zero plot for the *z*-transform of the convolution of the sequences $u[n]$ and $a^n u[n]$ (assuming $|a| < 1$).

The *z*-transform is particularly useful in the analysis of LTI systems described by difference equations. Recall that in Section 2.5, we showed that difference equations of the form

$$y[n] = -\sum_{k=1}^{N}\left(\frac{a_k}{a_0}\right)y[n-k] + \sum_{k=0}^{M}\left(\frac{b_k}{a_0}\right)x[n-k], \tag{3.66}$$

behave as causal LTI systems when the input is zero prior to $n = 0$ and initial rest conditions are imposed prior to the time when the input becomes nonzero; i.e.,

$$y[-N], y[-N+1], \ldots, y[-1]$$

are all assumed to be zero. The difference equation with assumed initial rest conditions defines the LTI system, but it is also of interest to know the system function. If we apply the linearity property (Section 3.4.1) and the time-shift property (Section 3.4.2) to Eq. (3.66), we obtain

$$Y(z) = -\sum_{k=1}^{N}\left(\frac{a_k}{a_0}\right)z^{-k}Y(z) + \sum_{k=0}^{M}\left(\frac{b_k}{a_0}\right)z^{-k}X(z). \tag{3.67}$$

Solving for $Y(z)$ in terms of $X(z)$ and the parameters of the difference equation yields

$$Y(z) = \left(\frac{\displaystyle\sum_{k=0}^{M} b_k z^{-k}}{\displaystyle\sum_{k=0}^{N} a_k z^{-k}} \right) X(z), \tag{3.68}$$

and from a comparison of Eqs. (3.65) and (3.68) it follows that for the LTI system described by Eq. (3.66), the system function is

$$H(z) = \frac{\displaystyle\sum_{k=0}^{M} b_k z^{-k}}{\displaystyle\sum_{k=0}^{N} a_k z^{-k}}. \tag{3.69}$$

Since the system defined by the difference equation of Eq. (3.66) is a causal system, our discussion in Section 3.2 leads to the conclusion that $H(z)$ in Eq. (3.69) must have an ROC of the form $|z| > r_R$, and since the ROC can contain no poles, r_R must be equal to the magnitude of pole of $H(z)$ that is farthest from the origin. Furthermore, the discussion in Section 3.2 also confirms that if $r_R < 1$, i.e., all poles are inside the unit circle, then the system is stable and the frequency response of the system is obtained by setting $z = e^{j\omega}$ in Eq. (3.69).

Note that if Eq. (3.66) is expressed in the equivalent form

$$\sum_{k=0}^{N} a_k y[n-k] = \sum_{k=0}^{M} b_k x[n-k] \tag{3.70}$$

then Eq. (3.69), which gives the system function (and frequency response for stable systems) as a ratio of polynomials in the variable z^{-1}, can be written down directly by observing that the numerator is the z-transform representation of the coefficient and delay terms involving the input, whereas the denominator represents the coefficients and delays of the terms involving the output. Similarly, given the system function as a ratio of polynomials in z^{-1} as in Eq. (3.69), it is straightforward to write down the difference equation in the form of Eq. (3.70) and then write it in the form of Eq. (3.66) for recursive implementation.

Example 3.21   1st-Order System

Suppose that a causal LTI system is described by the difference equation
$$y[n] = ay[n-1] + x[n]. \tag{3.71}$$
By inspection, it follows that the system function for this system is
$$H(z) = \frac{1}{1 - az^{-1}}, \tag{3.72}$$
with ROC $|z| > |a|$. from which it follows from entry 5 of Table 3.1 that the impulse response of the system is
$$h[n] = a^n u[n]. \tag{3.73}$$

Finally, if $x[n]$ is a sequence with a rational z-transform such as $x[n] = Au[n]$, we can find the output of the system in three distinct ways. (1) We can iterate the difference equation in Eq. (3.71). In general, this approach could be used with any input and would generally be used to implement the system, but it would not lead directly to a closed-form solution valid for all n even if such expression exists. (2) We could evaluate the convolution of $x[n]$ and $h[n]$ explicitly using the techniques illustrated in Section 2.3. (3) Since the z-transforms of both $x[n]$ and $h[n]$ are rational functions of z, we can use the partial fraction method of Section 3.3.2 to find a closed-form expression for the output valid for all n. In fact, this was done in Example 3.20.

We shall have much more use for the z-transform in Chapter 5 and subsequent chapters. For example, in Section 5.2.3, we shall obtain general expressions for the impulse response of an LTI system with rational system function, and we shall show how the frequency response of the system is related to the locations of the poles and zeros of $H(z)$.

3.6 THE UNILATERAL z-TRANSFORM

The z-transform, as defined by Eq. (3.2), and as considered so far in this chapter, is more explicitly referred to as the bilateral z-transform or the two-sided z-transform. In contrast, the *unilateral* or *one-sided z-transform* is defined as

$$\mathcal{X}(z) = \sum_{n=0}^{\infty} x[n]z^{-n}. \tag{3.74}$$

The unilateral z-transform differs from the bilateral z-transform in that the lower limit of the sum is always fixed at zero, regardless of the values of $x[n]$ for $n < 0$. If $x[n] = 0$ for $n < 0$, the unilateral and bilateral z-transforms are identical, whereas, if $x[n]$ is not zero for all $n < 0$, they will be different. A simple example illustrates this.

Example 3.22   Unilateral Transform of an Impulse

Suppose that $x_1[n] = \delta[n]$. Then it is clear from Eq. (3.74) that $\mathcal{X}_1(z) = 1$, which is identical to the bilateral z-transform of the impulse. However, consider $x_2[n] = \delta[n + 1] = x_1[n + 1]$. This time using Eq. (3.74) we find that $\mathcal{X}_2(z) = 0$, whereas the bilateral z-transform would be $X_2(z) = zX_1(z) = z$.

Because the unilateral transform in effect ignores any left-sided part, the properties of the ROC of the unilateral z-transform will be the same as those of the bilateral transform of a right-sided sequence obtained by assuming that the sequence values are zero for $n < 0$. That is, the ROC for all unilateral z-transforms will be of the form $|z| > r_R$, and for rational unilateral z-transforms, the boundary of the ROC will be defined by the pole that is farthest from the origin of the z-plane.

In digital signal processing applications, difference equations of the form of Eq. (3.66) are generally employed with initial rest conditions. However, in some situations, noninitial rest conditions may occur. In such cases, the linearity and time-shifting properties of the unilateral z-transform are particularly useful tools. The linearity property is identical to that of the bilateral z-transform (Property 1 in Table 3.2). The time-shifting property is different in the unilateral case because the lower limit in the unilateral transform definition is fixed at zero. To illustrate how to develop this property, consider a sequence $x[n]$ with unilateral z-transform $\mathcal{X}(z)$ and let $y[n] = x[n-1]$. Then, by definition

$$\mathcal{Y}(z) = \sum_{n=0}^{\infty} x[n-1]z^{-n}.$$

With the substitution of summation index $m = n - 1$, we can write $\mathcal{Y}(z)$ as

$$\mathcal{Y}(z) = \sum_{m=-1}^{\infty} x[m]z^{-(m+1)} = x[-1] + z^{-1}\sum_{m=0}^{\infty} x[m]z^{-m},$$

so that

$$\mathcal{Y}(z) = x[-1] + z^{-1}\mathcal{X}(z). \tag{3.75}$$

Thus, to determine the unilateral z-transform of a delayed sequence, we must provide sequence values that are ignored in computing $\mathcal{X}(z)$. By a similar analysis, it can be shown that if $y[n] = x[n-k]$, where $k > 0$, then

$$\mathcal{Y}(z) = x[-k] + x[-k+1]z^{-1} + \ldots + x[-1]z^{-k+1} + z^{-k}\mathcal{X}(z)$$

$$= \sum_{m=1}^{k} x[m-k-1]z^{-m+1} + z^{-k}\mathcal{X}(z). \tag{3.76}$$

The use of the unilateral z-transform to solve for the output of a difference equation with nonzero initial conditions is illustrated by the following example.

Example 3.23   Effect of Nonzero Initial Conditions

Consider a system described by the linear constant-coefficient difference equation

$$y[n] - ay[n-1] = x[n], \tag{3.77}$$

which is the same as the system in Examples 3.20 and 3.21. Assume that $x[n] = 0$ for $n < 0$ and the initial condition at $n = -1$ is denoted $y[-1]$. Applying the unilateral z-transform to Eq. (3.77) and using the linearity property as well as the time-shift property in Eq. (3.75), we have

$$\mathcal{Y}(z) - ay[-1] - az^{-1}\mathcal{Y}(z) = \mathcal{X}(z).$$

Solving for $\mathcal{Y}(z)$ we obtain

$$\mathcal{Y}(z) = \frac{ay[-1]}{1 - az^{-1}} + \frac{1}{1 - az^{-1}}\mathcal{X}(z). \tag{3.78}$$

Note that if $y[-1] = 0$ the first term disappears, and we are left with $\mathcal{Y}(z) = H(z)\mathcal{X}(z)$, where

$$H(z) = \frac{1}{1 - az^{-1}}, \qquad |z| > |a|$$

is the system function of the LTI system corresponding to the difference equation in Eq. (3.77) when iterated with initial rest conditions. This confirms that initial rest conditions are necessary for the iterated difference equation to behave as an LTI system. Furthermore, note that if $x[n] = 0$ for all n, the output will be equal to

$$y[n] = y[-1]a^{n+1} \qquad n \geq -1.$$

This shows that if $y[-1] \neq 0$, the system does not behave linearly because the scaling property for linear systems [Eq. (2.23b)] requires that when the input is zero for all n, the output must likewise be zero for all n.

To be more specific, suppose that $x[n] = Au[n]$ as in Example 3.20. We can determine an equation for $y[n]$ for $n \geq -1$ by noting that the unilateral z-transform of $x[n] = Au[n]$ is

$$\mathcal{X}(z) = \frac{A}{1 - z^{-1}}, \qquad |z| > 1$$

so that Eq. (3.78) becomes

$$\mathcal{Y}(z) = \frac{ay[-1]}{1 - az^{-1}} + \frac{A}{(1 - az^{-1})(1 - z^{-1})}. \tag{3.79}$$

Applying the partial fraction expansion technique to Eq. (3.79) gives

$$\mathcal{Y}(z) = \frac{ay[-1]}{1 - az^{-1}} + \frac{\dfrac{A}{1 - a}}{1 - z^{-1}} + \frac{-\dfrac{aA}{1 - a}}{1 - az^{-1}},$$

from which it follows that the complete solution is

$$y[n] = \begin{cases} y[-1] & n = -1 \\ \underbrace{y[-1]a^{n+1}}_{\text{ZIR}} + \underbrace{\frac{A}{1 - a}\left(1 - a^{n+1}\right)}_{\text{ZICR}} & n \geq 0 \end{cases} \tag{3.80}$$

Equation (3.80) shows that the system response is composed of two parts. The zero input response (ZIR) is the response when the input is zero (in this case when $A = 0$). The zero initial conditions response (ZICR) is the part that is directly proportional to the input (as required for linearity). This part remains when $y[-1] = 0$. In Problem 3.49, this decomposition into ZIR and ZICR components is shown to hold for any difference equation of the form of Eq. (3.66).

3.7 SUMMARY

In this chapter, we have defined the z-transform of a sequence and shown that it is a generalization of the Fourier transform. The discussion focused on the properties of the z-transform and techniques for obtaining the z-transform of a sequence and vice versa. Specifically, we showed that the defining power series of the z-transform may converge when the Fourier transform does not. We explored in detail the dependence of the shape of the ROC on the properties of the sequence. A full understanding of the properties of the ROC is essential for successful use of the z-transform. This is particularly true in developing techniques for finding the sequence that corresponds to a given z-transform, i.e., finding inverse z-transforms. Much of the discussion focused on z-transforms that are rational functions in their region of convergence. For such functions, we described a technique of inverse transformation based on the partial fraction expansion of $X(z)$. We also discussed other techniques for inverse transformation, such as the use of tabulated power series expansions and long division.

An important part of the chapter was a discussion of some of the many properties of the z-transform that make it useful in analyzing discrete-time signals and systems. A variety of examples demonstrated how these properties can be used to find direct and inverse z-transforms.

Problems

Basic Problems with Answers

3.1. Determine the z-transform, including the ROC, for each of the following sequences:

(a) $\left(\frac{1}{2}\right)^n u[n]$

(b) $-\left(\frac{1}{2}\right)^n u[-n-1]$

(c) $\left(\frac{1}{2}\right)^n u[-n]$

(d) $\delta[n]$

(e) $\delta[n-1]$

(f) $\delta[n+1]$

(g) $\left(\frac{1}{2}\right)^n (u[n] - u[n-10])$.

3.2. Determine the z-transform of the sequence

$$x[n] = \begin{cases} n, & 0 \le n \le N-1, \\ N, & N \le n. \end{cases}$$

3.3. Determine the z-transform of each of the following sequences. Include with your answer the ROC in the z-plane and a sketch of the pole–zero plot. Express all sums in closed form; α can be complex.

(a) $x_a[n] = \alpha^{|n|}, \qquad 0 < |\alpha| < 1.$

(b) $x_b[n] = \begin{cases} 1, & 0 \le n \le N-1, \\ 0, & \text{otherwise.} \end{cases}$

(c) $x_c[n] = \begin{cases} n+1, & 0 \le n \le N-1, \\ 2N-1-n, & N \le n \le 2(N-1), \\ 0, & \text{otherwise.} \end{cases}$

Hint: Note that $x_b[n]$ is a rectangular sequence and $x_c[n]$ is a triangular sequence. First, express $x_c[n]$ in terms of $x_b[n]$.

3.4. Consider the z-transform $X(z)$ whose pole–zero plot is as shown in Figure P3.4.

(a) Determine the ROC of $X(z)$ if it is known that the Fourier transform exists. For this case, determine whether the corresponding sequence $x[n]$ is right sided, left sided, or two sided.

(b) How many possible two-sided sequences have the pole–zero plot shown in Figure P3.4?

(c) Is it possible for the pole–zero plot in Figure P3.4 to be associated with a sequence that is both stable and causal? If so, give the appropriate ROC.

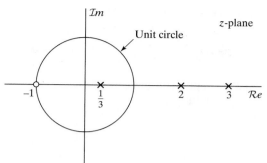

Figure P3.4

3.5. Determine the sequence $x[n]$ with z-transform

$$X(z) = (1 + 2z)(1 + 3z^{-1})(1 - z^{-1}).$$

3.6. Following are several z-transforms. For each, determine the inverse z-transform using both methods—partial fraction expansion and power series expansion—discussed in Section 3.3. In addition, indicate in each case whether the Fourier transform exists.

(a) $X(z) = \dfrac{1}{1 + \frac{1}{2}z^{-1}}, \qquad |z| > \dfrac{1}{2}$

(b) $X(z) = \dfrac{1}{1 + \frac{1}{2}z^{-1}}, \qquad |z| < \dfrac{1}{2}$

(c) $X(z) = \dfrac{1 - \frac{1}{2}z^{-1}}{1 + \frac{3}{4}z^{-1} + \frac{1}{8}z^{-2}}, \qquad |z| > \dfrac{1}{2}$

(d) $X(z) = \dfrac{1 - \frac{1}{2}z^{-1}}{1 - \frac{1}{4}z^{-2}}, \qquad |z| > \dfrac{1}{2}$

(e) $X(z) = \dfrac{1 - az^{-1}}{z^{-1} - a}, \qquad |z| > |1/a|$

3.7. The input to a causal LTI system is

$$x[n] = u[-n - 1] + \left(\frac{1}{2}\right)^n u[n].$$

The z-transform of the output of this system is

$$Y(z) = \frac{-\frac{1}{2}z^{-1}}{\left(1 - \frac{1}{2}z^{-1}\right)\left(1 + z^{-1}\right)}.$$

(a) Determine $H(z)$, the z-transform of the system impulse response. Be sure to specify the ROC.

(b) What is the ROC for $Y(z)$?

(c) Determine $y[n]$.

3.8. The system function of a causal LTI system is

$$H(z) = \frac{1 - z^{-1}}{1 + \frac{3}{4}z^{-1}}.$$

The input to this system is

$$x[n] = \left(\frac{1}{3}\right)^n u[n] + u[-n-1].$$

(a) Find the impulse response of the system, $h[n]$.
(b) Find the output $y[n]$.
(c) Is the system stable? That is, is $h[n]$ absolutely summable?

3.9. A causal LTI system has impulse response $h[n]$, for which the z-transform is

$$H(z) = \frac{1 + z^{-1}}{\left(1 - \frac{1}{2}z^{-1}\right)\left(1 + \frac{1}{4}z^{-1}\right)}.$$

(a) What is the ROC of $H(z)$?
(b) Is the system stable? Explain.
(c) Find the z-transform $X(z)$ of an input $x[n]$ that will produce the output

$$y[n] = -\frac{1}{3}\left(-\frac{1}{4}\right)^n u[n] - \frac{4}{3}(2)^n u[-n-1].$$

(d) Find the impulse response $h[n]$ of the system.

3.10. Without explicitly solving for $X(z)$, find the ROC of the z-transform of each of the following sequences, and determine whether the Fourier transform converges:

(a) $x[n] = \left[\left(\frac{1}{2}\right)^n + \left(\frac{3}{4}\right)^n\right]u[n-10]$

(b) $x[n] = \begin{cases} 1, & -10 \le n \le 10, \\ 0, & \text{otherwise}, \end{cases}$

(c) $x[n] = 2^n u[-n]$

(d) $x[n] = \left[\left(\frac{1}{4}\right)^{n+4} - (e^{j\pi/3})^n\right]u[n-1]$

(e) $x[n] = u[n+10] - u[n+5]$

(f) $x[n] = \left(\frac{1}{2}\right)^{n-1} u[n] + (2+3j)^{n-2}u[-n-1].$

3.11. Following are four z-transforms. Determine which ones *could* be the z-transform of a *causal* sequence. Do not evaluate the inverse transform. You should be able to give the answer by inspection. Clearly state your reasons in each case.

(a) $\dfrac{(1 - z^{-1})^2}{\left(1 - \frac{1}{2}z^{-1}\right)}$

(b) $\dfrac{(z - 1)^2}{\left(z - \frac{1}{2}\right)}$

(c) $\dfrac{\left(z - \frac{1}{4}\right)^5}{\left(z - \frac{1}{2}\right)^6}$

(d) $\dfrac{\left(z - \frac{1}{4}\right)^6}{\left(z - \frac{1}{2}\right)^5}$

3.12. Sketch the pole–zero plot for each of the following z-transforms and shade the ROC:

(a) $X_1(z) = \dfrac{1 - \frac{1}{2}z^{-1}}{1 + 2z^{-1}}$,   ROC: $|z| < 2$

(b) $X_2(z) = \dfrac{1 - \frac{1}{3}z^{-1}}{\left(1 + \frac{1}{2}z^{-1}\right)\left(1 - \frac{2}{3}z^{-1}\right)}$, $x_2[n]$ causal

(c) $X_3(z) = \dfrac{1 + z^{-1} - 2z^{-2}}{1 - \frac{13}{6}z^{-1} + z^{-2}}$, $x_3[n]$ absolutely summable.

3.13. A causal sequence $g[n]$ has the z-transform

$$G(z) = \sin(z^{-1})(1 + 3z^{-2} + 2z^{-4}).$$

Find $g[11]$.

3.14. If $H(z) = \dfrac{1}{1 - \frac{1}{4}z^{-2}}$ and $h[n] = A_1\alpha_1^n u[n] + A_2\alpha_2^n u[n]$, determine the values of A_1, A_2, α_1, and α_2.

3.15. If $H(z) = \dfrac{1 - \frac{1}{1024}z^{-10}}{1 - \frac{1}{2}z^{-1}}$ for $|z| > 0$, is the corresponding LTI system causal? Justify your answer.

3.16. When the input to an LTI system is

$$x[n] = \left(\frac{1}{3}\right)^n u[n] + (2)^n u[-n-1],$$

the corresponding output is

$$y[n] = 5\left(\frac{1}{3}\right)^n u[n] - 5\left(\frac{2}{3}\right)^n u[n].$$

(a) Find the system function $H(z)$ of the system. Plot the pole(s) and zero(s) of $H(z)$ and indicate the ROC.

(b) Find the impulse response $h[n]$ of the system.

(c) Write a difference equation that is satisfied by the given input and output.

(d) Is the system stable? Is it causal?

3.17. Consider an LTI system with input $x[n]$ and output $y[n]$ that satisfies the difference equation

$$y[n] - \frac{5}{2}y[n-1] + y[n-2] = x[n] - x[n-1].$$

Determine all possible values for the system's impulse response $h[n]$ at $n = 0$.

3.18. A causal LTI system has the system function

$$H(z) = \frac{1 + 2z^{-1} + z^{-2}}{\left(1 + \frac{1}{2}z^{-1}\right)(1 - z^{-1})}.$$

(a) Find the impulse response of the system, $h[n]$.

(b) Find the output of this system, $y[n]$, for the input

$$x[n] = 2^n.$$

3.19. For each of the following pairs of input z-transform $X(z)$ and system function $H(z)$, determine the ROC for the output z-transform $Y(z)$:

(a)

$$X(z) = \frac{1}{1 + \frac{1}{2}z^{-1}}, \qquad |z| > \frac{1}{2}$$

$$H(z) = \frac{1}{1 - \frac{1}{4}z^{-1}}, \qquad |z| > \frac{1}{4}$$

(b)

$$X(z) = \frac{1}{1 - 2z^{-1}}, \qquad |z| < 2$$

$$H(z) = \frac{1}{1 - \frac{1}{3}z^{-1}}, \qquad |z| > \frac{1}{3}$$

(c)

$$X(z) = \frac{1}{\left(1 - \frac{1}{5}z^{-1}\right)\left(1 + 3z^{-1}\right)}, \qquad \frac{1}{5} < |z| < 3$$

$$H(z) = \frac{1 + 3z^{-1}}{1 + \frac{1}{3}z^{-1}}, \qquad |z| > \frac{1}{3}$$

3.20. For each of the following pairs of input and output z-transforms $X(z)$ and $Y(z)$, determine the ROC for the system function $H(z)$:

(a)

$$X(z) = \frac{1}{1 - \frac{3}{4}z^{-1}}, \qquad |z| > \frac{3}{4}$$

$$Y(z) = \frac{1}{1 + \frac{2}{3}z^{-1}}, \qquad |z| > \frac{2}{3}$$

(b)

$$X(z) = \frac{1}{1 + \frac{1}{3}z^{-1}}, \qquad |z| < \frac{1}{3}$$

$$Y(z) = \frac{1}{\left(1 - \frac{1}{6}z^{-1}\right)\left(1 + \frac{1}{3}z^{-1}\right)}, \qquad \frac{1}{6} < |z| < \frac{1}{3}$$

Basic Problems

3.21. A causal LTI system has the following system function:

$$H(z) = \frac{4 + 0.25z^{-1} - 0.5z^{-2}}{(1 - 0.25z^{-1})(1 + 0.5z^{-1})}$$

(a) What is the ROC for $H(z)$?

(b) Determine if the system is stable or not.

(c) Determine the difference equation that is satisfied by the input $x[n]$ and the output $y[n]$.

(d) Use a partial fraction expansion to determine the impulse response $h[n]$.

(e) Find $Y(z)$, the z-transform of the output, when the input is $x[n] = u[-n-1]$. Be sure to specify the ROC for $Y(z)$.

(f) Find the output sequence $y[n]$ when the input is $x[n] = u[-n-1]$.

3.22. A causal LTI system has system function

$$H(z) = \frac{1 - 4z^{-2}}{1 + 0.5z^{-1}}.$$

The input to this system is

$$x[n] = u[n] + 2\cos\left(\frac{\pi}{2}n\right) \qquad -\infty < n < \infty,$$

Determine the output $y[n]$ for large positive n; i.e., find an expression for $y[n]$ that is asymptotically correct as n gets large. (*Of course, one approach is to find an expression for $y[n]$ that is valid for all n, but you should see an easier way.*)

3.23. Consider an LTI system with impulse response

$$h[n] = \begin{cases} a^n, & n \geq 0, \\ 0, & n < 0, \end{cases}$$

and input

$$x[n] = \begin{cases} 1, & 0 \leq n \leq (N-1), \\ 0, & \text{otherwise.} \end{cases}$$

(a) Determine the output $y[n]$ by explicitly evaluating the discrete convolution of $x[n]$ and $h[n]$.

(b) Determine the output $y[n]$ by computing the inverse z-transform of the product of the z-transforms of $x[n]$ and $h[n]$.

3.24. Consider an LTI system that is stable and for which $H(z)$, the z-transform of the impulse response, is given by

$$H(z) = \frac{3}{1 + \frac{1}{3}z^{-1}}.$$

Suppose $x[n]$, the input to the system, is a unit step sequence.

(a) Determine the output $y[n]$ by evaluating the discrete convolution of $x[n]$ and $h[n]$.

(b) Determine the output $y[n]$ by computing the inverse z-transform of $Y(z)$.

3.25. Sketch each of the following sequences and determine their z-transforms, including the ROC:

(a) $\displaystyle\sum_{k=-\infty}^{\infty} \delta[n - 4k]$

(b) $\dfrac{1}{2}\left[e^{j\pi n} + \cos\left(\frac{\pi}{2}n\right) + \sin\left(\frac{\pi}{2} + 2\pi n\right) \right] u[n]$

3.26. Consider a right-sided sequence $x[n]$ with z-transform

$$X(z) = \frac{1}{(1 - az^{-1})(1 - bz^{-1})} = \frac{z^2}{(z - a)(z - b)}.$$

In Section 3.3, we considered the determination of $x[n]$ by carrying out a partial fraction expansion, with $X(z)$ considered as a ratio of polynomials in z^{-1}. Carry out a partial fraction expansion of $X(z)$, considered as a ratio of polynomials in z, and determine $x[n]$ from this expansion.

3.27. Determine the unilateral z-transform, including the ROC, for each of the following sequences:

(a) $\delta[n]$
(b) $\delta[n - 1]$
(c) $\delta[n + 1]$
(d) $\left(\frac{1}{2}\right)^n u[n]$
(e) $-\left(\frac{1}{2}\right)^n u[-n - 1]$
(f) $\left(\frac{1}{2}\right)^n u[-n]$
(g) $\{\left(\frac{1}{2}\right)^n + \left(\frac{1}{4}\right)^n\}u[n]$
(h) $\left(\frac{1}{2}\right)^{n-1} u[n - 1]$

3.28. If $\mathcal{X}(z)$ denotes the unilateral z-transform of $x[n]$, determine, in terms of $\mathcal{X}(z)$, the unilateral z-transform of the following:

(a) $x[n - 2]$
(b) $x[n + 1]$
(c) $\sum_{m=-\infty}^{n} x[m]$

3.29. For each of the following difference equations and associated input and initial conditions, determine the response $y[n]$ for $n \geq 0$ by using the unilateral z-transform.

(a) $y[n] + 3y[n - 1] = x[n]$
$x[n] = \left(\frac{1}{2}\right)^n u[n]$
$y[-1] = 1$
(b) $y[n] - \frac{1}{2}y[n - 1] = x[n] - \frac{1}{2}x[n - 1]$
$x[n] = u[n]$
$y[-1] = 0$
(c) $y[n] - \frac{1}{2}y[n - 1] = x[n] - \frac{1}{2}x[n - 1]$
$x[n] = \left(\frac{1}{2}\right)^n u[n]$
$y[-1] = 1$

Advanced Problems

3.30. A causal LTI system has system function

$$H(z) = \frac{1 - z^{-1}}{1 - 0.25z^{-2}} = \frac{1 - z^{-1}}{(1 - 0.5z^{-1})(1 + 0.5z^{-1})}.$$

(a) Determine the output of the system when the input is $x[n] = u[n]$.

(b) Determine the input $x[n]$ so that the corresponding output of the above system is $y[n] = \delta[n] - \delta[n - 1]$.

(c) Determine the output $y[n]$ when the input is $x[n] = \cos(0.5\pi n)$ for $-\infty < n < \infty$. You may leave your answer in any convenient form.

3.31. Determine the inverse z-transform of each of the following. In parts (a)–(c), use the methods specified. (In part (d), use any method you prefer.)

(a) Long division:

$$X(z) = \frac{1 - \frac{1}{3}z^{-1}}{1 + \frac{1}{3}z^{-1}}, \qquad x[n] \text{ a right-sided sequence}$$

(b) Partial fraction:

$$X(z) = \frac{3}{z - \frac{1}{4} - \frac{1}{8}z^{-1}}, \qquad x[n] \text{ stable}$$

(c) Power series:

$$X(z) = \ln(1 - 4z), \qquad |z| < \frac{1}{4}$$

(d) $X(z) = \dfrac{1}{1 - \frac{1}{3}z^{-3}}, \qquad |z| > (3)^{-1/3}$

3.32. Using any method, determine the inverse z-transform for each of the following:

(a) $X(z) = \dfrac{1}{\left(1 + \frac{1}{2}z^{-1}\right)^2 (1 - 2z^{-1})(1 - 3z^{-1})}$,
($x[n]$ is a stable sequence)

(b) $X(z) = e^{z^{-1}}$

(c) $X(z) = \dfrac{z^3 - 2z}{z - 2}, \qquad (x[n] \text{ is a left-sided sequence})$

3.33. Determine the inverse z-transform of each of the following. You should find the z-transform properties in Section 3.4 helpful.

(a) $X(z) = \dfrac{3z^{-3}}{\left(1 - \frac{1}{4}z^{-1}\right)^2}, \qquad x[n] \text{ left sided}$

(b) $X(z) = \sin(z), \qquad \text{ROC includes } |z| = 1$

(c) $X(z) = \dfrac{z^7 - 2}{1 - z^{-7}}, \qquad |z| > 1$

3.34. Determine a sequence $x[n]$ whose z-transform is $X(z) = e^z + e^{1/z}, z \neq 0$.

3.35. Determine the inverse z-transform of

$$X(z) = \log(1 - 2z), \qquad |z| < \frac{1}{2},$$

by

(a) using the power series

$$\log(1 - x) = -\sum_{m=1}^{\infty} \frac{x^m}{m}, \qquad |x| < 1;$$

(b) first differentiating $X(z)$ and then using the derivative to recover $x[n]$.

3.36. For each of the following sequences, determine the z-transform and ROC, and sketch the pole–zero diagram:

(a) $x[n] = a^n u[n] + b^n u[n] + c^n u[-n - 1], \qquad |a| < |b| < |c|$

(b) $x[n] = n^2 a^n u[n]$

(c) $x[n] = e^{n^4}\left[\cos\left(\frac{\pi}{12}n\right)\right]u[n] - e^{n^4}\left[\cos\left(\frac{\pi}{12}n\right)\right]u[n - 1]$

3.37. The pole–zero diagram in Figure P3.37 corresponds to the z-transform $X(z)$ of a causal sequence $x[n]$. Sketch the pole–zero diagram of $Y(z)$, where $y[n] = x[-n+3]$. Also, specify the ROC for $Y(z)$.

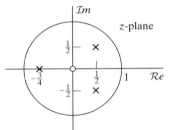

Figure P3.37

3.38. Let $x[n]$ be the sequence with the pole–zero plot shown in Figure P3.38. Sketch the pole–zero plot for:

(a) $y[n] = \left(\frac{1}{2}\right)^n x[n]$

(b) $w[n] = \cos\left(\frac{\pi n}{2}\right)x[n]$

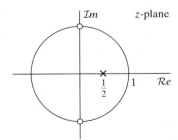

Figure P3.38

3.39. Determine the unit step response of the causal system for which the z-transform of the impulse response is

$$H(z) = \frac{1 - z^3}{1 - z^4}.$$

3.40. If the input $x[n]$ to an LTI system is $x[n] = u[n]$, the output is

$$y[n] = \left(\frac{1}{2}\right)^{n-1} u[n + 1].$$

(a) Find $H(z)$, the z-transform of the system impulse response, and plot its pole–zero diagram.
(b) Find the impulse response $h[n]$.
(c) Is the system stable?
(d) Is the system causal?

3.41. Consider a sequence $x[n]$ for which the z-transform is

$$X(z) = \frac{\frac{1}{3}}{1 - \frac{1}{2}z^{-1}} + \frac{\frac{1}{4}}{1 - 2z^{-1}}$$

and for which the ROC includes the unit circle. Determine $x[0]$ using the initial-value theorem (see Problem 3.57).

3.42. In Figure P3.42, $H(z)$ is the system function of a causal LTI system.

(a) Using z-transforms of the signals shown in the figure, obtain an expression for $W(z)$ in the form

$$W(z) = H_1(z)X(z) + H_2(z)E(z),$$

where both $H_1(z)$ and $H_2(z)$ are expressed in terms of $H(z)$.
(b) For the special case $H(z) = z^{-1}/(1 - z^{-1})$, determine $H_1(z)$ and $H_2(z)$.
(c) Is the system $H(z)$ stable? Are the systems $H_1(z)$ and $H_2(z)$ stable?

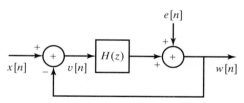

Figure P3.42

3.43. In Figure P3.43, $h[n]$ is the impulse response of the LTI system within the inner box. The input to system $h[n]$ is $v[n]$, and the output is $w[n]$. The z-transform of $h[n]$, $H(z)$, exists in the following ROC:

$$0 < r_{min} < |z| < r_{max} < \infty.$$

(a) Can the LTI system with impulse response $h[n]$ be bounded input, bounded output stable? If so, determine inequality constraints on r_{min} and r_{max} such that it is stable. If not, briefly explain why.
(b) Is the overall system (in the large box, with input $x[n]$ and output $y[n]$) LTI? If so, find its impulse response $g[n]$. If not, briefly explain why.

(c) Can the overall system be BIBO stable? If so, determine inequality constraints relating α, r_{min}, and r_{max} such that it is stable. If not, briefly explain why.

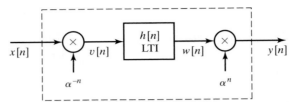

Figure P3.43

3.44. A causal and stable LTI system S has its input $x[n]$ and output $y[n]$ related by the linear constant-coefficient difference equation

$$y[n] + \sum_{k=1}^{10} \alpha_k y[n-k] = x[n] + \beta x[n-1].$$

Let the impulse response of S be the sequence $h[n]$.

(a) Show that $h[0]$ must be nonzero.
(b) Show that α_1 can be determined from knowledge of β, $h[0]$, and $h[1]$.
(c) If $h[n] = (0.9)^n \cos(\pi n/4)$ for $0 \le n \le 10$, sketch the pole–zero plot for the system function of S, and indicate the ROC.

3.45. When the input to an LTI system is

$$x[n] = \left(\frac{1}{2}\right)^n u[n] + 2^n u[-n-1],$$

the output is

$$y[n] = 6\left(\frac{1}{2}\right)^n u[n] - 6\left(\frac{3}{4}\right)^n u[n].$$

(a) Find the system function $H(z)$ of the system. Plot the poles and zeros of $H(z)$, and indicate the ROC.
(b) Find the impulse response $h[n]$ of the system.
(c) Write the difference equation that characterizes the system.
(d) Is the system stable? Is it causal?

3.46. The following information is known about an LTI system:

(i) The system is causal.
(ii) When the input is

$$x[n] = -\frac{1}{3}\left(\frac{1}{2}\right)^n u[n] - \frac{4}{3}(2)^n u[-n-1],$$

then the z-transform of the output is

$$Y(z) = \frac{1 - z^{-2}}{(1 - \frac{1}{2}z^{-1})(1 - 2z^{-1})}.$$

(a) Find the z-transform of $x[n]$.

(b) What are the possible choices for the ROC of $Y(z)$?

(c) What are the possible choices for a linear constant-coefficient difference equation used to describe the system?

(d) What are the possible choices for the impulse response of the system?

3.47. Let $x[n]$ be a discrete-time signal with $x[n] = 0$ for $n \leq 0$ and z-transform $X(z)$. Furthermore, given $x[n]$, let the discrete-time signal $y[n]$ be defined by

$$y[n] = \begin{cases} \frac{1}{n}x[n], & n > 0, \\ 0, & \text{otherwise.} \end{cases}$$

(a) Compute $Y(z)$ in terms of $X(z)$.

(b) Using the result of part (a), find the z-transform of

$$w[n] = \frac{1}{n + \delta[n]} u[n - 1].$$

3.48. The signal $y[n]$ is the output of an LTI system with impulse response $h[n]$ for a given input $x[n]$. Throughout the problem, assume that $y[n]$ is stable and has a z-transform $Y(z)$ with the pole–zero diagram shown in Figure P3.48-1. The signal $x[n]$ is stable and has the pole–zero diagram shown in Figure P3.48-2.

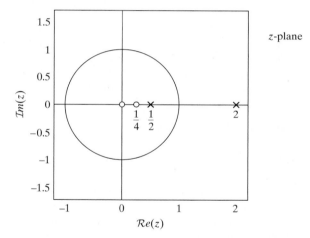

Figure P3.48-1

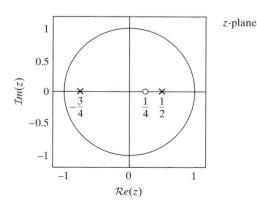

Figure P3.48-2

(a) What is the ROC, $Y(z)$?

(b) Is $y[n]$ left sided, right sided, or two sided?

(c) What is the ROC of $X(z)$?

(d) Is $x[n]$ a causal sequence? That is, does $x[n] = 0$ for $n < 0$?

(e) What is $x[0]$?

(f) Draw the pole–zero plot of $H(z)$, and specify its ROC.

(g) Is $h[n]$ anticausal? That is, does $h[n] = 0$ for $n > 0$?

3.49. Consider the difference equation of Eq. (3.66).

(a) Show that with nonzero initial conditions the unilateral z-transform of the output of the difference equation is

$$
\mathcal{Y}(z) = -\frac{\displaystyle\sum_{k=1}^{N} a_k \left(\sum_{m=1}^{k} y[m-k-1]z^{-m+1} \right)}{\displaystyle\sum_{k=0}^{N} a_k z^{-k}} + \frac{\displaystyle\sum_{k=0}^{M} b_k z^{-k}}{\displaystyle\sum_{k=0}^{N} a_k z^{-k}} \mathcal{X}(z).
$$

(b) Use the result of (a) to show that the output has the form
$$y[n] = y_{\text{ZIR}}[n] + y_{\text{ZICR}}[n]$$
where $y_{\text{ZIR}}[n]$ is the output when the input is zero for all n and $y_{\text{ZICR}}[n]$ is the output when the initial conditions are all zero.

(c) Show that when the initial conditions are all zero, the result reduces to the result that is obtained with the bilateral z-transform.

Extension Problems

3.50. Let $x[n]$ denote a causal sequence; i.e., $x[n] = 0, n < 0$. Furthermore, assume that $x[0] \neq 0$ and that the z-transform is a rational function.

(a) Show that there are no poles or zeros of $X(z)$ at $z = \infty$, i.e., that $\lim_{z \to \infty} X(z)$ is nonzero and finite.

(b) Show that the number of poles in the finite z-plane equals the number of zeros in the finite z-plane. (The finite z-plane excludes $z = \infty$.)

3.51. Consider a sequence with z-transform $X(z) = P(z)/Q(z)$, where $P(z)$ and $Q(z)$ are polynomials in z. If the sequence is absolutely summable and if all the roots of $Q(z)$ are inside the unit circle, is the sequence necessarily causal? If your answer is yes, clearly explain. If your answer is no, give a counterexample.

3.52. Let $x[n]$ be a causal stable sequence with z-transform $X(z)$. The *complex cepstrum* $\hat{x}[n]$ is defined as the inverse transform of the logarithm of $X(z)$; i.e.,

$$\hat{X}(z) = \log X(z) \overset{\mathcal{Z}}{\longleftrightarrow} \hat{x}[n],$$

where the ROC of $\hat{X}(z)$ includes the unit circle. (Strictly speaking, taking the logarithm of a complex number requires some careful considerations. Furthermore, the logarithm of a valid z-transform may not be a valid z-transform. For now, we assume that this operation is valid.)

Determine the complex cepstrum for the sequence
$$x[n] = \delta[n] + a\delta[n-N], \qquad \text{where } |a| < 1.$$

3.53. Assume that $x[n]$ is real and even; i.e., $x[n] = x[-n]$. Further, assume that z_0 is a zero of $X(z)$; i.e., $X(z_0) = 0$.

(a) Show that $1/z_0$ is also a zero of $X(z)$.

(b) Are there other zeros of $X(z)$ implied by the information given?

3.54. Using the definition of the z-transform in Eq. (3.2), show that if $X(z)$ is the z-transform of $x[n] = x_R[n] + jx_I[n]$, then

(a) $x^*[n] \overset{Z}{\longleftrightarrow} X^*(z^*)$

(b) $x[-n] \overset{Z}{\longleftrightarrow} X(1/z)$

(c) $x_R[n] \overset{Z}{\longleftrightarrow} \frac{1}{2}[X(z) + X^*(z^*)]$

(d) $x_I[n] \overset{Z}{\longleftrightarrow} \frac{1}{2j}[X(z) - X^*(z^*)]$.

3.55. Consider a *real* sequence $x[n]$ that has all the poles and zeros of its z-transform inside the unit circle. Determine, in terms of $x[n]$, a *real* sequence $x_1[n]$ not equal to $x[n]$, but for which $x_1[0] = x[0]$, $|x_1[n]| = |x[n]|$, and the z-transform of $x_1[n]$ has all its poles and zeros inside the unit circle.

3.56. A real finite-duration sequence whose z-transform has no zeros at conjugate reciprocal pair locations and no zeros on the unit circle is uniquely specified to within a positive scale factor by its Fourier transform phase (Hayes et al., 1980).

An example of zeros at conjugate reciprocal pair locations is $z = a$ and $(a^*)^{-1}$. Even though we can generate sequences that do not satisfy the preceding set of conditions, almost any sequence of practical interest satisfies the conditions and therefore is uniquely specified to within a positive scale factor by the phase of its Fourier transform.

Consider a sequence $x[n]$ that is real, that is zero outside $0 \leq n \leq N - 1$, and whose z-transform has no zeros at conjugate reciprocal pair locations and no zeros on the unit circle. We wish to develop an algorithm that reconstructs $cx[n]$ from $\angle X(e^{j\omega})$, the Fourier transform phase of $x[n]$, where c is a positive scale factor.

(a) Specify a set of $(N-1)$ linear equations, the solution to which will provide the recovery of $x[n]$ to within a positive or negative scale factor from $\tan\{\angle X(e^{j\omega})\}$. You do not have to prove that the set of $(N-1)$ linear equations has a unique solution. Further, show that if we know $\angle X(e^{j\omega})$ rather than just $\tan\{\angle X(e^{j\omega})\}$, the sign of the scale factor can also be determined.

(b) Suppose

$$x[n] = \begin{cases} 0, & n < 0, \\ 1, & n = 0, \\ 2, & n = 1, \\ 3, & n = 2, \\ 0, & n \geq 3. \end{cases}$$

Using the approach developed in part (a), demonstrate that $cx[n]$ can be determined from $\angle X(e^{j\omega})$, where c is a positive scale factor.

3.57. For a sequence $x[n]$ that is zero for $n < 0$, use Eq. (3.2) to show that

$$\lim_{z \to \infty} X(z) = x[0].$$

This result is called the *initial value theorem*. What is the corresponding theorem if the sequence is zero for $n > 0$?

3.58. The aperiodic autocorrelation function for a real-valued stable sequence $x[n]$ is defined as

$$c_{xx}[n] = \sum_{k=-\infty}^{\infty} x[k]x[n+k].$$

(a) Show that the z-transform of $c_{xx}[n]$ is

$$C_{xx}(z) = X(z)X(z^{-1}).$$

Determine the ROC for $C_{xx}(z)$.

(b) Suppose that $x[n] = a^n u[n]$. Sketch the pole–zero plot for $C_{xx}(z)$, including the ROC. Also, find $c_{xx}[n]$ by evaluating the inverse z-transform of $C_{xx}(z)$.

(c) Specify another sequence, $x_1[n]$, that is not equal to $x[n]$ in part (b), but that has the same autocorrelation function, $c_{xx}[n]$, as $x[n]$ in part (b).

(d) Specify a third sequence, $x_2[n]$, that is not equal to $x[n]$ or $x_1[n]$, but that has the same autocorrelation function as $x[n]$ in part (b).

3.59. Determine whether or not the function $X(z) = z^*$ can correspond to the z-transform of a sequence. Clearly explain your reasoning.

3.60. Let $X(z)$ denote a ratio of polynomials in z; i.e.,

$$X(z) = \frac{B(z)}{A(z)}.$$

Show that if $X(z)$ has a 1st-order pole at $z = z_0$, then the residue of $X(z)$ at $z = z_0$ is equal to

$$\frac{B(z_0)}{A'(z_0)},$$

where $A'(z_0)$ denotes the derivative of $A(z)$ evaluated at $z = z_0$.

4

Sampling of
Continuous-Time
Signals

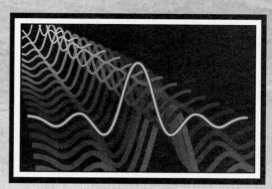

4.0 INTRODUCTION

Discrete-time signals can arise in many ways, but they occur most commonly as representations of sampled continuous-time signals. While sampling will no doubt be familiar to many readers, we shall review many of the basic issues such as the phenomenon of aliasing and the important fact that continuous-time signal processing can be implemented through a process of sampling, discrete-time processing, and reconstruction of a continuous-time signal. After a thorough discussion of these basic issues, we discuss multirate signal processing, A/D conversion, and the use of oversampling in A/D conversion.

4.1 PERIODIC SAMPLING

Discrete representations of signals can take many forms including basis expansions of various types, parametric models for signal modeling (Chapter 11), and nonuniform sampling (see for example Yen (1956), Yao and Thomas (1967) and Eldar and Oppenheim (2000)). Such representations are often based on prior knowledge of properties of the signal that can be exploited to obtain more efficient representations. However, even these alternative representations generally begin with a discrete-time representation of a continuous-time signal obtained through periodic sampling; i.e., a sequence of samples, $x[n]$, is obtained from a continuous-time signal $x_c(t)$ according to the relation

$$x[n] = x_c(nT), \qquad -\infty < n < \infty. \tag{4.1}$$

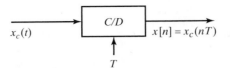

Figure 4.1   Block diagram representation of an ideal continuous-to-discrete-time (C/D) converter.

In Eq. (4.1), T is the *sampling period,* and its reciprocal, $f_s = 1/T$, is the *sampling frequency,* in samples per second. We also express the sampling frequency as $\Omega_s = 2\pi/T$ when we want to use frequencies in radians per second. Since sampling representations rely only on the assumption of a bandlimited Fourier transform, they are applicable to a wide class of signals that arise in many practical applications.

We refer to a system that implements the operation of Eq. (4.1) as an *ideal continuous-to-discrete-time (C/D) converter,* and we depict it in block diagram form as indicated in Figure 4.1. As an example of the relationship between $x_c(t)$ and $x[n]$, in Figure 2.2 we illustrated a continuous-time speech waveform and the corresponding sequence of samples.

In a practical setting, the operation of sampling is implemented by an analog-to-digital (A/D) converter. Such systems can be viewed as approximations to the ideal C/D converter. In addition to sampling rate, which is sufficient to define the ideal C/D converter, important considerations in the implementation or choice of an A/D converter include quantization of the output samples, linearity of quantization steps, the need for sample-and-hold circuits, and limitations on the sampling rate. The effects of quantization are discussed in Sections 4.8.2 and 4.8.3. Other practical issues of A/D conversion are electronic circuit concerns that are outside the scope of this text.

The sampling operation is generally not invertible; i.e., given the output $x[n]$, it is not possible in general to reconstruct $x_c(t)$, the input to the sampler, since many continuous-time signals can produce the same output sequence of samples. The inherent ambiguity in sampling is a fundamental issue in signal processing. However, it is possible to remove the ambiguity by restricting the frequency content of input signals that go into the sampler.

It is convenient to represent the sampling process mathematically in the two stages depicted in Figure 4.2(a). The stages consist of an impulse train modulator, followed by conversion of the impulse train to a sequence. The periodic impulse train is

$$s(t) = \sum_{n=-\infty}^{\infty} \delta(t - nT),\tag{4.2}$$

where $\delta(t)$ is the unit impulse function, or Dirac delta function. The product of $s(t)$ and $x_c(t)$ is therefore

$$x_s(t) = x_c(t)s(t)$$

$$= x_c(t) \sum_{n=-\infty}^{\infty} \delta(t - nT) = \sum_{n=-\infty}^{\infty} x_c(t)\delta(t - nT).\tag{4.3}$$

Using the property of the continuous-time impulse function, $x(t)\delta(t) = x(0)\delta(t)$, sometimes called the "sifting property" of the impulse function, (see e.g., Oppenheim and

Willsky, 1997), $x_s(t)$ can be expressed as

$$x_s(t) = \sum_{n=-\infty}^{\infty} x_c(nT)\delta(t - nT), \tag{4.4}$$

i.e., the size (area) of the impulse at sample time nT is equal to the value of the continuous-time signal at that time. In this sense, the impulse train modulation of Eq. (4.3) is a mathematical representation of sampling.

Figure 4.2(b) shows a continuous-time signal $x_c(t)$ and the results of impulse train sampling for two different sampling rates. Note that the impulses $x_c(nT)\delta(t - nT)$ are represented by arrows with length proportional to their area. Figure 4.2(c) depicts the corresponding output sequences. The essential difference between $x_s(t)$ and $x[n]$ is that $x_s(t)$ is, in a sense, a continuous-time signal (specifically, an impulse train) that is zero,

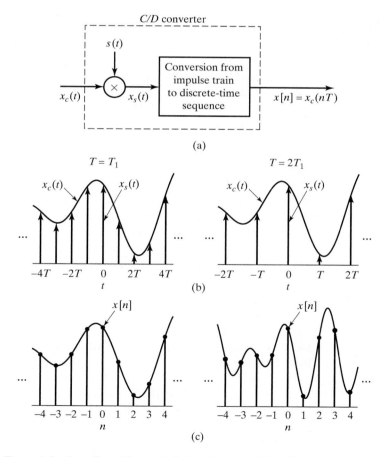

Figure 4.2 Sampling with a periodic impulse train, followed by conversion to a discrete-time sequence. (a) Overall system. (b) $x_s(t)$ for two sampling rates. (c) The output sequence for the two different sampling rates.

except at integer multiples of T. The sequence $x[n]$, on the other hand, is indexed on the integer variable n, which, in effect, introduces a time normalization; i.e., the sequence of numbers $x[n]$ contains no explicit information about the sampling period T. Furthermore, the samples of $x_c(t)$ are represented by finite numbers in $x[n]$ rather than as the areas of impulses, as with $x_s(t)$.

It is important to emphasize that Figure 4.2(a) is strictly a mathematical representation convenient for gaining insight into sampling in both the time domain and frequency domain. It is not a close representation of any physical circuits or systems designed to implement the sampling operation. Whether a piece of hardware can be construed to be an approximation to the block diagram of Figure 4.2(a) is a secondary issue at this point. We have introduced this representation of the sampling operation because it leads to a simple derivation of a key result and because the approach leads to a number of important insights that are difficult to obtain from a more formal derivation based on manipulation of Fourier transform formulas.

4.2 FREQUENCY-DOMAIN REPRESENTATION OF SAMPLING

To derive the frequency-domain relation between the input and output of an ideal C/D converter, consider the Fourier transform of $x_s(t)$. Since, from Eq. (4.3), $x_s(t)$ is the product of $x_c(t)$ and $s(t)$, the Fourier transform of $x_s(t)$ is the convolution of the Fourier transforms $X_c(j\Omega)$ and $S(j\Omega)$ scaled by $\frac{1}{2\pi}$. The Fourier transform of the periodic impulse train $s(t)$ is the periodic impulse train

$$S(j\Omega) = \frac{2\pi}{T} \sum_{k=-\infty}^{\infty} \delta(\Omega - k\,\Omega_s), \tag{4.5}$$

where $\Omega_s = 2\pi/T$ is the sampling frequency in radians/s (see Oppenheim and Willsky, 1997 or McClellan, Schafer and Yoder, 2003). Since

$$X_s(j\Omega) = \frac{1}{2\pi} X_c(j\Omega) * S(j\Omega),$$

where $*$ denotes the operation of continuous-variable convolution, it follows that

$$X_s(j\Omega) = \frac{1}{T} \sum_{k=-\infty}^{\infty} X_c(j(\Omega - k\,\Omega_s)). \tag{4.6}$$

Equation (4.6) is the desired relationship between the Fourier transforms of the input and the output of the impulse train modulator in Figure 4.2(a). Equation (4.6) states that the Fourier transform of $x_s(t)$ consists of periodically repeated copies of $X_c(j\Omega)$, the Fourier transform of $x_c(t)$. These copies are shifted by integer multiples of the sampling frequency, and then superimposed to produce the periodic Fourier transform of the impulse train of samples. Figure 4.3 depicts the frequency-domain representation of impulse train sampling. Figure 4.3(a) represents a bandlimited Fourier transform having the property that $X_c(j\Omega) = 0$ for $|\Omega| \geq \Omega_N$. Figure 4.3(b) shows the periodic impulse train $S(j\Omega)$, and Figure 4.3(c) shows $X_s(j\Omega)$, the result of convolving $X_c(j\Omega)$ with $S(j\Omega)$ and scaling by $\frac{1}{2\pi}$. It is evident that when

$$\Omega_s - \Omega_N \geq \Omega_N, \quad \text{or} \quad \Omega_s \geq 2\Omega_N, \tag{4.7}$$

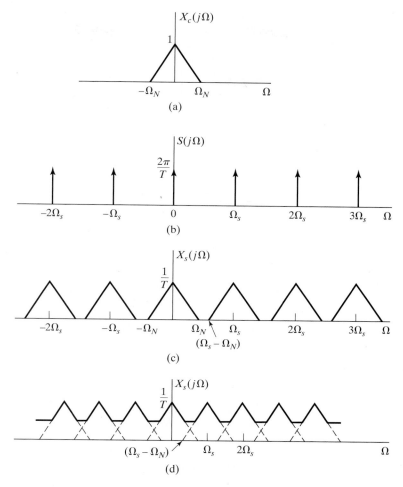

Figure 4.3 Frequency-domain representation of sampling in the time domain.
(a) Spectrum of the original signal. (b) Fourier transform of the sampling function.
(c) Fourier transform of the sampled signal with $\Omega_S > 2\Omega_N$. (d) Fourier transform
of the sampled signal with $\Omega_S < 2\Omega_N$.

as in Figure 4.3(c), the replicas of $X_c(j\Omega)$ do not overlap, and therefore, when they are
added together in Eq. (4.6), there remains (to within a scale factor of $1/T$) a replica
of $X_c(j\Omega)$ at each integer multiple of Ω_s. Consequently, $x_c(t)$ can be recovered from
$x_s(t)$ with an ideal lowpass filter. This is depicted in Figure 4.4(a), which shows the
impulse train modulator followed by an LTI system with frequency response $H_r(j\Omega)$.
For $X_c(j\Omega)$ as in Figure 4.4(b), $X_s(j\Omega)$ would be as shown in Figure 4.4(c), where it is
assumed that $\Omega_s > 2\Omega_N$. Since

$$X_r(j\Omega) = H_r(j\Omega)X_s(j\Omega), \tag{4.8}$$

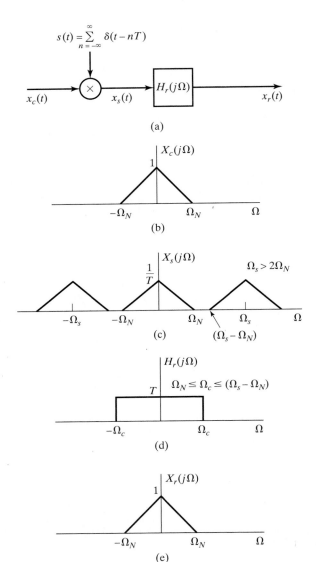

Figure 4.4 Exact recovery of a continuous-time signal from its samples using an ideal lowpass filter.

it follows that if $H_r(j\Omega)$ is an ideal lowpass filter with gain T and cutoff frequency Ω_c such that

$$\Omega_N \leq \Omega_c \leq (\Omega_s - \Omega_N), \tag{4.9}$$

then

$$X_r(j\Omega) = X_c(j\Omega), \tag{4.10}$$

as depicted in Figure 4.4(e) and therefore $x_r(t) = x_c(t)$.

If the inequality of Eq. (4.7) does not hold, i.e., if $\Omega_s < 2\Omega_N$, the copies of $X_c(j\Omega)$ overlap, so that when they are added together, $X_c(j\Omega)$ is no longer recoverable by lowpass filtering. This is illustrated in Figure 4.3(d). In this case, the reconstructed output $x_r(t)$ in Figure 4.4(a) is related to the original continuous-time input through a distortion referred to as *aliasing distortion,* or, more simply, *aliasing.* Figure 4.5 illustrates aliasing in the frequency domain for the simple case of a cosine signal of the form

$$x_c(t) = \cos \Omega_0 t, \tag{4.11a}$$

whose Fourier transform is

$$X_c(j\Omega) = \pi\delta(\Omega - \Omega_0) + \pi\delta(\Omega + \Omega_0) \tag{4.11b}$$

as depicted in Figure 4.5(a). Note that the impulse at $-\Omega_0$ is dashed. It will be helpful to observe its effect in subsequent plots. Figure 4.5(b) shows the Fourier transform of $x_s(t)$ with $\Omega_0 < \Omega_s/2$, and Figure 4.5(c) shows the Fourier transform of $x_s(t)$ with $\frac{\Omega_s}{2} < \Omega_0 < \Omega_s$. Figures 4.5(d) and (e) correspond to the Fourier transform of the

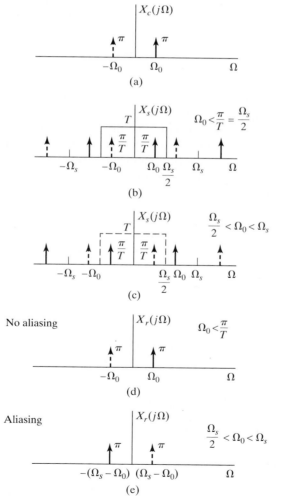

Figure 4.5 The effect of aliasing in the sampling of a cosine signal.

lowpass filter output for $\Omega_0 < \Omega_s/2 = \pi/T$ and $\Omega_s/2 < \Omega_0 < \Omega_s$, respectively, with $\Omega_c = \Omega_s/2$. Figures 4.5(c) and (e) correspond to the case of aliasing. With no aliasing [Figures 4.5(b) and (d)], the reconstructed output is

$$x_r(t) = \cos \Omega_0 t. \tag{4.12}$$

With aliasing, the reconstructed output is

$$x_r(t) = \cos(\Omega_s - \Omega_0)t; \tag{4.13}$$

i.e., the higher frequency signal $\cos \Omega_0 t$ has taken on the identity (alias) of the lower frequency signal $\cos(\Omega_s - \Omega_0)t$ as a consequence of the sampling and reconstruction. This discussion is the basis for the Nyquist sampling theorem (Nyquist 1928; Shannon, 1949), stated as follows.

Nyquist-Shannon Sampling Theorem: Let $x_c(t)$ be a bandlimited signal with

$$X_c(j\Omega) = 0 \qquad \text{for } |\Omega| \geq \Omega_N. \tag{4.14a}$$

Then $x_c(t)$ is uniquely determined by its samples $x[n] = x_c(nT)$, $n = 0, \pm 1, \pm 2, \ldots$, if

$$\Omega_s = \frac{2\pi}{T} \geq 2\Omega_N. \tag{4.14b}$$

The frequency Ω_N is commonly referred to as the *Nyquist frequency*, and the frequency $2\Omega_N$ as the *Nyquist rate*.

Thus far, we have considered only the impulse train modulator in Figure 4.2(a). Our eventual objective is to express $X(e^{j\omega})$, the discrete-time Fourier transform (DTFT) of the sequence $x[n]$, in terms of $X_s(j\Omega)$ and $X_c(j\Omega)$. Toward this end, let us consider an alternative expression for $X_s(j\Omega)$. Applying the continuous-time Fourier transform to Eq. (4.4), we obtain

$$X_s(j\Omega) = \sum_{n=-\infty}^{\infty} x_c(nT)e^{-j\Omega Tn}. \tag{4.15}$$

Since

$$x[n] = x_c(nT) \tag{4.16}$$

and

$$X(e^{j\omega}) = \sum_{n=-\infty}^{\infty} x[n]e^{-j\omega n}, \tag{4.17}$$

it follows that

$$X_s(j\Omega) = X(e^{j\omega})|_{\omega=\Omega T} = X(e^{j\Omega T}). \tag{4.18}$$

Consequently, from Eqs. (4.6) and (4.18),

$$X(e^{j\Omega T}) = \frac{1}{T} \sum_{k=-\infty}^{\infty} X_c(j(\Omega - k\Omega_s)), \tag{4.19}$$

or equivalently,

$$X(e^{j\omega}) = \frac{1}{T} \sum_{k=-\infty}^{\infty} X_c\left[j\left(\frac{\omega}{T} - \frac{2\pi k}{T}\right)\right]. \tag{4.20}$$

From Eqs. (4.18)–(4.20), we see that $X(e^{j\omega})$ is a frequency-scaled version of $X_s(j\Omega)$ with the frequency scaling specified by $\omega = \Omega T$. This scaling can alternatively be thought of as a normalization of the frequency axis so that the frequency $\Omega = \Omega_s$ in $X_s(j\Omega)$ is normalized to $\omega = 2\pi$ for $X(e^{j\omega})$. The frequency scaling or normalization in the transformation from $X_s(j\Omega)$ to $X(e^{j\omega})$ is directly a result of the time normalization in the transformation from $x_s(t)$ to $x[n]$. Specifically, as we see in Figure 4.2, $x_s(t)$ retains a spacing between samples equal to the sampling period T. In contrast, the "spacing" of sequence values $x[n]$ is always unity; i.e., the time axis is normalized by a factor of T. Correspondingly, in the frequency domain the frequency axis is normalized by $f_s = 1/T$.

For a sinusoid of the form $x_c(t) = \cos(\Omega_0 t)$, the highest (and only) frequency is Ω_0. Since the signal is described by a simple equation, it is easy to compute the samples of the signal. The next two examples use sinusoidal signals to illustrate some important points about sampling.

Example 4.1   Sampling and Reconstruction of a Sinusoidal Signal

If we sample the continuous-time signal $x_c(t) = \cos(4000\pi t)$ with sampling period $T = 1/6000$, we obtain $x[n] = x_c(nT) = \cos(4000\pi Tn) = \cos(\omega_0 n)$, where $\omega_0 = 4000\pi T = 2\pi/3$. In this case, $\Omega_s = 2\pi/T = 12000\pi$, and the highest frequency of the signal is $\Omega_0 = 4000\pi$, so the conditions of the Nyquist sampling theorem are satisfied and there is no aliasing. The Fourier transform of $x_c(t)$ is

$$X_c(j\Omega) = \pi\delta(\Omega - 4000\pi) + \pi\delta(\Omega + 4000\pi).$$

Figure 4.6(a) shows

$$X_s(j\Omega) = \frac{1}{T} \sum_{k=-\infty}^{\infty} X_c[j(\Omega - k\Omega_s)] \tag{4.21}$$

for $\Omega_s = 12000\pi$. Note that $X_c(j\Omega)$ is a pair of impulses at $\Omega = \pm 4000\pi$, and we see shifted copies of this Fourier transform centered on $\pm\Omega_s, \pm 2\Omega_s$, etc. Plotting $X(e^{j\omega}) = X_s(j\omega/T)$ as a function of the normalized frequency $\omega = \Omega T$ results in Figure 4.6(b), where we have used the fact that scaling the independent variable of an impulse also scales its area, i.e., $\delta(\omega/T) = T\delta(\omega)$ (Oppenheim and Willsky, 1997). Note that the original frequency $\Omega_0 = 4000\pi$ corresponds to the normalized frequency $\omega_0 = 4000\pi T = 2\pi/3$, which satisfies the inequality $\omega_0 < \pi$, corresponding to the fact that $\Omega_0 = 4000\pi < \pi/T = 6000\pi$. Figure 4.6(a) also shows the frequency response of an ideal reconstruction filter $H_r(j\Omega)$ for the given sampling rate of $\Omega_s = 12000\pi$. This figure shows that the reconstructed signal would have frequency $\Omega_0 = 4000\pi$, which is the frequency of the original signal $x_c(t)$.

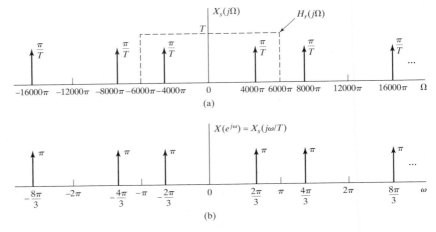

Figure 4.6 (a) Continuous-time and (b) discrete-time Fourier transforms for sampled cosine signal with frequency $\Omega_0 = 4000\pi$ and sampling period $T = 1/6000$.

Example 4.2   Aliasing in Sampling a Sinusoidal Signal

Now suppose that the continuous-time signal is $x_c(t) = \cos(16000\pi t)$, but the sampling period is $T = 1/6000$, as it was in Example 4.1. This sampling period fails to satisfy the Nyquist criterion, since $\Omega_s = 2\pi/T = 12000\pi < 2\Omega_0 = 32000\pi$. Consequently, we expect to see aliasing. The Fourier transform $X_s(j\Omega)$ for this case is identical to that of Figure 4.6(a). However, now the impulse located at $\Omega = -4000\pi$ is from $X_c[j(\Omega - \Omega_s)]$ in Eq. (4.21) rather than from $X_c(j\Omega,)$ and the impulse at $\Omega = 4000\pi$ is from $X_c[j(\Omega + \Omega_s)]$. That is, the frequencies $\pm 4000\pi$ are alias frequencies. Plotting $X(e^{j\omega}) = X_s(j\omega/T)$ as a function of ω yields the same graph as shown in Figure 4.6(b), since we are normalizing by the same sampling period. The fundamental reason for this is that the sequence of samples is the same in both cases; i.e.,

$$\cos(16000\pi n/6000) = \cos(2\pi n + 4000\pi n/6000) = \cos(2\pi n/3).$$

(Recall that we can add any integer multiple of 2π to the argument of the cosine without changing its value.) Thus, we have obtained the same sequence of samples, $x[n] = \cos(2\pi n/3)$, by sampling two different continuous-time signals with the same sampling frequency. In one case, the sampling frequency satisfied the Nyquist criterion, and in the other case it did not. As before, Figure 4.6(a) shows the frequency response of an ideal reconstruction filter $H_r(j\Omega)$ for the given sampling rate of $\Omega_s = 12000\pi$. It is clear from this figure that the signal that would be reconstructed would have the frequency $\Omega_0 = 4000\pi$, which is the alias frequency of the original frequency 16000π with respect to the sampling frequency $\Omega_s = 12000\pi$.

Examples 4.1 and 4.2 use sinusoidal signals to illustrate some of the ambiguities that are inherent in the sampling operation. Example 4.1 verifies that if the conditions of the sampling theorem hold, the original signal can be reconstructed from the samples. Example 4.2 illustrates that if the sampling frequency violates the sampling theorem, we cannot reconstruct the original signal using an ideal lowpass reconstruction filter with cutoff frequency at one-half the sampling frequency. The signal that is reconstructed is one of the alias frequencies of the original signal with respect to the sampling rate used in sampling the original continuous-time signal. In both examples, the sequence of samples was $x[n] = \cos(2\pi n/3)$, but the original continuous-time signal was different. As suggested by these two examples, there are unlimited ways of obtaining this same set of samples by periodic sampling of a continuous-time sinusoid. All ambiguity is removed, however, if we choose $\Omega_s > 2\Omega_0$.

4.3 RECONSTRUCTION OF A BANDLIMITED SIGNAL FROM ITS SAMPLES

According to the sampling theorem, samples of a continuous-time bandlimited signal taken frequently enough are sufficient to represent the signal exactly, in the sense that the signal can be recovered from the samples and with knowledge of the sampling period. Impulse train modulation provides a convenient means for understanding the process of reconstructing the continuous-time bandlimited signal from its samples.

In Section 4.2, we saw that if the conditions of the sampling theorem are met and if the modulated impulse train is filtered by an appropriate lowpass filter, then the Fourier transform of the filter output will be identical to the Fourier transform of the original continuous-time signal $x_c(t)$, and thus, the output of the filter will be $x_c(t)$. If we are given a sequence of samples, $x[n]$, we can form an impulse train $x_s(t)$ in which successive impulses are assigned an area equal to successive sequence values, i.e.,

$$x_s(t) = \sum_{n=-\infty}^{\infty} x[n]\delta(t - nT). \tag{4.22}$$

The nth sample is associated with the impulse at $t = nT$, where T is the sampling period associated with the sequence $x[n]$. If this impulse train is the input to an ideal lowpass continuous-time filter with frequency response $H_r(j\Omega)$ and impulse response $h_r(t)$, then the output of the filter will be

$$x_r(t) = \sum_{n=-\infty}^{\infty} x[n]h_r(t - nT). \tag{4.23}$$

A block diagram representation of this signal reconstruction process is shown in Figure 4.7(a). Recall that the ideal reconstruction filter has a gain of T [to compensate for the factor of $1/T$ in Eq. (4.19) or (4.20)] and a cutoff frequency Ω_c between Ω_N and $\Omega_s - \Omega_N$. A convenient and commonly used choice of the cutoff frequency is $\Omega_c = \Omega_s/2 = \pi/T$. This choice is appropriate for any relationship between Ω_s and Ω_N that avoids aliasing (i.e., so long as $\Omega_s \geq 2\Omega_N$). Figure 4.7(b) shows the frequency response of the ideal reconstruction filter. The corresponding impulse response, $h_r(t)$, is the inverse Fourier transform of $H_r(j\Omega)$, and for cutoff frequency π/T it is given by

$$h_r(t) = \frac{\sin(\pi t/T)}{\pi t/T}. \tag{4.24}$$

This impulse response is shown in Figure 4.7(c). Substituting Eq. (4.24) into Eq. (4.23) leads to

$$x_r(t) = \sum_{n=-\infty}^{\infty} x[n]\frac{\sin[\pi(t - nT)/T]}{\pi(t - nT)/T}. \tag{4.25}$$

Ideal reconstruction system

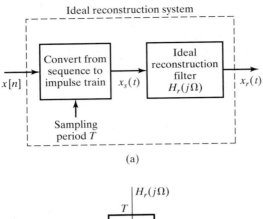

(a)

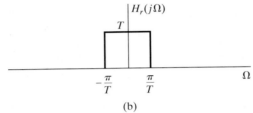

(b)

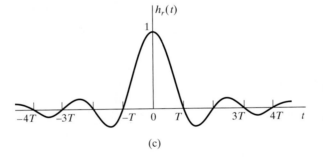

(c)

Figure 4.7 (a) Block diagram of an ideal bandlimited signal reconstruction system. (b) Frequency response of an ideal reconstruction filter. (c) Impulse response of an ideal reconstruction filter.

Equations (4.23) and (4.25) express the continuous-time signal in terms of a linear combination of basis functions $h_r(t - nT)$ with the samples $x[n]$ playing the role of coefficients. Other choices of the basis functions and corresponding coefficients could be used to represent other classes of continuous-time functions [see, for example Unser (2000)]. However, the functions in Eq. (4.24) and the samples $x[n]$ are the natural basis functions and coefficients for representing bandlimited continuous-time signals.

From the frequency-domain argument of Section 4.2, we saw that if $x[n] = x_c(nT)$, where $X_c(j\Omega) = 0$ for $|\Omega| \geq \pi/T$, then $x_r(t)$ is equal to $x_c(t)$. It is not immediately obvious that this is true by considering Eq. (4.25) alone. However, useful insight is gained by looking at that equation more closely. First, let us consider the function $h_r(t)$ given by Eq. (4.24). We note that

$$h_r(0) = 1. \tag{4.26a}$$

This follows from l'Hôpital's rule or the small angle approximation for the sine function. In addition,

$$h_r(nT) = 0 \qquad \text{for } n = \pm1, \pm2, \ldots. \tag{4.26b}$$

It follows from Eqs. (4.26a) and (4.26b) and Eq. (4.23) that if $x[n] = x_c(nT)$, then

$$x_r(mT) = x_c(mT) \tag{4.27}$$

for all integer values of m. That is, the signal that is reconstructed by Eq. (4.25) has the same values at the sampling times as the original continuous-time signal, independently of the sampling period T.

In Figure 4.8, we show a continuous-time signal $x_c(t)$ and the corresponding modulated impulse train. Figure 4.8(c) shows several of the terms

$$x[n]\frac{\sin[\pi(t - nT)/T]}{\pi(t - nT)/T}$$

and the resulting reconstructed signal $x_r(t)$. As suggested by this figure, the ideal lowpass filter *interpolates* between the impulses of $x_s(t)$ to construct a continuous-time signal $x_r(t)$. From Eq. (4.27), the resulting signal is an exact reconstruction of $x_c(t)$ at the sampling times. The fact that, if there is no aliasing, the lowpass filter interpolates the

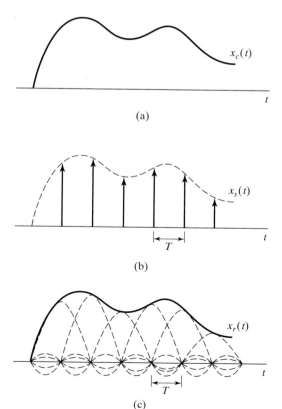

Figure 4.8 Ideal bandlimited interpolation.

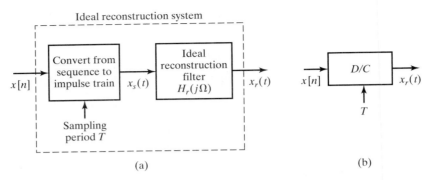

Figure 4.9 (a) Ideal bandlimited signal reconstruction. (b) Equivalent representation as an ideal D/C converter.

correct reconstruction between the samples follows from our frequency-domain analysis of the sampling and reconstruction process.

It is useful to formalize the preceding discussion by defining an ideal system for reconstructing a bandlimited signal from a sequence of samples. We will call this system the *ideal discrete-to-continuous-time (D/C) converter*. The desired system is depicted in Figure 4.9. As we have seen, the ideal reconstruction process can be represented as the conversion of the sequence to an impulse train, as in Eq. (4.22), followed by filtering with an ideal lowpass filter, resulting in the output given by Eq. (4.25). The intermediate step of conversion to an impulse train is a mathematical convenience in deriving Eq. (4.25) and in understanding the signal reconstruction process. However, once we are familiar with this process, it is useful to define a more compact representation, as depicted in Figure 4.9(b), where the input is the sequence $x[n]$ and the output is the continuous-time signal $x_r(t)$ given by Eq. (4.25).

The properties of the ideal D/C converter are most easily seen in the frequency domain. To derive an input/output relation in this domain, consider the Fourier transform of Eq. (4.23) or Eq. (4.25), which is

$$X_r(j\Omega) = \sum_{n=-\infty}^{\infty} x[n]H_r(j\Omega)e^{-j\Omega Tn}.$$

Since $H_r(j\Omega)$ is common to all the terms in the sum, we can write

$$X_r(j\Omega) = H_r(j\Omega)X(e^{j\Omega T}). \tag{4.28}$$

Equation (4.28) provides a frequency-domain description of the ideal D/C converter. According to Eq. (4.28), $X(e^{j\omega})$ is frequency scaled (in effect, going from the sequence to the impulse train causes ω to be replaced by ΩT). Then the ideal lowpass filter $H_r(j\Omega)$ selects the base period of the resulting periodic Fourier transform $X(e^{j\Omega T})$ and compensates for the $1/T$ scaling inherent in sampling. Thus, if the sequence $x[n]$ has been obtained by sampling a bandlimited signal at the Nyquist rate or higher, the reconstructed signal $x_r(t)$ will be equal to the original bandlimited signal. In any case, it is also clear from Eq. (4.28) that the output of the ideal D/C converter is always bandlimited to at most the cutoff frequency of the lowpass filter, which is typically taken to be one-half the sampling frequency.

4.4 DISCRETE-TIME PROCESSING OF CONTINUOUS-TIME SIGNALS

A major application of discrete-time systems is in the processing of continuous-time signals. This is accomplished by a system of the general form depicted in Figure 4.10. The system is a cascade of a C/D converter, followed by a discrete-time system, followed by a D/C converter. Note that the overall system is equivalent to a continuous-time system, since it transforms the continuous-time input signal $x_c(t)$ into the continuous-time output signal $y_r(t)$. The properties of the overall system are dependent on the choice of the discrete-time system and the sampling rate. We assume in Figure 4.10 that the C/D and D/C converters have the same sampling rate. This is not essential, and later sections of this chapter and some of the problems at the end of the chapter consider systems in which the input and output sampling rates are not the same.

The previous sections of the chapter have been devoted to understanding the C/D and D/C conversion operations in Figure 4.10. For convenience, and as a first step in understanding the overall system of Figure 4.10, we summarize the mathematical representations of these operations.

The C/D converter produces a discrete-time signal

$$x[n] = x_c(nT), \tag{4.29}$$

i.e., a sequence of samples of the continuous-time input signal $x_c(t)$. The DTFT of this sequence is related to the continuous-time Fourier transform of the continuous-time input signal by

$$X(e^{j\omega}) = \frac{1}{T} \sum_{k=-\infty}^{\infty} X_c\left[j\left(\frac{\omega}{T} - \frac{2\pi k}{T}\right)\right]. \tag{4.30}$$

The D/C converter creates a continuous-time output signal of the form

$$y_r(t) = \sum_{n=-\infty}^{\infty} y[n] \frac{\sin[\pi(t-nT)/T]}{\pi(t-nT)/T}, \tag{4.31}$$

where the sequence $y[n]$ is the output of the discrete-time system when the input to the system is $x[n]$. From Eq. (4.28), $Y_r(j\Omega)$, the continuous-time Fourier transform of $y_r(t)$, and $Y(e^{j\omega})$, the DTFT of $y[n]$, are related by

$$Y_r(j\Omega) = H_r(j\Omega)Y(e^{j\Omega T}) = \begin{cases} TY(e^{j\Omega T}), & |\Omega| < \pi/T, \\ 0, & \text{otherwise.} \end{cases} \tag{4.32}$$

Next, let us relate the output sequence $y[n]$ to the input sequence $x[n]$, or equivalently, $Y(e^{j\omega})$ to $X(e^{j\omega})$. A simple example is the identity system, i.e., $y[n] = x[n]$. This

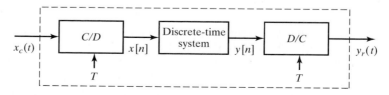

Figure 4.10   Discrete-time processing of continuous-time signals.

is in effect the case that we have studied in detail so far. We know that if $x_c(t)$ has a band-limited Fourier transform such that $X_c(j\Omega) = 0$ for $|\Omega| \geq \pi/T$ and if the discrete-time system in Figure 4.10 is the identity system such that $y[n] = x[n] = x_c(nT)$, then the output will be $y_r(t) = x_c(t)$. Recall that, in proving this result, we utilized the frequency-domain representations of the continuous-time and discrete-time signals, since the key concept of aliasing is most easily understood in the frequency domain. Likewise, when we deal with systems more complicated than the identity system, we generally carry out the analysis in the frequency domain. If the discrete-time system is nonlinear or time varying, it is usually difficult to obtain a general relationship between the Fourier transforms of the input and the output of the system. (In Problem 4.51, we consider an example of the system of Figure 4.10 in which the discrete-time system is nonlinear.) However, the LTI case leads to a rather simple and generally useful result.

4.4.1 Discrete-Time LTI Processing of Continuous-Time Signals

If the discrete-time system in Figure 4.10 is linear and time invariant, we have

$$Y(e^{j\omega}) = H(e^{j\omega})X(e^{j\omega}), \tag{4.33}$$

where $H(e^{j\omega})$ is the frequency response of the system or, equivalently, the Fourier transform of the unit sample response, and $X(e^{j\omega})$ and $Y(e^{j\omega})$ are the Fourier transforms of the input and output, respectively. Combining Eqs. (4.32) and (4.33), we obtain

$$Y_r(j\Omega) = H_r(j\Omega)H(e^{j\Omega T})X(e^{j\Omega T}). \tag{4.34}$$

Next, using Eq. (4.30) with $\omega = \Omega T$, we have

$$Y_r(j\Omega) = H_r(j\Omega)H(e^{j\Omega T})\frac{1}{T}\sum_{k=-\infty}^{\infty} X_c\left[j\left(\Omega - \frac{2\pi k}{T}\right)\right]. \tag{4.35}$$

If $X_c(j\Omega) = 0$ for $|\Omega| \geq \pi/T$, then the ideal lowpass reconstruction filter $H_r(j\Omega)$ cancels the factor $1/T$ and selects only the term in Eq. (4.35) for $k = 0$; i.e.,

$$Y_r(j\Omega) = \begin{cases} H(e^{j\Omega T})X_c(j\Omega), & |\Omega| < \pi/T, \\ 0, & |\Omega| \geq \pi/T. \end{cases} \tag{4.36}$$

Thus, if $X_c(j\Omega)$ is bandlimited and the sampling rate is at or above the Nyquist rate, the output is related to the input through an equation of the form

$$Y_r(j\Omega) = H_{\text{eff}}(j\Omega)X_c(j\Omega), \tag{4.37}$$

where

$$H_{\text{eff}}(j\Omega) = \begin{cases} H(e^{j\Omega T}), & |\Omega| < \pi/T, \\ 0, & |\Omega| \geq \pi/T. \end{cases} \tag{4.38}$$

That is, the overall continuous-time system is equivalent to an LTI system whose *effective* frequency response is given by Eq. (4.38).

It is important to emphasize that the linear and time-invariant behavior of the system of Figure 4.10 depends on two factors. First, the discrete-time system must be linear and time invariant. Second, the input signal must be bandlimited, and the sampling rate must be high enough so that any aliased components are removed by the discrete-time system. As a simple illustration of this second condition being violated, consider the case when $x_c(t)$ is a single finite-duration unit-amplitude pulse whose duration is less than the sampling period. If the pulse is unity at $t = 0$, then $x[n] = \delta[n]$. However, it is clearly possible to shift the pulse so that it is not aligned with any of the sampling times, i.e., $x[n] = 0$ for all n. Such a pulse, being limited in time, is not bandlimited, and the conditions of the sampling theorem cannot hold. Even if the discrete-time system is the identity system, such that $y[n] = x[n]$, the overall system will not be time invariant if aliasing occurs in sampling the input. In general, if the discrete-time system in Figure 4.10 is linear and time invariant, and if the sampling frequency is at or above the Nyquist rate associated with the bandwidth of the input $x_c(t)$, then the overall system will be equivalent to an LTI continuous-time system with an effective frequency response given by Eq. (4.38). Furthermore, Eq. (4.38) is valid even if some aliasing occurs in the C/D converter, as long as $H(e^{j\omega})$ does not pass the aliased components. Example 4.3 is a simple illustration of this.

Example 4.3   Ideal Continuous-Time Lowpass Filtering Using a Discrete-Time Lowpass Filter

Consider Figure 4.10, with the LTI discrete-time system having frequency response

$$H(e^{j\omega}) = \begin{cases} 1, & |\omega| < \omega_c, \\ 0, & \omega_c < |\omega| \le \pi. \end{cases} \tag{4.39}$$

This frequency response is periodic with period 2π, as shown in Figure 4.11(a). For bandlimited inputs sampled at or above the Nyquist rate, it follows from Eq. (4.38) that the overall system of Figure 4.10 will behave as an LTI continuous-time system with frequency response

$$H_{\text{eff}}(j\Omega) = \begin{cases} 1, & |\Omega T| < \omega_c \text{ or } |\Omega| < \omega_c/T, \\ 0, & |\Omega T| \ge \omega_c \text{ or } |\Omega| \ge \omega_c/T. \end{cases} \tag{4.40}$$

As shown in Figure 4.11(b), this effective frequency response is that of an ideal lowpass filter with cutoff frequency $\Omega_c = \omega_c/T$

The graphical illustration given in Figure 4.12 provides an interpretation of how this effective response is achieved. Figure 4.12(a) represents the Fourier transform

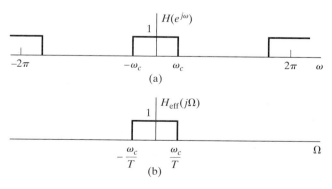

Figure 4.11   (a) Frequency response of discrete-time system in Figure 4.10. (b) Corresponding effective continuous-time frequency response for bandlimited inputs.

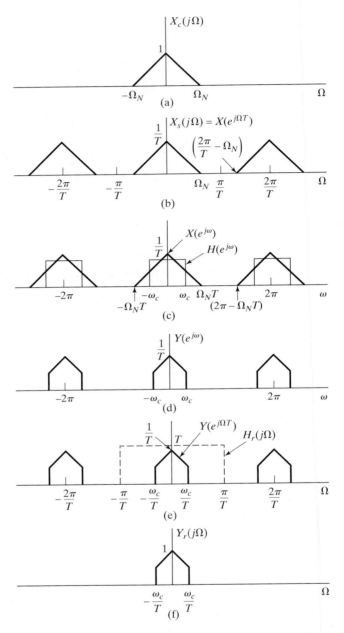

Figure 4.12   (a) Fourier transform of a bandlimited input signal. (b) Fourier transform of sampled input plotted as a function of continuous-time frequency Ω. (c) Fourier transform $X(e^{j\omega})$ of sequence of samples and frequency response $H(e^{j\omega})$ of discrete-time system plotted versus ω. (d) Fourier transform of output of discrete-time system. (e) Fourier transform of output of discrete-time system and frequency response of ideal reconstruction filter plotted versus Ω. (f) Fourier transform of output.

of a bandlimited signal. Figure 4.12(b) shows the Fourier transform of the intermediate modulated impulse train, which is identical to $X(e^{j\Omega T})$, the DTFT of the sequence of samples evaluated for $\omega = \Omega T$. In Figure 4.12(c), the DTFT of the sequence of samples and the frequency response of the discrete-time system are both plotted as a function of the normalized discrete-time frequency variable ω. Figure 4.12(d) shows $Y(e^{j\omega}) = H(e^{j\omega})X(e^{j\omega})$, the Fourier transform of the output of the discrete-time system. Figure 4.12(e) illustrates the Fourier transform of the output of the discrete-time system as a function of the continuous-time frequency Ω, together with the frequency response of the ideal reconstruction filter $H_r(j\Omega)$ of the D/C converter. Finally, Figure 4.12(f) shows the resulting Fourier transform of the output of the D/C converter. By comparing Figures 4.12(a) and 4.12(f), we see that the system behaves as an LTI system with frequency response given by Eq. (4.40) and plotted in Figure 4.11(b).

Several important points are illustrated in Example 4.3. First, note that the ideal lowpass discrete-time filter with discrete-time cutoff frequency ω_c has the effect of an ideal lowpass filter with cutoff frequency $\Omega_c = \omega_c/T$ when used in the configuration of Figure 4.10. This cutoff frequency depends on both ω_c and T. In particular, by using a fixed discrete-time lowpass filter, but varying the sampling period T, an equivalent continuous-time lowpass filter with a variable cutoff frequency can be implemented. For example, if T were chosen so that $\Omega_N T < \omega_c$, then the output of the system of Figure 4.10 would be $y_r(t) = x_c(t)$. Also, as illustrated in Problem 4.31, Eq. (4.40) will be valid even if some aliasing is present in Figures 4.12(b) and (c), as long as these distorted (aliased) components are eliminated by the filter $H(e^{j\omega})$. In particular, from Figure 4.12(c), we see that for no aliasing to be present in the output, we require that

$$(2\pi - \Omega_N T) \geq \omega_c, \tag{4.41}$$

compared with the Nyquist requirement that

$$(2\pi - \Omega_N T) \geq \Omega_N T. \tag{4.42}$$

As another example of continuous-time processing using a discrete-time system, let us consider the implementation of an ideal differentiator for bandlimited signals.

Example 4.4   Discrete-Time Implementation of an Ideal Continuous-Time Bandlimited Differentiator

The ideal continuous-time differentiator system is defined by

$$y_c(t) = \frac{d}{dt}[x_c(t)], \tag{4.43}$$

with corresponding frequency response

$$H_c(j\Omega) = j\Omega. \tag{4.44}$$

Since we are considering a realization in the form of Figure 4.10, the inputs are restricted to be bandlimited. For processing bandlimited signals, it is sufficient that

$$H_{\text{eff}}(j\Omega) = \begin{cases} j\Omega, & |\Omega| < \pi/T, \\ 0, & |\Omega| \geq \pi/T, \end{cases} \tag{4.45}$$

as depicted in Figure 4.13(a). The corresponding discrete-time system has frequency response

$$H(e^{j\omega}) = \frac{j\omega}{T}, \qquad |\omega| < \pi, \tag{4.46}$$

and is periodic with period 2π. This frequency response is plotted in Figure 4.13(b). The corresponding impulse response can be shown to be

$$h[n] = \frac{1}{2\pi} \int_{-\pi}^{\pi} \left(\frac{j\omega}{T}\right) e^{j\omega n} d\omega = \frac{\pi n \cos \pi n - \sin \pi n}{\pi n^2 T}, \qquad -\infty < n < \infty,$$

or equivalently,

$$h[n] = \begin{cases} 0, & n = 0, \\ \dfrac{\cos \pi n}{nT}, & n \neq 0. \end{cases} \tag{4.47}$$

Thus, if a discrete-time system with this impulse response was used in the configuration of Figure 4.10, the output for every appropriately bandlimited input would be the derivative of the input. Problem 4.22 concerns the verification of this for a sinusoidal input signal.

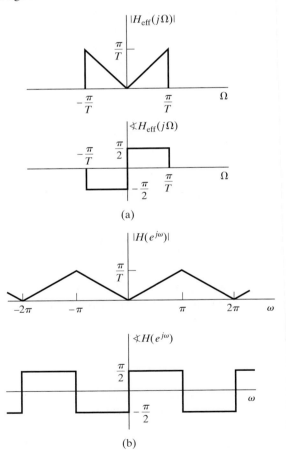

(a)

(b)

Figure 4.13 (a) Frequency response of a continuous-time ideal bandlimited differentiator $H_c(j\Omega) = j\Omega$, $|\Omega| < \pi/T$. (b) Frequency response of a discrete-time filter to implement a continuous-time bandlimited differentiator.

4.4.2 Impulse Invariance

We have shown that the cascade system of Figure 4.10 can be equivalent to an LTI system for bandlimited input signals. Let us now assume that, as depicted in Figure 4.14, we are given a desired continuous-time system that we wish to implement in the form of Figure 4.10. With $H_c(j\Omega)$ bandlimited, Eq. (4.38) specifies how to choose $H(e^{j\omega})$ so that $H_{\text{eff}}(j\Omega) = H_c(j\Omega)$. Specifically,

$$H(e^{j\omega}) = H_c(j\omega/T), \qquad |\omega| < \pi, \tag{4.48}$$

with the further requirement that T be chosen such that

$$H_c(j\Omega) = 0, \qquad |\Omega| \geq \pi/T. \tag{4.49}$$

Under the constraints of Eqs. (4.48) and (4.49), there is also a straightforward and useful relationship between the continuous-time impulse response $h_c(t)$ and the discrete-time impulse response $h[n]$. In particular, as we shall verify shortly,

$$h[n] = Th_c(nT); \tag{4.50}$$

i.e., the impulse response of the discrete-time system is a scaled, sampled version of $h_c(t)$. When $h[n]$ and $h_c(t)$ are related through Eq. (4.50), the discrete-time system is said to be an *impulse-invariant* version of the continuous-time system.

Equation (4.50) is a direct consequence of the discussion in Section 4.2. Specifically, with $x[n]$ and $x_c(t)$ respectively replaced by $h[n]$ and $h_c(t)$ in Eq. (4.16), i.e.,

$$h[n] = h_c(nT), \tag{4.51}$$

Eq. (4.20) becomes

$$H(e^{j\omega}) = \frac{1}{T} \sum_{k=-\infty}^{\infty} H_c\left(j\left(\frac{\omega}{T} - \frac{2\pi k}{T}\right)\right), \tag{4.52}$$

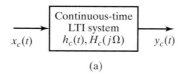

(a)

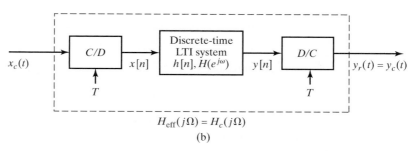

$$H_{\text{eff}}(j\Omega) = H_c(j\Omega)$$

(b)

Figure 4.14   (a) Continuous-time LTI system. (b) Equivalent system for bandlimited inputs.

or, if Eq. (4.49) is satisfied,

$$H(e^{j\omega}) = \frac{1}{T} H_c\left(j\frac{\omega}{T}\right), \qquad |\omega| < \pi. \tag{4.53}$$

Modifying Eqs. (4.51) and (4.53) to account for the scale factor of T in Eq. (4.50), we have

$$h[n] = T h_c(nT), \tag{4.54}$$

$$H(e^{j\omega}) = H_c\left(j\frac{\omega}{T}\right), \qquad |\omega| < \pi. \tag{4.55}$$

Example 4.5   A Discrete-Time Lowpass Filter Obtained by Impulse Invariance

Suppose that we wish to obtain an ideal lowpass discrete-time filter with cutoff frequency $\omega_c < \pi$. We can do this by sampling a continuous-time ideal lowpass filter with cutoff frequency $\Omega_c = \omega_c/T < \pi/T$ defined by

$$H_c(j\Omega) = \begin{cases} 1, & |\Omega| < \Omega_c, \\ 0, & |\Omega| \geq \Omega_c. \end{cases}$$

The impulse response of this continuous-time system is

$$h_c(t) = \frac{\sin(\Omega_c t)}{\pi t},$$

so we define the impulse response of the discrete-time system to be

$$h[n] = T h_c(nT) = T\frac{\sin(\Omega_c nT)}{\pi nT} = \frac{\sin(\omega_c n)}{\pi n},$$

where $\omega_c = \Omega_c T$. We have already shown that this sequence corresponds to the DTFT

$$H(e^{j\omega}) = \begin{cases} 1, & |\omega| < \omega_c, \\ 0, & \omega_c \leq |\omega| \leq \pi, \end{cases}$$

which is identical to $H_c(j\omega/T)$, as predicted by Eq. (4.55).

Example 4.6   Impulse Invariance Applied to Continuous-Time Systems with Rational System Functions

Many continuous-time systems have impulse responses composed of a sum of exponential sequences of the form

$$h_c(t) = A e^{s_0 t} u(t).$$

Such time functions have Laplace transforms

$$H_c(s) = \frac{A}{s - s_0} \qquad \mathcal{R}e(s) > \mathcal{R}e(s_0).$$

If we apply the impulse invariance concept to such a continuous-time system, we obtain the impulse response

$$h[n] = T h_c(nT) = A T e^{s_0 Tn} u[n],$$

which has z-transform system function

$$H(z) = \frac{A\,T}{1 - e^{s_0 T} z^{-1}} \qquad |z| > |e^{s_0 T}|$$

and, assuming $\mathcal{R}e(s_0) < 0$, the frequency response

$$H(e^{j\omega}) = \frac{A\,T}{1 - e^{s_0 T} e^{-j\omega}}.$$

In this case, Eq. (4.55) does not hold exactly, because the original continuous-time system did not have a strictly bandlimited frequency response, and therefore, the resulting discrete-time frequency response is an *aliased* version of $H_c(j\Omega)$. Even though aliasing occurs in such a case as this, the effect may be small. Higher-order systems whose impulse responses are sums of complex exponentials may in fact have frequency responses that fall off rapidly at high frequencies, so that aliasing is minimal if the sampling rate is high enough. Thus, one approach to the discrete-time simulation of continuous-time systems and also to the design of digital filters is through sampling of the impulse response of a corresponding analog filter.

4.5 CONTINUOUS-TIME PROCESSING OF DISCRETE-TIME SIGNALS

In Section 4.4, we discussed and analyzed the use of discrete-time systems for processing continuous-time signals in the configuration of Figure 4.10. In this section, we consider the complementary situation depicted in Figure 4.15, which is appropriately referred to as continuous-time processing of discrete-time signals. Although the system of Figure 4.15 is not typically used to implement discrete-time systems, it provides a useful interpretation of certain discrete-time systems that have no simple interpretation in the discrete domain.

From the definition of the ideal D/C converter, $X_c(j\Omega)$ and therefore also $Y_c(j\Omega)$, will necessarily be zero for $|\Omega| \geq \pi/T$. Thus, the C/D converter samples $y_c(t)$ without aliasing, and we can express $x_c(t)$ and $y_c(t)$ respectively as

$$x_c(t) = \sum_{n=-\infty}^{\infty} x[n] \frac{\sin[\pi(t - nT)/T]}{\pi(t - nT)/T} \qquad (4.56)$$

and

$$y_c(t) = \sum_{n=-\infty}^{\infty} y[n] \frac{\sin[\pi(t - nT)/T]}{\pi(t - nT)/T}, \qquad (4.57)$$

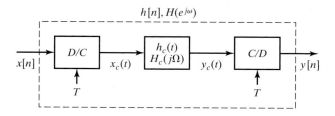

Figure 4.15 Continuous-time processing of discrete-time signals.

where $x[n] = x_c(nT)$ and $y[n] = y_c(nT)$. The frequency-domain relationships for Figure 4.15 are

$$X_c(j\Omega) = TX(e^{j\Omega T}), \qquad |\Omega| < \pi/T, \qquad (4.58a)$$

$$Y_c(j\Omega) = H_c(j\Omega)X_c(j\Omega), \qquad (4.58b)$$

$$Y(e^{j\omega}) = \frac{1}{T}Y_c\left(j\frac{\omega}{T}\right), \qquad |\omega| < \pi. \qquad (4.58c)$$

Therefore, by substituting Eqs. (4.58a) and (4.58b) into Eq. (4.58c), it follows that the overall system behaves as a discrete-time system whose frequency response is

$$H(e^{j\omega}) = H_c\left(j\frac{\omega}{T}\right), \qquad |\omega| < \pi, \qquad (4.59)$$

or equivalently, the overall frequency response of the system in Figure 4.15 will be equal to a given $H(e^{j\omega})$ if the frequency response of the continuous-time system is

$$H_c(j\Omega) = H(e^{j\Omega T}), \qquad |\Omega| < \pi/T. \qquad (4.60)$$

Since $X_c(j\Omega) = 0$ for $|\Omega| \geq \pi/T$, $H_c(j\Omega)$ may be chosen arbitrarily above π/T. A convenient—but arbitrary—choice is $H_c(j\Omega) = 0$ for $|\Omega| \geq \pi/T$.

With this representation of a discrete-time system, we can focus on the equivalent effect of the continuous-time system on the bandlimited continuous-time signal $x_c(t)$. This is illustrated in Examples 4.7 and 4.8.

Example 4.7   Noninteger Delay

Let us consider a discrete-time system with frequency response

$$H(e^{j\omega}) = e^{-j\omega\Delta}, \qquad |\omega| < \pi. \qquad (4.61)$$

When Δ is an integer, this system has a straightforward interpretation as a delay of Δ, i.e.,

$$y[n] = x[n - \Delta]. \qquad (4.62)$$

When Δ is not an integer, Eq. (4.62) has no formal meaning, because we cannot shift the sequence $x[n]$ by a noninteger amount. However, with the use of the system of Figure 4.15, a useful time-domain interpretation can be applied to the system specified by Eq. (4.61). Let $H_c(j\Omega)$ in Figure 4.15 be chosen to be

$$H_c(j\Omega) = H(e^{j\Omega T}) = e^{-j\Omega T\Delta}. \qquad (4.63)$$

Then, from Eq. (4.59), the overall discrete-time system in Figure 4.15 will have the frequency response given by Eq. (4.61), whether or not Δ is an integer. To interpret the system of Eq. (4.61), we note that Eq. (4.63) represents a time delay of $T\Delta$ seconds. Therefore,

$$y_c(t) = x_c(t - T\Delta). \qquad (4.64)$$

Furthermore, $x_c(t)$ is the bandlimited interpolation of $x[n]$, and $y[n]$ is obtained by sampling $y_c(t)$. For example, if $\Delta = \frac{1}{2}$, $y[n]$ would be the values of the bandlimited interpolation halfway between the input sequence values. This is illustrated in

Figure 4.16. We can also obtain a direct convolution representation for the system defined by Eq. (4.61). From Eqs. (4.64) and (4.56), we obtain

$$y[n] = y_c(nT) = x_c(nT - T\Delta)$$

$$= \sum_{k=-\infty}^{\infty} x[k] \left. \frac{\sin[\pi(t - T\Delta - kT)/T]}{\pi(t - T\Delta - kT)/T} \right|_{t=nT} \quad (4.65)$$

$$= \sum_{k=-\infty}^{\infty} x[k] \frac{\sin \pi(n - k - \Delta)}{\pi(n - k - \Delta)},$$

which is, by definition, the convolution of $x[n]$ with

$$h[n] = \frac{\sin \pi(n - \Delta)}{\pi(n - \Delta)}, \quad -\infty < n < \infty.$$

When Δ is not an integer, $h[n]$ has infinite extent. However, when $\Delta = n_0$ is an integer, it is easily shown that $h[n] = \delta[n - n_0]$, which is the impulse response of the ideal integer delay system.

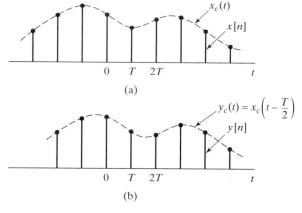

Figure 4.16   (a) Continuous-time processing of the discrete-time sequence (b) can produce a new sequence with a "half-sample" delay.

The noninteger delay represented by Eq. (4.65) has considerable practical significance, since such a factor often arises in the frequency-domain representation of systems. When this kind of term is found in the frequency response of a causal discrete-time system, it can be interpreted in the light of this example. This interpretation is illustrated in Example 4.8.

Example 4.8   Moving-Average System with Noninteger Delay

In Example 2.16, we considered the general moving-average system and obtained its frequency response. For the case of the causal $(M + 1)$-point moving-average system,

$M_1 = 0$ and $M_2 = M$, and the frequency response is

$$H(e^{j\omega}) = \frac{1}{(M+1)} \frac{\sin[\omega(M+1)/2]}{\sin(\omega/2)} e^{-j\omega M/2}, \qquad |\omega| < \pi. \tag{4.66}$$

This representation of the frequency response suggests the interpretation of the $(M+1)$-point moving-average system as the cascade of two systems, as indicated in Figure 4.17. The first system imposes a frequency-domain amplitude weighting. The second system represents the linear-phase term in Eq. (4.66). If M is an even integer (meaning the moving average of an odd number of samples), then the linear-phase term corresponds to an integer delay, i.e.,

$$y[n] = w[n - M/2]. \tag{4.67}$$

However, if M is odd, the linear-phase term corresponds to a noninteger delay, specifically, an integer-plus-one-half sample interval. This noninteger delay can be interpreted in terms of the discussion in Example 4.7; i.e., $y[n]$ is equivalent to bandlimited interpolation of $w[n]$, followed by a continuous-time delay of $MT/2$ (where T is the assumed, but arbitrary, sampling period associated with the D/C interpolation of $w[n]$), followed by C/D conversion again with sampling period T. This fractional delay is illustrated in Figure 4.18. Figure 4.18(a) shows a discrete-time sequence $x[n] = \cos(0.25\pi n)$. This sequence is the input to a six-point ($M = 5$) moving-average filter. In this example, the input is "turned on" far enough in the past so that the output consists only of the steady-state response for the time interval shown. Figure 4.18(b) shows the corresponding output sequence, which is given by

$$y[n] = H(e^{j0.25\pi}) \frac{1}{2} e^{j0.25\pi n} + H(e^{-j0.25\pi}) \frac{1}{2} e^{-j0.25\pi n}$$

$$= \frac{1}{2} \frac{\sin[3(0.25\pi)]}{6\sin(0.125\pi)} e^{-j(0.25\pi)5/2} e^{j0.25\pi n} + \frac{1}{2} \frac{\sin[3(-0.25\pi)]}{6\sin(-0.125\pi)} e^{j(0.25\pi)5/2} e^{-j0.25\pi n}$$

$$= 0.308 \cos[0.25\pi(n - 2.5)].$$

Thus, the six-point moving-average filter reduces the amplitude of the cosine signal and introduces a phase shift that corresponds to 2.5 samples of delay. This is apparent in Figure 4.18, where we have plotted the continuous-time cosines that would be interpolated by the ideal D/C converter for both the input and the output sequence. Note in Figure 4.18(b) that the six-point moving-average filtering gives a sampled cosine signal such that the sample points have been shifted by 2.5 samples with respect to the sample points of the input. This can be seen from Figure 4.18 by comparing the positive peak at 8 in the interpolated cosine for the input to the positive peak at 10.5 in the interpolated cosine for the output. Thus, the six-point moving-average filter is seen to have a delay of $5/2 = 2.5$ samples.

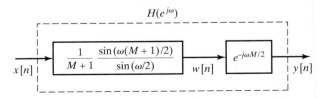

Figure 4.17 The moving-average system represented as a cascade of two systems.

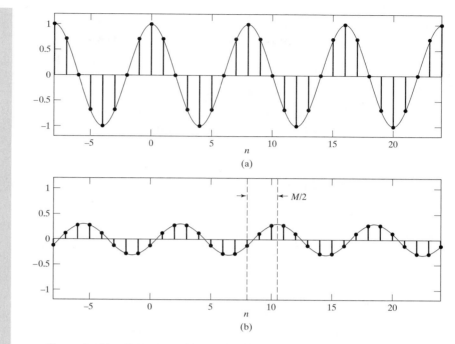

Figure 4.18 Illustration of moving-average filtering. (a) Input signal $x[n] = \cos(0.25\pi n)$. (b) Corresponding output of six-point moving-average filter.

4.6 CHANGING THE SAMPLING RATE USING DISCRETE-TIME PROCESSING

We have seen that a continuous-time signal $x_c(t)$ can be represented by a discrete-time signal consisting of a sequence of samples

$$x[n] = x_c(nT). \tag{4.68}$$

Alternatively, our previous discussion has shown that, even if $x[n]$ was not obtained originally by sampling, we can always use the bandlimited interpolation formula of Eq. (4.25) to reconstruct a continuous-time bandlimited signal $x_r(t)$ whose samples are $x[n] = x_r(nT) = x_c(nT)$, i.e., the samples of $x_c(t)$ and $x_r(t)$ are identical at the sampling times even when $x_r(t) \neq x_c(t)$.

It is often necessary to change the sampling rate of a discrete-time signal, i.e., to obtain a new discrete-time representation of the underlying continuous-time signal of the form

$$x_1[n] = x_c(nT_1), \tag{4.69}$$

where $T_1 \neq T$. This operation is often called *resampling*. Conceptually, $x_1[n]$ can be obtained from $x[n]$ by reconstructing $x_c(t)$ from $x[n]$ using Eq. (4.25) and then resampling $x_c(t)$ with period T_1 to obtain $x_1[n]$. However, this is not usually a practical approach,

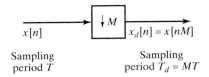

Figure 4.19 Representation of a compressor or discrete-time sampler.

because of the nonideal analog reconstruction filter, D/A converter, and A/D converter that would be used in a practical implementation. Thus, it is of interest to consider methods of changing the sampling rate that involve only discrete-time operations.

4.6.1 Sampling Rate Reduction by an Integer Factor

The sampling rate of a sequence can be reduced by "sampling" it, i.e., by defining a new sequence

$$x_d[n] = x[nM] = x_c(nMT). \tag{4.70}$$

Equation (4.70) defines the system depicted in Figure 4.19, which is called a *sampling rate compressor* (see Crochiere and Rabiner, 1983 and Vaidyanathan, 1993) or simply a *compressor*. From Eq. (4.70), it follows that $x_d[n]$ is identical to the sequence that would be obtained from $x_c(t)$ by sampling with period $T_d = MT$. Furthermore, if $X_c(j\Omega) = 0$ for $|\Omega| \geq \Omega_N$, then $x_d[n]$ is an exact representation of $x_c(t)$ if $\pi/T_d = \pi/(MT) \geq \Omega_N$. That is, the sampling rate can be reduced to π/M without aliasing if the original sampling rate is at least M times the Nyquist rate or if the bandwidth of the sequence is first reduced by a factor of M by discrete-time filtering. In general, the operation of reducing the sampling rate (including any prefiltering) is called *downsampling*.

As in the case of sampling a continuous-time signal, it is useful to obtain a frequency-domain relation between the input and output of the compressor. This time, however, it will be a relationship between DTFTs. Although several methods can be used to derive the desired result, we will base our derivation on the results already obtained for sampling continuous-time signals. First, recall that the DTFT of $x[n] = x_c(nT)$ is

$$X(e^{j\omega}) = \frac{1}{T} \sum_{k=-\infty}^{\infty} X_c \left[j \left(\frac{\omega}{T} - \frac{2\pi k}{T} \right) \right]. \tag{4.71}$$

Similarly, the DTFT of $x_d[n] = x[nM] = x_c(nT_d)$ with $T_d = MT$ is

$$X_d(e^{j\omega}) = \frac{1}{T_d} \sum_{r=-\infty}^{\infty} X_c \left[j \left(\frac{\omega}{T_d} - \frac{2\pi r}{T_d} \right) \right]. \tag{4.72}$$

Now, since $T_d = MT$, we can write Eq. (4.72) as

$$X_d(e^{j\omega}) = \frac{1}{MT} \sum_{r=-\infty}^{\infty} X_c \left[j \left(\frac{\omega}{MT} - \frac{2\pi r}{MT} \right) \right]. \tag{4.73}$$

To see the relationship between Eqs. (4.73) and (4.71), note that the summation index r in Eq. (4.73) can be expressed as

$$r = i + kM, \tag{4.74}$$

where k and i are integers such that $-\infty < k < \infty$ and $0 \leq i \leq M - 1$. Clearly, r is still an integer ranging from $-\infty$ to ∞, but now Eq. (4.73) can be expressed as

$$X_d(e^{j\omega}) = \frac{1}{M} \sum_{i=0}^{M-1} \left\{ \frac{1}{T} \sum_{k=-\infty}^{\infty} X_c\left[j\left(\frac{\omega}{MT} - \frac{2\pi k}{T} - \frac{2\pi i}{MT} \right) \right] \right\}. \qquad (4.75)$$

The term inside the square brackets in Eq. (4.75) is recognized from Eq. (4.71) as

$$X(e^{j(\omega - 2\pi i)/M}) = \frac{1}{T} \sum_{k=-\infty}^{\infty} X_c\left[j\left(\frac{\omega - 2\pi i}{MT} - \frac{2\pi k}{T} \right) \right]. \qquad (4.76)$$

Thus, we can express Eq. (4.75) as

$$X_d(e^{j\omega}) = \frac{1}{M} \sum_{i=0}^{M-1} X(e^{j(\omega/M - 2\pi i/M)}). \qquad (4.77)$$

There is a strong analogy between Eqs. (4.71) and (4.77): Equation (4.71) expresses the Fourier transform of the sequence of samples, $x[n]$ (with period T), in terms of the Fourier transform of the continuous-time signal $x_c(t)$; Equation (4.77) expresses the Fourier transform of the discrete-time sampled sequence $x_d[n]$ (with sampling period M) in terms of the Fourier transform of the sequence $x[n]$. If we compare Eqs. (4.72) and (4.77), we see that $X_d(e^{j\omega})$ can be thought of as being composed of the superposition of either an infinite set of amplitude-scaled copies of $X_c(j\Omega)$, frequency scaled through $\omega = \Omega T_d$ and shifted by integer multiples of 2π [Eq. (4.72)], or M amplitude-scaled copies of the periodic Fourier transform $X(e^{j\omega})$, frequency scaled by M and shifted by integer multiples of 2π [Eq. (4.77)]. Either interpretation makes it clear that $X_d(e^{j\omega})$ is periodic with period 2π (as are all DTFTs) and that aliasing can be avoided by ensuring that $X(e^{j\omega})$ is bandlimited, i.e.,

$$X(e^{j\omega}) = 0, \qquad \omega_N \leq |\omega| \leq \pi, \qquad (4.78)$$

and $2\pi/M \geq 2\omega_N$.

Downsampling is illustrated in Figure 4.20 for $M = 2$. Figure 4.20(a) shows the Fourier transform of a bandlimited continuous-time signal, and Figure 4.20(b) shows the Fourier transform of the impulse train of samples when the sampling period is T. Figure 4.20(c) shows $X(e^{j\omega})$ and is related to Figure 4.20(b) through Eq. (4.18). As we have already seen, Figures 4.20(b) and (c) differ only in a scaling of the frequency variable. Figure 4.20(d) shows the DTFT of the downsampled sequence when $M = 2$. We have plotted this Fourier transform as a function of the normalized frequency $\omega = \Omega T_d$. Finally, Figure 4.20(e) shows the DTFT of the downsampled sequence plotted as a function of the continuous-time frequency variable Ω. Figure 4.20(e) is identical to Figure 4.20(d), except for the scaling of the frequency axis through the relation $\Omega = \omega/T_d$.

In this example, $2\pi/T = 4\Omega_N$; i.e., the original sampling rate is exactly twice the minimum rate to avoid aliasing. Thus, when the original sampled sequence is downsampled by a factor of $M = 2$, no aliasing results. If the downsampling factor is more than 2 in this case, aliasing will result, as illustrated in Figure 4.21.

Figure 4.21(a) shows the continuous-time Fourier transform of $x_c(t)$, and Figure 4.21(b) shows the DTFT of the sequence $x[n] = x_c(nT)$, when $2\pi/T = 4\Omega_N$. Thus,

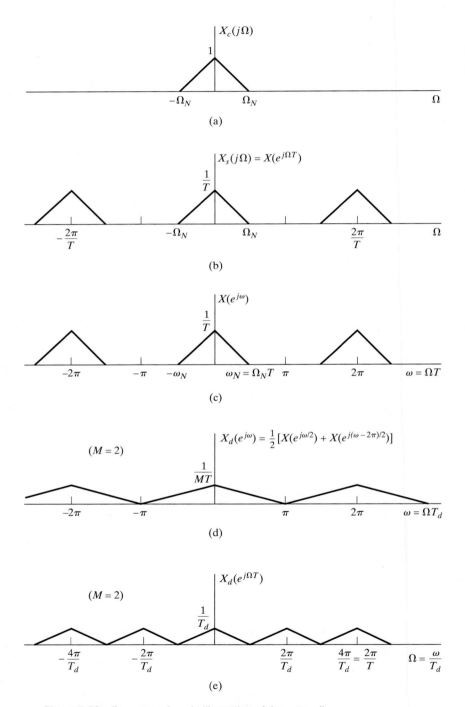

Figure 4.20 Frequency-domain illustration of downsampling.

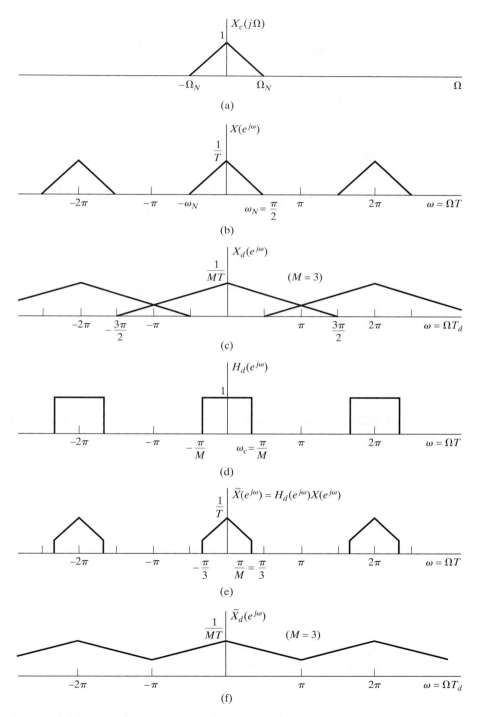

Figure 4.21 (a)–(c) Downsampling with aliasing. (d)–(f) Downsampling with prefiltering to avoid aliasing.

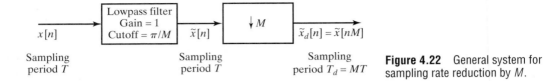

Figure 4.22   General system for sampling rate reduction by M.

$\omega_N = \Omega_N T = \pi/2$. Now, if we downsample by a factor of $M = 3$, we obtain the sequence $x_d[n] = x[3n] = x_c(n3T)$ whose DTFT is plotted in Figure 4.21(c) with normalized frequency $\omega = \Omega T_d$. Note that because $M\omega_N = 3\pi/2$, which is greater than π, aliasing occurs. In general, to avoid aliasing in downsampling by a factor of M requires that

$$\omega_N M \le \pi \qquad \text{or} \qquad \omega_N \le \pi/M. \tag{4.79}$$

If this condition does not hold, aliasing occurs, but it may be tolerable for some applications. In other cases, downsampling can be done without aliasing if we are willing to reduce the bandwidth of the signal $x[n]$ before downsampling. Thus, if $x[n]$ is filtered by an ideal lowpass filter with cutoff frequency π/M, then the output $\tilde{x}[n]$ can be downsampled without aliasing, as illustrated in Figures 4.21(d), (e), and (f). Note that the sequence $\tilde{x}_d[n] = \tilde{x}[nM]$ no longer represents the original underlying continuous-time signal $x_c(t)$. Rather, $\tilde{x}_d[n] = \tilde{x}_c(nT_d)$, where $T_d = MT$, and $\tilde{x}_c(t)$ is obtained from $x_c(t)$ by lowpass filtering with cutoff frequency $\Omega_c = \pi/T_d = \pi/(MT)$.

From the preceding discussion, we see that a general system for downsampling by a factor of M is the one shown in Figure 4.22. Such a system is called a *decimator*, and downsampling by lowpass filtering followed by compression has been termed *decimation* (Crochiere and Rabiner, 1983 and Vaidyanathan, 1993).

4.6.2 Increasing the Sampling Rate by an Integer Factor

We have seen that the reduction of the sampling rate of a discrete-time signal by an integer factor involves sampling the sequence in a manner analogous to sampling a continuous-time signal. Not surprisingly, increasing the sampling rate involves operations analogous to D/C conversion. To see this, consider a signal $x[n]$ whose sampling rate we wish to increase by a factor of L. If we consider the underlying continuous-time signal $x_c(t)$, the objective is to obtain samples

$$x_i[n] = x_c(nT_i), \tag{4.80}$$

where $T_i = T/L$, from the sequence of samples

$$x[n] = x_c(nT). \tag{4.81}$$

We will refer to the operation of increasing the sampling rate as *upsampling*.

From Eqs. (4.80) and (4.81), it follows that

$$x_i[n] = x[n/L] = x_c(nT/L), \qquad n = 0, \pm L, \pm 2L, \dots. \tag{4.82}$$

Figure 4.23 shows a system for obtaining $x_i[n]$ from $x[n]$ using only discrete-time processing. The system on the left is called a *sampling rate expander* (see Crochiere and Rabiner, 1983 and Vaidyanathan, 1993) or simply an *expander*. Its output is

$$x_e[n] = \begin{cases} x[n/L], & n = 0, \pm L, \pm 2L, \dots, \\ 0, & \text{otherwise,} \end{cases} \tag{4.83}$$

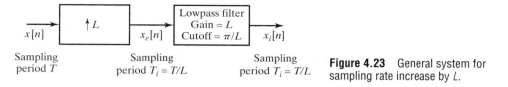

Sampling period T

Sampling period $T_i = T/L$

Sampling period $T_i = T/L$

Figure 4.23   General system for sampling rate increase by L.

or equivalently,

$$x_e[n] = \sum_{k=-\infty}^{\infty} x[k]\delta[n - kL]. \tag{4.84}$$

The system on the right is a lowpass discrete-time filter with cutoff frequency π/L and gain L. This system plays a role similar to the ideal D/C converter in Figure 4.9(b). First, we create a discrete-time impulse train $x_e[n]$, and we then use a lowpass filter to reconstruct the sequence.

The operation of the system in Figure 4.23 is most easily understood in the frequency domain. The Fourier transform of $x_e[n]$ can be expressed as

$$\begin{aligned} X_e(e^{j\omega}) &= \sum_{n=-\infty}^{\infty} \left(\sum_{k=-\infty}^{\infty} x[k]\delta[n - kL] \right) e^{-j\omega n} \\ &= \sum_{k=-\infty}^{\infty} x[k]e^{-j\omega L k} = X(e^{j\omega L}). \end{aligned} \tag{4.85}$$

Thus, the Fourier transform of the output of the expander is a frequency-scaled version of the Fourier transform of the input; i.e., ω is replaced by ωL so that ω is now normalized by

$$\omega = \Omega T_i. \tag{4.86}$$

This effect is illustrated in Figure 4.24. Figure 4.24(a) shows a bandlimited continuous-time Fourier transform, and Figure 4.24(b) shows the DTFT of the sequence $x[n] = x_c(nT)$, where $\pi/T = \Omega_N$. Figure 4.24(c) shows $X_e(e^{j\omega})$ according to Eq. (4.85), with $L = 2$, and Figure 4.24(e) shows the Fourier transform of the desired signal $x_i[n]$. We see that $X_i(e^{j\omega})$ can be obtained from $X_e(e^{j\omega})$ by correcting the amplitude scale from $1/T$ to $1/T_i$ and by removing all the frequency-scaled images of $X_c(j\Omega)$ except at integer multiples of 2π. For the case depicted in Figure 4.24, this requires a lowpass filter with a gain of 2 and cutoff frequency $\pi/2$, as shown in Figure 4.24(d). In general, the required gain would be L, since $L(1/T) = [1/(T/L)] = 1/T_i$, and the cutoff frequency would be π/L.

This example shows that the system of Figure 4.23 does indeed give an output satisfying Eq. (4.80) if the input sequence $x[n] = x_c(nT)$ was obtained by sampling without aliasing. Therefore, that system is called an *interpolator*, since it fills in the missing samples, and the operation of upsampling is consequently considered to be synonymous with *interpolation*.

As in the case of the D/C converter, it is possible to obtain an interpolation formula for $x_i[n]$ in terms of $x[n]$. First, note that the impulse response of the lowpass filter in Figure 4.23 is

$$h_i[n] = \frac{\sin(\pi n/L)}{\pi n/L}. \tag{4.87}$$

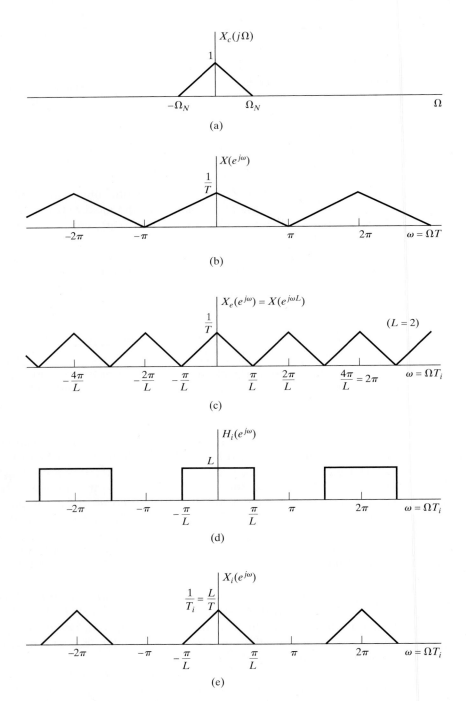

Figure 4.24 Frequency-domain illustration of interpolation.

Using Eq. (4.84), we obtain

$$x_i[n] = \sum_{k=-\infty}^{\infty} x[k] \frac{\sin[\pi(n-kL)/L]}{\pi(n-kL)/L}. \tag{4.88}$$

The impulse response $h_i[n]$ has the properties

$$\begin{aligned} h_i[0] &= 1, \\ h_i[n] &= 0, \qquad n = \pm L, \pm 2L, \ldots. \end{aligned} \tag{4.89}$$

Thus, for the ideal lowpass interpolation filter, we have

$$x_i[n] = x[n/L] = x_c(nT/L) = x_c(nT_i), \qquad n = 0, \pm L, \pm 2L, \ldots, \tag{4.90}$$

as desired. The fact that $x_i[n] = x_c(nT_i)$ for all n follows from our frequency-domain argument.

4.6.3 Simple and Practical Interpolation Filters

Although ideal lowpass filters for interpolation cannot be implemented exactly, very good approximations can be designed using techniques to be discussed in Chapter 7. However, in some cases, very simple interpolation procedures are adequate or are forced on us by computational limitations. Since linear interpolation is often used (even though it is often not very accurate), it is worthwhile to examine this process within the general framework that we have just developed.

Linear interpolation corresponds to interpolation so that the samples between two original samples lie on a straight line connecting the two original sample values. Linear interpolation can be accomplished with the system of Figure 4.23 with the filter having the triangularly shaped impulse response

$$h_{\text{lin}}[n] = \begin{cases} 1 - |n|/L, & |n| \le L, \\ 0, & \text{otherwise}, \end{cases} \tag{4.91}$$

as shown in Figure 4.25 for $L = 5$. With this filter, the interpolated output will be

$$x_{\text{lin}}[n] = \sum_{k=n-L+1}^{n+L-1} x_e[k] h_{\text{lin}}[n-k]. \tag{4.92}$$

Figure 4.26(a) depicts $x_e[k]$ (with the envelope of $h_{\text{lin}}[n-k]$ shown dashed for a particular value $n = 18$) and the corresponding output $x_{\text{lin}}[n]$ for the case $L = 5$. In this case, $x_{\text{lin}}[n]$ for $n = 18$ depends only on original samples $x[3]$ and $x[4]$. From this figure, we see that $x_{\text{lin}}[n]$ is identical to the sequence obtained by connecting the two original samples on either side of n by a straight line and then resampling at the $L - 1$ desired points in

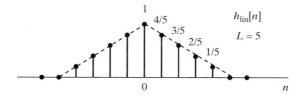

Figure 4.25 Impulse response for linear interpolation.

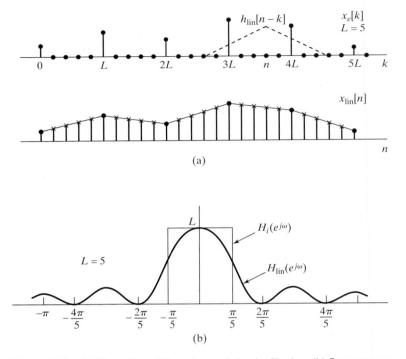

Figure 4.26 (a) Illustration of linear interpolation by filtering. (b) Frequency response of linear interpolator compared with ideal lowpass interpolation filter.

between. Also, note that the original sample values are preserved because $h_{\text{lin}}[0] = 1$ and $h_{\text{lin}}[n] = 0$ for $|n| \geq L$.

The nature of the distortion in the intervening samples can be better understood by comparing the frequency response of the linear interpolator with that of the ideal lowpass interpolator for a factor of L interpolation. It can be shown (see Problem 4.56) that

$$H_{\text{lin}}(e^{j\omega}) = \frac{1}{L} \left[\frac{\sin(\omega L/2)}{\sin(\omega/2)} \right]^2. \tag{4.93}$$

This function is plotted in Figure 4.26(b) for $L = 5$ together with the ideal lowpass interpolation filter. From the figure, we see that if the original signal is sampled at just the Nyquist rate, i.e., not oversampled, linear interpolation will not be very accurate, since the output of the filter will contain considerable energy in the band $\pi/L < |\omega| \leq \pi$ due to the frequency-scaled images of $X_c(j\Omega)$ at multiples of $2\pi/L$ that are not removed by the linear interpolation filter. However, if the original sampling rate is much higher than the Nyquist rate, then the linear interpolator will be more successful in removing these images because $H_{\text{lin}}(e^{j\omega})$ is small in a narrow region around these normalized frequencies, and at higher sampling rates, the increased frequency scaling causes the shifted copies of $X_c(j\Omega)$ to be more localized at multiples of $2\pi/L$. This is intuitively reasonable from a time domain perspective too, since, if the original sampling rate greatly exceeds the Nyquist rate, the signal will not vary significantly between samples, and thus, linear interpolation should be more accurate for oversampled signals.

Because of its double-sided infinite-length impulse response, the ideal bandlimited interpolator involves *all* of the original samples in the computation of each interpolated sample. In contrast, linear interpolation involves only *two* of the original samples in the computation of each interpolated sample. To better approximate ideal bandlimited interpolation, it is necessary to use filters with longer impulse responses. For this purpose FIR filters have many advantages. The impulse response $\tilde{h}_i[n]$ of an FIR filter for interpolation by a factor L usually is designed to have the following properties:

$$\tilde{h}_i[n] = 0 \qquad |n| \geq KL \tag{4.94a}$$

$$\tilde{h}_i[n] = \tilde{h}_i[-n] \qquad |n| \leq KL \tag{4.94b}$$

$$\tilde{h}_i[0] = 1 \qquad n = 0 \tag{4.94c}$$

$$\tilde{h}_i[n] = 0 \qquad n = \pm L, \pm 2L, \ldots, \pm KL. \tag{4.94d}$$

The interpolated output will therefore be

$$\tilde{x}_i[n] = \sum_{k=n-KL+1}^{n+KL-1} x_e[k]\tilde{h}_i[n-k]. \tag{4.95}$$

Note that the impulse response for linear interpolation satisfies Eqs. (4.94a)–(4.94d) with $K = 1$.

It is important to understand the motivation for the constraints of Eqs. (4.94a)–(4.94d). Equation (4.94a) states that the length of the FIR filter is $2KL - 1$ samples. Furthermore, this constraint ensures that only $2K$ original samples are involved in the computation of each sample of $\tilde{x}_i[n]$. This is because, even though $\tilde{h}_i[n]$ has $2KL - 1$ nonzero samples, the input $x_e[k]$ has only $2K$ nonzero samples within the region of support of $\tilde{h}_i[n - k]$ for any n between two of the original samples. Equation (4.94b) ensures that the filter will not introduce any phase shift into the interpolated samples since the corresponding frequency response is a real function of ω. The system could be made causal by introducing a delay of at least $KL - 1$ samples. In fact, the impulse response $\tilde{h}_i[n - KL]$ would yield an interpolated output delayed by KL samples, which would correspond to a delay of K samples at the original sampling rate. We might want to insert other amounts of delay so as to equalize delay among parts of a larger system that involves subsystems operating at different sampling rates. Finally, Eqs. (4.94c) and (4.94d) guarantee that the original signal samples will be preserved in the output, i.e.,

$$\tilde{x}_i[n] = x[n/L] \quad \text{at} \quad n = 0, \pm L, \pm 2L, \ldots. \tag{4.96}$$

Thus, if the sampling rate of $\tilde{x}_i[n]$ is subsequently reduced back to the original rate (with no intervening delay or a delay by a multiple of L samples) then $\tilde{x}_i[nL] = x[n]$; i.e., the original signal is recovered exactly. If this consistency is not required, the conditions of Eqs. (4.94c) and (4.94d) could be relaxed in the design of $\tilde{h}_i[n]$.

Figure 4.27 shows $x_e[k]$ and $\tilde{h}_i[n - k]$ with $K = 2$. The figure shows that each interpolated value depends on $2K = 4$ samples of the original input signal. Also note that computation of each interpolated sample requires only $2K$ multiplications and $2K - 1$ additions since there are always $L - 1$ zero samples in $x_e[k]$ between each of the original samples.

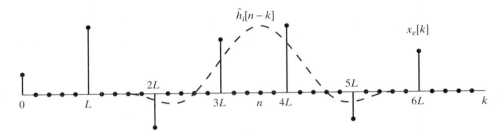

Figure 4.27   Illustration of interpolation involving $2K = 4$ samples when $L = 5$.

Interpolation is a much-studied problem in numerical analysis. Much of the development in this field is based on interpolation formulas that exactly interpolate polynomials of a certain degree. For example, the linear interpolator gives exact results for a constant signal and one whose samples vary along a straight line. Just as in the case of linear interpolation, higher-order Lagrange interpolation formulas (Schafer and Rabiner, 1973) and cubic spline interpolation formulas (Keys, 1981 and Unser, 2000) can be cast into our linear filtering framework to provide longer filters for interpolation. For example, the equation

$$\tilde{h}_i[n] = \begin{cases} (a + 2)|n/L|^3 - (a + 3)|n/L|^2 + 1 & 0 \leq n \leq L \\ a|n/L|^3 - 5|n/L|^2 + 8a|n/L| - 4a & L \leq n \leq 2L \\ 0 & \text{otherwise} \end{cases} \qquad (4.97)$$

defines a convenient family of interpolation filter impulse responses that involve four ($K = 2$) original samples in the computation of each interpolated sample. Figure 4.28(a) shows the impulse response of a cubic filter for $a = -0.5$ and $L = 5$ along with the filter (dashed triangle) for linear ($K = 1$) interpolation. The corresponding frequency responses are shown in Figure 4.28(b) on a logarithmic amplitude (dB) scale. Note that the cubic filter has much wider regions around the frequencies $2\pi/L$ and $4\pi/L$ (0.4π and 0.8π in this case) but lower sidelobes than the linear interpolator, which is shown as the dashed line.

4.6.4 Changing the Sampling Rate by a Noninteger Factor

We have shown how to increase or decrease the sampling rate of a sequence by an integer factor. By combining decimation and interpolation, it is possible to change the sampling rate by a noninteger factor. Specifically, consider Figure 4.29(a), which

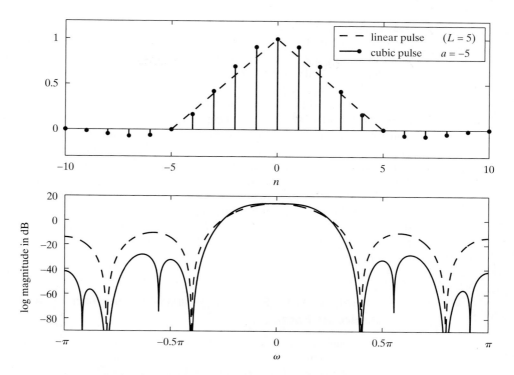

Figure 4.28 Impulse responses and frequency responses for linear and cubic interpolation.

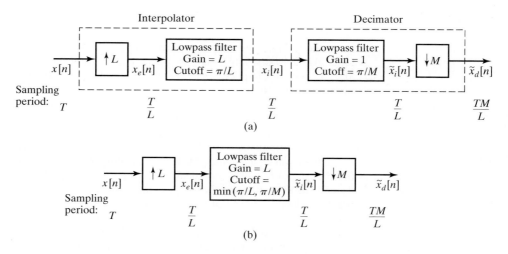

Figure 4.29 (a) System for changing the sampling rate by a noninteger factor. (b) Simplified system in which the decimation and interpolation filters are combined.

shows an interpolator that decreases the sampling period from T to T/L, followed by a decimator that increases the sampling period by M, producing an output sequence $\tilde{x}_d[n]$ that has an effective sampling period of (TM/L). By choosing L and M appropriately, we can approach arbitrarily close to any desired ratio of sampling periods. For example, if $L = 100$ and $M = 101$, then the effective sampling period is $1.01T$.

If $M > L$, there is a net increase in the sampling period (a decrease in the sampling rate), and if $M < L$, the opposite is true. Since the interpolation and decimation filters in Figure 4.29(a) are in cascade, they can be combined as shown in Figure 4.29(b) into one lowpass filter with gain L and cutoff equal to the minimum of π/L and π/M. If $M > L$, then π/M is the dominant cutoff frequency, and there is a net reduction in sampling rate. As pointed out in Section 4.6.1, if $x[n]$ was obtained by sampling at the Nyquist rate, the sequence $\tilde{x}_d[n]$ will correspond to a lowpass-filtered version of the original underlying bandlimited signal if we are to avoid aliasing. On the other hand, if $M < L$, then π/L is the dominant cutoff frequency, and there will be no need to further limit the bandwidth of the signal below the original Nyquist frequency.

Example 4.9   Sampling Rate Conversion by a Noninteger Rational Factor

Figure 4.30 illustrates sampling rate conversion by a rational factor. Suppose that a bandlimited signal with $X_c(j\Omega)$ as given in Figure 4.30(a) is sampled at the Nyquist rate; i.e., $2\pi/T = 2\Omega_N$. The resulting DTFT

$$X(e^{j\omega}) = \frac{1}{T} \sum_{k=-\infty}^{\infty} X_c\left(j\left(\frac{\omega}{T} - \frac{2\pi k}{T}\right)\right)$$

is plotted in Figure 4.30(b). An effective approach to changing the sampling period to $(3/2)T$, is to first interpolate by a factor $L = 2$ and then decimate by a factor of $M = 3$. Since this implies a net decrease in sampling rate, and the original signal was sampled at the Nyquist rate, we must incorporate additional bandlimiting to avoid aliasing.

Figure 4.30(c) shows the DTFT of the output of the $L = 2$ upsampler. If we were interested only in interpolating by a factor of 2, we could choose the lowpass filter to have a cutoff frequency of $\omega_c = \pi/2$ and a gain of $L = 2$. However, since the output of the filter will be decimated by $M = 3$, we must use a cutoff frequency of $\omega_c = \pi/3$, but the gain of the filter should still be 2 as in Figure 4.30(d). The Fourier transform $\tilde{X}_i(e^{j\omega})$ of the output of the lowpass filter is shown in Figure 4.30(e). The shaded regions indicate the part of the signal spectrum that is removed owing to the lower cutoff frequency for the interpolation filter. Finally, Figure 4.30(f) shows the DTFT of the output of the downsampler by $M = 3$. Note that the shaded regions show the aliasing that would have occurred if the cutoff frequency of the interpolation lowpass filter had been $\pi/2$ instead of $\pi/3$.

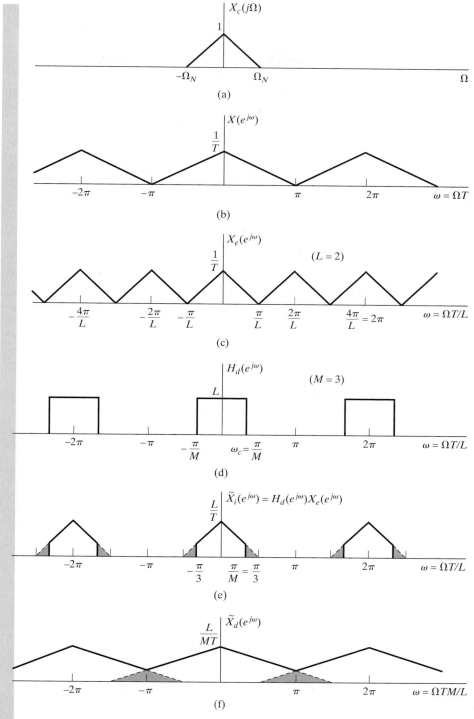

Figure 4.30   Illustration of changing the sampling rate by a noninteger factor.

4.7 MULTIRATE SIGNAL PROCESSING

As we have seen, it is possible to change the sampling rate of a discrete-time signal by a combination of interpolation and decimation. For example, if we want a new sampling period of $1.01T$, we can first interpolate by $L = 100$ using a lowpass filter that cuts off at $\omega_c = \pi/101$ and then decimate by $M = 101$. These large intermediate changes in sampling rate would require large amounts of computation for each output sample if we implement the filtering in a straightforward manner at the high intermediate sampling rate that is required. Fortunately, it is possible to greatly reduce the amount of computation required by taking advantage of some basic techniques broadly characterized as *multirate signal processing*. These multirate techniques refer in general to utilizing upsampling, downsampling, compressors, and expanders in a variety of ways to increase the efficiency of signal-processing systems. Besides their use in sampling rate conversion, they are exceedingly useful in A/D and D/A systems that exploit oversampling and noise shaping. Another important class of signal-processing algorithms that relies increasingly on multirate techniques is filter banks for the analysis and/or processing of signals.

Because of their widespread applicability, there is a large body of results on multirate signal processing techniques. In this section, we will focus on two basic results and show how a combination of these results can greatly improve the efficiency of sampling rate conversion. The first result is concerned with the interchange of filtering and downsampling or upsampling operations. The second is the polyphase decomposition. We shall also give two examples of how multirate techniques are used.

4.7.1 Interchange of Filtering with Compressor/Expander

First, we will derive two identities that aid in manipulating and understanding the operation of multirate systems. It is straightforward to show that the two systems in Figure 4.31 are equivalent. To see the equivalence, note that in Figure 4.31(b),

$$X_b(e^{j\omega}) = H(e^{j\omega M})X(e^{j\omega}), \tag{4.98}$$

and from Eq. (4.77),

$$Y(e^{j\omega}) = \frac{1}{M} \sum_{i=0}^{M-1} X_b(e^{j(\omega/M - 2\pi i/M)}). \tag{4.99}$$

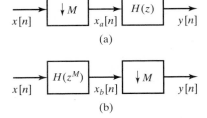

Figure 4.31 Two equivalent systems based on downsampling identities.

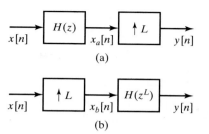

Figure 4.32   Two equivalent systems based on upsampling identities.

Substituting Eq. (4.98) into Eq. (4.99) gives

$$Y(e^{j\omega}) = \frac{1}{M} \sum_{i=0}^{M-1} X\left(e^{j(\omega/M - 2\pi i/M)}\right) H\left(e^{j(\omega - 2\pi i)}\right). \tag{4.100}$$

Since $H(e^{j(\omega - 2\pi i)}) = H(e^{j\omega})$, Eq. (4.100) reduces to

$$Y(e^{j\omega}) = H(e^{j\omega}) \frac{1}{M} \sum_{i=0}^{M-1} X\left(e^{j(\omega/M - 2\pi i/M)}\right)$$

$$= H(e^{j\omega}) X_a(e^{j\omega}), \tag{4.101}$$

which corresponds to Figure 4.31(a). Therefore, the systems in Figure 4.31(a) and 4.31(b) are completely equivalent.

A similar identity applies to upsampling. Specifically, using Eq. (4.85) in Section 4.6.2, it is also straightforward to show the equivalence of the two systems in Figure 4.32. We have, from Eq. (4.85) and Figure 4.32(a),

$$Y(e^{j\omega}) = X_a(e^{j\omega L})$$

$$= X(e^{j\omega L}) H(e^{j\omega L}). \tag{4.102}$$

Since, from Eq. (4.85),

$$X_b(e^{j\omega}) = X(e^{j\omega L}),$$

it follows that Eq. (4.102) is, equivalently,

$$Y(e^{j\omega}) = H(e^{j\omega L}) X_b(e^{j\omega}),$$

which corresponds to Figure 4.32(b).

In summary, we have shown that the operations of linear filtering and downsampling or upsampling can be interchanged if we modify the linear filter.

4.7.2 Multistage Decimation and Interpolation

When decimation or interpolation ratios are large, it is necessary to use filters with very long impulse responses to achieve adequate approximations to the required lowpass filters. In such cases, there can be significant reduction in computation through the use of multistage decimation or interpolation. Figure 4.33(a) shows a two-stage decimation

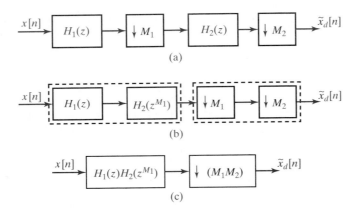

(a)

(b)

(c)

Figure 4.33   Multistage decimation:
(a) Two-stage decimation system.
(b) Modification of (a) using
downsampling identity of Figure 4.31.
(c) Equivalent one-stage decimation.

system where the overall decimation ratio is $M = M_1 M_2$. In this case, two lowpass filters are required; $H_1(z)$ corresponds to a lowpass filter with nominal cutoff frequency π/M_1 and likewise, $H_2(z)$ has nominal cutoff frequency π/M_2. Note that for single-stage decimation, the required nominal cutoff frequency would be $\pi/M = \pi/(M_1 M_2)$, which would be much smaller than that of either of the two filters. In Chapter 7 we will see that narrowband filters generally require high-order system functions to achieve sharp cutoff approximations to frequency-selective filter characteristics. Because of this effect, the two-stage implementation is often much more efficient than a single-stage implementation.

The single-stage system that is equivalent to Figure 4.33(a) can be derived using the downsampling identity of Figure 4.31. Figure 4.33(b) shows the result of replacing the system $H_2(z)$ and its preceding downsampler (by M_1) by the system $H_2(z^{M_1})$ followed by a downsampler by M_1. Figure 4.33(c) shows the result of combining the cascaded linear systems and cascaded downsamplers into corresponding single-stage systems. From this, we see that the system function of the equivalent single-stage lowpass filter is the product

$$H(z) = H_1(z) H_2(z^{M_1}). \qquad (4.103)$$

This equation, which can be generalized to any number of stages if M has many factors, is a useful representation of the overall effective frequency response of the two-stage decimator. Since it explicitly shows the effects of the two filters, it can be used as an aid in designing effective multistage decimators that minimize computation. (See Crochiere and Rabiner, 1983, Vaidyanathan, 1993, and Bellanger, 2000.) The factorization in Eq. (4.103) has also been used directly to design lowpass filters (Neuvo et al., 1984). In this context, the filter with system function represented by Eq. (4.103) is called an *interpolated FIR filter*. This is because the corresponding impulse response can be seen to be the convolution of $h_1[n]$ with the second impulse response expanded by M_1; i.e.,

$$h[n] = h_1[n] * \sum_{k=-\infty}^{\infty} h_2[k]\delta[n - kM_1]. \qquad (4.104)$$

The same multistage principles can be applied to interpolation, where, in this case, the upsampling identity of Figure 4.32 is used to relate the two-stage interpolator to an equivalent one-stage system. This is depicted in Figure 4.34.

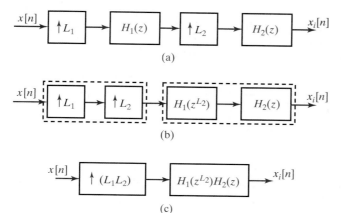

Figure 4.34 Multistage interpolation:
(a) Two-stage interpolation system.
(b) Modification of (a) using upsampling
identity of Figure 4.32. (c) Equivalent
one-stage interpolation.

4.7.3 Polyphase Decompositions

The polyphase decomposition of a sequence is obtained by representing it as a superposition of M subsequences, each consisting of every Mth value of successively delayed versions of the sequence. When this decomposition is applied to a filter impulse response, it can lead to efficient implementation structures for linear filters in several contexts. Specifically, consider an impulse response $h[n]$ that we decompose into M subsequences $h_k[n]$ with $k = 0, 1, \ldots, M-1$ as follows:

$$h_k[n] = \begin{cases} h[n+k], & n = \text{integer multiple of } M, \\ 0, & \text{otherwise.} \end{cases} \qquad (4.105)$$

By successively delaying these subsequences, we can reconstruct the original impulse response $h[n]$; i.e.,

$$h[n] = \sum_{k=0}^{M-1} h_k[n-k]. \qquad (4.106)$$

This decomposition can be represented with the block diagram in Figure 4.35. If we create a chain of advance elements at the input and a chain of delay elements at the output, the block diagram in Figure 4.36 is equivalent to that of Figure 4.35. In the decomposition in Figures 4.35 and 4.36, the sequences $e_k[n]$ are

$$e_k[n] = h[nM+k] = h_k[nM] \qquad (4.107)$$

and are referred to in general as the polyphase components of $h[n]$. There are several other ways to derive the polyphase components, and there are other ways to index them for notational convenience (Bellanger, 2000 and Vaidyanathan, 1993), but the definition in Eq. (4.107) is adequate for our purpose in this section.

Figures 4.35 and 4.36 are not realizations of the filter, but they show how the filter can be decomposed into M parallel filters. We see this by noting that Figures 4.35 and

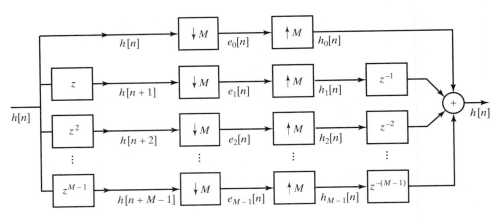

Figure 4.35   Polyphase decomposition of filter $h[n]$ using components $e_k[n]$.

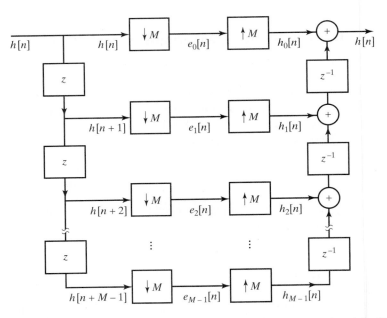

Figure 4.36   Polyphase decomposition of filter $h[n]$ using components $e_k[n]$ with chained delays.

4.36 show that, in the frequency or z-transform domain, the polyphase representation corresponds to expressing $H(z)$ as

$$H(z) = \sum_{k=0}^{M-1} E_k(z^M)z^{-k}. \tag{4.108}$$

Equation (4.108) expresses the system function $H(z)$ as a sum of delayed polyphase component filters. For example, from Eq. (4.108), we obtain the filter structure shown in Figure 4.37.

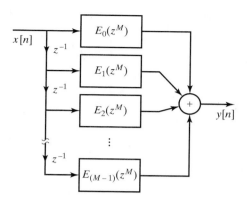

Figure 4.37 Realization structure based on polyphase decomposition of $h[n]$.

4.7.4 Polyphase Implementation of Decimation Filters

One of the important applications of the polyphase decomposition is in the implementation of filters whose output is then downsampled as indicated in Figure 4.38.

In the most straightforward implementation of Figure 4.38, the filter computes an output sample at each value of n, but then only one of every M output samples is retained. Intuitively, we might expect that it should be possible to obtain a more efficient implementation, which does not compute the samples that are thrown away.

To obtain a more efficient implementation, we can exploit a polyphase decomposition of the filter. Specifically, suppose we express $h[n]$ in polyphase form with polyphase components

$$e_k[n] = h[nM + k]. \tag{4.109}$$

From Eq. (4.108),

$$H(z) = \sum_{k=0}^{M-1} E_k(z^M)z^{-k}. \tag{4.110}$$

With this decomposition and the fact that downsampling commutes with addition, Figure 4.38 can be redrawn as shown in Figure 4.39. Applying the identity in Figure 4.31 to the system in Figure 4.39, we see that the latter then becomes the system shown in Figure 4.40.

To illustrate the advantage of Figure 4.40 compared with Figure 4.38, suppose that the input $x[n]$ is clocked at a rate of one sample per unit time and that $H(z)$ is an N-point FIR filter. In the straightforward implementation of Figure 4.38, we require N multiplications and $(N - 1)$ additions per unit time. In the system of Figure 4.40, each of the filters $E_k(z)$ is of length N/M, and their inputs are clocked at a rate of 1 per M units of time. Consequently, each filter requires $\frac{1}{M}\left(\frac{N}{M}\right)$ multiplications per unit time and $\frac{1}{M}\left(\frac{N}{M} - 1\right)$ additions per unit time. Since there are M polyphase components, the entire system therefore requires (N/M) multiplications and $\left(\frac{N}{M} - 1\right) + (M - 1)$ additions per unit time. Thus, we can achieve a significant savings for some values of M and N.

Figure 4.38 Decimation system.

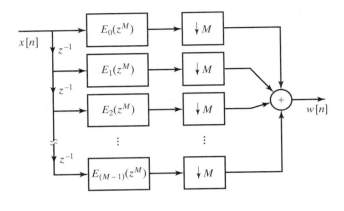

Figure 4.39   Implementation of decimation filter using polyphase decomposition.

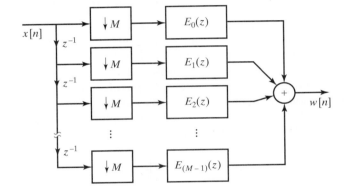

Figure 4.40   Implementation of decimation filter after applying the downsampling identity to the polyphase decomposition.

4.7.5 Polyphase Implementation of Interpolation Filters

A savings similar to that just discussed for decimation can be achieved by applying the polyphase decomposition to systems in which a filter is preceded by an upsampler as shown in Figure 4.41. Since only every Lth sample of $w[n]$ is nonzero, the most straightforward implementation of Figure 4.41 would involve multiplying filter coefficients by sequence values that are known to be zero. Intuitively, here again we would expect that a more efficient implementation was possible.

To implement the system in Figure 4.41 more efficiently, we again utilize the polyphase decomposition of $H(z)$. For example, we can express $H(z)$ as in the form of Eq. (4.110) and represent Figure 4.41 as shown in Figure 4.42. Applying the identity in Figure 4.32, we can rearrange Figure 4.42 as shown in Figure 4.43.

To illustrate the advantage of Figure 4.43 compared with Figure 4.41, we note that in Figure 4.41 if $x[n]$ is clocked at a rate of one sample per unit time, then $w[n]$ is clocked at a rate of L samples per unit time. If $H(z)$ is an FIR filter of length N, we then

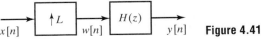

Figure 4.41   Interpolation system.

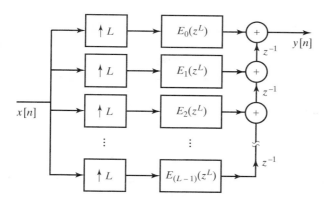

Figure 4.42 Implementation of interpolation filter using polyphase decomposition.

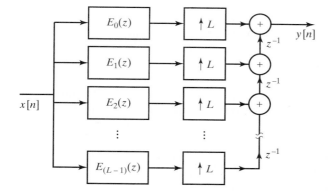

Figure 4.43 Implementation of interpolation filter after applying the upsampling identity to the polyphase decomposition.

require NL multiplications and $(NL - 1)$ additions per unit time. Figure 4.43, on the other hand, requires $L\,(N/L)$ multiplications and $L\left(\frac{N}{L} - 1\right)$ additions per unit time for the set of polyphase filters, plus $(L - 1)$ additions, to obtain $y[n]$. Thus, we again have the possibility of significant savings in computation for some values of L and N.

For both decimation and interpolation, gains in computational efficiency result from rearranging the operations so that the filtering is done at the low sampling rate. Combinations of interpolation and decimation systems for noninteger rate changes lead to significant savings when high intermediate rates are required.

4.7.6 Multirate Filter Banks

Polyphase structures for decimation and interpolation are widely used in filter banks for analysis and synthesis of audio and speech signals. For example, Figure 4.44 shows the block diagram of a two-channel analysis and synthesis filter bank commonly used in speech coding applications. The purpose of the analysis part of the system is to split the frequency spectrum of the input $x[n]$ into a lowpass band represented by the downsampled signal $v_0[n]$ and a highpass band represented by $v_1[n]$. In speech and audio coding applications, the channel signals are quantized for transmission and/or storage. Since the original band is nominally split into two equal parts of width $\pi/2$ radians, the

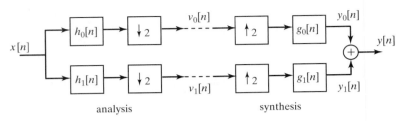

Figure 4.44   Two-channel analysis and synthesis filter bank.

sampling rates of the filter outputs can be 1/2 that of the input so that the total number of samples per second remains the same.[1] Note that downsampling the output of the lowpass filter expands the low-frequency band to the entire range $|\omega| < \pi$. On the other hand, downsampling the output of the highpass filter down-shifts the high-frequency band and expands it to the full range $|\omega| < \pi$.

The decomposition requires that $h_0[n]$ and $h_1[n]$ be impulse responses of lowpass and highpass filters respectively. A common approach is to derive the highpass filter from the lowpass filter by $h_1[n] = e^{j\pi n}h_0[n]$. This implies that $H_1(e^{j\omega}) = H_0(e^{j(\omega-\pi)})$ so that if $H_0(e^{j\omega})$ is a lowpass filter with nominal passband $0 \le |\omega| \le \pi/2$, then $H_1(e^{j\omega})$ will be a highpass filter with nominal passband $\pi/2 < |\omega| \le \pi$. The purpose of the righthand (synthesis) part of Figure 4.44 is to reconstitute an approximation to $x[n]$ from the two channel signals $v_0[n]$ and $v_1[n]$. This is achieved by upsampling both signals and passing them through a lowpass filter $g_0[n]$ and highpass filter $g_1[n]$ respectively. The resulting interpolated signals are added to produce the full-band output signal $y[n]$ sampled at the input sampling rate.

Applying the frequency-domain results for downsampling and upsampling to the system in Figure 4.44 leads to the following result:

$$Y(e^{j\omega}) = \frac{1}{2}\left[G_0(e^{j\omega})H_0(e^{j\omega}) + G_1(e^{j\omega})H_1(e^{j\omega})\right]X(e^{j\omega}) \qquad (4.111\text{a})$$

$$+\frac{1}{2}\Big[G_0(e^{j\omega})H_0(e^{j(\omega-\pi)})$$

$$+G_1(e^{j\omega})H_1(e^{j(\omega-\pi)})\Big]X(e^{j(\omega-\pi)}). \qquad (4.111\text{b})$$

If the analysis and synthesis filters are ideal so that they exactly split the band $0 \le |\omega| \le \pi$ into two equal segments without overlapping, then it is straightforward to verify that $Y(e^{j\omega}) = X(e^{j\omega})$; i.e., the synthesis filter bank reconstructs the input signal exactly. However, perfect or nearly perfect reconstruction also can be achieved with nonideal filters for which aliasing will occur in the downsampling operations of the analysis filter bank. To see this, note that the second term in the expression for $Y(e^{j\omega})$ (line labeled Eq. (4.111b)), which represents potential aliasing distortion from the downsampling operation, can be eliminated by choosing the filters such that

$$G_0(e^{j\omega})H_0(e^{j(\omega-\pi)}) + G_1(e^{j\omega})H_1(e^{j(\omega-\pi)}) = 0. \qquad (4.112)$$

---

[1]Filter banks that conserve the total number of samples per second are termed *maximally decimated* filter banks.

This condition is called the *alias cancellation condition*. One set of conditions that satisfy Eq. (4.112) is

$$h_1[n] = e^{j\pi n}h_0[n] \iff H_1(e^{j\omega}) = H_0(e^{j(\omega-\pi)}) \tag{4.113a}$$

$$g_0[n] = 2h_0[n] \iff G_0(e^{j\omega}) = 2H_0(e^{j\omega}) \tag{4.113b}$$

$$g_1[n] = -2h_1[n] \iff G_1(e^{j\omega}) = -2H_0(e^{j(\omega-\pi)}). \tag{4.113c}$$

The filters $h_0[n]$ and $h_1[n]$ are termed *quadrature mirror filters* since Eq. (4.113a) imposes mirror symmetry about $\omega = \pi/2$. Substituting these relations into Eq. (4.111a) leads to the relation

$$Y(e^{j\omega}) = \left[H_0^2(e^{j\omega}) - H_0^2(e^{j(\omega-\pi)}) \right] X(e^{j\omega}), \tag{4.114}$$

from which it follows that perfect reconstruction (with possible delay of M samples) requires

$$H_0^2(e^{j\omega}) - H_0^2(e^{j(\omega-\pi)}) = e^{-j\omega M}. \tag{4.115}$$

It can be shown (Vaidyanathan, 1993) that the only computationally realizable filters satisfying Eq. (4.115) exactly are systems with impulse responses of the form $h_0[n] = c_0\delta[n - 2n_0] + c_1\delta[n - 2n_1 - 1]$ where n_0 and n_1 are arbitrarily chosen integers and $c_0c_1 = \frac{1}{4}$. Such systems cannot provide the sharp frequency selective properties needed in speech and audio coding applications, but to illustrate that such systems can achieve exact reconstruction, consider the simple two-point moving average lowpass filter

$$h_0[n] = \frac{1}{2}(\delta[n] + \delta[n - 1]), \tag{4.116a}$$

which has frequency response

$$H_0(e^{j\omega}) = \cos(\omega/2)e^{-j\omega/2}. \tag{4.116b}$$

For this filter, $Y(e^{j\omega}) = e^{-j\omega}X(e^{j\omega})$ as can be verified by substituting Eq. (4.116b) into Eq. (4.114).

Either FIR or IIR filters can be used in the analysis/synthesis system of Figure 4.44 with the filters related as in Eq. (4.113a)–(4.113c) to provide nearly perfect reconstruction. The design of such filters is based on finding a design for $H_0(e^{j\omega})$ that is an acceptable lowpass filter approximation while satisfying Eq. (4.115) to within an acceptable approximation error. A set of such filters and an algorithm for their design was given by Johnston (1980). Smith and Barnwell (1984) and Mintzer (1985) showed that perfect reconstruction is possible with the two-channel filter bank of Figure 4.44 if the filters have a different relationship to one another than is specified by Eq. (4.113a)–(4.113c). The different relationship leads to filters called conjugate quadrature filters (CQF).

Polyphase techniques can be employed to save computation in the implementation of the analysis/synthesis system of Figure 4.44. Applying the polyphase downsampling result depicted in Figure 4.40 to the two channels leads to the block diagram in Figure 4.45(a), where

$$e_{00}[n] = h_0[2n] \tag{4.117a}$$

$$e_{01}[n] = h_0[2n + 1] \tag{4.117b}$$

$$e_{10}[n] = h_1[2n] = e^{j2\pi n}h_0[2n] = e_{00}[n] \tag{4.117c}$$

$$e_{11}[n] = h_1[2n + 1] = e^{j2\pi n}e^{j\pi}h_0[2n + 1] = -e_{01}[n]. \tag{4.117d}$$

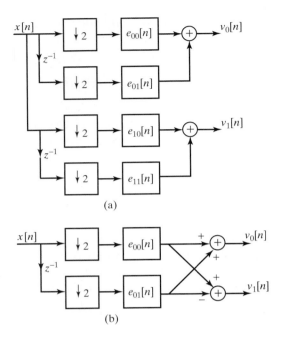

(a)

(b)

Figure 4.45 Polyphase representation of the two-channel analysis filter bank of Figure 4.44.

Equations (4.117c) and (4.117d) show that the polyphase filters for $h_1[n]$ are the same (except for sign) as those for $h_0[n]$. Therefore, only one set, $e_{00}[n]$ and $e_{01}[n]$ need be implemented. Figure 4.45(b) shows how both $v_0[n]$ and $v_1[n]$ can be formed from the outputs of the two polyphase filters. This equivalent structure, which requires only half the computation of Figure 4.45(a), is, of course, owing entirely to the simple relation between the two filters.

The polyphase technique can likewise be applied to the synthesis filter bank, by recognizing that the two interpolators can be replaced by their polyphase implementations and then the polyphase structures can be combined because $g_1[n] = -e^{j\pi n} g_0[n] = -e^{j\pi n} 2h_0[n]$. The resulting polyphase synthesis system can be represented in terms of the polyphase filters $f_{00}[n] = 2e_{00}[n]$ and $f_{01}[n] = 2e_{01}[n]$ as in Figure 4.46. As in the case of the analysis filter bank, the synthesis polyphase filters can be shared between the two channels thereby halving the computation.

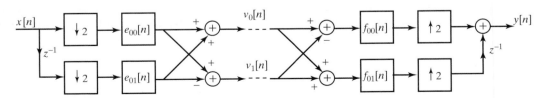

Figure 4.46 Polyphase representation of the two-channel analysis and synthesis filter bank of Figure 4.44.

This two-band analysis/synthesis system can be generalized to N equal width channels to obtain a finer decomposition of the spectrum. Such systems are used in audio coding, where they facilitate exploitation of the characteristics of human auditory perception in compression of the digital information rate. (See MPEG audio coding standard and Spanias, Painter, and Atti, 2007.) Also, the two-band system can be incorporated into a tree structure to obtain an analysis/synthesis system with either uniformly or nonuniformly spaced channels. When the CQF filters of Smith and Barnwell, and Mintzer are used, exact reconstruction is possible, and the resulting analysis synthesis system is essentially the discrete wavelet transform. (See Vaidyanathan, 1993 and Burrus, Gopinath and Guo, 1997.)

4.8 DIGITAL PROCESSING OF ANALOG SIGNALS

So far, our discussions of the representation of continuous-time signals by discrete-time signals have focused on idealized models of periodic sampling and bandlimited interpolation. We have formalized those discussions in terms of an idealized sampling system that we have called the *ideal continuous-to-discrete* (C/D) *converter* and an idealized bandlimited interpolator system called the *ideal discrete-to-continuous* (D/C) *converter*. These idealized conversion systems allow us to concentrate on the essential mathematical details of the relationship between a bandlimited signal and its samples. For example, in Section 4.4 we used the idealized C/D and D/C conversion systems to show that LTI discrete-time systems can be used in the configuration of Figure 4.47(a) to implement LTI continuous-time systems if the input is bandlimited and the sampling rate is at or above the Nyquist rate. In a practical setting, continuous-time signals are not precisely bandlimited, ideal filters cannot be realized, and the ideal C/D and D/C converters can only be approximated by devices that are called analog-to-digital (A/D) and digital-to-analog (D/A) converters, respectively. The block diagram of Figure 4.47(b) shows a

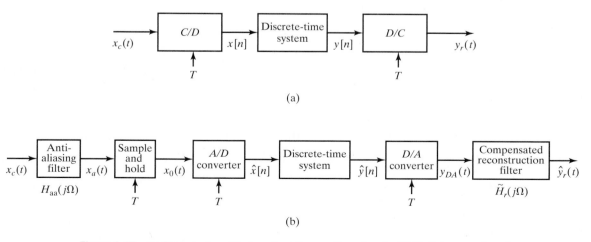

(a)

(b)

Figure 4.47   (a) Discrete-time filtering of continuous-time signals. (b) Digital processing of analog signals.

more realistic model for digital processing of continuous-time (analog) signals. In this section, we will examine some of the considerations introduced by each of the components of the system in Figure 4.47(b).

4.8.1 Prefiltering to Avoid Aliasing

In processing analog signals using discrete-time systems, it is generally desirable to minimize the sampling rate. This is because the amount of arithmetic processing required to implement the system is proportional to the number of samples to be processed. If the input is not bandlimited or if the Nyquist frequency of the input is too high, prefiltering may be necessary. An example of such a situation occurs in processing speech signals, where often only the low-frequency band up to about 3 to 4 kHz is required for intelligibility, even though the speech signal may have significant frequency content in the 4 kHz to 20 kHz range. Also, even if the signal is naturally bandlimited, wideband additive noise may fill in the higher frequency range, and as a result of sampling, these noise components would be aliased into the low-frequency band. If we wish to avoid aliasing, the input signal must be forced to be bandlimited to frequencies below one-half the desired sampling rate. This can be accomplished by lowpass filtering the continuous-time signal prior to C/D conversion, as shown in Figure 4.48. In this context, the lowpass filter that precedes the C/D converter is called an *antialiasing filter*. Ideally, the frequency response of the antialiasing filter would be

$$H_{aa}(j\Omega) = \begin{cases} 1, & |\Omega| < \Omega_c \leq \pi/T, \\ 0, & |\Omega| \geq \Omega_c. \end{cases} \tag{4.118}$$

From the discussion of Section 4.4.1, it follows that the overall system, from the output of the antialiasing filter $x_a(t)$ to the output $y_r(t)$, will always behave as an LTI system, since the input to the C/D converter, $x_a(t)$, is forced by the antialiasing filter to be bandlimited to frequencies below π/T radians/s. Thus, the overall effective frequency response of Figure 4.48 will be the product of $H_{aa}(j\Omega)$ and the effective frequency response from $x_a(t)$ to $y_r(t)$. Combining Eqs. (4.118) and (4.38) gives

$$H_{eff}(j\Omega) = \begin{cases} H(e^{j\Omega T}), & |\Omega| < \Omega_c, \\ 0, & |\Omega| \geq \Omega_c. \end{cases} \tag{4.119}$$

Thus, for an ideal lowpass antialiasing filter, the system of Figure 4.48 behaves as an LTI system with frequency response given by Eq. (4.119), even when $X_c(j\Omega)$ is not bandlimited. In practice, the frequency response $H_{aa}(j\Omega)$ cannot be ideally bandlimited, but $H_{aa}(j\Omega)$ can be made small for $|\Omega| > \pi/T$ so that aliasing is minimized. In this case, the overall frequency response of the system in Figure 4.48 should be approximately

$$H_{eff}(j\Omega) \approx H_{aa}(j\Omega)H(e^{j\Omega T}). \tag{4.120}$$

To achieve a negligibly small frequency response above π/T, it would be necessary for $H_{aa}(j\Omega)$ to begin to "roll off," i.e., begin to introduce attenuation, at frequencies below π/T, Eq. (4.120) suggests that the roll-off of the antialiasing filter (and other LTI distortions to be discussed later) could be at least partially compensated for by taking them into account in the design of the discrete-time system. This is illustrated in Problem 4.62.

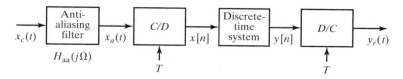

Figure 4.48   Use of prefiltering to avoid aliasing.

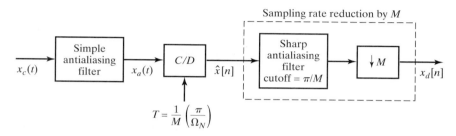

Figure 4.49   Using oversampled A/D conversion to simplify a continuous-time antialiasing filter.

The preceding discussion requires sharp-cutoff antialiasing filters. Such sharp-cutoff analog filters can be realized using active networks and integrated circuits. However, in applications involving powerful, but inexpensive, digital processors, these continuous-time filters may account for a major part of the cost of a system for discrete-time processing of analog signals. Sharp-cutoff filters are difficult and expensive to implement, and if the system is to operate with a variable sampling rate, adjustable filters would be required. Furthermore, sharp-cutoff analog filters generally have a highly nonlinear phase response, particularly at the passband edge. Thus, it is desirable for several reasons to eliminate the continuous-time filters or simplify the requirements on them.

One approach is depicted in Figure 4.49. With Ω_N denoting the highest frequency component to eventually be retained after the antialiasing filtering is completed, we first apply a very simple antialiasing filter that has a gradual cutoff with significant attenuation at $M\Omega_N$. Next, implement the C/D conversion at a sampling rate much higher than $2\Omega_N$, e.g., at $2M\Omega_N$. After that, sampling rate reduction by a factor of M that includes sharp antialiasing filtering is implemented in the discrete-time domain. Subsequent discrete-time processing can then be done at the low sampling rate to minimize computation.

This use of oversampling followed by sampling rate conversion is illustrated in Figure 4.50. Figure 4.50(a) shows the Fourier transform of a signal that occupies the band $|\Omega| < \Omega_N$, plus the Fourier transform of what might correspond to high-frequency "noise" or unwanted components that we eventually want to eliminate with the antialiasing filter. Also shown (dotted line) is the frequency response of an antialiasing filter that does not cut off sharply but gradually falls to zero at frequencies above the frequency Ω_N. Figure 4.50(b) shows the Fourier transform of the output of this filter. If the signal $x_a(t)$ is sampled with period T such that $(2\pi/T - \Omega_c) \geq \Omega_N$, then the DTFT

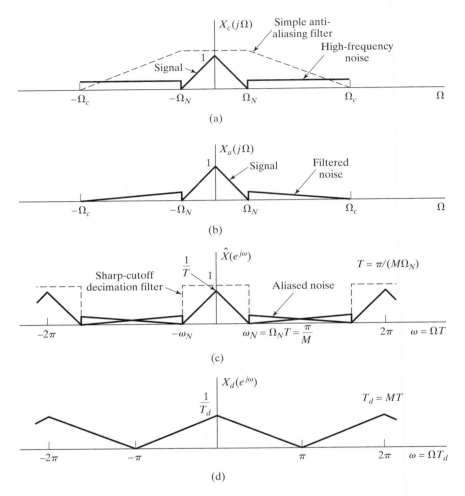

Figure 4.50   Use of oversampling followed by decimation in C/D conversion.

of the sequence $\hat{x}[n]$ will be as shown in Figure 4.50(c). Note that the "noise" will be aliased, but aliasing will not affect the signal band $|\omega| < \omega_N = \Omega_N T$. Now, if T and T_d are chosen so that $T_d = MT$ and $\pi/T_d = \Omega_N$, then $\hat{x}[n]$ can be filtered by a sharp-cutoff discrete-time filter (shown idealized in Figure 4.50(c)) with unity gain and cutoff frequency π/M. The output of the discrete-time filter can be downsampled by M to obtain the sampled sequence $x_d[n]$ whose Fourier transform is shown in Figure 4.50(d). Thus, all the sharp-cutoff filtering has been done by a discrete-time system, and only nominal continuous-time filtering is required. Since discrete-time FIR filters can have an exactly linear phase, it is possible using this oversampling approach to implement antialiasing filtering with virtually no phase distortion. This can be a significant advantage in situations where it is critical to preserve not only the frequency spectrum, but the waveshape as well.

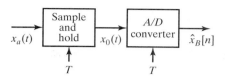

Figure 4.51   Physical configuration for A/D conversion.

4.8.2 A/D Conversion

An ideal C/D converter converts a continuous-time signal into a discrete-time signal, where each sample is known with infinite precision. As an approximation to this for digital signal processing, the system of Figure 4.51 converts a continuous-time (analog) signal into a digital signal, i.e., a sequence of finite-precision or quantized samples. The two systems in Figure 4.51 are available as physical devices. The A/D converter is a physical device that converts a voltage or current amplitude at its input into a binary code representing a quantized amplitude value closest to the amplitude of the input. Under the control of an external clock, the A/D converter can be caused to start and complete an A/D conversion every T seconds. However, the conversion is not instantaneous, and for this reason, a high-performance A/D system typically includes a sample-and-hold, as in Figure 4.51. The ideal sample-and-hold system is the system whose output is

$$x_0(t) = \sum_{n=-\infty}^{\infty} x[n]h_0(t - nT), \tag{4.121}$$

where $x[n] = x_a(nT)$ are the ideal samples of $x_a(t)$ and $h_0(t)$ is the impulse response of the zero-order-hold system, i.e.,

$$h_0(t) = \begin{cases} 1, & 0 < t < T, \\ 0, & \text{otherwise.} \end{cases} \tag{4.122}$$

If we note that Eq. (4.121) has the equivalent form

$$x_0(t) = h_0(t) * \sum_{n=-\infty}^{\infty} x_a(nT)\delta(t - nT), \tag{4.123}$$

we see that the ideal sample-and-hold is equivalent to impulse train modulation followed by linear filtering with the zero-order-hold system, as depicted in Figure 4.52(a). The relationship between the Fourier transform of $x_0(t)$ and the Fourier transform of $x_a(t)$ can be worked out following the style of analysis of Section 4.2, and we will do a similar analysis when we discuss the D/A converter. However, the analysis is unnecessary at this point, since everything we need to know about the behavior of the system can be seen from the time-domain expression. Specifically, the output of the zero-order hold is a staircase waveform where the sample values are held constant during the sampling period of T seconds. This is illustrated in Figure 4.52(b). Physical sample-and-hold circuits are designed to sample $x_a(t)$ as nearly instantaneously as possible and to hold the sample value as nearly constant as possible until the next sample is taken. The purpose of this is to provide the constant input voltage (or current) required by the A/D converter. The details of the wide variety of A/D conversion processes and the details of sample-and-hold and A/D circuit implementations are outside the scope of this book. Many practical issues arise in obtaining a sample-and-hold that samples

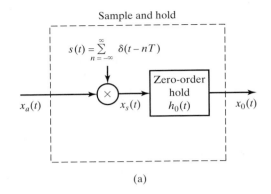

Sample and hold

(a)

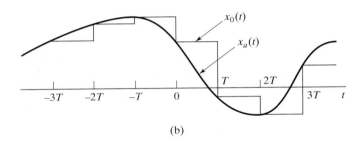

(b)

Figure 4.52 (a) Representation of an ideal sample-and-hold. (b) Representative input and output signals for the sample-and-hold.

quickly and holds the sample value constant with no decay or "glitches." Likewise, many practical concerns dictate the speed and accuracy of conversion of A/D converter circuits. Such questions are considered in Hnatek (1988) and Schmid (1976), and details of the performance of specific products are available in manufacturers' specification and data sheets. Our concern in this section is the analysis of the quantization effects in A/D conversion.

Since the purpose of the sample-and-hold in Figure 4.51 is to implement ideal sampling and to hold the sample value for quantization by the A/D converter, we can represent the system of Figure 4.51 by the system of Figure 4.53, where the ideal C/D converter represents the sampling performed by the sample-and-hold and, as we will describe later, the quantizer and coder together represent the operation of the A/D converter.

The quantizer is a nonlinear system whose purpose is to transform the input sample $x[n]$ into one of a finite set of prescribed values. We represent this operation as

$$\hat{x}[n] = Q(x[n]) \tag{4.124}$$

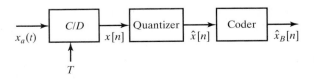

Figure 4.53 Conceptual representation of the system in Figure 4.51.

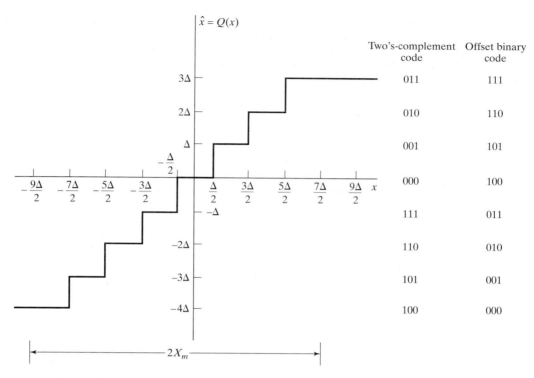

Figure 4.54   Typical quantizer for A/D conversion.

and refer to $\hat{x}[n]$ as the quantized sample. Quantizers can be defined with either uniformly or nonuniformly spaced quantization levels; however, when numerical calculations are to be done on the samples, the quantization steps usually are uniform. Figure 4.54 shows a typical uniform quantizer characteristic,[2] in which the sample values are *rounded* to the nearest quantization level.

Several features of Figure 4.54 should be emphasized. First, note that this quantizer would be appropriate for a signal whose samples are both positive and negative (bipolar). If it is known that the input samples are always positive (or negative), then a different distribution of the quantization levels would be appropriate. Next, observe that the quantizer of Figure 4.54 has an even number of quantization levels. With an even number of levels, it is not possible to have a quantization level at zero amplitude and also have an equal number of positive and negative quantization levels. Generally, the number of quantization levels will be a power of two, but the number will be much greater than eight, so this difference is usually inconsequential.

Figure 4.54 also depicts coding of the quantization levels. Since there are eight quantization levels, we can label them by a binary code of 3 bits. (In general, 2^{B+1} levels can be coded with a $(B+1)$-bit binary code.) In principle, any assignment of symbols

---

[2]Such quantizers are also called *linear* quantizers because of the linear progression of quantization steps.

can be used, and many binary coding schemes exist, each with its own advantages and disadvantages, depending on the application. For example, the right-hand column of binary numbers in Figure 4.54 illustrates the *offset binary* coding scheme, in which the binary symbols are assigned in numeric order, starting with the most negative quantization level. However, in digital signal processing, we generally wish to use a binary code that permits us to do arithmetic directly with the code words as scaled representations of the quantized samples.

The left-hand column in Figure 4.54 shows an assignment according to the two's complement binary number system. This system for representing signed numbers is used in most computers and microprocessors; thus, it is perhaps the most convenient labeling of the quantization levels. Note, incidentally, that the offset binary code can be converted to two's-complement code simply by complementing the most significant bit.

In the two's-complement system, the leftmost, or most significant, bit is considered as the sign bit, and we take the remaining bits as representing either binary integers or fractions. We will assume the latter; i.e., we assume a binary fraction point between the two most significant bits. Then, for the two's-complement interpretation, the binary symbols have the following meaning for $B = 2$:

Binary symbol	Numeric value, $\hat{x}_B$
$0_\diamond1\,1$	$3/4$
$0_\diamond1\,0$	$1/2$
$0_\diamond0\,1$	$1/4$
$0_\diamond0\,0$	0
$1_\diamond1\,1$	$-1/4$
$1_\diamond1\,0$	$-1/2$
$1_\diamond0\,1$	$-3/4$
$1_\diamond0\,0$	-1

In general, if we have a $(B + 1)$-bit binary two's-complement fraction of the form

$$a_0{}_\diamond a_1 a_2 \ldots a_B,$$

then its value is

$$-a_0 2^0 + a_1 2^{-1} + a_2 2^{-2} + \cdots + a_B 2^{-B}.$$

Note that the symbol $\diamond$ denotes the "binary point" of the number. The relationship between the code words and the quantized signal levels depends on the parameter X_m in Figure 4.54. This parameter determines the full-scale level of the A/D converter. From Figure 4.54, we see that the step size of the quantizer would in general be

$$\Delta = \frac{2X_m}{2^{B+1}} = \frac{X_m}{2^B}. \tag{4.125}$$

The smallest quantization levels ($\pm\Delta$) correspond to the least significant bit of the binary code word. Furthermore, the numeric relationship between the code words and the quantized samples is

$$\hat{x}[n] = X_m \hat{x}_B[n], \tag{4.126}$$

since we have assumed that $\hat{x}_B[n]$ is a binary number such that $-1 \leq \hat{x}_B[n] < 1$ (for two's complement). In this scheme, the binary coded samples $\hat{x}_B[n]$ are directly proportional

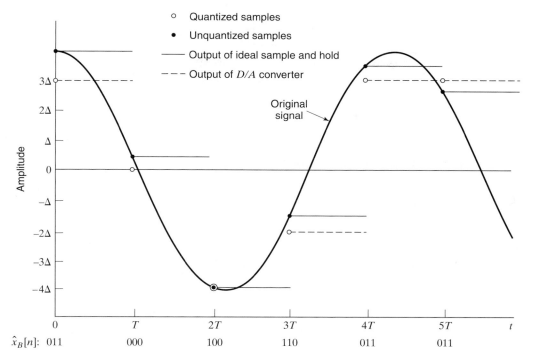

Figure 4.55   Sampling, quantization, coding, and D/A conversion with a 3-bit quantizer.

to the quantized samples (in two's-complement binary); therefore, they can be used as a numeric representation of the amplitude of the samples. Indeed, it is generally appropriate to assume that the input signal is normalized, so that the numeric values of $\hat{x}[n]$ and $\hat{x}_B[n]$ are identical and there is no need to distinguish between the quantized samples and the binary coded samples.

Figure 4.55 shows a simple example of quantization and coding of the samples of a sine wave using a 3-bit quantizer. The unquantized samples $x[n]$ are illustrated with solid dots, and the quantized samples $\hat{x}[n]$ are illustrated with open circles. Also shown is the output of an ideal sample-and-hold. The dotted lines labeled "output of D/A converter" will be discussed later. Figure 4.55 shows, in addition, the 3-bit code words that represent each sample. Note that, since the analog input $x_a(t)$ exceeds the full-scale value of the quantizer, some of the positive samples are "clipped."

Although much of the preceding discussion pertains to two's-complement coding of the quantization levels, the basic principles of quantization and coding in A/D conversion are the same regardless of the binary code used to represent the samples. A more detailed discussion of the binary arithmetic systems used in digital computing can be found in texts on computer arithmetic. (See, for example, Knuth, 1998.) We now turn to an analysis of the effects of quantization. Since this analysis does not depend on the assignment of binary code words, it will lead to rather general conclusions.

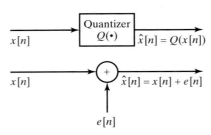

Figure 4.56   Additive noise model for quantizer.

4.8.3 Analysis of Quantization Errors

From Figures 4.54 and 4.55, we see that the quantized sample $\hat{x}[n]$ will generally be different from the true sample value $x[n]$. The difference between them is the *quantization error,* defined as

$$e[n] = \hat{x}[n] - x[n]. \tag{4.127}$$

For example, for the 3-bit quantizer of Figure 4.54, if $\Delta/2 < x[n] \le 3\Delta/2$, then $\hat{x}[n] = \Delta$, and it follows that

$$-\Delta/2 \le e[n] < \Delta/2. \tag{4.128}$$

In the case of Figure 4.54, Eq. (4.128) holds whenever

$$-9\Delta/2 < x[n] \le 7\Delta/2. \tag{4.129}$$

In the general case of a $(B+1)$-bit quantizer with Δ given by Eq. (4.125), the quantization error satisfies Eq. (4.128) whenever

$$(-X_m - \Delta/2) < x[n] \le (X_m - \Delta/2). \tag{4.130}$$

If $x[n]$ is outside this range, as it is for the sample at $t = 0$ in Figure 4.55, then the quantization error may be larger in magnitude than $\Delta/2$, and such samples are said to be *clipped,* and the quantizer is said to be *overloaded.*

A simplified, but useful, model of the quantizer is depicted in Figure 4.56. In this model, the quantization error samples are thought of as an additive noise signal. The model is exactly equivalent to the quantizer if we know $e[n]$. In most cases, however, $e[n]$ is not known, and a statistical model based on Figure 4.56 is then often useful in representing the effects of quantization. We will also use such a model in Chapters 6 and 9 to describe the effects of quantization in signal-processing algorithms. The statistical representation of quantization errors is based on the following assumptions:

1. The error sequence $e[n]$ is a sample sequence of a stationary random process.
2. The error sequence is uncorrelated with the sequence $x[n]$.[3]
3. The random variables of the error process are uncorrelated; i.e., the error is a white-noise process.
4. The probability distribution of the error process is uniform over the range of quantization error.

---

[3]This does not, of course, imply statistical independence, since the error is directly determined by the input signal.

As we will see, the preceding assumptions lead to a rather simple, but effective, analysis of quantization effects that can yield useful predictions of system performance. It is easy to find situations where these assumptions are not valid. For example, if $x_a(t)$ is a step function, the assumptions would not be justified. However, when the signal is a complicated signal, such as speech or music, where the signal fluctuates rapidly in a somewhat unpredictable manner, the assumptions are more realistic. Experimental measurements and theoretical analyses for random signal inputs have shown that, when the quantization step size (and therefore the error) is small and when the signal varies in a complicated manner, the measured correlation between the signal and the quantization error decreases, and the error samples also become uncorrelated. (See Bennett, 1948; Widrow, 1956, 1961; Sripad and Snyder, 1977; and Widrow and Kollár, 2008.) In a heuristic sense, the assumptions of the statistical model appear to be valid when the quantizer is not overloaded and when the signal is sufficiently complex, and the quantization steps are sufficiently small, so that the amplitude of the signal is likely to traverse many quantization steps from sample to sample.

Example 4.10   Quantization Error for a Sinusoidal Signal

As an illustration, Figure 4.57(a) shows the sequence of unquantized samples of the cosine signal $x[n] = 0.99 \cos(n/10)$. Figure 4.57(b) shows the quantized sample sequence $\hat{x}[n] = Q\{x[n]\}$ for a 3-bit quantizer ($B + 1 = 3$), assuming that $X_m = 1$. The dashed lines in this figure show the eight possible quantization levels. Figures 4.57(c) and 4.57(d) show the quantization error $e[n] = \hat{x}[n] - x[n]$ for 3- and 8-bit quantization, respectively. In each case, the scale of the quantization error is adjusted so that the range $\pm\Delta/2$ is indicated by the dashed lines.

Notice that in the 3-bit case, the error signal is highly correlated with the unquantized signal. For example, around the positive and negative peaks of the cosine, the quantized signal remains constant over many consecutive samples, so that the error has the shape of the input sequence during these intervals. Also, note that during the intervals around the positive peaks, the error is greater than $\Delta/2$ in magnitude because the signal level is too large for this setting of the quantizer parameters. On the other hand, the quantization error for 8-bit quantization has no apparent patterns.[4] Visual inspection of these figures supports the preceding assertions about the quantization-noise properties in the finely quantized (8-bit) case; i.e., the error samples appear to vary randomly, with no apparent correlation with the unquantized signal, and they range between $-\Delta/2$ and $+\Delta/2$.

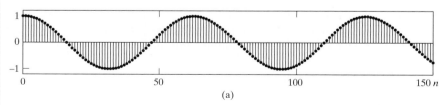

(a)

Figure 4.57 Example of quantization noise. (a) Unquantized samples of the signal $x[n] = 0.99 \cos(n/10)$.

---

[4]For periodic cosine signals, the quantization error would, of course, be periodic, too; and therefore, its power spectrum would be concentrated at multiples of the frequency of the input signal. We used the frequency $\omega_0 = 1/10$ to avoid this case in the example.

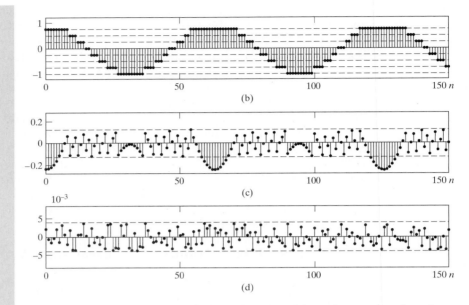

Figure 4.57 (*continued*) (b) Quantized samples of the cosine waveform in part (a) with a 3-bit quantizer. (c) Quantization error sequence for 3-bit quantization of the signal in (a). (d) Quantization error sequence for 8-bit quantization of the signal in (a).

For quantizers that round the sample value to the nearest quantization level, as shown in Figure 4.54, the amplitude of the quantization noise is in the range

$$-\Delta/2 \le e[n] < \Delta/2. \tag{4.131}$$

For small Δ, it is reasonable to assume that $e[n]$ is a random variable uniformly distributed from $-\Delta/2$ to $\Delta/2$. Therefore, the 1st-order probability density assumed for the quantization noise is as shown in Figure 4.58. (If truncation rather than rounding is used in implementing quantization, then the error would always be negative, and we would assume a uniform probability density from $-\Delta$ to 0.) To complete the statistical model for the quantization noise, we assume that successive noise samples are uncorrelated with each other and that $e[n]$ is uncorrelated with $x[n]$. Thus, $e[n]$ is assumed to be a uniformly distributed white-noise sequence. The mean value of $e[n]$ is zero, and its

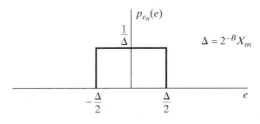

Figure 4.58 Probability density function of quantization error for a rounding quantizer such as that of Figure 4.54.

variance is

$$\sigma_e^2 = \int_{-\Delta/2}^{\Delta/2} e^2 \frac{1}{\Delta} de = \frac{\Delta^2}{12}. \tag{4.132}$$

For a $(B+1)$-bit quantizer with full-scale value X_m, the noise variance, or power, is

$$\sigma_e^2 = \frac{2^{-2B} X_m^2}{12}. \tag{4.133}$$

Equation (4.133) completes the white-noise model of quantization noise since the autocorrelation function would be $\phi_{ee}[m] = \sigma_e^2 \delta[m]$ and the corresponding power density spectrum would be

$$P_{ee}(e^{j\omega}) = \sigma_e^2 = \frac{2^{-2B} X_m^2}{12} \qquad |\omega| \leq \pi. \tag{4.134}$$

Example 4.11   Measurements of Quantization Noise

To confirm and illustrate the validity of the model for quantization noise, consider again quantization of the signal $x[n] = .99 \cos(n/10)$ which can be computed with 64-bit floating-point precisions (for all practical purposes unquantized) and then quantized to $B + 1$ bits. The quantization noise sequence can also be computed since we know both the input and the output of the quantizer. An amplitude histogram, which gives a count of the number of samples lying in each of a set of contiguous amplitude intervals or "bins," is often used as an estimate of the probability distribution of a random signal. Figure 4.59 shows histograms of the quantization noise for 16- and 8-bit quantization

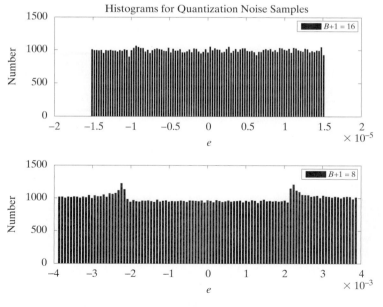

Figure 4.59   Histograms of quantization noise for (a) $B + 1 = 16$ and (b) $B + 1 = 8$.

with $X_m = 1$. Since the total number of samples was 101000, and the number of bins was 101, we should expect approximately 1000 samples in each bin if the noise is uniformly distributed. Furthermore the total range of samples should be $\pm 1/2^{16} = 1.53 \times 10^{-5}$ for 16-bit quantization and $\pm 1/2^8 = 3.9 \times 10^{-3}$ for 8-bit quantization. The histograms of Figure 4.59 are consistent with these values, although the 8-bit case shows some obvious deviation from the uniform distribution.

In Chapter 10, we show how to calculate estimates of the power density spectrum. Figure 4.60 shows such spectrum estimates for quantization noise signals where $B+1 = 16, 12, 8,$ and 4 bits. Observe that in this example, when the number of bits is 8 or greater, the spectrum is quite flat over the entire frequency range $0 \leq \omega \leq \pi$, and the spectrum level (in dB) is quite close to the value

$$10 \log_{10}(P_{ee}(e^{j\omega})) = 10 \log_{10}\left(\frac{1}{12(2^{2B})}\right) = -(10.79 + 6.02B),$$

which is predicted by the white-noise uniform-distribution model. Note that the curves for $B = 7, 11,$ and 15 differ at all frequencies by about 24 dB. Observe, however, that when $B + 1 = 4$, the model fails to predict the shape of the power spectrum of the noise.

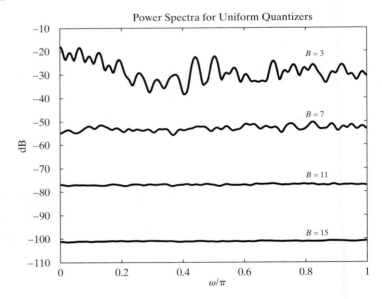

Figure 4.60   Spectra of quantization noise for several values of B.

This example demonstrates that the assumed model for quantization noise is useful in predicting the performance of uniform quantizers. A common measure of the amount of degradation of a signal by additive noise in general and quantization noise in particular is the signal-to-noise ratio (SNR), defined as the ratio of signal variance (power) to noise variance. Expressed in dB, the signal-to-quantization-noise ratio of a

$(B + 1)$-bit uniform quantizer is

$$\text{SNR}_Q = 10 \log_{10}\left(\frac{\sigma_x^2}{\sigma_e^2}\right) = 10 \log_{10}\left(\frac{12 \cdot 2^{2B}\sigma_x^2}{X_m^2}\right)$$

$$= 6.02B + 10.8 - 20 \log_{10}\left(\frac{X_m}{\sigma_x}\right).$$

(4.135)

From Eq. (4.135), we see that the SNR increases approximately 6 dB for each bit added to the word length of the quantized samples, i.e., for each doubling of the number of quantization levels. It is important to consider the term

$$-20 \log_{10}\left(\frac{X_m}{\sigma_x}\right)$$

(4.136)

in Eq. (4.135). First, recall that X_m is a parameter of the quantizer, and it would usually be fixed in a practical system. The quantity σ_x is the rms value of the signal amplitude, and it would necessarily be less than the peak amplitude of the signal. For example, if $x_a(t)$ is a sine wave of peak amplitude X_p, then $\sigma_x = X_p/\sqrt{2}$. If σ_x is too large, the peak signal amplitude will exceed the full-scale amplitude X_m of the A/D converter. In this case Eq. (4.135) is no longer valid, and severe distortion results. On the other hand, if σ_x is too small, then the term in Eq. (4.136) will become large and negative, thereby decreasing the SNR in Eq. (4.135). In fact, it is easily seen that when σ_x is halved, the SNR decreases by 6 dB. Thus, it is very important that the signal amplitude be carefully matched to the full-scale amplitude of the A/D converter.

Example 4.12   SNR for Sinusoidal Signal

Using the signal $x[n] = A\cos(n/10)$, we can compute the quantization error for different values of $B + 1$ with $X_m = 1$ and A varying. Figure 4.61 shows estimates of SNR as a function of X_m/σ_x obtained by computing the average power over many samples of the signal and dividing by the corresponding estimate of the average power of the noise; i.e.,

$$\text{SNR}_Q = 10 \log_{10}\left(\frac{\frac{1}{N}\sum_{n=0}^{N-1}(x[n])^2}{\frac{1}{N}\sum_{n=0}^{N-1}(e[n])^2}\right),$$

where in the case of Figure 4.61, $N = 101000$.

Observe that the curves in Figure 4.61 closely follow Eq. (4.135) over a wide range of values of B. In particular, the curves are straight lines as a function of $\log(X_m/\sigma_x)$, and they are offset from one another by 12 dB because the values of B differ by 2. SNR increases as X_m/σ_x decreases since increasing σ_x with X_m fixed

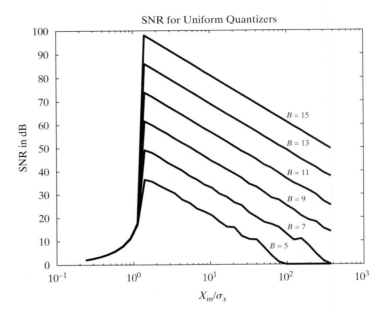

Figure 4.61   Signal-to-quantization-noise ratio as a function of X_m/σ_x for several values of B.

means that the signal uses more of the available quantization levels. However, note the precipitous fall of the curves as $X_m/\sigma_x \rightarrow 1$. Since $\sigma_x = .707A$ for a sine wave, this means that the amplitude A becomes greater than $X_m = 1$ and severe clipping occurs. Thus, the SNR decreases rapidly after the amplitude exceeds X_m.

For analog signals such as speech or music, the distribution of amplitudes tends to be concentrated about zero and falls off rapidly with increasing amplitude. In such cases, the probability that the magnitude of a sample will exceed three or four times the rms value is very low. For example, if the signal amplitude has a Gaussian distribution, only 0.064 percent of the samples would have an amplitude greater than $4\sigma_x$. Thus, to avoid clipping the peaks of the signal (as is assumed in our statistical model), we might set the gain of filters and amplifiers preceding the A/D converter so that $\sigma_x = X_m/4$. Using this value of σ_x in Eq. (4.135) gives

$$\text{SNR}_Q \approx 6B - 1.25 \text{ dB}. \tag{4.137}$$

For example, obtaining a SNR of about 90–96 dB for use in high-quality music recording and playback requires 16-bit quantization, but it should be remembered that such performance is obtained only if the input signal is carefully matched to the full-scale range of the A/D converter.

This trade-off between peak signal amplitude and absolute size of the quantization noise is fundamental to any quantization process. We will see its importance again in Chapter 6 when we discuss round-off noise in implementing discrete-time linear systems.

4.8.4 D/A Conversion

In Section 4.3, we discussed how a bandlimited signal can be reconstructed from a sequence of samples using ideal lowpass filtering. In terms of Fourier transforms, the reconstruction is represented as

$$X_r(j\Omega) = X(e^{j\Omega T})H_r(j\Omega), \tag{4.138}$$

where $X(e^{j\omega})$ is the DTFT of the sequence of samples and $X_r(j\Omega)$ is the Fourier transform of the reconstructed continuous-time signal. The ideal reconstruction filter is

$$H_r(j\Omega) = \begin{cases} T, & |\Omega| < \pi/T, \\ 0, & |\Omega| \geq \pi/T. \end{cases} \tag{4.139}$$

For this choice of $H_r(j\Omega)$, the corresponding relation between $x_r(t)$ and $x[n]$ is

$$x_r(t) = \sum_{n=-\infty}^{\infty} x[n]\frac{\sin[\pi(t-nT)/T]}{\pi(t-nT)/T}. \tag{4.140}$$

The system that takes the sequence $x[n]$ as input and produces $x_r(t)$ as output is called the *ideal D/C converter*. A physically realizable counterpart to the ideal D/C converter is a *digital-to-analog converter* (D/A converter) followed by an analog lowpass filter. As depicted in Figure 4.62, a D/A converter takes a sequence of binary code words $\hat{x}_B[n]$ as its input and produces a continuous-time output of the form

$$\begin{aligned} x_{DA}(t) &= \sum_{n=-\infty}^{\infty} X_m\hat{x}_B[n]h_0(t-nT) \\ &= \sum_{n=-\infty}^{\infty} \hat{x}[n]h_0(t-nT), \end{aligned} \tag{4.141}$$

where $h_0(t)$ is the impulse response of the zero-order hold given by Eq. (4.122). The dotted lines in Figure 4.55 show the output of a D/A converter for the quantized examples of the sine wave. Note that the D/A converter holds the quantized sample for one sample period in the same way that the sample-and-hold holds the unquantized input sample. If we use the additive-noise model to represent the effects of quantization, Eq. (4.141) becomes

$$x_{DA}(t) = \sum_{n=-\infty}^{\infty} x[n]h_0(t-nT) + \sum_{n=-\infty}^{\infty} e[n]h_0(t-nT). \tag{4.142}$$

To simplify our discussion, we define

$$x_0(t) = \sum_{n=-\infty}^{\infty} x[n]h_0(t-nT), \tag{4.143}$$

$$e_0(t) = \sum_{n=-\infty}^{\infty} e[n]h_0(t-nT), \tag{4.144}$$

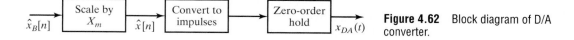

Figure 4.62   Block diagram of D/A converter.

so that Eq. (4.142) can be written as

$$x_{DA}(t) = x_0(t) + e_0(t). \tag{4.145}$$

The signal component $x_0(t)$ is related to the input signal $x_a(t)$, since $x[n] = x_a(nT)$. The noise signal $e_0(t)$ depends on the quantization-noise samples $e[n]$ in the same way that $x_0(t)$ depends on the unquantized signal samples. The Fourier transform of Eq. (4.143) is

$$
\begin{aligned}
X_0(j\Omega) &= \sum_{n=-\infty}^{\infty} x[n]H_0(j\Omega)e^{-j\Omega nT} \\
&= \left(\sum_{n=-\infty}^{\infty} x[n]e^{-j\Omega Tn} \right) H_0(j\Omega) \\
&= X(e^{j\Omega T})H_0(j\Omega).
\end{aligned} \tag{4.146}
$$

Now, since

$$X(e^{j\Omega T}) = \frac{1}{T} \sum_{k=-\infty}^{\infty} X_a\left(j\left(\Omega - \frac{2\pi k}{T}\right)\right), \tag{4.147}$$

it follows that

$$X_0(j\Omega) = \left[\frac{1}{T} \sum_{k=-\infty}^{\infty} X_a\left(j\left(\Omega - \frac{2\pi k}{T}\right)\right)\right] H_0(j\Omega). \tag{4.148}$$

If $X_a(j\Omega)$ is bandlimited to frequencies below π/T, the shifted copies of $X_a(j\Omega)$ do not overlap in Eq. (4.148), and if we define a compensated reconstruction filter as

$$\tilde{H}_r(j\Omega) = \frac{H_r(j\Omega)}{H_0(j\Omega)}, \tag{4.149}$$

then the output of the filter will be $x_a(t)$ if the input is $x_0(t)$. The frequency response of the zero-order-hold filter is easily shown to be

$$H_0(j\Omega) = \frac{2\sin(\Omega T/2)}{\Omega}e^{-j\Omega T/2}. \tag{4.150}$$

Therefore, the compensated reconstruction filter is

$$\tilde{H}_r(j\Omega) = \begin{cases} \frac{\Omega T/2}{\sin(\Omega T/2)}e^{j\Omega T/2}, & |\Omega| < \pi/T, \\ 0, & |\Omega| \geq \pi/T. \end{cases} \tag{4.151}$$

Figure 4.63(a) shows $|H_0(j\Omega)|$ as given by Eq. (4.150), compared with the magnitude of the ideal interpolation filter $|H_r(j\Omega)|$ as given by Eq. (4.139). Both filters have a gain of T at $\Omega = 0$, but the zero-order-hold, although lowpass in nature, does not cut off sharply at $\Omega = \pi/T$. Figure 4.63(b) shows the magnitude of the frequency response of the ideal compensated reconstruction filter to be used following a zero-order-hold reconstruction system such as a D/A converter. The phase response would ideally correspond to an advance time shift of $T/2$ seconds to compensate for the delay of that amount introduced by the zero-order hold. Since such a time advance cannot be realized in practical real-time approximations to the ideal compensated reconstruction

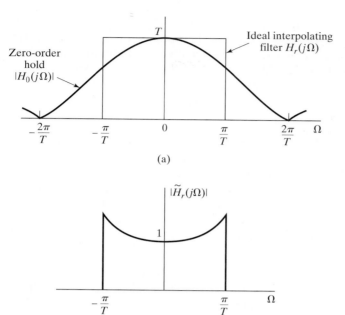

(a)

(b)

Figure 4.63   (a) Frequency response of zero-order hold compared with ideal interpolating filter. (b) Ideal compensated reconstruction filter for use with a zero-order-hold output.

filter, only the magnitude response would normally be compensated, and often even this compensation is neglected, since the gain of the zero-order hold drops only to $2/\pi$ (or -4 dB) at $\Omega = \pi/T$.

Figure 4.64 shows a D/A converter followed by an ideal compensated reconstruction filter. As can be seen from the preceding discussion, with the ideal compensated reconstruction filter following the D/A converter, the reconstructed output signal would be

$$\hat{x}_r(t) = \sum_{n=-\infty}^{\infty} \hat{x}[n]\frac{\sin[\pi(t-nT)/T]}{\pi(t-nT)/T}$$

$$= \sum_{n=-\infty}^{\infty} x[n]\frac{\sin[\pi(t-nT)/T]}{\pi(t-nT)/T} + \sum_{n=-\infty}^{\infty} e[n]\frac{\sin[\pi(t-nT)/T]}{\pi(t-nT)/T}.$$

(4.152)

In other words, the output would be

$$\hat{x}_r(t) = x_a(t) + e_a(t),$$

(4.153)

where $e_a(t)$ would be a bandlimited white-noise signal.

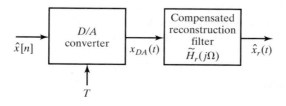

Figure 4.64   Physical configuration for D/A conversion.

Returning to a consideration of Figure 4.47(b), we are now in a position to understand the behavior of systems for digital processing of analog signals. If we assume that the output of the antialiasing filter is bandlimited to frequencies below π/T, that $\tilde{H}_r(j\Omega)$ is similarly bandlimited, and that the discrete-time system is linear and time invariant, then the output of the overall system will be of the form

$$\hat{y}_r(t) = y_a(t) + e_a(t), \tag{4.154}$$

where

$$T Y_a(j\Omega) = \tilde{H}_r(j\Omega)H_0(j\Omega)H(e^{j\Omega T})H_{aa}(j\Omega)X_c(j\Omega), \tag{4.155}$$

in which $H_{aa}(j\Omega)$, $H_0(j\Omega)$, and $\tilde{H}_r(j\Omega)$ are the frequency responses of the antialiasing filter, the zero-order hold of the D/A converter, and the reconstruction lowpass filter, respectively. $H(e^{j\Omega T})$ is the frequency response of the discrete-time system. Similarly, assuming that the quantization noise introduced by the A/D converter is white noise with variance $\sigma_e^2 = \Delta^2/12$, it can be shown that the power spectrum of the output noise is

$$P_{e_a}(j\Omega) = |\tilde{H}_r(j\Omega)H_0(j\Omega)H(e^{j\Omega T})|^2\sigma_e^2, \tag{4.156}$$

i.e., the input quantization noise is changed by the successive stages of discrete- and continuous-time filtering. From Eq. (4.155), it follows that, under the model for the quantization error and the assumption of negligible aliasing, the overall effective frequency response from $x_c(t)$ to $\hat{y}_r(t)$ is

$$T H_{\text{eff}}(j\Omega) = \tilde{H}_r(j\Omega)H_0(j\Omega)H(e^{j\Omega T})H_{aa}(j\Omega). \tag{4.157}$$

If the antialiasing filter is ideal, as in Eq. (4.118), and if the compensation of the reconstruction filter is ideal, as in Eq. (4.151), then the effective frequency response is as given in Eq. (4.119). Otherwise Eq. (4.157) provides a reasonable model for the effective response. Note that Eq. (4.157) suggests that compensation for imperfections in any of the four terms can, in principle, be included in any of the other terms; e.g., the discrete-time system can include appropriate compensation for the antialiasing filter or the zero-order hold or the reconstruction filter or all of these.

In addition to the filtering supplied by Eq. (4.157), Eq. (4.154) reminds us that the output will also be contaminated by the filtered quantization noise. In Chapter 6 we will see that noise can be introduced as well in the implementation of the discrete-time linear system. This internal noise will, in general, be filtered by parts of the discrete-time system implementation, by the zero-order hold of the D/A converter, and by the reconstruction filter.

4.9 OVERSAMPLING AND NOISE SHAPING IN A/D AND D/A CONVERSION

In Section 4.8.1, we showed that oversampling can make it possible to implement sharp-cutoff antialiasing filtering by incorporating digital filtering and decimation. As we discuss in Section 4.9.1, oversampling and subsequent discrete-time filtering and downsampling also permit an increase in the step size Δ of the quantizer or, equivalently, a reduction in the number of bits required in the A/D conversion. In Section 4.9.2 we show how the step size can be reduced even further by using oversampling together with quantization-noise feedback, and in Section 4.9.3 we show how the oversampling principle can be applied in D/A conversion.

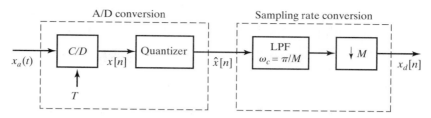

Figure 4.65 Oversampled A/D conversion with simple quantization and down-sampling.

4.9.1 Oversampled A/D Conversion with Direct Quantization

To explore the relation between oversampling and the quantization step size, we consider the system in Figure 4.65. To analyze the effect of oversampling in this system, we consider $x_a(t)$ to be a zero-mean, wide-sense-stationary, random process with power-spectral density denoted by $\Phi_{x_a x_a}(j\Omega)$ and autocorrelation function denoted by $\phi_{x_a x_a}(\tau)$. To simplify our discussion, we assume initially that $x_a(t)$ is already bandlimited to Ω_N, i.e.,

$$\Phi_{x_a x_a}(j\Omega) = 0, \qquad |\Omega| \geq \Omega_N, \qquad (4.158)$$

and we assume that $2\pi/T = 2M\Omega_N$. The constant M, which is assumed to be an integer, is called the *oversampling ratio*. Using the additive noise model discussed in detail in Section 4.8.3, we can replace Figure 4.65 by Figure 4.66. The decimation filter in Figure 4.66 is an ideal lowpass filter with unity gain and cutoff frequency $\omega_c = \pi/M$. Because the entire system of Figure 4.66 is linear, its output $x_d[n]$ has two components, one due to the signal input $x_a(t)$ and one due to the quantization noise input $e[n]$. We denote these components by $x_{da}[n]$ and $x_{de}[n]$, respectively.

Our goal is to determine the ratio of signal power $\mathcal{E}\{x_{da}^2[n]\}$ to quantization-noise power $\mathcal{E}\{x_{de}^2[n]\}$ in the output $x_d[n]$ as a function of the quantizer step size Δ and the oversampling ratio M. Since the system of Figure 4.66 is linear, and since the noise is assumed to be uncorrelated with the signal, we can treat the two sources separately in computing the respective powers of the signal and noise components at the output.

First, we will consider the signal component of the output. We begin by relating the power spectral density, autocorrelation function, and signal power of the sampled signal $x[n]$ to the corresponding functions for the continuous-time analog signal $x_a(t)$.

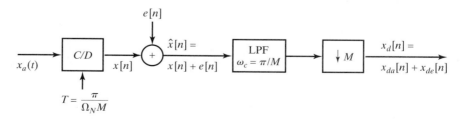

Figure 4.66 System of Figure 4.65 with quantizer replaced by linear noise model.

Let $\phi_{xx}[m]$ and $\Phi_{xx}(e^{j\omega})$ respectively denote the autocorrelation and power spectral density of $x[n]$. Then, by definition, $\phi_{xx}[m] = \mathcal{E}\{x[n+m]x[n]\}$, and since $x[n] = x_a(nT)$ and $x[n+m] = x_a(nT+mT)$,

$$\mathcal{E}\{x[n+m]x[n]\} = \mathcal{E}\{x_a((n+m)T)x_a(nT)\}. \tag{4.159}$$

Therefore,

$$\phi_{xx}[m] = \phi_{x_a x_a}(mT); \tag{4.160}$$

i.e., the autocorrelation function of the sequence of samples is a sampled version of the autocorrelation function of the corresponding continuous-time signal. In particular the wide-sense-stationarity assumption implies that $\mathcal{E}\{x_a^2(t)\}$ is a constant independent of t. It then follows that

$$\mathcal{E}\{x^2[n]\} = \mathcal{E}\{x_a^2(nT)\} = \mathcal{E}\{x_a^2(t)\} \qquad \text{for all } n \text{ or } t. \tag{4.161}$$

Since the power spectral densities are the Fourier transforms of the autocorrelation functions, as a consequence of Eq. (4.160),

$$\Phi_{xx}(e^{j\Omega T}) = \frac{1}{T}\sum_{k=-\infty}^{\infty}\Phi_{x_a x_a}\left[j\left(\Omega - \frac{2\pi k}{T}\right)\right]. \tag{4.162}$$

Assuming that the input is bandlimited as in Eq. (4.158), and assuming oversampling by a factor of M so that $2\pi/T = 2M\Omega_N$, we obtain, by substituting $\Omega = \omega/T$ into Eq. (4.162)

$$\Phi_{xx}(e^{j\omega}) = \begin{cases} \dfrac{1}{T}\Phi_{x_a x_a}\left(j\dfrac{\omega}{T}\right), & |\omega| < \pi/M, \\[2mm] 0, & \pi/M < \omega \le \pi. \end{cases} \tag{4.163}$$

For example, if $\Phi_{x_a x_a}(j\Omega)$ is as depicted in Figure 4.67(a), and if we choose the sampling rate to be $2\pi/T = 2M\Omega_N$, then $\Phi_{xx}(e^{j\omega})$ will be as depicted in Figure 4.67(b).

It is instructive to demonstrate that Eq. (4.161) is true by utilizing the power spectrum. The total power of the original analog signal is given by

$$\mathcal{E}\{x_a^2(t)\} = \frac{1}{2\pi}\int_{-\Omega_N}^{\Omega_N}\Phi_{x_a x_a}(j\Omega)d\Omega.$$

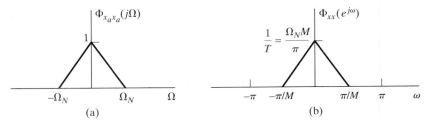

Figure 4.67   Illustration of frequency and amplitude scaling between $\Phi_{x_a x_a}(j\Omega)$ and $\Phi_{xx}(e^{j\omega})$.

From Eq. (4.163), the total power of the sampled signal is

$$\mathcal{E}\{x^2[n]\} = \frac{1}{2\pi} \int_{-\pi}^{\pi} \Phi_{xx}(e^{j\omega})d\omega \tag{4.164}$$

$$= \frac{1}{2\pi} \int_{-\pi/M}^{\pi/M} \frac{1}{T} \Phi_{x_a x_a}\left(j\frac{\omega}{T}\right) d\omega. \tag{4.165}$$

Using the fact that $\Omega_N T = \pi/M$ and making the substitution $\Omega = \omega/T$ in Eq. (4.165) gives

$$\mathcal{E}\{x^2[n]\} = \frac{1}{2\pi} \int_{-\Omega_N}^{\Omega_N} \Phi_{x_a x_a}(j\Omega)d\Omega = \mathcal{E}\{x_a^2(t)\}.$$

Thus, the total power of the sampled signal and the total power of the original analog signal are exactly the same as was also shown in Eq. (4.161). Since the decimation filter is an ideal lowpass filter with cutoff $\omega_c = \pi/M$, the signal $x[n]$ passes unaltered through the filter. Therefore, the downsampled signal component at the output, $x_{da}[n] = x[nM] = x_a(nMT)$, also has the same total power. This can be seen from the power spectrum by noting that, since $\Phi_{xx}(e^{j\omega})$ is bandlimited to $|\omega| < \pi/M$,

$$\Phi_{x_{da} x_{da}}(e^{j\omega}) = \frac{1}{M} \sum_{k=0}^{M-1} \Phi_{xx}(e^{j(\omega-2\pi k)/M})$$

$$= \frac{1}{M} \Phi_{xx}(e^{j\omega/M}) \qquad |\omega| < \pi. \tag{4.166}$$

Using Eq. (4.166), we obtain

$$\mathcal{E}\{x_{da}^2[n]\} = \frac{1}{2\pi} \int_{-\pi}^{\pi} \Phi_{x_{da} x_{da}}(e^{j\omega})d\omega$$

$$= \frac{1}{2\pi} \int_{-\pi}^{\pi} \frac{1}{M} \Phi_{xx}(e^{j\omega/M})d\omega$$

$$= \frac{1}{2\pi} \int_{-\pi/M}^{\pi/M} \Phi_{xx}(e^{j\omega}) d\omega = \mathcal{E}\{x^2[n]\},$$

which shows that the power of the signal component stays the same as it traverses the entire system from the input $x_a(t)$ to the corresponding output component $x_{da}[n]$. In terms of the power spectrum, this occurs because, for each scaling of the frequency axis that results from sampling, we have a counterbalancing inverse scaling of the amplitude, so that the area under the power spectrum remains the same as we go from $\Phi_{x_a x_a}(j\Omega)$ to $\Phi_{xx}(e^{j\omega})$ to $\Phi_{x_{da} x_{da}}(e^{j\omega})$ by sampling.

Now let us consider the noise component that is generated by quantization. According to the model in Section 4.8.3, we assume that $e[n]$ is a wide-sense-stationary white-noise process with zero mean and variance[5]

$$\sigma_e^2 = \frac{\Delta^2}{12}.$$

---

[5]Since the random process has zero mean, the average power and the variance are the same.

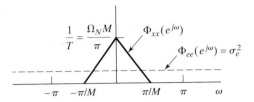

Figure 4.68 Power spectral density of signal and quantization noise with an oversampling factor of M.

Consequently, the autocorrelation function and power density spectrum for $e[n]$ are, respectively,

$$\phi_{ee}[m] = \sigma_e^2 \delta[m] \tag{4.167}$$

and

$$\Phi_{ee}(e^{j\omega}) = \sigma_e^2 \qquad |\omega| < \pi. \tag{4.168}$$

In Figure 4.68, we show the power density spectrum of $e[n]$ and of $x[n]$. The power density spectrum of the quantized signal $\hat{x}[n]$ is the sum of these, since the signal and quantization-noise samples are assumed to be uncorrelated in our model.

Although we have shown that the power in either $x[n]$ or $e[n]$ does not depend on M, we note that as the oversampling ratio M increases, less of the quantization-noise spectrum overlaps with the signal spectrum. It is this effect of the oversampling that lets us improve the signal-to-quantization-noise ratio by sampling-rate reduction. Specifically, the ideal lowpass filter removes the quantization noise in the band $\pi/M < |\omega| \leq \pi$, while it leaves the signal component unaltered. The noise power at the output of the ideal lowpass filter is

$$\mathcal{E}\{e^2[n]\} = \frac{1}{2\pi} \int_{-\pi/M}^{\pi/M} \sigma_e^2 d\omega = \frac{\sigma_e^2}{M}.$$

Next, the lowpass filtered signal is downsampled, and, as we have seen, the signal power in the downsampled output remains the same. In Figure 4.69, we show the resulting power density spectrum of both $x_{da}[n]$ and $x_{de}[n]$. Comparing Figures 4.68 and 4.69, we can see that the area under the power density spectrum for the signal has not changed, since the frequency axis and amplitude axis scaling have been inverses of each other. On the other hand, the noise power in the decimated output is the same as at the output of the lowpass filter; i.e.,

$$\mathcal{E}\{x_{de}^2[n]\} = \frac{1}{2\pi} \int_{-\pi}^{\pi} \frac{\sigma_e^2}{M} d\omega = \frac{\sigma_e^2}{M} = \frac{\Delta^2}{12M}. \tag{4.169}$$

Thus, the quantization-noise power $\mathcal{E}\{x_{de}^2[n]\}$ has been reduced by a factor of M through the filtering and downsampling, while the signal power has remained the same.

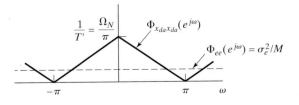

Figure 4.69 Power spectral density of signal and quantization noise after downsampling.

From Eq. (4.169), we see that for a given quantization noise power, there is a clear trade-off between the oversampling factor M and the quantizer step size Δ. Equation (4.125) states that for a quantizer with $(B+1)$ bits and maximum input signal level between plus and minus X_m, the step size is

$$\Delta = X_m/2^B,$$

and therefore,

$$\mathcal{E}\{x_{de}^2[n]\} = \frac{1}{12M}\left(\frac{X_m}{2^B}\right)^2. \tag{4.170}$$

Equation (4.170) shows that for a fixed quantizer, the noise power can be decreased by increasing the oversampling ratio M. Since the signal power is independent of M, increasing M will increase the signal-to-quantization-noise ratio. Alternatively, for a fixed quantization noise power $P_{de} = \mathcal{E}\{x_{de}^2[n]\}$, the required value for B is

$$B = -\frac{1}{2}\log_2 M - \frac{1}{2}\log_2 12 - \frac{1}{2}\log_2 P_{de} + \log_2 X_m. \tag{4.171}$$

From Eq. (4.171), we see that for every doubling of the oversampling ratio M, we need $1/2$ bit less to achieve a given signal-to-quantization-noise ratio, or, in other words, if we oversample by a factor $M = 4$, we need one less bit to achieve a desired accuracy in representing the signal.

4.9.2 Oversampled A/D Conversion with Noise Shaping

In the previous section, we showed that oversampling and decimation can improve the signal-to-quantization-noise ratio. This seems to be a somewhat remarkable result. It implies that we can, in principle, use very crude quantization in our initial sampling of the signal, and if the oversampling ratio is high enough, we can still obtain an accurate representation of the original samples by doing digital computation on the noisy samples. The problem with what we have seen so far is that, to make a significant reduction in the required number of bits, we need very large oversampling ratios. For example, to reduce the number of bits from 16 to 12 would require $M = 4^4 = 256$. This seems to be a rather high cost. However, the basic oversampling principle can lead to much higher gains if we combine it with the concept of noise spectrum shaping by feedback.

As was indicated in Figure 4.68, with direct quantization the power density spectrum of the quantization noise is constant over the entire frequency band. The basic concept in noise shaping is to modify the A/D conversion procedure so that the power density spectrum of the quantization noise is no longer uniform, but rather, is shaped such that most of the noise power is outside the band $|\omega| < \pi/M$. In that way, the subsequent filtering and downsampling removes more of the quantization-noise power.

The noise-shaping quantizer, generally referred to as a sampled-data Delta-Sigma modulator, is shown in Figure 4.70. (See Candy and Temes, 1992 and Schreier and Temes, 2005.) Figure 4.70(a) shows a block diagram of how the system is implemented with integrated circuits. The integrator is a switched-capacitor discrete-time integrator. The A/D converter can be implemented in many ways, but generally, it is a simple 1-bit quantizer or comparator. The D/A converter converts the digital output back to an analog pulse that is subtracted from the input signal at the input to the integrator. This system can

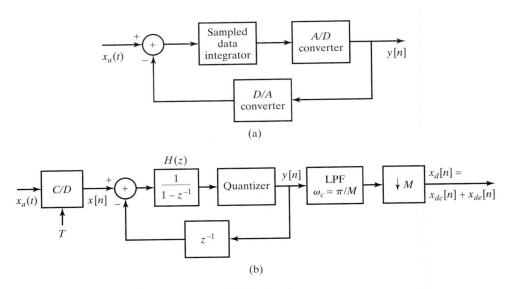

(a)

(b)

Figure 4.70   Oversampled quantizer with noise shaping.

be represented by the discrete-time equivalent system shown in Figure 4.70(b). The switched-capacitor integrator is represented by an accumulator system, and the delay in the feedback path represents the delay introduced by the D/A converter.

As before, we model the quantization error as an additive noise source so that the system in Figure 4.70 can be replaced by the linear model in Figure 4.71. In this system, the output $y[n]$ is the sum of two components: $y_x[n]$ due to the input $x[n]$ alone and $\hat{e}[n]$ due to the noise $e[n]$ alone.

We denote the transfer function from $x[n]$ to $y[n]$ as $H_x(z)$ and from $e[n]$ to $y[n]$ as $H_e(z)$. These transfer functions can both be calculated in a straightforward manner and are

$$H_x(z) = 1, \tag{4.172a}$$

$$H_e(z) = (1 - z^{-1}). \tag{4.172b}$$

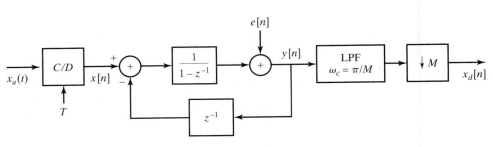

Figure 4.71   System of Figure 4.70 from $x_a(t)$ to $x_d[n]$ with quantizer replaced by a linear noise model.

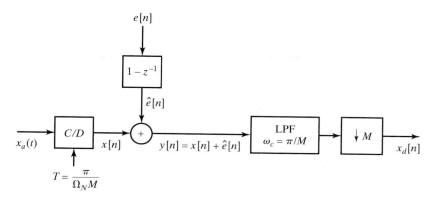

Figure 4.72   Equivalent representation of Figure 4.71.

Consequently,

$$y_x[n] = x[n], \tag{4.173a}$$

and

$$\hat{e}[n] = e[n] - e[n-1]. \tag{4.173b}$$

Therefore, the output $y[n]$ can be represented equivalently as $y[n] = x[n] + \hat{e}[n]$, where $x[n]$ appears unmodified at the output and the quantization noise $e[n]$ is modified by the first-difference operator $H_e(z)$. This is depicted in the block diagram in Figure 4.72. With the power density spectrum for $e[n]$ given by Eq. (4.168), the power density spectrum of the quantization noise $\hat{e}[n]$ that is present in $y[n]$ is

$$\begin{aligned}\Phi_{\hat{e}\hat{e}}(e^{j\omega}) &= \sigma_e^2 |H_e(e^{j\omega})|^2 \\ &= \sigma_e^2 [2\sin(\omega/2)]^2. \end{aligned} \tag{4.174}$$

In Figure 4.73, we show the power density spectrum of $\hat{e}[n]$, the power spectrum of $e[n]$, and the same signal power spectrum that was shown in Figure 4.67(b) and Figure 4.68. It is interesting to observe that the *total* noise power is increased from $\mathcal{E}\{e^2[n]\} = \sigma_e^2$

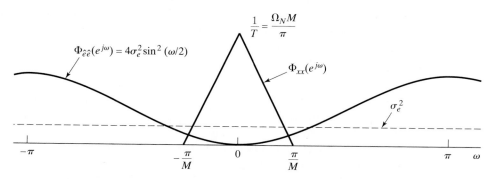

Figure 4.73   The power spectral density of the quantization noise and the signal.

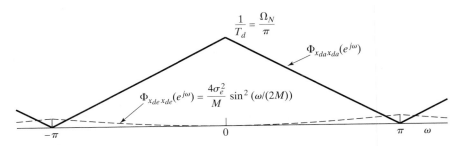

Figure 4.74   Power spectral density of the signal and quantization noise after downsampling.

at the quantizer to $\mathcal{E}\{\hat{e}^2[n]\} = 2\sigma_e^2$ at the output of the noise-shaping system. However, note that in comparison with Figure 4.68, the quantization noise has been shaped in such a way that more of the noise power is outside the signal band $|\omega| < \pi/M$ than in the direct oversampled case, where the noise spectrum is flat.

In the system of Figure 4.70, this out-of-band noise power is removed by the low-pass filter. Specifically, in Figure 4.74 we show the power density spectrum of $\Phi_{x_{da}x_{da}}(e^{j\omega})$ superimposed on the power density spectrum of $\Phi_{x_{de}x_{de}}(e^{j\omega})$. Since the downsampler does not remove any of the signal power, the signal power in $x_{da}[n]$ is

$$P_{da} = \mathcal{E}\{x_{da}^2[n]\} = \mathcal{E}\{x^2[n]\} = \mathcal{E}\{x_a^2(t)\}.$$

The quantization-noise power in the final output is

$$P_{de} = \frac{1}{2\pi} \int_{-\pi}^{\pi} \Phi_{x_{de}x_{de}}(e^{j\omega})d\omega = \frac{1}{2\pi} \frac{\Delta^2}{12M} \int_{-\pi}^{\pi} \left(2\sin\left(\frac{\omega}{2M}\right)\right)^2 d\omega. \qquad (4.175)$$

To compare this approximately with the results in Section 4.9.1, assume that M is sufficiently large so that

$$\sin\left(\frac{\omega}{2M}\right) \approx \frac{\omega}{2M}.$$

With this approximation, Eq. (4.175) is easily evaluated to obtain

$$P_{de} = \frac{1}{36} \frac{\Delta^2 \pi^2}{M^3}. \qquad (4.176)$$

From Eq. (4.176), we see again a trade-off between the oversampling ratio M and the quantizer step size Δ. For a $(B+1)$-bit quantizer and maximum input signal level between plus and minus X_m, $\Delta = X_m/2^B$. Therefore, to achieve a given quantization-noise power P_{de}, we must have

$$B = -\frac{3}{2}\log_2 M + \log_2(\pi/6) - \frac{1}{2}\log_2 P_{de} + \log_2 X_m. \qquad (4.177)$$

Comparing Eq. (4.177) with Eq. (4.171), we see that, whereas with direct quantization a doubling of the oversampling ratio M gained 1/2 bit in quantization, the use of noise shaping results in a gain of 1.5 bits.

Table 4.1 gives the equivalent savings in quantizer bits over direct quantization with no oversampling ($M=1$) for (a) direct quantization with oversampling, as discussed in Section 4.9.1, and (b) oversampling with noise shaping, as examined in this section.

TABLE 4.1   EQUIVALENT SAVINGS IN QUANTIZER BITS RELATIVE TO $M = 1$ FOR DIRECT QUANTIZATION AND 1st-ORDER NOISE SHAPING

M	Direct quantization	Noise shaping
4	1	2.2
8	1.5	3.7
16	2	5.1
32	2.5	6.6
64	3	8.1

The noise-shaping strategy in Figure 4.70 can be extended by incorporating a second stage of accumulation as shown in Figure 4.75. In this case, with the quantizer again modeled as an additive noise source $e[n]$, it can be shown that

$$y[n] = x[n] + \hat{e}[n]$$

where, in the two-stage case, $\hat{e}[n]$ is the result of processing the quantization noise $e[n]$ through the transfer function

$$H_e(z) = (1 - z^{-1})^2. \tag{4.178}$$

The corresponding power density spectrum of the quantization noise now present in $y[n]$ is

$$\Phi_{\hat{e}\hat{e}}(e^{j\omega}) = \sigma_e^2 [2\sin(\omega/2)]^4, \tag{4.179}$$

with the result that, although the total noise power at the output of the two-stage noise-shaping system is greater than for the one-stage case, even more of the noise lies outside the signal band. More generally, p stages of accumulation and feedback can be used, with corresponding noise shaping given by

$$\Phi_{\hat{e}\hat{e}}(e^{j\omega}) = \sigma_e^2 [2\sin(\omega/2)]^{2p}. \tag{4.180}$$

In Table 4.2, we show the equivalent reduction in quantizer bits as a function of the order p of the noise shaping and the oversampling ratio M. Note that with $p = 2$ and $M = 64$, we obtain almost 13 bits of increase in accuracy, suggesting that a 1-bit quantizer could achieve about 14-bit accuracy at the output of the decimator.

Although multiple feedback loops such as the one shown in Figure 4.75 promise greatly increased noise reduction, they are not without problems. Specifically, for large values of p, there is an increased potential for instability and oscillations to occur. An alternative structure known as multistage noise shaping (MASH) is considered in Problem 4.68.

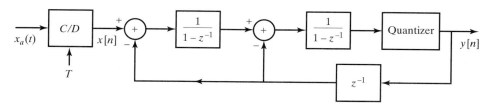

Figure 4.75   Oversampled quantizer with 2nd-order noise shaping.

TABLE 4.2   REDUCTION IN QUANTIZER
BITS AS ORDER p OF NOISE SHAPING

Quantizer order p	Oversampling factor M				
	4	8	16	32	64
0	1.0	1.5	2.0	2.5	3.0
1	2.2	3.7	5.1	6.6	8.1
2	2.9	5.4	7.9	10.4	12.9
3	3.5	7.0	10.5	14.0	17.5
4	4.1	8.5	13.0	17.5	22.0
5	4.6	10.0	15.5	21.0	26.5

4.9.3 Oversampling and Noise Shaping in D/A Conversion

In Sections 4.9.1 and 4.9.2, we discussed the use of oversampling to simplify the process of A/D conversion. As we mentioned, the signal is initially oversampled to simplify antialias filtering and improve accuracy, but the final output $x_d[n]$ of the A/D converter is sampled at the Nyquist rate for $x_a(t)$. The minimum sampling rate is, of course, highly desirable for digital processing or for simply representing the analog signal in digital form, as in the CD audio recording system. It is natural to apply the same principles in reverse to achieve improvements in the D/A conversion process.

The basic system, which is the counterpart to Figure 4.65, is shown in Figure 4.76. The sequence $y_d[n]$, which is to be converted to a continuous-time signal, is first up-sampled to produce the sequence $\hat{y}[n]$, which is then requantized before sending it to a D/A converter that accepts binary samples with the number of bits produced by the requantization process. We can use a simple D/A converter with few bits if we can be assured that the quantization noise does not occupy the signal band. Then the noise can be removed by inexpensive analog filtering.

In Figure 4.77, we show a structure for the quantizer that shapes the quantization noise in a similar manner to the 1st-order noise shaping provided by the system in Figure 4.70. In our analysis we assume that $y_d[n]$ is effectively unquantized or so finely quantized relative to $y[n]$ that the primary source of quantizer error is the quantizer in Figure 4.76. To analyze the system in Figures 4.76 and 4.77, we replace the quantizer in Figure 4.77 by an additive white-noise source $e[n]$, so that Figure 4.77 is replaced by Figure 4.78. The transfer function from $\hat{y}[n]$ to $y[n]$ is unity, i.e., the upsampled signal

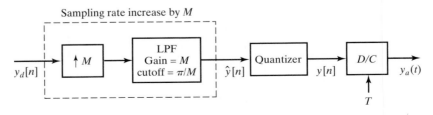

Figure 4.76   Oversampled D/A conversion.

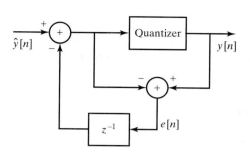

Figure 4.77   1st-order noise-shaping system for oversampled D/A quantization.

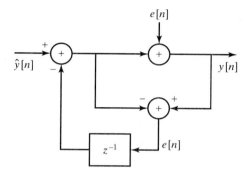

Figure 4.78   System of Figure 4.77 with quantizer replaced by linear noise model.

$\hat{y}[n]$ appears at the output unaltered. The transfer function $H_e(z)$ from $e[n]$ to $y[n]$ is

$$H_e(z) = 1 - z^{-1}.$$

Therefore, the quantization noise component $\hat{e}[n]$ that appears at the output of the noise-shaping system in Figure 4.78 has the power density spectrum

$$\Phi_{\hat{e}\hat{e}}(e^{j\omega}) = \sigma_e^2 (2\sin\omega/2)^2, \tag{4.181}$$

where, again, $\sigma_e^2 = \Delta^2/12$.

An illustration of this approach to D/A conversion is given in Figure 4.79. Figure 4.79(a) shows the power spectrum $\Phi_{y_d y_d}(e^{j\omega})$ of the input $y_d[n]$ in Figure 4.76. Note that we assume that the signal $y_d[n]$ is sampled at the Nyquist rate. Figure 4.79(b) shows the corresponding power spectrum at the output of the upsampler (by M), and Figure 4.79(c) shows the quantization noise spectrum at the output of the quantizer/noise-shaper system. Finally, Figure 4.79(d) shows the power spectrum of the signal component superimposed on the power spectrum of the noise component at the analog output of the D/C converter of Figure 4.76. In this case, we assume that the D/C converter has an ideal lowpass reconstruction filter with cutoff frequency $\pi/(MT)$, which will remove as much of the quantization noise as possible.

In a practical setting, we would like to avoid sharp-cutoff analog reconstruction filters. From Figure 4.79(d), it is clear that if we can tolerate somewhat more quantization noise, then the D/C reconstruction filter need not roll off so sharply. Furthermore, if we use multistage techniques in the noise shaping, we can obtain an output noise spectrum of the form

$$\Phi_{\hat{e}\hat{e}}(e^{j\omega}) = \sigma_e^2 (2\sin\omega/2)^{2p},$$

which would push more of the noise to higher frequencies. In this case, the analog reconstruction filter specifications could be relaxed even further.

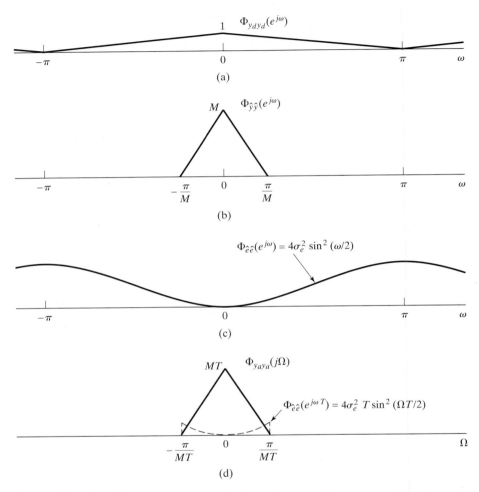

Figure 4.79   (a) Power spectral density of signal $y_d[n]$. (b) Power spectral density of signal $\hat{y}[n]$. (c) Power spectral density of quantization noise. (d) Power spectral density of the continuous-time signal and the quantization noise.

4.10 SUMMARY

In this chapter, we developed and explored the relationship between continuous-time signals and the discrete-time sequences obtained by periodic sampling. The fundamental theorem that allows the continuous-time signal to be represented by a sequence of samples is the Nyquist-Shannon theorem, which states that, for a bandlimited signal, periodic samples are a sufficient representation, as long as the sampling rate is sufficiently high relative to the highest frequency in the continuous-time signal. Under this condition, the continuous-time signal can be reconstructed by lowpass filtering from knowledge of only the original bandwidth, the sampling rate and the samples. This corresponds to bandlimited interpolation. If the sampling rate is too low relative to the bandwidth of the signal, then aliasing distortion occurs and the original signal cannot be reconstructed by bandlimited interpolation.

The ability to represent signals by sampling permits the discrete-time processing of continuous-time signals. This is accomplished by first sampling, then applying discrete-time processing, and, finally, reconstructing a continuous-time signal from the result. Examples given were lowpass filtering and differentiation.

Sampling rate changes are an important class of digital signal processing operations. Downsampling a discrete-time signal corresponds in the frequency domain to an amplitude-scaled replication of the discrete-time spectrum and rescaling of the frequency axis, which may require additional bandlimiting to avoid aliasing. Upsampling corresponds to effectively increasing the sampling rate and is also represented in the frequency domain by a rescaling of the frequency axis. By combining upsampling and downsampling by integer amounts, noninteger sampling rate conversion can be achieved. We also showed how this can be efficiently done using multirate techniques.

In the final sections of the chapter, we explored a number of practical considerations associated with the discrete-time processing of continuous-time signals, including the use of prefiltering to avoid aliasing, quantization error in A/D conversion, and some issues associated with the filtering used in sampling and reconstructing the continuous-time signals. Finally, we showed how discrete-time decimation and interpolation and noise shaping can be used to simplify the analog side of A/D and D/A conversion.

The focus of this chapter has been on periodic sampling as a process for obtaining a discrete representation of a continuous-time signal. While such representations are by far the most common and are the basis for almost all of the topics to be discussed in the remainder of this text, there are other approaches to obtaining discrete representations that may lead to more compact representations for signals where other information (besides bandwidth) is known about the signal. Some examples can be found in Unser (2000).

Problems

Basic Problems with Answers

4.1. The signal

$$x_c(t) = \sin(2\pi(100)t)$$

was sampled with sampling period $T = 1/400$ second to obtain a discrete-time signal $x[n]$. What is the resulting sequence $x[n]$?

4.2. The sequence

$$x[n] = \cos\left(\frac{\pi}{4}n\right), \qquad -\infty < n < \infty,$$

was obtained by sampling the continuous-time signal

$$x_c(t) = \cos(\Omega_0 t), \qquad -\infty < t < \infty,$$

at a sampling rate of 1000 samples/s. What are two possible positive values of Ω_0 that could have resulted in the sequence $x[n]$?

4.3. The continuous-time signal
$$x_c(t) = \cos(4000\pi t)$$
is sampled with a sampling period T to obtain the discrete-time signal
$$x[n] = \cos\left(\frac{\pi n}{3}\right).$$

(a) Determine a choice for T consistent with this information.
(b) Is your choice for T in part (a) unique? If so, explain why. If not, specify another choice of T consistent with the information given.

4.4. The continuous-time signal
$$x_c(t) = \sin(20\pi t) + \cos(40\pi t)$$
is sampled with a sampling period T to obtain the discrete-time signal
$$x[n] = \sin\left(\frac{\pi n}{5}\right) + \cos\left(\frac{2\pi n}{5}\right).$$

(a) Determine a choice for T consistent with this information.
(b) Is your choice for T in part (a) unique? If so, explain why. If not, specify another choice of T consistent with the information given.

4.5. Consider the system of Figure 4.10, with the discrete-time system an ideal lowpass filter with cutoff frequency $\pi/8$ radians/s.
(a) If $x_c(t)$ is bandlimited to 5 kHz, what is the maximum value of T that will avoid aliasing in the C/D converter?
(b) If $1/T = 10$ kHz, what will the cutoff frequency of the effective continuous-time filter be?
(c) Repeat part (b) for $1/T = 20$ kHz.

4.6. Let $h_c(t)$ denote the impulse response of an LTI continuous-time filter and $h_d[n]$ the impulse response of an LTI discrete-time filter.
(a) If
$$h_c(t) = \begin{cases} e^{-at}, & t \geq 0, \\ 0, & t < 0, \end{cases}$$
where a is a positive real constant, determine the continuous-time filter frequency response and sketch its magnitude.
(b) If $h_d[n] = Th_c(nT)$ with $h_c(t)$ as in part (a), determine the discrete-time filter frequency response and sketch its magnitude.
(c) For a given value of a, determine, as a function of T, the minimum magnitude of the discrete-time filter frequency response.

4.7. A simple model of a multipath communication channel is indicated in Figure P4.7-1. Assume that $s_c(t)$ is bandlimited such that $S_c(j\Omega) = 0$ for $|\Omega| \geq \pi/T$ and that $x_c(t)$ is sampled with a sampling period T to obtain the sequence
$$x[n] = x_c(nT).$$

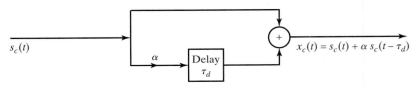

Figure P4.7-1

(a) Determine the Fourier transform of $x_c(t)$ and the Fourier transform of $x[n]$ in terms of $S_c(j\Omega)$.

(b) We want to simulate the multipath system with a discrete-time system by choosing $H(e^{j\omega})$ in Figure P4.7-2 so that the output $r[n] = x_c(nT)$ when the input is $s[n] = s_c(nT)$. Determine $H(e^{j\omega})$ in terms of T and τ_d.

(c) Determine the impulse response $h[n]$ in Figure P4.7 when (i) $\tau_d = T$ and (ii) $\tau_d = T/2$.

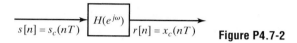

$$s[n] = s_c(nT) \quad\boxed{H(e^{j\omega})}\quad r[n] = x_c(nT)$$ **Figure P4.7-2**

4.8. Consider the system in Figure P4.8 with the following relations:

$$X_c(j\Omega) = 0, \qquad |\Omega| \geq 2\pi \times 10^4,$$

$$x[n] = x_c(nT),$$

$$y[n] = T \sum_{k=-\infty}^{n} x[k].$$

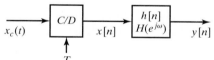

Figure P4.8

(a) For this system, what is the maximum allowable value of T if aliasing is to be avoided, i.e., so that $x_c(t)$ can be recovered from $x[n]$.

(b) Determine $h[n]$.

(c) In terms of $X(e^{j\omega})$, what is the value of $y[n]$ for $n \to \infty$?

(d) Determine whether there is any value of T for which

$$y[n]\bigg|_{n=\infty} = \int_{-\infty}^{\infty} x_c(t)dt. \qquad (P4.8-1)$$

If there is such a value for T, determine the maximum value. If there is not, explain and specify how T would be chosen so that the equality in Eq. (P4.8-1) is best approximated.

4.9. Consider a stable discrete-time signal $x[n]$ whose discrete-time Fourier transform $X(e^{j\omega})$ satisfies the equation

$$X(e^{j\omega}) = X\left(e^{j(\omega-\pi)}\right)$$

and has even symmetry, i.e., $x[n] = x[-n]$.

(a) Show that $X(e^{j\omega})$ is periodic with a period π.

(b) Find the value of $x[3]$. (*Hint:* Find values for all odd-indexed points.)

(c) Let $y[n]$ be the decimated version of $x[n]$, i.e., $y[n] = x[2n]$. Can you reconstruct $x[n]$ from $y[n]$ for all n. If yes, how? If no, justify your answer.

4.10. Each of the following continuous-time signals is used as the input $x_c(t)$ for an ideal C/D converter as shown in Figure 4.1 with the sampling period T specified. In each case, find the resulting discrete-time signal $x[n]$.

(a) $x_c(t) = \cos(2\pi(1000)t)$, $T = (1/3000)$ sec
(b) $x_c(t) = \sin(2\pi(1000)t)$, $T = (1/1500)$ sec
(c) $x_c(t) = \sin(2\pi(1000)t)/(\pi t)$, $T = (1/5000)$ sec

4.11. The following continuous-time input signals $x_c(t)$ and corresponding discrete-time output signals $x[n]$ are those of an ideal C/D as shown in Figure 4.1. Specify a choice for the sampling period T that is consistent with each pair of $x_c(t)$ and $x[n]$. In addition, indicate whether your choice of T is unique. If not, specify a second possible choice of T consistent with the information given.

(a) $x_c(t) = \sin(10\pi t)$, $x[n] = \sin(\pi n/4)$
(b) $x_c(t) = \sin(10\pi t)/(10\pi t)$, $x[n] = \sin(\pi n/2)/(\pi n/2)$.

4.12. In the system of Figure 4.10, assume that

$$H(e^{j\omega}) = j\omega/T,   -\pi \le \omega < \pi,$$

and $T = 1/10$ sec.

(a) For each of the following inputs $x_c(t)$, find the corresponding output $y_c(t)$.

(i) $x_c(t) = \cos(6\pi t)$.
(ii) $x_c(t) = \cos(14\pi t)$.

(b) Are the outputs $y_c(t)$ those you would expect from a differentiator?

4.13. In the system shown in Figure 4.15, $h_c(t) = \delta(t - T/2)$.

(a) Suppose the input $x[n] = \sin(\pi n/2)$ and $T = 10$. Find $y[n]$.
(b) Suppose you use the same $x[n]$ as in part (a), but halve T to be 5. Find the resulting $y[n]$.
(c) In general, how does the continuous-time LTI system $h_c(t)$ limit the range of the sampling period T that can be used without changing $y[n]$?

4.14. Which of the following signals can be downsampled by a factor of 2 using the system in Figure 4.19 without any loss of information?

(a) $x[n] = \delta[n - n_0]$, for n_0 some unknown integer
(b) $x[n] = \cos(\pi n/4)$
(c) $x[n] = \cos(\pi n/4) + \cos(3\pi n/4)$
(d) $x[n] = \sin(\pi n/3)/(\pi n/3)$
(e) $x[n] = (-1)^n \sin(\pi n/3)/(\pi n/3)$.

4.15. Consider the system shown in Figure P4.15. For each of the following input signals $x[n]$, indicate whether the output $x_r[n] = x[n]$.

(a) $x[n] = \cos(\pi n/4)$
(b) $x[n] = \cos(\pi n/2)$
(c)

$$x[n] = \left[\frac{\sin(\pi n/8)}{\pi n}\right]^2$$

Hint: Use the modulation property of the Fourier transform to find $X(e^{j\omega})$.

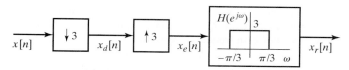

Figure P4.15

4.16. Consider the system in Figure 4.29. The input $x[n]$ and corresponding output $\tilde{x}_d[n]$ are given for a specific choice of M/L in each of the following parts. Determine a choice for M/L based on the information given, and specify whether your choice is unique.

   (a) $x[n] = \sin\left(\pi n/3\right)/\left(\pi n/3\right)$, $\tilde{x}_d[n] = \sin\left(5\pi n/6\right)/\left(5\pi n/6\right)$
   (b) $x[n] = \cos\left(3\pi n/4\right)$, $\tilde{x}_d[n] = \cos(\pi n/2)$.

4.17. Each of the following parts lists an input signal $x[n]$ and the upsampling and downsampling rates L and M for the system in Figure 4.29. Determine the corresponding output $\tilde{x}_d[n]$.

   (a) $x[n] = \sin(2\pi n/3)/\pi n$, $L = 4$, $M = 3$
   (b) $x[n] = \sin(3\pi n/4)$, $L = 6$, $M = 7$.

4.18. For the system shown in Figure 4.29, $X\left(e^{j\omega}\right)$, the Fourier transform of the input signal $x[n]$, is shown in Figure P4.18. For each of the following choices of L and M, specify the maximum possible value of ω_0 such that $\tilde{X}_d(e^{j\omega}) = aX\left(e^{jM\omega/L}\right)$ for some constant a.

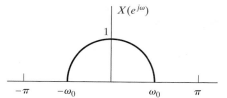

Figure P4.18

   (a) $M = 3$, $L = 2$
   (b) $M = 5$, $L = 3$
   (c) $M = 2$, $L = 3$.

4.19. The continuous-time signal $x_c(t)$ with the Fourier transform $X_c(j\Omega)$ shown in Figure P4.19-1 is passed through the system shown in Figure P4.19-2. Determine the range of values for T for which $x_r(t) = x_c(t)$.

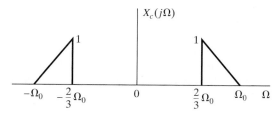

Figure P4.19-1

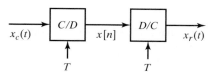

Figure P4.19-2

4.20. Consider the system in Figure 4.10. The input signal $x_c(t)$ has the Fourier transform shown in Figure P4.20 with $\Omega_0 = 2\pi(1000)$ radians/second. The discrete-time system is an ideal lowpass filter with frequency response

$$H(e^{j\omega}) = \begin{cases} 1, & |\omega| < \omega_c, \\ 0, & \text{otherwise.} \end{cases}$$

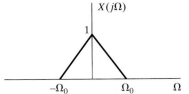

Figure P4.20

(a) What is the minimum sampling rate $F_s = 1/T$ such that no aliasing occurs in sampling the input?

(b) If $\omega_c = \pi/2$, what is the minimum sampling rate such that $y_r(t) = x_c(t)$?

Basic Problems

4.21. Consider a continuous-time signal $x_c(t)$ with Fourier transform $X_c(j\Omega)$ shown in Figure P4.21-1.

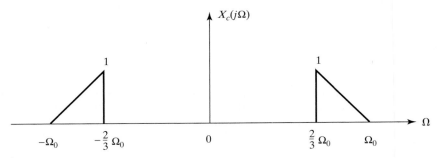

Figure P4.21-1   Fourier transform $X_c(j\Omega)$

(a) A continuous-time signal $x_r(t)$ is obtained through the process shown in Figure P4.21-2. First, $x_c(t)$ is multiplied by an impulse train of period T_1 to produce the waveform $x_s(t)$, i.e.,

$$x_s(t) = \sum_{n=-\infty}^{+\infty} x[n]\delta(t - nT_1).$$

Next, $x_s(t)$ is passed through a low pass filter with frequency response $H_r(j\Omega)$. $H_r(j\Omega)$ is shown in Figure P4.21-3.

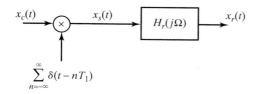

Figure P4.21-2   Conversion system for part (a)

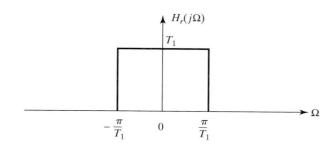

Figure P4.21-3   Frequency response $H_r(j\Omega)$

Determine the range of values for T_1 for which $x_r(t) = x_c(t)$.

(b) Consider the system in Figure P4.21-4. The system in this case is the same as the one in part (a), except that the sampling period is now T_2. The system $H_s(j\Omega)$ is some continuous-time ideal LTI filter. We want $x_o(t)$ to be equal to $x_c(t)$ for all t, i.e., $x_o(t) = x_c(t)$ for some choice of $H_s(j\Omega)$. Find all values of T_2 for which $x_o(t) = x_c(t)$ is possible. For the largest T_2 you determined that would still allow recovery of $x_c(t)$, choose $H_s(j\Omega)$ so that $x_o(t) = x_c(t)$. Sketch $H_s(j\Omega)$.

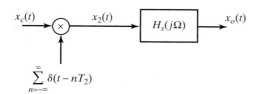

Figure P4.21-4   Conversion system for part (b)

4.22. Suppose that the bandlimited differentiator of Example 4.4 has input $x_c(t) = \cos(\Omega_0 t)$ with $\Omega_0 < \pi/T$. In this problem, we wish to verify that the continuous-time signal reconstructed from the output of the bandlimited differentiator is indeed the derivative of $x_c(t)$.

(a) The sampled input will be $x[n] = \cos(\omega_0 n)$, where $\omega_0 = \Omega_0 T < \pi$. Determine an expression for $X(e^{j\omega})$ that is valid for $|\omega| \leq \pi$.

(b) Now use Eq. (4.46) to determine the DTFT of $Y(e^{j\omega})$, the output of the discrete-time system.

(c) From Eq. (4.32) determine $Y_r(j\Omega)$, the continuous-time Fourier transform of the output of the D/C converter.

(d) Use the result of (c) to show that

$$y_r(t) = -\Omega_0 \sin(\Omega_0 t) = \frac{d}{dt}\left[x_c(t)\right].$$

4.23. Figure P4.23-1 shows a continuous-time filter that is implemented using an LTI discrete-time filter ideal lowpass filter with frequency response over $-\pi \leq \omega \leq \pi$ as

$$H(e^{j\omega}) = \begin{cases} 1 & |\omega| < \omega_c \\ 0 & \omega_c < |\omega| \leq \pi. \end{cases}$$

(a) If the continuous-time Fourier transform of $x_c(t)$, namely $X_c(j\Omega)$, is as shown in Figure P4.23-2 and $\omega_c = \frac{\pi}{5}$, sketch and label $X(e^{j\omega})$, $Y(e^{j\omega})$ and $Y_c(j\Omega)$ for each of the following cases:

　(i) $1/T_1 = 1/T_2 = 2 \times 10^4$
　(ii) $1/T_1 = 4 \times 10^4$, $1/T_2 = 10^4$
　(iii) $1/T_1 = 10^4$, $1/T_2 = 3 \times 10^4$.

(b) For $1/T_1 = 1/T_2 = 6 \times 10^3$, and for input signals $x_c(t)$ whose spectra are bandlimited to $|\Omega| < 2\pi \times 5 \times 10^3$ (but otherwise unconstrained), what is the maximum choice of the cutoff frequency ω_c of the filter $H(e^{j\omega})$ for which the overall system is LTI? For this maximum choice of ω_c, specify $H_c(j\Omega)$.

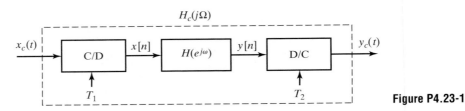

Figure P4.23-1

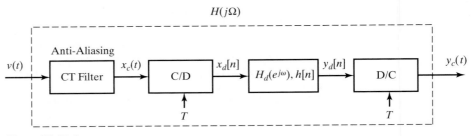

Figure P4.23-2

4.24. Consider the system shown in Figure P4.24-1.

$H(j\Omega)$

Figure P4.24-1

The anti-aliasing filter is a continuous-time filter with the frequency response $L(j\Omega)$ shown in Figure P4.24-2.

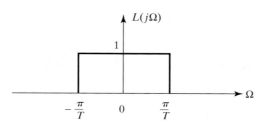

Figure P4.24-2

The frequency response of the LTI discrete-time system between the converters is given by:

$$H_d(e^{j\omega}) = e^{-j\frac{\omega}{3}}, \quad |\omega| < \pi$$

(a) What is the effective continuous-time frequency response of the overall system, $H(j\Omega)$?

(b) Choose the most accurate statement:

(i) $y_c(t) = \frac{d}{dt}x_c(3t)$.

(ii) $y_c(t) = x_c(t - \frac{T}{3})$.

(iii) $y_c(t) = \frac{d}{dt}x_c(t - 3T)$.

(iv) $y_c(t) = x_c(t - \frac{1}{3})$.

(a) Express $y_d[n]$ in terms of $y_c(t)$.

(b) Determine the impulse response $h[n]$ of the discrete-time LTI system.

4.25. Two bandlimited signals, $x_1(t)$ and $x_2(t)$, are multiplied, producing the product signal $w(t) = x_1(t)x_2(t)$. This signal is sampled by a periodic impulse train yielding the signal

$$w_p(t) = w(t) \sum_{n=-\infty}^{\infty} \delta(t - nT) = \sum_{n=-\infty}^{\infty} w(nT)\delta(t - nT).$$

Assume that $x_1(t)$ is bandlimited to Ω_1, and $x_2(t)$ is bandlimited to Ω_2; that is,

$$X_1(j\Omega) = 0, \quad |\Omega| \geq \Omega_1$$

$$X_2(j\Omega) = 0, \quad |\Omega| \geq \Omega_2.$$

Determine the *maximum* sampling interval T such that $w(t)$ is recoverable from $w_p(t)$ through the use of an ideal lowpass filter.

4.26. The system of Figure P4.26 is to be used to filter continuous time music signals using a sampling rate of 16kHz.

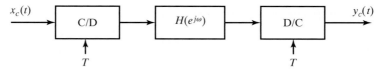

Figure P4.26

$H(e^{j\omega})$ is an ideal lowpass filter with a cutoff of $\pi/2$. If the input has been bandlimited such that $X_c(j\Omega) = 0$ for $|\Omega| > \Omega_c$, how should Ω_c be chosen so that the overall system in Figure P4.26 is LTI?

4.27. The system shown in Figure P4.27 is intended to approximate a differentiator for bandlimited continuous-time input waveforms.

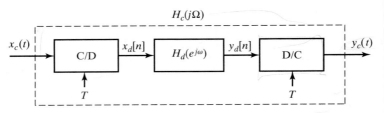

$H_c(j\Omega)$

$x_c(t)$ ──▶ C/D ──$x_d[n]$──▶ $H_d(e^{j\omega})$ ──$y_d[n]$──▶ D/C ──▶ $y_c(t)$

T T

Figure P4.27

• The continuous-time input signal $x_c(t)$ is bandlimited to $|\Omega| < \Omega_M$.
• The C/D converter has sampling rate $T = \dfrac{\pi}{\Omega_M}$, and produces the signal
  $x_d[n] = x_c(nT)$.
• The discrete-time filter has frequency response

$$H_d(e^{j\omega}) = \frac{e^{j\omega/2} - e^{-j\omega/2}}{T}, \qquad |\omega| \le \pi.$$

• The ideal D/C converter is such that $y_d[n] = y_c(nT)$.

(a) Find the continuous-time frequency response $H_c(j\Omega)$ of the end-to-end system.
(b) Find $x_d[n]$, $y_c(t)$, and $y_d[n]$, when the input signal is

$$x_c(t) = \frac{\sin(\Omega_M t)}{\Omega_M t}.$$

4.28. Consider the representation of the process of sampling followed by reconstruction shown in Figure P4.28.

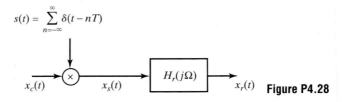

$$s(t) = \sum_{n=-\infty}^{\infty} \delta(t - nT)$$

$x_c(t)$ ──▶ ⊗ ──$x_s(t)$──▶ $H_r(j\Omega)$ ──▶ $x_r(t)$ **Figure P4.28**

Assume that the input signal is

$$x_c(t) = 2\cos(100\pi t - \pi/4) + \cos(300\pi t + \pi/3) \qquad -\infty < t < \infty$$

The frequency response of the reconstruction filter is

$$H_r(j\Omega) = \begin{cases} T & |\Omega| \le \pi/T \\ 0 & |\Omega| > \pi/T \end{cases}$$

(a) Determine the continuous-time Fourier transform $X_c(j\Omega)$ and plot it as a function of Ω.
(b) Assume that $f_s = 1/T = 500$ samples/sec and plot the Fourier transform $X_s(j\Omega)$ as a function of Ω for $-2\pi/T \le \Omega \le 2\pi/T$. What is the output $x_r(t)$ in this case? (You should be able to give an exact equation for $x_r(t)$.)
(c) Now, assume that $f_s = 1/T = 250$ samples/sec. Repeat part (b) for this condition.

(d) Is it possible to choose the sampling rate so that

$$x_r(t) = A + 2\cos(100\pi t - \pi/4)$$

where A is a constant? If so, what is the sampling rate $f_s = 1/T$, and what is the numerical value of A?

4.29. In Figure P4.29, assume that $X_c(j\Omega) = 0$, $|\Omega| \geq \pi/T_1$. For the general case in which $T_1 \neq T_2$ in the system, express $y_c(t)$ in terms of $x_c(t)$. Is the basic relationship different for $T_1 > T_2$ and $T_1 < T_2$?

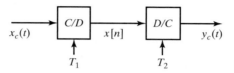

Figure P4.29

4.30. In the system of Figure P4.30, $X_c(j\Omega)$ and $H(e^{j\omega})$ are as shown. Sketch and label the Fourier transform of $y_c(t)$ for each of the following cases:

(a) $1/T_1 = 1/T_2 = 10^4$
(b) $1/T_1 = 1/T_2 = 2 \times 10^4$
(c) $1/T_1 = 2 \times 10^4$, $\quad 1/T_2 = 10^4$
(d) $1/T_1 = 10^4$, $\quad 1/T_2 = 2 \times 10^4$.

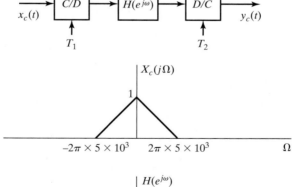

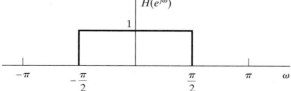

Figure P4.30

4.31. Figure P4.31-1 shows the overall system for filtering a continuous-time signal using a discrete-time filter. The frequency responses of the reconstruction filter $H_r(j\Omega)$ and the discrete-time filter $H(e^{j\omega})$ are shown in Figure P4.31-2.

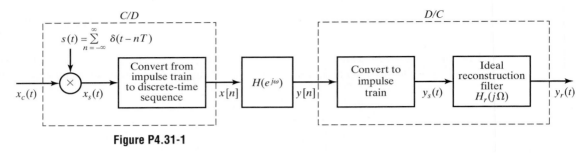

Figure P4.31-1

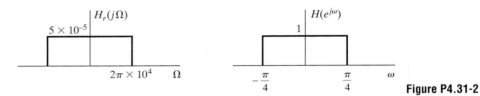

Figure P4.31-2

(a) For $X_c(j\Omega)$ as shown in Figure P4.31-3 and $1/T = 20\,\text{kHz}$, sketch $X_s(j\Omega)$ and $X(e^{j\omega})$.

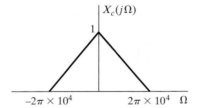

Figure P4.31-3

For a certain range of values of T, the overall system, with input $x_c(t)$ and output $y_c(t)$, is equivalent to a continuous-time lowpass filter with frequency response $H_{eff}(j\Omega)$ sketched in Figure P4.31-4.

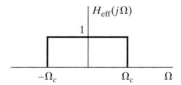

Figure P4.31-4

(b) Determine the range of values of T for which the information presented in (a) is true when $X_c(j\Omega)$ is bandlimited to $|\Omega| \leq 2\pi \times 10^4$ as shown in Figure P4.31-3.

(c) For the range of values determined in (b), sketch Ω_c as a function of $1/T$.

Note: This is one way of implementing a variable-cutoff continuous-time filter using fixed continuous-time and discrete-time filters and a variable sampling rate.

4.32. Consider the discrete-time system shown in Figure P4.32-1

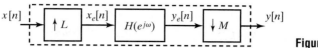

Figure P4.32-1

where

(i) L and M are positive integers.

(ii) $x_e[n] = \begin{cases} x[n/L] & n = kL, \quad k \text{ is any integer} \\ 0 & \text{otherwise.} \end{cases}$

(iii) $y[n] = y_e[nM]$.

(iv) $H(e^{j\omega}) = \begin{cases} M & |\omega| \le \frac{\pi}{4} \\ 0 & \frac{\pi}{4} < |\omega| \le \pi \ . \end{cases}$

(a) Assume that $L = 2$ and $M = 4$, and that $X(e^{j\omega})$, the DTFT of $x[n]$, is real and is as shown in Figure P4.32-2. Make an appropriately labeled sketch of $X_e(e^{j\omega})$, $Y_e(e^{j\omega})$, and $Y(e^{j\omega})$, the DTFTs of $x_e[n]$, $y_e[n]$, and $y[n]$, respectively. Be sure to clearly label salient amplitudes and frequencies.

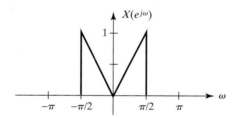

Figure P4.32-2

(b) Now assume $L = 2$ and $M = 8$. Determine $y[n]$ in this case.

Hint: See which diagrams in your answer to part (a) change.

4.33. For the system shown in Figure P4.33, find an expression for $y[n]$ in terms of $x[n]$. Simplify the expression as much as possible.

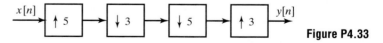

Figure P4.33

Advanced Problems

4.34. In the system shown in Figure P4.34, the individual blocks are defined as indicated.

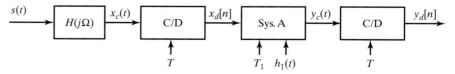

Figure P4.34

$$H(j\Omega): \quad H(j\Omega) = \begin{cases} 1, & |\Omega| < \pi \cdot 10^{-3} \text{ rad/sec} \\ 0, & |\Omega| > \pi \cdot 10^{-3} \text{ rad/sec} \end{cases}$$

$$\text{System A:} \quad y_c(t) = \sum_{k=-\infty}^{\infty} x_d[k] h_1(t - kT_1)$$

$$\text{Second C/D:} \quad y_d[n] = y_c(nT)$$

(a) Specify a choice for T, T_1, and $h_1(t)$ so that $y_c(t)$ and $x_c(t)$ are guaranteed to be equal for any choice of $s(t)$.

(b) State whether your choice in (a) is unique or whether there are other choices for T, T_1, and $h_1(t)$ that will guarantee that $y_c(t)$ and $x_c(t)$ are equal. As usual, clearly show your reasoning.

(c) For this part, we are interested in what is often referred to as *consistent resampling*. Specifically, the system A constructs a continuous-time signal $y_c(t)$ from $x_d[n]$ the sequence of samples of $x_c(t)$ and is then resampled to obtain $y_d[n]$. The resampling is referred to as consistent if $y_d[n] = x_d[n]$. Determine the most general conditions you can on T, T_1, and $h_1(t)$ so that $y_d[n] = x_d[n]$.

4.35. Consider the system shown in Figure P4.35-1.

For parts (a) and (b) only, $X_c(j\Omega) = 0$ for $|\Omega| > 2\pi \times 10^3$ and $H(e^{j\omega})$ is as shown in Figure P4.35-2 (and of course periodically repeats).

(a) Determine the most general condition on T, if any, so that the overall continuous-time system from $x_c(t)$ to $y_c(t)$ is LTI.

(b) Sketch and clearly label the overall equivalent continuous-time frequency response $H_{\text{eff}}(j\Omega)$ that results when the condition determined in (a) holds.

(c) **For this part only** assume that $X_c(j\Omega)$ in Figure P4.35-1 is bandlimited to avoid aliasing, i.e., $X_c(j\Omega) = 0$ for $|\Omega| \geq \frac{\pi}{T}$. For a general sampling period T, we would like to choose the system $H(e^{j\omega})$ in Figure P4.35-1 so that the overall continuous-time system from $x_c(t)$ to $y_c(t)$ is LTI for any input $x_c(t)$ bandlimited as above. Determine the most general conditions on $H(e^{j\omega})$, if any, so that the overall CT system is LTI. Assuming that these conditions hold, also specify in terms of $H(e^{j\omega})$ the overall equivalent continuous-time frequency response $H_{\text{eff}}(j\Omega)$.

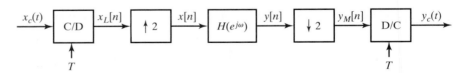

Figure P4.35-1

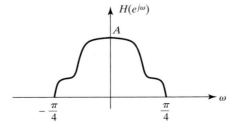

Figure P4.35-2

4.36. We have a discrete-time signal, $x[n]$, arriving from a source at a rate of $\frac{1}{T_1}$ samples per second. We want to digitally resample it to create a signal $y[n]$ that has $\frac{1}{T_2}$ samples per second, where $T_2 = \frac{3}{5}T_1$.

   (a) Draw a block diagram of a discrete-time system to perform the resampling. Specify the input/output relationship for all the boxes in the Fourier domain.

   (b) For an input signal $x[n] = \delta[n] = \begin{cases} 1, & n = 0 \\ 0, & \text{otherwise,} \end{cases}$ determine $y[n]$.

4.37. Consider the decimation filter structure shown in Figure P4.37-1:

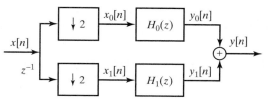

Figure P4.37-1

where $y_0[n]$ and $y_1[n]$ are generated according to the following difference equations:

$$y_0[n] = \frac{1}{4}y_0[n-1] - \frac{1}{3}x_0[n] + \frac{1}{8}x_0[n-1]$$

$$y_1[n] = \frac{1}{4}y_1[n-1] + \frac{1}{12}x_1[n]$$

   (a) How many multiplies per output sample does the implementation of the filter structure require? Consider a divide to be equivalent to a multiply.

The decimation filter can also be implemented as shown in Figure P4.37-2,

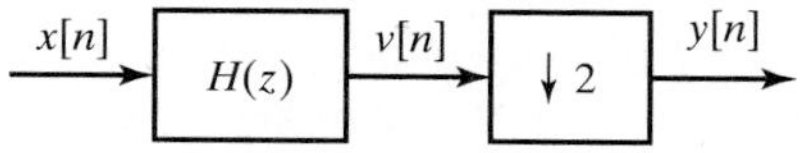

$$x[n] \longrightarrow \boxed{H(z)} \xrightarrow{v[n]} \boxed{\downarrow 2} \xrightarrow{y[n]}$$

Figure P4.37-2

where $v[n] = av[n-1] + bx[n] + cx[n-1]$.

   (b) Determine a, b, and c.

   (c) How many multiplies per output sample does this second implementation require?

4.38. Consider the two systems of Figure P4.38.

   (a) For $M = 2$, $L = 3$, and any arbitrary $x[n]$, will $y_A[n] = y_B[n]$? If your answer is yes, justify your answer. If your answer is no, clearly explain or give a counterexample.

   (b) How must M and L be related to guarantee $y_A[n] = y_B[n]$ for arbitrary $x[n]$?

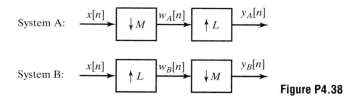

Figure P4.38

4.39. In system A, a continuous-time signal $x_c(t)$ is processed as indicated in Figure P4.39-1.

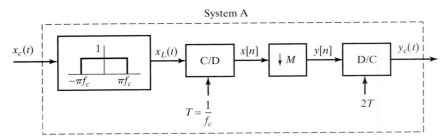

Figure P4.39-1

(a) If $M = 2$ and $x_c(t)$ has the Fourier transform shown in Figure P4.39-2, determine $y[n]$. Clearly show your work on this part.

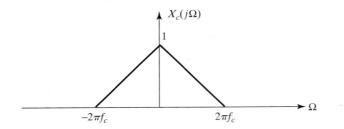

Figure P4.39-2

We would now like to modify system A by appropriately placing additional processing modules in the cascade chain of system A (i.e., blocks can be added at any point in the cascade chain—at the beginning, at the end, or even in between existing blocks). All of the current blocks in system A must be kept. We would like the modified system to be an ideal LTI lowpass filter, as indicated in Figure P4.39-3.

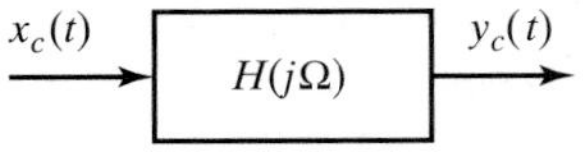

$$\xrightarrow{x_c(t)} \boxed{H(j\Omega)} \xrightarrow{y_c(t)}$$

Figure P4.39-3

$$H(j\Omega) = \begin{cases} 1 & |\Omega| < \frac{2\pi f_c}{5} \\ 0 & \text{otherwise} \end{cases}$$

We have available an unlimited number of the six modules specified in the table given in Figure P4.39-4. The per unit cost for each module is indicated, and we would like the final cost to be as low as possible. **Note that the D/C converter is running at a rate of "$2T$".**

(b) Design the lowest-cost modified system if $M = 2$ in System A. Specify the parameters for all the modules used.

(c) Design the lowest-cost modified system if $M = 4$ in System A. Specify the parameters for all the modules used.

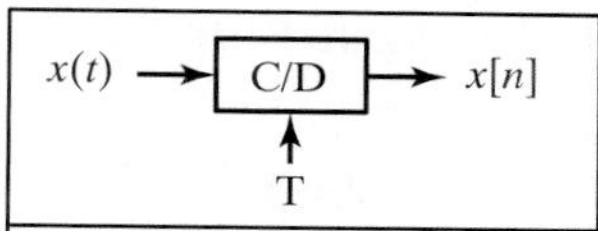 $x(t) \rightarrow$ C/D $\rightarrow x[n]$ T	**Continuous to Discrete-Time Converter** Parameters: T Cost : 10
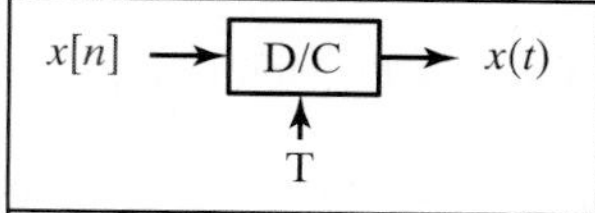 $x[n] \rightarrow$ D/C $\rightarrow x(t)$ T	**Discrete to Continuous Time-Converter** Parameters: T Cost : 10
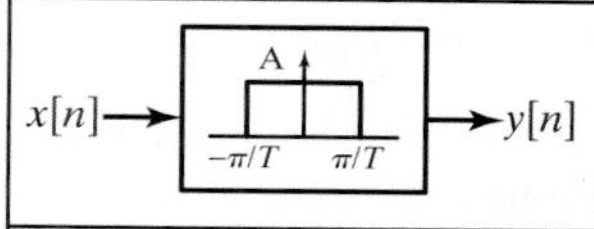 $x[n] \rightarrow$ A, $-\pi/T$, π/T $\rightarrow y[n]$	**Discrete-Time Lowpass Filter** Parameters: A, T Cost : 10
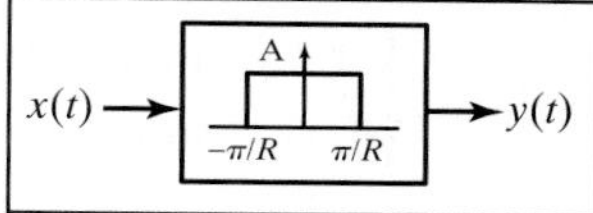 $x(t) \rightarrow$ A, $-\pi/R$, π/R $\rightarrow y(t)$	**Continuous-Time Lowpass Filter** Parameters: A, R Cost : 20
 $\rightarrow \uparrow L \rightarrow$	**Expander** Parameters: L Cost : 5
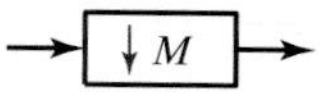 $\rightarrow \downarrow M \rightarrow$	**Compressor** Parameters: M Cost : 5

Figure P4.39-4

4.40. Consider the discrete-time system shown in Figure P4.40-1.

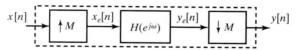

$x[n] \rightarrow \uparrow M \rightarrow x_e[n] \rightarrow H(e^{j\omega}) \rightarrow y_e[n] \rightarrow \downarrow M \rightarrow y[n]$

Figure P4.40-1

where

(i) M is an integer.

(ii) $x_e[n] = \begin{cases} x[n/M] & n = kM, \quad k \text{ is any integer} \\ 0 & \text{otherwise.} \end{cases}$

(iii) $y[n] = y_e[nM]$.

(iv) $H(e^{j\omega}) = \begin{cases} M & |\omega| \leq \frac{\pi}{4} \\ 0 & \frac{\pi}{4} < |\omega| \leq \pi \,. \end{cases}$

(a) Assume that $M = 2$ and that $X(e^{j\omega})$, the DTFT of $x[n]$, is real and is as shown in Figure P4.40-2. Make an appropriately labeled sketch of $X_e(e^{j\omega})$, $Y_e(e^{j\omega})$, and $Y(e^{j\omega})$, the DTFTs of $x_e[n]$, $y_e[n]$, and $y[n]$, respectively. Be sure to clearly label salient amplitudes and frequencies.

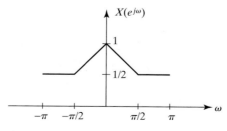

Figure P4.40-2

(b) For $M = 2$ and $X(e^{j\omega})$ as given in Figure P4.40-2, find the value of

$$\varepsilon = \sum_{n=-\infty}^{\infty} |x[n] - y[n]|^2 .$$

(c) For $M = 2$, the overall system is LTI. Determine and sketch the magnitude of the frequency response of the overall system $|H_{\text{eff}}(e^{j\omega})|$.

(d) For $M = 6$, the overall system is <u>still</u> LTI. Determine and sketch the magnitude of the overall system's frequency response $|H_{\text{eff}}(e^{j\omega})|$.

4.41. (a) Consider the system in Figure P4.41-1 where a filter $H(z)$ is followed by a compressor. Suppose that $H(z)$ has an impulse response given by:

$$h[n] = \begin{cases} (\frac{1}{2})^n, & 0 \le n \le 11 \\ 0, & \text{otherwise.} \end{cases} \qquad \text{(P4.41-1)}$$

Figure P4.41-1

The efficiency of this system can be improved by implementing the filter $H(z)$ and the compressor using a polyphase decomposition. Draw an efficient polyphase structure for this system with two polyphase components. Please specify the filters you use.

(b) Now consider the system in Figure P4.41-2 where a filter $H(z)$ is preceded by an expander. Suppose that $H(z)$ has the impulse response as given in Eq. (P4.41-1).

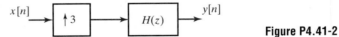

Figure P4.41-2

The efficiency of this system can be improved by implementing the expander and filter $H(z)$ using a polyphase decomposition. Draw an efficient polyphase structure for this system with three polyphase components. Please specify the filters you use.

4.42. For the systems shown in Figure P4.42-1 and Figure P4.42-2, determine whether or not it is possible to specify a choice for $H_2(z)$ in System 2 so that $y_2[n] = y_1[n]$ when $x_2[n] = x_1[n]$ and $H_1(z)$ is as specified. If it is possible, specify $H_2(z)$. If it is not possible, clearly explain.

System 1:

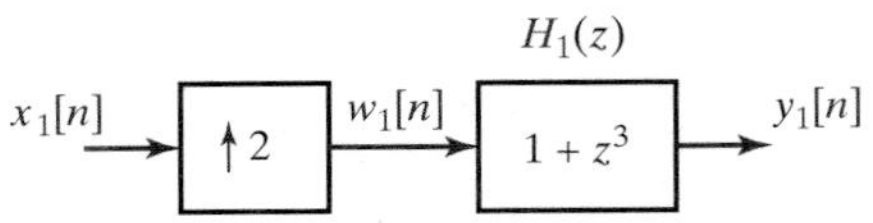

$$H_1(z)$$

$$x_1[n] \longrightarrow \boxed{\uparrow 2} \xrightarrow{w_1[n]} \boxed{1 + z^3} \xrightarrow{y_1[n]}$$

$$w_1[n] = \begin{cases} x_1[n/2] & , \ n/2 \text{ integer} \\ 0 & , \ \text{otherwise} \end{cases}$$

Figure P4.42-1

System 2:

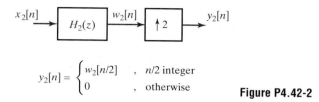

$$y_2[n] = \begin{cases} w_2[n/2] & , \quad n/2 \text{ integer} \\ 0 & , \quad \text{otherwise} \end{cases}$$

Figure P4.42-2

4.43. The block diagram in Figure P4.43 represents a system that we would like to implement. Determine a block diagram of an equivalent system consisting of a cascade of LTI systems, compressor blocks, and expander blocks which results in the minimum number of multiplications per output sample.

Note: By "equivalent system," we mean that it produces the same output sequence for any given input sequence.

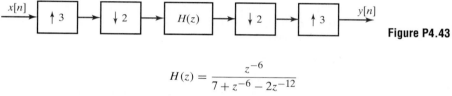

Figure P4.43

$$H(z) = \frac{z^{-6}}{7 + z^{-6} - 2z^{-12}}$$

4.44. Consider the two systems shown in Figure P4.44.

System A:

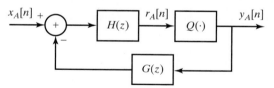

System B:

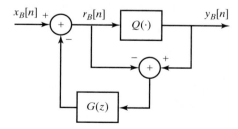

Figure P4.44

where $Q(\cdot)$ represents a quantizer which is the same in both systems. For any given $G(z)$, can $H(z)$ always be specified so that the two systems are equivalent (i.e., $y_A[n] = y_B[n]$ when $x_A[n] = x_B[n]$) for any arbitrary quantizer $Q(\cdot)$? If so, specify $H(z)$. If not, clearly explain your reasoning.

4.45. The quantizer $Q(\cdot)$ in the system S_1 (Figure P4.45-1) can be modeled with an additive noise. Figure P4.45-2 shows system S_2, which is a model for system S_1

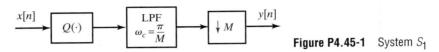

Figure P4.45-1   System S_1

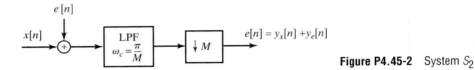

Figure P4.45-2   System S_2

The input $x[n]$ is a zero-mean, wide-sense stationary random process with power spectral density $\Phi_{xx}(e^{j\omega})$ which is bandlimited to π/M and we have $E\left[x^2[n]\right] = 1$. The additive noise $e[n]$ is wide-sense stationary white noise with zero mean and variance σ_e^2. Input and additive noise are uncorrelated. The frequency response of the low-pass filter in all the diagrams has a unit gain.

(a) For system S_2 find the signal to noise ratio: $SNR = 10\log\frac{E[y_x^2[n]]}{E[y_e^2[n]]}$. Note that $y_x[n]$ is the output due to $x[n]$ alone and $y_e[n]$ is the output due to $e[n]$ alone.

(b) To improve the SNR owing to quantization, the system of Figure P4.45-3 is proposed:

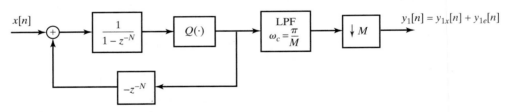

Figure P4.45-3

where $N > 0$ is an integer such that $\pi N << M$. Replace the quantizer with the additive model, as in Figure P4.45-4. Express $y_{1x}[n]$ in terms of $x[n]$ and $y_{1e}[n]$ in terms of $e[n]$.

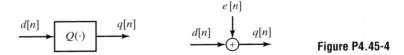

Figure P4.45-4

(c) Assume that $e[n]$ is a zero mean wide-sense stationary white noise that is uncorrelated with input $x[n]$. Is $y_{1e}[n]$ a wide-sense stationary signal? How about $y_1[n]$? Explain.

(d) Is the proposed method in part (b) improving the SNR? For which value of N is the SNR of the system in part (b) maximized?

4.46. The following are three proposed identities involving compressors and expanders. For each, state whether or not the proposed identity is valid. If your answer is that it is valid, explicitly show why. If your answer is no, explicitly give a simple counterexample.

(a) Proposed identity (a):

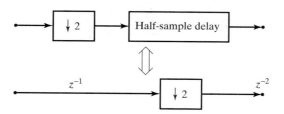

Figure P4.46-1

(b) Proposed identity (b):

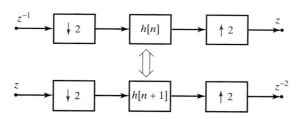

Figure P4.46-2

(c) Proposed identity (c):

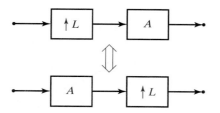

Figure P4.46-3

where L is a positive integer, and A is defined in terms of $X(e^{j\omega})$ and $Y(e^{j\omega})$ (the respective DTFTs of A's input and output) as:

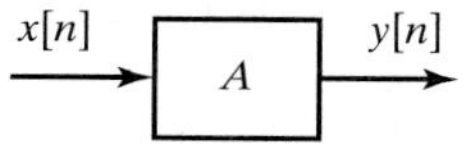

$$x[n] \longrightarrow \boxed{A} \longrightarrow y[n]$$

$$Y(e^{j\omega}) = \left(X(e^{j\omega})\right)^L$$

Figure P4.46-4

4.47. Consider the system shown in Figure P4.47-1 for discrete-time processing of the continuous-time input signal $g_c(t)$.

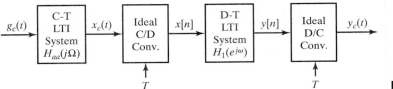

Figure P4.47-1

The continuous-time input signal to the overall system is of the form $g_c(t) = f_c(t) + e_c(t)$ where $f_c(t)$ is considered to be the "signal" component and $e_c(t)$ is considered to be an "additive noise" component. The Fourier transforms of $f_c(t)$ and $e_c(t)$ are as shown in Figure P4.47-2.

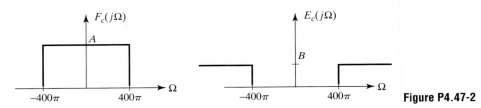

Figure P4.47-2

Since the total input signal $g_c(t)$ does not have a bandlimited Fourier transform, a zero-phase continuous-time antialiasing filter is used to combat aliasing distortion. Its frequency response is given in Figure P4.47-3.

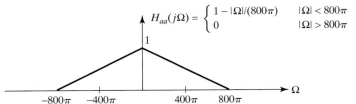

$$H_{aa}(j\Omega) = \begin{cases} 1 - |\Omega|/(800\pi) & |\Omega| < 800\pi \\ 0 & |\Omega| > 800\pi \end{cases}$$

Figure P4.47-3

(a) If in Figure P4.47-1 the sampling rate is $2\pi/T = 1600\pi$, and the discrete-time system has frequency response

$$H_1(e^{j\omega}) = \begin{cases} 1 & |\omega| < \pi/2 \\ 0 & \pi/2 < |\omega| \le \pi \end{cases}$$

sketch the Fourier transform of the continuous-time output signal for the input whose Fourier transform is defined in Figure P4.47-2.

(b) If the sampling rate is $2\pi/T = 1600\pi$, determine the magnitude and phase of $H_1(e^{j\omega})$ (the frequency response of the discrete-time system) so that the output of the system in Figure P4.47-1 is $y_c(t) = f_c(t - 0.1)$. You may use any combination of equations or carefully labeled plots to express your answer.

(c) It turns out that since we are only interested in obtaining $f_c(t)$ at the output, we can use a lower sampling rate than $2\pi/T = 1600\pi$ while still using the antialiasing filter in Figure P4.47-3. Determine the minimum sampling rate that will avoid aliasing distortion of $F_c(j\Omega)$ and determine the frequency response of the filter $H_1(e^{j\omega})$ that can be used so that $y_c(t) = f_c(t)$ at the output of the system in Figure P4.47-1.

(d) Now consider the system shown in Figure P4.47-4, where $2\pi/T = 1600\pi$, and the input signal is defined in Figure P4.47-2 and the antialiasing filter is as shown in Figure P4.47-3.

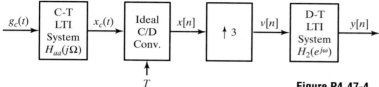

Figure P4.47-4   Another System Block Diagram

where

$$v[n] = \begin{cases} x[n/3] & n = 0, \pm 3, \pm 6, \dots \\ 0 & \text{otherwise} \end{cases}$$

What should $H_2(e^{j\omega})$ be if it is desired that $y[n] = f_c(nT/3)$?

4.48. **(a)** A finite sequence $b[n]$ is such that:

$$B(z) + B(-z) = 2c, \quad c \neq 0.$$

Explain the structure of $b[n]$. Is there any constraint on the length of $b[n]$?

(b) Is it possible to have $B(z) = H(z)H(z^{-1})$? Explain.

(c) A length-N filter $H(z)$ is such that,

$$H(z)H(z^{-1}) + H(-z)H(-z^{-1}) = c. \tag{P4.48-1}$$

Find $G_0(z)$ and $G_1(z)$ such that the filter shown in Figure P4.48 is LTI:

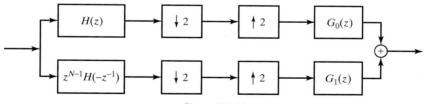

Figure P4.48

(d) For $G_0(z)$ and $G_1(z)$ given in part (c), does the overall system perfectly reconstruct the input? Explain.

4.49. Consider the multirate system shown in Figure P4.49-1 with input $x[n]$ and output $y[n]$:

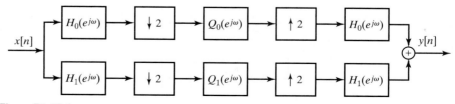

Figure P4.49-1

where $Q_0(e^{j\omega})$ and $Q_1(e^{j\omega})$ are the frequency responses of two LTI systems. $H_0(e^{j\omega})$ and $H_1(e^{j\omega})$ are ideal lowpass and highpass filters, respectively, with cutoff frequency at $\pi/2$ as shown in Figure P4.49-2:

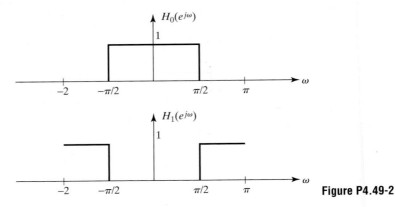

Figure P4.49-2

The overall system is LTI if $Q_0(e^{j\omega})$ and $Q_1(e^{j\omega})$ are as shown in Figure P4.49-3:

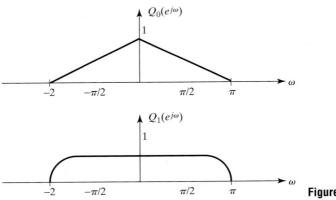

Figure P4.49-3

For these choices of $Q_0(e^{j\omega})$ and $Q_1(e^{j\omega})$, sketch the frequency response

$$G(e^{j\omega}) = \frac{Y(e^{j\omega})}{X(e^{j\omega})}$$

of the overall system.

4.50. Consider the QMF filterbank shown in Figure P4.50:

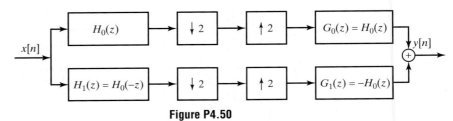

Figure P4.50

The input–output relationship is $Y(z) = T(z)X(z)$, where

$$T(z) = \frac{1}{2}(H_0^2(z) - H_0^2(-z)) = 2z^{-1}E_0(z^2)E_1(z^2)$$

and $E_0(z^2)$, $E_1(z^2)$ are the polyphase components of $H_0(z)$.

Parts (a) and (b) are independent.

(a) Explain whether the following two statements are correct:

(a1) If $H_0(z)$ is linear phase, then $T(z)$ is linear phase.
(a2) If $E_0(z)$ and $E_1(z)$ are linear phase, then $T(z)$ is linear phase.

(b) The prototype filter is known, $h_0[n] = \delta[n] + \delta[n-1] + \frac{1}{4}\delta[n-2]$:

(b1) What are $h_1[n]$, $g_0[n]$ and $g_1[n]$?
(b2) What are $e_0[n]$ and $e_1[n]$?
(b3) What are $T(z)$ and $t[n]$?

4.51. Consider the system in Figure 4.10 with $X_c(j\Omega) = 0$ for $|\Omega| \geq 2\pi(1000)$ and the discrete-time system a squarer, i.e., $y[n] = x^2[n]$. What is the largest value of T such that $y_c(t) = x_c^2(t)$?

4.52. In the system of Figure P4.52,

$$X_c(j\Omega) = 0, \qquad |\Omega| \geq \pi/T,$$

and

$$H(e^{j\omega}) = \begin{cases} e^{-j\omega}, & |\omega| < \pi/L, \\ 0, & \pi/L < |\omega| \leq \pi. \end{cases}$$

How is $y[n]$ related to the input signal $x_c(t)$?

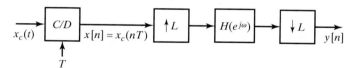

Figure P4.52

Extension Problems

4.53. In many applications, discrete-time random signals arise through periodic sampling of continuous-time random signals. We are concerned in this problem with a derivation of the sampling theorem for random signals. Consider a continuous-time, stationary, random process defined by the random variables $\{x_a(t)\}$, where t is a continuous variable. The autocorrelation function is defined as

$$\phi_{x_cx_c}(\tau) = \mathcal{E}\{x(t)x^*(t+\tau)\},$$

and the power density spectrum is

$$P_{x_cx_c}(\Omega) = \int_{-\infty}^{\infty} \phi_{x_cx_c}(\tau)e^{-j\Omega\tau}\,d\tau.$$

A discrete-time random process obtained by periodic sampling is defined by the set of random variables $\{x[n]\}$, where $x[n] = x_a(nT)$ and T is the sampling period.

(a) What is the relationship between $\phi_{xx}[n]$ and $\phi_{x_c x_c}(\tau)$?

(b) Express the power density spectrum of the discrete-time process in terms of the power density spectrum of the continuous-time process.

(c) Under what condition is the discrete-time power density spectrum a faithful representation of the continuous-time power density spectrum?

4.54. Consider a continuous-time random process $x_c(t)$ with a bandlimited power density spectrum $P_{x_c x_c}(\Omega)$ as depicted in Figure P4.54-1. Suppose that we sample $x_c(t)$ to obtain the discrete-time random process $x[n] = x_c(nT)$.

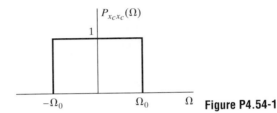

Figure P4.54-1

(a) What is the autocorrelation sequence of the discrete-time random process?

(b) For the continuous-time power density spectrum in Figure P4.54-1, how should T be chosen so that the discrete-time process is white, i.e., so that the power spectrum is constant for all ω?

(c) If the continuous-time power density spectrum is as shown in Figure P4.54-2, how should T be chosen so that the discrete-time process is white?

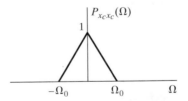

Figure P4.54-2

(d) What is the general requirement on the continuous-time process and the sampling period such that the discrete-time process is white?

4.55. This problem explores the effect of interchanging the order of two operations on a signal, namely, sampling and performing a memoryless nonlinear operation.

(a) Consider the two signal-processing systems in Figure P4.55-1, where the C/D and D/C converters are ideal. The mapping $g[x] = x^2$ represents a memoryless nonlinear device. For the two systems in the figure, sketch the signal spectra at points 1, 2, and 3 when the sampling rate is selected to be $1/T = 2f_m$ Hz and $x_c(t)$ has the Fourier transform shown in Figure P4.55-2. Is $y_1(t) = y_2(t)$? If not, why not? Is $y_1(t) = x^2(t)$? Explain your answer.

System 1:

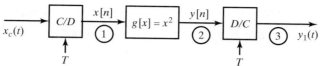

System 2:

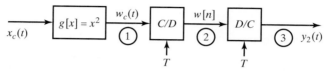

Figure P4.55-1

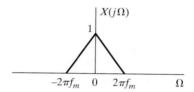

Figure P4.55-2

(b) Consider System 1, and let $x(t) = A \cos(30\pi t)$. Let the sampling rate be $1/T = 40$ Hz. Is $y_1(t) = x_c^2(t)$? Explain why or why not.

(c) Consider the signal-processing system shown in Figure P4.55-3, where $g[x] = x^3$ and $g^{-1}[v]$ is the (unique) inverse, i.e., $g^{-1}[g(x)] = x$. Let $x(t) = A \cos(30\pi t)$ and $1/T = 40$ Hz. Express $v[n]$ in terms of $x[n]$. Is there spectral aliasing? Express $y[n]$ in terms of $x[n]$. What conclusion can you reach from this example? You may find the following identity helpful:

$$\cos^3 \Omega_0 t = \tfrac{3}{4} \cos \Omega_0 t + \tfrac{1}{4} \cos 3\Omega_0 t.$$

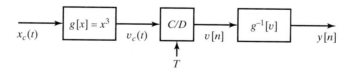

Figure P4.55-3

(d) One practical problem is that of digitizing a signal having a large dynamic range. Suppose we compress the dynamic range by passing the signal through a memoryless nonlinear device prior to A/D conversion and then expand it back after A/D conversion. What is the impact of the nonlinear operation prior to the A/D converter in our choice of the sampling rate?

4.56. Figure 4.23 depicts a system for interpolating a signal by a factor of L, where

$$x_e[n] = \begin{cases} x[n/L], & n = 0, \pm L, \pm 2L, \text{ etc} \ldots, \\ 0, & \text{otherwise}, \end{cases}$$

and the lowpass filter interpolates between the nonzero values of $x_e[n]$ to generate the upsampled or interpolated signal $x_i[n]$. When the lowpass filter is ideal, the interpolation is referred to as bandlimited interpolation. As indicated in Section 4.6.3, simple interpolation procedures are adequate in many applications. Two simple procedures often used are zero-order-hold and linear interpolation. For zero-order-hold interpolation, each value of $x[n]$ is simply repeated L times; i.e.,

$$x_i[n] = \begin{cases} x_e[0], & n = 0, \ 1, \ldots, L-1, \\ x_e[L], & n = L, \ L+1, \ldots, 2L-1, \\ x_e[2L], & n = 2L, \ 2L+1, \ldots, \\ \vdots & \end{cases}$$

Linear interpolation is described in Section 4.6.2.

(a) Determine an appropriate choice for the impulse response of the lowpass filter in Figure 4.23 to implement zero-order-hold interpolation. Also, determine the corresponding frequency response.

(b) Equation (4.91) specifies the impulse response for linear interpolation. Determine the corresponding frequency response. (You may find it helpful to use the fact that $h_{\text{lin}}[n]$ is triangular and consequently corresponds to the convolution of two rectangular sequences.)

(c) Sketch the magnitude of the filter frequency response for zero-order-hold and linear interpolation. Which is a better approximation to ideal bandlimited interpolation?

4.57. We wish to compute the autocorrelation function of an upsampled signal, as indicated in Figure P4.57-1. It is suggested that this can equivalently be accomplished with the system of Figure P4.57-2. Can $H_2(e^{j\omega})$ be chosen so that $\phi_3[m] = \phi_1[m]$? If not, why not? If so, specify $H_2(e^{j\omega})$.

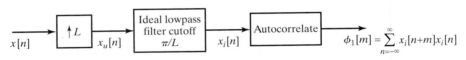

Figure P4.57-1

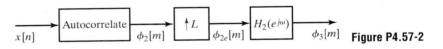

Figure P4.57-2

4.58. We are interested in upsampling a sequence by a factor of 2, using a system of the form of Figure 4.23. However, the lowpass filter in that figure is to be approximated by a five-point filter with impulse response $h[n]$ indicated in Figure P4.58-1. In this system, the output $y_1[n]$ is obtained by direct convolution of $h[n]$ with $w[n]$.

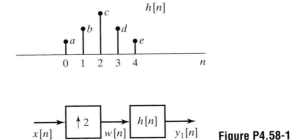

Figure P4.58-1

(a) A proposed implementation of the system with the preceding choice of $h[n]$ is shown in Figure P4.58-2. The three impulse responses $h_1[n], h_2[n]$, and $h_3[n]$ are all restricted to be zero outside the range $0 \leq n \leq 2$. Determine and clearly justify a choice for $h_1[n], h_2[n]$, and $h_3[n]$ so that $y_1[n] = y_2[n]$ for any $x[n]$, i.e., so that the two systems are identical.

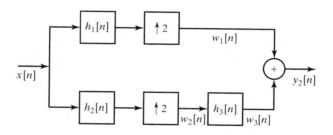

Figure P4.58-2

(b) Determine the number of multiplications per output point required in the system of Figure P4.58-1 and in the system of Figure P4.58-2. You should find that the system of Figure P4.58-2 is more efficient.

4.59. Consider the analysis–synthesis system shown in Figure P4.59-1. The lowpass filter $h_0[n]$ is identical in the analyzer and synthesizer, and the highpass filter $h_1[n]$ is identical in the analyzer and synthesizer. The Fourier transforms of $h_0[n]$ and $h_1[n]$ are related by

$$H_1(e^{j\omega}) = H_0(e^{j(\omega+\pi)}).$$

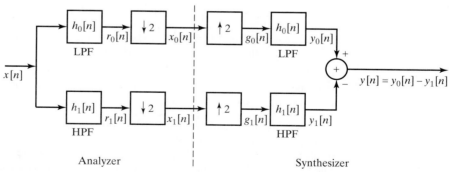

Figure P4.59-1

(a) If $X(e^{j\omega})$ and $H_0(e^{j\omega})$ are as shown in Figure P4.59-2, sketch (to within a scale factor) $X_0(e^{j\omega}), G_0(e^{j\omega})$, and $Y_0(e^{j\omega})$.

(b) Write a general expression for $G_0(e^{j\omega})$ in terms of $X(e^{j\omega})$ and $H_0(e^{j\omega})$. Do *not* assume that $X(e^{j\omega})$ and $H_0(e^{j\omega})$ are as shown in Figure P4.59-2.

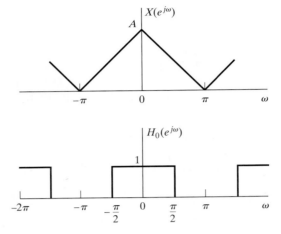

Figure P4.59-2

(c) Determine a set of conditions on $H_0(e^{j\omega})$ that is as general as possible and that will guarantee that $|Y(e^{j\omega})|$ is proportional to $|X(e^{j\omega})|$ for any stable input $x[n]$.

Note: Analyzer–synthesizer filter banks of the form developed in this problem are very similar to quadrature mirror filter banks. (For further reading, see Crochiere and Rabiner (1983), pp. 378–392.)

4.60. Consider a real-valued sequence $x[n]$ for which

$$X(e^{j\omega}) = 0, \qquad \frac{\pi}{3} \le |\omega| \le \pi.$$

One value of $x[n]$ may have been corrupted, and we would like to approximately or exactly recover it. With $\hat{x}[n]$ denoting the corrupted signal,

$$\hat{x}[n] = x[n] \text{ for } n \ne n_0,$$

and $\hat{x}[n_0]$ is real but not related to $x[n_0]$. In each of the following three cases, specify a practical algorithm for exactly or approximately recovering $x[n]$ from $\hat{x}[n]$:

(a) The value of n_0 is known.
(b) The exact value of n_0 is *not* known, but we know that n_0 is an even number.
(c) Nothing about n_0 is known.

4.61. Communication systems often require conversion from time-division multiplexing (TDM) to frequency-division multiplexing (FDM). In this problem, we examine a simple example of such a system. The block diagram of the system to be studied is shown in Figure P4.61-1. The TDM input is assumed to be the sequence of interleaved samples

$$w[n] = \begin{cases} x_1[n/2] & \text{for } n \text{ an even integer,} \\ x_2[(n-1)/2] & \text{for } n \text{ an odd integer.} \end{cases}$$

Assume that the sequences $x_1[n] = x_{c1}(nT)$ and $x_2[n] = x_{c2}(nT)$ have been obtained by sampling the continuous-time signals $x_{c1}(t)$ and $x_{c2}(t)$, respectively, without aliasing.

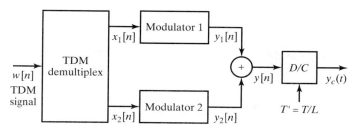

Figure P4.61-1

Assume also that these two signals have the same highest frequency, Ω_N, and that the sampling period is $T = \pi / \Omega_N$.

(a) Draw a block diagram of a system that produces $x_1[n]$ and $x_2[n]$ as outputs; i.e., obtain a system for demultiplexing a TDM signal using simple operations. State whether or not your system is linear, time invariant, causal, and stable.

The kth modulator system ($k = 1$ or 2) is defined by the block diagram in Figure P4.61-2. The lowpass filter $H_i(e^{j\omega})$, which is the same for both channels, has gain L and cutoff frequency π/L, and the highpass filters $H_k(e^{j\omega})$ have unity gain and cutoff frequency ω_k. The modulator frequencies are such that

$$\omega_2 = \omega_1 + \pi/L \quad\text{and}\quad \omega_2 + \pi/L \le \pi \quad (\text{assume } \omega_1 > \pi/2).$$

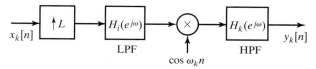

Figure P4.61-2

(b) Assume that $\Omega_N = 2\pi \times 5 \times 10^3$. Find ω_1 and L so that, after ideal D/C conversion with sampling period T/L, the Fourier transform of $y_c(t)$ is zero, except in the band of frequencies

$$2\pi \times 10^5 \le |\omega| \le 2\pi \times 10^5 + 2\Omega_N.$$

(c) Assume that the continuous-time Fourier transforms of the two original input signals are as sketched in Figure P4.61-3. Sketch the Fourier transforms at each point in the system.

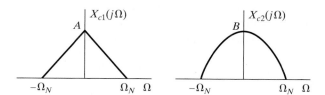

Figure P4.61-3

(d) Based on your solution to parts (a)–(c), discuss how the system could be generalized to handle M equal-bandwidth channels.

4.62. In Section 4.8.1, we considered the use of prefiltering to avoid aliasing. In practice, the antialiasing filter cannot be ideal. However, the nonideal characteristics can be at least partially compensated for with a discrete-time system applied to the sequence $x[n]$ that is the output of the C/D converter.

Consider the two systems in Figure P4.62-1. The antialiasing filters $H_{\text{ideal}}(j\Omega)$ and $H_{\text{aa}}(j\Omega)$ are shown in Figure P4.62-2. $H(e^{j\omega})$ in Figure P4.62-1 is to be specified to compensate for the nonideal characteristics of $H_{\text{aa}}(j\Omega)$.

Sketch $H(e^{j\omega})$ so that the two sequences $x[n]$ and $w[n]$ are identical.

System 1:

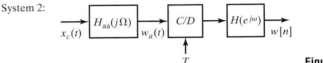

System 2:

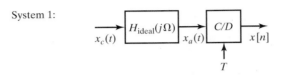

Figure P4.62-1

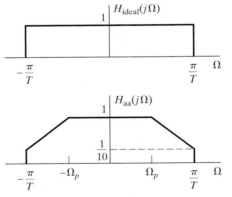

Figure P4.62-2

4.63. As discussed in Section 4.8.2, to process sequences on a digital computer, we must quantize the amplitude of the sequence to a set of discrete levels. This quantization can be expressed in terms of passing the input sequence $x[n]$ through a quantizer $Q(x)$ that has an input–output relation as shown in Figure 4.54.

As discussed in Section 4.8.3, if the quantization interval Δ is small compared with changes in the level of the input sequence, we can assume that the output of the quantizer is of the form

$$y[n] = x[n] + e[n],$$

where $e[n] = Q(x[n]) - x[n]$ and $e[n]$ is a stationary random process with a 1st-order probability density uniform between $-\Delta/2$ and $\Delta/2$, uncorrelated from sample to sample and uncorrelated with $x[n]$, so that $\mathcal{E}\{e[n]x[m]\} = 0$ for all m and n.

Let $x[n]$ be a stationary white-noise process with zero mean and variance σ_x^2.

(a) Find the mean, variance, and autocorrelation sequence of $e[n]$.

(b) What is the signal-to-quantizing-noise ratio σ_x^2/σ_e^2?

(c) The quantized signal $y[n]$ is to be filtered by a digital filter with impulse response $h[n] = \frac{1}{2}[a^n + (-a)^n]u[n]$. Determine the variance of the noise produced at the output due to the input quantization noise, and determine the SNR at the output.

In some cases we may want to use nonlinear quantization steps, for example, logarithmically spaced quantization steps. This can be accomplished by applying uniform quantization to the logarithm of the input as depicted in Figure P4.63, where $Q[\cdot]$ is a uniform quantizer as specified in Figure 4.54. In this case, if we assume that Δ is small compared with changes in the sequence $\ln(x[n])$, then we can assume that the output of the quantizer is

$$\ln(y[n]) = \ln(x[n]) + e[n].$$

Thus,

$$y[n] = x[n] \cdot \exp(e[n]).$$

For small e, we can approximate $\exp(e[n])$ by $(1 + e[n])$, so that

$$y[n] \approx x[n](1 + e[n]) = x[n] + f[n]. \qquad \text{(P4.63-1)}$$

This equation will be used to describe the effect of logarithmic quantization. We assume $e[n]$ to be a stationary random process, uncorrelated from sample to sample, independent of the signal $x[n]$, and with 1st-order probability density uniform between $\pm\Delta/2$.

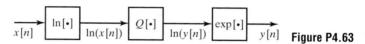

$x[n] \quad \boxed{\ln[\cdot]} \quad \ln(x[n]) \quad \boxed{Q[\cdot]} \quad \ln(y[n]) \quad \boxed{\exp[\cdot]} \quad y[n]$ **Figure P4.63**

(d) Determine the mean, variance, and autocorrelation sequence of the *additive* noise $f[n]$ defined in Eq. (4.57).

(e) What is the signal-to-quantizing-noise ratio σ_x^2/σ_f^2? Note that in this case σ_x^2/σ_f^2 is independent of σ_x^2. Within the limits of our assumption, therefore, the signal-to-quantizing-noise ratio is independent of the input signal level, whereas, for linear quantization, the ratio σ_x^2/σ_e^2 depends directly on σ_x^2.

(f) The quantized signal $y[n]$ is to be filtered by means of a digital filter with impulse response $h[n] = \frac{1}{2}[a^n + (-a)^n]u[n]$. Determine the variance of the noise produced at the output due to the input quantization noise, and determine the SNR at the output.

4.64. Figure P4.64-1 shows a system in which two continuous-time signals are multiplied and a discrete-time signal is then obtained from the product by sampling the product at the Nyquist rate; i.e., $y_1[n]$ is samples of $y_c(t)$ taken at the Nyquist rate. The signal $x_1(t)$ is bandlimited to 25 kHz ($X_1(j\Omega) = 0$ for $|\Omega| \geq 5\pi \times 10^4$), and $x_2(t)$ is limited to 2.5 kHz ($X_2(j\Omega) = 0$ for $|\Omega| \geq (\pi/2) \times 10^4$).

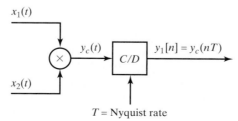

T = Nyquist rate   **Figure P4.64-1**

In some situations (digital transmission, for example), the continuous-time signals have already been sampled at their individual Nyquist rates, and the multiplication is to be carried out in the discrete-time domain, perhaps with some additional processing before and after multiplication, as indicated in Figure P4.64-2. Each of the systems A, B, and C either is an identity or can be implemented using one or more of the modules shown in Figure P4.64-3.

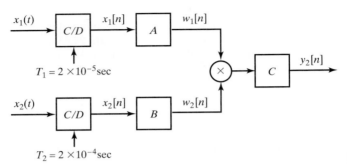

Figure P4.64-2

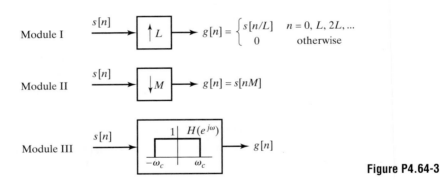

Figure P4.64-3

For each of the three systems A, B, and C, either specify that the system is an identity system or specify an appropriate interconnection of one or more of the modules shown in Figure P4.64-3. Also, specify all relevant parameters L, M, and ω_c. The systems A, B, and C should be constructed such that $y_2[n]$ is proportional to $y_1[n]$, i.e.,

$$y_2[n] = ky_1[n] = ky_c(nT) = kx_1(nT) \times x_2(nT),$$

and these samples are at the Nyquist rate, i.e., $y_2[n]$ does not represent oversampling or undersampling of $y_c(t)$.

4.65. Suppose $s_c(t)$ is a speech signal with the continuous-time Fourier transform $S_c(j\Omega)$ shown in Figure P4.65-1. We obtain a discrete-time sequence $s_r[n]$ from the system shown in Figure P4.65-2, where $H(e^{j\omega})$ is an ideal discrete-time lowpass filter with cutoff frequency ω_c and a gain of L throughout the passband, as shown in Figure 4.29(b). The signal $s_r[n]$ will be used as an input to a speech coder, which operates correctly only on discrete-time samples representing speech sampled at an 8-kHz rate. Choose values of L, M, and ω_c that produce the correct input signal $s_r[n]$ for the speech coder.

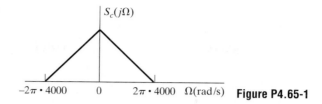

Figure P4.65-1

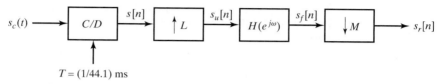

Figure P4.65-2

4.66. In many audio applications, it is necessary to sample a continuous-time signal $x_c(t)$ at a sampling rate $1/T = 44$ kHz. Figure P4.66-1 shows a straightforward system, including a continuous-time antialias filter $H_{a0}(j\Omega)$, to acquire the desired samples. In many applications, the "4x oversampling" system shown in Figure P4.66-2 is used instead of the conventional system shown in Figure P4.66-1. In the system in Figure P4.66-2,

$$H(e^{j\omega}) = \begin{cases} 1, & |\omega| \leq \pi/4, \\ 0, & \text{otherwise}, \end{cases}$$

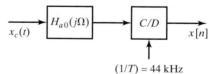

Figure P4.66-1

is an ideal lowpass filter, and

$$H_{a1}(j\Omega) = \begin{cases} 1, & |\Omega| \leq \Omega_p, \\ 0, & |\Omega| > \Omega_s, \end{cases}$$

for some $0 \leq \Omega_p \leq \Omega_s \leq \infty$.

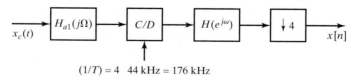

Figure P4.66-2

Assuming that $H(e^{j\omega})$ is ideal, find the minimal set of specifications on the antialias filter $H_{a1}(j\Omega)$, i.e., the smallest Ω_p and the largest Ω_s, such that the overall system of Figure P4.66-2 is equivalent to the system in Figure P4.66-1.

4.67. In this problem, we will consider the "double integration" system for quantization with noise shaping shown in Figure P4.67. In this system,

$$H_1(z) = \frac{1}{1 - z^{-1}} \quad \text{and} \quad H_2(z) = \frac{z^{-1}}{1 - z^{-1}},$$

and the frequency response of the decimation filter is

$$H_3(e^{j\omega}) = \begin{cases} 1, & |\omega| < \pi/M, \\ 0, & \pi/M \le |\omega| \le \pi. \end{cases}$$

The noise source $e[n]$, which represents a quantizer, is assumed to be a zero-mean white-noise (constant power spectrum) signal that is uniformly distributed in amplitude and has noise power $\sigma_e^2 = \Delta^2/12$.

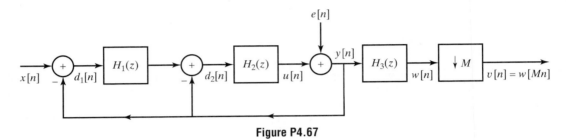

Figure P4.67

(a) Determine an equation for $Y(z)$ in terms of $X(z)$ and $E(z)$. Assume for this part that $E(z)$ exists. From the z-transform relation, show that $y[n]$ can be expressed in the form $y[n] = x[n-1] + f[n]$, where $f[n]$ is the output owing to the noise source $e[n]$. What is the time-domain relation between $f[n]$ and $e[n]$?

(b) Now assume that $e[n]$ is a white-noise signal as described prior to part (a). Use the result from part (a) to show that the power spectrum of the noise $f[n]$ is

$$P_{ff}(e^{j\omega}) = 16\sigma_e^2 \sin^4(\omega/2).$$

What is the *total* noise power (σ_f^2) in the noise component of the signal $y[n]$? On the same set of axes, sketch the power spectra $P_{ee}(e^{j\omega})$ and $P_{ff}(e^{j\omega})$ for $0 \le \omega \le \pi$.

(c) Now assume that $X(e^{j\omega}) = 0$ for $\pi/M < \omega \le \pi$. Argue that the output of $H_3(z)$ is $w[n] = x[n-1] + g[n]$. State in words what $g[n]$ is.

(d) Determine an expression for the noise power σ_g^2 at the output of the decimation filter. Assume that $\pi/M \ll \pi$, i.e., M is large, so that you can use a small-angle approximation to simplify the evaluation of the integral.

(e) After the decimator, the output is $v[n] = w[Mn] = x[Mn-1] + q[n]$, where $q[n] = g[Mn]$. Now suppose that $x[n] = x_c(nT)$ (i.e., $x[n]$ was obtained by sampling a continuous-time signal). What condition must be satisfied by $X_c(j\Omega)$ so that $x[n-1]$ will pass through the filter unchanged? Express the "signal component" of the output $v[n]$ in terms of $x_c(t)$. What is the total power σ_q^2 of the noise at the output? Give an expression for the power spectrum of the noise at the output, and, on the same set of axes, sketch the power spectra $P_{ee}(e^{j\omega})$ and $P_{qq}(e^{j\omega})$ for $0 \le \omega \le \pi$.

4.68. For sigma-delta oversampled A/D converters with high-order feedback loops, stability becomes a significant consideration. An alternative approach referred to as multi-stage noise shaping (MASH) achieves high-order noise shaping with only 1st-order feedback. The structure for 2nd-order MASH noise shaping is shown in Figure P4.68-2 and analyzed in this problem.

Figure P4.68-1 is a 1st-order sigma-delta ($\Sigma - \Delta$) noise shaping system, where the effect of the quantizer is represented by the additive noise signal $e[n]$. The noise $e[n]$ is explicitly shown in the diagram as a second output of the system. Assume that the input $x[n]$ is a zero-mean wide-sense stationary random process. Assume also that $e[n]$ is zero-mean, white, wide-sense stationary, and has variance σ_e^2. $e[n]$ is uncorrelated with $x[n]$.

(a) For the system in Figure P4.68-1, the output $y[n]$ has a component $y_x[n]$ due only to $x[n]$ and a component $y_e[n]$ due only to $e[n]$, i.e., $y[n] = y_x[n] + y_e[n]$.

    (i) Determine $y_x[n]$ in terms of $x[n]$.

    (ii) Determine $P_{y_e}(\omega)$, the power spectral density of $y_e[n]$.

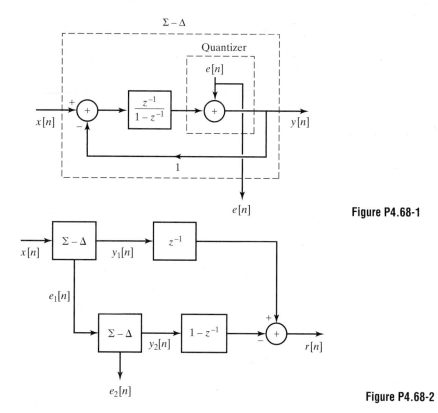

Figure P4.68-1

Figure P4.68-2

(a) The system of Figure P4.68 is now connected in the configuration shown in Figure P4.68, which shows the structure of the MASH system. Notice that $e_1[n]$ and $e_2[n]$ are the noise signals resulting from the quantizers in the sigma-delta noise shaping systems. The output of the system $r[n]$ has a component $r_x[n]$ owing only to $x[n]$, and a component $r_e[n]$ due only to the quantization noise, i.e., $r[n] = r_x[n] + r_e[n]$. Assume that $e_1[n]$ and $e_2[n]$ are zero-mean, white, wide-sense stationary, each with variance σ_e^2. Also assume that $e_1[n]$ is uncorrelated with $e_2[n]$.

    (i) Determine $r_x[n]$ in terms of $x[n]$.

    (ii) Determine $P_{r_e}(\omega)$, the power spectral density of $r_e[n]$.

5

Transform Analysis of Linear Time-Invariant Systems

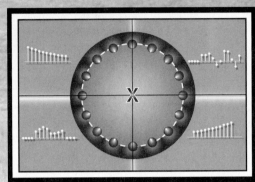

5.0 INTRODUCTION

In Chapter 2, we developed the Fourier transform representation of discrete-time signals and systems, and in Chapter 3 we extended that representation to the z-transform. In both chapters, the emphasis was on the transforms and their properties, with only a brief preview of the details of their use in the analysis of linear time-invariant (LTI) systems. In this chapter, we develop in more detail the representation and analysis of LTI systems using the Fourier and z-transforms. The material is essential background for our discussion in Chapter 6 of the implementation of LTI systems and in Chapter 7 of the design of such systems.

As discussed in Chapter 2, an LTI system can be completely characterized in the time domain by its impulse response $h[n]$, with the output $y[n]$ due to a given input $x[n]$ specified through the convolution sum

$$y[n] = \sum_{k=-\infty}^{\infty} x[k]h[n-k]. \tag{5.1}$$

Alternatively, since the frequency response and impulse response are directly related through the Fourier transform, the frequency response, assuming it exists (i.e., $H(z)$ has an ROC that includes $z = e^{j\omega}$), provides an equally complete characterization of LTI systems. In Chapter 3, we developed the z-transform as a generalization of the Fourier transform. The z-transform of the output of an LTI system is related to the z-transform of the input and the z-transform of the system impulse response by

$$Y(z) = H(z)X(z), \tag{5.2}$$

where $Y(z)$, $X(z)$, and $H(z)$ denote the z-transforms of $y[n]$, $x[n]$ and $h[n]$ respectively and have appropriate regions of convergence. $H(z)$ is typically referred to as the *system function.* Since a sequence and its z-transform form a unique pair, it follows that any LTI system is completely characterized by its system function, again assuming convergence.

Both the frequency response, which corresponds to the system function evaluated on the unit circle, and the system function more generally as a function of the complex variable z, are extremely useful in the analysis and representation of LTI systems, because we can readily infer many properties of the system response from them.

5.1 THE FREQUENCY RESPONSE OF LTI SYSTEMS

The frequency response $H(e^{j\omega})$ of an LTI system was defined in Section 2.6 as the complex gain (eigenvalue) that the system applies to a complex exponential input (eigenfunction) $e^{j\omega n}$. Furthermore, as discussed in Section 2.9.6, since the Fourier transform of a sequence represents a decomposition as a linear combination of complex exponentials, the Fourier transforms of the system input and output are related by

$$Y(e^{j\omega}) = H(e^{j\omega})X(e^{j\omega}), \tag{5.3}$$

where $X(e^{j\omega})$ and $Y(e^{j\omega})$ are the Fourier transforms of the system input and output, respectively.

5.1.1 Frequency Response Phase and Group Delay

The frequency response is in general a complex number at each frequency. With the frequency response expressed in polar form, the magnitude and phase of the Fourier transforms of the system input and output are related by

$$|Y(e^{j\omega})| = |H(e^{j\omega})| \cdot |X(e^{j\omega})|, \tag{5.4a}$$

$$\angle Y(e^{j\omega}) = \angle H(e^{j\omega}) + \angle X(e^{j\omega}), \tag{5.4b}$$

where $|H(e^{j\omega})|$ represents the *magnitude response* or the *gain* of the system, and $\angle H(e^{j\omega})$ the *phase response* or *phase shift* of the system.

The magnitude and phase effects represented by Eqs. (5.4a) and (5.4b) can be either desirable, if the input signal is modified in a useful way, or undesirable, if the input signal is changed in a deleterious manner. In the latter, we often refer to the effects of an LTI system on a signal, as represented by Eqs. (5.4a) and (5.4b), as *magnitude* and *phase distortions*, respectively.

The phase angle of any complex number is not uniquely defined, since any integer multiple of 2π can be added without affecting the complex number. When the phase is numerically computed with the use of an arctangent subroutine, the principal value is typically obtained. We will denote the principal value of the phase of $H(e^{j\omega})$ as $\mathrm{ARG}[H(e^{j\omega})]$, where

$$-\pi < \mathrm{ARG}[H(e^{j\omega})] \leq \pi. \tag{5.5}$$

Any other angle that gives the correct complex value of the function $H(e^{j\omega})$ can be represented in terms of the principal value as

$$\angle H(e^{j\omega}) = \text{ARG}[H(e^{j\omega})] + 2\pi r(\omega), \tag{5.6}$$

where $r(\omega)$ is a positive or negative integer that can be different at each value of ω. We will in general use the angle notation on the left side of Eq. (5.6) to denote ambiguous phase, since $r(\omega)$ is somewhat arbitrary.

In many cases, the principal value will exhibit discontinuities of 2π radians when viewed as a function of ω. This is illustrated in Figure 5.1, which shows a continuous-phase function $\arg[H(e^{j\omega})]$ and its principal value $\text{ARG}[H(e^{j\omega})]$ plotted over the range

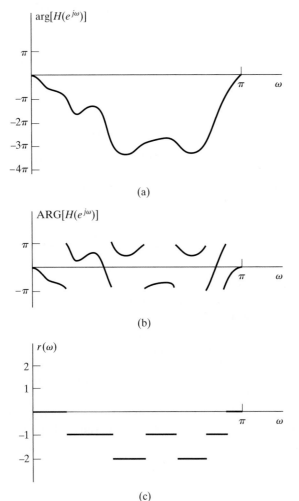

(a)

(b)

(c)

Figure 5.1   (a) Continuous-phase curve for a system function evaluated on the unit circle. (b) Principal value of the phase curve in part (a). (c) Integer multiples of 2π to be added to $\text{ARG}[H(e^{j\omega})]$ to obtain $\arg[H(e^{j\omega})]$.

$0 \leq \omega \leq \pi$. The phase function plotted in Figure 5.1(a) exceeds the range $-\pi$ to $+\pi$. The principal value, shown in Figure 5.1(b), has jumps of 2π, owing to the integer multiples of 2π that must be subtracted in certain regions to bring the phase curve within the range of the principal value. Figure 5.1(c) shows the corresponding value of $r(\omega)$ in Eq. (5.6).

Throughout this text, in our discussion of phase, we refer to $\mathrm{ARG}[H(e^{j\omega})]$ as the "wrapped" phase, since the evaluation modulo 2π can be thought of as wrapping the phase on a circle. In an amplitude and phase representation (in which the amplitude is real-valued but can be positive or negative), $\mathrm{ARG}[H(e^{j\omega})]$ can be "unwrapped" to a phase curve that is continuous in ω. The continuous (unwrapped) phase curve is denoted as $\arg[H(e^{j\omega})]$. Another particularly useful representation of phase is through the group delay $\tau(\omega)$ defined as

$$\tau(\omega) = \mathrm{grd}[H(e^{j\omega})] = -\frac{d}{d\omega}\{\arg[H(e^{j\omega})]\}. \tag{5.7}$$

It is worth noting that since the derivative of $\arg[H(e^{j\omega})]$ and $\mathrm{ARG}[H(e^{j\omega})]$ will be identical except for the presence of impulses in the derivative of $\mathrm{ARG}[H(e^{j\omega})]$ at the discontinuities, the group delay can be obtained from the principal value by differentiating, except at the discontinuities. Similarly, we can express the group delay in terms of the ambiguous phase $\angle H(e^{j\omega})$ as

$$\mathrm{grd}[H(e^{j\omega})] = -\frac{d}{d\omega}\{\angle H(e^{j\omega})\}, \tag{5.8}$$

with the interpretation that impulses caused by discontinuities of size 2π in $\angle H(e^{j\omega})$ are ignored.

To understand the effect of the phase and specifically the group delay of a linear system, let us first consider the ideal delay system. The impulse response is

$$h_{\mathrm{id}}[n] = \delta[n - n_d], \tag{5.9}$$

and the frequency response is

$$H_{\mathrm{id}}(e^{j\omega}) = e^{-j\omega n_d}, \tag{5.10}$$

or

$$|H_{\mathrm{id}}(e^{j\omega})| = 1, \tag{5.11a}$$

$$\angle H_{\mathrm{id}}(e^{j\omega}) = -\omega n_d, \qquad |\omega| < \pi, \tag{5.11b}$$

with periodicity 2π in ω assumed. From Eq. (5.11b) we note that time delay (or advance if $n_d < 0$) is associated with phase that is linear with frequency.

In many applications, delay distortion would be considered a rather mild form of phase distortion, since its effect is only to shift the sequence in time. Often this would be inconsequential, or it could easily be compensated for by introducing delay in other parts of a larger system. Thus, in designing approximations to ideal filters and other LTI systems, we frequently are willing to accept a linear-phase response rather than a zero-phase response as our ideal. For example, an ideal lowpass filter with linear phase would have frequency response

$$H_{\text{lp}}(e^{j\omega}) = \begin{cases} e^{-j\omega n_d}, & |\omega| < \omega_c, \\ 0, & \omega_c < |\omega| \leq \pi. \end{cases} \tag{5.12}$$

The corresponding impulse response is

$$h_{\text{lp}}[n] = \frac{\sin \omega_c(n - n_d)}{\pi(n - n_d)}, \qquad -\infty < n < \infty. \tag{5.13}$$

The group delay represents a convenient measure of the linearity of the phase. Specifically, consider the output of a system with frequency response $H(e^{j\omega})$ for a narrowband input of the form $x[n] = s[n]\cos(\omega_0 n)$. Since it is assumed that $X(e^{j\omega})$ is nonzero only around $\omega = \omega_0$, the effect of the phase of the system can be approximated in a narrow band around $\omega = \omega_0$ with the linear approximation

$$\arg[H(e^{j\omega})] \simeq -\phi_0 - \omega n_d, \tag{5.14}$$

where n_d now represents the group delay. With this approximation, it can be shown (see Problem 5.63) that the response $y[n]$ to $x[n] = s[n]\cos(\omega_0 n)$ is approximately $y[n] = |H(e^{j\omega_0})||s[n - n_d]\cos(\omega_0 n - \phi_0 - \omega_0 n_d)$. Consequently, the time delay of the envelope $s[n]$ of the narrowband signal $x[n]$ with Fourier transform centered at ω_0 is given by the negative of the slope of the phase at ω_0. In general, we can think of a broadband signal as a superposition of narrowband signals with different center frequencies. If the group delay is constant with frequency then each narrowband component will undergo identical delay. If the group delay is not constant, there will be different delays applied to different frequency packets resulting in a dispersion in time of the output signal energy. Thus, nonlinearity of the phase or equivalently nonconstant group delay results in time dispersion.

5.1.2 Illustration of Effects of Group Delay and Attenuation

As an illustration of the effects of phase, group delay, and attenuation, consider the specific system having system function

$$H(z) = \underbrace{\left(\frac{(1 - .98e^{j.8\pi}z^{-1})(1 - .98e^{-j.8\pi}z^{-1})}{(1 - .8e^{j.4\pi}z^{-1})(1 - .8e^{-j.4\pi}z^{-1})}\right)}_{H_1(z)} \underbrace{\prod_{k=1}^{4}\left(\frac{(c_k^* - z^{-1})(c_k - z^{-1})}{(1 - c_k z^{-1})(1 - c_k^* z^{-1})}\right)^2}_{H_2(z)} \tag{5.15}$$

with $c_k = 0.95e^{j(.15\pi + .02\pi k)}$ for $k = 1, 2, 3, 4$ and $H_1(z)$ and $H_2(z)$ defined as indicated. The pole-zero plot for the overall system function $H(z)$ is shown in Figure 5.2, where the factor $H_1(z)$ in Eq. (5.15) contributes the complex conjugate pair of poles at $z = 0.8e^{\pm j.4\pi}$ as well as the pair of zeros close to the unit circle at $z = .98e^{\pm j.8\pi}$.

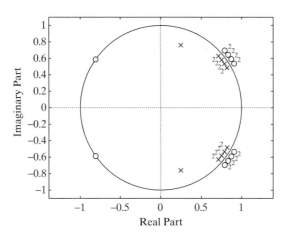

Figure 5.2   Pole-zero plot for the filter in the example of Section 5.1.2. (The number 2 indicates double-order poles and zeroes.)

The factor $H_2(z)$ in Eq. (5.15) contributes the groups of double-order poles at $z = c_k = 0.95e^{\pm j(.15\pi + .02\pi k)}$ and double-order zeros at $z = 1/c_k = 1/0.95e^{\mp j(.15\pi + .02\pi k)}$ for $k = 1, 2, 3, 4$. By itself, $H_2(z)$ represents an allpass system (see Section 5.5), i.e., $|H_2(e^{j\omega})| = 1$ for all ω. As we will see, $H_2(z)$ introduces a large amount of group delay over a narrow band of frequencies.

The frequency response functions for the overall system are shown in Figures 5.3 and 5.4. These figures illustrate several important points. First observe in Figure 5.3(a) that the principal value phase response exhibits multiple discontinuities of size 2π. These are due to the modulo 2π computation of the phase. Figure 5.3(b) shows the unwrapped (continuous) phase curve obtained by appropriately removing the jumps of size 2π.

Figure 5.4 shows the group delay and magnitude response of the overall system. Observe that, since the unwrapped phase is monotonically decreasing except around $\omega = \pm.8\pi$, the group delay is positive everywhere except in that region. Also, the group delay has a large positive peak in the bands of frequencies $.17\pi < |\omega| < .23\pi$ where the continuous phase has maximum negative slope. This frequency band corresponds to the angular location of the clusters of poles and reciprocal zeros in Figure 5.2. Also note the negative dip around $\omega = \pm.8\pi$, where the phase has positive slope. Since $H_2(z)$ represents an allpass filter, the magnitude response of the overall filter is entirely controlled by the poles and zeros of $H_1(z)$. Thus, since the frequency response is $H(z)$ evaluated for $z = e^{j\omega}$, the zeros at $z = .98e^{\pm j.8\pi}$ cause the overall frequency response to be very small in a band around frequencies $\omega = \pm.8\pi$.

In Figure 5.5(a) we show an input signal $x[n]$ consisting of three narrowband pulses separated in time. Figure 5.5(b) shows the corresponding DTFT magnitude $|X(e^{j\omega})|$. The pulses are given by

$$x_1[n] = w[n]\cos(0.2\pi n), \tag{5.16a}$$

$$x_2[n] = w[n]\cos(0.4\pi n - \pi/2), \tag{5.16b}$$

$$x_3[n] = w[n]\cos(0.8\pi n + \pi/5). \tag{5.16c}$$

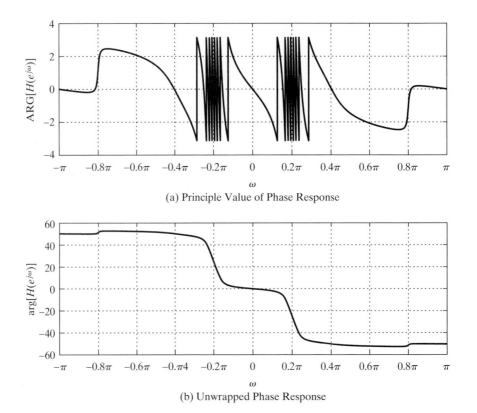

Figure 5.3 Phase response functions for system in the example of Section 5.1.2; (a) Principal value phase, ARG[$H(e^{j\omega})$], (b) Continuous phase arg [$H(e^{j\omega})$].

where each sinusoid is shaped into a finite-duration pulse by the 61-point envelope sequence

$$w[n] = \begin{cases} 0.54 - 0.46\cos(2\pi n/M), & 0 \le n \le M, \\ 0, & \text{otherwise} \end{cases} \tag{5.17}$$

with $M = 60$.[1] The complete input sequence shown in Figure 5.5(a) is

$$x[n] = x_3[n] + x_1[n - M - 1] + x_2[n - 2M - 2], \tag{5.18}$$

i.e., the highest frequency pulse comes first, then the lowest, followed by the mid-frequency pulse. From the windowing or modulation theorem for discrete-time Fourier transforms (Section 2.9.7), the DTFT of a windowed (truncated-in-time) sinusoid is the convolution of the DTFT of the infinite-duration sinusoid (comprised of impulses at ± the frequency of the sinusoid) with the DTFT of the window. The three sinusoidal frequencies are $\omega_1 = 0.2\pi$, $\omega_2 = 0.4\pi$, and $\omega_3 = 0.8\pi$. Correspondingly, in the Fourier transform magnitude in Figure 5.5(b) we see significant energy centered and concen-

---

[1] In Chapters 7 and 10, we will see that this envelope sequence is called a Hamming window when used in filter design and spectrum analysis respectively.

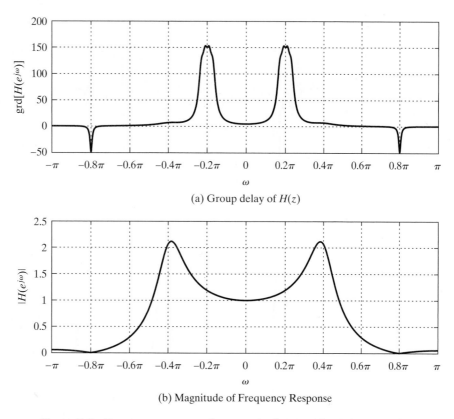

(a) Group delay of $H(z)$

(b) Magnitude of Frequency Response

Figure 5.4   Frequency response of system in the example of Section 5.1.2;
(a) Group delay function, grd[$H(e^{j\omega})$], (b) Magnitude of frequency response,
$|H(e^{j\omega})|$.

trated around each of the three frequencies. Each pulse contributes (in the frequency domain) a band of frequencies centered at the frequency of the sinusoid and with a shape and width corresponding to the Fourier transform of the time window applied to the sinusoid.[2]

When used as input to the system with system function $H(z)$, each of the frequency packets or groups associated with each of the narrowband pulses will be affected by the filter response magnitude and group delay over the frequency band of that group. From the filter frequency response magnitude, we see that the frequency group centered and concentrated around $\omega = \omega_1 = 0.2\pi$ will experience a slight amplitude gain, and the one around $\omega = \omega_2 = 0.4\pi$ will experience a gain of about 2. Since the magnitude of the frequency response is very small around frequency $\omega = \omega_3 = 0.8\pi$, the highest-frequency pulse will be significantly attenuated. It will not be totally eliminated, of course, since the frequency content of that group extends below and above frequency $\omega = \omega_3 = 0.8\pi$ because of the windowing applied to the sinusoid. Examining the plot of

---

[2] As we will see later in Chapters 7 and 10, the width of the frequency bands is approximately inversely proportional to the length of the window $M + 1$.

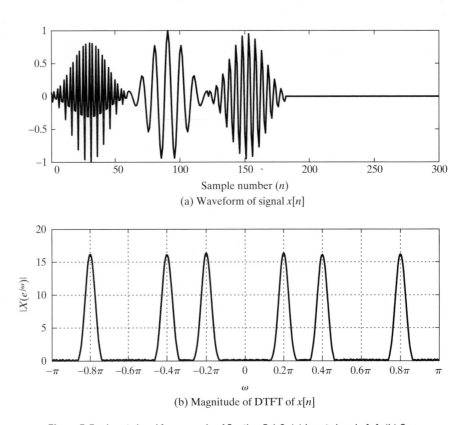

(a) Waveform of signal $x[n]$

(b) Magnitude of DTFT of $x[n]$

Figure 5.5 Input signal for example of Section 5.1.2; (a) Input signal $x[n]$, (b) Corresponding DTFT magnitude $|X(e^{j\omega})|$.

the system group delay in Figure 5.4(a), we see that the group delay around frequency $\omega = \omega_1 = 0.2\pi$ is significantly larger than for either $\omega = \omega_2 = 0.4\pi$ or $\omega = \omega_3 = 0.8\pi$, and consequently the lowest-frequency pulse will experience the most delay through the system.

The system output is shown in Figure 5.6. The pulse at frequency $\omega = \omega_3 = 0.8\pi$ has been essentially eliminated, which is consistent with the low values of the frequency response magnitude around that frequency. The two other pulses have been increased in amplitude and delayed; the pulse at $\omega = 0.2\pi$ is slightly larger and delayed by about 150 samples, and the pulse at $\omega = 0.4\pi$ has about twice the amplitude and is delayed by about 10 samples. This is consistent with the magnitude response and group delay at those frequencies. In fact, since the low-frequency pulse is delayed by 140 samples more than the mid-frequency pulse and the pulses are each only 61 samples long, these two pulses are interchanged in time order in the output.

The example that we have presented in this subsection was designed to illustrate how LTI systems can modify signals through the combined effects of amplitude scaling and phase shift. For the specific signal that we chose, consisting of a sum of narrowband components, it was possible to trace the effects on the individual pulses. This is because the frequency response functions were smooth and varied only slightly across the narrow

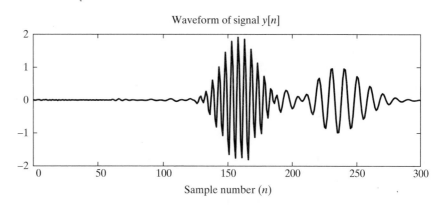

Figure 5.6   Output signal for the example of Section 5.1.2.

frequency bands occupied by the individual components. Therefore, all the frequencies corresponding to a given pulse were subject to approximately the same gain and were delayed by approximately the same amount, resulting in the pulse shape being replicated with only scaling and delay at the output. For wideband signals, this would generally not be the case, because different parts of the spectrum would be modified differently by the system. In such cases, recognizable features of the input such as pulse shape generally would not be obvious in the output signal, and individual pulses separated in time in the input might cause overlapping contributions to the output.

This example has illustrated a number of important concepts that will be further elaborated on in this and subsequent chapters. After completing a thorough study of this chapter, it would be worthwhile to study the example of this subsection again carefully to gain a greater appreciation of its nuances. To fully appreciate this example, it also would be useful to duplicate it with variable parameters in a convenient programming system such as MATLAB. Before testing the computer program, the reader should attempt to predict what would happen, for example, when the window length is either increased or decreased or when the frequencies of the sinusoids are changed.

5.2 SYSTEMS CHARACTERIZED BY LINEAR CONSTANT-COEFFICIENT DIFFERENCE EQUATIONS

While ideal filters are useful conceptually, discrete-time filters are most typically realized through the implementation of a linear constant-coefficient difference equation of the form of Eq. (5.19).

$$\sum_{k=0}^{N} a_k y[n-k] = \sum_{k=0}^{M} b_k x[n-k]. \tag{5.19}$$

In Chapter 6, we discuss various computational structures for realizing such systems, and in Chapter 7, we discuss various procedures for obtaining the parameters of the difference equation to approximate a desired frequency response. In this section, with the aid of the z-transform, we examine the properties and characteristics of LTI systems represented by Eq. (5.19). The results and insights will play an important role in many of the later chapters.

As we saw in Section 3.5, applying the z-transform to both sides of Eq. (5.19) and using the linearity property (Section 3.4.1) and the time-shifting property (Section 3.4.2), it follows that, for a system whose input and output satisfy a difference equation of the form of Eq. (5.19), the system function has the algebraic form

$$H(z) = \frac{Y(z)}{X(z)} = \frac{\displaystyle\sum_{k=0}^{M} b_k z^{-k}}{\displaystyle\sum_{k=0}^{N} a_k z^{-k}}. \tag{5.20}$$

In Eq. (5.20) $H(z)$ takes the form of a ratio of polynomials in z^{-1}, because Eq. (5.19) consists of two linear combinations of delay terms. Although Eq. (5.20) can, of course, be rewritten so that the polynomials are expressed as powers of z rather than of z^{-1}, it is common practice not to do so. Also, it is often convenient to express Eq. (5.20) in factored form as

$$H(z) = \left(\frac{b_0}{a_0}\right) \frac{\displaystyle\prod_{k=1}^{M}(1 - c_k z^{-1})}{\displaystyle\prod_{k=1}^{N}(1 - d_k z^{-1})}. \tag{5.21}$$

Each of the factors $(1 - c_k z^{-1})$ in the numerator contributes a zero at $z = c_k$ and a pole at $z = 0$. Similarly, each of the factors $(1 - d_k z^{-1})$ in the denominator contributes a zero at $z = 0$ and a pole at $z = d_k$.

There is a straightforward relationship between the difference equation and the corresponding algebraic expression for the system function. Specifically, the numerator polynomial in Eq. (5.20) has the same coefficients and algebraic structure as the right-hand side of Eq. (5.19) (the terms of the form $b_k z^{-k}$ correspond to $b_k x[n - k]$), whereas the denominator polynomial in Eq. (5.20) has the same coefficients and algebraic structure as the left-hand side of Eq. (5.19) (the terms of the form $a_k z^{-k}$ correspond to $a_k y[n - k]$). Thus, given either the system function in the form of Eq. (5.20) or the difference equation in the form of Eq. (5.19), it is straightforward to obtain the other. This is illustrated in the following example.

Example 5.1   2nd-Order System

Suppose that the system function of an LTI system is

$$H(z) = \frac{(1 + z^{-1})^2}{\left(1 - \frac{1}{2}z^{-1}\right)\left(1 + \frac{3}{4}z^{-1}\right)}. \tag{5.22}$$

To find the difference equation that is satisfied by the input and output of this system, we express $H(z)$ in the form of Eq. (5.20) by multiplying the numerator and denominator factors to obtain the ratio of polynomials

$$H(z) = \frac{1 + 2z^{-1} + z^{-2}}{1 + \frac{1}{4}z^{-1} - \frac{3}{8}z^{-2}} = \frac{Y(z)}{X(z)}. \tag{5.23}$$

Thus,

$$\left(1 + \tfrac{1}{4}z^{-1} - \tfrac{3}{8}z^{-2}\right) Y(z) = (1 + 2z^{-1} + z^{-2})X(z),$$

and the difference equation is

$$y[n] + \tfrac{1}{4}y[n-1] - \tfrac{3}{8}y[n-2] = x[n] + 2x[n-1] + x[n-2]. \tag{5.24}$$

5.2.1 Stability and Causality

To obtain Eq. (5.20) from Eq. (5.19), we assumed that the system was linear and time invariant, so that Eq. (5.2) applied, but we made no further assumption about stability or causality. Correspondingly, from the difference equation, we can obtain the algebraic expression for the system function, but not the region of convergence (ROC). Specifically, the ROC of $H(z)$ is not determined from the derivation leading to Eq. (5.20), since all that is required for Eq. (5.20) to hold is that $X(z)$ and $Y(z)$ have overlapping ROCs. This is consistent with the fact that, as we saw in Chapter 2, the difference equation does not uniquely specify the impulse response of an LTI system. For the system function of Eq. (5.20) or (5.21), there are a number of choices for the ROC. For a given ratio of polynomials, each possible choice for the ROC will lead to a different impulse response, but they will all correspond to the same difference equation. However, if we assume that the system is causal, it follows that $h[n]$ must be a right-sided sequence, and therefore, the ROC of $H(z)$ must be outside the outermost pole. Alternatively, if we assume that the system is stable, then, from the discussion in Section 2.4, the impulse response must be absolutely summable, i.e.,

$$\sum_{n=-\infty}^{\infty} |h[n]| < \infty. \tag{5.25}$$

Since Eq. (5.25) is identical to the condition that

$$\sum_{n=-\infty}^{\infty} |h[n]z^{-n}| < \infty \tag{5.26}$$

for $|z| = 1$, the condition for stability is equivalent to the condition that the ROC of $H(z)$ includes the unit circle. Determining the ROC to associate with the system function obtained from the difference equation is illustrated in the following example.

Example 5.2   Determining the ROC

Consider the LTI system with input and output related through the difference equation
$$y[n] - \tfrac{5}{2}y[n-1] + y[n-2] = x[n]. \tag{5.27}$$
From the previous discussions, the algebraic expression for $H(z)$ is given by
$$H(z) = \frac{1}{1 - \tfrac{5}{2}z^{-1} + z^{-2}} = \frac{1}{\left(1 - \tfrac{1}{2}z^{-1}\right)(1 - 2z^{-1})}. \tag{5.28}$$
The corresponding pole–zero plot for $H(z)$ is indicated in Figure 5.7. There are three possible choices for the ROC. If the system is assumed to be causal, then the ROC

is outside the outermost pole, i.e., $|z| > 2$. In this case, the system will not be stable, since the ROC does not include the unit circle. If we assume that the system is stable, then the ROC will be $\frac{1}{2} < |z| < 2$, and $h[n]$ will be a two-sided sequence. For the third possible choice of ROC, $|z| < \frac{1}{2}$, the system will be neither stable nor causal.

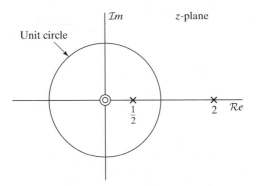

Figure 5.7   Pole–zero plot for Example 5.2.

As Example 5.2 suggests, causality and stability are not necessarily compatible requirements. For an LTI system whose input and output satisfy a difference equation of the form of Eq. (5.19) to be both causal and stable, the ROC of the corresponding system function must be outside the outermost pole and include the unit circle. Clearly, this requires that all the poles of the system function be inside the unit circle.

5.2.2 Inverse Systems

For a given LTI system with system function $H(z)$, the corresponding inverse system is defined to be the system with system function $H_i(z)$ such that if it is cascaded with $H(z)$, the overall effective system function is unity; i.e.,

$$G(z) = H(z)H_i(z) = 1. \tag{5.29}$$

This implies that

$$H_i(z) = \frac{1}{H(z)}. \tag{5.30}$$

The time-domain condition equivalent to Eq. (5.29) is

$$g[n] = h[n] * h_i[n] = \delta[n]. \tag{5.31}$$

From Eq. (5.30), the frequency response of the inverse system, if it exists, is

$$H_i(e^{j\omega}) = \frac{1}{H(e^{j\omega})}; \tag{5.32}$$

i.e., $H_i(e^{j\omega})$ is the reciprocal of $H(e^{j\omega})$. Equivalently, the log magnitude, phase, and group delay of the inverse system are negatives of the corresponding functions for the original system. Not all systems have an inverse. For example, the ideal lowpass filter does not. There is no way to recover the frequency components above the cutoff frequency that are set to zero by such a filter.

Many systems do have inverses, and the class of systems with rational system functions provides a very useful and interesting example. Specifically, consider

$$H(z) = \left(\frac{b_0}{a_0}\right) \frac{\prod\limits_{k=1}^{M}(1 - c_k z^{-1})}{\prod\limits_{k=1}^{N}(1 - d_k z^{-1})}, \tag{5.33}$$

with zeros at $z = c_k$ and poles at $z = d_k$, in addition to possible zeros and/or poles at $z = 0$ and $z = \infty$. Then

$$H_i(z) = \left(\frac{a_0}{b_0}\right) \frac{\prod\limits_{k=1}^{N}(1 - d_k z^{-1})}{\prod\limits_{k=1}^{M}(1 - c_k z^{-1})}; \tag{5.34}$$

i.e., the poles of $H_i(z)$ are the zeros of $H(z)$ and vice versa. The question arises as to what ROC to associate with $H_i(z)$. The answer is provided by the convolution theorem, expressed in this case by Eq. (5.31). For Eq. (5.31) to hold, the ROC of $H(z)$ and $H_i(z)$ must overlap. If $H(z)$ is causal, its ROC is

$$|z| > \max_k |d_k|. \tag{5.35}$$

Thus, any appropriate ROC for $H_i(z)$ that overlaps with the region specified by Eq. (5.35) is a valid ROC for $H_i(z)$. Examples 5.3 and 5.4 will illustrate some of the possibilities.

Example 5.3   Inverse System for 1st-Order System

Let $H(z)$ be

$$H(z) = \frac{1 - 0.5z^{-1}}{1 - 0.9z^{-1}}$$

with ROC $|z| > 0.9$. Then $H_i(z)$ is

$$H_i(z) = \frac{1 - 0.9z^{-1}}{1 - 0.5z^{-1}}.$$

Since $H_i(z)$ has only one pole, there are only two possibilities for its ROC, and the only choice for the ROC of $H_i(z)$ that overlaps with $|z| > 0.9$ is $|z| > 0.5$. Therefore, the impulse response of the inverse system is

$$h_i[n] = (0.5)^n u[n] - 0.9(0.5)^{n-1} u[n-1].$$

In this case, the inverse system is both causal and stable.

Example 5.4   Inverse for System with a Zero in the ROC

Suppose that $H(z)$ is

$$H(z) = \frac{z^{-1} - 0.5}{1 - 0.9z^{-1}}, \qquad |z| > 0.9.$$

The inverse system function is

$$H_i(z) = \frac{1 - 0.9z^{-1}}{z^{-1} - 0.5} = \frac{-2 + 1.8z^{-1}}{1 - 2z^{-1}}.$$

As before, there are two possible ROCs that could be associated with this algebraic expression for $H_i(z)$: $|z| < 2$ and $|z| > 2$. In this case, however, both regions overlap with $|z| > 0.9$, so both are valid inverse systems. The corresponding impulse response for an ROC $|z| < 2$ is

$$h_{i1}[n] = 2(2)^n u[-n-1] - 1.8(2)^{n-1} u[-n]$$

and, for an ROC $|z| > 2$, is

$$h_{i2}[n] = -2(2)^n u[n] + 1.8(2)^{n-1} u[n-1].$$

We see that $h_{i1}[n]$ is stable and noncausal, while $h_{i2}[n]$ is unstable and causal. Theoretically, either system cascaded with $H(z)$ will result in the identity system.

A generalization from Examples 5.3 and 5.4 is that if $H(z)$ is a causal system with zeros at $c_k, k = 1, \ldots, M$, then its inverse system will be causal if and only if we associate the ROC,

$$|z| > \max_k |c_k|,$$

with $H_i(z)$. If we also require that the inverse system be stable, then the ROC of $H_i(z)$ must include the unit circle, in which case

$$\max_k |c_k| < 1;$$

i.e., all the zeros of $H(z)$ must be inside the unit circle. Thus, an LTI system is stable and causal and also has a stable and causal inverse if and only if both the poles and the zeros of $H(z)$ are inside the unit circle. Such systems are referred to as *minimum-phase* systems and will be discussed in more detail in Section 5.6.

5.2.3 Impulse Response for Rational System Functions

The discussion of the partial fraction expansion technique for finding inverse z-transforms (Section 3.3.2) can be applied to the system function $H(z)$ to obtain a general expression for the impulse response of a system that has a rational system function as in Eq. (5.21). Recall that any rational function of z^{-1} with only 1st-order poles can be expressed in the form

$$H(z) = \sum_{r=0}^{M-N} B_r z^{-r} + \sum_{k=1}^{N} \frac{A_k}{1 - d_k z^{-1}}, \qquad (5.36)$$

where the terms in the first summation would be obtained by long division of the denominator into the numerator and would be present only if $M \geq N$. The coefficients A_k in the second set of terms are obtained using Eq. (3.43). If $H(z)$ has a multiple-order pole, its partial fraction expansion would have the form of Eq. (3.46). If the system is assumed to be causal, then the ROC is outside all of the poles in Eq. (5.36), and it follows that

$$h[n] = \sum_{r=0}^{M-N} B_r \delta[n-r] + \sum_{k=1}^{N} A_k d_k^n u[n], \tag{5.37}$$

where the first summation is included only if $M \geq N$.

In discussing LTI systems, it is useful to identify two classes. In the first class, at least one nonzero pole of $H(z)$ is not canceled by a zero. In this case, $h[n]$ will have at least one term of the form $A_k(d_k)^n u[n]$, and $h[n]$ will not be of finite length, i.e., will not be zero outside a finite interval. Consequently, systems of this class are infinite impulse response (IIR) systems.

For the second class of systems, $H(z)$ has no poles except at $z = 0$; i.e., $N = 0$ in Eqs. (5.19) and (5.20). Thus, a partial fraction expansion is not possible, and $H(z)$ is simply a polynomial in z^{-1} of the form

$$H(z) = \sum_{k=0}^{M} b_k z^{-k}. \tag{5.38}$$

(We assume, without loss of generality, that $a_0 = 1$.) In this case, $H(z)$ is determined to within a constant multiplier by its zeros. From Eq. (5.38), $h[n]$ is seen by inspection to be

$$h[n] = \sum_{k=0}^{M} b_k \, \delta[n-k] = \begin{cases} b_n, & 0 \leq n \leq M, \\ 0, & \text{otherwise.} \end{cases} \tag{5.39}$$

In this case, the impulse response is finite in length; i.e., it is zero outside a finite interval. Consequently, these systems are finite impulse response (FIR) systems. Note that for FIR systems, the difference equation of Eq. (5.19) is identical to the convolution sum, i.e.,

$$y[n] = \sum_{k=0}^{M} b_k x[n-k]. \tag{5.40}$$

Example 5.5 gives a simple example of an FIR system.

Example 5.5   A Simple FIR System

Consider an impulse response that is a truncation of the impulse response of an IIR system with system function

$$G(z) = \frac{1}{1 - az^{-1}}, \qquad |z| > |a|,$$

i.e.,

$$h[n] = \begin{cases} a^n, & 0 \leq n \leq M, \\ 0 & \text{otherwise.} \end{cases}$$

Then, the system function is

$$H(z) = \sum_{n=0}^{M} a^n z^{-n} = \frac{1 - a^{M+1} z^{-M-1}}{1 - az^{-1}}. \tag{5.41}$$

Since the zeros of the numerator are at z-plane locations

$$z_k = ae^{j2\pi k/(M+1)}, \qquad k = 0, 1, \dots, M, \tag{5.42}$$

where a is assumed real and positive, the pole at $z = a$ is canceled by the zero denoted z_0. The pole–zero plot for the case $M = 7$ is shown in Figure 5.8.

The difference equation satisfied by the input and output of the LTI system is the discrete convolution

$$y[n] = \sum_{k=0}^{M} a^k x[n - k]. \tag{5.43}$$

However, Eq. (5.41) suggests that the input and output also satisfy the difference equation

$$y[n] - ay[n-1] = x[n] - a^{M+1}x[n - M - 1]. \tag{5.44}$$

These two equivalent difference equations result from the two equivalent forms of $H(z)$ in Eq. (5.41).

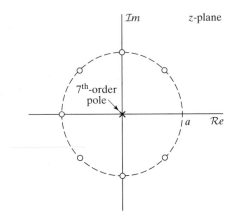

Figure 5.8   Pole–zero plot for Example 5.5.

5.3 FREQUENCY RESPONSE FOR RATIONAL SYSTEM FUNCTIONS

If a stable LTI system has a rational system function, i.e., if its input and output satisfy a difference equation of the form of Eq. (5.19), then its frequency response (the system function of Eq. (5.20) evaluated on the unit circle) has the form

$$H(e^{j\omega}) = \frac{\displaystyle\sum_{k=0}^{M} b_k e^{-j\omega k}}{\displaystyle\sum_{k=0}^{N} a_k e^{-j\omega k}}. \tag{5.45}$$

That is, $H(e^{j\omega})$ is a ratio of polynomials in the variable $e^{-j\omega}$. To determine the magnitude, phase, and group delay associated with the frequency response of such systems, it is useful to express $H(e^{j\omega})$ in terms of the poles and zeros of $H(z)$. Such an expression results from substituting $z = e^{j\omega}$ into Eq. (5.21) to obtain

$$H(e^{j\omega}) = \left(\frac{b_0}{a_0}\right) \frac{\displaystyle\prod_{k=1}^{M}(1 - c_k e^{-j\omega})}{\displaystyle\prod_{k=1}^{N}(1 - d_k e^{-j\omega})}. \tag{5.46}$$

From Eq. (5.46), it follows that

$$|H(e^{j\omega})| = \left|\frac{b_0}{a_0}\right| \frac{\displaystyle\prod_{k=1}^{M}|1 - c_k e^{-j\omega}|}{\displaystyle\prod_{k=1}^{N}|1 - d_k e^{-j\omega}|}. \tag{5.47}$$

Correspondingly, the magnitude-squared function is

$$|H(e^{j\omega})|^2 = H(e^{j\omega})H^*(e^{j\omega}) = \left(\frac{b_0}{a_0}\right)^2 \frac{\displaystyle\prod_{k=1}^{M}(1 - c_k e^{-j\omega})(1 - c_k^* e^{j\omega})}{\displaystyle\prod_{k=1}^{N}(1 - d_k e^{-j\omega})(1 - d_k^* e^{j\omega})}. \tag{5.48}$$

From Eq. (5.47), we note that $|H(e^{j\omega})|$ is the product of the magnitudes of all the zero factors of $H(z)$ evaluated on the unit circle, divided by the product of the magnitudes of all the pole factors evaluated on the unit circle. Expressed in decibels (dB), the gain is defined as

$$\text{Gain in dB} = 20\log_{10}|H(e^{j\omega})| \tag{5.49}$$

$$\text{Gain in dB} = 20\log_{10}\left|\frac{b_0}{a_0}\right| + \sum_{k=1}^{M} 20\log_{10}|1 - c_k e^{-j\omega}| \\ - \sum_{k=1}^{N} 20\log_{10}|1 - d_k e^{-j\omega}|. \tag{5.50}$$

The phase response for a rational system function has the form

$$\arg\left[H(e^{j\omega})\right] = \arg\left[\frac{b_0}{a_0}\right] + \sum_{k=1}^{M}\arg\left[1 - c_k e^{-j\omega}\right] - \sum_{k=1}^{N}\arg\left[1 - d_k e^{-j\omega}\right], \tag{5.51}$$

where $\arg[\ ]$ represents the continuous (unwrapped) phase.

The corresponding group delay for a rational system function is

$$\text{grd}[H(e^{j\omega})] = \sum_{k=1}^{N} \frac{d}{d\omega}(\arg[1 - d_k e^{-j\omega}]) - \sum_{k=1}^{M} \frac{d}{d\omega}(\arg[1 - c_k e^{-j\omega}]). \tag{5.52}$$

An equivalent expression is

$$\text{grd}[H(e^{j\omega})] = \sum_{k=1}^{N} \frac{|d_k|^2 - \mathcal{R}e\{d_k e^{-j\omega}\}}{1 + |d_k|^2 - 2\mathcal{R}e\{d_k e^{-j\omega}\}} - \sum_{k=1}^{M} \frac{|c_k|^2 - \mathcal{R}e\{c_k e^{-j\omega}\}}{1 + |c_k|^2 - 2\mathcal{R}e\{c_k e^{-j\omega}\}}. \quad (5.53)$$

In Eq. (5.51), as written, the phase of each of the terms is ambiguous; i.e., any integer multiple of 2π can be added to each term at each value of ω without changing the value of the complex number. The expression for the group delay, on the other hand, is defined in terms of the derivative of the unwrapped phase.

Equations (5.50), (5.51), and (5.53) represent the magnitude in dB, the phase, and the group delay, respectively, as a sum of contributions from each of the poles and zeros of the system function. Consequently, to gain an understanding of how the pole and zero locations of higher-order stable systems impact the frequency response, it is useful to consider in detail the frequency response of 1st-order and 2nd-order systems in relation to their pole and zero locations.

5.3.1 Frequency Response of 1st-Order Systems

In this section, we examine the properties of a single factor of the form $(1 - re^{j\theta}e^{-j\omega})$, where r is the radius and θ is the angle of the pole or zero in the z-plane. This factor is typical of either a pole or a zero at a radius r and angle θ in the z-plane.

The square of the magnitude of such a factor is

$$|1 - re^{j\theta}e^{-j\omega}|^2 = (1 - re^{j\theta}e^{-j\omega})(1 - re^{-j\theta}e^{j\omega}) = 1 + r^2 - 2r\cos(\omega - \theta). \quad (5.54)$$

The gain in dB associated with this factor is

$$(+/-)20\log_{10}|1 - re^{j\theta}e^{-j\omega}| = (+/-)10\log_{10}[1 + r^2 - 2r\cos(\omega - \theta)], \quad (5.55)$$

with a positive sign if the factor represents a zero and a negative sign if it represents a pole.

The contribution to the principal value of the phase for such a factor is

$$(+/-)\text{ARG}[1 - re^{j\theta}e^{-j\omega}] = (+/-)\arctan\left[\frac{r\sin(\omega - \theta)}{1 - r\cos(\omega - \theta)}\right]. \quad (5.56)$$

Differentiating the right-hand side of Eq. (5.56) (except at discontinuities) gives the contribution to the group delay of the factor as

$$(+/-)\text{grd}[1 - re^{j\theta}e^{-j\omega}] = (+/-)\frac{r^2 - r\cos(\omega - \theta)}{1 + r^2 - 2r\cos(\omega - \theta)} = (+/-)\frac{r^2 - r\cos(\omega - \theta)}{|1 - re^{j\theta}e^{-j\omega}|^2}. \quad (5.57)$$

again, with the positive sign for a zero and a negative sign for a pole. The functions in Eqs. (5.54)–(5.57) are, of course, periodic in ω with period 2π. Figure 5.9(a) shows a plot of Eq. (5.55) as a function of ω over one period ($0 \le \omega < 2\pi$) for several values of θ with $r = 0.9$.

Figure 5.9(b) shows the phase function in Eq. (5.56) as a function of ω for $r = 0.9$ and several values of θ. Note that the phase is zero at $\omega = \theta$ and that, for fixed r, the function simply shifts with θ. Figure 5.9(c) shows the group delay function in Eq. (5.57) for the same conditions on r and θ. Note that the high positive slope of the phase around $\omega = \theta$ corresponds to a large negative peak in the group delay function at $\omega = \theta$.

In inferring frequency response characteristics from pole–zero plots of either continuous-time or discrete-time systems, the associated vector diagrams in the

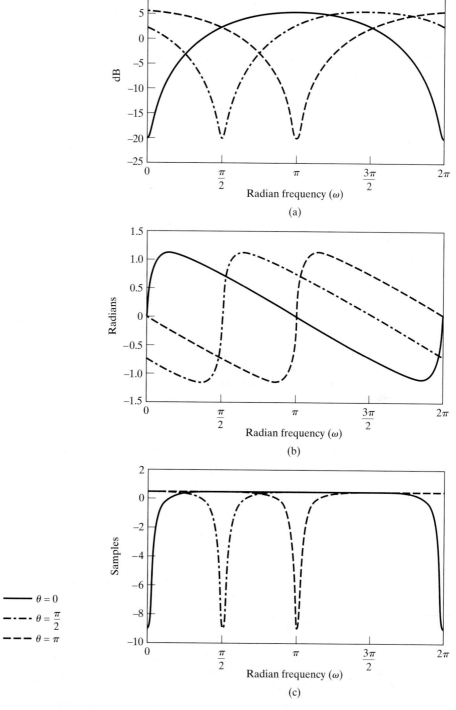

Figure 5.9 Frequency response for a single zero, with $r = 0.9$ and the three values of θ shown. (a) Log magnitude. (b) Phase. (c) Group delay.

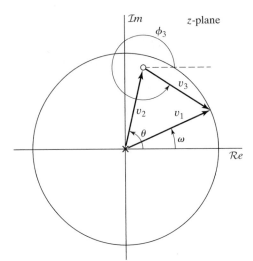

Figure 5.10   z-plane vectors for a 1st-order system function evaluated on the unit circle, with $r < 1$.

complex plane are typically helpful. In this construction, each pole and zero factor can be represented by a vector in the z-plane from the pole or zero to a point on the unit circle. For a 1st-order system function of the form

$$H(z) = (1 - re^{j\theta}z^{-1}) = \frac{(z - re^{j\theta})}{z}, \qquad r < 1, \tag{5.58}$$

the pole–zero pattern is illustrated in Figure 5.10. Also indicated in this figure are the vectors v_1, v_2, and $v_3 = v_1 - v_2$, representing the complex numbers $e^{j\omega}$, $re^{j\theta}$, and $(e^{j\omega} - re^{j\theta})$, respectively. In terms of these vectors, the magnitude of the complex number

$$\frac{e^{j\omega} - re^{j\theta}}{e^{j\omega}}$$

is the ratio of the magnitudes of the vectors v_3 and v_1, i.e.,

$$|1 - re^{j\theta}e^{-j\omega}| = \left| \frac{e^{j\omega} - re^{j\theta}}{e^{j\omega}} \right| = \frac{|v_3|}{|v_1|}, \tag{5.59}$$

or, since $|v_1| = 1$, Eq. (5.59) is just equal to $|v_3|$. The corresponding phase is

$$\angle(1 - re^{j\theta}e^{-j\omega}) = \angle(e^{j\omega} - re^{j\theta}) - \angle(e^{j\omega}) = \angle(v_3) - \angle(v_1)$$

$$= \phi_3 - \phi_1 = \phi_3 - \omega. \tag{5.60}$$

Thus, the contribution of a single factor $(1 - re^{j\theta}z^{-1})$ to the magnitude function at frequency ω is the length of the vector v_3 from the zero to the point $z = e^{j\omega}$ on the unit circle. The vector has minimum length when $\omega = \theta$. This accounts for the sharp dip in the magnitude function at $\omega = \theta$ in Figure 5.9(a). The vector v_1 from the pole at $z = 0$ to $z = e^{j\omega}$ always has unit length. Thus, it does not have any effect on the magnitude response. Equation (5.60) states that the phase function is equal to the difference between the angle of the vector from the zero at $re^{j\theta}$ to the point $z = e^{j\omega}$ and the angle of the vector from the pole at $z = 0$ to the point $z = e^{j\omega}$.

The dependence of the frequency-response contributions of a single factor $(1 - re^{j\theta}e^{-j\omega})$ on the radius r is shown in Figure 5.11 for $\theta = \pi$ and several values of r.

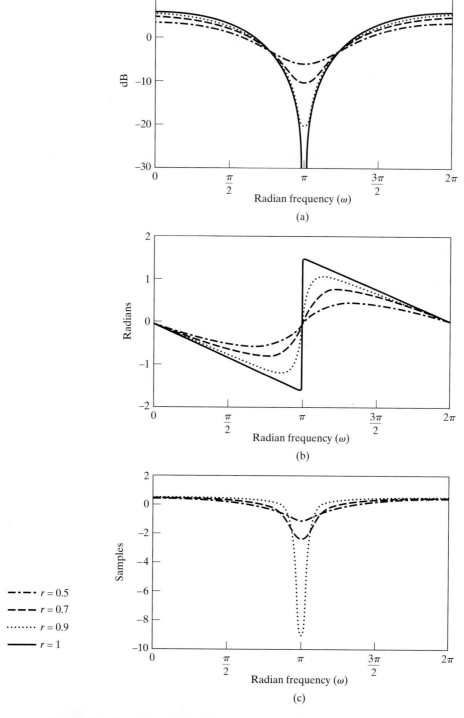

Figure 5.11 Frequency response for a single zero, with $\theta = \pi$, $r = 1$, 0.9, 0.7, and 0.5. (a) Log magnitude. (b) Phase. (c) Group delay for $r = 0.9$, 0.7, and 0.5.

295

Note that the log magnitude function plotted in Figure 5.11(a) dips more sharply as r becomes closer to 1; indeed, the magnitude in dB approaches $-\infty$ at $\omega = \theta$ as r approaches 1. The phase function plotted in Figure 5.11(b) has positive slope around $\omega = \theta$, which becomes infinite as r approaches 1. Thus, for $r = 1$, the phase function is discontinuous, with a jump of π radians at $\omega = \theta$. Away from $\omega = \theta$, the slope of the phase function is negative. Since the group delay is the negative of the slope of the phase curve, the group delay is negative around $\omega = \theta$, and it dips sharply as r approaches 1 becoming an impulse (not shown) when $r = 1$. Figure 5.11(c) shows that as we move away from $\omega = \theta$, the group delay becomes positive and relatively flat.

5.3.2 Examples with Multiple Poles and Zeros

In this section, we utilize and expand the discussion of Section 5.3.1 to determine the frequency response of systems with rational system functions.

Example 5.6   2nd-Order IIR System

Consider the 2nd-order system

$$H(z) = \frac{1}{(1 - re^{j\theta}z^{-1})(1 - re^{-j\theta}z^{-1})} = \frac{1}{1 - 2r\cos\theta z^{-1} + r^2 z^{-2}}. \tag{5.61}$$

The difference equation satisfied by the input and output of the system is

$$y[n] - 2r\cos\theta \, y[n-1] + r^2 y[n-2] = x[n].$$

Using the partial fraction expansion technique, we can show that the impulse response of a causal system with this system function is

$$h[n] = \frac{r^n \sin[\theta(n+1)]}{\sin\theta} u[n]. \tag{5.62}$$

The system function in Eq. (5.61) has a pole at $z = re^{j\theta}$ and at the conjugate location, $z = re^{-j\theta}$, and two zeros at $z = 0$. The pole–zero plot is shown in Figure 5.12.

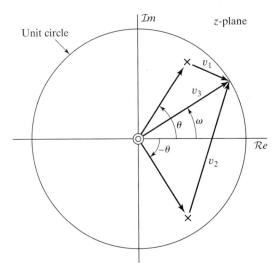

Figure 5.12   Pole–zero plot for Example 5.6.

From our discussion in Section 5.3.1,

$$20 \log_{10} |H(e^{j\omega})| = -10 \log_{10}[1 + r^2 - 2r \cos(\omega - \theta)] \\ -10 \log_{10}[1 + r^2 - 2r \cos(\omega + \theta)], \tag{5.63a}$$

$$\angle H(e^{j\omega}) = -\arctan\left[\frac{r \sin(\omega - \theta)}{1 - r \cos(\omega - \theta)}\right] - \arctan\left[\frac{r \sin(\omega + \theta)}{1 - r \cos(\omega + \theta)}\right], \tag{5.63b}$$

and

$$\text{grd}[H(e^{j\omega})] = -\frac{r^2 - r\cos(\omega - \theta)}{1 + r^2 - 2r\cos(\omega - \theta)} - \frac{r^2 - r\cos(\omega + \theta)}{1 + r^2 - 2r\cos(\omega + \theta)}. \tag{5.63c}$$

These functions are plotted in Figure 5.13 for $r = 0.9$ and $\theta = \pi/4$.

Figure 5.12 shows the pole and zero vectors v_1, v_2, and v_3. The magnitude response is the product of the lengths of the zero vectors (which in this case are always unity), divided by the product of the lengths of the pole vectors. That is,

$$|H(e^{j\omega})| = \frac{|v_3|^2}{|v_1| \cdot |v_2|} = \frac{1}{|v_1| \cdot |v_2|}. \tag{5.64}$$

When $\omega \approx \theta$, the length of the vector $v_1 = e^{j\omega} - re^{j\theta}$ becomes small and changes significantly as ω varies about θ, whereas the length of the vector $v_2 = e^{j\omega} - re^{-j\theta}$ changes only slightly as ω varies around $\omega = \theta$. Thus, the pole at angle θ dominates the frequency response around $\omega = \theta$, as is evident from Figure 5.13. By symmetry, the pole at angle $-\theta$ dominates the frequency response around $\omega = -\theta$.

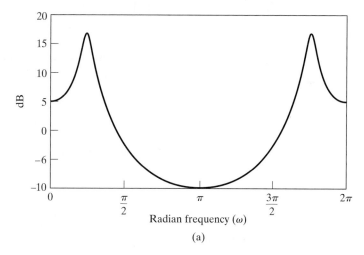

Figure 5.13 Frequency response for a complex-conjugate pair of poles as in Example 5.6, with $r = 0.9$, $\theta = \pi/4$. (a) Log magnitude.

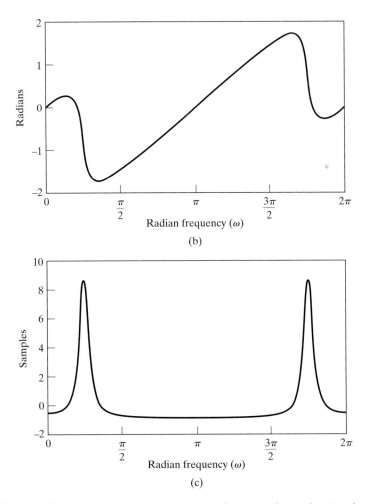

Figure 5.13 (*continued*) Frequency response for a complex-conjugate pair of poles as in Example 5.6, with $r = 0.9$, $\theta = \pi/4$. (b) Phase. (c) Group delay.

Example 5.7   2nd-Order FIR System

In this example we consider an FIR system whose impulse response is

$$h[n] = \delta[n] - 2r \cos\theta \delta[n - 1] + r^2 \delta[n - 2]. \qquad (5.65)$$

The corresponding system function is

$$H(z) = 1 - 2r \cos\theta z^{-1} + r^2 z^{-2}, \qquad (5.66)$$

which is the reciprocal of the system function in Example 5.6. Therefore, the frequency-response plots for this FIR system are simply the negative of the plots in Figure 5.13. Note that the pole and zero locations are interchanged in the reciprocal.

Example 5.8   3ʳᵈ–Order IIR System

In this example, we consider a lowpass filter designed using one of the approximation methods to be described in Chapter 7. The system function to be considered is

$$H(z) = \frac{0.05634(1 + z^{-1})(1 - 1.0166z^{-1} + z^{-2})}{(1 - 0.683z^{-1})(1 - 1.4461z^{-1} + 0.7957z^{-2})}, \tag{5.67}$$

and the system is specified to be stable. The zeros of this system function are at the following locations:

Radius	Angle
1	π rad
1	± 1.0376 rad (59.45°)

The poles are at the following locations:

Radius	Angle
0.683	0
0.892	± 0.6257 rad (35.85°)

The pole–zero plot for this system is shown in Figure 5.14. Figure 5.15 shows the log magnitude, phase, and group delay of the system. The effect of the zeros that are on the unit circle at $\omega = \pm 1.0376$ and π is clearly evident. However, the poles are placed so that, rather than peaking for frequencies close to their angles, the total log magnitude remains close to 0 dB over a band from $\omega = 0$ to $\omega = 0.2\pi$ (and, by symmetry, from $\omega = 1.8\pi$ to $\omega = 2\pi$), and then it drops abruptly and remains below -25 dB from about $\omega = 0.3\pi$ to 1.7π. As suggested by this example, useful approximations to frequency-selective filter responses can be achieved using poles to build up the magnitude response and zeros to suppress it.

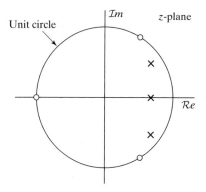

Figure 5.14   Pole–zero plot for the lowpass filter of Example 5.8.

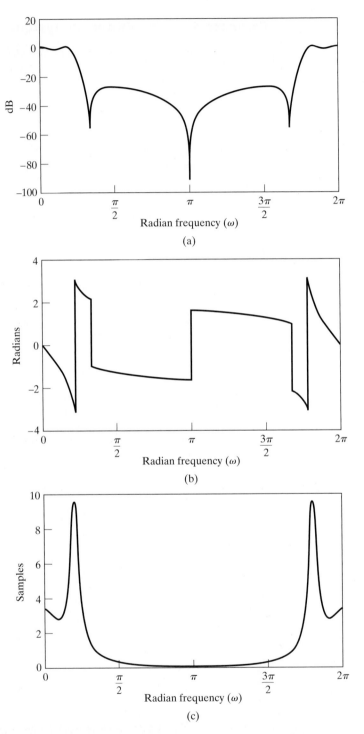

Figure 5.15 Frequency response for the lowpass filter of Example 5.8. (a) Log magnitude. (b) Phase. (c) Group delay.

In this example, we see both types of discontinuities in the plotted phase curve. At $\omega \approx 0.22\pi$, there is a discontinuity of 2π owing to the use of the principal value in plotting. At $\omega = \pm 1.0376$ and $\omega = \pi$, the discontinuities of π are due to the zeros on the unit circle.

5.4 RELATIONSHIP BETWEEN MAGNITUDE AND PHASE

In general, knowledge about the magnitude of the frequency response of an LTI system provides no information about the phase, and vice versa. However, for systems described by linear constant-coefficient difference equations, i.e., rational system functions, there is some constraint between magnitude and phase. In particular, as we discuss in this section, if the magnitude of the frequency response and the number of poles and zeros are known, then there are only a finite number of choices for the associated phase. Similarly, if the number of poles and zeros and the phase are known, then, to within a scale factor, there are only a finite number of choices for the magnitude. Furthermore, under a constraint referred to as minimum phase, the frequency-response magnitude specifies the phase uniquely, and the frequency-response phase specifies the magnitude to within a scale factor.

To explore the possible choices of system function, given the square of the magnitude of the system frequency response, we consider $|H(e^{j\omega})|^2$ expressed as

$$
\begin{aligned}
|H(e^{j\omega})|^2 &= H(e^{j\omega})H^*(e^{j\omega}) \\
&= H(z)H^*(1/z^*)|_{z=e^{j\omega}}.
\end{aligned}
\tag{5.68}
$$

Restricting the system function $H(z)$ to be rational in the form of Eq. (5.21), i.e.,

$$
H(z) = \left(\frac{b_0}{a_0}\right)\frac{\displaystyle\prod_{k=1}^{M}(1 - c_k z^{-1})}{\displaystyle\prod_{k=1}^{N}(1 - d_k z^{-1})},
\tag{5.69}
$$

we see that $H^*(1/z^*)$ in Eq. (5.68) is

$$
H^*\left(\frac{1}{z^*}\right) = \left(\frac{b_0}{a_0}\right)\frac{\displaystyle\prod_{k=1}^{M}(1 - c_k^* z)}{\displaystyle\prod_{k=1}^{N}(1 - d_k^* z)},
\tag{5.70}
$$

wherein we have assumed that a_0 and b_0 are real. Therefore, Eq. (5.68) states that the square of the magnitude of the frequency response is the evaluation on the unit circle of the z-transform

$$C(z) = H(z)H^*(1/z^*) \tag{5.71}$$

$$= \left(\frac{b_0}{a_0}\right)^2 \frac{\displaystyle\prod_{k=1}^{M}(1 - c_k z^{-1})(1 - c_k^* z)}{\displaystyle\prod_{k=1}^{N}(1 - d_k z^{-1})(1 - d_k^* z)}. \tag{5.72}$$

If we know $|H(e^{j\omega})|^2$ expressed as a function of $e^{j\omega}$, then by replacing $e^{j\omega}$ by z, we can construct $C(z)$. From $C(z)$, we would like to infer as much as possible about $H(z)$. We first note that for each pole d_k of $H(z)$, there is a pole of $C(z)$ at d_k and $(d_k^*)^{-1}$. Similarly, for each zero c_k of $H(z)$, there is a zero of $C(z)$ at c_k and $(c_k^*)^{-1}$. Consequently, the poles and zeros of $C(z)$ occur in conjugate reciprocal pairs, with one element of each pair associated with $H(z)$ and one element of each pair associated with $H^*(1/z^*)$. Furthermore, if one element of each pair is inside the unit circle, then the other (i.e., the conjugate reciprocal) will be outside the unit circle. The only other alternative is for both to be on the unit circle in the same location.

If $H(z)$ is assumed to correspond to a causal, stable system, then all its poles must lie inside the unit circle. With this constraint, the poles of $H(z)$ can be identified from the poles of $C(z)$. However, with this constraint alone, the zeros of $H(z)$ cannot be uniquely identified from the zeros of $C(z)$. This can be seen from the following example.

Example 5.9   Different Systems with the Same C(z)

Consider two different stable systems with system functions

$$H_1(z) = \frac{2(1 - z^{-1})(1 + 0.5z^{-1})}{(1 - 0.8e^{j\pi/4}z^{-1})(1 - 0.8e^{-j\pi/4}z^{-1})} \tag{5.73}$$

and

$$H_2(z) = \frac{(1 - z^{-1})(1 + 2z^{-1})}{(1 - 0.8e^{j\pi/4}z^{-1})(1 - 0.8e^{-j\pi/4}z^{-1})}. \tag{5.74}$$

The pole–zero plots for these systems are shown in Figures 5.16(a) and 5.16(b), respectively. The two systems have identical pole locations and both have a zero at $z = 1$ but differ in the location of the second zero.

Now,

$$C_1(z) = H_1(z)H_1^*(1/z^*)$$

$$= \frac{2(1 - z^{-1})(1 + 0.5z^{-1})2(1 - z)(1 + 0.5z)}{(1 - 0.8e^{j\pi/4}z^{-1})(1 - 0.8e^{-j\pi/4}z^{-1})(1 - 0.8e^{-j\pi/4}z)(1 - 0.8e^{j\pi/4}z)} \tag{5.75}$$

and

$$C_2(z) = H_2(z)H_2^*(1/z^*)$$

$$= \frac{(1 - z^{-1})(1 + 2z^{-1})(1 - z)(1 + 2z)}{(1 - 0.8e^{j\pi/4}z^{-1})(1 - 0.8e^{-j\pi/4}z^{-1})(1 - 0.8e^{-j\pi/4}z)(1 - 0.8e^{j\pi/4}z)}. \tag{5.76}$$

Using the fact that

$$4(1 + 0.5z^{-1})(1 + 0.5z) = (1 + 2z^{-1})(1 + 2z), \tag{5.77}$$

we see that $C_1(z) = C_2(z)$. The pole–zero plot for $C_1(z)$ and $C_2(z)$, which are identical, is shown in Figure 5.16(c).

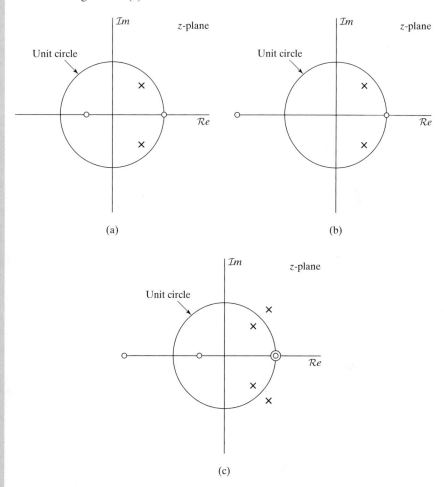

Figure 5.16   Pole–zero plots for two system functions and their common magnitude-squared function. (a) $H_1(z)$. (b) $H_2(z)$. (c) $C_1(z)$, $C_2(z)$.

The system functions $H_1(z)$ and $H_2(z)$ in Example 5.9 differ only in the location of one of the zeros. In the example, the factor $2(1 + 0.5z^{-1}) = (z^{-1} + 2)$ contributes the same to the square of the magnitude of the frequency response as the factor $(1 + 2z^{-1})$, and consequently, $|H_1(e^{j\omega})|$ and $|H_2(e^{j\omega})|$ are equal. However, the phase functions for these two frequency responses are different.

Example 5.10   Determination of *H(z)* from *C(z)*

Suppose we are given the pole–zero plot for $C(z)$ in Figure 5.17 and want to determine the poles and zeros to associate with $H(z)$. The conjugate reciprocal pairs of poles and zeros for which one element of each is associated with $H(z)$ and one with $H^*(1/z^*)$ are as follows:

$$\text{Pole pair 1}: \quad (p_1, p_4)$$

$$\text{Pole pair 2}: \quad (p_2, p_5)$$

$$\text{Pole pair 3}: \quad (p_3, p_6)$$

$$\text{Zero pair 1}: \quad (z_1, z_4)$$

$$\text{Zero pair 2}: \quad (z_2, z_5)$$

$$\text{Zero pair 3}: \quad (z_3, z_6)$$

Knowing that $H(z)$ corresponds to a stable, causal system, we must choose the poles from each pair that are inside the unit circle, i.e., p_1, p_2, and p_3. No such constraint is imposed on the zeros. However, if we assume that the coefficients a_k and b_k are real in Eqs. (5.19) and (5.20), the zeros (and poles) either are real or occur in complex conjugate pairs. Consequently, the zeros to associate with $H(z)$ are

$$z_3 \quad \text{or} \quad z_6$$

and

$$(z_1, z_2) \quad \text{or} \quad (z_4, z_5).$$

Therefore, there are a total of four different stable, causal systems with three poles and three zeros for which the pole–zero plot of $C(z)$ is that shown in Figure 5.17 and, equivalently, for which the frequency-response magnitude is the same. If we had not assumed that the coefficients a_k and b_k were real, the number of choices would be greater. Furthermore, if the number of poles and zeros of $H(z)$ were not restricted, the number of choices for $H(z)$ would be unlimited. To see this, assume that $H(z)$ has a factor of the form

$$\frac{z^{-1} - a^*}{1 - az^{-1}},$$

i.e.,

$$H(z) = H_1(z)\frac{z^{-1} - a^*}{1 - az^{-1}}. \tag{5.78}$$

Factors of this form represent *all-pass factors*, since they have unity magnitude on the unit circle; they are discussed in more detail in Section 5.5. It is easily verified that for $H(z)$ in Eq. (5.78),

$$C(z) = H(z)H^*(1/z^*) = H_1(z)H_1^*(1/z^*); \tag{5.79}$$

i.e., all-pass factors cancel in $C(z)$ and therefore would not be identifiable from the pole–zero plot of $C(z)$. Consequently, if the number of poles and zeros of $H(z)$ is unspecified, then, given $C(z)$, any choice for $H(z)$ can be cascaded with an arbitrary number of all-pass factors with poles inside the unit circle (i.e., $|a| < 1$).

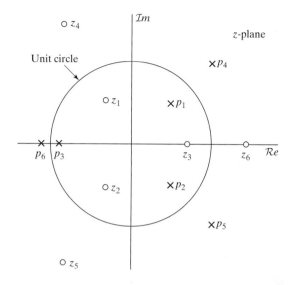

Figure 5.17   Pole–zero plot for the magnitude-squared function in Example 5.10.

5.5 ALL-PASS SYSTEMS

As indicated in the discussion of Example 5.10, a stable system function of the form

$$H_{ap}(z) = \frac{z^{-1} - a^*}{1 - az^{-1}} \tag{5.80}$$

has a frequency-response magnitude that is independent of ω. This can be seen by writing $H_{ap}(e^{j\omega})$ in the form

$$H_{ap}(e^{j\omega}) = \frac{e^{-j\omega} - a^*}{1 - ae^{-j\omega}}$$

$$= e^{-j\omega} \frac{1 - a^* e^{j\omega}}{1 - ae^{-j\omega}}. \tag{5.81}$$

In Eq. (5.81), the term $e^{-j\omega}$ has unity magnitude, and the remaining numerator and denominator factors are complex conjugates of each other and therefore have the same magnitude. Consequently, $|H_{ap}(e^{j\omega})| = 1$. A system for which the frequency-response magnitude is a constant, referred to as an all-pass system, passes all of the frequency components of its input with constant gain or attenuation.[3]
    The most general form for the system function of an all-pass system with a real-valued impulse response is a product of factors like Eq. (5.80), with complex poles being

---

[3] In some discussions, an all-pass system is defined to have gain of unity. In this text, we use the term all-pass system to refer to a system that passes all frequencies with a constant gain A that is not restricted to be unity.

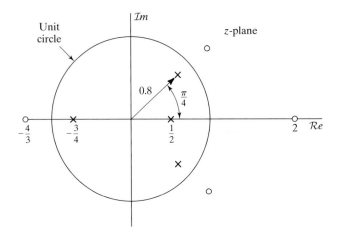

Figure 5.18 Typical pole–zero plot for an all-pass system.

paired with their conjugates; i.e.,

$$H_{\text{ap}}(z) = A \prod_{k=1}^{M_r} \frac{z^{-1} - d_k}{1 - d_k z^{-1}} \prod_{k=1}^{M_c} \frac{(z^{-1} - e_k^*)(z^{-1} - e_k)}{(1 - e_k z^{-1})(1 - e_k^* z^{-1})}, \tag{5.82}$$

where A is a positive constant and the d_ks are the real poles, and the e_ks the complex poles, of $H_{\text{ap}}(z)$. For causal and stable all-pass systems, $|d_k| < 1$ and $|e_k| < 1$. In terms of our general notation for system functions, all-pass systems have $M = N = 2M_c + M_r$ poles and zeros. Figure 5.18 shows a typical pole–zero plot for an all-pass system. In this case $M_r = 2$ and $M_c = 1$. Note that each pole of $H_{\text{ap}}(z)$ is paired with a conjugate reciprocal zero.

The frequency response for a general all-pass system can be expressed in terms of the frequency responses of 1st-order all-pass systems like that specified in Eq. (5.80). For a causal all-pass system, each of these terms consists of a single pole inside the unit circle and a zero at the conjugate reciprocal location. The magnitude response for such a term is unity, as we have shown. Thus, the log magnitude in dB is zero. With a expressed in polar form as $a = re^{j\theta}$, the phase function for Eq. (5.80) is

$$\angle \left[\frac{e^{-j\omega} - re^{-j\theta}}{1 - re^{j\theta} e^{-j\omega}} \right] = -\omega - 2 \arctan \left[\frac{r \sin(\omega - \theta)}{1 - r \cos(\omega - \theta)} \right]. \tag{5.83}$$

Likewise, the phase of a 2nd-order all-pass system with poles at $z = re^{j\theta}$ and $z = re^{-j\theta}$ is

$$\angle \left[\frac{(e^{-j\omega} - re^{-j\theta})(e^{-j\omega} - re^{j\theta})}{(1 - re^{j\theta} e^{-j\omega})(1 - re^{-j\theta} e^{-j\omega})} \right] = -2\omega - 2 \arctan \left[\frac{r \sin(\omega - \theta)}{1 - r \cos(\omega - \theta)} \right]$$

$$-2 \arctan \left[\frac{r \sin(\omega + \theta)}{1 - r \cos(\omega + \theta)} \right]. \tag{5.84}$$

Example 5.11   1st- and 2nd-Order All-Pass Systems

Figure 5.19 shows plots of the log magnitude, phase, and group delay for two 1st-order all-pass systems, one with a pole at $z = 0.9$ ($\theta = 0, r = 0.9$) and another with a

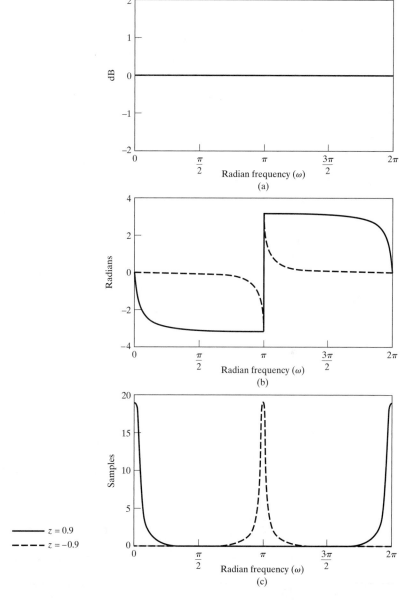

Figure 5.19 Frequency response for all-pass filters with real poles at $z = 0.9$ (solid line) and $z = -0.9$ (dashed line). (a) Log magnitude. (b) Phase (principal value). (c) Group delay.

pole at $z = -0.9$ ($\theta = \pi, r = 0.9$). For both systems, the radii of the poles are $r = 0.9$. Likewise, Figure 5.20 shows the same functions for a 2nd-order all-pass system with poles at $z = 0.9e^{j\pi/4}$ and $z = 0.9e^{-j\pi/4}$.

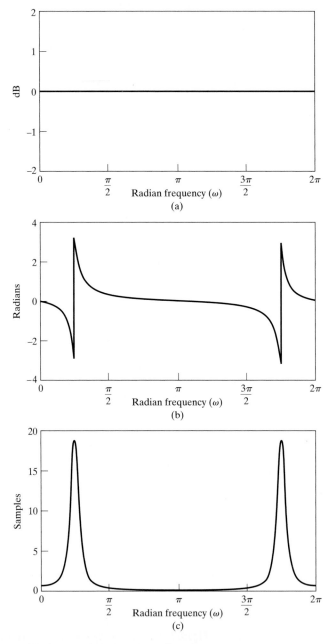

Figure 5.20   Frequency response of 2nd-order all-pass system with poles at $z = 0.9e^{\pm j\pi/4}$. (a) Log magnitude. (b) Phase (principal value). (c) Group delay.

Example 5.11 illustrates a general property of causal all-pass systems. In Figure 5.19(b), we see that the phase is nonpositive for $0 < \omega < \pi$. Similarly, in Figure 5.20(b), if the discontinuity of 2π resulting from the computation of the principal value is removed, the resulting continuous-phase curve is nonpositive for $0 < \omega < \pi$. Since the more general all-pass system given by Eq. (5.82) is a product of only such 1st- and 2nd-order factors, we can conclude that the (unwrapped) phase, $\arg[H_{ap}(e^{j\omega})]$, of a causal all-pass system is always nonpositive for $0 < \omega < \pi$. This may not appear to be true if the principal value is plotted, as is illustrated in Figure 5.21, which shows the log magnitude, phase, and group delay for an all-pass system with poles and zeros as in Figure 5.18. However, we can establish this result by first considering the group delay.

The group delay of the simple one-pole all-pass system of Eq. (5.80) is the negative derivative of the phase given by Eq. (5.83). With a small amount of algebra, it can be shown that

$$\text{grd}\left[\frac{e^{-j\omega} - re^{-j\theta}}{1 - re^{j\theta}e^{-j\omega}}\right] = \frac{1 - r^2}{1 + r^2 - 2r\cos(\omega - \theta)} = \frac{1 - r^2}{|1 - re^{j\theta}e^{-j\omega}|^2}. \quad (5.85)$$

Since $r < 1$ for a stable and causal all-pass system, from Eq. (5.85) the group delay contributed by a single causal all-pass factor is always positive. Since the group delay of a higher-order all-pass system will be a sum of positive terms, as in Eq. (5.85), it is true in general that the group delay of a causal rational all-pass system is always positive. This is confirmed by Figures 5.19(c), 5.20(c), and 5.21(c), which show the group delay for 1st-order, 2nd-order, and 3rd-order all-pass systems, respectively.

The positivity of the group delay of a causal all-pass system is the basis for a simple proof of the negativity of the phase of such a system. First, note that

$$\arg[H_{ap}(e^{j\omega})] = -\int_0^\omega \text{grd}[H_{ap}(e^{j\phi})]d\phi + \arg[H_{ap}(e^{j0})] \quad (5.86)$$

for $0 \le \omega \le \pi$. From Eq. (5.82), it follows that

$$H_{ap}(e^{j0}) = A\prod_{k=1}^{M_r}\frac{1 - d_k}{1 - d_k}\prod_{k=1}^{M_c}\frac{|1 - e_k|^2}{|1 - e_k|^2} = A. \quad (5.87)$$

Therefore, $\arg[H_{ap}(e^{j0})] = 0$, and since

$$\text{grd}[H_{ap}(e^{j\omega})] \ge 0, \quad (5.88)$$

it follows from Eq. (5.86) that

$$\arg[H_{ap}(e^{j\omega})] \le 0 \quad \text{for } 0 \le \omega < \pi. \quad (5.89)$$

The positivity of the group delay and the nonpositivity of the unwrapped phase are important properties of causal all-pass systems.

All-pass systems have importance in many contexts. They can be used as compensators for phase (or group delay) distortion, as we will see in Chapter 7, and they are useful in the theory of minimum-phase systems, as we will see in Section 5.6. They are also useful in transforming frequency-selective lowpass filters into other frequency-selective forms and in obtaining variable-cutoff frequency-selective filters. These applications are discussed in Chapter 7 and applied in the problems in that chapter.

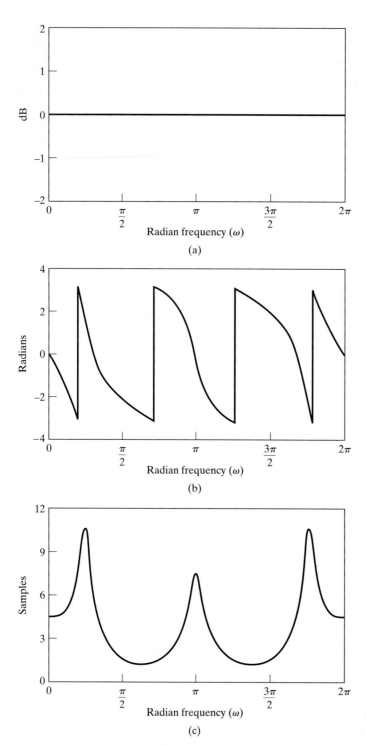

Figure 5.21 Frequency response for an all-pass system with the pole–zero plot in Figure 5.18. (a) Log magnitude. (b) Phase (principal value). (c) Group delay.

5.6 MINIMUM-PHASE SYSTEMS

In Section 5.4, we showed that the frequency-response magnitude for an LTI system with rational system function does not uniquely characterize the system. If the system is stable and causal, the poles must be inside the unit circle, but stability and causality place no such restriction on the zeros. For certain classes of problems, it is useful to impose the additional restriction that the inverse system (one with system function $1/H(z)$) also be stable and causal. As discussed in Section 5.2.2, this then restricts the zeros, as well as the poles, to be inside the unit circle, since the poles of $1/H(z)$ are the zeros of $H(z)$. Such systems are commonly referred to as *minimum-phase* systems. The name minimum-phase comes from a property of the phase response, which is not obvious from the preceding definition. This and other fundamental properties that we discuss are unique to this class of systems, and therefore, any one of them could be taken as the definition of the class. These properties are developed in Section 5.6.3.

If we are given a magnitude-squared function in the form of Eq. (5.72), and we know that both the system and its inverse are causal and stable (i.e., is a minimum-phase system), then $H(z)$ is uniquely determined and will consist of all the poles and zeros of $C(z) = H(z)H^*(1/z^*)$ that lie inside the unit circle.[4] This approach is often followed in filter design when only the magnitude response is determined by the design method used. (See Chapter 7.)

5.6.1 Minimum-Phase and All-Pass Decomposition

In Section 5.4, we saw that, from the square of the magnitude of the frequency response alone, we could not uniquely determine the system function $H(z)$, since any choice that had the given frequency-response magnitude could be cascaded with arbitrary all-pass factors without affecting the magnitude. A related observation is that any rational system function[5] can be expressed as

$$H(z) = H_{\min}(z)H_{\mathrm{ap}}(z), \tag{5.90}$$

where $H_{\min}(z)$ is a minimum-phase system and $H_{\mathrm{ap}}(z)$ is an all-pass system.

To show this, suppose that $H(z)$ has one zero outside the unit circle at $z = 1/c^*$, where $|c| < 1$, and the remaining poles and zeros are inside the unit circle. Then $H(z)$ can be expressed as

$$H(z) = H_1(z)(z^{-1} - c^*), \tag{5.91}$$

where, by definition, $H_1(z)$ is minimum phase. An equivalent expression for $H(z)$ is

$$H(z) = H_1(z)(1 - cz^{-1})\frac{z^{-1} - c^*}{1 - cz^{-1}}. \tag{5.92}$$

Since $|c| < 1$, the factor $H_1(z)(1 - cz^{-1})$ also is minimum phase, and it differs from $H(z)$ only in that the zero of $H(z)$ that was outside the unit circle at $z = 1/c^*$ is reflected inside the unit circle to the conjugate reciprocal location $z = c$. The term $(z^{-1} - c^*)/(1 - cz^{-1})$

---

[4]We have assumed that $C(z)$ has no poles or zeros on the unit circle. Strictly speaking, systems with poles on the unit circle are unstable and are generally to be avoided in practice. Zeros on the unit circle, however, often occur in practical filter designs. By our definition, such systems are nonminimum phase, but many of the properties of minimum-phase systems hold even in this case.

[5]Somewhat for convenience, we will restrict the discussion to stable, causal systems, although the observation applies more generally.

is all-pass. This example can be generalized in a straightforward way to include more zeros outside the unit circle, thereby showing that, in general, any system function can be expressed as

$$H(z) = H_{\min}(z) H_{\mathrm{ap}}(z), \tag{5.93}$$

where $H_{\min}(z)$ contains all the poles and zeros of $H(z)$ that lie inside the unit circle, together with zeros that are the conjugate reciprocals of the zeros of $H(z)$ that lie outside the unit circle. The system function $H_{\mathrm{ap}}(z)$ is comprised of all the zeros of $H(z)$ that lie outside the unit circle, together with poles to cancel the reflected conjugate reciprocal zeros in $H_{\min}(z)$.

Using Eq. (5.93), we can form a nonminimum-phase system from a minimum-phase system by reflecting one or more zeros lying inside the unit circle to their conjugate reciprocal locations outside the unit circle, or, conversely, we can form a minimum-phase system from a nonminimum-phase system by reflecting all the zeros lying outside the unit circle to their conjugate reciprocal locations inside. In either case, both the minimum-phase and the nonminimum-phase systems will have the same frequency-response magnitude.

Example 5.12   Minimum-Phase/All-Pass Decomposition

To illustrate the decomposition of a stable, causal system into the cascade of a minimum-phase and an all-pass system, consider the two stable, causal systems specified by the system functions

$$H_1(z) = \frac{(1 + 3z^{-1})}{1 + \frac{1}{2}z^{-1}}$$

and

$$H_2(z) = \frac{\left(1 + \frac{3}{2}e^{+j\pi/4}z^{-1}\right)\left(1 + \frac{3}{2}e^{-j\pi/4}z^{-1}\right)}{\left(1 - \frac{1}{3}z^{-1}\right)}.$$

The first system function, $H_1(z)$, has a pole inside the unit circle at $z = -\frac{1}{2}$, and a zero outside at $z = -3$. We will need to choose the appropriate all-pass system to reflect this zero inside the unit circle. From Eq. (5.91), we have $c = -\frac{1}{3}$. Therefore, from Eqs. (5.92) and (5.93), the all-pass component will be

$$H_{\mathrm{ap}}(z) = \frac{z^{-1} + \frac{1}{3}}{1 + \frac{1}{3}z^{-1}},$$

and the minimum-phase component will be

$$H_{\min}(z) = 3\frac{1 + \frac{1}{3}z^{-1}}{1 + \frac{1}{2}z^{-1}};$$

i.e.,

$$H_1(z) = \left(3\frac{1 + \frac{1}{3}z^{-1}}{1 + \frac{1}{2}z^{-1}}\right)\left(\frac{z^{-1} + \frac{1}{3}}{1 + \frac{1}{3}z^{-1}}\right).$$

The second system function, $H_2(z)$, has two complex zeros outside the unit circle and a real pole inside. We can express $H_2(z)$ in the form of Eq. (5.91) by factoring

$\frac{3}{2}e^{j\pi/4}$ and $\frac{3}{2}e^{-j\pi/4}$ out of the numerator terms to obtain

$$H_2(z) = \frac{9}{4} \frac{\left(z^{-1} + \frac{2}{3}e^{-j\pi/4}\right)\left(z^{-1} + \frac{2}{3}e^{j\pi/4}\right)}{1 - \frac{1}{3}z^{-1}}.$$

Factoring as in Eq. (5.92) yields

$$H_2(z) = \left[\frac{9}{4} \frac{\left(1 + \frac{2}{3}e^{-j\pi/4}z^{-1}\right)\left(1 + \frac{2}{3}e^{j\pi/4}z^{-1}\right)}{1 - \frac{1}{3}z^{-1}}\right]$$

$$\times \left[\frac{\left(z^{-1} + \frac{2}{3}e^{-j\pi/4}\right)\left(z^{-1} + \frac{2}{3}e^{j\pi/4}\right)}{\left(1 + \frac{2}{3}e^{j\pi/4}z^{-1}\right)\left(1 + \frac{2}{3}e^{-j\pi/4}z^{-1}\right)}\right].$$

The first term in square brackets is a minimum-phase system, while the second term is an all-pass system.

5.6.2 Frequency-Response Compensation of Non-Minimum-Phase Systems

In many signal-processing contexts, a signal has been distorted by an LTI system with an undesirable frequency response. It may then be of interest to process the distorted signal with a compensating system, as indicated in Figure 5.22. This situation may arise, for example, in transmitting signals over a communication channel. If perfect compensation is achieved, then $s_c[n] = s[n]$, i.e., $H_c(z)$ is the inverse of $H_d(z)$. However, if we assume that the distorting system is stable and causal and require the compensating system to be stable and causal, then perfect compensation is possible only if $H_d(z)$ is a minimum-phase system, so that it has a stable, causal inverse.

Based on the previous discussions, assuming that $H_d(z)$ is known or approximated as a rational system function, we can form a minimum-phase system $H_{d\min}(z)$ by reflecting all the zeros of $H_d(z)$ that are outside the unit circle to their conjugate reciprocal locations inside the unit circle. $H_d(z)$ and $H_{d\min}(z)$ have the same frequency-response magnitude and are related through an all-pass system $H_{\mathrm{ap}}(z)$, i.e.,

$$H_d(z) = H_{d\min}(z)H_{\mathrm{ap}}(z). \tag{5.94}$$

Choosing the compensating filter to be

$$H_c(z) = \frac{1}{H_{d\min}(z)}, \tag{5.95}$$

we find that the overall system function relating $s[n]$ and $s_c[n]$ is

$$G(z) = H_d(z)H_c(z) = H_{\mathrm{ap}}(z); \tag{5.96}$$

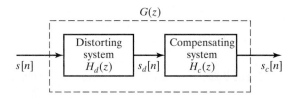

Figure 5.22 Illustration of distortion compensation by linear filtering.

i.e., $G(z)$ corresponds to an all-pass system. Consequently, the frequency-response magnitude is exactly compensated for, whereas the phase response is modified to $\angle H_{ap}(e^{j\omega})$.

The following example illustrates compensation of the frequency response magnitude when the system to be compensated for is a nonminimum-phase FIR system.

Example 5.13   Compensation of an FIR System

Consider the distorting system function to be

$$H_d(z) = (1 - 0.9e^{j0.6\pi}z^{-1})(1 - 0.9e^{-j0.6\pi}z^{-1})$$

$$\times (1 - 1.25e^{j0.8\pi}z^{-1})(1 - 1.25e^{-j0.8\pi}z^{-1}). \qquad (5.97)$$

The pole–zero plot is shown in Figure 5.23. Since $H_d(z)$ has only zeros (all poles are at $z = 0$), it follows that the system has a finite-duration impulse response. Therefore, the system is stable; and since $H_d(z)$ is a polynomial with only negative powers of z, the system is causal. However, since two of the zeros are outside the unit circle, the system is nonminimum phase. Figure 5.24 shows the log magnitude, phase, and group delay for $H_d(e^{j\omega})$.

The corresponding minimum-phase system is obtained by reflecting the zeros that occur at $z = 1.25e^{\pm j0.8\pi}$ to their conjugate reciprocal locations inside the unit circle. If we express $H_d(z)$ as

$$H_d(z) = (1 - 0.9e^{j0.6\pi}z^{-1})(1 - 0.9e^{-j0.6\pi}z^{-1})(1.25)^2$$

$$\times (z^{-1} - 0.8e^{-j0.8\pi})(z^{-1} - 0.8e^{j0.8\pi}), \qquad (5.98)$$

then

$$H_{min}(z) = (1.25)^2(1 - 0.9e^{j0.6\pi}z^{-1})(1 - 0.9e^{-j0.6\pi}z^{-1})$$

$$\times (1 - 0.8e^{-j0.8\pi}z^{-1})(1 - 0.8e^{j0.8\pi}z^{-1}), \qquad (5.99)$$

and the all-pass system that relates $H_{min}(z)$ and $H_d(z)$ is

$$H_{ap}(z) = \frac{(z^{-1} - 0.8e^{-j0.8\pi})(z^{-1} - 0.8e^{j0.8\pi})}{(1 - 0.8e^{j0.8\pi}z^{-1})(1 - 0.8e^{-j0.8\pi}z^{-1})}. \qquad (5.100)$$

The log magnitude, phase, and group delay of $H_{min}(e^{j\omega})$ are shown in Figure 5.25. Figures 5.24(a) and 5.25(a) are, of course, identical. The log magnitude, phase, and group delay for $H_{ap}(e^{j\omega})$ are plotted in Figure 5.26.

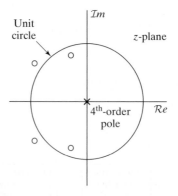

Figure 5.23   Pole–zero plot of FIR system in Example 5.13.

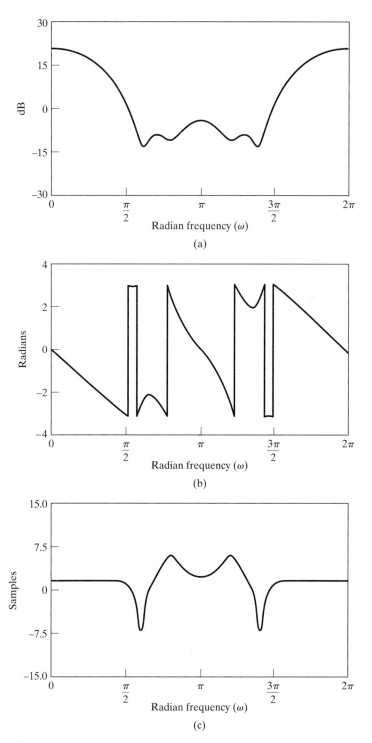

Figure 5.24   Frequency response for FIR system with pole–zero plot in Figure 5.23. (a) Log magnitude. (b) Phase (principal value). (c) Group delay.

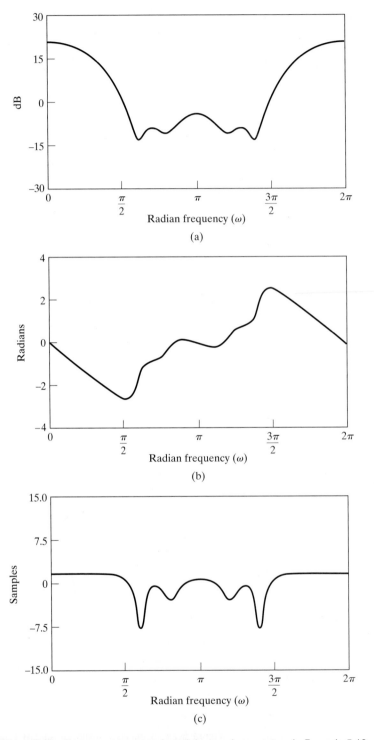

Figure 5.25   Frequency response for minimum-phase system in Example 5.13.
(a) Log magnitude. (b) Phase. (c) Group delay.

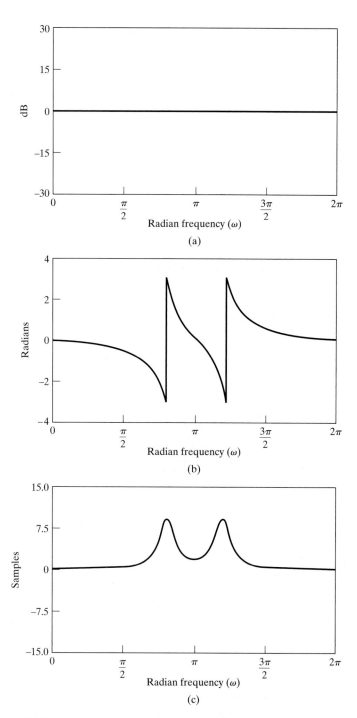

Figure 5.26   Frequency response of all-pass system of Example 5.13. (The sum of corresponding curves in Figures 5.25 and 5.26 equals the corresponding curve in Figure 5.24 with the sum of the phase curves taken modulo 2π.) (a) Log magnitude. (b) Phase (principal value). (c) Group delay.

Note that the inverse system for $H_d(z)$ would have poles at $z = 1.25e^{\pm j0.8\pi}$ and at $z = 0.9e^{\pm j0.6\pi}$, and thus, the causal inverse would be unstable. The minimum-phase inverse would be the reciprocal of $H_{min}(z)$, as given by Eq. (5.99), and if this inverse were used in the cascade system of Figure 5.22, the overall effective system function would be $H_{ap}(z)$, as given in Eq. (5.100).

5.6.3 Properties of Minimum-Phase Systems

We have been using the term "minimum phase" to refer to systems that are causal and stable and that have a causal and stable inverse. This choice of name is motivated by a property of the phase function that, while not obvious, follows from our chosen definition. In this section, we develop a number of interesting and important properties of minimum-phase systems relative to all other systems that have the same frequency-response magnitude.

The Minimum Phase-Lag Property

The use of the terminology "minimum phase" as a descriptive name for a system having all its poles and zeros inside the unit circle is suggested by Example 5.13. Recall that, as a consequence of Eq. (5.90), the unwrapped phase, i.e., $\arg[H(e^{j\omega})]$, of any nonminimum-phase system can be expressed as

$$\arg[H(e^{j\omega})] = \arg[H_{min}(e^{j\omega})] + \arg[H_{ap}(e^{j\omega})]. \qquad (5.101)$$

Therefore, the continuous phase that would correspond to the principal-value phase of Figure 5.24(b) is the sum of the unwrapped phase associated with the minimum-phase function of Figure 5.25(b) and the unwrapped phase of the all-pass system associated with the principal-value phase shown in Figure 5.26(b). As was shown in Section 5.5 and as indicated by the principal-value phase curves of Figures 5.19(b), 5.20(b), 5.21(b), and 5.26(b), the unwrapped-phase curve of an all-pass system is negative for $0 \le \omega \le \pi$. Thus, the reflection of zeros of $H_{min}(z)$ from inside the unit circle to conjugate reciprocal locations outside always decreases the (unwrapped) phase or increases the negative of the phase, which is called the *phase-lag* function. Hence, the causal, stable system that has $|H_{min}(e^{j\omega})|$ as its magnitude response and also has all its zeros (and, of course, poles) inside the unit circle has the minimum phase-lag function (for $0 \le \omega < \pi$) of all the systems having that same magnitude response. Therefore, a more precise terminology is *minimum phase-lag* system, but *minimum phase* is historically the established terminology.

To make the interpretation of minimum phase-lag systems more precise, it is necessary to impose the additional constraint that $H(e^{j\omega})$ be positive at $\omega = 0$, i.e.,

$$H(e^{j0}) = \sum_{n=-\infty}^{\infty} h[n] > 0. \qquad (5.102)$$

Note that $H(e^{j0})$ will be real if we restrict $h[n]$ to be real. The condition of Eq. (5.102) is necessary because a system with impulse response $-h[n]$ has the same poles and zeros for its system function as a system with impulse response $h[n]$. However, multiplying by -1 would alter the phase by π radians. Thus, to remove this ambiguity, we impose the condition of Eq. (5.102) to ensure that a system with all its poles and zeros inside the unit circle also has the minimum phase-lag property. However, this constraint is often of little significance, and our definition at the beginning of Section 5.6, which does not include it, is the generally accepted definition of the class of minimum-phase systems.

The Minimum Group-Delay Property

Example 5.13 illustrates another property of systems whose poles and zeros are all inside the unit circle. First, note that the group delay for the systems that have the same magnitude response is

$$\text{grd}[H(e^{j\omega})] = \text{grd}[H_{\min}(e^{j\omega})] + \text{grd}[H_{ap}(e^{j\omega})]. \tag{5.103}$$

The group delay for the minimum-phase system shown in Figure 5.25(c) is always less than the group delay for the nonminimum-phase system shown in Figure 5.24(c). This is because, as Figure 5.26(c) shows, the all-pass system that converts the minimum-phase system into the nonminimum-phase system has a positive group delay. In Section 5.5, we showed this to be a general property of all-pass systems; they always have positive group delay for all ω. Thus, if we again consider all the systems that have a given magnitude response $|H_{\min}(e^{j\omega})|$, the one that has all its poles and zeros inside the unit circle has the minimum group delay. An equally appropriate name for such systems would therefore be *minimum group-delay* systems, but this terminology is not generally used.

The Minimum Energy-Delay Property

In Example 5.13, there are a total of four causal FIR systems with real impulse responses that have the same frequency-response magnitude as the system in Eq. (5.97). The associated pole–zero plots are shown in Figure 5.27, where Figure 5.27(d) corresponds to Eq. (5.97) and Figure 5.27(a) to the minimum-phase system of Eq. (5.99). The impulse responses for these four cases are plotted in Figure 5.28. If we compare the four sequences in this figure, we observe that the minimum-phase sequence appears to have larger samples at its left-hand end than do all the other sequences. Indeed, it is true for this example and, in general, that

$$|h[0]| \leq |h_{\min}[0]| \tag{5.104}$$

for any causal, stable sequence $h[n]$ for which

$$|H(e^{j\omega})| = |H_{\min}(e^{j\omega})|. \tag{5.105}$$

A proof of this property is suggested in Problem 5.71.

All the impulse responses whose frequency-response magnitude is equal to $|H_{\min}(e^{j\omega})|$ have the same total energy as $h_{\min}[n]$, since, by Parseval's theorem,

$$\sum_{n=0}^{\infty} |h[n]|^2 = \frac{1}{2\pi} \int_{-\pi}^{\pi} |H(e^{j\omega})|^2 d\omega = \frac{1}{2\pi} \int_{-\pi}^{\pi} |H_{\min}(e^{j\omega})|^2 d\omega$$

$$= \sum_{n=0}^{\infty} |h_{\min}[n]|^2. \tag{5.106}$$

If we define the *partial energy* of the impulse response as

$$E[n] = \sum_{m=0}^{n} |h[m]|^2, \tag{5.107}$$

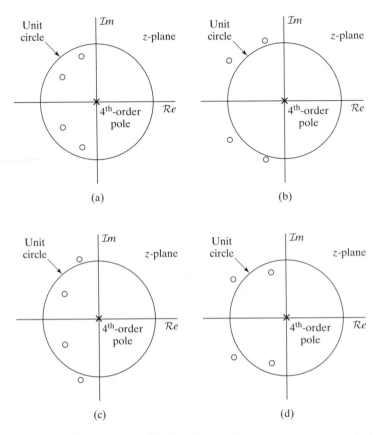

Figure 5.27   Four systems, all having the same frequency-response magnitude. Zeros are at all combinations of the complex conjugate zero pairs $0.9e^{\pm j0.6\pi}$ and $0.8e^{\pm j0.8\pi}$ and their reciprocals.

then it can be shown that (see Problem 5.72)

$$\sum_{m=0}^{n} |h[m]|^2 \leq \sum_{m=0}^{n} |h_{\min}[m]|^2 \tag{5.108}$$

for all impulse responses $h[n]$ belonging to the family of systems that have magnitude response given by Eq. (5.105). According to Eq. (5.108), the partial energy of the minimum-phase system is most concentrated around $n = 0$; i.e., the energy of the minimum-phase system is delayed the least of all systems having the same magnitude response function. For this reason, minimum-phase (lag) systems are also called *minimum energy-delay systems*, or simply, *minimum-delay systems*. This delay property is illustrated by Figure 5.29, which shows plots of the partial energy for the four sequences in Figure 5.28. We note for this example—and it is true in general—that the minimum energy delay occurs for the system that has all its zeros inside the unit circle (i.e., the minimum-phase system) and the maximum energy delay occurs for the system that has all its zeros outside the unit circle. Maximum energy-delay systems are also often called *maximum-phase systems*.

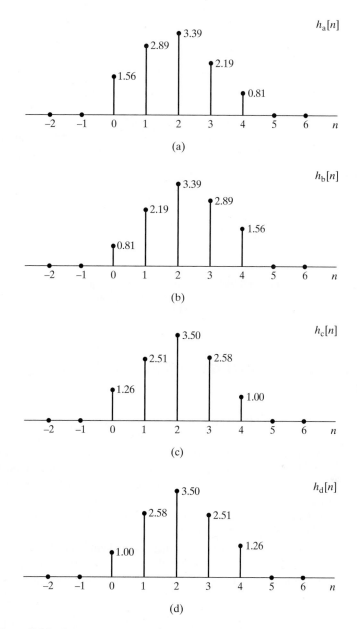

Figure 5.28 Sequences corresponding to the pole–zero plots of Figure 5.27.

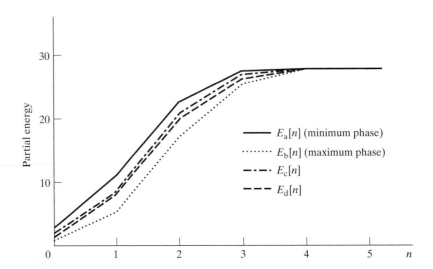

Figure 5.29   Partial energies for the four sequences of Figure 5.28. (Note that $E_a[n]$ is for the minimum-phase sequence $h_a[n]$ and $E_b[n]$ is for the maximum-phase sequence $h_b[n]$.)

5.7 LINEAR SYSTEMS WITH GENERALIZED LINEAR PHASE

In designing filters and other signal-processing systems that pass some portion of the frequency band undistorted, it is desirable to have approximately constant frequency-response magnitude and zero phase in that band. For causal systems, zero phase is not attainable, consequently, some phase distortion must be allowed. As we saw in Section 5.1, the effect of linear phase with integer slope is a simple time shift. A nonlinear phase, on the other hand, can have a major effect on the shape of a signal, even when the frequency-response magnitude is constant. Thus, in many situations it is particularly desirable to design systems to have exactly or approximately linear phase. In this section, we consider a formalization and generalization of the notions of linear phase and ideal time delay by considering the class of systems that have constant group delay. We begin by reconsidering the concept of delay in a discrete-time system.

5.7.1 Systems with Linear Phase

Consider an LTI system whose frequency response over one period is

$$H_{\mathrm{id}}(e^{j\omega}) = e^{-j\omega\alpha}, \qquad |\omega| < \pi, \tag{5.109}$$

where α is a real number, not necessarily an integer. Such a system is an "ideal delay" system, where α is the delay introduced by the system. Note that this system has constant magnitude response, linear phase, and constant group delay; i.e.,

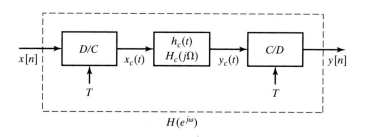

Figure 5.30 Interpretation of noninteger delay in discrete-time systems.

$$|H_{id}(e^{j\omega})| = 1, \tag{5.110a}$$

$$\angle H_{id}(e^{j\omega}) = -\omega\alpha, \tag{5.110b}$$

$$\text{grd}[H_{id}(e^{j\omega})] = \alpha. \tag{5.110c}$$

The inverse Fourier transform of $H_{id}(e^{j\omega})$ is the impulse response

$$h_{id}[n] = \frac{\sin \pi (n - \alpha)}{\pi (n - \alpha)}, \qquad -\infty < n < \infty. \tag{5.111}$$

The output of this system for an input $x[n]$ is

$$y[n] = x[n] * \frac{\sin \pi (n - \alpha)}{\pi (n - \alpha)} = \sum_{k=-\infty}^{\infty} x[k] \frac{\sin \pi (n - k - \alpha)}{\pi (n - k - \alpha)}. \tag{5.112}$$

If $\alpha = n_d$, where n_d is an integer, then, as mentioned in Section 5.1,

$$h_{id}[n] = \delta[n - n_d] \tag{5.113}$$

and

$$y[n] = x[n] * \delta[n - n_d] = x[n - n_d]. \tag{5.114}$$

That is, if $\alpha = n_d$ is an integer, the system with linear phase and unity gain in Eq. (5.109) simply shifts the input sequence by n_d samples. If α is not an integer, the most straightforward interpretation is the one developed in Example 4.7 in Chapter 4.

Specifically, a representation of the system of Eq. (5.109) is that shown in Figure 5.30, with $h_c(t) = \delta(t - \alpha T)$ and $H_c(j\Omega) = e^{-j\Omega\alpha T}$, so that

$$H(e^{j\omega}) = e^{-j\omega\alpha}, \qquad |\omega| < \pi. \tag{5.115}$$

In this representation, the choice of T is irrelevant and could simply be normalized to unity. It is important to stress again that the representation is valid whether or not $x[n]$ was originally obtained by sampling a continuous-time signal. According to the representation in Figure 5.30, $y[n]$ is the sequence of samples of the time-shifted, band-limited interpolation of the input sequence $x[n]$; i.e., $y[n] = x_c(nT - \alpha T)$. The system of Eq. (5.109) is said to have a time shift of α samples, even if α is not an integer. If the group delay α is positive, the time shift is a time delay. If α is negative, the time shift is a time advance.

This discussion also provides a useful interpretation of linear phase when it is associated with a nonconstant magnitude response. For example, consider a more general frequency response with linear phase, i.e.,

$$H(e^{j\omega}) = |H(e^{j\omega})|e^{-j\omega\alpha}, \qquad |\omega| < \pi. \tag{5.116}$$

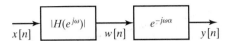

Figure 5.31   Representation of a linear-phase LTI system as a cascade of a magnitude filter and a time shift.

Equation (5.116) suggests the interpretation of Figure 5.31. The signal $x[n]$ is filtered by the zero-phase frequency response $|H(e^{j\omega})|$, and the filtered output is then "time shifted" by the (integer or noninteger) amount α. Suppose, for example, that $H(e^{j\omega})$ is the linear-phase ideal lowpass filter

$$H_{\text{lp}}(e^{j\omega}) = \begin{cases} e^{-j\omega\alpha}, & |\omega| < \omega_c, \\ 0, & \omega_c < |\omega| \leq \pi. \end{cases} \tag{5.117}$$

The corresponding impulse response is

$$h_{\text{lp}}[n] = \frac{\sin\omega_c(n-\alpha)}{\pi(n-\alpha)}. \tag{5.118}$$

Note that Eq. (5.111) is obtained if $\omega_c = \pi$.

Example 5.14   Ideal Lowpass with Linear Phase

The impulse response of the ideal lowpass filter illustrates some interesting properties of linear-phase systems. Figure 5.32(a) shows $h_{\text{lp}}[n]$ for $\omega_c = 0.4\pi$ and $\alpha = n_d = 5$. Note that when α is an integer, the impulse response is symmetric about $n = n_d$; i.e.,

$$h_{\text{lp}}[2n_d - n] = \frac{\sin\omega_c(2n_d - n - n_d)}{\pi(2n_d - n - n_d)}$$

$$= \frac{\sin\omega_c(n_d - n)}{\pi(n_d - n)} \tag{5.119}$$

$$= h_{\text{lp}}[n].$$

In this case, we could define a *zero-phase system*

$$\hat{H}_{\text{lp}}(e^{j\omega}) = H_{\text{lp}}(e^{j\omega})e^{j\omega n_d} = |H_{\text{lp}}(e^{j\omega})|, \tag{5.120}$$

wherein the impulse response is shifted to the left by n_d samples, yielding an even sequence

$$\hat{h}_{\text{lp}}[n] = \frac{\sin\omega_c n}{\pi n} = \hat{h}_{\text{lp}}[-n]. \tag{5.121}$$

Figure 5.32(b) shows $h_{\text{lp}}[n]$ for $\omega_c = 0.4\pi$ and $\alpha = 4.5$. This is typical of the case when the linear phase corresponds to an integer plus one-half. As in the case of the integer delay, it is easily shown that if α is an integer plus one-half (or 2α is an integer), then

$$h_{\text{lp}}[2\alpha - n] = h_{\text{lp}}[n]. \tag{5.122}$$

In this case, the point of symmetry is α, which is not an integer. Therefore, since the symmetry is not about a point of the sequence, it is not possible to shift the sequence to obtain an even sequence that has zero phase. This is similar to the case of Example 4.8 with M odd.

Figure 5.32(c) represents a third case, in which there is no symmetry at all. In this case, $\omega_c = 0.4\pi$ and $\alpha = 4.3$.

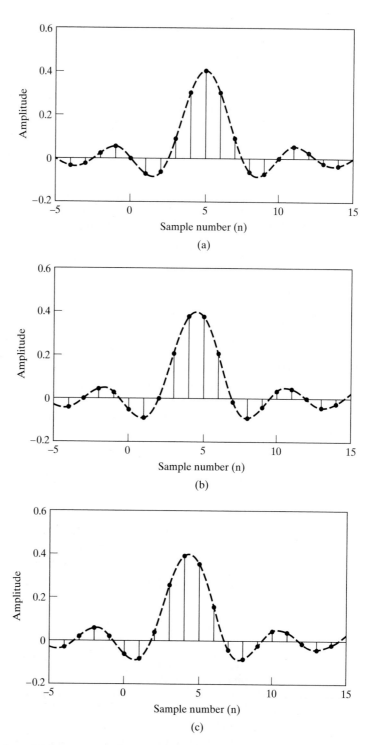

Figure 5.32   Ideal lowpass filter impulse responses, with $\omega_c = 0.4\pi$. (a) Delay $= \alpha = 5$. (b) Delay $= \alpha = 4.5$. (c) Delay $= \alpha = 4.3$.

In general a linear-phase system has frequency response

$$H(e^{j\omega}) = |H(e^{j\omega})|e^{-j\omega\alpha}. \tag{5.123}$$

As illustrated in Example 5.14, if 2α is an integer (i.e., if α is an integer or an integer plus one-half), the corresponding impulse response has even symmetry about α; i.e.,

$$h[2\alpha - n] = h[n]. \tag{5.124}$$

If 2α is not an integer, then the impulse response will not have symmetry. This is illustrated in Figure 5.32(c), which shows an impulse response that is not symmetric, but that has linear phase, or equivalently, constant group delay.

5.7.2 Generalized Linear Phase

In the discussion in Section 5.7.1, we considered a class of systems whose frequency response is of the form of Eq. (5.116), i.e., a real-valued nonnegative function of ω multiplied by a linear-phase term $e^{-j\omega\alpha}$. For a frequency response of this form, the phase of $H(e^{j\omega})$ is entirely associated with the linear-phase factor $e^{-j\omega\alpha}$, i.e., $\arg[H(e^{j\omega})] = -\omega\alpha$, and consequently, systems in this class are referred to as linear-phase systems. In the moving average of Example 4.8, the frequency response in Eq. (4.66) is a real-valued function of ω multiplied by a linear-phase term, but the system is not, strictly speaking, a linear-phase system, since, at frequencies for which the factor

$$\frac{1}{M+1} \frac{\sin[\omega(M+1)/2]}{\sin(\omega/2)}$$

is negative, this term contributes an additional phase of π radians to the total phase.

Many of the advantages of linear-phase systems apply to systems with frequency response having the form of Eq. (4.66) as well, and consequently, it is useful to generalize somewhat the definition and concept of linear phase. Specifically, a system is referred to as a *generalized linear-phase system* if its frequency response can be expressed in the form

$$H(e^{j\omega}) = A(e^{j\omega})e^{-j\alpha\omega + j\beta}, \tag{5.125}$$

where α and β are constants and $A(e^{j\omega})$ is a real (possibly bipolar) function of ω. For the linear-phase system of Eq. (5.117) and the moving-average filter of Example 4.8, $\alpha = -M/2$ and $\beta = 0$. We see, however, that the bandlimited differentiator of Example 4.4 has the form of Eq. (5.125) with $\alpha = 0$, $\beta = \pi/2$, and $A(e^{j\omega}) = \omega/T$.

A system whose frequency response has the form of Eq. (5.125) is called a generalized linear-phase system because the phase of such a system consists of constant terms added to the linear function $-\omega\alpha$; i.e., $-\omega\alpha + \beta$ is the equation of a straight line.

However, if we ignore any discontinuities that result from the addition of constant phase over all or part of the band $|\omega| < \pi$, then such a system can be characterized by constant group delay. That is, the class of systems such that

$$\tau(\omega) = \text{grd}[H(e^{j\omega})] = -\frac{d}{d\omega}\{\arg[H(e^{j\omega})]\} = \alpha \tag{5.126}$$

have linear phase of the more general form

$$\arg[H(e^{j\omega})] = \beta - \omega\alpha, \qquad 0 < \omega < \pi, \tag{5.127}$$

where β and α are both real constants.

Recall that we showed in Section 5.7.1 that the impulse responses of linear-phase systems may have symmetry about α if 2α is an integer. To see the implication of this for generalized linear-phase systems, it is useful to derive an equation that must be satisfied by $h[n]$, α, and β for constant group-delay systems. This equation is derived by noting that, for such systems, the frequency response can be expressed as

$$H(e^{j\omega}) = A(e^{j\omega})e^{j(\beta-\alpha\omega)}$$
$$= A(e^{j\omega})\cos(\beta - \omega\alpha) + jA(e^{j\omega})\sin(\beta - \omega\alpha), \tag{5.128}$$

or equivalently, as

$$H(e^{j\omega}) = \sum_{n=-\infty}^{\infty} h[n]e^{-j\omega n}$$
$$= \sum_{n=-\infty}^{\infty} h[n]\cos\omega n - j\sum_{n=-\infty}^{\infty} h[n]\sin\omega n, \tag{5.129}$$

where we have assumed that $h[n]$ is real. The tangent of the phase angle of $H(e^{j\omega})$ can be expressed as

$$\tan(\beta - \omega\alpha) = \frac{\sin(\beta - \omega\alpha)}{\cos(\beta - \omega\alpha)} = \frac{-\displaystyle\sum_{n=-\infty}^{\infty} h[n]\sin\omega n}{\displaystyle\sum_{n=-\infty}^{\infty} h[n]\cos\omega n}.$$

Cross multiplying and combining terms with a trigonometric identity leads to the equation

$$\sum_{n=-\infty}^{\infty} h[n]\sin[\omega(n-\alpha) + \beta] = 0 \quad \text{for all } \omega. \tag{5.130}$$

This equation is a necessary condition on $h[n]$, α, and β for the system to have constant group delay. It is not a sufficient condition, however, and, owing to its implicit nature, it does not tell us how to find a linear-phase system.

A class of examples of generalized linear-phase systems are those for which

$$\beta = 0 \quad \text{or} \quad \pi, \tag{5.131a}$$
$$2\alpha = M = \text{an integer}, \tag{5.131b}$$
$$h[2\alpha - n] = h[n]. \tag{5.131c}$$

With $\beta = 0$ or π, Eq. (5.130) becomes

$$\sum_{n=-\infty}^{\infty} h[n] \sin[\omega(n - \alpha)] = 0, \tag{5.132}$$

from which it can be shown that if 2α is an integer, terms in Eq. (5.132) can be paired so that each pair of terms is identically zero for all ω. These conditions in turn imply that the corresponding frequency response has the form of Eq. (5.125) with $\beta = 0$ or π and $A(e^{j\omega})$ an even (and, of course, real) function of ω.

Another class of examples of generalized linear-phase systems are those for which

$$\beta = \pi/2 \quad \text{or} \quad 3\pi/2, \tag{5.133a}$$

$$2\alpha = M = \text{an integer}, \tag{5.133b}$$

and

$$h[2\alpha - n] = -h[n] \tag{5.133c}$$

Equations (5.133) imply that the frequency response has the form of Eq. (5.125) with $\beta = \pi/2$ and $A(e^{j\omega})$ an odd function of ω. For these cases Eq. (5.130) becomes

$$\sum_{n=-\infty}^{\infty} h[n] \cos[\omega(n - \alpha)] = 0, \tag{5.134}$$

and is satisfied for all ω.

Note that Eqs. (5.131) and (5.133) give two sets of sufficient conditions that guarantee generalized linear phase or constant group delay, but as we have already seen in Figure 5.32(c), there are other systems that satisfy Eq. (5.125) without these symmetry conditions.

5.7.3 Causal Generalized Linear-Phase Systems

If the system is causal, then Eq. (5.130) becomes

$$\sum_{n=0}^{\infty} h[n] \sin[\omega(n - \alpha) + \beta] = 0 \qquad \text{for all } \omega. \tag{5.135}$$

Causality and the conditions in Eqs. (5.131) and (5.133) imply that

$$h[n] = 0, \qquad n < 0 \quad \text{and} \quad n > M;$$

i.e., causal FIR systems have generalized linear phase if they have impulse response length $(M + 1)$ and satisfy either Eq. (5.131c) or (5.133c). Specifically, it can be shown that if

$$h[n] = \begin{cases} h[M - n], & 0 \le n \le M, \\ 0, & \text{otherwise}, \end{cases} \tag{5.136a}$$

then

$$H(e^{j\omega}) = A_e(e^{j\omega})e^{-j\omega M/2}, \tag{5.136b}$$

where $A_e(e^{j\omega})$ is a real, even, periodic function of ω. Similarly, if

$$h[n] = \begin{cases} -h[M-n], & 0 \leq n \leq M, \\ 0, & \text{otherwise,} \end{cases} \tag{5.137a}$$

then it follows that

$$H(e^{j\omega}) = jA_o(e^{j\omega})e^{-j\omega M/2} = A_o(e^{j\omega})e^{-j\omega M/2 + j\pi/2}, \tag{5.137b}$$

where $A_o(e^{j\omega})$ is a real, odd, periodic function of ω. Note that in both cases the length of the impulse response is $(M+1)$ samples.

The conditions in Eqs. (5.136a) and (5.137a) are sufficient to guarantee a causal system with generalized linear phase. However, they are not necessary conditions. Clements and Pease (1989) have shown that causal infinite-duration impulse responses can also have Fourier transforms with generalized linear phase. The corresponding system functions, however, are not rational, and thus, the systems cannot be implemented with difference equations.

Expressions for the frequency response of FIR linear-phase systems are useful in filter design and in understanding some of the properties of such systems. In deriving these expressions, it turns out that significantly different expressions result, depending on the type of symmetry and whether M is an even or odd integer. For this reason, it is generally useful to define four types of FIR generalized linear-phase systems.

Type I FIR Linear-Phase Systems

A type I system is defined as a system that has a symmetric impulse response

$$h[n] = h[M-n], \qquad 0 \leq n \leq M, \tag{5.138}$$

with M an even integer. The delay $M/2$ is an integer. The frequency response is

$$H(e^{j\omega}) = \sum_{n=0}^{M} h[n]e^{-j\omega n}. \tag{5.139}$$

By applying the symmetry condition, Eq. (5.138), the sum in Eq. (5.139) can be rewritten in the form

$$H(e^{j\omega}) = e^{-j\omega M/2} \left(\sum_{k=0}^{M/2} a[k] \cos \omega k \right), \tag{5.140a}$$

where

$$a[0] = h[M/2], \tag{5.140b}$$

$$a[k] = 2h[(M/2) - k], \qquad k = 1, 2, \ldots, M/2. \tag{5.140c}$$

Thus, from Eq. (5.140a), we see that $H(e^{j\omega})$ has the form of Eq. (5.136b), and in particular, β in Eq. (5.125) is either 0 or π.

Type II FIR Linear-Phase Systems

A type II system has a symmetric impulse response as in Eq. (5.138), with M an odd integer. $H(e^{j\omega})$ for this case can be expressed as

$$H(e^{j\omega}) = e^{-j\omega M/2} \left\{ \sum_{k=1}^{(M+1)/2} b[k] \cos\left[\omega\left(k - \tfrac{1}{2}\right)\right] \right\}, \qquad (5.141a)$$

where

$$b[k] = 2h[(M+1)/2 - k], \qquad k = 1, 2, \ldots, (M+1)/2. \qquad (5.141b)$$

Again, $H(e^{j\omega})$ has the form of Eq. (5.136b) with a time delay of $M/2$, which in this case is an integer plus one-half, and β in Eq. (5.125) is either 0 or π.

Type III FIR Linear-Phase Systems

If the system has an antisymmetric impulse response

$$h[n] = -h[M - n], \qquad 0 \le n \le M, \qquad (5.142)$$

with M an even integer, then $H(e^{j\omega})$ has the form

$$H(e^{j\omega}) = je^{-j\omega M/2} \left[\sum_{k=1}^{M/2} c[k] \sin \omega k \right], \qquad (5.143a)$$

where

$$c[k] = 2h[(M/2) - k], \qquad k = 1, 2, \ldots, M/2. \qquad (5.143b)$$

In this case, $H(e^{j\omega})$ has the form of Eq. (5.137b) with a delay of $M/2$, which is an integer, and β in Eq. (5.125) is $\pi/2$ or $3\pi/2$.

Type IV FIR Linear-Phase Systems

If the impulse response is antisymmetric as in Eq. (5.142) and M is odd, then

$$H(e^{j\omega}) = je^{-j\omega M/2} \left[\sum_{k=1}^{(M+1)/2} d[k] \sin\left[\omega\left(k - \tfrac{1}{2}\right)\right] \right], \qquad (5.144a)$$

where

$$d[k] = 2h[(M+1)/2 - k], \qquad k = 1, 2, \ldots, (M+1)/2. \qquad (5.144b)$$

As in the case of type III systems, $H(e^{j\omega})$ has the form of Eq. (5.137b) with delay $M/2$, which is an integer plus one-half, and β in Eq. (5.125) is $\pi/2$ or $3\pi/2$.

Examples of FIR Linear-Phase Systems

Figure 5.33 shows an example of each of the four types of FIR linear-phase impulse responses. The associated frequency responses are given in Examples 5.15–5.18.

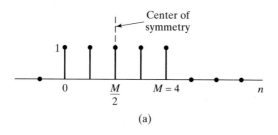

(a)

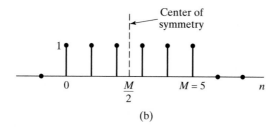

(b)

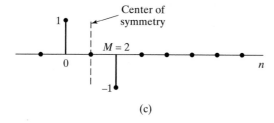

(c)

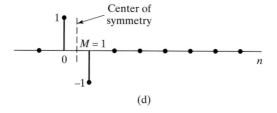

(d)

Figure 5.33 Examples of FIR linear-phase systems. (a) Type I, M even, $h[n] = h[M - n]$. (b) Type II, M odd, $h[n] = h[M - n]$. (c) Type III, M even, $h[n] = -h[M - n]$. (d) Type IV, M odd, $h[n] = -h[M - n]$.

Example 5.15   Type I Linear-Phase System

If the impulse response is

$$h[n] = \begin{cases} 1, & 0 \le n \le 4, \\ 0, & \text{otherwise}, \end{cases} \tag{5.145}$$

as shown in Figure 5.33(a), the system satisfies the condition of Eq. (5.138). The frequency response is

$$H(e^{j\omega}) = \sum_{n=0}^{4} e^{-j\omega n} = \frac{1 - e^{-j\omega 5}}{1 - e^{-j\omega}}$$

$$= e^{-j\omega 2} \frac{\sin(5\omega/2)}{\sin(\omega/2)}. \tag{5.146}$$

The magnitude, phase, and group delay of the system are shown in Figure 5.34. Since $M = 4$ is even, the group delay is an integer, i.e., $\alpha = 2$.

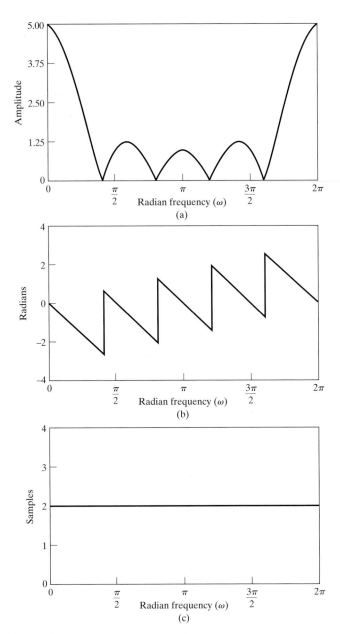

Figure 5.34   Frequency response of type I system of Example 5.15. (a) Magnitude. (b) Phase. (c) Group delay.

Example 5.16   Type II Linear-Phase System

If the length of the impulse response of the previous example is extended by one sample, we obtain the impulse response of Figure 5.33(b), which has frequency response

$$H(e^{j\omega}) = e^{-j\omega 5/2} \frac{\sin(3\omega)}{\sin(\omega/2)}. \tag{5.147}$$

The frequency-response functions for this system are shown in Figure 5.35. Note that the group delay in this case is constant with $\alpha = 5/2$.

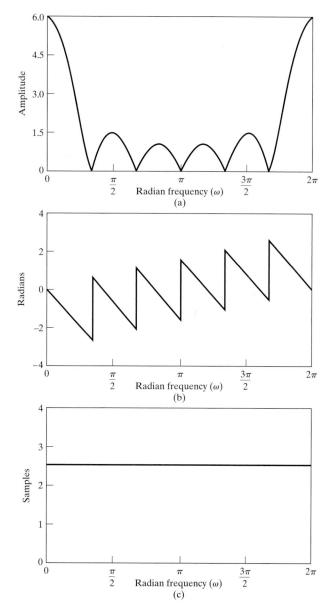

Figure 5.35   Frequency response of type II system of Example 5.16. (a) Magnitude. (b) Phase. (c) Group delay.

Example 5.17   Type III Linear-Phase System

If the impulse response is

$$h[n] = \delta[n] - \delta[n-2],$$ (5.148)

as in Figure 5.33(c), then

$$H(e^{j\omega}) = 1 - e^{-j2\omega} = j[2\sin(\omega)]e^{-j\omega}. \tag{5.149}$$

The frequency-response plots for this example are given in Figure 5.36. Note that the group delay in this case is constant with $\alpha = 1$.

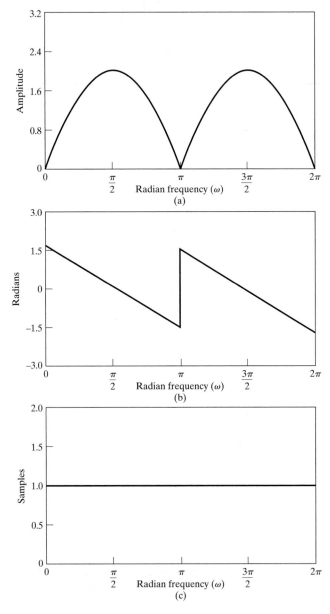

Figure 5.36   Frequency response of type III system of Example 5.17. (a) Magnitude. (b) Phase. (c) Group delay.

Example 5.18   Type IV Linear-Phase System

In this case (Figure 5.33(d)), the impulse response is

$$h[n] = \delta[n] - \delta[n-1],\tag{5.150}$$

for which the frequency response is

$$H(e^{j\omega}) = 1 - e^{-j\omega}$$
$$= j[2\sin(\omega/2)]e^{-j\omega/2}.\tag{5.151}$$

The frequency response for this system is shown in Figure 5.37. Note that the group delay is equal to $\frac{1}{2}$ for all ω.

Locations of Zeros for FIR Linear-Phase Systems

The preceding examples illustrate the properties of the impulse response and the frequency response for all four types of FIR linear-phase systems. It is also instructive to consider the locations of the zeros of the system function for FIR linear-phase systems. The system function is

$$H(z) = \sum_{n=0}^{M} h[n]z^{-n}.\tag{5.152}$$

In the symmetric cases (types I and II), we can use Eq. (5.138) to express $H(z)$ as

$$H(z) = \sum_{n=0}^{M} h[M-n]z^{-n} = \sum_{k=M}^{0} h[k]z^k z^{-M}$$
$$= z^{-M} H(z^{-1}).\tag{5.153}$$

From Eq. (5.153), we conclude that if z_0 is a zero of $H(z)$, then

$$H(z_0) = z_0^{-M} H(z_0^{-1}) = 0.\tag{5.154}$$

This implies that if $z_0 = re^{j\theta}$ is a zero of $H(z)$, then $z_0^{-1} = r^{-1}e^{-j\theta}$ is also a zero of $H(z)$. When $h[n]$ is real and z_0 is a zero of $H(z)$, $z_0^* = re^{-j\theta}$ will also be a zero of $H(z)$, and by the preceding argument, so will $(z_0^*)^{-1} = r^{-1}e^{j\theta}$. Therefore, when $h[n]$ is real, each complex zero not on the unit circle will be part of a set of four conjugate reciprocal zeros of the form

$$(1 - re^{j\theta}z^{-1})(1 - re^{-j\theta}z^{-1})(1 - r^{-1}e^{j\theta}z^{-1})(1 - r^{-1}e^{-j\theta}z^{-1}).$$

If a zero of $H(z)$ is on the unit circle, i.e., $z_0 = e^{j\theta}$, then $z_0^{-1} = e^{-j\theta} = z_0^*$, so zeros on the unit circle come in pairs of the form

$$(1 - e^{j\theta}z^{-1})(1 - e^{-j\theta}z^{-1}).$$

If a zero of $H(z)$ is real and not on the unit circle, the reciprocal will also be a zero of $H(z)$, and $H(z)$ will have factors of the form

$$(1 \pm rz^{-1})(1 \pm r^{-1}z^{-1}).$$

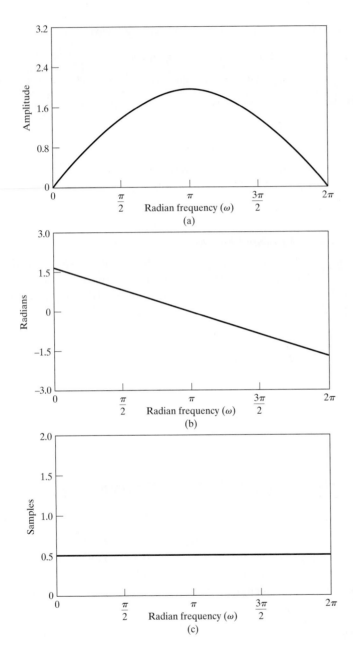

Figure 5.37   Frequency response of type IV system of Example 5.18. (a) Magnitude. (b) Phase. (c) Group delay.

Finally, a zero of $H(z)$ at $z = \pm 1$ can appear by itself, since ± 1 is its own reciprocal and its own conjugate. Thus, we may also have factors of $H(z)$ of the form

$$(1 \pm z^{-1}).$$

The case of a zero at $z = -1$ is particularly important. From Eq. (5.153),

$$H(-1) = (-1)^M H(-1).$$

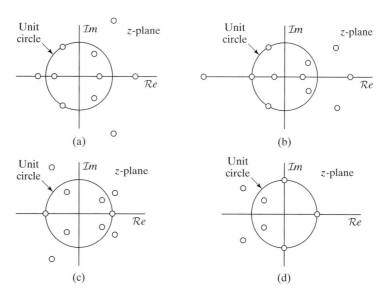

Figure 5.38   Typical plots of zeros for linear-phase systems. (a) Type I. (b) Type II. (c) Type III. (d) Type IV.

If M is even, we have a simple identity, but if M is odd, $H(-1) = -H(-1)$, so $H(-1)$ must be zero. Thus, for symmetric impulse responses with M odd, the system function *must* have a zero at $z = -1$. Figures 5.38(a) and 5.38(b) show typical locations of zeros for type I (M even) and type II (M odd) systems, respectively.

If the impulse response is antisymmetric (types III and IV), then, following the approach used to obtain Eq. (5.153), we can show that

$$H(z) = -z^{-M} H(z^{-1}). \tag{5.155}$$

This equation can be used to show that the zeros of $H(z)$ for the antisymmetric case are constrained in the same way as the zeros for the symmetric case. In the antisymmetric case, however, both $z = 1$ and $z = -1$ are of special interest. If $z = 1$, Eq. (5.155) becomes

$$H(1) = -H(1). \tag{5.156}$$

Thus, $H(z)$ *must* have a zero at $z = 1$ for both M even and M odd. If $z = -1$, Eq. (5.155) gives

$$H(-1) = (-1)^{-M+1} H(-1). \tag{5.157}$$

In this case, if $(M - 1)$ is odd (i.e., if M is even), $H(-1) = -H(-1)$, so $z = -1$ *must* be a zero of $H(z)$ if M is even. Figures 5.38(c) and 5.38(d) show typical zero locations for type III and IV systems, respectively.

These constraints on the zeros are important in designing FIR linear-phase systems, since they impose limitations on the types of frequency responses that can be achieved. For example, we note that, in approximating a highpass filter using a symmetric impulse response, M should not be odd, since the frequency response is constrained to be zero at $\omega = \pi (z = -1)$.

5.7.4 Relation of FIR Linear-Phase Systems to Minimum-Phase Systems

The previous discussion shows that all FIR linear-phase systems with real impulse responses have zeros either on the unit circle or at conjugate reciprocal locations. Thus, it is easily shown that the system function of any FIR linear-phase system can be factored into a minimum-phase term $H_{\min}(z)$, a maximum-phase term $H_{\max}(z)$, and a term $H_{\mathrm{uc}}(z)$ containing only zeros on the unit circle; i.e.,

$$H(z) = H_{\min}(z)H_{\mathrm{uc}}(z)H_{\max}(z), \tag{5.158a}$$

where

$$H_{\max}(z) = H_{\min}(z^{-1})z^{-M_i} \tag{5.158b}$$

and M_i is the number of zeros of $H_{\min}(z)$. In Eq. (5.158a), $H_{\min}(z)$ has all M_i of its zeros *inside* the unit circle, and $H_{\mathrm{uc}}(z)$ has all M_o of its zeros *on* the unit circle. $H_{\max}(z)$ has all M_i of its zeros *outside* the unit circle, and, from Eq. (5.158b), its zeros are the reciprocals of the M_i zeros of $H_{\min}(z)$. The order of the system function $H(z)$ is therefore $M = 2M_i + M_o$.

Example 5.19   Decomposition of a Linear-Phase System

As a simple example of the use of Eqs. (5.158), consider the minimum-phase system function of Eq. (5.99), for which the frequency response is plotted in Figure 5.25. The system obtained by applying Eq. (5.158b) to $H_{\min}(z)$ in Eq. (5.99) is

$$H_{\max}(z) = (0.9)^2(1 - 1.1111e^{j0.6\pi}z^{-1})(1 - 1.1111e^{-j0.6\pi}z^{-1})$$

$$\times (1 - 1.25e^{-j0.8\pi}z^{-1})(1 - 1.25e^{j0.8\pi}z^{-1}).$$

$H_{\max}(z)$ has the frequency response shown in Figure 5.39. Now, if these two systems are cascaded, it follows from Eq. (5.158b) that the overall system

$$H(z) = H_{\min}(z)H_{\max}(z)$$

has linear phase. The frequency response of the composite system would be obtained by adding the respective log magnitude, phase, and group-delay functions. Therefore,

$$20\log_{10}|H(e^{j\omega})| = 20\log_{10}|H_{\min}(e^{j\omega})| + 20\log_{10}|H_{\max}(e^{j\omega})|$$

$$= 40\log_{10}|H_{\min}(e^{j\omega})|. \tag{5.159}$$

Similarly,

$$\angle H(e^{j\omega}) = \angle H_{\min}(e^{j\omega}) + \angle H_{\max}(e^{j\omega}). \tag{5.160}$$

From Eq. (5.158b), it follows that

$$\angle H_{\max}(e^{j\omega}) = -\omega M_i - \angle H_{\min}(e^{j\omega}). \tag{5.161}$$

and therefore

$$\angle H(e^{j\omega}) = -\omega M_i,$$

where $M_i = 4$ is the number of zeros of $H_{\min}(z)$. In like manner, the group-delay functions of $H_{\min}(e^{j\omega})$ and $H_{\max}(e^{j\omega})$ combine to give

$$\mathrm{grd}[H(e^{j\omega})] = M_i = 4.$$

The frequency-response plots for the composite system are given in Figure 5.40. Note that the curves are sums of the corresponding functions in Figures 5.25 and 5.39.

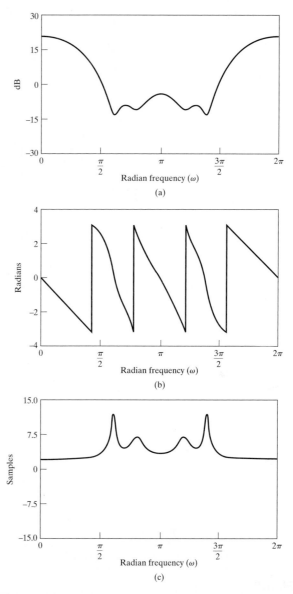

(a)

(b)

(c)

Figure 5.39 Frequency response of maximum-phase system having the same magnitude as the system in Figure 5.25. (a) Log magnitude. (b) Phase (principal value). (c) Group delay.

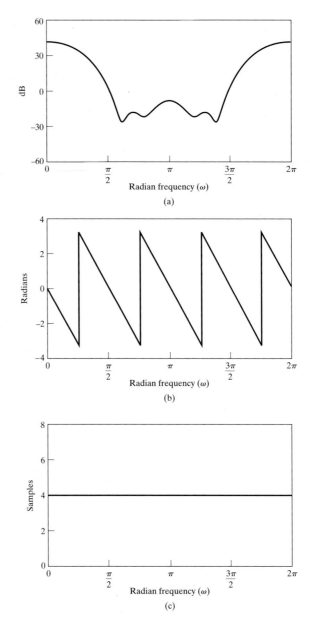

Figure 5.40   Frequency response of cascade of maximum-phase and minimum-phase systems, yielding a linear-phase system. (a) Log magnitude. (b) Phase (principal value). (c) Group delay.

5.8 SUMMARY

In this chapter, we developed and explored the representation and analysis of LTI systems using the Fourier and z-transforms. The importance of transform analysis for LTI systems stems directly from the fact that complex exponentials are eigenfunctions of such systems and the associated eigenvalues correspond to the system function or frequency response.

A particularly important class of LTI systems is that characterized by linear constant-coefficient difference equations. Systems characterized by difference equations may have an impulse response that is infinite in duration (IIR) or finite in duration (FIR). Transform analysis is particularly useful for analyzing these systems, since the Fourier transform or z-transform converts a difference equation to an algebraic equation. In particular, the system function is a ratio of polynomials, the coefficients of which correspond directly to the coefficients in the difference equation. The roots of these polynomials provide a useful system representation in terms of the pole–zero plot.

The frequency response of LTI systems is often characterized in terms of magnitude and phase or group delay, which is the negative of the derivative of the phase. Linear phase is often a desirable characteristic of a system frequency response, since it is a relatively mild form of phase distortion, corresponding to a time shift. The importance of FIR systems lies in part in the fact that such systems can be easily designed to have exactly linear phase (or generalized linear phase), whereas, for a given set of frequency response magnitude specifications, IIR systems are more efficient. These and other trade-offs will be discussed in detail in Chapter 7.

While, in general, for LTI systems, the frequency-response magnitude and phase are independent, for minimum-phase systems the magnitude uniquely specifies the phase and the phase uniquely specifies the magnitude to within a scale factor. Nonminimum-phase systems can be represented as the cascade combination of a minimum-phase system and an all-pass system. Relations between Fourier transform magnitude and phase will be discussed in considerably more detail in Chapter 12.

Problems

Basic Problems with Answers

5.1. In the system shown in Figure P5.1-1, $H(e^{j\omega})$ is an ideal lowpass filter. Determine whether for some choice of input $x[n]$ and cutoff frequency ω_c, the output can be the pulse

$$y[n] = \begin{cases} 1, & 0 \le n \le 10, \\ 0, & \text{otherwise,} \end{cases}$$

shown in Figure P5.1-2.

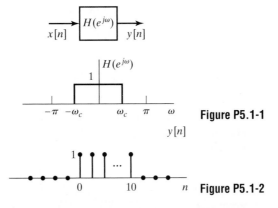

Figure P5.1-1

Figure P5.1-2

5.2. Consider a stable LTI system with input $x[n]$ and output $y[n]$. The input and output satisfy the difference equation

$$y[n-1] - \tfrac{10}{3}y[n] + y[n+1] = x[n].$$

(a) Plot the poles and zeros of the system function in the z-plane.
(b) Determine the impulse response $h[n]$.

5.3. Consider an LTI discrete-time system for which the input $x[n]$ and output $y[n]$ are related by the 2nd-order difference equation

$$y[n-1] + \tfrac{1}{3}y[n-2] = x[n].$$

From the following list, choose *two* possible impulse responses for the system:

(a) $\left(-\tfrac{1}{3}\right)^{n+1} u[n+1]$
(b) $3^{n+1}u[n+1]$
(c) $3(-3)^{n+2}u[-n-2]$
(d) $\tfrac{1}{3}\left(-\tfrac{1}{3}\right)^{n} u[-n-2]$
(e) $\left(-\tfrac{1}{3}\right)^{n+1} u[-n-2]$
(f) $\left(\tfrac{1}{3}\right)^{n+1} u[n+1]$
(g) $(-3)^{n+1}u[n]$
(h) $n^{1/3}u[n]$.

5.4. When the input to an LTI system is

$$x[n] = \left(\tfrac{1}{2}\right)^{n} u[n] + (2)^{n}u[-n-1],$$

the output is

$$y[n] = 6 \left(\tfrac{1}{2}\right)^{n} u[n] - 6 \left(\tfrac{3}{4}\right)^{n} u[n].$$

(a) Determine the system function $H(z)$ of the system. Plot the poles and zeros of $H(z)$, and indicate the ROC.
(b) Determine the impulse response $h[n]$ of the system for all values of n.
(c) Write the difference equation that characterizes the system.
(d) Is the system stable? Is it causal?

5.5. Consider a system described by a linear constant-coefficient difference equation with initial-rest conditions. The step response of the system is given by

$$y[n] = \left(\tfrac{1}{3}\right)^{n} u[n] + \left(\tfrac{1}{4}\right)^{n} u[n] + u[n].$$

(a) Determine the difference equation.
(b) Determine the impulse response of the system.
(c) Determine whether or not the system is stable.

5.6. The following information is known about an LTI system:

(1) The system is causal.
(2) When the input is

$$x[n] = -\tfrac{1}{3}\left(\tfrac{1}{2}\right)^{n} u[n] - \tfrac{4}{3}(2)^{n}u[-n-1],$$

the z-transform of the output is

$$Y(z) = \frac{1 - z^{-2}}{\left(1 - \frac{1}{2}z^{-1}\right)(1 - 2z^{-1})}.$$

(a) Determine the z-transform of $x[n]$.

(b) What are the possible choices for the ROC of $Y(z)$?

(c) What are the possible choices for the impulse response of the system?

5.7. When the input to an LTI system is

$$x[n] = 5u[n],$$

the output is

$$y[n] = \left[2\left(\tfrac{1}{2}\right)^n + 3\left(-\tfrac{3}{4}\right)^n\right]u[n].$$

(a) Determine the system function $H(z)$ of the system. Plot the poles and zeros of $H(z)$, and indicate the ROC.

(b) Determine the impulse response of the system for all values of n.

(c) Write the difference equation that characterizes the system.

5.8. A causal LTI system is described by the difference equation

$$y[n] = \tfrac{3}{2}y[n-1] + y[n-2] + x[n-1].$$

(a) Determine the system function $H(z) = Y(z)/X(z)$ for this system. Plot the poles and zeros of $H(z)$, and indicate the ROC.

(b) Determine the impulse response of the system.

(c) You should have found the system to be unstable. Determine a stable (noncausal) impulse response that satisfies the difference equation.

5.9. Consider an LTI system with input $x[n]$ and output $y[n]$ for which

$$y[n-1] - \tfrac{5}{2}y[n] + y[n+1] = x[n].$$

The system may or may not be stable or causal. By considering the pole–zero pattern associated with this difference equation, determine three possible choices for the impulse response of the system. Show that each choice satisfies the difference equation. Indicate which choice corresponds to a stable system and which choice corresponds to a causal system.

5.10. If the system function $H(z)$ of an LTI system has a pole–zero diagram as shown in Figure P5.10 and the system is causal, can the inverse system $H_i(z)$, where $H(z)H_i(z) = 1$, be both causal and stable? Clearly justify your answer.

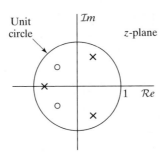

Figure P5.10

5.11. The system function of an LTI system has the pole–zero plot shown in Figure P5.11 Specify whether each of the following statements is true, is false, or cannot be determined from the information given.

(a) The system is stable.
(b) The system is causal.
(c) If the system is causal, then it must be stable.
(d) If the system is stable, then it must have a two-sided impulse response.

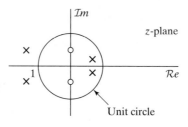

Figure P5.11

5.12. A discrete-time causal LTI system has the system function

$$H(z) = \frac{(1+0.2z^{-1})(1-9z^{-2})}{(1+0.81z^{-2})}.$$

(a) Is the system stable?
(b) Determine expressions for a minimum-phase system $H_1(z)$ and an all-pass system $H_{ap}(z)$ such that

$$H(z) = H_1(z)H_{ap}(z).$$

5.13. Figure P5.13 shows the pole–zero plots for four different LTI systems. Based on these plots, state whether or not each system is an all-pass system.

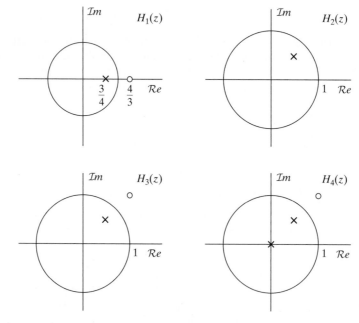

Figure P5.13

5.14. Determine the group delay for $0 < \omega < \pi$ for each of the following sequences:

(a)

$$x_1[n] = \begin{cases} n - 1, & 1 \le n \le 5, \\ 9 - n, & 5 < n \le 9, \\ 0, & \text{otherwise.} \end{cases}$$

(b)

$$x_2[n] = \left(\frac{1}{2}\right)^{|n-1|} + \left(\frac{1}{2}\right)^{|n|}.$$

5.15. Consider the class of discrete-time filters whose frequency response has the form

$$H(e^{j\omega}) = |H(e^{j\omega})|e^{-j\alpha\omega},$$

where $|H(e^{j\omega})|$ is a real and nonnegative function of ω and α is a real constant. As discussed in Section 5.7.1, this class of filters is referred to as *linear-phase* filters.

Consider also the class of discrete-time filters whose frequency response has the form

$$H(e^{j\omega}) = A(e^{j\omega})e^{-j\alpha\omega+j\beta},$$

where $A(e^{j\omega})$ is a real function of ω, α is a real constant, and β is a real constant. As discussed in Section 5.7.2, filters in this class are referred to as *generalized linear-phase* filters.

For each of the filters in Figure P5.15, determine whether it is a generalized linear-phase filter. If it is, then find $A(e^{j\omega})$, α, and β. In addition, for each filter you determine to be a generalized linear-phase filter, indicate whether it also meets the more stringent criterion for being a linear-phase filter.

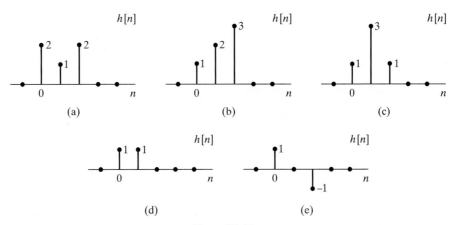

Figure P5.15

5.16. Figure P5.16 plots the continuous-phase $\arg[H(e^{j\omega})]$ for the frequency response of a specific LTI system, where

$$\arg[H(e^{j\omega})] = -\alpha\omega$$

for $|\omega| < \pi$ and α is a positive integer.

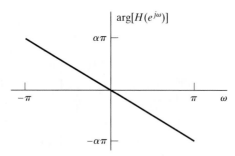

Figure P5.16

Is the impulse response $h[n]$ of this system a causal sequence? If the system is definitely causal, or if it is definitely not causal, give a proof. If the causality of the system cannot be determined from Figure P5.16, give examples of a noncausal sequence and a causal sequence that both have the foregoing phase response $\arg[H(e^{j\omega})]$.

5.17. For each of the following system functions, state whether or not it is a minimum-phase system. Justify your answers:

$$H_1(z) = \frac{(1 - 2z^{-1})\left(1 + \frac{1}{2}z^{-1}\right)}{\left(1 - \frac{1}{3}z^{-1}\right)\left(1 + \frac{1}{3}z^{-1}\right)},$$

$$H_2(z) = \frac{\left(1 + \frac{1}{4}z^{-1}\right)\left(1 - \frac{1}{4}z^{-1}\right)}{\left(1 - \frac{2}{3}z^{-1}\right)\left(1 + \frac{2}{3}z^{-1}\right)},$$

$$H_3(z) = \frac{1 - \frac{1}{3}z^{-1}}{\left(1 - \frac{j}{2}z^{-1}\right)\left(1 + \frac{j}{2}z^{-1}\right)},$$

$$H_4(z) = \frac{z^{-1}\left(1 - \frac{1}{3}z^{-1}\right)}{\left(1 - \frac{j}{2}z^{-1}\right)\left(1 + \frac{j}{2}z^{-1}\right)}.$$

5.18. For each of the following system functions $H_k(z)$, specify a minimum-phase system function $H_{\min}(z)$ such that the frequency-response magnitudes of the two systems are equal, i.e., $|H_k(e^{j\omega})| = |H_{\min}(e^{j\omega})|$.

(a)

$$H_1(z) = \frac{1 - 2z^{-1}}{1 + \frac{1}{3}z^{-1}}$$

(b)

$$H_2(z) = \frac{(1 + 3z^{-1})\left(1 - \frac{1}{2}z^{-1}\right)}{z^{-1}\left(1 + \frac{1}{3}z^{-1}\right)}$$

(c)

$$H_3(z) = \frac{(1 - 3z^{-1})\left(1 - \frac{1}{4}z^{-1}\right)}{\left(1 - \frac{3}{4}z^{-1}\right)\left(1 - \frac{4}{3}z^{-1}\right)}.$$

5.19. Figure P5.19 shows the impulse responses for several different LTI systems. Determine the group delay associated with each system.

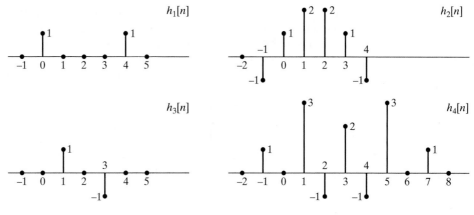

Figure P5.19

5.20. Figure P5.20 shows just the zero locations for several different system functions. For each plot, state whether the system function could be a generalized linear-phase system implemented by a linear constant-coefficient difference equation with real coefficients.

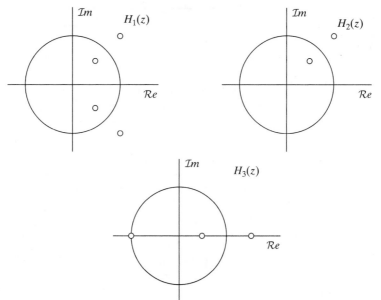

Figure P5.20

Basic Problems

5.21. Let $h_{lp}[n]$ denote the impulse response of an ideal lowpass filter with unity passband gain and cutoff frequency $\omega_c = \pi/4$. Figure P5.21 shows five systems, each of which is equivalent to an ideal LTI frequency-selective filter. For each system shown, sketch the equivalent frequency response, indicating explicitly the band-edge frequencies in terms of ω_c. In each case, specify whether the system is a lowpass, highpass, bandpass, bandstop, or multiband filter.

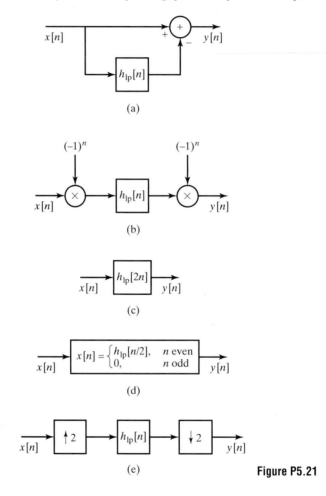

(a)

(b)

(c)

(d)

(e)   **Figure P5.21**

5.22. Many properties of a discrete-time sequence $h[n]$ or an LTI system with impulse response $h[n]$ can be discerned from a pole–zero plot of $H(z)$. In this problem, we are concerned only with causal systems. Clearly describe the z-plane characteristic that corresponds to each of the following properties:

(a) Real-valued impulse response

(b) Finite impulse response

(c) $h[n] = h[2\alpha - n]$ where 2α is an integer

(d) Minimum phase

(e) All-pass.

5.23. For all parts of this problem, $H(e^{j\omega})$ is the frequency response of a DT filter and can be expressed in polar coordinates as

$$H(e^{j\omega}) = A(\omega)e^{j\theta(\omega)}$$

where $A(\omega)$ is even and real-valued and $\theta(\omega)$ is a continuous, odd function of ω for $-\pi < \omega < \pi$, i.e., $\theta(\omega)$ is what we have referred to as the *unwrapped phase*. Recall:

- The *group delay* $\tau(\omega)$ associated with the filter is defined as

$$\tau(\omega) = -\frac{d\theta(\omega)}{d\omega} \qquad \text{for} |\omega| < \pi.$$

- An LTI filter is called *minimum phase* if it is stable and causal and has a stable and causal inverse.

For each of the following statements, state whether it is TRUE or FALSE. If you state that it is TRUE, give a clear, brief justification. If you state that it is FALSE, give a simple counterexample with a clear, brief explanation of why it is a counterexample.

(a) "If the filter is causal, its group delay must be nonnegative at all frequencies in the range $|\omega| < \pi$."

(b) "If the group delay of the filter is a positive constant integer for $|\omega| < \pi$ the filter must be a simple integer delay."

(c) "If the filter is minimum phase and all the poles and zeros are on the real axis then $\int_0^\pi \tau(\omega)d\omega = 0$."

5.24. A stable system with system function $H(z)$ has the pole–zero diagram shown in Figure P5.24. It can be represented as the cascade of a stable minimum-phase system $H_{min}(z)$ and a stable all-pass system $H_{ap}(z)$.

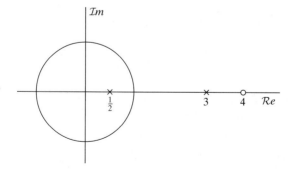

Figure P5.24   Pole–zero diagram for $H(z)$.

Determine a choice for $H_{min}(z)$ and $H_{ap}(z)$ (up to a scale factor) and draw their corresponding pole–zero plots. Indicate whether your decomposition is unique up to a scale factor.

5.25. (a) An ideal lowpass filter with impulse response $h[n]$ is designed with zero phase, a cutoff frequency of $\omega_c = \pi/4$, a passband gain of 1, and a stopband gain of 0. ($H(e^{j\omega})$ is shown in Figure P5.21.) Sketch the discrete-time Fourier transform of $(-1)^n h[n]$.

(b) A complex-valued filter with impulse response $g[n]$ has the pole–zero diagram shown in Figure P5.25. Sketch the pole–zero diagram for $(-1)^n g[n]$. If there is not sufficient information provided, explain why.

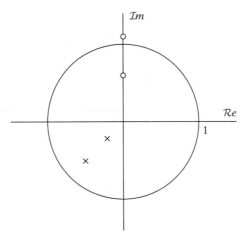

Figure P5.25

5.26. Consider a discrete-time LTI system for which the frequency response $H(e^{j\omega})$ is described by:

$$H(e^{j\omega}) = -j, \quad 0 < \omega < \pi$$

$$H(e^{j\omega}) = j, \quad -\pi < \omega < 0$$

(a) Is the impulse response of the system $h[n]$ real-valued? (i.e., is $h[n] = h^*[n]$ for all n)

(b) Calculate the following:

$$\sum_{n=-\infty}^{\infty} |h[n]|^2$$

(c) Determine the response of the system to the input $x[n] = s[n]\cos(\omega_c n)$, where $0 < \omega_c < \pi/2$ and $S(e^{j\omega}) = 0$ for $\omega_c/3 \leq |\omega| \leq \pi$.

5.27. We process the signal $x[n] = \cos(0.3\pi n)$ with a unity-gain all-pass LTI system, with frequency response $w = H(e^{j\omega})$ and a group delay of 4 samples at frequency $\omega = 0.3\pi$, to get the output $y[n]$. We also know that $\angle H(e^{j0.3\pi}) = \theta$ and $\angle H(e^{-j0.3\pi}) = -\theta$. Choose the most accurate statement:

(a) $y[n] = \cos(0.3\pi n + \theta)$
(b) $y[n] = \cos(0.3\pi(n-4) + \theta)$
(c) $y[n] = \cos(0.3\pi(n-4-\theta))$
(d) $y[n] = \cos(0.3\pi(n-4))$
(e) $y[n] = \cos(0.3\pi(n-4+\theta))$.

5.28. A causal LTI system has the system function

$$H(z) = \frac{(1 - e^{j\pi/3}z^{-1})(1 - e^{-j\pi/3}z^{-1})(1 + 1.1765z^{-1})}{(1 - 0.9e^{j\pi/3}z^{-1})(1 - 0.9e^{-j\pi/3}z^{-1})(1 + 0.85z^{-1})}.$$

(a) Write the difference equation that is satisfied by the input $x[n]$ and output $y[n]$ of this system.

(b) Plot the pole–zero diagram and indicate the ROC for the system function.

(c) Make a carefully labeled sketch of $|H(e^{j\omega})|$. Use the pole–zero locations to explain why the frequency response looks as it does.

(d) State whether the following are true or false about the system:

   (i) The system is stable.
   (ii) The impulse response approaches a nonzero constant for large n.
   (iii) Because the system function has a pole at angle $\pi/3$, the magnitude of the frequency response has a peak at approximately $\omega = \pi/3$.
   (iv) The system is a minimum-phase system.
   (v) The system has a causal and stable inverse.

5.29. Consider the cascade of an LTI system with its inverse system shown in Figure P5.29.

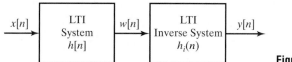

Figure P5.29

The impulse response of the first system is $h[n] = \delta[n] + 2\delta[n - 1]$.

(a) Determine the impulse response $h_i[n]$ of a stable inverse system for $h[n]$. Is the inverse system causal?

(b) Now consider the more general case where $h[n] = \delta[n] + \alpha\delta[n - 1]$. Under what conditions on α will there exist an inverse system that is both stable and causal?

5.30. In each of the following parts, state whether the statement is always TRUE or FALSE. Justify each of your answers.

(a) "An LTI discrete-time system consisting of the cascade connection of two minimum-phase systems is also minimum-phase."

(b) "An LTI discrete-time system consisting of the parallel connection of two minimum-phase systems is also minimum-phase."

5.31. Consider the system function

$$H(z) = \frac{rz^{-1}}{1 - (2r\cos\omega_0)z^{-1} + r^2z^{-2}}, \qquad |z| > r.$$

Assume first that $\omega_0 \neq 0$.

(a) Draw a labeled pole–zero diagram and determine $h[n]$.

(b) Repeat part (a) when $\omega_0 = 0$. This is known as a critically damped system.

Advanced Problems

5.32. Suppose that a causal LTI system has an impulse response of length 6 as shown in Figure P5.32, where c is a real-valued constant (positive or negative).

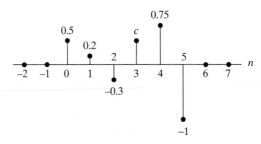

Figure P5.32

Which of the following statements is true:

(a) This system must be minimum phase.
(b) This system cannot be minimum phase.
(c) This system may or may not be minimum phase, depending on the value of c.

Justify your answer.

5.33. $H(z)$ is the system function for a stable LTI system and is given by:

$$H(z) = \frac{(1 - 2z^{-1})(1 - 0.75z^{-1})}{z^{-1}(1 - 0.5z^{-1})}.$$

(a) $H(z)$ can be represented as a cascade of a minimum-phase system $H_{min1}(z)$ and a unity-gain all-pass system $H_{ap}(z)$, i.e.,

$$H(z) = H_{min1}(z)H_{ap}(z).$$

Determine a choice for $H_{min1}(z)$ and $H_{ap}(z)$ and specify whether or not they are unique up to a scale factor.

(b) $H(z)$ can be expressed as a cascade of a minimum-phase system $H_{min2}(z)$ and a generalized linear-phase FIR system $H_{lp}(z)$:

$$H(z) = H_{min2}(z)H_{lp}(z).$$

Determine a choice for $H_{min2}(z)$ and $H_{lp}(z)$ and specify whether or not these are unique up to a scale factor.

5.34. A discrete-time LTI system with input $x[n]$ and output $y[n]$ has the frequency response magnitude and group delay functions shown in Figure P5.34-1. The signal $x[n]$, also shown in Figure P5.34-1, is the sum of three narrowband pulses. In particular, Figure P5.34-1 contains the following plots:

- $x[n]$
- $|X(e^{j\omega})|$, the Fourier transform magnitude of a particular input $x[n]$
- Frequency response magnitude plot for the system
- Group delay plot for the system

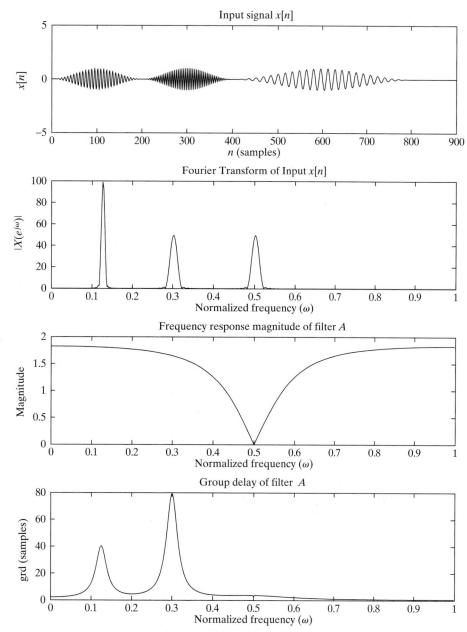

Figure P5.34-1 The input signal and the filter frequency response

In Figure P5.34-2 you are given four possible output signals, $y_i[n]$ $i = 1, 2, \ldots, 4$. Determine which one of the possible output signals is the output of the system when the input is $x[n]$. Provide a justification for your choice.

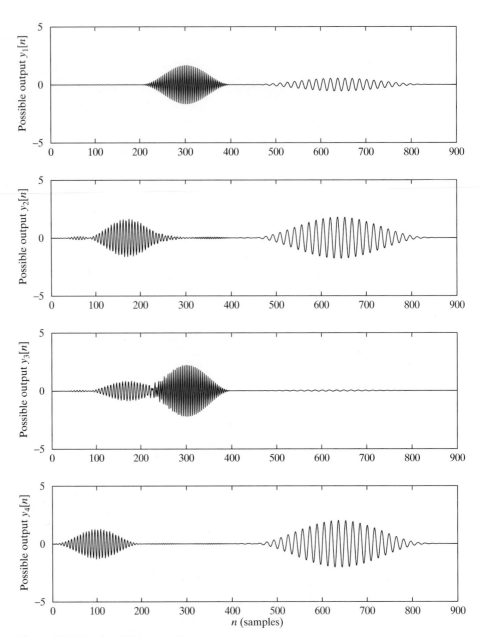

Figure P5.34-2   Possible output signals

5.35. Suppose that a discrete-time filter has group delay $\tau(\omega)$. Does the condition $\tau(\omega) > 0$ for $-\pi < \omega \le \pi$ imply that the filter is necessarily causal? Clearly explain your reasoning.

5.36. Consider the stable LTI system with system function

$$H(z) = \frac{1 + 4z^{-2}}{1 - \frac{1}{4}z^{-1} - \frac{3}{8}z^{-2}}.$$

The system function $H(z)$ can be factored such that

$$H(z) = H_{min}(z)H_{ap}(z),$$

where $H_{min}(z)$ is a minimum-phase system, and $H_{ap}(z)$ is an all-pass system, i.e.,

$$|H_{ap}(e^{j\omega})| = 1.$$

Sketch the pole–zero diagrams for $H_{min}(z)$ and $H_{ap}(z)$. Be sure to label the positions of all the poles and zeros. Also, indicate the ROC for $H_{min}(z)$ and $H_{ap}(z)$.

5.37. An LTI system has generalized linear phase and system function $H(z) = a + bz^{-1} + cz^{-2}$. The impulse response has unit energy, $a \geq 0$, and $H(e^{j\pi}) = H(e^{j0}) = 0$.

(a) Determine the impulse response $h[n]$.
(b) Plot $|H(e^{j\omega})|$.

5.38. $H(z)$ is the system function for a stable LTI system and is given by:

$$H(z) = \frac{(1 - 9z^{-2})(1 + \frac{1}{3}z^{-1})}{1 - \frac{1}{3}z^{-1}}.$$

(a) $H(z)$ can be represented as a cascade of a minimum-phase system $H_{min}(z)$ and a unity-gain all-pass system $H_{ap}(z)$. Determine a choice for $H_{min}(z)$ and $H_{ap}(z)$ and specify whether or not they are unique up to a scale factor.
(b) Is the minimum-phase system, $H_{min}(z)$, an FIR system? Explain.
(c) Is the minimum-phase system, $H_{min}(z)$, a generalized linear-phase system? If not, can $H(z)$ be represented as a cascade of a generalized linear-phase system $H_{lin}(z)$ and an all-pass system $H_{ap2}(z)$? If your answer is yes, determine $H_{lin}(z)$ and $H_{ap2}(z)$. If your answer is no, explain why such representation does not exist.

5.39. $H(z)$ is the transfer function of a stable LTI system and is given by:

$$H(z) = \frac{z - 2}{z(z - 1/3)}.$$

(a) Is the system causal? Clearly justify your answer.
(b) $H(z)$ can also be expressed as $H(z) = H_{min}(z)H_{lin}(z)$ where $H_{min}(z)$ is a minimum-phase system and $H_{lin}(z)$ is a generalized linear-phase system. Determine a choice for $H_{min}(z)$ and $H_{lin}(z)$.

5.40. System S_1 has a real impulse response $h_1[n]$ and a real-valued frequency response $H_1(e^{j\omega})$.

(a) Does the impulse response $h_1[n]$ have any symmetry? Explain.
(b) System S_2 is a linear-phase system with the same magnitude response as system S_1. What is the relationship between $h_2[n]$, the impulse response of system S_2, and $h_1[n]$?
(c) Can a causal IIR filter have a linear phase? Explain. If your answer is yes, provide an example.

5.41. Consider a discrete-time LTI filter whose impulse response $h[n]$ is nonzero only over five consecutive time samples; the filter's frequency response is $H(e^{j\omega})$. Let signals $x[n]$ and $y[n]$ denote the filter's input and output, respectively.

Moreover, you are given the following information about the filter:

(i) $\int_{-\pi}^{\pi} H(e^{j\omega}) \, d\omega = 4\pi.$

(ii) There exists a signal $a[n]$ that has a real and even DTFT $A(e^{j\omega})$ given by

$$A(e^{j\omega}) = H(e^{j\omega}) e^{j2\omega}.$$

(iii) $A(e^{j0}) = 8$, and $A(e^{j\pi}) = 12$.

Completely specify the impulse response $h[n]$, i.e., specify the impulse response at each time instant where it takes a nonzero value. Plot $h[n]$, carefully and accurately labeling its salient features.

5.42. A bounded-input bounded-output stable discrete-time LTI system has impulse response $h[n]$ corresponding to a rational system function $H(z)$ with the pole–zero diagram shown in Figure P5.42.

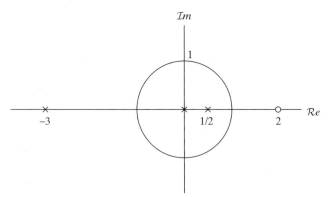

Figure P5.42

In addition, we know that $\sum_{n=-\infty}^{\infty} (-1)^n h[n] = -1$.

(a) Determine $H(z)$ and its ROC.

(b) Consider a new system having an impulse response $g[n] = h[n + n_0]$, where n_0 is an integer. Given that $G(z)|_{z=0} = 0$, and $\lim_{z \to \infty} G(z) < \infty$, determine the values of n_0 and $g[0]$.

(c) A new system has an impulse response, $f[n] = h[n] * h[-n]$.
Determine $F(z)$ and its ROC.

(d) Is there a right-sided signal $e[n]$ such that $e[n] * h[n] = u[n]$, where $u[n]$ is the unit-step sequence? If so, is $e[n]$ causal?

5.43. Consider an LTI system with system function:

$$H(z) = \frac{z^{-2}(1 - 2z^{-1})}{2(1 - \frac{1}{2}z^{-1})}, \quad |z| > \frac{1}{2}.$$

(a) Is $H(z)$ an all-pass system? Explain.

(b) The system is to be implemented as the cascade of three systems $H_{min}(z)$, $H_{max}(z)$, and $H_d(z)$, denoting minimum-phase, maximum-phase, and integer time shift, respectively. Determine the impulse responses $h_{min}[n]$, $h_{max}[n]$, and $h_d[n]$, corresponding to each of the three systems.

5.44. The impulse responses of four linear-phase FIR filters $h_1[n]$, $h_2[n]$, $h_3[n]$, and $h_4[n]$ are given below. Moreover, four magnitude response plots, A, B, C, and D, that potentially correspond to these impulse responses are shown in Figure P5.44. For each impulse response $h_i[n]$, $i = 1, \ldots, 4$, specify which of the four magnitude response plots, if any, corresponds to it. If none of the magnitude response plots matches a given $h_i[n]$, then specify "none" as the answer for that $h_i[n]$.

$$h_1[n] = 0.5\delta[n] + 0.7\delta[n - 1] + 0.5\delta[n - 2]$$

$$h_2[n] = 1.5\delta[n] + \delta[n - 1] + \delta[n - 2] + 1.5\delta[n - 3]$$

$$h_3[n] = -0.5\delta[n] - \delta[n - 1] + \delta[n - 3] + 0.5\delta[n - 4]$$

$$h_4[n] = -\delta[n] + 0.5\delta[n - 1] - 0.5\delta[n - 2] + \delta[n - 3].$$

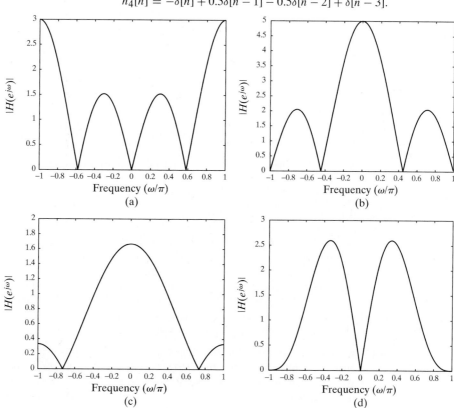

Figure P5.44

5.45. The pole–zero plots in Figure P5.45 describe six different causal LTI systems.

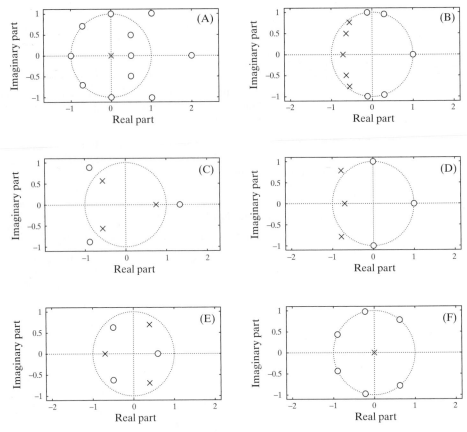

Figure P5.45

Answer the following questions about the systems having the above pole–zero plots. In each case, an acceptable answer could be *none* or *all*.

(a) Which systems are IIR systems?
(b) Which systems are FIR systems?
(c) Which systems are stable systems?
(d) Which systems are minimum-phase systems?
(e) Which systems are generalized linear-phase systems?
(f) Which systems have $|H(e^{j\omega})|$=constant for all ω?
(g) Which systems have corresponding stable and causal inverse systems?
(h) Which system has the shortest (least number of nonzero samples) impulse response?
(i) Which systems have lowpass frequency responses?
(j) Which systems have minimum group delay?

5.46. Assume that the two linear systems in the cascade shown in Figure P5.46 are linear-phase FIR filters. Suppose that $H_1(z)$ has order M_1 (impulse response length $M_1 + 1$) and $H_2(z)$ has order M_2. Suppose that the frequency responses are of the form $H_1(e^{j\omega}) = A_1(e^{j\omega})e^{-j\omega M_1/2}$ and $H_2(e^{j\omega}) = jA_2(e^{j\omega})e^{-j\omega M_2/2}$, where M_1 is an even integer and M_2 is an odd integer.

(a) Determine the overall frequency response $H(e^{j\omega})$.
(b) Determine the length of the impulse response of the overall system.
(c) Determine the group delay of the overall system.
(d) Is the overall system a Type I, Type II, Type III, or Type-IV generalized linear-phase system?

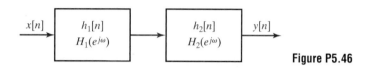

Figure P5.46

5.47. A linear-phase FIR system has a real impulse response $h[n]$ whose z-transform is known to have the form

$$H(z) = (1 - az^{-1})(1 - e^{j\pi/2}z^{-1})(1 - bz^{-1})(1 - 0.5z^{-1})(1 - cz^{-1})$$

where a, b, and c are zeros of $H(z)$ that you are to find. It is also known that $H(e^{j\omega}) = 0$ for $\omega = 0$. This information and knowledge of the properties of linear-phase systems are sufficient to completely determine the system function (and therefore the impulse response) and to answer the following questions:

(a) Determine the length of the impulse response (i.e., the number of nonzero samples).
(b) Is this a Type I, Type II, Type III, or Type IV system?
(c) Determine the group delay of the system in samples.
(d) Determine the unknown zeros a, b, and c. (The labels are arbitrary, but there are three more zeros to find.)
(e) Determine the values of the impulse response and sketch it as a stem plot.

5.48. The system function $H(z)$ of a causal LTI system has the pole–zero configuration shown in Figure P5.48. It is also known that $H(z) = 6$ when $z = 1$.

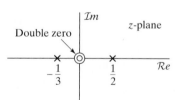

Figure P5.48

(a) Determine $H(z)$.
(b) Determine the impulse response $h[n]$ of the system.
(c) Determine the response of the system to the following input signals:

(i) $x[n] = u[n] - \frac{1}{2}u[n-1]$
(ii) The sequence $x[n]$ obtained from sampling the continuous-time signal

$$x(t) = 50 + 10\cos 20\pi t + 30\cos 40\pi t$$

at a sampling frequency $\Omega_s = 2\pi(40)$ rad/s.

5.49. The system function of an LTI system is given by

$$H(z) = \frac{21}{\left(1 - \frac{1}{2}z^{-1}\right)(1 - 2z^{-1})(1 - 4z^{-1})}.$$

It is known that the system is not stable and that the impulse response is two sided.

(a) Determine the impulse response $h[n]$ of the system.
(b) The impulse response found in part (a) can be expressed as the sum of a causal impulse response $h_1[n]$ and an anticausal impulse response $h_2[n]$. Determine the corresponding system functions $H_1(z)$ and $H_2(z)$.

5.50. The Fourier transform of a stable LTI system is purely real and is shown in Figure P5.50. Determine whether this system has a stable inverse system.

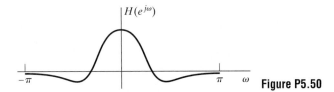

Figure P5.50

5.51. A causal LTI system has the system function

$$H(z) = \frac{(1 - 1.5z^{-1} - z^{-2})(1 + 0.9z^{-1})}{(1 - z^{-1})(1 + 0.7jz^{-1})(1 - 0.7jz^{-1})}.$$

(a) Write the difference equation that is satisfied by the input and the output of the system.
(b) Plot the pole–zero diagram and indicate the ROC for the system function.
(c) Sketch $|H(e^{j\omega})|$.
(d) State whether the following are true or false about the system:

(i) The system is stable.
(ii) The impulse response approaches a constant for large n.
(iii) The magnitude of the frequency response has a peak at approximately $\omega = \pm\pi/4$.
(iv) The system has a stable and causal inverse.

5.52. Consider a causal sequence $x[n]$ with the z-transform

$$X(z) = \frac{\left(1 - \frac{1}{2}z^{-1}\right)\left(1 - \frac{1}{4}z^{-1}\right)\left(1 - \frac{1}{5}z\right)}{\left(1 - \frac{1}{6}z\right)}.$$

For what values of α is $\alpha^n x[n]$ a real, minimum-phase sequence?

5.53. Consider the LTI system whose system function is

$$H(z) = (1 - 0.9e^{j0.6\pi}z^{-1})(1 - 0.9e^{-j0.6\pi}z^{-1})(1 - 1.25e^{j0.8\pi}z^{-1})(1 - 1.25e^{-j0.8\pi}z^{-1}).$$

(a) Determine all causal system functions that result in the same frequency-response magnitude as $H(z)$ and for which the impulse responses are real valued and of the same length as the impulse response associated with $H(z)$. (There are four different such system functions.) Identify which system function is minimum phase and which, to within a time shift, is maximum phase.

(b) Determine the impulse responses for the system functions in part (a).

(c) For each of the sequences in part (b), compute and plot the quantity

$$E[n] = \sum_{m=0}^{n} (h[m])^2$$

for $0 \le n \le 5$. Indicate explicitly which plot corresponds to the minimum-phase system.

5.54. Shown in Figure P5.54 are eight different finite-duration sequences. Each sequence is four points long. The magnitude of the Fourier transform is the same for all sequences. Which of the sequences has all the zeros of its z-transform *inside* the unit circle?

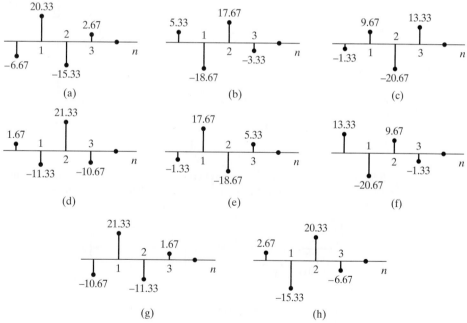

Figure P5.54

5.55. Each of the pole–zero plots in Figure P5.55, together with the specification of the ROC, describes an LTI system with system function $H(z)$. In each case, determine whether any of the following statements are true. Justify your answer with a brief statement or a counterexample.

(a) The system is a zero-phase or a generalized linear-phase system.

(b) The system has a stable inverse $H_i(z)$.

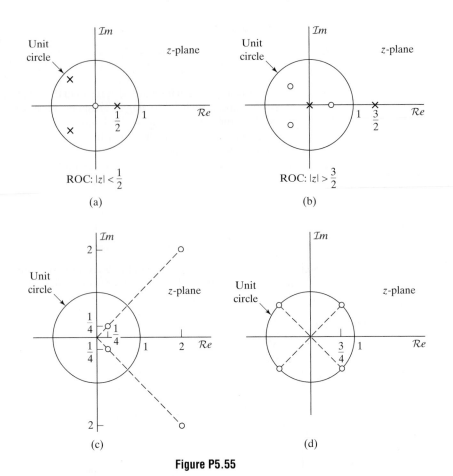

Figure P5.55

5.56. Assuming ideal D/C amd C/D converters, the overall system of Figure P5.56 is a discrete-time LTI system with frequency response $H(e^{j\omega})$ and impulse response $h[n]$.

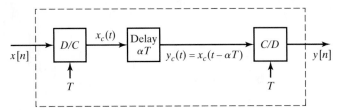

Figure P5.56

(a) $H(e^{j\omega})$ can be expressed in the form
$$H(e^{j\omega}) = A(e^{j\omega})e^{j\phi(\omega)},$$
with $A(e^{j\omega})$ real. Determine and sketch $A(e^{j\omega})$ and $\phi(\omega)$ for $|\omega| < \pi$.
(b) Sketch $h[n]$ for the following:
   (i) $\alpha = 3$
   (ii) $\alpha = 3\frac{1}{2}$
   (iii) $\alpha = 3\frac{1}{4}$.

(c) Consider a discrete-time LTI system for which

$$H(e^{j\omega}) = A\,(e^{j\omega})e^{j\alpha\omega}, \qquad |\omega| < \pi,$$

with $A\,(e^{j\omega})$ real. What can be said about the symmetry of $h[n]$ for the following?

(i) $\alpha = $ integer

(ii) $\alpha = M/2$, where M is an odd integer

(iii) General α.

5.57. Consider the class of FIR filters that have $h[n]$ real, $h[n] = 0$ for $n < 0$ and $n > M$, and one of the following symmetry properties:

$$\text{Symmetric:} \quad h[n] = h[M - n]$$

$$\text{Antisymmetric:} \quad h[n] = -h[M - n]$$

All filters in this class have generalized linear phase, i.e., have frequency response of the form

$$H(e^{j\omega}) = A\,(e^{j\omega})e^{-j\alpha\omega + j\beta},$$

where $A\,(e^{j\omega})$ is a real function of ω, α is a real constant, and β is a real constant.

For the following table, show that $A\,(e^{j\omega})$ has the indicated form, and find the values of α and β.

Type	Symmetry	$(M + 1)$	Form of $A\,(e^{j\omega})$	α	β
I	Symmetric	Odd	$\displaystyle\sum_{n=0}^{M/2} a[n]\cos\omega n$		
II	Symmetric	Even	$\displaystyle\sum_{n=1}^{(M+1)/2} b[n]\cos\omega(n - 1/2)$		
III	Antisymmetric	Odd	$\displaystyle\sum_{n=1}^{M/2} c[n]\sin\omega n$		
IV	Antisymmetric	Even	$\displaystyle\sum_{n=1}^{(M+1)/2} d[n]\sin\omega(n - 1/2)$		

Here are several helpful suggestions.

- For type I filters, first show that $H(e^{j\omega})$ can be written in the form

$$H(e^{j\omega}) = \sum_{n=0}^{(M-2)/2} h[n]e^{-j\omega n} + \sum_{n=0}^{(M-2)/2} h[M - n]e^{-j\omega[M-n]} + h[M/2]e^{-j\omega(M/2)}.$$

- The analysis for type III filters is very similar to that for type I, with the exception of a sign change and removal of one of the preceding terms.

- For type II filters, first write $H(e^{j\omega})$ in the form

$$H(e^{j\omega}) = \sum_{n=0}^{(M-1)/2} h[n]e^{-j\omega n} + \sum_{n=0}^{(M-1)/2} h[M - n]e^{-j\omega[M-n]},$$

and then pull out a common factor of $e^{-j\omega(M/2)}$ from both sums.

- The analysis for type IV filters is very similar to that for type II filters.

5.58. Let $h_{lp}[n]$ denote the impulse response of an FIR generalized linear-phase lowpass filter. The impulse response $h_{hp}[n]$ of an FIR generalized linear-phase highpass filter can be obtained by the transformation

$$h_{hp}[n] = (-1)^n h_{lp}[n].$$

If we decide to design a highpass filter using this transformation and we wish the resulting highpass filter to be symmetric, which of the four types of generalized linear-phase FIR filters can we use for the design of the lowpass filter? Your answer should consider *all* the possible types.

5.59. **(a)** A specific minimum-phase system has system function $H_{min}(z)$ such that

$$H_{min}(z) H_{ap}(z) = H_{lin}(z),$$

where $H_{ap}(z)$ is an all-pass system function and $H_{lin}(z)$ is a causal generalized linear-phase system. What does this information tell you about the poles and zeros of $H_{min}(z)$?

(b) A generalized linear-phase FIR system has an impulse response with real values and $h[n] = 0$ for $n < 0$ and for $n \geq 8$, and $h[n] = -h[7 - n]$. The system function of this system has a zero at $z = 0.8e^{j\pi/4}$ and another zero at $z = -2$. What is $H(z)$?

5.60. This problem concerns a discrete-time filter with a real-valued impulse response $h[n]$. Determine whether the following statement is true or false:

Statement: If the group delay of the filter is a constant for $0 < \omega < \pi$, then the impulse response must have the property that either

$$h[n] = h[M - n]$$

or

$$h[n] = -h[M - n],$$

where M is an integer.

If the statement is true, show why it is true. If it is false, provide a counterexample.

5.61. The system function $H_{II}(z)$ represents a type II FIR generalized linear-phase system with impulse response $h_{II}[n]$. This system is cascaded with an LTI system whose system function is $(1 - z^{-1})$ to produce a third system with system function $H(z)$ and impulse response $h[n]$. Prove that the overall system is a generalized linear-phase system, and determine what type of linear-phase system it is.

5.62. Let S_1 be a causal and stable LTI system with impulse response $h_1[n]$ and frequency response $H_1(e^{j\omega})$. The input $x[n]$ and output $y[n]$ for S_1 are related by the difference equation

$$y[n] - y[n - 1] + \tfrac{1}{4}y[n - 2] = x[n].$$

(a) If an LTI system S_2 has a frequency response given by $H_2(e^{j\omega}) = H_1(-e^{j\omega})$, would you characterize S_2 as being a lowpass filter, a bandpass filter, or a highpass filter? Justify your answer.

(b) Let S_3 be a causal LTI system whose frequency response $H_3(e^{j\omega})$ has the property that

$$H_3(e^{j\omega}) H_1(e^{j\omega}) = 1.$$

Is S_3 a minimum-phase filter? Could S_3 be classified as one of the four types of FIR filters with generalized linear phase? Justify your answers.

(c) Let S_4 be a stable and noncausal LTI system whose frequency response is $H_4(e^{j\omega})$ and whose input $x[n]$ and output $y[n]$ are related by the difference equation:

$$y[n] + \alpha_1 y[n - 1] + \alpha_2 y[n - 2] = \beta_0 x[n],$$

where α_1, α_2, and β_0 are all real and nonzero constants. Specify a value for α_1, a value for α_2, and a value for β_0 such that $|H_4(e^{j\omega})| = |H_1(e^{j\omega})|$.

(d) Let S_5 be an FIR filter whose impulse response is $h_5[n]$ and whose frequency response, $H_5(e^{j\omega})$, has the property that $H_5(e^{j\omega}) = |A(e^{j\omega})|^2$ for some DTFT $A(e^{j\omega})$ (i.e., S_5 is a zero-phase filter). Determine $h_5[n]$ such that $h_5[n] * h_1[n]$ is the impulse response of a noncausal FIR filter.

Extension Problems

5.63. In the system shown in Figure P5.63-1, assume that the input can be expressed in the form

$$x[n] = s[n]\cos(\omega_0 n).$$

Assume also that $s[n]$ is lowpass and relatively narrowband; i.e., $S(e^{j\omega}) = 0$ for $|\omega| > \Delta$, with Δ very small and $\Delta \ll \omega_0$, so that $X(e^{j\omega})$ is narrowband around $\omega = \pm\omega_0$.

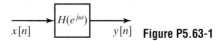

$x[n]$ $y[n]$ **Figure P5.63-1**

(a) If $|H(e^{j\omega})| = 1$ and $\angle H(e^{j\omega})$ is as illustrated in Figure P5.63-2, show that $y[n] = s[n]\cos(\omega_0 n - \phi_0)$.

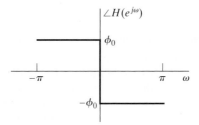

Figure P5.63-2

(b) If $|H(e^{j\omega})| = 1$ and $\angle H(e^{j\omega})$ is as illustrated in Figure P5.63-3, show that $y[n]$ can be expressed in the form

$$y[n] = s[n - n_d]\cos(\omega_0 n - \phi_0 - \omega_0 n_d).$$

Show also that $y[n]$ can be equivalently expressed as

$$y[n] = s[n - n_d]\cos(\omega_0 n - \phi_1),$$

where $-\phi_1$ is the phase of $H(e^{j\omega})$ at $\omega = \omega_0$.

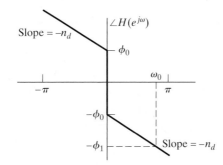

Figure P5.63-3

(c) The group delay associated with $H(e^{j\omega})$ is defined as

$$\tau_{gr}(\omega) = -\frac{d}{d\omega}\arg[H(e^{j\omega})],$$

and the phase delay is defined as $\tau_{ph}(\omega) = -(1/\omega)\angle H(e^{j\omega})$. Assume that $|H(e^{j\omega})|$ is unity over the bandwidth of $x[n]$. Based on your results in parts (a) and (b) and on the assumption that $x[n]$ is narrowband, show that if $\tau_{gr}(\omega_0)$ and $\tau_{ph}(\omega_0)$ are both integers, then

$$y[n] = s[n - \tau_{gr}(\omega_0)]\cos\{\omega_0[n - \tau_{ph}(\omega_0)]\}.$$

This equation shows that, for a narrowband signal $x[n]$, $\angle H(e^{j\omega})$ effectively applies a delay of $\tau_{gr}(\omega_0)$ to the envelope $s[n]$ of $x[n]$ and a delay of $\tau_{ph}(\omega_0)$ to the carrier $\cos\omega_0 n$.

(d) Referring to the discussion in Section 4.5 associated with noninteger delays of a sequence, how would you interpret the effect of group delay and phase delay if $\tau_{gr}(\omega_0)$ or $\tau_{ph}(\omega_0)$ (or both) is not an integer?

5.64. The signal $y[n]$ is the output of an LTI system with input $x[n]$, which is zero-mean white noise. The system is described by the difference equation

$$y[n] = \sum_{k=1}^{N} a_k y[n - k] + \sum_{k=0}^{M} b_k x[n - k], \qquad b_0 = 1.$$

(a) What is the z-transform $\Phi_{yy}(z)$ of the autocorrelation function $\phi_{yy}[n]$?

Sometimes it is of interest to process $y[n]$ with a linear filter such that the power spectrum of the linear filter's output will be flat when the input to the linear filter is $y[n]$. This procedure is known as "whitening" $y[n]$, and the linear filter that accomplishes the task is said to be the "whitening filter" for the signal $y[n]$. Suppose that we know the autocorrelation function $\phi_{yy}[n]$ and its z-transform $\Phi_{yy}(z)$, but not the values of the coefficients a_k and b_k.

(b) Describe a procedure for finding a system function $H_w(z)$ of the whitening filter.

(c) Is the whitening filter unique?

5.65. In many practical situations, we are faced with the problem of recovering a signal that has been "blurred" by a convolution process. We can model this blurring process as a linear filtering operation, as depicted in Figure P5.65-1, where the blurring impulse response is as shown in Figure P5.65-2. This problem will consider ways to recover $x[n]$ from $y[n]$.

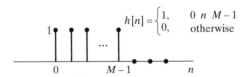

$x[n]$ $y[n]$
Desired signal   Blurred signal   **Figure P5.65–1**

$$h[n] = \begin{cases} 1, & 0 \leq n \leq M-1 \\ 0, & \text{otherwise} \end{cases}$$

Figure P5.65–2

(a) One approach to recovering $x[n]$ from $y[n]$ is to use an inverse filter; i.e., $y[n]$ is filtered by a system whose frequency response is

$$H_i(e^{j\omega}) = \frac{1}{H(e^{j\omega})},$$

where $H(e^{j\omega})$ is the Fourier transform of $h[n]$. For the impulse response $h[n]$ shown in Figure P5.65-2, discuss the practical problems involved in implementing the inverse filtering approach. Be complete, but also be brief and to the point.

(b) Because of the difficulties involved in inverse filtering, the following approach is suggested for recovering $x[n]$ from $y[n]$: The blurred signal $y[n]$ is processed by the system shown in Figure P5.65-3, which produces an output $w[n]$ from which we can extract an improved replica of $x[n]$. The impulse responses $h_1[n]$ and $h_2[n]$ are shown in Figure P5.65-4. Explain in detail the working of this system. In particular, state precisely the conditions under which we can recover $x[n]$ exactly from $w[n]$. *Hint:* Consider the impulse response of the overall system from $x[n]$ to $w[n]$.

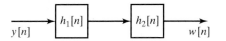

Figure P5.65-3

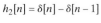

$$h_1[n] = \sum_{k=0}^{q} \delta[n-kM]$$

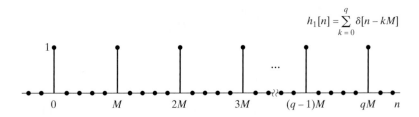

$$h_2[n] = \delta[n] - \delta[n-1]$$

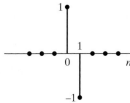

Figure P5.65-4

(c) Let us now attempt to generalize this approach to arbitrary finite-length blurring impulse responses $h[n]$; i.e., assume only that $h[n] = 0$ for $n < 0$ or $n \geq M$. Further, assume that $h_1[n]$ is the same as in Figure P5.65-4. How must $H_2(z)$ and $H(z)$ be related for the system to work as in part (b)? What condition must $H(z)$ satisfy in order that it be possible to implement $H_2(z)$ as a causal system?

5.66. In this problem, we demonstrate that, for a rational z-transform, a factor of the form $(z - z_0)$ and a factor of the form $z/(z - 1/z_0^*)$ contribute the same phase.

(a) Let $H(z) = z - 1/a$, where a is real and $0 < a < 1$. Sketch the poles and zeros of the system, including an indication of those at $z = \infty$. Determine $\angle H(e^{j\omega})$, the phase of the system.

(b) Let $G(z)$ be specified such that it has poles at the conjugate-reciprocal locations of zeros of $H(z)$ and zeros at the conjugate-reciprocal locations of poles of $H(z)$, including those at zero and ∞. Sketch the pole–zero diagram of $G(z)$. Determine $\angle G(e^{j\omega})$, the phase of the system, and show that it is identical to $\angle H(e^{j\omega})$.

5.67. Prove the validity of the following two statements:

(a) The convolution of two minimum-phase sequences is also a minimum-phase sequence.

(b) The sum of two minimum-phase sequences is not necessarily a minimum-phase sequence. Specifically, give an example of both a minimum-phase and a nonminimum-phase sequence that can be formed as the sum of two minimum-phase sequences.

5.68. A sequence is defined by the relationship

$$r[n] = \sum_{m=-\infty}^{\infty} h[m]h[n+m] = h[n] * h[-n],$$

where $h[n]$ is a minimum-phase sequence and

$$r[n] = \tfrac{4}{3}\left(\tfrac{1}{2}\right)^n u[n] + \tfrac{4}{3}2^n u[-n-1].$$

(a) Determine $R(z)$ and sketch the pole–zero diagram.

(b) Determine the minimum-phase sequence $h[n]$ to within a scale factor of ± 1. Also, determine the z-transform $H(z)$ of $h[n]$.

5.69. A *maximum-phase* sequence is a stable sequence whose z-transform has all its poles and zeros *outside* the unit circle.

(a) Show that maximum-phase sequences are necessarily anti-causal, i.e., that they are zero for $n > 0$.

FIR maximum-phase sequences can be made causal by including a finite amount of delay. A finite-duration causal maximum-phase sequence having a Fourier transform of a given magnitude can be obtained by reflecting all the zeros of the z-transform of a minimum-phase sequence to conjugate-reciprocal positions outside the unit circle. That is, we can express the z-transform of a maximum-phase causal finite-duration sequence as

$$H_{\max}(z) = H_{\min}(z)H_{ap}(z).$$

Obviously, this process ensures that $|H_{\max}(e^{j\omega})| = |H_{\min}(e^{j\omega})|$. Now, the z-transform of a finite-duration minimum-phase sequence can be expressed as

$$H_{\min}(z) = h_{\min}[0] \prod_{k=1}^{M} (1 - c_k z^{-1}), \qquad |c_k| < 1.$$

(b) Obtain an expression for the all-pass system function required to reflect all the zeros of $H_{\min}(z)$ to positions outside the unit circle.

(c) Show that $H_{\max}(z)$ can be expressed as

$$H_{\max}(z) = z^{-M} H_{\min}(z^{-1}).$$

(d) Using the result of part (c), express the maximum-phase sequence $h_{\max}[n]$ in terms of $h_{\min}[n]$.

5.70. It is not possible to obtain a causal and stable inverse system (a perfect compensator) for a nonminimum-phase system. In this problem, we study an approach to compensating for only the magnitude of the frequency response of a nonminimum-phase system.

Suppose that a stable nonminimum-phase LTI discrete-time system with a rational system function $H(z)$ is cascaded with a compensating system $H_c(z)$ as shown in Figure P5.70.

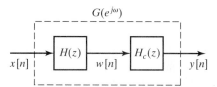

Figure P5.70

(a) How should $H_c(z)$ be chosen so that it is stable and causal and so that the magnitude of the overall effective frequency response is unity? (Recall that $H(z)$ can always be represented as $H(z) = H_{ap}(z)H_{min}(z)$.)

(b) What are the corresponding system functions $H_c(z)$ and $G(z)$?

(c) Assume that

$$H(z) = (1 - 0.8e^{j0.3\pi}z^{-1})(1 - 0.8e^{-j0.3\pi}z^{-1})(1 - 1.2e^{j0.7\pi}z^{-1})(1 - 1.2e^{-j0.7\pi}z^{-1}).$$

Determine $H_{min}(z)$, $H_{ap}(z)$, $H_c(z)$, and $G(z)$ for this case, and construct the pole–zero plots for each system function.

5.71. Let $h_{min}[n]$ denote a minimum-phase sequence with z-transform $H_{min}(z)$. If $h[n]$ is a causal nonminimum-phase sequence whose Fourier transform magnitude is equal to $|H_{min}(e^{j\omega})|$, show that

$$|h[0]| < |h_{min}[0]|.$$

(Use the initial-value theorem together with Eq. (5.93).)

5.72. One of the interesting and important properties of minimum-phase sequences is the minimum-energy delay property; i.e., of all the causal sequences having the same Fourier transform magnitude function $|H(e^{j\omega})|$, the quantity

$$E[n] = \sum_{m=0}^{n} |h[m]|^2$$

is maximum for all $n \geq 0$ when $h[n]$ is the minimum-phase sequence. This result is proved as follows: Let $h_{min}[n]$ be a minimum-phase sequence with z-transform $H_{min}(z)$. Furthermore, let z_k be a zero of $H_{min}(z)$ so that we can express $H_{min}(z)$ as

$$H_{min}(z) = Q(z)(1 - z_k z^{-1}), \qquad |z_k| < 1,$$

where $Q(z)$ is again minimum phase. Now consider another sequence $h[n]$ with z-transform $H(z)$ such that

$$|H(e^{j\omega})| = |H_{min}(e^{j\omega})|$$

and such that $H(z)$ has a zero at $z = 1/z_k^*$ instead of at z_k.

(a) Express $H(z)$ in terms of $Q(z)$.

(b) Express $h[n]$ and $h_{min}[n]$ in terms of the minimum-phase sequence $q[n]$ that has z-transform $Q(z)$.

(c) To compare the distribution of energy of the two sequences, show that

$$\varepsilon = \sum_{m=0}^{n} |h_{\min}[m]|^2 - \sum_{m=0}^{n} |h[m]|^2 = (1 - |z_k|^2)|q[n]|^2.$$

(d) Using the result of part (c), argue that

$$\sum_{m=0}^{n} |h[m]|^2 \le \sum_{m=0}^{n} |h_{\min}[m]|^2 \qquad \text{for all } n.$$

5.73. A causal all-pass system $H_{\text{ap}}(z)$ has input $x[n]$ and output $y[n]$.

(a) If $x[n]$ is a real minimum-phase sequence (which also implies that $x[n] = 0$ for $n < 0$), using Eq. (5.108), show that

$$\sum_{k=0}^{n} |x[k]|^2 \ge \sum_{k=0}^{n} |y[k]|^2. \qquad (\text{P5.73-1})$$

(b) Show that Eq. (P5.73-1) holds even if $x[n]$ is not minimum phase, but is zero for $n < 0$.

5.74. In the design of either continuous-time or discrete-time filters, we often approximate a specified magnitude characteristic without particular regard to the phase. For example, standard design techniques for lowpass and bandpass filters are derived from a consideration of the magnitude characteristics only.

   In many filtering problems, we would prefer that the phase characteristics be zero or linear. For causal filters, it is impossible to have zero phase. However, for many filtering applications, it is not necessary that the impulse response of the filter be zero for $n < 0$ if the processing is not to be carried out in real time.

   One technique commonly used in discrete-time filtering when the data to be filtered are of finite duration and are stored, for example, in computer memory is to process the data forward and then backward through the same filter.

   Let $h[n]$ be the impulse response of a causal filter with an arbitrary phase characteristic. Assume that $h[n]$ is real, and denote its Fourier transform by $H(e^{j\omega})$. Let $x[n]$ be the data that we want to filter.

(a) *Method A:* The filtering operation is performed as shown in Figure P5.74-1.

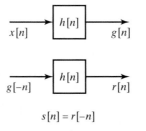

$$s[n] = r[-n]$$ **Figure P5.74-1**

   1. Determine the overall impulse response $h_1[n]$ that relates $x[n]$ to $s[n]$, and show that it has a zero-phase characteristic.
   2. Determine $|H_1(e^{j\omega})|$, and express it in terms of $|H(e^{j\omega})|$ and $\angle H(e^{j\omega})$.

(b) *Method B:* As depicted in Figure P5.74-2, process $x[n]$ through the filter $h[n]$ to get $g[n]$. Also, process $x[n]$ backward through $h[n]$ to get $r[n]$. The output $y[n]$ is then taken as the sum of $g[n]$ and $r[-n]$. This composite set of operations can be represented by a filter with input $x[n]$, output $y[n]$, and impulse response $h_2[n]$.

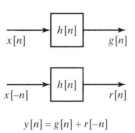

$$y[n] = g[n] + r[-n]$$ **Figure P5.74-2**

    1. Show that the composite filter $h_2[n]$ has a zero-phase characteristic.
    2. Determine $|H_2(e^{j\omega})|$, and express it in terms of $|H(e^{j\omega})|$ and $\angle H(e^{j\omega})$.

(c) Suppose that we are given a sequence of finite duration on which we would like to perform a bandpass zero-phase filtering operation. Furthermore, assume that we are given the bandpass filter $h[n]$, with frequency response as specified in Figure P5.74-3, which has the magnitude characteristic that we desire, but has linear phase. To achieve zero phase, we could use either method A or B. Determine and sketch $|H_1(e^{j\omega})|$ and $|H_2(e^{j\omega})|$. From these results, which method would you use to achieve the desired bandpass filtering operation? Explain why. More generally, if $h[n]$ has the desired magnitude, but a nonlinear phase characteristic, which method is preferable to achieve a zero-phase characteristic?

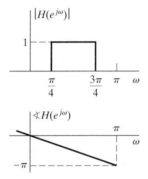

Figure P5.74-3

5.75. Determine whether the following statement is true or false. If it is true, concisely state your reasoning. If it is false, give a counterexample.
Statement: If the system function $H(z)$ has poles anywhere other than at the origin or infinity, then the system cannot be a zero-phase or a generalized linear-phase system.

5.76. Figure P5.76 shows the zeros of the system function $H(z)$ for a real causal linear-phase FIR filter. All of the indicated zeros represent factors of the form $(1 - az^{-1})$. The corresponding poles at $z = 0$ for these factors are not shown in the figure. The filter has approximately unity gain in its passband.

    (a) One of the zeros has magnitude 0.5 and angle 153 degrees. Determine the exact location of as many other zeros as you can from this information.
    (b) The system function $H(z)$ is used in the system for discrete-time processing of continuous-time signals shown in Figure 4.10, with the sampling period $T = 0.5$ msec. Assume that the continuous-time input $X_c(j\Omega)$ is bandlimited and that the sampling rate is high enough to avoid aliasing. What is the time delay (in msec) through the entire system, assuming that both C/D and D/C conversion require negligible amounts of time?

(c) For the system in part (b), sketch the overall effective continuous-time frequency response $20 \log_{10} |H_{\text{eff}}(j\Omega)|$ for $0 \leq \Omega \leq \pi/T$ as accurately as possible using the given information. From the information in Figure P5.76 estimate the frequencies at which $H_{\text{eff}}(j\Omega) = 0$, and mark them on your plot.

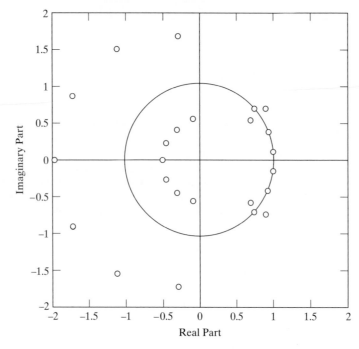

Figure P5.76

5.77. A signal $x[n]$ is processed through an LTI system $H(z)$ and then downsampled by a factor of 2 to yield $y[n]$ as indicated in Figure P5.77. Also, as shown in the same figure, $x[n]$ is first downsampled and then processed through an LTI system $G(z)$ to obtain $r[n]$.

(a) Specify a choice for $H(z)$ (other than a constant) and $G(z)$ so that $r[n] = y[n]$ for an arbitrary $x[n]$.

(b) Specify a choice for $H(z)$ so that there is no choice for $G(z)$ that will result in $r[n] = y[n]$ for an arbitrary $x[n]$.

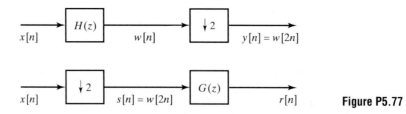

Figure P5.77

(c) Determine as general a set of conditions as you can on $H(z)$ such that $G(z)$ can be chosen so that $r[n] = y[n]$ for an arbitrary $x[n]$. The conditions should not depend on $x[n]$. If you first develop the conditions in terms of $h[n]$, restate them in terms of $H(z)$.

(d) For the conditions determined in part (c), what is $g[n]$ in terms of $h[n]$ so that $r[n] = y[n]$.

5.78. Consider a discrete-time LTI system with a real-valued impulse response $h[n]$. We want to find $h[n]$, or equivalently, the system function $H(z)$ from the autocorrelation $c_{hh}[\ell]$ of the impulse response. The definition of the autocorrelation is

$$c_{hh}[\ell] = \sum_{k=-\infty}^{\infty} h[k]h[k + \ell].$$

(a) If the system $h[n]$ is causal and stable, can you uniquely recover $h[n]$ from $c_{hh}[\ell]$? Justify your answer.

(b) Assume that $h[n]$ is causal and stable and that, in addition, you know that the system function has the form

$$H(z) = \cfrac{1}{1 - \displaystyle\sum_{k=1}^{N} a_k z^{-k}}$$

for some finite a_k. Can you uniquely recover $h[n]$ from $c_{hh}[\ell]$? Clearly justify your answer.

5.79. Let $h[n]$ and $H(z)$ denote the impulse response and system function of a stable all-pass LTI system. Let $h_i[n]$ denote the impulse response of the (stable) LTI inverse system. Assume that $h[n]$ is real. Show that $h_i[n] = h[-n]$.

5.80. Consider a real-valued sequence $x[n]$ for which $X(e^{j\omega}) = 0$ for $\frac{\pi}{4} \le |\omega| \le \pi$. One sequence value of $x[n]$ may have been corrupted, and we would like to recover it approximately or exactly. With $g[n]$ denoting the corrupted signal,

$$g[n] = x[n] \qquad \text{for } n \neq n_0,$$

and $g[n_0]$ is real but not related to $x[n_0]$. In each of the following two cases, specify a practical algorithm for recovering $x[n]$ from $g[n]$ exactly or approximately.

(a) The exact value of n_0 is not known, but we know that n_0 is an odd number.

(b) Nothing about n_0 is known.

5.81. Show that if $h[n]$ is an $(M + 1)$-point FIR filter such that $h[n] = h[M - n]$ and $H(z_0) = 0$, then $H(1/z_0) = 0$. This shows that even symmetric linear-phase FIR filters have zeros that are reciprocal images. (If $h[n]$ is real, the zeros also will be real or will occur in complex conjugates.)

6

Structures for
Discrete-Time
Systems

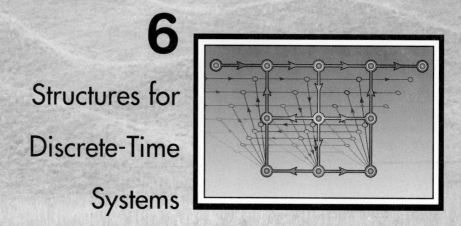

6.0 INTRODUCTION

As we saw in Chapter 5, an LTI system with a rational system function has the property that the input and output sequences satisfy a linear constant-coefficient difference equation. Since the system function is the z-transform of the impulse response, and since the difference equation satisfied by the input and output can be determined by inspection of the system function, it follows that the difference equation, the impulse response, and the system function are equivalent characterizations of the input–output relation of an LTI discrete-time system. When such systems are implemented with discrete-time analog or digital hardware, the difference equation or the system function representation must be converted to an algorithm or structure that can be realized in the desired technology. As we will see in this chapter, systems described by linear constant-coefficient difference equations can be represented by structures consisting of an interconnection of the basic operations of addition, multiplication by a constant, and delay, the exact implementation of which is dictated by the technology to be used.

As an illustration of the computation associated with a difference equation, consider the system described by the system function

$$H(z) = \frac{b_0 + b_1 z^{-1}}{1 - az^{-1}}, \qquad |z| > |a|. \qquad (6.1)$$

The impulse response of this system is

$$h[n] = b_0 a^n u[n] + b_1 a^{n-1} u[n-1], \qquad (6.2)$$

and the 1st-order difference equation that is satisfied by the input and output sequences is

$$y[n] - ay[n-1] = b_0 x[n] + b_1 x[n-1]. \qquad (6.3)$$

Equation (6.2) gives a formula for the impulse response for this system. However, since the system impulse response has infinite duration, even if we only wanted to compute the output over a finite interval, it would not be efficient to do so by discrete convolution since the amount of computation required to compute $y[n]$ would grow with n. However, rewriting Eq. (6.3) in the form

$$y[n] = ay[n-1] + b_0 x[n] + b_1 x[n-1] \tag{6.4}$$

provides the basis for an algorithm for recursive computation of the output at any time n in terms of the previous output $y[n-1]$, the current input sample $x[n]$, and the previous input sample $x[n-1]$. As discussed in Section 2.5, if we further assume initial-rest conditions (i.e., if $x[n] = 0$ for $n < 0$, then $y[n] = 0$ for $n < 0$), and if we use Eq. (6.4) as a recurrence formula for computing the output from past values of the output and present and past values of the input, the system will be linear and time invariant. A similar procedure can be applied to the more general case of an Nth-order difference equation. However, the algorithm suggested by Eq. (6.4), and its generalization for higher-order difference equations is not the only computational algorithm for implementing a particular system, and often, it is not the best choice. As we will see, an unlimited variety of computational structures result in the same relation between the input sequence $x[n]$ and the output sequence $y[n]$.

In the remainder of this chapter, we consider the important issues in the implementation of LTI discrete-time systems. We first present the block diagram and signal flow graph descriptions of computational structures for linear constant-coefficient difference equations representing LTI causal systems.[1] Using a combination of algebraic manipulations and manipulations of block diagram representations, we derive a number of basic equivalent structures for implementing a causal LTI system including lattice structures. Although two structures may be equivalent with regard to their input–output characteristics for infinite-precision representations of coefficients and variables, they may have vastly different behavior when the numerical precision is limited. This is the major reason that it is of interest to study different implementation structures. The effects of finite-precision representation of the system coefficients and the effects of truncation or rounding of intermediate computations are considered in the latter sections of the chapter.

6.1 BLOCK DIAGRAM REPRESENTATION OF LINEAR CONSTANT-COEFFICIENT DIFFERENCE EQUATIONS

The implementation of an LTI discrete-time system by iteratively evaluating a recurrence formula obtained from a difference equation requires that delayed values of the output, input, and intermediate sequences be available. The delay of sequence values implies the need for storage of past sequence values. Also, we must provide means for multiplication of the delayed sequence values by the coefficients, as well as means for adding the resulting products. Therefore, the basic elements required for the implementation of an LTI discrete-time system are adders, multipliers, and memory for storing

---

[1]Such flow graphs are also called "networks" in analogy to electrical circuit diagrams. We shall use the terms flow graph, structure, and network interchangeably with respect to graphic representations of difference equations.

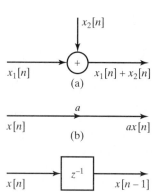

Figure 6.1 Block diagram symbols.
(a) Addition of two sequences.
(b) Multiplication of a sequence by a
constant. (c) Unit delay.

delayed sequence values and coefficients. The interconnection of these basic elements
is conveniently depicted by block diagrams composed of the basic pictorial symbols
shown in Figure 6.1. Figure 6.1(a) represents the addition of two sequences. In general
block diagram notation, an adder may have any number of inputs. However, in almost
all practical implementations, adders have only two inputs. In all the diagrams of this
chapter, we indicate this explicitly by limiting the number of inputs as in Figure 6.1(a).
Figure 6.1(b) depicts multiplication of a sequence by a constant, and Figure 6.1(c) depicts delaying a sequence by one sample. In digital implementations, the delay operation
can be implemented by providing a storage register for each unit delay that is required.
For this reason, we sometimes refer to the operator of Figure 6.1(c) as a *delay register*.
In analog discrete-time implementations such as switched-capacitor filters, the delays
are implemented by charge storage devices. The unit delay system is represented in Figure 6.1(c) by its system function, z^{-1}. Delays of more than one sample can be denoted
as in Figure 6.1(c), with a system function of z^{-M}, where M is the number of samples
of delay; however, the actual implementation of M samples of delay would generally
be done by cascading M unit delays. In an integrated-circuit implementation, these unit
delays might form a shift register that is clocked at the sampling rate of the input signal.
In a software implementation, M cascaded unit delays would be implemented as M
consecutive memory registers.

Example 6.1   Block Diagram Representation of a Difference Equation

As an example of the representation of a difference equation in terms of the elements
in Figure 6.1, consider the 2nd-order difference equation

$$y[n] = a_1 y[n-1] + a_2 y[n-2] + b_0 x[n]. \qquad (6.5)$$

The corresponding system function is

$$H(z) = \frac{b_0}{1 - a_1 z^{-1} - a_2 z^{-2}}. \qquad (6.6)$$

The block diagram representation of the system realization based on Eq. (6.5) is shown
in Figure 6.2. Such diagrams give a pictorial representation of a computational algorithm for implementing the system. When the system is implemented on either a

general-purpose computer or a digital signal processing (DSP) chip, network structures such as the one shown in Figure 6.2 serve as the basis for a program that implements the system. If the system is implemented with discrete components or as a complete system with very large-scale integration (VLSI) technology, the block diagram is the basis for determining a hardware architecture for the system. In both cases, diagrams such as Figure 6.2 show explicitly that we must provide storage for the delayed variables (in this case, $y[n-1]$ and $y[n-2]$) and also the coefficients of the difference equation (in this case, $a_1, a_2,$ and b_0). Furthermore, we see from Figure 6.2 that an output sequence value $y[n]$ is computed by first forming the products $a_1 y[n-1]$ and $a_2 y[n-2]$, then adding them, and, finally, adding the result to $b_0 x[n]$. Thus, Figure 6.2 conveniently depicts the complexity of the associated computational algorithm, the steps of the algorithm, and the amount of hardware required to realize the system.

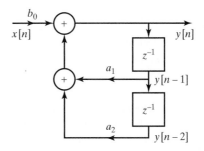

Figure 6.2   Example of a block diagram representation of a difference equation.

Example 6.1 can be generalized to higher-order difference equations of the form[2]

$$y[n] - \sum_{k=1}^{N} a_k y[n-k] = \sum_{k=0}^{M} b_k x[n-k], \tag{6.7}$$

with the corresponding system function

$$H(z) = \frac{\displaystyle\sum_{k=0}^{M} b_k z^{-k}}{1 - \displaystyle\sum_{k=1}^{N} a_k z^{-k}}. \tag{6.8}$$

Rewriting Eq. (6.7) as a recurrence formula for $y[n]$ in terms of a linear combination of past values of the output sequence and current and past values of the input sequence

---

[2]The form used in previous chapters for a general N^{th}-order difference equation was

$$\sum_{k=0}^{N} a_k y[n-k] = \sum_{k=0}^{M} b_k x[n-k].$$

In the remainder of the book, it will be more convenient to use the form in Eq. (6.7), where the coefficient of $y[n]$ is normalized to unity and the coefficients associated with the delayed output appear with a positive sign after they have been moved to the right-hand side of the equation. (See Eq. (6.9).)

leads to the relation

$$y[n] = \sum_{k=1}^{N} a_k y[n-k] + \sum_{k=0}^{M} b_k x[n-k]. \tag{6.9}$$

The block diagram of Figure 6.3 is an explicit pictorial representation of Eq. (6.9). More precisely, it represents the pair of difference equations

$$v[n] = \sum_{k=0}^{M} b_k x[n-k], \tag{6.10a}$$

$$y[n] = \sum_{k=1}^{N} a_k y[n-k] + v[n]. \tag{6.10b}$$

The assumption of a two-input adder implies that the additions are done in a specified order. That is, Figure 6.3 shows that the products $a_N y[n-N]$ and $a_{N-1} y[n-N+1]$ must be computed, then added, and the resulting sum added to $a_{N-2} y[n-N+2]$, and so on. After $y[n]$ has been computed, the delay variables must be updated by moving $y[n-N+1]$ into the register holding $y[n-N]$, and so on, with the newly computed $y[n]$ becoming $y[n-1]$ for the next iteration.

A block diagram can be rearranged or modified in a variety of ways without changing the overall system function. Each appropriate rearrangement represents a *different* computational algorithm for implementing the *same* system. For example, the block diagram of Figure 6.3 can be viewed as a cascade of two systems, the first representing the computation of $v[n]$ from $x[n]$ and the second representing the computation of $y[n]$ from $v[n]$. Since each of the two systems is an LTI system (assuming initial-rest conditions for the delay registers), the order in which the two systems are cascaded can be reversed, as shown in Figure 6.4, without affecting the overall system function. In Figure 6.4, for convenience, we have assumed that $M = N$. Clearly, there is no loss of generality, since if $M \neq N$, some of the coefficients a_k or b_k in the figure would be zero, and the diagram could be simplified accordingly.

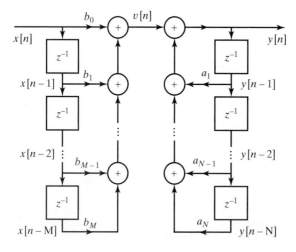

Figure 6.3   Block diagram representation for a general N^{th}-order difference equation.

In terms of the system function $H(z)$ in Eq. (6.8), Figure 6.3 can be viewed as an implementation of $H(z)$ through the decomposition

$$H(z) = H_2(z)H_1(z) = \left(\frac{1}{1 - \sum\limits_{k=1}^{N} a_k z^{-k}} \right) \left(\sum\limits_{k=0}^{M} b_k z^{-k} \right) \tag{6.11}$$

or, equivalently, through the pair of equations

$$V(z) = H_1(z)X(z) = \left(\sum\limits_{k=0}^{M} b_k z^{-k} \right) X(z), \tag{6.12a}$$

$$Y(z) = H_2(z)V(z) = \left(\frac{1}{1 - \sum\limits_{k=1}^{N} a_k z^{-k}} \right) V(z). \tag{6.12b}$$

Figure 6.4, on the other hand, represents $H(z)$ as

$$H(z) = H_1(z)H_2(z) = \left(\sum\limits_{k=0}^{M} b_k z^{-k} \right) \left(\frac{1}{1 - \sum\limits_{k=1}^{N} a_k z^{-k}} \right) \tag{6.13}$$

or, equivalently, through the equations

$$W(z) = H_2(z)X(z) = \left(\frac{1}{1 - \sum\limits_{k=1}^{N} a_k z^{-k}} \right) X(z), \tag{6.14a}$$

$$Y(z) = H_1(z)W(z) = \left(\sum\limits_{k=0}^{M} b_k z^{-k} \right) W(z). \tag{6.14b}$$

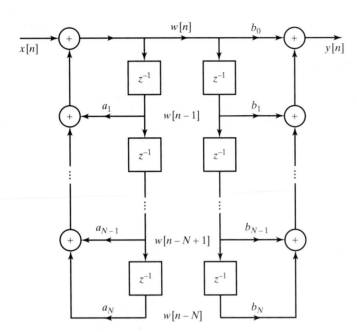

Figure 6.4 Rearrangement of block diagram of Figure 6.3. We assume for convenience that $N = M$. If $N \neq M$, some of the coefficients will be zero.

In the time domain, Figure 6.4 and, equivalently, Eqs. (6.14a) and (6.14b) can be represented by the pair of difference equations

$$w[n] = \sum_{k=1}^{N} a_k w[n-k] + x[n], \tag{6.15a}$$

$$y[n] = \sum_{k=0}^{M} b_k w[n-k]. \tag{6.15b}$$

The block diagrams of Figures 6.3 and 6.4 have several important differences. In Figure 6.3, the zeros of $H(z)$, represented by $H_1(z)$, are implemented first, followed by the poles, represented by $H_2(z)$. In Figure 6.4, the poles are implemented first, followed by the zeros. Theoretically, the order of implementation does not affect the overall system function. However, as we will see, when a difference equation is implemented with finite-precision arithmetic, there can be a significant difference between two systems that are equivalent with the assumption of infinite precision arithmetic in the real number system. Another important point concerns the number of delay elements in the two systems. As drawn, the systems in Figures 6.3 and 6.4 each have a total of $(N + M)$ delay elements. However, the block diagram of Figure 6.4 can be redrawn by noting that exactly the same signal, $w[n]$, is stored in the two chains of delay elements in the figure. Consequently, the two can be collapsed into one chain, as indicated in Figure 6.5.

The total number of delay elements in Figure 6.5 is less than or equal to the number required in either Figure 6.3 or Figure 6.4, and in fact it is the minimum number required to implement a system with system function given by Eq. (6.8). Specifically, the minimum number of delay elements required is, in general, $\max(N, M)$. An implementation with the minimum number of delay elements is commonly referred to as a *canonic form*

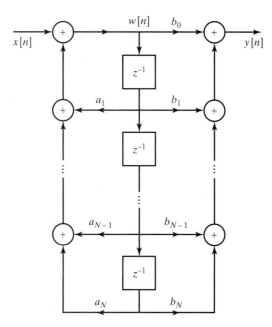

Figure 6.5   Combination of delays in Figure 6.4.

implementation. The noncanonic block diagram in Figure 6.3 is referred to as the *direct form I* implementation of the general Nth-order system because it is a direct realization of the difference equation satisfied by the input $x[n]$ and the output $y[n]$, which in turn can be written directly from the system function by inspection. Figure 6.5 is often referred to as the *direct form II* or *canonic direct form* implementation. Knowing that Figure 6.5 is an appropriate realization structure for $H(z)$ given by Eq. (6.8), we can go directly back and forth in a straightforward manner between the system function and the block diagram (or the equivalent difference equation).

Example 6.2   Direct Form I and Direct Form II Implementation of an LTI System

Consider the LTI system with system function

$$H(z) = \frac{1 + 2z^{-1}}{1 - 1.5z^{-1} + 0.9z^{-2}}. \tag{6.16}$$

Comparing this system function with Eq. (6.8), we find $b_0 = 1$, $b_1 = 2$, $a_1 = +1.5$, and $a_2 = -0.9$, so it follows from Figure 6.3 that we can implement the system in a direct form I block diagram as shown in Figure 6.6. Referring to Figure 6.5, we can also implement the system function in direct form II, as shown in Figure 6.7. In both cases, note that the coefficients in the feedback branches in the block diagram have opposite signs from the corresponding coefficients of z^{-1} and z^{-2} in Eq. (6.16). Although this change of sign is sometimes confusing, it is essential to remember that the feedback coefficients $\{a_k\}$ always have the opposite sign in the difference equation from their sign in the system function. Note also that the direct form II requires only two delay elements to implement $H(z)$, one less than the direct form I implementation.

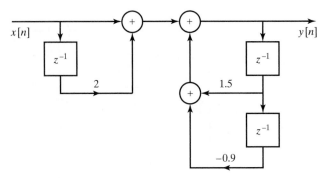

Figure 6.6   Direct form I implementation of Eq. (6.16).

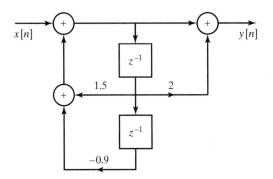

Figure 6.7   Direct form II implementation of Eq. (6.16).

In the preceding discussion, we developed two equivalent block diagrams for implementing an LTI system with system function given by Eq. (6.8). These block diagrams, which represent different computational algorithms for implementing the system, were obtained by manipulations based on the linearity of the system and the algebraic properties of the system function. Indeed, since the basic difference equations that represent an LTI system are linear, equivalent sets of difference equations can be obtained simply by linear transformations of the variables of the difference equations. Thus, there are an unlimited number of equivalent realizations of any given system. In Section 6.3, using an approach similar to that employed in this section, we will develop a number of other important and useful equivalent structures for implementing a system with system function as in Eq. (6.8). Before discussing these other forms, however, it is convenient to introduce signal flow graphs as an alternative to block diagrams for representing difference equations.

6.2 SIGNAL FLOW GRAPH REPRESENTATION OF LINEAR CONSTANT-COEFFICIENT DIFFERENCE EQUATIONS

A signal flow graph representation of a difference equation is essentially the same as a block diagram representation, except for a few notational differences. Formally, a

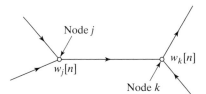

Figure 6.8   Example of nodes and branches in a signal flow graph.

signal flow graph is a network of directed branches that connect at nodes. Associated with each node is a variable or node value. The value associated with node k might be denoted w_k, or, since node variables for digital filters are generally sequences, we often indicate this explicitly with the notation $w_k[n]$. Branch (j, k) denotes a branch originating at node j and terminating at node k, with the direction from j to k being indicated by an arrowhead on the branch. This is shown in Figure 6.8. Each branch has an input signal and an output signal. The input signal from node j to branch (j, k) is the node value $w_j[n]$. In a linear signal flow graph, which is the only class we will consider, the output of a branch is a linear transformation of the input to the branch. The simplest example is a constant gain, i.e., when the output of the branch is simply a constant multiple of the input to the branch. The linear operation represented by the branch is typically indicated next to the arrowhead showing the direction of the branch. For the case of a constant multiplier, the constant is simply shown next to the arrowhead. When an explicit indication of the branch operation is omitted, this indicates a branch transmittance of unity, or the identity transformation. By definition, the value at each node in a graph is the sum of the outputs of all the branches entering the node.

To complete the definition of signal flow graph notation, we define two special types of nodes. *Source nodes* are nodes that have no entering branches. Source nodes are used to represent the injection of external inputs or signal sources into a graph. *Sink nodes* are nodes that have only entering branches. Sink nodes are used to extract outputs from a graph. Source nodes, sink nodes, and simple branch gains are illustrated in the signal flow graph of Figure 6.9. The linear equations represented by the figure are as follows:

$$w_1[n] = x[n] + aw_2[n] + bw_2[n],$$

$$w_2[n] = cw_1[n], \tag{6.17}$$

$$y[n] = dx[n] + ew_2[n].$$

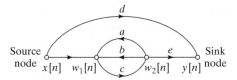

Figure 6.9   Example of a signal flow graph showing source and sink nodes.

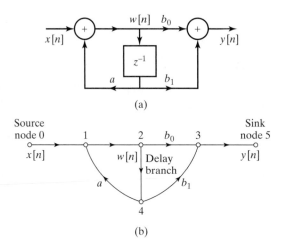

(a)

(b)

Figure 6.10 (a) Block diagram representation of a 1st-order digital filter. (b) Structure of the signal flow graph corresponding to the block diagram in (a).

Addition, multiplication by a constant, and delay are the basic operations required to implement a linear constant-coefficient difference equation. Since these are all linear operations, it is possible to use signal flow graph notation to depict algorithms for implementing LTI discrete-time systems. As an example of how the flow graph concepts just discussed can be applied to the representation of a difference equation, consider the block diagram in Figure 6.10(a), which is the direct form II realization of the system whose system function is given by Eq. (6.1). A signal flow graph corresponding to this system is shown in Figure 6.10(b). In the signal flow graph representation of difference equations, the node variables are sequences. In Figure 6.10(b), node 0 is a source node whose value is determined by the input sequence $x[n]$, and node 5 is a sink node whose value is denoted $y[n]$. Notice that the source and sink nodes are connected to the rest of the graph by unity-gain branches to clearly denote the input and output of the system. Obviously, nodes 3 and 5 have identical values. The extra branch with unity gain is simply used to highlight the fact that node 3 is the output of the system. In Figure 6.10(b), all branches except one (the delay branch (2, 4)) can be represented by a simple branch gain; i.e., the output signal is a constant multiple of the branch input. A delay cannot be represented in the time domain by a branch gain. However, the z-transform representation of a unit delay is multiplication by the factor z^{-1}. If we represented the difference equations by their corresponding z-transform equations, all the branches would be characterized by their system functions. In this case, each branch gain would be a function of z; e.g., a unit delay branch would have a gain of z^{-1}. By convention, we represent the variables in a signal flow graph as sequences rather than as z-transforms of sequences. However, to simplify the notation, we normally indicate a delay branch by showing its branch gain as z^{-1}, but it is understood that the output of such a branch is the branch input delayed by one sequence value. That is, the use of z^{-1} in a signal flow graph is in the sense of an operator that produces a delay of one sample. The graph of Figure 6.10(b) is shown in Figure 6.11 with this convention. The

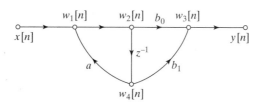

Figure 6.11   Signal flow graph of Figure 6.10(b) with the delay branch indicated by z^{-1}.

equations represented by Figure 6.11 are as follows:

$$w_1[n] = aw_4[n] + x[n], \tag{6.18a}$$

$$w_2[n] = w_1[n], \tag{6.18b}$$

$$w_3[n] = b_0 w_2[n] + b_1 w_4[n], \tag{6.18c}$$

$$w_4[n] = w_2[n-1], \tag{6.18d}$$

$$y[n] = w_3[n]. \tag{6.18e}$$

A comparison of Figure 6.10(a) and Figure 6.11 shows that there is a direct correspondence between branches in the block diagram and branches in the flow graph. In fact, the important difference between the two is that nodes in the flow graph represent both branching points and adders, whereas in the block diagram a special symbol is used for adders. A branching point in the block diagram is represented in the flow graph by a node that has only one incoming branch and one or more outgoing branches. An adder in the block diagram is represented in the signal flow graph by a node that has two (or more) incoming branches. In general, we will draw flow graphs with at most two inputs to each node, since most hardware implementations of addition have only two inputs. Signal flow graphs are therefore totally equivalent to block diagrams as pictorial representations of difference equations, but they are simpler to draw. Like block diagrams, they can be manipulated graphically to gain insight into the properties of a given system. A large body of signal flow graph theory exists that can be directly applied to discrete-time systems when they are represented in this form. (See Mason and Zimmermann, 1960; Chow and Cassignol, 1962; and Phillips and Nagle, 1995.) Although we will use flow graphs primarily for their pictorial value, we will use certain theorems relating to signal flow graphs in examining alternative structures for implementing linear systems.

Equations (6.18a)–(6.18e) define a multistep algorithm for computing the output of the LTI system from the input sequence $x[n]$. This example illustrates the kind of data precedence relations that generally arise in the implementation of IIR systems. Equations (6.18a)–(6.18e) cannot be computed in arbitrary order. Equations (6.18a) and (6.18c) require multiplications and additions, but Eqs. (6.18b) and (6.18e) simply rename variables. Equation (6.18d) represents the "updating" of the memory of the system. It would be implemented simply by replacing the contents of the memory register representing $w_4[n]$ by the value of $w_2[n]$, but this would have to be done consistently either *before* or *after* the evaluation of all the other equations. Initial-rest conditions would be imposed in this case by defining $w_2[-1] = 0$ or $w_4[0] = 0$. Clearly, Eqs. (6.18a)–(6.18e) must be computed in the order given, except that the last two could be interchanged or Eq. (6.18d) could be consistently evaluated first.

The flow graph represents a set of difference equations, with one equation being written at each node of the network. In the case of the flow graph of Figure 6.11, we can eliminate some of the variables rather easily to obtain the pair of equations

$$w_2[n] = aw_2[n-1] + x[n], \tag{6.19a}$$

$$y[n] = b_0 w_2[n] + b_1 w_2[n-1], \tag{6.19b}$$

which are in the form of Eqs. (6.15a) and (6.15b); i.e., in direct form II. Often, the manipulation of the difference equations of a flow graph is difficult when dealing with the time-domain variables, owing to feedback of delayed variables. In such cases, it is always possible to work with the z-transform representation, wherein all branches are simple gains since delay is represented in the z-transform by multiplication by z^{-1}. Problems 6.1–6.28 illustrate the utility of z-transform analysis of flow graphs for obtaining equivalent sets of difference equations.

Example 6.3   Determination of the System Function from a Flow Graph

To illustrate the use of the z-transform in determining the system function from a flow graph, consider Figure 6.12. The flow graph in this figure is not in direct form. Therefore, the system function cannot be written down by inspection of the graph. However, the set of difference equations represented by the graph can be written down by writing an equation for the value of each node variable in terms of the other node variables. The five equations are

$$w_1[n] = w_4[n] - x[n], \tag{6.20a}$$

$$w_2[n] = \alpha w_1[n], \tag{6.20b}$$

$$w_3[n] = w_2[n] + x[n], \tag{6.20c}$$

$$w_4[n] = w_3[n-1], \tag{6.20d}$$

$$y[n] = w_2[n] + w_4[n]. \tag{6.20e}$$

These are the equations that would be used to implement the system in the form described by the flow graph. Equations (6.20a)–(6.20e) can be represented by the z-transform equations

$$W_1(z) = W_4(z) - X(z), \tag{6.21a}$$

$$W_2(z) = \alpha W_1(z), \tag{6.21b}$$

$$W_3(z) = W_2(z) + X(z), \tag{6.21c}$$

$$W_4(z) = z^{-1} W_3(z), \tag{6.21d}$$

$$Y(z) = W_2(z) + W_4(z). \tag{6.21e}$$

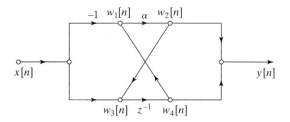

Figure 6.12   Flow graph not in standard direct form.

We can eliminate $W_1(z)$ and $W_3(z)$ from this set of equations by substituting Eq. (6.21a) into Eq. (6.21b) and Eq. (6.21c) into Eq. (6.21d), obtaining

$$W_2(z) = \alpha(W_4(z) - X(z)), \tag{6.22a}$$

$$W_4(z) = z^{-1}(W_2(z) + X(z)), \tag{6.22b}$$

$$Y(z) = W_2(z) + W_4(z). \tag{6.22c}$$

Equations (6.22a) and (6.22b) can be solved for $W_2(z)$ and $W_4(z)$, yielding

$$W_2(z) = \frac{\alpha(z^{-1} - 1)}{1 - \alpha z^{-1}} X(z), \tag{6.23a}$$

$$W_4(z) = \frac{z^{-1}(1 - \alpha)}{1 - \alpha z^{-1}} X(z), \tag{6.23b}$$

and substituting Eqs. (6.23a) and (6.23b) into Eq. (6.22c) leads to

$$Y(z) = \left(\frac{\alpha(z^{-1} - 1) + z^{-1}(1 - \alpha)}{1 - \alpha z^{-1}} \right) X(z) = \left(\frac{z^{-1} - \alpha}{1 - \alpha z^{-1}} \right) X(z). \tag{6.24}$$

Therefore, the system function of the flow graph of Figure 6.12 is

$$H(z) = \frac{z^{-1} - \alpha}{1 - \alpha z^{-1}}, \tag{6.25}$$

from which it follows that the impulse response of the system is

$$h[n] = \alpha^{n-1} u[n-1] - \alpha^{n+1} u[n]$$

and the direct form I flow graph is as shown in Figure 6.13.

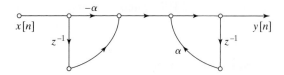

Figure 6.13   Direct form I equivalent of Figure 6.12.

Example 6.3 shows how the z-transform converts the time-domain expressions, which involve feedback and thus are difficult to solve, into linear equations that can be solved by algebraic techniques. The example also illustrates that different flow graph representations define computational algorithms that require different amounts of computational resources. By comparing Figures 6.12 and 6.13, we see that the original implementation requires only one multiplication and one delay (memory) element, whereas the direct form I implementation would require two multiplications and two delay elements. The direct form II implementation would require one less delay, but it still would require two multiplications.

6.3 BASIC STRUCTURES FOR IIR SYSTEMS

In Section 6.1, we introduced two alternative structures for implementing an LTI system with system function as in Eq. (6.8). In this section we present the signal flow graph representations of those systems, and we also develop several other commonly used equivalent flow graph network structures. Our discussion will make it clear that, for any given rational system function, a wide variety of equivalent sets of difference equations or network structures exists. One consideration in the choice among these different structures is computational complexity. For example, in some digital implementations, structures with the fewest constant multipliers and the fewest delay branches are often most desirable. This is because multiplication is generally a time-consuming and costly operation in digital hardware and because each delay element corresponds to a memory register. Consequently, a reduction in the number of constant multipliers means an increase in speed, and a reduction in the number of delay elements means a reduction in memory requirements.

Other, more subtle, trade-offs arise in VLSI implementations, in which the area of a chip is often an important measure of efficiency. Modularity and simplicity of data transfer on the chip are also frequently very desirable in such implementations. In multiprocessor implementations, the most important considerations are often related to partitioning of the algorithm and communication requirements between processors. Other major considerations are the effects of a finite register length and finite-precision arithmetic. These effects depend on the way in which the computations are organized, i.e., on the structure of the signal flow graph. Sometimes it is desirable to use a structure that does not have the minimum number of multipliers and delay elements if that structure is less sensitive to finite register length effects.

In this section, we develop several of the most commonly used forms for implementing an LTI IIR system and obtain their flow graph representations.

6.3.1 Direct Forms

In Section 6.1, we obtained block diagram representations of the direct form I (Figure 6.3) and direct form II, or canonic direct form (Figure 6.5), structures for an LTI system whose input and output satisfy a difference equation of the form

$$y[n] - \sum_{k=1}^{N} a_k y[n-k] = \sum_{k=0}^{M} b_k x[n-k], \tag{6.26}$$

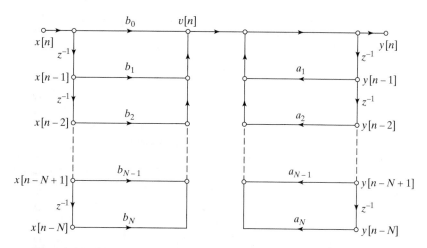

Figure 6.14   Signal flow graph of direct form I structure for an Nth-order system.

with the corresponding rational system function

$$H(z) = \frac{\displaystyle\sum_{k=0}^{M} b_k z^{-k}}{1 - \displaystyle\sum_{k=1}^{N} a_k z^{-k}}. \tag{6.27}$$

In Figure 6.14, the direct form I structure of Figure 6.3 is shown using signal flow graph conventions, and Figure 6.15 shows the signal flow graph representation of the direct form II structure of Figure 6.5. Again, we have assumed for convenience that $N = M$. Note that we have drawn the flow graph so that each node has no more than two inputs. A node in a signal flow graph may have any number of inputs, but, as indicated earlier, this two-input convention results in a graph that is more closely related to programs and architectures for implementing the computation of the difference equations represented by the graph.

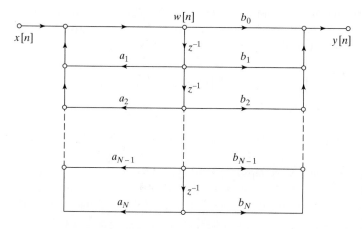

Figure 6.15   Signal flow graph of direct form II structure for an Nth-order system.

Example 6.4   Illustration of Direct Form I and Direct Form II Structures

Consider the system function

$$H(z) = \frac{1 + 2z^{-1} + z^{-2}}{1 - 0.75z^{-1} + 0.125z^{-2}}.$$ (6.28)

Since the coefficients in the direct form structures correspond directly to the coefficients of the numerator and denominator polynomials (taking into account the minus sign in the denominator of Eq. (6.27)), we can draw these structures by inspection with reference to Figures 6.14 and 6.15. The direct form I and direct form II structures for this example are shown in Figures 6.16 and 6.17, respectively.

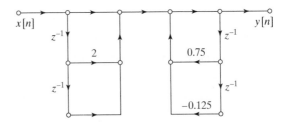

Figure 6.16   Direct form I structure for Example 6.4.

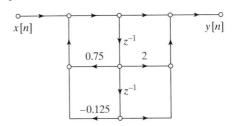

Figure 6.17   Direct form II structure for Example 6.4.

6.3.2 Cascade Form

The direct form structures were obtained directly from the system function $H(z)$, written as a ratio of polynomials in the variable z^{-1} as in Eq. (6.27). If we factor the numerator and denominator polynomials, we can express $H(z)$ in the form

$$H(z) = A \frac{\displaystyle\prod_{k=1}^{M_1}(1 - f_k z^{-1})\prod_{k=1}^{M_2}(1 - g_k z^{-1})(1 - g_k^* z^{-1})}{\displaystyle\prod_{k=1}^{N_1}(1 - c_k z^{-1})\prod_{k=1}^{N_2}(1 - d_k z^{-1})(1 - d_k^* z^{-1})},$$ (6.29)

where $M = M_1 + 2M_2$ and $N = N_1 + 2N_2$. In this expression, the 1st-order factors represent real zeros at f_k and real poles at c_k, and the 2nd-order factors represent complex conjugate pairs of zeros at g_k and g_k^* and complex conjugate pairs of poles

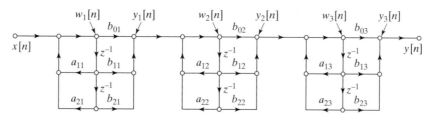

Figure 6.18   Cascade structure for a 6th-order system with a direct form II realization of each 2nd-order subsystem.

at d_k and d_k^*. This represents the most general distribution of poles and zeros when all the coefficients in Eq. (6.27) are real. Equation (6.29) suggests a class of structures consisting of a cascade of 1st- and 2nd-order systems. There is considerable freedom in the choice of composition of the subsystems and in the order in which the subsystems are cascaded. In practice, however, it is often desirable to implement the cascade realization using a minimum of storage and computation. A modular structure that is advantageous for many types of implementations is obtained by combining pairs of real factors and complex conjugate pairs into 2nd-order factors so that Eq. (6.29) can be expressed as

$$H(z) = \prod_{k=1}^{N_s} \frac{b_{0k} + b_{1k}z^{-1} + b_{2k}z^{-2}}{1 - a_{1k}z^{-1} - a_{2k}z^{-2}}, \tag{6.30}$$

where $N_s = \lfloor (N+1)/2 \rfloor$ is the largest integer contained in $(N+1)/2$. In writing $H(z)$ in this form, we have assumed that $M \leq N$ and that the real poles and zeros have been combined in pairs. If there are an odd number of real zeros, one of the coefficients b_{2k} will be zero. Likewise, if there are an odd number of real poles, one of the coefficients a_{2k} will be zero. The individual 2nd-order sections can be implemented using either of the direct form structures; however, the previous discussion shows that we can implement a cascade structure with a minimum number of multiplications and a minimum number of delay elements if we use the direct form II structure for each 2nd-order section. A cascade structure for a 6th-order system using three direct form II 2nd-order sections is shown in Figure 6.18. The difference equations represented by a general cascade of direct form II 2nd-order sections are of the form

$$y_0[n] = x[n], \tag{6.31a}$$

$$w_k[n] = a_{1k}w_k[n-1] + a_{2k}w_k[n-2] + y_{k-1}[n], \quad k = 1, 2, \ldots, N_s, \tag{6.31b}$$

$$y_k[n] = b_{0k}w_k[n] + b_{1k}w_k[n-1] + b_{2k}w_k[n-2], \quad k = 1, 2, \ldots, N_s, \tag{6.31c}$$

$$y[n] = y_{N_s}[n]. \tag{6.31d}$$

It is easy to see that a variety of theoretically equivalent systems can be obtained by simply pairing the poles and zeros in different ways and by ordering the 2nd-order sections in different ways. Indeed, if there are N_s 2nd-order sections, there are $N_s!$ (N_s factorial) pairings of the poles with zeros and $N_s!$ orderings of the resulting 2nd-order sections, or a total of $(N_s!)^2$ different pairings and orderings. Although these all have the same overall system function and corresponding input–output relation when infinite-precision arithmetic is used, their behavior with finite-precision arithmetic can be quite different, as we will see in Sections 6.8–6.10.

Example 6.5   Illustration of Cascade Structures

Let us again consider the system function of Eq. (6.28). Since this is a 2nd-order system, a cascade structure with direct form II 2nd-order sections reduces to the structure of Figure 6.17. Alternatively, to illustrate the cascade structure, we can use 1st-order systems by expressing $H(z)$ as a product of 1st-order factors, as in

$$H(z) = \frac{1 + 2z^{-1} + z^{-2}}{1 - 0.75z^{-1} + 0.125z^{-2}} = \frac{(1 + z^{-1})(1 + z^{-1})}{(1 - 0.5z^{-1})(1 - 0.25z^{-1})}. \tag{6.32}$$

Since all of the poles and zeros are real, a cascade structure with 1st-order sections has real coefficients. If the poles and/or zeros were complex, only a 2nd-order section would have real coefficients. Figure 6.19 shows two equivalent cascade structures, each of which has the system function in Eq. (6.32). The difference equations represented by the flow graphs in the figure can be written down easily. Problem 6.22 is concerned with finding other, equivalent system configurations.

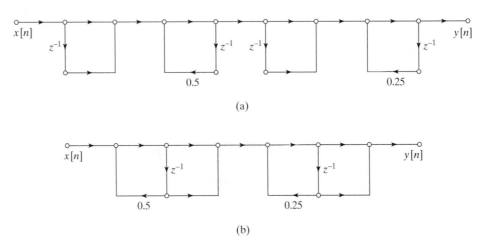

Figure 6.19   Cascade structures for Example 6.5. (a) Direct form I subsections. (b) Direct form II subsections.

A final comment should be made about our definition of the system function for the cascade form. As defined in Eq. (6.30), each 2nd-order section has five constant multipliers. For comparison, let us assume that $M = N$ in $H(z)$ as given by Eq. (6.27), and furthermore, assume that N is an even integer, so that $N_s = N/2$. Then, the direct form I and II structures have $2N + 1$ constant multipliers, while the cascade form structure suggested by Eq. (6.30) has $5N/2$ constant multipliers. For the 6th-order system in Figure 6.18, we require a total of 15 multipliers, while the equivalent direct forms would require a total of 13 multipliers. Another definition of the cascade form is

$$H(z) = b_0 \prod_{k=1}^{N_s} \frac{1 + \tilde{b}_{1k}z^{-1} + \tilde{b}_{2k}z^{-2}}{1 - a_{1k}z^{-1} - a_{2k}z^{-2}}, \tag{6.33}$$

where b_0 is the leading coefficient in the numerator polynomial of Eq. (6.27) and $\tilde{b}_{ik} = b_{ik}/b_{0k}$ for $i = 1, 2$ and $k = 1, 2, \ldots, N_s$. This form for $H(z)$ suggests a cascade of four-multiplier 2nd-order sections, with a single overall gain constant b_0. This cascade form has the same number of constant multipliers as the direct form structures. As discussed in Section 6.9, the five-multiplier 2nd-order sections are commonly used when implemented with fixed-point arithmetic, because they make it possible to distribute the overall gain of the system and thereby control the size of signals at various critical points in the system. When floating-point arithmetic is used and dynamic range is not a problem, the four-multiplier 2nd-order sections can be used to decrease the amount of computation. Further simplification results for zeros on the unit circle. In this case, $\tilde{b}_{2k} = 1$, and we require only three multipliers per 2nd-order section.

6.3.3 Parallel Form

As an alternative to factoring the numerator and denominator polynomials of $H(z)$, we can express a rational system function as given by Eq. (6.27) or (6.29) as a partial fraction expansion in the form

$$H(z) = \sum_{k=0}^{N_p} C_k z^{-k} + \sum_{k=1}^{N_1} \frac{A_k}{1 - c_k z^{-1}} + \sum_{k=1}^{N_2} \frac{B_k(1 - e_k z^{-1})}{(1 - d_k z^{-1})(1 - d_k^* z^{-1})}, \tag{6.34}$$

where $N = N_1 + 2N_2$. If $M \geq N$, then $N_p = M - N$; otherwise, the first summation in Eq. (6.34) is not included. If the coefficients a_k and b_k are real in Eq. (6.27), then the quantities A_k, B_k, C_k, c_k, and e_k are all real. In this form, the system function can be interpreted as representing a parallel combination of 1st- and 2nd-order IIR systems, with possibly N_p simple scaled delay paths. Alternatively, we may group the real poles in pairs, so that $H(z)$ can be expressed as

$$H(z) = \sum_{k=0}^{N_p} C_k z^{-k} + \sum_{k=1}^{N_s} \frac{e_{0k} + e_{1k} z^{-1}}{1 - a_{1k} z^{-1} - a_{2k} z^{-2}}, \tag{6.35}$$

where, as in the cascade form, $N_s = \lfloor (N+1)/2 \rfloor$ is the largest integer contained in $(N+1)/2$, and if $N_p = M - N$ is negative, the first sum is not present. A typical example for $N = M = 6$ is shown in Figure 6.20. The general difference equations for the parallel form with 2nd-order direct form II sections are

$$w_k[n] = a_{1k} w_k[n-1] + a_{2k} w_k[n-2] + x[n], \quad k = 1, 2, \ldots, N_s, \tag{6.36a}$$

$$y_k[n] = e_{0k} w_k[n] + e_{1k} w_k[n-1], \quad\quad\quad k = 1, 2, \ldots, N_s, \tag{6.36b}$$

$$y[n] = \sum_{k=0}^{N_p} C_k x[n-k] + \sum_{k=1}^{N_s} y_k[n]. \tag{6.36c}$$

If $M < N$, then the first summation in Eq. (6.36c) is not included.

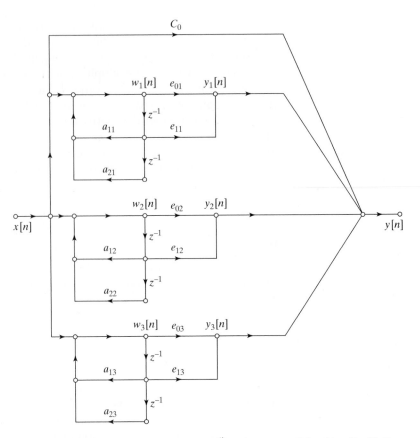

Figure 6.20 Parallel form structure for 6th-order system ($M = N = 6$) with the real and complex poles grouped in pairs.

Example 6.6   Illustration of Parallel Form Structures

Consider again the system function used in Examples 6.4 and 6.5. For the parallel form, we must express $H(z)$ in the form of either Eq. (6.34) or Eq. (6.35). If we use 2nd-order sections,

$$H(z) = \frac{1 + 2z^{-1} + z^{-2}}{1 - 0.75z^{-1} + 0.125z^{-2}} = 8 + \frac{-7 + 8z^{-1}}{1 - 0.75z^{-1} + 0.125z^{-2}}. \tag{6.37}$$

The parallel form realization for this example with a 2nd-order section is shown in Figure 6.21.

Since all the poles are real, we can obtain an alternative parallel form realization by expanding $H(z)$ as

$$H(z) = 8 + \frac{18}{1 - 0.5z^{-1}} - \frac{25}{1 - 0.25z^{-1}}. \tag{6.38}$$

The resulting parallel form with 1st-order sections is shown in Figure 6.22. As in the general case, the difference equations represented by both Figures 6.21 and 6.22 can be written down by inspection.

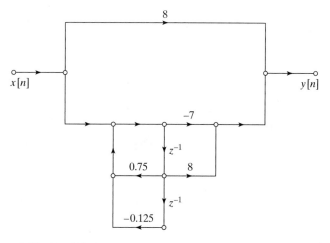

Figure 6.21   Parallel form structure for Example 6.6 using a 2nd-order system.

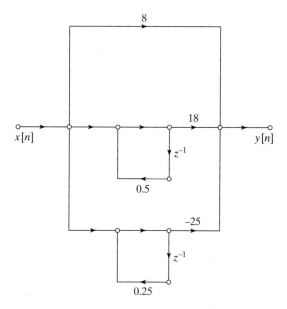

Figure 6.22   Parallel form structure for Example 6.6 using 1st-order systems.

6.3.4 Feedback in IIR Systems

All the flow graphs of this section have feedback loops; i.e., they have closed paths that begin at a node and return to that node by traversing branches only in the direction of their arrowheads. Such a structure in the flow graph implies that a node variable in a loop depends directly or indirectly on itself. A simple example is shown in Figure 6.23(a), which represents the difference equation

$$y[n] = ay[n-1] + x[n]. \tag{6.39}$$

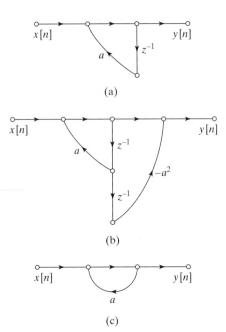

Figure 6.23 (a) System with feedback loop. (b) FIR system with feedback loop. (c) Noncomputable system.

Such loops are necessary (but not sufficient) to generate infinitely long impulse responses. This can be seen if we consider a network with no feedback loops. In such a case, any path from the input to the output can pass through each delay element only once. Therefore, the longest delay between the input and output would occur for a path that passes through all of the delay elements in the network. Thus, for a network with no loops, the impulse response is no longer than the total number of delay elements in the network. From this, we conclude that if a network has no loops, then the system function has only zeros (except for poles at $z = 0$), and the number of zeros can be no more than the number of delay elements in the network.

Returning to the simple example of Figure 6.23(a), we see that when the input is the impulse sequence $\delta[n]$, the single-input sample continually recirculates in the feedback loop with either increasing (if $|a| > 1$) or decreasing (if $|a| < 1$) amplitude owing to multiplication by the constant a, so that the impulse response is $h[n] = a^n u[n]$. This illustrates how feedback can create an infinitely long impulse response.

If a system function has poles, a corresponding block diagram or signal flow graph will have feedback loops. On the other hand, neither poles in the system function nor loops in the network are sufficient for the impulse response to be infinitely long. Figure 6.23(b) shows a network with a feedback loop, but with an impulse response of finite length. This is because the pole of the system function cancels with a zero; i.e., for Figure 6.23(b),

$$H(z) = \frac{1 - a^2 z^{-2}}{1 - az^{-1}} = \frac{(1 - az^{-1})(1 + az^{-1})}{1 - az^{-1}} = 1 + az^{-1}. \qquad (6.40)$$

The impulse response of this system is $h[n] = \delta[n] + a\delta[n - 1]$. The system is a simple example of a general class of FIR systems called *frequency-sampling systems*. This class of systems is considered in more detail in Problems 6.39 and 6.51.

Loops in a network pose special problems in implementing the computations implied by the network. As we have discussed, it must be possible to compute the node variables in a network in sequence such that all necessary values are available when needed. In some cases, there is no way to order the computations so that the node variables of a flow graph can be computed in sequence. Such a network is called *noncomputable* (see Crochiere and Oppenheim, 1975). A simple noncomputable network is shown in Figure 6.23(c). The difference equation for this network is

$$y[n] = ay[n] + x[n]. \tag{6.41}$$

In this form, we cannot compute $y[n]$ because the right-hand side of the equation involves the quantity we wish to compute. The fact that a flow graph is noncomputable does *not* mean the equations represented by the flow graph cannot be solved; indeed, the solution to Eq. (6.41) is $y[n] = x[n]/(1-a)$. It simply means that the flow graph does not represent a set of difference equations that can be solved successively for the node variables. The key to the computability of a flow graph is that all loops must contain at least one unit delay element. Thus, in manipulating flow graphs representing implementations of LTI systems, we must be careful not to create delay-free loops. Problem 6.37 deals with a system having a delay-free loop. Problem 7.51 shows how a delay-free loop can be introduced.

6.4 TRANSPOSED FORMS

The theory of linear signal flow graphs provides a variety of procedures for transforming such graphs into different forms while leaving the overall system function between input and output unchanged. One of these procedures, called *flow graph reversal* or *transposition*, leads to a set of transposed system structures that provide some useful alternatives to the structures discussed in the previous section.

Transposition of a flow graph is accomplished by reversing the directions of all branches in the network while keeping the branch transmittances as they were and reversing the roles of the input and output so that source nodes become sink nodes and vice versa. For single-input, single-output systems, the resulting flow graph has the same system function as the original graph if the input and output nodes are interchanged. Although we will not formally prove this result here,[3] we will demonstrate that it is valid with two examples.

Example 6.7   Transposed Form for a 1st-Order System with No Zeros

The 1st-order system corresponding to the flow graph in Figure 6.24(a) has system function

$$H(z) = \frac{1}{1 - az^{-1}}. \tag{6.42}$$

---

[3]The theorem follows directly from Mason's gain formula of signal flow graph theory. (See Mason and Zimmermann, 1960; Chow and Cassignol, 1962; or Phillips and Nagle, 1995.)

To obtain the transposed form for this system, we reverse the directions of all the branch arrows, taking the output where the input was and injecting the input where the output was. The result is shown in Figure 6.24(b). It is usually convenient to draw the transposed network with the input on the left and the output on the right, as shown in Figure 6.24(c). Comparing Figures 6.24(a) and 6.24(c) we note that the only difference is that in Figure 6.24(a), we multiply the *delayed* output sequence $y[n-1]$ by the coefficient a, whereas in Figure 6.24(c) we multiply the output $y[n]$ by the coefficient a and then delay the resulting product. Since the two operations can be interchanged, we can conclude by inspection that the original system in Figure 6.24(a) and the corresponding transposed system in Figure 6.24(c) have the same system function.

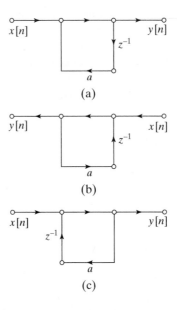

Figure 6.24 (a) Flow graph of simple 1st-order system. (b) Transposed form of (a). (c) Structure of (b) redrawn with input on left.

In Example 6.7, it is straightforward to see that the original system and its transpose have the same system function. However, for more complicated graphs, the result is often not so obvious. This is illustrated by the next example.

Example 6.8   Transposed Form for a Basic 2nd-Order Section

Consider the basic 2nd-order section depicted in Figure 6.25. The corresponding difference equations for this system are

$$w[n] = a_1 w[n-1] + a_2 w[n-2] + x[n], \tag{6.43a}$$

$$y[n] = b_0 w[n] + b_1 w[n-1] + b_2 w[n-2]. \tag{6.43b}$$

The transposed flow graph is shown in Figure 6.26; its corresponding difference equations are

$$v_0[n] = b_0 x[n] + v_1[n-1], \tag{6.44a}$$

$$y[n] = v_0[n], \tag{6.44b}$$

$$v_1[n] = a_1 y[n] + b_1 x[n] + v_2[n-1], \tag{6.44c}$$

$$v_2[n] = a_2 y[n] + b_2 x[n]. \tag{6.44d}$$

Equations (6.43a)–(6.43b) and Eqs. (6.44a)–(6.44d) are different ways to organize the computation of the output samples $y[n]$ from the input samples $x[n]$, and it is not immediately clear that the two sets of difference equations are equivalent. One way to show this equivalence is to use the z-transform representations of both sets of equations, solve for the ratio $Y(z)/X(z) = H(z)$ in both cases, and compare the results. Another way is to substitute Eq. (6.44d) into Eq. (6.44c), substitute the result into Eq. (6.44a), and finally, substitute that result into Eq. (6.44b). The final result is

$$y[n] = a_1 y[n-1] + a_2 y[n-2] + b_0 x[n] + b_1 x[n-1] + b_2 x[n-2]. \tag{6.45}$$

Since the network of Figure 6.25 is a direct form II structure, it is easily seen that the input and output of the system in Figure 6.25 also satisfies the difference Eq. (6.45). Therefore, for initial-rest conditions, the systems in Figures 6.25 and 6.26 are equivalent.

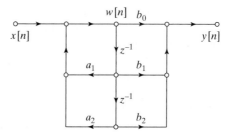

Figure 6.25   Direct form II structure for Example 6.8.

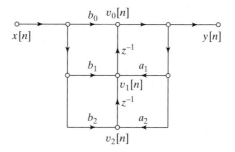

Figure 6.26   Transposed direct form II structure for Example 6.8.

The transposition theorem can be applied to any of the structures that we have discussed so far. For example, the result of applying the theorem to the direct form I structure of Figure 6.14 is shown in Figure 6.27, and similarly, the structure obtained by

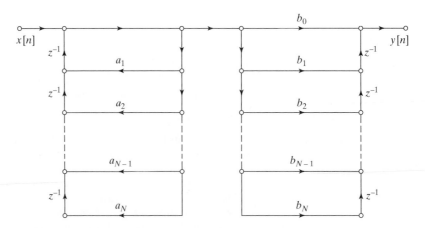

Figure 6.27   General flow graph resulting from applying the transposition theorem to the direct form I structure of Figure 6.14.

transposing the direct form II structure of Figure 6.15 is shown in Figure 6.28. If a signal flow graph configuration is transposed, the number of delay branches and the number of coefficients remain the same. Thus, the transposed direct form II structure is also a canonic structure. Transposed structures derived from direct forms are also "direct" in the sense that they can be constructed by inspection of the numerator and denominator of the system function.

An important point becomes evident through a comparison of Figures 6.15 and 6.28. Whereas the direct form II structure implements the poles first and then the zeros, the transposed direct form II structure implements the zeros first and then the poles. These differences can become important in the presence of quantization in finite-precision digital implementations or in the presence of noise in discrete-time analog implementations.

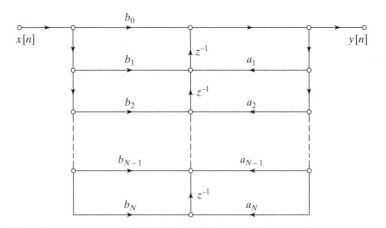

Figure 6.28   General flow graph resulting from applying the transposition theorem to the direct form II structure of Figure 6.15.

When the transposition theorem is applied to cascade or parallel structures, the individual 2nd-order systems are replaced by transposed structures. For example, applying the transposition theorem to Figure 6.18 results in a cascade of three transposed direct form II sections (each like the one in Example 6.8) with the same coefficients as in Figure 6.18, but with the order of the cascade reversed. A similar statement can be made about the transposition of Figure 6.20.

The transposition theorem further emphasizes that an infinite variety of implementation structures exists for any given rational system function. The transposition theorem provides a simple procedure for generating new structures. The problems of implementing systems with finite-precision arithmetic have motivated the development of many more classes of equivalent structures than we can discuss here. Thus, we concentrate only on the most commonly used structures.

6.5 BASIC NETWORK STRUCTURES FOR FIR SYSTEMS

The direct, cascade, and parallel form structures discussed in Sections 6.3 and 6.4 are the most common basic structures for IIR systems. These structures were developed under the assumption that the system function had both poles and zeros. Although the direct and cascade forms for IIR systems include FIR systems as a special case, there are additional specific forms for FIR systems.

6.5.1 Direct Form

For causal FIR systems, the system function has only zeros (except for poles at $z = 0$), and since the coefficients a_k are all zero, the difference equation of Eq. (6.9) reduces to

$$y[n] = \sum_{k=0}^{M} b_k x[n - k]. \tag{6.46}$$

This can be recognized as the discrete convolution of $x[n]$ with the impulse response

$$h[n] = \begin{cases} b_n & n = 0, 1, \ldots, M, \\ 0 & \text{otherwise.} \end{cases} \tag{6.47}$$

In this case, the direct form I and direct form II structures in Figures 6.14 and 6.15 both reduce to the direct form FIR structure as redrawn in Figure 6.29. Because of the chain of delay elements across the top of the diagram, this structure is also referred to as a *tapped delay line* structure or a *transversal filter* structure. As seen from Figure 6.29, the signal at each tap along this chain is weighted by the appropriate coefficient (impulse-response value), and the resulting products are summed to form the output $y[n]$.

The transposed direct form for the FIR case is obtained by applying the transposition theorem to Figure 6.29 or, equivalently, by setting the coefficients a_k to zero in Figure 6.27 or Figure 6.28. The result is shown in Figure 6.30.

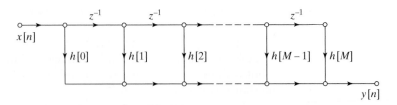

Figure 6.29   Direct form realization of an FIR system.

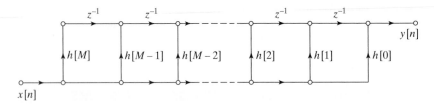

Figure 6.30   Transposition of the network of Figure 6.29.

6.5.2 Cascade Form

The cascade form for FIR systems is obtained by factoring the polynomial system function. That is, we represent $H(z)$ as

$$H(z) = \sum_{n=0}^{M} h[n]z^{-n} = \prod_{k=1}^{M_s}(b_{0k} + b_{1k}z^{-1} + b_{2k}z^{-2}), \qquad (6.48)$$

where $M_s = \lfloor (M+1)/2 \rfloor$ is the largest integer contained in $(M+1)/2$. If M is odd, one of the coefficients b_{2k} will be zero, since $H(z)$ in that case would have an odd number of real zeros. The flow graph representing Eq. (6.48) is shown in Figure 6.31, which is identical in form to Figure 6.18 with the coefficients a_{1k} and a_{2k} all zero. Each of the 2nd-order sections in Figure 6.31 uses the direct form shown in Figure 6.29. Another alternative is to use transposed direct form 2nd-order sections or, equivalently, to apply the transposition theorem to Figure 6.31.

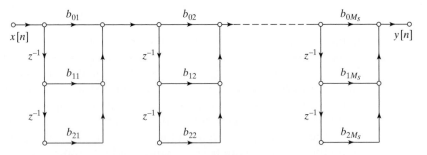

Figure 6.31   Cascade form realization of an FIR system.

6.5.3 Structures for Linear-Phase FIR Systems

In Chapter 5, we showed that causal FIR systems have generalized linear phase if the impulse response satisfies the symmetry condition

$$h[M - n] = h[n] \qquad n = 0, 1, \ldots, M \qquad (6.49a)$$

or

$$h[M - n] = -h[n] \qquad n = 0, 1, \ldots, M. \qquad (6.49b)$$

With either of these conditions, the number of coefficient multipliers can be essentially halved. To see this, consider the following manipulations of the discrete convolution equation, assuming that M is an even integer corresponding to type I or type III systems:

$$y[n] = \sum_{k=0}^{M} h[k]x[n - k]$$

$$= \sum_{k=0}^{M/2-1} h[k]x[n - k] + h[M/2]x[n - M/2] + \sum_{k=M/2+1}^{M} h[k]x[n - k]$$

$$= \sum_{k=0}^{M/2-1} h[k]x[n - k] + h[M/2]x[n - M/2] + \sum_{k=0}^{M/2-1} h[M - k]x[n - M + k].$$

For type I systems, we use Eq. (6.49a) to obtain

$$y[n] = \sum_{k=0}^{M/2-1} h[k](x[n - k] + x[n - M + k]) + h[M/2]x[n - M/2]. \qquad (6.50)$$

For type III systems, we use Eq. (6.49b) to obtain

$$y[n] = \sum_{k=0}^{M/2-1} h[k](x[n - k] - x[n - M + k]). \qquad (6.51)$$

For the case of M an odd integer, the corresponding equations are, for type II systems,

$$y[n] = \sum_{k=0}^{(M-1)/2} h[k](x[n - k] + x[n - M + k]) \qquad (6.52)$$

and, for type IV systems,

$$y[n] = \sum_{k=0}^{(M-1)/2} h[k](x[n - k] - x[n - M + k]). \qquad (6.53)$$

Equations (6.50)–(6.53) imply structures with either $M/2 + 1$, $M/2$, or $(M + 1)/2$ coefficient multipliers, rather than the M coefficient multipliers of the general direct form structure of Figure 6.29. Figure 6.32 shows the structure implied by Eq. (6.50), and Figure 6.33 shows the structure implied by Eq. (6.52).

In our discussion of linear-phase systems in Section 5.7.3, we showed that the symmetry conditions of Eqs. (6.49a) and (6.49b) cause the zeros of $H(z)$ to occur in mirror-image pairs. That is, if z_0 is a zero of $H(z)$, then $1/z_0$ is also a zero of $H(z)$. Furthermore, if $h[n]$ is real, then the zeros of $H(z)$ occur in complex-conjugate pairs.

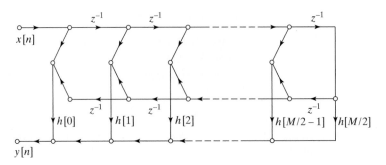

Figure 6.32 Direct form structure for an FIR linear-phase system when M is an even integer.

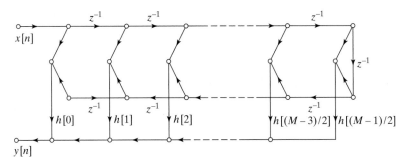

Figure 6.33 Direct form structure for an FIR linear-phase system when M is an odd integer.

As a consequence, real zeros not on the unit circle occur in reciprocal pairs. Complex zeros not on the unit circle occur in groups of four, corresponding to the complex conjugates and reciprocals. If a zero is on the unit circle, its reciprocal is also its conjugate. Consequently, complex zeros on the unit circle are conveniently grouped into pairs. Zeros at $z = \pm 1$ are their own reciprocal and complex conjugate. The four cases are summarized in Figure 6.34, where the zeros at z_1, z_1^*, $1/z_1$, and $1/z_1^*$ are considered as a group of four. The zeros at z_2 and $1/z_2$ are considered as a group of two, as are the zeros at z_3 and z_3^*. The zero at z_4 is considered singly. If $H(z)$ has the zeros shown in Figure 6.34, it can be factored into a product of 1st-, 2nd-, and 4th-order factors. Each of these factors is a polynomial whose coefficients have the same symmetry as the coefficients of $H(z)$; i.e., each factor is a linear-phase polynomial in z^{-1}. Therefore, the system can be implemented as a cascade of 1st-, 2nd-, and 4th-order systems. For example, the system function corresponding to the zeros of Figure 6.34 can be expressed as

$$H(z) = h[0](1 + z^{-1})(1 + az^{-1} + z^{-2})(1 + bz^{-1} + z^{-2})$$
$$\times (1 + cz^{-1} + dz^{-2} + cz^{-3} + z^{-4}), \tag{6.54}$$

where

$$a = (z_2 + 1/z_2), \qquad b = 2\mathcal{R}e\{z_3\}, \qquad c = -2\mathcal{R}e\{z_1 + 1/z_1\}, \qquad d = 2 + |z_1 + 1/z_1|^2.$$

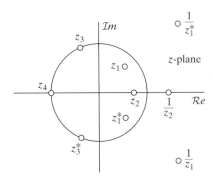

Figure 6.34   Symmetry of zeros for a linear-phase FIR filter.

This representation suggests a cascade structure consisting of linear-phase elements. It can be seen that the order of the system function polynomial is $M = 9$ and the number of different coefficient multipliers is five. This is the same number $((M + 1)/2 = 5)$ of constant multipliers required for implementing the system in the linear-phase direct form of Figure 6.32. Thus, with no additional multiplications, we obtain a modular structure in terms of a cascade of short linear-phase FIR systems.

6.6 LATTICE FILTERS

In Sections 6.3.2 and 6.5.2, we discussed cascade forms for both IIR and FIR systems obtained by factoring their system functions into 1st- and 2nd-order sections. Another interesting and useful cascade structure is based on a cascade (output to input) connection of the basic structure shown in Figure 6.35(a). In the case of Figure 6.35(a) the basic building block system has two inputs and two outputs, and is called a two-port flow graph. Figure 6.35(b) shows the equivalent flow graph representation. Figure 6.36 shows a cascade of M of these basic elements with a "termination" at each end of the cascade so that the overall system is a single-input single-output system with input $x[n]$ feeding both inputs of two-port building block (1) and output $y[n]$ defined to be $a^{(M)}[n]$, the upper output of the last two-port building block M. (The lower output of the Mth stage is generally ignored.) Although such a structure could take many different forms

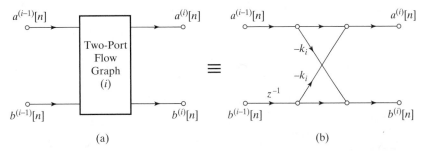

Figure 6.35   One section of the lattice structure for FIR lattice filters. (a) Block diagram representation of a two-port building block (b) Equivalent flow graph.

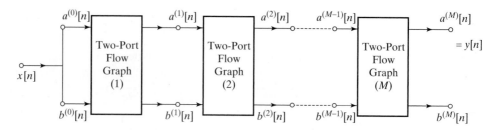

Figure 6.36   Cascade connection of M basic building block sections.

depending on the definition of the basic building block, we will limit our attention to the particular choice in Figure 6.35(b), which leads to a widely used class of FIR and IIR filter structures known as *lattice filters*.

6.6.1 FIR Lattice Filters

If the basic butterfly-shaped two-port building block in Figure 6.35(b) is used in the cascade of Figure 6.36, we obtain a flow graph like the one shown in Figure 6.37, whose lattice shape motivates the name *lattice filter*. The coefficients $k_1, k_2, \ldots, k_M$, are referred to generally as the *k-parameters* of the lattice structure. In Chapter 11, we will see that the *k*-parameters have a special meaning in the context of all-pole modeling of signals, and the lattice filter of Figure 6.37 is an implementation structure for a linear prediction of signal samples. In the current chapter, our focus is only on the use of lattice filters to implement FIR and all-pole IIR transfer functions.

The node variables $a^{(i)}[n]$ and $b^{(i)}[n]$ in Figure 6.37 are intermediate sequences that depend upon the input $x[n]$ through the set of difference equations

$$a^{(0)}[n] = b^{(0)}[n] = x[n] \tag{6.55a}$$

$$a^{(i)}[n] = a^{(i-1)}[n] - k_i b^{(i-1)}[n-1] \quad i = 1, 2, \ldots, M \tag{6.55b}$$

$$b^{(i)}[n] = b^{(i-1)}[n-1] - k_i a^{(i-1)}[n] \quad i = 1, 2, \ldots, M \tag{6.55c}$$

$$y[n] = a^{(M)}[n]. \tag{6.55d}$$

As we can see, the *k*-parameters are coefficients in this set of M coupled difference equations represented by Figure 6.37 and Eqs. (6.55a)–(6.55d). It should be clear that

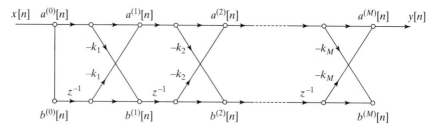

Figure 6.37   Lattice flow graph for an FIR system based on a cascade of M two-port building blocks of Figure 6.35(b).

these equations must be computed in the order shown ($i = 0, 1, \ldots, M$) since the output of stage ($i - 1$) is needed as input to stage (i), etc.

The lattice structure in Figure 6.37 is clearly an LTI system, since it is a linear signal flow graph with only delays and constant branch coefficients. Furthermore, note that there are no feedback loops, which implies that the system has a finite-duration impulse response. In fact, a straightforward argument is sufficient to show that the impulse response from the input to any internal node has finite length. Specifically, consider the impulse response from $x[n]$ to the node variable $a^{(i)}[n]$, i.e., from the input to the ith upper node. It is clear that if $x[n] = \delta[n]$, then $a^{(i)}[0] = 1$ for every i, since the impulse propagates with no delay through the top branch of all the stages. All other paths to any node variable $a^{(i)}[n]$ or $b^{(i)}[n]$ pass through at least one delay, with the greatest delay being along the bottom path and then up to node variable $a^{(i)}[n]$ through the coefficient $-k_i$. This will be the last impulse that arrives at $a^{(i)}[n]$, so the impulse response will have length $i + 1$ samples. All other paths to an internal node zigzag between the top and bottom of the graph, thereby passing through at least one, but not all, of the delays occurring before the outputs of section (i).

Note that in our introduction to lattice filters, $a^{(i)}[n]$ and $b^{(i)}[n]$ were used in Figure 6.37 and Eqs. (6.55a)–(6.55d) to denote the node variables of building block (i) for *any* input $x[n]$. However, for the remainder of our discussion, it is convenient to assume specifically that $x[n] = \delta[n]$ so that $a^{(i)}[n]$ and $b^{(i)}[n]$ are the resulting impulse responses at the associated nodes, and that the corresponding z-transforms $A^{(i)}(z)$ and $B^{(i)}(z)$ are the transfer functions between the input and the ith nodes. Consequently, the transfer function between the input and the upper ith node is

$$A^{(i)}(z) = \sum_{n=0}^{i} a^{(i)}[n]z^{-n} = 1 - \sum_{m=1}^{i} \alpha_m^{(i)} z^{-m}, \tag{6.56}$$

where in the second form, the coefficients $\alpha_m^{(i)}$ for $m \le i$ are composed of sums of products of the coefficients k_j for $j \le m$. As we have shown, the coefficient for the longest delay from the input to the upper node i is $\alpha_i^{(i)} = k_i$. In this notation, the impulse response from $x[n]$ to node variable $a^{(i)}[n]$ is

$$a^{(i)}[n] = \begin{cases} 1 & n = 0 \\ -\alpha_n^{(i)} & 1 \le n \le i \\ 0 & \text{otherwise} \end{cases} \tag{6.57}$$

Similarly, the transfer function from the input to the lower node i is denoted $B^{(i)}(z)$. Therefore, from Figure 6.35(b) or Eqs. (6.55b) and (6.55c), we see that

$$A^{(i)}(z) = A^{(i-1)}(z) - k_i z^{-1} B^{(i-1)}(z) \tag{6.58a}$$

$$B^{(i)}(z) = -k_i A^{(i-1)}(z) + z^{-1} B^{(i-1)}(z). \tag{6.58b}$$

Also, we note that at the input end ($i = 0$)

$$A_0(z) = B_0(z) = 1. \tag{6.59}$$

Using Eqs. (6.58a) and (6.58b) and starting with Eq. (6.59), we can calculate $A^{(i)}(z)$ and $B^{(i)}(z)$ recursively up to any value of i. If we continue, the pattern that emerges in the relationship between $B^{(i)}(z)$ and $A^{(i)}(z)$ is

$$B^{(i)}(z) = z^{-i} A^{(i)}(1/z) \tag{6.60a}$$

or by replacing z by $1/z$ in Eq. (6.60a) we have the equivalent relation

$$A^{(i)}(z) = z^{-i}B^{(i)}(1/z). \tag{6.60b}$$

We can verify these equivalent relationships formally by induction, i.e., by verifying that if they are true for some value $i - 1$ then they will be true for i. Specifically, it is straightforward to see from Eq. (6.59) that Eqs. (6.60a) and (6.60b) are true for $i = 0$. Now note that for $i = 1$,

$$A^{(1)}(z) = A^{(0)}(z) - k_1 z^{-1}B^{(0)}(z) = 1 - k_1 z^{-1}$$
$$B^{(1)}(z) = -k_1 A^{(0)}(z) + z^{-1}B^{(0)}(z) = -k_1 + z^{-1}$$
$$= z^{-1}(1 - k_1 z)$$
$$= z^{-1}A^{(1)}(1/z)$$

and for $i = 2$,

$$A^{(2)}(z) = A^{(1)}(z) - k_2 z^{-1}B^{(1)}(z) = 1 - k_1 z^{-1} - k_2 z^{-2}(1 - k_1 z)$$
$$= 1 - k_1(1 - k_2)z^{-1} - k_2 z^{-2}$$
$$B^{(2)}(z) = -k_2 A^{(1)}(z) + z^{-1}B^{(1)}(z) = -k_2(1 - k_1 z^{-1}) + z^{-2}(1 - k_1 z)$$
$$= z^{-2}(1 - k_1(1 - k_2)z - k_2 z^2)$$
$$= z^{-2}A^{(2)}(1/z).$$

We can prove the general result by assuming that Eq. (6.60a) and Eq. (6.60b) are true for $i - 1$, and then substitute into Eq. (6.58b) to obtain

$$B^{(i)}(z) = -k_i z^{-(i-1)}B^{(i-1)}(1/z) + z^{-1}z^{-(i-1)}A^{(i-1)}(1/z)$$
$$= z^{-i}\left[A^{(i-1)}(1/z) - k_i z B^{(i-1)}(1/z)\right].$$

From Eq. (6.58a) it follows that the term in brackets is $A^{(i)}(1/z)$, so that in general,

$$B^{(i)}(z) = z^{-i}A^{(i)}(1/z),$$

as in Eq. (6.60a). Thus, we have shown that Eqs. (6.60a) and (6.60b) hold for any $i \geq 0$.

As indicated earlier, the transfer functions $A^{(i)}(z)$ and $B^{(i)}(z)$ can be computed recursively using Eq. (6.58a) and (6.58b). These transfer functions are i^{th}-order polynomials, and it is particularly useful to obtain a direct relationship among the coefficients of the polynomials. Toward this end, the right side of Eq. (6.57) defines the coefficients of $A^{(i)}(z)$ to be $-\alpha_m^{(i)}$, for $m = 1, 2, \ldots, i$ with the leading coefficient equal to one; i.e., as in Eq. (6.56),

$$A^{(i)}(z) = 1 - \sum_{m=1}^{i} \alpha_m^{(i)} z^{-m}, \tag{6.61}$$

and similarly,

$$A^{(i-1)}(z) = 1 - \sum_{m=1}^{i-1} \alpha_m^{(i-1)} z^{-m}. \tag{6.62}$$

To obtain a direct recursive relationship for the coefficients $\alpha_m^{(i)}$ in terms of $\alpha_m^{(i-1)}$ and k_i, we combine Eqs. (6.60a) and (6.62) from which it follows that

$$B^{(i-1)}(z) = z^{-(i-1)} A^{(i-1)}(1/z) = z^{-(i-1)} \left[1 - \sum_{m=1}^{i-1} \alpha_m^{(i-1)} z^{+m} \right]. \tag{6.63}$$

Substituting Eqs. (6.62) and (6.63) into Eq. (6.58a), $A^{(i)}(z)$ can also be expressed as

$$A^{(i)}(z) = \left(1 - \sum_{m=1}^{i-1} \alpha_m^{(i-1)} z^{-m} \right) - k_i z^{-1} \left(z^{-(i-1)} \left[1 - \sum_{m=1}^{i-1} \alpha_m^{(i-1)} z^{+m} \right] \right). \tag{6.64}$$

Re-indexing the second summation by reversing the ordering of the terms (i.e., replacing m by $i - m$ and re-summing) and combining terms in Eq. (6.64) leads to

$$A^{(i)}(z) = 1 - \sum_{m=1}^{i-1} \left[\alpha_m^{(i-1)} - k_i \alpha_{i-m}^{(i-1)} \right] z^{-m} - k_i z^{-i}, \tag{6.65}$$

where we see that, as indicated earlier, the coefficient of z^{-i} is $-k_i$. Comparing Eqs. (6.65) and (6.61) shows that

$$\alpha_m^{(i)} = \left[\alpha_m^{(i-1)} - k_i \alpha_{i-m}^{(i-1)} \right] \quad m = 1, ..., i - 1 \tag{6.66a}$$

$$\alpha_i^{(i)} = k_i. \tag{6.66b}$$

Equations (6.66) are the desired direct recursion between the coefficients of $A^{(i)}(z)$ and the coefficients of $A^{(i-1)}(z)$. These equations, together with Eq. (6.60a) also determine the transfer function $B^{(i)}(z)$.

The recursion of Eqs. (6.66) can also be expressed compactly in matrix form. We denote by $\boldsymbol{\alpha}_{i-1}$ the vector of transfer function coefficients for $A^{(i-1)}(z)$ and by $\check{\boldsymbol{\alpha}}_{i-1}$ these coefficients in reverse order, i.e.,

$$\boldsymbol{\alpha}_{i-1} = \left[\alpha_1^{(i-1)} \ \alpha_2^{(i-1)} \ \cdots \ \alpha_{i-1}^{(i-1)} \right]^T$$

and

$$\check{\boldsymbol{\alpha}}_{i-1} = \left[\alpha_{i-1}^{(i-1)} \ \alpha_{i-2}^{(i-1)} \ \cdots \ \alpha_1^{(i-1)} \right]^T.$$

Then Eqs. (6.66) can be expressed as the matrix equation

$$\boldsymbol{\alpha}_i = \begin{bmatrix} \boldsymbol{\alpha}_{i-1} \\ \cdots \cdots \\ 0 \end{bmatrix} - k_i \begin{bmatrix} \check{\boldsymbol{\alpha}}_{i-1} \\ \cdots \cdots \\ -1 \end{bmatrix}. \tag{6.67}$$

The recursion in Eqs. (6.66) or Eqs. (6.67) is the basis for an algorithm for analyzing an FIR lattice structure to obtain its transfer function. We begin with the flow graph specified as in Figure 6.37 by the set of k-parameters $\{k_1, k_2, \ldots, k_M\}$. Then we can use Eqs. (6.66) recursively to compute the transfer functions of successively higher-order FIR filters until we come to end of the cascade giving us

$$A(z) = 1 - \sum_{m=1}^{M} \alpha_m z^{-m} = \frac{Y(z)}{X(z)}, \tag{6.68a}$$

k-Parameters-to-Coefficients Algorithm

Given $k_1, k_2, \ldots, k_M$
for $i = 1, 2, \ldots, M$
$\quad \alpha_i^{(i)} = k_i$ Eq. (6.66b)
$\quad$ if $i > 1$ then for $j = 1, 2, \ldots, i - 1$
$\quad\quad \alpha_j^{(i)} = \alpha_j^{(i-1)} - k_i \alpha_{i-j}^{(i-1)}$ Eq. (6.66a)
$\quad$ end
end
$\alpha_j = \alpha_j^{(M)} \quad j = 1, 2, \ldots, M$ Eq. (6.68b)

Figure 6.38 Algorithm for converting from k-parameters to FIR filter coefficients.

where

$$\alpha_m = \alpha_m^{(M)} \quad m = 1, 2, \ldots, M. \tag{6.68b}$$

The steps of this algorithm are represented in Figure 6.38.

It is also of interest to obtain the k-parameters in the FIR lattice structure that realize a given desired transfer function from input $x[n]$ to the output $y[n] = a^{(M)}[n]$; i.e., we wish to go from $A(z)$ specified as a polynomial by Eqs. (6.68a) and (6.68b) to the set of k-parameters for the lattice structure in Figure 6.37. This can be done by reversing the recursion of Eqs. (6.66) or (6.67) to obtain successively the transfer function $A^{(i-1)}(z)$ in terms of $A^{(i)}(z)$ for $i = M, M-1, \ldots, 2$. The k-parameters are obtained as a by-product of this recursion.

Specifically, we assume that the coefficients $\alpha_m^{(M)} = \alpha_m$ for $m = 1, \ldots, M$ are specified and we want to obtain the k-parameters $k_1, \ldots, k_M$ to realize this transfer function in lattice form. We start with the last stage of the FIR lattice, i.e., with $i = M$. From Eq. (6.66b),

$$k_M = \alpha_M^{(M)} = \alpha_M \tag{6.69}$$

with $A^{(M)}(z)$ defined in terms of the specified coefficients as

$$A^{(M)}(z) = 1 - \sum_{m=1}^{M} \alpha_m^{(M)} z^{-m} = 1 - \sum_{m=1}^{M} \alpha_m z^{-m}. \tag{6.70}$$

Inverting Eqs. (6.66) or equivalently Eq. (6.67), with $i = M$ and $k_M = \alpha_M^{(M)}$ then determines $\boldsymbol{\alpha}_{M-1}$, the vector of transform coefficients at the next to last stage $i = M - 1$. This process is repeated until we reach $A^{(1)}(z)$.

To obtain a general recursion formula for $\alpha_m^{(i-1)}$ in terms of $\alpha_m^{(i)}$ from Eq. (6.66a) note that $\alpha_{i-m}^{(i-1)}$ must be eliminated. To do this, replace m by $i - m$ in Eq. (6.66a) and multiply both sides of the resulting equation by k_i thereby obtaining

$$k_i \alpha_{i-m}^{(i)} = k_i \alpha_{i-m}^{(i-1)} - k_i^2 \alpha_m^{(i-1)}.$$

Adding this equation to Eq. (6.66a) results in

$$\alpha_m^{(i)} + k_i \alpha_{i-m}^{(i)} = \alpha_m^{(i-1)} - k_i^2 \alpha_m^{(i-1)}$$

from which it follows that

$$\alpha_m^{(i-1)} = \frac{\alpha_m^{(i)} + k_i \alpha_{i-m}^{(i)}}{1 - k_i^2} \quad m = 1, 2, \ldots, i - 1. \tag{6.71a}$$

With $\alpha_m^{(i-1)}$ calculated for $m = 1, 2, \ldots, i - 1$ we note from Eq. (6.66b) that

$$k_{i-1} = \alpha_{i-1}^{(i-1)}. \tag{6.71b}$$

Thus, starting with $\alpha_m^{(M)} = \alpha_m$, $m = 1, 2, \ldots M$ we can use Eqs. (6.71a) and (6.71b) to compute $\alpha_m^{(M-1)}$, for $m = 1, 2, \ldots, M - 1$ and k_{M-1}, and then repeat this process recursively to obtain all of the transfer functions $A^{(i)}(z)$ and, as a by-product, all of the k-parameters needed for the lattice structure. The algorithm is represented in Figure 6.39.

Coefficients-to-k-Parameters Algorithm

Given $\alpha_j^{(M)} = \alpha_j$ $j = 1, 2, \ldots, M$
$k_M = \alpha_M^{(M)}$ Eq. (6.69)
for $i = M, M - 1, \ldots, 2$
   for $j = 1, 2, \ldots, i - 1$

$$\alpha_j^{(i-1)} = \frac{\alpha_j^{(i)} + k_i \alpha_{i-j}^{(i)}}{1 - k_i^2} \qquad \text{Eq. (6.71a)}$$

   end
$$k_{i-1} = \alpha_{i-1}^{(i-1)} \qquad\qquad\qquad \text{Eq. (6.71b)}$$
end

Figure 6.39   Algorithm for converting from FIR filter coefficients to k-parameters.

Example 6.9 k-Parameters for a 3rd-Order FIR System

Consider the FIR system shown in Figure 6.40a whose system function is
$$A(z) = 1 - 0.9z^{-1} + 0.64z^{-2} - 0.576z^{-3}.$$
Consequently, $M = 3$ and the coefficients $\alpha_k^{(3)}$ in Eq. (6.70), are
$$\alpha_1^{(3)} = 0.9 \quad \alpha_2^{(3)} = 0.64 \quad \alpha_3^{(3)} = 0.576.$$
We begin by observing that $k_3 = \alpha_3^{(3)} = 0.576$.

Next we want to calculate the coefficients for transfer function $A^{(2)}(z)$ using Eq. (6.71a). Specifically, applying Eq. (6.71a), we obtain (rounded to three decimal places):

$$\alpha_1^{(2)} = \frac{\alpha_1^{(3)} + k_3 \alpha_2^{(3)}}{1 - k_3^2} = 0.795$$

$$\alpha_2^{(2)} = \frac{\alpha_2^{(3)} + k_3 \alpha_1^{(3)}}{1 - k_3^2} = -0.182$$

From Eq. (6.71b) we then identify $k_2 = \alpha_2^{(2)} = -0.182$

To obtain $A^{(1)}(z)$ we again apply Eq. (6.71a) obtaining

$$\alpha_1^{(1)} = \frac{\alpha_1^{(2)} + k_2 \alpha_1^{(2)}}{1 - k_2^2} = 0.673.$$

We then identify $k_1 = \alpha_1^{(1)} = 0.673$. The resulting lattice structure is shown in Figure 6.40b.

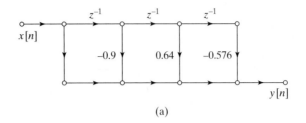

(a)

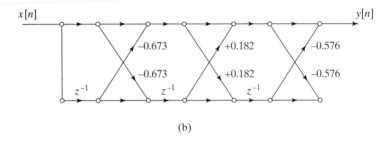

(b)

Figure 6.40 Flow graphs for example. (a) Direct form. (b) Lattice form (coefficients rounded).

6.6.2 All-Pole Lattice Structure

A lattice structure for implementing the all-pole system function $H(z) = 1/A(z)$ can be developed from the FIR lattice of the previous section by recognizing that $H(z)$ is the inverse filter for the FIR system function $A(z)$. To derive the all-pole lattice structure, assume that we are given $y[n] = a^{(M)}[n]$, and we wish to compute the input $a^{(0)}[n] = x[n]$. This can be done by working from right to left to invert the computations in Figure 6.37. More specifically, if we solve Eq. (6.58a) for $A^{(i-1)}(z)$ in terms of $A^{(i)}(z)$ and $B^{(i-1)}(z)$ and leave Eq. (6.58b) as it is, we obtain the pair of equations

$$A^{(i-1)}(z) = A^{(i)}(z) + k_i z^{-1} B^{(i-1)}(z) \tag{6.72a}$$

$$B^{(i)}(z) = -k_i A^{(i-1)}(z) + z^{-1} B^{(i-1)}(z), \tag{6.72b}$$

which have the flow graph representation shown in Figure 6.41. Note that in this case, the signal flow is from i to $i - 1$ along the top of the diagram and from $i - 1$ to i along the bottom. Successive connection of M stages of Figure 6.41 with the appropriate k_i in each section takes the input $a^{(M)}[n]$ to the output $a^{(0)}[n]$ as shown in the flow graph of Figure 6.42. Finally, the condition $x[n] = a^{(0)}[n] = b^{(0)}[n]$ at the terminals of the last stage in Figure 6.42 causes a feedback connection that provides the sequences $b^{(i)}[n]$ that propagate in the reverse direction. Such feedback is, of course, necessary for an IIR system.

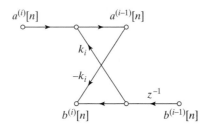

Figure 6.41 One stage of computation for an all-pole lattice system.

The set of difference equations represented by Figure 6.42 is[4]

$$a^{(M)}[n] = y[n] \tag{6.73a}$$

$$a^{(i-1)}[n] = a^{(i)}[n] + k_i b^{(i-1)}[n-1] \quad i = M, M-1, \ldots, 1 \tag{6.73b}$$

$$b^{(i)}[n] = b^{(i-1)}[n-1] - k_i a^{(i-1)}[n] \quad i = M, M-1, \ldots, 1 \tag{6.73c}$$

$$x[n] = a^{(0)}[n] = b^{(0)}[n]. \tag{6.73d}$$

Because of the feedback inherent in Figure 6.42 and these corresponding equations, initial conditions must be specified for all of the node variables associated with delays. Typically, we would specify $b^{(i)}[-1] = 0$ for initial rest conditions. Then, if Eq. (6.73b) is evaluated first, $a^{(i-1)}[n]$ will be available at times $n \geq 0$ for evaluation of Eq. (6.73c) with the values of $b^{(i-1)}[n-1]$ having been provided by the previous iteration.

Now we can state that all the analysis of Section 6.6.1 applies to the all-pole lattice system of Figure 6.42. If we wish to obtain a lattice implementation of an all-pole system with system function $H(z) = 1/A(z)$, we can simply use the algorithms in Figures 6.39 and 6.38 to obtain k-parameters from denominator polynomial coefficients or vice-versa.

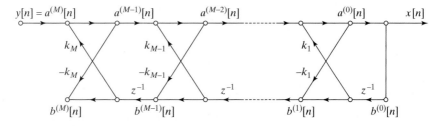

Figure 6.42 All-pole lattice system.

---

[4]Note that by basing our derivation of the all-pole lattice on the FIR lattice in Figure 6.37, we have ended up with the input denoted $y[n]$ and the output $x[n]$ in opposition to our normal convention. This labeling is, of course, arbitrary once the derivation has been completed.

Example 6.10   Lattice Implementation of an IIR System

As an example of an IIR system, consider the system function

$$H(z) = \frac{1}{1 - 0.9z^{-1} + 0.64z^{-2} - 0.576z^{-3}} \tag{6.74a}$$

$$= \frac{1}{(1 - 0.8jz^{-1})(1 + 0.8jz^{-1})(1 - 0.9z^{-1})} \tag{6.74b}$$

which is the inverse system for the system in Example 6.9. Figure 6.43(a) shows the direct form realization of this system, whereas Figure 6.43(b) shows the equivalent IIR lattice system using the k-parameters computed as in Example 6.9. Note that the lattice structure has the same number of delays (memory registers) as the direct form structure. However, the number of multipliers is twice the number of the direct form. This is obviously true for any order M.

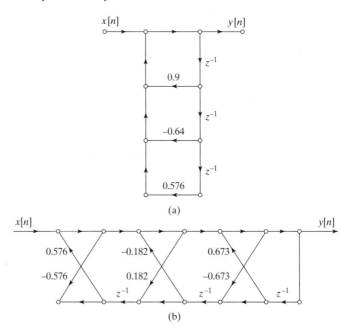

Figure 6.43   Signal flow graph of IIR filter; (a) direct form, (b) lattice form.

Since the lattice structure of Figure 6.42 is an IIR system, we must be concerned about its stability. We will see in Chapter 13 that a necessary and sufficient condition for all the zeros of a polynomial $A(z)$ to be inside the unit circle is $|k_i| < 1$, $i = 1, 2, \ldots, M$. (See Markel and Gray, 1976.) Example 6.10 confirms this fact since, as shown in Eq. (6.74b) the poles of $H(z)$ (zeros of $A(z)$) are located inside the unit circle of the z-plane and all the k-parameters have magnitude less than one. For IIR systems, the guarantee of stability inherent in the condition $|k_i| < 1$ is particularly important. Even though the lattice structure requires twice the number of multiplications per output sample as the direct form, it is insensitive to quantization of the k-parameters. This property accounts for the popularity of lattice filters in speech synthesis applications. (See Quatieri, 2002 and Rabiner and Schafer, 1978.)

6.6.3 Generalization of Lattice Systems

We have shown that FIR systems and all-pole IIR systems have a lattice structure representation. When the system function has both poles and zeros, it is still possible to find a lattice structure based upon a modification of the all-pole structure of Figure 6.42. The derivation will not be provided here (See Gray and Markel, 1973, 1976.), but it is outlined in Problem 11.27.

6.7 OVERVIEW OF FINITE-PRECISION NUMERICAL EFFECTS

We have seen that a particular LTI discrete-time system can be implemented by a variety of computational structures. One motivation for considering alternatives to the simple direct form structures is that different structures that are equivalent for infinite-precision arithmetic may behave differently when implemented with finite numerical precision. In this section, we give a brief introduction to the major numerical problems that arise in implementing discrete-time systems. A more detailed analysis of these finite word-length effects is given in Sections 6.8–6.10.

6.7.1 Number Representations

In theoretical analyses of discrete-time systems, we generally assume that signal values and system coefficients are represented in the real-number system. However, with analog discrete-time systems, the limited precision of the components of a circuit makes it difficult to realize coefficients exactly. Similarly, when implementing digital signal-processing systems, we must represent signals and coefficients in some digital number system that must always have finite precision.

The problem of finite numerical precision has already been discussed in Section 4.8.2 in the context of A/D conversion. We showed there that the output samples from an A/D converter are quantized and thus can be represented by fixed-point binary numbers. For compactness and simplicity in implementing arithmetic, one of the bits of the binary number is assumed to indicate the algebraic sign of the number. Formats such as *sign and magnitude, one's complement,* and *two's complement* are possible, but two's complement is most common.[5] A real number can be represented with infinite precision in two's-complement form as

$$x = X_m \left(-b_0 + \sum_{i=1}^{\infty} b_i 2^{-i} \right), \tag{6.75}$$

where X_m is an arbitrary scale factor and the b_is are either 0 or 1. The quantity b_0 is referred to as the *sign bit.* If $b_0 = 0$, then $0 \le x \le X_m$, and if $b_0 = 1$, then $-X_m \le x < 0$. Thus, any real number whose magnitude is less than or equal to X_m can be represented by Eq. (6.75). An arbitrary real number x would require an infinite number of bits for its exact binary representation. As we saw in the case of A/D conversion, if we use only

---

[5]A detailed description of binary number systems and corresponding arithmetic is given by Knuth (1997).

a finite number of bits $(B + 1)$, then the representation of Eq. (6.75) must be modified to

$$\hat{x} = Q_B[x] = X_m \left(-b_0 + \sum_{i=1}^{B} b_i 2^{-i} \right) = X_m \hat{x}_B. \qquad (6.76)$$

The resulting binary representation is quantized, so that the smallest difference between numbers is

$$\Delta = X_m 2^{-B}. \qquad (6.77)$$

In this case, the quantized numbers are in the range $-X_m \leq \hat{x} < X_m$. The fractional part of $\hat{x}$ can be represented with the positional notation

$$\hat{x}_B = b_0 \diamond b_1 b_2 b_3 \cdots b_B, \qquad (6.78)$$

where $\diamond$ represents the binary point.

The operation of quantizing a number to $(B + 1)$ bits can be implemented by rounding or by truncation, but in either case, quantization is a nonlinear memory-less operation. Figures 6.44(a) and 6.44(b) show the input–output relation for two's-complement rounding and truncation, respectively, for the case $B = 2$. In considering the effects of quantization, we often define the *quantization error* as

$$e = Q_B[x] - x. \qquad (6.79)$$

For the case of two's-complement rounding, $-\Delta/2 < e \leq \Delta/2$, and for two's-complement truncation, $-\Delta < e \leq 0$.[6]

If a number is larger than X_m (a situation called an overflow), we must implement some method of determining the quantized result. In the two's-complement arithmetic system, this need arises when we add two numbers whose sum is greater than X_m. For example, consider the 4-bit two's-complement number 0111, which in decimal form is 7. If we add the number 0001, the carry propagates all the way to the sign bit, so that the result is 1000, which in decimal form is -8. Thus, the resulting error can be very large when overflow occurs. Figure 6.45(a) shows the two's-complement rounding quantizer, including the effect of regular two's-complement arithmetic overflow. An alternative, which is called *saturation overflow* or *clipping*, is shown in Figure 6.45(b). This method of handling overflow is generally implemented for A/D conversion, and it sometimes is implemented in specialized DSP microprocessors for the addition of two's-complement numbers. With saturation overflow, the size of the error does not increase abruptly when overflow occurs; however, a disadvantage of the method is that it voids the following interesting and useful property of two's-complement arithmetic: If several two's-complement numbers whose sum would not overflow are added, then the result of two's-complement accumulation of these numbers is correct, even though intermediate sums might overflow.

Both quantization and overflow introduce errors in digital representations of numbers. Unfortunately, to minimize overflow while keeping the number of bits the same, we must increase X_m and thus increase the size of quantization errors proportionately. Hence, to simultaneously achieve wider dynamic range and lower quantization error, we must increase the number of bits in the binary representation.

---

[6]Note that Eq. (6.76) also represents the result of rounding or truncating any $(B_1 + 1)$-bit binary representation, where $B_1 > B$. In this case Δ would be replaced by $(\Delta - X_m 2^{-B_1})$ in the bounds on the size of the quantization error.

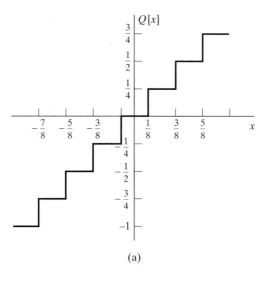

(a)

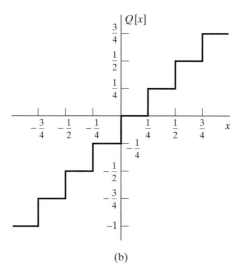

(b)

Figure 6.44 Nonlinear relationships representing two's-complement (a) rounding and (b) truncation for $B = 2$.

So far, we have simply stated that the quantity X_m is an arbitrary scale factor; however, this factor has several useful interpretations. In A/D conversion, we considered X_m to be the full-scale amplitude of the A/D converter. In this case, X_m would probably represent a voltage or current in the analog part of the system. Thus, X_m serves as a calibration constant for relating binary numbers in the range $-1 \le \hat{x}_B < 1$ to analog signal amplitudes.

In digital signal-processing implementations, it is common to assume that all signal variables and all coefficients are binary fractions. Thus, if we multiply a $(B+1)$-bit signal variable by a $(B + 1)$-bit coefficient, the result is a $(2B + 1)$-bit fraction that can be conveniently reduced to $(B+1)$ bits by rounding or truncating the least significant bits. With this convention, the quantity X_m can be thought of as a scale factor that allows the

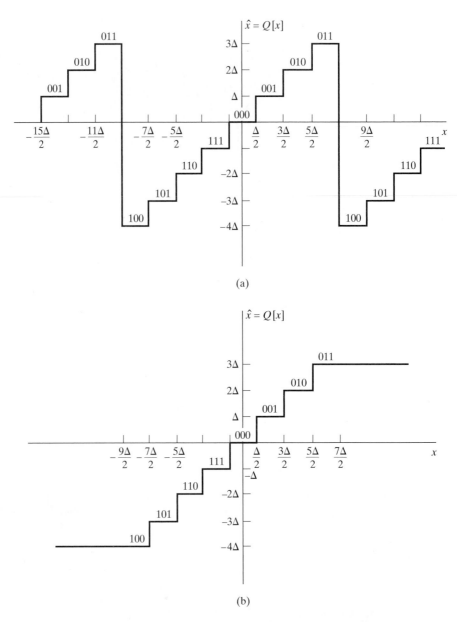

Figure 6.45 Two's-complement rounding. (a) Natural overflow. (b) Saturation.

representation of numbers that are greater than unity in magnitude. For example, in fixed-point computations, it is common to assume that each binary number has a scale factor of the form $X_m = 2^c$. Accordingly, a value $c = 2$ implies that the binary point is actually located between b_2 and b_3 of the binary word in Eq. (6.78). Often, this scale factor is not explicitly represented; instead, it is implicit in the implementation program or hardware architecture.

Still another way of thinking about the scale factor X_m leads to the *floating-point representations,* in which the exponent c of the scale factor is called the *characteristic* and the fractional part $\hat{x}_B$ is called the *mantissa.* The characteristic and the mantissa are each represented explicitly as binary numbers in floating-point arithmetic systems. Floating-point representations provide a convenient means for maintaining both a wide dynamic range and a small quantization noise; however, quantization error manifests itself in a somewhat different way.

6.7.2 Quantization in Implementing Systems

Numerical quantization affects the implementation of LTI discrete-time systems in several ways. As a simple illustration, consider Figure 6.46(a), which shows a block diagram

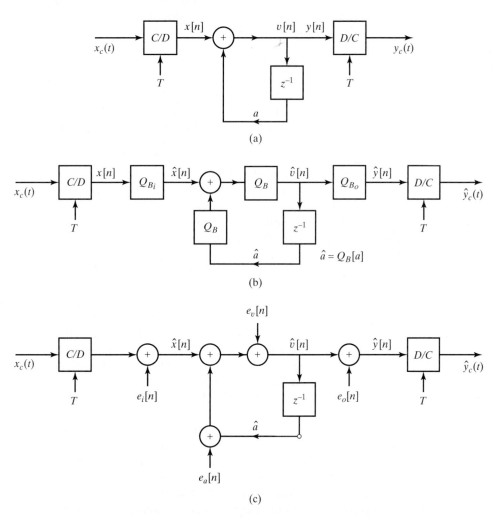

(a)

(b)

(c)

Figure 6.46   Implementation of discrete-time filtering of an analog signal. (a) Ideal system. (b) Nonlinear model. (c) Linearized model.

for a system in which a bandlimited continuous-time signal $x_c(t)$ is sampled to obtain the sequence $x[n]$, which is the input to an LTI system whose system function is

$$H(z) = \frac{1}{1 - az^{-1}}. \tag{6.80}$$

The output of this system, $y[n]$, is converted by ideal bandlimited interpolation to the bandlimited signal $y_c(t)$.

A more realistic model is shown in Figure 6.46(b). In a practical setting, sampling would be done with an A/D converter with finite precision of $(B_i + 1)$ bits. The system would be implemented with binary arithmetic of $(B + 1)$ bits. The coefficient a in Figure 6.46(a) would be represented with $(B + 1)$ bits of precision. Also, the delayed variable $\hat{v}[n - 1]$ would be stored in a $(B + 1)$-bit register, and when the $(B + 1)$-bit number $\hat{v}[n - 1]$ is multiplied by the $(B + 1)$-bit number $\hat{a}$, the resulting product would be $(2B + 1)$ bits in length. If we assume that a $(B + 1)$-bit adder is used, the product $\hat{a}\hat{v}[n - 1]$ must be quantized (i.e., rounded or truncated) to $(B + 1)$ bits before it can be added to the $(B_i + 1)$-bit input sample $\hat{x}[n]$. When $B_i < B$, the $(B_i + 1)$ bits of the input samples can be placed anywhere in the $(B + 1)$-bit binary word with appropriate extension of the sign. Different choices correspond to different scalings of the input. The coefficient a has been quantized, so leaving aside the other quantization errors, the system response cannot in general be the same as in Figure 6.46(a). Finally, the $(B + 1)$-bit samples $\hat{v}[n]$, computed by iterating the difference equation represented by the block diagram, would be converted to an analog signal by a $(B_o + 1)$-bit D/A converter. When $B_o < B$, the output samples must be quantized further before D/A conversion.

Although the model of Figure 6.46(b) could be an accurate representation of a real system, it would be difficult to analyze. The system is nonlinear owing to the quantizers and the possibility of overflow at the adder. Also, quantization errors are introduced at many points in the system. The effects of these errors are impossible to analyze precisely, since they depend on the input signal, which we generally consider to be unknown. Thus, we are forced to adopt several different approximation approaches to simplify the analysis of such systems.

The effect of quantizing the system parameters, such as the coefficient a in Figure 6.46(a), is generally determined separately from the effect of quantization in data conversion or in implementing difference equations. That is, the ideal coefficients of a system function are replaced by their quantized values, and the resulting response functions are tested to see whether, in the absence of other quantization in the arithmetic, quantization of the filter coefficients has degraded the performance of the system to unacceptable levels. For the example of Figure 6.46, if the real number a is quantized to $(B + 1)$ bits, we must consider whether the resulting system with system function

$$\hat{H}(z) = \frac{1}{1 - \hat{a}z^{-1}} \tag{6.81}$$

is close enough to the desired system function $H(z)$ given by Eq. (6.80). Since there are only 2^{B+1} different $(B + 1)$-bit binary numbers, the pole of $H(z)$ can occur only at 2^{B+1} locations on the real axis of the z-plane, and, while it is possible that $\hat{a} = a$, in most cases some deviation from the ideal response would result. This type of analysis is discussed in more general terms in Section 6.8.

The nonlinearity of the system of Figure 6.46(b) causes behavior that cannot occur in a linear system. Specifically, systems such as this can exhibit zero-input limit cycles, where the output oscillates periodically when the input becomes zero after having been nonzero. Limit cycles are caused both by quantization and by overflow. Although the analysis of such phenomena is difficult, some useful approximate results have been developed. Limit cycles are discussed briefly in Section 6.10.

If care is taken in the design of a digital implementation, we can ensure that overflow occurs only rarely and quantization errors are small. Under these conditions, the system of Figure 6.46(b) behaves very much like a linear system (with quantized coefficients) in which quantization errors are injected at the input and output and at internal points in the structure where rounding or truncation occurs. Therefore, we can replace the model of Figure 6.46(b) by the linearized model of Figure 6.46(c), where the quantizers are replaced by additive noise sources (see Gold and Rader, 1969; Jackson, 1970a, 1970b). Figure 6.46(c) is equivalent to Figure 6.46(b) if we know each of the noise sources exactly. However, as discussed in Section 4.8.3, useful results are obtained if we assume a random noise model for the quantization noise in A/D conversion. This same approach can be used in analyzing the effects of arithmetic quantization in digital implementations of linear systems. As seen in Figure 6.46(c), each noise source injects a random signal that is processed by a different part of the system, but since we assume that all parts of the system are linear, we can compute the overall effect by superposition. In Section 6.9, we illustrate this style of analysis for several important systems.

In the simple example of Figure 6.46, there is little flexibility in the choice of structure. However, for higher-order systems, we have seen that there is a wide variety of choices. Some of the structures are less sensitive to coefficient quantization than others. Similarly, because different structures have different quantization noise sources and because these noise sources are filtered in different ways by the system, we will find that structures that are theoretically equivalent sometimes perform quite differently when finite-precision arithmetic is used to implement them.

6.8 THE EFFECTS OF COEFFICIENT QUANTIZATION

LTI discrete-time systems are generally used to perform a filtering operation. Methods for designing FIR and IIR filters, which are discussed in Chapter 7, typically assume a particular form for the system function. The result of the filter design process is a system function for which we must choose an implementation structure (a set of difference equations) from an unlimited number of theoretically equivalent implementations. Although we are almost always interested in implementations that require the least hardware or software complexity, it is not always possible to base the choice of implementation structure on this criterion alone. As we will see in Section 6.9, the implementation structure determines the quantization noise generated internally in the system. Also, some structures are more sensitive than others to perturbations of the coefficients. As we pointed out in Section 6.7, the standard approach to the study of coefficient quantization and round-off noise is to treat them independently. In this section, we consider the effects of quantizing the system parameters.

6.8.1 Effects of Coefficient Quantization in IIR Systems

When the parameters of a rational system function or corresponding difference equation are quantized, the poles and zeros of the system function move to new positions in the z-plane. Equivalently, the frequency response is perturbed from its original value. If the system implementation structure is highly sensitive to perturbations of the coefficients, the resulting system may no longer meet the original design specifications, or an IIR system might even become unstable.

A detailed sensitivity analysis for the general case is complicated and usually of limited value in specific cases of digital filter implementation. Using powerful simulation tools, it is usually easy to simply quantize the coefficients of the difference equations employed in implementing the system and then compute the corresponding frequency response and compare it with the desired frequency-response function. Even though simulation of the system is generally necessary in specific cases, it is still worthwhile to consider, in general, how the system function is affected by quantization of the coefficients of the difference equations. For example, the system function representation corresponding to both direct forms (and their corresponding transposed versions) is the ratio of polynomials

$$H(z) = \frac{\displaystyle\sum_{k=0}^{M} b_k z^{-k}}{1 - \displaystyle\sum_{k=1}^{N} a_k z^{-k}}. \tag{6.82}$$

The sets of coefficients $\{a_k\}$ and $\{b_k\}$ are the ideal infinite-precision coefficients in both direct form implementation structures (and corresponding transposed structures). If we quantize these coefficients, we obtain the system function

$$\hat{H}(z) = \frac{\displaystyle\sum_{k=0}^{M} \hat{b}_k z^{-k}}{1 - \displaystyle\sum_{k=1}^{N} \hat{a}_k z^{-k}}, \tag{6.83}$$

where $\hat{a}_k = a_k + \Delta a_k$ and $\hat{b}_k = b_k + \Delta b_k$ are the quantized coefficients that differ from the original coefficients by the quantization errors Δa_k and Δb_k.

Now consider how the roots of the denominator and numerator polynomials (the poles and zeros of $H(z)$) are affected by the errors in the coefficients. Each polynomial root is affected by *all* of the errors in the coefficients of the polynomial since each root is a function of all the coefficients of the polynomial. Thus, each pole and zero will be affected by all of the quantization errors in the denominator and numerator polynomials, respectively. More specifically, Kaiser (1966) showed that if the poles (or zeros) are tightly clustered, it is possible that small errors in the denominator (numerator) coefficients can cause large shifts of the poles (zeros) for the direct form structures. Thus, if the poles (zeros) are tightly clustered, corresponding to a narrow-bandpass filter or a narrow-bandwidth lowpass filter, then we can expect the poles of the direct form structure to be quite sensitive to quantization errors in the coefficients. Furthermore, Kaiser's analysis showed that the larger the number of clustered poles (zeros), the greater is the sensitivity.

The cascade and parallel form system functions, which are given by Eqs. (6.30) and (6.35), respectively, consist of combinations of 2nd-order direct form systems. However, in both cases, each pair of complex-conjugate poles is realized independently of all the other poles. Thus, the error in a particular pole pair is independent of its distance from the other poles of the system function. For the cascade form, the same argument holds for the zeros, since they are realized as independent 2nd-order factors. Thus, the cascade form is generally much less sensitive to coefficient quantization than the equivalent direct form realization.

As seen in Eq. (6.35), the zeros of the parallel form system function are realized implicitly, through combining the quantized 2nd-order sections to obtain a common denominator. Thus, a particular zero is affected by quantization errors in the numerator and denominator coefficients of *all* the 2nd-order sections. However, for most practical filter designs, the parallel form is also found to be much less sensitive to coefficient quantization than the equivalent direct forms because the 2nd-order subsystems are not extremely sensitive to quantization. In many practical filters, the zeros are often widely distributed around the unit circle, or in some cases they may all be located at $z = \pm 1$. In the latter situation, the zeros mainly provide much higher attenuation around frequencies $\omega = 0$ and $\omega = \pi$ than is specified, and thus, movements of zeros away from $z = \pm 1$ do not significantly degrade the performance of the system.

6.8.2 Example of Coefficient Quantization in an Elliptic Filter

As an illustration of the effect of coefficient quantization, consider the example of an IIR bandpass elliptic filter designed using approximation techniques to be discussed in Chapter 7. The filter was designed to meet the following specifications:

$$0.99 \leq |H(e^{j\omega})| \leq 1.01, \qquad 0.3\pi \leq |\omega| \leq 0.4\pi,$$

$$|H(e^{j\omega})| \leq 0.01 \,(\text{i.e.,} -40\,\text{dB}), \qquad |\omega| \leq 0.29\pi,$$

$$|H(e^{j\omega})| \leq 0.01 \,(\text{i.e.,} -40\,\text{dB}), \qquad 0.41\pi \leq |\omega| \leq \pi.$$

That is, the filter should approximate one in the passband, $0.3\pi \leq |\omega| \leq 0.4\pi$, and zero elsewhere in the base interval $0 \leq |\omega| \leq \pi$. As a concession to computational realizability, a transition (do not care) region of 0.01π is allowed on either side of the passband. In Chapter 7, we will see that specifications for frequency-selective filter design algorithms are often represented in this form. The MATLAB function for elliptic filter design produces the coefficients of a 12th-order direct form representation of the system function of the form of Eq. (6.82), where the coefficients a_k and b_k were computed with 64-bit floating-point arithmetic and are shown in Table 6.1 with full 15-decimal-digit precision. We shall refer to this representation of the filter as "unquantized."

The frequency response $20 \log_{10} |H(e^{j\omega})|$ of the unquantized filter is shown in Figure 6.47(a), which shows that the filter meets the specifications in the stopbands (at least 40 dB attenuation). Also, the solid line in Figure 6.47(b), which is a blow-up of the passband region $0.3\pi \leq |\omega| \leq 0.4\pi$ for the unquantized filter, shows that the filter also meets the specifications in the passband.

TABLE 6.1 UNQUANTIZED DIRECT-FORM COEFFICIENTS FOR A 12TH-ORDER ELLIPTIC FILTER

k	b_k	a_k
0	0.01075998066934	1.00000000000000
1	-0.05308642937079	-5.22581881365349
2	0.16220359377307	16.78472670299535
3	-0.34568964826145	-36.88325765883139
4	0.57751602647909	62.39704677556246
5	-0.77113336470234	-82.65403268814103
6	0.85093484466974	88.67462886449437
7	-0.77113336470234	-76.47294840588104
8	0.57751602647909	53.41004513122380
9	-0.34568964826145	-29.20227549870331
10	0.16220359377307	12.29074563512827
11	-0.05308642937079	-3.53766014466313
12	0.01075998066934	0.62628586102551

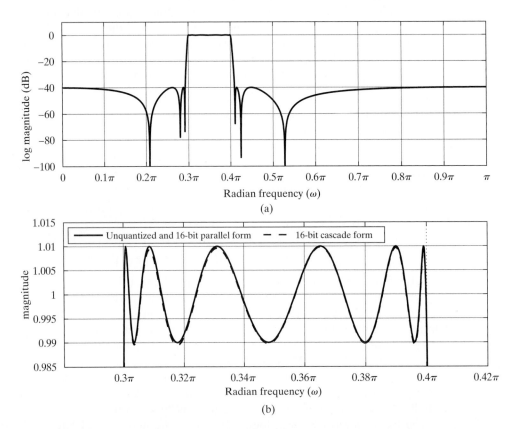

Figure 6.47 IIR coefficient quantization example. (a) Log magnitude for unquantized elliptic bandpass filter. (b) Magnitude in passband for unquantized (solid line) and 16-bit quantized cascade form (dashed line).

TABLE 6.2   ZEROS AND POLES OF UNQUANTIZED 12TH-ORDER ELLIPTIC FILTER.

k	$|c_k|$	$\angle c_k$	$|d_k|$	$\angle d_{1k}$
1	1.0	± 1.65799617112574	0.92299356261936	± 1.15956955465354
2	1.0	± 0.65411612347125	0.92795010695052	± 1.02603244134180
3	1.0	± 1.33272553462313	0.96600955362927	± 1.23886921536789
4	1.0	± 0.87998582176421	0.97053510266510	± 0.95722682653782
5	1.0	± 1.28973944928129	0.99214245914242	± 1.26048962626170
6	1.0	± 0.91475122405407	0.99333628602629	± 0.93918174153968

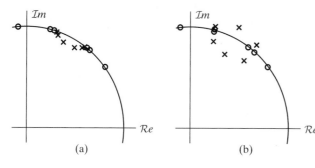

(a)   (b)

Figure 6.48   IIR coefficient quantization example. (a) Poles and zeros of $H(z)$ for unquantized coefficients. (b) Poles and zeros for 16-bit quantization of the direct form coefficients.

Factoring the numerator and denominator polynomials corresponding to the coefficients in Table 6.1 in Eq. (6.82) yields a representation

$$H(z) = \prod_{k=1}^{12} \frac{b_0(1 - c_k z^{-1})}{(1 - d_k z^{-1})}. \tag{6.84}$$

in terms of the zeros and poles, which are given in Table 6.2.

The poles and zeros of the unquantized filter that lie in the upper half of the z-plane are plotted in Figure 6.48(a). Note that the zeros are all on the unit circle, with their angular locations corresponding to the deep nulls in Figure 6.47. The zeros are strategically placed by the filter design method on either side of the passband to provide the desired stopband attenuation and sharp cutoff. Also note that the poles are clustered in the narrow passband, with two of the complex conjugate pole pairs having radii greater than 0.99. This finely tuned arrangement of zeros and poles is required to produce the narrowband sharp-cutoff bandpass filter frequency response shown in Figure 6.47(a).

A glance at the coefficients in Table 6.1 suggests that quantization of the direct form may present significant problems. Recall that with a fixed quantizer, the quantization error size is the same, regardless of the size of the number being quantized; i.e., the quantization error for coefficient $a_{12} = 0.62628586102551$ can be as large as the error for coefficient $a_6 = 88.67462886449437$, if we use the same number of bits and the same scale factor for both. For this reason, when the direct form coefficients in Table 6.1 were quantized with 16-bit precision, each coefficient was quantized independently of the other coefficients so as to maximize the accuracy for each coefficient; i.e., each 16-bit coefficient requires its own scale factor.[7] With this conservative approach, the

---

[7]To simplify implementation, it would be desirable, but far less accurate, if each coefficient had the same scale factor.

resulting poles and zeros are as depicted in Figure 6.48(b). Note that the zeros have shifted noticeably, but not dramatically. In particular, the closely-spaced pair of zeros toward the top of the circle has remained at about the same angle, but they have moved off of the unit circle into a group of four complex conjugate reciprocal zeros, whereas the other zeros are shifted angularly but remain on the unit circle. This constrained movement is a result of the symmetry of the coefficients of the numerator polynomial, which is preserved under quantization. However, the tightly clustered poles, having no symmetry constraints, have moved to much different positions, and, as is easily observed, some of the poles have moved outside the unit circle. Therefore, the direct form system cannot be implemented with 16-bit coefficients because it would be unstable.

On the other hand, the cascade form is much less sensitive to coefficient quantization. The cascade form of the present example can be obtained by grouping the complex conjugate pairs of poles and zeros in Eq. (6.84) and Table 6.2, to form six 2nd-order factors as in

$$H(z) = \prod_{k=1}^{6} \frac{b_{0k}(1 - c_k z^{-1})(1 - c_k^* z^{-1})}{(1 - d_k z^{-1})(1 - d_k^* z^{-1})} = \prod_{k=1}^{6} \frac{b_{0k} + b_{1k} z^{-1} + b_{2k} z^{-2}}{1 - a_{1k} z^{-1} - a_{2k} z^{-2}}. \qquad (6.85)$$

The zeros c_k and poles d_k and coefficients b_{ik} and a_{ik} of the cascade form can be computed with 64-bit floating-point accuracy so these coefficients can still considered to be unquantized. Table 6.3 gives the coefficients of the six 2nd-order sections (as defined in Eq. (6.85). The pairing and ordering of the poles and zeros follows a procedure to be discussed in Section 6.9.3.

TABLE 6.3   UNQUANTIZED CASCADE-FORM COEFFICIENTS FOR A 12TH-ORDER ELLIPTIC FILTER

k	a_{1k}	a_{2k}	b_{0k}	b_{1k}	b_{2k}
1	0.737904	-0.851917	0.137493	0.023948	0.137493
2	0.961757	-0.861091	0.281558	-0.446881	0.281558
3	0.629578	-0.933174	0.545323	-0.257205	0.545323
4	1.117648	-0.941938	0.706400	-0.900183	0.706400
5	0.605903	-0.984347	0.769509	-0.426879	0.769509
6	1.173028	-0.986717	0.937657	-1.143918	0.937657

To illustrate how coefficients are quantized and represented as fixed-point numbers, the coefficients in Table 6.3 were quantized to 16-bit accuracy. The resulting coefficients are presented in Table 6.4. The fixed-point coefficients are shown as a decimal integer times a power-of-2 scale factor. The binary representation would be obtained by converting the decimal integer to a binary number. In a fixed-point implementation, the scale factor would be represented only implicitly in the data shifts that would be necessary to line up the binary points of products prior to their addition to other products. Notice that binary points of the coefficients are not all in the same location. For example, all the coefficients with scale factor 2^{-15} have their binary points between the sign bit, b_0, and the highest fractional bit, b_1, as shown in Eq. (6.78). However, numbers whose magnitudes do not exceed 0.5, such as the coefficient b_{02}, can be shifted left by one or more bit positions.[8] Thus, the binary point for b_{02} is actually to the left of the

---

[8]The use of different binary point locations retains greater accuracy in the coefficients, but it complicates the programming or system architecture.

sign bit as if the word length is extended to 17 bits. On the other hand, numbers whose magnitudes exceed 1 but are less than 2, such as a_{16}, must have their binary points moved one position to the right, i.e., between b_1 and b_2 in Eq. (6.78).

TABLE 6.4   SIXTEEN-BIT QUANTIZED CASCADE-FORM COEFFICIENTS FOR A 12TH-ORDER ELLIPTIC FILTER

k	a_{1k}	a_{2k}	b_{0k}	b_{1k}	b_{2k}
1	24196×2^{-15}	-27880×2^{-15}	17805×2^{-17}	3443×2^{-17}	17805×2^{-17}
2	31470×2^{-15}	-28180×2^{-15}	18278×2^{-16}	-29131×2^{-16}	18278×2^{-16}
3	20626×2^{-15}	-30522×2^{-15}	17556×2^{-15}	-8167×2^{-15}	17556×2^{-15}
4	18292×2^{-14}	-30816×2^{-15}	22854×2^{-15}	-29214×2^{-15}	22854×2^{-15}
5	19831×2^{-15}	-32234×2^{-15}	25333×2^{-15}	-13957×2^{-15}	25333×2^{-15}
6	19220×2^{-14}	-32315×2^{-15}	15039×2^{-14}	-18387×2^{-14}	15039×2^{-14}

The dashed line in Figure 6.47(b) shows the magnitude response in the passband for the quantized cascade form implementation. The frequency response is only slightly degraded in the passband region and negligibly in the stopband.

To obtain other equivalent structures, the cascade form system function must be rearranged into a different form. For example, if a parallel form structure is determined (by partial fraction expansion of the unquantized system function), and the resulting coefficients are quantized to 16 bits as before, the frequency response in the passband is so close to the unquantized frequency response that the difference is not observable in Figure 6.47(a) and barely observable in Figure 6.47(b).

The example just discussed illustrates the robustness of the cascade and parallel forms to the effects of coefficient quantization, and it also illustrates the extreme sensitivity of the direct forms for high-order filters. Because of this sensitivity, the direct forms are rarely used for implementing anything other than 2nd-order systems.[9] Since the cascade and parallel forms can be configured to require the same amount of memory and the same or only slightly more computation as the canonic direct form, these modular structures are the most commonly used. More complex structures such as lattice structures may be more robust for very short word lengths, but they require significantly more computation for systems of the same order.

6.8.3 Poles of Quantized 2nd-Order Sections

Even for the 2nd-order systems that are used to implement the cascade and parallel forms, there remains some flexibility to improve the robustness to coefficient quantization. Consider a complex-conjugate pole pair implemented using the direct form, as in Figure 6.49. With infinite-precision coefficients, this flow graph has poles at $z = re^{j\theta}$ and $z = re^{-j\theta}$. However, if the coefficients $2r\cos\theta$ and $-r^2$ are quantized, only a finite number of different pole locations is possible. The poles must lie on a grid in the z-plane defined by the intersection of concentric circles (corresponding to the quantization of r^2) and vertical lines (corresponding to the quantization of $2r\cos\theta$). Such a grid is

---

[9]An exception is in speech synthesis, where systems of 10th-order and higher are routinely implemented using the direct form. This is possible because in speech synthesis the poles of the system function are widely separated (see Rabiner and Schafer, 1978).

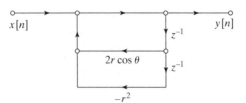

Figure 6.49 Direct form implementation of a complex-conjugate pole pair.

illustrated in Figure 6.50(a) for 4-bit quantization (3 bits plus 1 bit for the sign); i.e., r^2 is restricted to seven positive values and zero, whereas $2r\cos\theta$ is restricted to seven positive values, eight negative values, and zero. Figure 6.50(b) shows a denser grid obtained with 7-bit quantization (6 bits plus 1 bit for the sign). The plots of Figure 6.50 are, of course, symmetrically mirrored into each of the other quadrants of the z-plane.

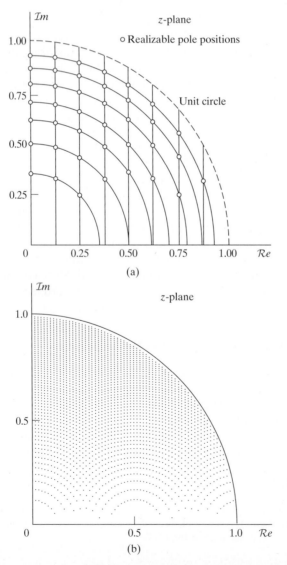

Figure 6.50 Pole-locations for the 2nd-order IIR direct form system of Figure 6.49. (a) Four-bit quantization of coefficients. (b) Seven-bit quantization.

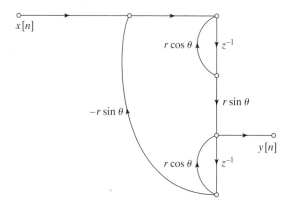

Figure 6.51 Coupled form implementation of a complex-conjugate pole pair.

Notice that for the direct form, the grid is rather sparse around the real axis. Thus, poles located around $\theta = 0$ or $\theta = \pi$ may be shifted more than those around $\theta = \pi/2$. Of course, it is always possible that the infinite-precision pole location is very close to one of the allowed quantized poles. In this case, quantization causes no problem whatsoever, but in general, quantization can be expected to degrade performance.

An alternative 2nd-order structure for realizing poles at $z = re^{j\theta}$ and $z = re^{-j\theta}$ is shown in Figure 6.51. This structure is referred to as the *coupled form* for the 2nd-order system (see Rader and Gold, 1967). It is easily verified that the systems of Figures 6.49 and 6.51 have the same poles for infinite-precision coefficients. To implement the system of Figure 6.51, we must quantize $r\cos\theta$ and $r\sin\theta$. Since these quantities are the real and imaginary parts, respectively, of the pole locations, the quantized pole locations are at intersections of evenly spaced horizontal and vertical lines in the z-plane. Figures 6.52(a) and 6.52(b) show the possible pole locations for 4-bit and 7-bit quantization, respectively. In this case, the density of pole locations is uniform throughout the interior of the unit circle. Twice as many constant multipliers are required to achieve this more uniform density. In some situations, the extra computation might be justified to achieve more accurate pole location with reduced word length.

6.8.4 Effects of Coefficient Quantization in FIR Systems

For FIR systems, we need only be concerned with the locations of the zeros of the system function, since, for causal FIR systems, all the poles are at $z = 0$. Although we have just seen that the direct form structure should be avoided for high-order IIR systems, it turns out that the direct form structure is commonly used for FIR systems. To understand why this is so, we express the system function for a direct form FIR system in the form

$$H(z) = \sum_{n=0}^{M} h[n]z^{-n}. \tag{6.86}$$

Now, suppose that the coefficients $\{h[n]\}$ are quantized, resulting in a new set of coefficients $\{\hat{h}[n] = h[n] + \Delta h[n]\}$. The system function for the quantized system is then

$$\hat{H}(z) = \sum_{n=0}^{M} \hat{h}[n]z^{-n} = H(z) + \Delta H(z), \tag{6.87}$$

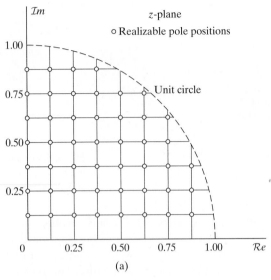

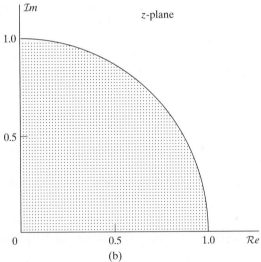

Figure 6.52   Pole locations for coupled form 2nd-order IIR system of Figure 6.51. (a) Four-bit quantization of coefficients. (b) Seven-bit quantization.

where

$$\Delta H(z) = \sum_{n=0}^{M} \Delta h[n] z^{-n}. \tag{6.88}$$

Thus, the system function (and therefore, also the frequency response) of the quantized system is linearly related to the quantization errors in the impulse-response coefficients. For this reason, the quantized system can be represented as in Figure 6.53, which shows the unquantized system in parallel with an error system whose impulse response is the sequence of quantization error samples $\{\Delta h[n]\}$ and whose system function is the corresponding z-transform, $\Delta H(z)$.

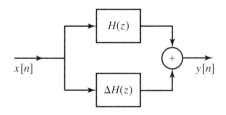

Figure 6.53   Representation of coefficient quantization in FIR systems.

TABLE 6.5   UNQUANTIZED AND QUANTIZED COEFFICIENTS FOR AN OPTIMUM FIR LOWPASS FILTER ($M = 27$)

Coefficient	Unquantized	16 bits	14 bits	13 bits	8 bits
$h[0] = h[27]$	1.359657×10^{-3}	45×2^{-15}	11×2^{-13}	6×2^{-12}	0×2^{-7}
$h[1] = h[26]$	-1.616993×10^{-3}	-53×2^{-15}	-13×2^{-13}	-7×2^{-12}	0×2^{-7}
$h[2] = h[25]$	-7.738032×10^{-3}	-254×2^{-15}	-63×2^{-13}	-32×2^{-12}	-1×2^{-7}
$h[3] = h[24]$	-2.686841×10^{-3}	-88×2^{-15}	-22×2^{-13}	-11×2^{-12}	0×2^{-7}
$h[4] = h[23]$	1.255246×10^{-2}	411×2^{-15}	103×2^{-13}	51×2^{-12}	2×2^{-7}
$h[5] = h[22]$	6.591530×10^{-3}	216×2^{-15}	54×2^{-13}	27×2^{-12}	1×2^{-7}
$h[6] = h[21]$	-2.217952×10^{-2}	-727×2^{-15}	-182×2^{-13}	-91×2^{-12}	-3×2^{-7}
$h[7] = h[20]$	-1.524663×10^{-2}	-500×2^{-15}	-125×2^{-13}	-62×2^{-12}	-2×2^{-7}
$h[8] = h[19]$	3.720668×10^{-2}	1219×2^{-15}	305×2^{-13}	152×2^{-12}	5×2^{-7}
$h[9] = h[18]$	3.233332×10^{-2}	1059×2^{-15}	265×2^{-13}	132×2^{-12}	4×2^{-7}
$h[10] = h[17]$	-6.537057×10^{-2}	-2142×2^{-15}	-536×2^{-13}	-268×2^{-12}	-8×2^{-7}
$h[11] = h[16]$	-7.528754×10^{-2}	-2467×2^{-15}	-617×2^{-13}	-308×2^{-12}	-10×2^{-7}
$h[12] = h[15]$	1.560970×10^{-1}	5115×2^{-15}	1279×2^{-13}	639×2^{-12}	20×2^{-7}
$h[13] = h[14]$	4.394094×10^{-1}	14399×2^{-15}	3600×2^{-13}	1800×2^{-12}	56×2^{-7}

Another approach to studying the sensitivity of the direct form FIR structure would be to examine the sensitivity of the zeros to quantization errors in the impulse-response coefficients, which are, of course the coefficients of the polynomial $H(z)$. If the zeros of $H(z)$ are tightly clustered, then their locations will be highly sensitive to quantization errors in the impulse-response coefficients. The reason that the direct form FIR system is widely used is that, for most linear phase FIR filters, the zeros are more or less uniformly spread in the z-plane. We demonstrate this by the following example.

6.8.5 Example of Quantization of an Optimum FIR Filter

As an example of the effect of coefficient quantization in the FIR case, consider a linear-phase lowpass filter designed to meet the following specifications:

$$0.99 \le |H(e^{j\omega})| \le 1.01, \qquad 0 \le |\omega| \le 0.4\pi,$$

$$|H(e^{j\omega})| \le 0.001 \,(\text{i.e.,} -60\,\text{dB}), \qquad 0.6\pi \le |\omega| \le \pi.$$

This filter was designed using the Parks–McClellan design technique, which will be discussed in Section 7.7.3. The details of the design for this example are considered in Section 7.8.1.

Table 6.5 shows the unquantized impulse-response coefficients for the system, along with quantized coefficients for 16-, 14-, 13-, and 8-bit quantization. Figure 6.54

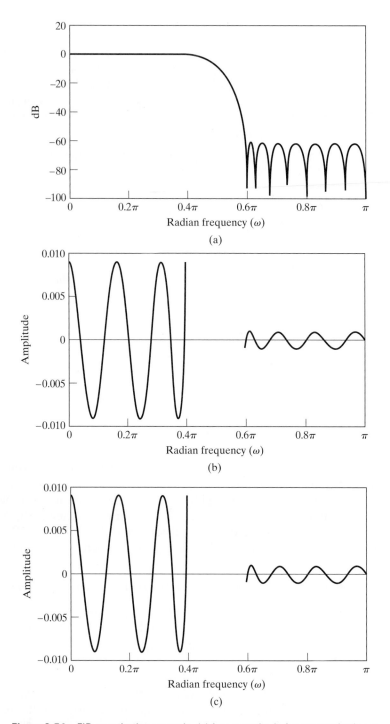

Figure 6.54 FIR quantization example. (a) Log magnitude for unquantized case. (b) Approximation error for unquantized case. (Error not defined in transition band.) (c) Approximation error for 16-bit quantization.

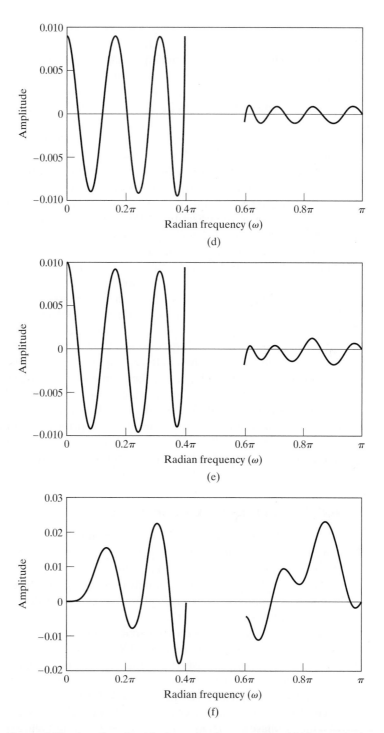

Figure 6.54 (*continued*) (d) Approximation error for 14-bit quantization. (e) Approximation error for 13-bit quantization. (f) Approximation error for 8-bit quantization.

gives a comparison of the frequency responses of the various systems. Figure 6.54(a) shows the log magnitude in dB of the frequency response for unquantized coefficients. Figures 6.54(b), (c), (d), (e), and (f) show the passband and stopband approximation errors (errors in approximating unity in the passband and zero in the stopband) for the unquantized, 16-, 14-, 13-, and 8-bit quantized cases, respectively. From Figure 6.54, we see that the system meets the specifications for the unquantized case and both the 16-bit and 14-bit quantized cases. However, with 13-bit quantization the stopband approximation error becomes greater than 0.001, and with 8-bit quantization the stopband approximation error is over 10 times as large as specified. Thus, we see that at least 14-bit coefficients are required for a direct form implementation of the system. However, this is not a serious limitation, since 16-bit or 14-bit coefficients are well matched to many of the technologies that might be used to implement such a filter.

The effect of quantization of the filter coefficients on the locations of the zeros of the filter is shown in Figure 6.55. Note that in the unquantized case, shown in Figure 6.55(a), the zeros are spread around the z-plane, although there is some clustering on the unit circle. The zeros on the unit circle are primarily responsible for developing the stopband attenuation, whereas those at conjugate reciprocal locations off the unit circle are primarily responsible for forming the passband. Note that little difference is observed in Figure 6.55(b) for 16-bit quantization, but in Figure 6.55(c), showing 13-bit quantization, the zeros on the unit circle have moved significantly. Finally, in Figure 6.55(d), we see that 8-bit quantization causes several of the zeros on the unit circle to pair up and move off the circle to conjugate reciprocal locations. This behavior of the zeros explains the behavior of the frequency response shown in Figure 6.54.

A final point about this example is worth mentioning. All of the unquantized coefficients have magnitudes less than 0.5. Consequently, if all of the coefficients (and therefore, the impulse response) are doubled prior to quantization, more efficient use of the available bits will result, corresponding in effect to increasing B by 1. In Table 6.5 and Figure 6.54, we did not take this potential for increased accuracy into account.

6.8.6 Maintaining Linear Phase

So far, we have not made any assumptions about the phase response of the FIR system. However, the possibility of generalized linear phase is one of the major advantages of an FIR system. Recall that a linear-phase FIR system has either a symmetric ($h[M - n] = h[n]$) or an antisymmetric ($h[M - n] = -h[n]$) impulse response. These linear-phase conditions are easily preserved for the direct form quantized system. Thus, all the systems discussed in the example of the previous subsection have a precisely linear phase, regardless of the coarseness of the quantization. This can be seen in the way in which the conjugate reciprocal locations are preserved in Figure 6.55.

Figure 6.55(d) suggests that, in situations where quantization is very coarse or for high-order systems with closely spaced zeros, it may be worthwhile to realize smaller sets of zeros independently with a cascade form FIR system. To maintain linear phase, each of the sections in the cascade must also have linear phase. Recall that the zeros of a linear-phase system must occur as illustrated in Figure 6.34. For example, if we use 2nd-order sections of the form $(1 + az^{-1} + z^{-2})$ for each complex-conjugate pair of zeros on the unit circle, the zero can move only on the unit circle when the coefficient a is

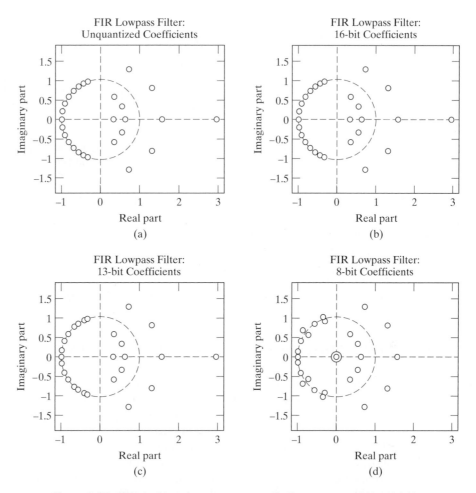

Figure 6.55 Effect of impulse response quantization on zeros of $H(z)$. (a) Un-quantized. (b) 16-bit quantization. (c) 13-bit quantization. (d) 8-bit quantization.

quantized. This prevents zeros from moving away from the unit circle, thereby lessening their attenuating effect. Similarly, real zeros inside the unit circle and at the reciprocal location outside the unit circle would remain real. Also, zeros at $z = \pm 1$ can be realized exactly by 1st-order systems. If a pair of complex-conjugate zeros inside the unit circle is realized by a 2nd-order system rather than a 4th-order system, then we must ensure that, for each complex zero inside the unit circle, there is a conjugate reciprocal zero outside the unit circle. This can be done by expressing the 4th-order factor corresponding to zeros at $z = re^{j\theta}$ and $z = r^{-1}e^{-j\theta}$ as

$$1 + cz^{-1} + dz^{-2} + cz^{-3} + z^{-4}$$
$$= (1 - 2r \cos\theta z^{-1} + r^2 z^{-2})\frac{1}{r^2}(r^2 - 2r \cos\theta z^{-1} + z^{-2}). \tag{6.89}$$

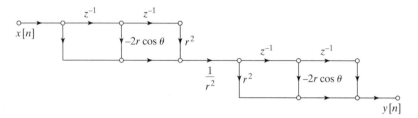

Figure 6.56 Subsystem to implement 4th-order factors in a linear-phase FIR system such that linearity of the phase is maintained independently of parameter quantization.

This condition corresponds to the subsystem shown in Figure 6.56. This system uses the same coefficients, $-2r\cos\theta$ and r^2, to realize both the zeros inside the unit circle and the conjugate reciprocal zeros outside the unit circle. Thus, the linear-phase condition is preserved under quantization. Notice that the factor $(1-2r\cos\theta z^{-1}+r^2z^{-2})$ is identical to the denominator of the 2nd-order direct form IIR system of Figure 6.49. Therefore, the set of quantized zeros is as depicted in Figure 6.50. More details on cascade realizations of FIR systems are given by Herrmann and Schüssler (1970b).

6.9 EFFECTS OF ROUND-OFF NOISE IN DIGITAL FILTERS

Difference equations realized with finite-precision arithmetic are nonlinear systems. Although it is important in general to understand how this nonlinearity affects the performance of discrete-time systems, a precise analysis of arithmetic quantization effects is generally not required in practical applications, where we are typically concerned with the performance of a specific system. Indeed, just as with coefficient quantization, the most effective approach is often to simulate the system and measure its performance. For example, a common objective in quantization error analysis is to choose the digital word length such that the digital system is a sufficiently accurate realization of the desired linear system and at the same time requires a minimum of hardware or software complexity. The digital word length can, of course, be changed only in steps of 1 bit, and as we have already seen in Section 4.8.2, the addition of 1 bit to the word length reduces the size of quantization errors by a factor of 2. Thus, the choice of word length is insensitive to inaccuracies in the quantization error analysis; an analysis that is correct to within 30 to 40 percent is often adequate. For this reason, many of the important effects of quantization can be studied using linear additive noise approximations. We develop such approximations in this section and illustrate their use with several examples. An exception is the phenomenon of zero-input limit cycles, which are strictly nonlinear phenomena. We restrict our study of nonlinear models for digital filters to a brief introduction to zero-input limit cycles in Section 6.10.

6.9.1 Analysis of the Direct Form IIR Structures

To introduce the basic ideas, let us consider the direct form structure for an LTI discrete-time system. The flow graph of a direct form I 2nd-order system is shown in

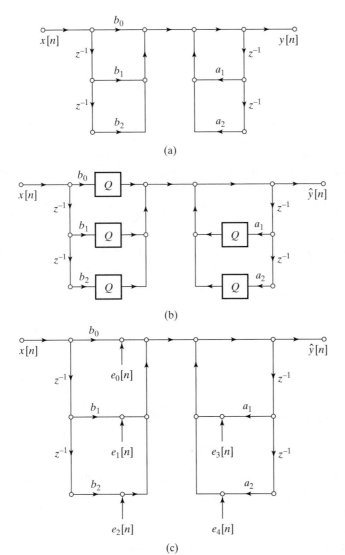

Figure 6.57   Models for direct form I system. (a) Infinite-precision model. (b) Nonlinear quantized model. (c) Linear-noise model.

Figure 6.57(a). The general N^{th}-order difference equation for the direct form I structure is

$$y[n] = \sum_{k=1}^{N} a_k y[n-k] + \sum_{k=0}^{M} b_k x[n-k], \qquad (6.90)$$

and the system function is

$$H(z) = \frac{\displaystyle\sum_{k=0}^{M} b_k z^{-k}}{1 - \displaystyle\sum_{k=1}^{N} a_k z^{-k}} = \frac{B(z)}{A(z)}. \tag{6.91}$$

Let us assume that all signal values and coefficients are represented by $(B+1)$-bit fixed-point binary numbers. Then, in implementing Eq. (6.90) with a $(B+1)$-bit adder, it would be necessary to reduce the length of the $(2B+1)$-bit products resulting from multiplying two $(B+1)$-bit numbers back to $(B+1)$ bits. Since all numbers are treated as fractions, we would discard the least significant B bits by either rounding or truncation. This is represented by replacing each constant multiplier branch in Figure 6.57(a) by a constant multiplier followed by a quantizer, as in the nonlinear model of Figure 6.57(b). The difference equation corresponding to Figure 6.57(b) is the nonlinear equation

$$\hat{y}[n] = \sum_{k=1}^{N} Q[a_k \hat{y}[n-k]] + \sum_{k=0}^{M} Q[b_k x[n-k]]. \tag{6.92}$$

Figure 6.57(c) shows an alternative representation in which the quantizers are replaced by noise sources that are equal to the quantization error at the output of each quantizer. For example, rounding or truncation of a product $bx[n]$ is represented by a noise source of the form

$$e[n] = Q[bx[n]] - bx[n]. \tag{6.93}$$

If the noise sources are known exactly, then Figure 6.57(c) is exactly equivalent to Figure 6.57(b). However, Figure 6.57(c) is most useful when we assume that each quantization noise source has the following properties:

1. Each quantization noise source $e[n]$ is a wide-sense-stationary white-noise process.
2. Each quantization noise source has a uniform distribution of amplitudes over one quantization interval.
3. Each quantization noise source is *uncorrelated* with the input to the corresponding quantizer, all other quantization noise sources, and the input to the system.

These assumptions are identical to those made in the analysis of A/D conversion in Section 4.8. Strictly speaking, our assumptions here cannot be valid, since the quantization error depends directly on the input to the quantizer. This is readily apparent for constant and sinusoidal signals. However, experimental and theoretical analyses have shown (see Bennett, 1948; Widrow, 1956, 1961; Widrow and Kollár, 2008) that in many situations the approximation just described leads to accurate predictions of measured statistical averages such as the mean, variance, and correlation function. This is true when the input signal is a complicated wideband signal such as speech, in which the signal fluctuates rapidly among all the quantization levels and traverses many of those levels in going from sample to sample (see Gold and Rader, 1969). The simple linear-noise approximation presented here allows us to characterize the noise generated in the system by averages such as the mean and variance and to determine how these averages are modified by the system.

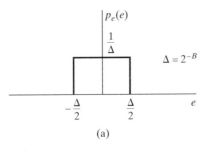

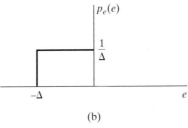

Figure 6.58 Probability density function for quantization errors. (a) Rounding. (b) Truncation.

For $(B+1)$-bit quantization, we showed in Section 6.7.1 that, for rounding,

$$-\frac{1}{2}2^{-B} < e[n] \le \frac{1}{2}2^{-B}, \tag{6.94a}$$

and for two's-complement truncation,

$$-2^{-B} < e[n] \le 0. \tag{6.94b}$$

Thus, according to our second assumption, the probability density functions for the random variables representing quantization error are the uniform densities shown in Figure 6.58(a) for rounding and in Figure 6.58(b) for truncation. The mean and variance for rounding are, respectively,

$$m_e = 0, \tag{6.95a}$$

$$\sigma_e^2 = \frac{2^{-2B}}{12}. \tag{6.95b}$$

For two's-complement truncation, the mean and variance are

$$m_e = -\frac{2^{-B}}{2}, \tag{6.96a}$$

$$\sigma_e^2 = \frac{2^{-2B}}{12}. \tag{6.96b}$$

In general, the autocorrelation sequence of a quantization noise source is, according to the first assumption,

$$\phi_{ee}[n] = \sigma_e^2 \,\delta[n] + m_e^2. \tag{6.97}$$

In the case of rounding, which we will assume for convenience henceforth, $m_e = 0$, so the autocorrelation function is $\phi_{ee}[n] = \sigma_e^2 \,\delta[n]$, and the power spectrum is $\Phi_{ee}(e^{j\omega}) = \sigma_e^2$ for $|\omega| \le \pi$. In this case, the variance and the average power are identical. In the

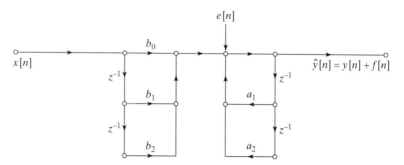

Figure 6.59   Linear-noise model for direct form I with noise sources combined.

case of truncation, the mean is not zero, so average-power results derived for rounding must be corrected by computing the mean of the signal and adding its square to the average-power results for rounding.

With this model for each of the noise sources in Figure 6.57(c), we can now proceed to determine the effect of the quantization noise on the output of the system. To aid us in doing this, it is helpful to observe that all of the noise sources in that figure are effectively injected between the part of the system that implements the zeros and the part that implements the poles. Thus, Figure 6.59 is equivalent to Figure 6.57(c) if $e[n]$ in Figure 6.59 is

$$e[n] = e_0[n] + e_1[n] + e_2[n] + e_3[n] + e_4[n]. \tag{6.98}$$

Since we assume that all the noise sources are independent of the input and independent of each other, the variance of the combined noise sources for the 2nd-order direct form I case is

$$\sigma_e^2 = \sigma_{e_0}^2 + \sigma_{e_1}^2 + \sigma_{e_2}^2 + \sigma_{e_3}^2 + \sigma_{e_4}^2 = 5 \cdot \frac{2^{-2B}}{12}, \tag{6.99}$$

and for the general direct form I case, it is

$$\sigma_e^2 = (M + 1 + N)\frac{2^{-2B}}{12}. \tag{6.100}$$

To obtain an expression for the output noise, we note from Figure 6.59 that the system has two inputs, $x[n]$ and $e[n]$, and since the system is now assumed to be linear, the output can be represented as $\hat{y}[n] = y[n] + f[n]$, where $y[n]$ is the response of the ideal unquantized system to the input sequence $x[n]$ and $f[n]$ is the response of the system to the input $e[n]$. The output $y[n]$ is given by the difference equation (6.90), but since $e[n]$ is injected after the zeros and before the poles, the output noise satisfies the difference equation

$$f[n] = \sum_{k=1}^{N} a_k f[n-k] + e[n]; \tag{6.101}$$

i.e., the properties of the output noise in the direct form I implementation depend only on the poles of the system.

To determine the mean and variance of the output noise sequence, we can use some results from Section 2.10. Consider a linear system with system function $H_{ef}(z)$

with a white-noise input $e[n]$ and corresponding output $f[n]$. Then, from Eqs. (2.184) and (2.185), the mean of the output is

$$m_f = m_e \sum_{n=-\infty}^{\infty} h_{ef}[n] = m_e H_{ef}(e^{j0}). \qquad (6.102)$$

Since $m_e = 0$ for rounding, the mean of the output will be zero, so we need not be concerned with the mean value of the noise if we assume rounding. From Eqs. (6.97) and (2.190), it follows that, because, for rounding, $e[n]$ is a zero-mean white-noise sequence, the power density spectrum of the output noise is simply

$$P_{ff}(\omega) = \Phi_{ff}(e^{j\omega}) = \sigma_e^2 |H_{ef}(e^{j\omega})|^2. \qquad (6.103)$$

From Eq. (2.192), the variance of the output noise can be shown to be

$$\sigma_f^2 = \frac{1}{2\pi} \int_{-\pi}^{\pi} P_{ff}(\omega)d\omega = \sigma_e^2 \frac{1}{2\pi} \int_{-\pi}^{\pi} |H_{ef}(e^{j\omega})|^2 d\omega. \qquad (6.104)$$

Using Parseval's theorem in the form of Eq. (2.162), we can also express σ_f^2 as

$$\sigma_f^2 = \sigma_e^2 \sum_{n=-\infty}^{\infty} |h_{ef}[n]|^2. \qquad (6.105)$$

When the system function corresponding to $h_{ef}[n]$ is a rational function, as it will always be for difference equations of the type considered in this chapter, we can use Eq. (A.66) in Appendix A for evaluating infinite sums of squares of the form of Eq. (6.105).

We will use the results summarized in Eqs. (6.102)–(6.105) often in our analysis of quantization noise in linear systems. For example, for the direct form I system of Figure 6.59, $H_{ef}(z) = 1/A(z)$; i.e., the system function from the point where all the noise sources are injected to the output consists only of the poles of the system function $H(z)$ in Eq. (6.91). Thus, we conclude that, in general, the total output variance owing to internal round-off or truncation is

$$\sigma_f^2 = (M+1+N)\frac{2^{-2B}}{12}\frac{1}{2\pi}\int_{-\pi}^{\pi}\frac{d\omega}{|A(e^{j\omega})|^2}$$
$$= (M+1+N)\frac{2^{-2B}}{12}\sum_{n=-\infty}^{\infty}|h_{ef}[n]|^2, \qquad (6.106)$$

where $h_{ef}[n]$ is the impulse response corresponding to $H_{ef}(z) = 1/A(z)$. The use of the preceding results is illustrated by the following examples.

Example 6.11   Round-off Noise in a 1st-Order System

Suppose that we wish to implement a stable system having the system function

$$H(z) = \frac{b}{1-az^{-1}}, \qquad |a| < 1. \qquad (6.107)$$

Figure 6.60 shows the flow graph of the linear-noise model for the implementation in which products are quantized before addition. Each noise source is filtered by the system from $e[n]$ to the output, for which the impulse response is $h_{ef}[n] = a^n u[n]$.

Since $M = 0$ and $N = 1$ for this example, from Eq. (6.103), the power spectrum of the output noise is

$$P_{ff}(\omega) = 2\frac{2^{-2B}}{12}\left(\frac{1}{1 + a^2 - 2a\cos\omega}\right), \tag{6.108}$$

and the total noise variance at the output is

$$\sigma_f^2 = 2\frac{2^{-2B}}{12}\sum_{n=0}^{\infty}|a|^{2n} = 2\frac{2^{-2B}}{12}\left(\frac{1}{1 - |a|^2}\right). \tag{6.109}$$

From Eq. (6.109), we see that the output noise variance increases as the pole at $z = a$ approaches the unit circle. Thus, to maintain the noise variance below a specified level as $|a|$ approaches unity, we must use longer word lengths. The following example also illustrates this point.

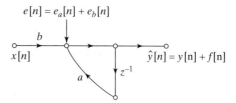

Figure 6.60   1st-order linear noise model.

Example 6.12   Round-off Noise in a 2nd-Order System

Consider a stable 2nd-order direct form I system with system function

$$H(z) = \frac{b_0 + b_1 z^{-1} + b_2 z^{-2}}{(1 - re^{j\theta}z^{-1})(1 - re^{-j\theta}z^{-1})}. \tag{6.110}$$

The linear-noise model for this system is shown in Figure 6.57(c), or equivalently, Figure 6.59, with $a_1 = 2r\cos\theta$ and $a_2 = -r^2$. In this case, the total output noise power can be expressed in the form

$$\sigma_f^2 = 5\frac{2^{-2B}}{12}\frac{1}{2\pi}\int_{-\pi}^{\pi}\frac{d\omega}{|(1 - re^{j\theta}e^{-j\omega})(1 - re^{-j\theta}e^{-j\omega})|^2}. \tag{6.111}$$

Using Eq. (A.66) in Appendix A, the output noise power is found to be

$$\sigma_f^2 = 5\frac{2^{-2B}}{12}\left(\frac{1 + r^2}{1 - r^2}\right)\frac{1}{r^4 + 1 - 2r^2\cos 2\theta}. \tag{6.112}$$

As in Example 6.11, we see that as the complex conjugate poles approach the unit circle ($r \rightarrow 1$), the total output noise variance increases, thus requiring longer word lengths to maintain the variance below a prescribed level.

The techniques of analysis developed so far for the direct form I structure can also be applied to the direct form II structure. The nonlinear difference equations for the direct form II structure are of the form

$$\hat{w}[n] = \sum_{k=1}^{N} Q[a_k \hat{w}[n-k]] + x[n], \tag{6.113a}$$

$$\hat{y}[n] = \sum_{k=0}^{M} Q[b_k \hat{w}[n-k]]. \tag{6.113b}$$

Figure 6.61(a) shows the linear-noise model for a 2nd-order direct form II system. A noise source has been introduced after each multiplication, indicating that the products are quantized to $(B+1)$ bits before addition. Figure 6.61(b) shows an equivalent linear model, wherein we have moved the noise sources resulting from implementation of the poles and combined them into a single noise source $e_a[n] = e_3[n] + e_4[n]$ at the input. Likewise, the noise sources due to implementation of the zeros are combined into the single noise source $e_b[n] = e_0[n] + e_1[n] + e_2[n]$ that is added directly to the output. From this equivalent model, it follows that for M zeros and N poles and rounding ($m_e = 0$), the power spectrum of the output noise is

$$P_{ff}(\omega) = N \frac{2^{-2B}}{12} |H(e^{j\omega})|^2 + (M+1) \frac{2^{-2B}}{12}, \tag{6.114}$$

and the output noise variance is

$$\sigma_f^2 = N \frac{2^{-2B}}{12} \frac{1}{2\pi} \int_{-\pi}^{\pi} |H(e^{j\omega})|^2 d\omega + (M+1) \frac{2^{-2B}}{12}$$

$$= N \frac{2^{-2B}}{12} \sum_{n=-\infty}^{\infty} |h[n]|^2 + (M+1) \frac{2^{-2B}}{12}. \tag{6.115}$$

That is, the white noise produced in implementing the poles is filtered by the entire system, whereas the white noise produced in implementing the zeros is added directly to the output of the system. In writing Eq. (6.115), we have assumed that the N noise sources at the input are independent, so that their sum has N times the variance of a single quantization noise source. The same assumption was made about the $(M+1)$ noise sources at the output. These results are easily modified for two's-complement truncation. Recall from Eqs. (6.95a)–(6.95b) and Eqs. (6.96a)–(6.96b) that the variance of a truncation noise source is the same as that of a rounding noise source, but the mean of a truncation noise source is not zero. Consequently, the formulas in Eqs. (6.106) and (6.115) for the total output noise variance also hold for truncation. However, the output noise will have a nonzero average value that can be computed using Eq. (6.102).

A comparison of Eq. (6.106) with Eq. (6.115) shows that the direct form I and direct form II structures are affected differently by the quantization of products in implementing the corresponding difference equations. In general, other equivalent structures such as cascade, parallel, and transposed forms will have a total output noise variance different from that of either of the direct form structures. However, even though Eqs. (6.106) and (6.115) are different, we cannot say which system will have the smaller output noise variance unless we know specific values for the coefficients of the system. In other words, it is not possible to state that a particular structural form will always produce the least output noise.

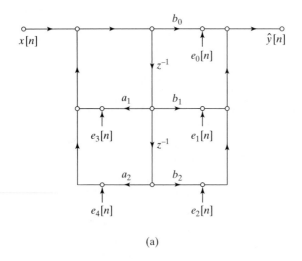

(a)

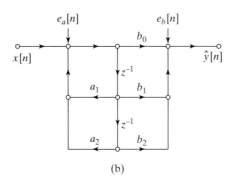

(b)

Figure 6.61 Linear-noise models for direct form II. (a) Showing quantization of individual products. (b) With noise sources combined.

It is possible to improve the noise performance of the direct form systems (and therefore cascade and parallel forms as well) by using a $(2B+1)$-bit adder to accumulate the sum of products required in both direct form systems. For example, for the direct form I implementation, we could use a difference equation of the form

$$\hat{y}[n] = Q\left[\sum_{k=1}^{N} a_k \hat{y}[n-k] + \sum_{k=0}^{M} b_k x[n-k]\right]; \tag{6.116}$$

i.e., the sums of products are accumulated with $(2B+1)$- or $(2B+2)$-bit accuracy, and the result is quantized to $(B+1)$ bits for output and storage in the delay memory. In the direct form I case, this means that the quantization noise is still filtered by the poles, but the factor $(M+1+N)$ in Eq. (6.106) is replaced by unity. Similarly, for the direct form II realization, the difference equations (6.113a)–(6.113b) can respectively be replaced by

$$\hat{w}[n] = Q\left[\sum_{k=1}^{N} a_k \hat{w}[n-k] + x[n]\right] \tag{6.117a}$$

and

$$\hat{y}[n] = Q\left[\sum_{k=0}^{M} b_k \hat{w}[n-k]\right]. \tag{6.117b}$$

This implies a single noise source at both the input and output, so the factors N and $(M+1)$ in Eq. (6.115) are each replaced by unity. Thus, the double-length accumulator provided in most DSP chips can be used to significantly reduce quantization noise in direct form systems.

6.9.2 Scaling in Fixed-Point Implementations of IIR Systems

The possibility of overflow is another important consideration in the implementation of IIR systems using fixed-point arithmetic. If we follow the convention that each fixed-point number represents a fraction (possibly times a known scale factor), each node in the structure must be constrained to have a magnitude less than unity to avoid overflow. If $w_k[n]$ denotes the value of the kth node variable, and $h_k[n]$ denotes the impulse response from the input $x[n]$ to the node variable $w_k[n]$, then

$$|w_k[n]| = \left| \sum_{m=-\infty}^{\infty} x[n-m]h_k[m] \right|. \qquad (6.118)$$

The bound

$$|w_k[n]| \le x_{\max} \sum_{m=-\infty}^{\infty} |h_k[m]| \qquad (6.119)$$

is obtained by replacing $x[n-m]$ by its maximum value $x_{\max}$ and using the fact that the magnitude of a sum is less than or equal to the sum of the magnitudes of the summands. Therefore, a sufficient condition for $|w_k[n]| < 1$ is

$$x_{\max} < \frac{1}{\displaystyle\sum_{m=-\infty}^{\infty} |h_k[m]|} \qquad (6.120)$$

for all nodes in the flow graph. If $x_{\max}$ does not satisfy Eq. (6.120), then we can multiply $x[n]$ by a scaling multiplier s at the input to the system so that $sx_{\max}$ satisfies Eq. (6.120) for all nodes in the flow graph; i.e.,

$$sx_{\max} < \frac{1}{\displaystyle\max_k \left[\sum_{m=-\infty}^{\infty} |h_k[m]| \right]}. \qquad (6.121)$$

Scaling the input in this way guarantees that overflow never occurs at any of the nodes in the flow graph. Equation (6.120) is necessary as well as sufficient, since an input always exists such that Eq. (6.119) is satisfied with equality. (See Eq. (2.70) in the discussion of stability in Section 2.4.) However, Eq. (6.120) leads to a very conservative scaling of the input for most signals.

Another approach to scaling is to assume that the input is a narrowband signal, modeled as $x[n] = x_{\max} \cos \omega_0 n$. In this case, the node variables will have the form

$$w_k[n] = |H_k(e^{j\omega_0})|x_{\max} \cos(\omega_0 n + \angle H_k(e^{j\omega_0})). \qquad (6.122)$$

Therefore, overflow is avoided for *all* sinusoidal signals if

$$\max_{k, |\omega| \le \pi} |H_k(e^{j\omega})|x_{\max} < 1 \qquad (6.123)$$

or if the input is scaled by the scale factor s such that

$$sx_{\max} < \frac{1}{\max\limits_{k,|\omega|\leq\pi} |H_k(e^{j\omega})|}. \tag{6.124}$$

Still another scaling approach is based on the energy $E = \Sigma_n|x[n]|^2$ of the input signal. We can derive the scale factor in this case by applying the Schwarz inequality (see Bartle, 2000) to obtain the following inequality relating the square of the node signal to the energies of the input signal and the node impulse response:

$$|w_k[n]|^2 = \left| \frac{1}{2\pi} \int_{-\pi}^{\pi} H_k(e^{j\omega})X(e^{j\omega})e^{j\omega n}d\omega \right|^2$$

$$\leq \left(\frac{1}{2\pi} \int_{-\pi}^{\pi} |H_k(e^{j\omega})|^2d\omega \right) \left(\frac{1}{2\pi} \int_{-\pi}^{\pi} |X(e^{j\omega})|^2d\omega \right). \tag{6.125}$$

Therefore, if we scale the input sequence values by s and apply Parseval's theorem, we see that $|w_k[n]|^2 < 1$ for all nodes k if

$$s^2 \left(\sum_{n=-\infty}^{\infty} |x[n]|^2 \right) = s^2 E < \frac{1}{\max\limits_k \left[\sum\limits_{n=-\infty}^{\infty} |h_k[n]|^2 \right]}. \tag{6.126}$$

Since it can be shown that for the kth node,

$$\left\{ \sum_{n=-\infty}^{\infty} |h_k[n]|^2 \right\}^{1/2} \leq \max_\omega |H_k(e^{j\omega})| \leq \sum_{n=-\infty}^{\infty} |h_k[n]|, \tag{6.127}$$

it follows that (for most input signals) Eqs. (6.121), (6.124), and (6.126) give three decreasingly conservative ways of scaling the input to a digital filter (equivalently decreasing the gain of the filter). Of the three, Eq. (6.126) is generally the easiest to evaluate analytically because the partial fraction method of Appendix A can be used; however use of Eq. (6.126) requires an assumption about the mean-squared value of the signal, E. On the other hand, Eq. (6.121) is difficult to evaluate analytically, except for the simplest systems. Of course, if the filter coefficients are fixed numbers, the scale factors can be estimated by computing the impulse response or frequency response numerically.

If the input must be scaled down ($s < 1$), the signal-to-noise ratio (SNR) at the output of the system will be reduced because the signal power is reduced, but the noise power is dependent only on the rounding operation. Figure 6.62 shows 2nd-order direct form I and direct form II systems with scaling multipliers at the input. In determining the scaling multiplier for these systems, it is not necessary to examine each node in the flow graph. Some nodes do not represent addition and thus cannot overflow. Other nodes represent partial sums. If we use nonsaturation two's-complement arithmetic, such nodes are permitted to overflow if certain key nodes do not. For example, in Figure 6.62(a), we can focus on the node enclosed by the dashed circle. In the figure, the scaling multiplier is shown combined with the b_ks, so that the noise source is the same

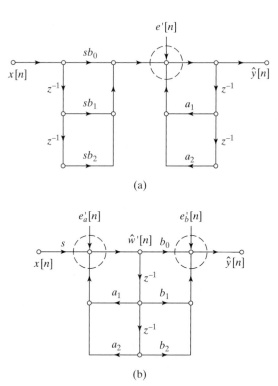

Figure 6.62 Scaling of direct form systems. (a) Direct form I. (b) Direct form II.

as in Figure 6.59; i.e., it has five times the power of a single quantization noise source.[10] Since the noise source is again filtered only by the poles, the output noise power is the same in Figures 6.59 and 6.62(a). However, the overall system function of the system in Figure 6.62(a) is $sH(z)$ instead of $H(z)$, so the unquantized component of the output $\hat{y}[n]$ is $sy[n]$ instead of $y[n]$. Since the noise is injected after the scaling, the ratio of signal power to noise power in the scaled system is s^2 times the SNR for Figure 6.59. Because $s < 1$ if scaling is required to avoid overflow, the SNR is reduced by scaling.

The same is true for the direct form II system of Figure 6.62(b). In this case, we must determine the scaling multiplier to avoid overflow at both of the circled nodes. Again, the overall gain of the system is s times the gain of the system in Figure 6.61(b), but it may be necessary to implement the scaling multiplier explicitly in this case to avoid overflow at the node on the left. This scaling multiplier adds an additional noise component to $e_a[n]$, so the noise power at the input is, in general, $(N + 1)2^{-2B}/12$. Otherwise, the noise sources are filtered by the system in exactly the same way in both Figure 6.61(b) and Figure 6.62(b). Therefore, the signal power is multiplied by s^2, and the noise power at the output is again given by Eq. (6.115), with N replaced by $(N+1)$. The SNR is again reduced if scaling is required to avoid overflow.

---

[10]This eliminates a separate scaling multiplication and quantization noise source. However, scaling (and quantizing) the b_ks can change the frequency response of the system. If a separate input scaling multiplier precedes the implementation of the zeros in Figure 6.62(a), then an additional quantization noise source would contribute to the output noise after going through the entire system $H(z)$.

Example 6.13   Interaction Between Scaling and Round-off Noise

To illustrate the interaction of scaling and round-off noise, consider the system of Example 6.11 with system function given by Eq. (6.107). If the scaling multiplier is combined with the coefficient b, we obtain the flow graph of Figure 6.63 for the scaled system. Suppose that the input is white noise with amplitudes uniformly distributed between -1 and $+1$. Then the total signal variance is $\sigma_x^2 = 1/3$. To guarantee no overflow in computing $\hat{y}[n]$, we use Eq. (6.121) to compute the scale factor

$$s = \frac{1}{\displaystyle\sum_{n=0}^{\infty} |b|\,|a|^n} = \frac{1 - |a|}{|b|}. \tag{6.128}$$

The output noise variance was determined in Example 6.11 to be

$$\sigma_f^2 = 2\frac{2^{-2B}}{12}\frac{1}{1 - a^2} \tag{6.129}$$

and since we again have two $(B+1)$-bit rounding operations, the noise power at the output is the same, i.e., $\sigma_{f'}^2 = \sigma_f^2$. The variance of the output $y'[n]$ due to the scaled input $sx[n]$ is

$$\sigma_{y'}^2 = \left(\frac{1}{3}\right)\frac{s^2 b^2}{1 - a^2} = s^2 \sigma_y^2. \tag{6.130}$$

Therefore, the SNR at the output is

$$\frac{\sigma_{y'}^2}{\sigma_{f'}^2} = s^2\frac{\sigma_y^2}{\sigma_f^2} = \left(\frac{1 - |a|}{|b|}\right)^2 \frac{\sigma_y^2}{\sigma_f^2}. \tag{6.131}$$

As the pole of the system approaches the unit circle, the SNR decreases because the quantization noise is amplified by the system and because the high gain of the system forces the input to be scaled down to avoid overflow. Again, we see that overflow and quantization noise work in opposition to decrease the performance of the system.

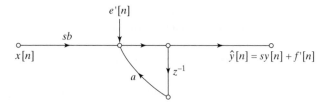

Figure 6.63   Scaled 1st-order system.

6.9.3 Example of Analysis of a Cascade IIR Structure

The previous results of this section can be applied directly to the analysis of either parallel or cascade structures composed of 2nd-order direct form subsystems. The interaction of scaling and quantization is particularly interesting in the cascade form. Our general comments on cascade systems will be interwoven with a specific example.

An elliptic lowpass filter was designed to meet the following specifications:

$$0.99 \le |H(e^{j\omega})| \le 1.01, \qquad |\omega| \le 0.5\pi,$$

$$|H(e^{j\omega})| \le 0.01, \qquad 0.56\pi \le |\omega| \le \pi.$$

The system function of the resulting system is

$$H(z) = 0.079459 \prod_{k=1}^{3} \left(\frac{1 + b_{1k}z^{-1} + z^{-2}}{1 - a_{1k}z^{-1} - a_{2k}z^{-2}} \right) = 0.079459 \prod_{k=1}^{3} H_k(z), \qquad (6.132)$$

where the coefficients are given in Table 6.6. Notice that all the zeros of $H(z)$ are on the unit circle in this example; however, that need not be the case in general.

Figure 6.64(a) shows a flow graph of a possible implementation of this system as a cascade of 2nd-order transposed direct form II subsystems. The gain constant, 0.079459, is such that the overall gain of the system is approximately unity in the passband, and it is assumed that this guarantees no overflow at the output of the system. Figure 6.64(a) shows the gain constant placed at the input to the system. This approach reduces the amplitude of the signal immediately, with the result that the subsequent filter sections must have high gain to produce an overall gain of unity. Since the quantization noise sources are introduced after the gain of 0.079459 but are likewise amplified by the rest of the system, this is not a good approach. Ideally, the overall gain constant, being less than unity, should be placed at the very end of the cascade, so that the signal and noise will be attenuated by the same amount. However, this creates the possibility of overflow along the cascade. Therefore, a better approach is to distribute the gain among the three stages of the system, so that overflow is just avoided at each stage of the cascade. This distribution is represented by

$$H(z) = s_1 H_1(z) s_2 H_2(z) s_3 H_3(z), \qquad (6.133)$$

where $s_1 s_2 s_3 = 0.079459$. The scaling multipliers can be incorporated into the coefficients of the numerators of the individual system functions $H_k'(z) = s_k H_k(z)$, as in

$$H(z) = \prod_{k=1}^{3} \left(\frac{b_{0k}' + b_{1k}'z^{-1} + b_{2k}'z^{-2}}{1 - a_{1k}z^{-1} - a_{2k}z^{-2}} \right) = \prod_{k=1}^{3} H_k'(z), \qquad (6.134)$$

where $b_{0k}' = b_{2k}' = s_k$ and $b_{1k}' = s_k b_{1k}$. The resulting scaled system is depicted in Figure 6.64(b).

Also shown in Figure 6.64(b) are quantization noise sources representing the quantization of the products before addition. Figure 6.64(c) shows an equivalent noise model,

TABLE 6.6 COEFFICIENTS FOR ELLIPTIC LOWPASS FILTER IN CASCADE FORM

k	a_{1k}	a_{2k}	b_{1k}
1	0.478882	−0.172150	1.719454
2	0.137787	−0.610077	0.781109
3	−0.054779	−0.902374	0.411452

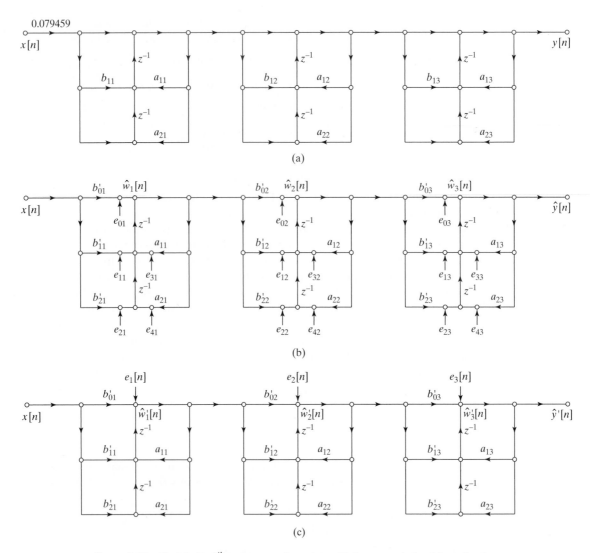

Figure 6.64   Models for 6th-order cascade system with transposed direct form II subsystems. (a) Infinite-precision model. (b) Linear-noise model for scaled system, showing quantization of individual multiplications. (c) Linear-noise model with noise sources combined.

for which it is recognized that all the noise sources in a particular section are filtered only by the poles of that section (and the subsequent subsystems). Figure 6.64(c) also uses the fact that delayed white-noise sources are still white noise and are independent of all the other noise sources, so that all five sources in a subsection can be combined into a single noise source having five times the variance of a single quantization noise source.[11] Since the noise sources are assumed independent, the variance of the output

---

[11]This discussion can be generalized to show that the transposed direct form II has the same noise behavior as the direct form I system.

noise is the sum of the variances owing to the three noise sources in Figure 6.64(c). Therefore, for rounding, the power spectrum of the output noise is

$$P_{f'f'}(\omega) = 5\frac{2^{-2B}}{12}\left[\frac{s_2^2|H_2(e^{j\omega})|^2 s_3^2|H_3(e^{j\omega})|^2}{|A_1(e^{j\omega})|^2} + \frac{s_3^2|H_3(e^{j\omega})|^2}{|A_2(e^{j\omega})|^2} + \frac{1}{|A_3(e^{j\omega})|^2}\right], \quad (6.135)$$

and the total output noise variance is

$$\sigma_{f'}^2 = 5\frac{2^{-2B}}{12}\left[\frac{1}{2\pi}\int_{-\pi}^{\pi}\frac{s_2^2|H_2(e^{j\omega})|^2 s_3^2|H_3(e^{j\omega})|^2}{|A_1(e^{j\omega})|^2}d\omega\right.$$

$$\left. + \frac{1}{2\pi}\int_{-\pi}^{\pi}\frac{s_3^2|H_3(e^{j\omega})|^2}{|A_2(e^{j\omega})|^2}d\omega + \frac{1}{2\pi}\int_{-\pi}^{\pi}\frac{1}{|A_3(e^{j\omega})|^2}d\omega\right]. \quad (6.136)$$

If a double-length accumulator is available, it would be necessary to quantize only the sums that are the inputs to the delay elements in Figure 6.64(b). In this case the factor of 5 in Eqs. (6.135) and (6.136) would be changed to 3. Furthermore, if a double-length register were used to implement the delay elements, only the variables $\hat{w}_k[n]$ would have to be quantized, and there would be only one quantization noise source per subsystem. In that case, the factor of 5 in Eqs. (6.135) and (6.136) would be changed to unity.

The scale factors s_k are chosen to avoid overflow at points along the cascade system. We will use the scaling convention of Eq. (6.124). Therefore, the scaling constants are chosen to satisfy

$$s_1 \max_{|\omega|\leq\pi}|H_1(e^{j\omega})| < 1, \quad (6.137a)$$

$$s_1 s_2 \max_{|\omega|\leq\pi}|H_1(e^{j\omega})H_2(e^{j\omega})| < 1, \quad (6.137b)$$

$$s_1 s_2 s_3 = 0.079459. \quad (6.137c)$$

The last condition ensures that there will be no overflow at the output of the system for unit-amplitude sinusoidal inputs, because the maximum overall gain of the filter is unity. For the coefficients of Table 6.6, the resulting scale factors are $s_1 = 0.186447$, $s_2 = 0.529236$, and $s_3 = 0.805267$.

Equations (6.135) and (6.136) show that the shape of the output noise power spectrum and the total output noise variance depends on the way that zeros and poles are paired to form the 2nd-order sections and on the order of the 2nd-order sections in the cascade form realization. Indeed, it is easily seen that, for N sections, there are ($N!$) ways to pair the poles and zeros, and there are likewise ($N!$) ways to order the resulting 2nd-order sections, a total of ($N!$)^2 different systems. In addition, we can choose either direct form I or direct form II (or their transposes) for the implementation of the 2nd-order sections. In our example, this implies that there are 144 different cascade systems to consider, if we wish to find the system with the lowest output noise variance. For five cascaded sections, there would be 57,600 different systems. Clearly, the complete analysis of even low-order systems is a tedious task, since an expression like Eq. (6.136) must be evaluated for each pairing and ordering. Hwang (1974) used dynamic programming and Liu and Peled (1975) used a heuristic approach to reduce the amount of computation.

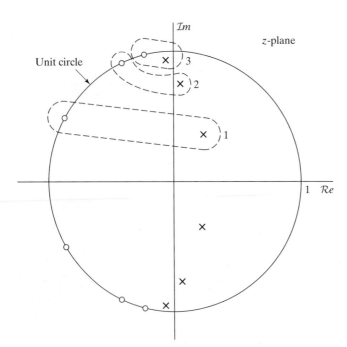

Figure 6.65   Pole–zero plot for 6th-order system of Figure 6.64, showing pairing of poles and zeros.

Even though finding the best pairing and ordering may require computer optimization, Jackson (1970a, 1970b, 1996) found that good results are almost always obtained by applying simple rules of the following form:

1. The pole that is closest to the unit circle should be paired with the zero that is closest to it in the z-plane.
2. Rule 1 should be repeatedly applied until all the poles and zeros have been paired.
3. The resulting 2nd-order sections should be ordered according to either increasing closeness to the unit circle or decreasing closeness to the unit circle.

The pairing rules are based on the observation that subsystems with high peak gain are undesirable because they can cause overflow and because they can amplify quantization noise. Pairing a pole that is close to the unit circle with an adjacent zero tends to reduce the peak gain of that section. These heuristic rules are implemented in design and analysis tools such as the MATLAB function zp2sos.

One motivation for rule 3 is suggested by Eq. (6.135). We see that the frequency responses of some of the subsystems appear more than once in the equation for the power spectrum of the output noise. If we do not want the output noise variance spectrum to have a high peak around a pole that is close to the unit circle, then it is advantageous to have the frequency-response component owing to that pole not appear frequently in Eq. (6.135). This suggests moving such "high Q" poles to the beginning of the cascade. On the other hand, the frequency response from the input to a particular node in the flow graph will involve a product of the frequency responses of the subsystems that precede the node. Thus, to avoid excessive reduction of the signal level in the early stages of the cascade, we should place the poles that are close to the unit circle last in order.

Clearly then, the question of ordering hinges on a variety of considerations, including total output noise variance and the shape of the output noise spectrum. Jackson (1970a, 1970b) used L_p norms to quantify the analysis of the pairing-and-ordering problem and gave a much more detailed set of "rules of thumb" for obtaining good results without having to evaluate all possibilities.

The pole–zero plot for the system in our example is shown in Figure 6.65. The paired poles and zeros are circled. In this case, we have chosen to order the sections from least peaked to most peaked frequency response. Figure 6.66 illustrates how the frequency responses of the individual sections combine to form the overall frequency response. Figures 6.66(a)–(c) show the frequency responses of the individual unscaled subsystems. Figures 6.66(d)–(f) show how the overall frequency response is built up. Notice that Figures 6.66(d)–(f) demonstrate that the scaling Eqs. (6.137a)–(6.137c) ensure that the maximum gain from the input to the output of any subsystem is less than unity. The solid curve in Figure 6.67 shows the power spectrum of the output noise for the ordering 123 (least peaked to most peaked). We assume that $B + 1 = 16$ for the plot. Note that the spectrum peaks in the vicinity of the pole that is closest to the unit circle. The dotted curve shows the power spectrum of the output noise when the section order is reversed (i.e., 321). Since section 1 has high gain at low frequencies, the noise spectrum is appreciably larger at low frequencies and slightly lower around the peak. The high Q pole still filters the noise sources of the first section in the cascade, so it still tends to dominate the spectrum. The total noise power for the two orderings turns out to be almost the same in this case.

The example we have just presented shows the complexity of the issues that arise in fixed-point implementations of cascade IIR systems. The parallel form is somewhat simpler because the issue of pairing and ordering does not arise. However, scaling is still required to avoid overflow in individual 2nd-order subsystems and when the outputs of the subsystems are summed to produce the overall output. The techniques that we have developed must therefore be applied for the parallel form as well. Jackson (1996) discusses the analysis of the parallel form in detail and concludes that its total output noise power is typically comparable to that of the best pairings and orderings of the cascade form. Even so, the cascade form is more common, because, for widely used IIR filters such that the zeros of the system function are on the unit circle, the cascade form can be implemented with fewer multipliers and with more control over the locations of the zeros.

6.9.4 Analysis of Direct-Form FIR Systems

Since the direct form I and direct form II IIR systems include the direct form FIR system as a special case (i.e., the case where all coefficients a_k in Figures 6.14 and 6.15 are zero), the results and analysis techniques of Sections 6.9.1 and 6.9.2 apply to FIR systems if we eliminate all reference to the poles of the system function and eliminate the feedback paths in all the signal flow graphs.

The direct form FIR system is simply the discrete convolution

$$y[n] = \sum_{k=0}^{M} h[k]x[n-k]. \tag{6.138}$$

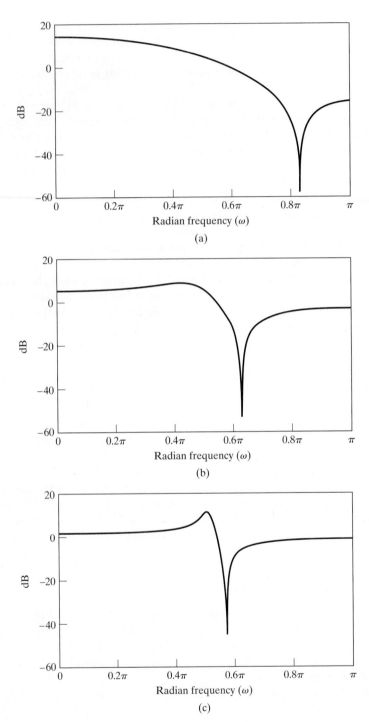

Figure 6.66 Frequency-response functions for example system.
(a) $20 \log_{10} |H_1(e^{j\omega})|$.
(b) $20 \log_{10} |H_2(e^{j\omega})|$.
(c) $20 \log_{10} |H_3(e^{j\omega})|$.

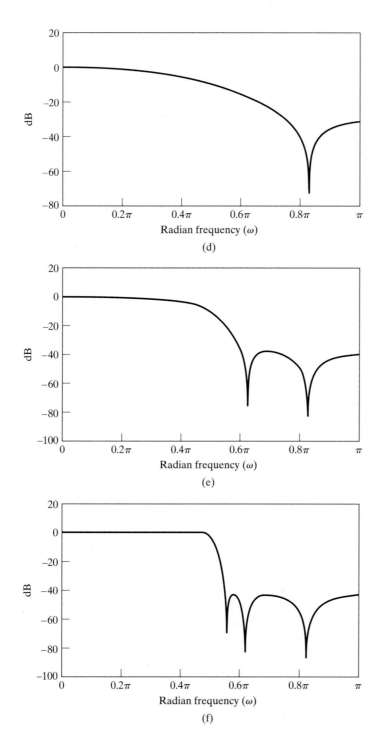

Figure 6.66 *(continued)*
(d) $20 \log_{10} |H_1'(e^{j\omega})|$.
(e) $20 \log_{10} |H_1'(e^{j\omega}) H_2'(e^{j\omega})|$.
(f) $20 \log_{10} |H_1'(e^{j\omega}) H_2'(e^{j\omega}) H_3'(e^{j\omega})|$
$= 20 \log_{10} |H'(e^{j\omega})|$.

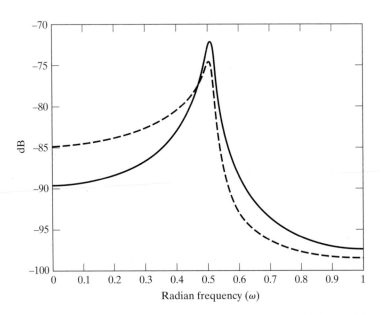

Figure 6.67 Output noise power spectrum for 123 ordering (solid line) and 321 ordering (dashed line) of 2nd-order sections.

Figure 6.68(a) shows the ideal unquantized direct form FIR system, and Figure 6.68(b) shows the linear-noise model for the system, assuming that all products are quantized before additions are performed. The effect is to inject $(M + 1)$ white-noise sources directly at the output of the system, so that the total output noise variance is

$$\sigma_f^2 = (M + 1)\frac{2^{-2B}}{12}. \tag{6.139}$$

This is exactly the result we would obtain by setting $N = 0$ and $h_{ef}[n] = \delta[n]$ in Eqs. (6.106) and (6.115). When a double-length accumulator is available, we would need to quantize only the output. Therefore, the factor $(M+1)$ in Eq. (6.139) would be replaced by unity. This makes the double-length accumulator a very attractive hardware feature for implementing FIR systems.

Overflow is also a problem for fixed-point realizations of FIR systems in direct form. For two's-complement arithmetic, we need to be concerned only about the size of the output, since all the other sums in Figure 6.68(b) are partial sums. Thus, the impulse-response coefficients can be scaled to reduce the possibility of overflow. Scaling multipliers can be determined using any of the alternatives discussed in Section 6.9.2. Of course, scaling the impulse response reduces the gain of the system, and therefore the SNR at the output is reduced as discussed in that section.

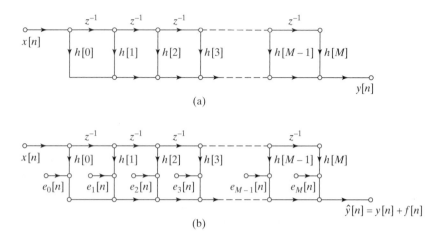

Figure 6.68   Direct form realization of an FIR system. (a) Infinite-precision model. (b) Linear-noise model.

Example 6.14   Scaling Considerations for the FIR System in Section 6.8.5

The impulse-response coefficients for the system in Section 6.8.5 are given in Table 6.5. Simple calculations show, and from Figure 6.54(b) we see, that

$$\sum_{n=0}^{27} |h[n]| = 1.751352,$$

$$\left(\sum_{n=0}^{27} |h[n]|^2 \right)^{1/2} = 0.679442,$$

$$\max_{|\omega| \le \pi} |H(e^{j\omega})| \approx 1.009.$$

These numbers satisfy the ordering relationship in Eq. (6.127). Thus, the system, as given, is scaled so that overflow is theoretically possible for a sinusoidal signal whose amplitude is greater than $1/1.009 = 0.9911$, but even so, overflow is unlikely for most signals. Indeed, since the filter has a linear phase, we can argue that, for wideband signals, since the gain in the passband is approximately unity, and the gain elsewhere is less than unity, the output signal should be smaller than the input signal.

In Section 6.5.3, we showed that linear-phase systems like the one in Example 6.14 can be implemented with about half the number of multiplications of the general FIR system. This is evident from the signal flow graphs of Figures 6.32 and 6.33. In these cases, it should be clear that the output noise variance would be halved if products were quantized before addition. However, the utilization of such structures involves a more complicated indexing algorithm than the direct form. The architecture of most DSP chips combines a double-length accumulator with an efficient pipelined multiply–

accumulate operation and simple looping control to optimize for the case of the direct form FIR system. For this reason, direct form FIR implementations are often most attractive, even compared with IIR filters that meet frequency-response specifications with fewer multiplications, since cascade or parallel structures do not permit long sequences of multiply-accumulate operations.

In Section 6.5.3, we discussed cascade realizations of FIR systems. The results and analysis techniques of Section 6.9.3 apply to these realizations; but for FIR systems with no poles, the pairing and ordering problem reduces to just an ordering problem. As in the case of IIR cascade systems, the analysis of all possible orderings can be very difficult if the system is composed of many subsystems. Chan and Rabiner (1973a, 1973b) studied this problem and found experimentally that the noise performance is relatively insensitive to the ordering. Their results suggest that a good ordering is an ordering for which the frequency response from each noise source to the output is relatively flat and for which the peak gain is small.

6.9.5 Floating-Point Realizations of Discrete-Time Systems

From the preceding discussion, it is clear that the limited dynamic range of fixed-point arithmetic makes it necessary to carefully scale the input and intermediate signal levels in fixed-point digital realizations of discrete-time systems. The need for such scaling can be essentially eliminated by using floating-point numeric representations and floating-point arithmetic.

In floating-point representations, a real number x is represented by the binary number $2^c \hat{x}_M$, where the exponent c of the scale factor is called the *characteristic* and $\hat{x}_M$ is a fractional part called the *mantissa*. Both the characteristic and the mantissa are represented explicitly as fixed-point binary numbers in floating-point arithmetic systems. Floating-point representations provide a convenient means for maintaining both a wide dynamic range and low quantization noise; however, quantization error manifests itself in a somewhat different way. Floating-point arithmetic generally maintains its high accuracy and wide dynamic range by adjusting the characteristic and normalizing the mantissa so that $0.5 \leq \hat{x}_M < 1$. When floating-point numbers are multiplied, their characteristics are added and their mantissas are multiplied. Thus, the mantissa must be quantized. When two floating-point numbers are added, their characteristics must be adjusted to be the same by moving the binary point of the mantissa of the smaller number. Hence, addition results in quantization, too. If we assume that the range of the characteristic is sufficient so that no numbers become larger than 2^c, then quantization affects only the mantissa, but the error in the mantissa is also scaled by 2^c. Thus, a quantized floating-point number is conveniently represented as

$$\hat{x} = x(1 + \varepsilon) = x + \varepsilon x. \tag{6.140}$$

By representing the quantization error as a fraction ε of x, we automatically represent the fact that the quantization error is scaled up and down with the signal level.

The aforementioned properties of floating-point arithmetic complicate the quantization error analysis of floating-point implementations of discrete-time systems. First, noise sources must be inserted both after each multiplication and after each addition. An important consequence is that, in contrast to fixed-point arithmetic, the *order* in which multiplications and additions are performed can sometimes make a big difference. More important for analysis, we can no longer justify the assumption that the quantization noise sources are white noise and are independent of the signal. In fact, in Eq. (6.140), the noise is expressed explicitly in terms of the signal. Therefore, we can no longer analyze the noise without making assumptions about the nature of the input signal. If the input is assumed to be known (e.g., white noise), a reasonable assumption is that the relative error ε is independent of x and is uniformly distributed white noise.

With these types of assumptions, useful results have been obtained by Sandberg (1967), Liu and Kaneko (1969), Weinstein and Oppenheim (1969), and Kan and Aggarwal (1971). In particular, Weinstein and Oppenheim, comparing floating-point and fixed-point realizations of 1st- and 2nd-order IIR systems, showed that if the number of bits representing the floating-point mantissa is equal to the length of the fixed-point word, then floating-point arithmetic leads to higher SNR at the output. Not surprisingly, the difference was found to be greater for poles close to the unit circle. However, additional bits are required to represent the characteristic, and the greater the desired dynamic range, the more bits are required for the characteristic. Also, the hardware to implement floating-point arithmetic is much more complex than that for fixed-point arithmetic. Therefore, the use of floating-point arithmetic entails an increased word length and increased complexity in the arithmetic unit. Its major advantage is that it essentially eliminates the problem of overflow, and if a sufficiently long mantissa is used, quantization also becomes much less of a problem. This translates into greater simplicity in system design and implementation.

Nowadays, digital filtering of multi-media signals is often implemented on personal computers or workstations that have very accurate floating point numerical representations and high speed arithmetic units. In such cases, the quantization issues discussed in Sections 6.7–6.9 are generally of little or no concern. However, in high volume systems, fixed point arithmetic is generally required to achieve low cost.

6.10 ZERO-INPUT LIMIT CYCLES IN FIXED-POINT REALIZATIONS OF IIR DIGITAL FILTERS

For stable IIR discrete-time systems implemented with infinite-precision arithmetic, if the excitation becomes zero and remains zero for n greater than some value n_0, the output for $n > n_0$ will decay asymptotically toward zero. For the same system, implemented with finite-register-length arithmetic, the output may continue to oscillate indefinitely with a periodic pattern while the input remains equal to zero. This effect is often referred to as *zero-input limit cycle behavior* and is a consequence either of the nonlinear quantizers in the feedback loop of the system or of overflow of additions. The limit cycle behavior of a digital filter is complex and difficult to analyze, and we will not attempt to treat the topic in any general sense. To illustrate the point, however, we will give two simple examples that will show how such limit cycles can arise.

6.10.1 Limit Cycles Owing to Round-off and Truncation

Successive round-off or truncation of products in an iterated difference equation can create repeating patterns. This is illustrated in the following example.

Example 6.15   Limit Cycle Behavior in a 1st-Order System

Consider the 1st-order system characterized by the difference equation

$$y[n] = ay[n-1] + x[n], \qquad |a| < 1. \tag{6.141}$$

The signal flow graph of this system is shown in Figure 6.69(a). Let us assume that the register length for storing the coefficient a, the input $x[n]$, and the filter node variable $y[n-1]$ is 4 bits (i.e., a sign bit to the left of the binary point and 3 bits to the right of the binary point). Because of the finite-length registers, the product $ay[n-1]$ must be rounded or truncated to 4 bits before being added to $x[n]$. The flow graph representing the actual realization based on Eq. (6.141) is shown in Figure 6.69(b). Assuming rounding of the product, the actual output $\hat{y}[n]$ satisfies the nonlinear difference equation

$$\hat{y}[n] = Q[a\hat{y}[n-1]] + x[n], \tag{6.142}$$

where $Q[\cdot]$ represents the rounding operation. Let us assume that $a = 1/2 = 0_\diamond 100$ and that the input is $x[n] = (7/8)\delta[n] = (0_\diamond 111)\delta[n]$. Using Eq. (6.142), we see that for $n = 0$, $\hat{y}[0] = 7/8 = 0_\diamond 111$. To obtain $\hat{y}[1]$, we multiply $\hat{y}[0]$ by a, obtaining the result $a\hat{y}[0] = 0_\diamond 011100$, a 7-bit number that must be rounded to 4 bits. This number, 7/16, is exactly halfway between the two 4-bit quantization levels 4/8 and 3/8. If we choose always to round upward in such cases, then $0_\diamond 011100$ rounded to 4 bits is $0_\diamond 100 = 1/2$. Since $x[1] = 0$, it follows that $\hat{y}[1] = 0_\diamond 100 = 1/2$. Continuing to iterate the difference equation gives $\hat{y}[2] = Q[a\hat{y}[1]] = 0_\diamond 010 = 1/4$ and $\hat{y}[3] = 0_\diamond 001 = 1/8$. In both these cases, no rounding is necessary. However, to obtain $\hat{y}[4]$, we must round the 7-bit number $a\hat{y}[3] = 0_\diamond 000100$ to $0_\diamond 001$. The same result is obtained for all values of $n \geq 3$. The output sequence for this example is shown in Figure 6.70(a). If $a = -1/2$, we can carry out the preceding computation again to demonstrate that the output is as shown in Figure 6.70(b). Thus, because of rounding of the product $a\hat{y}[n-1]$, the output reaches a constant value of 1/8 when $a = 1/2$ and a periodic steady-state oscillation between +1/8 and -1/8 when $a = -1/2$. These are periodic outputs similar to those that would be obtained from a 1st-order pole at $z = \pm 1$ instead of at $z = \pm 1/2$.

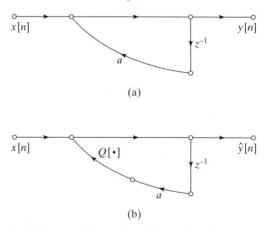

Figure 6.69   1st-order IIR system. (a) Infinite-precision linear system. (b) Nonlinear system due to quantization.

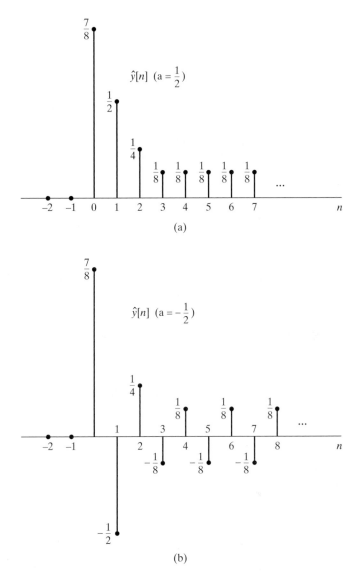

Figure 6.70 Response of the 1st-order system of Figure 6.69 to an impulse.
(a) $a = \frac{1}{2}$. (b) $a = -\frac{1}{2}$.

When $a = +1/2$, the period of the oscillation is 1, and when $a = -1/2$, the period of oscillation is 2. Such steady-state periodic outputs are called *limit cycles*, and their existence was first noted by Blackman (1965), who referred to the amplitude intervals to which such limit cycles are confined as *dead bands*. In this case, the dead band is $-2^{-B} \leq \hat{y}[n] \leq +2^{-B}$, where $B = 3$.

The foregoing example has illustrated that a zero-input limit cycle can result from rounding in a 1st-order IIR system. Similar results can be demonstrated for truncation. 2nd-order systems can also exhibit limit cycle behavior. In the case of parallel realizations of higher-order systems, the outputs of the individual 2nd-order systems are indepen-dent when the input is zero. In this case, one or more of the 2nd-order sections could contribute a limit cycle to the output sum. In the case of cascade realizations, only the first section has zero input; succeeding sections may exhibit their own characteristic limit cycle behavior, or they may appear to be simply filtering the limit cycle output of a previous section. For higher-order systems realized by other filter structures, the limit cycle behavior becomes more complex, as does its analysis.

In addition to giving an understanding of limit cycle effects in digital filters, the preceding results are useful when the zero-input limit cycle response of a system is the desired output. This is the case, for example, when one is concerned with digital sine wave oscillators for signal generation or for the generation of coefficients for calculation of the discrete Fourier transform.

6.10.2 Limit Cycles Owing to Overflow

In addition to the classes of limit cycles discussed in the preceding section, a more severe type of limit cycle can occur owing to overflow. The effect of overflow is to insert a gross error in the output, and in some cases the filter output thereafter oscillates between large-amplitude limits. Such limit cycles have been referred to as *overflow oscillation*. The problem of oscillations caused by overflow is discussed in detail by Ebert et al. (1969). Overflow oscillations are illustrated by the following example.

Example 6.16   Overflow Oscillations in a 2nd-Order System

Consider a 2nd-order system realized by the difference equation

$$\hat{y}[n] = x[n] + Q[a_1 \hat{y}[n-1]] + Q[a_2 \hat{y}[n-2]], \tag{6.143}$$

where $Q[\cdot]$ represents two's-complement rounding with a word length of 3 bits plus 1 bit for the sign. Overflow can occur with two's-complement addition of the rounded products. Suppose that $a_1 = 3/4 = 0_\diamond 110$ and $a_2 = -3/4 = 1_\diamond 010$, and assume that $x[n]$ remains equal to zero for $n \geq 0$. Furthermore, assume that $\hat{y}[-1] = 3/4 = 0_\diamond 110$ and $\hat{y}[-2] = -3/4 = 1_\diamond 010$. Now the output at sample $n = 0$ is

$$\hat{y}[0] = 0_\diamond 110 \times 0_\diamond 110 + 1_\diamond 010 \times 1_\diamond 010.$$

If we evaluate the products using two's-complement arithmetic, we obtain

$$\hat{y}[0] = 0_\diamond 100100 + 0_\diamond 100100,$$

and if we choose to round upward when a number is halfway between two quantization levels, the result of two's-complement addition is

$$\hat{y}[0] = 0_\diamond 101 + 0_\diamond 101 = 1_\diamond 010 = -\tfrac{3}{4}.$$

In this case the binary carry overflows into the sign bit, thus changing the positive sum into a negative number. Repeating the process gives

$$\hat{y}[1] = 1_\diamond 011 + 1_\diamond 011 = 0_\diamond 110 = \tfrac{3}{4}.$$

The binary carry resulting from the sum of the sign bits is lost, and the negative sum is mapped into a positive number. Clearly, $\hat{y}[n]$ will continue to oscillate between $+3/4$ and $-3/4$ until an input is applied. Thus, $\hat{y}[n]$ has entered a periodic limit cycle with a period of 2 and an amplitude of almost the full-scale amplitude of the implementation.

The preceding example illustrates how overflow oscillations occur. Much more complex behavior can be exhibited by higher-order systems, and other frequencies can occur. Some results are available for predicting when overflow oscillations can be supported by a difference equation (see Ebert et al., 1969). Overflow oscillations can be avoided by using the saturation overflow characteristic of Figure 6.45(b) (see Ebert et al., 1969).

6.10.3 Avoiding Limit Cycles

The possible existence of a zero-input limit cycle is important in applications where a digital filter is to be in continuous operation, since it is generally desired that the output approach zero when the input is zero. For example, suppose that a speech signal is sampled, filtered by a digital filter, and then converted back to an acoustic signal using a D/A converter. In such a situation it would be very undesirable for the filter to enter a periodic limit cycle whenever the input is zero, since the limit cycle would produce an audible tone.

One approach to the general problem of limit cycles is to seek structures that do not support limit cycle oscillations. Such structures have been derived by using state-space representations (see Barnes and Fam, 1977; Mills, Mullis and Roberts, 1978) and concepts analogous to passivity in analog systems (see Rao and Kailath, 1984; Fettweis, 1986). However, these structures generally require more computation than an equivalent cascade or parallel form implementation. By adding more bits to the computational wordlength, we can generally avoid overflow. Similarly, since round-off limit cycles usually are limited to the least significant bits of the binary word, additional bits can be used to reduce the effective amplitude of the limit cycle. Also, Claasen et al. (1973) showed that if a double-length accumulator is used so that quantization occurs after the accumulation of products, then limit cycles owing to round-off are much less likely to occur in 2nd-order systems. Thus, the trade-off between word length and computational algorithm complexity arises for limit cycles just as it does for coefficient quantization and round-off noise.

Finally, it is important to point out that zero-input limit cycles due to both overflow and round-off are a phenomenon unique to IIR systems: FIR systems cannot support zero-input limit cycles, because they have no feedback paths. The output of an FIR system will be zero no later than $(M + 1)$ samples after the input goes to zero and remains there. This is a major advantage of FIR systems in applications wherein limit cycle oscillations cannot be tolerated.

6.11 SUMMARY

In this chapter, we have considered many aspects of the problem of implementing an LTI discrete-time system. The first half of the chapter was devoted to basic implementation structures. After introducing block diagram and signal flow graphs as pictorial representations of difference equations, we discussed a number of basic structures for

IIR and FIR discrete-time systems. These included the direct form I, direct form II, cascade form, parallel form, lattice form, and transposed version of all the basic forms. We showed that these forms are all equivalent when implemented with infinite-precision arithmetic. However, the different structures are most significant in the context of finite-precision implementations. Therefore, the remainder of the chapter addressed problems associated with finite precision or quantization in fixed-point digital implementations of the basic structures.

We began the discussion of finite precision effects with a brief review of digital number representation and an overview showing that the quantization effects that are important in sampling (discussed in Chapter 4) are also important in representing the coefficients of a discrete-time system and in implementing systems using finite-precision arithmetic. We illustrated the effect of quantization of the coefficients of a difference equation through several examples. This issue was treated independently of the effects of finite-precision arithmetic, which we showed introduces nonlinearity into the system. We demonstrated that in some cases this nonlinearity was responsible for limit cycles that may persist after the input to a system has become zero. We also showed that quantization effects can be modeled in terms of independent random white-noise sources that are injected internally into the flow graph. Such linear-noise models were developed for the direct form structures and for the cascade structure. In all of our discussion of quantization effects, the underlying theme was the conflict between the desire for fine quantization and the need to maintain a wide range of signal amplitudes. We saw that in fixed-point implementations, one can be improved at the expense of the other, but to improve one while leaving the other unaffected requires that we increase the number of bits used to represent coefficients and signal amplitudes. This can be done either by increasing the fixed-point word length or by adopting a floating-point representation.

Our discussion of quantization effects serves two purposes. First, we developed several results that can be useful in guiding the design of practical implementations. We found that quantization effects depend greatly on the structure used and on the specific parameters of the system to be implemented, and even though simulation of the system is generally necessary to evaluate its performance, many of the results discussed are useful in making intelligent decisions in the design process. A second, equally important purpose of this part of the chapter was to illustrate a style of analysis that can be applied in studying quantization effects in a variety of digital signal-processing algorithms. The examples of the chapter indicate the types of assumptions and approximations that are commonly made in studying quantization effects. In Chapter 9, we will apply the analysis techniques developed here to the study of quantization in the computation of the discrete Fourier transform.

Problems

Basic Problems with Answers

6.1. Determine the system function of the two flow graphs in Figure P6.1, and show that they have the same poles.

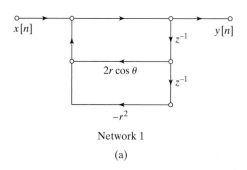

Network 1

(a)

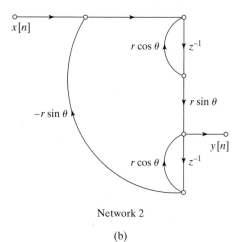

Network 2

(b)

Figure P6.1

6.2. The signal flow graph of Figure P6.2 represents a linear difference equation with constant coefficients. Determine the difference equation that relates the output $y[n]$ to the input $x[n]$.

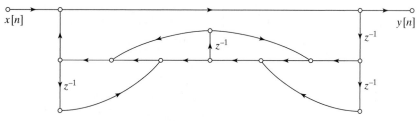

Figure P6.2

6.3. Figure P6.3 shows six systems. Determine which one of the last five, (b)–(f), has the same system function as (a). You should be able to eliminate some of the possibilities by inspection.

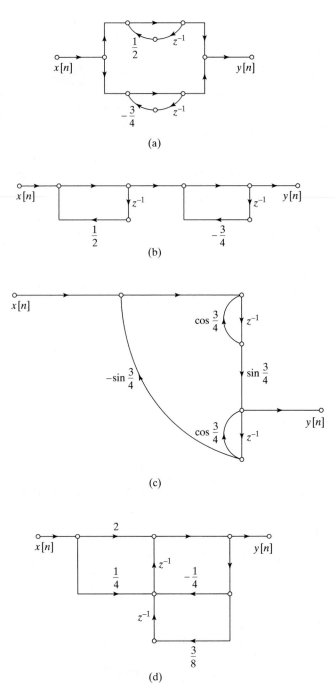

(a)

(b)

(c)

(d)

Figure P6.3

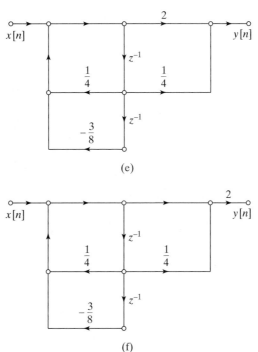

(e)

(f)

Figure P6.3 (*continued*)

6.4. Consider the system in Figure P6.3(d).

(a) Determine the system function relating the z-transforms of the input and output.

(b) Write the difference equation that is satisfied by the input sequence $x[n]$ and the output sequence $y[n]$.

6.5. An LTI system is realized by the flow graph shown in Figure P6.5.

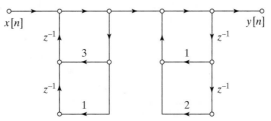

Figure P6.5

(a) Write the difference equation relating $x[n]$ and $y[n]$ for this flow graph.

(b) What is the system function of the system?

(c) In the realization of Figure P6.5, how many real multiplications and real additions are required to compute each sample of the output? (Assume that $x[n]$ is real, and assume that multiplication by 1 does not count in the total.)

(d) The realization of Figure P6.5 requires four storage registers (delay elements). Is it possible to reduce the number of storage registers by using a different structure? If so, draw the flow graph; if not, explain why the number of storage registers cannot be reduced.

6.6. Determine the impulse response of each of the systems in Figure P6.6.

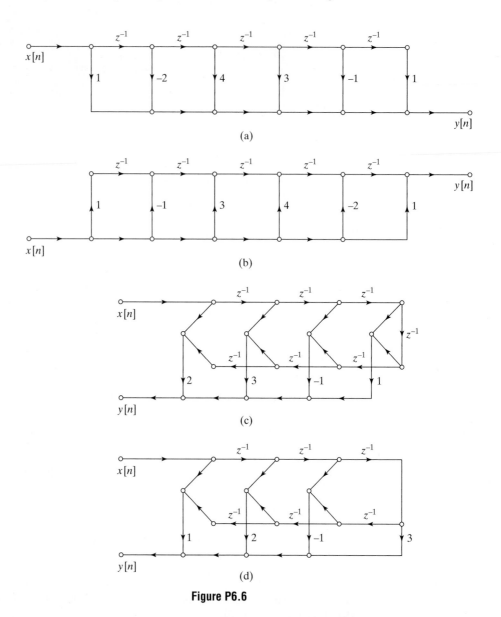

Figure P6.6

6.7. Let $x[n]$ and $y[n]$ be sequences related by the following difference equation:

$$y[n] - \frac{1}{4}y[n-2] = x[n-2] - \frac{1}{4}x[n].$$

Draw a direct form II signal flow graph for the causal LTI system corresponding to this difference equation.

6.8. The signal flow graph in Figure P6.8 represents an LTI system. Determine a difference equation that gives a relationship between the input $x[n]$ and the output $y[n]$ of this system. As usual, all branches of the signal flow graph have unity gain unless specifically indicated otherwise.

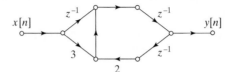

Figure P6.8

6.9. Figure P6.9 shows the signal flow graph for a causal discrete-time LTI system. Branches without gains explicitly indicated have a gain of unity.

(a) By tracing the path of an impulse through the flowgraph, determine $h[1]$, the impulse response at $n = 1$.

(b) Determine the difference equation relating $x[n]$ and $y[n]$.

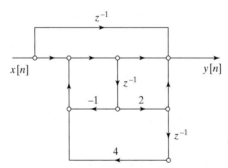

Figure P6.9

6.10. Consider the signal flow graph shown in Figure P6.10.

(a) Using the node variables indicated, write the set of difference equations represented by this flow graph.

(b) Draw the flow graph of an equivalent system that is the cascade of two 1st-order systems.

(c) Is the system stable? Explain.

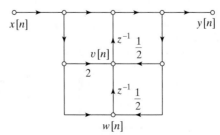

Figure P6.10

6.11. Consider a causal LTI system with impulse response $h[n]$ and system function

$$H(z) = \frac{(1 - 2z^{-1})(1 - 4z^{-1})}{z\left(1 - \frac{1}{2}z^{-1}\right)}$$

(a) Draw a direct form II flow graph for the system.
(b) Draw the transposed form of the flow graph in part (a).

6.12. For the LTI system described by the flow graph in Figure P6.12, determine the difference equation relating the input $x[n]$ to the output $y[n]$.

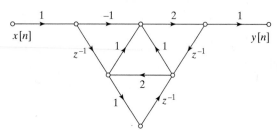

Figure P6.12

6.13. Draw the signal flow graph for the direct form I implementation of the LTI system with system function

$$H(z) = \frac{1 - \frac{1}{2}z^{-2}}{1 - \frac{1}{4}z^{-1} - \frac{1}{8}z^{-2}}.$$

6.14. Draw the signal flow graph for the direct form II implementation of the LTI system with system function

$$H(z) = \frac{1 + \frac{5}{6}z^{-1} + \frac{1}{6}z^{-2}}{1 - \frac{1}{2}z^{-1} - \frac{1}{2}z^{-2}}.$$

6.15. Draw the signal flow graph for the transposed direct form II implementation of the LTI system with system function

$$H(z) = \frac{1 - \frac{7}{6}z^{-1} + \frac{1}{6}z^{-2}}{1 + z^{-1} + \frac{1}{2}z^{-2}}.$$

6.16. Consider the signal flow graph shown in Figure P6.16.

(a) Draw the signal flow graph that results from applying the transposition theorem to this signal flow graph.
(b) Confirm that the transposed signal flow graph that you found in (a) has the same system function $H(z)$ as the original system in the figure.

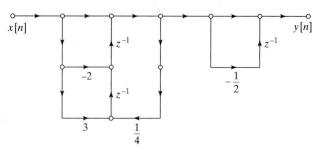

Figure P6.16

6.17. Consider the causal LTI system with system function

$$H(z) = 1 - \frac{1}{3}z^{-1} + \frac{1}{6}z^{-2} + z^{-3}.$$

(a) Draw the signal flow graph for the direct form implementation of this system.
(b) Draw the signal flow graph for the transposed direct form implementation of the system.

6.18. For some nonzero choices of the parameter a, the signal flow graph in Figure P6.18 can be replaced by a 2nd-order direct form II signal flow graph implementing the same system function. Give one such choice for a and the system function $H(z)$ that results.

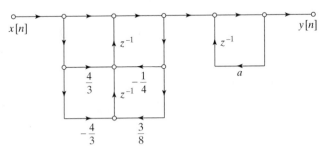

Figure P6.18

6.19. Consider the causal LTI system with the system function

$$H(z) = \frac{2 - \frac{8}{3}z^{-1} - 2z^{-2}}{\left(1 - \frac{1}{3}z^{-1}\right)\left(1 + \frac{2}{3}z^{-1}\right)}.$$

Draw a signal flow graph that implements this system as a parallel combination of 1st-order transposed direct form II sections.

6.20. Draw a signal flow graph implementing the system function

$$H(z) = \frac{(1 + (1 - j/2)z^{-1})(1 + (1 + j/2)z^{-1})}{(1 + (j/2)z^{-1})(1 - (j/2)z^{-1})(1 - (1/2)z^{-1})(1 - 2z^{-1})}$$

as a cascade of 2nd-order transposed direct form II sections with real coefficients.

Basic Problems

6.21. For many applications, it is useful to have a system that will generate a sinusoidal sequence. One possible way to do this is with a system whose impulse response is $h[n] = e^{j\omega_0 n}u[n]$. The real and imaginary parts of $h[n]$ are therefore $h_r[n] = (\cos \omega_0 n)u[n]$ and $h_i[n] = (\sin \omega_0 n)u[n]$, respectively.

In implementing a system with a complex impulse response, the real and imaginary parts are distinguished as separate outputs. By first writing the complex difference equation required to produce the desired impulse response and then separating it into its real and imaginary parts, draw a flow graph that will implement this system. The flow graph that you draw should have only real coefficients. This implementation is sometimes called the *coupled form oscillator,* since, when the input is excited by an impulse, the outputs are sinusoidal.

6.22. For the system function

$$H(z) = \frac{1 + 2z^{-1} + z^{-2}}{1 - 0.75z^{-1} + 0.125z^{-2}},$$

draw the flow graphs of all possible realizations for this system as cascades of 1st-order systems.

6.23. We want to implement a causal system $H(z)$ with the pole–zero diagram shown in Figure P6.23. For all parts of this problem, z_1, z_2, p_1, and p_2 are real, and a gain constant that is independent of frequency can be absorbed into a gain coefficient in the output branch of each flow graph.

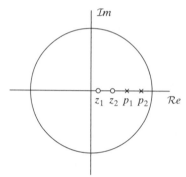

Figure P6.23

(a) Draw the flow graph of the direct form II implementation. Determine an expression for each of the branch gains in terms of the variables z_1, z_2, p_1, and p_2.

(b) Draw the flow graph of an implementation as a cascade of 2nd-order direct form II sections. Determine an expression for each of the branch gains in terms of the variables z_1, z_2, p_1, and p_2.

(c) Draw the flow graph of a parallel form implementation with 1st-order direct form II sections. Specify a system of linear equations that can be solved to express the branch gains in terms of the variables z_1, z_2, p_1, and p_2.

6.24. Consider a causal LTI system whose system function is

$$H(z) = \frac{1 - \frac{3}{10}z^{-1} + \frac{1}{3}z^{-2}}{\left(1 - \frac{4}{5}z^{-1} + \frac{2}{3}z^{-2}\right)\left(1 + \frac{1}{5}z^{-1}\right)} = \frac{\frac{1}{2}}{1 - \frac{4}{5}z^{-1} + \frac{2}{3}z^{-2}} + \frac{\frac{1}{2}}{1 + \frac{1}{5}z^{-1}}$$

(a) Draw the signal flow graphs for implementations of the system in each of the following forms:

    (i) Direct form I
    (ii) Direct form II
    (iii) Cascade form using 1st- and 2nd-order direct form II sections
    (iv) Parallel form using 1st- and 2nd-order direct form I sections
    (v) Transposed direct form II.

(b) Write the difference equations for the flow graph of part (v) in (a) and show that this system has the correct system function.

6.25. A causal LTI system is defined by the signal flow graph shown in Figure P6.25, which represents the system as a cascade of a 2nd-order system with a 1st-order system.

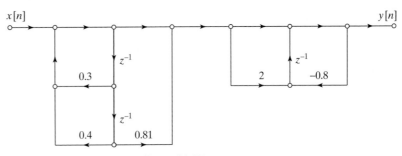

Figure P6.25

(a) What is the system function of the overall cascade system?
(b) Is the overall system stable? Explain briefly.
(c) Is the overall system a minimum-phase system? Explain briefly.
(d) Draw the signal flow graph of a transposed direct form II implementation of this system.

6.26. A causal LTI system has system function given by the following expression:

$$H(z) = \frac{1}{1-z^{-1}} + \frac{1-z^{-1}}{1-z^{-1}+0.8z^{-2}}.$$

(a) Is this system stable? Explain briefly.
(b) Draw the signal flow graph of a parallel form implementation of this system.
(c) Draw the signal flow graph of a cascade form implementation of this system as a cascade of a 1st-order system and a 2nd-order system. Use a transposed direct form II implementation for the 2nd-order system.

6.27. An LTI system with system function

$$H(z) = \frac{0.2(1+z^{-1})^6}{\left(1-2z^{-1}+\frac{7}{8}z^{-2}\right)\left(1+z^{-1}+\frac{1}{2}z^{-2}\right)\left(1-\frac{1}{2}z^{-1}+z^{-2}\right)}$$

is to be implemented using a flow graph of the form shown in Figure P6.27.

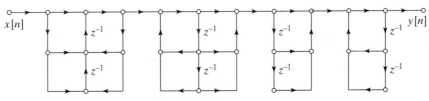

Figure P6.27

(a) Fill in all the coefficients in the diagram of Figure P6.27. Is your solution unique?

(b) Define appropriate node variables in Figure P6.27, and write the set of difference equations that is represented by the flow graph.

6.28. (a) Determine the system function, $H(z)$, from $x[n]$ to $y[n]$ for the flow graph shown in Figure P6.28-1 (note that the location where the diagonal lines criss-cross is not a single node).

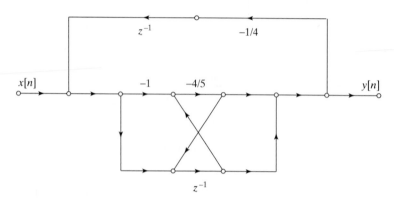

Figure P6.28-1

(b) Draw the direct form (I and II) flow graph of systems having the system function $H(z)$.

(c) Design $H_1(z)$ such that $H_2(z)$ in Figure P6.28-2 has a causal stable inverse and $|H_2(e^{j\omega})| = |H(e^{j\omega})|$. Note: Zero-pole cancellation is permitted.

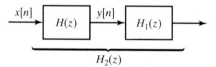

Figure P6.28-2

(d) Draw the transposed direct form II flow graph for $H_2(z)$.

6.29. (a) Determine the system function $H(z)$ relating the input $x[n]$ to the output $y[n]$ for the FIR lattice filter depicted in Figure P6.29.

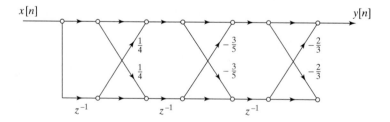

Figure P6.29

(b) Draw the lattice filter structure for the all-pole filter $1/H(z)$.

6.30. Determine and draw the lattice filter implementation of the following causal all-pole system function:

$$H(z) = \frac{1}{1 + \frac{3}{2}z^{-1} - z^{-2} + \frac{3}{4}z^{-3} + 2z^{-4}}$$

Is the system stable?

6.31. An IIR lattice filter is shown in Figure P6.31.

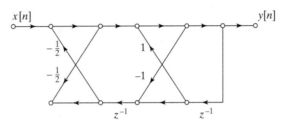

Figure P6.31

(a) By tracing the path of an impulse through the flowgraph, determine $y[1]$ for input $x[n] = \delta[n]$.

(b) Determine a flow graph for the corresponding inverse filter.

(c) Determine the transfer function for the IIR filter in Figure P6.31.

6.32. The flow graph shown in Figure P6.32 is an implementation of a causal, LTI system.

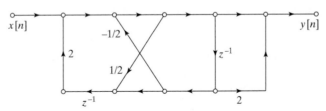

Figure P6.32

(a) Draw the transpose of the signal flow graph.

(b) For either the original system or its transpose, determine the difference equation relating the input $x[n]$ to the output $y[n]$. (*Note:* The difference equations will be the same for both structures.)

(c) Is the system BIBO stable?

(d) Determine $y[2]$ if $x[n] = (1/2)^n u[n]$.

Advanced Problems

6.33. Consider the LTI system represented by the FIR lattice structure in Figure P6.33-1.

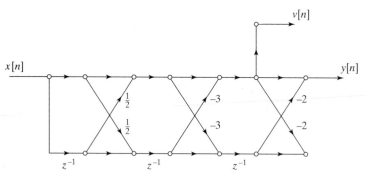

Figure P6.33-1

(a) Determine the system function from the input $x[n]$ to the output $v[n]$ (NOT $y[n]$).

(b) Let $H(z)$ be the system function from the input $x[n]$ to the output $y[n]$, and let $g[n]$ be the result of expanding the associated impulse response $h[n]$ by 2 as shown in Figure P6.33-2.

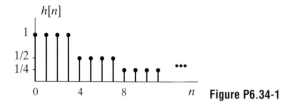

Figure P6.33-2

The impulse response $g[n]$ defines a new system with system function $G(z)$. We would like to implement $G(z)$ using an FIR lattice structure. Determine the k-parameters necessary for an FIR lattice implementation of $G(z)$. *Note:* You should think carefully before diving into a long calculation.

6.34. Figure P6.34-1 shows an impulse response $h[n]$, specified as

$$h[n] = \begin{cases} \left(\frac{1}{2}\right)^{n/4} u[n], & \text{for } n \text{ an integer multiple of 4} \\ \text{constant in between as indicated} \end{cases}.$$

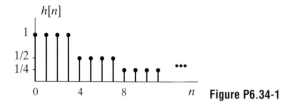

Figure P6.34-1

(a) Determine a choice for $h_1[n]$ and $h_2[n]$ such that

$$h[n] = h_1[n] * h_2[n],$$

where $h_1[n]$ is an FIR filter and where $h_2[n] = 0$ for $n/4$ not an integer. Is $h_2[n]$ an FIR or IIR filter?

(b) The impulse response $h[n]$ is to be used in a downsampling system as indicated in Figure P6.34-2.

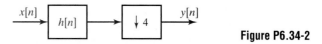

Figure P6.34-2

Draw a flow graph implementation of the system in Figure P6.34-2 that requires the minimum number of nonzero and nonunity coefficient multipliers. You may use unit delay elements, coefficient multipliers, adders and compressors. (Multiplication by a zero or a one does not require a multiplier.)

(c) For your system, state how many multiplications per input sample and per output sample are required, giving a brief explanation.

6.35. Consider the system shown in Figure P6.35-1.

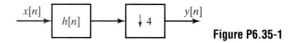

Figure P6.35-1

We want to implement this system using the polyphase structure shown in Figure P6.35-2.

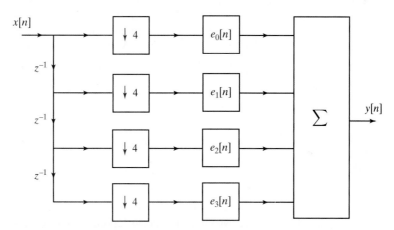

Figure P6.35-2   Polyphase structure of the system.

For parts (a) and (b) only, assume $h[n]$ is defined in Figure P6.35-3

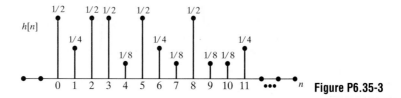

Figure P6.35-3

($h[n] = 0$ for all $n < 0$ and $n \geq 12$).

(a) Give the sequences $e_0[n], e_1[n], e_2[n],$ and $e_3[n]$ that result in a correct implementation.

(b) We want to minimize the total number of multiplies per output sample for the implementation of the structure in Figure P6.35-2. Using the appropriate choice of $e_0[n]$, $e_1[n]$, $e_2[n]$, and $e_3[n]$ from part (a), determine the minimum number of multiplies per output sample for the overall system. Also, determine the minimum number of multiplies per input sample for the overall system. Explain.

(c) Instead of using the sequences $e_0[n], e_1[n], e_2[n],$ and $e_3[n]$ identified in part (a), now assume that $E_0(e^{j\omega})$ and $E_2(e^{j\omega})$ the DTFTs of $e_0[n]$ and $e_2[n]$, respectively, are as given in Figure P6.35-4, and $E_1(e^{j\omega}) = E_3(e^{j\omega}) = 0$.

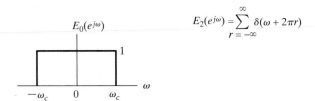

Figure P6.35-4

Sketch and label $H(e^{j\omega})$ from $(-\pi, \pi)$.

6.36. Consider a general flow graph (denoted Network A) consisting of coefficient multipliers and delay elements, as shown in Figure P6.36-1. If the system is initially at rest, its behavior is completely specified by its impulse response $h[n]$. We wish to modify the system to create a new flow graph (denoted Network A_1) with impulse response $h_1[n] = (-1)^n h[n]$.

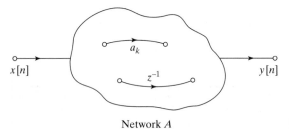

Network A Figure P6.36-1

(a) If $H(e^{j\omega})$ is as given in Figure P6.36-2, sketch $H_1(e^{j\omega})$.

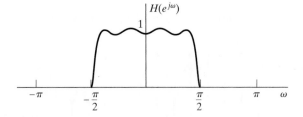

Figure P6.36-2

(b) Explain how to modify Network A by simple modifications of its coefficient multipliers and/or the delay branches to form the new Network A_1 whose impulse response is $h_1[n]$.

(c) If Network A is as given in Figure P6.36-3, show how to modify it by simple modifications to *only the coefficient multipliers* so that the resulting Network A_1 has impulse response $h_1[n]$.

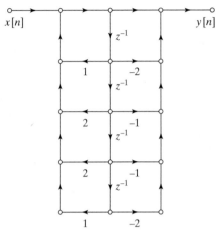

Figure P6.36-3

6.37. The flow graph shown in Figure P6.37 is noncomputable; i.e., it is not possible to compute the output using the difference equations represented by the flow graph because it contains a closed loop having no delay elements.

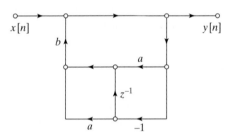

Figure P6.37

(a) Write the difference equations for the system of Figure P6.37, and from them, find the system function of the flow graph.

(b) From the system function, obtain a flow graph that is computable.

6.38. The impulse response of an LTI system is

$$h[n] = \begin{cases} a^n, & 0 \le n \le 7, \\ 0, & \text{otherwise.} \end{cases}$$

(a) Draw the flow graph of a direct form nonrecursive implementation of the system.

(b) Show that the corresponding system function can be expressed as

$$H(z) = \frac{1 - a^8 z^{-8}}{1 - az^{-1}}, \quad |z| > |a|.$$

(c) Draw the flow graph of an implementation of $H(z)$, as expressed in part (b), corresponding to a cascade of an FIR system (numerator) with an IIR system (denominator).

(d) Is the implementation in part (c) recursive or nonrecursive? Is the overall system FIR or IIR?

(e) Which implementation of the system requires

  (i) the most storage (delay elements)?

  (ii) the most arithmetic (multiplications and additions per output sample)?

6.39. Consider an FIR system whose impulse response is

$$h[n] = \begin{cases} \frac{1}{15}(1 + \cos[(2\pi/15)(n - n_0)]), & 0 \le n \le 14, \\ 0, & \text{otherwise.} \end{cases}$$

This system is an example of a class of filters known as frequency-sampling filters. Problem 6.51 discusses these filters in detail. In this problem, we consider just one specific case.

(a) Sketch the impulse response of the system for the cases $n_0 = 0$ and $n_0 = 15/2$.

(b) Show that the system function of the system can be expressed as

$$H(z) = (1 - z^{-15}) \cdot \frac{1}{15} \left[\frac{1}{1 - z^{-1}} + \frac{\frac{1}{2}e^{-j2\pi n_0/15}}{1 - e^{j2\pi/15}z^{-1}} + \frac{\frac{1}{2}e^{j2\pi n_0/15}}{1 - e^{-j2\pi/15}z^{-1}} \right].$$

(c) Show that if $n_0 = 15/2$, the frequency response of the system can be expressed as

$$H(e^{j\omega}) = \frac{1}{15}e^{-j\omega 7} \left\{ \frac{\sin(\omega 15/2)}{\sin(\omega/2)} + \frac{1}{2}\frac{\sin[(\omega - 2\pi/15)15/2]}{\sin[(\omega - 2\pi/15)/2]} \right.$$

$$\left. + \frac{1}{2}\frac{\sin[(\omega + 2\pi/15)15/2]}{\sin[(\omega + 2\pi/15)/2]} \right\}.$$

Use this expression to sketch the magnitude of the frequency response of the system for $n_0 = 15/2$. Obtain a similar expression for $n_0 = 0$. Sketch the magnitude response for $n_0 = 0$. For which choices of n_0 does the system have a generalized linear phase?

(d) Draw a signal flow graph of an implementation of the system as a cascade of an FIR system whose system function is $1 - z^{-15}$ and a parallel combination of a 1st- and 2nd-order IIR system.

6.40. Consider the discrete-time system depicted in Figure P6.40-1.

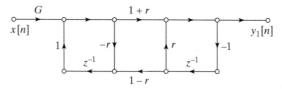

Figure P6.40-1

(a) Write the set of difference equations represented by the flow graph of Figure P6.40-1.

(b) Determine the system function $H_1(z) = Y_1(z)/X(z)$ of the system in Figure P6.40-1, and determine the magnitudes and angles of the poles of $H_1(z)$ as a function of r for $-1 < r < 1$.

(c) Figure P6.40-2 shows a different flow graph obtained from the flow graph of Figure P6.40-1 by moving the delay elements to the opposite top branch. How is the system function $H_2(z) = Y_2(z)/X(z)$ related to $H_1(z)$?

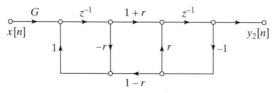

Figure P6.40-2

6.41. The three flow graphs in Figure P6.41 are all equivalent implementations of the same two-input, two-output LTI system.

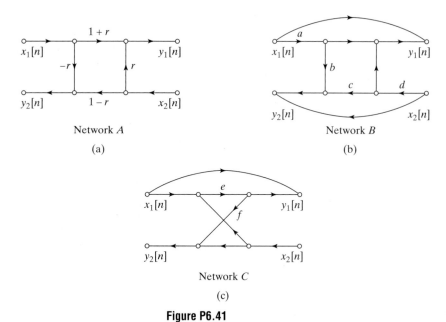

Network A

(a)

Network B

(b)

Network C

(c)

Figure P6.41

(a) Write the difference equations for Network A.

(b) Determine values of a, b, c, and d for Network B in terms of r in Network A such that the two systems are equivalent.

(c) Determine values of e and f for Network C in terms of r in Network A such that the two systems are equivalent.

(d) Why might Network B or C be preferred over Network A? What possible advantage could Network A have over Network B or C?

6.42. Consider an all-pass system with system function

$$H(z) = -0.54 \frac{1 - (1/0.54)z^{-1}}{1 - 0.54z^{-1}}.$$

A flow graph for an implementation of this system is shown in Figure P6.42.

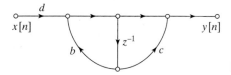

Figure P6.42

(a) Determine the coefficients b, c, and d such that the flow graph in Figure P6.42 is a direct realization of $H(z)$.

(b) In a practical implementation of the network in Figure P6.42, the coefficients b, c, and d might be quantized by rounding the exact value to the nearest tenth (e.g., 0.54 will be rounded to 0.5 and $1/0.54 = 1.8518\ldots$ will be rounded to 1.9). Would the resulting system still be an all-pass system?

(c) Show that the difference equation relating the input and output of the all-pass system with system function $H(z)$ can be expressed as

$$y[n] = 0.54(y[n-1] - x[n]) + x[n-1].$$

Draw the flow graph of a network that implements this difference equation with two delay elements, but only one multiplication by a constant other than ± 1.

(d) With quantized coefficients, would the flow graph of part (c) be an all-pass system?

The primary disadvantage of the implementation in part (c) compared with the implementation in part (a) is that it requires two delay elements. However, for higher-order systems, it is necessary to implement a cascade of all-pass systems. For N all-pass sections in cascade, it is possible to use all-pass sections in the form determined in part (c) while requiring only $(N+1)$ delay elements. This is accomplished by sharing a delay element between sections.

(e) Consider the all-pass system with system function

$$H(z) = \left(\frac{z^{-1} - a}{1 - az^{-1}}\right)\left(\frac{z^{-1} - b}{1 - bz^{-1}}\right).$$

Draw the flow graph of a "cascade" realization composed of two sections of the form obtained in part (c) with one delay element shared between the sections. The resulting flow graph should have only three delay elements.

(f) With quantized coefficients a and b, would the flow graph in part (e) be an all-pass system?

6.43. All branches of the signal flow graphs in this problem have unity gain unless specifically indicated otherwise.

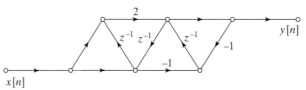

Figure P6.43-1

(a) The signal flow graph of System A, shown in Figure P6.43-1, represents a causal LTI system. Is it possible to implement the same input–output relationship using fewer delays? If it is possible, what is the minimum number of delays required to implement an equivalent system? If it is not possible, explain why not.

(b) Does the System B shown in Figure P6.43-2 represent the same input–output relationship as System A in Figure P6.43-1? Explain clearly.

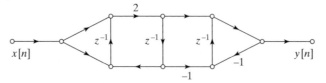

Figure P6.43-2

6.44. Consider an all-pass system whose system function is

$$H(z) = \frac{z^{-1} - \frac{1}{3}}{1 - \frac{1}{3}z^{-1}}.$$

(a) Draw the direct form I signal flow graph for the system. How many delays and multipliers do you need? (Do not count multiplying by ± 1.)

(b) Draw a signal flow graph for the system that uses one multiplier. Minimize the number of delays.

(c) Now consider another all-pass system whose system function is

$$H(z) = \frac{(z^{-1} - \frac{1}{3})(z^{-1} - 2)}{(1 - \frac{1}{3}z^{-1})(1 - 2z^{-1})}.$$

Determine and draw a signal flow graph for the system with two multipliers and three delays.

6.45. With infinite-precision arithmetic, the flow graphs shown in Figure P6.45 have the same system function, but with quantized fixed-point arithmetic they behave differently. Assume that a and b are real numbers and $0 < a < 1$.

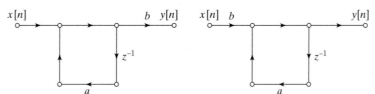

Figure P6.45

(a) Determine x_{max}, the maximum amplitude of the input samples so that the maximum value of the output $y[n]$ of either of the two systems is guaranteed to be less than one.

(b) Assume that the above systems are implemented with two's-complement fixed-point arithmetic, and that in both cases all products are immediately rounded to $B + 1$ bits (*before* any additions are done). Insert round-off noise sources at appropriate locations in the above diagrams to model the rounding error. Assume that each of the noise sources inserted has average power equal to $\sigma_B^2 = 2^{-2B}/12$.

(c) If the products are rounded as described in part (b), the outputs of the two systems will differ; i.e., the output of the first system will be $y_1[n] = y[n] + f_1[n]$ and the output of the second system will be $y_2[n] = y[n] + f_2[n]$, where $f_1[n]$ and $f_2[n]$ are the outputs due to the noise sources. Determine the power density spectra $\Phi_{f_1 f_1}(e^{j\omega})$ and $\Phi_{f_2 f_2}(e^{j\omega})$ of the output noise for both systems.

(d) Determine the total noise powers $\sigma_{f_1}^2$ and $\sigma_{f_2}^2$ at the output for both systems.

6.46. An allpass system is to be implemented with fixed-point arithmetic. Its system function is

$$H(z) = \frac{(z^{-1} - a^*)(z^{-1} - a)}{(1 - az^{-1})(1 - a^*z^{-1})}$$

where $a = re^{j\theta}$.

(a) Draw the signal flow graphs for both the direct form I and direct form II implementations of this system as a 2nd-order system using only real coefficients.

(b) Assuming that the products are each rounded *before* additions are performed, insert appropriate noise sources into the networks drawn in part (a), combining noise sources where possible, and indicating the power of each noise source in terms of σ_B^2, the power of a single rounding noise source.

(c) Circle the nodes in your network diagrams where overflow may occur.

(d) Specify whether or not the output noise power of the direct form II system is independent of r, while the output noise power for the direct form I system increases as $r \to 1$. Give a convincing argument to support your answer. Try to answer the question *without* computing the output noise power of either system. Of course, such a computation would answer the question, but you should be able to see the answer without computing the noise power.

(e) Now determine the output noise power for both systems.

6.47. Assume that a in the flow graphs shown in Figure P6.47 is a real number and $0 < a < 1$. Note that under infinite-precision arithmetic, the two systems are equivalent.

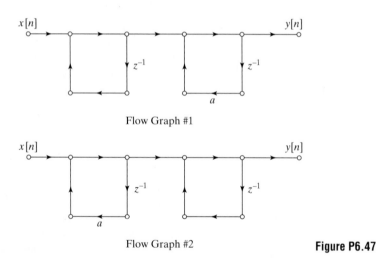

Flow Graph #1

Flow Graph #2   **Figure P6.47**

(a) Assume that the two systems are implemented with two's-complement fixed-point arithmetic, and that in both cases all products are immediately rounded (*before* any additions are done). Insert round-off noise sources at appropriate locations in both flow graphs to model the rounding error (multiplications by unity do not introduce noise). Assume that each of the noise sources inserted has average power equal to $\sigma_B^2 = 2^{-2B}/12$.

(b) If the products are rounded as described in part (a), the outputs of the two systems will differ; i.e., the output of the first system will be $y_1[n] = y[n] + f_1[n]$ and the output of the second system will be $y_2[n] = y[n] + f_2[n]$, where $y[n]$ is the output owing to $x[n]$ acting alone, and $f_1[n]$ and $f_2[n]$ are the outputs owing to the noise sources. Determine the power density spectrum of the output noise $\Phi_{f_1 f_1}(e^{j\omega})$. Also determine the total noise power of the output of flow graph #1; i.e., determine $\sigma_{f_1}^2$.

(c) Without actually computing the output noise power for flow graph #2, you should be able to determine which system has the largest total noise power at the output. Give a brief explanation of your answer.

6.48. Consider the parallel form flow graph shown in Figure P6.48

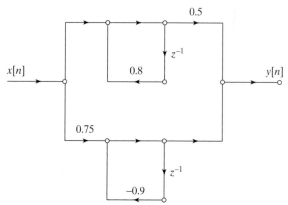

Figure P6.48

(a) Assume that the system is implemented with two's-complement fixed-point arithmetic, and that all products (multiplications by 1 do not introduce noise) are immediately rounded (*before* any additions are done). Insert round-off noise sources at appropriate locations in the flow graph to model the rounding error. Indicate the size (average power) of each noise source in terms of σ_B^2, the average power of one $(B + 1)$-bit rounding operation.

(b) If the products are rounded as described in part (a), the output can be represented as $\hat{y}[n] = y[n] + f[n]$ where $y[n]$ is the output owing to $x[n]$ acting alone, and $f[n]$ is the total output due to all the noise sources acting independently. Determine the power density spectrum of the output noise $\Phi_{ff}(e^{j\omega})$.

(c) Also determine the total noise power σ_f^2 of the noise component of the output.

6.49. Consider the system shown in Figure P6.49, which consists of a 16-bit A/D converter whose output is the input to an FIR digital filter that is implemented with 16-bit fixed-point arithmetic.

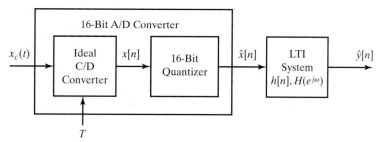

Figure P6.49

The impulse response of the digital filter is

$$h[n] = -.375\delta[n] + .75\delta[n-1] - .375\delta[n-2].$$

This system is implemented with 16-bit two's-complement arithmetic. The products are rounded to 16-bits *before* they are accumulated to produce the output. In anticipation of using the linear noise model to analyze this system, we define $\hat{x}[n] = x[n] + e[n]$ and $\hat{y}[n] = y[n] + f[n]$, where $e[n]$ is the quantization error introduced by the A/D converter and $f[n]$ is the *total* quantization noise at the output of the filter.

(a) Determine the maximum magnitude of $\hat{x}[n]$ such that no overflow can possibly occur in implementing the digital filter; i.e., determine x_{max} such that $\hat{y}[n] < 1$ for all $-\infty < n < \infty$ when $\hat{x}[n] < x_{max}$ for all $-\infty < n < \infty$.

(b) Draw the linear noise model for the complete system (including the linear noise model of the A/D). Include a detailed flow-graph for the digital filter including all noise sources due to quantization.

(c) Determine the total noise power at the output. Denote this σ_f^2.

(d) Determine the power spectrum of the noise at the output of the filter; i.e., determine $\Phi_{ff}(e^{j\omega})$. Plot your result.

Extension Problems

6.50. In this problem, we consider the implementation of a causal filter with system function

$$H(z) = \frac{1}{(1-.63z^{-1})(1-.83z^{-1})} = \frac{1}{1-1.46z^{-1}+0.5229z^{-2}}$$

This system is to be implemented with $(B+1)$-bit two's-complement rounding arithmetic with products rounded before additions are performed. The input to the system is a zero-mean, white, wide-sense stationary random process, with values uniformly distributed between $-x_{max}$ and $+x_{max}$.

(a) Draw the direct form flow graph implementation for the filter, with all coefficient multipliers rounded to the nearest tenth.

(b) Draw a flow graph implementation of this system as a cascade of two 1st-order systems, with all coefficient multipliers rounded to the nearest tenth.

(c) Only one of the implementations from parts (a) and (b) above is usable. Which one? Explain.

(d) To prevent overflow at the output node, we must carefully choose the parameter x_{max}. For the implementation selected in part (c), determine a value for x_{max} that guarantees the output will stay between -1 and 1. (Ignore any potential overflow at nodes other than the output.)

(e) Redraw the flow graph selected in part (c), this time including linearized noise models representing quantization round-off error.

(f) Whether you chose the direct form or cascade implementation for part (c), there is still at least one more design alternative:

  (i) If you chose the direct form, you could also use a transposed direct form.

  (ii) If you chose the cascade form, you could implement the smaller pole first or the larger pole first.

For the system chosen in part (c), which alternative (if any) has lower output quantization noise power? Note you do not need to explicitly calculate the total output quantization noise power, but you must justify your answer with some analysis.

6.51. In this problem, we will develop some of the properties of a class of discrete-time systems called frequency-sampling filters. This class of filters has system functions of the form

$$H(z) = (1 - z^{-N}) \cdot \sum_{k=0}^{N-1} \frac{\tilde{H}[k]/N}{1 - z_k z^{-1}},$$

where $z_k = e^{j(2\pi/N)k}$ for $k = 0, 1, \ldots, N - 1$.

(a) System functions such as $H(z)$ can be implemented as a cascade of an FIR system whose system function is $(1 - z^{-N})$ with a parallel combination of 1st-order IIR systems. Draw the signal flow graph of such an implementation.

(b) Show that $H(z)$ is an $(N - 1)$st-degree polynomial in z^{-1}. To do this, it is necessary to show that $H(z)$ has no poles other than $z = 0$ and that it has no powers of z^{-1} higher than $(N - 1)$. What do these conditions imply about the length of the impulse response of the system?

(c) Show that the impulse response is given by the expression

$$h[n] = \left(\frac{1}{N} \sum_{k=0}^{N-1} \tilde{H}[k] e^{j(2\pi/N)kn} \right) (u[n] - u[n - N]).$$

Hint: Determine the impulse responses of the FIR and the IIR parts of the system, and convolve them to find the overall impulse response.

(d) Use l'Hôpital's rule to show that

$$H(z_m) = H(e^{j(2\pi/N)m}) = \tilde{H}[m], \qquad m = 0, 1, \ldots, N - 1;$$

i.e., show that the constants $\tilde{H}[m]$ are samples of the frequency response of the system, $H(e^{j\omega})$, at equally spaced frequencies $\omega_m = (2\pi/N)m$ for $m = 0, 1, \ldots, N - 1$. It is this property that accounts for the name of this class of FIR systems.

(e) In general, both the poles z_k of the IIR part and the samples of the frequency response $\tilde{H}[k]$ will be complex. However, if $h[n]$ is real, we can find an implementation involving only real quantities. Specifically, show that if $h[n]$ is real and N is an even integer, then $H(z)$ can be expressed as

$$H(z) = (1 - z^{-N}) \left\{ \frac{H(1)/N}{1 - z^{-1}} + \frac{H(-1)/N}{1 + z^{-1}} \right.$$

$$\left. + \sum_{k=1}^{(N/2)-1} \frac{2|H(e^{j(2\pi/N)k})|}{N} \cdot \frac{\cos[\theta(2\pi k/N)] - z^{-1} \cos[\theta(2\pi k/N) - 2\pi k/N]}{1 - 2\cos(2\pi k/N)z^{-1} + z^{-2}} \right\},$$

where $H(e^{j\omega}) = |H(e^{j\omega})|e^{j\theta(\omega)}$. Draw the signal flow graph representation of such a system when $N = 16$ and $H(e^{j\omega_k}) = 0$ for $k = 3, 4, \ldots, 14$.

6.52. In Chapter 4, we showed that, in general, the sampling rate of a discrete-time signal can be reduced by a combination of linear filtering and time compression. Figure P6.52 shows a block diagram of an M-to-1 decimator that can be used to reduce the sampling rate by an integer factor M. According to the model, the linear filter operates at the high sampling rate. However, if M is large, most of the output samples of the filter will be discarded by the compressor. In some cases, more efficient implementations are possible.

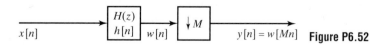

Figure P6.52

(a) Assume that the filter is an FIR system with impulse response such that $h[n] = 0$ for $n < 0$ and for $n > 10$. Draw the system in Figure P6.52, but replace the filter $h[n]$ with an equivalent signal flow graph based on the given information. Note that it is not possible to implement the M-to-1 compressor using a signal flow graph, so you must leave this as a box, as shown in Figure P6.52.

(b) Note that some of the branch operations can be commuted with the compression operation. Using this fact, draw the flow graph of a more efficient realization of the system of part (a). By what factor has the total number of computations required in obtaining the output $y[n]$ been decreased?

(c) Now suppose that the filter in Figure P6.52 has system function

$$H(z) = \frac{1}{1 - \frac{1}{2}z^{-1}}, \qquad |z| > \frac{1}{2}.$$

Draw the flow graph of the direct form realization of the complete system in the figure. With this system for the linear filter, can the total computation per output sample be reduced? If so, by what factor?

(d) Finally, suppose that the filter in Figure P6.52 has system function

$$H(z) = \frac{1 + \frac{7}{8}z^{-1}}{1 - \frac{1}{2}z^{-1}}, \qquad |z| > \frac{1}{2}.$$

Draw the flow graph for the complete system of the figure, using each of the following forms for the linear filter:

(i) direct form I
(ii) direct form II
(iii) transposed direct form I
(iv) transposed direct form II.

For which of the four forms can the system of Figure P6.52 be more efficiently implemented by commuting operations with the compressor?

6.53. Speech production can be modeled by a linear system representing the vocal cavity, which is excited by puffs of air released through the vibrating vocal cords. One approach to synthesizing speech involves representing the vocal cavity as a connection of cylindrical acoustic tubes of equal length, but with varying cross-sectional areas, as depicted in Figure P6.53. Let us assume that we want to simulate this system in terms of the volume velocity representing airflow. The input is coupled into the vocal tract through a small constriction, the vocal cords. We will assume that the input is represented by a change in volume velocity at the left end, but that the boundary condition for traveling waves at the left end is that the net volume velocity must be zero. This is analogous to an electrical transmission line driven by a current source at one end and with an open circuit at the far end. Current in the transmission line is then analogous to volume velocity in the acoustic tube, whereas voltage is analogous to acoustic pressure. The output of the acoustic tube is the volume velocity at the right end. We assume that each section is a lossless acoustic transmission line.

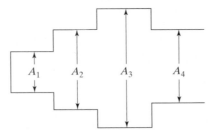

Figure P6.53

At each interface between sections, a forward-traveling wave f^+ is transmitted to the next section with one coefficient and reflected as a backward-traveling wave f^- with a different coefficient. Similarly, a backward-traveling wave f^- arriving at an interface is transmitted with one coefficient and reflected with a different coefficient. Specifically, if we consider a forward-traveling wave f^+ in a tube with cross-sectional area A_1 arriving at the interface with a tube of cross-sectional area A_2, then the forward-traveling wave transmitted is $(1+r)f^+$ and the reflected wave is rf^+, where

$$r = \frac{A_2 - A_1}{A_2 + A_1}.$$

Consider the length of each section to be 3.4 cm, with the velocity of sound in air equal to 34,000 cm/s. Draw a flow graph that will implement the four-section model in Figure P6.53, with the output sampled at 20,000 samples/s.

In spite of the lengthy introduction, this a reasonably straightforward problem. If you find it difficult to think in terms of acoustic tubes, think in terms of transmission-line sections with different characteristic impedances. Just as with transmission lines, it is difficult to express the impulse response in closed form. Therefore, draw the flow graph directly from physical considerations, in terms of forward- and backward-traveling pulses in each section.

6.54. In modeling the effects of round-off and truncation in digital filter implementations, quantized variables are represented as

$$\hat{x}[n] = Q[x[n]] = x[n] + e[n],$$

where $Q[\cdot]$ denotes either rounding or truncation to $(B+1)$ bits and $e[n]$ is the *quantization error*. We assume that the quantization noise sequence is a stationary white-noise sequence such that

$$\mathcal{E}\{(e[n] - m_e)(e[n+m] - m_e)\} = \sigma_e^2\, \delta[m]$$

and that the amplitudes of the noise sequence values are uniformly distributed over the quantization step $\Delta = 2^{-B}$. The 1st-order probability densities for rounding and truncation are shown in Figures P6.54(a) and (b), respectively.

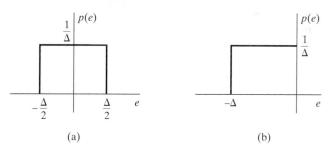

(a)   (b)   **Figure P6.54**

(a) Determine the mean m_e and the variance σ_e^2 for the noise owing to rounding.
(b) Determine the mean m_e and the variance σ_e^2 for the noise owing to truncation.

6.55. Consider an LTI system with two inputs, as depicted in Figure P6.55. Let $h_1[n]$ and $h_2[n]$ be the impulse responses from nodes 1 and 2, respectively, to the output, node 3. Show that if $x_1[n]$ and $x_2[n]$ are uncorrelated, then their corresponding outputs $y_1[n]$ and $y_2[n]$ are also uncorrelated.

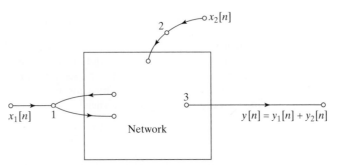

Figure P6.55

6.56. The flow graphs in Figure P6.56 all have the same system function. Assume that the systems in the figure are implemented using $(B + 1)$-bit fixed-point arithmetic in all the computations. Assume also that all products are rounded to $(B + 1)$ bits *before* additions are performed.

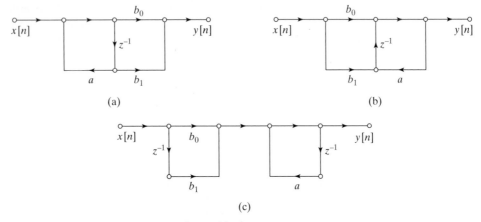

Figure P6.56

(a) Draw linear-noise models for each of the systems in Figure P6.56.

(b) Two of the flow graphs in Figure P6.56 have the *same* total output noise power owing to arithmetic round-off. Without explicitly computing the output noise power, determine which two have the same output noise power.

(c) Determine the output noise power for each of the flow graphs in Figure P6.56. Express your answer in terms of σ_B^2, the power of a single source of round-off noise.

6.57. The flow graph of a 1st-order system is shown in Figure P6.57.

Figure P6.57

(a) Assuming infinite-precision arithmetic, find the response of the system to the input

$$x[n] = \begin{cases} \frac{1}{2}, & n \geq 0, \\ 0, & n < 0. \end{cases}$$

What is the response of the system for large n?

Now suppose that the system is implemented with fixed-point arithmetic. The coefficient and all variables in the flow graph are represented in sign-and-magnitude notation with 5-bit registers. That is, all numbers are to be considered signed fractions represented as

$$b_0 b_1 b_2 b_3 b_4,$$

where b_0, b_1, b_2, b_3, and b_4 are either 0 or 1 and

$$|\text{Register value}| = b_1 2^{-1} + b_2 2^{-2} + b_3 2^{-3} + b_4 2^{-4}.$$

If $b_0 = 0$, the fraction is positive, and if $b_0 = 1$, the fraction is negative. The result of a multiplication of a sequence value by a coefficient is truncated before additions occur; i.e., only the sign bit and the most significant four bits are retained.

(b) Compute the response of the quantized system to the input of part (a), and plot the responses of both the quantized and unquantized systems for $0 \le n \le 5$. How do the responses compare for large n?

(c) Now consider the system depicted in Figure P6.57, where

$$x[n] = \begin{cases} \frac{1}{2}(-1)^n, & n \ge 0, \\ 0, & n < 0. \end{cases}$$

Repeat parts (a) and (b) for this system and input.

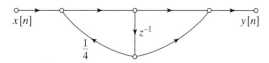

Figure P6.57

6.58. A causal LTI system has a system function

$$H(z) = \frac{1}{1 - 1.04z^{-1} + 0.98z^{-2}}.$$

(a) Is this system stable?

(b) If the coefficients are rounded to the nearest tenth, would the resulting system be stable?

6.59. When implemented with infinite-precision arithmetic, the flow graphs in Figure P6.59 have the same system function.

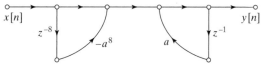

Network 1

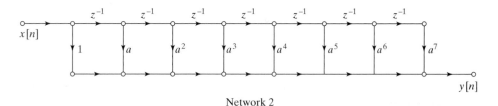

Network 2

Figure P6.59

(a) Show that the two systems have the same overall system function from input $x[n]$ to output $y[n]$.

(b) Assume that the preceding systems are implemented with two's complement fixed-point arithmetic and that products are rounded *before* additions are performed. Draw signal flow graphs that insert round-off noise sources at appropriate locations in the signal flow graphs of Figure P6.59.

(c) Circle the nodes in your figure from part (b) where overflow can occur.

(d) Determine the maximum size of the input samples such that overflow cannot occur in either of the two systems.

(e) Assume that $|a| < 1$. Determine the total noise power at the output of each system, and determine the maximum value of $|a|$ such that Network 1 has lower output noise power than Network 2.

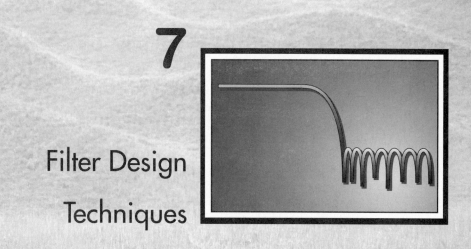

7

Filter Design
Techniques

7.0 INTRODUCTION

Filters are a particularly important class of LTI systems. Strictly speaking, the term *frequency-selective filter* suggests a system that passes certain frequency components of an input signal and totally rejects all others, but in a broader context, any system that modifies certain frequencies relative to others is also called a filter. While the primary emphasis in this chapter is on the design of frequency-selective filters, some of the techniques are more broadly applicable. We concentrate on the design of causal filters, although in many contexts, filters need not be restricted to causal designs. Very often, noncausal filters are designed and implemented by modifying causal designs.

The design of discrete-time filters corresponds to determining the parameters of a transfer function or difference equation that approximates a desired impulse response or frequency response within specified tolerances. As discussed in Chapter 2, discrete-time systems implemented with difference equations fall into two basic categories: infinite impulse response (IIR) systems and finite impulse response (FIR) systems. Designing IIR filters implies obtaining an approximating transfer function that is a rational function of z, whereas designing FIR filters implies polynomial approximation. The commonly used design techniques for these two classes take different forms. When discrete-time filters first came into common use, their designs were based on mapping well-formulated and well-understood continuous-time filter designs to discrete-time designs through techniques such as impulse invariance and the bilinear transformation, as we will discuss in Sections 7.2.1 and 7.2.2. These always resulted in IIR filters and remain at the core of the design of frequency selective discrete-time IIR filters. In contrast, since there is not a body of FIR design techniques in continuous time that could be adapted to the discrete-time case, design techniques for that class of filters emerged only after they became important in practical systems. The most prevalent approaches to designing FIR filters are the use of windowing, as we will discuss in Section 7.5 and the class of iterative algorithms discussed in Section 7.7 and collectively referred to as the Parks–McClellan algorithm.

The design of filters involves the following stages: the specification of the desired properties of the system, the approximation of the specifications using a causal discrete-time system, and the realization of the system. Although these three steps are certainly not independent, we focus our attention primarily on the second step, the first being highly dependent on the application and the third dependent on the technology to be used for the implementation. In a practical setting, the desired filter is generally implemented with digital hardware and often used to filter a signal that is derived from a continuous-time signal by means of periodic sampling followed by A/D conversion. For this reason, it has become common to refer to discrete-time filters as *digital filters,* even though the underlying design techniques most often relate only to the discrete-time nature of the signals and systems. The issues associated with quantization of filter coefficients and signals inherent in digital representations is handled separately, as already discussed in Chapter 6.

In this chapter, we will discuss a wide range of methods for designing both IIR and FIR filters. In any practical context, there are a variety of trade offs between these two classes of filters, and many factors that need to be considered in choosing a specific design procedure or class of filters. Our goal in this chapter is to discuss and illustrate some of the most widely used design techniques and to suggest some of the trade offs involved. The projects and problems on the companion website provide an opportunity to explore in more depth the characteristics of the various filter types and classes and the associated issues and trade offs.

7.1 FILTER SPECIFICATIONS

In our discussion of filter design techniques, we will focus primarily on frequency-selective lowpass filters, although many of the techniques and examples generalize to other types of filters. Furthermore, as discussed in Section 7.4, lowpass filter designs are easily transformed into other types of frequency-selective filters.

Figure 7.1 depicts the typical representation of the tolerance limits associated with approximating a discrete-time lowpass filter that ideally has unity gain in the passband and zero gain in the stopband. We refer to a plot such as Figure 7.1 as a "tolerance scheme."

Since the approximation cannot have an abrupt transition from passband to stopband, a transition region from the passband edge frequency ω_p to the beginning of the stopband at ω_s is allowed, in which the filter gain is unconstrained.

Depending somewhat on the application, and the historical basis for the design technique, the passband tolerance limits may vary symmetrically around unity gain in which case $\delta_{p_1} = \delta_{p_2}$, or the passband may be constrained to have maximum gain of unity, in which case $\delta_{p_1} = 0$.

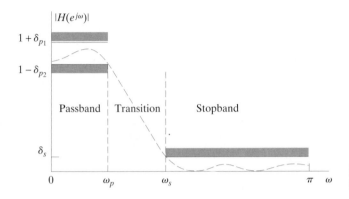

Figure 7.1   Lowpass filter tolerance scheme.

Many of the filters used in practice are specified by a tolerance scheme similar to that which is presented below in Example 7.1, with no constraints on the phase response other than those imposed implicitly by requirements of stability and causality. For example, the poles of the system function for a causal and stable IIR filter must lie inside the unit circle. In designing FIR filters, we often impose the constraint of linear phase. This removes the phase of the signal from consideration in the design process.

Example 7.1   Determining Specifications for a Discrete-Time Filter

Consider a discrete-time lowpass filter that is to be used to filter a continuous-time signal using the basic configuration of Figure 7.2. As shown in Section 4.4, if an LTI discrete-time system is used as in Figure 7.2, and if the input is bandlimited and the sampling frequency is high enough to avoid aliasing, then the overall system behaves as an LTI continuous-time system with frequency response

$$H_{\text{eff}}(j\Omega) = \begin{cases} H(e^{j\Omega T}), & |\Omega| < \pi/T, \\ 0, & |\Omega| \geq \pi/T. \end{cases} \tag{7.1a}$$

In such cases, it is straightforward to convert from specifications on the effective continuous-time filter to specifications on the discrete-time filter through the relation $\omega = \Omega T$. That is, $H(e^{j\omega})$ is specified over one period by the equation

$$H(e^{j\omega}) = H_{\text{eff}}\left(j\frac{\omega}{T}\right), \qquad |\omega| < \pi. \tag{7.1b}$$

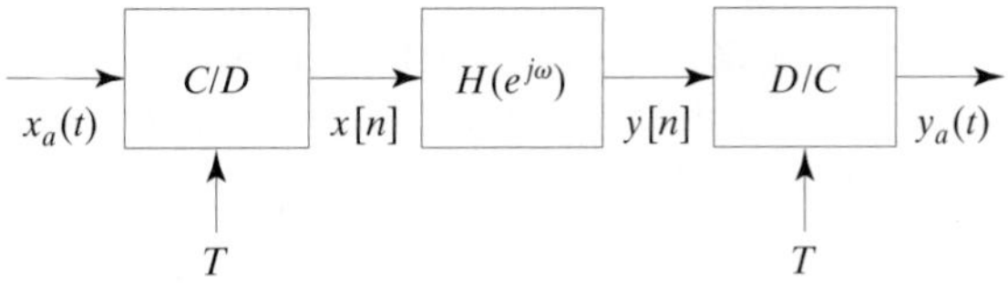

Figure 7.2   Basic system for discrete-time filtering of continuous-time signals.

For this example, the overall system of Figure 7.2 is to have the following properties when the sampling rate is 10^4 samples/s ($T = 10^{-4}$ s):

1. The gain $|H_{\text{eff}}(j\Omega)|$ should be within ± 0.01 of unity in the frequency band $0 \leq \Omega \leq 2\pi(2000)$.

2. The gain should be no greater than 0.001 in the frequency band $2\pi(3000) \leq \Omega$.

Since Eq. (7.1a) is a mapping between the continuous-time and discrete-time frequencies, it only affects the passband and stopband edge frequencies and not the tolerance limits on frequency response magnitude. For this specific example, the parameters would be

$$\delta_{p1} = \delta_{p2} = 0.01$$

$$\delta_s = 0.001$$

$$\omega_p = 0.4\pi \text{ radians}$$

$$\omega_s = 0.6\pi \text{ radians}.$$

Therefore, in this case, the ideal passband magnitude is unity and is allowed to vary between $(1+\delta_{p1})$ and $(1-\delta_{p2})$, and the stopband magnitude is allowed to vary between 0 and δ_s. Expressed in units of decibels,

$$
\begin{aligned}
\text{ideal passband gain in decibels} &= 20\log_{10}(1) &&= 0 \text{ dB} \\
\text{maximum passband gain in decibels} &= 20\log_{10}(1.01) &&= 0.0864 \text{ dB} \\
\text{minimum passband gain at passband edge in decibels} &= 20\log_{10}(0.99) &&= -0.873 \text{ dB} \\
\text{maximum stopband gain in decibels} &= 20\log_{10}(0.001) &&= -60 \text{ dB}
\end{aligned}
$$

Example 7.1 was phrased in the context of using a discrete-time filter to process a continuous-time signal after periodic sampling. There are many applications in which a discrete-time signal to be filtered is not derived from a continuous-time signal, and there are a variety of means besides periodic sampling for representing continuous-time signals in terms of sequences. Also, in most of the design techniques that we discuss, the sampling period plays no role whatsoever in the approximation procedure. For these reasons, we take the point of view that the filter design problem begins from a set of desired specifications in terms of the discrete-time frequency variable ω. Depending on the specific application or context, these specifications may or may not have been obtained from a consideration of filtering in the framework of Figure 7.2.

7.2 DESIGN OF DISCRETE-TIME IIR FILTERS FROM CONTINUOUS-TIME FILTERS

Historically, as the field of digital signal processing was emerging, techniques for the design of discrete-time IIR filters relied on the transformation of a continuous-time filter into a discrete-time filter meeting prescribed specifications. This was and still is a reasonable approach for several reasons:

- The art of continuous-time IIR filter design is highly advanced, and since useful results can be achieved, it is advantageous to use the design procedures already developed for continuous-time filters.

- Many useful continuous-time IIR design methods have relatively simple closed-form design formulas. Therefore, discrete-time IIR filter design methods based on such standard continuous-time design formulas are simple to carry out.

- The standard approximation methods that work well for continuous-time IIR filters do not lead to simple closed-form design formulas when these methods are applied directly to the discrete-time IIR case, because the frequency response of a discrete-time filter is periodic, and that of a continuous-time filter is not.

The fact that continuous-time filter designs can be mapped to discrete-time filter designs is totally unrelated to, and independent of, whether the discrete-time filter is to be used in the configuration of Figure 7.2 for processing continuous-time signals. We emphasize again that the design procedure for the discrete-time system begins from a set of discrete-time specifications. Henceforth, we assume that these specifications have been appropriately determined. We will use continuous-time filter approximation methods only as a convenience in determining the discrete-time filter that meets the desired specifications. Indeed, the continuous-time filter on which the approximation is based may have a frequency response that is vastly different from the effective frequency response when the discrete-time filter is used in the configuration of Figure 7.2.

In designing a discrete-time filter by transforming a prototype continuous-time filter, the specifications for the continuous-time filter are obtained by a transformation of the specifications for the desired discrete-time filter. The system function $H_c(s)$ or impulse response $h_c(t)$ of the continuous-time filter is then obtained through one of the established approximation methods used for continuous-time filter design, such as those which are discussed in Appendix B. Next, the system function $H(z)$ or impulse response $h[n]$ for the discrete-time filter is obtained by applying to $H_c(s)$ or $h_c(t)$ a transformation of the type discussed in this section.

In such transformations, we generally require that the essential properties of the continuous-time frequency response be preserved in the frequency response of the resulting discrete-time filter. Specifically, this implies that we want the imaginary axis of the s-plane to map onto the unit circle of the z-plane. A second condition is that a stable continuous-time filter should be transformed to a stable discrete-time filter. This means that if the continuous-time system has poles only in the left half of the s-plane, then the discrete-time filter must have poles only inside the unit circle in the z-plane. These constraints are basic to all the techniques discussed in this section.

7.2.1 Filter Design by Impulse Invariance

In Section 4.4.2, we discussed the concept of *impulse invariance*, wherein a discrete-time system is defined by sampling the impulse response of a continuous-time system. We showed that impulse invariance provides a direct means of computing samples of the output of a bandlimited continuous-time system for bandlimited input signals. In some contexts, it is particularly appropriate and convenient to design a discrete-time filter by sampling the impulse response of a continuous-time filter. For example, if the overall objective is to simulate a continuous-time system in a discrete-time setting, we might typically carry out the simulation in the configuration of Figure 7.2, with the discrete-time system design such that its impulse response corresponds to samples of the continuous-time filter to be simulated. In other contexts, it might be desirable to maintain, in a discrete-time setting, certain time-domain characteristics of well-developed continuous-time filters, such as desirable time-domain overshoot, energy compaction, controlled time-domain ripple, and so on. Alternatively, in the context of filter design, we can think of impulse invariance as a method for obtaining a discrete-time system whose frequency response is determined by the frequency response of a continuous-time system.

In the impulse invariance design procedure for transforming continuous-time filters into discrete-time filters, the impulse response of the discrete-time filter is chosen proportional to equally spaced samples of the impulse response of the continuous-time filter; i.e.,

$$h[n] = T_d h_c(nT_d), \tag{7.2}$$

where T_d represents a sampling interval. As we will see, because we begin the design problem with the discrete-time filter specifications, the parameter T_d in Eq. (7.2) in fact has no role whatsoever in the design process or the resulting discrete-time filter. However, since it is customary to specify this parameter in defining the procedure, we include it in the following discussion. Even if the filter is used in the basic configuration of Figure 7.2, the design sampling period T_d need not be the same as the sampling period T associated with the C/D and D/C conversion.

When impulse invariance is used as a means for designing a discrete-time filter with a specified frequency response, we are especially interested in the relationship between the frequency responses of the discrete-time and continuous-time filters. From the discussion of sampling in Chapter 4, it follows that the frequency response of the discrete-time filter obtained through Eq. (7.2) is related to the frequency response of the continuous-time filter by

$$H(e^{j\omega}) = \sum_{k=-\infty}^{\infty} H_c\left(j\frac{\omega}{T_d} + j\frac{2\pi}{T_d}k\right). \tag{7.3}$$

If the continuous-time filter is bandlimited, so that

$$H_c(j\Omega) = 0, \qquad |\Omega| \geq \pi/T_d, \tag{7.4}$$

then

$$H(e^{j\omega}) = H_c\left(j\frac{\omega}{T_d}\right), \qquad |\omega| \leq \pi; \tag{7.5}$$

i.e., the discrete-time and continuous-time frequency responses are related by a linear scaling of the frequency axis, namely, $\omega = \Omega T_d$ for $|\omega| < \pi$. Unfortunately, any practical continuous-time filter cannot be exactly bandlimited, and consequently, interference between successive terms in Eq. (7.3) occurs, causing aliasing, as illustrated in Figure 7.3. However, if the continuous-time filter approaches zero at high frequencies, the aliasing may be negligibly small, and a useful discrete-time filter can result from sampling the impulse response of a continuous-time filter.

When the impulse invariance design procedure is used to utilize continuous-time filter design procedures for the design of a discrete-time filter with given frequency response specifications, the discrete-time filter specifications are first transformed to continuous-time filter specifications through the use of Eq. (7.5). Assuming that the

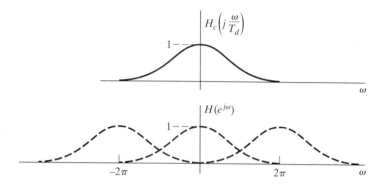

Figure 7.3   Illustration of aliasing in the impulse invariance design technique.

aliasing involved in the transformation from $H_c(j\Omega)$ to $H(e^{j\omega})$ is negligible, we obtain the specifications on $H_c(j\Omega)$ by applying the relation

$$\Omega = \omega/T_d \tag{7.6}$$

to obtain the continuous-time filter specifications from the specifications on $H(e^{j\omega})$. After obtaining a continuous-time filter that meets these specifications, the continuous-time filter with system function $H_c(s)$ is transformed to the desired discrete-time filter with system function $H(z)$. We develop the algebraic details of the transformation from $H_c(s)$ to $H(z)$ shortly. Note, however, that in the transformation back to discrete-time frequency, $H(e^{j\omega})$ will be related to $H_c(j\Omega)$ through Eq. (7.3), which again applies the transformation of Eq. (7.6) to the frequency axis. As a consequence, the "sampling" parameter T_d cannot be used to control aliasing. Since the basic specifications are in terms of discrete-time frequency, if the sampling rate is increased (i.e., if T_d is made smaller), then the cutoff frequency of the continuous-time filter must increase in proportion. In practice, to compensate for aliasing that might occur in the transformation from $H_c(j\Omega)$ to $H(e^{j\omega})$, the continuous-time filter may be somewhat overdesigned, i.e., designed to exceed the specifications, particularly in the stopband.

While the impulse invariance transformation from continuous time to discrete time is defined in terms of time-domain sampling, it is easy to carry out as a transformation on the system function. To develop this transformation, we consider the system function of a causal continuous-time filter expressed in terms of a partial fraction expansion, so that[1]

$$H_c(s) = \sum_{k=1}^{N} \frac{A_k}{s - s_k}. \tag{7.7}$$

The corresponding impulse response is

$$h_c(t) = \begin{cases} \sum_{k=1}^{N} A_k e^{s_k t}, & t \geq 0, \\ 0, & t < 0. \end{cases} \tag{7.8}$$

---

[1] For simplicity, we assume in the discussion that all poles of $H(s)$ are single order. In Problem 7.41, we consider the modifications required for multiple-order poles.

The impulse response of the causal discrete-time filter obtained by sampling $T_d h_c(t)$ is

$$
h[n] = T_d h_c(nT_d) = \sum_{k=1}^{N} T_d A_k e^{s_k n T_d} u[n]
$$

$$
= \sum_{k=1}^{N} T_d A_k (e^{s_k T_d})^n u[n].
$$

(7.9)

The system function of the causal discrete-time filter is therefore given by

$$
H(z) = \sum_{k=1}^{N} \frac{T_d A_k}{1 - e^{s_k T_d} z^{-1}}.
$$

(7.10)

In comparing Eqs. (7.7) and (7.10), we observe that a pole at $s = s_k$ in the s-plane transforms to a pole at $z = e^{s_k T_d}$ in the z-plane and the coefficients in the partial fraction expansions of $H_c(s)$ and $H(z)$ are equal, except for the scaling multiplier T_d. If the continuous-time causal filter is stable, corresponding to the real part of s_k being less than zero, then the magnitude of $e^{s_k T_d}$ will be less than unity, so that the corresponding pole in the discrete-time filter is inside the unit circle. Therefore, the causal discrete-time filter is also stable. Although the poles in the s-plane map to poles in the z-plane according to the relationship $z_k = e^{s_k T_d}$, it is important to recognize that the impulse invariance design procedure does not correspond to a simple mapping of the s-plane to the z-plane by that relationship. In particular, the zeros in the discrete-time system function are a function of the poles $e^{s_k T_d}$ and the coefficients $T_d A_k$ in the partial fraction expansion, and they will not in general be mapped in the same way the poles are mapped. We illustrate the impulse invariance design procedure of a lowpass filter with the following example.

Example 7.2   Impulse Invariance with a Butterworth Filter

In this example we consider the design of a lowpass discrete-time filter by applying impulse invariance to an appropriate continuous-time filter. The class of filters that we choose for this example is referred to as Butterworth filters, which we discuss in more detail in Section 7.3 and in Appendix B.[2] The specifications for the discrete-time filter correspond to passband gain between 0 dB and -1 dB, and stopband attenuation of at least -15 dB, i.e.,

$$
0.89125 \leq |H(e^{j\omega})| \leq 1, \qquad 0 \leq |\omega| \leq 0.2\pi, \tag{7.11a}
$$

$$
|H(e^{j\omega})| \leq 0.17783, \qquad 0.3\pi \leq |\omega| \leq \pi. \tag{7.11b}
$$

Since the parameter T_d cancels in the impulse invariance procedure, we can just as well choose $T_d = 1$, so that $\omega = \Omega$. In Problem 7.2, this same example is considered, but with the parameter T_d explicitly included to illustrate how and where it cancels.

In designing the filter using impulse invariance on a continuous-time Butterworth filter, we must first transform the discrete-time specifications to specifications on the continuous-time filter. For this example, we will assume that the effect of aliasing in Eq. (7.3) is negligible. After the design is complete, we can evaluate the resulting frequency response against the specifications in Eqs. (7.11a) and (7.11b).

---

[2] Continuous-time Butterworth and Chebyshev filters are discussed in Appendix B.

Because of the preceding considerations, we want to design a continuous-time Butterworth filter with magnitude function $|H_c(j\Omega)|$ for which

$$0.89125 \le |H_c(j\Omega)| \le 1, \qquad 0 \le |\Omega| \le 0.2\pi, \tag{7.12a}$$

$$|H_c(j\Omega)| \le 0.17783, \qquad 0.3\pi \le |\Omega| \le \pi. \tag{7.12b}$$

Since the magnitude response of an analog Butterworth filter is a monotonic function of frequency, Eqs. (7.12a) and (7.12b) will be satisfied if $H_c(j0) = 1$,

$$|H_c(j0.2\pi)| \ge 0.89125 \tag{7.13a}$$

and

$$|H_c(j0.3\pi)| \le 0.17783. \tag{7.13b}$$

The magnitude-squared function of a Butterworth filter is of the form

$$|H_c(j\Omega)|^2 = \frac{1}{1 + (\Omega/\Omega_c)^{2N}}, \tag{7.14}$$

so that the filter design process consists of determining the parameters N and Ω_c to meet the desired specifications. Using Eq. (7.14) in Eqs. (7.13) with equality leads to the equations

$$1 + \left(\frac{0.2\pi}{\Omega_c}\right)^{2N} = \left(\frac{1}{0.89125}\right)^2 \tag{7.15a}$$

and

$$1 + \left(\frac{0.3\pi}{\Omega_c}\right)^{2N} = \left(\frac{1}{0.17783}\right)^2. \tag{7.15b}$$

The simultaneous solution of these two equations is $N = 5.8858$ and $\Omega_c = 0.70474$. The parameter N, however, must be an integer. In order that the specifications are met or exceeded, we must round N up to the nearest integer, $N = 6$, in which case the filter will not exactly satisfy both Eqs. (7.15a) and (7.15b) simultaneously. With $N = 6$, the filter parameter Ω_c can be chosen to exceed the specified requirements (i.e., have lower approximation error) in either the passband, the stopband, or both. Specifically, as the value of Ω_c varies, there is a trade-off in the amount by which the stopband and passband specifications are exceeded. If we substitute $N = 6$ into Eq. (7.15a), we obtain ($\Omega_c = 0.7032$). With this value, the passband specifications (of the continuous-time filter) will be met exactly, and the stopband specifications (of the continuous-time filter) will be exceeded. This allows some margin for aliasing in the discrete-time filter. With ($\Omega_c = 0.7032$) and with $N = 6$, the 12 poles of the magnitude-squared function $H_c(s)H_c(-s) = 1/[1+(s/j\Omega_c)^{2N}]$ are uniformly distributed in angle on a circle of radius ($\Omega_c = 0.7032$), as indicated in Figure 7.4. Consequently, the poles of $H_c(s)$ are the three pole pairs in the left half of the s-plane with the following coordinates:

Pole pair 1: $-0.182 \pm j(0.679)$,

Pole pair 2: $-0.497 \pm j(0.497)$,

Pole pair 3: $-0.679 \pm j(0.182)$.

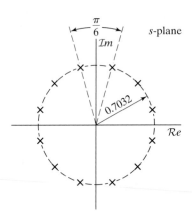

Figure 7.4 s-plane locations for poles of $H_c(s)H_c(-s)$ for 6th-order Butterworth filter in Example 7.2.

Therefore,

$$H_c(s) = \frac{0.12093}{(s^2 + 0.3640s + 0.4945)(s^2 + 0.9945s + 0.4945)(s^2 + 1.3585s + 0.4945)}.$$

$$(7.16)$$

If we express $H_c(s)$ as a partial fraction expansion, perform the transformation of Eq. (7.10), and then combine complex-conjugate terms, the resulting system function of the discrete-time filter is

$$H(z) = \frac{0.2871 - 0.4466z^{-1}}{1 - 1.2971z^{-1} + 0.6949z^{-2}} + \frac{-2.1428 + 1.1455z^{-1}}{1 - 1.0691z^{-1} + 0.3699z^{-2}}$$

$$+ \frac{1.8557 - 0.6303z^{-1}}{1 - 0.9972z^{-1} + 0.2570z^{-2}}.$$

$$(7.17)$$

As is evident from Eq. (7.17), the system function resulting from the impulse invariance design procedure may be realized directly in parallel form. If either the cascade or direct form is desired, the separate 2nd-order terms are first combined in an appropriate way.

The frequency-response functions of the discrete-time system are shown in Figure 7.5. The prototype continuous-time filter had been designed to meet the specifications exactly at the passband edge and to exceed the specifications at the stopband edge, and this turns out to be true for the resulting discrete-time filter. This is an indication that the continuous-time filter was sufficiently bandlimited so that aliasing presented no problem. Indeed, the difference between $20\log_{10}|H(e^{j\omega})|$ and $20\log_{10}|H_c(j\Omega)|$ would not be visible on this plotting scale, except for a slight deviation around $\omega = \pi$. (Recall that $T_d = 1$, so $\Omega = \omega$.) Sometimes, aliasing is much more of a problem. If the resulting discrete-time filter fails to meet the specifications because of aliasing, there is no alternative with impulse invariance but to try again with a higher-order filter or with different filter parameters, holding the order fixed.

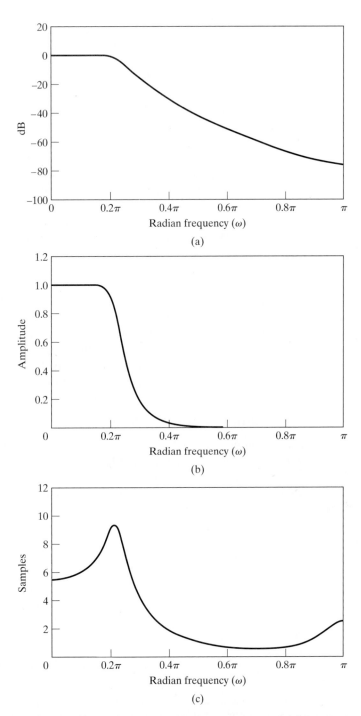

Figure 7.5 Frequency response of 6th-order Butterworth filter transformed by impulse invariance. (a) Log magnitude in dB. (b) Magnitude. (c) Group delay.

The basis for impulse invariance is to choose an impulse response for the discrete-time filter that is similar in some sense to the impulse response of the continuous-time filter. The use of this procedure may be motivated by a desire to maintain the shape of the impulse response or by the knowledge that if the continuous-time filter is bandlimited, consequently the discrete-time filter frequency response will closely approximate the continuous-time frequency response. When the primary objective is to control some aspect of the time response, such as the impulse response or the step response, a natural approach might be to design the discrete-time filter by impulse invariance or by step invariance. In the latter case, the response of the filter to a sampled unit step function is defined to be the sequence obtained by sampling the continuous-time step response. If the continuous-time filter has good step response characteristics, such as a small rise time and low peak overshoot, these characteristics will be preserved in the discrete-time filter. Clearly, this concept of waveform invariance can be extended to the preservation of the output waveshape for a variety of inputs, as illustrated in Problem 7.1. The problem points out the fact that transforming the same continuous-time filter by impulse invariance and also by step invariance (or some other waveform invariance criterion) does not lead to the same discrete-time filter in the two cases.

In the impulse invariance design procedure, the relationship between continuous-time and discrete-time frequency is linear; consequently, except for aliasing, the shape of the frequency response is preserved. This is in contrast to the procedure discussed next, which is based on an algebraic transformation. In concluding this subsection we iterate that the impulse invariance technique is appropriate only for bandlimited filters; highpass or bandstop continuous-time filters, for example, would require additional bandlimiting to avoid severe aliasing distortion if impulse invariance design is used.

7.2.2 Bilinear Transformation

The technique discussed in this subsection uses the bilinear transformation, an algebraic transformation between the variables s and z that maps the entire $j\Omega$-axis in the s-plane to one revolution of the unit circle in the z-plane. Since with this approach, $-\infty \leq \Omega \leq \infty$ maps onto $-\pi \leq \omega \leq \pi$, the transformation between the continuous-time and discrete-time frequency variables is necessarily nonlinear. Therefore, the use of this technique is restricted to situations in which the corresponding nonlinear warping of the frequency axis is acceptable.

With $H_c(s)$ denoting the continuous-time system function and $H(z)$ the discrete-time system function, the bilinear transformation corresponds to replacing s by

$$s = \frac{2}{T_d}\left(\frac{1-z^{-1}}{1+z^{-1}}\right); \tag{7.18}$$

that is,

$$H(z) = H_c\left(\frac{2}{T_d}\left(\frac{1-z^{-1}}{1+z^{-1}}\right)\right). \tag{7.19}$$

As in impulse invariance, a "sampling" parameter T_d is often included in the definition of the bilinear transformation. Historically, this parameter has been included, because the difference equation corresponding to $H(z)$ can be obtained by applying the trapezoidal integration rule to the differential equation corresponding to $H_c(s)$, with T_d

representing the step size of the numerical integration. (See Kaiser, 1966, and Problem 7.49.) However, in filter design, our use of the bilinear transformation is based on the properties of the algebraic transformation given in Eq. (7.18). As with impulse invariance, the parameter T_d is of no consequence in the design procedure, since we assume that the design problem always begins with specifications on the discrete-time filter $H(e^{j\omega})$. When these specifications are mapped to continuous-time specifications, and the continuous-time filter is then mapped back to a discrete-time filter, the effect of T_d will cancel. We will retain the parameter T_d in our discussion for historical reasons; in specific problems and examples, any convenient value can be chosen.

To develop the properties of the algebraic transformation specified in Eq. (7.18), we solve for z to obtain

$$z = \frac{1 + (T_d/2)s}{1 - (T_d/2)s},\qquad (7.20)$$

and, substituting $s = \sigma + j\Omega$ into Eq. (7.20), we obtain

$$z = \frac{1 + \sigma T_d/2 + j\Omega T_d/2}{1 - \sigma T_d/2 - j\Omega T_d/2}.\qquad (7.21)$$

If $\sigma < 0$, then, from Eq. (7.21), it follows that $|z| < 1$ for any value of Ω. Similarly, if $\sigma > 0$, then $|z| > 1$ for all Ω. That is, if a pole of $H_c(s)$ is in the left-half s-plane, its image in the z-plane will be inside the unit circle. Therefore, causal stable continuous-time filters map into causal stable discrete-time filters.

Next, to show that the $j\Omega$-axis of the s-plane maps onto the unit circle, we substitute $s = j\Omega$ into Eq. (7.20), obtaining

$$z = \frac{1 + j\Omega T_d/2}{1 - j\Omega T_d/2}.\qquad (7.22)$$

From Eq. (7.22), it is clear that $|z| = 1$ for all values of s on the $j\Omega$-axis. That is, the $j\Omega$-axis maps onto the unit circle, so Eq. (7.22) takes the form

$$e^{j\omega} = \frac{1 + j\Omega T_d/2}{1 - j\Omega T_d/2}.\qquad (7.23)$$

To derive a relationship between the variables ω and Ω, it is useful to return to Eq. (7.18) and substitute $z = e^{j\omega}$. We obtain

$$s = \frac{2}{T_d}\left(\frac{1 - e^{-j\omega}}{1 + e^{-j\omega}}\right),\qquad (7.24)$$

or, equivalently,

$$s = \sigma + j\Omega = \frac{2}{T_d}\left[\frac{2e^{-j\omega/2}(j\sin\omega/2)}{2e^{-j\omega/2}(\cos\omega/2)}\right] = \frac{2j}{T_d}\tan(\omega/2).\qquad (7.25)$$

Equating real and imaginary parts on both sides of Eq. (7.25) leads to the relations $\sigma = 0$ and

$$\Omega = \frac{2}{T_d}\tan(\omega/2),\qquad (7.26)$$

or

$$\omega = 2\arctan(\Omega T_d/2).\qquad (7.27)$$

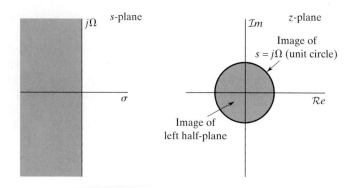

Figure 7.6   Mapping of the s-plane onto the z-plane using the bilinear transformation.

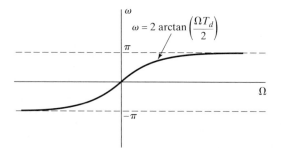

Figure 7.7   Mapping of the continuous-time frequency axis onto the discrete-time frequency axis by bilinear transformation.

These properties of the bilinear transformation as a mapping from the s-plane to the z-plane are summarized in Figures 7.6 and 7.7. From Eq. (7.27) and Figure 7.7, we see that the range of frequencies $0 \leq \Omega \leq \infty$ maps to $0 \leq \omega \leq \pi$, while the range $-\infty \leq \Omega \leq 0$ maps to $-\pi \leq \omega \leq 0$. The bilinear transformation avoids the problem of aliasing encountered with the use of impulse invariance, because it maps the entire imaginary axis of the s-plane onto the unit circle in the z-plane. The price paid for this, however, is the nonlinear compression of the frequency axis, as depicted in Figure 7.7. Consequently, the design of discrete-time filters using the bilinear transformation is useful only when this compression can be tolerated or compensated for, as in the case of filters that approximate ideal piecewise-constant magnitude-response characteristics. This is illustrated in Figure 7.8, wherein we show how a continuous-time frequency response and tolerance scheme maps to a corresponding discrete-time frequency response and tolerance scheme through the frequency warping of Eqs. (7.26) and (7.27). If the critical frequencies (such as the passband and stopband edge frequencies) of the continuous-time filter are prewarped according to Eq. (7.26) then, when the continuous-time filter is transformed to the discrete-time filter using Eq. (7.19), the discrete-time filter will meet the desired specifications.

Although the bilinear transformation can be used effectively in mapping a piecewise-constant magnitude-response characteristic from the s-plane to the z-plane, the distortion in the frequency axis also manifests itself as a warping of the phase response of the filter. For example, Figure 7.9 shows the result of applying the bilinear transformation to an ideal linear-phase factor $e^{-s\alpha}$. If we substitute Eq. (7.18) for s and evaluate the result on the unit circle, the phase angle is $-(2\alpha/T_d)\tan(\omega/2)$. In Figure 7.9, the

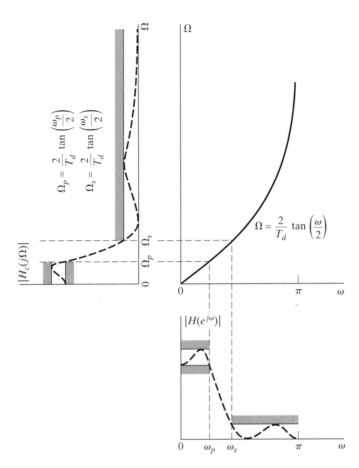

Figure 7.8 Frequency warping inherent in the bilinear transformation of a continuous-time lowpass filter into a discrete-time lowpass filter. To achieve the desired discrete-time cutoff frequencies, the continuous-time cutoff frequencies must be prewarped as indicated.

solid curve shows the function $-(2\alpha/T_d)\tan(\omega/2)$, and the dotted curve is the periodic linear-phase function $-(\omega\alpha/T_d)$, which is obtained by using the small-angle approximation $\omega/2 \approx \tan(\omega/2)$. From this, it should be evident that if we desire a discrete-time lowpass filter with a linear-phase characteristic, we cannot obtain such a filter by applying the bilinear transformation to a continuous-time lowpass filter with a linear-phase characteristic.

As mentioned previously, because of the frequency warping, the bilinear transformation is most useful in the design of approximations to filters with piecewise-constant frequency magnitude characteristics, such as highpass, lowpass and bandpass filters. As demonstrated in Example 7.2, impulse invariance can also be used to design lowpass filters. However, impulse invariance cannot be used to map highpass continuous-time designs to highpass discrete-time designs, since highpass continuous-time filters are not bandlimited.

In Example 4.4, we discussed a class of filters often referred to as discrete-time differentiators. A significant feature of the frequency response of this class of filters is that it is linear with frequency. The nonlinear warping of the frequency axis introduced by the bilinear transformation will not preserve that property. Consequently, the

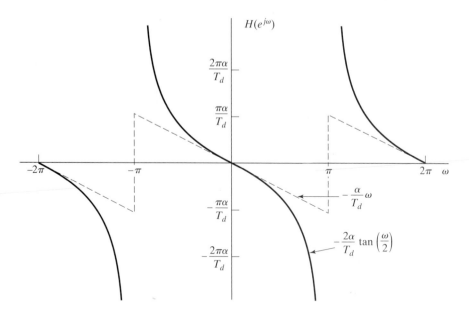

Figure 7.9 Illustration of the effect of the bilinear transformation on a linear-phase characteristic. (Dashed line is linear phase and solid line is phase resulting from bilinear transformation.)

bilinear transformation applied to a continuous-time differentiator will not result in a discrete-time differentiator. However, impulse invariance applied to an appropriately bandlimited continuous-time differentiator will result in a discrete-time differentiator.

7.3 DISCRETE-TIME BUTTERWORTH, CHEBYSHEV AND ELLIPTIC FILTERS

Historically, the most widely used classes of frequency-selective continuous-time filters are those referred to as Butterworth, Chebyshev and elliptic filter designs. In Appendix B we briefly summarize the characteristics of these three classes of continuous-time filters. The associated closed-form design formulas make the design procedure relatively straightforward. As discussed in Appendix B, the magnitude of the frequency response of a Butterworth continuous-time filter is monotonic in the passband and the stopband. A type I Chebyshev filter has an equiripple frequency response in the passband and varies monotonically in the stopband. A type II Chebyshev filter is monotonic in the passband and equiripple in the stopband. An elliptic filter is equiripple in both the passband and the stopband. Clearly, these properties will be preserved when the filter is mapped to a digital filter with the bilinear transformation. This is illustrated by the dashed approximation shown in Figure 7.8. The filters resulting from applying the bilinear transformation to these classes of continuous-time filters, referred to respectively as discrete-time Butterworth, Chebyshev and elliptic filters have similarly become widely used as discrete-time frequency selective filters.

As a first step in the design procedure for any of these classes of filters, the critical frequencies, i.e., the band edge frequencies, must be prewarped to the continuous-time frequencies using Eq. (7.26) so that the frequency distortion inherent in the bilinear transformation will map the continuous-time frequencies back to the correct discrete-time frequencies. This prewarping will be illustrated in more detail in Example 7.3. The allowed tolerances in the passbands and stopbands will be the same for the discrete-time and continuous-time filters since the bilinear mapping only distorts the frequency axis, not the amplitude scale. In using a discrete-time filter design package such as found in MATLAB and LabVIEW, the typical inputs would be the desired tolerances and the discrete-time critical frequencies. The design program explicitly or implicitly handles any necessary prewarping of the frequencies.

In advance of illustrating these classes of filters with several examples, it is worth commenting on some general characteristics to expect. We have noted above that we expect the discrete-time Butterworth, Chebyshev and elliptic filter frequency responses to retain the monotonicity and ripple characteristics of the corresponding continuous-time filters. The Nth-order continuous-time lowpass Butterworth filter has N zeros at $\Omega = \infty$. Since the bilinear transformation maps $s = \infty$ to $z = -1$, we would expect any Butterworth design utilizing the bilinear transformation to result in N zeros at $z = -1$. The same is also true for the Chebyshev type I lowpass filter.

7.3.1 Examples of IIR Filter Design

In the following discussion, we present a number of examples to illustrate IIR filter design. The purpose of Example 7.3 is to illustrate the steps in the design of a Butterworth filter using the bilinear transformation, in comparison with the use of impulse invariance. Example 7.4 presents a set of examples comparing the design of a Butterworth, Chebyshev I, Chebyshev II, and elliptic filter. Example 7.5 illustrates, with a different set of specifications, the design of a Butterworth, Chebyshev I, Chebyshev II and elliptic filter. These designs will be compared in Section 7.8.1 with FIR designs. For both Example 7.4 and 7.5 the filter design package in the signal processing toolbox of MATLAB was used.

Example 7.3   Bilinear Transformation of a Butterworth Filter

Consider the discrete-time filter specifications of Example 7.2, in which we illustrated the impulse invariance technique for the design of a discrete-time filter. The specifications for the discrete-time filter are

$$0.89125 \leq |H(e^{j\omega})| \leq 1, \qquad 0 \leq \omega \leq 0.2\pi, \qquad (7.28a)$$

$$|H(e^{j\omega})| \leq 0.17783, \qquad 0.3\pi \leq \omega \leq \pi. \qquad (7.28b)$$

In carrying out the design using the bilinear transformation applied to a continuous-time design, the critical frequencies of the discrete-time filter are first prewarped to the corresponding continuous-time frequencies using Eq. (7.26) so that the frequency distortion inherent in the bilinear transformation will map the continuous-time frequencies back to the correct discrete-time critical frequencies. For this specific filter, with $|H_c(j\Omega)|$ representing the magnitude-response function of the continuous-time filter, we require that

$$0.89125 \leq |H_c(j\Omega)| \leq 1, \qquad 0 \leq \Omega \leq \frac{2}{T_d} \tan\left(\frac{0.2\pi}{2}\right), \tag{7.29a}$$

$$|H_c(j\Omega)| \leq 0.17783, \qquad \frac{2}{T_d} \tan\left(\frac{0.3\pi}{2}\right) \leq \Omega \leq \infty. \tag{7.29b}$$

For convenience, we choose $T_d = 1$. Also, as with Example 7.2, since a continuous-time Butterworth filter has a monotonic magnitude response, we can equivalently require that

$$|H_c(j2\tan(0.1\pi))| \geq 0.89125 \tag{7.30a}$$

and

$$|H_c(j2\tan(0.15\pi))| \leq 0.17783. \tag{7.30b}$$

The form of the magnitude-squared function for the Butterworth filter is

$$|H_c(j\Omega)|^2 = \frac{1}{1 + (\Omega/\Omega_c)^{2N}}. \tag{7.31}$$

Solving for N and Ω_c with the equality sign in Eqs. (7.30a) and (7.30b), we obtain

$$1 + \left(\frac{2\tan(0.1\pi)}{\Omega_c}\right)^{2N} = \left(\frac{1}{0.89}\right)^2 \tag{7.32a}$$

and

$$1 + \left(\frac{2\tan(0.15\pi)}{\Omega_c}\right)^{2N} = \left(\frac{1}{0.178}\right)^2, \tag{7.32b}$$

and solving for N in Eqs. (7.32a) and (7.32b) gives

$$N = \frac{\log\left[\left(\left(\frac{1}{0.178}\right)^2 - 1\right) \Big/ \left(\left(\frac{1}{0.89}\right)^2 - 1\right)\right]}{2\log[\tan(0.15\pi)/\tan(0.1\pi)]} \tag{7.33}$$

$$= 5.305.$$

Since N must be an integer, we choose $N = 6$. Substituting $N = 6$ into Eq. (7.32b), we obtain $\Omega_c = 0.766$. For this value of Ω_c, the passband specifications are exceeded and the stopband specifications are met exactly. This is reasonable for the bilinear transformation, since we do not have to be concerned with aliasing. That is, with proper prewarping, we can be certain that the resulting discrete-time filter will meet the specifications exactly at the desired stopband edge.

In the s-plane, the 12 poles of the magnitude-squared function are uniformly distributed in angle on a circle of radius 0.766, as shown in Figure 7.10. The system function of the causal continuous-time filter obtained by selecting the left half-plane poles is

$$H_c(s) = \frac{0.20238}{(s^2 + 0.3996s + 0.5871)(s^2 + 1.0836s + 0.5871)(s^2 + 1.4802s + 0.5871)}.$$

(7.34)

The system function for the discrete-time filter is then obtained by applying the bilinear transformation to $H_c(s)$ with $T_d = 1$. The result is

$$H(z) = \frac{0.0007378(1 + z^{-1})^6}{(1 - 1.2686z^{-1} + 0.7051z^{-2})(1 - 1.0106z^{-1} + 0.3583z^{-2})}$$

(7.35)

$$\times \frac{1}{(1 - 0.9044z^{-1} + 0.2155z^{-2})}.$$

The magnitude, log magnitude, and group delay of the frequency response of the discrete-time filter are shown in Figure 7.11. At $\omega = 0.2\pi$ the log magnitude is -0.56 dB, and at $\omega = 0.3\pi$ the log magnitude is exactly -15 dB.

Since the bilinear transformation maps the entire $j\Omega$-axis of the s-plane onto the unit circle in the z-plane, the magnitude response of the discrete-time filter falls off much more rapidly than that of the continuous-time filter or the Butterworth discrete-time filter designed by impulse invariance. In particular, the behavior of $H(e^{j\omega})$ at $\omega = \pi$ corresponds to the behavior of $H_c(j\Omega)$ at $\Omega = \infty$. Therefore, since the continuous-time Butterworth filter has a 6th-order zero at $s = \infty$, the resulting discrete-time filter has a 6th-order zero at $z = -1$.

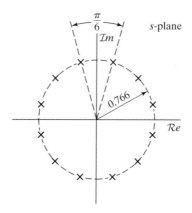

Figure 7.10 s-plane locations for poles of $H_c(s)H_c(-s)$ for 6th-order Butterworth filter in Example 7.3.

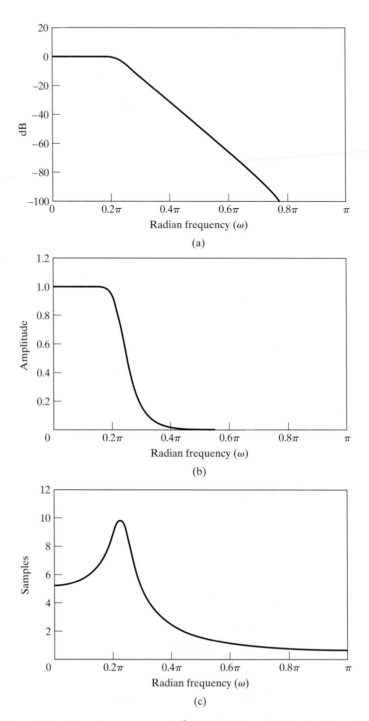

Figure 7.11   Frequency response of 6th-order Butterworth filter transformed by bilinear transform. (a) Log magnitude in dB. (b) Magnitude. (c) Group delay.

Since the general form of the magnitude-squared of the Nth-order Butterworth continuous-time filter is as given by Eq. (7.31), and since ω and Ω are related by Eq. (7.26), it follows that the general N^{th}-order Butterworth discrete-time filter has the magnitude-squared function

$$|H(e^{j\omega})|^2 = \frac{1}{1 + \left(\dfrac{\tan(\omega/2)}{\tan(\omega_c/2)}\right)^{2N}}, \qquad (7.36)$$

where $\tan(\omega_c/2) = \Omega_c T_d/2$. The frequency-response function of Eq. (7.36) has the same properties as the continuous-time Butterworth response; i.e., it is maximally flat[3] and $|H(e^{j\omega_c})|^2 = 0.5$. However, the function in Eq. (7.36) is periodic with period 2π and falls off more sharply than the continuous-time Butterworth response.

Discrete-time Butterworth filters are not typically designed directly by starting with Eq. (7.36), because it is not straightforward to determine the z-plane locations of the poles (all the zeros are at $z = -1$) associated with the magnitude-squared function of Eq. (7.36). It is necessary to determine the poles so as to factor the magnitude-squared function into $H(z)H(z^{-1})$ and thereby determine $H(z)$. It is much easier to factor the continuous-time system function, and then transform the left half-plane poles by the bilinear transformation as we did in Example 7.3.

Equations of the form of Eq. (7.36) may also be obtained for discrete-time Chebyshev and elliptic filters. However, the details of the design computations for these commonly used classes of filters are best carried out by computer programs that incorporate the appropriate closed-form design equations.

In the next example, we compare the design of a lowpass filter based on Butterworth, Chebyshev I, Chebyshev II and elliptic filter designs. There are some specific characteristics of the frequency response magnitude and the pole–zero patterns for each of these four discrete-time lowpass filter types, and these characteristics will be evident in the designs in Example 7.4 and Example 7.5 that follow.

For a Butterworth lowpass filter, the frequency response magnitude decreases monotonically in both the passband and stopband, and all the zeros of the transfer function are at $z = -1$. For a Chebyshev Type I lowpass filter, the frequency response magnitude will always be equiripple in the passband, i.e., will oscillate with equal maximum error on either side of the desired gain and will be monotonic in the stopband. All the zeros of the corresponding transfer function will be at $z = -1$. For a Chebyshev Type II lowpass filter, the frequency response magnitude will be monotonic in the passband and equiripple in the stopband, i.e., oscillates around zero gain. Because of this equiripple stopband behavior, the zeros of the transfer function will correspondingly be distributed on the unit circle.

In both cases of Chebyshev approximation, the monotonic behavior in either the stopband or the passband suggests that perhaps a lower-order system might be obtained if an equiripple approximation were used in both the passband and the stopband. Indeed, it can be shown (see Papoulis, 1957) that for fixed values of δ_{p_1}, δ_{p_2}, δ_s, ω_p, and ω_s in the tolerance scheme of Figure 7.1, the lowest order filter is obtained when the approximation error ripples equally between the extremes of the two approximation bands. This equiripple behavior is achieved with the class of filters referred to as elliptic

---

[3] The first $(2N - 1)$ derivatives of $|H(e^{j\omega})|^2$ are zero at $\omega = 0$.

filters. Elliptic filters, like the Chebyshev type II filter, has its zeros arrayed in the stopband region of the unit circle. These properties of Butterworth, Chebyshev, and elliptic filters are illustrated by the following example.

Example 7.4   Design Comparisons

For the four filter designs that follow, the signal processing toolbox in MATLAB was used. This and other typical design programs for IIR lowpass filter design, assume tolerance specifications as indicated in Figure 7.1 with $\delta_{p1} = 0$. Although the resulting designs correspond to what would result from applying the bilinear transformation to appropriate continuous-time designs, any required frequency prewarping and incorporation of the bilinear transformation, are internal to these design programs and transparent to the user. Consequently the specifications are given to the design program directly in terms of the discrete-time parameters. For this example, the filter has been designed to meet or exceed the following specifications.:

$$\begin{aligned}
\text{passband edge frequency } \omega_p &= 0.5\pi \\
\text{stopband edge frequency } \omega_s &= 0.6\pi \\
\text{maximum passband gain} &= 0 \text{ dB} \\
\text{minimum passband gain} &= -0.3 \text{ dB} \\
\text{maximum stopband gain} &= -30 \text{ dB}
\end{aligned}$$

Referring to Figure 7.1, the corresponding passband and stopband tolerance limits are

$$\begin{aligned}
20 \log_{10}(1 + \delta_{p_1}) &= 0 & &\text{or equivalently } \delta_{p_1} = 0 \\
20 \log_{10}(1 - \delta_{p_2}) &= -0.3 & &\text{or equivalently } \delta_{p_2} = 0.0339 \\
20 \log_{10}(\delta_s) &= -30 & &\text{or equivalently } \delta_s = 0.0316.
\end{aligned}$$

Note that the specifications are only on the magnitudes of the frequency response. The phase is implicitly determined by the nature of the approximating functions.

Using the filter design program, it is determined that for a Butterworth design, the minimum (integer) filter order that meets or exceeds the given specifications is a 15th-order filter. The resulting frequency response magnitude, group delay, and pole–zero plot are shown in Figure 7.12. As expected, all of the zeros of the Butterworth filter are at $z = -1$.

For a Chebyshev type I design, the minimum filter order is 7. The resulting frequency response magnitude and group delay, and the corresponding pole–zero plot are shown in Figure 7.13. As expected, all of the zeros of the transfer function are at $z = -1$ and the frequency response magnitude is equiripple in the passband and monotonic in the stopband.

For a Chebyshev type II design, the minimum filter order is again 7. The resulting frequency response magnitude, group delay and pole–zero plot are shown in Figure 7.14. Again as expected, the frequency response magnitude is monotonic in the passband and equiripple in the stopband. The zeros of the transfer function are arrayed on the unit circle in the stopband.

In comparing the Chebyshev I and Chebyshev II designs it is worth noting that for both, the order of the denominator polynomial in the transfer function corresponding to the poles is 7, and the order of the numerator polynomial is also 7. In the implementation of the difference equation for both the Chebyshev I design and the Butterworth design, significant advantage can be taken of the fact that all the zeros

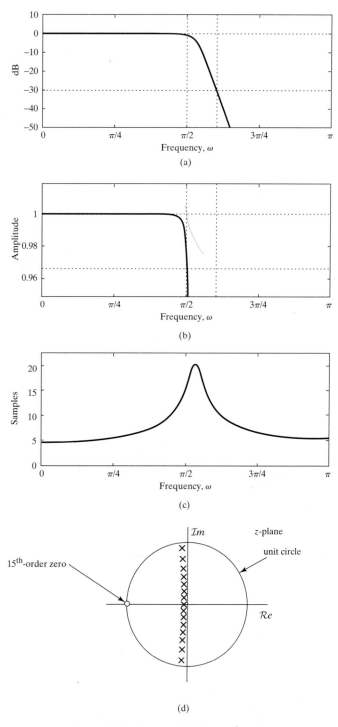

Figure 7.12 Butterworth filter, 15th-order.

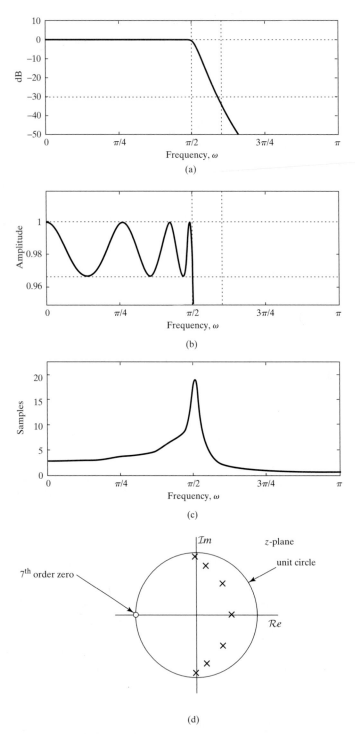

Figure 7.13   Chebyshev Type I filter, 7th-order.

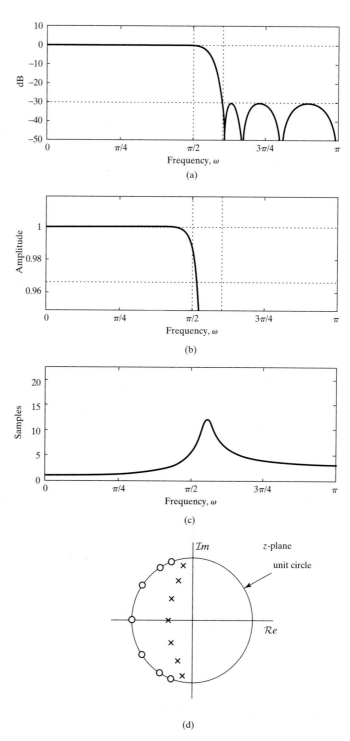

Figure 7.14   Chebyshev Type II filter, 7th-order.

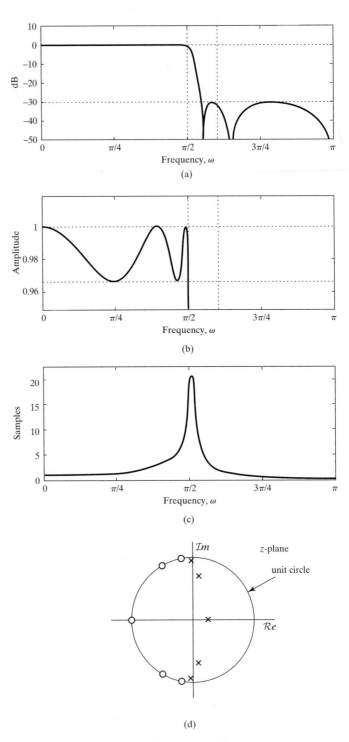

Figure 7.15   Elliptic filter, 5th-order, exceeds design specifications.

occur at $z = -1$. This is not the case for the Chebyshev II filter. Consequently, in an implementation of the filter, the Chebyshev II design will require more multiplications than the Chebyshev I design. For the Butterworth design, while advantage can be taken of the clustered zeros at $z = -1$, the filter order is more than twice that of the Chebyshev designs and consequently requires more multiplications.

For the design of an elliptic filter to meet the given specifications, a filter of at least 5th-order is required. Figure 7.15 shows the resulting design. As with previous examples, in designing a filter with given specifications, the minimum specifications are likely to be exceeded, since the filter order is necessarily an integer. Depending on the application, the designer may choose which of the specifications to exactly meet and which to exceed. For example, with the elliptic filter design we may choose to exactly meet the passband and stopband edge frequencies and the passband variation and minimize the stopband gain. The resulting filter, which achieves 43 dB of attenuation in the stopband, is shown in Figure 7.16. Alternately, the added flexibility can be used to narrow the transition band or reduce the deviation from 0 dB gain in the passband. Again as expected, the frequency response of the elliptic filter is equiripple in both the passband and the stopband.

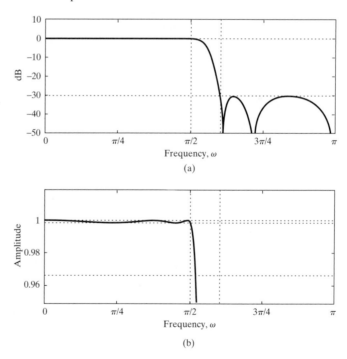

Figure 7.16   Elliptic filter, 5th-order, minimizing the passband ripple.

Example 7.5   Design Example for Comparison with FIR Designs

In this example we return to the specifications of Example 7.1 and illustrate the realization of this filter specification with a Butterworth, Chebyshev I, Chebyshev II,

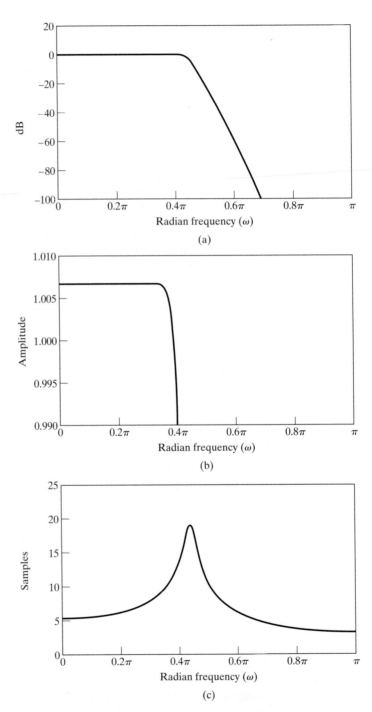

Figure 7.17   Frequency response of 14th-order Butterworth filter in Example 7.5. (a) Log magnitude in dB. (b) Detailed plot of magnitude in passband. (c) Group delay.

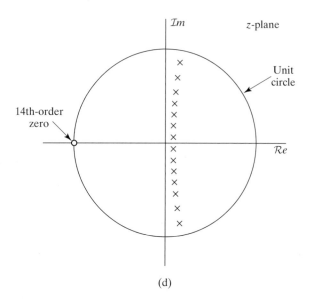

(d)

Figure 7.17 (*continued*) (d) Pole–zero plot of 14th-order Butterworth filter in Example 7.5.

and elliptic designs. The designs are again carried out using the filter design program in the MATLAB signal processing toolbox. In Section 7.8.1 we will compare these IIR designs with FIR designs with the same specifications. Typical design programs for FIR filters require the passband tolerance limits in Figure 7.1 to be specified with $\delta_{p1} = \delta_{p2}$, whereas for IIR filters, it is typically assumed that $\delta_{p1} = 0$. Consequently to carry out a comparison of IIR and FIR designs, some renormalization of the passband and stopband specifications may need to be carried out (see, for example, Problem 7.3), as will be done in Example 7.5.

The lowpass discrete-time filter specifications as used for this example are:

$$0.99 \le |H(e^{j\omega})| \le 1.01, \qquad |\omega| \le 0.4\pi, \tag{7.37a}$$

and

$$|H(e^{j\omega})| \le 0.001, \qquad 0.6\pi \le |\omega| \le \pi. \tag{7.37b}$$

In terms of the tolerance scheme of Figure 7.1, $\delta_{p_1} = \delta_{p_2} = 0.01$, $\delta_s = 0.001$, $\omega_p = 0.4\pi$, and $\omega_s = 0.6\pi$. Rescaling these specifications so that $\delta_{p1} = 0$ corresponds to scaling the filter by $1/(1 + \delta_{p1})$ to obtain: $\delta_{p1} = 0$, $\delta_{p2} = 0.0198$ and $\delta_s = .00099$.

The filters are first designed using the filter design program with these specifications and the filter designs returned by the filter design program are then rescaled by a factor of 1.01 to satisfy the specifications in Eqs. (7.37a) and (7.37b).

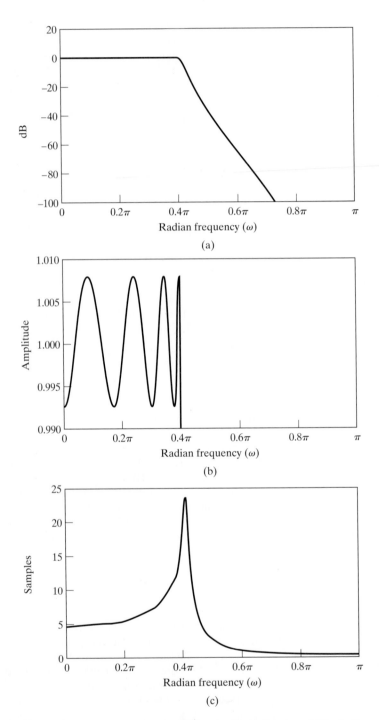

Figure 7.18   Frequency response of 8th-order Chebyshev type I filter in Example 7.5. (a) Log magnitude in dB. (b) Detailed plot of magnitude in passband. (c) Group delay.

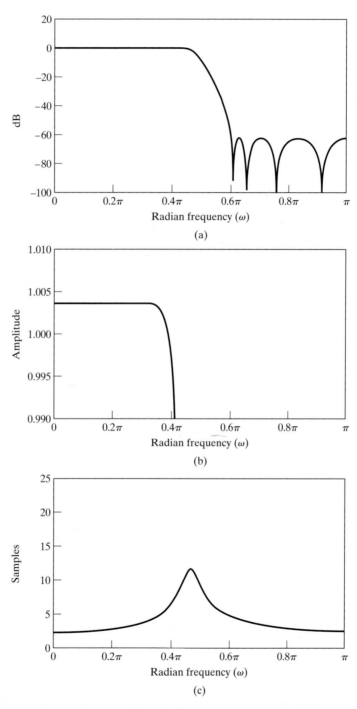

Figure 7.19 Frequency response of 8th-order Chebyshev type II filter in Example 7.5. (a) Log magnitude in dB. (b) Detailed plot of magnitude in passband. (c) Group delay.

For the specifications in this example, the Butterworth approximation method requires a system of 14th-order. The frequency response of the discrete-time filter that results from the bilinear transformation of the appropriate prewarped Butterworth filter is shown in Figure 7.17. Figure 7.17(a) shows the log magnitude in dB, Figure 7.17(b) shows the magnitude of $H(e^{j\omega})$ in the passband only, and Figure 7.17(c) shows the group delay of the filter. From these plots, we see that as expected, the Butterworth frequency response decreases monotonically with frequency, and the gain of the filter becomes very small above about $\omega = 0.7\pi$.

Both Chebyshev designs I and II lead to the same order for a given set of specifications. For our specifications the required order is 8 rather than 14, as was required for the Butterworth approximation. Figure 7.18 shows the log magnitude, passband magnitude, and group delay for the type I approximation to the specifications of Eqs. (7.37a) and (7.37b). Note that as expected, the frequency response oscillates with equal maximum error on either side of the desired gain of unity in the passband.

Figure 7.19 shows the frequency-response functions for the Chebyshev type II approximation. In this case, the equiripple approximation behavior is in the stopband. The pole–zero plots for the Chebyshev filters are shown in Figure 7.20. Note that the Chebyshev type I filter is similar to the Butterworth filter in that it has all eight of its zeros at $z = -1$. On the other hand, the type II filter has its zeros arrayed on the unit circle. These zeros are naturally positioned by the design equations so as to achieve the equiripple behavior in the stopband.

The specifications of Eqs. (7.37a) and (7.37b) are met by an elliptic filter of order six. This is the lowest order rational function approximation to the specifications. Figure 7.21 clearly shows the equiripple behavior in both approximation bands. Figure 7.22 shows that the elliptic filter, like the Chebyshev type II, has its zeros arrayed in the stopband region of the unit circle.

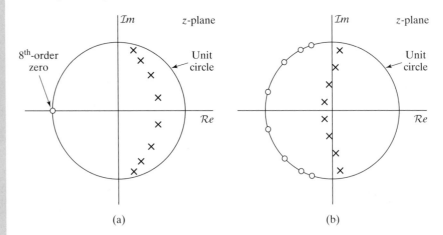

(a)   (b)

Figure 7.20   Pole–zero plot of 8th-order Chebyshev filters in Example 7.5. (a) Type I. (b) Type II.

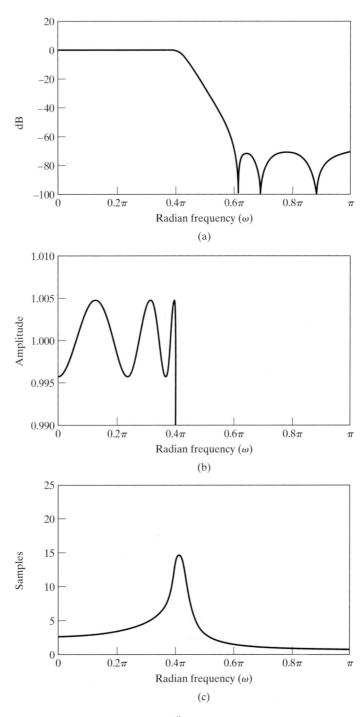

Figure 7.21   Frequency response of 6th-order elliptic filter in Example 7.5. (a) Log magnitude in dB. (b) Detailed plot of magnitude in passband. (c) Group delay.

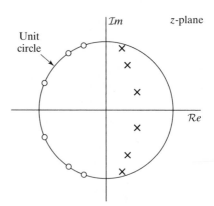

Figure 7.22   Pole–zero plot of 6th-order elliptic filter in Example 7.5.

7.4 FREQUENCY TRANSFORMATIONS OF LOWPASS IIR FILTERS

Our discussion and examples of IIR filter design have focused on the design of frequency-selective lowpass filters. Other types of frequency-selective filters such as highpass, bandpass, bandstop, and multiband filters are equally important. As with lowpass filters, these other classes are characterized by one or several passbands and stopbands, each specified by passband and stopband edge frequencies. Generally the desired filter gain is unity in the passbands and zero in the stopbands, but as with lowpass filters, the filter design specifications include tolerance limits by which the ideal gains or attenuation in the pass- and stopbands can be exceeded. A typical tolerance scheme for a multiband filter with two passbands and one stopband is shown in Figure 7.23.

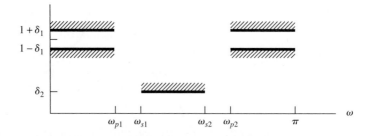

Figure 7.23   Tolerance scheme for a multiband filter.

The traditional approach to the design of many continuous-time frequency-selective filters is to first design a frequency-normalized prototype lowpass filter and then, using an algebraic transformation, derive the desired filter from the prototype lowpass filter (see Guillemin, 1957 and Daniels, 1974). In the case of discrete-time frequency-selective filters, we could design a continuous-time frequency-selective filter of the desired type and then transform it to a discrete-time filter. This procedure would be acceptable with the bilinear transformation, but impulse invariance clearly could not be used to transform highpass and bandstop continuous-time filters into corresponding discrete-time filters because of the aliasing that results from sampling. An alternative procedure that works with either the bilinear transformation or impulse invariance is to design a discrete-time prototype lowpass filter and then perform an algebraic transformation on it to obtain the desired frequency-selective discrete-time filter.

Frequency-selective filters of the lowpass, highpass, bandpass, and bandstop types can be obtained from a lowpass discrete-time filter by use of transformations very similar to the bilinear transformation used to transform continuous-time system functions into discrete-time system functions. To see how this is done, assume that we are given a lowpass system function $H_{lp}(Z)$ that we wish to transform to a new system function $H(z)$, which has either lowpass, highpass, bandpass, or bandstop characteristics when evaluated on the unit circle. Note that we associate the complex variable Z with the prototype lowpass filter and the complex variable z with the transformed filter. Then, we define a mapping from the Z-plane to the z-plane of the form

$$Z^{-1} = G(z^{-1}) \tag{7.38}$$

such that

$$H(z) = H_{lp}(Z)\big|_{Z^{-1}=G(z^{-1})} \tag{7.39}$$

Instead of expressing Z as a function of z, we have assumed in Eq. (7.38) that Z^{-1} is expressed as a function of z^{-1}. Thus, according to Eq. (7.39), in obtaining $H(z)$ from $H_{lp}(z)$ we replace Z^{-1} everywhere in $H_{lp}(Z)$ by the function $G(z^{-1})$. This is a convenient representation, because $H_{lp}(Z)$ is normally expressed as a rational function of Z^{-1}.

If $H_{lp}(Z)$ is the rational system function of a causal and stable system, we naturally require that the transformed system function $H(z)$ be a rational function of z^{-1} and that the system also be causal and stable. This places the following constraints on the transformation $Z^{-1} = G(z^{-1})$:

1. $G(z^{-1})$ must be a rational function of z^{-1}.
2. The inside of the unit circle of the Z-plane must map to the inside of the unit circle of the z-plane.
3. The unit circle of the Z-plane must map onto the unit circle of the z-plane.

Let θ and ω be the frequency variables (angles) in the Z-plane and z-plane, respectively, i.e., on the respective unit circles $Z = e^{j\theta}$ and $z = e^{j\omega}$. Then, for condition 3 to hold, it must be true that

$$e^{-j\theta} = |G(e^{-j\omega})|e^{j\angle G(e^{-j\omega})}, \tag{7.40}$$

and thus,

$$|G(e^{-j\omega})| = 1. \tag{7.41}$$

Therefore, the relationship between the frequency variables is

$$-\theta = \angle G(e^{-j\omega}). \tag{7.42}$$

Constantinides (1970) showed that the most general form of the function $G(z^{-1})$ that satisfies all the above requirements is

$$Z^{-1} = G(z^{-1}) = \pm \prod_{k=1}^{N} \frac{z^{-1} - \alpha_k}{1 - \alpha_k z^{-1}}. \tag{7.43}$$

From our discussion of allpass systems in Chapter 5, it should be clear that $G(z^{-1})$ as given in Eq. (7.43) satisfies Eq. (7.41), and it is easily shown that Eq. (7.43) maps the inside of the unit circle of the Z-plane to the inside of the unit circle of the z-plane if and only if $|\alpha_k| < 1$. By choosing appropriate values for N and the constants α_k, a variety of mappings can be obtained. The simplest is the one that transforms a lowpass filter into another lowpass filter with different passband and stopband edge frequencies. For this case,

$$Z^{-1} = G(z^{-1}) = \frac{z^{-1} - \alpha}{1 - \alpha z^{-1}}. \tag{7.44}$$

If we substitute $Z = e^{j\theta}$ and $z = e^{j\omega}$, we obtain

$$e^{-j\theta} = \frac{e^{-j\omega} - \alpha}{1 - \alpha e^{-j\omega}}, \tag{7.45}$$

from which it follows that

$$\omega = \arctan\left[\frac{(1-\alpha^2)\sin\theta}{2\alpha + (1+\alpha^2)\cos\theta}\right]. \tag{7.46}$$

This relationship is plotted in Figure 7.24 for different values of α. Although a warping of the frequency scale is evident in Figure 7.24 (except in the case $\alpha = 0$, which corresponds to $Z^{-1} = z^{-1}$), if the original system has a piecewise-constant lowpass frequency response with cutoff frequency θ_p, then the transformed system will likewise have a similar lowpass response with cutoff frequency ω_p determined by the choice of α.

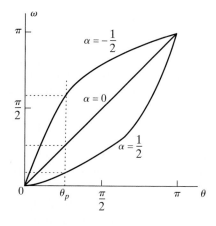

Figure 7.24 Warping of the frequency scale in lowpass-to-lowpass transformation.

TABLE 7.1 TRANSFORMATIONS FROM A LOWPASS DIGITAL FILTER PROTOTYPE OF CUTOFF FREQUENCY θ_p TO HIGHPASS, BANDPASS, AND BANDSTOP FILTERS

Filter Type	Transformations	Associated Design Formulas
Lowpass	$Z^{-1} = \dfrac{z^{-1} - \alpha}{1 - az^{-1}}$	$\alpha = \dfrac{\sin\left(\frac{\theta_p - \omega_p}{2}\right)}{\sin\left(\frac{\theta_p + \omega_p}{2}\right)}$ $\omega_p =$ desired cutoff frequency
Highpass	$Z^{-1} = -\dfrac{z^{-1} + \alpha}{1 + \alpha z^{-1}}$	$\alpha = -\dfrac{\cos\left(\frac{\theta_p + \omega_p}{2}\right)}{\cos\left(\frac{\theta_p - \omega_p}{2}\right)}$ $\omega_p =$ desired cutoff frequency
Bandpass	$Z^{-1} = -\dfrac{z^{-2} - \frac{2\alpha k}{k+1} z^{-1} + \frac{k-1}{k+1}}{\frac{k-1}{k+1} z^{-2} - \frac{2\alpha k}{k+1} z^{-1} + 1}$	$\alpha = \dfrac{\cos\left(\frac{\omega_{p2} + \omega_{p1}}{2}\right)}{\cos\left(\frac{\omega_{p2} - \omega_{p1}}{2}\right)}$ $k = \cot\left(\frac{\omega_{p2} - \omega_{p1}}{2}\right)\tan\left(\frac{\theta_p}{2}\right)$ $\omega_{p1} =$ desired lower cutoff frequency $\omega_{p2} =$ desired upper cutoff frequency
Bandstop	$Z^{-1} = \dfrac{z^{-2} - \frac{2\alpha}{1+k} z^{-1} + \frac{1-k}{1+k}}{\frac{1-k}{1+k} z^{-2} - \frac{2\alpha}{1+k} z^{-1} + 1}$	$\alpha = \dfrac{\cos\left(\frac{\omega_{p2} + \omega_{p1}}{2}\right)}{\cos\left(\frac{\omega_{p2} - \omega_{p1}}{2}\right)}$ $k = \tan\left(\frac{\omega_{p2} - \omega_{p1}}{2}\right)\tan\left(\frac{\theta_p}{2}\right)$ $\omega_{p1} =$ desired lower cutoff frequency $\omega_{p2} =$ desired upper cutoff frequency

Solving for α in terms of θ_p and ω_p, we obtain

$$\alpha = \frac{\sin[(\theta_p - \omega_p)/2]}{\sin[(\theta_p + \omega_p)/2]}. \tag{7.47}$$

Thus, to use these results to obtain a lowpass filter $H(z)$ with cutoff frequency ω_p from an already available lowpass filter $H_{lp}(Z)$ with cutoff frequency θ_p, we would use Eq. (7.47) to determine α in the expression

$$H(z) = H_{lp}(Z)\big|_{Z^{-1}=(z^{-1}-\alpha)/(1-\alpha z^{-1})}. \tag{7.48}$$

(Problem 7.51 explores how the lowpass–lowpass transformation can be used to obtain a network structure for a variable cutoff frequency filter where the cutoff frequency is determined by a single parameter α.)

Transformations from a lowpass filter to highpass, bandpass, and bandstop filters can be derived in a similar manner. These transformations are summarized in Table 7.1. In the design formulas, all of the cutoff frequencies are assumed to be between zero and π radians. The following example illustrates the use of such transformations.

Example 7.6   Transformation of a Lowpass Filter to a Highpass Filter

Consider a Type I Chebyshev lowpass filter with system function

$$H_{lp}(Z) = \frac{0.001836(1 + Z^{-1})^4}{(1 - 1.5548Z^{-1} + 0.6493Z^{-2})(1 - 1.4996Z^{-1} + 0.8482Z^{-2})}. \tag{7.49}$$

This 4th-order system was designed to meet the specifications

$$0.89125 \le |H_{lp}(e^{j\theta})| \le 1, \quad 0 \le \theta \le 0.2\pi, \tag{7.50a}$$

$$|H_{lp}(e^{j\theta})| \le 0.17783, \quad 0.3\pi \le \theta \le \pi. \tag{7.50b}$$

The frequency response of this filter is shown in Figure 7.25.

To transform this filter to a highpass filter with passband cutoff frequency $\omega_p = 0.6\pi$, we obtain from Table 7.1

$$\alpha = -\frac{\cos[(0.2\pi + 0.6\pi)/2]}{\cos[(0.2\pi - 0.6\pi)/2]} = -0.38197. \tag{7.51}$$

Thus, using the lowpass–highpass transformation indicated in Table 7.1, we obtain

$$H(z) = H_{lp}(Z)\big|_{Z^{-1}=-[(z^{-1}-0.38197)/(1-0.38197z^{-1})]}$$

$$= \frac{0.02426(1 - z^{-1})^4}{(1 + 1.0416z^{-1} + 0.4019z^{-2})(1 + 0.5661z^{-1} + 0.7657z^{-2})}. \tag{7.52}$$

The frequency response of this system is shown in Figure 7.26. Note that except for some distortion of the frequency scale, the highpass frequency response appears very much as if the lowpass frequency response were shifted in frequency by π. Also note that the 4th-order zero at $Z = -1$ for the lowpass filter now appears at $z = 1$ for the highpass filter. This example also verifies that the equiripple passband and stopband behavior is preserved by frequency transformations of this type. Also note that the group delay in Figure 7.26(c) is not simply a stretched and shifted version of Figure 7.25(c). This is because the phase variations are stretched and shifted, so that the derivative of the phase is smaller for the highpass filter.

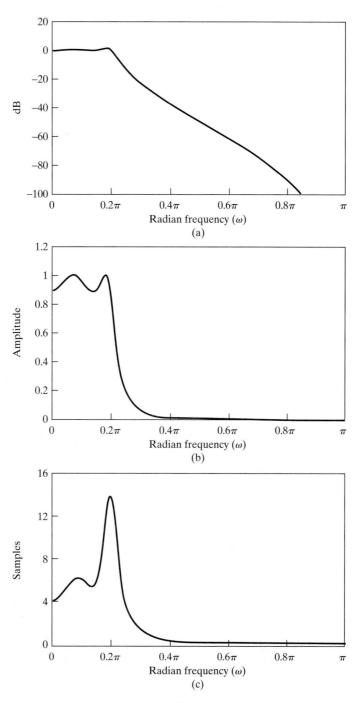

Figure 7.25 Frequency response of 4th-order Chebyshev lowpass filter. (a) Log magnitude in dB. (b) Magnitude. (c) Group delay.

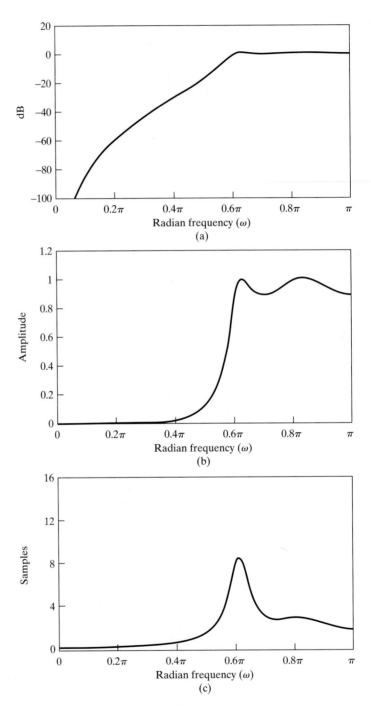

Figure 7.26   Frequency response of 4th-order Chebyshev highpass filter obtained by frequency transformation. (a) Log magnitude in dB. (b) Magnitude. (c) Group delay.

7.5 DESIGN OF FIR FILTERS BY WINDOWING

As discussed in Section 7.2, commonly used techniques for the design of IIR filters have evolved from applying transformations of continuous-time IIR systems into discrete-time IIR systems. In contrast, the design techniques for FIR filters are based on directly approximating the desired frequency response or impulse response of the discrete-time system.

The simplest method of FIR filter design is called the *window method*. This method generally begins with an ideal desired frequency response that can be represented as

$$H_d(e^{j\omega}) = \sum_{n=-\infty}^{\infty} h_d[n]e^{-j\omega n}, \tag{7.53}$$

where $h_d[n]$ is the corresponding impulse response sequence, which can be expressed in terms of $H_d(e^{j\omega})$ as

$$h_d[n] = \frac{1}{2\pi} \int_{-\pi}^{\pi} H_d(e^{j\omega})e^{j\omega n}\, d\omega. \tag{7.54}$$

Many idealized systems are defined by piecewise-constant or piecewise-smooth frequency responses with discontinuities at the boundaries between bands. As a result, these systems have impulse responses that are noncausal and infinitely long. The most straightforward approach to obtaining an FIR approximation to such systems is to truncate the ideal impulse response through the process referred to as windowing. Equation (7.53) can be thought of as a Fourier series representation of the periodic frequency response $H_d(e^{j\omega})$, with the sequence $h_d[n]$ playing the role of the Fourier coefficients. Thus, the approximation of an ideal filter by truncation of the ideal impulse response is identical to the issue of the convergence of Fourier series, a subject that has received a great deal of study. A particularly important concept from this theory is the Gibbs phenomenon, which was discussed in Example 2.18. In the following discussion, we will see how this effect of nonuniform convergence manifests itself in the design of FIR filters.

A particularly simple way to obtain a causal FIR filter from $h_d[n]$ is to truncate $h_d[n]$, i.e., to define a new system with impulse response $h[n]$ given by[4]

$$h[n] = \begin{cases} h_d[n], & 0 \le n \le M, \\ 0, & \text{otherwise.} \end{cases} \tag{7.55}$$

More generally, we can represent $h[n]$ as the product of the desired impulse response and a finite-duration "window" $w[n]$; i.e.,

$$h[n] = h_d[n]w[n], \tag{7.56}$$

---

[4]The notation for FIR systems was established in Chapter 5. That is, M is the order of the system function polynomial. Thus, $(M+1)$ is the length, or duration, of the impulse response. Often in the literature, N is used for the length of the impulse response of an FIR filter; however, we have used N to denote the order of the denominator polynomial in the system function of an IIR filter. Thus, to avoid confusion and maintain consistency throughout this book, we will consider the length of the impulse response of an FIR filter to be $(M+1)$.

where, for simple truncation as in Eq. (7.55), the window is the *rectangular window*

$$w[n] = \begin{cases} 1, & 0 \leq n \leq M, \\ 0, & \text{otherwise.} \end{cases} \tag{7.57}$$

It follows from the modulation, or windowing, theorem (Section 2.9.7) that

$$H(e^{j\omega}) = \frac{1}{2\pi} \int_{-\pi}^{\pi} H_d(e^{j\theta}) W(e^{j(\omega-\theta)}) d\theta. \tag{7.58}$$

That is, $H(e^{j\omega})$ is the periodic convolution of the desired ideal frequency response with the Fourier transform of the window. Thus, the frequency response $H(e^{j\omega})$ will be a "smeared" version of the desired response $H_d(e^{j\omega})$. Figure 7.27(a) depicts typical functions $H_d(e^{j\theta})$ and $W(e^{j(\omega-\theta)})$ as a function of θ, as required in Eq. (7.58).

If $w[n] = 1$ for all n (i.e., if we do not truncate at all), $W(e^{j\omega})$ is a periodic impulse train with period 2π, and therefore, $H(e^{j\omega}) = H_d(e^{j\omega})$. This interpretation suggests that if $w[n]$ is chosen so that $W(e^{j\omega})$ is concentrated in a narrow band of frequencies around $\omega = 0$, i.e., it approximates an impulse, then $H(e^{j\omega})$ will "look like" $H_d(e^{j\omega})$, except where $H_d(e^{j\omega})$ changes very abruptly. Consequently, the choice of window is governed by the desire to have $w[n]$ as short as possible in duration, so as to minimize computation in the implementation of the filter, while having $W(e^{j\omega})$ approximate an impulse; that is, we want $W(e^{j\omega})$ to be highly concentrated in frequency so that the convolution of Eq. (7.58) faithfully reproduces the desired frequency response. These are conflicting requirements, as can be seen in the case of the rectangular window of Eq. (7.57), where

$$W(e^{j\omega}) = \sum_{n=0}^{M} e^{-j\omega n} = \frac{1 - e^{-j\omega(M+1)}}{1 - e^{-j\omega}} = e^{-j\omega M/2} \frac{\sin[\omega(M+1)/2]}{\sin(\omega/2)}. \tag{7.59}$$

The magnitude of the function $\sin[\omega(M+1)/2]/\sin(\omega/2)$ is plotted in Figure 7.28 for the case $M = 7$. Note that $W(e^{j\omega})$ for the rectangular window has a generalized linear phase. As M increases, the width of the "main lobe" decreases. The main lobe is usually defined as the region between the first zero-crossings on either side of the origin. For the rectangular window, the width of the main lobe is $\Delta\omega_m = 4\pi/(M+1)$. However, for the rectangular window, the side lobes are large, and in fact, as M increases, the peak amplitudes of the main lobe and the side lobes grow in a manner such that the area under each lobe is a constant while the width of each lobe decreases with M. Consequently, as $W(e^{j(\omega-\theta)})$ "slides by" a discontinuity of $H_d(e^{j\theta})$ as ω varies, the integral of $W(e^{j(\omega-\theta)})H_d(e^{j\theta})$ will oscillate as each side lobe of $W(e^{j(\omega-\theta)})$ moves past the discontinuity. This result is depicted in Figure 7.27(b). Since the area under each lobe remains constant with increasing M, the oscillations occur more rapidly, but do not decrease in amplitude as M increases.

In the theory of Fourier series, it is well known that this nonuniform convergence, the *Gibbs phenomenon,* can be moderated through the use of a less abrupt truncation of the Fourier series. By tapering the window smoothly to zero at each end, the height of the side lobes can be diminished; however, this is achieved at the expense of a wider main lobe and thus a wider transition at the discontinuity.

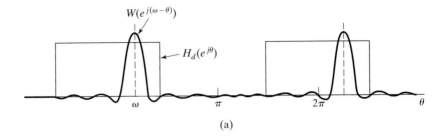

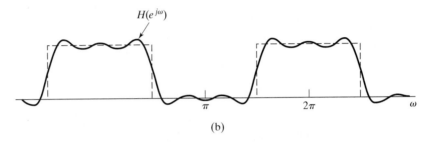

Figure 7.27   (a) Convolution process implied by truncation of the ideal impulse response. (b) Typical approximation resulting from windowing the ideal impulse response.

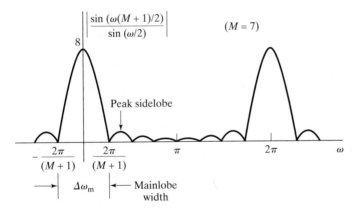

Figure 7.28   Magnitude of the Fourier transform of a rectangular window $(M = 7)$.

7.5.1 Properties of Commonly Used Windows

Some commonly used windows are shown in Figure 7.29. These windows are defined by the following equations:

Rectangular

$$w[n] = \begin{cases} 1, & 0 \leq n \leq M, \\ 0, & \text{otherwise} \end{cases} \qquad (7.60a)$$

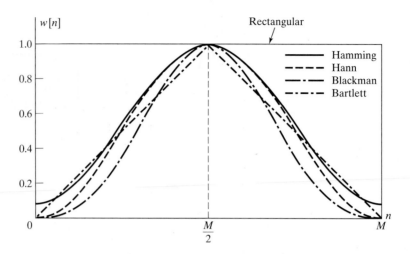

Figure 7.29   Commonly used windows.

Bartlett (triangular)

$$w[n] = \begin{cases} 2n/M, & 0 \le n \le M/2, \ M \text{ even} \\ 2 - 2n/M, & M/2 < n \le M, \\ 0, & \text{otherwise} \end{cases} \quad (7.60\text{b})$$

Hann

$$w[n] = \begin{cases} 0.5 - 0.5\cos(2\pi n/M), & 0 \le n \le M, \\ 0, & \text{otherwise} \end{cases} \quad (7.60\text{c})$$

Hamming

$$w[n] = \begin{cases} 0.54 - 0.46\cos(2\pi n/M), & 0 \le n \le M, \\ 0, & \text{otherwise} \end{cases} \quad (7.60\text{d})$$

Blackman

$$w[n] = \begin{cases} 0.42 - 0.5\cos(2\pi n/M) + 0.08\cos(4\pi n/M), & 0 \le n \le M, \\ 0, & \text{otherwise} \end{cases} \quad (7.60\text{e})$$

(For convenience, Figure 7.29 shows these windows plotted as functions of a continuous variable; however, as specified in Eq. (7.60), the window sequence is defined only at integer values of n.)

The Bartlett, Hann, Hamming, and Blackman windows are all named after their originators. The Hann window is associated with Julius von Hann, an Austrian meteorologist. The term "hanning" was used by Blackman and Tukey (1958) to describe the operation of applying this window to a signal and has since become the most widely used name for the window, with varying preferences for the choice of "Hanning" or "hanning." There is some slight variation in the definition of the Bartlett and Hann windows. As we have defined them, $w[0] = w[M] = 0$, so that it would be reasonable to assert that with this definition, the window length is really only $M - 1$ samples. Other definitions of the Bartlett and Hann windows are related to our definitions by a shift of one sample and redefinition of the window length.

As will be discussed in Chapter 10, the windows defined in Eq. (7.60) are commonly used for spectrum analysis as well as for FIR filter design. They have the desirable property that their Fourier transforms are concentrated around $\omega = 0$, and they have a simple functional form that allows them to be computed easily. The Fourier transform of the Bartlett window can be expressed as a product of Fourier transforms of rectangular windows, and the Fourier transforms of the other windows can be expressed as sums of frequency-shifted Fourier transforms of the rectangular window, as given by Eq. (7.59). (See Problem 7.43.)

The function $20 \log_{10} |W(e^{j\omega})|$ is plotted in Figure 7.30 for each of these windows with $M = 50$. The rectangular window clearly has the narrowest main lobe, and thus, for a given length, it should yield the sharpest transitions of $H(e^{j\omega})$ at a discontinuity of $H_d(e^{j\omega})$. However, the first side lobe is only about 13 dB below the main peak, resulting in oscillations of $H(e^{j\omega})$ of considerable size around discontinuities of $H_d(e^{j\omega})$. Table 7.2, which compares the windows of Eq. (7.60), shows that, by tapering the window smoothly to zero, as with the Bartlett, Hamming, Hann, and Blackman windows, the side lobes (second column) are greatly reduced in amplitude; however, the price paid is a much wider main lobe (third column) and thus wider transitions at discontinuities of $H_d(e^{j\omega})$. The other columns of Table 7.2 will be discussed later.

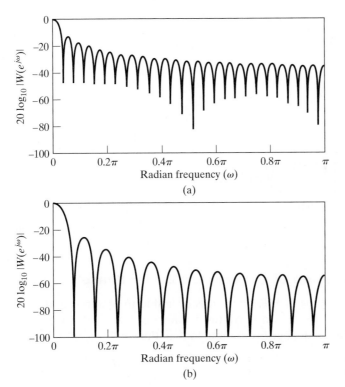

Figure 7.30   Fourier transforms (log magnitude) of windows of Figure 7.29 with $M = 50$. (a) Rectangular. (b) Bartlett.

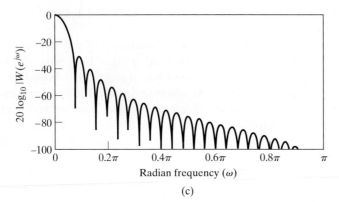

(c)

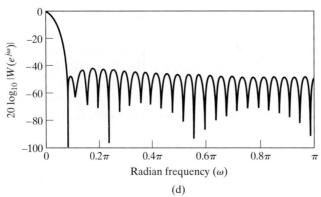

(d)

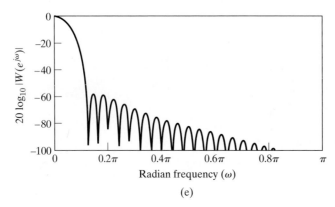

(e)

Figure 7.30 (*continued*) (c) Hann. (d) Hamming. (e) Blackman.

7.5.2 Incorporation of Generalized Linear Phase

In designing many types of FIR filters, it is desirable to obtain causal systems with a generalized linear-phase response. All the windows of Eq. (7.60) have been defined in anticipation of this need. Specifically, note that all the windows have the property that

$$
w[n] = \begin{cases} w[M - n], & 0 \le n \le M, \\ 0, & \text{otherwise;} \end{cases} \tag{7.61}
$$

TABLE 7.2   COMPARISON OF COMMONLY USED WINDOWS

Type of Window	Peak Side-Lobe Amplitude (Relative)	Approximate Width of Main Lobe	Peak Approximation Error, $20 \log_{10} \delta$ (dB)	Equivalent Kaiser Window, β	Transition Width of Equivalent Kaiser Window
Rectangular	−13	$4\pi/(M+1)$	−21	0	$1.81\pi/M$
Bartlett	−25	$8\pi/M$	−25	1.33	$2.37\pi/M$
Hann	−31	$8\pi/M$	−44	3.86	$5.01\pi/M$
Hamming	−41	$8\pi/M$	−53	4.86	$6.27\pi/M$
Blackman	−57	$12\pi/M$	−74	7.04	$9.19\pi/M$

i.e., they are symmetric about the point $M/2$. As a result, their Fourier transforms are of the form

$$W(e^{j\omega}) = W_e(e^{j\omega})e^{-j\omega M/2}, \tag{7.62}$$

where $W_e(e^{j\omega})$ is a real, even function of ω. This is illustrated by Eq. (7.59). The convention of Eq. (7.61) leads to causal filters in general, and if the desired impulse response is also symmetric about $M/2$, i.e., if $h_d[M-n] = h_d[n]$, then the windowed impulse response will also have that symmetry, and the resulting frequency response will have a generalized linear phase; that is,

$$H(e^{j\omega}) = A_e(e^{j\omega})e^{-j\omega M/2}, \tag{7.63}$$

where $A_e(e^{j\omega})$ is real and is an even function of ω. Similarly, if the desired impulse response is antisymmetric about $M/2$, i.e., if $h_d[M-n] = -h_d[n]$, then the windowed impulse response will also be antisymmetric about $M/2$, and the resulting frequency response will have a generalized linear phase with a constant phase shift of ninety degrees; i.e.,

$$H(e^{j\omega}) = jA_o(e^{j\omega})e^{-j\omega M/2}, \tag{7.64}$$

where $A_o(e^{j\omega})$ is real and is an odd function of ω.

Although the preceding statements are straightforward if we consider the product of the symmetric window with the symmetric (or antisymmetric) desired impulse response, it is useful to consider the frequency-domain representation. Suppose $h_d[M-n] = h_d[n]$. Then,

$$H_d(e^{j\omega}) = H_e(e^{j\omega})e^{-j\omega M/2}, \tag{7.65}$$

where $H_e(e^{j\omega})$ is real and even.

If the window is symmetric, we can substitute Eqs. (7.62) and (7.65) into Eq. (7.58) to obtain

$$H(e^{j\omega}) = \frac{1}{2\pi} \int_{-\pi}^{\pi} H_e(e^{j\theta})e^{-j\theta M/2} W_e(e^{j(\omega-\theta)})e^{-j(\omega-\theta)M/2}d\theta. \tag{7.66}$$

A simple manipulation of the phase factors leads to

$$H(e^{j\omega}) = A_e(e^{j\omega})e^{-j\omega M/2}, \tag{7.67}$$

where

$$A_e(e^{j\omega}) = \frac{1}{2\pi} \int_{-\pi}^{\pi} H_e(e^{j\theta}) W_e(e^{j(\omega-\theta)}) d\theta. \tag{7.68}$$

Thus, we see that the resulting system has a generalized linear phase and, moreover, the real function $A_e(e^{j\omega})$ is the result of the periodic convolution of the real functions $H_e(e^{j\omega})$ and $W_e(e^{j\omega})$.

The detailed behavior of the convolution of Eq. (7.68) determines the magnitude response of the filter that results from windowing. The following example illustrates this for a linear-phase lowpass filter.

Example 7.7   Linear-Phase Lowpass Filter

The desired frequency response is defined as

$$H_{\text{lp}}(e^{j\omega}) = \begin{cases} e^{-j\omega M/2}, & |\omega| < \omega_c, \\ 0, & \omega_c < |\omega| \leq \pi, \end{cases} \tag{7.69}$$

where the generalized linear-phase factor has been incorporated into the definition of the ideal lowpass filter. The corresponding ideal impulse response is

$$h_{\text{lp}}[n] = \frac{1}{2\pi} \int_{-\omega_c}^{\omega_c} e^{-j\omega M/2} e^{j\omega n} d\omega = \frac{\sin[\omega_c(n - M/2)]}{\pi(n - M/2)} \tag{7.70}$$

for $-\infty < n < \infty$. It is easily shown that $h_{\text{lp}}[M - n] = h_{\text{lp}}[n]$, so if we use a symmetric window in the equation

$$h[n] = \frac{\sin[\omega_c(n - M/2)]}{\pi(n - M/2)} w[n], \tag{7.71}$$

then a linear-phase system will result.

The upper part of Figure 7.31 depicts the character of the amplitude response that would result for all the windows of Eq. (7.60), except the Bartlett window, which is rarely used for filter design. (For M even, the Bartlett window would produce a monotonic function $A_e(e^{j\omega})$, because $W_e(e^{j\omega})$ is a positive function.) The figure displays the important properties of window method approximations to desired frequency responses that have step discontinuities. It applies accurately when ω_c is not close to zero or to π and when the width of the main lobe is smaller than $2\omega_c$. At the bottom of the figure is a typical Fourier transform for a symmetric window (except for the linear phase). This function should be visualized in different positions as an aid in understanding the shape of the approximation $A_e(e^{j\omega})$ in the vicinity of ω_c.

When $\omega = \omega_c$, the symmetric function $W_e(e^{j(\omega-\theta)})$ is centered on the discontinuity, and about one-half its area contributes to $A_e(e^{j\omega})$. Similarly, we can see that the peak overshoot occurs when $W_e(e^{j(\omega-\theta)})$ is shifted such that the first negative side lobe on the right is just to the right of ω_c. Similarly, the peak negative undershoot occurs when the first negative side lobe on the left is just to the left of ω_c. This means that the distance between the peak ripples on either side of the discontinuity is approximately the main-lobe width $\Delta\omega_m$, as shown in Figure 7.31. The transition width $\Delta\omega$ as defined in the figure is therefore somewhat less than the main-lobe width. Finally, owing to the symmetry of $W_e(e^{j(\omega-\theta)})$, the approximation tends to be symmetric around ω_c; i.e., the approximation overshoots by an amount δ in the passband and undershoots by the same amount in the stopband.

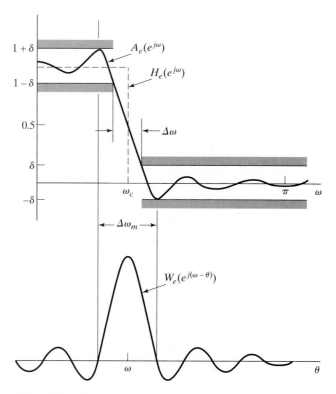

Figure 7.31   Illustration of type of approximation obtained at a discontinuity of the ideal frequency response.

The fourth column of Table 7.2 shows the peak approximation error at a discontinuity (in dB) for the windows of Eq. (7.60). Clearly, the windows with the smaller side lobes yield better approximations of the ideal response at a discontinuity. Also, the third column, which shows the width of the main lobe, suggests that narrower transition regions can be achieved by increasing M. Thus, through the choice of the shape and duration of the window, we can control the properties of the resulting FIR filter. However, trying different windows and adjusting lengths by trial and error is not a very satisfactory way to design filters. Fortunately, a simple formalization of the window method has been developed by Kaiser (1974).

7.5.3 The Kaiser Window Filter Design Method

The trade-off between the main-lobe width and side-lobe area can be quantified by seeking the window function that is maximally concentrated around $\omega = 0$ in the frequency domain. The issue was considered in depth in a series of classic papers by Slepian et al. (1961). The solution found in this work involves prolate spheroidal wave functions, which are difficult to compute and therefore unattractive for filter design. However, Kaiser (1966, 1974) found that a near-optimal window could be formed using the zero[th]-order modified Bessel function of the first kind, a function that is much easier to compute. The Kaiser window is defined as

$$
w[n] = \begin{cases} \dfrac{I_0[\beta(1 - [(n - \alpha)/\alpha]^2)^{1/2}]}{I_0(\beta)}, & 0 \le n \le M, \\ 0, & \text{otherwise,} \end{cases} \tag{7.72}
$$

where $\alpha = M/2$, and $I_0(\cdot)$ represents the zeroth-order modified Bessel function of the first kind. In contrast to the other windows in Eqs. (7.60), the Kaiser window has two parameters: the length $(M + 1)$ and a shape parameter β. By varying $(M + 1)$ and β, the window length and shape can be adjusted to trade side-lobe amplitude for main-lobe width. Figure 7.32(a) shows continuous envelopes of Kaiser windows of length $M + 1 = 21$ for $\beta = 0, 3$, and 6. Notice from Eq. (7.72) that the case $\beta = 0$ reduces to the rectangular window. Figure 7.32(b) shows the corresponding Fourier transforms of the Kaiser windows in Figure 7.32(a). Figure 7.32(c) shows Fourier transforms of Kaiser windows with $\beta = 6$ and $M = 10, 20$, and 40. The plots in Figures 7.32(b) and (c) clearly show that the desired trade-off can be achieved. If the window is tapered more, the side lobes of the Fourier transform become smaller, but the main lobe becomes wider. Figure 7.32(c) shows that increasing M while holding β constant causes the main lobe to decrease in width, but it does not affect the peak amplitude of the side lobes. In fact, through extensive numerical experimentation, Kaiser obtained a pair of formulas that permit the filter designer to predict in advance the values of M and β needed to meet a given frequency-selective filter specification. The upper plot of Figure 7.31 is also typical of approximations obtained using the Kaiser window, and Kaiser (1974) found that, over a usefully wide range of conditions, the peak approximation error (δ in Figure 7.31) is determined by the choice of β. Given that δ is fixed, the passband cutoff frequency ω_p of the lowpass filter is defined to be the highest frequency such that $|H(e^{j\omega})| \ge 1 - \delta$. The stopband cutoff frequency ω_s is defined to be the lowest frequency such that $|H(e^{j\omega})| \le \delta$. Therefore, the transition region has width

$$
\Delta\omega = \omega_s - \omega_p \tag{7.73}
$$

for the lowpass filter approximation. Defining

$$
A = -20 \log_{10} \delta, \tag{7.74}
$$

Kaiser determined empirically that the value of β needed to achieve a specified value of A is given by

$$
\beta = \begin{cases} 0.1102(A - 8.7), & A > 50, \\ 0.5842(A - 21)^{0.4} + 0.07886(A - 21), & 21 \le A \le 50, \\ 0.0, & A < 21. \end{cases} \tag{7.75}
$$

(Recall that the case $\beta = 0$ is the rectangular window for which $A = 21$.) Furthermore, Kaiser found that to achieve prescribed values of A and $\Delta\omega$, M must satisfy

$$
M = \frac{A - 8}{2.285\Delta\omega}. \tag{7.76}
$$

Equation (7.76) predicts M to within ± 2 over a wide range of values of $\Delta\omega$ and A. Thus, with these formulas, the Kaiser window design method requires almost no iteration or trial and error. The examples in Section 7.6 outline and illustrate the procedure.

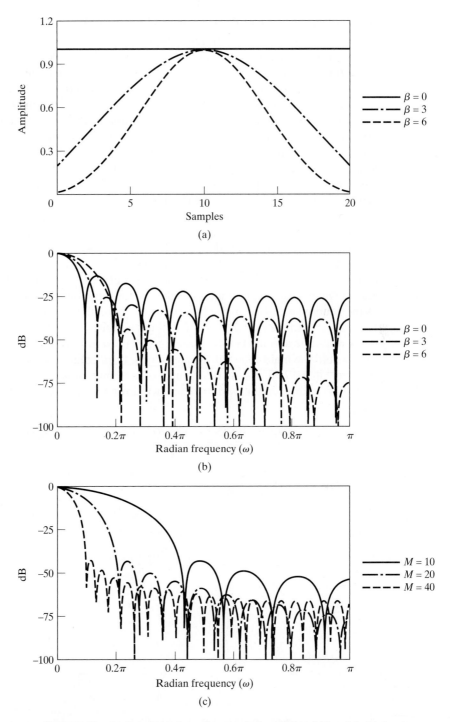

Figure 7.32 (a) Kaiser windows for $\beta = 0$, 3, and 6 and $M = 20$. (b) Fourier transforms corresponding to windows in (a). (c) Fourier transforms of Kaiser windows with $\beta = 6$ and $M = 10$, 20, and 40.

Relationship of the Kaiser Window to Other Windows

The basic principle of the window design method is to truncate the ideal impulse response with a finite-length window such as one of those discussed in this section. The corresponding effect in the frequency domain is that the ideal frequency response is convolved with the Fourier transform of the window. If the ideal filter is a lowpass filter, the discontinuity in its frequency response is smeared as the main lobe of the Fourier transform of the window moves across the discontinuity in the convolution process. To a first approximation, the width of the resulting transition band is determined by the width of the main lobe of the Fourier transform of the window, and the passband and stopband ripples are determined by the side lobes of the Fourier transform of the window. Because the passband and stopband ripples are produced by integration of the symmetric window side lobes, the ripples in the passband and the stopband are approximately the same. Furthermore, to a very good approximation, the maximum passband and stopband deviations are not dependent on M and can be changed only by changing the shape of the window used. This is illustrated by Kaiser's formula, Eq. (7.75), for the window shape parameter, which is independent of M. The last two columns of Table 7.2 compare the Kaiser window with the windows of Eqs. (7.60). The fifth column gives the Kaiser window shape parameter (β) that yields the same peak approximation error (δ) as the window indicated in the first column. The sixth column shows the corresponding transition width [from Eq. (7.76)] for filters designed with the Kaiser window. This formula would be a much better predictor of the transition width for the other windows than would the main-lobe width given in the third column of the table.

In Figure 7.33 is shown a comparison of maximum approximation error versus transition width for the various fixed windows and the Kaiser window for different

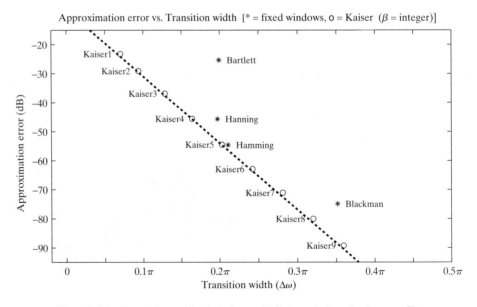

Figure 7.33   Comparison of fixed windows with Kaiser windows in a lowpass filter design application ($M = 32$ and $\omega_c = \pi/2$). (Note that the designation "Kaiser 6" means Kaiser window with $\beta = 6$, etc.)

values of β. The dashed line obtained from Eq. (7.76), shows that Kaiser's formula is an accurate representation of approximation error as a function of transition width for the Kaiser window.

7.6 EXAMPLES OF FIR FILTER DESIGN BY THE KAISER WINDOW METHOD

In this section, we give several examples that illustrate the use of the Kaiser window to obtain FIR approximations to several filter types including lowpass filters. These examples also serve to point out some important properties of FIR systems.

7.6.1 Lowpass Filter

With the use of the design formulas for the Kaiser window, it is straightforward to design an FIR lowpass filter to meet prescribed specifications. The procedure is as follows:

1. First, the specifications must be established. This means selecting the desired ω_p and ω_s and the maximum tolerable approximation error. For window design, the resulting filter will have the same peak error δ in both the passband and the stopband. For this example, we use the same specifications as in Example 7.5, $\omega_p = 0.4\pi$, $\omega_s = 0.6\pi$, $\delta_1 = 0.01$, and $\delta_2 = 0.001$. Since filters designed by the window method inherently have $\delta_1 = \delta_2$, we must set $\delta = 0.001$.

2. The cutoff frequency of the underlying ideal lowpass filter must be found. Owing to the symmetry of the approximation at the discontinuity of $H_d(e^{j\omega})$, we would set

$$\omega_c = \frac{\omega_p + \omega_s}{2} = 0.5\pi.$$

3. To determine the parameters of the Kaiser window, we first compute

$$\Delta\omega = \omega_s - \omega_p = 0.2\pi, \qquad A = -20\log_{10}\delta = 60.$$

We substitute these two quantities into Eqs. (7.75) and (7.76) to obtain the required values of β and M. For this example the formulas predict

$$\beta = 5.653, \qquad M = 37.$$

4. The impulse response of the filter is computed using Eqs. (7.71) and (7.72). We obtain

$$h[n] = \begin{cases} \dfrac{\sin\omega_c(n-\alpha)}{\pi(n-\alpha)} \cdot \dfrac{I_0[\beta(1-[(n-\alpha)/\alpha]^2)^{1/2}]}{I_0(\beta)}, & 0 \le n \le M, \\ 0, & \text{otherwise,} \end{cases}$$

where $\alpha = M/2 = 37/2 = 18.5$. Since $M = 37$ is an odd integer, the resulting linear-phase system would be of type II. (See Section 5.7.3 for the definitions of the four types of FIR systems with generalized linear phase.) The response characteristics of the filter are shown in Figure 7.34. Figure 7.34(a), which shows the impulse response, displays the characteristic symmetry of a type II system.

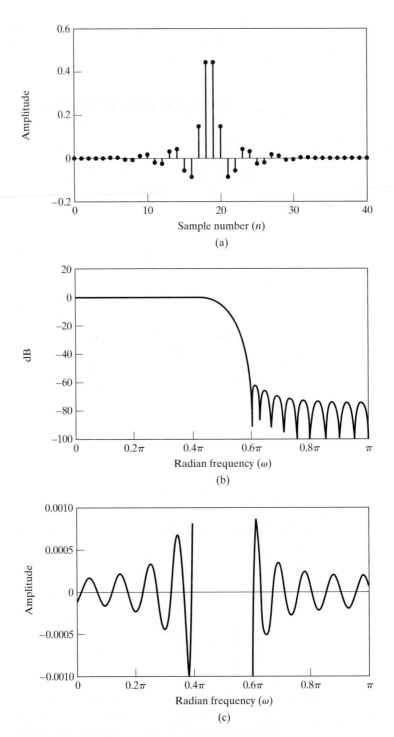

Figure 7.34 Response functions for the lowpass filter designed with a Kaiser window. (a) Impulse response ($M = 37$). (b) Log magnitude. (c) Approximation error for $A_e(e^{j\omega})$.

Figure 7.34(b), which shows the log magnitude response in dB, indicates that $H(e^{j\omega})$ is zero at $\omega = \pi$ or, equivalently, that $H(z)$ has a zero at $z = -1$, as required for a type II FIR system. Figure 7.34(c) shows the approximation error in the passband and stopbands. This error function is defined as

$$E_A(\omega) = \begin{cases} 1 - A_e(e^{j\omega}), & 0 \le \omega \le \omega_p, \\ 0 - A_e(e^{j\omega}), & \omega_s \le \omega \le \pi. \end{cases} \tag{7.77}$$

(The error is not defined in the transition region, $0.4\pi < \omega < 0.6\pi$.) Note the slight asymmetry of the approximation error, and note also that the peak approximation error is $\delta = 0.00113$ instead of the desired value of 0.001. In this case it is necessary to increase M to 40 in order to meet the specifications.

5. Finally, observe that it is not necessary to plot either the phase or the group delay, since we know that the phase is precisely linear and the delay is $M/2 = 18.5$ samples.

7.6.2 Highpass Filter

The ideal highpass filter with generalized linear phase has the frequency response

$$H_{hp}(e^{j\omega}) = \begin{cases} 0, & |\omega| < \omega_c, \\ e^{-j\omega M/2}, & \omega_c < |\omega| \le \pi. \end{cases} \tag{7.78}$$

The corresponding impulse response can be found by evaluating the inverse transform of $H_{hp}(e^{j\omega})$, or we can observe that

$$H_{hp}(e^{j\omega}) = e^{-j\omega M/2} - H_{lp}(e^{j\omega}), \tag{7.79}$$

where $H_{lp}(e^{j\omega})$ is given by Eq. (7.69). Thus, $h_{hp}[n]$ is

$$h_{hp}[n] = \frac{\sin \pi(n - M/2)}{\pi(n - M/2)} - \frac{\sin \omega_c(n - M/2)}{\pi(n - M/2)}, \qquad -\infty < n < \infty. \tag{7.80}$$

To design an FIR approximation to the highpass filter, we can proceed in a manner similar to that in Section 7.6.1.

Suppose that we wish to design a filter to meet the highpass specifications

$$|H(e^{j\omega})| \le \delta_2, \qquad |\omega| \le \omega_s$$

$$1 - \delta_1 \le |H(e^{j\omega})| \le 1 + \delta_1, \qquad \omega_p \le |\omega| \le \pi$$

where $\omega_s = 0.35\pi$, $\omega_p = 0.5\pi$, and $\delta_1 = \delta_2 = \delta = 0.02$. Since the ideal response also has a discontinuity, we can apply Kaiser's formulas in Eqs. (7.75) and (7.76) with $A = 33.98$ and $\Delta\omega = 0.15\pi$ to estimate the required values of $\beta = 2.65$ and $M = 24$. Figure 7.35 shows the response characteristics that result when a Kaiser window with these parameters is applied to $h_{hp}[n]$ with $\omega_c = (0.35\pi + 0.5\pi)/2$. Note that, since M is an even integer, the filter is a type I FIR system with linear phase, and the delay is precisely $M/2 = 12$ samples. In this case, the actual peak approximation error is $\delta = 0.0209$ rather than 0.02, as specified. Since the error is less than 0.02 everywhere except at the stopband edge, it is tempting to simply increase M to 25, keeping β the same, thereby narrowing the transition region. This type II filter, which is shown in Figure 7.36, is highly unsatisfactory, owing to the zero of $H(z)$ that is forced by the

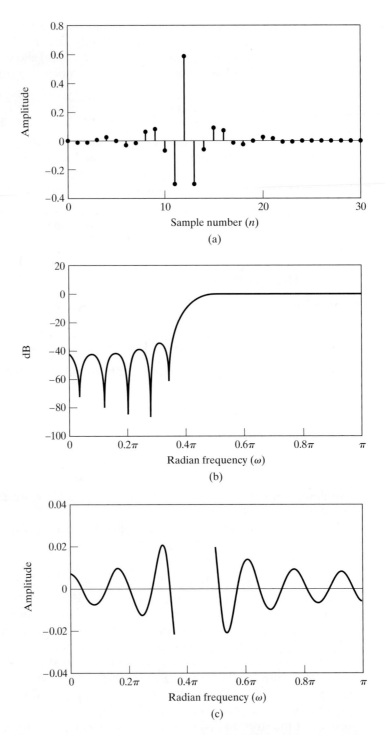

Figure 7.35 Response functions for type I FIR highpass filter. (a) Impulse response ($M = 24$). (b) Log magnitude. (c) Approximation error for $A_e(e^{j\omega})$.

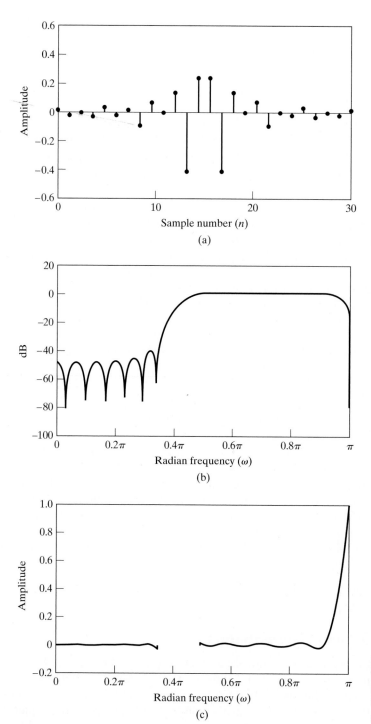

Figure 7.36 Response functions for type II FIR highpass filter. (a) Impulse response ($M = 25$). (b) Log magnitude of Fourier transform. (c) Approximation error for $A_e(e^{j\omega})$.

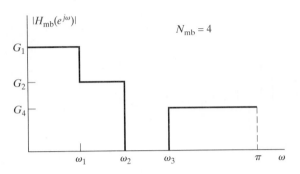

Figure 7.37   Ideal frequency response for multiband filter.

linear-phase constraint to be at $z = -1$, i.e., $\omega = \pi$. Although increasing the order by 1 leads to a worse result, increasing M to 26 would, of course, lead to a type I system that would exceed the specifications. Clearly, type II FIR linear-phase systems are generally not appropriate approximations for either highpass or bandstop filters.

The previous discussion of highpass filter design can be generalized to the case of multiple passbands and stopbands. Figure 7.37 shows an ideal multiband frequency-selective frequency response. This generalized multiband filter includes lowpass, highpass, bandpass, and bandstop filters as special cases. If such a magnitude function is multiplied by a linear-phase factor $e^{-j\omega M/2}$, the corresponding ideal impulse response is

$$h_{\mathrm{mb}}[n] = \sum_{k=1}^{N_{\mathrm{mb}}} (G_k - G_{k+1}) \frac{\sin \omega_k (n - M/2)}{\pi (n - M/2)}, \qquad (7.81)$$

where N_{mb} is the number of bands and $G_{N_{\mathrm{mb}}+1} = 0$. If $h_{\mathrm{mb}}[n]$ is multiplied by a Kaiser window, the type of approximations that we have observed at the single discontinuity of the lowpass and highpass systems will occur at *each* of the discontinuities. The behavior will be the same at each discontinuity, provided that the discontinuities are far enough apart. Thus, Kaiser's formulas for the window parameters can be applied to this case to predict approximation errors and transition widths. Note that the approximation errors will be scaled by the size of the jump that produces them. That is, if a discontinuity of unity produces a peak error of δ, then a discontinuity of one-half will have a peak error of $\delta/2$.

7.6.3 Discrete-Time Differentiators

As illustrated in Example 4.4, sometimes it is of interest to obtain samples of the derivative of a bandlimited signal from samples of the signal itself. Since the Fourier transform of the derivative of a continuous-time signal is $j\Omega$ times the Fourier transform of the signal, it follows that, for bandlimited signals, a discrete-time system with frequency response $(j\omega/T)$ for $-\pi < \omega < \pi$ (and that is periodic, with period 2π) will yield output samples that are equal to samples of the derivative of the continuous-time signal. A system with this property is referred to as a discrete-time differentiator.

For an ideal discrete-time differentiator with linear phase, the appropriate frequency response is

$$H_{\text{diff}}(e^{j\omega}) = (j\omega)e^{-j\omega M/2}, \qquad -\pi < \omega < \pi. \tag{7.82}$$

(We have omitted the factor $1/T$.) The corresponding ideal impulse response is

$$h_{\text{diff}}[n] = \frac{\cos \pi(n - M/2)}{(n - M/2)} - \frac{\sin \pi(n - M/2)}{\pi(n - M/2)^2}, \qquad -\infty < n < \infty. \tag{7.83}$$

If $h_{\text{diff}}[n]$ is multiplied by a symmetric window of length $(M+1)$, then it is easily shown that $h[n] = -h[M - n]$. Thus, the resulting system is either a type III or a type IV generalized linear-phase system.

Since Kaiser's formulas were developed for frequency responses with simple magnitude discontinuities, it is not straightforward to apply them to differentiators, wherein the discontinuity in the ideal frequency response is introduced by the phase. Nevertheless, as we show in the next example, the window method is very effective in designing such systems.

Kaiser Window Design of a Differentiator

To illustrate the window design of a differentiator, suppose $M = 10$ and $\beta = 2.4$. The resulting response characteristics are shown in Figure 7.38. Figure 7.38(a) shows the antisymmetric impulse response. Since M is even, the system is a type III linear-phase system, which implies that $H(z)$ has zeros at *both* $z = +1$ ($\omega = 0$) and $z = -1$ ($\omega = \pi$). This is clearly displayed in the magnitude response shown in Figure 7.38(b). The phase is exact, since type III systems have a $\pi/2$-radian constant phase shift plus a linear phase corresponding in this case to $M/2 = 5$ samples delay. Figure 7.38(c) shows the amplitude approximation error

$$E_{\text{diff}}(\omega) = \omega - A_o(e^{j\omega}), \qquad 0 \le \omega \le 0.8\pi, \tag{7.84}$$

where $A_o(e^{j\omega})$ is the amplitude of the approximation. (Note that the error is large around $\omega = \pi$ and is not plotted for frequencies above $\omega = 0.8\pi$.) Clearly, the linearly increasing magnitude is not achieved over the whole band, and, obviously, the relative error (i.e., $E_{\text{diff}}(\omega)/\omega$) is very large for low frequencies or high frequencies (around $\omega = \pi$).

Type IV linear-phase systems do not constrain $H(z)$ to have a zero at $z = -1$. This type of system leads to much better approximations to the amplitude function, as shown in Figure 7.39, for $M = 5$ and $\beta = 2.4$. In this case, the amplitude approximation error is very small up to and beyond $\omega = 0.8\pi$. The phase for this system is again a $\pi/2$-radian constant phase shift plus a linear phase corresponding to a delay of $M/2 = 2.5$ samples. This noninteger delay is the price paid for the exceedingly good amplitude approximation. Instead of obtaining samples of the derivative of the continuous-time signal at the original sampling times $t = nT$, we obtain samples of the derivative at times $t = (n - 2.5)T$. However, in many applications, this noninteger delay may not cause a problem, or it could be compensated for by other noninteger delays in a more complex system involving other linear-phase filters.

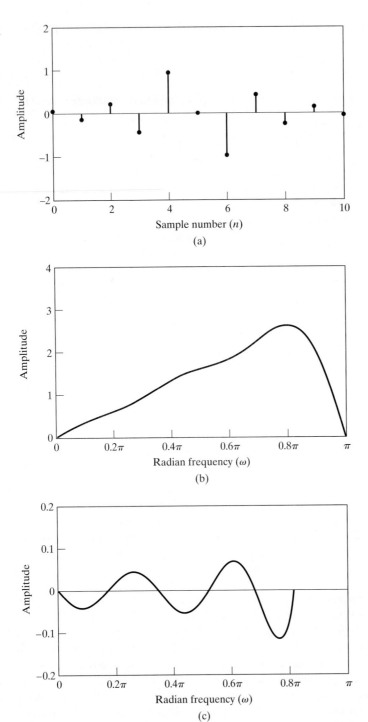

Figure 7.38 Response functions for type III FIR discrete-time differentiator. (a) Impulse response ($M = 10$). (b) Magnitude. (c) Approximation error for $A_0(e^{j\omega})$.

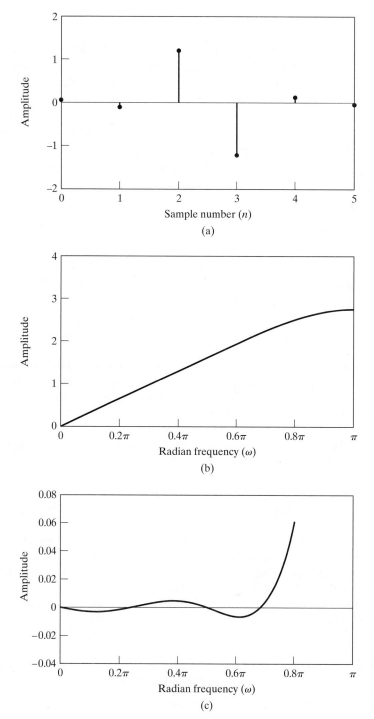

Figure 7.39 Response functions for type IV FIR discrete-time differentiator. (a) Impulse response ($M = 5$). (b) Magnitude. (c) Approximation error for $A_0(e^{j\omega})$.

7.7 OPTIMUM APPROXIMATIONS OF FIR FILTERS

The design of FIR filters by windowing is straightforward and is quite general, even though it has a number of limitations as discussed below. However, we often wish to design a filter that is the "best" that can be achieved for a given value of M. It is meaningless to discuss this question in the absence of an approximation criterion. For example, in the case of the window design method, it follows from the theory of Fourier series that the rectangular window provides the best mean-square approximation to a desired frequency response for a given value of M. That is,

$$h[n] = \begin{cases} h_d[n], & 0 \leq n \leq M, \\ 0, & \text{otherwise}, \end{cases} \tag{7.85}$$

minimizes the expression

$$\varepsilon^2 = \frac{1}{2\pi} \int_{-\pi}^{\pi} |H_d(e^{j\omega}) - H(e^{j\omega})|^2 d\omega. \tag{7.86}$$

(See Problem 7.25.) However, as we have seen, this approximation criterion leads to adverse behavior at discontinuities of $H_d(e^{j\omega})$. Furthermore, the window method does not permit individual control over the approximation errors in different bands. For many applications, better filters result from a minimax strategy (minimization of the maximum errors) or a frequency-weighted error criterion. Such designs can be achieved using algorithmic techniques.

As the previous examples show, frequency-selective filters designed by windowing often have the property that the error is greatest on either side of a discontinuity of the ideal frequency response, and the error becomes smaller for frequencies away from the discontinuity. Furthermore, as suggested by Figure 7.31, such filters typically result in approximately equal errors in the passband and stopband. (See Figures 7.34(c) and 7.35(c), for example.) We have already seen that, for IIR filters, if the approximation error is spread out uniformly in frequency and if the passband and stopband ripples are adjusted separately, a given design specification can be met with a lower-order filter than if the approximation just meets the specification at one frequency and far exceeds it at others. This intuitive notion is confirmed for FIR systems by a theorem to be discussed later in the section.

In the following discussion, we consider a particularly effective and widely used algorithmic procedure for the design of FIR filters with a generalized linear phase. Although we consider only type I filters in detail, we indicate where appropriate, how the results apply to types II, III, and IV generalized linear-phase filters.

In designing a causal type I linear-phase FIR filter, it is convenient first to consider the design of a zero-phase filter, i.e., one for which

$$h_e[n] = h_e[-n], \tag{7.87}$$

and then to insert a delay sufficient to make it causal. Consequently, we consider $h_e[n]$ satisfying the condition of Eq. (7.87). The corresponding frequency response is given by

$$A_e(e^{j\omega}) = \sum_{n=-L}^{L} h_e[n]e^{-j\omega n}, \tag{7.88}$$

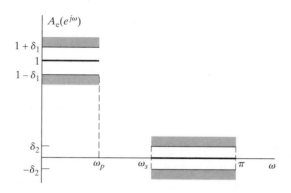

Figure 7.40 Tolerance scheme and ideal response for lowpass filter.

with $L = M/2$ an integer, or, because of Eq. (7.87),

$$A_e(e^{j\omega}) = h_e[0] + \sum_{n=1}^{L} 2h_e[n]\cos(\omega n). \qquad (7.89)$$

Note that $A_e(e^{j\omega})$ is a real, even, and periodic function of ω. A causal system can be obtained from $h_e[n]$ by delaying it by $L = M/2$ samples. The resulting system has impulse response

$$h[n] = h_e[n - M/2] = h[M - n] \qquad (7.90)$$

and frequency response

$$H(e^{j\omega}) = A_e(e^{j\omega})e^{-j\omega M/2}. \qquad (7.91)$$

Figure 7.40 shows a tolerance scheme for an approximation to a lowpass filter with a real function such as $A_e(e^{j\omega})$. Unity is to be approximated in the band $0 \le |\omega| \le \omega_p$ with maximum absolute error δ_1, and zero is to be approximated in the band $\omega_s \le |\omega| \le \pi$ with maximum absolute error δ_2. An algorithmic technique for designing a filter to meet these specifications must, in effect, systematically vary the $(L + 1)$ unconstrained impulse response values $h_e[n]$, where $0 \le n \le L$. Design algorithms have been developed in which some of the parameters $L, \delta_1, \delta_2, \omega_p$, and ω_s are fixed and an iterative procedure is used to obtain optimum adjustments of the remaining parameters. Two distinct approaches have been developed. Herrmann (1970), Herrmann and Schüssler (1970a), and Hofstetter, Oppenheim and Siegel (1971) developed procedures in which L, δ_1, and δ_2 are fixed, and ω_p and ω_s are variable. Parks and McClellan (1972a, 1972b), McClellan and Parks (1973), and Rabiner (1972a, 1972b) developed procedures in which L, ω_p, ω_s, and the ratio δ_1/δ_2 are fixed and δ_1 (or δ_2) is variable. Since the time when these different approaches were developed, the Parks–McClellan algorithm has become the dominant method for optimum design of FIR filters. This is because it is the most flexible and the most computationally efficient. Thus, we will discuss only that algorithm here.

The Parks–McClellan algorithm is based on reformulating the filter design problem as a problem in polynomial approximation. Specifically, the terms $\cos(\omega n)$ in Eq. (7.89) can be expressed as a sum of powers of $\cos \omega$ in the form

$$\cos(\omega n) = T_n(\cos \omega), \qquad (7.92)$$

where $T_n(x)$ is an nth-order polynomial.[5] Consequently, Eq. (7.89) can be rewritten as an Lth-order polynomial in $\cos \omega$, namely,

$$A_e(e^{j\omega}) = \sum_{k=0}^{L} a_k (\cos \omega)^k, \qquad (7.93)$$

where the a_ks are constants that are related to $h_e[n]$, the values of the impulse response. With the substitution $x = \cos \omega$, we can express Eq. (7.93) as

$$A_e(e^{j\omega}) = P(x)|_{x=\cos \omega}, \qquad (7.94)$$

where $P(x)$ is the Lth-order polynomial

$$P(x) = \sum_{k=0}^{L} a_k x^k. \qquad (7.95)$$

We will see that it is not necessary to know the relationship between the a_ks and $h_e[n]$ (although a formula can be obtained); it is enough to know that $A_e(e^{j\omega})$ can be expressed as the Lth-order trigonometric polynomial of Eq. (7.93).

The key to gaining control over ω_p and ω_s is to fix them at their desired values and let δ_1 and δ_2 vary. Parks and McClellan (1972a, 1972b) showed that with L, ω_p, and ω_s fixed, the frequency-selective filter design problem becomes a problem in Chebyshev approximation over disjoint sets, an important problem in approximation theory and one for which several useful theorems and procedures have been developed. (See Cheney, 1982.) To formalize the approximation problem in this case, let us define an approximation error function

$$E(\omega) = W(\omega)[H_d(e^{j\omega}) - A_e(e^{j\omega})], \qquad (7.96)$$

where the weighting function $W(\omega)$ incorporates the approximation error parameters into the design process. In this design method, the error function $E(\omega)$, the weighting function $W(\omega)$, and the desired frequency response $H_d(e^{j\omega})$ are defined only over closed subintervals of $0 \le \omega \le \pi$. For example, to approximate a lowpass filter, these functions are defined for $0 \le \omega \le \omega_p$ and $\omega_s \le \omega \le \pi$. The approximating function $A_e(e^{j\omega})$ is not constrained in the transition region(s) (e.g., $\omega_p < \omega < \omega_s$), and it may take any shape necessary to achieve the desired response in the other subintervals.

For example, suppose that we wish to obtain an approximation as in Figure 7.40, where L, ω_p, and ω_s are fixed design parameters. For this case,

$$H_d(e^{j\omega}) = \begin{cases} 1, & 0 \le \omega \le \omega_p, \\ 0, & \omega_s \le \omega \le \pi. \end{cases} \qquad (7.97)$$

The weighting function $W(\omega)$ allows us to weight the approximation errors differently in the different approximation intervals. For the lowpass filter approximation problem, the weighting function is

$$W(\omega) = \begin{cases} \dfrac{1}{K}, & 0 \le \omega \le \omega_p, \\ 1, & \omega_s \le \omega \le \pi, \end{cases} \qquad (7.98)$$

---

[5]More specifically, $T_n(x)$ is the nth-order Chebyshev polynomial, defined as $T_n(x) = \cos(n \cos^{-1} x)$.

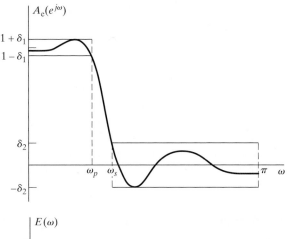

Figure 7.41 Typical frequency response meeting the specifications of Figure 7.40.

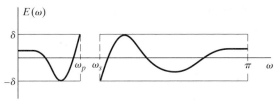

Figure 7.42 Weighted error for the approximation of Figure 7.41.

where $K = \delta_1/\delta_2$. If $A_e(e^{j\omega})$ is as shown in Figure 7.41, the weighted approximation error, $E(\omega)$ in Eq. (7.96), would be as indicated in Figure 7.42. Note that with this weighting, the maximum weighted absolute approximation error is $\delta = \delta_2$ in both bands.

The particular criterion used in this design procedure is the so-called minimax or Chebyshev criterion, where, within the frequency intervals of interest (the passband and stopband for a lowpass filter), we seek a frequency response $A_e(e^{j\omega})$ that *minimizes* the *maximum* weighted approximation error of Eq. (7.96). Stated more compactly, the best approximation is to be found in the sense of

$$\min_{\{h_e[n]:0 \le n \le L\}} \left(\max_{\omega \in F} |E(\omega)| \right),$$

where F is the closed subset of $0 \le \omega \le \pi$ such that $0 \le \omega \le \omega_p$ or $\omega_s \le \omega \le \pi$. Thus, we seek the set of impulse response values that minimizes δ in Figure 7.42.

Parks and McClellan (1972a, 1972b) applied the following theorem of approximation theory to this filter design problem.

Alternation Theorem: Let F_P denote the closed subset consisting of the disjoint union of closed subsets of the real axis x. Furthermore,

$$P(x) = \sum_{k=0}^{r} a_k x^k$$

is an rth-order polynomial, and $D_P(x)$ denotes a given desired function of x that is continuous on F_P; $W_P(x)$ is a positive function, continuous on F_P, and

$$E_P(x) = W_P(x)[D_P(x) - P(x)]$$

is the weighted error. The maximum error is defined as

$$\|E\| = \max_{x \in F_P} |E_P(x)|.$$

A necessary and sufficient condition that $P(x)$ be the unique r^{th}-order polynomial that minimizes $\|E\|$ is that $E_P(x)$ exhibit *at least* $(r + 2)$ alternations; i.e., there must exist at least $(r + 2)$ values x_i in F_p such that $x_1 < x_2 < \cdots < x_{r+2}$ and such that $E_P(x_i) = -E_P(x_{i+1}) = \pm \|E\|$ for $i = 1, 2, \ldots, (r + 1)$.

At first glance, it may seem to be difficult to relate this formal theorem to the problem of filter design. However, in the discussion that follows, all of the elements of the theorem will be shown to be important in developing the design algorithm. To aid in understanding the alternation theorem, in Section 7.7.1 we will interpret it specifically for the design of a type I lowpass filter. Before proceeding to apply the alternation theorem to filter design, however, we illustrate in Example 7.8 how the theorem is applied to polynomials.

Example 7.8   Alternation Theorem and Polynomials

The alternation theorem provides a necessary and sufficient condition that a polynomial must satisfy in order that it be the polynomial that minimizes the maximum weighted error for a given order. To illustrate how the theorem is applied, suppose we want to examine polynomials $P(x)$ that approximate unity for $-1 \leq x \leq -0.1$ and zero for $0.1 \leq x \leq 1$. Consider three such polynomials, as shown in Figure 7.43. Each of these polynomials is of 5th-order, and we would like to determine which, if any, satisfy the alternation theorem. The closed subsets of the real axis x referred to in the theorem are the regions $-1 \leq x \leq -0.1$ and $0.1 \leq x \leq 1$. We will weight errors equally in both regions, i.e., $W_p(x) = 1$. To begin, it will be useful for the reader to carefully construct sketches of the approximation error function for each polynomial in Figure 7.43.

According to the alternation theorem, the optimal 5th-order polynomial must exhibit *at least* seven alternations of the error in the regions corresponding to the closed subset F_p. $P_1(x)$ has only five alternations—three in the region $-1 \leq x \leq -0.1$ and two in the region $0.1 \leq x \leq 1$. The points x at which the polynomial attains the maximum approximation error $\|E\|$ within the set F_p are called extremal points (or simply extremals). All alternations occur at extremals, but not all extremal points are alternations, as we will see. For example, the point with zero slope close to $x = 1$ that does not touch the dotted line is a local maximum, but is not an alternation, because the corresponding error function does not reach the negative extreme value.[6] The alternation theorem specifies that adjacent alternations must alternate sign, so the extremal value at $x = 1$ cannot be an alternation either, since the previous alternation was a positive extremal value at the first point with zero slope in $0.1 \leq x \leq 1$. The locations of the alternations are indicated by the symbol $\circ$ on the polynomials in Figure 7.43.

$P_2(x)$ also has only five alternations and thus is not optimal. Specifically, $P_2(x)$ has three alternations in $-1 \leq x \leq -0.1$, but again, only two alternations in $0.1 \leq x \leq 1$. The difficulty occurs because $x = 0.1$ is not a negative extremal value. The previous alternation at $x = -0.1$ is a positive extremal value, so we need a negative extremal value for the next alternation. The first point with zero slope inside $0.1 \leq x \leq 1$ also cannot be counted, since it is a positive extremal value, like $x = -0.1$, and does not alternate sign. We can count the second point with zero slope in this region and $x = 1$, giving two alternations in $0.1 \leq x \leq 1$ and a total of five.

---

[6] In this discussion, we refer to positive and negative extremals of the error function. Since the polynomial is subtracted from a constant to form the error, the extremal points are easily located on the polynomial curves in Figure 7.43, but the sign is opposite of the variation above and below the desired constant values.

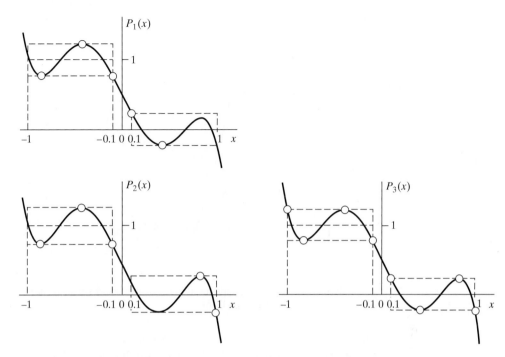

Figure 7.43 5th-order polynomials for Example 7.8. Alternation points are indicated by ∘.

$P_3(x)$ has eight alternations; all points of zero slope, $x = -1$, $x = -0.1$, $x = 0.1$ and $x = 1$. Since eight alternations satisfies the alternation theorem, which specifies a minimum of seven, $P_3(x)$ is the unique optimal 5th-order polynomial approximation for this region.

7.7.1 Optimal Type I Lowpass Filters

For type I filters, the polynomial $P(x)$ is the cosine polynomial $A_e(e^{j\omega})$ in Eq. (7.93), with the transformation of variable $x = \cos\omega$ and $r = L$:

$$P(\cos\omega) = \sum_{k=0}^{L} a_k (\cos\omega)^k. \tag{7.99}$$

$D_P(x)$ is the desired lowpass filter frequency response in Eq. (7.97), with $x = \cos\omega$:

$$D_P(\cos\omega) = \begin{cases} 1, & \cos\omega_p \le \cos\omega \le 1, \\ 0, & -1 \le \cos\omega \le \cos\omega_s. \end{cases} \tag{7.100}$$

$W_P(\cos\omega)$ is given by Eq. (7.98), rephrased in terms of $\cos\omega$:

$$W_P(\cos\omega) = \begin{cases} \dfrac{1}{K}, & \cos\omega_p \le \cos\omega \le 1, \\ 1, & -1 \le \cos\omega \le \cos\omega_s. \end{cases} \tag{7.101}$$

And the weighted approximation error is

$$E_P(\cos\omega) = W_P(\cos\omega)[D_P(\cos\omega) - P(\cos\omega)]. \tag{7.102}$$

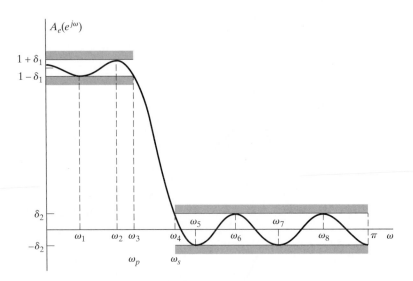

Figure 7.44 Typical example of a lowpass filter approximation that is optimal according to the alternation theorem for $L = 7$.

The closed subset F_P is made up of the union of the intervals $0 \le \omega \le \omega_p$ and $\omega_s \le \omega \le \pi$, or, in terms of $\cos \omega$, of the corresponding intervals $\cos \omega_p \le \cos \omega \le 1$ and $-1 \le \cos \omega \le \cos \omega_s$. The alternation theorem then states that a set of coefficients a_k in Eq. (7.99) will correspond to the filter representing the unique best approximation to the ideal lowpass filter, with the ratio δ_1/δ_2 fixed at K and with passband and stopband edges ω_p and ω_s, if and only if $E_P(\cos \omega)$ exhibits at least $(L+2)$ alternations on F_P, i.e., if and only if $E_P(\cos \omega)$ alternately equals plus and minus its maximum value at least $(L + 2)$ times. We have previously seen such *equiripple approximations* in the case of elliptic IIR filters.

Figure 7.44 shows a filter frequency response that is optimal according to the alternation theorem for $L = 7$. In this figure, $A_e(e^{j\omega})$ is plotted against ω. To formally test the alternation theorem, we should first redraw $A_e(e^{j\omega})$ as a function of $x = \cos \omega$. Furthermore, we want to explicitly examine the alternations of $E_P(x)$. Consequently, in Figure 7.45(a), (b), and (c), we show $P(x)$, $W_P(x)$, and $E_P(x)$, respectively, as a function of $x = \cos \omega$. In this example, where $L = 7$, we see that there are nine alternations of the error. Consequently, the alternation theorem is satisfied. An important point is that, in counting alternations, we include the points $\cos \omega_p$ and $\cos \omega_s$, since, according to the alternation theorem, the subsets (or subintervals) included in F_P are closed, i.e., the endpoints of the intervals are counted. Although this might seem to be a small issue, it is in fact very significant, as we will see.

Comparing Figures 7.44 and 7.45 suggests that when the desired filter is a lowpass filter (or any piecewise-constant filter) we could easily count the alternations by direct examination of the frequency response, keeping in mind that the maximum error is different (in the ratio $K = \delta_1/\delta_2$) in the passband and stopband.

The alternation theorem states that the optimum filter must have a minimum of $(L + 2)$ alternations, but it does not exclude the possibility of more than $(L + 2)$ alternations. In fact, we will show that for a lowpass filter, the maximum possible number of alternations is $(L + 3)$. First, however, we illustrate this in Figure 7.46 for $L = 7$.

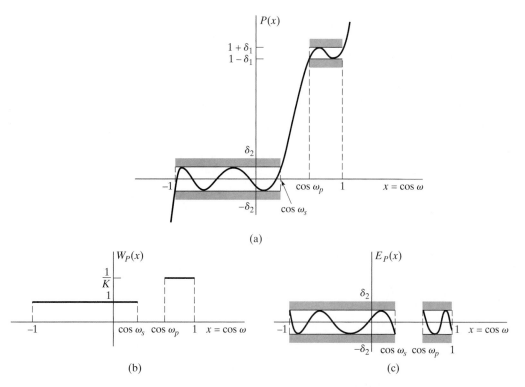

Figure 7.45 Equivalent polynomial approximation functions as a function of $x = \cos \omega$. (a) Approximating polynomial. (b) Weighting function. (c) Approximation error.

Figure 7.46(a) has $L + 3 = 10$ alternations, whereas Figures 7.46(b), (c), and (d) each have $L+2 = 9$ alternations. The case of $L+3$ alternations (Figure 7.46a) is often referred to as the *extraripple case*. Note that for the extraripple filter, there are alternations at $\omega = 0$ and π, as well as at $\omega = \omega_p$ and $\omega = \omega_s$, i.e., at all the band edges. For Figures 7.46(b) and (c), there are again alternations at ω_p and ω_s, but not at both $\omega = 0$ and $\omega = \pi$. In Figure 7.46(d), there are alternations at $0, \pi, \omega_p$, and ω_s, but there is one less point of zero slope inside the stopband. We also observe that all of these cases are equiripple inside the passband and stopband; i.e., all points of zero slope inside the interval $0 < \omega < \pi$ are frequencies at which the magnitude of the weighted error is maximal. Finally, because all of the filters in Figure 7.46 satisfy the alternation theorem for $L = 7$ and for the same value of $K = \delta_1/\delta_2$, it follows that ω_p and/or ω_s must be different for each, since the alternation theorem states that the optimum filter under the conditions of the theorem is unique.

The properties referred to in the preceding paragraph for the filters in Figure 7.46 result from the alternation theorem. Specifically, we will show that for type I lowpass filters:

- The maximum possible number of alternations of the error is $(L + 3)$.
- Alternations will always occur at ω_p and ω_s.
- All points with zero slope inside the passband and all points with zero slope inside the stopband (for $0 < \omega < \omega_p$ and $\omega_s < \omega < \pi$) will correspond to alternations; i.e., the filter will be equiripple, except possibly at $\omega = 0$ and $\omega = \pi$.

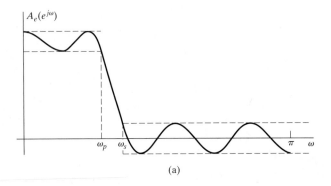

(a)

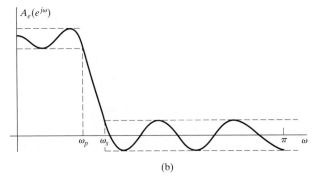

(b)

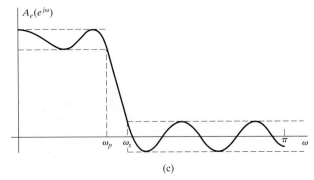

(c)

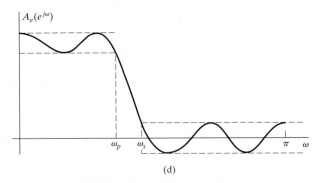

(d)

Figure 7.46 Possible optimum lowpass filter approximations for $L = 7$. (a) $L + 3$ alternations (extraripple case). (b) $L + 2$ alternations (extremum at $\omega = \pi$). (c) $L + 2$ alternations (extremum at $\omega = 0$). (d) $L + 2$ alternations (extremum at both $\omega = 0$ and $\omega = \pi$).

The maximum possible number of alternations is $(L + 3)$

Reference to Figure 7.44 or Figure 7.46 suggests that the maximum possible number of locations for alternations, are the four band edges ($\omega = 0, \pi, \omega_p, \omega_s$) and the frequencies at which $A_e(e^{j\omega})$ has zero slope. Since an Lth-order polynomial can have at most $(L-1)$ points with zero slope in an open interval, the maximum possible number of locations for alternations are the $(L-1)$ local maxima or minima of the polynomial plus the four band edges, a total of $(L + 3)$. In considering points with zero slope for trigonometric polynomials, it is important to observe that the trigonometric polynomial

$$P(\cos \omega) = \sum_{k=0}^{L} a_k (\cos \omega)^k, \tag{7.103}$$

when considered as a function of ω, will always have zero slope at $\omega = 0$ and $\omega = \pi$, even though $P(x)$ considered as a function of x may not have zero slope at the corresponding points $x = 1$ and $x = -1$. This is because

$$\frac{dP(\cos \omega)}{d\omega} = -\sin \omega \left(\sum_{k=0}^{L} k a_k (\cos \omega)^{k-1} \right)$$

$$= -\sin \omega \left(\sum_{k=0}^{L-1} (k+1) a_{k+1} (\cos \omega)^k \right), \tag{7.104}$$

which is always zero at $\omega = 0$ and $\omega = \pi$, as well as at the $(L-1)$ roots of the $(L-1)$st-order polynomial represented by the sum. This behavior at $\omega = 0$ and $\omega = \pi$ is evident in Figure 7.46. In Figure 7.46(d), it happens that the polynomial $P(x)$ also has zero slope at $x = -1 = \cos \pi$.

Alternations always occur at ω_p and ω_s

For all of the frequency responses in Figure 7.46, $A_e(e^{j\omega})$ is exactly equal to $1 - \delta_1$ at the passband edge ω_p and exactly equal to $+\delta_2$ at the stopband edge ω_s. To suggest why this must always be the case, let us consider whether the filter in Figure 7.46(a) could also be optimal if we redefined ω_p as indicated in Figure 7.47 leaving the polynomial unchanged. The frequencies at which the magnitude of the maximum weighted error are equal are the frequencies $\omega = 0, \omega_1, \omega_2, \omega_s, \omega_3, \omega_4, \omega_5, \omega_6$, and $\omega = \pi$, for a total of $(L + 2) = 9$. However, not all of the frequencies are alternations, since, to be counted in the alternation theorem, the error must *alternate* between $\delta = \pm \|E\|$ at these frequencies. Therefore, because the error is negative at both ω_2 and ω_s, the frequencies counted in the alternation theorem are $\omega = 0, \omega_1, \omega_2, \omega_3, \omega_4; \omega_5, \omega_6$, and π, for a total of 8. Since $(L + 2) = 9$, the conditions of the alternation theorem are not

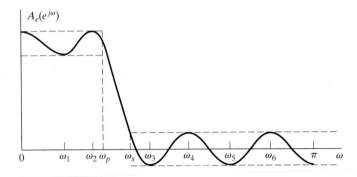

Figure 7.47   Illustration that the passband edge ω_p must be an alternation frequency.

satisfied, and the frequency response of Figure 7.47 is not optimal with ω_p and ω_s as indicated. In other words, the removal of ω_p as an alternation frequency removes two alternations. Since the maximum number is $(L + 3)$, this leaves at most $(L + 1)$, which is not a sufficient number. An identical argument would hold if ω_s were removed as an alternation frequency. A similar argument can be constructed for highpass filters, but this is not necessarily the case for bandpass or multiband filters. (See Problem 7.63.)

The filter will be equiripple except possibly at $\omega = 0$ or $\omega = \pi$

The argument here is very similar to the one used to show that both ω_p and ω_s must be alternations. Suppose, for example, that the filter in Figure 7.46(a) was modified as indicated in Figure 7.48, so that one point with zero slope did not achieve the maximum error. Although the maximum error occurs at nine frequencies, only eight of these can be counted as alternations. Consequently, eliminating one ripple as a point of maximum error reduces the number of alternations by two, leaving $(L+1)$ as the maximum possible number.

The foregoing represent only a few of many properties that can be inferred from the alternation theorem. A variety of others are discussed in Rabiner and Gold (1975). Furthermore, we have considered only type I lowpass filters. While a much broader and detailed discussion of type II, III, and IV filters or filters with more general desired frequency responses is beyond the scope of this book, we briefly consider type II lowpass filters to further emphasize a number of aspects of the alternation theorem.

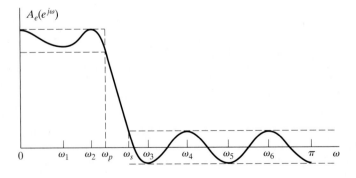

Figure 7.48   Illustration that the frequency response must be equiripple in the approximation bands.

7.7.2 Optimal Type II Lowpass Filters

A type II causal filter is a filter for which $h[n] = 0$ outside the range $0 \leq n \leq M$, with the filter length $(M + 1)$ even, i.e., M odd, and with the symmetry property

$$h[n] = h[M - n]. \tag{7.105}$$

Consequently, the frequency response $H(e^{j\omega})$ can be expressed in the form

$$H(e^{j\omega}) = e^{-j\omega M/2} \sum_{n=0}^{(M-1)/2} 2h[n] \cos\left[\omega\left(\frac{M}{2} - n\right)\right]. \tag{7.106}$$

Letting $b[n] = 2h[(M+1)/2 - n]$, $n = 1, 2, \ldots, (M+1)/2$, we can rewrite Eq. (7.106) as

$$H(e^{j\omega}) = e^{-j\omega M/2} \left\{ \sum_{n=1}^{(M+1)/2} b[n] \cos\left[\omega\left(n - \frac{1}{2}\right)\right] \right\}. \tag{7.107}$$

To apply the alternation theorem to the design of type II filters, we must be able to identify the problem as one of polynomial approximation. To accomplish this, we express the summation in Eq. (7.107) in the form

$$\sum_{n=1}^{(M+1)/2} b[n] \cos\left[\omega\left(n - \frac{1}{2}\right)\right] = \cos(\omega/2) \left[\sum_{n=0}^{(M-1)/2} \tilde{b}[n] \cos(\omega n) \right]. \tag{7.108}$$

(See Problem 7.58.) The summation on the right-hand side of Eq. (7.108) can now be represented as a trigonometric polynomial $P(\cos\omega)$ so that

$$H(e^{j\omega}) = e^{-j\omega M/2} \cos(\omega/2) P(\cos\omega), \tag{7.109a}$$

where

$$P(\cos\omega) = \sum_{k=0}^{L} a_k (\cos\omega)^k \tag{7.109b}$$

and $L = (M - 1)/2$. The coefficients a_k in Eq. (7.109b) are related to the coefficients $\tilde{b}[n]$ in Eq. (7.108), which in turn are related to the coefficients $b[n] = 2h[(M+1)/2 - n]$ in Eq. (7.107). As in the type I case, it is not necessary to obtain an explicit relationship between the impulse response and the a_ks. We now can apply the alternation theorem to the weighted error between $P(\cos\omega)$ and the desired frequency response. For a type I lowpass filter with a specified ratio K of passband to stopband ripple, the desired function is given by Eq. (7.97), and the weighting function for the error is given by Eq. (7.98). For type II lowpass filters, because of the presence of the factor $\cos(\omega/2)$ in Eq. (7.109a), the function to be approximated by the polynomial $P(\cos\omega)$ is defined as

$$H_d(e^{j\omega}) = D_P(\cos\omega) = \begin{cases} \dfrac{1}{\cos(\omega/2)}, & 0 \leq \omega \leq \omega_p, \\[2mm] 0, & \omega_s \leq \omega \leq \pi, \end{cases} \tag{7.110}$$

and the weighting function to be applied to the error is

$$W(\omega) = W_P(\cos\omega) = \begin{cases} \dfrac{\cos(\omega/2)}{K}, & 0 \leq \omega \leq \omega_p, \\[2mm] \cos(\omega/2), & \omega_s \leq \omega \leq \pi. \end{cases} \tag{7.111}$$

Consequently, type II filter design is a different polynomial approximation problem than type I filter design.

In this section, we have only outlined the design of type II filters, principally to highlight the requirement that the design problem first be formulated as a polynomial approximation problem. A similar set of issues arises in the design of type III and type IV linear-phase FIR filters. Specifically, these classes also can be formulated as polynomial approximation problems, but in each class, the weighting function applied to the error has a trigonometric form, just as it does for type II filters. (See Problem 7.58.) A detailed discussion of the design and properties of these classes of filters can be found in Rabiner and Gold (1975).

The details of the formulation of the problem for type I and type II linear-phase systems have been illustrated for the case of the lowpass filter. However, the discussion of type II systems in particular should suggest that there is great flexibility in the choice of both the desired response function $H_d(e^{j\omega})$ and the weighting function $W(\omega)$. For example, the weighting function can be defined in terms of the desired function so as to yield equiripple percentage error approximation. This approach is valuable in designing type III and type IV differentiator systems.

7.7.3 The Parks–McClellan Algorithm

The alternation theorem gives necessary and sufficient conditions on the error for optimality in the Chebyshev or minimax sense. Although the theorem does not state explicitly how to find the optimum filter, the conditions that are presented serve as the basis for an efficient algorithm for finding it. While our discussion is phrased in terms of type I lowpass filters, the algorithm easily generalizes.

From the alternation theorem, we know that the optimum filter $A_e(e^{j\omega})$ will satisfy the set of equations

$$W(\omega_i)[H_d(e^{j\omega_i}) - A_e(e^{j\omega_i})] = (-1)^{i+1}\delta, \qquad i = 1, 2, \ldots, (L+2), \qquad (7.112)$$

where δ is the optimum error and $A_e(e^{j\omega})$ is given by either Eq. (7.89) or Eq. (7.93). Using Eq. (7.93) for $A_e(e^{j\omega})$, we can write these equations as

$$\begin{bmatrix} 1 & x_1 & x_1^2 & \cdots & x_1^L & \dfrac{1}{W(\omega_1)} \\ 1 & x_2 & x_2^2 & \cdots & x_2^L & \dfrac{-1}{W(\omega_2)} \\ \vdots & \vdots & \vdots & & \vdots & \vdots \\ 1 & x_{L+2} & x_{L+2}^2 & \cdots & x_{L+2}^L & \dfrac{(-1)^{L+1}}{W(\omega_{L+2})} \end{bmatrix} \begin{bmatrix} a_0 \\ a_1 \\ \vdots \\ \delta \end{bmatrix} = \begin{bmatrix} H_d(e^{j\omega_1}) \\ H_d(e^{j\omega_2}) \\ \vdots \\ H_d(e^{j\omega_{L+2}}) \end{bmatrix}, \qquad (7.113)$$

where $x_i = \cos \omega_i$. This set of equations serves as the basis for an iterative algorithm for finding the optimum $A_e(e^{j\omega})$. The procedure begins by guessing a set of alternation frequencies ω_i for $i = 1, 2, \ldots, (L+2)$. Note that ω_p and ω_s are fixed and, based on our discussion in Section 7.7.1, are necessarily members of the set of alternation frequencies. Specifically, if $\omega_\ell = \omega_p$, then $\omega_{\ell+1} = \omega_s$. The set of Eqs. (7.113) could be solved for the set of coefficients a_k and δ. However, a more efficient alternative is to use polynomial interpolation. In particular, Parks and McClellan (1972a, 1972b) found that, for the given set of the extremal frequencies,

$$\delta = \frac{\displaystyle\sum_{k=1}^{L+2} b_k H_d(e^{j\omega_k})}{\displaystyle\sum_{k=1}^{L+2} \frac{b_k(-1)^{k+1}}{W(\omega_k)}}, \tag{7.114}$$

where

$$b_k = \prod_{\substack{i=1 \\ i \neq k}}^{L+2} \frac{1}{(x_k - x_i)} \tag{7.115}$$

and, as before, $x_i = \cos\omega_i$. That is, if $A_e(e^{j\omega})$ is determined by the set of coefficients a_k that satisfy Eq. (7.113), with δ given by Eq. (7.114), then the error function goes through $\pm\delta$ at the $(L+2)$ frequencies ω_i, or, equivalently, $A_e(e^{j\omega})$ has values $1 \pm K\delta$ if $0 \leq \omega_i \leq \omega_p$ and $\pm\delta$ if $\omega_s \leq \omega_i \leq \pi$. Now, since $A_e(e^{j\omega})$ is known to be an Lth-order trigonometric polynomial, we can interpolate a trigonometric polynomial through $(L+1)$ of the $(L+2)$ known values $E(\omega_i)$ (or equivalently, $A_e(e^{j\omega_i})$). Parks and McClellan used the Lagrange interpolation formula to obtain

$$A_e(e^{j\omega}) = P(\cos\omega) = \frac{\displaystyle\sum_{k=1}^{L+1} [d_k/(x - x_k)]C_k}{\displaystyle\sum_{k=1}^{L+1} [d_k/(x - x_k)]}, \tag{7.116a}$$

where $x = \cos\omega$, $x_i = \cos\omega_i$,

$$C_k = H_d(e^{j\omega_k}) - \frac{(-1)^{k+1}\delta}{W(\omega_k)}, \tag{7.116b}$$

and

$$d_k = \prod_{\substack{i=1 \\ i \neq k}}^{L+1} \frac{1}{(x_k - x_i)} = b_k(x_k - x_{L+2}). \tag{7.116c}$$

Although only the frequencies $\omega_1, \omega_2, \ldots, \omega_{L+1}$ are used in fitting the Lth-order polynomial, we can be assured that the polynomial also takes on the correct value at ω_{L+2} because Eqs. (7.113) are satisfied by the resulting $A_e(e^{j\omega})$.

Now $A_e(e^{j\omega})$ is available at any desired frequency, without the need to solve the set of equations (7.113) for the coefficients a_k. The polynomial of Eq. (7.116a) can be used to evaluate $A_e(e^{j\omega})$ and also $E(\omega)$ on a dense set of frequencies in the passband and stopband. If $|E(\omega)| \leq \delta$ for all ω in the passband and stopband, then the optimum approximation has been found. Otherwise, we must find a new set of extremal frequencies.

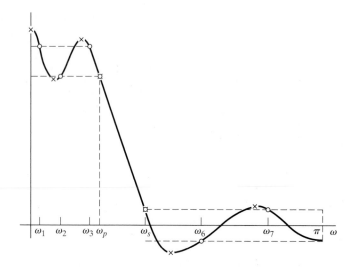

Figure 7.49   Illustration of the Parks–McClellan algorithm for equiripple approximation.

Figure 7.49 shows a typical example for a type I lowpass filter before the optimum has been found. Clearly, the set of frequencies ω_i used to find δ (as represented by open circles in the figure) was such that δ was too small. Adopting the philosophy of the Remez exchange method (see Cheney, 2000), the extremal frequencies are exchanged for a completely new set defined by the $(L+2)$ largest peaks of the error curve. The points marked with $\times$ would be the new set of frequencies for the example shown in the figure. As before, ω_p and ω_s must be selected as extremal frequencies. Recall that there are at most $(L-1)$ local minima and maxima in the open intervals $0 < \omega < \omega_p$ and $\omega_s < \omega < \pi$. The remaining extremal frequency can be at either $\omega = 0$ or $\omega = \pi$. If there is a maximum of the error function at both 0 and π, then the frequency at which the greatest error occurs is taken as the new estimate of the frequency of the remaining extremum. The cycle—computing the value of δ, fitting a polynomial to the assumed error peaks, and then locating the actual error peaks—is repeated until δ does not change from its previous value by more than a prescribed small amount. This value of δ is then the desired minimum maximum weighted approximation error.

A flowchart for the Parks–McClellan algorithm is shown in Figure 7.50. In this algorithm, all the impulse response values $h_e[n]$ are implicitly varied on each iteration to obtain the desired optimal approximation, but the values of $h_e[n]$ are never explicitly computed. After the algorithm has converged, the impulse response can be computed from samples of the polynomial representation using the discrete Fourier transform, as will be discussed in Chapter 8.

7.7.4 Characteristics of Optimum FIR Filters

Optimum lowpass FIR filters have the smallest maximum weighted approximation error δ for prescribed passband and stopband edge frequencies ω_p and ω_s. For the weighting function of Eq. (7.98), the resulting maximum stopband approximation error is $\delta_2 = \delta$, and the maximum passband approximation error is $\delta_1 = K\delta$. In Figure 7.51, we illustrate how δ varies with the order of the filter and the passband cutoff frequency. For this

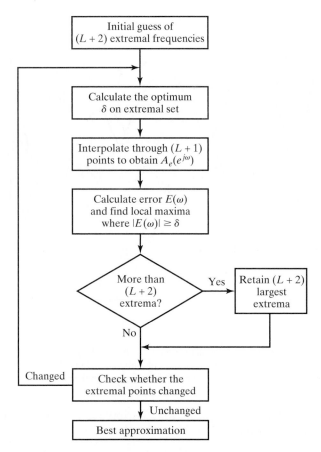

Figure 7.50 Flowchart of Parks–McClellan algorithm.

example, $K = 1$ and the transition width is fixed at $(\omega_s - \omega_p) = 0.2\pi$. The curves show that as ω_p increases, the error δ attains local minima. These minima on the curves correspond to the extraripple ($L + 3$ extrema) filters. All points between the minima correspond to filters that are optimal according to the alternation theorem. The filters for $M = 8$ and $M = 10$ are type I filters, while $M = 9$ and $M = 11$ correspond to a type II filter. It is interesting to note that, for some choices of parameters, a shorter filter ($M = 9$) may be better (i.e., it yields a smaller error) than a longer filter ($M = 10$). This may at first seem surprising and even contradictory. However, the cases $M = 9$ and $M = 10$ represent fundamentally different types of filters. Interpreted another way, filters for $M = 9$ cannot be considered to be special cases of $M = 10$ with one point set to zero, since this would violate the linear-phase symmetry requirement. On the other hand, $M = 8$ could always be thought of as a special case of $M = 10$ with the first and last samples set to zero. For that reason, an optimal filter for $M = 8$ cannot be better than one for $M = 10$. This restriction can be seen in Figure 7.51, where the curve for $M = 8$ is always above or equal to the one for $M = 10$. The points at which the two curves touch correspond to identical impulse responses, with the $M = 10$ filter having the first and last points equal to zero.

Herrmann et al. (1973) did an extensive computational study of the relationships

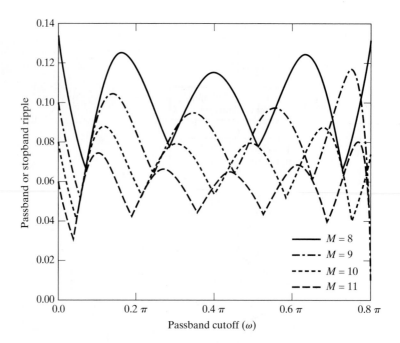

Figure 7.51   Illustration of the dependence of passband and stopband error on cutoff frequency for optimal approximations of a lowpass filter. For this example, $K = 1$ and $(\omega_s - \omega_p) = 0.2\pi$. (After Herrmann, Rabiner and Chan, 1973.)

among the parameters $M, \delta_1, \delta_2, \omega_p$, and ω_s for equiripple lowpass approximations, and Kaiser (1974) subsequently obtained the simplified formula

$$M = \frac{-10\log_{10}(\delta_1\delta_2) - 13}{2.324\Delta\omega}, \tag{7.117}$$

where $\Delta\omega = \omega_s - \omega_p$, as a fit to their data. By comparing Eq. (7.117) with the design formula of Eq. (7.76) for the Kaiser window method, we can see that, for the comparable case ($\delta_1 = \delta_2 = \delta$), the optimal approximations provide about 5 dB better approximation error for a given value of M. Another important advantage of the equiripple filters is that δ_1 and δ_2 need not be equal, as must be the case for the window method.

7.8 EXAMPLES OF FIR EQUIRIPPLE APPROXIMATION

The Parks–McClellan algorithm for optimum equiripple approximation of FIR filters can be used to design a wide variety of such filters. In this section, we give several examples that illustrate some of the properties of the optimum approximation and suggest the great flexibility that is afforded by the design method.

7.8.1 Lowpass Filter

For the lowpass filter case, we again approximate the set of specifications used in Example 7.5 and Section 7.6.1 so that we can compare all the major design methods on the same lowpass filter specifications. These specifications call for $\omega_p = 0.4\pi$, $\omega_s = 0.6\pi$, $\delta_1 = 0.01$, and $\delta_2 = 0.001$. In contrast to the window method, the Parks–McClellan algorithm can accommodate the different approximation error in the passband versus that in the stopband by fixing the weighting function parameter at $K = \delta_1/\delta_2 = 10$.

Substituting the foregoing specifications into Eq. (7.117) and rounding up yields the estimate $M = 26$ for the value of M that is necessary to achieve the specifications. Figures 7.52(a), (b), and (c) show the impulse response, log magnitude, and approximation error, respectively, for the optimum filter with $M = 26$, $\omega_p = 0.4\pi$, and $\omega_s = 0.6\pi$. Figure 7.52(c) shows the *unweighted* approximation error

$$E_A(\omega) = \frac{E(\omega)}{W(\omega)} = \begin{cases} 1 - A_e(e^{j\omega}), & 0 \le \omega \le \omega_p, \\ 0 - A_e(e^{j\omega}), & \omega_s \le \omega \le \pi, \end{cases} \tag{7.118}$$

rather than the weighted error used in the formulation of the design algorithm. The weighted error would be identical to Figure 7.52(c), except that the error would be divided by 10 in the passband.[7] The alternations of the approximation error are clearly in evidence in Figure 7.52(c). There are seven alternations in the passband and eight in the stopband, for a total of fifteen alternations. Since $L = M/2$ for type I (M even) systems, and $M = 26$, the minimum number of alternations is $(L + 2) = (26/2 + 2) = 15$. Thus, the filter of Figure 7.52 is the optimum filter for $M = 26$, $\omega_p = 0.4\pi$, and $\omega_s = 0.6\pi$. However, Figure 7.52(c) shows that the filter fails to meet the original specifications on passband and stopband error. (The maximum errors in the passband and stopband are 0.0116 and 0.00116, respectively.) To meet or exceed the specifications, we must increase M.

The filter response functions for the case $M = 27$ are shown in Figure 7.53. Now the passband and stopband approximation errors are slightly less than the specified values. (The maximum errors in the passband and stopband are 0.0092 and 0.00092, respectively.) In this case, there are again seven alternations in the passband and eight alternations in the stopband, for a total of fifteen. Note that, since $M = 27$, this is a type II system, and for type II systems, the order of the implicit approximating polynomial is $L = (M - 1)/2 = (27 - 1)/2 = 13$. Thus, the minimum number of alternations is still 15. Note also that in the type II case, the system is constrained to have a zero of its system function at $z = -1$ or $\omega = \pi$. This is clearly shown in Figures 7.53(b) and (c).

If we compare the results of this example with the results of Section 7.6.1, we find that the Kaiser window method requires a value $M = 40$ to meet or exceed the specifications, whereas the Parks–McClellan method requires $M = 27$. This disparity is accentuated because the window method produces approximately equal maximum errors in the passband and stopband, while the Parks–McClellan method can weight the errors differently.

7.8.2 Compensation for Zero-Order Hold

In many cases, a discrete-time filter is designed to be used in a system such as that depicted in Figure 7.54; i.e., the filter is used to process a sequence of samples $x[n]$ to obtain

---

[7]For frequency-selective filters, the unweighted approximation error also conveniently displays the passband and stopband behavior, since $A_e(e^{j\omega}) = 1 - E(\omega)$ in the passband and $A_e(e^{j\omega}) = -E(\omega)$ in the stopband.

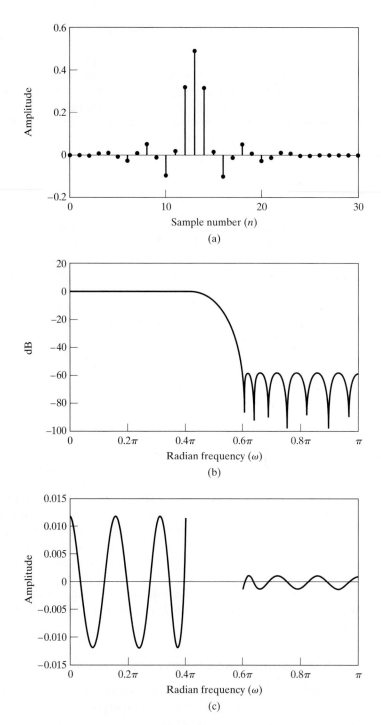

Figure 7.52 Optimum type I FIR lowpass filter for $\omega_p = 0.4\pi$, $\omega_s = 0.6\pi$, $K = 10$, and $M = 26$. (a) Impulse response. (b) Log magnitude of the frequency response. (c) Approximation error (unweighted).

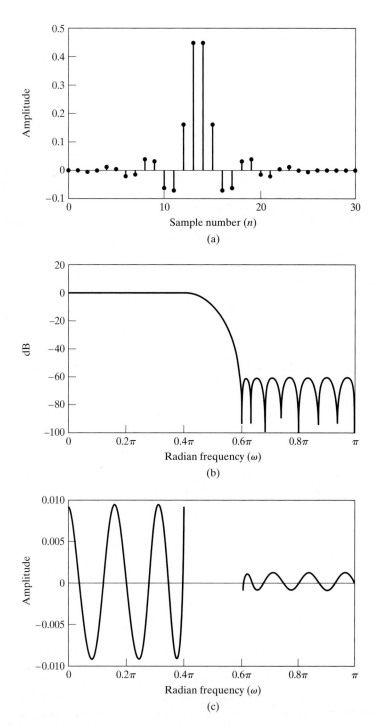

Figure 7.53 Optimum type II FIR lowpass filter for $\omega_p = 0.4\pi$, $\omega_S = 0.6\pi$, $K = 10$, and $M = 27$. (a) Impulse response. (b) Log magnitude of frequency response. (c) Approximation error (unweighted).

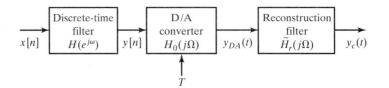

Figure 7.54 Precompensation of a discrete-time filter for the effects of a D/A converter.

a sequence $y[n]$, which is then the input to a D/A converter and continuous-time lowpass filter (as an approximation to the ideal D/C converter) used for the reconstruction of a continuous-time signal $y_c(t)$. Such a system arises as part of a system for discrete-time filtering of a continuous-time signal, as discussed in Section 4.8. If the D/A converter holds its output constant for the entire sampling period T, the Fourier transform of the output $y_c(t)$ is

$$Y_c(j\Omega) = \tilde{H}_r(j\Omega)H_o(j\Omega)H(e^{j\Omega T})X(e^{j\Omega T}), \tag{7.119}$$

where $\tilde{H}_r(j\Omega)$ is the frequency response of an appropriate lowpass reconstruction filter and

$$H_o(j\Omega) = \frac{\sin(\Omega T/2)}{\Omega/2}e^{-j\Omega T/2} \tag{7.120}$$

is the frequency response of the zero-order hold of the D/A converter. In Section 4.8.4, we suggested that compensation for $H_o(j\Omega)$ could be incorporated into the continuous-time reconstruction filter; i.e., $\tilde{H}_r(j\Omega)$ could be chosen as

$$\tilde{H}_r(j\Omega) = \begin{cases} \dfrac{\Omega T/2}{\sin(\Omega T/2)} & |\Omega| < \dfrac{\pi}{T} \\ 0 & \text{otherwise} \end{cases} \tag{7.121}$$

so that the effect of the discrete-time filter $H(e^{j\Omega T})$ would be undistorted by the zero-order hold. Another approach is to build the compensation into the discrete-time filter by designing a filter $\tilde{H}(e^{j\Omega T})$ such that

$$\tilde{H}(e^{j\Omega T}) = \frac{\Omega T/2}{\sin(\Omega T/2)}H(e^{j\Omega T}). \tag{7.122}$$

A D/A-compensated lowpass filter can be readily designed by the Parks–McClellan algorithm if we simply define the desired response as

$$\tilde{H}_d(e^{j\omega}) = \begin{cases} \dfrac{\omega/2}{\sin(\omega/2)}, & 0 \le \omega \le \omega_p, \\ 0, & \omega_s \le \omega \le \pi. \end{cases} \tag{7.123}$$

Figure 7.55 shows the response functions for such a filter, wherein the specifications are again $\omega_p = 0.4\pi$, $\omega_s = 0.6\pi$, $\delta_1 = 0.01$, and $\delta_2 = 0.001$. In this case, the specifications are met with $M = 28$ rather than $M = 27$ as in the previous constant-gain case. Thus, for essentially no penalty, we have incorporated compensation for the D/A converter into the discrete-time filter so that the effective passband of the filter will be flat. (To emphasize the sloping nature of the passband, Figure 7.55(c) shows the magnitude response in the passband, rather than the approximation error, as in the frequency response plots for the other FIR examples.)

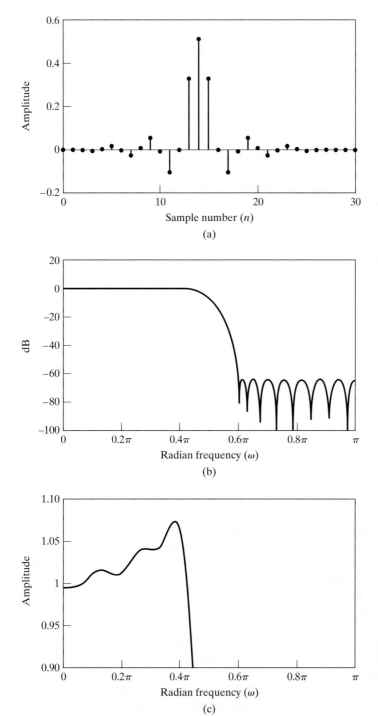

Figure 7.55 Optimum D/A-compensated lowpass filter for $\omega_p = 0.4\pi$, $\omega_S = 0.6\pi$, $K = 10$, and $M = 28$. (a) Impulse response. (b) Log magnitude of the frequency response. (c) Magnitude response in passband.

7.8.3 Bandpass Filter

Section 7.7 focused entirely on the lowpass optimal FIR, for which there are only two approximation bands. However, bandpass and bandstop filters require three approximation bands. To design such filters, it is necessary to generalize the discussion of Section 7.7 to the multiband case. This requires that we explore the implications of the alternation theorem and the properties of the approximating polynomial in the more general context. First, recall that, as stated, the alternation theorem does not assume any limit on the number of disjoint approximation intervals. Therefore, the *minimum* number of alternations for the optimum approximation is still $(L+2)$. However, multiband filters can have more than $(L+3)$ alternations, because there are more band edges. (Problem 7.63 explores this issue.) This means that some of the statements proved in Section 7.7.1 are not true in the multiband case. For example, it is *not* necessary for all the local maxima or minima of $A_e(e^{j\omega})$ to lie inside the approximation intervals. Thus, local extrema can occur in the transition regions, and the approximation need not be equiripple in the approximation regions.

To illustrate this, consider the desired response

$$H_d(e^{j\omega}) = \begin{cases} 0, & 0 \le \omega \le 0.3\pi, \\ 1, & 0.35\pi \le \omega \le 0.6\pi, \\ 0, & 0.7\pi \le \omega \le \pi, \end{cases} \tag{7.124}$$

and the error weighting function

$$W(\omega) = \begin{cases} 1, & 0 \le \omega \le 0.3\pi, \\ 1, & 0.35\pi \le \omega \le 0.6\pi, \\ 0.2, & 0.7\pi \le \omega \le \pi. \end{cases} \tag{7.125}$$

A value of $M + 1 = 75$ was chosen for the length of the impulse response of the filter. Figure 7.56 shows the response functions for the resulting filter. Note that the transition region from the second approximation band to the third is no longer monotonic. However, the use of two local extrema in this unconstrained region does not violate the alternation theorem. Since $M = 74$, the filter is a type I system, and the order of the implicit approximating polynomial is $L = M/2 = 74/2 = 37$. Thus, the alternation theorem requires at least $L+2 = 39$ alternations. It can be readily seen in Figure 7.56(c), which shows the unweighted approximation error, that there are 13 alternations in each band, for a total of 39.

Such approximations as shown in Figure 7.56 are optimal in the sense of the alternation theorem, but they would probably be unacceptable in a filtering application. In general, there is no guarantee that the transition regions of a multiband filter will be monotonic, because the Parks–McClellan algorithm leaves these regions completely unconstrained. When this kind of response results for a particular choice of the filter parameters, acceptable transition regions can usually be obtained by systematically changing one or more of the band edge frequencies, the impulse-response length, or the error-weighting function and redesigning the filter.

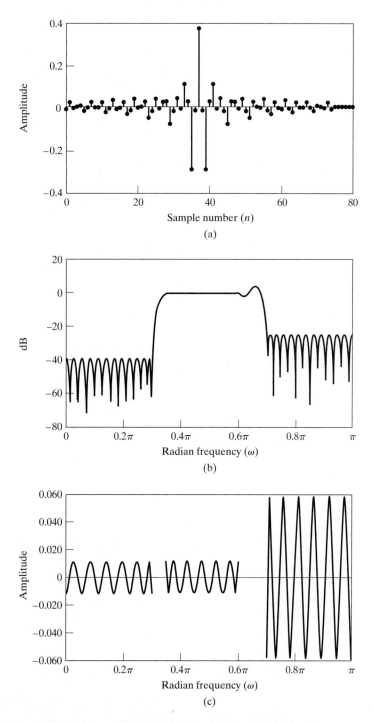

Figure 7.56 Optimum FIR bandpass filter for $M = 74$. (a) Impulse response. (b) Log magnitude of the frequency response. (c) Approximation error (unweighted).

7.9 COMMENTS ON IIR AND FIR DISCRETE-TIME FILTERS

This chapter has been concerned with design methods for LTI discrete-time systems. We have discussed a wide range of methods of designing both infinite-duration and finite-duration impulse-response filters.

The choice between an FIR filter and an IIR filter depends on the importance to the design problem of the advantages of each type. IIR filters, for example, have the advantage that a variety of frequency-selective filters can be designed using closed-form design formulas. That is, once the problem has been specified in terms appropriate for a given approximation method (e.g., Butterworth, Chebyshev, or elliptic), then the order of the filter that will meet the specifications can be computed, and the coefficients (or poles and zeros) of the discrete-time filter can be obtained by straightforward substitution into a set of design equations. This kind of simplicity of the design procedure makes it feasible to design IIR filters by manual computation if necessary, and it leads to straightforward noniterative computer programs for IIR filter design. These methods are limited to frequency-selective filters, and they permit only the magnitude response to be specified. If other magnitude shapes are desired, or if it is necessary to approximate a prescribed phase- or group-delay response, an algorithmic procedure will be required.

In contrast, FIR filters can have a precisely (generalized) linear phase. However, closed-form design equations do not exist for FIR filters. Although the window method is straightforward to apply, some iteration may be necessary to meet a prescribed specification. The Parks–McClellan algorithm leads to lower-order filters than the window method and filter design programs are readily available for both methods. Also, the window method and most of the algorithmic methods afford the possibility of approximating rather arbitrary frequency-response characteristics with little more difficulty than is encountered in the design of lowpass filters. In addition, the design problem for FIR filters is much more under control than the IIR design problem, because of the existence of an optimality theorem for FIR filters that is meaningful in a wide range of practical situations. Design techniques for FIR filters without linear phase have been given by Chen and Parks (1987), Parks and Burrus (1987), Schüssler and Steffen (1988), and Karam and McClellan (1995).

Questions of economics also arise in implementing a discrete-time filter. Economic concerns are usually measured in terms of hardware complexity, chip area, or computational speed. These factors are more or less directly related to the order of the filter required to meet a given specification. In applications where the efficiencies of polyphase implementations cannot be exploited, it is generally true that a given magnitude-response specification can be met most efficiently with an IIR filter. However, in many cases, the linear phase available with an FIR filter may be well worth the extra cost.

In any specific practical setting, the choice of class of filters and design method will be highly dependent on the context, constraints, specifications, and implementation platform. In this section, we conclude the chapter with one specific example to illustrate some of the trade offs and issues that can arise. However, it is only one of many scenarios, each of which can result in different choices and conclusions.

7.10 DESIGN OF AN UPSAMPLING FILTER

We conclude this chapter with a comparison, in the context of upsampling, of IIR and FIR filter designs. As discussed in Chapter 4, Sections 4.6.2 and 4.9.3, integer upsampling and oversampled D/A conversion employ an expander-by-L followed by a discrete-time lowpass filter. Because the sampling rate at the output of the expander is L times the rate at the input, the lowpass filter operates at a rate which is L-times the rate of the input to the upsampler or D/A converter. As we illustrate in this example, the order of the lowpass filter is very dependent on whether the filter is designed as an IIR or FIR filter and also within those classes, which filter design method is chosen. While the order of the resulting IIR filter might be significantly less than the order of the FIR filter, the FIR filter can exploit the efficiencies of a polyphase implementation. For the IIR designs, polyphase can be exploited for the implementation of the zeros of the transfer function but not for the poles.

The system to be implemented is an upsampler-by-four, i.e., $L = 4$. As discussed in Chapter 4, the ideal filter for 1:4 interpolation is an ideal lowpass filter with gain of 4 and cutoff frequency $\pi/4$. To approximate this filter we set the specifications as follows:[8]

$$
\begin{aligned}
\text{passband edge frequency } \omega_p &= 0.22\pi \\
\text{stopband edge frequency } \omega_s &= 0.29\pi \\
\text{maximum passband gain} &= 0 \text{ dB} \\
\text{minimum passband gain} &= -1 \text{ dB} \\
\text{maximum stopband gain} &= -40 \text{ dB}.
\end{aligned}
$$

Six different filters were designed to meet these specifications: the four IIR filter designs discussed in Section 7.3 (Butterworth, Chebyshev I, Chebyshev II, elliptic) and two FIR filter designs (a Kaiser window design and an optimal filter designed using the Parks–McClellan algorithm). The designs were done using the signal processing toolbox in MATLAB. Since the FIR design program used requires passband tolerance limits that are symmetric about unity, the specifications above were first scaled appropriately for the FIR designs and the resulting FIR filter was then rescaled for a maximum of 0 dB gain in the passband. (See Problem 7.3.)

The resulting filter orders for the six filters are shown in Table 7.3 and the corresponding pole–zero plots are shown in Figure 7.57(a)–(f). For the two FIR designs only the zero locations are shown in Figure 7.57. If these filters are implemented as causal

TABLE 7.3 ORDERS OF DESIGNED FILTERS.

Filter design	Order
Butterworth	18
Chebyshev I	8
Chebyshev II	8
Elliptic	5
Kaiser	63
Parks–McClellan	44

---

[8]The gain was normalized to unity in the passband. In all cases the filters can be scaled by 4 for use in interpolation.

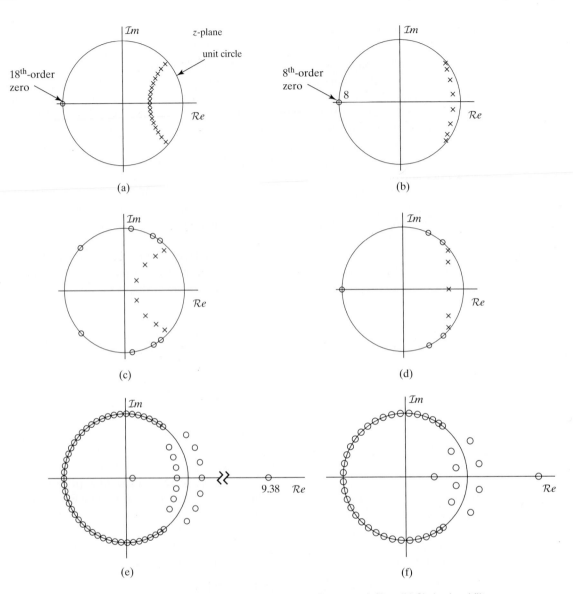

Figure 7.57 Pole–zero plots for the six designs. (a) Butterworth filter. (b) Chebyshev I filter. (c) Chebyshev II filter. (d) Elliptic filter. (e) Kaiser filter. (f) Parks–McClellan filter.

filters there will be a multiple-order pole at the origin to match the total number of zeros of the transfer function.

Without exploiting available efficiencies, such as the use of a polyphase implementation, the two FIR designs require significantly more multiplications per output sample than any of the IIR designs. In the IIR designs, the number of multiplications per output sample will be dependent on specifically how the zeros are implemented. A discussion of how to efficiently implement each of the six designs follows below with a summary in Table 7.4 comparing the required number of multiplications per output sample. The four IIR designs can be considered as a cascade of an FIR filter (implementing the zeros of the transfer function) and an IIR filter (implementing the poles). We first discuss the two FIR designs since efficiencies that can be exploited for those can also be utilized with the FIR component of the IIR filters.

Parks–McClellan and Kaiser window designs: Without exploiting symmetry of the impulse response or a polyphase implementation, the required number of multiplications per *output* sample is equal to the length of the filter. If a polyphase implementation is used as discussed in Section 4.7.5, then the number of multiplications per *input* sample is equal to the length of the filter. Alternatively, since both filters are symmetric, the folded structure discussed in Section 6.5.3 (Figures 6.32 and 6.33) can be used to reduce the number of multiplications at the input rate by approximately a factor of 2.[9]

Butterworth design: As is characteristic of discrete-time Butterworth filters, all of the zeros occur at $z = -1$ and the poles are, of course, in complex conjugate pairs. By implementing the zeros as a cascade of 18 1st-order terms of the form $(1 + z^{-1})$ no multiplications are required for implementing the zeros. The 18 poles require a total of 18 multiplications per output sample.

Chebyshev I design: The Chebyshev I filter has order 8 with the zeros at $z = -1$ and consequently the zeros can be implemented with no multiplications. The 8 poles require 8 multiplies per output sample.

Chebyshev II design: In this design, the filter order is again 8. Since the zeros are now distributed around the unit circle, their implementation will require some multiplications. However, since all the zeros are on the unit circle, the associated FIR impulse response will be symmetric, and folding and/or polyphase efficiencies can be exploited for implementing the zeros.

Elliptic lter design: The elliptic filter has the lowest (order 5) of the four IIR designs. From the pole–zero plot we note that it has all its zeros on the unit circle. Consequently the zeros can be implemented efficiently exploiting symmetry as well as polyphase implementation.

Table 7.4 summarizes the number of multiplications required per output sample for each of the six designs with several different implementation structures. The direct form implementation assumes that both the poles and zeros are implemented in direct form, i.e., it does not take advantage of the possibility of cascade implementation of multiple zeroes at $z = -1$. Exploiting a polyphase implementation but not also the symmetry of the impulse response, the FIR designs are slightly less efficient than the most efficient IIR designs, although they are also the only ones that have linear phase. Exploiting both symmetry and polyphase together in implementing the Parks–McClellan design, it and the elliptic filter are the most efficient.

---

[9]It is possible to combine both folding and polyphase efficiencies in implementing symmetric FIR filters (see Baran and Oppenheim, 2007). The resulting number of multiplications is approximately half the filter length and at the rate of the input samples rather than at the rate of the output samples. However, the resulting structure is significantly more complex.

TABLE 7.4   AVERAGE NUMBER OF REQUIRED
MULTIPLICATIONS PER OUTPUT SAMPLE FOR
EACH OF THE DESIGNED FILTERS.

Filter design	Direct form	Symmetric	Polyphase
Butterworth	37	18	18
Chebyshev I	17	8	8
Chebyshev II	17	13	10.25
Elliptic	11	8	6.5
Kaiser	64	32	16
Parks–McClellan	45	23	11.25

7.11 SUMMARY

In this chapter, we have considered a variety of design techniques for both infinite-duration and finite-duration impulse-response discrete-time filters. Our emphasis was on the frequency-domain specification of the desired filter characteristics, since this is most common in practice. Our objective was to give a general picture of the wide range of possibilities available for discrete-time filter design, while also giving sufficient detail about some of the techniques so that they may be applied directly, without further reference to the extensive literature on discrete-time filter design. In the FIR case, considerable detail was presented on both the window method and the Parks–McClellan algorithmic method of filter design.

The chapter concluded with some remarks on the choice between the two classes of digital filters. The main point of that discussion was that the choice is not always clear cut and may depend on a multitude of factors that are often difficult to quantify or discuss in general terms. However, it should be clear from this chapter and Chapter 6 that digital filters are characterized by great flexibility in design and implementation. This flexibility makes it possible to implement rather sophisticated signal-processing schemes that in many cases would be difficult, if not impossible, to implement by analog means.

Problems

Basic Problems with Answers

7.1. Consider a causal continuous-time system with impulse response $h_c(t)$ and system function

$$H_c(s) = \frac{s + a}{(s + a)^2 + b^2}.$$

(a) Use impulse invariance to determine $H_1(z)$ for a discrete-time system such that $h_1[n] = h_c(nT)$.

(b) Use step invariance to determine $H_2(z)$ for a discrete-time system such that $s_2[n] = s_c(nT)$, where

$$s_2[n] = \sum_{k=-\infty}^{n} h_2[k] \quad \text{and} \quad s_c(t) = \int_{-\infty}^{t} h_c(\tau)d\tau.$$

(c) Determine the step response $s_1[n]$ of system 1 and the impulse response $h_2[n]$ of system 2. Is it true that $h_2[n] = h_1[n] = h_c(nT)$? Is it true that $s_1[n] = s_2[n] = s_c(nT)$?

7.2. A discrete-time lowpass filter is to be designed by applying the impulse invariance method to a continuous-time Butterworth filter having magnitude-squared function

$$|H_c(j\Omega)|^2 = \frac{1}{1 + (\Omega/\Omega_c)^{2N}}.$$

The specifications for the discrete-time system are those of Example 7.2, i.e.,

$$0.89125 \le |H(e^{j\omega})| \le 1, \quad 0 \le |\omega| \le 0.2\pi,$$

$$|H(e^{j\omega})| \le 0.17783, \quad 0.3\pi \le |\omega| \le \pi.$$

Assume, as in that example, that aliasing will not be a problem; i.e., design the continuous-time Butterworth filter to meet passband and stopband specifications as determined by the desired discrete-time filter.

(a) Sketch the tolerance bounds on the magnitude of the frequency response, $|H_c(j\Omega)|$, of the continuous-time Butterworth filter such that after application of the impulse invariance method (i.e., $h[n] = T_d h_c(nT_d)$), the resulting discrete-time filter will satisfy the given design specifications. Do not assume that $T_d = 1$ as in Example 7.2.

(b) Determine the integer order N and the quantity $T_d\Omega_c$ such that the continuous-time Butterworth filter exactly meets the specifications determined in part (a) at the passband edge.

(c) Note that if $T_d = 1$, your answer in part (b) should give the values of N and Ω_c obtained in Example 7.2. Use this observation to determine the system function $H_c(s)$ for $T_d \ne 1$ and to argue that the system function $H(z)$ which results from impulse invariance design with $T_d \ne 1$ is the same as the result for $T_d = 1$ given by Eq. (7.17).

7.3. We wish to use impulse invariance or the bilinear transformation to design a discrete-time filter that meets specifications of the following form:

$$1 - \delta_1 \le |H(e^{j\omega})| \le 1 + \delta_1, \quad 0 \le |\omega| \le \omega_p,$$

$$|H(e^{j\omega})| \le \delta_2, \quad \omega_s \le |\omega| \le \pi. \tag{P7.3-1}$$

For historical reasons, most of the design formulas, tables, or charts for continuous-time filters are normally specified with a peak gain of unity in the passband; i.e.,

$$1 - \hat{\delta}_1 \le |H_c(j\Omega)| \le 1, \quad 0 \le |\Omega| \le \Omega_p,$$

$$|H_c(\Omega)| \le \hat{\delta}_2, \quad \Omega_s \le |\Omega|. \tag{P7.3-2}$$

Useful design charts for continuous-time filters specified in this form were given by Rabiner, Kaiser, Herrmann, and Dolan (1974).

(a) To use such tables and charts to design discrete-time systems with a peak gain of $(1+\delta_1)$, it is necessary to convert the discrete-time specifications into specifications of the form of Eq. (P7.3-2). This can be done by dividing the discrete-time specifications by $(1+\delta_1)$. Use this approach to obtain an expression for $\hat{\delta}_1$ and $\hat{\delta}_2$ in terms of δ_1 and δ_2.

(b) In Example 7.2, we designed a discrete-time filter with a maximum passband gain of unity. This filter can be converted to a filter satisfying a set of specifications such as those in Eq. (P7.3-1) by multiplying by a constant of the form $(1+\delta_1)$. Find the required value of δ_1 and the corresponding value of δ_2 for this example, and use Eq. (7.17) to determine the coefficients of the system function of the new filter.

(c) Repeat part (b) for the filter in Example 7.3.

7.4. The system function of a discrete-time system is

$$H(z) = \frac{2}{1 - e^{-0.2}z^{-1}} - \frac{1}{1 - e^{-0.4}z^{-1}}.$$

(a) Assume that this discrete-time filter was designed by the impulse invariance method with $T_d = 2$; i.e., $h[n] = 2h_c(2n)$, where $h_c(t)$ is real. Find the system function $H_c(s)$ of a continuous-time filter that could have been the basis for the design. Is your answer unique? If not, find another system function $H_c(s)$.

(b) Assume that $H(z)$ was obtained by the bilinear transform method with $T_d = 2$. Find the system function $H_c(s)$ that could have been the basis for the design. Is your answer unique? If not, find another $H_c(s)$.

7.5. We wish to use the Kaiser window method to design a discrete-time filter with generalized linear phase that meets specifications of the following form:

$$|H(e^{j\omega})| \le 0.01, \qquad 0 \le |\omega| \le 0.25\pi,$$
$$0.95 \le |H(e^{j\omega})| \le 1.05, \qquad 0.35\pi \le |\omega| \le 0.6\pi,$$
$$|H(e^{j\omega})| \le 0.01, \qquad 0.65\pi \le |\omega| \le \pi.$$

(a) Determine the minimum length $(M + 1)$ of the impulse response and the value of the Kaiser window parameter β for a filter that meets the preceding specifications.

(b) What is the delay of the filter?

(c) Determine the ideal impulse response $h_d[n]$ to which the Kaiser window should be applied.

7.6. We wish to use the Kaiser window method to design a symmetric real-valued FIR filter with zero phase that meets the following specifications:

$$0.9 < H(e^{j\omega}) < 1.1, \qquad 0 \le |\omega| \le 0.2\pi,$$
$$-0.06 < H(e^{j\omega}) < 0.06, \qquad 0.3\pi \le |\omega| \le 0.475\pi,$$
$$1.9 < H(e^{j\omega}) < 2.1, \qquad 0.525\pi \le |\omega| \le \pi.$$

This specification is to be met by applying the Kaiser window to the ideal real-valued impulse response associated with the ideal frequency response $H_d(e^{j\omega})$ given by

$$H_d(e^{j\omega}) = \begin{cases} 1, & 0 \le |\omega| \le 0.25\pi, \\ 0, & 0.25\pi \le |\omega| \le 0.5\pi, \\ 2, & 0.5\pi \le |\omega| \le \pi. \end{cases}$$

(a) What is the maximum value of δ that can be used to meet this specification? What is the corresponding value of β? Clearly explain your reasoning.

(b) What is the maximum value of $\Delta\omega$ that can be used to meet the specification? What is the corresponding value of $M + 1$, the length of the impulse response? Clearly explain your reasoning.

7.7. We are interested in implementing a continuous-time LTI lowpass filter $H(j\Omega)$ using the system shown in Figure 4.10 when the discrete-time system has frequency response $H_d(e^{j\omega})$. The sampling time $T = 10^{-4}$ second and the input signal $x_c(t)$ is appropriately bandlimited with $X_c(j\Omega) = 0$ for $|\Omega| \ge 2\pi(5000)$.

Let the specifications on $|H(j\Omega)|$ be

$$0.99 \le |H(j\Omega)| \le 1.01, \qquad |\Omega| \le 2\pi(1000),$$
$$|H(j\Omega)| \le 0.01, \qquad |\Omega| \ge 2\pi(1100).$$

Determine the corresponding specifications on the discrete-time frequency response $H_d(e^{j\omega})$.

7.8. We wish to design an optimal (Parks–McClellan) zero-phase Type I FIR lowpass filter with passband frequency $\omega_p = 0.3\pi$ and stopband frequency $\omega_s = 0.6\pi$ with equal error weighting in the passband and stopband. The impulse response of the desired filter has length 11; i.e., $h[n] = 0$ for $n < -5$ or $n > 5$. Figure P7.8 shows the frequency response $H(e^{j\omega})$ for two different filters. For each filter, specify how many alternations the filter has, and state whether it satisfies the alternation theorem as the optimal filter in the minimax sense meeting the preceding specifications.

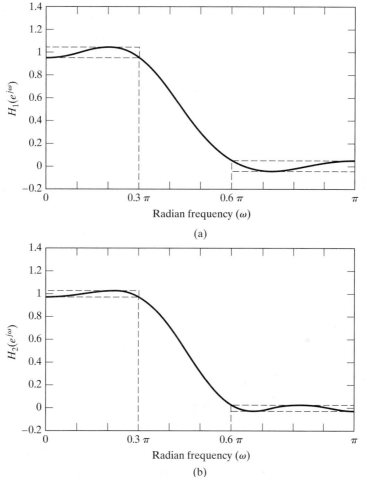

(a)

(b)   **Figure P7.8**

7.9. Suppose we design a discrete-time filter using the impulse invariance technique with an ideal continuous-time lowpass filter as a prototype. The prototype filter has a cutoff frequency of $\Omega_c = 2\pi(1000)$ rad/s, and the impulse invariance transformation uses $T = 0.2$ ms. What is the cutoff frequency ω_c for the resulting discrete-time filter?

7.10. We wish to design a discrete-time lowpass filter using the bilinear transformation on a continuous-time ideal lowpass filter. Assume that the continuous-time prototype filter has cutoff frequency $\Omega_c = 2\pi(2000)$ rad/s, and we choose the bilinear transformation parameter $T = 0.4$ ms. What is the cutoff frequency ω_c for the resulting discrete-time filter?

7.11. Suppose that we have an ideal discrete-time lowpass filter with cutoff frequency $\omega_c = \pi/4$. In addition, we are told that this filter resulted from applying impulse invariance to a continuous-time prototype lowpass filter using $T = 0.1$ ms. What was the cutoff frequency Ω_c for the prototype continuous-time filter?

7.12. An ideal discrete-time highpass filter with cutoff frequency $\omega_c = \pi/2$ was designed using the bilinear transformation with $T = 1$ ms. What was the cutoff frequency Ω_c for the prototype continuous-time ideal highpass filter?

7.13. An ideal discrete-time lowpass filter with cutoff frequency $\omega_c = 2\pi/5$ was designed using impulse invariance from an ideal continuous-time lowpass filter with cutoff frequency $\Omega_c = 2\pi(4000)$ rad/s. What was the value of T? Is this value unique? If not, find another value of T consistent with the information given.

7.14. The bilinear transformation is used to design an ideal discrete-time lowpass filter with cutoff frequency $\omega_c = 3\pi/5$ from an ideal continuous-time lowpass filter with cutoff frequency $\Omega_c = 2\pi(300)$ rad/s. Give a choice for the parameter T that is consistent with this information. Is this choice unique? If not, give another choice that is consistent with the information.

7.15. We wish to design an FIR lowpass filter satisfying the specifications

$$0.95 < H(e^{j\omega}) < 1.05, \qquad 0 \le |\omega| \le 0.25\pi,$$

$$-0.1 < H(e^{j\omega}) < 0.1, \qquad 0.35\pi \le |\omega| \le \pi,$$

by applying a window $w[n]$ to the impulse response $h_d[n]$ for the ideal discrete-time lowpass filter with cutoff $\omega_c = 0.3\pi$. Which of the windows listed in Section 7.5.1 can be used to meet this specification? For each window that you claim will satisfy this specification, give the minimum length $M + 1$ required for the filter.

7.16. We wish to design an FIR lowpass filter satisfying the specifications

$$0.98 < H(e^{j\omega}) < 1.02, \qquad 0 \le |\omega| \le 0.63\pi,$$

$$-0.15 < H(e^{j\omega}) < 0.15, \qquad 0.65\pi \le |\omega| \le \pi,$$

by applying a Kaiser window to the impulse response $h_d[n]$ for the ideal discrete-time lowpass filter with cutoff $\omega_c = 0.64\pi$. Find the values of β and M required to satisfy this specification.

7.17. Suppose that we wish to design a bandpass filter satisfying the following specification:

$$-0.02 < |H(e^{j\omega})| < 0.02, \qquad 0 \le |\omega| \le 0.2\pi,$$

$$0.95 < |H(e^{j\omega})| < 1.05, \qquad 0.3\pi \le |\omega| \le 0.7\pi,$$

$$-0.001 < |H(e^{j\omega})| < 0.001, \qquad 0.75\pi \le |\omega| \le \pi.$$

The filter will be designed by applying impulse invariance with $T = 5$ ms to a prototype continuous-time filter. State the specifications that should be used to design the prototype continuous-time filter.

7.18. Suppose that we wish to design a highpass filter satisfying the following specification:

$$-0.04 < |H(e^{j\omega})| < 0.04, \qquad 0 \le |\omega| \le 0.2\pi,$$

$$0.995 < |H(e^{j\omega})| < 1.005, \qquad 0.3\pi \le |\omega| \le \pi.$$

The filter will be designed using the bilinear transformation and $T = 2$ ms with a prototype continuous-time filter. State the specifications that should be used to design the prototype continuous-time filter to ensure that the specifications for the discrete-time filter are met.

7.19. We wish to design a discrete-time ideal bandpass filter that has a passband $\pi/4 \leq \omega \leq \pi/2$ by applying impulse invariance to an ideal continuous-time bandpass filter with passband $2\pi(300) \leq \Omega \leq 2\pi(600)$. Specify a choice for T that will produce the desired filter. Is your choice of T unique?

7.20. Specify whether the following statement is true or false. Justify your answer.
Statement: If the bilinear transformation is used to transform a continuous-time all-pass system to a discrete-time system, the resulting discrete-time system will also be an all-pass system.

Basic Problems

7.21. An engineer is asked to evaluate the signal processing system shown in Figure P7.21-1 and improve it if necessary. The input $x[n]$ is obtained by sampling a continuous-time signal at a sampling rate of $1/T = 100$ Hz.

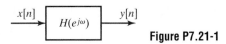

Figure P7.21-1

The goal is for $H(e^{j\omega})$ to be a linear-phase FIR filter, and ideally it should have the following amplitude response (so it can function as a bandlimited differentiator):

$$\text{amplitude of } H_{id}(e^{j\omega}) = \begin{cases} -\omega/T & \omega < 0 \\ \omega/T & \omega \geq 0 \end{cases}$$

(a) For one implementation of $H(e^{j\omega})$, referred to as $H_1(e^{j\omega})$, the designer, motivated by the definition

$$\frac{d\,(x(t))}{dt} = \lim_{\Delta t \to 0} \frac{x(t) - x(t - \Delta t)}{\Delta t},$$

chooses the system impulse response $h_1[n]$ so that the input–output relationship is

$$y[n] = \frac{x[n] - x[n-1]}{T}$$

Plot the amplitude response of $H_1(e^{j\omega})$ and discuss how well it matches the ideal response. You may find the following expansions helpful:

$$\sin(\theta) = \theta - \frac{1}{3!}\theta^3 + \frac{1}{5!}\theta^5 - \frac{1}{7!}\theta^7 + \cdots$$

$$\cos(\theta) = 1 - \frac{1}{2!}\theta^2 + \frac{1}{4!}\theta^4 - \frac{1}{6!}\theta^6 + \cdots$$

(b) We want to cascade $H_1(e^{j\omega})$ with another *linear-phase* FIR filter $G(e^{j\omega})$, to ensure that for the combination of the two filters, the group delay is an integer number of samples. Should the length of the impulse response $g[n]$ be an even or an odd integer? Explain.

(c) Another method for designing the discrete-time H filter is the method of impulse invariance. In this method, the ideal bandlimited continuous-time impulse response, as given in Eq. (P7.21-1), is sampled.

$$h(t) = \frac{\Omega_c \pi t \cos(\Omega_c t) - \pi \sin(\Omega_c t)}{\pi^2 t^2} \qquad \text{(P7.21-1)}$$

(In a typical application, Ω_c might be slightly less than π/T, making $h(t)$ the impulse response of a differentiator which is bandlimited to $|\Omega| < \pi/T$.) Based on this impulse response, we would have to create a new filter H_2 which is also FIR and linear phase. Therefore, the impulse response, $h_2[n]$, should preserve the odd symmetry of $h(t)$ about $t = 0$. Using the plot in Figure P7.21-2, indicate the location of samples that result if the impulse response is sampled at 100 Hz, and an impulse response of length 9 is obtained using a rectangular window.

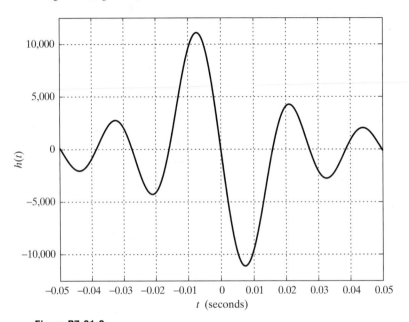

Figure P7.21-2

(d) Again using the plot in Figure P7.21-2, indicate the location of samples if the impulse response $h_2[n]$ is designed to have length 8, again preserving the odd symmetry of $h(t)$ about $t = 0$.

(e) Since the desired magnitude response of $H(e^{j\omega})$ is large near $\omega = \pi$, you do not want H_2 to have a zero at $\omega = \pi$. Would you use an impulse response with an even or an odd number of samples? Explain.

7.22. In the system shown in Figure P7.22, the discrete-time system is a linear-phase FIR lowpass filter designed by the Parks–McClellan algorithm with $\delta_1 = 0.01$, $\delta_2 = 0.001$, $\omega_p = 0.4\pi$, and $\omega_s = 0.6\pi$. The length of the impulse response is 28 samples. The sampling rate for the ideal C/D and D/C converters is $1/T = 10000$ samples/sec.

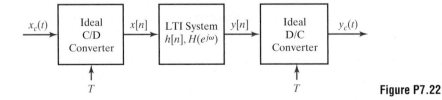

Figure P7.22

(a) What property should the input signal have so that the overall system behaves as an LTI system with $Y_c(j\Omega) = H_{eff}(j\Omega)X_c(j\Omega)$?

(b) For the conditions found in (a), determine the approximation error specifications satisfied by $|H_{eff}(j\Omega)|$. Give your answer as either an equation or a plot as a function of Ω.

(c) What is the overall delay from the continuous-time input to the continuous-time output (in seconds) of the system in Figure P7.22?

7.23. Consider a continuous-time system with system function

$$H_C(s) = \frac{1}{s}.$$

This system is called an *integrator*, since the output $y_C(t)$ is related to the input $x_C(t)$ by

$$y_C(t) = \int_{-\infty}^{t} x_C(\tau)d\tau.$$

Suppose a discrete-time system is obtained by applying the bilinear transformation to $H_C(s)$.

(a) What is the system function $H(z)$ of the resulting discrete-time system? What is the impulse response $h[n]$?

(b) If $x[n]$ is the input and $y[n]$ is the output of the resulting discrete-time system, write the difference equation that is satisfied by the input and output. What problems do you anticipate in implementing the discrete-time system using this difference equation?

(c) Obtain an expression for the frequency response $H(e^{j\omega})$ of the system. Sketch the magnitude and phase of the discrete-time system for $0 \le |\omega| \le \pi$. Compare them with the magnitude and phase of the frequency response $H_C(j\Omega)$ of the continuous-time integrator. Under what conditions could the discrete-time "integrator" be considered a good approximation to the continuous-time integrator?

Now consider a continuous-time system with system function

$$G_C(s) = s.$$

This system is a *differentiator*; i.e., the output is the derivative of the input. Suppose a discrete-time system is obtained by applying the bilinear transformation to $G_C(s)$.

(d) What is the system function $G(z)$ of the resulting discrete-time system? What is the impulse response $g[n]$?

(e) Obtain an expression for the frequency response $G(e^{j\omega})$ of the system. Sketch the magnitude and phase of the discrete-time system for $0 \le |\omega| \le \pi$. Compare them with the magnitude and phase of the frequency response $G_C(j\Omega)$ of the continuous-time differentiator. Under what conditions could the discrete-time "differentiator" be considered a good approximation to the continuous-time differentiator?

(f) The continuous-time integrator and differentiator are exact inverses of one another. Is the same true of the discrete-time approximations obtained by using the bilinear transformation?

7.24. Suppose we have an even-symmetric FIR filter $h[n]$ of length $2L + 1$, i.e.,

$$h[n] = 0 \text{ for } |n| > L,$$

$$h[n] = h[-n].$$

The frequency response $H(e^{j\omega})$, i.e., the DTFT of $h[n]$, is plotted over $-\pi \le \omega \le \pi$ in Figure P7.24.

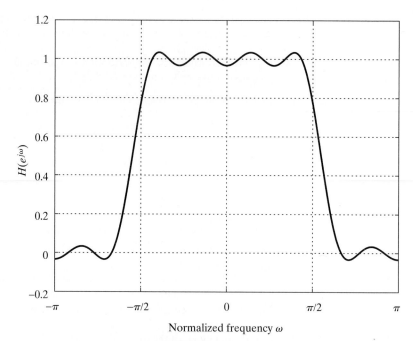

Figure P7.24

What can be inferred from Figure P7.24 about the possible range of values of L? Clearly explain the reason(s) for your answer. Do not make any assumptions about the design procedure that might have been used to obtain $h[n]$.

7.25. Let $h_d[n]$ denote the impulse response of an ideal desired system with corresponding frequency response $H_d(e^{j\omega})$, and let $h[n]$ and $H(e^{j\omega})$ denote the impulse response and frequency response, respectively, of an FIR approximation to the ideal system. Assume that $h[n] = 0$ for $n < 0$ and $n > M$. We wish to choose the $(M+1)$ samples of the impulse response so as to minimize the mean-square error of the frequency response defined as

$$\varepsilon^2 = \frac{1}{2\pi} \int_{-\pi}^{\pi} |H_d(e^{j\omega}) - H(e^{j\omega})|^2 d\omega.$$

   (a) Use Parseval's relation to express the error function in terms of the sequences $h_d[n]$ and $h[n]$.
   (b) Using the result of part (a), determine the values of $h[n]$ for $0 \le n \le M$ that minimize ε^2.
   (c) The FIR filter determined in part (b) could have been obtained by a windowing operation. That is, $h[n]$ could have been obtained by multiplying the desired infinite-length sequence $h_d[n]$ by a certain finite-length sequence $w[n]$. Determine the necessary window $w[n]$ such that the optimal impulse response is $h[n] = w[n]h_d[n]$.

Advanced Problems

7.26. *Impulse invariance* and the *bilinear transformation* are two methods for designing discrete-time filters. Both methods transform a continuous-time system function $H_c(s)$ into a discrete-time system function $H(z)$. Answer the following questions by indicating which method(s) will yield the desired result:

(a) A minimum-phase continuous-time system has all its poles and zeros in the left-half s-plane. If a minimum-phase continuous-time system is transformed into a discrete-time system, which method(s) will result in a minimum-phase discrete-time system?

(b) If the continuous-time system is an all-pass system, its poles will be at locations s_k in the left-half s-plane, and its zeros will be at corresponding locations $-s_k$ in the right-half s-plane. Which design method(s) will result in an all-pass discrete-time system?

(c) Which design method(s) will guarantee that

$$H(e^{j\omega})|_{\omega=0} = H_c(j\Omega)|_{\Omega=0}?$$

(d) If the continuous-time system is a bandstop filter, which method(s) will result in a discrete-time bandstop filter?

(e) Suppose that $H_1(z)$, $H_2(z)$, and $H(z)$ are transformed versions of $H_{c1}(s)$, $H_{c2}(s)$, and $H_c(s)$, respectively. Which design method(s) will guarantee that $H(z) = H_1(z)H_2(z)$ whenever $H_c(s) = H_{c1}(s)H_{c2}(s)$?

(f) Suppose that $H_1(z)$, $H_2(z)$, and $H(z)$ are transformed versions of $H_{c1}(s)$, $H_{c2}(s)$, and $H_c(s)$, respectively. Which design method(s) will guarantee that $H(z) = H_1(z)+H_2(z)$ whenever $H_c(s) = H_{c1}(s) + H_{c2}(s)$?

(g) Assume that two continuous-time system functions satisfy the condition

$$\frac{H_{c1}(j\Omega)}{H_{c2}(j\Omega)} = \begin{cases} e^{-j\pi/2}, & \Omega > 0, \\ e^{j\pi/2}, & \Omega < 0. \end{cases}$$

If $H_1(z)$ and $H_2(z)$ are transformed versions of $H_{c1}(s)$ and $H_{c2}(s)$, respectively, which design method(s) will result in discrete-time systems such that

$$\frac{H_1(e^{j\omega})}{H_2(e^{j\omega})} = \begin{cases} e^{-j\pi/2}, & 0 < \omega < \pi, \\ e^{j\pi/2}, & -\pi < \omega < 0? \end{cases}$$

(Such systems are called "90-degree phase splitters.")

7.27. Suppose that we are given an ideal lowpass discrete-time filter with frequency response

$$H(e^{j\omega}) = \begin{cases} 1, & |\omega| < \pi/4, \\ 0, & \pi/4 < |\omega| \le \pi. \end{cases}$$

We wish to derive new filters from this prototype by manipulations of the impulse response $h[n]$.

(a) Plot the frequency response $H_1(e^{j\omega})$ for the system whose impulse response is $h_1[n] = h[2n]$.

(b) Plot the frequency response $H_2(e^{j\omega})$ for the system whose impulse response is

$$h_2[n] = \begin{cases} h[n/2], & n = 0, \pm 2, \pm 4, \dots, \\ 0, & \text{otherwise.} \end{cases}$$

(c) Plot the frequency response $H_3(e^{j\omega})$ for the system whose impulse response is $h_3[n] = e^{j\pi n}h[n] = (-1)^n h[n]$.

7.28. Consider a continuous-time lowpass filter $H_c(s)$ with passband and stopband specifications

$$1 - \delta_1 \le |H_c(j\Omega)| \le 1 + \delta_1, \qquad |\Omega| \le \Omega_p,$$
$$|H_c(j\Omega)| \le \delta_2, \qquad \Omega_s \le |\Omega|.$$

This filter is transformed to a lowpass discrete-time filter $H_1(z)$ by the transformation

$$H_1(z) = H_c(s)|_{s=(1-z^{-1})/(1+z^{-1})},$$

and the same continuous-time filter is transformed to a highpass discrete-time filter by the transformation

$$H_2(z) = H_c(s)\big|_{s=(1+z^{-1})/(1-z^{-1})}.$$

(a) Determine a relationship between the passband cutoff frequency Ω_p of the continuous-time lowpass filter and the passband cutoff frequency ω_{p1} of the discrete-time lowpass filter.

(b) Determine a relationship between the passband cutoff frequency Ω_p of the continuous-time lowpass filter and the passband cutoff frequency ω_{p2} of the discrete-time highpass filter.

(c) Determine a relationship between the passband cutoff frequency ω_{p1} of the discrete-time lowpass filter and the passband cutoff frequency ω_{p2} of the discrete-time highpass filter.

(d) The network in Figure P7.28 depicts an implementation of the discrete-time lowpass filter with system function $H_1(z)$. The coefficients $A, B, C,$ and D are real. How should these coefficients be modified to obtain a network that implements the discrete-time highpass filter with system function $H_2(z)$?

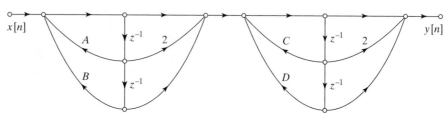

Figure P7.28

7.29. A discrete-time system with system function $H(Z)$ and impulse response $h[n]$ has frequency response

$$H(e^{j\theta}) = \begin{cases} A, & |\theta| < \theta_c, \\ 0, & \theta_c < |\theta| \leq \pi, \end{cases}$$

where $0 < \theta_c < \pi$. This filter is transformed into a new filter by the transformation $Z = -z^2$; i.e.,

$$H_1(z) = H(Z)\big|_{Z=-z^2} = H(-z^2).$$

(a) Obtain a relationship between the frequency variable θ for the original lowpass system $H(Z)$ and the frequency variable ω for the new system $H_1(z)$.

(b) Sketch and carefully label the frequency response $H_1(e^{j\omega})$ for the new filter.

(c) Obtain a relationship expressing $h_1[n]$ in terms of $h[n]$.

(d) Assume that $H(Z)$ can be realized by the set of difference equations

$$g[n] = x[n] - a_1 g[n-1] - b_1 f[n-2],$$

$$f[n] = a_2 g[n-1] + b_2 f[n-1],$$

$$y[n] = c_1 f[n] - c_2 g[n-1],$$

where $x[n]$ is the input and $y[n]$ is the output of the system. Determine a set of difference equations that will realize the transformed system $H_1(z) = H(-z^2)$.

7.30. Consider designing a discrete-time filter with system function $H(z)$ from a continuous-time filter with rational system function $H_c(s)$ by the transformation

$$H(z) = H_c(s)\big|_{s=\beta[(1-z^{-\alpha})/(1+z^{-\alpha})]},$$

where α is a nonzero integer and β is real.

(a) If $\alpha > 0$, for what values of β does a stable, causal continuous-time filter with rational $H_c(s)$ always lead to a stable, causal discrete-time filter with rational $H(z)$?
(b) If $\alpha < 0$, for what values of β does a stable, causal continuous-time filter with rational $H_c(s)$ always lead to a stable, causal discrete-time filter with rational $H(z)$?
(c) For $\alpha = 2$ and $\beta = 1$, determine to what contour in the z-plane the $j\Omega$-axis of the s-plane maps.
(d) Suppose that the continuous-time filter is a stable lowpass filter with passband frequency response such that

$$1 - \delta_1 \le |H_c(j\Omega)| \le 1 + \delta_1 \qquad \text{for} \quad |\Omega| \le 1.$$

If the discrete-time system $H(z)$ is obtained by the transformation set forth at the beginning of this problem, with $\alpha = 2$ and $\beta = 1$, determine the values of ω in the interval $|\omega| \le \pi$ for which

$$1 - \delta_1 \le |H(e^{j\omega})| \le 1 + \delta_1.$$

7.31. Suppose that we have used the Parks–McClellan algorithm to design a causal FIR linear-phase lowpass filter. The system function of this system is denoted $H(z)$. The *length* of the impulse response is 25 samples, i.e., $h[n] = 0$ for $n < 0$ and for $n > 24$, and $h[0] \ne 0$. The desired response and weighting function used were

$$H_d(e^{j\omega}) = \begin{cases} 1 & |\omega| \le 0.3\pi \\ 0 & 0.4\pi \le |\omega| \le \pi \end{cases} \qquad W(e^{j\omega}) = \begin{cases} 1 & |\omega| \le 0.3\pi \\ 2 & 0.4\pi \le |\omega| \le \pi. \end{cases}$$

In each case below, determine whether the statement is true or false or that insufficient information is given. Justify your conclusions.

(a) $h[n + 12] = h[12 - n]$ or $h[n + 12] = -h[12 - n]$ for $-\infty < n < \infty$.
(b) The system has a stable and causal inverse.
(c) We know that $H(-1) = 0$.
(d) The maximum weighted approximation error is the same in all approximation bands.
(e) If z_0 is a zero of $H(z)$, then $1/z_0$ is a pole of $H(z)$.
(f) The system can be implemented by a network (flow graph) that has no feedback paths.
(g) The group delay is equal to 24 for $0 < \omega < \pi$.
(h) If the coefficients of the system function are quantized to 10 bits each, the system is still optimum in the Chebyshev sense for the original desired response and weighting function.
(i) If the coefficients of the system function are quantized to 10 bits each, the system is still guaranteed to be a linear-phase filter.
(j) If the coefficients of the system function are quantized to 10 bits each, the system may become unstable.

7.32. You are required to design an FIR filter, $h[n]$, with the following magnitude specifications:

- Passband edge: $\omega_p = \pi/100$.
- Stopband edge: $\omega_s = \pi/50$.
- Maximum stopband gain: $\delta_s \leq -60$ dB relative to passband.

It is suggested that you try using a Kaiser window. The Kaiser window design rules for shape parameter β and filter length M are provided in Section 7.5.3.

(a) What values of β and M are necessary to meet the required specifications?

You show the resulting filter to your boss, and he is unsatisfied. He asks you to reduce the computations required for the filter. You bring in a consultant who suggests that you design the filter as a cascade of two stages: $h'[n] = p[n] * q[n]$. To design $p[n]$ he suggests first designing a filter, $g[n]$, with passband edge $\omega'_p = 10\omega_p$, stopband edge $\omega'_s = 10\omega_s$ and stopband gain $\delta'_s = \delta_s$. The filter $p[n]$ is then obtained by expanding $g[n]$ by a factor of 10:

$$p[n] = \begin{cases} g[n/10], & \text{when } n/10 \text{ is an integer,} \\ 0, & \text{otherwise.} \end{cases}$$

(b) What values of β' and M' are necessary to meet the required specifications for $g[n]$?
(c) Sketch $P(e^{j\omega})$ from $\omega = 0$ to $\omega = \pi/4$. You do not need to draw the exact shape of the frequency response; instead, you should show which regions of the frequency response are near 0 dB, and which regions are at or below -60 dB. Label all band edges in your sketch.
(d) What specifications should be used in designing $q[n]$ to guarantee that $h'[n] = p[n] * q[n]$ meets or exceeds the original requirements? Specify the passband edge, ω''_p, stopband edge, ω''_s, and stopband attenuation, δ''_s, required for $q[n]$.
(e) What values of β'' and M'' are necessary to meet the required specifications for $q[n]$? How many nonzero samples will $h'[n] = q[n] * p[n]$ have?
(f) The filter $h'[n]$ from parts (b)–(e) is implemented by first directly convolving the input with $q[n]$ and then directly convolving the results with $p[n]$. The filter $h[n]$ from part (a) is implemented by directly convolving the input with $h[n]$. Which of these two implementations requires fewer multiplications? Explain. Note: you should not count multiplications by 0 as an operation.

7.33. Consider a real, bandlimited signal $x_a(t)$ whose Fourier transform $X_a(j\Omega)$ has the following property:

$$X_a(j\Omega) = 0 \quad \text{for} \quad |\Omega| > 2\pi \cdot 10000 \, .$$

That is, the signal is bandlimited to 10 kHz.

We wish to process $x_a(t)$ with a highpass analog filter whose magnitude response satisfies the following specifications (see Figure P7.33):

$$\begin{cases} 0 \leq |H_a(j\Omega)| \leq 0.1 & \text{for } 0 \leq |\Omega| \leq 2\pi \cdot 4000 = \Omega_s \\ 0.9 \leq |H_a(j\Omega)| \leq 1 & \text{for } \Omega_p = 2\pi \cdot 8000 \leq |\Omega|, \end{cases}$$

where Ω_s and Ω_p denote the stopband and passband frequencies, respectively.

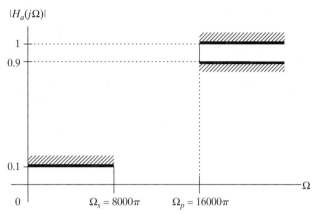

Figure P7.33

(a) Suppose the analog filter $H_a(j\Omega)$ is implemented by discrete-time processing, according to the diagram shown in Figure 7.2.

The sampling frequency $f_s = \dfrac{1}{T}$ is 24 kHz for both the ideal C/D and D/C converters. Determine the appropriate filter specification for $|H(e^{j\omega})|$, the magnitude response of the digital filter.

(b) Using the bilinear transformation $s = \dfrac{1 - z^{-1}}{1 + z^{-1}}$, we want to design a digital filter whose magnitude response specifications were found in part (a). Find the specifications of $|G_{HP}(j\Omega_1)|$, the magnitude response of the highpass analog filter that is related to the digital filter through the bilinear transformation. Again, provide a fully labelled sketch of the magnitude response specifications on $|G_{HP}(j\Omega_1)|$.

(c) Using the frequency transformation $s_1 = \dfrac{1}{s_2}$, (i.e., replacing the Laplace transform variable s by its reciprocal), design the highpass analog filter $G_{HP}(j\Omega_1)$ from the lowest-order Butterworth filter, whose magnitude-squared frequency response is given below:

$$|G(j\Omega_2)|^2 = \frac{1}{1 + \left(\Omega_2/\Omega_c\right)^{2N}}.$$

In particular, find the lowest filter order N and its corresponding cutoff frequency Ω_c, such that the original filter's passband specification ($|H_a(j\Omega_p)| = 0.9$) is met *exactly*. In a diagram, label the salient features of the Butterworth filter magnitude response that you have designed.

(d) Draw the pole–zero diagram of the (lowpass) Butterworth filter $G(s_2)$, and find an expression for its transfer function.

7.34. A zero-phase FIR filter $h[n]$ has associated DTFT $H(e^{j\omega})$, shown in Figure P7.34.

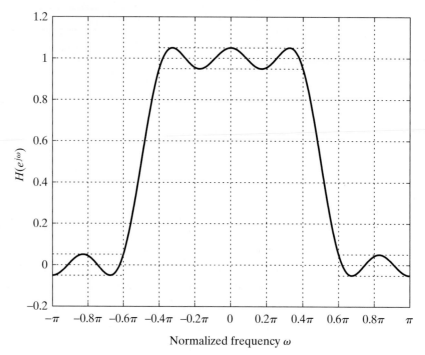

Normalized frequency ω

Figure P7.34

The filter is known to have been designed using the Parks–McClellan (PM) algorithm. The input parameters to the PM algorithm are known to have been:

- Passband edge: $\omega_p = 0.4\pi$
- Stopband edge: $\omega_s = 0.6\pi$
- Ideal passband gain: $G_p = 1$
- Ideal stopband gain: $G_s = 0$
- Error weighting function $W(\omega) = 1$

The length of the impulse response $h[n]$, is $M + 1 = 2L + 1$ and

$$h[n] = 0 \ \text{ for } \ |n| > L.$$

The value of L is not known.

It is claimed that there are two filters, each with frequency response identical to that shown in Figure P7.34, and each having been designed by the Parks–McClellan algorithm with *different* values for the input parameter L.

- **Filter 1:** $L = L_1$
- **Filter 2:** $L = L_2 > L_1$.

Both filters were designed using exactly the same Parks–McClellan algorithm and input parameters, *except* for the value of L.

(a) What are possible values for L_1?
(b) What are possible values for $L_2 > L_1$?

(c) Are the impulse responses $h_1[n]$ and $h_2[n]$ of the two filters identical?

(d) The alternation theorem guarantees "*uniqueness* of the rth-order polynomial." If your answer to (c) is yes, explain why the alternation theorem is not violated. If your answer is no, show how the two filters, $h_1[n]$ and $h_2[n]$, are related.

7.35. We are given an FIR bandpass filter $h[n]$ that is zero phase, i.e., $h[n] = h[-n]$. Its associated DTFT $H(e^{j\omega})$ is shown in Figure P7.35.

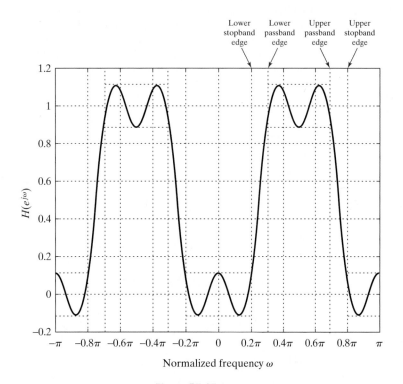

Figure P7.35

The filter is known to have been designed using the Parks–McClellan algorithm. The input parameters to the Parks–McClellan algorithm are known to have been:

- Lower stopband edge: $\omega_1 = 0.2\pi$
- Lower passband edge: $\omega_2 = 0.3\pi$
- Upper passband edge: $\omega_3 = 0.7\pi$
- Upper stopband edge: $\omega_4 = 0.8\pi$

- Ideal passband gain: $G_p = 1$
- Ideal stopband gain: $G_s = 0$
- Error weighting function $W(\omega) = 1$

The value of the input parameter $M+1$, which represents the maximum number of nonzero impulse response values (equivalently the filter length), is not known.

It is claimed that there are two filters, each with a frequency response identical to that shown in Figure P7.35, but having *different* impulse response lengths $M + 1 = 2L + 1$.

- **Filter 1:** $M = M_1 = 14$
- **Filter 2:** $M = M_2 \neq M_1$

Both filters were designed using exactly the same Parks–McClellan algorithm and input parameters, *except* for the value of M.

(a) What are possible values for M_2?

(b) The alternation theorem guarantees "*uniqueness* of the rth-order polynomial." Explain why the alternation theorem is not violated.

7.36. The graphs in Figure P7.36 depict four frequency-response magnitude plots of linear-phase FIR filters, labelled $|A_e^i(e^{j\omega})|$, $i = 1, 2, 3, 4$. One or more of these plots may belong to equiripple linear-phase FIR filters designed by the Parks–McClellan algorithm. The maximum approximation errors in the passband and the stopband, as well as the desired cutoff frequencies of those bands, are also shown in the plots. Please note that the approximation error and filter length specifications may have been chosen differently to ensure that the cutoff frequencies are the same in each design.

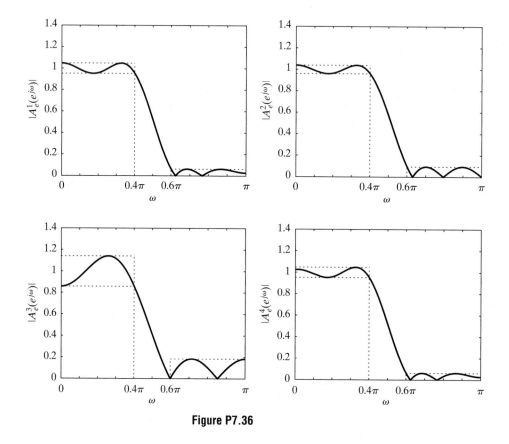

Figure P7.36

(a) What type(s) (I, II, III, IV) of linear-phase FIR filters can $|A_e^i(e^{j\omega})|$ correspond to, for $i = 1, 2, 3, 4$? Please note that there may be more than one linear-phase FIR filter type corresponding to each $|A_e^i(e^{j\omega})|$. If you feel this is the case, list all possible choices.

(b) How many alternations does each $|A_e^i(e^{j\omega})|$ exhibit, for $i = 1, 2, 3, 4$?

(c) For each i, $i = 1, 2, 3, 4$, can $|A_e^i(e^{j\omega})|$ belong to an output of the Parks–McClellan algorithm?

(d) If you claimed that a given $|A_e^i(e^{j\omega})|$ could correspond to an output of the Parks–McClellan algorithm, and that it could be type I, what is the length of the impulse response of $|A_e^i(e^{j\omega})|$?

7.37. Consider the two-stage system shown in Figure P7.37 for interpolating a sequence $x[n] = x_c(nT)$ to a sampling rate that is 15 times as high as the input sampling rate; i.e., we desire $y[n] = x_c(nT/15)$.

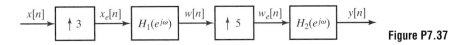

Figure P7.37

Assume that the input sequence $x[n] = x_c(nT)$ was obtained by sampling a band-limited continuous-time signal whose Fourier transform satisfies the following condition: $|X_c(j\Omega)| = 0$ for $|\Omega| \geq 2\pi(3600)$. Assume that the original sampling period was $T = 1/8000$.

(a) Make a sketch of the Fourier transform $X_c(j\Omega)$ of a "typical" bandlimited input signal and the corresponding discrete-time Fourier transforms $X(e^{j\omega})$ and $X_e(e^{j\omega})$.

(b) To implement the interpolation system, we must, of course, use nonideal filters. Use your plot of $X_e(e^{j\omega})$ obtained in part (a) to determine the passband and stopband cutoff frequencies (ω_{p1} and ω_{s1}) required to preserve the original band of frequencies essentially unmodified while significantly attenuating the images of the baseband spectrum. (That is, we desire that $w[n] \approx x_c(nT/3)$.) Assuming that this can be achieved with passband approximation error $\delta_1 = 0.005$ (for filter passband gain of 1) and stopband approximation error $\delta_2 = 0.01$, plot the specifications for the design of the filter $H_1(e^{j\omega})$ for $-\pi \leq \omega \leq \pi$.

(c) Assuming that $w[n] = x_c(nT/3)$, make a sketch of $W_e(e^{j\omega})$ and use it to determine the passband and stopband cutoff frequencies ω_{p2} and ω_{s2} required for the second filter.

(d) Use the formula of Eq. (7.117) to determine the filter orders M_1 and M_2 for Parks–McClellan filters that have the passband and stopband cutoff frequencies determined in parts (b) and (c) with $\delta_1 = 0.005$ and $\delta_2 = 0.01$ for both filters.

(e) Determine how many multiplications are required to compute 15 samples of the output for this case.

7.38. The system of Figure 7.2 is used to perform filtering of continuous-time signals with a digital filter. The sampling rate of the C/D and D/C converters is $f_s = 1/T = 10,000$ samples/sec.

A Kaiser window $w_K[n]$ of length $M + 1 = 23$ and $\beta = 3.395$ is used to design a linear-phase lowpass filter with frequency response $H_{\text{lp}}(e^{j\omega})$. When used in the system of Figure 7.1 so that $H(e^{j\omega}) = H_{\text{lp}}(e^{j\omega})$, the overall effective frequency response (from input $x_a(t)$ to output $y_a(t)$) meets the following specifications:

$$0.99 \leq |H_{\text{eff}}(j\Omega)| \leq 1.01, \quad 0 \leq |\Omega| \leq 2\pi(2000)$$
$$|H_{\text{eff}}(j\Omega)| \leq 0.01 \quad 2\pi(3000) \leq |\Omega| \leq 2\pi(5000).$$

(a) The linear phase of the FIR filter introduces a time delay t_d. Find the time delay through the system (in milliseconds).

(b) Now a highpass filter is designed with the *same* Kaiser window by applying it to the ideal impulse response $h_d[n]$ whose corresponding frequency response is

$$H_d(e^{j\omega}) = \begin{cases} 0 & |\omega| < 0.25\pi \\ 2e^{-j\omega n_d} & 0.25\pi < |\omega| \leq \pi. \end{cases}$$

That is, a linear-phase FIR highpass filter with impulse response $h_{hp}[n] = w_K[n]h_d[n]$ and frequency response $H_{hp}(e^{j\omega})$ was obtained by multiplying $h_d[n]$ by the same Kaiser window $w_K[n]$ that was used to design the first mentioned lowpass filter. The resulting FIR highpass discrete-time filter meets a set of specifications of the following form:

$$|H_{hp}(e^{j\omega})| \leq \delta_1 \qquad 0 \leq |\omega| \leq \omega_1$$
$$G - \delta_2 \leq |H_{hp}(e^{j\omega})| \leq G + \delta_2 \quad \omega_2 \leq |\omega| \leq \pi$$

Use information from the lowpass filter specifications to determine the values of ω_1, ω_2, δ_1, δ_2, and G.

7.39. Figure P7.39 is the ideal, desired frequency response amplitude for a bandpass filter to be designed as a Type I FIR filter $h[n]$, with DTFT $H(e^{j\omega})$ that approximates $H_d(e^{j\omega})$ and meets the following constraints:

$$-\delta_1 \leq H(e^{j\omega}) \leq \delta_1, \ \ 0 \leq |\omega| \leq \omega_1$$

$$1 - \delta_2 \leq H(e^{j\omega}) \leq 1 + \delta_2, \ \ \omega_2 \leq |\omega| \leq \omega_3$$

$$-\delta_3 \leq H(e^{j\omega}) \leq \delta_3, \ \ \omega_4 \leq |\omega| \leq \pi$$

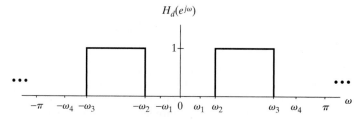

Figure P7.39

The resulting filter $h[n]$ is to minimize the maximum weighted error and therefore must satisfy the alternation theorem.

Determine and sketch an appropriate choice for the weighting function to use with the Parks–McClellan algorithm.

7.40. (a) Figure P7.40-1 shows the frequency response $A_e(e^{j\omega})$ of a lowpass Type I Parks–McClellan filter based on the following specifications. Consequently it satisfies the alternation theorem.

$$\text{Passband edge:} \quad \omega_p = 0.45\pi$$
$$\text{Stopband edge:} \quad \omega_s = 0.50\pi$$
$$\text{Desired passband magnitude:} \quad 1$$
$$\text{Desired stopband magnitude:} \quad 0$$

The weighting function used in both the passband and the stopband is $W(\omega) = 1$.

What can you conclude about the maximum possible number of nonzero values in the impulse response of the filter?

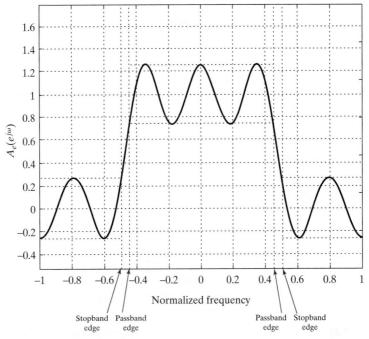

Figure P7.40-1

(b) Figure P7.40-2 shows another frequency response $B_e(e^{j\omega})$ for a Type I FIR filter. $B_e(e^{j\omega})$ is obtained from $A_e(e^{j\omega})$ from part (a) as follows:

$$B_e(e^{j\omega}) = k_1 \left(A_e(e^{j\omega}) \right)^2 + k_2,$$

where k_1 and k_2 are constants. Observe that $B_e(e^{j\omega})$ displays equiripple behavior, with different maximum error in the passband and stopband.

Does this filter satisfy the alternation theorem with the passband and stopband edge frequencies indicated and with passband ripple and stopband ripple indicated by the dashed lines?

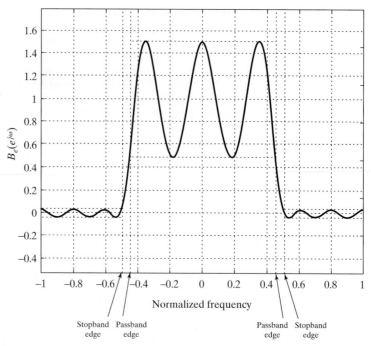

Figure P7.40-2

7.41. Assume that $H_c(s)$ has an rth-order pole at $s = s_0$, so that $H_c(s)$ can be expressed as

$$H_c(s) = \sum_{k=1}^{r} \frac{A_k}{(s - s_0)^k} + G_c(s),$$

where $G_c(s)$ has only 1st-order poles. Assume $H_c(s)$ is causal.

   (a) Give a formula for determining the constants A_k from $H_c(s)$.

   (b) Obtain an expression for the impulse response $h_c(t)$ in terms of s_0 and $g_c(t)$, the inverse Laplace transform of $G_c(s)$.

7.42. As discussed in Chapter 12, an *ideal discrete-time Hilbert transformer* is a system that introduces -90 degrees ($-\pi/2$ radians) of phase shift for $0 < \omega < \pi$ and $+90$ degrees ($+\pi/2$ radians) of phase shift for $-\pi < \omega < 0$. The magnitude of the frequency response is constant (unity) for $0 < \omega < \pi$ and for $-\pi < \omega < 0$. Such systems are also called *ideal 90-degree phase shifters*.

   (a) Give an equation for the ideal desired frequency response $H_d(e^{j\omega})$ of an ideal discrete-time Hilbert transformer that also includes constant (nonzero) group delay. Plot the phase response of this system for $-\pi < \omega < \pi$.

   (b) What type(s) of FIR linear-phase systems (I, II, III, or IV) can be used to approximate the ideal Hilbert transformer in part (a)?

   (c) Suppose that we wish to use the window method to design a linear-phase approximation to the ideal Hilbert transformer. Use $H_d(e^{j\omega})$ given in part (a) to determine the ideal impulse response $h_d[n]$ if the FIR system is to be such that $h[n] = 0$ for $n < 0$ and $n > M$.

   (d) What is the delay of the system if $M = 21$? Sketch the magnitude of the frequency response of the FIR approximation for this case, assuming a rectangular window.

(e) What is the delay of the system if $M = 20$? Sketch the magnitude of the frequency response of the FIR approximation for this case, assuming a rectangular window.

7.43. The commonly used windows presented in Section 7.5.1 can all be expressed in terms of rectangular windows. This fact can be used to obtain expressions for the Fourier transforms of the Bartlett window and the raised-cosine family of windows, which includes the Hanning, Hamming, and Blackman windows.

(a) Show that the $(M+1)$-point Bartlett window, defined by Eq. (7.60b), can be expressed as the convolution of two smaller rectangular windows. Use this fact to show that the Fourier transform of the $(M+1)$-point Bartlett window is

$$W_B(e^{j\omega}) = e^{-j\omega M/2}(2/M)\left(\frac{\sin(\omega M/4)}{\sin(\omega/2)}\right)^2 \qquad \text{for } M \text{ even,}$$

or

$$W_B(e^{j\omega}) = e^{-j\omega M/2}(2/M)\left(\frac{\sin[\omega(M+1)/4]}{\sin(\omega/2)}\right)\left(\frac{\sin[\omega(M-1)/4]}{\sin(\omega/2)}\right) \qquad \text{for } M \text{ odd.}$$

(b) It can easily be seen that the $(M+1)$-point raised-cosine windows defined by Eqs. (7.60c)–(7.60e) can all be expressed in the form

$$w[n] = [A + B\cos(2\pi n/M) + C\cos(4\pi n/M)]w_R[n],$$

where $w_R[n]$ is an $(M+1)$-point rectangular window. Use this relation to find the Fourier transform of the general raised-cosine window.

(c) Using appropriate choices for A, B, and C and the result determined in part (b), sketch the magnitude of the Fourier transform of the Hanning window.

7.44. Consider the following ideal frequency response for a multiband filter:

$$H_d(e^{j\omega}) = \begin{cases} e^{-j\omega M/2}, & 0 \le |\omega| < 0.3\pi, \\ 0, & 0.3\pi < |\omega| < 0.6\pi, \\ 0.5e^{-j\omega M/2}, & 0.6\pi < |\omega| \le \pi. \end{cases}$$

The impulse response $h_d[n]$ is multiplied by a Kaiser window with $M = 48$ and $\beta = 3.68$, resulting in a linear-phase FIR system with impulse response $h[n]$.

(a) What is the delay of the filter?
(b) Determine the ideal desired impulse response $h_d[n]$.
(c) Determine the set of approximation error specifications that is satisfied by the FIR filter; i.e., determine the parameters $\delta_1, \delta_2, \delta_3, B, C, \omega_{p1}, \omega_{s1}, \omega_{s2},$ and ω_{p2} in

$$B - \delta_1 \le |H(e^{j\omega})| \le B + \delta_1, \qquad 0 \le \omega \le \omega_{p1},$$
$$|H(e^{j\omega})| \le \delta_2, \qquad \omega_{s1} \le \omega \le \omega_{s2},$$
$$C - \delta_3 \le |H(e^{j\omega})| \le C + \delta_3, \qquad \omega_{p2} \le \omega \le \pi.$$

7.45. The frequency response of a desired filter $h_d[n]$ is shown in Figure P7.45. In this problem, we wish to design an $(M+1)$-point causal linear-phase FIR filter $h[n]$ that minimizes the integral-squared error

$$\epsilon_d^2 = \frac{1}{2\pi}\int_{-\pi}^{\pi} |A(e^{j\omega}) - H_d(e^{j\omega})|^2 d\omega,$$

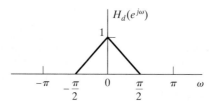

Figure P7.45

where the frequency response of the filter $h[n]$ is

$$H(e^{j\omega}) = A(e^{j\omega})e^{-j\omega M/2}$$

and M is an even integer.

(a) Determine $h_d[n]$.

(b) What symmetry should $h[n]$ have in the range $0 \leq n \leq M$? Briefly explain your reasoning.

(c) Determine $h[n]$ in the range $0 \leq n \leq M$.

(d) Determine an expression for the minimum integral-squared error ϵ^2 as a function of $h_d[n]$ and M.

7.46. Consider a type I linear-phase FIR lowpass filter with impulse response $h_{LP}[n]$ of length $(M + 1)$ and frequency response

$$H_{LP}(e^{j\omega}) = A_e(e^{j\omega})e^{-j\omega M/2}.$$

The system has the amplitude function $A_e(e^{j\omega})$ shown in Figure P7.46.

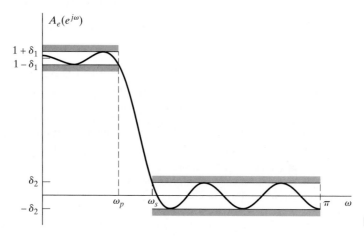

Figure P7.46

This amplitude function is the optimal (in the Parks–McClellan sense) approximation to unity in the band $0 \leq \omega \leq \omega_p$, where $\omega_p = 0.27\pi$, and the optimal approximation to zero in the band $\omega_s \leq \omega \leq \pi$, wherein $\omega_s = 0.4\pi$.

(a) What is the value of M?

Suppose now that a highpass filter is derived from this lowpass filter by defining

$$h_{HP}[n] = (-1)^{n+1}h_{LP}[n] = -e^{j\pi n}h_{LP}[n].$$

(b) Show that the resulting frequency response is of the form $H_{HP}(e^{j\omega}) = B_e(e^{j\omega})e^{-j\omega M/2}$.

(c) Sketch $B_e(e^{j\omega})$ for $0 \le \omega \le \pi$.

(d) It is asserted that for the given value of M (as found in part (a)), the resulting highpass filter is the optimum approximation to zero in the band $0 \le \omega \le 0.6\pi$ and to unity in the band $0.73\pi \le \omega \le \pi$. Is this assertion correct? Justify your answer.

7.47. Design a three-point optimal (in the minimax sense) causal lowpass filter with $\omega_s = \pi/2$, $\omega_p = \pi/3$, and $K = 1$. Specify the impulse response $h[n]$ of the filter you design. *Note:* $\cos(\pi/2) = 0$ and $\cos(\pi/3) = 0.5$.

Extension Problems

7.48. If an LTI continuous-time system has a rational system function, then its input and output satisfy an ordinary linear differential equation with constant coefficients. A standard procedure in the simulation of such systems is to use finite-difference approximations to the derivatives in the differential equations. In particular, since, for continuous differentiable functions $y_c(t)$,

$$\frac{dy_c(t)}{dt} = \lim_{T \to 0} \left[\frac{y_c(t) - y_c(t - T)}{T} \right],$$

it seems plausible that if T is "small enough," we should obtain a good approximation if we replace $dy_c(t)/dt$ by $[y_c(t) - y_c(t - T)]/T$.

While this simple approach may be useful for simulating continuous-time systems, it is *not* generally a useful method for designing discrete-time systems for filtering applications. To understand the effect of approximating differential equations by difference equations, it is helpful to consider a specific example. Assume that the system function of a continuous-time system is

$$H_c(s) = \frac{A}{s + c},$$

where A and c are constants.

(a) Show that the input $x_c(t)$ and the output $y_c(t)$ of the system satisfy the differential equation

$$\frac{dy_c(t)}{dt} + cy_c(t) = Ax_c(t).$$

(b) Evaluate the differential equation at $t = nT$, and substitute

$$\left. \frac{dy_c(t)}{dt} \right|_{t=nT} \approx \frac{y_c(nT) - y_c(nT - T)}{T},$$

i.e., replace the first derivative by the *first backward difference*.

(c) Define $x[n] = x_c(nT)$ and $y[n] = y_c(nT)$. With this notation and the result of part (b), obtain a difference equation relating $x[n]$ and $y[n]$, and determine the system function $H(z) = Y(z)/X(z)$ of the resulting discrete-time system.

(d) Show that, for this example,

$$H(z) = H_c(s)\big|_{s=(1-z^{-1})/T};$$

i.e., show that $H(z)$ can be obtained directly from $H_c(s)$ by the mapping

$$s = \frac{1 - z^{-1}}{T}.$$

(It can be demonstrated that if higher-order derivatives are approximated by repeated application of the first backward difference, then the result of part (d) holds for higher-order systems as well.)

(e) For the mapping of part (d), determine the contour in the z-plane to which the $j\Omega$-axis of the s-plane maps. Also, determine the region of the z-plane that corresponds to the left half of the s-plane. If the continuous-time system with system function $H_c(s)$ is stable, will the discrete-time system obtained by first backward difference approximation also be stable? Will the frequency response of the discrete-time system be a faithful reproduction of the frequency response of the continuous-time system? How will the stability and frequency response be affected by the choice of T?

(f) Assume that the first derivative is approximated by the *first forward difference;* i.e.,

$$\left.\frac{dy_c(t)}{dt}\right|_{t=nT} \approx \frac{y_c(nT+T) - y_c(nT)}{T}.$$

Determine the corresponding mapping from the s-plane to the z-plane, and repeat part (e) for this mapping.

7.49. Consider an LTI continuous-time system with rational system function $H_c(s)$. The input $x_c(t)$ and the output $y_c(t)$ satisfy an ordinary linear differential equation with constant coefficients. One approach to simulating such systems is to use numerical techniques to integrate the differential equation. In this problem, we demonstrate that if the trapezoidal integration formula is used, this approach is equivalent to transforming the continuous-time system function $H_c(s)$ to a discrete-time system function $H(z)$ using the bilinear transformation.

To demonstrate this statement, consider the continuous-time system function

$$H_c(s) = \frac{A}{s+c},$$

where A and c are constants. The corresponding differential equation is

$$\dot{y}_c(t) + cy_c(t) = Ax_c(t),$$

where

$$\dot{y}_c(t) = \frac{dy_c(t)}{dt}.$$

(a) Show that $y_c(nT)$ can be expressed in terms of $\dot{y}_c(t)$ as

$$y_c(nT) = \int_{(nT-T)}^{nT} \dot{y}_c(\tau)d\tau + y_c(nT-T).$$

The definite integral in this equation represents the area beneath the function $\dot{y}_c(t)$ for the interval from $(nT-T)$ to nT. Figure P7.49 shows a function $\dot{y}_c(t)$ and a shaded trapezoid-shaped region whose area approximates the area beneath the curve. This approximation to the integral is known as the *trapezoidal approximation.* Clearly, as T approaches zero, the approximation improves. Use the trapezoidal approximation to obtain an expression for $y_c(nT)$ in terms of $y_c(nT-T)$, $\dot{y}_c(nT)$, and $\dot{y}_c(nT-T)$.

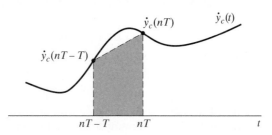

Figure P7.49

(b) Use the differential equation to obtain an expression for $\dot{y}_c(nT)$, and substitute this expression into the expression obtained in part (a).

(c) Define $x[n] = x_c(nT)$ and $y[n] = y_c(nT)$. With this notation and the result of part (b), obtain a difference equation relating $x[n]$ and $y[n]$, and determine the system function $H(z) = Y(z)/X(z)$ of the resulting discrete-time system.

(d) Show that, for this example,

$$H(z) = H_c(s)\big|_{s=(2/T)[(1-z^{-1})/(1+z^{-1})]};$$

i.e., show that $H(z)$ can be obtained directly from $H_c(s)$ by the bilinear transformation. (For higher-order differential equations, repeated trapezoidal integration applied to the highest order derivative of the output will result in the same conclusion for a general continuous-time system with rational system function.)

7.50. In this problem, we consider a method of filter design that might be called *autocorrelation invariance*. Consider a stable continuous-time system with impulse response $h_c(t)$ and system function $H_c(s)$. The autocorrelation function of the system impulse response is defined as

$$\phi_c(\tau) = \int_{-\infty}^{\infty} h_c(t)h_c(t+\tau)d\tau,$$

and for a real impulse response, it is easily shown that the Laplace transform of $\phi_c(\tau)$ is $\Phi_c(s) = H_c(s)H_c(-s)$. Similarly, consider a discrete-time system with impulse response $h[n]$ and system function $H(z)$. The autocorrelation function of a discrete-time system impulse response is defined as

$$\phi[m] = \sum_{n=-\infty}^{\infty} h[n]h[n+m],$$

and for a real impulse response, $\Phi(z) = H(z)H(z^{-1})$.

Autocorrelation invariance implies that a discrete-time filter is defined by equating the autocorrelation function of the discrete-time system to the sampled autocorrelation function of a continuous-time system; i.e.,

$$\phi[m] = T_d\phi_c(mT_d), \qquad -\infty < m < \infty.$$

The following design procedure is proposed for autocorrelation invariance when $H_c(s)$ is a rational function having N 1st-order poles at $s_k, k = 1, 2, \ldots, N$, and $M < N$ zeros:

1. Obtain a partial fraction expansion of $\Phi_c(s)$ in the form

$$\Phi_c(s) = \sum_{k=1}^{N}\left(\frac{A_k}{s-s_k} + \frac{B_k}{s+s_k}\right).$$

2. Form the z-transform

$$\Phi(z) = \sum_{k=1}^{N}\left(\frac{T_d A_k}{1 - e^{s_k T_d}z^{-1}} + \frac{T_d B_k}{1 - e^{-s_k T_d}z^{-1}}\right).$$

3. Find the poles and zeros of $\Phi(z)$, and form a minimum-phase system function $H(z)$ from the poles and zeros of $\Phi(z)$ that are *inside* the unit circle.

(a) Justify each step in the proposed design procedure; i.e., show that the autocorrelation function of the resulting discrete-time system is a sampled version of the autocorrelation function of the continuous-time system. To verify the procedure, it may be helpful to try it out on the 1st-order system with impulse response

$$h_c(t) = e^{-\alpha t}u(t)$$

and corresponding system function

$$H_c(s) = \frac{1}{s + \alpha}.$$

(b) What is the relationship between $|H(e^{j\omega})|^2$ and $|H_c(j\Omega)|^2$? What types of frequency-response functions would be appropriate for autocorrelation invariance design?

(c) Is the system function obtained in Step 3 unique? If not, describe how to obtain additional autocorrelation-invariant discrete-time systems.

7.51. Let $H_{lp}(Z)$ denote the system function for a discrete-time lowpass filter. The implementations of such a system can be represented by linear signal flow graphs consisting of adders, gains, and unit delay elements as in Figure P7.51-1. We want to implement a lowpass filter for which the cutoff frequency can be varied by changing a single parameter. The proposed strategy is to replace each unit delay element in a flow graph representing $H_{lp}(Z)$ by the network shown in Figure P7.51-2, where α is real and $|\alpha| < 1$.

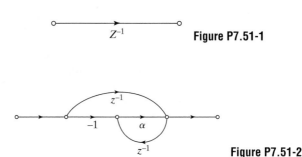

$$Z^{-1} \qquad \textbf{Figure P7.51-1}$$

$$\textbf{Figure P7.51-2}$$

(a) Let $H(z)$ denote the system function for the filter that results when the network of Figure P7.51-2 is substituted for each unit delay branch in the network that implements $H_{lp}(Z)$. Show that $H(z)$ and $H_{lp}(Z)$ are related by a mapping of the Z-plane into the z-plane.

(b) If $H(e^{j\omega})$ and $H_{lp}(e^{j\theta})$ are the frequency responses of the two systems, determine the relationship between the frequency variables ω and θ. Sketch ω as a function of θ for $\alpha = 0.5$, and -0.5, and show that $H(e^{j\omega})$ is a lowpass filter. Also, if θ_p is the passband cutoff frequency for the original lowpass filter $H_{lp}(Z)$, obtain an equation for ω_p, the cutoff frequency of the new filter $H(z)$, as a function of α and θ_p.

(c) Assume that the original lowpass filter has the system function

$$H_{lp}(Z) = \frac{1}{1 - 0.9Z^{-1}}.$$

Draw the flow graph of an implementation of $H_{lp}(Z)$, and also draw the flow graph of the implementation of $H(z)$ obtained by replacing the unit delay elements in the first flow graph by the network in Figure P7.51-2. Does the resulting network correspond to a computable difference equation?

(d) If $H_{lp}(Z)$ corresponds to an FIR system implemented in direct form, would the flow graph manipulation lead to a computable difference equation? If the FIR system $H_{lp}(Z)$ was a linear-phase system, would the resulting system $H(z)$ also be a linear-phase system? If the FIR system has an impulse response of length $M + 1$ samples what would be the length of the impulse response of the transformed system?

(e) To avoid the difficulties that arose in part (c), it is suggested that the network of Figure P7.51-2 be cascaded with a unit delay element, as depicted in Figure P7.51-3. Repeat the analysis of part (a) when the network of Figure P7.51-3 is substituted for each unit delay element. Determine an equation that expresses θ as a function of ω, and show that if $H_{lp}(e^{j\theta})$ is a lowpass filter, then $H(e^{j\omega})$ is not a lowpass filter.

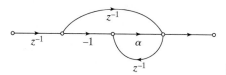

Figure P7.51-3

7.52. If we are given a basic filter module (a hardware or computer subroutine), it is sometimes possible to use it repetitively to implement a new filter with sharper frequency-response characteristics. One approach is to cascade the filter with itself two or more times, but it can easily be shown that, while stopband errors are squared (thereby reducing them if they are less than 1), this approach will increase the passband approximation error. Another approach, suggested by Tukey (1977), is shown in the block diagram of Figure P7.52-1. Tukey called this approach "twicing."

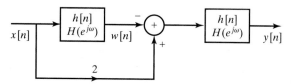

Figure P7.52-1

(a) Assume that the basic system has a symmetric finite-duration impulse response; i.e.,

$$h[n] = \begin{cases} h[-n], & -L \le n \le L, \\ 0 & \text{otherwise.} \end{cases}$$

Determine whether the overall impulse response $g[n]$ is (i) FIR and (ii) symmetric.

(b) Suppose that $H(e^{j\omega})$ satisfies the following approximation error specifications:

$$(1 - \delta_1) \le H(e^{j\omega}) \le (1 + \delta_1), \qquad 0 \le \omega \le \omega_p,$$
$$-\delta_2 \le H(e^{j\omega}) \le \delta_2, \qquad \omega_s \le \omega \le \pi.$$

It can be shown that if the basic system has these specifications, the overall frequency response $G(e^{j\omega})$ (from $x[n]$ to $y[n]$) satisfies specifications of the form

$$A \le G(e^{j\omega}) \le B, \qquad 0 \le \omega \le \omega_p,$$
$$C \le G(e^{j\omega}) \le D, \qquad \omega_s \le \omega \le \pi.$$

Determine A, B, C, and D in terms of δ_1 and δ_2. If $\delta_1 \ll 1$ and $\delta_2 \ll 1$, what are the approximate maximum passband and stopband approximation errors for $G(e^{j\omega})$?

(c) As determined in part (b), Tukey's twicing method improves the passband approximation error, but increases the stopband error. Kaiser and Hamming (1977) generalized the twicing method so as to improve *both* the passband and the stopband. They called their approach "sharpening." The simplest sharpening system that improves both passband and stopband is shown in Figure P7.52-2. Assume again that the impulse response of the basic system is as given in part (a). Repeat part (b) for the system of Figure P7.52-2.

(d) The basic system was assumed to be noncausal. If the impulse response of the basic system is a causal linear-phase FIR system such that

$$h[n] = \begin{cases} h[M - n], & 0 \le n \le M, \\ 0, & \text{otherwise,} \end{cases}$$

how should the systems of Figures P7.52-1 and P7.52-2 be modified? What type(s) (I, II, III, or IV) of causal linear-phase FIR system(s) can be used? What are the lengths of the impulse responses $g[n]$ for the systems in Figures P7.52-1 and P7.52-2 (in terms of L)?

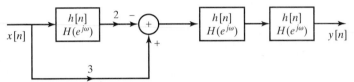

Figure P7.52-2

7.53. Consider the design of a lowpass linear-phase FIR filter by means of the Parks–McClellan algorithm. Use the alternation theorem to argue that the approximation must decrease monotonically in the "don't care" region between the passband and the stopband approximation intervals. *Hint*: Show that all the local maxima and minima of the trigonometric polynomial must be in either the passband or the stopband to satisfy the alternation theorem.

7.54. Figure P7.54 shows the frequency response $A_e(e^{j\omega})$ of a discrete-time FIR system for which the impulse response is

$$h_e[n] = \begin{cases} h_e[-n], & -L \le n \le L, \\ 0, & \text{otherwise.} \end{cases}$$

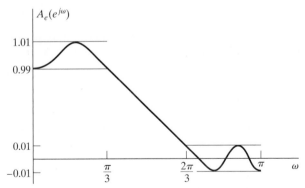

Figure P7.54

(a) Show that $A_e(e^{j\omega})$ cannot correspond to an FIR filter generated by the Parks–McClellan algorithm with a passband edge frequency of $\pi/3$, a stopband edge frequency of $2\pi/3$, and an error-weighting function of unity in the passband and stopband. Clearly explain your reasoning. *Hint*: The alternation theorem states that the best approximation is unique.

(b) Based on Figure P7.54 and the statement that $A_e(e^{j\omega})$ cannot correspond to an optimal filter, what can be concluded about the value of L?

7.55. Consider the system in Figure P7.55.

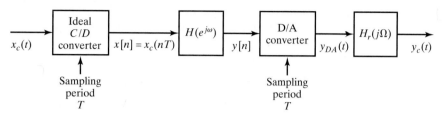

Figure P7.55

1. Assume that $X_c(j\Omega) = 0$ for $|\Omega| \geq \pi/T$ and that

$$H_r(j\Omega) = \begin{cases} 1, & |\Omega| < \pi/T, \\ 0, & |\Omega| > \pi/T, \end{cases}$$

   denotes an ideal lowpass reconstruction filter.
2. The D/A converter has a built-in zero-order-hold circuit, so that

$$Y_{DA}(t) = \sum_{n=-\infty}^{\infty} y[n]h_0(t - nT),$$

   where

$$h_0(t) = \begin{cases} 1, & 0 \leq t < T, \\ 0, & \text{otherwise.} \end{cases}$$

   (We neglect quantization in the D/A converter.)
3. The second system in Figure P7.55 is a linear-phase FIR discrete-time system with frequency response $H(e^{j\omega})$.

We wish to design the FIR system using the Parks–McClellan algorithm to compensate for the effects of the zero-order-hold system.

(a) The Fourier transform of the output is $Y_c(j\Omega) = H_{\text{eff}}(j\Omega)X_c(j\Omega)$. Determine an expression for $H_{\text{eff}}(j\Omega)$ in terms of $H(e^{j\Omega T})$ and T.
(b) If the linear-phase FIR system is such that $h[n] = 0$ for $n < 0$ and $n > 51$, and $T = 10^{-4}$ s, what is the overall time delay (in ms) between $x_c(t)$ and $y_c(t)$?
(c) Suppose that when $T = 10^{-4}$ s, we want the effective frequency response to be equiripple (in both the passband and the stopband) within the following tolerances:

$$0.99 \leq |H_{\text{eff}}(j\Omega)| \leq 1.01, \qquad |\Omega| \leq 2\pi(1000),$$

$$|H_{\text{eff}}(j\Omega)| \leq 0.01, \qquad 2\pi(2000) \leq |\Omega| \leq 2\pi(5000).$$

We want to achieve this by designing an optimum linear-phase filter (using the Parks–McClellan algorithm) that includes compensation for the zero-order hold. Give an equation for the ideal response $H_d(e^{j\omega})$ that should be used. Find and sketch the weighting function $W(\omega)$ that should be used. Sketch a "typical" frequency response $H(e^{j\omega})$ that might result.
(d) How would you modify your results in part (c) to include magnitude compensation for a reconstruction filter $H_r(j\Omega)$ with zero gain above $\Omega = 2\pi(5000)$, but with sloping passband?

7.56. After a discrete-time signal is lowpass filtered, it is often downsampled or decimated, as depicted in Figure P7.56-1. Linear-phase FIR filters are frequently desirable in such applications, but if the lowpass filter in the figure has a narrow transition band, an FIR system will have a long impulse response and thus will require a large number of multiplications and additions per output sample.

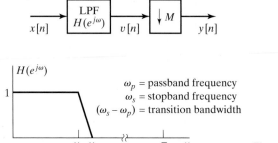

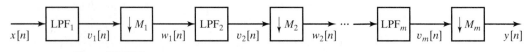

ω_p = passband frequency
ω_s = stopband frequency
$(\omega_s - \omega_p)$ = transition bandwidth

Figure P7.56-1

In this problem, we will study the merits of a multistage implementation of the system in Figure P7.56-1. Such implementations are particularly useful when ω_s is small and the decimation factor M is large. A general multistage implementation is depicted in Figure P7.56-2. The strategy is to use a wider transition band in the lowpass filters of the earlier stages, thereby reducing the length of the required filter impulse responses in those stages. As decimation occurs, the number of signal samples is reduced, and we can progressively decrease the widths of the transition bands of the filters that operate on the decimated signal. In this manner, the overall number of computations required to implement the decimator may be reduced.

Figure P7.56-2

(a) If no aliasing is to occur as a result of the decimation in Figure P7.56-1, what is the maximum allowable decimation factor M in terms of ω_s?

(b) Let $M = 100$, $\omega_s = \pi/100$, and $\omega_p = 0.9\pi/100$ in the system of Figure P7.56-2. If $x[n] = \delta[n]$, sketch $V(e^{j\omega})$ and $Y(e^{j\omega})$.

Now consider a two-stage implementation of the decimator for $M = 100$, as depicted in Figure P7.56-3, where $M_1 = 50$, $M_2 = 2$, $\omega_{p1} = 0.9\pi/100$, $\omega_{p2} = 0.9\pi/2$, and $\omega_{s2} = \pi/2$. We must choose ω_{s1} or, equivalently, the transition band of LPF_1, $(\omega_{s1} - \omega_{p1})$, such that the two-stage implementation yields the same equivalent passband and stopband frequencies as the single-stage decimator. (We are not concerned about the detailed shape of the frequency response in the transition band, except that both systems should have a monotonically decreasing response in the transition band.)

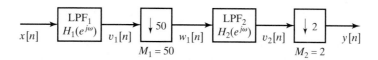

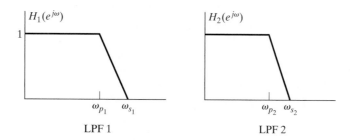

Figure P7.56-3

(c) For an arbitrary ω_{s1} and the input $x[n] = \delta[n]$, sketch $V_1(e^{j\omega})$, $W_1(e^{j\omega})$, $V_2(e^{j\omega})$, and $Y(e^{j\omega})$ for the two-stage decimator of Figure P7.56-3.

(d) Find the *largest* value of ω_{s1} such that the two-stage decimator yields the same equivalent passband and stopband cutoff frequencies as the single-stage system in part (b).

In addition to possessing a nonzero transition bandwidth, the lowpass filters must differ from the ideal by passband and stopband approximation errors of δ_p and δ_s, respectively. Assume that linear-phase equiripple FIR approximations are used. It follows from Eq. (7.117) that, for optimum lowpass filters,

$$N \approx \frac{-10\log_{10}(\delta_p\delta_s) - 13}{2.324\Delta\omega} + 1, \tag{P7.56-1}$$

where N is the length of the impulse response and $\Delta\omega = \omega_s - \omega_p$ is the transition band of the lowpass filter. Equation P7.56-1 provides the basis for comparing the two implementations of the decimator. Equation (7.76) could be used in place of Eq. (P7.56-1) to estimate the impulse-response length if the filters are designed by the Kaiser window method.

(e) Assume that $\delta_p = 0.01$ and $\delta_s = 0.001$ for the lowpass filter in the single-stage implementation. Compute the length N of the impulse response of the lowpass filter, and determine the number of multiplications required to compute each sample of the output. Take advantage of the symmetry of the impulse response of the linear-phase FIR system. (Note that in this decimation application, only every M^{th} sample of the output need be computed; i.e., the compressor commutes with the multiplications of the FIR system.)

(f) Using the value of ω_{s1} found in part (d), compute the impulse response lengths N_1 and N_2 of LPF_1 and LPF_2, respectively, in the two-stage decimator of Figure P7.56-3. Determine the total number of multiplications required to compute each sample of the output in the two-stage decimator.

(g) If the approximation error specifications $\delta_p = 0.01$ and $\delta_s = 0.001$ are used for both filters in the two-stage decimator, the overall passband ripple may be greater than 0.01, since the passband ripples of the two stages can reinforce each other; e.g., $(1 + \delta_p)(1 + \delta_p) > (1 + \delta_p)$. To compensate for this, the filters in the two-stage implementation can each be designed to have only one-half the passband ripple of the single-stage implementation. Therefore, assume that $\delta_p = 0.005$ and $\delta_s = 0.001$ for each filter in the two-stage decimator. Calculate the impulse response lengths N_1 and N_2 of LPF_1 and LPF_2, respectively, and determine the total number of multiplications required to compute each sample of the output.

(h) Should we also reduce the specification on the stopband approximation error for the filters in the two-stage decimator?

(i) *Optional.* The combination of $M_1 = 50$ and $M_2 = 2$ may not yield the smallest total number of multiplications per output sample. Other integer choices for M_1 and M_2 are possible such that $M_1 M_2 = 100$. Determine the values of M_1 and M_2 that minimize the number of multiplications per output sample.

7.57. In this problem, we develop a technique for designing discrete-time filters with minimum phase. Such filters have all their poles and zeros inside (or on) the unit circle. (We will allow zeros on the unit circle.) Let us first consider the problem of converting a type I linear-phase FIR equiripple lowpass filter to a minimum-phase system. If $H(e^{j\omega})$ is the frequency response of a type I linear-phase filter, then

1. The corresponding impulse response

$$h[n] = \begin{cases} h[M-n], & 0 \le n \le M, \\ 0, & \text{otherwise,} \end{cases}$$

is real and M is an even integer.

2. It follows from part 1 that $H(e^{j\omega}) = A_e(e^{j\omega})e^{-j\omega n_0}$, where $A_e(e^{j\omega})$ is real and $n_0 = M/2$ is an integer.

3. The passband ripple is δ_1; i.e., in the passband, $A_e(e^{j\omega})$ oscillates between $(1+\delta_1)$ and $(1-\delta_1)$. (See Figure P7.57-1.)

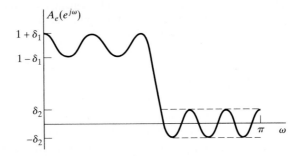

Figure P7.57-1

4. The stopband ripple is δ_2; i.e., in the stopband, $-\delta_2 \le A_e(e^{j\omega}) \le \delta_2$, and $A_e(e^{j\omega})$ oscillates between $-\delta_2$ and $+\delta_2$. (See Figure P7.57-1.)

The following technique was proposed by Herrmann and Schüssler (1970a) for converting this linear-phase system into a minimum-phase system that has a system function $H_{\min}(z)$ and unit sample response $h_{\min}[n]$ (in this problem, we assume that minimum-phase systems can have zeros *on* the unit circle):

Step 1. Create a new sequence

$$h_1[n] = \begin{cases} h[n], & n \ne n_0, \\ h[n_0] + \delta_2, & n = n_0. \end{cases}$$

Step 2. Recognize that $H_1(z)$ can be expressed in the form

$$H_1(z) = z^{-n_0} H_2(z) H_2(1/z) = z^{-n_0} H_3(z)$$

for some $H_2(z)$, where $H_2(z)$ has all its poles and zeros inside or on the unit circle and $h_2[n]$ is real.

Step 3. Define

$$H_{\min}(z) = \frac{H_2(z)}{a}.$$

The denominator constant where $a = (\sqrt{1 - \delta_1 + \delta_2} + \sqrt{1 + \delta_1 + \delta_2})/2$ normalizes the passband so that the resulting frequency response $H_{\min}(e^{j\omega})$ will oscillate about a value of unity.

(a) Show that if $h_1[n]$ is chosen as in Step 1, then $H_1(e^{j\omega})$ can be written as

$$H_1(e^{j\omega}) = e^{-j\omega n_0} H_3(e^{j\omega}),$$

where $H_3(e^{j\omega})$ is real and nonnegative for all values of ω.

(b) If $H_3(e^{j\omega}) \geq 0$, as was shown in part (a), show that there exists an $H_2(z)$ such that

$$H_3(z) = H_2(z)H_2(1/z),$$

where $H_2(z)$ is a minimum-phase system function and $h_2[n]$ is real (i.e., justify Step 2).

(c) Demonstrate that the new filter $H_{\min}(e^{j\omega})$ is an equiripple lowpass filter (i.e., that its magnitude characteristic is of the form shown in Figure P7.57-2) by evaluating δ_1' and δ_2'. What is the length of the new impulse response $h_{\min}[n]$?

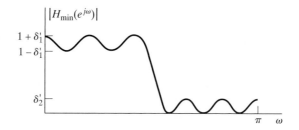

Figure P7.57-2

(d) In parts (a), (b), and (c), we assumed that we started with a type I FIR linear-phase filter. Will this technique work if we remove the linear-phase constraint? Will it work if we use a type II FIR linear-phase system?

7.58. Suppose that we have a program that finds the set of coefficients $a[n]$, $n = 0, 1, \ldots, L$, that minimizes

$$\max_{\omega \in F} \left\{ \left| W(\omega) \left[H_d(e^{j\omega}) - \sum_{n=0}^{L} a[n] \cos \omega n \right] \right| \right\},$$

given L, F, $W(\omega)$, and $H_d(e^{j\omega})$. We have shown that the solution to this optimization problem implies a noncausal FIR zero-phase system with impulse response satisfying $h_e[n] = h_e[-n]$. By delaying $h_e[n]$ by L samples, we obtain a causal type I FIR linear-phase system with frequency response

$$H(e^{j\omega}) = e^{-j\omega M/2} \sum_{n=0}^{L} a[n] \cos \omega n = \sum_{n=0}^{2L} h[n]e^{-j\omega n},$$

where the impulse response is related to the coefficients $a[n]$ by

$$a[n] = \begin{cases} 2h[M/2 - n] & \text{for } 1 \leq n \leq L, \\ h[M/2] & \text{for } n = 0, \end{cases}$$

and $M = 2L$ is the order of the system function polynomial. (The length of the impulse response is $M + 1$.)

The other three types (II, III, and IV) of linear-phase FIR filters can be designed by the available program if we make suitable modifications to the weighting function $W(\omega)$

and the desired frequency response $H_d(e^{j\omega})$. To see how to do this, it is necessary to manipulate the expressions for the frequency response into the standard form assumed by the program.

(a) Assume that we wish to design a causal type II FIR linear-phase system such that $h[n] = h[M - n]$ for $n = 0, 1, \ldots, M$, where M is an odd integer. Show that the frequency response of this type of system can be expressed as

$$H(e^{j\omega}) = e^{-j\omega M/2} \sum_{n=1}^{(M+1)/2} b[n] \cos\omega\left(n - \tfrac{1}{2}\right),$$

and determine the relationship between the coefficients $b[n]$ and $h[n]$.

(b) Show that the summation

$$\sum_{n=1}^{(M+1)/2} b[n] \cos\omega\left(n - \tfrac{1}{2}\right)$$

can be written as

$$\cos(\omega/2) \sum_{n=0}^{(M-1)/2} \tilde{b}[n] \cos\omega n$$

by obtaining an expression for $b[n]$ for $n = 1, 2, \ldots, (M+1)/2$ in terms of $\tilde{b}[n]$ for $n = 0, 1, \ldots, (M-1)/2$. *Hint:* Note carefully that $b[n]$ is to be expressed in terms of $\tilde{b}[n]$. Also, use the trigonometric identity $\cos\alpha\cos\beta = \tfrac{1}{2}\cos(\alpha + \beta) + \tfrac{1}{2}\cos(\alpha - \beta)$.

(c) If we wish to use the given program to design type II systems (M odd) for a given F, $W(\omega)$, and $H_d(e^{j\omega})$, show how to obtain $\tilde{L}$, $\tilde{F}$, $\tilde{W}(\omega)$, and $\tilde{H}_d(e^{j\omega})$ in terms of M, F, $W(\omega)$, and $H_d(e^{j\omega})$ such that if we run the program using $\tilde{L}$, $\tilde{F}$, $\tilde{W}(\omega)$, and $\tilde{H}_d(e^{j\omega})$, we may use the resulting set of coefficients to determine the impulse response of the desired type II system.

(d) Parts (a)–(c) can be repeated for types III and IV causal linear-phase FIR systems where $h[n] = -h[M - n]$. For these cases, you must show that, for type III systems (M even), the frequency response can be expressed as

$$H(e^{j\omega}) = e^{-j\omega M/2} \sum_{n=1}^{M/2} c[n] \sin\omega n$$

$$= e^{-j\omega M/2} \sin\omega \sum_{n=0}^{(M-2)/2} \tilde{c}[n] \cos\omega n,$$

and for type IV systems (M odd),

$$H(e^{j\omega}) = e^{-j\omega M/2} \sum_{n=1}^{(M+1)/2} d[n] \sin\omega\left(n - \tfrac{1}{2}\right)$$

$$= e^{-j\omega M/2} \sin(\omega/2) \sum_{n=0}^{(M-1)/2} \tilde{d}[n] \cos\omega n.$$

As in part (b), it is necessary to express $c[n]$ in terms of $\tilde{c}[n]$ and $d[n]$ in terms of $\tilde{d}[n]$ using the trigonometric identity $\sin\alpha\cos\beta = \tfrac{1}{2}\sin(\alpha + \beta) + \tfrac{1}{2}\sin(\alpha - \beta)$. McClellan and Parks (1973) and Rabiner and Gold (1975) give more details on issues raised in this problem.

7.59. In this problem, we consider a method of obtaining an implementation of a variable-cutoff linear-phase filter. Assume that we are given a zero-phase filter designed by the Parks–McClellan method. The frequency response of this filter can be represented as

$$A_e(e^{j\theta}) = \sum_{k=0}^{L} a_k (\cos\theta)^k,$$

and its system function can therefore be represented as

$$A_e(Z) = \sum_{k=0}^{L} a_k \left(\frac{Z + Z^{-1}}{2}\right)^k,$$

with $e^{j\theta} = Z$. (We use Z for the original system and z for the system to be obtained by transformation of the original system.)

(a) Using the preceding expression for the system function, draw a block diagram or flow graph of an implementation of the system that utilizes multiplications by the coefficients a_k, additions, and elemental systems having system function $(Z + Z^{-1})/2$.

(b) What is the length of the impulse response of the system? The overall system can be made causal by cascading the system with a delay of L samples. Distribute this delay as unit delays so that all parts of the network will be causal.

(c) Suppose that we obtain a new system function from $A_e(Z)$ by the substitution

$$B_e(z) = A_e(Z)\big|_{(Z+Z^{-1})/2=\alpha_0+\alpha_1[(z+z^{-1})/2]}.$$

Using the flow graph obtained in part (a), draw the flow graph of a system that implements the system function $B_e(z)$. What is the length of the impulse response of this system? Modify the network as in part (b) to make the overall system and all parts of the network causal.

(d) If $A_e(e^{j\theta})$ is the frequency response of the original filter and $B_e(e^{j\omega})$ is the frequency response of the transformed filter, determine the relationship between θ and ω.

(e) The frequency response of the original optimal filter is shown in Figure P7.59. For the case $\alpha_1 = 1 - \alpha_0$ and $0 \leq \alpha_0 < 1$, describe how the frequency response $B_e(e^{j\omega})$ changes as α_0 varies. *Hint:* Plot $A_e(e^{j\theta})$ and $B_e(e^{j\omega})$ as functions of $\cos\theta$ and $\cos\omega$. Are the resulting transformed filters also optimal in the sense of having the minimum maximum weighted approximation errors in the transformed passband and stopband?

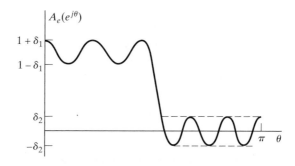

Figure P7.59

(f) *Optional.* Repeat part (e) for the case $\alpha_1 = 1 + \alpha_0$ and $-1 < \alpha_0 \leq 0$.

7.60. In this problem, we consider the effect of mapping continuous-time filters to discrete-time filters by replacing derivatives in the differential equation for a continuous-time filter by central differences to obtain a difference equation. The first central difference of a sequence $x[n]$ is defined as

$$\Delta^{(1)}\{x[n]\} = x[n+1] - x[n-1],$$

and the kth central difference is defined recursively as

$$\Delta^{(k)}\{x[n]\} = \Delta^{(1)}\{\Delta^{(k-1)}\{x[n]\}\}.$$

For consistency, the zeroth central difference is defined as

$$\Delta^{(0)}\{x[n]\} = x[n].$$

(a) If $X(z)$ is the z-transform of $x[n]$, determine the z-transform of $\Delta^{(k)}\{x[n]\}$.

The mapping of an LTI continuous-time filter to an LTI discrete-time filter is as follows: Let the continuous-time filter with input $x(t)$ and output $y(t)$ be specified by a differential equation of the form

$$\sum_{k=0}^{N} a_k \frac{d^k y(t)}{dt^k} = \sum_{r=0}^{M} b_r \frac{d^r x(t)}{dt^r}.$$

Then the corresponding discrete-time filter with input $x[n]$ and output $y[n]$ is specified by the difference equation

$$\sum_{k=0}^{N} a_k \Delta^{(k)}\{y[n]\} = \sum_{r=0}^{M} b_r \Delta^{(r)}\{x[n]\}.$$

(b) If $H_c(s)$ is a rational continuous-time system function and $H_d(z)$ is the discrete-time system function obtained by mapping the differential equation to a difference equation as indicated in part (a), then

$$H_d(z) = H_c(s)\big|_{s=m(z)}.$$

Determine $m(z)$.

(c) Assume that $H_c(s)$ approximates a continuous-time lowpass filter with a cutoff frequency of $\Omega = 1$; i.e.,

$$H(j\Omega) \approx \begin{cases} 1, & |\Omega| < 1, \\ 0, & \text{otherwise.} \end{cases}$$

This filter is mapped to a discrete-time filter using central differences as discussed in part (a). Sketch the approximate frequency response that you would expect for the discrete-time filter, assuming that it is stable.

7.61. Let $h[n]$ be the optimal type I equiripple lowpass filter shown in Figure P7.61, designed with weighting function $W(e^{j\omega})$ and desired frequency response $H_d(e^{j\omega})$. For simplicity, assume that the filter is zero phase (i.e., noncausal). We will use $h[n]$ to design five different FIR filters as follows:

$$h_1[n] = h[-n],$$

$$h_2[n] = (-1)^n h[n],$$

$$h_3[n] = h[n] * h[n],$$

$$h_4[n] = h[n] - K\delta[n], \text{ where } K \text{ is a constant,}$$

$$h_5[n] = \begin{cases} h[n/2] & \text{for } n \text{ even,} \\ 0 & \text{otherwise.} \end{cases}$$

For each filter $h_i[n]$, determine whether $h_i[n]$ is optimal in the minimax sense. That is, determine whether

$$h_i[n] = \min_{h_i[n]} \max_{\omega \in F} \left(W(e^{j\omega})|H_d(e^{j\omega}) - H_i(e^{j\omega})| \right)$$

for some choices of a piecewise-constant $H_d(e^{j\omega})$ and a piecewise-constant $W(e^{j\omega})$, where F is a union of disjoint closed intervals on $0 \le \omega \le \pi$. If $h_i[n]$ is optimal, determine the corresponding $H_d(e^{j\omega})$ and $W(e^{j\omega})$. If $h_i[n]$ is not optimal, explain why.

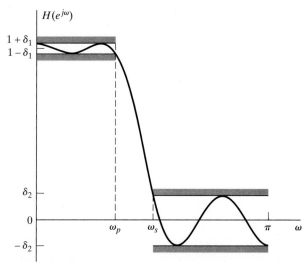

Figure P7.61

7.62. Suppose that you have used the Parks–McClellan algorithm to design a causal FIR linear-phase system. The system function of this system is denoted $H(z)$. The length of the impulse response is 25 samples, $h[n] = 0$ for $n < 0$ and for $n > 24$, and $h[0] \ne 0$. For each of the following questions, answer "true," "false," or "insufficient information given":

(a) $h[n + 12] = h[12 - n]$ or $h[n + 12] = -h[12 - n]$ for $-\infty < n < \infty$.
(b) The system has a stable and causal inverse.
(c) We know that $H(-1) = 0$.
(d) The maximum weighted approximation error is the same in all approximation bands.
(e) The system can be implemented by a signal flow graph that has no feedback paths.
(f) The group delay is positive for $0 < \omega < \pi$.

7.63. Consider the design of a type I bandpass linear-phase FIR filter using the Parks–McClellan algorithm. The impulse response length is $M+1 = 2L+1$. Recall that for type I systems, the frequency response is of the form $H(e^{j\omega}) = A_e(e^{j\omega})e^{-j\omega M/2}$, and the Parks–McClellan algorithm finds the function $A_e(e^{j\omega})$ that minimizes the maximum value of the error function

$$E(\omega) = W(\omega)[H_d(e^{j\omega}) - A_e(e^{j\omega})], \qquad \omega \in F,$$

where F is a closed subset of the interval $0 \le \omega \le \pi$, $W(\omega)$ is a weighting function, and $H_d(e^{j\omega})$ defines the desired frequency response in the approximation intervals F. The tolerance scheme for a bandpass filter is shown in Figure P7.63.

(a) Give the equation for the desired response $H_d(e^{j\omega})$ for the tolerance scheme in Figure P7.63.

(b) Give the equation for the weighting function $W(\omega)$ for the tolerance scheme in Figure P7.63.

(c) What is the *minimum* number of alternations of the error function for the optimum filter?

(d) What is the *maximum* number of alternations of the error function for the optimum filter?

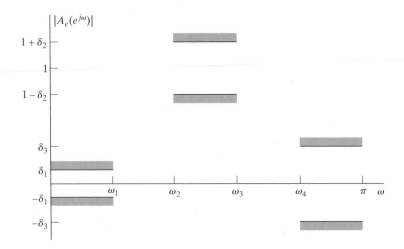

Figure P7.63

(e) Sketch a "typical" weighted error function $E(\omega)$ that could be the error function for an optimum bandpass filter if $M = 14$. Assume the *maximum* number of alternations.

(f) Now suppose that $M, \omega_1, \omega_2, \omega_3$, the weighting function, and the desired function are kept the same, but ω_4 is *increased,* so that the transition band ($\omega_4 - \omega_3$) is increased. Will the optimum filter for these new specifications *necessarily* have a *smaller* value of the maximum approximation error than the optimum filter associated with the original specifications? Clearly show your reasoning.

(g) In the lowpass filter case, all local minima and maxima of $A_e(e^{j\omega})$ must occur in the approximation bands $\omega \in F$; they *cannot* occur in the "don't care" bands. Also, in the lowpass case, the local minima and maxima that occur in the approximation bands must be alternations of the error. Show that this is not necessarily true in the bandpass filter case. Specifically, use the alternation theorem to show (i) that local maxima and minima of $A_e(e^{j\omega})$ are not restricted to the approximation bands and (ii) that local maxima and minima in the approximation bands need not be alternations.

7.64. It is often desirable to transform a prototype discrete-time lowpass filter to another kind of discrete-time frequency-selective filter. In particular, the impulse invariance approach cannot be used to convert continuous-time highpass or bandstop filters to discrete-time highpass or bandstop filters. Consequently, the traditional approach has been to design a prototype lowpass discrete-time filter using either impulse invariance or the bilinear transformation and then to use an algebraic transformation to convert the discrete-time lowpass filter into the desired frequency-selective filter.

To see how this is done, assume that we are given a lowpass system function $H_{lp}(Z)$ that we wish to transform to a new system function $H(z)$, which has either lowpass, highpass, bandpass, or bandstop characteristics when it is evaluated on the unit circle. Note that we associate the complex variable Z with the prototype lowpass filter and the complex variable z with the transformed filter. Then, we define a mapping from the Z-plane to the z-plane of the form

$$Z^{-1} = G(z^{-1}) \qquad \text{(P7.64-1)}$$

such that

$$H(z) = H_{lp}(Z)\big|_{Z^{-1}=G(z^{-1})}. \qquad \text{(P7.64-2)}$$

Instead of expressing Z as a function of z, we have assumed in Eq. (P7.64-1) that Z^{-1} is expressed as a function of z^{-1}. Thus, according to Eq. (P7.64-2), in obtaining $H(z)$ from $H_{lp}(Z)$, we simply replace Z^{-1} everywhere in $H_{lp}(Z)$ by the function $G(z^{-1})$. This is a convenient representation, because $H_{lp}(Z)$ is normally expressed as a rational function of z^{-1}.

   If $H_{lp}(Z)$ is the rational system function of a causal and stable system, we naturally require that the transformed system function $H(z)$ be a rational function of z^{-1} and that the system also be causal and stable. This places the following constraints on the transformation $Z^{-1} = G(z^{-1})$:

   1. $G(z^{-1})$ must be a rational function of z^{-1}.
   2. The inside of the unit circle of the Z-plane must map to the inside of the unit circle of the z-plane.
   3. The unit circle of the Z-plane must map onto the unit circle of the z-plane.

In this problem, you will derive and characterize the algebraic transformations necessary to convert a discrete-time lowpass filter into another lowpass filter with a different cutoff frequency or to a discrete-time highpass filter.

(a) Let θ and ω be the frequency variables (angles) in the Z-plane and z-plane, respectively, i.e., on the respective unit circles $Z = e^{j\theta}$ and $z = e^{j\omega}$. Show that, for Condition 3 to hold, $G(z^{-1})$ must be an all-pass system, i.e.,

$$|G(e^{-j\omega})| = 1. \qquad \text{(P7.64-3)}$$

(b) It is possible to show that the most general form of $G(z^{-1})$ that satisfies all of the preceding three conditions is

$$Z^{-1} = G(z^{-1}) = \pm \prod_{k=1}^{N} \frac{z^{-1} - \alpha_k}{1 - \alpha_k z^{-1}}. \qquad \text{(P7.64-4)}$$

From our discussion of all-pass systems in Chapter 5, it should be clear that $G(z^{-1})$, as given in Eq. (P7.64-4), satisfies Eq. (P7.64-3), i.e., is an allpass system, and thus meets Condition 3. Eq. (P7.64-4) also clearly meets Condition 1. Demonstrate that Condition 2 is satisfied if and only if $|\alpha_k| < 1$.

(c) A simple 1st-order $G(z^{-1})$ can be used to map a prototype lowpass filter $H_{lp}(Z)$ with cutoff θ_p to a new filter $H(z)$ with cutoff ω_p. Demonstrate that

$$G(z^{-1}) = \frac{z^{-1} - \alpha}{1 - \alpha z^{-1}}$$

will produce the desired mapping for some value of α. Solve for α as a function of θ_p and ω_p. Problem 7.51 uses this approach to design lowpass filters with adjustable cutoff frequencies.

(d) Consider the case of a prototype lowpass filter with $\theta_p = \pi/2$. For each of the following choices of α, specify the resulting cutoff frequency ω_p for the transformed filter:
   (i) $\alpha = -0.2679$.
   (ii) $\alpha = 0$.
   (iii) $\alpha = 0.4142$.

(e) It is also possible to find a 1st-order all-pass system for $G(z^{-1})$ such that the prototype lowpass filter is transformed to a discrete-time highpass filter with cutoff ω_p. Note that such a transformation must map $Z^{-1} = e^{j\theta_p} \rightarrow z^{-1} = e^{j\omega_p}$ and also map $Z^{-1} = 1 \rightarrow z^{-1} = -1$; i.e., $\theta = 0$ maps to $\omega = \pi$. Find $G(z^{-1})$ for this transformation, and also, find an expression for α in terms of θ_p and ω_p.

(f) Using the same prototype filter and values for α as in part (d), sketch the frequency responses for the highpass filters resulting from the transformation you specified in part (e).

Similar, but more complicated, transformations can be used to convert the prototype lowpass filter $H_{lp}(Z)$ into bandpass and bandstop filters. Constantinides (1970) describes these transformations in more detail.

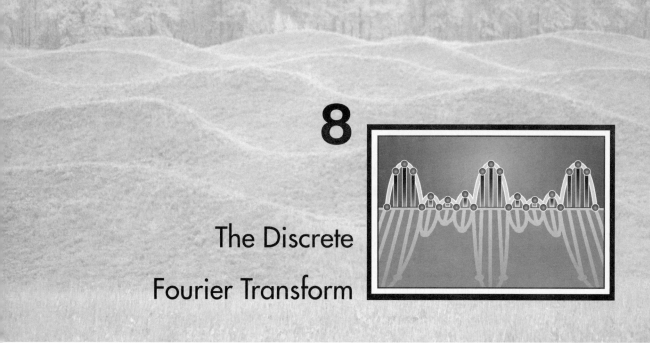

8 The Discrete Fourier Transform

8.0 INTRODUCTION

In Chapters 2 and 3, we discussed the representation of sequences and LTI systems in terms of the discrete-time Fourier and z-transforms, respectively. For finite-duration sequences, there is an alternative discrete-time Fourier representation, referred to as the *discrete Fourier transform* (DFT). The DFT is itself a sequence rather than a function of a continuous variable, and it corresponds to samples, equally spaced in frequency, of the DTFT of the signal. In addition to its theoretical importance as a Fourier representation of sequences, the DFT plays a central role in the implementation of a variety of digital signal-processing algorithms. This is because efficient algorithms exist for the computation of the DFT. These algorithms will be discussed in detail in Chapter 9. The application of the DFT to spectrum analysis will be described in Chapter 10.

Although several points of view can be taken toward the derivation and interpretation of the DFT representation of a finite-duration sequence, we have chosen to base our presentation on the relationship between periodic sequences and finite-length sequences. We begin by considering the Fourier series representation of periodic sequences. Although this representation is important in its own right, we are most often interested in the application of Fourier series results to the representation of finite-length sequences. We accomplish this by constructing a periodic sequence for which each period is identical to the finite-length sequence. The Fourier series representation of the periodic sequence then corresponds to the DFT of the finite-length sequence. Thus, our approach is to define the Fourier series representation for periodic sequences and to study the properties of such representations. Then, we repeat essentially the same derivations, assuming that the sequence to be represented is a finite-length sequence.

This approach to the DFT emphasizes the fundamental inherent periodicity of the DFT representation and ensures that this periodicity is not overlooked in applications of the DFT.

8.1 REPRESENTATION OF PERIODIC SEQUENCES: THE DISCRETE FOURIER SERIES

Consider a sequence $\tilde{x}[n]$ that is periodic[1] with period N, so that $\tilde{x}[n] = \tilde{x}[n + rN]$ for any integer values of n and r. As with continuous-time periodic signals, such a sequence can be represented by a Fourier series corresponding to a sum of harmonically related complex exponential sequences, i.e., complex exponentials with frequencies that are integer multiples of the fundamental frequency $(2\pi/N)$ associated with the periodic sequence $\tilde{x}[n]$. These periodic complex exponentials are of the form

$$e_k[n] = e^{j(2\pi/N)kn} = e_k[n + rN], \tag{8.1}$$

where k is any integer, and the Fourier series representation then has the form[2]

$$\tilde{x}[n] = \frac{1}{N} \sum_k \tilde{X}[k] e^{j(2\pi/N)kn}. \tag{8.2}$$

The Fourier series representation of a continuous-time periodic signal generally requires infinitely many harmonically related complex exponentials, whereas the Fourier series for any discrete-time signal with period N requires only N harmonically related complex exponentials. To see this, note that the harmonically related complex exponentials $e_k[n]$ in Eq. (8.1) are identical for values of k separated by N; i.e., $e_0[n] = e_N[n]$, $e_1[n] = e_{N+1}[n]$, and, in general,

$$e_{k+\ell N}[n] = e^{j(2\pi/N)(k+\ell N)n} = e^{j(2\pi/N)kn} e^{j2\pi \ell n} = e^{j(2\pi/N)kn} = e_k[n], \tag{8.3}$$

where ℓ is any integer. Consequently, the set of N periodic complex exponentials $e_0[n]$, $e_1[n], \ldots, e_{N-1}[n]$ defines all the distinct periodic complex exponentials with frequencies that are integer multiples of $(2\pi/N)$. Thus, the Fourier series representation of a periodic sequence $\tilde{x}[n]$ need contain only N of these complex exponentials. For notational convenience, we choose k in the range of 0 to $N - 1$; hence, Eq. (8.2) has the form

$$\tilde{x}[n] = \frac{1}{N} \sum_{k=0}^{N-1} \tilde{X}[k] e^{j(2\pi/N)kn}. \tag{8.4}$$

However, choosing k to range over any full period of $\tilde{X}[k]$ would be equally valid.

To obtain the sequence of Fourier series coefficients $\tilde{X}[k]$ from the periodic sequence $\tilde{x}[n]$, we exploit the orthogonality of the set of complex exponential sequences.

---

[1] Henceforth, we will use the tilde (˜) to denote periodic sequences whenever it is important to clearly distinguish between periodic and aperiodic sequences.

[2] The multiplicative constant $1/N$ is included in Eq. (8.2) for convenience. It could also be absorbed into the definition of $\tilde{X}[k]$.

After multiplying both sides of Eq. (8.4) by $e^{-j(2\pi/N)rn}$ and summing from $n = 0$ to $n = N - 1$, we obtain

$$\sum_{n=0}^{N-1} \tilde{x}[n]e^{-j(2\pi/N)rn} = \sum_{n=0}^{N-1} \frac{1}{N} \sum_{k=0}^{N-1} \tilde{X}[k]e^{j(2\pi/N)(k-r)n}. \tag{8.5}$$

After interchanging the order of summation on the right-hand side, Eq. (8.5) becomes

$$\sum_{n=0}^{N-1} \tilde{x}[n]e^{-j(2\pi/N)rn} = \sum_{k=0}^{N-1} \tilde{X}[k] \left[\frac{1}{N} \sum_{n=0}^{N-1} e^{j(2\pi/N)(k-r)n} \right]. \tag{8.6}$$

The following identity expresses the orthogonality of the complex exponentials:

$$\frac{1}{N} \sum_{n=0}^{N-1} e^{j(2\pi/N)(k-r)n} = \begin{cases} 1, & k - r = mN, \quad m \text{ an integer,} \\ 0, & \text{otherwise.} \end{cases} \tag{8.7}$$

This identity can easily be proved (see Problem 8.54), and when it is applied to the summation in brackets in Eq. (8.6), the result is

$$\sum_{n=0}^{N-1} \tilde{x}[n]e^{-j(2\pi/N)rn} = \tilde{X}[r]. \tag{8.8}$$

Thus, the Fourier series coefficients $\tilde{X}[k]$ in Eq. (8.4) are obtained from $\tilde{x}[n]$ by the relation

$$\tilde{X}[k] = \sum_{n=0}^{N-1} \tilde{x}[n]e^{-j(2\pi/N)kn}. \tag{8.9}$$

Note that the sequence $\tilde{X}[k]$ defined in Eq. (8.9) is also periodic with period N if Eq. (8.9) is evaluated outside the range $0 \le k \le N - 1$; i.e., $\tilde{X}[0] = \tilde{X}[N]$, $\tilde{X}[1] = \tilde{X}[N+1]$, and, more generally,

$$\tilde{X}[k + N] = \sum_{n=0}^{N-1} \tilde{x}[n]e^{-j(2\pi/N)(k+N)n}$$

$$= \left(\sum_{n=0}^{N-1} \tilde{x}[n]e^{-j(2\pi/N)kn} \right) e^{-j2\pi n} = \tilde{X}[k],$$

for any integer k.

The Fourier series coefficients can be interpreted to be a sequence of finite length, given by Eq. (8.9) for $k = 0, \ldots, (N-1)$, and zero otherwise, or as a periodic sequence defined for all k by Eq. (8.9). Clearly, both of these interpretations are acceptable, since in Eq. (8.4) we use only the values of $\tilde{X}[k]$ for $0 \le k \le (N-1)$. An advantage to interpreting the Fourier series coefficients $\tilde{X}[k]$ as a periodic sequence is that there is then a duality between the time and frequency domains for the Fourier series representation of periodic sequences. Equations (8.9) and (8.4) together are an analysis–synthesis pair and will be referred to as the *discrete Fourier series* (DFS) representation of a periodic sequence.

For convenience in notation, these equations are often written in terms of the complex quantity

$$W_N = e^{-j(2\pi/N)}. \tag{8.10}$$

With this notation, the DFS analysis–synthesis pair is expressed as follows:

$$\text{Analysis equation:} \quad \tilde{X}[k] = \sum_{n=0}^{N-1} \tilde{x}[n]W_N^{kn}. \tag{8.11}$$

$$\text{Synthesis equation:} \quad \tilde{x}[n] = \frac{1}{N}\sum_{k=0}^{N-1} \tilde{X}[k]W_N^{-kn}. \tag{8.12}$$

In both of these equations, $\tilde{X}[k]$ and $\tilde{x}[n]$ are periodic sequences. We will sometimes find it convenient to use the notation

$$\tilde{x}[n] \overset{\mathcal{DFS}}{\longleftrightarrow} \tilde{X}[k] \tag{8.13}$$

to signify the relationships of Eqs. (8.11) and (8.12). The following examples illustrate the use of those equations.

Example 8.1   DFS of a Periodic Impulse Train

We consider the periodic impulse train

$$\tilde{x}[n] = \sum_{r=-\infty}^{\infty} \delta[n-rN] = \begin{cases} 1, & n=rN, \quad r \text{ any integer}, \\ 0, & \text{otherwise.} \end{cases} \tag{8.14}$$

Since $\tilde{x}[n] = \delta[n]$ for $0 \le n \le N-1$, the DFS coefficients are found, using Eq. (8.11), to be

$$\tilde{X}[k] = \sum_{n=0}^{N-1} \delta[n]W_N^{kn} = W_N^0 = 1. \tag{8.15}$$

In this case, $\tilde{X}[k] = 1$ for all k. Thus, substituting Eq. (8.15) into Eq. (8.12) leads to the representation

$$\tilde{x}[n] = \sum_{r=-\infty}^{\infty} \delta[n-rN] = \frac{1}{N}\sum_{k=0}^{N-1} W_N^{-kn} = \frac{1}{N}\sum_{k=0}^{N-1} e^{j(2\pi/N)kn}. \tag{8.16}$$

Example 8.1 produced a useful representation of a periodic impulse train in terms of a sum of complex exponentials, wherein all the complex exponentials have the same magnitude and phase and add to unity at integer multiples of N and to zero for all other integers. If we look closely at Eqs. (8.11) and (8.12), we see that the two equations are very similar, differing only in a constant multiplier and the sign of the exponents. This duality between the periodic sequence $\tilde{x}[n]$ and its DFS coefficients $\tilde{X}[k]$ is illustrated in the following example.

Example 8.2   Duality in the DFS

In this example, the DFS coefficients are a periodic impulse train:

$$\tilde{Y}[k] = \sum_{r=-\infty}^{\infty} N\delta[k - rN].$$

Substituting $\tilde{Y}[k]$ into Eq. (8.12) gives

$$\tilde{y}[n] = \frac{1}{N} \sum_{k=0}^{N-1} N\delta[k]W_N^{-kn} = W_N^{-0} = 1.$$

In this case, $\tilde{y}[n] = 1$ for all n. Comparing this result with the results for $\tilde{x}[n]$ and $\tilde{X}[k]$ of Example 8.1, we see that $\tilde{Y}[k] = N\tilde{x}[k]$ and $\tilde{y}[n] = \tilde{X}[n]$. In Section 8.2.3, we will show that this example is a special case of a more general duality property.

If the sequence $\tilde{x}[n]$ is equal to unity over only part of one period, we can also obtain a closed-form expression for the DFS coefficients. This is illustrated by the following example.

Example 8.3   The DFS of a Periodic Rectangular Pulse Train

For this example, $\tilde{x}[n]$ is the sequence shown in Figure 8.1, whose period is $N = 10$. From Eq. (8.11),

$$\tilde{X}[k] = \sum_{n=0}^{4} W_{10}^{kn} = \sum_{n=0}^{4} e^{-j(2\pi/10)kn}. \tag{8.17}$$

This finite sum has the closed form

$$\tilde{X}[k] = \frac{1 - W_{10}^{5k}}{1 - W_{10}^{k}} = e^{-j(4\pi k/10)} \frac{\sin(\pi k/2)}{\sin(\pi k/10)}. \tag{8.18}$$

The magnitude and phase of the periodic sequence $\tilde{X}[k]$ are shown in Figure 8.2.

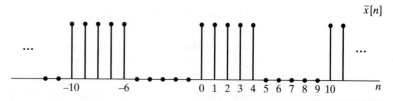

Figure 8.1 Periodic sequence with period $N = 10$ for which the Fourier series representation is to be computed.

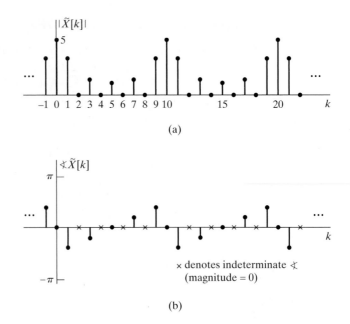

Figure 8.2   Magnitude and phase of the Fourier series coefficients of the sequence of Figure 8.1.

We have shown that any periodic sequence can be represented as a sum of complex exponential sequences. The key results are summarized in Eqs. (8.11) and (8.12). As we will see, these relationships are the basis for the DFT, which focuses on finite-length sequences. Before discussing the DFT, however, we will consider some of the basic properties of the DFS representation of periodic sequences in Section 8.2, and then, in Section 8.3, we will show how we can use the DFS representation to obtain a DTFT representation of periodic signals.

8.2 PROPERTIES OF THE DFS

Just as with Fourier series and Fourier and Laplace transforms for continuous-time signals, and with discrete-time Fourier and z-transforms for nonperiodic sequences, certain properties of the DFS are of fundamental importance to its successful use in signal-processing problems. In this section, we summarize these important properties. It is not surprising that many of the basic properties are analogous to properties of the z-transform and DTFT. However, we will be careful to point out where the periodicity of both $\tilde{x}[n]$ and $\tilde{X}[k]$ results in some important distinctions. Furthermore, an exact duality exists between the time and frequency domains in the DFS representation that does not exist in the DTFT and z-transform representation of sequences.

8.2.1 Linearity

Consider two periodic sequences $\tilde{x}_1[n]$ and $\tilde{x}_2[n]$, both with period N, such that

$$\tilde{x}_1[n] \overset{\mathcal{DFS}}{\longleftrightarrow} \tilde{X}_1[k], \tag{8.19a}$$

and

$$\tilde{x}_2[n] \overset{\mathcal{DFS}}{\longleftrightarrow} \tilde{X}_2[k]. \tag{8.19b}$$

Then

$$a\tilde{x}_1[n] + b\tilde{x}_2[n] \overset{\mathcal{DFS}}{\longleftrightarrow} a\tilde{X}_1[k] + b\tilde{X}_2[k]. \tag{8.20}$$

This linearity property follows immediately from the form of Eqs. (8.11) and (8.12).

8.2.2 Shift of a Sequence

If a periodic sequence $\tilde{x}[n]$ has Fourier coefficients $\tilde{X}[k]$, then $\tilde{x}[n - m]$ is a shifted version of $\tilde{x}[n]$, and

$$\tilde{x}[n - m] \overset{\mathcal{DFS}}{\longleftrightarrow} W_N^{km}\tilde{X}[k]. \tag{8.21}$$

The proof of this property is considered in Problem 8.55. Note that any shift that is greater than or equal to the period (i.e., $m \geq N$) cannot be distinguished in the time domain from a shorter shift m_1 such that $m = m_1 + m_2 N$, where m_1 and m_2 are integers and $0 \leq m_1 \leq N - 1$. (Another way of stating this is that $m_1 = m$ modulo N or, equivalently, m_1 is the remainder when m is divided by N.) It is easily shown that with this representation of m, $W_N^{km} = W_N^{km_1}$; i.e., as it must be, the ambiguity of the shift in the time domain is also manifest in the frequency-domain representation.

Because the sequence of Fourier series coefficients of a periodic sequence is a periodic sequence, a similar result applies to a shift in the Fourier coefficients by an integer ℓ. Specifically,

$$W_N^{-n\ell}\tilde{x}[n] \overset{\mathcal{DFS}}{\longleftrightarrow} \tilde{X}[k - \ell]. \tag{8.22}$$

Note the difference in the sign of the exponents in Eqs. (8.21) and (8.22).

8.2.3 Duality

Because of the strong similarity between the Fourier analysis and synthesis equations in continuous time, there is a duality between the time domain and frequency domain. However, for the DTFT of aperiodic signals, no similar duality exists, since aperiodic signals and their Fourier transforms are very different kinds of functions: Aperiodic discrete-time signals are, of course, aperiodic sequences, whereas their DTFTs are always periodic functions of a continuous frequency variable.

From Eqs. (8.11) and (8.12), we see that the DFS analysis and synthesis equations differ only in a factor of $1/N$ and in the sign of the exponent of W_N. Furthermore, a periodic sequence and its DFS coefficients are the same kinds of functions; they are both periodic sequences. Specifically, taking account of the factor $1/N$ and the difference in sign in the exponent between Eqs. (8.11) and (8.12), it follows from Eq. (8.12) that

$$N\tilde{x}[-n] = \sum_{k=0}^{N-1} \tilde{X}[k]W_N^{kn} \tag{8.23}$$

or, interchanging the roles of n and k in Eq. (8.23),

$$N\tilde{x}[-k] = \sum_{n=0}^{N-1} \tilde{X}[n]W_N^{nk}. \tag{8.24}$$

We see that Eq. (8.24) is similar to Eq. (8.11). In other words, the sequence of DFS coefficients of the periodic sequence $\tilde{X}[n]$ is $N\tilde{x}[-k]$, i.e., the original periodic sequence in reverse order and multiplied by N. This duality property is summarized as follows: If

$$\tilde{x}[n] \overset{\mathcal{DFS}}{\longleftrightarrow} \tilde{X}[k], \tag{8.25a}$$

then

$$\tilde{X}[n] \overset{\mathcal{DFS}}{\longleftrightarrow} N\tilde{x}[-k]. \tag{8.25b}$$

8.2.4 Symmetry Properties

As we discussed in Section 2.8, the Fourier transform of an aperiodic sequence has a number of useful symmetry properties. The same basic properties also hold for the DFS representation of a periodic sequence. The derivation of these properties, which is similar in style to the derivations in Chapter 2, is left as an exercise. (See Problem 8.56.) The resulting properties are summarized for reference as properties 9–17 in Table 8.1 in Section 8.2.6.

8.2.5 Periodic Convolution

Let $\tilde{x}_1[n]$ and $\tilde{x}_2[n]$ be two periodic sequences, each with period N and with DFS coefficients denoted by $\tilde{X}_1[k]$ and $\tilde{X}_2[k]$, respectively. If we form the product

$$\tilde{X}_3[k] = \tilde{X}_1[k]\tilde{X}_2[k], \tag{8.26}$$

then the periodic sequence $\tilde{x}_3[n]$ with Fourier series coefficients $\tilde{X}_3[k]$ is

$$\tilde{x}_3[n] = \sum_{m=0}^{N-1} \tilde{x}_1[m]\tilde{x}_2[n-m]. \tag{8.27}$$

This result is not surprising, since our previous experience with transforms suggests that multiplication of frequency-domain functions corresponds to convolution of time-domain functions and Eq. (8.27) looks very much like a convolution sum. Equation (8.27) involves the summation of values of the product of $\tilde{x}_1[m]$ with $\tilde{x}_2[n-m]$, which is a time-reversed and time-shifted version of $\tilde{x}_2[m]$, just as in aperiodic discrete convolution. However, the sequences in Eq. (8.27) are all periodic with period N, and the summation is over only one period. A convolution in the form of Eq. (8.27) is referred to as a *periodic convolution*. Just as with aperiodic convolution, periodic convolution is commutative; i.e.,

$$\tilde{x}_3[n] = \sum_{m=0}^{N-1} \tilde{x}_2[m]\tilde{x}_1[n-m]. \tag{8.28}$$

To demonstrate that $\tilde{X}_3[k]$, given by Eq. (8.26), is the sequence of Fourier coefficients corresponding to $\tilde{x}_3[n]$ given by Eq. (8.27), let us first apply Eq. (8.11), the DFS analysis equation, to Eq. (8.27) to obtain

$$\tilde{X}_3[k] = \sum_{n=0}^{N-1} \left(\sum_{m=0}^{N-1} \tilde{x}_1[m]\tilde{x}_2[n-m] \right) W_N^{kn}, \tag{8.29}$$

which, after we interchange the order of summation, becomes

$$\tilde{X}_3[k] = \sum_{m=0}^{N-1} \tilde{x}_1[m] \left(\sum_{n=0}^{N-1} \tilde{x}_2[n-m] W_N^{kn} \right). \tag{8.30}$$

The inner sum on the index n is the DFS for the shifted sequence $\tilde{x}_2[n-m]$. Therefore, from the shifting property of Section 8.2.2, we obtain

$$\sum_{n=0}^{N-1} \tilde{x}_2[n-m] W_N^{kn} = W_N^{km}\tilde{X}_2[k],$$

which can be substituted into Eq. (8.30) to yield

$$\tilde{X}_3[k] = \sum_{m=0}^{N-1} \tilde{x}_1[m] W_N^{km}\tilde{X}_2[k] = \left(\sum_{m=0}^{N-1} \tilde{x}_1[m] W_N^{km} \right) \tilde{X}_2[k] = \tilde{X}_1[k]\tilde{X}_2[k]. \tag{8.31}$$

In summary,

$$\sum_{m=0}^{N-1} \tilde{x}_1[m]\tilde{x}_2[n-m] \stackrel{\mathcal{DFS}}{\longleftrightarrow} \tilde{X}_1[k]\tilde{X}_2[k]. \tag{8.32}$$

The periodic convolution of periodic sequences thus corresponds to multiplication of the corresponding periodic sequences of Fourier series coefficients.

Since periodic convolutions are somewhat different from aperiodic convolutions, it is worthwhile to consider the mechanics of evaluating Eq. (8.27). First, note that Eq. (8.27) calls for the product of sequences $\tilde{x}_1[m]$ and $\tilde{x}_2[n-m] = \tilde{x}_2[-(m-n)]$ viewed as functions of m with n fixed. This is the same as for an aperiodic convolution, but with the following two major differences:

1. The sum is over the finite interval $0 \le m \le N-1$.
2. The values of $\tilde{x}_2[n-m]$ in the interval $0 \le m \le N-1$ repeat periodically for m outside of that interval.

These details are illustrated by the following example.

Example 8.4   Periodic Convolution

An illustration of the procedure for forming the periodic convolution of two periodic sequences corresponding to Eq. (8.27) is given in Figure 8.3, wherein we have illustrated the sequences $\tilde{x}_2[m]$, $\tilde{x}_1[m]$, $\tilde{x}_2[-m]$, $\tilde{x}_2[1-m] = \tilde{x}_2[-(m-1)]$, and $\tilde{x}_2[2-m] = \tilde{x}_2[-(m-2)]$. To evaluate $\tilde{x}_3[n]$ in Eq. (8.27) for $n = 2$, for example, we multiply $\tilde{x}_1[m]$ by $\tilde{x}_2[2-m]$ and then sum the product terms $\tilde{x}_1[m]\tilde{x}_2[2-m]$ for $0 \leq m \leq N-1$, obtaining $\tilde{x}_3[2]$. As n changes, the sequence $\tilde{x}_2[n-m]$ shifts appropriately, and Eq. (8.27) is evaluated for each value of $0 \leq n \leq N-1$. Note that as the sequence $\tilde{x}_2[n-m]$ shifts to the right or left, values that leave the interval between the dotted lines at one end reappear at the other end because of the periodicity. Because of the periodicity of $\tilde{x}_3[n]$, there is no need to continue to evaluate Eq. (8.27) outside the interval $0 \leq n \leq N-1$.

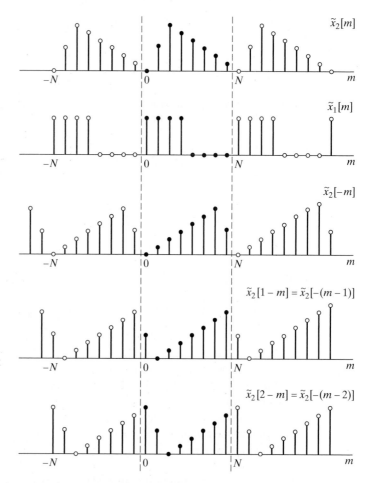

Figure 8.3   Procedure for forming the periodic convolution of two periodic sequences.

The duality theorem in Section 8.2.3 suggests that if the roles of time and frequency are interchanged, we will obtain a result almost identical to the previous result. That is, the periodic sequence

$$\tilde{x}_3[n] = \tilde{x}_1[n]\tilde{x}_2[n], \tag{8.33}$$

where $\tilde{x}_1[n]$ and $\tilde{x}_2[n]$ are periodic sequences, each with period N, has the DFS coefficients given by

$$\tilde{X}_3[k] = \frac{1}{N}\sum_{\ell=0}^{N-1}\tilde{X}_1[\ell]\tilde{X}_2[k-\ell], \tag{8.34}$$

corresponding to $1/N$ times the periodic convolution of $\tilde{X}_1[k]$ and $\tilde{X}_2[k]$. This result can also be verified by substituting $\tilde{X}_3[k]$, given by Eq. (8.34), into the Fourier series relation of Eq. (8.12) to obtain $\tilde{x}_3[n]$.

8.2.6 Summary of Properties of the DFS Representation of Periodic Sequences

The properties of the DFS representation discussed in this section are summarized in Table 8.1.

8.3 THE FOURIER TRANSFORM OF PERIODIC SIGNALS

As discussed in Section 2.7, uniform convergence of the Fourier transform of a sequence requires that the sequence be absolutely summable, and mean-square convergence requires that the sequence be square summable. Periodic sequences satisfy neither condition. However, as we discussed briefly in Section 2.7, sequences that can be expressed as a sum of complex exponentials can be considered to have a Fourier transform representation in the form of Eq. (2.147), i.e., as a train of impulses. Similarly, it is often useful to incorporate the DFS representation of periodic signals within the framework of the discrete-time Fourier transform. This can be done by interpreting the discrete-time Fourier transform of a periodic signal to be an impulse train in the frequency domain with the impulse values proportional to the DFS coefficients for the sequence. Specifically, if $\tilde{x}[n]$ is periodic with period N and the corresponding DFS coefficients are $\tilde{X}[k]$, then the Fourier transform of $\tilde{x}[n]$ is defined to be the impulse train

$$\tilde{X}(e^{j\omega}) = \sum_{k=-\infty}^{\infty}\frac{2\pi}{N}\tilde{X}[k]\delta\left(\omega - \frac{2\pi k}{N}\right). \tag{8.35}$$

Note that $\tilde{X}(e^{j\omega})$ has the necessary periodicity with period 2π since $\tilde{X}[k]$ is periodic with period N, and the impulses are spaced at integer multiples of $2\pi/N$, where N is an

TABLE 8.1   SUMMARY OF PROPERTIES OF THE DFS

Periodic Sequence (Period N)	DFS Coefficients (Period N)
1. $\tilde{x}[n]$	$\tilde{X}[k]$ periodic with period N
2. $\tilde{x}_1[n], \tilde{x}_2[n]$	$\tilde{X}_1[k], \tilde{X}_2[k]$ periodic with period N
3. $a\tilde{x}_1[n] + b\tilde{x}_2[n]$	$a\tilde{X}_1[k] + b\tilde{X}_2[k]$
4. $\tilde{X}[n]$	$N\tilde{x}[-k]$
5. $\tilde{x}[n-m]$	$W_N^{km}\tilde{X}[k]$
6. $W_N^{-\ell n}\tilde{x}[n]$	$\tilde{X}[k-\ell]$
7. $\displaystyle\sum_{m=0}^{N-1}\tilde{x}_1[m]\tilde{x}_2[n-m]$ (periodic convolution)	$\tilde{X}_1[k]\tilde{X}_2[k]$
8. $\tilde{x}_1[n]\tilde{x}_2[n]$	$\displaystyle\frac{1}{N}\sum_{\ell=0}^{N-1}\tilde{X}_1[\ell]\tilde{X}_2[k-\ell]$ (periodic convolution)
9. $\tilde{x}^*[n]$	$\tilde{X}^*[-k]$
10. $\tilde{x}^*[-n]$	$\tilde{X}^*[k]$
11. $\mathcal{R}e\{\tilde{x}[n]\}$	$\tilde{X}_e[k] = \frac{1}{2}(\tilde{X}[k] + \tilde{X}^*[-k])$
12. $j\mathcal{I}m\{\tilde{x}[n]\}$	$\tilde{X}_o[k] = \frac{1}{2}(\tilde{X}[k] - \tilde{X}^*[-k])$
13. $\tilde{x}_e[n] = \frac{1}{2}(\tilde{x}[n] + \tilde{x}^*[-n])$	$\mathcal{R}e\{\tilde{X}[k]\}$
14. $\tilde{x}_o[n] = \frac{1}{2}(\tilde{x}[n] - \tilde{x}^*[-n])$	$j\mathcal{I}m\{\tilde{X}[k]\}$
Properties 15–17 apply only when $x[n]$ is real.	
15. Symmetry properties for $\tilde{x}[n]$ real.	$\begin{cases} \tilde{X}[k] = \tilde{X}^*[-k] \\ \mathcal{R}e\{\tilde{X}[k]\} = \mathcal{R}e\{\tilde{X}[-k]\} \\ \mathcal{I}m\{\tilde{X}[k]\} = -\mathcal{I}m\{\tilde{X}[-k]\} \\ |\tilde{X}[k]| = |\tilde{X}[-k]| \\ \angle\tilde{X}[k] = -\angle\tilde{X}[-k] \end{cases}$
16. $\tilde{x}_e[n] = \frac{1}{2}(\tilde{x}[n] + \tilde{x}[-n])$	$\mathcal{R}e\{\tilde{X}[k]\}$
17. $\tilde{x}_0[n] = \frac{1}{2}(\tilde{x}[n] - \tilde{x}[-n])$	$j\mathcal{I}m\{\tilde{X}[k]\}$

integer. To show that $\tilde{X}(e^{j\omega})$ as defined in Eq. (8.35) is a Fourier transform representation of the periodic sequence $\tilde{x}[n]$, we substitute Eq. (8.35) into the inverse Fourier transform Eq. (2.130); i.e.,

$$\frac{1}{2\pi}\int_{0-\epsilon}^{2\pi-\epsilon}\tilde{X}(e^{j\omega})e^{j\omega n}d\omega = \frac{1}{2\pi}\int_{0-\epsilon}^{2\pi-\epsilon}\sum_{k=-\infty}^{\infty}\frac{2\pi}{N}\tilde{X}[k]\delta\left(\omega-\frac{2\pi k}{N}\right)e^{j\omega n}d\omega, \quad (8.36)$$

where ϵ satisfies the inequality $0 < \epsilon < (2\pi/N)$. Recall that in evaluating the inverse Fourier transform, we can integrate over any interval of length 2π, since the integrand $\tilde{X}(e^{j\omega})e^{j\omega n}$ is periodic with period 2π. In Eq. (8.36) the integration limits are denoted $0-\epsilon$ and $2\pi-\epsilon$, which means that the integration is from just before $\omega = 0$ to just before $\omega = 2\pi$. These limits are convenient, because they include the impulse at $\omega = 0$ and exclude the impulse at $\omega = 2\pi$.[3] Interchanging the order of integration and summation leads to

$$\frac{1}{2\pi} \int_{0-\epsilon}^{2\pi-\epsilon} \tilde{X}(e^{j\omega}) e^{j\omega n} d\omega = \frac{1}{N} \sum_{k=-\infty}^{\infty} \tilde{X}[k] \int_{0-\epsilon}^{2\pi-\epsilon} \delta\left(\omega - \frac{2\pi k}{N}\right) e^{j\omega n} d\omega$$

$$= \frac{1}{N} \sum_{k=0}^{N-1} \tilde{X}[k] e^{j(2\pi/N)kn}.$$

(8.37)

The final form of Eq. (8.37) results because only the impulses corresponding to $k = 0, 1, \ldots, (N-1)$ are included in the interval between $\omega = 0 - \epsilon$ and $\omega = 2\pi - \epsilon$.

Comparing Eq. (8.37) and Eq. (8.12), we see that the final right-hand side of Eq. (8.37) is exactly equal to the Fourier series representation for $\tilde{x}[n]$, as specified by Eq. (8.12). Consequently, the inverse Fourier transform of the impulse train in Eq. (8.35) is the periodic signal $\tilde{x}[n]$, as desired.

Although the Fourier transform of a periodic sequence does not converge in the normal sense, the introduction of impulses permits us to include periodic sequences formally within the framework of Fourier transform analysis. This approach was also used in Chapter 2 to obtain a Fourier transform representation of other nonsummable sequences, such as the two-sided constant sequence (Example 2.19) or the complex exponential sequence (Example 2.20). Although the DFS representation is adequate for most purposes, the Fourier transform representation of Eq. (8.35) sometimes leads to simpler or more compact expressions and simplified analysis.

Example 8.5   The Fourier Transform of a Periodic Discrete-Time Impulse Train

Consider the periodic discrete-time impulse train

$$\tilde{p}[n] = \sum_{r=-\infty}^{\infty} \delta[n - rN],$$

(8.38)

which is the same as the periodic sequence $\tilde{x}[n]$ considered in Example 8.1. From the results of that example, it follows that

$$\tilde{P}[k] = 1, \qquad \text{for all } k.$$

(8.39)

Therefore, the DTFT of $\tilde{p}[n]$ is

$$\tilde{P}(e^{j\omega}) = \sum_{k=-\infty}^{\infty} \frac{2\pi}{N} \delta\left(\omega - \frac{2\pi k}{N}\right).$$

(8.40)

The result of Example 8.5 is the basis for a useful interpretation of the relation between a periodic signal and a finite-length signal. Consider a finite-length signal $x[n]$ such that $x[n] = 0$ except in the interval $0 \leq n \leq N - 1$, and consider the convolution

---

[3] The limits 0 to 2π would present a problem since the impulses at both 0 and 2π would require special handling.

Figure 8.4 Periodic sequence $\tilde{x}[n]$ formed by repeating a finite-length sequence, $x[n]$, periodically. Alternatively, $x[n] = \tilde{x}[n]$ over one period and is zero otherwise.

of $x[n]$ with the periodic impulse train $\tilde{p}[n]$ of Example 8.5:

$$\tilde{x}[n] = x[n] * \tilde{p}[n] = x[n] * \sum_{r=-\infty}^{\infty} \delta[n - rN] = \sum_{r=-\infty}^{\infty} x[n - rN]. \qquad (8.41)$$

Equation (8.41) states that $\tilde{x}[n]$ consists of a set of periodically repeated copies of the finite-length sequence $x[n]$. Figure 8.4 illustrates how a periodic sequence $\tilde{x}[n]$ can be formed from a finite-length sequence $x[n]$ through Eq. (8.41). The Fourier transform of $x[n]$ is $X(e^{j\omega})$, and the Fourier transform of $\tilde{x}[n]$ is

$$\tilde{X}(e^{j\omega}) = X(e^{j\omega})\tilde{P}(e^{j\omega})$$

$$= X(e^{j\omega}) \sum_{k=-\infty}^{\infty} \frac{2\pi}{N} \delta\left(\omega - \frac{2\pi k}{N}\right) \qquad (8.42)$$

$$= \sum_{k=-\infty}^{\infty} \frac{2\pi}{N} X(e^{j(2\pi/N)k}) \delta\left(\omega - \frac{2\pi k}{N}\right).$$

Comparing Eq. (8.42) with Eq. (8.35), we conclude that

$$\tilde{X}[k] = X(e^{j(2\pi/N)k}) = X(e^{j\omega})\Big|_{\omega=(2\pi/N)k}. \qquad (8.43)$$

In other words, the periodic sequence $\tilde{X}[k]$ of DFS coefficients in Eq. (8.11) has an discrete-time interpretation as equally spaced samples of the DTFT of the finite-length sequence obtained by extracting one period of $\tilde{x}[n]$; i.e.,

$$x[n] = \begin{cases} \tilde{x}[n], & 0 \le n \le N-1, \\ 0, & \text{otherwise.} \end{cases} \qquad (8.44)$$

This is also consistent with Figure 8.4, where it is clear that $x[n]$ can be obtained from $\tilde{x}[n]$ using Eq. (8.44). We can verify Eq. (8.43) in yet another way. Since $x[n] = \tilde{x}[n]$ for $0 \le n \le N-1$ and $x[n] = 0$ otherwise,

$$X(e^{j\omega}) = \sum_{n=0}^{N-1} x[n]e^{-j\omega n} = \sum_{n=0}^{N-1} \tilde{x}[n]e^{-j\omega n}. \qquad (8.45)$$

Comparing Eq. (8.45) and Eq. (8.11), we see again that

$$\tilde{X}[k] = X(e^{j\omega})|_{\omega=2\pi k/N}. \qquad (8.46)$$

This corresponds to sampling the Fourier transform at N equally spaced frequencies between $\omega = 0$ and $\omega = 2\pi$ with a frequency spacing of $2\pi/N$.

Example 8.6   Relationship Between the Fourier Series Coefficients and the Fourier Transform of One Period

We again consider the sequence $\tilde{x}[n]$ of Example 8.3, which is shown in Figure 8.1. One period of $\tilde{x}[n]$ for the sequence in Figure 8.1 is

$$x[n] = \begin{cases} 1, & 0 \le n \le 4, \\ 0, & \text{otherwise.} \end{cases} \tag{8.47}$$

The Fourier transform of one period of $\tilde{x}[n]$ is given by

$$X(e^{j\omega}) = \sum_{n=0}^{4} e^{-j\omega n} = e^{-j2\omega} \frac{\sin(5\omega/2)}{\sin(\omega/2)}. \tag{8.48}$$

Equation (8.46) can be shown to be satisfied for this example by substituting $\omega = 2\pi k/10$ into Eq. (8.48), giving

$$\tilde{X}[k] = e^{-j(4\pi k/10)} \frac{\sin(\pi k/2)}{\sin(\pi k/10)},$$

which is identical to the result in Eq. (8.18). The magnitude and phase of $X(e^{j\omega})$ are sketched in Figure 8.5. Note that the phase is discontinuous at the frequencies where $X(e^{j\omega}) = 0$. That the sequences in Figures 8.2(a) and (b) correspond to samples of Figures 8.5(a) and (b), respectively, is demonstrated in Figure 8.6, where Figures 8.2 and 8.5 have been superimposed.

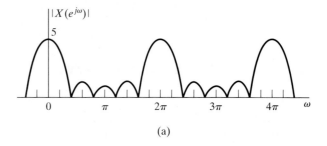

(a)

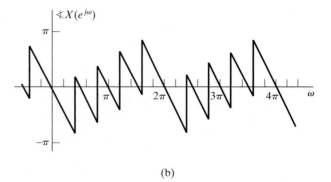

(b)

Figure 8.5   Magnitude and phase of the Fourier transform of one period of the sequence in Figure 8.1.

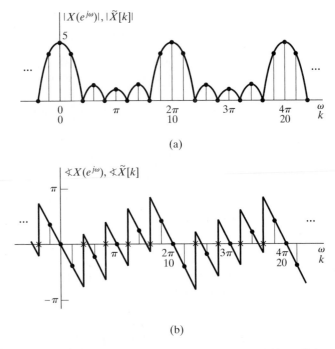

Figure 8.6 Overlay of Figures 8.2 and 8.5 illustrating the DFS coefficients of a periodic sequence as samples of the Fourier transform of one period.

8.4 SAMPLING THE FOURIER TRANSFORM

In this section, we discuss with more generality the relationship between an aperiodic sequence with Fourier transform $X(e^{j\omega})$ and the periodic sequence for which the DFS coefficients correspond to samples of $X(e^{j\omega})$ equally spaced in frequency. We will find this relationship to be particularly important when we discuss the discrete Fourier transform and its properties later in the chapter.

Consider an aperiodic sequence $x[n]$ with Fourier transform $X(e^{j\omega})$, and assume that a sequence $\tilde{X}[k]$ is obtained by sampling $X(e^{j\omega})$ at frequencies $\omega_k = 2\pi k/N$; i.e.,

$$\tilde{X}[k] = X(e^{j\omega})|_{\omega=(2\pi/N)k} = X(e^{j(2\pi/N)k}). \tag{8.49}$$

Since the Fourier transform is periodic in ω with period 2π, the resulting sequence is periodic in k with period N. Also, since the Fourier transform is equal to the z-transform evaluated on the unit circle, it follows that $\tilde{X}[k]$ can also be obtained by sampling $X(z)$ at N equally spaced points on the unit circle. Thus,

$$\tilde{X}[k] = X(z)|_{z=e^{j(2\pi/N)k}} = X(e^{j(2\pi/N)k}). \tag{8.50}$$

These sampling points are depicted in Figure 8.7 for $N = 8$. The figure makes it clear that the sequence of samples is periodic, since the N points are equally spaced starting with zero angle. Therefore, the same sequence repeats as k varies outside the range $0 \le k \le N - 1$ since we simply continue around the unit circle visiting the same set of N points.

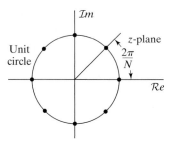

Figure 8.7   Points on the unit circle at which $X(z)$ is sampled to obtain the periodic sequence $\tilde{X}[k]$ ($N = 8$).

Note that the sequence of samples $\tilde{X}[k]$, being periodic with period N, could be the sequence of DFS coefficients of a sequence $\tilde{x}[n]$. To obtain that sequence, we can simply substitute $\tilde{X}[k]$ obtained by sampling into Eq. (8.12):

$$\tilde{x}[n] = \frac{1}{N} \sum_{k=0}^{N-1} \tilde{X}[k] W_N^{-kn}. \tag{8.51}$$

Since we have made no assumption about $x[n]$ other than that the Fourier transform exists, we can use infinite limits to indicate that the sum is

$$X(e^{j\omega}) = \sum_{m=-\infty}^{\infty} x[m] e^{-j\omega m} \tag{8.52}$$

is over all nonzero values of $x[m]$.

Substituting Eq. (8.52) into Eq. (8.49) and then substituting the resulting expression for $\tilde{X}[k]$ into Eq. (8.51) gives

$$\tilde{x}[n] = \frac{1}{N} \sum_{k=0}^{N-1} \left[\sum_{m=-\infty}^{\infty} x[m] e^{-j(2\pi/N)km} \right] W_N^{-kn}, \tag{8.53}$$

which, after we interchange the order of summation, becomes

$$\tilde{x}[n] = \sum_{m=-\infty}^{\infty} x[m] \left[\frac{1}{N} \sum_{k=0}^{N-1} W_N^{-k(n-m)} \right] = \sum_{m=-\infty}^{\infty} x[m] \tilde{p}[n-m]. \tag{8.54}$$

The term in brackets in Eq. (8.54) can be seen from either Eq. (8.7) or Eq. (8.16) to be the Fourier series representation of the periodic impulse train of Examples 8.1 and 8.2. Specifically,

$$\tilde{p}[n-m] = \frac{1}{N} \sum_{k=0}^{N-1} W_N^{-k(n-m)} = \sum_{r=-\infty}^{\infty} \delta[n-m-rN] \tag{8.55}$$

and therefore,

$$\tilde{x}[n] = x[n] * \sum_{r=-\infty}^{\infty} \delta[n-rN] = \sum_{r=-\infty}^{\infty} x[n-rN], \tag{8.56}$$

where $*$ denotes aperiodic convolution. That is, $\tilde{x}[n]$ is the periodic sequence that results from the aperiodic convolution of $x[n]$ with a periodic unit-impulse train. Thus, the

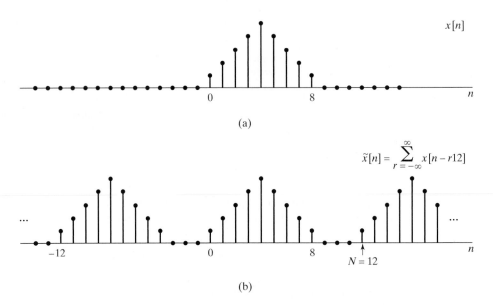

(a)

$$\tilde{x}[n] = \sum_{r=-\infty}^{\infty} x[n-r12]$$

(b)

Figure 8.8 (a) Finite-length sequence $x[n]$. (b) Periodic sequence $\tilde{x}[n]$ corresponding to sampling the Fourier transform of $x[n]$ with $N = 12$.

periodic sequence $\tilde{x}[n]$, corresponding to $\tilde{X}[k]$ obtained by sampling $X(e^{j\omega})$, is formed from $x[n]$ by adding together an infinite number of shifted replicas of $x[n]$. The shifts are all the positive and negative integer multiples of N, the period of the sequence $\tilde{X}[k]$. This is illustrated in Figure 8.8, where the sequence $x[n]$ is of length 9 and the value of N in Eq. (8.56) is $N = 12$. Consequently, the delayed replications of $x[n]$ do not overlap, and one period of the periodic sequence $\tilde{x}[n]$ is recognizable as $x[n]$. This is consistent with the discussion in Section 8.3 and Example 8.6, wherein we showed that the Fourier series coefficients for a periodic sequence are samples of the Fourier transform of one period. In Figure 8.9 the same sequence $x[n]$ is used, but the value of N is now $N = 7$. In this case, the replicas of $x[n]$ overlap and one period of $\tilde{x}[n]$ is no longer identical to $x[n]$. In both cases, however, Eq. (8.49) still holds; i.e., in both cases, the DFS coefficients of $\tilde{x}[n]$ are samples of the Fourier transform of $x[n]$ spaced

$$\tilde{x}[n] = \sum_{r=-\infty}^{\infty} x[n-r7]$$

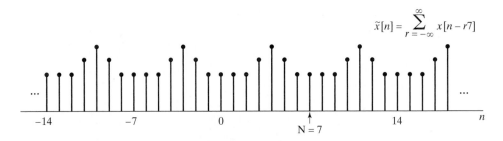

Figure 8.9 Periodic sequence $\tilde{x}[n]$ corresponding to sampling the Fourier transform of $x[n]$ in Figure 8.8(a) with $N = 7$.

in frequency at integer multiples of $2\pi/N$. This discussion should be reminiscent of our discussion of sampling in Chapter 4. The difference is that here we are sampling in the frequency domain rather than in the time domain. However, the general outlines of the mathematical representations are very similar.

For the example in Figure 8.8, the original sequence $x[n]$ can be recovered from $\tilde{x}[n]$ by extracting one period. Equivalently, the Fourier transform $X(e^{j\omega})$ can be recovered from the samples spaced in frequency by $2\pi/12$. In contrast, in Figure 8.9, $x[n]$ cannot be recovered by extracting one period of $\tilde{x}[n]$, and, equivalently, $X(e^{j\omega})$ cannot be recovered from its samples if the sample spacing is only $2\pi/7$. In effect, for the case illustrated in Figure 8.8, the Fourier transform of $x[n]$ has been sampled at a sufficiently small spacing (in frequency) to be able to recover it from these samples, whereas Figure 8.9 represents a case for which the Fourier transform has been undersampled. The relationship between $x[n]$ and one period of $\tilde{x}[n]$ in the undersampled case can be thought of as a form of aliasing in the time domain, essentially identical to the frequency-domain aliasing (discussed in Chapter 4) that results from undersampling in the time domain. Obviously, time-domain aliasing can be avoided only if $x[n]$ has finite length, just as frequency-domain aliasing can be avoided only for signals that have bandlimited Fourier transforms.

This discussion highlights several important concepts that will play a central role in the remainder of the chapter. We have seen that samples of the Fourier transform of an aperiodic sequence $x[n]$ can be thought of as DFS coefficients of a periodic sequence $\tilde{x}[n]$ obtained through summing periodic replicas of $x[n]$. If $x[n]$ is finite length and we take a sufficient number of equally spaced samples of its Fourier transform (specifically, a number greater than or equal to the length of $x[n]$), then the Fourier transform is recoverable from these samples, and, equivalently, $x[n]$ is recoverable from the corresponding periodic sequence $\tilde{x}[n]$. Specifically, if $x[n] = 0$ outside the interval $n = 0$, $n = N - 1$, then

$$x[n] = \begin{cases} \tilde{x}[n], & 0 \le n \le N - 1, \\ 0, & \text{otherwise.} \end{cases} \tag{8.57}$$

If the interval of support of $x[n]$ is different than $0, N - 1$ then Eq. (8.57) would be appropriately modified.

A direct relationship between $X(e^{j\omega})$ and its samples $\tilde{X}[k]$, i.e., an interpolation formula for $X(e^{j\omega})$, can be derived (see Problem 8.57). However, the essence of our previous discussion is that to represent or to recover $x[n]$, it is not necessary to know $X(e^{j\omega})$ at all frequencies if $x[n]$ has finite length. Given a finite-length sequence $x[n]$, we can form a periodic sequence using Eq. (8.56), which in turn can be represented by a DFS. Alternatively, given the sequence of Fourier coefficients $\tilde{X}[k]$, we can find $\tilde{x}[n]$ and then use Eq. (8.57) to obtain $x[n]$. When the Fourier series is used in this way to represent finite-length sequences, it is called the discrete Fourier transform or DFT. In developing, discussing, and applying the DFT, it is always important to remember that the representation through samples of the Fourier transform is in effect a representation of the finite-duration sequence by a periodic sequence, one period of which is the finite-duration sequence that we wish to represent.

8.5 FOURIER REPRESENTATION OF FINITE-DURATION SEQUENCES: THE DFT

In this section, we formalize the point of view suggested at the end of the previous section. We begin by considering a finite-length sequence $x[n]$ of length N samples such that $x[n] = 0$ outside the range $0 \leq n \leq N - 1$. In many instances, we will want to assume that a sequence has length N, even if its length is $M \leq N$. In such cases, we simply recognize that the last $(N - M)$ samples are zero. To each finite-length sequence of length N, we can always associate a periodic sequence

$$\tilde{x}[n] = \sum_{r=-\infty}^{\infty} x[n - rN]. \qquad (8.58a)$$

The finite-length sequence $x[n]$ can be recovered from $\tilde{x}[n]$ through Eq. (8.57), i.e.,

$$x[n] = \begin{cases} \tilde{x}[n], & 0 \leq n \leq N - 1, \\ 0, & \text{otherwise.} \end{cases} \qquad (8.58b)$$

Recall from Section 8.4 that the DFS coefficients of $\tilde{x}[n]$ are samples (spaced in frequency by $2\pi/N$) of the Fourier transform of $x[n]$. Since $x[n]$ is assumed to have finite length N, there is no overlap between the terms $x[n - rN]$ for different values of r. Thus, Eq. (8.58a) can alternatively be written as

$$\tilde{x}[n] = x[(n \text{ modulo } N)]. \qquad (8.59)$$

For convenience, we will use the notation $((n))_N$ to denote (n modulo N); with this notation, Eq. (8.59) is expressed as

$$\tilde{x}[n] = x[((n))_N]. \qquad (8.60)$$

Note that Eq. (8.60) is equivalent to Eq. (8.58a) only when $x[n]$ has length less than or equal to N. The finite-duration sequence $x[n]$ is obtained from $\tilde{x}[n]$ by extracting one period, as in Eq. (8.58b).

One informal and useful way of visualizing Eq. (8.59) is to think of wrapping a plot of the finite-duration sequence $x[n]$ around a cylinder with a circumference equal to the length of the sequence. As we repeatedly traverse the circumference of the cylinder, we see the finite-length sequence periodically repeated. With this interpretation, representation of the finite-length sequence by a periodic sequence corresponds to wrapping the sequence around the cylinder; recovering the finite-length sequence from the periodic sequence using Eq. (8.58b) can be visualized as unwrapping the cylinder and laying it flat so that the sequence is displayed on a linear time axis rather than a circular (modulo N) time axis.

As defined in Section 8.1, the sequence of DFS coefficients $\tilde{X}[k]$ of the periodic sequence $\tilde{x}[n]$ is itself a periodic sequence with period N. To maintain a duality between the time and frequency domains, we will choose the Fourier coefficients that we associate with a finite-duration sequence to be a finite-duration sequence corresponding to one period of $\tilde{X}[k]$. This finite-duration sequence, $X[k]$, will be referred to as the DFT. Thus, the DFT, $X[k]$, is related to the DFS coefficients, $\tilde{X}[k]$, by

$$X[k] = \begin{cases} \tilde{X}[k], & 0 \le k \le N-1, \\ 0, & \text{otherwise}, \end{cases} \tag{8.61}$$

and

$$\tilde{X}[k] = X[(k \text{ modulo } N)] = X[((k))_N]. \tag{8.62}$$

From Section 8.1, $\tilde{X}[k]$ and $\tilde{x}[n]$ are related by

$$\tilde{X}[k] = \sum_{n=0}^{N-1} \tilde{x}[n] W_N^{kn}, \tag{8.63}$$

$$\tilde{x}[n] = \frac{1}{N} \sum_{k=0}^{N-1} \tilde{X}[k] W_N^{-kn}. \tag{8.64}$$

where $W_N = e^{-j(2\pi/N)}$.

Since the summations in Eqs. (8.63) and (8.64) involve only the interval between zero and $(N-1)$, it follows from Eqs. (8.58b) to (8.64) that

$$X[k] = \begin{cases} \sum_{n=0}^{N-1} x[n] W_N^{kn}, & 0 \le k \le N-1, \\ 0, & \text{otherwise}, \end{cases} \tag{8.65}$$

$$x[n] = \begin{cases} \dfrac{1}{N} \sum_{k=0}^{N-1} X[k] W_N^{-kn}, & 0 \le n \le N-1, \\ 0, & \text{otherwise}. \end{cases} \tag{8.66}$$

Generally, the DFT analysis and synthesis equations are written as follows:

$$\text{Analysis equation:}\quad X[k] = \sum_{n=0}^{N-1} x[n] W_N^{kn}, \qquad 0 \le k \le N-1, \tag{8.67}$$

$$\text{Synthesis equation:}\quad x[n] = \frac{1}{N} \sum_{k=0}^{N-1} X[k] W_N^{-kn}, \qquad 0 \le n \le N-1. \tag{8.68}$$

That is, the fact that $X[k] = 0$ for k outside the interval $0 \le k \le N-1$ and that $x[n] = 0$ for n outside the interval $0 \le n \le N-1$ is implied, but not always stated explicitly. The relationship between $x[n]$ and $X[k]$ implied by Eqs. (8.67) and (8.68) will sometimes be denoted as

$$x[n] \overset{\mathcal{DFT}}{\longleftrightarrow} X[k]. \tag{8.69}$$

In recasting Eqs. (8.11) and (8.12) in the form of Eqs. (8.67) and (8.68) for finite-duration sequences, we have not eliminated the inherent periodicity. As with the DFS, the DFT $X[k]$ is equal to samples of the periodic Fourier transform $X(e^{j\omega})$, and if Eq. (8.68) is evaluated for values of n outside the interval $0 \le n \le N-1$, the result will not be zero, but rather a periodic extension of $x[n]$. The inherent periodicity is always present. Sometimes, it causes us difficulty, and sometimes we can exploit it, but to totally ignore it is to invite trouble. In defining the DFT representation, we are simply recognizing that we are interested in values of $x[n]$ only in the interval $0 \le n \le N-1$, because $x[n]$ is really zero outside that interval, and we are interested in values of $X[k]$ only in the interval $0 \le k \le N-1$ because these are the only values needed in Eq. (8.68) to reconstruct $X[n]$.

Example 8.7   The DFT of a Rectangular Pulse

To illustrate the DFT of a finite-duration sequence, consider $x[n]$ shown in Figure 8.10(a). In determining the DFT, we can consider $x[n]$ as a finite-duration sequence with any length greater than or equal to $N = 5$. Considered as a sequence of length $N = 5$, the periodic sequence $\tilde{x}[n]$ whose DFS corresponds to the DFT of $x[n]$ is shown in Figure 8.10(b). Since the sequence in Figure 8.10(b) is constant over the interval $0 \le n \le 4$, it follows that

$$\tilde{X}[k] = \sum_{n=0}^{4} e^{-j(2\pi k/5)n} = \frac{1 - e^{-j2\pi k}}{1 - e^{-j(2\pi k/5)}} \tag{8.70}$$

$$= \begin{cases} 5, & k = 0, \pm 5, \pm 10, \ldots, \\ 0, & \text{otherwise;} \end{cases}$$

i.e., the only nonzero DFS coefficients $\tilde{X}[k]$ are at $k = 0$ and integer multiples of $k = 5$ (all of which represent the same complex exponential frequency). The DFS coefficients are shown in Figure 8.10(c). Also shown is the magnitude of the DTFT, $|X(e^{j\omega})|$. Clearly, $\tilde{X}[k]$ is a sequence of samples of $X(e^{j\omega})$ at frequencies $\omega_k = 2\pi k/5$. According to Eq. (8.61), the five-point DFT of $x[n]$ corresponds to the finite-length sequence obtained by extracting one period of $\tilde{X}[k]$. Consequently, the five-point DFT of $x[n]$ is shown in Figure 8.10(d).

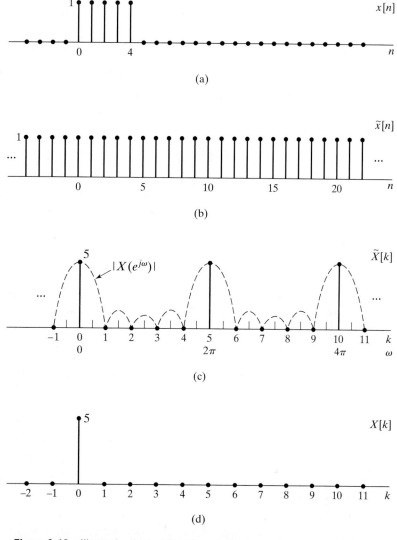

Figure 8.10 Illustration of the DFT. (a) Finite-length sequence $x[n]$. (b) Periodic sequence $\tilde{x}[n]$ formed from $x[n]$ with period $N = 5$. (c) Fourier series coefficients $\tilde{X}[k]$ for $\tilde{x}[n]$. To emphasize that the Fourier series coefficients are samples of the Fourier transform, $|X(e^{j\omega})|$ is also shown. (d) DFT of $x[n]$.

If, instead, we consider $x[n]$ to be of length $N = 10$, then the underlying periodic sequence is that shown in Figure 8.11(b), which is the periodic sequence considered in Example 8.3. Therefore, $\tilde{X}[k]$ is as shown in Figures 8.2 and 8.6, and the 10-point DFT $X[k]$ shown in Figures 8.11(c) and 8.11(d) is one period of $\tilde{X}[k]$.

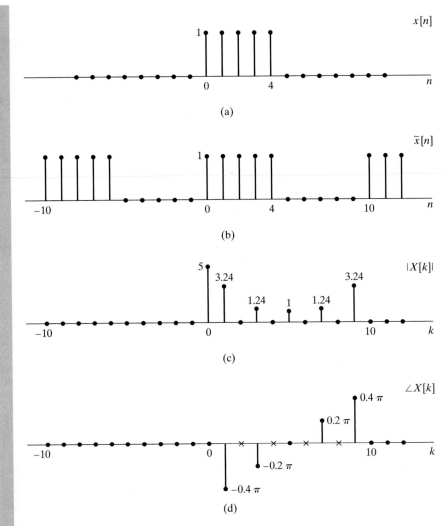

Figure 8.11   Illustration of the DFT. (a) Finite-length sequence $x[n]$. (b) Periodic sequence $\tilde{x}[n]$ formed from $x[n]$ with period $N = 10$. (c) DFT magnitude. (d) DFT phase. (x's indicate indeterminate values.)

The distinction between the finite-duration sequence $x[n]$ and the periodic sequence $\tilde{x}[n]$ related through Eqs. (8.57) and (8.60) may seem minor, since, by using these equations, it is straightforward to construct one from the other. However, the distinction becomes important in considering properties of the DFT and in considering the effect on $x[n]$ of modifications to $X[k]$. This will become evident in the next section, where we discuss the properties of the DFT representation.

8.6 PROPERTIES OF THE DFT

In this section, we consider a number of properties of the DFT for finite-duration sequences. Our discussion parallels the discussion of Section 8.2 for periodic sequences. However, particular attention is paid to the interaction of the finite-length assumption and the implicit periodicity of the DFT representation of finite-length sequences.

8.6.1 Linearity

If two finite-duration sequences $x_1[n]$ and $x_2[n]$ are linearly combined, i.e., if

$$x_3[n] = ax_1[n] + bx_2[n], \tag{8.71}$$

then the DFT of $x_3[n]$ is

$$X_3[k] = aX_1[k] + bX_2[k]. \tag{8.72}$$

Clearly, if $x_1[n]$ has length N_1 and $x_2[n]$ has length N_2, then the maximum length of $x_3[n]$ will be $N_3 = \max(N_1, N_2)$. Thus, in order for Eq. (8.72) to be meaningful, both DFTs must be computed with the same length $N \geq N_3$. If, for example, $N_1 < N_2$, then $X_1[k]$ is the DFT of the sequence $x_1[n]$ augmented by $(N_2 - N_1)$ zeros. That is, the N_2-point DFT of $x_1[n]$ is

$$X_1[k] = \sum_{n=0}^{N_1-1} x_1[n]W_{N_2}^{kn}, \qquad 0 \leq k \leq N_2 - 1, \tag{8.73}$$

and the N_2-point DFT of $x_2[n]$ is

$$X_2[k] = \sum_{n=0}^{N_2-1} x_2[n]W_{N_2}^{kn}, \qquad 0 \leq k \leq N_2 - 1. \tag{8.74}$$

In summary, if

$$x_1[n] \overset{\mathcal{DFS}}{\longleftrightarrow} X_1[k] \tag{8.75a}$$

and

$$x_2[n] \overset{\mathcal{DFS}}{\longleftrightarrow} X_2[k], \tag{8.75b}$$

then

$$ax_1[n] + bx_2[n] \overset{\mathcal{DFJ}}{\longleftrightarrow} aX_1[k] + bX_2[k], \tag{8.76}$$

where the lengths of the sequences and their DFTs are all equal to at least the maximum of the lengths of $x_1[n]$ and $x_2[n]$. Of course, DFTs of greater length can be computed by augmenting both sequences with zero-valued samples.

8.6.2 Circular Shift of a Sequence

According to Section 2.9.2 and property 2 in Table 2.2, if $X(e^{j\omega})$ is the discrete-time Fourier transform of $x[n]$, then $e^{-j\omega m} X(e^{j\omega})$ is the Fourier transform of the time-shifted sequence $x[n - m]$. In other words, a shift in the time domain by m points (with positive m corresponding to a time delay and negative m to a time advance) corresponds in the frequency domain to multiplication of the Fourier transform by the linear-phase factor $e^{-j\omega m}$. In Section 8.2.2, we discussed the corresponding property for the DFS coefficients of a periodic sequence; specifically, if a periodic sequence $\tilde{x}[n]$ has Fourier series coefficients $\tilde{X}[k]$, then the shifted sequence $\tilde{x}[n - m]$ has Fourier series coefficients $e^{-j(2\pi k/N)m} \tilde{X}[k]$. Now we will consider the operation in the time domain that corresponds to multiplying the DFT coefficients of a finite-length sequence $x[n]$ by the linear-phase factor $e^{-j(2\pi k/N)m}$. Specifically, let $x_1[n]$ denote the finite-length sequence for which the DFT is $e^{-j(2\pi k/N)m} X[k]$; i.e., if

$$x[n] \overset{\mathcal{DFJ}}{\longleftrightarrow} X[k], \tag{8.77}$$

then we are interested in $x_1[n]$ such that

$$x_1[n] \overset{\mathcal{DFJ}}{\longleftrightarrow} X_1[k] = e^{-j(2\pi k/N)m} X[k] = W_N^{mk} X[k]. \tag{8.78}$$

Since the N-point DFT represents a finite-duration sequence of length N, both $x[n]$ and $x_1[n]$ must be zero outside the interval $0 \le n \le N - 1$, and consequently, $x_1[n]$ cannot result from a simple time shift of $x[n]$. The correct result follows directly from the result of Section 8.2.2 and the interpretation of the DFT as the Fourier series coefficients of the periodic sequence $x_1[((n))_N]$. In particular, from Eqs. (8.59) and (8.62) it follows that

$$\tilde{x}[n] = x[((n))_N] \overset{\mathcal{DFS}}{\longleftrightarrow} \tilde{X}[k] = X[((k))_N], \tag{8.79}$$

and similarly, we can define a periodic sequence $\tilde{x}_1[n]$ such that

$$\tilde{x}_1[n] = x_1[((n))_N] \overset{\mathcal{DFS}}{\longleftrightarrow} \tilde{X}_1[k] = X_1[((k))_N], \tag{8.80}$$

where, by assumption,

$$X_1[k] = e^{-j(2\pi k/N)m} X[k]. \tag{8.81}$$

Therefore, the DFS coefficients of $\tilde{x}_1[n]$ are

$$\tilde{X}_1[k] = e^{-j[2\pi((k))_N/N]m} X[((k))_N]. \tag{8.82}$$

Note that

$$e^{-j[2\pi((k))_N/N]m} = e^{-j(2\pi k/N)m}. \tag{8.83}$$

That is, since $e^{-j(2\pi k/N)m}$ is periodic with period N in both k and m, we can drop the notation $((k))_N$. Hence, Eq. (8.82) becomes

$$\tilde{X}_1[k] = e^{-j(2\pi k/N)m}\, \tilde{X}[k], \tag{8.84}$$

so that it follows from Section 8.2.2 that

$$\tilde{x}_1[n] = \tilde{x}[n - m] = x[((n - m))_N]. \tag{8.85}$$

Thus, the finite-length sequence $x_1[n]$ whose DFT is given by Eq. (8.81) is

$$x_1[n] = \begin{cases} \tilde{x}_1[n] = x[((n - m))_N], & 0 \le n \le N - 1, \\ 0, & \text{otherwise.} \end{cases} \tag{8.86}$$

Equation (8.86) tells us how to construct $x_1[n]$ from $x[n]$.

Example 8.8   Circular Shift of a Sequence

The circular shift procedure is illustrated in Figure 8.12 for $m = -2$; i.e., we want to determine $x_1[n] = x[((n + 2))_N]$ for $N = 6$, which we have shown will have DFT $X_1[k] = W_6^{-2k} X[k]$. Specifically, from $x[n]$, we construct the periodic sequence $\tilde{x}[n] = x[((n))_6]$, as indicated in Figure 8.12(b). According to Eq. (8.85), we then shift $\tilde{x}[n]$ by 2 to the left, obtaining $\tilde{x}_1[n] = \tilde{x}[n + 2]$ as in Figure 8.12(c). Finally, using Eq. (8.86), we extract one period of $\tilde{x}_1[n]$ to obtain $x_1[n]$, as indicated in Figure 8.12(d).

A comparison of Figures 8.12(a) and (d) indicates clearly that $x_1[n]$ does not correspond to a linear shift of $x[n]$, and in fact, both sequences are confined to the interval between 0 and $(N - 1)$. By reference to Figure 8.12, we see that $x_1[n]$ can be formed by shifting $x[n]$, so that as a sequence value leaves the interval 0 to $(N - 1)$ at one end, it enters at the other end. Another interesting point is that, for the example shown in Figure 8.12(a), if we form $x_2[n] = x[((n - 4))_6]$ by shifting the sequence by 4 to the right modulo 6, we obtain the same sequence as $x_1[n]$. In terms of the DFT, this results because $W_6^{4k} = W_6^{-2k}$ or, more generally, $W_N^{mk} = W_N^{-(N-m)k}$, which implies that an N-point circular shift in one direction by m is the same as a circular shift in the opposite direction by $N - m$.

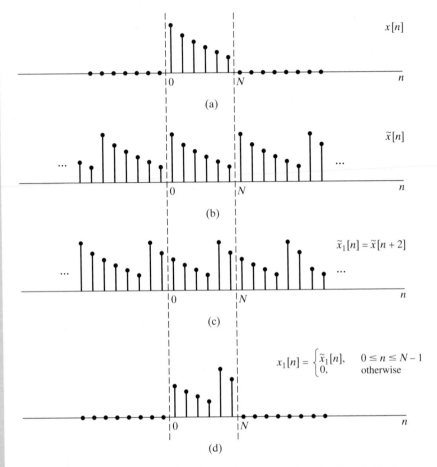

Figure 8.12 Circular shift of a finite-length sequence; i.e., the effect in the time domain of multiplying the DFT of the sequence by a linear-phase factor.

In Section 8.5, we suggested the interpretation of forming the periodic sequence $\tilde{x}[n]$ from the finite-length sequence $x[n]$ by displaying $x[n]$ around the circumference of a cylinder with a circumference of exactly N points. As we repeatedly traverse the circumference of the cylinder, the sequence that we see is the periodic sequence $\tilde{x}[n]$. A linear shift of this sequence corresponds, then, to a rotation of the cylinder. In the context of finite-length sequences and the DFT, such a shift is called a *circular* shift or a *rotation* of the sequence within the interval $0 \leq n \leq N - 1$.

In summary, the circular shift property of the DFT is

$$x[((n - m))_N], \quad 0 \leq n \leq N - 1 \overset{\mathcal{DFT}}{\longleftrightarrow} e^{-j(2\pi k/N)m} X[k] = W_N^m X[k]. \tag{8.87}$$

8.6.3 Duality

Since the DFT is so closely associated with the DFS, we would expect the DFT to exhibit a duality property similar to that of the DFS discussed in Section 8.2.3. In fact, from an examination of Eqs. (8.67) and (8.68), we see that the analysis and synthesis equations differ only in the factor $1/N$ and the sign of the exponent of the powers of W_N.

The DFT duality property can be derived by exploiting the relationship between the DFT and the DFS as in our derivation of the circular shift property. Toward this end, consider $x[n]$ and its DFT $X[k]$, and construct the periodic sequences

$$\tilde{x}[n] = x[((n))_N], \tag{8.88a}$$

$$\tilde{X}[k] = X[((k))_N], \tag{8.88b}$$

so that

$$\tilde{x}[n] \overset{\mathcal{DFS}}{\longleftrightarrow} \tilde{X}[k]. \tag{8.89}$$

From the duality property given in Eqs. (8.25),

$$\tilde{X}[n] \overset{\mathcal{DFS}}{\longleftrightarrow} N\tilde{x}[-k]. \tag{8.90}$$

If we define the periodic sequence $\tilde{x}_1[n] = \tilde{X}[n]$, one period of which is the finite-length sequence $x_1[n] = X[n]$, then the DFS coefficients of $\tilde{x}_1[n]$ are $\tilde{X}_1[k] = N\tilde{x}[-k]$. Therefore, the DFT of $x_1[n]$ is

$$X_1[k] = \begin{cases} N\tilde{x}[-k], & 0 \le k \le N-1, \\ 0, & \text{otherwise,} \end{cases} \tag{8.91}$$

or, equivalently,

$$X_1[k] = \begin{cases} Nx[((-k))_N], & 0 \le k \le N-1, \\ 0, & \text{otherwise.} \end{cases} \tag{8.92}$$

Consequently, the duality property for the DFT can be expressed as follows: If

$$x[n] \overset{\mathcal{DFT}}{\longleftrightarrow} X[k], \tag{8.93a}$$

then

$$X[n] \overset{\mathcal{DFT}}{\longleftrightarrow} Nx[((-k))_N], \qquad 0 \le k \le N-1. \tag{8.93b}$$

The sequence $Nx[((-k))_N]$ is $Nx[k]$ index reversed, modulo N. Index-reversing modulo N corresponds specifically to $((-k))_N = N-k$ for $1 \le k \le N-1$ and $((-k))_N = ((k))_N$ for $k = 0$. As in the case of shifting modulo N, the process of index-reversing modulo N is usually best visualized in terms of the underlying periodic sequences.

Example 8.9   The Duality Relationship for the DFT

To illustrate the duality relationship in Eqs. (8.93), let us consider the sequence $x[n]$ of Example 8.7. Figure 8.13(a) shows the finite-length sequence $x[n]$, and Figures 8.13(b) and 8.13(c) are the real and imaginary parts, respectively, of the corresponding 10-point DFT $X[k]$. By simply relabeling the horizontal axis, we obtain the complex sequence $x_1[n] = X[n]$, as shown in Figures 8.13(d) and 8.13(e). According to the duality relation in Eqs. (8.93), the 10-point DFT of the (complex-valued) sequence $X[n]$ is the sequence shown in Figure 8.13(f).

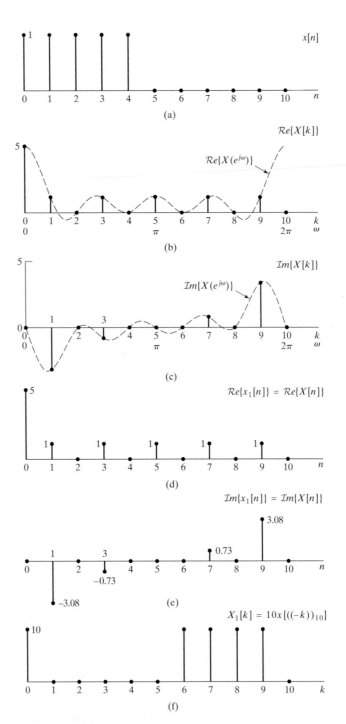

Figure 8.13 Illustration of duality. (a) Real finite-length sequence $x[n]$. (b) and (c) Real and imaginary parts of corresponding DFT $X[k]$. (d) and (e) The real and imaginary parts of the dual sequence $x_1[n] = X[n]$. (f) The DFT of $x_1[n]$.

8.6.4 Symmetry Properties

Since the DFT of $x[n]$ is identical to the DFS coefficients of the periodic sequence $\tilde{x}[n] = x[((n))_N]$, symmetry properties associated with the DFT can be inferred from the symmetry properties of the DFS summarized in Table 8.1 in Section 8.2.6. Specifically, using Eqs. (8.88) together with Properties 9 and 10 in Table 8.1, we have

$$x^*[n] \overset{\mathcal{DFS}}{\longleftrightarrow} X^*[((-k))_N], \qquad 0 \le n \le N-1, \tag{8.94}$$

and

$$x^*[((-n))_N] \overset{\mathcal{DFS}}{\longleftrightarrow} X^*[k], \qquad 0 \le n \le N-1. \tag{8.95}$$

Properties 11–14 in Table 8.1 refer to the decomposition of a periodic sequence into the sum of a conjugate-symmetric and a conjugate-antisymmetric sequence. This suggests the decomposition of the finite-duration sequence $x[n]$ into the two finite-duration sequences of duration N corresponding to one period of the conjugate-symmetric and one period of the conjugate-antisymmetric components of $\tilde{x}[n]$. We will denote these components of $x[n]$ as $x_{ep}[n]$ and $x_{op}[n]$. Thus, with

$$\tilde{x}[n] = x[((n))_N] \tag{8.96}$$

and the conjugate-symmetric part being

$$\tilde{x}_e[n] = \tfrac{1}{2}\{\tilde{x}[n] + \tilde{x}^*[-n]\}, \tag{8.97}$$

and the conjugate-antisymmetric part being

$$\tilde{x}_o[n] = \tfrac{1}{2}\{\tilde{x}[n] - \tilde{x}^*[-n]\}, \tag{8.98}$$

we define $x_{ep}[n]$ and $x_{op}[n]$ as

$$x_{ep}[n] = \tilde{x}_e[n], \qquad 0 \le n \le N-1, \tag{8.99}$$

$$x_{op}[n] = \tilde{x}_o[n], \qquad 0 \le n \le N-1, \tag{8.100}$$

or, equivalently,

$$x_{ep}[n] = \tfrac{1}{2}\{x[((n))_N] + x^*[((-n))_N]\}, \qquad 0 \le n \le N-1, \tag{8.101a}$$

$$x_{op}[n] = \tfrac{1}{2}\{x[((n))_N] - x^*[((-n))_N]\}, \qquad 0 \le n \le N-1, \tag{8.101b}$$

with both $x_{ep}[n]$ and $x_{op}[n]$ being finite-length sequences, i.e., both zero outside the interval $0 \le n \le N-1$. Since $((-n))_N = (N-n)$ and $((n))_N = n$ for $0 \le n \le N-1$, we can also express Eqs. (8.101) as

$$x_{ep}[n] = \tfrac{1}{2}\{x[n] + x^*[N-n]\}, \qquad 1 \le n \le N-1, \tag{8.102a}$$

$$x_{ep}[0] = \mathcal{Re}\{x[0]\}, \tag{8.102b}$$

$$x_{op}[n] = \tfrac{1}{2}\{x[n] - x^*[N-n]\}, \qquad 1 \le n \le N-1, \tag{8.102c}$$

$$x_{op}[0] = j\mathcal{Im}\{x[0]\}. \tag{8.102d}$$

This form of the equations is convenient, since it avoids the modulo N computation of indices.

Clearly, $x_{ep}[n]$ and $x_{op}[n]$ are not equivalent to $x_e[n]$ and $x_o[n]$ as defined by Eqs. (2.149a) and (2.149b). However, it can be shown (see Problem 8.59) that

$$x_{ep}[n] = \{x_e[n] + x_e[n - N]\}, \qquad 0 \le n \le N - 1, \qquad (8.103)$$

and

$$x_{op}[n] = \{x_o[n] + x_o[n - N]\}, \qquad 0 \le n \le N - 1. \qquad (8.104)$$

In other words, $x_{ep}[n]$ and $x_{op}[n]$ can be generated by time-aliasing $x_e[n]$ and $x_o[n]$ into the interval $0 \le n \le N - 1$. The sequences $x_{ep}[n]$ and $x_{op}[n]$ will be referred to as *the periodic conjugate-symmetric* and *periodic conjugate-antisymmetric* components, respectively, of $x[n]$. When $x_{ep}[n]$ and $x_{op}[n]$ are real, they will be referred to as the *periodic even* and *periodic odd* components, respectively. Note that the sequences $x_{ep}[n]$ and $x_{op}[n]$ are not periodic sequences; they are, however, finite-length sequences that are equal to one period of the periodic sequences $\tilde{x}_e[n]$ and $\tilde{x}_o[n]$, respectively.

Equations (8.101) and (8.102) define $x_{ep}[n]$ and $x_{op}[n]$ in terms of $x[n]$. The inverse relation, expressing $x[n]$ in terms of $x_{ep}[n]$ and $x_{op}[n]$, can be obtained by using Eqs. (8.97) and (8.98) to express $\tilde{x}[n]$ as

$$\tilde{x}[n] = \tilde{x}_e[n] + \tilde{x}_o[n]. \qquad (8.105)$$

Thus,

$$x[n] = \tilde{x}[n] = \tilde{x}_e[n] + \tilde{x}_o[n], \qquad 0 \le n \le N - 1. \qquad (8.106)$$

Combining Eqs. (8.106) with Eqs. (8.99) and (8.100), we obtain

$$x[n] = x_{ep}[n] + x_{op}[n]. \qquad (8.107)$$

Alternatively, Eqs. (8.102), when added, also lead to Eq. (8.107). The symmetry properties of the DFT associated with properties 11–14 in Table 8.1 now follow in a straightforward way:

$$\mathcal{R}e\{x[n]\} \overset{\mathcal{DFS}}{\longleftrightarrow} X_{ep}[k], \qquad (8.108)$$

$$j\mathcal{I}m\{x[n]\} \overset{\mathcal{DFS}}{\longleftrightarrow} X_{op}[k], \qquad (8.109)$$

$$x_{ep}[n] \overset{\mathcal{DFS}}{\longleftrightarrow} \mathcal{R}e\{X[k]\}, \qquad (8.110)$$

$$x_{op}[n] \overset{\mathcal{DFS}}{\longleftrightarrow} j\mathcal{I}m\{X[k]\}. \qquad (8.111)$$

8.6.5 Circular Convolution

In Section 8.2.5, we showed that multiplication of the DFS coefficients of two periodic sequences corresponds to a periodic convolution of the sequences. Here, we consider two *finite-duration* sequences $x_1[n]$ and $x_2[n]$, both of length N, with DFTs $X_1[k]$ and $X_2[k]$, respectively, and we wish to determine the sequence $x_3[n]$, for which the DFT is $X_3[k] = X_1[k]X_2[k]$. To determine $x_3[n]$, we can apply the results of Section 8.2.5. Specifically, $x_3[n]$ corresponds to one period of $\tilde{x}_3[n]$, which is given by Eq. (8.27). Thus,

$$x_3[n] = \sum_{m=0}^{N-1} \tilde{x}_1[m]\tilde{x}_2[n - m], \qquad 0 \le n \le N - 1, \qquad (8.112)$$

or, equivalently,

$$x_3[n] = \sum_{m=0}^{N-1} x_1[((m))_N]x_2[((n-m))_N], \qquad 0 \le n \le N-1. \tag{8.113}$$

Since $((m))_N = m$ for $0 \le m \le N-1$, Eq. (8.113) can be written

$$x_3[n] = \sum_{m=0}^{N-1} x_1[m]x_2[((n-m))_N], \qquad 0 \le n \le N-1. \tag{8.114}$$

Equation (8.114) differs from a linear convolution of $x_1[n]$ and $x_2[n]$ as defined by Eq. (2.49) in some important respects. In linear convolution, the computation of the sequence value $x_3[n]$ involves multiplying one sequence by a time-reversed and linearly shifted version of the other and then summing the values of the product $x_1[m]x_2[n-m]$ over all m. To obtain successive values of the sequence formed by the convolution operation, the two sequences are successively shifted relative to each other along a linear axis. In contrast, for the convolution defined by Eq. (8.114), the second sequence is circularly time reversed and circularly shifted with respect to the first. For this reason, the operation of combining two finite-length sequences according to Eq. (8.114) is called *circular convolution*. More specifically, we refer to Eq. (8.114) as an N-point circular convolution, explicitly identifying the fact that both sequences have length N (or less) and that the sequences are shifted modulo N. Sometimes, the operation of forming a sequence $x_3[n]$ for $0 \le n \le N-1$ using Eq. (8.114) will be denoted

$$x_3[n] = x_1[n] \, \text{\textcircled{N}} \, x_2[n], \tag{8.115}$$

i.e., the symbol $\text{\textcircled{N}}$ denotes N-point circular convolution.

Since the DFT of $x_3[n]$ is $X_3[k] = X_1[k]X_2[k]$ and since $X_1[k]X_2[k] = X_2[k]X_1[k]$, it follows with no further analysis that

$$x_3[n] = x_2[n] \, \text{\textcircled{N}} \, x_1[n], \tag{8.116}$$

or, more specifically,

$$x_3[n] = \sum_{m=0}^{N-1} x_2[m]x_1[((n-m))_N]. \tag{8.117}$$

That is, circular convolution, like linear convolution, is a commutative operation.

Since circular convolution is really just periodic convolution, Example 8.4 and Figure 8.3 are also illustrative of circular convolution. However, if we use the notion of circular shifting, it is not necessary to construct the underlying periodic sequences as in Figure 8.3. This is illustrated in the following examples.

Example 8.10   Circular Convolution with a Delayed Impulse Sequence

An example of circular convolution is provided by the result of Section 8.6.2. Let $x_2[n]$ be a finite-duration sequence of length N and

$$x_1[n] = \delta[n - n_0], \tag{8.118}$$

where $0 < n_0 < N$. Clearly, $x_1[n]$ can be considered as the finite-duration sequence

$$x_1[n] = \begin{cases} 0, & 0 \le n < n_0, \\ 1, & n = n_0, \\ 0, & n_0 < n \le N - 1. \end{cases} \tag{8.119}$$

as depicted in Figure 8.14 for $n_0 = 1$.

The DFT of $x_1[n]$ is

$$X_1[k] = W_N^{kn_0}. \tag{8.120}$$

If we form the product

$$X_3[k] = W_N^{kn_0} X_2[k], \tag{8.121}$$

we see from Section 8.6.2 that the finite-duration sequence corresponding to $X_3[k]$ is the sequence $x_2[n]$ rotated to the right by n_0 samples in the interval $0 \le n \le N - 1$. That is, the circular convolution of a sequence $x_2[n]$ with a single delayed unit impulse results in a rotation of $x_2[n]$ in the interval $0 \le n \le N - 1$. This example is illustrated in Figure 8.14 for $N = 5$ and $n_0 = 1$. Here, we show the sequences $x_2[m]$

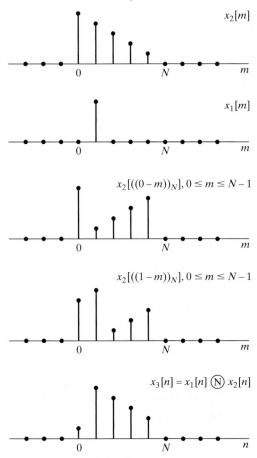

Figure 8.14   Circular convolution of a finite-length sequence $x_2[n]$ with a single delayed impulse, $x_1[n] = \delta[n - 1]$.

and $x_1[m]$ and then $x_2[((0-m))_N]$ and $x_2[((1-m))_N]$. It is clear from these two cases that the result of circular convolution of $x_2[n]$ with a single shifted unit impulse will be to circularly shift $x_2[n]$. The last sequence shown is $x_3[n]$, the result of the circular convolution of $x_1[n]$ and $x_2[n]$.

Example 8.11   Circular Convolution of Two Rectangular Pulses

As another example of circular convolution, let

$$x_1[n] = x_2[n] = \begin{cases} 1, & 0 \le n \le L-1, \\ 0, & \text{otherwise,} \end{cases} \tag{8.122}$$

where, in Figure 8.15, $L = 6$. If we let N denote the DFT length, then, for $N = L$, the N-point DFTs are

$$X_1[k] = X_2[k] = \sum_{n=0}^{N-1} W_N^{kn} = \begin{cases} N, & k = 0, \\ 0, & \text{otherwise.} \end{cases} \tag{8.123}$$

If we explicitly multiply $X_1[k]$ and $X_2[k]$, we obtain

$$X_3[k] = X_1[k]X_2[k] = \begin{cases} N^2, & k = 0, \\ 0, & \text{otherwise,} \end{cases} \tag{8.124}$$

from which it follows that

$$x_3[n] = N, \qquad 0 \le n \le N-1. \tag{8.125}$$

This result is depicted in Figure 8.15. Clearly, as the sequence $x_2[((n-m))_N]$ is rotated with respect to $x_1[m]$, the sum of products $x_1[m]x_2[((n-m))_N]$ will always be equal to N.

Of course, it is possible to consider $x_1[n]$ and $x_2[n]$ as $2L$-point sequences by augmenting them with L zeros. If we then perform a $2L$-point circular convolution of the augmented sequences, we obtain the sequence in Figure 8.16, which can be seen to be identical to the linear convolution of the finite-duration sequences $x_1[n]$ and $x_2[n]$. This important observation will be discussed in much more detail in Section 8.7.

Note that for $N = 2L$, as in Figure 8.16,

$$X_1[k] = X_2[k] = \frac{1 - W_N^{Lk}}{1 - W_N^k},$$

so the DFT of the triangular-shaped sequence $x_3[n]$ in Figure 8.16(e) is

$$X_3[k] = \left(\frac{1 - W_N^{Lk}}{1 - W_N^k} \right)^2,$$

with $N = 2L$.

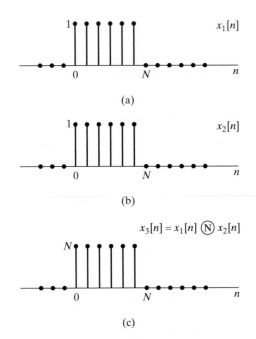

Figure 8.15   *N*-point circular convolution of two constant sequences of length *N*.

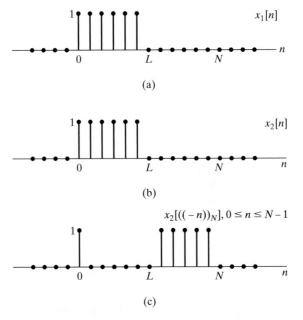

Figure 8.16   2*L*-point circular convolution of two constant sequences of length *L*.

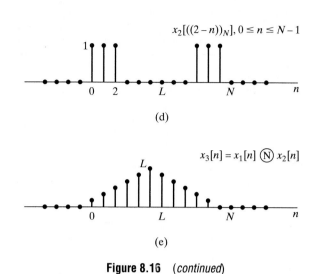

(d)

(e)

Figure 8.16   (*continued*)

The circular convolution property is represented as

$$x_1[n] \ \textcircled{N} \ x_2[n] \ \overset{\mathcal{DFT}}{\longleftrightarrow} \ X_1[k]X_2[k]. \tag{8.126}$$

In view of the duality of the DFT relations, it is not surprising that the DFT of a product of two N-point sequences is the circular convolution of their respective DFTs. Specifically, if $x_3[n] = x_1[n]x_2[n]$, then

$$X_3[k] = \frac{1}{N} \sum_{\ell=0}^{N-1} X_1[\ell]X_2[((k - \ell))_N] \tag{8.127}$$

or

$$x_1[n]x_2[n] \ \overset{\mathcal{DFT}}{\longleftrightarrow} \ \frac{1}{N}X_1[k] \ \textcircled{N} \ X_2[k]. \tag{8.128}$$

8.6.6 Summary of Properties of the DFT

The properties of the DFT that we discussed in Section 8.6 are summarized in Table 8.2. Note that for all of the properties, the expressions given specify $x[n]$ for $0 \le n \le N - 1$ and $X[k]$ for $0 \le k \le N - 1$. Both $x[n]$ and $X[k]$ are equal to zero outside those ranges.

TABLE 8.2   SUMMARY OF PROPERTIES OF THE DFT

Finite-Length Sequence (Length N)	N-point DFT (Length N)
1. $x[n]$	$X[k]$
2. $x_1[n], x_2[n]$	$X_1[k], X_2[k]$
3. $ax_1[n] + bx_2[n]$	$aX_1[k] + bX_2[k]$
4. $X[n]$	$Nx[((-k))_N]$
5. $x[((n-m))_N]$	$W_N^{km} X[k]$
6. $W_N^{-\ell n} x[n]$	$X[((k-\ell))_N]$
7. $\displaystyle\sum_{m=0}^{N-1} x_1[m]x_2[((n-m))_N]$	$X_1[k]X_2[k]$
8. $x_1[n]x_2[n]$	$\displaystyle\frac{1}{N}\sum_{\ell=0}^{N-1} X_1[\ell]X_2[((k-\ell))_N]$
9. $x^*[n]$	$X^*[((-k))_N]$
10. $x^*[((-n))_N]$	$X^*[k]$
11. $\mathcal{Re}\{x[n]\}$	$X_{ep}[k] = \frac{1}{2}\{X[((k))_N] + X^*[((-k))_N]\}$
12. $j\mathcal{Im}\{x[n]\}$	$X_{op}[k] = \frac{1}{2}\{X[((k))_N] - X^*[((-k))_N]\}$
13. $x_{ep}[n] = \frac{1}{2}\{x[n] + x^*[((-n))_N]\}$	$\mathcal{Re}\{X[k]\}$
14. $x_{op}[n] = \frac{1}{2}\{x[n] - x^*[((-n))_N]\}$	$j\mathcal{Im}\{X[k]\}$

Properties 15–17 apply only when $x[n]$ is real.

15. Symmetry properties	$\begin{cases} X[k] = X^*[((-k))_N] \\ \mathcal{Re}\{X[k]\} = \mathcal{Re}\{X[((-k))_N]\} \\ \mathcal{Im}\{X[k]\} = -\mathcal{Im}\{X[((-k))_N]\} \\ |X[k]| = |X[((-k))_N]| \\ \angle\{X[k]\} = -\angle\{X[((-k))_N]\} \end{cases}$
16. $x_{ep}[n] = \frac{1}{2}\{x[n] + x[((-n))_N]\}$	$\mathcal{Re}\{X[k]\}$
17. $x_{op}[n] = \frac{1}{2}\{x[n] - x[((-n))_N]\}$	$j\mathcal{Im}\{X[k]\}$

8.7 COMPUTING LINEAR CONVOLUTION USING THE DFT

We will show in Chapter 9 that efficient algorithms are available for computing the DFT of a finite-duration sequence. These are known collectively as FFT algorithms. Because these algorithms are available, it is computationally efficient to implement a convolution of two sequences by the following procedure:

(a) Compute the N-point DFTs $X_1[k]$ and $X_2[k]$ of the two sequences $x_1[n]$ and $x_2[n]$, respectively.

(b) Compute the product $X_3[k] = X_1[k]X_2[k]$ for $0 \le k \le N-1$.

(c) Compute the sequence $x_3[n] = x_1[n] \,\circledN\, x_2[n]$ as the inverse DFT of $X_3[k]$.

In many DSP applications, we are interested in implementing a linear convolution of two sequences; i.e., we wish to implement an LTI system. This is certainly true, for example, in filtering a sequence such as a speech waveform or a radar signal or in computing the autocorrelation function of such signals. As we saw in Section 8.6.5, the multiplication of DFTs corresponds to a circular convolution of the sequences. To obtain a linear convolution, we must ensure that circular convolution has the effect of linear convolution. The discussion at the end of Example 8.11 hints at how this might be done. We now present a more detailed analysis.

8.7.1 Linear Convolution of Two Finite-Length Sequences

Consider a sequence $x_1[n]$ whose length is L points and a sequence $x_2[n]$ whose length is P points, and suppose that we wish to combine these two sequences by linear convolution to obtain a third sequence

$$x_3[n] = \sum_{m=-\infty}^{\infty} x_1[m]x_2[n-m]. \tag{8.129}$$

Figure 8.17(a) shows a typical sequence $x_1[m]$ and Figure 8.17(b) shows a typical sequence $x_2[n-m]$ for the three cases $n = -1$, for $0 \leq n \leq L - 1$, and $n = L + P - 1$. Clearly, the product $x_1[m]x_2[n-m]$ is zero for all m whenever $n < 0$ and $n > L + P - 2$; i.e., $x_3[n] \neq 0$ for $0 \leq n \leq L + P - 2$. Therefore, $(L + P - 1)$ is the maximum length of the sequence $x_3[n]$ resulting from the linear convolution of a sequence of length L with a sequence of length P.

8.7.2 Circular Convolution as Linear Convolution with Aliasing

As Examples 8.10 and 8.11 show, whether a circular convolution corresponding to the product of two N-point DFTs is the same as the linear convolution of the corresponding finite-length sequences depends on the length of the DFT in relation to the length of the finite-length sequences. An extremely useful interpretation of the relationship between circular convolution and linear convolution is in terms of time aliasing. Since this interpretation is so important and useful in understanding circular convolution, we will develop it in several ways.

In Section 8.4, we observed that if the Fourier transform $X(e^{j\omega})$ of a sequence $x[n]$ is sampled at frequencies $\omega_k = 2\pi k/N$, then the resulting sequence corresponds to the DFS coefficients of the periodic sequence

$$\tilde{x}[n] = \sum_{r=-\infty}^{\infty} x[n-rN]. \tag{8.130}$$

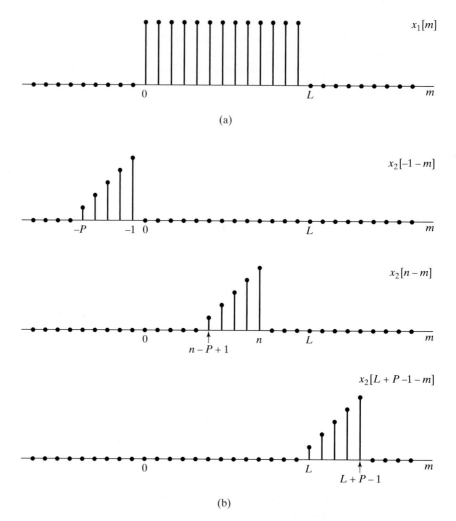

Figure 8.17 Example of linear convolution of two finite-length sequences showing that the result is such that $x_3[n] = 0$ for $n \leq -1$ and for $n \geq L + P - 1$.
(a) Finite-length sequence $x_1[m]$. (b) $x_2[n - m]$ for several values of n.

From our discussion of the DFT, it follows that the finite-length sequence

$$X[k] = \begin{cases} X(e^{j(2\pi k/N)}), & 0 \leq k \leq N - 1, \\ 0, & \text{otherwise,} \end{cases} \tag{8.131}$$

is the DFT of one period of $\tilde{x}[n]$, as given by Eq. (8.130); i.e.,

$$x_p[n] = \begin{cases} \tilde{x}[n], & 0 \leq n \leq N - 1, \\ 0, & \text{otherwise.} \end{cases} \tag{8.132}$$

Obviously, if $x[n]$ has length less than or equal to N, no time aliasing occurs and
$x_p[n] = x[n]$. However, if the length of $x[n]$ is greater than N, $x_p[n]$ may not be equal
to $x[n]$ for some or all values of n. We will henceforth use the subscript p to denote

that a sequence is one period of a periodic sequence resulting from an inverse DFT of a sampled Fourier transform. The subscript can be dropped if it is clear that time aliasing is avoided.

The sequence $x_3[n]$ in Eq. (8.129) has Fourier transform

$$X_3(e^{j\omega}) = X_1(e^{j\omega})X_2(e^{j\omega}). \tag{8.133}$$

If we define a DFT

$$X_3[k] = X_3(e^{j(2\pi k/N)}), \qquad 0 \le k \le N-1, \tag{8.134}$$

then it is clear from Eqs. (8.133) and (8.134) that, also

$$X_3[k] = X_1(e^{j(2\pi k/N)})X_2(e^{j(2\pi k/N)}), \qquad 0 \le k \le N-1, \tag{8.135}$$

and therefore,

$$X_3[k] = X_1[k]X_2[k]. \tag{8.136}$$

That is, the sequence resulting as the inverse DFT of $X_3[k]$ is

$$x_{3p}[n] = \begin{cases} \displaystyle\sum_{r=-\infty}^{\infty} x_3[n-rN], & 0 \le n \le N-1, \\ 0, & \text{otherwise,} \end{cases} \tag{8.137}$$

and from Eq. (8.136), it follows that

$$x_{3p}[n] = x_1[n] \,Ⓝ\, x_2[n]. \tag{8.138}$$

Thus, the circular convolution of two finite-length sequences is equivalent to linear convolution of the two sequences, followed by time aliasing according to Eq. (8.137).

Note that if N is greater than or equal to either L or P, $X_1[k]$ and $X_2[k]$ represent $x_1[n]$ and $x_2[n]$ exactly, but $x_{3p}[n] = x_3[n]$ for all n only if N is greater than or equal to the length of the sequence $x_3[n]$. As we showed in Section 8.7.1, if $x_1[n]$ has length L and $x_2[n]$ has length P, then $x_3[n]$ has maximum length $(L+P-1)$. Therefore, the circular convolution corresponding to $X_1[k]X_2[k]$ is identical to the linear convolution corresponding to $X_1(e^{j\omega})X_2(e^{j\omega})$ if N, the length of the DFTs, satisfies $N \ge L+P-1$.

Example 8.12   Circular Convolution as Linear Convolution with Aliasing

The results of Example 8.11 are easily understood in light of the interpretation just discussed. Note that $x_1[n]$ and $x_2[n]$ are identical constant sequences of length $L = P = 6$, as shown in Figure 8.18(a). The linear convolution of $x_1[n]$ and $x_2[n]$ is of length $L+P-1 = 11$ and has the triangular shape shown in Figure 8.18(b). In Figures 8.18(c) and (d) are shown two of the shifted versions $x_3[n-rN]$ in Eq. (8.137), $x_3[n-N]$ and $x_3[n+N]$ for $N = 6$. The N-point circular convolution of $x_1[n]$ and $x_2[n]$ can be formed by using Eq. (8.137). This is shown in Figure 8.18(e) for $N = L = 6$ and in Figure 8.18(f) for $N = 2L = 12$. Note that for $N = L = 6$, only $x_3[n]$ and $x_3[n+N]$ contribute to the result. For $N = 2L = 12$, only $x_3[n]$ contributes to the result. Since the length of the linear convolution is $(2L-1)$, the result of the circular convolution for $N = 2L$ is identical to the result of linear convolution for all $0 \le n \le N-1$. In fact, this would be true for $N = 2L - 1 = 11$ as well.

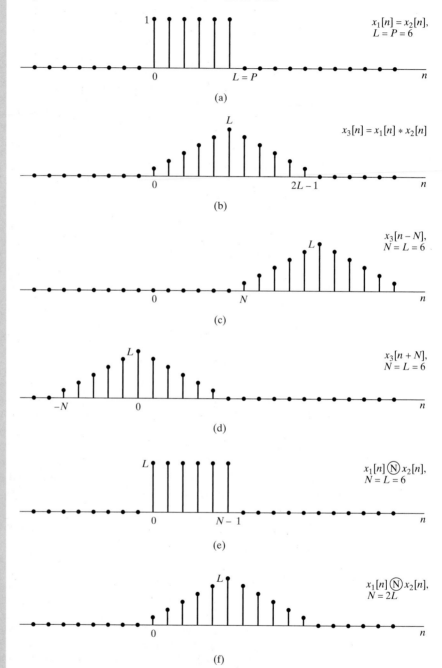

Figure 8.18 Illustration that circular convolution is equivalent to linear convolution followed by aliasing. (a) The sequences $x_1[n]$ and $x_2[n]$ to be convolved. (b) The linear convolution of $x_1[n]$ and $x_2[n]$. (c) $x_3[n - N]$ for $N = 6$. (d) $x_3[n + N]$ for $N = 6$. (e) $x_1[n]$ ⑥ $x_2[n]$, which is equal to the sum of (b), (c), and (d) in the interval $0 \le n \le 5$. (f) $x_1[n]$ ⑫ $x_2[n]$.

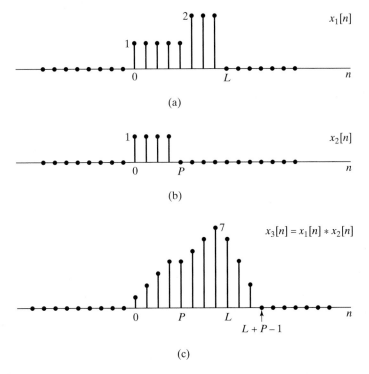

Figure 8.19   An example of linear convolution of two finite-length sequences.

As Example 8.12 points out, time aliasing in the circular convolution of two finite-length sequences can be avoided if $N \geq L + P - 1$. Also, it is clear that if $N = L = P$, all of the sequence values of the circular convolution may be different from those of the linear convolution. However, if $P < L$, some of the sequence values in an L-point circular convolution will be equal to the corresponding sequence values of the linear convolution. The time-aliasing interpretation is useful for showing this.

Consider two finite-duration sequences $x_1[n]$ and $x_2[n]$, with $x_1[n]$ of length L and $x_2[n]$ of length P, where $P < L$, as indicated in Figures 8.19(a) and 8.19(b), respectively. Let us first consider the L-point circular convolution of $x_1[n]$ and $x_2[n]$ and inquire as to which sequence values in the circular convolution are identical to values that would be obtained from a linear convolution and which are not. The linear convolution of $x_1[n]$ with $x_2[n]$ will be a finite-length sequence of length $(L + P - 1)$, as indicated in Figure 8.19(c). To determine the L-point circular convolution, we use Eqs. (8.137) and (8.138) so that

$$x_{3p}[n] = \begin{cases} x_1[n] \; ⓁΧ \; x_2[n] = \displaystyle\sum_{r=-\infty}^{\infty} x_3[n-rL], & 0 \leq n \leq L-1, \\ 0, & \text{otherwise.} \end{cases} \quad (8.139)$$

Figure 8.20(a) shows the term in Eq. (8.139) for $r = 0$, and Figures 8.20(b) and 8.20(c) show the terms for $r = -1$ and $r = +1$, respectively. From Figure 8.20, it should be clear that in the interval $0 \leq n \leq L-1$, $x_{3p}[n]$ is influenced only by $x_3[n]$ and $x_3[n+L]$.

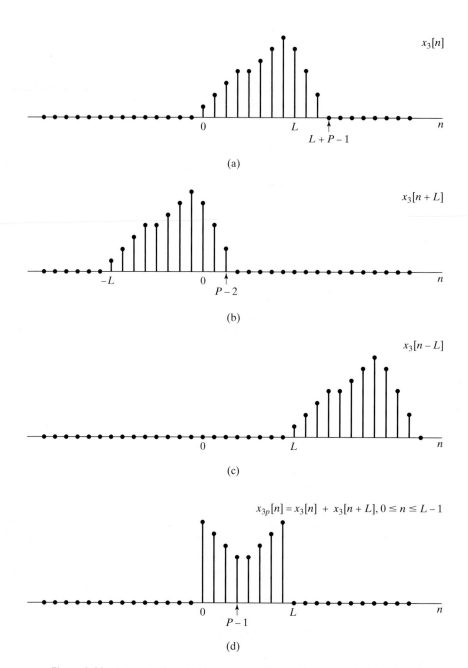

Figure 8.20 Interpretation of circular convolution as linear convolution followed by aliasing for the circular convolution of the two sequences $x_1[n]$ and $x_2[n]$ in Figure 8.19.

In general, whenever $P < L$, only the term $x_3[n + L]$ will alias into the interval $0 \le n \le L-1$. More specifically, when these terms are summed, the last $(P-1)$ points of $x_3[n+L]$, which extend from $n = 0$ to $n = P-2$, will be added to the first $(P-1)$ points of $x_3[n]$, and the last $(P-1)$ points of $x_3[n]$, extending from $n = L$ to $n = L+P-2$, will contribute only to the next period of the underlying periodic result $\tilde{x}_3[n]$. Then, $x_{3p}[n]$ is formed by extracting the portion for $0 \le n \le L - 1$. Since the last $(P - 1)$ points of $x_3[n + L]$ and the last $(P - 1)$ points of $x_3[n]$ are identical, we can alternatively view the process of forming the circular convolution $x_{3p}[n]$ through linear convolution plus aliasing, as taking the $(P - 1)$ values of $x_3[n]$ from $n = L$ to $n = L + P - 2$ and adding them to the first $(P - 1)$ values of $x_3[n]$. This process is illustrated in Figure 8.21 for the case $P = 4$ and $L = 8$. Figure 8.21(a) shows the linear convolution $x_3[n]$, with the points for $n \ge L$ denoted by open symbols. Note that only $(P - 1)$ points for $n \ge L$ are nonzero. Figure 8.21(b) shows the formation of $x_{3p}[n]$ by "wrapping $x_3[n]$ around on itself." The first $(P - 1)$ points are corrupted by the time aliasing, and the remaining points from $n = P - 1$ to $n = L - 1$ (i.e., the last $L - P + 1$ points) are not corrupted; that is, they are identical to what would be obtained with a linear convolution.

From this discussion, it should be clear that if the circular convolution is of sufficient length relative to the lengths of the sequences $x_1[n]$ and $x_2[n]$, then aliasing with nonzero values can be avoided, in which case the circular convolution and linear convolution will be identical. Specifically, if, for the case just considered, $x_3[n]$ is replicated with period $N \ge L + P - 1$, then no nonzero overlap will occur. Figures 8.21(c) and 8.21(d) illustrate this case, again for $P = 4$ and $L = 8$, with $N = 11$.

8.7.3 Implementing Linear Time-Invariant Systems Using the DFT

The previous discussion focused on ways of obtaining a linear convolution from a circular convolution. Since LTI systems can be implemented by convolution, this implies that circular convolution (implemented by the procedure suggested at the beginning of Section 8.7) can be used to implement these systems. To see how this can be done, let us first consider an L-point input sequence $x[n]$ and a P-point impulse response $h[n]$. The linear convolution of these two sequences, which will be denoted by $y[n]$, has finite duration with length $(L + P - 1)$. Consequently, as discussed in Section 8.7.2, for the circular convolution and linear convolution to be identical, the circular convolution must have a length of at least $(L + P - 1)$ points. The circular convolution can be achieved by multiplying the DFTs of $x[n]$ and $h[n]$. Since we want the product to represent the DFT of the linear convolution of $x[n]$ and $h[n]$, which has length $(L + P - 1)$, the DFTs that we compute must also be of at least that length, i.e., both $x[n]$ and $h[n]$ must be augmented with sequence values of zero amplitude. This process is often referred to as *zero-padding*.

This procedure permits the computation of the linear convolution of two finite-length sequences using the DFT; i.e., the output of an FIR system whose input also has finite length can be computed with the DFT. In many applications, such as filtering a speech waveform, the input signal is of indefinite duration. Theoretically, while we might be able to store the entire waveform and then implement the procedure just discussed using a DFT for a large number of points, such a DFT might be impractical to compute.

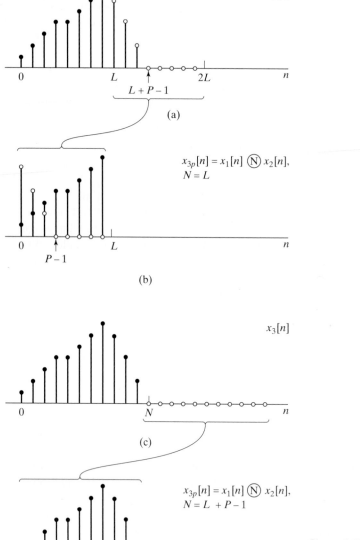

Figure 8.21 Illustration of how the result of a circular convolution "wraps around." (a) and (b) $N = L$, so the aliased "tail" overlaps the first $(P - 1)$ points. (c) and (d) $N = (L + P - 1)$, so no overlap occurs.

Another consideration is that for this method of filtering, no filtered samples can be computed until all the input samples have been collected. Generally, we would like to avoid such a large delay in processing. The solution to both of these problems is to use *block convolution*, in which the signal to be filtered is segmented into sections of length L. Each section can then be convolved with the finite-length impulse response and the

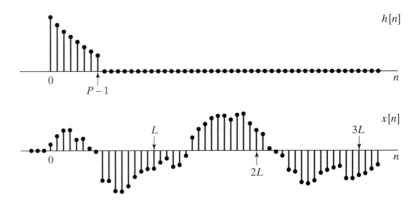

Figure 8.22   Finite-length impulse response $h[n]$ and indefinite-length signal $x[n]$ to be filtered.

filtered sections fitted together in an appropriate way. The linear filtering of each block can then be implemented using the DFT.

To illustrate the procedure and to develop the procedure for fitting the filtered sections together, consider the impulse response $h[n]$ of length P and the signal $x[n]$ depicted in Figure 8.22. Henceforth, we will assume that $x[n] = 0$ for $n < 0$ and that the length of $x[n]$ is much greater than P. The sequence $x[n]$ can be represented as a sum of shifted nonoverlapping finite-length segments of length L; i.e.,

$$x[n] = \sum_{r=0}^{\infty} x_r[n - rL], \tag{8.140}$$

where

$$x_r[n] = \begin{cases} x[n + rL], & 0 \le n \le L - 1, \\ 0, & \text{otherwise.} \end{cases} \tag{8.141}$$

Figure 8.23(a) illustrates this segmentation for $x[n]$ in Figure 8.22. Note that within each segment $x_r[n]$, the first sample is at $n = 0$; however, the zero[th] sample of $x_r[n]$ is the rL[th] sample of the sequence $x[n]$. This is shown in Figure 8.23(a) by plotting the segments in their shifted positions but with the redefined time origin indicated.

Because convolution is an LTI operation, it follows from Eq. (8.140) that

$$y[n] = x[n] * h[n] = \sum_{r=0}^{\infty} y_r[n - rL], \tag{8.142}$$

where

$$y_r[n] = x_r[n] * h[n]. \tag{8.143}$$

Since the sequences $x_r[n]$ have only L nonzero points and $h[n]$ is of length P, each of the terms $y_r[n] = x_r[n] * h[n]$ has length $(L + P - 1)$. Thus, the linear convolution $x_r[n] * h[n]$ can be obtained by the procedure described earlier using N-point DFTs, wherein $N \ge L + P - 1$. Since the beginning of each input section is separated from its neighbors by L points and each filtered section has length $(L + P - 1)$, the nonzero points in the filtered sections will overlap by $(P - 1)$ points, and these overlap samples

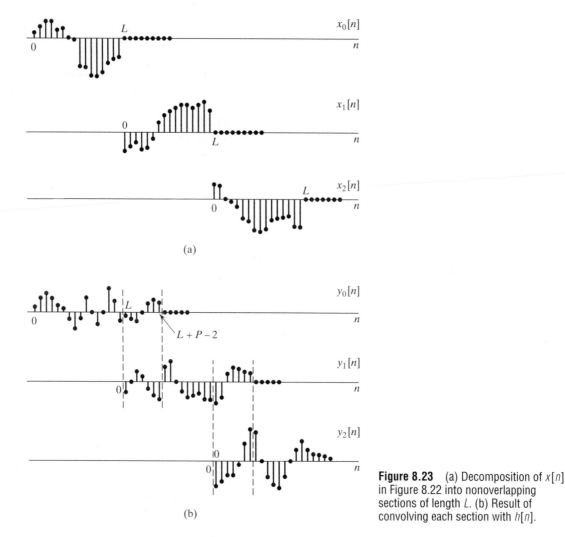

(a)

(b)

Figure 8.23   (a) Decomposition of $x[n]$ in Figure 8.22 into nonoverlapping sections of length L. (b) Result of convolving each section with $h[n]$.

must be added in carrying out the sum required by Eq. (8.142). This is illustrated in Figure 8.23(b), which illustrates the filtered sections, $y_r[n] = x_r[n] * h[n]$. Just as the input waveform is reconstructed by adding the delayed waveforms in Figure 8.23(a), the filtered result $x[n] * h[n]$ is constructed by adding the delayed filtered sections depicted in Figure 8.23(b). This procedure for constructing the filtered output from filtered sections is often referred to as the *overlap–add method*, because the filtered sections are overlapped and added to construct the output. The overlapping occurs because the linear convolution of each section with the impulse response is, in general, longer than the length of the section. The *overlap–add method* of block convolution is not tied to the DFT and circular convolution. Clearly, all that is required is that the smaller convolutions be computed and the results combined appropriately.

An alternative block convolution procedure, commonly called the *overlap–save method*, corresponds to implementing an L-point circular convolution of a P-point impulse response $h[n]$ with an L-point segment $x_r[n]$ and identifying the part of the circular convolution that corresponds to a linear convolution. The resulting output segments are then "patched together" to form the output. Specifically, we showed that if an L-point sequence is circularly convolved with a P-point sequence $(P < L)$, then the first $(P - 1)$ points of the result are incorrect due to time aliasing, whereas the remaining points are identical to those that would be obtained had we implemented a linear convolution. Therefore, we can divide $x[n]$ into sections of length L so that each input section overlaps the preceding section by $(P - 1)$ points. That is, we define the sections as

$$x_r[n] = x[n + r(L - P + 1) - P + 1], \qquad 0 \le n \le L - 1, \qquad (8.144)$$

wherein, as before, we have defined the time origin for each section to be at the beginning of that section rather than at the origin of $x[n]$. This method of sectioning is depicted in Figure 8.24(a). The circular convolution of each section with $h[n]$ is denoted $y_{rp}[n]$, the extra subscript p indicating that $y_{rp}[n]$ is the result of a circular convolution in which time aliasing has occurred. These sequences are depicted in Figure 8.24(b). The portion of each output section in the region $0 \le n \le P - 2$ is the part that must be discarded. The remaining samples from successive sections are then abutted to construct the final filtered output. That is,

$$y[n] = \sum_{r=0}^{\infty} y_r[n - r(L - P + 1) + P - 1], \qquad (8.145)$$

where

$$y_r[n] = \begin{cases} y_{rp}[n], & P - 1 \le n \le L - 1, \\ 0, & \text{otherwise.} \end{cases} \qquad (8.146)$$

This procedure is called the overlap–save method because the input segments overlap, so that each succeeding input section consists of $(L - P + 1)$ new points and $(P - 1)$ points saved from the previous section.

The utility of the overlap–add and the overlap–save methods of block convolution may not be immediately apparent. In Chapter 9, we consider highly efficient algorithms for computing the DFT. These algorithms, collectively called the FFT, are so efficient that, for FIR impulse responses of even modest length (on the order of 25 or 30), it may be more efficient to carry out block convolution using the DFT than to implement the linear convolution directly. The length P at which the DFT method becomes more efficient is, of course, dependent on the hardware and software available to implement the computations. (See Stockham, 1966, and Helms, 1967.)

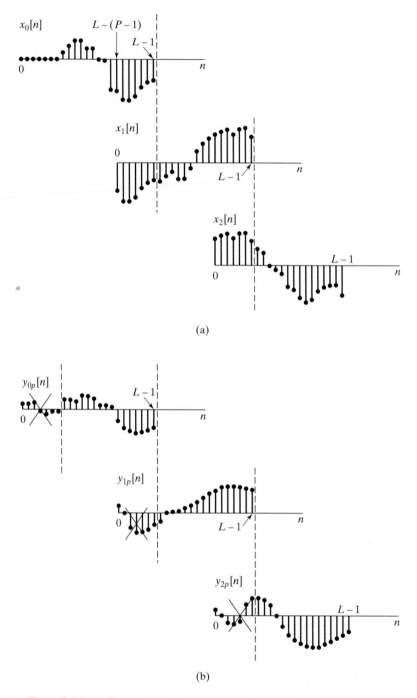

Figure 8.24 (a) Decomposition of $x[n]$ in Figure 8.22 into overlapping sections of length L. (b) Result of convolving each section with $h[n]$. The portions of each filtered section to be discarded in forming the linear convolution are indicated.

8.8 THE DISCRETE COSINE TRANSFORM (DCT)

The DFT is perhaps the most common example of a general class of finite-length transform representations of the form

$$A[k] = \sum_{n=0}^{N-1} x[n]\phi_k^*[n], \tag{8.147}$$

$$x[n] = \frac{1}{N} \sum_{k=0}^{N-1} A[k]\phi_k[n], \tag{8.148}$$

where the sequences $\phi_k[n]$, referred to as the *basis sequences,* are orthogonal to one another; i.e.,

$$\frac{1}{N} \sum_{n=0}^{N-1} \phi_k[n]\phi_m^*[n] = \begin{cases} 1, & m = k, \\ 0, & m \neq k. \end{cases} \tag{8.149}$$

In the case of the DFT, the basis sequences are the complex periodic sequences $e^{j2\pi kn/N}$, and the sequence $A[k]$ is, in general, complex even if the sequence $x[n]$ is real. It is natural to inquire as to whether there exist sets of real-valued basis sequences that would yield a real-valued transform sequence $A[k]$ when $x[n]$ is real. This has led to the definition of a number of other orthogonal transform representations, such as Haar transforms, Hadamard transforms (see Elliott and Rao, 1982), and Hartley transforms (Bracewell, 1983, 1984, 1989). (The definition and properties of the Hartley transform are explored in Problem 8.68.) Another orthogonal transform for real sequences is the discrete cosine transform (DCT). (See Ahmed, Natarajan and Rao, 1974 and Rao and Yip, 1990.) The DCT is closely related to the DFT and has become especially useful and important in a number of signal-processing applications, particularly speech and image compression. In this section, we conclude our discussion of the DFT by introducing the DCT and showing its relationship to the DFT.

8.8.1 Definitions of the DCT

The DCT is a transform in the form of Eqs. (8.147) and (8.148) with basis sequences $\phi_k[n]$ that are cosines. Since cosines are both periodic and have even symmetry, the extension of $x[n]$ outside the range $0 \leq n \leq (N-1)$ in the synthesis Eq. (8.148) will be both periodic and symmetric. In other words, just as the DFT involves an implicit assumption of periodicity, the DCT involves implicit assumptions of both periodicity and *even symmetry*.

In the development of the DFT, we represented finite-length sequences by first forming periodic sequences from which the finite-length sequence can be uniquely recovered and then using an expansion in terms of periodic complex exponentials. In a similar style, the DCT corresponds to forming a periodic, symmetric sequence from a finite-length sequence in such a way that the original finite-length sequence can be uniquely recovered. Because there are many ways to do this, there are many definitions of the DCT. In Figure 8.25, we show 17 samples for each of four examples of symmetric periodic extensions of a four-point sequence. The original finite-length sequence is shown in each subfigure as the samples with solid dots. These sequences are all periodic

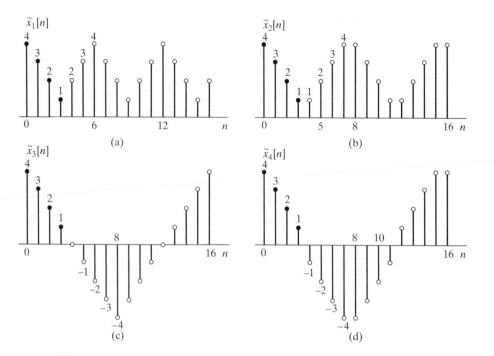

Figure 8.25 Four ways to extend a four-point sequence $x[n]$ both periodically and symmetrically. The finite-length sequence $x[n]$ is plotted with solid dots. (a) Type-1 periodic extension for DCT-1. (b) Type-2 periodic extension for DCT-2. (c) Type-3 periodic extension for DCT-3. (d) Type-4 periodic extension for DCT-4.

(with period 16 or less) and also have even symmetry. In each case, the finite-length sequence is easily extracted as the first four points of one period. For convenience, we denote the periodic sequences obtained by replicating with period 16 each of the four subsequences in Figure 8.25(a), (b), (c), and (d) as $\tilde{x}_1[n]$, $\tilde{x}_2[n]$, $\tilde{x}_3[n]$, and $\tilde{x}_4[n]$, respectively. We note that $\tilde{x}_1[n]$ has period $(2N - 2) = 6$ and has even symmetry about both $n = 0$ and $n = (N - 1) = 3$. The sequence $\tilde{x}_2[n]$ has period $2N = 8$ and has even symmetry about the "half sample" points $n = -\frac{1}{2}$ and $\frac{7}{2}$. The sequence $\tilde{x}_3[n]$ has period $4N = 16$ and has even symmetry about $n = 0$ and $n = 8$. The sequence $\tilde{x}_4[n]$ also has period $4N = 16$ and even symmetry about the "half sample" points $n = -\frac{1}{2}$ and $n = (2N - \frac{1}{2}) = \frac{15}{2}$.

The four different cases shown in Figure 8.25 illustrate the periodicity that is implicit in the four common forms of the DCT, which are referred to as DCT-1, DCT-2, DCT-3, and DCT-4 respectively. It can be shown (see Martucci, 1994) that there are four more ways to create an even periodic sequence from $x[n]$. This implies four other possible DCT representations. Furthermore, it is also possible to create eight odd-symmetric periodic real sequences from $x[n]$, leading to eight different versions of the *discrete sine transform* (DST), wherein the basis sequences in the orthonormal representation are sine functions. These transforms make up a family of 16 orthonormal transforms for real sequences. Of these, the DCT-1 and DCT-2 representations are the most used, and we shall focus on them in the remainder of our discussion.

8.8.2 Definition of the DCT-1 and DCT-2

All of the periodic extensions leading to different forms of the DCT can be thought of as a sum of shifted copies of the N-point sequences $\pm x[n]$ and $\pm x[-n]$. The differences between the extensions for the DCT-1 and DCT-2 depend on whether the endpoints overlap with shifted versions of themselves and, if so, which of the endpoints overlap. For the DCT-1, $x[n]$ is first modified at the endpoints and then extended to have period $2N - 2$. The resulting periodic sequence is

$$\tilde{x}_1[n] = x_\alpha[((n))_{2N-2}] + x_\alpha[((-n))_{2N-2}], \tag{8.150}$$

where $x_\alpha[n]$ is the modified sequence $x_\alpha[n] = \alpha[n]x[n]$, with

$$\alpha[n] = \begin{cases} \frac{1}{2}, & n = 0 \text{ and } N - 1, \\ 1, & 1 \leq n \leq N - 2. \end{cases} \tag{8.151}$$

The weighting of the endpoints compensates for the doubling that occurs when the two terms in Eq. (8.150) overlap at $n = 0$, $n = (N - 1)$, and at the corresponding points spaced from these by integer multiples of $(2N - 2)$. With this weighting, it is easily verified that $x[n] = \tilde{x}_1[n]$ for $n = 0, 1, \ldots, N - 1$. The resulting periodic sequence $\tilde{x}_1[n]$ has even periodic symmetry about the points $n = 0$ and $n = N-1, 2N-2$, etc., which we refer to as *Type-1* periodic symmetry. Figure 8.25 (a) is an example of Type-1 symmetry where $N = 4$ and the periodic sequence $\tilde{x}_1[n]$ has period $2N - 2 = 6$. The DCT-1 is defined by the transform pair

$$X^{c1}[k] = 2 \sum_{n=0}^{N-1} \alpha[n]x[n] \cos\left(\frac{\pi k n}{N - 1}\right), \qquad 0 \leq k \leq N - 1, \tag{8.152}$$

$$x[n] = \frac{1}{N - 1} \sum_{k=0}^{N-1} \alpha[k]X^{c1}[k] \cos\left(\frac{\pi k n}{N - 1}\right), \qquad 0 \leq n \leq N - 1, \tag{8.153}$$

where $\alpha[n]$ is defined in Eq. (8.151).

For the DCT-2, $x[n]$ is extended to have period $2N$, and the periodic sequence is given by

$$\tilde{x}_2[n] = x[((n))_{2N}] + x[((-n - 1))_{2N}], \tag{8.154}$$

Because the endpoints do not overlap, no modification of them is required to ensure that $x[n] = \tilde{x}_2[n]$ for $n = 0, 1, \ldots, N - 1$. In this case, which we call *Type-2* periodic symmetry, the periodic sequence $\tilde{x}_2[n]$ has even periodic symmetry about the "half sample" points $-1/2$, $N - 1/2$, $2N - 1/2$, etc. This is illustrated by Figure 8.25(b) for $N = 4$ and period $2N = 8$. The DCT-2 is defined by the transform pair

$$X^{c2}[k] = 2 \sum_{n=0}^{N-1} x[n] \cos\left(\frac{\pi k(2n + 1)}{2N}\right), \qquad 0 \leq k \leq N - 1, \tag{8.155}$$

$$x[n] = \frac{1}{N} \sum_{k=0}^{N-1} \beta[k]X^{c2}[k] \cos\left(\frac{\pi k(2n + 1)}{2N}\right), \qquad 0 \leq n \leq N - 1, \tag{8.156}$$

where the inverse DCT-2 involves the weighting function

$$\beta[k] = \begin{cases} \frac{1}{2}, & k = 0 \\ 1, & 1 \le k \le N - 1. \end{cases} \tag{8.157}$$

In many treatments, the DCT definitions include normalization factors that make the transforms *unitary*.[4] For example, the DCT-2 form is often defined as

$$\tilde{X}^{c2}[k] = \sqrt{\frac{2}{N}}\tilde{\beta}[k] \sum_{n=0}^{N-1} x[n]\cos\left(\frac{\pi k(2n+1)}{2N}\right), \qquad 0 \le k \le N-1, \tag{8.158}$$

$$x[n] = \sqrt{\frac{2}{N}} \sum_{k=0}^{N-1} \tilde{\beta}[k]\tilde{X}^{c2}[k]\cos\left(\frac{\pi k(2n+1)}{2N}\right), \qquad 0 \le n \le N-1, \tag{8.159}$$

where

$$\tilde{\beta}[k] = \begin{cases} \frac{1}{\sqrt{2}}, & k = 0, \\ 1, & k = 1, 2, \ldots, N-1. \end{cases} \tag{8.160}$$

Comparing these equations with Eqs. (8.155) and (8.156), we see that the multiplicative factors 2, $1/N$, and $\beta[k]$ have been redistributed between the direct and inverse transforms. (A similar normalization can be applied to define a normalized version of the DCT-1.) While this normalization creates a unitary transform representation, the definitions in Eqs. (8.152) and (8.153) and Eqs. (8.155) and (8.156) are simpler to relate to the DFT as we have defined it in this chapter. Therefore, in the following discussions, we use our definitions rather than the normalized definitions that are found, for example, in Rao and Yip (1990) and many other texts.

Although we normally evaluate the DCT only for $0 \le k \le N-1$, nothing prevents our evaluating the DCT equations outside that interval, as illustrated in Figure 8.26, wherein the DCT values for $0 \le k \le N - 1$ are shown as solid dots. These figures illustrate that the DCTs also are even periodic sequences. However, the symmetry of the transform sequence is not always the same as the symmetry of the implicit periodic input sequence. While $\tilde{x}_1[n]$ and the extension of $X^{c1}[k]$ both have Type-1 symmetry with the same period, we see from a comparison of Figures 8.25(c) and 8.26(b) that the extended $X^{c2}[k]$ has the same symmetry as $\tilde{x}_3[n]$ rather than $\tilde{x}_2[n]$. Furthermore, $X^{c2}[n]$ extends with period $4N$ while $\tilde{x}_2[n]$ has period $2N$.

Since the DCTs are orthogonal transform representations, they have properties similar in form to those of the DFT. These properties are elaborated on in some detail in Ahmed, Natarajan and Rao (1974) and Rao and Yip (1990).

8.8.3 Relationship between the DFT and the DCT-1

As might be expected, there is a close relationship between the DFT and the various classes of the DCT of a finite-length sequence. To develop this relationship, we note that,

---

[4]The DCT would be a unitary transform if it is orthonormal and also has the property that
$$\sum_{n=0}^{N-1}(x[n])^2 = \sum_{k=0}^{N-1}(X^{c2}[k])^2.$$

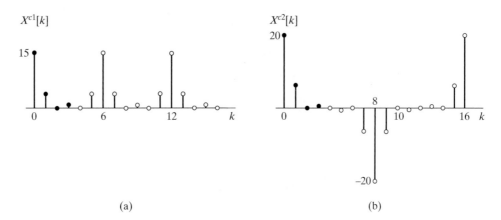

(a)   (b)

Figure 8.26   DCT-1 and DCT-2 for the four-point sequence used in Figure 8.25.
(a) DCT-1. (b) DCT-2.

since, for the DCT-1, $\tilde{x}_1[n]$ is constructed from $x_1[n]$ through Eqs. (8.150) and (8.151), one period of the periodic sequence $\tilde{x}_1[n]$ defines the finite-length sequence

$$x_1[n] = x_\alpha[((n))_{2N-2}] + x_\alpha[((-n))_{2N-2}] = \tilde{x}_1[n], \qquad n = 0, 1, \ldots, 2N - 3, \quad (8.161)$$

where $x_\alpha[n] = \alpha[n]x[n]$ is the N-point real sequence with endpoints divided by 2. From Eq. (8.161), it follows that the $(2N - 2)$-point DFT of the $(2N - 2)$-point sequence $x_1[n]$ is

$$X_1[k] = X_\alpha[k] + X_\alpha^*[k] = 2\mathcal{R}e\{X_\alpha[k]\}, \qquad k = 0, 1, \ldots, 2N - 3, \quad (8.162)$$

where $X_\alpha[k]$ is the $(2N - 2)$-point DFT of the N-point sequence $\alpha[n]x[n]$; i.e., $\alpha[n]x[n]$ is padded with $(N - 2)$ zero samples. Using the definition of the $(2N - 2)$-point DFT of the padded sequence, we obtain for $k = 0, 1 \ldots, N - 1$,

$$X_1[k] = 2\mathcal{R}e\{X_\alpha[k]\} = 2 \sum_{n=0}^{N-1} \alpha[n]x[n] \cos\left(\frac{2\pi kn}{2N - 2}\right) = X^{c1}[k]. \quad (8.163)$$

Therefore, the DCT-1 of an N-point sequence is identical to the first N points of $X_1[k]$, the $(2N - 2)$-point DFT of the symmetrically extended sequence $x_1[n]$, and it is also identical to twice the real part of the first N points of $X_\alpha[k]$, the $(2N - 2)$-point DFT of the weighted sequence $x_\alpha[n]$.

Since, as discussed in Chapter 9, fast computational algorithms exist for the DFT, they can be used to compute the DFTs $X_\alpha[k]$ or $X_1[k]$ in Eq. (8.163), thus providing a convenient and readily available fast computation of the DCT-1. Since the definition of the DCT-1 involves only real-valued coefficients, there are also efficient algorithms for computing the DCT-1 of real sequences more directly without requiring the use of complex multiplications and additions. (See Ahmed, Natarajan and Rao, 1974 and Chen and Fralick, 1977.)

The inverse DCT-1 can also be computed using the inverse DFT. It is only necessary to use Eq. (8.163) to construct $X_1[k]$ from $X^{c1}[k]$ and then compute the inverse $(2N - 2)$-point DFT. Specifically,

$$X_1[k] = \begin{cases} X^{c1}[k], & k = 0, \ldots, N - 1, \\ X^{c1}[2N - 2 - k], & k = N, \ldots, 2N - 3, \end{cases} \quad (8.164)$$

and, using the definition of the $(2N - 2)$-point inverse DFT, we can compute the symmetrically extended sequence

$$x_1[n] = \frac{1}{2N-2} \sum_{k=0}^{2N-3} X_1[k] e^{j2\pi kn/(2N-2)}, \qquad n = 0, 1, \ldots, 2N - 3, \qquad (8.165)$$

from which we can obtain $x[n]$ by extracting the first N points, i.e., $x[n] = x_1[n]$ for $n = 0, 1, \ldots, N - 1$. By substitution of Eq. (8.164) into Eq. (8.165), it also follows that the inverse DCT-1 relation can be expressed in terms of $X^{c1}[k]$ and cosine functions, as in Eq. (8.153). This is suggested as an exercise in Problem 8.71.

8.8.4 Relationship between the DFT and the DCT-2

It is also possible to express the DCT-2 of a finite-length sequence $x[n]$ in terms of the DFT. To develop this relationship, observe that one period of the periodic sequence $\tilde{x}_2[n]$ defines the $2N$-point sequence

$$x_2[n] = x[((n))_{2N}] + x[((-n-1))_{2N}] = \tilde{x}_2[n], \qquad n = 0, 1, \ldots, 2N - 1, \qquad (8.166)$$

where $x[n]$ is the original N-point real sequence. From Eq. (8.166), it follows that the $2N$-point DFT of the $2N$-point sequence $x_2[n]$ is

$$X_2[k] = X[k] + X^*[k] e^{j2\pi k/(2N)}, \qquad k = 0, 1, \ldots, 2N - 1, \qquad (8.167)$$

where $X[k]$ is the $2N$-point DFT of the N-point sequence $x[n]$; i.e., in this case, $x[n]$ is padded with N zero samples. From Eq. (8.167), we obtain

$$X_2[k] = X[k] + X^*[k] e^{j2\pi k/(2N)}$$
$$= e^{j\pi k/(2N)} \left(X[k] e^{-j\pi k/(2N)} + X^*[k] e^{j\pi k/(2N)} \right) \qquad (8.168)$$
$$= e^{j\pi k/(2N)} 2\mathcal{R}e\left\{ X[k] e^{-j\pi k/(2N)} \right\}.$$

From the definition of the $2N$-point DFT of the padded sequence, it follows that

$$\mathcal{R}e\left\{ X[k] e^{-j\pi k/(2N)} \right\} = \sum_{n=0}^{N-1} x[n] \cos\left(\frac{\pi k(2n+1)}{2N} \right). \qquad (8.169)$$

Therefore, using Eqs. (8.155), (8.167), and (8.169), we can express $X^{c2}[k]$ in terms of $X[k]$, the $2N$-point DFT of the N-point sequence $x[n]$, as

$$X^{c2}[k] = 2\mathcal{R}e\left\{ X[k] e^{-j\pi k/(2N)} \right\}, \qquad k = 0, 1, \ldots, N - 1, \qquad (8.170)$$

or in terms of the $2N$-point DFT of the $2N$-point symmetrically extended sequence $x_2[n]$ defined by Eq. (8.166) as

$$X^{c2}[k] = e^{-j\pi k/(2N)} X_2[k], \qquad k = 0, 1, \ldots, N - 1, \qquad (8.171)$$

and equivalently,

$$X_2[k] = e^{j\pi k/(2N)} X^{c2}[k], \qquad k = 0, 1, \ldots, N - 1. \qquad (8.172)$$

As in the case of the DCT-1, fast algorithms can be used to compute the $2N$-point DFTs $X[k]$ and $X_2[k]$ in Eqs. (8.170) and (8.171), respectively. Makhoul (1980) discusses other ways that the DFT can be used to compute the DCT-2. (See also Problem 8.72.) In addition, special fast algorithms for the computation of the DCT-2 have been developed (Rao and Yip, 1990).

The inverse DCT-2 can also be computed using the inverse DFT. The procedure utilizes Eq. (8.172) together with a symmetry property of the DCT-2. Specifically, it is easily verified by direct substitution into Eq. (8.155) that

$$X^{c2}[2N - k] = -X^{c2}[k], \qquad k = 0, 1, \ldots, 2N - 1, \tag{8.173}$$

from which it follows that

$$X_2[k] = \begin{cases} X^{c2}[0], & k = 0, \\ e^{j\pi k/(2N)} X^{c2}[k], & k = 1, \ldots, N - 1, \\ 0, & k = N, \\ -e^{j\pi k/(2N)} X^{c2}[2N - k], & k = N + 1, N + 2, \ldots, 2N - 1. \end{cases} \tag{8.174}$$

Using the inverse DFT, we can compute the symmetrically extended sequence

$$x_2[n] = \frac{1}{2N} \sum_{k=0}^{2N-1} X_2[k] e^{j2\pi kn/(2N)}, \qquad n = 0, 1, \ldots, 2N - 1, \tag{8.175}$$

from which we can obtain $x[n] = x_2[n]$ for $n = 0, 1, \ldots, N-1$. By substituting Eq. (8.174) into Eq. (8.175), we can easily show that the inverse DCT-2 relation is that given by Eq. (8.156). (See Problem 8.73.)

8.8.5 Energy Compaction Property of the DCT-2

The DCT-2 is used in many data compression applications in preference to the DFT because of a property that is frequently referred to as "energy compaction." Specifically, the DCT-2 of a finite-length sequence often has its coefficients more highly concentrated at low indices than the DFT does. The importance of this flows from Parseval's theorem, which, for the DCT-1, is

$$\sum_{n=0}^{N-1} \alpha[n]|x[n]|^2 = \frac{1}{2N - 2} \sum_{k=0}^{N-1} \alpha[k]|X^{c1}[k]|^2, \tag{8.176}$$

and, for the DCT-2, is

$$\sum_{n=0}^{N-1} |x[n]|^2 = \frac{1}{N} \sum_{k=0}^{N-1} \beta[k]|X^{c2}[k]|^2, \tag{8.177}$$

where $\beta[k]$ is defined in Eq. (8.157). The DCT can be said to be concentrated in the low indices of the DCT if the remaining DCT coefficients can be set to zero without a significant impact on the energy of the signal. We illustrate the energy compaction property in the following example.

Example 8.13   Energy Compaction in the DCT-2

Consider a test input of the form

$$x[n] = a^n \cos(\omega_0 n + \phi), \qquad n = 0, 1, \ldots, N - 1. \tag{8.178}$$

Such a signal is illustrated in Figure 8.27 for $a = .9$, $\omega_0 = 0.1\pi$, $\phi = 0$, and $N = 32$.

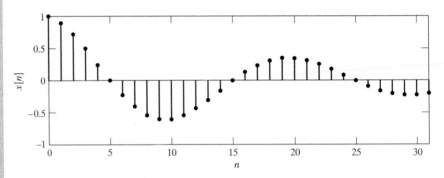

Figure 8.27   Test signal for comparing DFT and DCT.

The real and imaginary parts of the 32-point DFT of the 32-point sequence in Figure 8.27 are shown in Figures 8.28(a) and (b), respectively, and the DCT-2 of the sequence is shown in Figure 8.28(c). In the case of the DFT, the real and imaginary parts are shown for $k = 0, 1, \ldots, 16$. Since the signal is real, $X[0]$ and $X[16]$ are real. The remaining values are complex and conjugate symmetric. Thus, the 32 real numbers shown in Figures 8.28(a) and (b) completely specify the 32-point DFT. In the case of the DCT-2, we show all 32 of the real DCT-2 values. Clearly, the DCT-2 values are highly concentrated at low indices, so Parseval's theorem suggests that the energy of the sequence is more concentrated in the DCT-2 representation than in the DFT representation.

This energy concentration property can be quantified by truncating the two representations and comparing the mean-squared approximation error for the two representations when both use the same number of real coefficient values. To do this, we define

$$x_m^{\text{dft}}[n] = \frac{1}{N} \sum_{k=0}^{N-1} T_m[k] X[k] e^{j2\pi kn/N}, \qquad n = 0, 1, \ldots, N - 1, \tag{8.179}$$

where, in this case, $X[k]$ is the N-point DFT of $x[n]$ and

$$T_m[k] = \begin{cases} 1, & 0 \le k \le (N - 1 - m)/2, \\ 0, & (N + 1 - m)/2 \le k \le (N - 1 + m)/2, \\ 1, & (N + 1 + m)/2 \le k \le N - 1. \end{cases}$$

If $m = 1$, the term $X[N/2]$ is removed. If $m = 3$, then the terms $X[N/2]$ and $X[N/2-1]$ and its corresponding complex conjugate $X[N/2 + 1]$ are removed, and so forth; i.e., $x_m^{\text{dft}}[n]$ for $m = 1, 3, 5, \ldots, N - 1$ is the sequence that is synthesized by symmetrically omitting m DFT coefficients.[5] With the exception of the DFT value, $X[N/2]$, which is

---

[5] For simplicity, we assume that N is an even integer.

real, each omitted complex DFT value and its corresponding complex conjugate actually corresponds to omitting two real numbers. For example, $m = 5$ would correspond to setting the coefficients $X[14]$, $X[15]$, $X[16]$, $X[17]$, and $X[18]$ to zero in synthesizing $x_5^{\text{dft}}[n]$ from the 32-point DFT shown in Figures 8.28(a) and (b).

Likewise, we can truncate the DCT-2 representation, obtaining

$$x_m^{\text{dct}}[n] = \frac{1}{N} \sum_{k=0}^{N-1-m} \beta[k] X^{c2}[k] \cos\left(\frac{\pi k(2n+1)}{2N}\right), \qquad 0 \le n \le N-1. \qquad (8.180)$$

In this case, if $m = 5$, we omit the DCT-2 coefficients $X^{c2}[27], \ldots, X^{c2}[31]$ in the synthesis of $x_m^{\text{dct}}[n]$ from the DCT-2 shown in Figure 8.28(c). Since these coefficients are very small, $x_5^{\text{dct}}[n]$ should differ only slightly from $x[n]$.

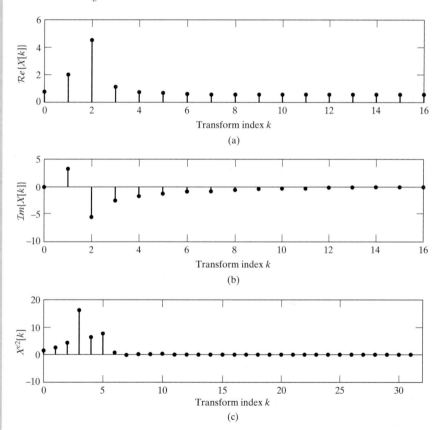

Figure 8.28   (a) Real part of 32-point DFT; (b) Imaginary part of 32-point DFT; (c) 32-point DCT-2 of the test signal plotted in Figure 8.27.

To show how the approximation errors depend on m for the DFT and the DCT-2, we define

$$E^{\text{dft}}[m] = \frac{1}{N} \sum_{n=0}^{N-1} |x[n] - x_m^{\text{dft}}[n]|^2$$

and

$$E^{\text{dct}}[m] = \frac{1}{N} \sum_{n=0}^{N-1} |x[n] - x_m^{\text{dct}}[n]|^2$$

to be the mean-squared approximation errors for the truncated DFT and DCT, respectively. These errors are plotted in Figure 8.29, with $E^{\text{dft}}[m]$ indicated with $\circ$ and $E^{\text{dct}}[m]$ shown with $\bullet$. For the special cases $m = 0$ (no truncation) and $m = N-1$ (only the DC value is retained), the DFT truncation function is $T_0[k] = 1$ for $0 \le k \le N-1$ and $T_{N-1}[k] = 0$ for $1 \le k \le N-1$ and $T_{N-1}[0] = 1$. In these cases, both representations give the same error. For values $1 \le m \le 30$, the DFT error grows steadily as m increases, whereas the DCT error remains very small—up to about $m = 25$—implying that the 32 numbers of the sequence $x[n]$ can be represented with slight error by only seven DCT-2 coefficients.

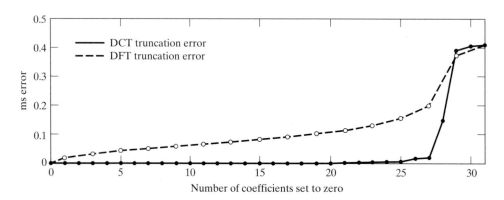

Figure 8.29   Comparison of truncation errors for DFT and DCT-2.

The signal in Example 8.13 is a low frequency exponentially decaying signal with zero phase. We have chosen this example very carefully to emphasize the energy compaction property. Not every choice of $x[n]$ will give such dramatic results. Highpass signals and even some signals of the form of Eq. (8.178) with different parameters do not show this dramatic difference. Nevertheless, in many cases of interest in data compression, the DCT-2 provides a distinct advantage over the DFT. It can be shown (Rao and Yip, 1990) that the DCT-2 is nearly optimum in the sense of minimum mean-squared truncation error for sequences with exponential correlation functions.

8.8.6 Applications of the DCT

The major application of the DCT-2 is in signal compression, where it is a key part of many standardized algorithms. (See Jayant and Noll, 1984, Pau, 1995, Rao and Hwang,

1996, Taubman and Marcellin, 2002, Bosi and Goldberg, 2003 and Spanias, Painter and Atti, 2007.) In this application, the blocks of the signal are represented by their cosine transforms. The popularity of the DCT in signal compression is mainly as a result of its energy concentration property, which we demonstrated by a simple example in the previous section.

The DCT representations, being orthogonal transforms like the DFT, have many properties similar to those of the DFT that make them very flexible for manipulating the signals that they represent. One of the most important properties of the DFT is that periodic convolution of two finite-length sequences corresponds to multiplication of their corresponding DFTs. We have seen in Section 8.7 that it is possible to exploit this property to compute linear convolutions by doing only DFT computations. In the case of the DCT, the corresponding result is that multiplication of DCTs corresponds to periodic convolution of the underlying symmetrically extended sequences. However, there are additional complications. For example, the periodic convolution of two Type-2 symmetric periodic sequences is not a Type-2 sequence, but rather, a Type-1 sequence. Alternatively, periodic convolution of a Type-1 sequence with a Type-2 sequence of the same implied period is a Type-2 sequence. Thus, a mixture of DCTs is required to effect periodic symmetric convolution by inverse transformation of the product of DCTs. There are many more ways to do this, because we have many different DCT definitions from which to choose. Each different combination would correspond to periodic convolution of a pair of symmetrically extended finite sequences. Martucci (1994) provides a complete discussion of the use of DCT and DST transforms in implementing symmetric periodic convolution.

Multiplication of DCTs corresponds to a special type of periodic convolution that has some features that may be useful in some applications. As we have seen for the DFT, periodic convolution is characterized by end effects, or "wrap around" effects. Indeed, even linear convolution of two finite-length sequences has end effects as the impulse response engages and disengages from the input. The end effects of periodic symmetric convolution are different from ordinary convolution and from periodic convolution as implemented by multiplying DFTs. The symmetric extension creates symmetry at the endpoints. The "smooth" boundaries that this implies often mitigate the end effects encountered in convolving finite-length sequences. One area in which symmetric convolution is particularly useful is image filtering, where objectionable edge effects are perceived as blocking artifacts. In such representations, the DCT may be superior to the DFT or even ordinary linear convolution. In doing periodic symmetric convolution by multiplication of DCTs, we can force the same result as ordinary convolution by extending the sequences with a sufficient number of zero samples placed at both the beginning and the end of each sequence.

8.9 SUMMARY

In this chapter, we have discussed discrete Fourier representations of finite-length sequences. Most of our discussion focused on the discrete Fourier transform (DFT), which is based on the DFS representation of periodic sequences. By defining a periodic sequence for which each period is identical to the finite-length sequence, the DFT becomes identical to one period of the DFS coefficients. Because of the importance of this underlying periodicity, we first examined the properties of DFS representations and then interpreted those properties in terms of finite-length sequences. An important result is that the DFT values are equal to samples of the z-transform at equally spaced points on the unit circle. This leads to the notion of time aliasing in the interpretation of DFT properties, a concept we used extensively in the study of circular convolution and its relation to linear convolution. We then used the results of this study to show how the DFT could be employed to implement the linear convolution of a finite-length impulse response with an indefinitely long input signal.

The chapter concluded with an introduction to the DCT. It was shown that the DCT and DFT are closely related and that they share an implicit assumption of periodicity. The energy compaction property, which is the main reason for the popularity of the DCT in data compression, was demonstrated with an example.

Problems

Basic Problems with Answers

8.1. Suppose $x_c(t)$ is a periodic continuous-time signal with period 1 ms and for which the Fourier series is

$$x_c(t) = \sum_{k=-9}^{9} a_k e^{j(2000\pi kt)}.$$

The Fourier series coefficients a_k are zero for $|k| > 9$. $x_c(t)$ is sampled with a sample spacing $T = \frac{1}{6} \times 10^{-3}$ s to form $x[n]$. That is,

$$x[n] = x_c\left(\frac{n}{6000}\right).$$

   (a) Is $x[n]$ periodic and, if so, with what period?
   (b) Is the sampling rate above the Nyquist rate? That is, is T sufficiently small to avoid aliasing?
   (c) Find the DFS coefficients of $x[n]$ in terms of a_k.

8.2. Suppose $\tilde{x}[n]$ is a periodic sequence with period N. Then $\tilde{x}[n]$ is also periodic with period $3N$. Let $\tilde{X}[k]$ denote the DFS coefficients of $\tilde{x}[n]$ considered as a periodic sequence with period N, and let $\tilde{X}_3[k]$ denote the DFS coefficients of $\tilde{x}[n]$ considered as a periodic sequence with period $3N$.

   (a) Express $\tilde{X}_3[k]$ in terms of $\tilde{X}[k]$.
   (b) By explicitly calculating $\tilde{X}[k]$ and $\tilde{X}_3[k]$, verify your result in part (a) when $\tilde{x}[n]$ is as given in Figure P8.2.

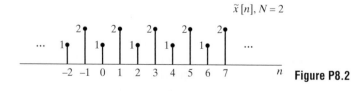

Figure P8.2

8.3. Figure P8.3 shows three periodic sequences $\tilde{x}_1[n]$ through $\tilde{x}_3[n]$. These sequences can be expressed in a Fourier series as

$$\tilde{x}[n] = \frac{1}{N} \sum_{k=0}^{N-1} \tilde{X}[k] e^{j(2\pi/N)kn}.$$

(a) For which sequences can the time origin be chosen such that all the $\tilde{X}[k]$ are real?
(b) For which sequences can the time origin be chosen such that all the $\tilde{X}[k]$ (except for k an integer multiple of N) are imaginary?
(c) For which sequences does $\tilde{X}[k] = 0$ for $k = \pm 2, \pm 4, \pm 6$?

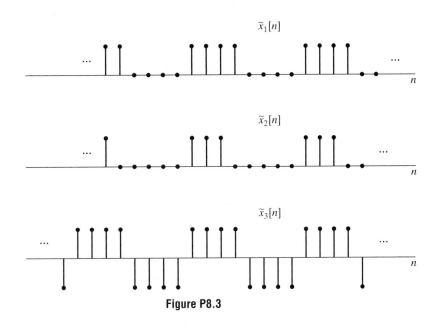

Figure P8.3

8.4. Consider the sequence $x[n]$ given by $x[n] = \alpha^n u[n]$. Assume $|\alpha| < 1$. A periodic sequence $\tilde{x}[n]$ is constructed from $x[n]$ in the following way:

$$\tilde{x}[n] = \sum_{r=-\infty}^{\infty} x[n + rN].$$

(a) Determine the Fourier transform $X(e^{j\omega})$ of $x[n]$.
(b) Determine the DFS coefficients $\tilde{X}[k]$ for the sequence $\tilde{x}[n]$.
(c) How is $\tilde{X}[k]$ related to $X(e^{j\omega})$?

8.5. Compute the DFT of each of the following finite-length sequences considered to be of length N (where N is even):

(a) $x[n] = \delta[n]$,
(b) $x[n] = \delta[n - n_0]$, $0 \le n_0 \le N - 1$,
(c) $x[n] = \begin{cases} 1, & n \text{ even}, & 0 \le n \le N - 1, \\ 0, & n \text{ odd}, & 0 \le n \le N - 1, \end{cases}$
(d) $x[n] = \begin{cases} 1, & 0 \le n \le N/2 - 1, \\ 0, & N/2 \le n \le N - 1, \end{cases}$

(e) $x[n] = \begin{cases} a^n, & 0 \le n \le N - 1, \\ 0, & \text{otherwise.} \end{cases}$

8.6. Consider the complex sequence

$$x[n] = \begin{cases} e^{j\omega_0 n}, & 0 \le n \le N - 1, \\ 0, & \text{otherwise.} \end{cases}$$

(a) Find the Fourier transform $X(e^{j\omega})$ of $x[n]$.
(b) Find the N-point DFT $X[k]$ of the finite-length sequence $x[n]$.
(c) Find the DFT of $x[n]$ for the case $\omega_0 = 2\pi k_0/N$, where k_0 is an integer.

8.7. Consider the finite-length sequence $x[n]$ in Figure P8.7. Let $X(z)$ be the z-transform of $x[n]$. If we sample $X(z)$ at $z = e^{j(2\pi/4)k}$, $k = 0, 1, 2, 3$, we obtain

$$X_1[k] = X(z)\big|_{z=e^{j(2\pi/4)k}}, \qquad k = 0, 1, 2, 3.$$

Sketch the sequence $x_1[n]$ obtained as the inverse DFT of $X_1[k]$.

Figure P8.7

8.8. Let $X(e^{j\omega})$ denote the Fourier transform of the sequence $x[n] = (0.5)^n u[n]$. Let $y[n]$ denote a finite-duration sequence of length 10; i.e., $y[n] = 0, n < 0$, and $y[n] = 0, n \ge 10$. The 10-point DFT of $y[n]$, denoted by $Y[k]$, corresponds to 10 equally spaced samples of $X(e^{j\omega})$; i.e., $Y[k] = X(e^{j2\pi k/10})$. Determine $y[n]$.

8.9. Consider a 20-point finite-duration sequence $x[n]$ such that $x[n] = 0$ outside $0 \le n \le 19$, and let $X(e^{j\omega})$ represent the discrete-time Fourier transform of $x[n]$.

(a) If it is desired to evaluate $X(e^{j\omega})$ at $\omega = 4\pi/5$ by computing one M-point DFT, determine the smallest possible M, and develop a method to obtain $X(e^{j\omega})$ at $\omega = 4\pi/5$ using the smallest M.
(b) If it is desired to evaluate $X(e^{j\omega})$ at $\omega = 10\pi/27$ by computing one L-point DFT, determine the smallest possible L, and develop a method to obtain $X(e^{j10\pi/27})$ using the smallest L.

8.10. The two eight-point sequences $x_1[n]$ and $x_2[n]$ shown in Figure P8.10 have DFTs $X_1[k]$ and $X_2[k]$, respectively. Determine the relationship between $X_1[k]$ and $X_2[k]$.

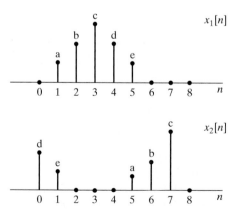

Figure P8.10

8.11. Figure P8.11 shows two finite-length sequences $x_1[n]$ and $x_2[n]$. Sketch their six-point circular convolution.

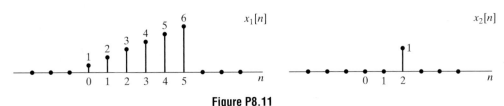

Figure P8.11

8.12. Suppose we have two four-point sequences $x[n]$ and $h[n]$ as follows:

$$x[n] = \cos\left(\frac{\pi n}{2}\right), \quad n = 0, 1, 2, 3,$$
$$h[n] = 2^n, \quad n = 0, 1, 2, 3.$$

(a) Calculate the four-point DFT $X[k]$.
(b) Calculate the four-point DFT $H[k]$.
(c) Calculate $y[n] = x[n] \textcircled{4} h[n]$ by doing the circular convolution directly.
(d) Calculate $y[n]$ of part (c) by multiplying the DFTs of $x[n]$ and $h[n]$ and performing an inverse DFT.

8.13. Consider the finite-length sequence $x[n]$ in Figure P8.13. The five-point DFT of $x[n]$ is
denoted by $X[k]$. Plot the sequence $y[n]$ whose DFT is

$$Y[k] = W_5^{-2k} X[k].$$

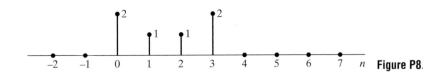

Figure P8.13

8.14. Two finite-length signals, $x_1[n]$ and $x_2[n]$, are sketched in Figure P8.14. Assume that $x_1[n]$
and $x_2[n]$ are zero outside of the region shown in the figure. Let $x_3[n]$ be the eight-point
circular convolution of $x_1[n]$ with $x_2[n]$; i.e., $x_3[n] = x_1[n] \textcircled{8} x_2[n]$. Determine $x_3[2]$.

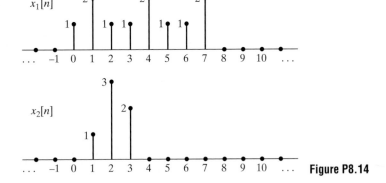

Figure P8.14

8.15. Figure P8.15-1 shows two sequences $x_1[n]$ and $x_2[n]$. The value of $x_2[n]$ at time $n = 3$ is not known, but is shown as a variable a. Figure P8.15-2 shows $y[n]$, the four-point circular convolution of $x_1[n]$ and $x_2[n]$. Based on the graph of $y[n]$, can you specify a uniquely? If so, what is a? If not, give two possible values of a that would yield the sequence $y[n]$ as shown.

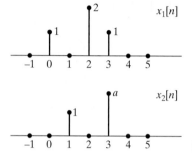

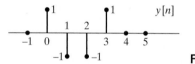

Figure P8.15-1

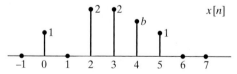

Figure P8.15-2

8.16. Figure P8.16-1 shows a six-point discrete-time sequence $x[n]$. Assume that $x[n] = 0$ outside the interval shown. The value of $x[4]$ is not known and is represented as b. Note that the sample shown for b in the figure is not necessarily to scale. Let $X(e^{j\omega})$ be the DTFT of $x[n]$ and $X_1[k]$ be samples of $X(e^{j\omega})$ every $\pi/2$; i.e.,

$$X_1[k] = X(e^{j\omega})|_{\omega=(\pi/2)k}, \qquad 0 \le k \le 3.$$

The four-point sequence $x_1[n]$ that results from taking the four-point inverse DFT of $X_1[k]$ is shown in Figure P8.16-2. Based on this figure, can you determine b uniquely? If so, give the value for b.

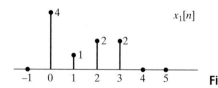

Figure P8.16-1

Figure P8.16-2

8.17. Figure P8.17 shows two finite-length sequences $x_1[n]$ and $x_2[n]$. What is the smallest N such that the N-point circular convolution of $x_1[n]$ and $x_2[n]$ are equal to the linear convolution of these sequences, i.e., such that $x_1[n] \,\circledN\, x_2[n] = x_1[n] * x_2[n]$?

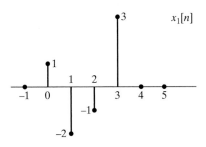

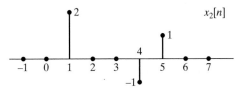

Figure P8.17

8.18. Figure P8.18-1 shows a sequence $x[n]$ for which the value of $x[3]$ is an unknown constant c. The sample with amplitude c is not necessarily drawn to scale. Let

$$X_1[k] = X[k]e^{j2\pi 3k/5},$$

where $X[k]$ is the five-point DFT of $x[n]$. The sequence $x_1[n]$ plotted in Figure P8.18-2 is the inverse DFT of $X_1[k]$. What is the value of c?

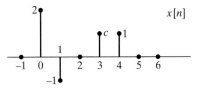

Figure P8.18-1

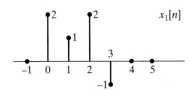

Figure P8.18-2

8.19. Two finite-length sequences $x[n]$ and $x_1[n]$ are shown in Figure P8.19. The DFTs of these sequences, $X[k]$ and $X_1[k]$, respectively, are related by the equation

$$X_1[k] = X[k]e^{-j(2\pi km/6)},$$

where m is an unknown constant. Can you determine a value of m consistent with Figure P8.19? Is your choice of m unique? If so, justify your answer. If not, find another choice of m consistent with the information given.

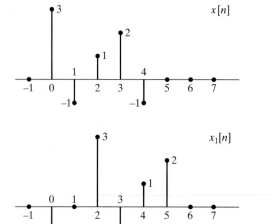

Figure P8.19

8.20. Two finite-length sequences $x[n]$ and $x_1[n]$ are shown in Figure P8.20. The N-point DFTs of these sequences, $X[k]$ and $X_1[k]$, respectively, are related by the equation

$$X_1[k] = X[k]e^{j2\pi k2/N},$$

where N is an unknown constant. Can you determine a value of N consistent with Figure P8.20? Is your choice for N unique? If so, justify your answer. If not, find another choice of N consistent with the information given.

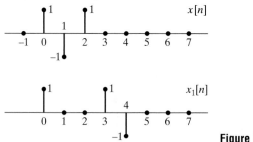

Figure P8.20

Basic Problems

8.21. **(a)** Figure P8.21-1 shows two periodic sequences, $\tilde{x}_1[n]$ and $\tilde{x}_2[n]$, with period $N = 7$. Find a sequence $\tilde{y}_1[n]$ whose DFS is equal to the product of the DFS of $\tilde{x}_1[n]$ and the DFS of $\tilde{x}_2[n]$, i.e.,

$$\tilde{Y}_1[k] = \tilde{X}_1[k]\tilde{X}_2[k].$$

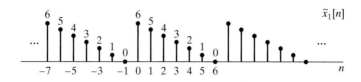

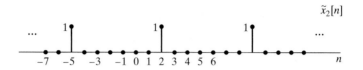

Figure P8.21-1

(b) Figure P8.21-2 shows a periodic sequence $\tilde{x}_3[n]$ with period $N = 7$. Find a sequence $\tilde{y}_2[n]$ whose DFS is equal to the product of the DFS of $\tilde{x}_1[n]$ and the DFS of $\tilde{x}_3[n]$, i.e.,

$$\tilde{Y}_2[k] = \tilde{X}_1[k]\tilde{X}_3[k].$$

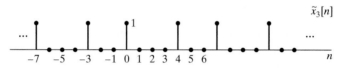

Figure P8.21-2

8.22. Consider an N-point sequence $x[n]$, i.e.,

$$x[n] = 0 \text{ for } n > N - 1 \text{ and } n < 0.$$

The discrete-time Fourier transform of $x[n]$ is $X(e^{j\omega})$, and the N-point DFT of $x[n]$ is $X[k]$.

If $\mathcal{R}e\,\{X[k]\} = 0$ for $k = 0, 1, \ldots, N - 1$, can we conclude that $\mathcal{R}e\left\{X(e^{j\omega})\right\} = 0$ for $-\pi \le \omega \le \pi$? If your answer is yes, explicitly show why. If not, give a simple counterexample.

8.23. Consider the real finite-length sequence $x[n]$ shown in Figure P8.23.

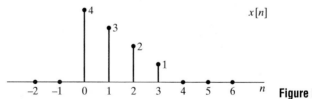

Figure P8.23

(a) Sketch the finite-length sequence $y[n]$ whose six-point DFT is

$$Y[k] = W_6^{5k} X[k],$$

where $X[k]$ is the six-point DFT of $x[n]$.

(b) Sketch the finite-length sequence $w[n]$ whose six-point DFT is

$$W[k] = \mathcal{I}m\{X[k]\}.$$

(c) Sketch the finite-length sequence $q[n]$ whose three-point DFT is

$$Q[k] = X[2k + 1], \qquad k = 0, 1, 2.$$

8.24. Figure P8.24 shows a finite-length sequence $x[n]$. Sketch the sequences

$$x_1[n] = x[((n-2))_4], \qquad 0 \le n \le 3,$$

and

$$x_2[n] = x[((-n))_4], \qquad 0 \le n \le 3.$$

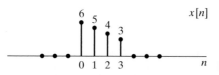

Figure P8.24

8.25. Consider the signal $x[n] = \delta[n-4] + 2\delta[n-5] + \delta[n-6]$.

    (a) Find $X(e^{j\omega})$ the discrete-time Fourier transform of $x[n]$. Write expressions for the magnitude and phase of $X(e^{j\omega})$, and sketch these functions.

    (b) Find all values of N for which the N-point DFT is a set of real numbers.

    (c) Can you find a three-point causal signal $x_1[n]$ (i.e., $x_1[n] = 0$ for $n < 0$ and $n > 2$) for which the three-point DFT of $x_1[n]$ is:

$$X_1[k] = |X[k]| \qquad k = 0, 1, 2$$

    where $X[k]$ is the three-point DFT of $x[n]$?

8.26. We have shown that the DFT $X[k]$ of a finite-length sequence $x[n]$ is identical to samples of the DTFT $X(e^{j\omega})$ of that sequence at frequencies $\omega_k = (2\pi/N)k$; i.e., $X[k] = X(e^{j(2\pi/N)k})$ for $k = 0, 1, \ldots, N-1$. Now consider a sequence $y[n] = e^{-j(\pi/N)n}x[n]$ whose DFT is $Y[k]$.

    (a) Determine the relationship between the DFT $Y[k]$ and the DTFT $X(e^{j\omega})$.

    (b) The result of part (a) shows that $Y[k]$ is a differently sampled version of $X(e^{j\omega})$. What are the frequencies at which $X(e^{j\omega})$ is sampled?

    (c) Given the modified DFT $Y[k]$, how would you recover the original sequence $x[n]$?

8.27. The 10-point DFT of a 10-point sequence $g[n]$ is

$$G[k] = 10\,\delta[k] .$$

Find $G(e^{j\omega})$, the DTFT of $g[n]$.

8.28. Consider the six-point sequence

$$x[n] = 6\delta[n] + 5\delta[n-1] + 4\delta[n-2] + 3\delta[n-3] + 2\delta[n-4] + \delta[n-5]$$

shown in Figure P8.28.

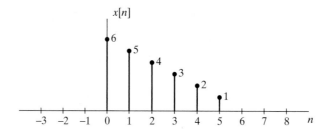

Figure P8.28

(a) Determine $X[k]$, the six-point DFT of $x[n]$. Express your answer in terms of $W_6 = e^{-j2\pi/6}$.

(b) Plot the sequence $w[n]$, $n = 0, 1, \ldots, 5$, that is obtained by computing the inverse six-point DFT of $W[k] = W_6^{-2k} X[k]$.

(c) Use any convenient method to evaluate the six-point circular convolution of $x[n]$ with the sequence $h[n] = \delta[n] + \delta[n-1] + \delta[n-2]$. Sketch the result.

(d) If we convolve the given $x[n]$ with the given $h[n]$ by N-point circular convolution, how should N be chosen so that the result of the circular convolution is identical to the result of linear convolution? That is, choose N so that

$$y_p[n] = x[n] \; \textcircled{N} \; h[n] = \sum_{m=0}^{N-1} x[m]h[((n-m))_N]$$

$$= x[n] * h[n] = \sum_{m=-\infty}^{\infty} x[m]h[n-m] \quad \text{for } 0 \le n \le N-1.$$

(e) In certain applications, such as multicarrier communication systems (see Starr et al, 1999), the linear convolution of a finite-length signal $x[n]$ of length L samples with a shorter finite-length impulse response $h[n]$ is required to be identical (over $0 \le n \le L-1$) to what would have been obtained by L-point circular convolution of $x[n]$ with $h[n]$. This can be achieved by augmenting the sequence $x[n]$ appropriately. Starting with the graph of Figure P8.28, where $L = 6$, add samples to the given sequence $x[n]$ to produce a new sequence $x_1[n]$ such that with the sequence $h[n]$ given in part (c), the ordinary convolution $y_1[n] = x_1[n] * h[n]$ satisfies the equation

$$y_1[n] = x_1[n] * h[n] = \sum_{m=-\infty}^{\infty} x_1[m]h[n-m]$$

$$= y_p[n] = x[n] \; \textcircled{L} \; h[n] = \sum_{m=0}^{5} x[m]h[((n-m))_6] \quad \text{for } 0 \le n \le 5.$$

(f) Generalize the result of part (e) for the case where $h[n]$ is nonzero for $0 \le n \le M$ and $x[n]$ is nonzero for $0 \le n \le L-1$, where $M < L$; i.e., show how to construct a sequence $x_1[n]$ from $x[n]$ such that the linear convolution $x_1[n] * h[n]$ is equal to the circular convolution $x[n] \; \textcircled{L} \; h[n]$ for $0 \le n \le L-1$.

8.29. Consider the real five-point sequence

$$x[n] = \delta[n] + \delta[n-1] + \delta[n-2] - \delta[n-3] + \delta[n-4].$$

The deterministic autocorrelation of this sequence is the inverse DTFT of

$$C(e^{j\omega}) = X(e^{j\omega})X^*(e^{j\omega}) = |X(e^{j\omega})|^2,$$

where $X^*(e^{j\omega})$ is the complex conjugate of $X(e^{j\omega})$. For the given $x[n]$, the autocorrelation can be found to be

$$c[n] = x[n] * x[-n].$$

(a) Plot the sequence $c[n]$. Observe that $c[-n] = c[n]$ for all n.

(b) Now assume that we compute the seven-point DFT ($N = 5$) of the sequence $x[n]$. Call this DFT $X_5[k]$. Then, we compute the inverse DFT of $C_5[k] = X_5[k]X_5^*[k]$. Plot the resulting sequence $c_5[n]$. How is $c_5[n]$ related to $c[n]$ from part (a)?

(c) Now assume that we compute the 10-point DFT ($N = 10$) of the sequence $x[n]$. Call this DFT $X_{10}[k]$. Then, we compute the inverse DFT of $C_{10}[k] = X_{10}[k]X_{10}^*[k]$. Plot the resulting sequence $c_{10}[n]$.

(d) Now suppose that we use $X_{10}[k]$ to form $D_{10}[k] = W_{10}^{5k}C_{10}[k] = W_{10}^{5k}X_{10}[k]X_{10}^*[k]$, where $W_{10} = e^{-j(2\pi/10)}$. Then, we compute the inverse DFT of $D_{10}[k]$. Plot the resulting sequence $d_{10}[n]$.

8.30. Consider two sequences $x[n]$ and $h[n]$, and let $y[n]$ denote their ordinary (linear) convolution, $y[n] = x[n] * h[n]$. Assume that $x[n]$ is zero outside the interval $21 \leq n \leq 31$, and $h[n]$ is zero outside the interval $18 \leq n \leq 31$.

(a) The signal $y[n]$ will be zero outside of an interval $N_1 \leq n \leq N_2$. Determine numerical values for N_1 and N_2.

(b) Now suppose that we compute the 32-point DFTs of

$$x_1[n] = \begin{cases} 0 & n = 0, 1, \ldots, 20 \\ x[n] & n = 21, 22, \ldots, 31 \end{cases}$$

and

$$h_1[n] = \begin{cases} 0 & n = 0, 1, \ldots, 17 \\ h[n] & n = 18, 19, \ldots, 31 \end{cases}$$

(i.e., the zero samples at the beginning of each sequence are included). Then, we form the product $Y_1[k] = X_1[k]H_1[k]$. If we define $y_1[n]$ to be the 32-point inverse DFT of $Y_1[k]$, how is $y_1[n]$ related to the ordinary convolution $y[n]$? That is, give an equation that expresses $y_1[n]$ in terms of $y[n]$ for $0 \leq n \leq 31$.

(c) Suppose that you are free to choose the DFT length (N) in part (b) so that the sequences are also zero-padded at their ends. What is the *minimum* value of N so that $y_1[n] = y[n]$ for $0 \leq n \leq N - 1$?

8.31. Consider the sequence $x[n] = 2\delta[n] + \delta[n-1] - \delta[n-2]$.

(a) Determine the DTFT $X(e^{j\omega})$ of $x[n]$ and the DTFT $Y(e^{j\omega})$ of the sequence $y[n] = x[-n]$.

(b) Using your results from part (a) find an expression for

$$W(e^{j\omega}) = X(e^{j\omega})Y(e^{j\omega}).$$

(c) Using the result of part (b) make a plot of $w[n] = x[n] * y[n]$.

(d) Now plot the sequence $y_p[n] = x[((-n))_4]$ as a function of n for $0 \leq n \leq 3$.

(e) Now use any convenient method to evaluate the four-point circular convolution of $x[n]$ with $y_p[n]$. Call your answer $w_p[n]$ and plot it.

(f) If we convolve $x[n]$ with $y_p[n] = x[((-n))_N]$, how should N be chosen to avoid time-domain aliasing?

8.32. Consider a finite-duration sequence $x[n]$ of length P such that $x[n] = 0$ for $n < 0$ and $n \geq P$. We want to compute samples of the Fourier transform at the N equally spaced frequencies

$$\omega_k = \frac{2\pi k}{N}, \qquad k = 0, 1, \ldots, N - 1.$$

Determine and justify procedures for computing the N samples of the Fourier transform using only one N-point DFT for the following two cases:

(a) $N > P$.

(b) $N < P$.

8.33. An FIR filter has a 10-point impulse response, i.e.,

$$h[n] = 0 \qquad \text{for } n < 0 \text{ and for } n > 9.$$

Given that the 10-point DFT of $h[n]$ is given by

$$H[k] = \frac{1}{5}\delta[k-1] + \frac{1}{3}\delta[k-7],$$

find $H(e^{j\omega})$, the DTFT of $h[n]$.

8.34. Suppose that $x_1[n]$ and $x_2[n]$ are two finite-length sequences of length N, i.e., $x_1[n] = x_2[n] = 0$ outside $0 \le n \le N-1$. Denote the z-transform of $x_1[n]$ by $X_1(z)$, and denote the N-point DFT of $x_2[n]$ by $X_2[k]$. The two transforms $X_1(z)$ and $X_2[k]$ are related by:

$$X_2[k] = X_1(z)\Big|_{z=\frac{1}{2}e^{-j\frac{2\pi k}{N}}} \quad , \quad k = 0, 1, \ldots, N-1$$

Determine the relationship between $x_1[n]$ and $x_2[n]$.

Advanced Problems

8.35. Figure P8.35-1 illustrates a six-point discrete-time sequence $x[n]$. Assume that $x[n]$ is zero outside the interval shown.

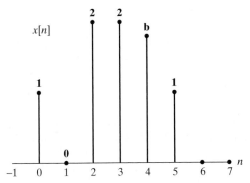

Figure P8.35-1

The value of $x[4]$ is not known and is represented as b. The sample in the figure is not shown to scale. Let $X(e^{j\omega})$ be the DTFT of $x[n]$ and $X_1[k]$ be samples of $X(e^{j\omega})$ at $\omega_k = 2\pi k/4$, i.e.,

$$X_1[k] = X(e^{j\omega})\big|_{\omega=\frac{\pi k}{2}}, \qquad 0 \le k \le 3.$$

The four-point sequence $x_1[n]$ that results from taking the four-point inverse DFT of $X_1[k]$ is shown in Figure P8.35-2. Based on the figure can you determine b uniquely? If so, give the value of b.

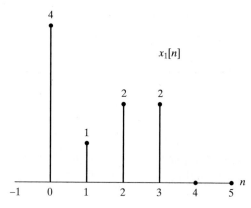

Figure P8.35-2

8.36. **(a)** $X(e^{j\omega})$ is the DTFT of the discrete-time signal

$$x[n] = (1/2)^n u[n].$$

Find a length-5 sequence $g[n]$ whose five-point DFT $G[k]$ is identical to samples of the DTFT of $x[n]$ at $\omega_k = 2\pi k/5$, i.e.,

$$g[n] = 0 \text{ for } n < 0, \ n > 4$$

and

$$G[k] = X(e^{j2\pi k/5}) \text{ for } k = 0, 1, \ldots, 4.$$

(b) Let $w[n]$ be a sequence that is strictly nonzero for $0 \le n \le 9$ and zero elsewhere, i.e.,

$$w[n] \neq 0, \quad 0 \le n \le 9$$
$$w[n] = 0 \quad \text{otherwise}$$

Determine a choice for $w[n]$ such that its DTFT $W(e^{j\omega})$ is equal to $X(e^{j\omega})$ at the frequencies $\omega = 2\pi k/5, \ k = 0, 1, \ldots, 4$, i.e.,

$$W(e^{j2\pi k/5}) = X(e^{j2\pi k/5}) \text{ for } k = 0, 1, \ldots, 4.$$

8.37. A discrete-time LTI filter S is to be implemented using the overlap–save method. In the overlap–save method, the input is divided into *overlapping* blocks, as opposed to the overlap–add method where the input blocks are nonoverlapping. For this implementation, the input signal $x[n]$ will be divided into overlapping 256-point blocks $x_r[n]$. Adjacent blocks will overlap by 255 points so that they differ by only one sample. This is represented by Eq. (P8.37-1) which is a relation between $x_r[n]$ and $x[n]$,

$$x_r[n] = \begin{cases} x[n+r] & 0 \le n \le 255 \\ 0 & \text{otherwise,} \end{cases} \qquad \text{(P8.37-1)}$$

where r ranges over all integers and we obtain a different block $x_r[n]$ for each value of r.

Each block is processed by computing the 256-point DFT of $x_r[n]$, multiplying the result with $H[k]$ given in Eq. (P8.37-2), and computing the 256-point inverse DFT of the product.

$$H[k] = \begin{cases} 1 & 0 \le k \le 31 \\ 0 & 32 \le k \le 224 \\ 1 & 225 \le k \le 255 \end{cases} \qquad \text{(P8.37-2)}$$

One sample from each block computation (in this case only a single sample per block) is then "saved" as part of the overall output.

(a) Is S an ideal frequency-selective filter? Justify your answer.
(b) Is the impulse response of S real valued? Justify your answer.
(c) Determine the impulse response of S.

8.38. $x[n]$ is a real-valued finite-length sequence of length 512, i.e.,

$$x[n] = 0 \qquad n < 0, n \ge 512$$

and has been stored in a 512-point data memory. It is known that $X[k]$ the 512-point DFT of $x[n]$ has the property

$$X[k] = 0 \qquad 250 \le k \le 262.$$

In storing the data, one data point at most may have been corrupted. Specifically, if $s[n]$ denotes the stored data, $s[n] = x[n]$ except possibly at one unknown memory location n_0. To test and possibly correct the data, you are able to examine $S[k]$, the 512-point DFT of $s[n]$.

(a) Specify whether, by examining $S[k]$, it is possible and if so, how, to *detect* whether an error has been made in one data point, i.e., whether or not $s[n] = x[n]$ for all n.

In parts (b) and (c), assume that you know for sure that one data point has been corrupted, i.e., that $s[n] = x[n]$ *except* at $n = n_0$.

(b) In this part, assume the value of n_0 is unknown. Specify a procedure for determining from $S[k]$ the value of n_0.
(c) In this part, assume that you know the value of n_0. Specify a procedure for determining $x[n_0]$ from $S[k]$.

8.39. In the system shown in the Figure P8.39, $x_1[n]$ and $x_2[n]$ are both causal, 32-point sequences, i.e., they are both zero outside the interval $0 \le n \le 31$. $y[n]$ denotes the linear convolution of $x_1[n]$ and $x_2[n]$, i.e., $y[n] = x_1[n] * x_2[n]$.

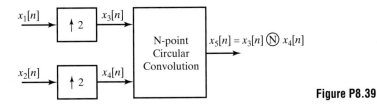

Figure P8.39

(a) Determine the values of N for which all the values of $y[n]$ can be completely recovered from $x_5[n]$.
(b) Specify explicitly how to recover $y[n]$ from $x_5[n]$ for the *smallest* value of N which you determined in part (a).

8.40. Three real-valued seven-point sequences ($x_1[n], x_2[n],$ and $x_3[n]$) are shown in Figure P8.40. For each of these sequences, specify whether the seven-point DFT can be written in the form

$$X_i[k] = A_i[k]e^{-j(2\pi k/7)k\alpha_i} \qquad k = 0, 1, \ldots, 6$$

where $A_i[k]$ is real-valued and $2\alpha_i$ is an integer. Include a brief explanation. For each sequence which can be written in this form, specify all corresponding values of α_i for $0 \le \alpha_i < 7$.

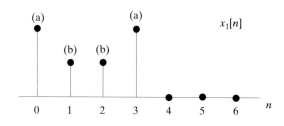

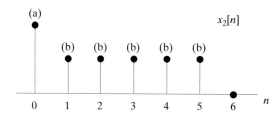

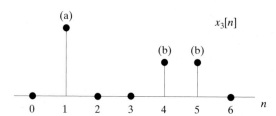

Figure P8.40

8.41. Suppose $x[n]$ is the eight-point complex-valued sequence with real part $x_r[n]$ and imaginary part $x_i[n]$ shown in Figure P8.41 (i.e., $x[n] = x_r[n] + jx_i[n]$). Let $y[n]$ be the four-point complex-valued sequence such that $Y[k]$, the four-point DFT of $y[n]$, is equal to the odd-indexed values of $X[k]$, the eight-point DFT of $x[n]$ (the odd-indexed values of $X[k]$ are those for which $k = 1, 3, 5, 7$).

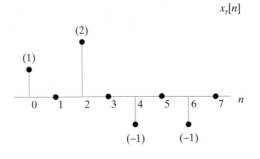

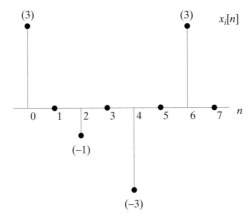

Figure P8.41

Determine the numerical values of $y_r[n]$ and $y_i[n]$, and the real and imaginary parts of $y[n]$.

8.42. $x[n]$ is a finite-length sequence of length 1024, i.e.,

$$x[n] = 0 \text{ for } n < 0, \ n > 1023.$$

The autocorrelation of $x[n]$ is defined as

$$c_{xx}[m] = \sum_{n=-\infty}^{\infty} x[n]x[n+m],$$

and $X_N[k]$ is defined as the N-point DFT of $x[n]$, with $N \geq 1024$.

We are interested in computing $c_{xx}[m]$. A proposed procedure begins by first computing the N-point inverse DFT of $|X_N[k]|^2$ to obtain an N-point sequence $g_N[n]$, i.e.,

$$g_N[n] = N\text{-point IDFT}\left\{|X_N[k]|^2\right\}.$$

(a) Determine the minimum value of N so that $c_{xx}[m]$ can be obtained from $g_N[n]$. Also specify how you would obtain $c_{xx}[m]$ from $g_N[n]$.

(b) Determine the minimum value of N so that $c_{xx}[m]$ for $|m| \leq 10$ can be obtained from $g_N[n]$. Also specify how you would obtain these values from $g_N[n]$.

8.43. In Figure P8.43, $x[n]$ is a finite-length sequence of length 1024. The sequence $R[k]$ is obtained by taking the 1024-point DFT of $x[n]$ and compressing the result by 2.

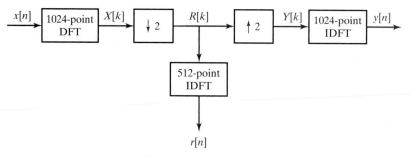

Figure P8.43

(a) Choose the most accurate statement for $r[n]$, the 512-point inverse DFT of $R[k]$. Justify your choice in a few concise sentences.

(i) $r[n] = x[n]$, $0 \leq n \leq 511$

(ii) $r[n] = x[2n]$, $0 \leq n \leq 511$

(iii) $r[n] = x[n] + x[n + 512]$, $0 \leq n \leq 511$

(iv) $r[n] = x[n] + x[-n + 512]$, $0 \leq n \leq 511$

(v) $r[n] = x[n] + x[1023 - n]$, $0 \leq n \leq 511$

In all cases $r[n] = 0$ outside $0 \leq n \leq 511$.

(b) The sequence $Y[k]$ is obtained by expanding $R[k]$ by 2. Choose the most accurate statement for $y[n]$, the 1024-point inverse DFT of $Y[k]$. Justify your choice in a few concise sentences.

(i) $y[n] = \begin{cases} \frac{1}{2}(x[n] + x[n + 512]), & 0 \leq n \leq 511 \\ \frac{1}{2}(x[n] + x[n - 512]), & 512 \leq n \leq 1023 \end{cases}$

(ii) $y[n] = \begin{cases} x[n], & 0 \leq n \leq 511 \\ x[n - 512], & 512 \leq n \leq 1023 \end{cases}$

(iii) $y[n] = \begin{cases} x[n], & n \text{ even} \\ 0, & n \text{ odd} \end{cases}$

(iv) $y[n] = \begin{cases} x[2n], & 0 \leq n \leq 511 \\ x[2(n - 512)], & 512 \leq n \leq 1023 \end{cases}$

(v) $y[n] = \frac{1}{2}(x[n] + x[1023 - n])$, $0 \leq n \leq 1023$

In all cases $y[n] = 0$ outside $0 \leq n \leq 1023$.

8.44. Figure P8.44 shows two finite-length sequences $x_1[n]$ and $x_2[n]$ of length 7. $X_i(e^{j\omega})$ denotes the DTFT of $x_i[n]$, and $X_i[k]$ denotes the seven-point DFT of $x_i[n]$.

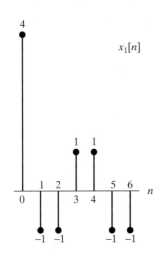
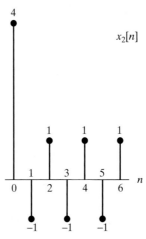

Figure P8.44

For each of the sequences $x_1[n]$ and $x_2[n]$, indicate whether each one of the following properties holds:

(a) $X_i(e^{j\omega})$ can be written in the form

$$X_i(e^{j\omega}) = A_i(\omega)e^{j\alpha_i\omega}, \quad \text{for } \omega \in (-\pi, \pi),$$

where $A_i(\omega)$ is real and α_i is a constant.

(b) $X_i[k]$ can be written in the form

$$X_i[k] = B_i[k]e^{j\beta_i k},$$

where $B_i[k]$ is real and β_i is a constant.

8.45. The sequence $x[n]$ is a 128-point sequence (i.e., $x[n] = 0$ for $n < 0$ and for $n > 127$), and $x[n]$ has at least one nonzero sample. The DTFT of $x[n]$ is denoted $X(e^{j\omega})$. What is the largest integer M such that it is possible for $X(e^{j2\pi k/M})$ to be zero for all integer values of k? Construct an example for the maximal M that you have found.

8.46. Each part of this problem may be solved independently. All parts use the signal $x[n]$ given by

$$x[n] = 3\delta[n] - \delta[n-1] + 2\delta[n-3] + \delta[n-4] - \delta[n-6].$$

(a) Let $X\left(e^{j\omega}\right)$ be the DTFT of $x[n]$. Define

$$R[k] = X\left(e^{j\omega}\right)\Big|_{\omega=\frac{2\pi k}{4}}, \quad 0 \le k \le 3$$

Plot the signal $r[n]$ which is the four-point inverse DFT of $R[k]$.

(b) Let $X[k]$ be the eight-point DFT of $x[n]$, and let $H[k]$ be the eight-point DFT of the impulse response $h[n]$ given by

$$h[n] = \delta[n] - \delta[n-4].$$

Define $Y[k] = X[k]H[k]$ for $0 \le k \le 7$. Plot $y[n]$, the eight-point DFT of $Y[k]$.

8.47. Consider a time-limited continuous-time signal $x_c(t)$ whose duration is 100 ms. Assume that this signal has a bandlimited Fourier transform such that $X_c(j\Omega) = 0$ for $|\Omega| \geq 2\pi(10,000)$ rad/s; i.e., assume that aliasing is negligible. We want to compute samples of $X_c(j\Omega)$ with 5 Hz spacing over the interval $0 \leq \Omega \leq 2\pi(10,000)$. This can be done with a 4000-point DFT. Specifically, we want to obtain a 4000-point sequence $x[n]$ for which the 4000-point DFT is related to $X_c(j\Omega)$ by:

$$X[k] = \alpha X_c(j2\pi \cdot 5 \cdot k), \qquad k = 0, 1, \ldots, 1999, \qquad \text{(P8.47-1)}$$

where α is a known scale factor. The following method is proposed to obtain a 4000-point sequence whose DFT gives the desired samples of $X_c(j\Omega)$. First, $x_c(t)$ is sampled with a sampling period of $T = 50\mu s$. Next, the resulting 2000-point sequence is used to form the sequence $\hat{x}[n]$ as follows:

$$\hat{x}[n] = \begin{cases} x_c(nT), & 0 \leq n \leq 1999, \\ x_c((n-2000)T), & 2000 \leq n \leq 3999, \\ 0, & \text{otherwise.} \end{cases} \qquad \text{(P8.47-2)}$$

Finally, the 4000-point DFT $\hat{X}[k]$ of this sequence is computed. For this method, determine how $\hat{X}[k]$ is related to $X_c(j\Omega)$. Indicate this relationship in a sketch for a "typical" Fourier transform $X_c(j\Omega)$. Explicitly state whether or not $\hat{X}[k]$ is the desired result, i.e., whether $\hat{X}[k]$ equals $X[k]$ as specified in Eq. (P8.47-1).

8.48. $x[n]$ is a real-valued finite length sequence of length 1024, i.e.,

$$x[n] = 0 \qquad n < 0, n \geq 1023.$$

Only the following samples of the 1024-point DFT of $x[n]$ are known

$$X[k] \qquad k = 0, 16, 16 \times 2, 16 \times 3, \ldots, 16 \times (64 - 1)$$

Also, we observe $s[n]$ which is a corrupted version of $x[n]$, with first 64 points corrupted, i.e., $s[n] = x[n]$ for $n \geq 64$, and $s[n] \neq x[n]$, for $0 \leq n \leq 63$. Describe a procedure to recover the first 64 samples of $x[n]$ using only 1024-point DFT and IDFT blocks, multipliers, and adders.

8.49. The deterministic crosscorrelation function between two real sequences is defined as

$$c_{xy}[n] = \sum_{m=-\infty}^{\infty} y[m]x[n+m] = \sum_{m=-\infty}^{\infty} y[-m]x[n-m] = y[-n] * x[n] \qquad -\infty < n < \infty$$

(a) Show that the DTFT of $c_{xy}[n]$ is $C_{xy}(e^{j\omega}) = X(e^{j\omega})Y^*(e^{j\omega})$.

(b) Suppose that $x[n] = 0$ for $n < 0$ and $n > 99$ and $y[n] = 0$ for $n < 0$ and $n > 49$. The corresponding crosscorrelation function $c_{xy}[n]$ will be nonzero only in a finite-length interval $N_1 \leq n \leq N_2$. What are N_1 and N_2?

(c) Suppose that we wish to compute values of $c_{xy}[n]$ in the interval $0 \leq n \leq 20$ using the following procedure:

    (i) Compute $X[k]$, the N-point DFT of $x[n]$

    (ii) Compute $Y[k]$, the N-point DFT of $y[n]$

    (iii) Compute $C[k] = X[k]Y^*[k]$ for $0 \leq k \leq N - 1$

    (iv) Compute $c[n]$, the inverse DFT of $C[k]$

What is the *minimum* value of N such that $c[n] = c_{xy}[n], 0 \leq n \leq 20$? Explain your reasoning.

8.50. The DFT of a finite-duration sequence corresponds to samples of its z-transform on the unit circle. For example, the DFT of a 10-point sequence $x[n]$ corresponds to samples of $X(z)$ at the 10 equally spaced points indicated in Figure P8.50-1. We wish to find the equally spaced samples of $X(z)$ on the contour shown in Figure P8.50-2; i.e., we wish to obtain

$$X(z)\big|_{z=0.5e^{j[(2\pi k/10)+(\pi/10)]}}.$$

Show how to modify $x[n]$ to obtain a sequence $x_1[n]$ such that the DFT of $x_1[n]$ corresponds to the desired samples of $X(z)$.

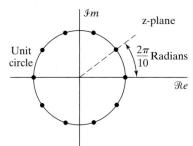

Figure P8.50-1

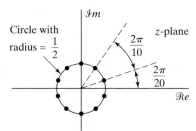

Figure P8.50-2

8.51. Let $w[n]$ denote the linear convolution of $x[n]$ and $y[n]$. Let $g[n]$ denote the 40-point circular convolution of $x[n]$ and $y[n]$:

$$w[n] = x[n] * y[n] = \sum_{k=-\infty}^{\infty} x[k]y[n-k],$$

$$g[n] = x[n] \textcircled{40}\, y[n] = \sum_{k=0}^{39} x[k]y[((n-k))_{40}].$$

(a) Determine the values of n for which $w[n]$ can be nonzero.

(b) Determine the values of n for which $w[n]$ can be obtained from $g[n]$. Explicitly specify at what index values n in $g[n]$ these values of $w[n]$ appear.

8.52. Let $x[n] = 0$, $n < 0, n > 7$, be a real eight-point sequence, and let $X[k]$ be its eight-point DFT.

(a) Evaluate

$$\left(\frac{1}{8}\sum_{k=0}^{7} X[k]e^{j(2\pi/8)kn}\right)\Bigg|_{n=9}$$

in terms of $x[n]$.

(b) Let $v[n] = 0$, $n < 0, n > 7$, be an eight-point sequence, and let $V[k]$ be its eight-point DFT.

If $V[k] = X(z)$ at $z = 2\exp(j(2\pi k + \pi)/8)$ for $k = 0, \ldots, 7$, where $X(z)$ is the z-transform of $x[n]$, express $v[n]$ in terms of $x[n]$.

(c) Let $w[n] = 0$, $n < 0, n > 3$, be a four-point sequence, and let $W[k]$ be its four-point DFT.

If $W[k] = X[k] + X[k + 4]$, express $w[n]$ in terms of $x[n]$.

(d) Let $y[n] = 0$, $n < 0, n > 7$, be an eight-point sequence, and let $Y[k]$ be its eight-point DFT.

If

$$Y[k] = \begin{cases} 2X[k], & k = 0, 2, 4, 6, \\ 0, & k = 1, 3, 5, 7, \end{cases}$$

express $y[n]$ in terms of $x[n]$.

8.53. Read each part of this problem carefully to note the differences among parts.

(a) Consider the signal

$$x[n] = \begin{cases} 1 + \cos(\pi n/4) - 0.5\cos(3\pi n/4), & 0 \le n \le 7, \\ 0, & \text{otherwise,} \end{cases}$$

which can be represented by the IDFT equation as

$$x[n] = \begin{cases} \dfrac{1}{8}\displaystyle\sum_{k=0}^{7} X_8[k]e^{j(2\pi k/8)n}, & 0 \le n \le 7, \\ 0, & \text{otherwise,} \end{cases}$$

where $X_8[k]$ is the eight-point DFT of $x[n]$. Plot $X_8[k]$ for $0 \le k \le 7$.

(b) Determine $V_{16}[k]$, the 16-point DFT of the 16-point sequence

$$v[n] = \begin{cases} 1 + \cos(\pi n/4) - 0.5\cos(3\pi n/4), & 0 \le n \le 15, \\ 0, & \text{otherwise.} \end{cases}$$

Plot $V_{16}[k]$ for $0 \le k \le 15$.

(c) Finally, consider $|X_{16}[k]|$, the magnitude of the 16-point DFT of the eight-point sequence

$$x[n] = \begin{cases} 1 + \cos(\pi n/4) - 0.5\cos(3\pi n/4), & 0 \le n \le 7, \\ 0, & \text{otherwise.} \end{cases}$$

Plot $|X_{16}[k]|$ for $0 \le k \le 15$ *without explicitly evaluating the DFT expression*. You will not be able to find all values of $|X_{16}[k]|$ by inspection as in parts (a) and (b), but you should be able to find some of the values exactly. Plot all the values you know exactly with a solid circle, and plot estimates of the other values with an open circle.

Extension Problems

8.54. In deriving the DFS analysis Eq. (8.11), we used the identity of Eq. (8.7). To verify this identity, we will consider the two conditions $k - r = mN$ and $k - r \neq mN$ separately.

(a) For $k - r = mN$, show that $e^{j(2\pi/N)(k-r)n} = 1$ and, from this, that

$$\frac{1}{N} \sum_{n=0}^{N-1} e^{j(2\pi/N)(k-r)n} = 1 \qquad \text{for } k - r = mN. \tag{P8.54-1}$$

(b) Since k and r are both integers in Eq. (8.7), we can make the substitution $k - r = \ell$ and consider the summation

$$\frac{1}{N} \sum_{n=0}^{N-1} e^{j(2\pi/N)\ell n} = \frac{1}{N} \sum_{n=0}^{N-1} [e^{j(2\pi/N)\ell}]^n. \tag{P8.54-2}$$

Because this is the sum of a finite number of terms in a geometric series, it can be expressed in closed form as

$$\frac{1}{N} \sum_{n=0}^{N-1} [e^{j(2\pi/N)\ell}]^n = \frac{1}{N} \frac{1 - e^{j(2\pi/N)\ell N}}{1 - e^{j(2\pi/N)\ell}}. \tag{P8.54-3}$$

For what values of ℓ is the right-hand side of Eq. (P8.54-3) equation indeterminate? That is, are the numerator and denominator both zero?

(c) From the result in part (b), show that if $k - r \neq mN$, then

$$\frac{1}{N} \sum_{n=0}^{N-1} e^{j(2\pi/N)(k-r)n} = 0. \tag{P8.54-4}$$

8.55. In Section 8.2, we stated the property that if

$$\tilde{x}_1[n] = \tilde{x}[n - m],$$

then

$$\tilde{X}_1[k] = W_N^{km} \tilde{X}[k],$$

where $\tilde{X}[k]$ and $\tilde{X}_1[k]$ are the DFS coefficients of $\tilde{x}[n]$ and $\tilde{x}_1[n]$, respectively. In this problem, we consider the proof of that property.

(a) Using Eq. (8.11) together with an appropriate substitution of variables, show that $\tilde{X}_1[k]$ can be expressed as

$$\tilde{X}_1[k] = W_N^{km} \sum_{r=-m}^{N-1-m} \tilde{x}[r] W_N^{kr}. \tag{P8.55-1}$$

(b) The summation in Eq. (P8.55-1) can be rewritten as

$$\sum_{r=-m}^{N-1-m} \tilde{x}[r] W_N^{kr} = \sum_{r=-m}^{-1} \tilde{x}[r] W_N^{kr} + \sum_{r=0}^{N-1-m} \tilde{x}[r] W_N^{kr}. \tag{P8.55-2}$$

Using the fact that $\tilde{x}[r]$ and W_N^{kr} are both periodic, show that

$$\sum_{r=-m}^{-1} \tilde{x}[r] W_N^{kr} = \sum_{r=N-m}^{N-1} \tilde{x}[r] W_N^{kr}. \tag{P8.55-3}$$

(c) From your results in parts (a) and (b), show that

$$\tilde{X}_1[k] = W_N^{km} \sum_{r=0}^{N-1} \tilde{x}[r] W_N^{kr} = W_N^{km} \tilde{X}[k].$$

8.56. (a) Table 8.1 lists a number of symmetry properties of the DFS for periodic sequences, several of which we repeat here. Prove that each of these properties is true. In carrying out your proofs, you may use the definition of the DFS and any previous property in the list. (For example, in proving property 3, you may use properties 1 and 2.)

Sequence	*DFS*
1. $\tilde{x}^*[n]$	$\tilde{X}^*[-k]$
2. $\tilde{x}^*[-n]$	$\tilde{X}^*[k]$
3. $\mathcal{Re}\{\tilde{x}[n]\}$	$\tilde{X}_e[k]$
4. $j\mathcal{Im}\{\tilde{x}[n]\}$	$\tilde{X}_o[k]$

(b) From the properties proved in part (a), show that for a real periodic sequence $\tilde{x}[n]$, the following symmetry properties of the DFS hold:

1. $\mathcal{Re}\{\tilde{X}[k]\} = \mathcal{Re}\{\tilde{X}[-k]\}$
2. $\mathcal{Im}\{\tilde{X}[k]\} = -\mathcal{Im}\{\tilde{X}[-k]\}$
3. $|\tilde{X}[k]| = |\tilde{X}[-k]|$
4. $\angle\tilde{X}[k] = -\angle\tilde{X}[-k]$

8.57. We stated in Section 8.4 that a direct relationship between $X(e^{j\omega})$ and $\tilde{X}[k]$ can be derived, where $\tilde{X}[k]$ is the DFS coefficients of a periodic sequence and $X(e^{j\omega})$ is the Fourier transform of one period. Since $\tilde{X}[k]$ corresponds to samples of $X(e^{j\omega})$, the relationship then corresponds to an interpolation formula.

One approach to obtaining the desired relationship is to rely on the discussion of Section 8.4, the relationship of Eq. (8.54), and the modulation property of Section 2.9.7. The procedure is as follows:

1. With $\tilde{X}[k]$ denoting the DFS coefficients of $\tilde{x}[n]$, express the Fourier transform $\tilde{X}(e^{j\omega})$ of $\tilde{x}[n]$ as an impulse train; i.e., scaled and shifted impulse functions $S(\omega)$.
2. From Eq. (8.57), $x[n]$ can be expressed as $x[n] = \tilde{x}[n]w[n]$, where $w[n]$ is an appropriate finite-length window.
3. Since $x[n] = \tilde{x}[n]w[n]$, from Section 2.9.7, $X(e^{j\omega})$ can be expressed as the (periodic) convolution of $\tilde{X}(e^{j\omega})$ and $W(e^{j\omega})$.

By carrying out the details in this procedure, show that $X(e^{j\omega})$ can be expressed as

$$X(e^{j\omega}) = \frac{1}{N} \sum_k \tilde{X}[k] \frac{\sin[(\omega N - 2\pi k)/2]}{\sin\{[\omega - (2\pi k/N)]/2\}} e^{-j[(N-1)/2](\omega - 2\pi k/N)}.$$

Specify explicitly the limits on the summation.

8.58. Let $X[k]$ denote the N-point DFT of the N-point sequence $x[n]$.

(a) Show that if

$$x[n] = -x[N - 1 - n],$$

then $X[0] = 0$. Consider separately the cases of N even and N odd.

(b) Show that if N is even and if

$$x[n] = x[N - 1 - n],$$

then $X[N/2] = 0$.

8.59. In Section 2.8, the conjugate-symmetric and conjugate-antisymmetric components of a sequence $x[n]$ were defined, respectively, as

$$x_e[n] = \frac{1}{2}(x[n] + x^*[-n]),$$

$$x_o[n] = \frac{1}{2}(x[n] - x^*[-n]).$$

In Section 8.6.4, we found it convenient to define respectively the periodic conjugate-symmetric and periodic conjugate-antisymmetric components of a sequence of finite duration N as

$$x_{ep}[n] = \tfrac{1}{2}\{x[((n))_N] + x^*[((-n))_N]\}, \qquad 0 \le n \le N - 1,$$

$$x_{op}[n] = \tfrac{1}{2}\{x[((n))_N] - x^*[((-n))_N]\}, \qquad 0 \le n \le N - 1.$$

(a) Show that $x_{ep}[n]$ can be related to $x_e[n]$ and that $x_{op}[n]$ can be related to $x_o[n]$ by the relations

$$x_{ep}[n] = (x_e[n] + x_e[n - N]), \qquad 0 \le n \le N - 1,$$

$$x_{op}[n] = (x_o[n] + x_o[n - N]), \qquad 0 \le n \le N - 1.$$

(b) $x[n]$ is considered to be a sequence of length N, and in general, $x_e[n]$ cannot be recovered from $x_{ep}[n]$, and $x_o[n]$ cannot be recovered from $x_{op}[n]$. Show that with $x[n]$ considered as a sequence of length N, but with $x[n] = 0, n > N/2$, $x_e[n]$ can be obtained from $x_{ep}[n]$, and $x_o[n]$ can be obtained from $x_{op}[n]$.

8.60. Show from Eqs. (8.65) and (8.66) that with $x[n]$ as an N-point sequence and $X[k]$ as its N-point DFT,

$$\sum_{n=0}^{N-1} |x[n]|^2 = \frac{1}{N} \sum_{k=0}^{N-1} |X[k]|^2.$$

This equation is commonly referred to as *Parseval's relation* for the DFT.

8.61. $x[n]$ is a real-valued, nonnegative, finite-length sequence of length N; i.e., $x[n]$ is real and nonnegative for $0 \le n \le N - 1$ and is zero otherwise. The N-point DFT of $x[n]$ is $X[k]$, and the Fourier transform of $x[n]$ is $X(e^{j\omega})$.

Determine whether each of the following statements is true or false. For each statement, if you indicate that it is true, clearly show your reasoning. If you state that it is false, construct a counterexample.

(a) If $X(e^{j\omega})$ is expressible in the form

$$X(e^{j\omega}) = B(\omega)e^{j\alpha\omega},$$

where $B(\omega)$ is real and α is a real constant, then $X[k]$ can be expressed in the form

$$X[k] = A[k]e^{j\gamma k},$$

where $A[k]$ is real and γ is a real constant.

(b) If $X[k]$ is expressible in the form

$$X[k] = A[k]e^{j\gamma k},$$

where $A[k]$ is real and γ is a real constant, then $X(e^{j\omega})$ can be expressed in the form

$$X(e^{j\omega}) = B(\omega)e^{j\alpha\omega},$$

where $B(\omega)$ is real and α is a real constant.

8.62. $x[n]$ and $y[n]$ are two real-valued, positive, finite-length sequences of length 256; i.e.,

$$x[n] > 0, \qquad 0 \le n \le 255,$$

$$y[n] > 0, \qquad 0 \le n \le 255,$$

$$x[n] = y[n] = 0, \qquad \text{otherwise.}$$

$r[n]$ denotes the *linear* convolution of $x[n]$ and $y[n]$. $R(e^{j\omega})$ denotes the Fourier transform of $r[n]$. $R_S[k]$ denotes 128 equally spaced samples of $R(e^{j\omega})$; i.e.,

$$R_S[k] = R(e^{j\omega})\Big|_{\omega=2\pi k/128}, \qquad k = 0, 1, \dots, 127.$$

Given $x[n]$ and $y[n]$, we want to obtain $R_S[k]$ as efficiently as possible. The *only* modules available are those shown in Figure P8.62. The costs associated with each module are as follows:

Modules I and II are free.
Module III costs 10 units.
Module IV costs 50 units.
Module V costs 100 units.

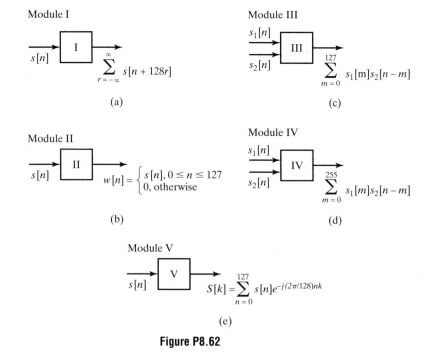

(a)

(b)

(c)

(d)

(e)

Figure P8.62

By appropriately connecting one or several of each module, construct a system for which the inputs are $x[n]$ and $y[n]$ and the output is $R_S[k]$. The important considerations are (a) whether the system works and (b) how efficient it is. The lower the *total* cost, the more efficient the system is.

8.63. $y[n]$ is the output of a stable LTI system with system function $H(z) = 1/(z - bz^{-1})$, where b is a known constant. We would like to recover the input signal $x[n]$ by operating on $y[n]$. The following procedure is proposed for recovering part of $x[n]$ from the data $y[n]$:

1. Using $y[n], 0 \le n \le N - 1$, calculate $Y[k]$, the N-point DFT of $y[n]$.

2. Form

$$V[k] = (W_N^{-k} - bW_N^{k})Y[k].$$

3. Calculate the inverse DFT of $V[k]$ to obtain $v[n]$.

For which values of the index n in the range $n = 0, 1, \ldots, N-1$ are we guaranteed that

$$x[n] = v[n]?$$

8.64. A modified discrete Fourier transform (MDFT) was proposed (Vernet, 1971) that computes samples of the z-transform on the unit circle offset from those computed by the DFT. In particular, with $X_M[k]$ denoting the MDFT of $x(n)$,

$$X_M[k] = X(z)\Big|_{z = e^{j[2\pi k/N + \pi/N]}}, \qquad k = 0, 1, 2, \ldots, N - 1.$$

Assume that N is even.

(a) The N-point MDFT of a sequence $x[n]$ corresponds to the N-point DFT of a sequence $x_M[n]$, which is easily constructed from $x[n]$. Determine $x_M[n]$ in terms of $x[n]$.

(b) If $x[n]$ is real, not all the points in the DFT are independent, since the DFT is conjugate symmetric; i.e., $X[k] = X^*[((-k))_N]$ for $0 \le k \le N - 1$. Similarly, if $x[n]$ is real, not all the points in the MDFT are independent. Determine, for $x[n]$ real, the relationship between points in $X_M[k]$.

(c) (i) Let $R[k] = X_M[2k]$; that is, $R[k]$ contains the even-numbered points in $X_M[k]$. From your answer in part (b), show that $X_M[k]$ can be recovered from $R[k]$.

(ii) $R[k]$ can be considered to be the $N/2$-point MDFT of an $N/2$-point sequence $r[n]$. Determine a simple expression relating $r[n]$ directly to $x[n]$.

According to parts (b) and (c), the N-point MDFT of a real sequence $x[n]$ can be computed by forming $r[n]$ from $x[n]$ and then computing the $N/2$-point MDFT of $r[n]$. The next two parts are directed at showing that the MDFT can be used to implement a linear convolution.

(d) Consider three sequences $x_1[n]$, $x_2[n]$, and $x_3[n]$, all of length N. Let $X_{1M}[k]$, $X_{2M}[k]$, and $X_{3M}[k]$, respectively, denote the MDFTs of the three sequences. If

$$X_{3M}[k] = X_{1M}[k]X_{2M}[k],$$

express $x_3[n]$ in terms of $x_1[n]$ and $x_2[n]$. Your expression must be of the form of a single summation over a "combination" of $x_1[n]$ and $x_2[n]$ in the same style as (but not identical to) a circular convolution.

(e) It is convenient to refer to the result in part (d) as a modified circular convolution. If the sequences $x_1[n]$ and $x_2[n]$ are both zero for $n \ge N/2$, show that the modified circular convolution of $x_1[n]$ and $x_2[n]$ is identical to the linear convolution of $x_1[n]$ and $x_2[n]$.

8.65. In some applications in coding theory, it is necessary to compute a 63-point circular convolution of two 63-point sequences $x[n]$ and $h[n]$. Suppose that the only computational
devices available are multipliers, adders, and processors that compute N-point DFTs, with
N restricted to be a power of 2.

    (a) It is possible to compute the 63-point circular convolution of $x[n]$ and $h[n]$ using a num-
ber of 64-point DFTs, inverse DFTs, and the overlap–add method. How many DFTs
are needed? *Hint*: Consider each of the 63-point sequences as the sum of a 32-point
sequence and 31-point sequence.

    (b) Specify an algorithm that computes the 63-point circular convolution of $x[n]$ and $h[n]$
using two 128-point DFTs and one 128-point inverse DFT.

    (c) We could also compute the 63-point circular convolution of $x[n]$ and $h[n]$ by computing
their linear convolution in the time domain and then aliasing the results. In terms of
multiplications, which of these three methods is most efficient? Which is least efficient?
(Assume that one complex multiplication requires four real multiplications and that
$x[n]$ and $h[n]$ are real.)

8.66. We want to filter a very long sequence with an FIR filter whose impulse response is 50 samples long. We wish to implement this filter with a DFT using the overlap–save technique.
The procedure is as follows:

    1. The input sections must be overlapped by V samples.

    2. From the output of each section, we must extract M samples such that when these
samples from each section are abutted, the resulting sequence is the desired filtered
output.

Assume that the input segments are 100 samples long and that the size of the DFT is
$128\ (= 2^7)$ points. Assume further that the output sequence from the circular convolution
is indexed from point 0 to point 127.

    (a) Determine V.

    (b) Determine M.

    (c) Determine the index of the beginning and the end of the M points extracted; i.e., deter-
mine which of the 128 points from the circular convolution are extracted to be abutted
with the result from the previous section.

8.67. A problem that often arises in practice is one in which a distorted signal $y[n]$ is the output
that results when a desired signal $x[n]$ has been filtered by an LTI system. We wish to recover the original signal $x[n]$ by processing $y[n]$. In theory, $x[n]$ can be recovered from $y[n]$
by passing $y[n]$ through an inverse filter having a system function equal to the reciprocal of
the system function of the distorting filter.

    Suppose that the distortion is caused by an FIR filter with impulse response

$$h[n] = \delta[n] - 0.5\delta[n - n_0],$$

where n_0 is a positive integer, i.e., the distortion of $x[n]$ takes the form of an echo at delay n_0.

    (a) Determine the z-transform $H(z)$ and the N-point DFT $H[k]$ of the impulse response
$h[n]$. Assume that $N = 4n_0$.

    (b) Let $H_i(z)$ denote the system function of the inverse filter, and let $h_i[n]$ be the corre-
sponding impulse response. Determine $h_i[n]$. Is this an FIR or an IIR filter? What is
the duration of $h_i[n]$?

(c) Suppose that we use an FIR filter of length N in an attempt to implement the inverse filter, and let the N-point DFT of the FIR filter be

$$G[k] = 1/H[k], \qquad k = 0, 1, \ldots, N - 1.$$

What is the impulse response $g[n]$ of the FIR filter?

(d) It might appear that the FIR filter with DFT $G[k] = 1/H[k]$ implements the inverse filter perfectly. After all, one might argue that the FIR distorting filter has an N-point DFT $H[k]$ and the FIR filter in cascade has an N-point DFT $G[k] = 1/H[k]$, and since $G[k]H[k] = 1$ for all k, we have implemented an all-pass, nondistorting filter. Briefly explain the fallacy in this argument.

(e) Perform the convolution of $g[n]$ with $h[n]$, and thus determine how well the FIR filter with N-point DFT $G[k] = 1/H[k]$ implements the inverse filter.

8.68. A sequence $x[n]$ of length N has a discrete Hartley transform (DHT) defined as

$$X_H[k] = \sum_{n=0}^{N-1} x[n]H_N[nk], \qquad k = 0, 1, \ldots, N - 1, \qquad \text{(P8.68-1)}$$

where

$$H_N[a] = C_N[a] + S_N[a],$$

with

$$C_N[a] = \cos(2\pi a/N), \qquad S_N[a] = \sin(2\pi a/N).$$

Originally proposed by R.V.L. Hartley in 1942 for the continuous-time case, the Hartley transform has properties that make it useful and attractive in the discrete-time case as well (Bracewell, 1983, 1984). Specifically, from Eq. (P8.68-1), it is apparent that the DHT of a real sequence is also a real sequence. In addition, the DHT has a convolution property, and fast algorithms exist for its computation.

In complete analogy with the DFT, the DHT has an implicit periodicity that must be acknowledged in its use. That is, if we consider $x[n]$ to be a finite-length sequence such that $x[n] = 0$ for $n < 0$ and $n > N - 1$, then we can form a periodic sequence

$$\tilde{x}[n] = \sum_{r=-\infty}^{\infty} x[n + rN]$$

such that $x[n]$ is simply one period of $\tilde{x}[n]$. The periodic sequence $\tilde{x}[n]$ can be represented by a discrete Hartley series (DHS), which in turn can be interpreted as the DHT by focusing attention on only one period of the periodic sequence.

(a) The DHS analysis equation is defined by

$$\tilde{X}_H[k] = \sum_{n=0}^{N-1} \tilde{x}[n]H_N[nk]. \qquad \text{(P8.68-2)}$$

Show that the DHS coefficients form a sequence that is also periodic with period N; i.e.,

$$\tilde{X}_H[k] = \tilde{X}_H[k + N] \qquad \text{for all } k.$$

(b) It can also be shown that the sequences $H_N[nk]$ are orthogonal; i.e.,

$$\sum_{k=0}^{N-1} H_N[nk]H_N[mk] = \begin{cases} N, & ((n))_N = ((m))_N, \\ 0, & \text{otherwise.} \end{cases}$$

Using this property and the DHS analysis formula of Eq. (P8.68-2), show that the DHS synthesis formula is

$$\tilde{x}[n] = \frac{1}{N}\sum_{k=0}^{N-1} \tilde{X}_H[k]H_N[nk]. \tag{P8.68-3}$$

Note that the DHT is simply one period of the DHS coefficients, and likewise, the DHT synthesis (inverse) equation is identical to the DHS synthesis Eq. (P8.68-3), except that we simply extract one period of $\tilde{x}[n]$; i.e., the DHT synthesis expression is

$$x[n] = \frac{1}{N}\sum_{k=0}^{N-1} X_H[k]H_N[nk], \qquad n = 0, 1, \dots, N-1. \tag{P8.68-4}$$

With Eqs. (P8.68-1) and (P8.68-4) as definitions of the analysis and synthesis relations, respectively, for the DHT, we may now proceed to derive the useful properties of this representation of a finite-length discrete-time signal.

(c) Verify that $H_N[a] = H_N[a+N]$, and verify the following useful property of $H_N[a]$:

$$H_N[a+b] = H_N[a]C_N[b] + H_N[-a]S_N[b]$$

$$= H_N[b]C_N[a] + H_N[-b]S_N[a].$$

(d) Consider a circularly shifted sequence

$$x_1[n] = \begin{cases} \tilde{x}[n-n_0] = x[((n-n_0))_N], & n = 0, 1, \dots, N-1, \\ 0, & \text{otherwise.} \end{cases} \tag{P8.68-5}$$

In other words, $x_1[n]$ is the sequence that is obtained by extracting one period from the shifted periodic sequence $\tilde{x}[n-n_0]$. Using the identity verified in part (c), show that the DHS coefficients for the shifted periodic sequence are

$$\tilde{x}[n-n_0] \overset{\mathcal{DHS}}{\longleftrightarrow} \tilde{X}_H[k]C_N[n_0k] + \tilde{X}_H[-k]S_N[n_0k]. \tag{P8.68-6}$$

From this, we conclude that the DHT of the finite-length circularly shifted sequence $x[((n-n_0))_N]$ is

$$x[((n-n_0))_N] \overset{\mathcal{DHT}}{\longleftrightarrow} X_H[k]C_N[n_0k] + X_H[((-k))_N]S_N[n_0k]. \tag{P8.68-7}$$

(e) Suppose that $x_3[n]$ is the N-point circular convolution of two N-point sequences $x_1[n]$ and $x_2[n]$; i.e.,

$$x_3[n] = x_1[n] \text{Ⓝ} x_2[n]$$

$$= \sum_{m=0}^{N-1} x_1[m]x_2[((n-m))_N], \qquad n = 0, 1, \dots, N-1. \tag{P8.68-8}$$

By applying the DHT to both sides of Eq. (P8.68-8) and using Eq. (P8.68-7), show that

$$X_{H3}[k] = \frac{1}{2}X_{H1}[k](X_{H2}[k] + X_{H2}[((-k))_N])$$
$$+\frac{1}{2}X_{H1}[((-k))_N](X_{H2}[k] - X_{H2}[((-k))_N]) \qquad \text{(P8.68-9)}$$

for $k = 0, 1, \ldots, N - 1$. This is the desired convolution property.

Note that a linear convolution can be computed using the DHT in the same way that the DFT can be used to compute a linear convolution. While computing $X_{H3}[k]$ from $X_{H1}[k]$ and $X_{H2}[k]$ requires the same amount of computation as computing $X_3[k]$ from $X_1[k]$ and $X_2[k]$, the computation of the DHT requires only half the number of real multiplications required to compute the DFT.

(f) Suppose that we wish to compute the DHT of an N-point sequence $x[n]$ and we have available the means to compute the N-point DFT. Describe a technique for obtaining $X_H[k]$ from $X[k]$ for $k = 0, 1, \ldots, N - 1$.

(g) Suppose that we wish to compute the DFT of an N-point sequence $x[n]$ and we have available the means to compute the N-point DHT. Describe a technique for obtaining $X[k]$ from $X_H[k]$ for $k = 0, 1, \ldots, N - 1$.

8.69. Let $x[n]$ be an N-point sequence such that $x[n] = 0$ for $n < 0$ and for $n > N - 1$. Let $\hat{x}[n]$ be the $2N$-point sequence obtained by repeating $x[n]$; i.e.,

$$\hat{x}[n] = \begin{cases} x[n], & 0 \leq n \leq N - 1, \\ x[n - N], & N \leq n \leq 2N - 1, \\ 0, & \text{otherwise.} \end{cases}$$

Consider the implementation of a discrete-time filter shown in Figure P8.69. This system has an impulse response $h[n]$ that is $2N$ points long; i.e., $h[n] = 0$ for $n < 0$ and for $n > 2N - 1$.

(a) In Figure P8.69-1, what is $\hat{X}[k]$, the $2N$-point DFT of $\hat{x}[n]$, in terms of $X[k]$, the N-point DFT of $x[n]$?

(b) The system shown in Figure P8.69-1 can be implemented using only N-point DFTs as indicated in Figure P8.69-2 for appropriate choices for System A and System B. Specify System A and System B so that $\hat{y}[n]$ in Figure P8.69-1 and $y[n]$ in Figure P8.69-2 are equal for $0 \leq n \leq 2N - 1$. Note that $h[n]$ and $y[n]$ in Figure P8.69-2 are $2N$-point sequences and $w[n]$ and $g[n]$ are N-point sequences.

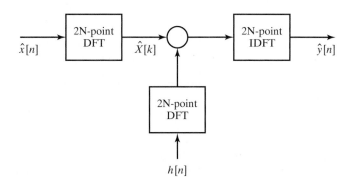

Figure P8.69-1

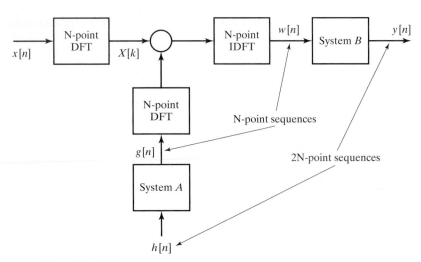

Figure P8.69-2

8.70. In this problem, you will examine the use of the DFT to implement the filtering necessary
for the discrete-time interpolation, or upsampling, of a signal. Assume that the discrete-time
signal $x[n]$ was obtained by sampling a continuous-time signal $x_c(t)$ with a sampling period
T. Moreover, the continuous-time signal is appropriately bandlimited; i.e., $X_c(j\Omega) = 0$ for
$|\Omega| \geq 2\pi/T$. For this problem, we will assume that $x[n]$ has length N; i.e., $x[n] = 0$ for
$n < 0$ or $n > N - 1$, where N is even. It is not strictly possible to have a signal that is both
perfectly bandlimited and of finite duration, but this is often assumed in practical systems
processing finite-length signals which have very little energy outside the band $|\Omega| \leq 2\pi/T$.

   We wish to implement a 1:4 interpolation, i.e., increase the sampling rate by a factor
of 4. As seen in Figure 4.23, we can perform this sampling rate conversion using a sampling
rate expander followed by an appropriate lowpass filter. In this chapter, we have seen that
the lowpass filter could be implemented using the DFT if the filter is an FIR impulse response. For this problem, assume that this filter has an impulse response $h[n]$ that is $N + 1$
points long. Figure P8.70-1 depicts such a system, where $H[k]$ is the $4N$-point DFT of the
impulse response of the lowpass filter. Note that both $v[n]$ and $y[n]$ are $4N$-point sequences.

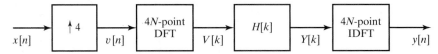

Figure P8.70-1

(a) Specify the DFT $H[k]$ such that the system in Figure P8.70-1 implements the desired
upsampling system. Think carefully about the phase of the values of $H[k]$.

(b) It is also possible to upsample $x[n]$ using the system in Figure P8.70-2. Specify System
A in the middle box so that the $4N$-point signal $y_2[n]$ in this figure is the same as $y[n]$
in Figure P8.70-2. Note that System A may consist of more than one operation.

(c) Is there a reason that the implementation in Figure P8.70-2 might be preferable to Figure P8.70-1?

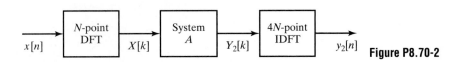

Figure P8.70-2

8.71. Derive Eq. (8.153) using Eqs. (8.164) and (8.165).

8.72. Consider the following procedure

(a) Form the sequence $v[n] = x_2[2n]$ where $x_2[n]$ is given by Eq. (8.166). This yields

$$v[n] = x[2n] \qquad n = 0, 1, \ldots, N/2 - 1$$
$$v[N - 1 - n] = x[2n + 1], \quad n = 0, 1, \ldots, N/2 - 1.$$

(b) Compute $V[k]$, the N-point DFT of $v[n]$.
Demonstrate that the following is true:

$$X^{c2}[k] = 2\mathcal{R}e\{e^{-j2\pi k/(4N)} V[k]\}, \qquad k = 0, 1, \ldots, N - 1,$$

$$= 2 \sum_{n=0}^{N-1} v[n]\cos\left[\frac{\pi k(4n + 1)}{2N}\right], \qquad k = 0, 1, \ldots, N - 1,$$

$$= 2 \sum_{n=0}^{N-1} x[n]\cos\left[\frac{\pi k(2n + 1)}{2N}\right], \qquad k = 0, 1, \ldots, N - 1.$$

Note that this algorithm uses N-point rather than $2N$-point DFTs as required in Eq. (8.167). In addition, since $v[n]$ is a real sequence, we can exploit even and odd symmetries to do the computation of $V[k]$ in one $N/4$-point complex DFT.

8.73. Derive Eq. (8.156) using Eqs. (8.174) and (8.157).

8.74. (a) Use Parseval's theorem for the DFT to derive a relationship between $\sum_k |X^{c1}[k]|^2$

and $\sum_n |x[n]|^2$.

(b) Use Parseval's theorem for the DFT to derive a relationship between $\sum_k |X^{c2}[k]|^2$

and $\sum_n |x[n]|^2$.

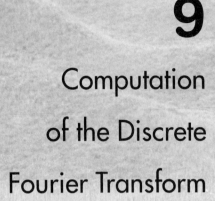

9

Computation
of the Discrete
Fourier Transform

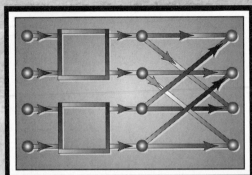

9.0 INTRODUCTION

The discrete Fourier transform (DFT) plays an important role in the analysis, design, and implementation of discrete-time signal-processing algorithms and systems because the basic properties of the discrete-time Fourier transform and discrete Fourier transform, discussed in Chapters 2 and 8, respectively, make it particularly convenient to analyze and design systems in the Fourier domain. It is equally important that efficient algorithms exist for explicitly computing the DFT. As a result, the DFT is an important component in many practical applications of discrete-time systems.

In this chapter, we discuss several methods for computing values of the DFT. The major focus of the chapter is a particularly efficient class of algorithms for the digital computation of the N-point DFT. Collectively, these efficient algorithms, which are discussed in Sections 9.2, 9.3, and 9.5, are called FFT algorithms. To achieve the highest efficiency, the FFT algorithms must compute all N values of the DFT. When we require values of the DFT at only a few frequencies in the range $0 \leq \omega < 2\pi$, other algorithms may be more efficient and flexible, even though they are less efficient than the FFT algorithms for computation of all the values of the DFT. Examples of such algorithms are the Goertzel algorithm, discussed in Section 9.1.2, and the chirp transform algorithm, discussed in Section 9.6.2.

There are many ways to measure the complexity and efficiency of an implementation or algorithm, and a final assessment depends on both the available implementation technology and the intended application. We will use the number of arithmetic multiplications and additions as a measure of computational complexity. This measure is simple to apply, and the number of multiplications and additions is directly related to the computational speed when algorithms are implemented on general-purpose digital computers or special-purpose processors. However, other measures are sometimes more appropriate. For example, in custom VLSI implementations, the area of the chip and power requirements are important considerations and may not be directly related to the number of arithmetic operations.

In terms of multiplications and additions, the class of FFT algorithms can be orders of magnitude more efficient than competing algorithms. The efficiency of FFT algorithms is so high, in fact, that in many cases the most efficient procedure for implementing a convolution is to compute the transform of the sequences to be convolved, multiply their transforms, and then compute the inverse transform of the product of transforms. The details of this technique were discussed in Section 8.7. In seeming contradiction to this, there is a set of algorithms (mentioned briefly in Section 9.6) for evaluation of the DFT (or a more general set of samples of the Fourier transform) that derive their efficiency from a reformulation of the Fourier transform in terms of a convolution and thereby implement the Fourier transform computation by using efficient procedures for evaluating the associated convolution. This suggests the possibility of implementing a convolution by multiplication of DFTs, where the DFTs have been implemented by first expressing them as convolutions and then taking advantage of efficient procedures for implementing the associated convolutions. While this seems on the surface to be a basic contradiction, we will see in Section 9.6 that in certain cases it is an entirely reasonable approach.

In the sections that follow, we consider a number of algorithms for computing the discrete Fourier transform. We begin in Section 9.1 with discussions of direct computation methods, i.e., methods based on direct use of the defining relation for the DFT as a computational formula. We include in this discussion the Goertzel algorithm (Goertzel, 1958), which requires computation proportional to N^2, but with a smaller constant of proportionality than that of the direct evaluation of the defining formula. One of the principal advantages of the direct evaluation method or the Goertzel algorithm is that they are not restricted to computation of the DFT, but can be used to compute any desired set of samples of the DTFT of a finite-length sequence.

In Sections 9.2 and 9.3 we present a detailed discussion of FFT algorithms for which computation is proportional to $N \log_2 N$. This class of algorithms is considerably more efficient in terms of arithmetic operations than the Goertzel algorithm, but is specifically oriented toward computation of *all* the values of the DFT. We do not attempt to be exhaustive in our coverage of that class of algorithms, but we illustrate the general principles common to all algorithms of this type by considering in detail only a few of the more commonly used schemes.

In Section 9.4, we consider some of the practical issues that arise in implementing the power-of-two-length FFT algorithms discussed in Sections 9.2 and 9.3. Section 9.5 provides a brief overview of algorithms for N a composite number including a brief reference to FFT algorithms that are optimized for a particular computer architecture. In Section 9.6, we discuss algorithms that rely on formulating the computation of the DFT in terms of a convolution. In Section 9.7, we consider effects of arithmetic round-off in FFT algorithms.

9.1 DIRECT COMPUTATION OF THE DISCRETE FOURIER TRANSFORM

As defined in Chapter 8, the DFT of a finite-length sequence of length N is

$$X[k] = \sum_{n=0}^{N-1} x[n] W_N^{kn}, \qquad k = 0, 1, \ldots, N - 1, \tag{9.1}$$

where $W_N = e^{-j(2\pi/N)}$. The inverse discrete Fourier transform is given by

$$x[n] = \frac{1}{N} \sum_{k=0}^{N-1} X[k] W_N^{-kn}, \qquad n = 0, 1, \ldots, N - 1. \tag{9.2}$$

In Eqs. (9.1) and (9.2), both $x[n]$ and $X[k]$ may be complex.[1] Since the expressions on the right-hand sides of those equations differ only in the sign of the exponent of W_N and in the scale factor $1/N$, a discussion of computational procedures for Eq. (9.1) applies with straightforward modifications to Eq. (9.2). (See Problem 9.1.)

Most approaches to improving the efficiency of computation of the DFT exploit the symmetry and periodicity properties of W_N^{kn}; specifically,

$$W_N^{k(N-n)} = W_N^{-kn} = (W_N^{kn})^* \text{ (complex conjugate symmetry)} \tag{9.3a}$$

$$W_N^{kn} = W_N^{k(n+N)} = W_N^{(k+N)n} \quad \text{(periodicity in } n \text{ and } k). \tag{9.3b}$$

(Since $W_N^{kn} = \cos(2\pi kn/N) - j \sin(2\pi kn/N)$ these properties are a direct consequence of the symmetry and periodicity of the underlying sine and cosine functions.) Because the complex numbers W_N^{kn} have the role of coefficients in Eqs. (9.1) and (9.2), the redundancy implied by these conditions can be used to advantage in reducing the amount of computation required for their evaluation.

9.1.1 Direct Evaluation of the Definition of the DFT

To create a frame of reference, consider first the direct evaluation of the defining DFT expression in Eq. (9.1). Since $x[n]$ may be complex, N complex multiplications and $(N - 1)$ complex additions are required to compute each value of the DFT if we use Eq. (9.1) directly as a formula for computation. To compute all N values therefore requires a total of N^2 complex multiplications and $N(N-1)$ complex additions. Expressing

---

[1] In discussing algorithms for computing the DFT of a finite-length sequence $x[n]$, it is worthwhile to recall from Chapter 8 that the DFT values defined by Eq. (9.1) can be thought of either as samples of the DTFT $X(e^{j\omega})$ at frequencies $\omega_k = 2\pi k/N$ or as coefficients in the discrete-time Fourier series for the periodic sequence

$$\tilde{x}[n] = \sum_{r=-\infty}^{\infty} x[n + rN].$$

It will be helpful to keep both interpretations in mind and to be able to switch focus from one to the other as is convenient.

Eq. (9.1) in terms of operations on real numbers, we obtain

$$
\begin{aligned}
X[k] = \sum_{n=0}^{N-1} \Big[&(\mathcal{R}e\{x[n]\}\mathcal{R}e\{W_N^{kn}\} - \mathcal{I}m\{x[n]\}\mathcal{I}m\{W_N^{kn}\}) \\
&+ j(\mathcal{R}e\{x[n]\}\mathcal{I}m\{W_N^{kn}\} + \mathcal{I}m\{x[n]\}\mathcal{R}e\{W_N^{kn}\}) \Big], \\
&\hspace{5cm} k = 0, 1, \ldots, N-1,
\end{aligned}
\tag{9.4}
$$

which shows that each complex multiplication $x[n] \cdot W_N^{kn}$ requires four real multiplications and two real additions, and each complex addition requires two real additions. Therefore, for each value of k, the direct computation of $X[k]$ requires $4N$ real multiplications and $(4N - 2)$ real additions.[2] Since $X[k]$ must be computed for N different values of k, the direct computation of the discrete Fourier transform of a sequence $x[n]$ requires $4N^2$ real multiplications and $N(4N - 2)$ real additions. Besides the multiplications and additions called for by Eq. (9.4), the digital computation of the DFT on a general-purpose digital computer or with special-purpose hardware also requires provision for storing and accessing the N complex input sequence values $x[n]$ and values of the complex coefficients W_N^{kn}. Since the amount of computation, and thus the computation time, is approximately proportional to N^2, it is evident that the number of arithmetic operations required to compute the DFT by the direct method becomes very large for large values of N. For this reason, we are interested in computational procedures that reduce the number of multiplications and additions.

As an illustration of how the properties of W_N^{kn} can be exploited, using the symmetry property in Eq. (9.3a), we can group terms in the summation in Eq. (9.4) for n and $(N - n)$. For example, the grouping

$$
\mathcal{R}e\{x[n]\}\mathcal{R}e\{W_N^{kn}\} + \mathcal{R}e\{x[N - n]\}\mathcal{R}e\{W_N^{k(N-n)}\}
$$

$$
= (\mathcal{R}e\{x[n]\} + \mathcal{R}e\{x[N - n]\})\mathcal{R}e\{W_N^{kn}\}
$$

eliminates one real multiplication, as does the grouping

$$
-\mathcal{I}m\{x[n]\}\mathcal{I}m\{W_N^{kn}\} - \mathcal{I}m\{x[N - n]\}\mathcal{I}m\{W_N^{k(N-n)}\}
$$

$$
= -(\mathcal{I}m\{x[n]\} - \mathcal{I}m\{x[N - n]\})\mathcal{I}m\{W_N^{kn}\}.
$$

Similar groupings can be used for the other terms in Eq. (9.4). In this way, the number of multiplications can be reduced by approximately a factor of 2. We can also take advantage of the fact that for certain values of the product kn, the implicit sine and cosine functions take on the value 1 or 0, thereby eliminating the need for multiplications. However, reductions of this type still leave us with an amount of computation that is proportional to N^2. Fortunately, the second property [Eq. (9.3b)], the periodicity of the complex sequence W_N^{kn}, can be exploited with recursion to achieve significantly greater reductions of the computation.

9.1.2 The Goertzel Algorithm

The Goertzel algorithm (Goertzel, 1958) is an example of how the periodicity of the sequence W_N^{kn} can be used to reduce computation. To derive the algorithm, we begin

---

[2]Throughout the discussion, the formula for the number of computations may be only approximate. Multiplication by W_N^0, for example, does not require a multiplication. Nevertheless, when N is large, the estimate of computational complexity obtained by including such multiplications is sufficiently accurate to permit comparisons between different classes of algorithms.

by noting that

$$W_N^{-kN} = e^{j(2\pi/N)Nk} = e^{j2\pi k} = 1,$$ (9.5)

since k is an integer. This is a result of the periodicity with period N of W_N^{-kn} in either n or k. Because of Eq. (9.5), we may multiply the right side of Eq. (9.1) by W_N^{-kN} without affecting the equation. Thus,

$$X[k] = W_N^{-kN} \sum_{r=0}^{N-1} x[r]W_N^{kr} = \sum_{r=0}^{N-1} x[r]W_N^{-k(N-r)}.$$ (9.6)

To suggest the final result, we define the sequence

$$y_k[n] = \sum_{r=-\infty}^{\infty} x[r]W_N^{-k(n-r)}u[n-r].$$ (9.7)

From Eqs. (9.6) and (9.7) and the fact that $x[n] = 0$ for $n < 0$ and $n \geq N$, it follows that

$$X[k] = y_k[n]\Big|_{n=N}.$$ (9.8)

Equation (9.7) can be interpreted as a discrete convolution of the finite-duration sequence $x[n]$, $0 \leq n \leq N - 1$, with the sequence $W_N^{-kn}u[n]$. Consequently, $y_k[n]$ can be viewed as the response of a system with impulse response $W_N^{-kn}u[n]$ to a finite-length input $x[n]$. In particular, $X[k]$ is the value of the output when $n = N$.

The signal flow graph of a system with impulse response $W_N^{-kn}u[n]$ is shown in Figure 9.1, which represents the difference equation

$$y_k[n] = W_N^{-k}y_k[n-1] + x[n],$$ (9.9)

where initial rest conditions are assumed. Since the general input $x[n]$ and the coefficient W_N^{-k} are both complex, the computation of each new value of $y_k[n]$ using the system of Figure 9.1 requires 4 real multiplications and 4 real additions. All the intervening values $y_k[1], y_k[2], \ldots, y_k[N-1]$ must be computed in order to compute $y_k[N] = X[k]$, so the use of the system in Figure 9.1 as a computational algorithm requires $4N$ real multiplications and $4N$ real additions to compute $X[k]$ for a particular value of k. Thus, this procedure is slightly less efficient than the direct method. However, it avoids the computation or storage of the coefficients W_N^{kn}, since these quantities are implicitly computed by the recursion implied by Figure 9.1.

It is possible to retain this simplification while reducing the number of multiplications by a factor of 2. To see how this may be done, note that the system function of the system of Figure 9.1 is

$$H_k(z) = \frac{1}{1 - W_N^{-k}z^{-1}}.$$ (9.10)

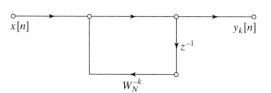

$x[n]$　　　　　　　　　　　　$y_k[n]$

z^{-1}

W_N^{-k}

Figure 9.1　Flow graph of 1st-order complex recursive computation of $X[k]$.

Multiplying both the numerator and the denominator of $H_k(z)$ by the factor $(1-W_N^k z^{-1})$, we obtain

$$H_k(z) = \frac{1 - W_N^k z^{-1}}{(1 - W_N^{-k} z^{-1})(1 - W_N^k z^{-1})}$$

$$= \frac{1 - W_N^k z^{-1}}{1 - 2\cos(2\pi k/N)z^{-1} + z^{-2}}. \qquad (9.11)$$

The signal flow graph of Figure 9.2 corresponds to the direct form II implementation of the system function of Eq. (9.11) for which the difference equation for the poles is

$$v_k[n] = 2\cos(2\pi k/N)v_k[n-1] - v_k[n-2] + x[n]. \qquad (9.12a)$$

After N iterations of Eq. (9.12a) starting with initial rest conditions $w_k[-2] = w_k[-1] = 0$, the desired DFT value can be obtained by implementing the zero as in

$$X[k] = y_k[n]\Big|_{n=N} = v_k[N] - W_N^k v_k[N-1]. \qquad (9.12b)$$

If the input is complex, only two real multiplications per sample are required to implement the poles of this system, since the coefficients are real and the factor -1 need not be counted as a multiplication. As in the case of the 1st-order system, for a complex input, four real additions per sample are required to implement the poles (if the input is complex). Since we only need to bring the system to a state from which $y_k[N]$ can be computed, the complex multiplication by $-W_N^k$ required to implement the zero of the system function need not be performed at every iteration of the difference equation, but only after the Nth iteration. Thus, the total computation is $2N$ real multiplications and $4N$ real additions for the poles,[3] plus 4 real multiplications and 4 real additions for the zero. The total computation is therefore $2(N+2)$ real multiplications and $4(N+1)$ real additions, about half the number of real multiplications required with the direct method. In this more efficient scheme, we still have the advantage that $\cos(2\pi k/N)$ and W_N^k are the only coefficients that must be computed and stored. The coefficients W_N^{kn} are again computed implicitly in the iteration of the recursion formula implied by Figure 9.2.

As an additional advantage of the use of this network, let us consider the computation of the DFT of $x[n]$ at the two symmetric frequencies $2\pi k/N$ and $2\pi(N-k)/N$,

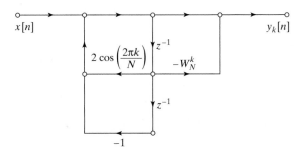

Figure 9.2   Flow graph of 2nd-order recursive computation of $X[k]$ (Goertzel algorithm).

---

[3]This assumes that $x[n]$ is complex. If $x[n]$ is real, the operation count is N real multiplications and $2N$ real additions for implementing the poles.

that is, the computation of $X[k]$ and $X[N - k]$. It is straightforward to verify that the network in the form of Figure 9.2 required to compute $X[N - k]$ has exactly the same poles as that in Figure 9.2, but the coefficient for the zero is the complex conjugate of that in Figure 9.2. (See Problem 9.21.) Since the zero is implemented only on the final iteration, the $2N$ multiplications and $4N$ additions required for the poles can be used for the computation of two DFT values. Thus, for the computation of all N values of the discrete Fourier transform using the Goertzel algorithm, the number of real multiplications required is approximately N^2 and the number of real additions is approximately $2N^2$. While this is more efficient than the direct computation of the discrete Fourier transform, the amount of computation is still proportional to N^2.

In either the direct method or the Goertzel algorithm we do not need to evaluate $X[k]$ at all N values of k. Indeed, we can evaluate $X[k]$ for any M values of k, with each DFT value being computed by a recursive system of the form of Figure 9.2 with appropriate coefficients. In this case, the total computation is proportional to NM. The Goertzel method and the direct method are attractive when M is small; however, as indicated previously, algorithms are available for which the computation is proportional to $N \log_2 N$ when N is a power of 2. Therefore, when M is less than $\log_2 N$, either the Goertzel algorithm or direct evaluation of the DFT may in fact be the most efficient method, but when all N values of $X[k]$ are required, the decimation-in-time algorithms, to be considered next, are roughly $(N/\log_2 N)$ times more efficient than either the direct method or the Goertzel algorithm.

As we have derived it, the Goertzel algorithm computes the DFT value $X[k]$, which is identical to the DTFT $X(e^{j\omega})$ evaluated at frequency $\omega = 2\pi k/N$. With only a minor modification of the above derivation, we can show that $X(e^{j\omega})$ can be evaluated at any frequency ω_a by iterating the difference equation

$$v_a[n] = 2\cos(\omega_0)v_a[n - 1] - v_a[n - 2] + x[n], \qquad (9.13a)$$

N times with the desired value of the DTFT obtained by

$$X(e^{j\omega_a}) = e^{-j\omega_a N}(v_a[N] - e^{-j\omega_a}v_a[N - 1]). \qquad (9.13b)$$

Note that in the case $\omega_a = 2\pi k/N$ Eqs. (9.13a) and (9.13b) reduce to Eqs. (9.12a) and (9.12b). Because Eq. (9.13b) must only be computed once, it is only slightly less efficient to compute the value of $X(e^{j\omega})$ at an arbitrarily chosen frequency than at a DFT frequency.

Still another advantage of the Goertzel algorithm in some real-time applications is that the computation can begin as soon as the first input sample is available. The computation then involves iterating the difference equation Eq. (9.12a) or Eq. (9.13a) as each new input sample becomes available. After N iterations, the desired value of $X(e^{j\omega})$ can be computed with either Eq. (9.12b) or Eq. (9.13b) as is appropriate.

9.1.3 Exploiting both Symmetry and Periodicity

Computational algorithms that exploit both the symmetry and the periodicity of the sequence W_N^{kn} were known long before the era of high-speed digital computation. At that time, any scheme that reduced manual computation by even a factor of 2 was welcomed. Heideman, Johnson and Burrus (1984) have traced the origins of the basic principles of the FFT back to Gauss, as early as 1805. Runge (1905) and later Danielson and Lanczos (1942) described algorithms for which computation was roughly proportional to $N \log_2 N$ rather than N^2. However, the distinction was not of great importance for the small values of N that were feasible for hand computation. The possibility of greatly reduced computation was generally overlooked until about 1965, when Cooley and Tukey (1965) published an algorithm for the computation of the discrete Fourier transform that is applicable when N is a composite number, i.e., the product of two or more integers. The publication of their paper touched off a flurry of activity in the application of the discrete Fourier transform to signal processing and resulted in the discovery of a number of highly efficient computational algorithms. Collectively, the entire set of such algorithms has come to be known as the *fast Fourier transform,* or the FFT.[4]

In contrast to the direct methods discussed above, FFT algorithms are based on the fundamental principle of decomposing the computation of the discrete Fourier transform of a sequence of length N into smaller-length discrete Fourier transforms that are combined to form the N-point transform. These smaller-length transforms may be evaluated by direct methods, or they may be further decomposed into even smaller transforms. The manner in which this principle is implemented leads to a variety of different algorithms, all with comparable improvements in computational speed. In this chapter, we are concerned with two basic classes of FFT algorithms. The first class, called decimation in time, derives its name from the fact that in the process of arranging the computation into smaller transformations, the sequence $x[n]$ (generally thought of as a time sequence) is decomposed into successively smaller subsequences. In the second general class of algorithms, the sequence of discrete Fourier transform coefficients $X[k]$ is decomposed into smaller subsequences—hence its name, decimation in frequency.

We discuss decimation-in-time algorithms in Section 9.2. Decimation-in-frequency algorithms are discussed in Section 9.3. This is an arbitrary ordering. The two sections are essentially independent and can therefore be read in either order.

9.2 DECIMATION-IN-TIME FFT ALGORITHMS

Dramatic efficiency in computing the DFT results from decomposing the computation into successively smaller DFT computations while exploiting both the symmetry and the periodicity of the complex exponential $W_N^{kn} = e^{-j(2\pi/N)kn}$. Algorithms in which the decomposition is based on decomposing the sequence $x[n]$ into successively smaller subsequences are called *decimation-in-time algorithms.*

The principle of decimation-in-time is conveniently illustrated by considering the special case of N an integer power of 2, i.e., $N = 2^\nu$. Since N is divisible by two, we can consider computing $X[k]$ by separating $x[n]$ into two $(N/2)$-point[5] sequences consisting of the even-numbered points $g[n] = x[2n]$ and the odd-numbered points $h[n] = x[2n+1]$. Figure 9.3 shows this decomposition and also the (somewhat obvious, but crucial) fact that the original sequence can be recovered simply by re-interleaving the two sequences.

---

[4]See Cooley, Lewis and Welch (1967) and Heideman, Johnson and Burrus (1984) for historical summaries of algorithmic developments related to the FFT.

[5]When discussing FFT algorithms, it is common to use the words *sample* and *point* interchangeably to mean *sequence value*, i.e., a single number. Also, we refer to a sequence of length N as an N-point sequence, and the DFT of a sequence of length N will be called an N-point DFT.

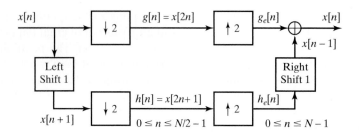

Figure 9.3   Illustration of the basic principle of decimation-in-time.

To understand the significance of Figure 9.3 as an organizing principle for computing the DFT, it is helpful to consider the frequency-domain equivalents of the operations
depicted in the block diagram. First, note that the time-domain operation labeled "Left
Shift 1" corresponds in the frequency domain to multiplying $X(e^{j\omega})$ by $e^{j\omega}$. As discussed
in Section 4.6.1, corresponding to the compression of the time sequences by 2, the DTFTs
$G(e^{j\omega})$ and $H(e^{j\omega})$ (and therefore $G[k]$ and $H[k]$) are obtained by frequency-domain
aliasing that occurs after expanding the frequency scale by the substitution $\omega \rightarrow \omega/2$ in
$X(e^{j\omega})$ and $e^{j\omega}X(e^{j\omega})$. That is, the DTFTs of the compressed sequences $g[n] = x[2n]$
and $h[n] = x[2n + 1]$ are respectively

$$G(e^{j\omega}) = \frac{1}{2}\left(X(e^{j\omega/2}) + X(e^{j(\omega-2\pi)/2})\right) \tag{9.14a}$$

$$H(e^{j\omega}) = \frac{1}{2}\left(X(e^{j\omega/2})e^{j\omega/2} + X(e^{j(\omega-2\pi)/2})e^{j(\omega-2\pi)/2}\right). \tag{9.14b}$$

The sequence-expansion-by-2 shown in the right half of the block diagram in Figure 9.3
results in the frequency-compressed DTFTs $G_e(e^{j\omega}) = G(e^{j2\omega})$ and $H_e(e^{j\omega}) = H(e^{j2\omega})$,
which, according to Figure 9.3, combine to form $X(e^{j\omega})$ through

$$X(e^{j\omega}) = G_e(e^{j\omega}) + e^{-j\omega}H_e(e^{j\omega})$$
$$= G(e^{j2\omega}) + e^{-j\omega}H(e^{j2\omega}). \tag{9.15}$$

Substituting Eqs. (9.14a) and (9.14b) into Eq. (9.15) will verify that the DTFT $X(e^{j\omega})$ of
the N-point sequence $x[n]$ can be represented as in Eq. (9.15) in terms of the DTFTs of
the $N/2$-point sequences $g[n] = x[2n]$ and $h[n] = x[2n + 1]$. Therefore, the DFT $X[k]$
can likewise be represented in terms of the DFTs of $g[n]$ and $h[n]$.
Specifically, $X[k]$ corresponds to evaluating $X(e^{j\omega})$ at frequencies $\omega_k = 2\pi k/N$
with $k = 0, 1, \ldots, N - 1$. Therefore, using Eq. (9.15) we obtain

$$X[k] = X(e^{j2\pi k/N}) = G(e^{j(2\pi k/N)2}) + e^{-j2\pi k/N}H(e^{(j2\pi k/N)2}). \tag{9.16}$$

From the definition of $g[n]$ and $G(e^{j\omega})$, it follows that

$$G(e^{j(2\pi k/N)2}) = \sum_{n=0}^{N/2-1} x[2n]e^{-j(2\pi k/N)2n}$$

$$= \sum_{n=0}^{N/2-1} x[2n]e^{-j(2\pi k/(N/2)n}$$

$$= \sum_{n=0}^{N/2-1} x[2n]W_{N/2}^{kn}, \qquad (9.17a)$$

and by a similar manipulation, it can be shown that

$$H(e^{j(2\pi k/N)2}) = \sum_{n=0}^{N/2-1} x[2n+1]W_{N/2}^{kn}. \qquad (9.17b)$$

Thus, from Eqs. (9.17a) and (9.17b) and Eq. (9.16), it follows that

$$X[k] = \sum_{n=0}^{N/2-1} x[2n]W_{N/2}^{kn} + W_N^k \sum_{n=0}^{N/2-1} x[2n+1]W_{N/2}^{kn} \qquad k = 0, 1, \ldots, N-1, \quad (9.18)$$

where the N-point DFT $X[k]$ is by definition

$$X[k] = \sum_{n=0}^{N-1} x[n]W_N^{nk}, \qquad k = 0, 1, \ldots, N-1. \qquad (9.19)$$

Likewise, by definition, the $(N/2)$-point DFTs of $g[n]$ and $h[n]$ are

$$G[k] = \sum_{n=0}^{N/2-1} x[2n]W_{N/2}^{nk}, \qquad k = 0, 1, \ldots, N/2-1 \qquad (9.20a)$$

$$H[k] = \sum_{n=0}^{N/2-1} x[2n+1]W_{N/2}^{nk}, \qquad k = 0, 1, \ldots, N/2-1. \qquad (9.20b)$$

Equation (9.18) shows that the N-point DFT $X[k]$ can be computed by evaluating the $(N/2)$-point DFTs $G[k]$ and $H[k]$ over $k = 0, 1, \ldots, N-1$ instead of $k = 0, 1, \ldots, N/2-1$ as we normally do for $(N/2)$-point DFTs. This is easily achieved even when $G[k]$ and $H[k]$ are computed only for $k = 0, 1, \ldots, N/2-1$, because the $(N/2)$-point transforms

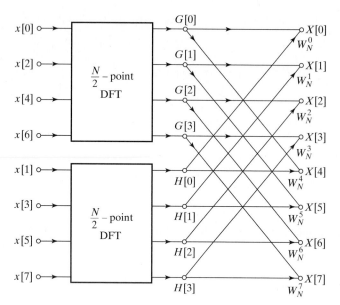

Figure 9.4   Flow graph of the decimation-in-time decomposition of an N-point DFT computation into two (N/2)-point DFT computations (N = 8).

are implicitly periodic with period $N/2$. With this observation, Eq. (9.18) can be rewritten as

$$X[k] = G[((k))_{N/2}] + W_N^k H[((k))_{N/2}] \qquad k = 0, 1, \ldots, N - 1. \tag{9.21}$$

The notation $((k))_{N/2}$ conveniently makes it explicit that even though $G[k]$ and $H[k]$ are only computed for $k = 0, 1, \ldots, N/2 - 1$, they are extended periodically (with no additional computation) by interpreting k modulo $N/2$.

After the two DFTs are computed, they are combined according to Eq. (9.21) to yield the N-point DFT $X[k]$. Figure 9.4 depicts this computation for $N = 8$. In this figure, we have used the signal flow graph conventions that were introduced in Chapter 6 for representing difference equations. That is, branches entering a node are summed to produce the node variable. When no coefficient is indicated, the branch transmittance is assumed to be unity. For other branches, the transmittance of a branch is an integer power of the complex number W_N.

In Figure 9.4, two 4-point DFTs are computed, with $G[k]$ designating the 4-point DFT of the even-numbered points and $H[k]$ designating the 4-point DFT of the odd-numbered points. According to Eq. (9.21), $X[0]$ is obtained by multiplying $H[0]$ by W_N^0 and adding the product to $G[0]$. The DFT value $X[1]$ is obtained by multiplying $H[1]$ by W_N^1 and adding that result to $G[1]$. Equation (9.21) states that, because of the implicit periodicity of $G[k]$ and $H[k]$, to compute $X[4]$, we should multiply $H[((4))_4]$ by W_N^4 and add the result to $G[((4))_4]$. Thus, $X[4]$ is obtained by multiplying $H[0]$ by W_N^4 and adding the result to $G[0]$. As shown in Figure 9.4, the values $X[5]$, $X[6]$, and $X[7]$ are obtained similarly.

With the computation restructured according to Eq. (9.21), we can compare the number of multiplications and additions required with those required for a direct computation of the DFT. Previously we saw that, for direct computation without exploiting symmetry, N^2 complex multiplications and additions were required.[6] By comparison, Eq. (9.21) requires the computation of two $(N/2)$-point DFTs, which in turn requires $2(N/2)^2$ complex multiplications and approximately $2(N/2)^2$ complex additions if we do the $(N/2)$-point DFTs by the direct method. Then the two $(N/2)$-point DFTs must be combined, requiring N complex multiplications, corresponding to multiplying the second sum by W_N^k, and N complex additions, corresponding to adding the product obtained to the first sum. Consequently, the computation of Eq. (9.21) for all values of k requires at most $N + 2(N/2)^2$ or $N + N^2/2$ complex multiplications and complex additions. It is easy to verify that for $N > 2$, the total $N + N^2/2$ will be less than N^2.

Equation (9.21) corresponds to breaking the original N-point computation into two $(N/2)$-point DFT computations. If $N/2$ is even, as it is when N is equal to a power of 2, then we can consider computing each of the $(N/2)$-point DFTs in Eq. (9.21) by breaking each of the sums in that equation into two $(N/4)$-point DFTs, which would then be combined to yield the $(N/2)$-point DFTs. Thus, $G[k]$ in Eq. (9.21) can be represented as

$$G[k] = \sum_{r=0}^{(N/2)-1} g[r]W_{N/2}^{rk} = \sum_{\ell=0}^{(N/4)-1} g[2\ell]W_{N/2}^{2\ell k} + \sum_{\ell=0}^{(N/4)-1} g[2\ell+1]W_{N/2}^{(2\ell+1)k}, \quad (9.22)$$

or

$$G[k] = \sum_{\ell=0}^{(N/4)-1} g[2\ell]W_{N/4}^{\ell k} + W_{N/2}^{k} \sum_{\ell=0}^{(N/4)-1} g[2\ell+1]W_{N/4}^{\ell k}. \quad (9.23)$$

Similarly, $H[k]$ can be represented as

$$H[k] = \sum_{\ell=0}^{(N/4)-1} h[2\ell]W_{N/4}^{\ell k} + W_{N/2}^{k} \sum_{\ell=0}^{(N/4)-1} h[2\ell+1]W_{N/4}^{\ell k}. \quad (9.24)$$

Consequently, the $(N/2)$-point DFT $G[k]$ can be obtained by combining the $(N/4)$-point DFTs of the sequences $g[2\ell]$ and $g[2\ell+1]$. Similarly, the $(N/2)$-point DFT $H[k]$ can be obtained by combining the $(N/4)$-point DFTs of the sequences $h[2\ell]$ and $h[2\ell+1]$. Thus, if the 4-point DFTs in Figure 9.4 are computed according to Eqs. (9.23) and (9.24), then that computation would be carried out as indicated in Figure 9.5. Inserting the computation of Figure 9.5 into the flow graph of Figure 9.4, we obtain the complete flow graph of Figure 9.6, where we have expressed the coefficients in terms of powers of W_N rather than powers of $W_{N/2}$, using the fact that $W_{N/2} = W_N^2$.

For the 8-point DFT that we have been using as an illustration, the computation has been reduced to a computation of 2-point DFTs. For example, the 2-point DFT of the sequence consisting of $x[0]$ and $x[4]$ is depicted in Figure 9.7. With the computation of Figure 9.7 inserted in the flow graph of Figure 9.6, we obtain the complete flow graph for computation of the 8-point DFT, as shown in Figure 9.9.

---

[6]For simplicity, we assume that N is large, so that $(N-1)$ can be approximated accurately by N.

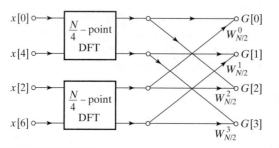

$x[0]$, $x[4]$ — $\frac{N}{4}$-point DFT

$x[2]$, $x[6]$ — $\frac{N}{4}$-point DFT

$G[0]$, $W_{N/2}^0$; $G[1]$, $W_{N/2}^1$; $G[2]$, $W_{N/2}^2$; $G[3]$, $W_{N/2}^3$

Figure 9.5   Flow graph of the decimation-in-time decomposition of an $(N/2)$-point DFT computation into two $(N/4)$-point DFT computations ($N = 8$).

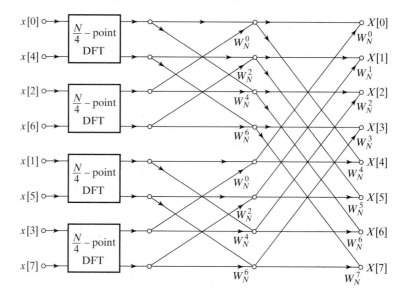

Figure 9.6   Result of substituting the structure of Figure 9.5 into Figure 9.4.

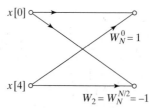

Figure 9.7   Flow graph of a 2-point DFT.

For the more general case, but with N still a power of 2, we would proceed by decomposing the $(N/4)$-point transforms in Eqs. (9.23) and (9.24) into $(N/8)$-point transforms and continue until we were left with only 2-point transforms. This requires $v = \log_2 N$ stages of computation. Previously, we found that in the original decomposition of an N-point transform into two $(N/2)$-point transforms, the number of complex multiplications and additions required was $N + 2(N/2)^2$. When the $(N/2)$-point transforms are decomposed into $(N/4)$-point transforms, the factor of $(N/2)^2$ is replaced by

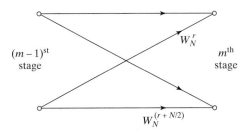

Figure 9.8   Flow graph of basic butterfly computation in Figure 9.9.

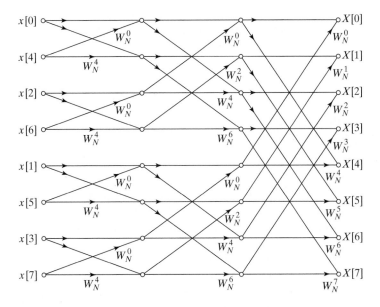

Figure 9.9   Flow graph of complete decimation-in-time decomposition of an 8-point DFT computation.

$N/2 + 2(N/4)^2$, so the overall computation then requires $N + N + 4(N/4)^2$ complex multiplications and additions. If $N = 2^\nu$, this can be done at most $\nu = \log_2 N$ times, so that after carrying out this decomposition as many times as possible, the number of complex multiplications and additions is equal to $N\nu = N \log_2 N$.

The flow graph of Figure 9.9 displays the operations explicitly. By counting branches with transmittances of the form W_N^r, we note that each stage has N complex multiplications and N complex additions. Since there are $\log_2 N$ stages, we have a total of $N \log_2 N$ complex multiplications and additions. This can be a substantial computational saving. For example, if $N = 2^{10} = 1024$, then $N^2 = 2^{20} = 1,048,576$, and $N \log_2 N = 10,240$, a reduction of more than two orders of magnitude!

The computation in the flow graph of Figure 9.9 can be reduced further by exploiting the symmetry and periodicity of the coefficients W_N^r. We first note that, in proceeding from one stage to the next in Figure 9.9, the basic computation is in the form of Figure 9.8, i.e., it involves obtaining a pair of values in one stage from a pair of values in the preceding stage, where the coefficients are always powers of W_N and the exponents are

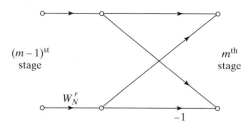

$(m-1)^{st}$
stage

m^{th}
stage

W_N^r

-1

Figure 9.10 Flow graph of simplified butterfly computation requiring only one complex multiplication.

separated by $N/2$. Because of the shape of the flow graph, this elementary computation is called a *butterfly*. Since

$$W_N^{N/2} = e^{-j(2\pi/N)N/2} = e^{-j\pi} = -1, \qquad (9.25)$$

the factor $W_N^{r+N/2}$ can be written as

$$W_N^{r+N/2} = W_N^{N/2} W_N^r = -W_N^r. \qquad (9.26)$$

With this observation, the butterfly computation of Figure 9.8 can be simplified to the form shown in Figure 9.10, which requires one complex addition and one complex subtraction, but only one complex multiplication instead of two. Using the basic flow graph of Figure 9.10 as a replacement for butterflies of the form of Figure 9.8, we obtain from Figure 9.9 the flow graph of Figure 9.11. In particular, the number of complex multiplications has been reduced by a factor of 2 over the number in Figure 9.9.

Figure 9.11 shows $\log_2 N$ stages of computation each involving a set of $N/2$ 2-point DFT computations (butterflies). Between the sets of 2-point transforms are com-

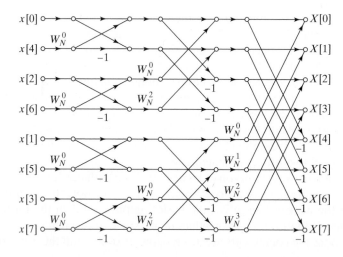

Figure 9.11 Flow graph of 8-point DFT using the butterfly computation of Figure 9.10.

plex multipliers of the form W_N^r. These complex multipliers have been called "twiddle factors" because they serve as adjustments in the process of converting the 2-point transforms into longer transforms.

9.2.1 Generalization and Programming the FFT

The flow graph of Figure 9.11, which describes an algorithm for computation of an 8-point discrete Fourier transform, is easily generalized to any $N = 2^v$, so it serves both as a proof that the computation requires on the order of $N \log N$ operations and as a graphical representation from which an implementation program could be written. While programs in high-level computer languages are widely available, it may be necessary in some cases to construct a program for a new machine architecture or to optimize a given program to take advantage of low-level features of a given machine architecture. A refined analysis of the diagram reveals many details that are important for programming or for designing special hardware for computing the DFT. We call attention to some of these details in Sections 9.2.2 and 9.2.3 for the decimation-in-time algorithms and in Sections 9.3.1 and 9.3.2 for the decimation-in-frequency algorithms. In Section 9.4 we discuss some additional practical considerations. While these sections are not essential for obtaining a basic understanding of FFT principles, they provide useful guidance for programming and system design.

9.2.2 In-Place Computations

The essential features of the flow graph of Figure 9.11 are the branches connecting the nodes and the transmittance of each of these branches. No matter how the nodes in the flow graph are rearranged, it will always represent the same computation, provided that the connections between the nodes and the transmittances of the connections are maintained. The particular form for the flow graph in Figure 9.11 arose out of deriving the algorithm by separating the original sequence into the even-numbered and odd-numbered points and then continuing to create smaller and smaller subsequences in the same way. An interesting by-product of this derivation is that this flow graph, in addition to describing an efficient procedure for computing the discrete Fourier transform, also suggests a useful way of storing the original data and storing the results of the computation in intermediate arrays.

   To see this, it is useful to note that according to Figure 9.11, each stage of the computation takes a set of N complex numbers and transforms them into another set of N complex numbers through basic butterfly computations of the form of Figure 9.10. This process is repeated $v = \log_2 N$ times, resulting in the computation of the desired discrete Fourier transform. When implementing the computations depicted in Figure 9.11, we can imagine the use of two arrays of (complex) storage registers, one for the array being computed and one for the data being used in the computation. For example, in computing the first array in Figure 9.11, one set of storage registers would contain the input data and the second set would contain the computed results for the first stage. While the validity of Figure 9.11 is not tied to the order in which the input data are stored, we can order the set of complex numbers in the same order that they appear in the figure (from top to bottom). We denote the sequence of complex numbers re-

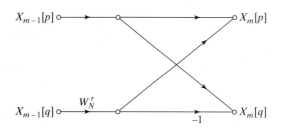

Figure 9.12   Flow graph of Eqs. (9.28).

sulting from the mth stage of computation as $X_m[\ell]$, where $\ell = 0, 1, \ldots, N - 1$, and $m = 1, 2, \ldots, \nu$. Furthermore, for convenience, we define the set of input samples as $X_0[\ell]$. We can think of $X_{m-1}[\ell]$ as the input array and $X_m[\ell]$ as the output array for the mth stage of the computations. Thus, for the case of $N = 8$, as in Figure 9.11,

$$X_0[0] = x[0],$$
$$X_0[1] = x[4],$$
$$X_0[2] = x[2],$$
$$X_0[3] = x[6],$$
$$X_0[4] = x[1], \tag{9.27}$$
$$X_0[5] = x[5],$$
$$X_0[6] = x[3],$$
$$X_0[7] = x[7].$$

Using this notation, we can label the input and output of the butterfly computation in Figure 9.10 as indicated in Figure 9.12, with the associated equations

$$X_m[p] = X_{m-1}[p] + W_N^r X_{m-1}[q], \tag{9.28a}$$

$$X_m[q] = X_{m-1}[p] - W_N^r X_{m-1}[q]. \tag{9.28b}$$

In Eqs. (9.28), p, q, and r vary from stage to stage in a manner that is readily inferred from Figure 9.11 and from Eqs. (9.21), (9.23), and (9.24) and. It is clear from Figures 9.11 and 9.12 that only the complex numbers in locations p and q of the $(m-1)^{\text{st}}$ array are required to compute the elements p and q of the mth array. Thus, only one complex array of N storage registers is physically necessary to implement the complete computation if $X_m[p]$ and $X_m[q]$ are stored in the same storage registers as $X_{m-1}[p]$ and $X_{m-1}[q]$, respectively. This kind of computation is commonly referred to as an *in-place* computation. The fact that the flow graph of Figure 9.11 (or Figure 9.9) represents an in-place computation is tied to the fact that we have associated nodes in the flow graph that are on the same horizontal line with the same storage location and the fact that the computation between two arrays consists of a butterfly computation in which the input nodes and the output nodes are horizontally adjacent.

In order that the computation may be done in place as just discussed, the input sequence must be stored (or at least accessed) in a nonsequential order, as shown in the flow graph of Figure 9.11. In fact, the order in which the input data are stored and accessed is referred to as *bit-reversed* order. To see what is meant by this terminology, we note that for the 8-point flow graph that we have been discussing, three binary digits

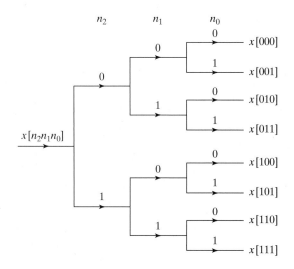

Figure 9.13   Tree diagram depicting normal-order sorting.

are required to index through the data. Writing the indices in Eqs. (9.27) in binary form, we obtain the following set of equations:

$$X_0[000] = x[000],$$
$$X_0[001] = x[100],$$
$$X_0[010] = x[010],$$
$$X_0[011] = x[110],$$
$$X_0[100] = x[001],$$
$$X_0[101] = x[101],$$
$$X_0[110] = x[011],$$
$$X_0[111] = x[111].$$

(9.29)

If (n_2, n_1, n_0) is the binary representation of the index of the sequence $x[n]$, then the sequence value $x[n_2, n_1, n_0]$ is stored in the array position $X_0[n_0, n_1, n_2]$. That is, in determining the position of $x[n_2, n_1, n_0]$ in the input array, we must reverse the order of the bits of the index n.

Consider the process depicted in Figure 9.13 for sorting a data sequence in normal order by successive examination of the bits representing the data index. If the most significant bit of the data index is zero, $x[n]$ belongs in the top half of the sorted array; otherwise it belongs in the bottom half. Next, the top half and bottom half subsequences can be sorted by examining the second most significant bit, and so on.

To see why bit-reversed order is necessary for in-place computation, recall the process that resulted in Figure 9.9 and Figure 9.11. The sequence $x[n]$ was first divided into the even-numbered samples, with the even-numbered samples occurring in the top half of Figure 9.4 and the odd-numbered samples occurring in the bottom half. Such a separation of the data can be carried out by examining the least significant bit $[n_0]$ in the index n. If the least significant bit is 0, the sequence value corresponds to an even-numbered sample and therefore will appear in the top half of the array $X_0[\ell]$, and if the least significant bit is 1, the sequence value corresponds to an odd-numbered

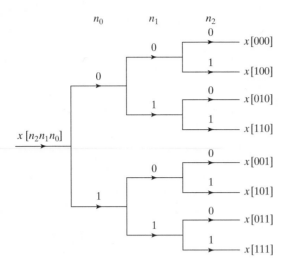

Figure 9.14   Tree diagram depicting bit-reversed sorting.

sample and consequently will appear in the bottom half of the array. Next, the even- and odd-indexed subsequences are sorted into their even- and odd-indexed parts, and this can be done by examining the second least significant bit in the index. Considering first the even-indexed subsequence, if the second least significant bit is 0, the sequence value is an even-numbered term in the subsequence, and if the second least significant bit is 1, then the sequence value has an odd-numbered index in this subsequence. The same process is carried out for the subsequence formed from the original odd-indexed sequence values. This process is repeated until N subsequences of length 1 are obtained. This sorting into even- and odd-indexed subsequences is depicted by the tree diagram of Figure 9.14.

The tree diagrams of Figures 9.13 and 9.14 are identical, except that for normal sorting, we examine the bits representing the index from left to right, whereas for the sorting leading naturally to Figure 9.9 or 9.11, we examine the bits in reverse order, right to left, resulting in bit-reversed sorting. Thus, the necessity for bit-reversed ordering of the sequence $x[n]$ results from the manner in which the DFT computation is decomposed into successively smaller DFT computations in arriving at Figures 9.9 and 9.11.

9.2.3 Alternative Forms

Although it is reasonable to store the results of each stage of the computation in the order in which the nodes appear in Figure 9.11, it is certainly not necessary to do so. No matter how the nodes of Figure 9.11 are rearranged, the result will always be a valid computation of the discrete Fourier transform of $x[n]$, as long as the branch transmittances are unchanged. Only the order in which data are accessed and stored will change. If we associate the nodes with indexing of an array of complex storage locations, it is clear from our previous discussion that a flow graph corresponding to an in-place computation results only if the rearrangement of nodes is such that the input and output nodes for each butterfly computation are horizontally adjacent. Otherwise two complex storage arrays will be required. Figure 9.11, is, of course, such an arrangement. Another is depicted in Figure 9.15. In this case, the input sequence is in normal order and the sequence of DFT values is in bit-reversed order. Figure 9.15 can be obtained from Figure 9.11 as follows: All the nodes that are horizontally adjacent to $x[4]$ in Figure 9.11 are interchanged with all the nodes horizontally adjacent to $x[1]$. Similarly, all the nodes that are horizontally adjacent to $x[6]$ in Figure 9.11 are interchanged with those that are horizontally adjacent to $x[3]$. The nodes horizontally adjacent to $x[0]$, $x[2]$, $x[5]$, and $x[7]$ are not disturbed. The resulting flow graph in Figure 9.15 corresponds to the form of the decimation-in-time algorithm originally given by Cooley and Tukey (1965).

The only difference between Figures 9.11 and 9.15 is in the ordering of the nodes. This implies that Figures 9.11 and 9.15 represent two different programs for carrying out the computations. The branch transmittances (powers of W_N) remain the same, and therefore the intermediate results will be exactly the same—they will be computed in a different order within each stage. There are, of course, a large variety of possible orderings. However, most do not make much sense from a computational viewpoint. As one example, suppose that the nodes are ordered such that the input and output both appear in normal order. A flow graph of this type is shown in Figure 9.16. In this case, however, the computation cannot be carried out in place because the butterfly structure does not continue past the first stage. Thus, two complex arrays of length N would be required to perform the computation depicted in Figure 9.16.

In realizing the computations depicted by Figures 9.11, 9.15, and 9.16, it is clearly necessary to access elements of intermediate arrays in non-sequential order. Thus, for greater computational speed, the complex numbers must be stored in random-access memory.[7] For example, in the computation of the first array in Figure 9.11 from the input array, the inputs to each butterfly computation are adjacent node variables and are thought of as being stored in adjacent storage locations. In the computation of the second intermediate array from the first, the inputs to a butterfly are separated by two storage locations; and in the computation of the third array from the second, the inputs to a butterfly computation are separated by four storage locations. If $N > 8$, the separation between butterfly inputs is 8 for the fourth stage, 16 for the fifth stage, and so on. The separation in the last (ν^{th}) stage is $N/2$.

In Figure 9.15 the situation is similar in that, to compute the first array from the input data we use data separated by 4, to compute the second array from the first array we use input data separated by 2, and then finally, to compute the last array we use adjacent data. It is straightforward to imagine simple algorithms for modifying index registers to access the data in the flow graph of either Figure 9.11 or Figure 9.15 if the data are stored in random-access memory. However, in the flow graph of Figure 9.16, the data are accessed non-sequentially, the computation is not in place, and a scheme for indexing the data is considerably more complicated than in either of the two previous cases. Even given the availability of large amounts of random-access memory, the overhead for index computations could easily nullify much of the computational advantage that is implied by eliminating multiplications and additions. Consequently, this structure has no apparent advantages.

---

[7]When the Cooley–Tukey algorithms first appeared in 1965, digital memory was expensive and of limited size. The size and availability of random access memory is no longer an issue except for exceedingly large values of N.

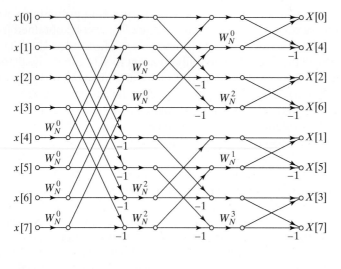

Figure 9.15 Rearrangement of Figure 9.11 with input in normal order and output in bit-reversed order.

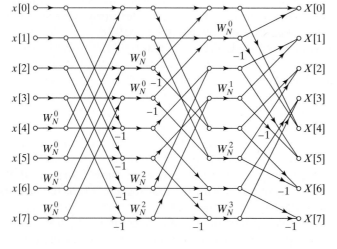

Figure 9.16 Rearrangement of Figure 9.11 with both input and output in normal order.

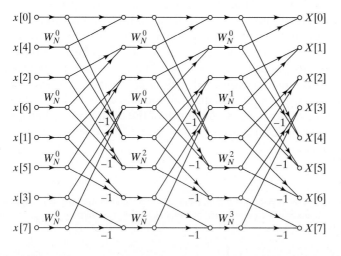

Figure 9.17 Rearrangement of Figure 9.11 having the same geometry for each stage, thereby simplifying data access.

Some forms have advantages even if they do not allow in-place computation. A rearrangement of the flow graph in Figure 9.11 that is particularly useful when an adequate amount of random-access memory is not available is shown in Figure 9.17. This flow graph represents the decimation-in-time algorithm originally given by Singleton (1969). Note first that in this flow graph the input is in bit-reversed order and the output is in normal order. The important feature of the flow graph is that the geometry is identical for each stage; only the branch transmittances change from stage to stage. This makes it possible to access data sequentially. Suppose, for example that we have four separate mass-storage files, and suppose that the first half of the input sequence (in bit-reversed order) is stored in one file and the second half is stored in a second file. Then the sequence can be accessed sequentially in files 1 and 2 and the results written sequentially on files 3 and 4, with the first half of the new array being written to file 3 and the second half to file 4. Then at the next stage of computation, files 3 and 4 are the input, and the output is written to files 1 and 2. This is repeated for each of the v stages.

Such an algorithm could be useful in computing the DFT of extremely long sequences. This could mean values of N on the order of hundreds of millions since random-access memories of giga-byte size are routinely available. Perhaps a more interesting feature of the diagram in Figure 9.17 is that the indexing is very simple and it is the same from stage-to-stage. With two banks of random-access memory, this algorithm would have very simple index calculations.

9.3 DECIMATION-IN-FREQUENCY FFT ALGORITHMS

The decimation-in-time FFT algorithms are based on structuring the DFT computation by forming smaller and smaller subsequences of the input sequence $x[n]$. Alternatively, we can consider dividing the DFT sequence $X[k]$ into smaller and smaller subsequences in the same manner. FFT algorithms based on this procedure are commonly called *decimation-in-frequency* algorithms.

To develop this class of FFT algorithms, we again restrict the discussion to N a power of 2 and consider computing separately the $N/2$ even-numbered frequency samples and the $N/2$ odd-numbered frequency samples. We have depicted this in the block diagram representation in Figure 9.18 where $X_0[k] = X[2k]$ and $X_1[k] = X[2k+1]$. In shifting left by 1 DFT sample so that the compressor selects the odd-indexed samples, it is important to remember that the DFT $X[k]$ is implicitly periodic with period N. This is denoted "Circular Left Shift 1" (and correspondingly "Circular Right Shift 1") in Figure 9.18. Observe that this diagram has a similar structure to Figure 9.3, where the same operations were applied to the time sequence $x[n]$ instead of the DFT $X[k]$. In this case, Figure 9.18 directly depicts the fact that the N-point transform $X[k]$ can be obtained by interleaving its even-indexed and odd-indexed samples after expansion by a factor of 2.

Figure 9.18 is a correct representation of $X[k]$, but in order to use it as the basis for computing $X[k]$, we first show that $X[2k]$ and $X[2k + 1]$ can be computed from the time-domain sequence $x[n]$. In Section 8.4 we saw that the DFT is related to the DTFT by sampling at frequencies $2\pi k/N$ with the result that the corresponding time-

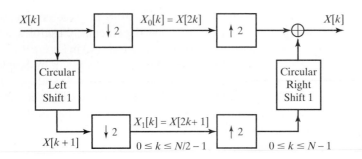

Figure 9.18   Illustration of the basic principle of decimation-in-frequency.

domain operation is time-aliasing with repetition length (period) N. As discussed in Section 8.4, if N is greater than or equal to the length of the sequence $x[n]$, the inverse DFT yields the original sequence over $0 \leq n \leq N - 1$ because the N-point copies of $x[n]$ do not overlap when time-aliased with repetition offset N. However, in Figure 9.18, the DFT is compressed by 2, which is equivalent to sampling the DTFT $X(e^{j\omega})$ at frequencies $2\pi k/(N/2)$. Thus, the implicit periodic time-domain signal represented by $X_0[k] = X[2k]$ is

$$\tilde{x}_0[n] = \sum_{m=-\infty}^{\infty} x[n + mN/2] \qquad -\infty < n < \infty. \tag{9.30}$$

Since $x[n]$ has length N, only two of the shifted copies of $x[n]$ overlap in the interval $0 \leq n \leq N/2 - 1$, so the corresponding finite-length sequence $x_0[n]$ is

$$x_0[n] = x[n] + x[n + N/2] \qquad 0 \leq n \leq N/2 - 1. \tag{9.31a}$$

To obtain the comparable result for the odd-indexed DFT samples, recall that the circularly shifted DFT $X[k + 1]$ corresponds to $W_N^n x[n]$ (see Property 6 of Table 8.2). Therefore the $N/2$-point sequence $x_1[n]$ corresponding to $X_1[k] = X[2k + 1]$ is

$$x_1[n] = x[n]W_N^n + x[n + N/2]W_N^{n+N/2}$$
$$= (x[n] - x[n + N/2])W_N^n \qquad 0 \leq n \leq N/2 - 1, \tag{9.31b}$$

since $W_N^{N/2} = -1$.

From Eqs. (9.31a) and (9.31b), it follows that

$$X_0[k] = \sum_{n=0}^{N/2-1} (x[n] + x[n + N/2])W_{N/2}^{kn} \tag{9.32a}$$

$$X_1[k] = \sum_{n=0}^{N/2-1} [(x[n] - x[n + N/2])W_N^n]W_{N/2}^{kn} \tag{9.32b}$$

$$k = 0, 1, \ldots, N/2 - 1.$$

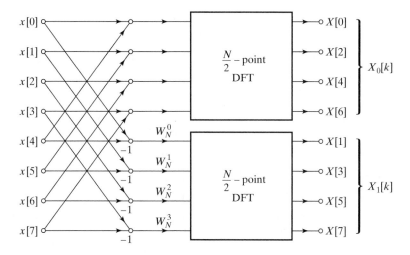

Figure 9.19   Flow graph of decimation-in-frequency decomposition of an N-point DFT computation into two $(N/2)$-point DFT computations ($N = 8$).

Equation (9.32a) is the $(N/2)$-point DFT of the sequence $x_0[n]$ obtained by adding the second half of the input sequence to the first half. Equation (9.32b) is the $(N/2)$-point DFT of the sequence $x_1[n]$ obtained by subtracting the second half of the input sequence from the first half and multiplying the resulting sequence by W_N^n.

Thus, using Eqs. (9.32a) and (9.32b), the even-numbered and odd-numbered output points of $X[k]$ can be computed since $X[2k] = X_0[k]$ and $X[2k + 1] = X_1[k]$, respectively. The procedure suggested by Eqs. (9.32a) and (9.32b) is illustrated for the case of an 8-point DFT in Figure 9.19.

Proceeding in a manner similar to that followed in deriving the decimation-in-time algorithm, we note that for N a power of 2, $N/2$ is divisible by 2 so the $(N/2)$-point DFTs can be computed by computing the even-numbered and odd numbered output points for those DFTs separately. As in the case of the procedure leading to Eqs. (9.32a) and (9.32b), this is accomplished by combining the first half and the last half of the input points for each of the $(N/2)$-point DFTs and then computing $(N/4)$-point DFTs. The flow graph resulting from taking this step for the 8-point example is shown in Figure 9.20. For the 8-point example, the computation has now been reduced to the computation of 2-point DFTs, which are implemented by adding and subtracting the input points, as discussed previously. Thus, the 2-point DFTs in Figure 9.20 can be replaced by the computation shown in Figure 9.21, so the computation of the 8-point DFT can be accomplished by the algorithm depicted in Figure 9.22. We again see $\log_2 N$ stages of 2-point transforms coupled together through twiddle factors that in this case occur at the output of the 2-point transforms.

By counting the arithmetic operations in Figure 9.22 and generalizing to $N = 2^v$, we see that the computation of Figure 9.22 requires $(N/2) \log_2 N$ complex multiplications and $N \log_2 N$ complex additions. Thus, the total number of computations is the same for the decimation-in-frequency and the decimation-in-time algorithms.

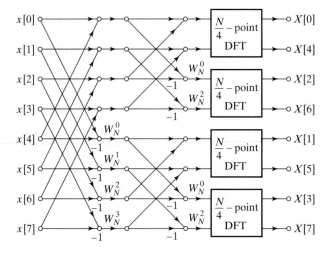

Figure 9.20 Flow graph of decimation-in-frequency decomposition of an 8-point DFT into four 2-point DFT computations.

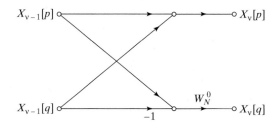

Figure 9.21 Flow graph of a typical 2-point DFT as required in the last stage of decimation-in-frequency decomposition.

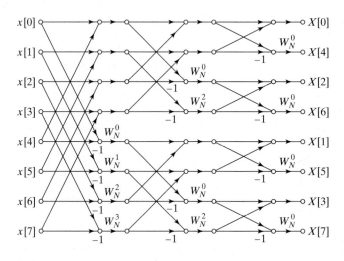

Figure 9.22 Flow graph of complete decimation-in-frequency decomposition of an 8-point DFT computation.

9.3.1 In-Place Computation

The flow graph in Figure 9.22 depicts one FFT algorithm based on decimation in frequency. We can observe a number of similarities and also a number of differences in comparing this graph with the flow graphs derived on the basis of decimation in time. As with decimation in time, of course, the flow graph of Figure 9.22 corresponds to a computation of the discrete Fourier transform, regardless of how the graph is drawn, as long as the same nodes are connected to each other with the proper branch transmittances. In other words, the flow graph of Figure 9.22 is not based on any assumption about the order in which the input sequence values are stored. However, as was done with the decimation-in-time algorithms, we can interpret successive vertical nodes in the flow graph of Figure 9.22 as corresponding to successive storage registers in a digital memory. In this case, the flow graph in Figure 9.22 begins with the input sequence in normal order and provides the output DFT in bit-reversed order. The basic computation again has the form of a butterfly computation, although the butterfly is different from that arising in the decimation-in-time algorithms. However, because of the butterfly nature of the computation, the flow graph of Figure 9.22 can be interpreted as an in-place computation of the discrete Fourier transform.

9.3.2 Alternative Forms

A variety of alternative forms for the decimation-in-frequency algorithm can be obtained by transposing the decimation-in-time forms developed in Section 9.2.3. If we denote the sequence of complex numbers resulting from the mth stage of the computation as $X_m[\ell]$, where $\ell = 0, 1, \ldots, N - 1$, and $m = 1, 2, \ldots, \nu$, then the basic butterfly computation shown in Figure 9.23 has the form

$$X_m[p] = X_{m-1}[p] + X_{m-1}[q], \tag{9.33a}$$

$$X_m[q] = (X_{m-1}[p] - X_{m-1}[q])W_N^r. \tag{9.33b}$$

Comparing Figures 9.12 and 9.23 or Eqs. (9.28) and (9.33), it appears that the butterfly computations are different for the two classes of FFT algorithms. However, the two butterfly flow graphs are, in the terminology of Chapter 6, transposes of one another. That is, if we reverse the direction of arrows and redefine the input and output nodes in Figure 9.12, we obtain Figure 9.23 and vice-versa. Since the FFT flow graphs consist of connected sets of butterflies, it is not surprising, therefore, that we also note

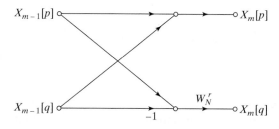

Figure 9.23   Flow graph of a typical butterfly computation required in Figure 9.22.

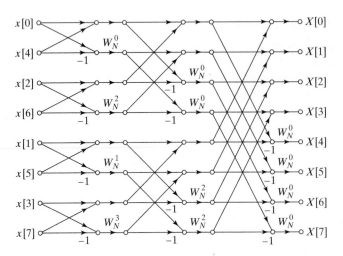

Figure 9.24 Flow graph of a decimation-in-frequency DFT algorithm obtained from Figure 9.22. Input in bit-reversed order and output in normal order. (Transpose of Figure 9.15.)

a resemblance between the FFT flow graphs of Figures 9.11 and 9.22. Specifically, Figure 9.22 can be obtained from Figure 9.11 by reversing the direction of signal flow and interchanging the input and output. That is, Figure 9.22 is the transpose of the flow graph in Figure 9.11. In Chapter 6 we stated a transposition theorem that applies only to single-input/single-output flow graphs. When viewed as flow graphs, however, FFT algorithms are multi-input/multi-output systems, which require a more general form of the transposition theorem. (See Claasen and Mecklenbräuker, 1978.) Nevertheless, it is intuitively clear that the input–output characteristics of the flow graphs in Figures 9.11 and 9.22 are the same based simply on the above observation that the butterflies are transposes of each other. This can be shown more formally by noting that the butterfly equations in Eqs. (9.33) can be solved backward, starting with the output array. (Problem 9.31 outlines a proof of this result.) More generally, it is true that for each decimation-in-time FFT algorithm, there exists a decimation-in-frequency FFT algorithm that corresponds to interchanging the input and output and reversing the direction of all the arrows in the flow graph.

This result implies that all the flow graphs of Section 9.2 have counterparts in the class of decimation-in-frequency algorithms. This, of course, also corresponds to the fact that, as before, it is possible to rearrange the nodes of a decimation-in-frequency flow graph without altering the final result.

Applying the transposition procedure to Figure 9.15 leads to Figure 9.24. In this flow graph, the output is in normal order and the input is in bit-reversed order. The transpose of the flow graph of Figure 9.16 would lead to a flow graph with both the input and output in normal order. An algorithm base on the resulting flow graph would suffer from the same limitations as for Figure 9.16.

The transpose of Figure 9.17 is shown in Figure 9.25. Each stage of Figure 9.25 has the same geometry, a property that simplifies data access, as discussed before.

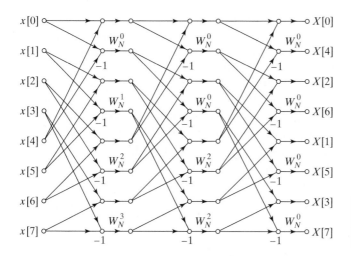

$x[0]$ ⟶ $X[0]$
$x[1]$ ⟶ $X[4]$
$x[2]$ ⟶ $X[2]$
$x[3]$ ⟶ $X[6]$
$x[4]$ ⟶ $X[1]$
$x[5]$ ⟶ $X[5]$
$x[6]$ ⟶ $X[3]$
$x[7]$ ⟶ $X[7]$

Figure 9.25   Rearrangement of Figure 9.22 having the same geometry for each stage, thereby simplifying data access. (Transpose of Figure 9.17.)

9.4 PRACTICAL CONSIDERATIONS

In Sections 9.2 and 9.3, we discussed the basic principles of efficient computation of the DFT when N is an integer power of 2. In these discussions, we favored the use of signal flow graph representations rather than explicitly writing out in detail the equations that such flow graphs represent. Of necessity, we have shown flow graphs for specific values of N. However, by considering a flow graph such as that in Figure 9.11, for a specific value of N, it is possible to see how to structure a general computational algorithm that would apply to any $N = 2^v$. While the discussion in Sections 9.2 and 9.3 is completely adequate for a basic understanding of the FFT principles, the material of this section is intended to provide useful guidance for programming and system design.

Although it is true that the flow graphs of the previous sections capture the essence of the FFT algorithms that they depict, a variety of details must be considered in the implementation of a given algorithm. In this section, we briefly suggest some of these. Specifically, in Section 9.4.1 we discuss issues associated with accessing and storing data in the intermediate arrays of the FFT. In Section 9.4.2 we discuss issues associated with computing or accessing the branch coefficients in the flow graph. Our emphasis is on algorithms for N a power of 2, but much of the discussion applies to the general case as well. For purposes of illustration, we focus primarily on the decimation-in-time algorithm of Figure 9.11.

9.4.1 Indexing

In the algorithm depicted in Figure 9.11, the input must be in bit-reversed order so that the computation can be performed in place. The resulting DFT is then in normal order. Generally, sequences do not originate in bit-reversed order, so the first step in the implementation of Figure 9.11 is to sort the input sequence into bit-reversed order. As can be seen from that figure and Eqs. (9.27) and (9.29), bit-reversed sorting can be done in place, since samples are only pairwise interchanged; i.e., a sample at a given index is interchanged with the sample in the location specified by the bit-reversed index. This is conveniently done in place by using two counters, one in normal order and the other in bit-reversed order. The data in the two positions specified by the two counters are simply interchanged. Once the input is in bit-reversed order, we can proceed with the first stage of computation. In this case, the inputs to the butterflies are adjacent elements of the array $X_0[\cdot]$. In the second stage, the inputs to the butterflies are separated by 2. In the mth stage, the butterfly inputs are separated by 2^{m-1}. The coefficients are powers of $W_N^{(N/2^m)}$ in the mth stage and are required in normal order if computation of butterflies begins at the top of the flow graph of Figure 9.11. The preceding statements define the manner in which data must be accessed at a given stage, which, of course, depends on the flow graph that is implemented. For example, in the mth stage of Figure 9.15, the butterfly spacing is 2^{v-m}, and in this case the coefficients are required in bit-reversed order. The input is in normal order; however, the output is in bit-reversed order, so it generally would be necessary to sort the output into normal order by using a normal-order counter and a bit-reversed counter, as discussed previously.

In general, if we consider all the flow graphs in Sections 9.2 and 9.3, we see that each algorithm has its own characteristic indexing issues. The choice of a particular algorithm depends on a number of factors. The algorithms utilizing an in-place computation have the advantage of making efficient use of memory. Two disadvantages, however, are that the kind of memory required is random-access rather than sequential memory and that either the input sequence or the output DFT sequence is in bit-reversed order. Furthermore, depending on whether a decimation-in-time or a decimation-in-frequency algorithm is chosen and whether the inputs or the outputs are in bit-reversed order, the coefficients are required to be accessed in either normal order or bit-reversed order. If non-random-access sequential memory is used, some fast Fourier transform algorithms utilize sequential memory, as we have shown, but either the inputs or the outputs must be in bit-reversed order. While the flow graph for the algorithm can be arranged so that the inputs, the outputs, and the coefficients are in normal order, the indexing structure required to implement these algorithms is complicated, and twice as much random access memory is required. Consequently, the use of such algorithms does not appear to be advantageous.

The in-place FFT algorithms of Figures 9.11, 9.15, 9.22, and 9.24 are among the most commonly used. If a sequence is to be transformed only once, then bit-reversed sorting must be implemented on either the input or the output. However, in some situations a sequence is transformed, the result is modified in some way, and then the inverse DFT is computed. For example, in implementing FIR digital filters by block convolution using the discrete Fourier transform, the DFT of a section of the input sequence is multiplied by the DFT of the filter impulse response, and the result is inverse transformed to obtain a segment of the output of the filter. Similarly, in computing an autocorrelation function or cross-correlation function using the discrete Fourier transform, a sequence will be transformed, the DFTs will be multiplied, and then the resulting product will be inverse transformed. When two transforms are cascaded in this way, it is possible, by appropriate choice of the FFT algorithms, to avoid the need for bit reversal. For example, in implementing an FIR digital filter using the DFT, we can choose an algorithm for the direct transform that utilizes the data in normal order and provides a DFT in bit-reversed order. Either the flow graph corresponding to Figure 9.15, based on decimation in time, or that of Figure 9.22, based on decimation in frequency, could be used in this way. The difference between these two forms is that the decimation-in-time form requires the coefficients in bit-reversed order, whereas the decimation-in-frequency form requires the coefficients in normal order.

Note that Figure 9.11 utilizes coefficients in normal order, whereas Figure 9.24 requires the coefficients in bit-reversed order. If the decimation-in-time form of the algorithm is chosen for the direct transform, then the decimation-in-frequency form of the algorithm should be chosen for the inverse transform, requiring coefficients in bit-reversed order. Likewise, the decimation-in-frequency algorithm for the direct transform should be paired with the decimation-in-time algorithm for the inverse transform, which would then utilize normally ordered coefficients.

9.4.2 Coefficients

We have observed that the coefficients W_N^r (twiddle factors) may be required in either bit-reversed order or in normal order. In either case we must store a table sufficient to look up all required values, or we must compute the values as needed. The first alternative has the advantage of speed, but of course requires extra storage. We observe from the flow graphs that we require W_N^r for $r = 0, 1, \ldots, (N/2) - 1$. Thus, we require $N/2$ complex storage registers for a complete table of values of W_N^r.[8] In the case of algorithms in which the coefficients are required in bit-reversed order, we can simply store the table in bit-reversed order.

The computation of the coefficients as they are needed saves storage, but is less efficient than storing a lookup table. If the coefficients are to be computed, it is generally most efficient to use a recursion formula. At any given stage, the required coefficients are all powers of a complex number of the form W_N^q, where q depends on the algorithm and the stage. Thus, if the coefficients are required in normal order, we can use the recursion formula

$$W_N^{q\ell} = W_N^q \cdot W_N^{q(\ell-1)} \tag{9.34}$$

to obtain the ℓth coefficient from the $(\ell-1)^{\text{st}}$ coefficient. Clearly, algorithms that require coefficients in bit-reversed order are not well suited to this approach. It should be noted that Eq. (9.34) is essentially the coupled-form oscillator of Problem 6.21. When using finite-precision arithmetic, errors can build up in the iteration of this difference equation. Therefore, it is generally necessary to reset the value at prescribed points (e.g., $W_N^{N/4} = -j$) so that errors do not become unacceptable.

9.5 MORE GENERAL FFT ALGORITHMS

The power-of-two algorithms discussed in detail in Sections 9.2 and 9.3 are straightforward, highly efficient and easy to program. However, there are many applications where efficient algorithms for other values of N are very useful.

---

[8]This number can be reduced using symmetry at the cost of greater complexity in accessing desired values.

9.5.1 Algorithms for Composite Values of N

Although the special case of N a power of 2 leads to algorithms that have a particularly simple structure, this is not the only restriction on N that can lead to reduced computation for the DFT. The same principles that were applied in the power-of-two decimation-in-time and decimation-in-frequency algorithms can be employed when N is a composite integer, i.e., the product of two or more integer factors. For example, if $N = N_1 N_2$, it is possible to express the N-point DFT as either a combination of N_1 N_2-point DFTs or as a combination of N_2 N_1-point DFTs, and thereby obtain reductions in the number of computations. To see this, the indices n and k are represented as follows:

$$n = N_2 n_1 + n_2 \qquad \begin{cases} n_1 = 0, 1, \ldots, N_1 - 1 \\ n_2 = 0, 1, \ldots, N_2 - 1 \end{cases} \tag{9.35a}$$

$$k = k_1 + N_1 k_2 \qquad \begin{cases} k_1 = 0, 1, \ldots, N_1 - 1 \\ k_2 = 0, 1, \ldots, N_2 - 1. \end{cases} \tag{9.35b}$$

Since $N = N_1 N_2$, these index decompositions ensure that n and k range over all the values $0, 1, \ldots, N - 1$. Substituting these representations of n and k into the definition of the DFT leads after a few manipulations to

$$X[k] = X[k_1 + N_1 k_2]$$

$$= \sum_{n_2=0}^{N_2-1} \left[\left(\sum_{n_1=0}^{N_1-1} x[N_2 n_1 + n_2] W_{N_1}^{k_1 n_1} \right) W_N^{k_1 n_2} \right] W_{N_2}^{k_2 n_2}, \tag{9.36}$$

where $k_1 = 0, 1, \ldots, N_1 - 1$ and $k_2 = 0, 1, \ldots, N_2 - 1$. The part of Eq. (9.36) inside the parentheses represents N_2 N_1-point DFTs, while the outer sum corresponds to N_1 N_2-point DFTs of the outputs of the first set of transforms occurring after modification by the twiddle factors $W_N^{k_1 n_2}$.

If $N_1 = 2$ and $N_2 = N/2$, Eq. (9.36) reduces to the first stage decomposition of the decimation-in-frequency power-of-two algorithm depicted in Figure 9.19 of Section 9.3, which consists of $N/2$ 2-point transforms followed by two $N/2$-point transforms. Conversely, if $N_1 = N/2$ and $N_2 = 2$, Eq. (9.36) reduces to the first stage decomposition of the decimation-in-time power-of-two algorithm depicted in Figure 9.4 Section 9.2, which consists of two $N/2$-point transforms followed by $N/2$ 2-point transforms.[9]

Cooley–Tukey algorithms for general composite N are obtained by first doing the N_1-point transforms and then again applying Eq. (9.36) to another remaining factor N_2 of N/N_1 until all the factors of N have been used. The repeated application of Eq. (9.36)

---

[9]For Figure 9.4 to be an exact representation of Eq. (9.36), the two-point butterflies of the last stage must be replaced by the butterflies of Figure 9.10.

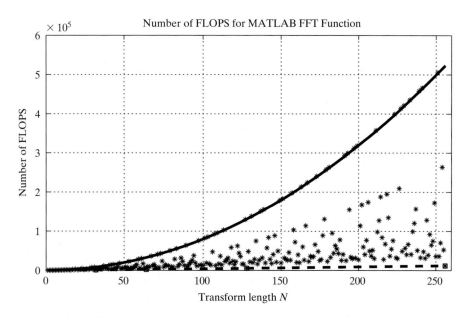

Figure 9.26   Number of floating-point operations as a function of N for MATLAB fft( ) function (revision 5.2).

leads to decompositions similar to the power-of-two algorithms. These algorithms re-quire only slightly more complicated indexing than the power of 2 case. If the factors of N are relatively prime, the number of multiplications can be further reduced at the cost of more complicated indexing. The "prime factor" algorithms use different index decompositions from those of Eqs. (9.35a) and (9.35b) so as to eliminate the twiddle factors in Eq. (9.36), and thus save a significant amount of computation. The details of the more general Cooley–Tukey and prime factor algorithms are discussed in Burrus and Parks (1985), Burrus (1988), and Blahut (1985).

As an illustration of what can be achieved using such prime factor algorithms, consider the measurements plotted in Figure 9.26. These measurements of the number of floating-point operations (FLOPS) as a function of N are for MATLAB's fft( ) function in Rev. 5.2 of MATLAB.[10] As we have discussed, the total number of floating point operations should be proportional to $N \log_2 N$ for N a power of two and propor-tional to N^2 for direct computation. For other values of N the total operation count will be dependent on the number (and cardinality) of the factors.

When N is a prime number, direct evaluation is required so the number of FLOPS will be proportional to N^2. The upper (solid) curve in Figure 9.26 shows the function

$$\text{FLOPS}(N) = 6N^2 + 2N(N - 1). \tag{9.37}$$

All the points falling on this curve are for values N a prime number. The lower dashed curve shows the function

$$\text{FLOPS}(N) = 6N \log_2 N. \tag{9.38}$$

---

[10]This graph was created with a modified version of a program written by C. S. Burrus. Since it is no longer possible to measure the number of floating-point operations in recent revisions of MATLAB, the reader may not be able to repeat this experiment.

The points falling on this curve are all for N a power of two. For other composite numbers the number of operations falls between the two curves. To see how efficiency varies from integer to integer, consider values of N from 199 to 202. The number 199 is a prime, so the number of operations (318004) falls on the maximum curve. The value $N = 200$ has the factorization $N = 2 \cdot 2 \cdot 2 \cdot 5 \cdot 5$, and the number of operations (27134) is near the minimum curve. For $N = 201 = 3 \cdot 67$, the number of FLOPS is 113788, and for $N = 202 = 2 \cdot 101$ the number is 167676. This wide difference between $N = 201$ and $N = 202$ is because a 101-point transform requires much more computation than a 67-point transform. Also note that when N has many small factors (such as $N = 200$) the efficiency is much greater.

9.5.2 Optimized FFT Algorithms

An FFT algorithm is based on the mathematical decomposition of the DFT into a combination of smaller transforms as we showed in detail in Sections 9.2 and 9.3. The FFT algorithm can be expressed in a high-level programming language that can be translated into machine-level instructions by compilers running on the target machine. In general, this will lead to implementations whose efficiency will vary with machine architecture. To address the issue of maximizing efficiency over a range of machines, Frigo and Johnson (1998 and 2005), developed a free-software library called FFTW ("Fastest Fourier Transform in the West"). FFTW uses a "planner" to adapt its generalized Coley–Tukey-type FFT algorithms to a given hardware platform, thereby maximizing efficiency. The system operates in two stages, the first being a planning stage in which the computations are organized so as to optimize performance on the given machine, and the second being a computation stage where the resulting plan (program) is executed. Once the plan is determined for a given machine, it can be executed on that machine as many times as needed. The details of FFTW are beyond our scope here. However, Frigo and Johnson, 2005 have shown that over a wide range of host machines, the FFTW algorithm is significantly faster than other implementations for values of N ranging from about 16 up to 8192. Above 8192, the performance of FFTW drops drastically due to memory cache issues.

9.6 IMPLEMENTATION OF THE DFT USING CONVOLUTION

Because of the dramatic efficiency of the FFT, convolution is often implemented by explicitly computing the inverse DFT of the product of the DFTs of each sequence to be convolved, where an FFT algorithm is used to compute both the forward and the inverse DFTs. In contrast, and even in apparent (but, of course, not actual) contradiction, it is sometimes preferable to compute the DFT by first reformulating it as a convolution. We have already seen an example of this in the Goertzel algorithm. A number of other, more sophisticated, procedures are based on this approach as discussed in the following sections.

9.6.1 Overview of the Winograd Fourier Transform Algorithm

One procedure proposed and developed by S. Winograd (1978), often referred to as the Winograd Fourier transform algorithm (WFTA), achieves its efficiency by expressing the DFT in terms of polynomial multiplication or, equivalently, convolution. The WFTA uses an indexing scheme corresponding to the decomposition of the DFT into a multiplicity of short-length DFTs where the lengths are relatively prime. Then the short DFTs are converted into periodic convolutions. A scheme for converting a DFT into a convolution when the number of input samples is prime was proposed by Rader (1968), but its application awaited the development of efficient methods for computing periodic convolutions. Winograd combined all of the foregoing procedures together with highly efficient algorithms for computing cyclic convolutions into a new approach to computing the DFT. The techniques for deriving efficient algorithms for computing short convolutions are based on relatively advanced number-theoretic concepts, such as the Chinese remainder theorem for polynomials, and consequently, we do not explore the details here. However, excellent discussions of the details of the WFTA are available in McClellan and Rader (1979), Blahut (1985), and Burrus (1988).

With the WFTA approach, the number of multiplications required for an N-point DFT is proportional to N rather than $N \log N$. Although this approach leads to algorithms that are optimal in terms of minimizing multiplications, the number of additions is significantly increased in comparison with the FFT. Therefore, the WFTA is most advantageous when multiplication is significantly slower than addition, as is often the case with fixed-point digital arithmetic. However, in processors where multiplication and accumulation are tied together, the Cooley–Tukey or prime factor algorithms are generally preferable. Additional difficulties with the WFTA are that indexing is more complicated, in-place computation is not possible, and there are major structural differences in algorithms for different values of N.

Thus, although the WFTA is extremely important as a benchmark for determining how efficient the DFT computation can be (in terms of number of multiplications), other factors often dominate in determining the speed and efficiency of a hardware or software implementation of the DFT computation.

9.6.2 The Chirp Transform Algorithm

Another algorithm based on expressing the DFT as a convolution is referred to as the chirp transform algorithm (CTA). This algorithm is not optimal in minimizing any measure of computational complexity, but it has been useful in a variety of applications, particularly when implemented in technologies that are well suited to doing convolution with a fixed, prespecified impulse response. The CTA is also more flexible than the FFT, since it can be used to compute *any* set of equally spaced samples of the Fourier transform on the unit circle.

To derive the CTA, we let $x[n]$ denote an N-point sequence and $X(e^{j\omega})$ its Fourier transform. We consider the evaluation of M samples of $X(e^{j\omega})$ that are equally spaced in angle on the unit circle, as indicated in Figure 9.27, i.e., at frequencies

$$\omega_k = \omega_0 + k\Delta\omega, \qquad k = 0, 1, \ldots, M-1, \tag{9.39}$$

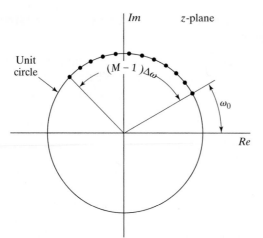

Figure 9.27 Frequency samples for chirp transform algorithm.

where the starting frequency ω_0 and the frequency increment $\Delta\omega$ can be chosen arbitrarily. (For the specific case of the DFT, $\omega_0 = 0$, $M = N$, and $\Delta\omega = 2\pi/N$.) The Fourier transform corresponding to this more general set of frequency samples is given by

$$X(e^{j\omega_k}) = \sum_{n=0}^{N-1} x[n]e^{-j\omega_k n}, \qquad k = 0, 1, \ldots, M-1, \tag{9.40}$$

or, with W defined as

$$W = e^{-j\Delta\omega} \tag{9.41}$$

and using Eq. (9.39),

$$X(e^{j\omega_k}) = \sum_{n=0}^{N-1} x[n]e^{-j\omega_0 n} W^{nk}. \tag{9.42}$$

To express $X(e^{j\omega_k})$ as a convolution, we use the identity

$$nk = \tfrac{1}{2}[n^2 + k^2 - (k-n)^2] \tag{9.43}$$

to express Eq. (9.42) as

$$X(e^{j\omega_k}) = \sum_{n=0}^{N-1} x[n]e^{-j\omega_0 n} W^{n^2/2} W^{k^2/2} W^{-(k-n)^2/2}. \tag{9.44}$$

Letting

$$g[n] = x[n]e^{-j\omega_0 n} W^{n^2/2}, \tag{9.45}$$

we can then write

$$X(e^{j\omega_k}) = W^{k^2/2} \left(\sum_{n=0}^{N-1} g[n] W^{-(k-n)^2/2} \right), \qquad k = 0, 1, \ldots, M-1. \tag{9.46}$$

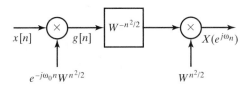

Figure 9.28   Block diagram of chirp transform algorithm.

In preparation for interpreting Eq. (9.46) as the output of a linear time-invariant system, we obtain more familiar notation by replacing k by n and n by k in Eq. (9.46):

$$X(e^{j\omega_n}) = W^{n^2/2}\left(\sum_{k=0}^{N-1} g[k]W^{-(n-k)^2/2}\right), \qquad n = 0, 1, \ldots, M-1. \qquad (9.47)$$

In the form of Eq. (9.47), $X(e^{j\omega_n})$ corresponds to the convolution of the sequence $g[n]$ with the sequence $W^{-n^2/2}$, followed by multiplication by the sequence $W^{n^2/2}$. The output sequence, indexed on the independent variable n, is the sequence of frequency samples $X(e^{j\omega_n})$. With this interpretation, the computation of Eq. (9.47) is as depicted in Figure 9.28. The sequence $W^{-n^2/2}$ can be thought of as a complex exponential sequence with linearly increasing frequency $n\Delta w$. In radar systems, such signals are called chirp signals—hence the name *chirp transform*. A system similar to Figure 9.28 is commonly used in radar and sonar signal processing for pulse compression (Skolnik, 2002).

For the evaluation of the Fourier transform samples specified in Eq. (9.47), we need only compute the output of the system in Figure 9.28 over a finite interval. In Figure 9.29, we depict illustrations of the sequences $g[n]$, $W^{-n^2/2}$, and $g[n] * W^{-n^2/2}$. Since $g[n]$ is of finite duration, only a finite portion of the sequence $W^{-n^2/2}$ is used in obtaining $g[n] * W^{-n^2/2}$ over the interval $n = 0, 1, \ldots, M-1$, specifically, that portion from $n = -(N-1)$ to $n = M-1$. Let us define

$$h[n] = \begin{cases} W^{-n^2/2}, & -(N-1) \leq n \leq M-1, \\ 0, & \text{otherwise}, \end{cases} \qquad (9.48)$$

as illustrated in Figure 9.30. It is easily verified by considering the graphical representation of the process of convolution that

$$g[n] * W^{-n^2/2} = g[n] * h[n], \qquad n = 0, 1, \ldots, M-1. \qquad (9.49)$$

Consequently, the infinite-duration impulse response $W^{-n^2/2}$ in the system of Figure 9.28 can be replaced by the finite-duration impulse response of Figure 9.30. The system is now as indicated in Figure 9.31, where $h[n]$ is specified by Eq. (9.48) and the frequency samples are given by

$$X(e^{j\omega_n}) = y[n], \qquad n = 0, 1, \ldots, M-1. \qquad (9.50)$$

Evaluation of frequency samples using the procedure indicated in Figure 9.31 has a number of potential advantages. In general, we do not require $N = M$ as in the FFT algorithms, and neither N nor M need be composite numbers. In fact, they may be prime numbers if desired. Furthermore, the parameter ω_0 is arbitrary. This increased flexibility over the FFT does not preclude efficient computation, since the convolution in Figure 9.31 can be implemented efficiently using an FFT algorithm with the technique

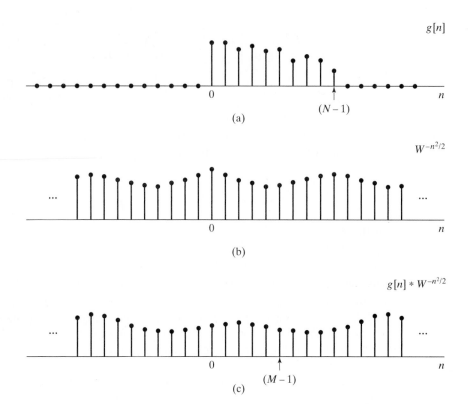

$g[n]$

(a)

$W^{-n^2/2}$

(b)

$g[n] * W^{-n^2/2}$

(c)

Figure 9.29 An illustration of the sequences used in the chirp transform algorithm. Note that the actual sequences involved are complex valued. (a) $g[n] = x[n]e^{-j\omega_0 n} W^{n^2/2}$. (b) $W^{-n^2/2}$. (c) $g[n] * W^{-n^2/2}$.

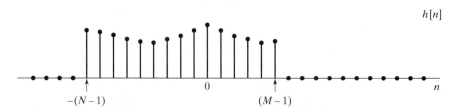

$h[n]$

Figure 9.30 An illustration of the region of support for the FIR chirp filter. Note that the actual values of $h[n]$ as given by Eq. (9.48) are complex.

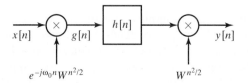

$x[n]$ $g[n]$ $h[n]$ $y[n]$

$e^{-j\omega_0 n} W^{n^2/2}$ $W^{n^2/2}$

Figure 9.31 Block diagram of chirp transform system for finite-length impulse response.

of Section 8.7 to compute the convolution. As discussed in that section, the FFT size must be greater than or equal to $(M + N - 1)$ in order that the circular convolution will be equal to $g[n] * h[n]$ for $0 \leq n \leq M - 1$. The FFT size is otherwise arbitrary and can, for example, be chosen to be a power of 2. It is interesting to note that the FFT algorithms used to compute the convolution implied by the CTA could be of the Winograd type. These algorithms themselves use convolution to implement the DFT computation.

In the system of Figure 9.31 $h[n]$ is noncausal, and for certain real-time implementations it must be modified to obtain a causal system. Since $h[n]$ is of finite duration, this modification is easily accomplished by delaying $h[n]$ by $(N - 1)$ to obtain a causal impulse response:

$$h_1[n] = \begin{cases} W^{-(n-N+1)^2/2}, & n = 0, 1, \ldots, M + N - 2, \\ 0, & \text{otherwise.} \end{cases} \tag{9.51}$$

Since both the chirp demodulation factor at the output and the output signal are also delayed by $(N - 1)$ samples, the Fourier transform values are

$$X(e^{j\omega_n}) = y_1[n + N - 1], \qquad n = 0, 1, \ldots, M - 1. \tag{9.52}$$

Modifying the system of Figure 9.31 to obtain a causal system results in the system of Figure 9.32. An advantage of this system stems from the fact that it involves the convolution of the input signal (modulated with a chirp) with a fixed, causal impulse response. Certain technologies, such as charge-coupled devices (CCD) and surface acoustic wave (SAW) devices, are particularly useful for implementing convolution with a fixed, pre-specified impulse response. These devices can be used to implement FIR filters, with the filter impulse response being specified at the time of fabrication by a geometric pattern of electrodes. A similar approach was followed by Hewes, Broderson and Buss (1979) in implementing the CTA with CCDs.

Further simplification of the CTA results when the frequency samples to be computed correspond to the DFT, i.e., when $\omega_0 = 0$ and $W = e^{-j2\pi/N}$, so that $\omega_n = 2\pi n/N$. In this case, it is convenient to modify the system of Figure 9.32. Specifically, with $\omega_0 = 0$ and $W = e^{-j2\pi/N} = W_N$, consider applying an additional unit of delay to the impulse response in Figure 9.32. With N even, $W_N^N = e^{j2\pi} = 1$, so

$$W_N^{-(n-N)^2/2} = W_N^{-n^2/2}. \tag{9.53}$$

Therefore, the system now is as shown in Figure 9.33, where

$$h_2[n] = \begin{cases} W_N^{-n^2/2}, & n = 1, 2, \ldots, M + N - 1, \\ 0, & \text{otherwise.} \end{cases} \tag{9.54}$$

In this case, the chirp signal modulating $x[n]$ and the chirp signal modulating the output of the FIR filter are identical, and

$$X(e^{j2\pi n/N}) = y_2[n + N], \qquad n = 0, 1, \ldots, M - 1. \tag{9.55}$$

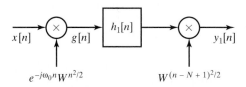

Figure 9.32   Block diagram of chirp transform system for causal finite-length impulse response.

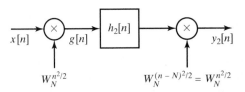

Figure 9.33 Block diagram of chirp transform system for obtaining DFT samples.

Example 9.1   Chirp Transform Parameters

Suppose we have a finite-length sequence $x[n]$ that is nonzero only on the interval $n = 0, \ldots, 25$, and we wish to compute 16 samples of the DTFT $X(e^{j\omega})$ at the frequencies $\omega_k = 2\pi/27 + 2\pi k/1024$ for $k = 0, \ldots, 15$. We can compute the desired frequency samples through convolution with a causal impulse response using the system in Figure 9.32 with an appropriate choice of parameters. We set $M = 16$, the number of samples desired, and $N = 26$, the length of the sequence. The frequency of the initial sample, ω_0, is $2\pi/27$, while the interval between adjacent frequency samples, $\Delta\omega$, is $2\pi/1024$. With these choices for the parameters, we know from Eq. (9.41) that $W = e^{-j\Delta\omega}$, and so the causal impulse response we desire is from Eq. (9.51)

$$h_1[n] = \begin{cases} [e^{-j2\pi/1024}]^{-(n-25)^2/2}, & n = 0, \ldots, 40, \\ 0, & \text{otherwise}. \end{cases}$$

For this causal impulse response, the output $y_1[n]$ will be the desired frequency samples beginning at $y_1[25]$, i.e.,

$$y_1[n + 25] = X(e^{j\omega_n})|_{\omega_n = 2\pi/27 + 2\pi n/1024}, \qquad n = 0, \ldots, 15.$$

An algorithm similar to the CTA was first proposed by Bluestein (1970), who showed that a recursive realization of Figure 9.32 can be obtained for the case $\Delta\omega = 2\pi/N$, N a perfect square. (See Problem 9.48.) Rabiner, Schafer and Rader (1969) generalized this algorithm to obtain samples of the z-transform equally spaced in angle on a spiral contour in the z-plane. This more general form of the CTA was called the chirp z-transform (CZT) algorithm. The algorithm that we have called the CTA is a special case of the CZT algorithm.

9.7 EFFECTS OF FINITE REGISTER LENGTH

Since the fast Fourier transform algorithm is widely used for digital filtering and spectrum analysis, it is important to understand the effects of finite register length in the computation. As in the case of digital filters, a precise analysis of the effects is difficult. However, a simplified analysis is often sufficient for the purpose of choosing the required register length. The analysis that we will present is similar in style to that carried out in Section 6.9. Specifically, we analyze arithmetic round-off by means of a linear-noise model obtained by inserting an additive noise source at each point in the computation algorithm where round-off occurs. Furthermore, we will make a number of assumptions to simplify the analysis. The results that we obtain lead to several simplified, but useful, estimates of the effect of arithmetic round-off. Although the analysis is for rounding, it is generally easy to modify the results for truncation.

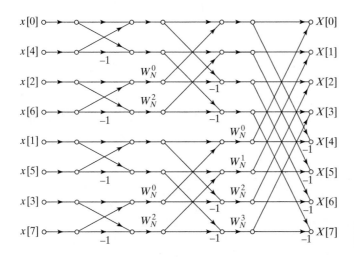

Figure 9.34 Flow graph for decimation-in-time FFT algorithm.

We have seen several different algorithmic structures for the FFT. However, the effects of round-off noise are very similar among the different classes of algorithms. Therefore, even though we consider only the radix-2 decimation-in-time algorithm, our results are representative of other forms as well.

The flow graph depicting a decimation-in-time algorithm for $N = 8$ was shown in Figure 9.11 and is reproduced in Figure 9.34. Some key aspects of this diagram are common to all standard radix-2 algorithms. The DFT is computed in $v = \log_2 N$ stages. At each stage a new array of N numbers is formed from the previous array by linear combinations of the elements, taken two at a time. The vth array contains the desired DFT. For radix-2 decimation-in-time algorithms, the basic 2-point DFT computation is of the form

$$X_m[p] = X_{m-1}[p] + W_N^r X_{m-1}[q], \qquad (9.56a)$$

$$X_m[q] = X_{m-1}[p] - W_N^r X_{m-1}[q]. \qquad (9.56b)$$

Here the subscripts m and $(m - 1)$ refer to the mth array and the $(m - 1)^{\text{st}}$ array, respectively, and p and q denote the location of the numbers in each array. (Note that $m = 0$ refers to the input array and $m = v$ refers to the output array.) A flow graph representing the butterfly computation is shown in Figure 9.35.

At each stage, $N/2$ separate butterfly computations are carried out to produce the next array. The integer r varies with p, q, and m in a manner that depends on the specific form of the FFT algorithm used. However, our analysis is not tied to the specific way

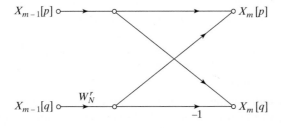

Figure 9.35 Butterfly computation for decimation-in-time.

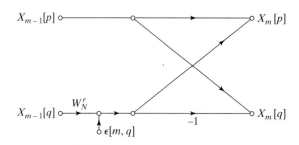

Figure 9.36   Linear-noise model for fixed-point round-off noise in a decimation-in-time butterfly computation.

in which r varies. Also, the specific relationship among p, q, and m, which determines how we index through the mth array, is not important for the analysis. The details of the analysis for decimation in time and decimation in frequency differ somewhat due to the different butterfly forms, but the basic results do not change significantly. In our analysis we assume a butterfly of the form of Eqs. (9.56a) and (9.56b), corresponding to decimation in time.

We model the round-off noise by associating an additive noise generator with each fixed-point multiplication. With this model, the butterfly of Figure 9.35 is replaced by that of Figure 9.36 for analyzing the round-off noise effects. The notation $\varepsilon[m, q]$ represents the complex-valued error introduced in computing the mth array from the $(m-1)^{\text{st}}$ array; specifically, it indicates the error resulting from quantization of multiplication of the qth element of the $(m-1)^{\text{st}}$ array by a complex coefficient.

Since we assume that, in general, the input to the FFT is a complex sequence, each of the multiplications is complex and thus consists of four real multiplications. We assume that the errors due to each real multiplication have the following properties:

1. The errors are uniformly distributed random variables over the range $-(1/2) \cdot 2^{-B}$ to $(1/2) \cdot 2^{-B}$, where, as defined in Section 6.7.1, numbers are represented as $(B + 1)$-bit signed fractions. Therefore, each error source has variance $2^{-2B}/12$.

2. The errors are uncorrelated with one another.

3. All the errors are uncorrelated with the input and, consequently, also with the output.

Since each of the four noise sequences is uncorrelated zero-mean white noise and all have the same variance,

$$\mathcal{E}\{|\varepsilon[m, q]|^2\} = 4 \cdot \frac{2^{-2B}}{12} = \tfrac{1}{3} \cdot 2^{-2B} = \sigma_B^2. \tag{9.57}$$

To determine the mean-square value of the output noise at any output node, we must account for the contribution from each of the noise sources that propagate to that node. We can make the following observations from the flow graph of Figure 9.34:

1. The transmission function from any node in the flow graph to any other node to which it is connected is multiplication by a complex constant of unity magnitude (because each branch transmittance is either unity or an integer power of W_N).

2. Each output node connects to seven butterflies in the flow graph. (In general, each output node would connect to $(N - 1)$ butterflies.) For example, Figure 9.37(a)

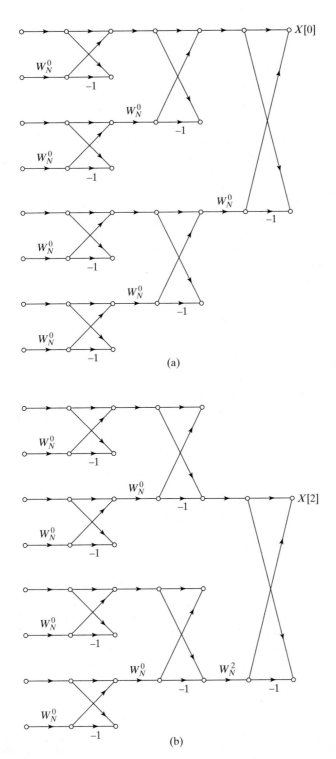

Figure 9.37 (a) Butterflies that affect $X[0]$; (b) butterflies that affect $X[2]$.

shows the flow graph with all the butterflies removed that do not connect to $X[0]$, and Figure 9.37(b) shows the flow graph with all the butterflies removed that do not connect to $X[2]$.

These observations can be generalized to the case of N an arbitrary power of 2.

As a consequence of the first observation, the mean-square value of the magnitude of the component of the output noise due to each elemental noise source is the same and equal to σ_B^2. The total output noise at each output node is equal to the sum of the noise propagated to that node. Since we assume that all the noise sources are uncorrelated, the mean-square value of the magnitude of the output noise is equal to σ_B^2 times the number of noise sources that propagate to that node. At most one complex noise source is introduced at each butterfly; consequently, from observation 2, at most $(N-1)$ noise sources propagate to each output node. In fact, not all the butterflies generate round-off noise, since some (for example, all those in the first and second stages for $N = 8$) involve only multiplication by unity. However, if for simplicity we assume that round-off occurs for each butterfly, we can consider the result as an upper bound on the output noise. With this assumption, then, the mean square value of the output noise in the kth DFT value, $F[k]$, is given by

$$\mathcal{E}\{|F[k]|^2\} = (N-1)\sigma_B^2, \tag{9.58}$$

which, for large N, we approximate as

$$\mathcal{E}\{|F[k]|^2\} \cong N\sigma_B^2. \tag{9.59}$$

According to this result, the mean-square value of the output noise is proportional to N, the number of points transformed. The effect of doubling N, or adding another stage in the FFT, is to double the mean-square value of the output noise. In Problem 9.52, we consider the modification of this result when we do not insert noise sources for those butterflies that involve only multiplication by unity or j. Note that for FFT algorithms, a double-length accumulator does not help us reduce round-off noise, since the outputs of the butterfly computation must be stored in $(B+1)$-bit registers at the output of each stage.

In implementing an FFT algorithm with fixed-point arithmetic, we must ensure against overflow. From Eqs. (9.56a) and (9.56b), it follows that

$$\max(|X_{m-1}[p]|, |X_{m-1}[q]|) \le \max(|X_m[p]|, |X_m[q]|) \tag{9.60}$$

and also

$$\max(|X_m[p]|, |X_m[q]|) \le 2\max(|X_{m-1}[p]|, |X_{m-1}[q]|). \tag{9.61}$$

(See Problem 9.51.) Equation (9.60) implies that the maximum magnitude is non-decreasing from stage to stage. If the magnitude of the output of the FFT is less than unity, then the magnitude of the points in each array must be less than unity, i.e., there will be no overflow in any of the arrays.[11]

To express this constraint as a bound on the input sequence, we note that the condition

$$|x[n]| < \frac{1}{N}, \qquad 0 \le n \le N-1, \tag{9.62}$$

---

[11] Actually, one should discuss overflow in terms of the real and imaginary parts of the data rather than the magnitude. However, $|x| < 1$ implies that $|\mathcal{R}e\{x\}| < 1$ and $|\mathcal{I}m\{x\}| < 1$, and only a slight increase in allowable signal level is achieved by scaling on the basis of real and imaginary parts.

is both necessary and sufficient to guarantee that

$$|X[k]| < 1, \qquad 0 \le k \le N - 1. \tag{9.63}$$

This follows from the definition of the DFT, since

$$|X[k]| = \left| \sum_{n=0}^{N-1} x[n] W_N^{kn} \right| \le \sum_{n=0}^{N-1} |x[n]| \quad k = 0, 1, \dots N - 1. \tag{9.64}$$

Thus, Eq. (9.62) is sufficient to guarantee that there will be no overflow for all stages of the algorithm.

To obtain an explicit expression for the noise-to-signal ratio at the output of the FFT algorithm, consider an input in which successive sequence values are uncorrelated, i.e., a white-noise input signal. Also, assume that the real and imaginary parts of the input sequence are uncorrelated and that each has an amplitude density that is uniform between $-1/(\sqrt{2}N)$ and $+1/(\sqrt{2}N)$. (Note that this signal satisfies Eq. (9.62).) Then the average squared magnitude of the complex input sequence is

$$\mathcal{E}\{|x[n]|^2\} = \frac{1}{3N^2} = \sigma_x^2. \tag{9.65}$$

The DFT of the input sequence is

$$X[k] = \sum_{n=0}^{N-1} x[n] W^{kn}, \tag{9.66}$$

from which it can be shown that, under the foregoing assumptions on the input,

$$\mathcal{E}\{|X[k]|^2\} = \sum_{n=0}^{N-1} \mathcal{E}\{|x[n]|^2\} |W^{kn}|^2$$

$$= N\sigma_x^2 = \frac{1}{3N}. \tag{9.67}$$

Combining Eqs. (9.59) and (9.67), we obtain

$$\frac{\mathcal{E}\{|F[k]|^2\}}{\mathcal{E}\{|X[k]|^2\}} = 3N^2 \sigma_B^2 = N^2 2^{-2B}. \tag{9.68}$$

According to Eq. (9.68), the noise-to-signal ratio increases as N^2, or 1 bit per stage. That is, if N is doubled, corresponding to adding one additional stage to the FFT, then to maintain the same noise-to-signal ratio, 1 bit must be added to the register length. The assumption of a white-noise input signal is, in fact, not critical here. For a variety of other inputs, the noise-to-signal ratio is still proportional to N^2, with only the constant of proportionality changing.

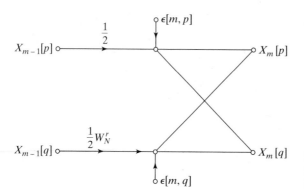

Figure 9.38   Butterfly showing scaling multipliers and associated fixed-point round-off noise.

Equation (9.61) suggests an alternative scaling procedure. Since the maximum magnitude increases by no more than a factor of 2 from stage to stage, we can prevent overflow by requiring that $|x[n]| < 1$ and incorporating an attenuation of $\frac{1}{2}$ at the input to each stage. In this case, the output will consist of the DFT scaled by $1/N$. Although the mean-square output signal will be $1/N$ times what it would be if no scaling were introduced, the input amplitude can be N times larger without causing overflow. For the white-noise input signal, this means that we can assume that the real and imaginary parts are uniformly distributed from $-1/\sqrt{2}$ to $1/\sqrt{2}$, so that $|x[n]| < 1$. Thus, with the ν divisions by 2, the maximum expected value of the magnitude squared of the DFT that can be attained (for the white input signal) is the same as that given in Eq. (9.67). However, the output noise level will be much less than in Eq. (9.59), since the noise introduced at early stages of the FFT will be attenuated by the scaling that takes place in the later arrays. Specifically, with scaling by $1/2$ introduced at the input to each butterfly, we modify the butterfly of Figure 9.36 to that of Figure 9.38, where, in particular, two noise sources are now associated with each butterfly. As before, we assume that the real and imaginary parts of these noise sources are uncorrelated and are also uncorrelated with the other noise sources and that the real and imaginary parts are uniformly distributed between $\pm(1/2) \cdot 2^{-B}$. Thus, as before,

$$\mathcal{E}\{|\varepsilon[m, q]|^2\} = \sigma_B^2 = \tfrac{1}{3} \cdot 2^{-2B} = \mathcal{E}\{|\varepsilon[m, p]|^2\}. \tag{9.69}$$

Because the noise sources are all uncorrelated, the mean-squared magnitude of the noise at each output node is again the sum of the mean-squared contributions of each noise source in the flow graph. However, unlike the previous case, the attenuation that each noise source experiences through the flow graph depends on the array at which it originates. A noise source originating at the mth array will propagate to the output with multiplication by a complex constant with magnitude $(1/2)^{\nu-m-1}$. By examination of Figure 9.34, we see that for the case $N = 8$, each output node connects to:

1 butterfly originating at the $(\nu - 1)^{\text{st}}$ array,

2 butterflies originating at the $(\nu - 2)^{\text{nd}}$ array,

4 butterflies originating at the $(\nu - 3)^{\text{rd}}$ array, etc.

For the general case with $N = 2^{\nu}$, each output node connects to $2^{\nu-m-1}$ butterflies and therefore to $2^{\nu-m}$ noise sources that originate at the mth array. Thus, at each output node, the mean-square magnitude of the noise is

$$\mathcal{E}\{|F[k]|^2\} = \sigma_B^2 \sum_{m=0}^{\nu-1} 2^{\nu-m} \cdot (0.5)^{2\nu-2m-2}$$

$$= \sigma_B^2 \sum_{m=0}^{\nu-1} (0.5)^{\nu-m-2} \tag{9.70}$$

$$= \sigma_B^2 \cdot 2 \sum_{k=0}^{\nu-1} 0.5^k$$

$$= 2\sigma_B^2 \frac{1-0.5^{\nu}}{1-0.5} = 4\sigma_B^2(1-0.5^{\nu}).$$

For large N, we assume that 0.5^{ν} (i.e., $1/N$) is negligible compared with unity, so

$$\mathcal{E}\{|F[k]|^2\} \cong 4\sigma_B^2 = \tfrac{4}{3} \cdot 2^{-2B}, \tag{9.71}$$

which is much less than the noise variance resulting when all the scaling is carried out on the input data.

Now we can combine Eq. (9.71) with Eq. (9.67) to obtain the output noise-to-signal ratio for the case of step-by-step scaling and white input. We obtain

$$\frac{\mathcal{E}\{|F[k]|^2\}}{\mathcal{E}\{|X[k]|^2\}} = 12N\sigma_B^2 = 4N \cdot 2^{-2B}, \tag{9.72}$$

a result proportional to N rather than to N^2. An interpretation of Eq. (9.72) is that the output noise-to-signal ratio increases as N, corresponding to half a bit per stage, a result first obtained by Welch (1969). It is important to note again that the assumption of a white-noise signal is not essential in the analysis. The basic result of an increase of half a bit per stage holds for a broad class of signals, with only the constant multiplier in Eq. (9.72) being dependent on the signal.

We should also note that the dominant factor that causes the increase of the noise-to-signal ratio with N is the decrease in signal level (required by the overflow constraint) as we pass from stage to stage. According to Eq. (9.71), very little noise (only a bit or two) is present in the final array. Most of the noise has been shifted out of the binary word by the scalings.

We have assumed straight fixed-point computation in the preceding discussion; i.e., only preset attenuations were allowed, and we were not permitted to rescale on the basis of an overflow test. Clearly, if the hardware or programming facility is such that straight fixed-point computation must be used, we should, if possible, incorporate attenuators of 1/2 at each array rather than use a large attenuation of the input array.

A third approach to avoiding overflow is the use of *block floating point*. In this procedure the original array is normalized to the far left of the computer word, with the restriction that $|x[n]| < 1$; the computation proceeds in a fixed-point manner, except that after every addition there is an overflow test. If overflow is detected, the entire array is divided by 2 and the computation continues. The number of necessary divisions by 2 are counted to determine a scale factor for the entire final array. The output noise-to-signal ratio depends strongly on how many overflows occur and at what stages of the computation they occur. The positions and timing of overflows are determined by the signal being transformed; thus, to analyze the noise-to-signal ratio in a block floating-point implementation of the FFT, we would need to know the input signal.

The preceding analysis shows that scaling to avoid overflow is the dominant factor in determining the noise-to-signal ratio of fixed-point implementations of FFT algorithms. Therefore, floating-point arithmetic should improve the performance of these algorithms. The effect of floating-point round-off on the FFT was analyzed both theoretically and experimentally by Gentleman and Sande (1966), Weinstein and Oppenheim (1969), and Kaneko and Liu (1970). These investigations show that, since scaling is no longer necessary, the decrease of noise-to-signal ratio with increasing N is much less dramatic than for fixed-point arithmetic.

For example, Weinstein (1969) showed theoretically that the noise-to-signal ratio is proportional to ν for $N = 2^\nu$, rather than proportional to N as in the fixed-point case. Therefore, quadrupling ν (raising N to the fourth power) increases the noise-to-signal ratio by only 1 bit.

9.8 SUMMARY

In this chapter we have considered techniques for computation of the discrete Fourier transform, and we have seen how the periodicity and symmetry of the complex factor $e^{-j(2\pi/N)kn}$ can be exploited to increase the efficiency of DFT computations.

We considered the Goertzel algorithm and the direct evaluation of the DFT expression because of the importance of these techniques when not all N of the DFT values are required. However, our major emphasis was on fast Fourier transform (FFT) algorithms. We described the decimation-in-time and decimation-in-frequency classes of FFT algorithms in some detail and some of the implementation considerations, such as indexing and coefficient quantization. Much of the detailed discussion concerned algorithms that require N to be a power of 2, since these algorithms are easy to understand, simple to program, and most often used.

The use of convolution as the basis for computing the DFT was briefly discussed. We presented a brief overview of the Winograd Fourier transform algorithm, and in somewhat more detail we discussed an algorithm called the chirp transform algorithm.

The final section of the chapter discussed effects of finite word length in DFT computations. We used linear-noise models to show that the noise-to-signal ratio of a DFT computation varies differently with the length of the sequence, depending on how scaling is done. We also commented briefly on the use of floating-point representations.

Problems

Basic Problems with Answers

9.1. Suppose that a computer program is available for computing the DFT

$$X[k] = \sum_{n=0}^{N-1} x[n]e^{-j(2\pi/N)kn}, \qquad k = 0, 1, \ldots, N-1;$$

i.e., the input to the program is the sequence $x[n]$ and the output is the DFT $X[k]$. Show how the input and/or output sequences may be rearranged such that the program can also be used to compute the inverse DFT

$$x[n] = \frac{1}{N} \sum_{k=0}^{N-1} X[k]e^{j(2\pi/N)kn}, \qquad n = 0, 1, \ldots, N-1;$$

i.e., the input to the program should be $X[k]$ or a sequence simply related to $X[k]$, and the output should be either $x[n]$ or a sequence simply related to $x[n]$. There are several possible approaches.

9.2. Computing the DFT generally requires complex multiplications. Consider the product $X + jY = (A + jB)(C + jD) = (AC - BD) + j(BC + AD)$. In this form, a complex multiplication requires four real multiplications and two real additions. Verify that a complex multiplication can be performed with three real multiplications and five additions using the algorithm

$$X = (A - B)D + (C - D)A,$$

$$Y = (A - B)D + (C + D)B.$$

9.3. Suppose that you time-reverse and delay a real-valued 32-point sequence $x[n]$ to obtain $x_1[n] = x[32-n]$. If $x_1[n]$ is used as the input for the system in Figure P9.4, find an expression for $y[32]$ in terms of $X(e^{j\omega})$, the DTFT of the original sequence $x[n]$.

9.4. Consider the system shown in Figure P9.4. If the input to the system, $x[n]$, is a 32-point sequence in the interval $0 \le n \le 31$, the output $y[n]$ at $n = 32$ is equal to $X(e^{j\omega})$ evaluated at a specific frequency ω_k. What is ω_k for the coefficients shown in Figure P9.4?

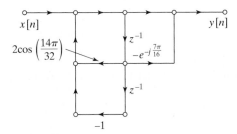

Figure P9.4

9.5. Consider the signal flow graph in Figure P9.5. Suppose that the input to the system $x[n]$ is an 8-point sequence. Choose the values of a and b such that $y[8] = X(e^{j6\pi/8})$.

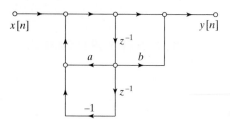

Figure P9.5

9.6. Figure P9.6 shows the graph representation of a decimation-in-time FFT algorithm for $N = 8$. The heavy line shows a path from sample $x[7]$ to DFT sample $X[2]$.

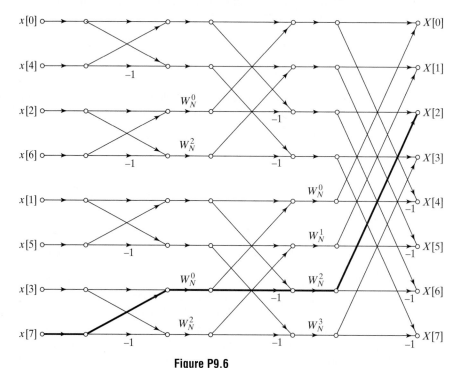

Figure P9.6

(a) What is the "gain" along the path that is emphasized in Figure P9.6?

(b) How many other paths in the flow graph begin at $x[7]$ and end at $X[2]$? Is this true in general? That is, how many paths are there between each input sample and each output sample?

(c) Now consider the DFT sample $X[2]$. By tracing paths in the flow graph of Figure P9.6, show that each input sample contributes the proper amount to the output DFT sample; i.e., verify that

$$X[2] = \sum_{n=0}^{N-1} x[n]e^{-j(2\pi/N)2n}.$$

9.7. Figure P9.7 shows the flow graph for an 8-point decimation-in-time FFT algorithm. Let $x[n]$ be the sequence whose DFT is $X[k]$. In the flow graph, $A[\cdot]$, $B[\cdot]$, $C[\cdot]$, and $D[\cdot]$ represent separate arrays that are indexed consecutively in the same order as the indicated nodes.

(a) Specify how the elements of the sequence $x[n]$ should be placed in the array $A[r]$, $r = 0, 1, \ldots, 7$. Also, specify how the elements of the DFT sequence should be extracted from the array $D[r]$, $r = 0, 1, \ldots, 7$.

(b) Without determining the values in the intermediate arrays, $B[\cdot]$ and $C[\cdot]$, determine and sketch the array sequence $D[r]$, $r = 0, 1, \ldots, 7$, if the input sequence is $x[n] = (-W_N)^n$, $n = 0, 1, \ldots, 7$.

(c) Determine and sketch the sequence $C[r]$, $r = 0, 1, \ldots, 7$, if the output Fourier transform is $X[k] = 1$, $k = 0, 1, \ldots, 7$.

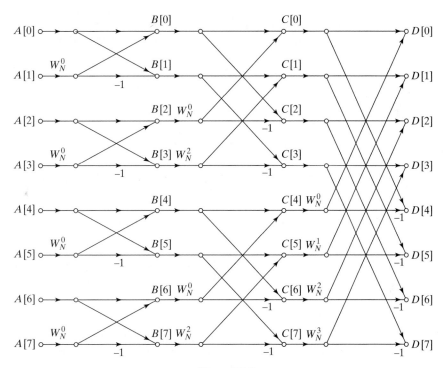

Figure P9.7

9.8. In implementing an FFT algorithm, it is sometimes useful to generate the powers of W_N with a recursive difference equation, or oscillator. In this problem we consider a radix-2 decimation-in-time algorithm for $N = 2^\nu$. Figure 9.11 depicts this type of algorithm for $N = 8$. To generate the coefficients efficiently, the frequency of the oscillator would change from stage to stage.

Assume that the arrays are numbered 0 through $\nu = \log_2 N$, so the array holding the initial input sequence is the zero[th] array and the DFT is in the νth array. In computing the butterflies in a given stage, all butterflies requiring the same coefficients W_N^r are evaluated before obtaining new coefficients. In indexing through the array, we assume that the data in the array are stored in consecutive complex registers numbered 0 through $(N-1)$. All the following questions are concerned with the computation of the mth array from the $(m-1)$[st] array, where $1 \le m \le \nu$. Answers should be expressed in terms of m.

(a) How many butterflies must be computed in the mth stage? How many different coefficients are required in the mth stage?

(b) Write a difference equation whose impulse response $h[n]$ contains the coefficients W_N^r required by the butterflies in the mth stage.

(c) The difference equation from part (b) should have the form of an oscillator, i.e., $h[n]$ should be periodic for $n \geq 0$. What is the period of $h[n]$? Based on this, write an expression for the frequency of this oscillator as a function of m.

9.9. Consider the butterfly in Figure P9.9. This butterfly was extracted from a signal flow graph implementing an FFT algorithm. Choose the most accurate statement from the following list:

1. The butterfly was extracted from a decimation-in-time FFT algorithm.
2. The butterfly was extracted from a decimation-in-frequency FFT algorithm.
3. It is not possible to say from the figure which kind of FFT algorithm the butterfly came from.

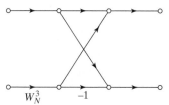

W_N^3 -1 **Figure P9.9**

9.10. A finite-length signal $x[n]$ is nonzero in the interval $0 \leq n \leq 19$. This signal is the input to the system shown in Figure P9.10, where

$$h[n] = \begin{cases} e^{j(2\pi/21)(n-19)^2/2}, & n = 0, 1, \ldots, 28, \\ 0, & \text{otherwise.} \end{cases}$$

$$W = e^{-j(2\pi/21)}$$

The output of the system, $y[n]$, for the interval $n = 19, \ldots, 28$ can be expressed in terms of the DTFT $X(e^{j\omega})$ for appropriate values of ω. Write an expression for $y[n]$ in this interval in terms of $X(e^{j\omega})$.

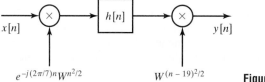

$e^{-j(2\pi/7)n}W^{n^2/2}$ $W^{(n-19)^2/2}$ **Figure P9.10**

9.11. The butterfly flow graph in Figure 9.10 can be used to compute the DFT of a sequence of length $N = 2^\nu$ "in-place," i.e., using a single array of complex-valued registers. Assume this array of registers $A[\ell]$ is indexed on $0 \leq l \leq N - 1$. The input sequence is initially stored in $A[\ell]$ in bit-reversed order. The array is then processed by ν stages of butterflies. Each butterfly takes two array elements $A[\ell_0]$ and $A[\ell_1]$ as inputs, then stores its outputs into those same array locations. The values of ℓ_0 and ℓ_1 depend on the stage number and the location of the butterfly in the signal flow graph. The stages of the computation are indexed by $m = 1, \ldots, \nu$.

(a) What is $|\ell_1 - \ell_0|$ as a function of the stage number m?

(b) Many stages contain butterflies with the same "twiddle" factor W_N^r. For these stages, how far apart are the values of ℓ_0 for the butterflies with the same W_N^r?

9.12. Consider the system shown in Figure P9.12, with

$$h[n] = \begin{cases} e^{j(2\pi/10)(n-11)^2/2}, & n = 0, 1, \ldots, 15, \\ 0, & \text{otherwise.} \end{cases}$$

It is desired that the output of the system, $y[n + 11] = X(e^{j\omega_n})$, where $\omega_n = (2\pi/19) + n(2\pi/10)$ for $n = 0, \ldots, 4$. Give the correct value for the sequence $r[n]$ in Figure P9.12 such that the output $y[n]$ provides the desired samples of the DTFT.

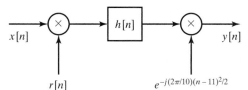

Figure P9.12

9.13. Assume that you wish to sort a sequence $x[n]$ of length $N = 16$ into bit-reversed order for input to an FFT algorithm. Give the new sample order for the bit-reversed sequence.

9.14. For the following statement, assume that the sequence $x[n]$ has length $N = 2^\nu$ and that $X[k]$ is the N-point DFT of $x[n]$. Indicate whether the statement is true or false, and justify your answer.

   Statement: It is impossible to construct a signal flow graph to compute $X[k]$ from $x[n]$ such that both $x[n]$ and $X[k]$ are in normal sequential (not bit-reversed) order.

9.15. The butterfly in Figure P9.15 was taken from a decimation-in-frequency FFT with $N = 16$, where the input sequence was arranged in normal order. Note that a 16-point FFT will have four stages, indexed $m = 1, \ldots, 4$. Which of the four stages have butterflies of this form? Justify your answer.

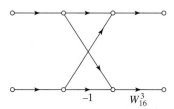

Figure P9.15

9.16. The butterfly in Figure P9.16 was taken from a decimation-in-time FFT with $N = 16$. Assume that the four stages of the signal flow graph are indexed by $m = 1, \ldots, 4$. What are the possible values of r for each of the four stages?

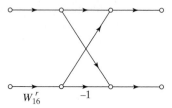

W_{16}^r -1 **Figure P9.16**

9.17. Suppose you have two programs for computing the DFT of a sequence $x[n]$ that has $N = 2^{\nu}$ nonzero samples. Program A computes the DFT by directly implementing the definition of the DFT sum from Eq. (8.67) and takes N^2 seconds to run. Program B implements the decimation-in-time FFT algorithm and takes $10N \log_2 N$ seconds to run. What is the shortest sequence N such that Program B runs faster than Program A?

9.18. The butterfly in Figure P9.18 was taken from a decimation-in-time FFT with $N = 16$. Assume that the four stages of the signal flow graph are indexed by $m = 1, \ldots, 4$. Which of the four stages have butterflies of this form?

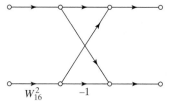

W_{16}^2 -1 **Figure P9.18**

9.19. Suppose you are told that an $N = 32$ FFT algorithm has a "twiddle" factor of W_{32}^2 for one of the butterflies in its fifth (last) stage. Is the FFT a decimation-in-time or decimation-in-frequency algorithm?

9.20. Suppose you have a signal $x[n]$ with 1021 nonzero samples whose DTFT you wish to estimate by computing the DFT. You find that it takes your computer 100 seconds to compute the 1021-point DFT of $x[n]$. You then add three zero-valued samples at the end of the sequence to form a 1024-point sequence $x_1[n]$. The same program on your computer requires only 1 second to compute $X_1[k]$. Reflecting, you realize that by using $x_1[n]$, you are able to compute more samples of $X(e^{j\omega})$ in a much shorter time by adding some zeros to the end of $x[n]$ and pretending that the sequence is longer. How do you explain this apparent paradox?

Basic Problems

9.21. In Section 9.1.2, we used the fact that $W_N^{-kN} = 1$ to derive a recurrence algorithm for computing a specific DFT value $X[k]$ for a finite-length sequence $x[n]$, $n = 0, 1, \ldots, N-1$.

(a) Using the fact that $W_N^{kN} = W_N^{Nn} = 1$, show that $X[N-k]$ can be obtained as the output after N iterations of the difference equation depicted in Figure P9.21-1. That is, show that

$$X[N-k] = y_k[N].$$

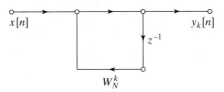

Figure P9.21-1

(b) Show that $X[N - k]$ is also equal to the output after N iterations of the difference equation depicted in Figure P9.21-2. Note that the system of Figure P9.21-2 has the same poles as the system in Figure 9.2, but the coefficient required to implement the complex zero in Figure P9.21-2 is the complex conjugate of the corresponding coefficient in Figure 9.2; i.e., $W_N^{-k} = (W_N^k)^*$.

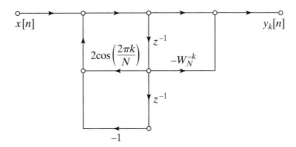

Figure P9.21-2

9.22. Consider the system shown in Figure P9.22. The subsystem from $x[n]$ to $y[n]$ is a causal, LTI system implementing the difference equation

$$y[n] = x[n] + ay[n - 1].$$

$x[n]$ is a finite length sequence of length 90, i.e.,

$$x[n] = 0 \quad \text{for } n < 0 \text{ and } n > 89.$$

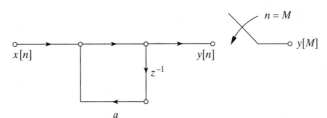

Figure P9.22

Determine a choice for the complex constant a and a choice for the sampling instant M so that

$$y[M] = X(e^{j\omega})\Big|_{\omega = 2\pi/60}.$$

9.23. Construct a flow graph for a 16-point radix-2 decimation-in-time FFT algorithm. Label all multipliers in terms of powers of W_{16}, and also label any branch transmittances that are equal to -1. Label the input and output nodes with the appropriate values of the input and DFT sequences, respectively. Determine the number of real multiplications and the number of real additions required to implement the flow graph.

9.24. It is suggested that if you have an FFT subroutine for computing a length-N DFT, the inverse DFT of an N-point sequence $X[k]$ can be implemented using this subroutine as follows:

1. Swap the real and imaginary parts of each DFT coefficient $X[k]$.
2. Apply the FFT routine to this input sequence.
3. Swap the real and imaginary parts of the output sequence.
4. Scale the resulting sequence by $\frac{1}{N}$ to obtain the sequence $x[n]$, corresponding to the inverse DFT of $X[k]$.

Determine whether this procedure works as claimed. If it doesn't, propose a simple modification that will make it work.

9.25. The DFT is a sampled version of the DTFT of a finite-length sequence; i.e.,

$$X[k] = X(e^{j(2\pi/N)k})$$

$$= X(e^{j\omega_k})\Big|_{\omega_k=(2\pi/N)k}$$

$$= \sum_{n=0}^{N-1} x[n]e^{-j(2\pi/N)kn} \qquad k = 0, 1, \ldots, N-1. \qquad \text{(P9.25-1)}$$

Furthermore, an FFT algorithm is an efficient way to compute the values $X[k]$.

Now consider a finite-length sequence $x[n]$ whose length is N samples. We want to evaluate $X(z)$, the z-transform of the finite-length sequence, at the following points in the z-plane

$$z_k = re^{j(2\pi/N)k} \qquad k = 0, 1, \ldots, N-1,$$

where r is a positive number. We have available an FFT algorithm.

(a) Plot the points z_k in the z-plane for the case $N = 8$ and $r = 0.9$.
(b) Write an equation [similar to Eq. (P9.25-1) above] for $X(z_k)$ that shows that $X(z_k)$ is the DFT of a modified sequence $\tilde{x}[n]$. What is $\tilde{x}[n]$?
(c) Describe an algorithm for computing $X(z_k)$ using the given FFT function. (*Direct evaluation is not an option.*) You may describe your algorithm using any combination of English text and equations, but you must give a step-by-step procedure that starts with the sequence $x[n]$ and ends with $X(z_k)$.

9.26. We are given a finite-length sequence $x[n]$ of length 627 (i.e., $x[n] = 0$ for $n < 0$ and $n > 626$), and we have available a program that will compute the DFT of a sequence of any length $N = 2^\nu$.

For the given sequence, we want to compute samples of the DTFT at frequencies

$$\omega_k = \frac{2\pi}{627} + \frac{2\pi k}{256}, \qquad k = 0, 1, \ldots, 255.$$

Specify how to obtain a new sequence $y[n]$ from $x[n]$ such that the desired frequency samples can be obtained by applying the available FFT program to $y[n]$ with ν *as small as possible.*

9.27. A finite-length signal of length $L = 500$ ($x[n] = 0$ for $n < 0$ and $n > L - 1$) is obtained by sampling a continuous-time signal with sampling rate 10,000 samples per second. We wish to compute samples of the z-transform of $x[n]$ at the N equally spaced points $z_k = (0.8)e^{j2\pi k/N}$, for $0 \le k \le N - 1$, with an effective frequency spacing of 50 Hz or less.

(a) Determine the minimum value for N if $N = 2^\nu$.
(b) Determine a sequence $y[n]$ of length N, where N is as determined in part (a), such that its DFT $Y[k]$ is equal to the desired samples of the z-transform of $x[n]$.

9.28. You are asked to build a system that computes the DFT of a 4-point sequence

$$x[0], x[1], x[2], x[3].$$

You can purchase any number of computational units at the per-unit cost shown in Table 9.1.

TABLE 9.1

Module	Per-Unit Cost
8-point DFT	$1
8-point IDFT	$1
adder	$10
multiplier	$100

Design a system of the lowest possible cost. Draw the associated block diagram and indicate the system cost.

Advanced Problems

9.29. Consider an N-point sequence $x[n]$ with DFT $X[k], k = 0, 1, \ldots, N-1$. The following algorithm computes the even-indexed DFT values $X[k], k = 0, 2, \ldots, N-2$, for N even, using only a single $N/2$-point DFT:

1. Form the sequence $y[n]$ by time aliasing, i.e.,

$$y[n] = \begin{cases} x[n] + x[n + N/2], & 0 \le n \le N/2 - 1, \\ 0, & \text{otherwise.} \end{cases}$$

2. Compute $Y[r], r = 0, 1, \ldots, (N/2) - 1$, the $N/2$-point DFT of $y[n]$.
3. Then the even-indexed values of $X[k]$ are $X[k] = Y[k/2]$, for $k = 0, 2, \ldots, N-2$.

(a) Show that the preceding algorithm produces the desired results.
(b) Now suppose that we form a finite-length sequence $y[n]$ from a sequence $x[n]$ by

$$y[n] = \begin{cases} \displaystyle\sum_{r=-\infty}^{\infty} x[n + rM], & 0 \le n \le M - 1, \\ 0, & \text{otherwise.} \end{cases}$$

Determine the relationship between the M-point DFT $Y[k]$ and $X(e^{j\omega})$, the Fourier transform of $x[n]$. Show that the result of part (a) is a special case of the result of part (b).
(c) Develop an algorithm similar to the one in part (a) to compute the odd-indexed DFT values $X[k], k = 1, 3, \ldots, N-1$, for N even, using only a single $N/2$-point DFT.

9.30. The system in Figure P9.30 computes an N-point (where N is an even number) DFT $X[k]$ of an N-point sequence $x[n]$ by decomposing $x[n]$ into two $N/2$-point sequences $g_1[n]$ and $g_2[n]$, computing the $N/2$-point DFT's $G_1[k]$ and $G_2[k]$, and then combining these to form $X[k]$.

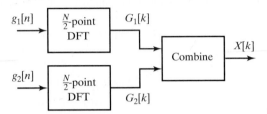

Figure P9.30

If $g_1[n]$ is the even-indexed values of $x[n]$ and $g_2[n]$ is the odd-indexed values of $x[n]$ i.e., $g_1[n] = x[2n]$ and $g_2[n] = x[2n+1]$ then $X[k]$ will be the DFT of $x[n]$.

In using the system in Figure P9.30 an error is made in forming $g_1[n]$ and $g_2[n]$, such that $g_1[n]$ is **incorrectly** chosen as the odd-indexed values and $g_2[n]$ as the even indexed values but $G_1[k]$ and $G_2[k]$ are still combined as in Figure P9.30 and the incorrect sequence $\hat{X}[k]$ results. Express $\hat{X}[k]$ in terms of $X[k]$.

9.31. In Section 9.3.2, it was asserted that the transpose of the flow graph of an FFT algorithm is also the flow graph of an FFT algorithm. The purpose of this problem is to develop that result for radix-2 FFT algorithms.

(a) The basic butterfly for the decimation-in-frequency radix-2 FFT algorithm is depicted in Figure P9.31-1. This flow graph represents the equations

$$X_m[p] = X_{m-1}[p] + X_{m-1}[q],$$

$$X_m[q] = (X_{m-1}[p] - X_{m-1}[q])W_N^r.$$

Starting with these equations, show that $X_{m-1}[p]$ and $X_{m-1}[q]$ can be computed from $X_m[p]$ and $X_m[q]$, respectively, using the butterfly shown in Figure P9.31-2.

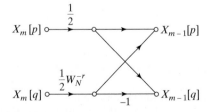

Figure P9.31-1

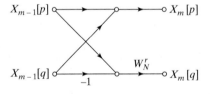

Figure P9.31-2

(b) In the decimation-in-frequency algorithm of Figure 9.22, $X_v[r], r = 0, 1, \ldots, N - 1$ is the DFT $X[k]$ arranged in bit-reversed order, and $X_0[r] = x[r], r = 0, 1, \ldots, N - 1$; i.e., the zeroth array is the input sequence arranged in normal order. If each butterfly in Figure 9.22 is replaced by the appropriate butterfly of the form of Figure P9.31, the result would be a flow graph for computing the sequence $x[n]$ (in normal order) from the DFT $X[k]$ (in bit-reversed order). Draw the resulting flow graph for $N = 8$.

(c) The flow graph obtained in part (b) represents an *inverse* DFT algorithm, i.e., an algorithm for computing

$$x[n] = \frac{1}{N} \sum_{n=0}^{N-1} X[k] W_N^{-kn}, \qquad n = 0, 1, \ldots, N - 1.$$

Modify the flow graph obtained in part (b) so that it computes the DFT

$$X[k] = \sum_{n=0}^{N-1} x[n] W_N^{kn}, \qquad k = 0, 1, \ldots, N - 1,$$

rather than the inverse DFT.

(d) Observe that the result in part (c) is the transpose of the decimation-in-frequency algorithm of Figure 9.22 and that it is identical to the decimation-in-time algorithm depicted in Figure 9.11. Does it follow that, to each decimation-in-time algorithm (e.g., Figures 9.15–9.17), there corresponds a decimation-in-frequency algorithm that is the transpose of the decimation-in-time algorithm and vice versa? Explain.

9.32. We want to implement a 6-point decimation-in-time FFT using a mixed radix approach. One option is to first take three 2-point DFTs, and then use the results to compute the 6-point DFT. For this option:

(a) Draw a flowgraph to show what a 2-point DFT calculates. Also, fill in the parts of the flowgraph in Figure P9.32-1 involved in calculating the DFT values X_0, X_1, and X_4.

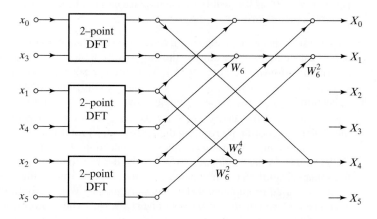

Figure P9.32-1

(b) How many complex multiplications does this option require? (Multiplying a number by -1 does not count as a complex multiplication.)

A second option is to start with two 3-point DFTs, and then use the results to compute the 6-point DFT.

(c) Draw a flowgraph to show what a 3-point DFT calculates. Also, fill in all of the flowgraph in Figure P9.32-2 and briefly explain how you derived your implementation:

(d) How many complex multiplications does this option require?

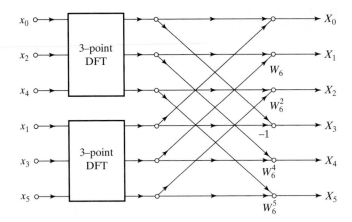

Figure P9.32-2

9.33. The decimation-in-frequency FFT algorithm was developed in Section 9.3 for radix 2, i.e., $N = 2^v$. A similar approach leads to a radix-3 algorithm when $N = 3^v$.

(a) Draw a flow graph for a 9-point decimation-in-frequency algorithm using a 3×3 decomposition of the DFT.

(b) For $N = 3^v$, how many complex multiplications by powers of W_N are needed to compute the DFT of an N-point complex sequence using a radix-3 decimation-in-frequency FFT algorithm?

(c) For $N = 3^v$, is it possible to use in-place computation for the radix-3 decimation-in-frequency algorithm?

9.34. We have seen that an FFT algorithm can be viewed as an interconnection of butterfly computational elements. For example, the butterfly for a radix-2 decimation-in-frequency FFT algorithm is shown in Figure P9.34-1. The butterfly takes two complex numbers as input and produces two complex numbers as output. Its implementation requires a complex multiplication by W_N^r, where r is an integer that depends on the location of the butterfly in the flow graph of the algorithm. Since the complex multiplier is of the form $W_N^r = e^{j\theta}$, the CORDIC (coordinate rotation digital computer) rotator algorithm (see Problem 9.46) can be used to implement the complex multiplication efficiently. Unfortunately, while the CORDIC rotator algorithm accomplishes the desired change of angle, it also introduces a fixed magnification that is independent of the angle θ. Thus, if the CORDIC rotator algorithm were used to implement the multiplications by W_N^r, the butterfly of Figure P9.34-1 would be replaced by the butterfly of Figure P9.34-2, where G represents the fixed magnification factor of the CORDIC rotator. (We assume no error in approximating the angle of rotation.) If each butterfly in the flow graph of the decimation-in-frequency FFT algorithm is replaced by the butterfly of Figure P9.34-2, we obtain a modified FFT algorithm for which the flow graph would be as shown in Figure P9.34-3 for $N = 8$. The output of this modified algorithm would not be the desired DFT.

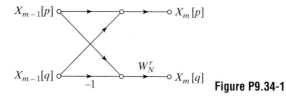

Figure P9.34-1

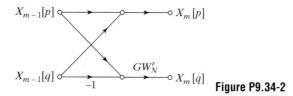

Figure P9.34-2

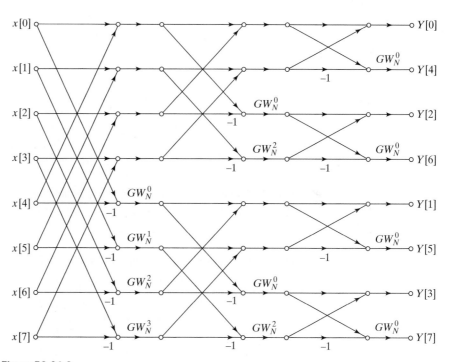

Figure P9.34-3

(a) Show that the output for the modified FFT algorithm is $Y[k] = W[k]X[k]$, where $X[k]$ is the correct DFT of the input sequence $x[n]$ and $W[k]$ is a function of G, N, and k.

(b) The sequence $W[k]$ can be described by a particularly simple rule. Find this rule and indicate its dependence on G, N, and k.

(c) Suppose that we wish to preprocess the input sequence $x[n]$ to compensate for the effect of the modified FFT algorithm. Determine a procedure for obtaining a sequence $\hat{x}[n]$ from $x[n]$ such that if $\hat{x}[n]$ is the input to the modified FFT algorithm, then the output will be $X[k]$, the correct DFT of the original sequence $x[n]$.

9.35. This problem deals with the efficient computation of samples of the z-transform of a finite-length sequence. Using the chirp transform algorithm, develop a procedure for computing values of $X(z)$ at 25 points spaced uniformly on an arc of a circle of radius 0.5, beginning at an angle of $-\pi/6$ and ending at an angle of $2\pi/3$. The length of the sequence is 100 samples.

9.36. Consider a 1024-point sequence $x[n]$ constructed by interleaving two 512-point sequences $x_e[n]$ and $x_o[n]$. Specifically,

$$x[n] = \begin{cases} x_e[n/2], & \text{if } n = 0,\, 2,\, 4,\, \dots,\, 1022; \\ x_o[(n-1)/2], & \text{if } n = 1,\, 3,\, 5,\, \dots,\, 1023; \\ 0, & \text{for } n \text{ outside of the range } 0 \le n \le 1023. \end{cases}$$

Let $X[k]$ denote the 1024-point DFT of $x[n]$ and $X_e[k]$ and $X_o[k]$ denote the 512-point DFTs of $x_e[n]$ and $x_o[n]$, respectively. Given $X[k]$ we would like to obtain $X_e[k]$ from $X[k]$ in a computationally efficient way where computational efficiency is measured in terms of the total number of complex multiplies and adds required. One not-very-efficient approach is as shown in Figure P9.36:

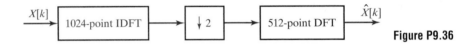

Figure P9.36

     Specify the most efficient algorithm that you can (certainly more efficient than the block diagram of Figure P9.36) to obtain $X_e[k]$ from $X[k]$.

9.37. Suppose that a program is available that computes the DFT of a complex sequence. If we wish to compute the DFT of a real sequence, we may simply specify the imaginary part to be zero and use the program directly. However, the symmetry of the DFT of a real sequence can be used to reduce the amount of computation.

(a) Let $x[n]$ be a real-valued sequence of length N, and let $X[k]$ be its DFT with real and imaginary parts denoted $X_R[k]$ and $X_I[k]$, respectively; i.e.,

$$X[k] = X_R[k] + jX_I[k].$$

Show that if $x[n]$ is real, then $X_R[k] = X_R[N-k]$ and $X_I[k] = -X_I[N-k]$ for $k = 1, \dots, N-1$.

(b) Now consider two real-valued sequences $x_1[n]$ and $x_2[n]$ with DFTs $X_1[k]$ and $X_2[k]$, respectively. Let $g[n]$ be the complex sequence $g[n] = x_1[n] + jx_2[n]$, with corresponding DFT $G[k] = G_R[k] + jG_I[k]$. Also, let $G_{OR}[k]$, $G_{ER}[k]$, $G_{OI}[k]$ and $G_{EI}[k]$ denote, respectively, the odd part of the real part, the even part of the real part, the odd part of the imaginary part, and the even part of the imaginary part of $G[k]$. Specifically, for $1 \le k \le N-1$,

$$G_{OR}[k] = \tfrac{1}{2}\{G_R[k] - G_R[N-k]\},$$

$$G_{ER}[k] = \tfrac{1}{2}\{G_R[k] + G_R[N-k]\},$$

$$G_{OI}[k] = \tfrac{1}{2}\{G_I[k] - G_I[N-k]\},$$

$$G_{EI}[k] = \tfrac{1}{2}\{G_I[k] + G_I[N-k]\},$$

and $G_{OR}[0] = G_{OI}[0] = 0$, $G_{ER}[0] = G_R[0]$, $G_{EI}[0] = G_I[0]$. Determine expressions for $X_1[k]$ and $X_2[k]$ in terms of $G_{OR}[k]$, $G_{ER}[k]$, $G_{OI}[k]$, and $G_{EI}[k]$.

(c) Assume that $N = 2^v$ and that a radix-2 FFT program is available to compute the DFT. Determine the number of real multiplications and the number of real additions required to compute both $X_1[k]$ and $X_2[k]$ by (i) using the program twice (with the imaginary part of the input set to zero) to compute the two complex N-point DFTs $X_1[k]$ and $X_2[k]$ separately and (ii) using the scheme suggested in part (b), which requires only one N-point DFT to be computed.

(d) Assume that we have only one real N-point sequence $x[n]$, where N is a power of 2. Let $x_1[n]$ and $x_2[n]$ be the two real $N/2$-point sequences $x_1[n] = x[2n]$ and $x_2[n] = x[2n + 1]$, where $n = 0, 1, \ldots, (N/2) - 1$. Determine $X[k]$ in terms of the $(N/2)$-point DFTs $X_1[k]$ and $X_2[k]$.

(e) Using the results of parts (b), (c), and (d), describe a procedure for computing the DFT of the real N-point sequence $x[n]$ utilizing only one $N/2$-point FFT computation. Determine the numbers of real multiplications and real additions required by this procedure, and compare these numbers with the numbers required if the $X[k]$ is computed using one N-point FFT computation with the imaginary part set to zero.

9.38. Let $x[n]$ and $h[n]$ be two real finite-length sequences such that

$$x[n] = 0 \quad \text{for } n \text{ outside the interval } 0 \le n \le L - 1,$$

$$h[n] = 0 \quad \text{for } n \text{ outside the interval } 0 \le n \le P - 1.$$

We wish to compute the sequence $y[n] = x[n]*h[n]$, where $*$ denotes ordinary convolution.

(a) What is the length of the sequence $y[n]$?

(b) For direct evaluation of the convolution sum, how many real multiplications are required to compute all of the nonzero samples of $y[n]$? The following identity may be useful:

$$\sum_{k=1}^{N} k = \frac{N(N+1)}{2}.$$

(c) State a procedure for using the DFT to compute all of the nonzero samples of $y[n]$. Determine the minimum size of the DFTs and inverse DFTs in terms of L and P.

(d) Assume that $L = P = N/2$, where $N = 2^v$ is the size of the DFT. Determine a formula for the number of real multiplications required to compute all the nonzero values of $y[n]$ using the method of part (c) if the DFTs are computed using a radix-2 FFT algorithm. Use this formula to determine the minimum value of N for which the FFT method requires fewer real multiplications than the direct evaluation of the convolution sum.

9.39. In Section 8.7.3, we showed that linear time-invariant filtering can be implemented by sectioning the input signal into finite-length segments and using the DFT to implement circular convolutions on these segments. The two methods discussed were called the overlap–add and the overlap–save methods. If the DFTs are computed using an FFT algorithm, these sectioning methods can require fewer complex multiplications per output sample than the direct evaluation of the convolution sum.

(a) Assume that the complex input sequence $x[n]$ is of infinite duration and that the complex impulse response $h[n]$ is of length P samples, so that $h[n] \ne 0$ only for $0 \le n \le P - 1$. Also, assume that the output is computed using the overlap–save method, with the DFTs of length $L = 2^v$, and suppose that these DFTs are computed using a radix-2 FFT algorithm. Determine an expression for the number of complex multiplications required per output sample as a function of v and P.

(b) Suppose that the length of the impulse response is $P = 500$. By evaluating the formula obtained in part (a), plot the number of multiplications per output sample as a function of v for the values of $v \leq 20$ such that the overlap–save method applies. For what value of v is the number of multiplications minimal? Compare the number of complex multiplications per output sample for the overlap–save method using the FFT with the number of complex multiplications per output sample required for direct evaluation of the convolution sum.

(c) Show that for large FFT lengths, the number of complex multiplications per output sample is approximately v. Thus, beyond a certain FFT length, the overlap–save method is less efficient than the direct method. If $P = 500$, for what value of v will the direct method be more efficient?

(d) Assume that the FFT length is twice the length of the impulse response (i.e., $L = 2P$), and assume that $L = 2^v$. Using the formula obtained in part (a), determine the smallest value of P such that the overlap–save method using the FFT requires fewer complex multiplications than the direct convolution method.

9.40. $x[n]$ is a 1024-point sequence that is nonzero only for $0 \leq n \leq 1023$. Let $X[k]$ be the 1024-point DFT of $x[n]$. Given $X[k]$, we want to compute $x[n]$ in the ranges $0 \leq n \leq 3$ and $1020 \leq n \leq 1023$ using the system in Figure P9.40. Note that the input to the system is the sequence of DFT coefficients. By selecting $m_1[n]$, $m_2[n]$, and $h[n]$, show how the system can be used to compute the desired samples of $x[n]$. Note that the samples $y[n]$ for $0 \leq n \leq 7$ must contain the desired samples of $x[n]$.

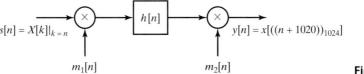

Figure P9.40

9.41. A system has been built for computing the 8-point DFT $Y[0]$, $Y[1]$, ..., $Y[7]$ of a sequence $y[0]$, $y[1]$, ..., $y[7]$. However, the system is not working properly: only the even DFT samples $Y[0]$, $Y[2]$, $Y[4]$, $Y[6]$ are being computed correctly. To help you solve the problem, the data you can access are:

- the (correct) even DFT samples, $Y[0]$, $Y[2]$, $Y[4]$, $Y[6]$;
- the first 4 input values $y[0]$, $y[1]$, $y[2]$, $y[3]$ (the other inputs are unavailable).

(a) If $y[0] = 1$, and $y[1] = y[2] = y[3] = 0$, and $Y[0] = Y[2] = Y[4] = Y[6] = 2$, what are the missing values $Y[1]$, $Y[3]$, $Y[5]$, $Y[7]$? Explain.

(b) You need to build an efficient system that computes the odd samples $Y[1]$, $Y[3]$, $Y[5]$, $Y[7]$ for any set of inputs. The computational modules you have available are one 4-point DFT and one 4-point IDFT. Both are free. You can purchase adders, subtracters, or multipliers for $10 each. Design a system of the lowest possible cost that takes as input

$$y[0], y[1], y[2], y[3], Y[0], Y[2], Y[4], Y[6]$$

and produces as output

$$Y[1], Y[3], Y[5], Y[7].$$

Draw the associated block diagram and indicate the total cost.

9.42. Consider a class of DFT-based algorithms for implementing a causal FIR filter with impulse response $h[n]$ that is zero outside the interval $0 \leq n \leq 63$. The input signal (for the FIR filter) $x[n]$ is segmented into an infinite number of possibly overlapping 128-point blocks $x_i[n]$, for i an integer and $-\infty \leq i \leq \infty$, such that

$$x_i[n] = \begin{cases} x[n], & iL \leq n \leq iL + 127, \\ 0, & \text{otherwise,} \end{cases}$$

where L is a positive integer.

Specify a method for computing

$$y_i[n] = x_i[n] * h[n]$$

for any i. Your answer should be in the form of a block diagram utilizing only the types of modules shown in Figures PP9.42-1 and PP9.42-2. A module may be used more than once or not at all.

The four modules in Figure P9.42-2 either use radix-2 FFTs to compute $X[k]$, the N-point DFT of $x[n]$, or use radix-2 inverse FFTs to compute $x[n]$ from $X[k]$.

Your specification must include the lengths of the FFTs and IFFTs used. For each "shift by n_0" module, you should also specify a value for n_0, the amount by which the input sequence is to be shifted.

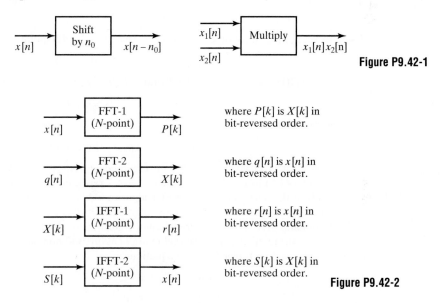

Figure P9.42-1

Figure P9.42-2

Extension Problems

9.43. In many applications (such as evaluating frequency responses and interpolation), it is of interest to compute the DFT of a short sequence that is "zero-padded." In such cases, a specialized "pruned" FFT algorithm can be used to increase the efficiency of computation (Markel, 1971). In this problem, we will consider pruning of the radix-2 decimation-in-frequency algorithm when the length of the input sequence is $M \leq 2^\mu$ and the length of the DFT is $N = 2^\nu$, where $\mu < \nu$.

(a) Draw the complete flow graph of a decimation-in-frequency radix-2 FFT algorithm for $N = 16$. Label all branches appropriately.

(b) Assume that the input sequence is of length $M = 2$; i.e., $x[n] \neq 0$ only for $N = 0$ and $N = 1$. Draw a new flow graph for $N = 16$ that shows how the nonzero input samples propagate to the output DFT; i.e., eliminate or prune all branches in the flow graph of part (a) that represent operations on zero-inputs.

(c) In part (b), all of the butterflies in the first three stages of computation should have been effectively replaced by a half-butterfly of the form shown in Figure P9.43, and in the last stage, all the butterflies should have been of the regular form. For the general case where the length of the input sequence is $M \leq 2^\mu$ and the length of the DFT is $N = 2^\nu$, where $\mu < \nu$, determine the number of stages in which the pruned butterflies can be used. Also, determine the number of complex multiplications required to compute the N-point DFT of an M-point sequence using the pruned FFT algorithm. Express your answers in terms of ν and μ.

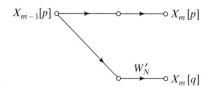

Figure P9.43

9.44. In Section 9.2, we showed that if N is divisible by 2, an N-point DFT may be expressed as

$$X[k] = G[((k))_{N/2}] + W_N^k H[((k))_{N/2}], \qquad 0 \leq k \leq N - 1. \qquad \text{(P9.44-1)}$$

where $G[k]$ is the $N/2$-point DFT of the sequence of even-indexed samples,

$$g[n] = x[2n], \qquad 0 \leq n \leq (N/2) - 1,$$

and $H[k]$ is the $N/2$-point DFT of the odd-indexed samples,

$$h[n] = x[2n + 1], \qquad 0 \leq n \leq (N/2) - 1.$$

Note that $G[k]$ and $H[k]$ must be repeated periodically for $N/2 \leq k \leq N - 1$ for Eq. (P9.44-1) to make sense. When $N = 2^\nu$, repeated application of this decomposition leads to the decimation-in-time FFT algorithm depicted for $N = 8$ in Figure 9.11. As we have seen, such algorithms require complex multiplications by the "twiddle" factors W_N^k. Rader and Brenner (1976) derived a new algorithm in which the multipliers are purely imaginary, thus requiring only two real multiplications and no real additions. In this algorithm, Eq. (P9.44-1) is replaced by the equations

$$X[0] = G[0] + F[0], \qquad \text{(P9.44-2)}$$

$$X[N/2] = G[0] - F[0], \qquad \text{(P9.44-3)}$$

$$X[k] = G[k] - \frac{1}{2}j\frac{F[k]}{\sin(2\pi k/N)}, \qquad k \neq 0, N/2. \qquad \text{(P9.44-4)}$$

Here, $F[k]$ is the $N/2$-point DFT of the sequence

$$f[n] = x[2n + 1] - x[2n - 1] + Q,$$

where

$$Q = \frac{2}{N} \sum_{n=0}^{(N/2)-1} x[2n + 1]$$

is a quantity that need be computed only once.

(a) Show that $F[0] = H[0]$ and therefore that Eqs. (P9.44-2) and (P9.44-3) give the same result as Eq. (P9.44-1) for $k = 0, N/2$.

(b) Show that

$$F[k] = H[k]W_N^k(W_N^{-k} - W_N^k)$$

for $k = 1, 2, \ldots, (N/2) - 1$. Use this result to obtain Eq. (P9.44-4). Why must we compute $X[0]$ and $X[N/2]$ using separate equations?

(c) When $N = 2^\nu$, we can apply Eqs. (P9.44-2)–(P9.44-4) repeatedly to obtain a complete decimation-in-time FFT algorithm. Determine formulas for the number of real multiplications and for the number of real additions as a function of N. In counting operations due to Eq. (P9.44-4), take advantage of any symmetries and periodicities, but do not exclude "trivial" multiplications by $\pm j/2$.

(d) Rader and Brenner (1976) state that FFT algorithms based on Eqs. (P9.44-2)–(P9.44-4) have "poor noise properties." Explain why this might be true.

9.45. A modified FFT algorithm called the *split-radix* FFT, or SRFFT, was proposed by Duhamel and Hollman (1984) and Duhamel (1986). The flow graph for the split-radix algorithm is similar to the radix-2 flow graph, but it requires fewer real multiplications. In this problem, we illustrate the principles of the SRFFT for computing the DFT $X[k]$ of a sequence $x[n]$ of length N.

(a) Show that the even-indexed terms of $X[k]$ can be expressed as the $N/2$-point DFT

$$X[2k] = \sum_{n=0}^{(N/2)-1} (x[n] + x[n + N/2])W_N^{2kn}$$

for $k = 0, 1, \ldots, (N/2) - 1$.

(b) Show that the odd-indexed terms of the DFT $X[k]$ can be expressed as the $N/4$-point DFTs

$$X[4k + 1]$$
$$= \sum_{n=0}^{(N/4)-1} \{(x[n] - x[n + N/2]) - j(x[n + N/4] - x[n + 3N/4])\}W_N^n W_N^{4kn}$$

for $k = 0, 1, \ldots, (N/4) - 1$, and

$$X[4k + 3]$$
$$= \sum_{n=0}^{(N/4)-1} \{(x[n] - x[n + N/2]) + j(x[n + N/4] - x[n + 3N/4])\}W_N^{3n} W_N^{4kn}$$

for $k = 0, 1, \ldots, (N/4) - 1$.

(c) The flow graph in Figure P9.45 represents the preceding decomposition of the DFT for a 16-point transform. Redraw this flow graph, labeling each branch with the appropriate multiplier coefficient.

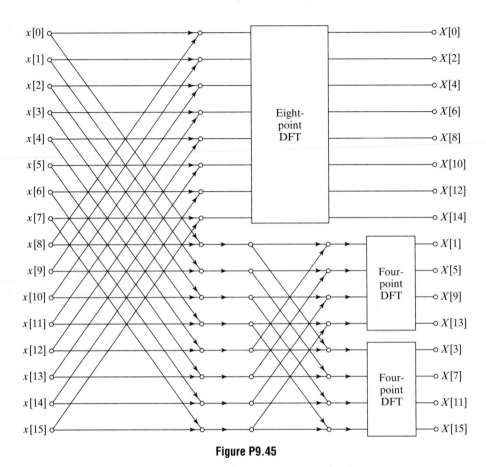

Figure P9.45

(d) Determine the number of real multiplications required to implement the 16-point transform when the SRFFT principle is applied to compute the other DFTs in Figure P9.45. Compare this number with the number of real multiplications required to implement a 16-point radix-2 decimation-in-frequency algorithm. In both cases, assume that multiplications by W_N^0 are not done.

9.46. In computing the DFT, it is necessary to multiply a complex number by another complex number whose magnitude is unity, i.e., $(X + jY)e^{j\theta}$. Clearly, such a complex multiplication changes only the angle of the complex number, leaving the magnitude unchanged. For this reason, multiplications by a complex number $e^{j\theta}$ are sometimes called *rotations*. In DFT or FFT algorithms, many different angles θ may be needed. However, it may be undesirable to store a table of all required values of $\sin\theta$ and $\cos\theta$, and computing these functions by a power series requires many multiplications and additions. With the CORDIC algorithm given by Volder (1959), the product $(X + jY)e^{j\theta}$ can be evaluated efficiently by a combination of additions, binary shifts, and table lookups from a small table.

(a) Define $\theta_i = \arctan(2^{-i})$. Show that any angle $0 < \theta < \pi/2$ can be represented as

$$\theta = \sum_{i=0}^{M-1} \alpha_i \theta_i + \epsilon = \hat{\theta} + \epsilon,$$

where $\alpha_i = \pm 1$ and the error ϵ is bounded by

$$|\epsilon| \leq \arctan(2^{-M}).$$

(b) The angles θ_i may be computed in advance and stored in a small table of length M. State an algorithm for obtaining the sequence $\{\alpha_i\}$ for $i = 0, 1, \ldots, M-1$, such that $\alpha_i = \pm 1$. Use your algorithm to determine the sequence $\{\alpha_i\}$ for representing the angle $\theta = 100\pi/512$ when $M = 11$.

(c) Using the result of part (a), show that the recursion

$$
\begin{aligned}
X_0 &= X, \\
Y_0 &= Y, \\
X_i &= X_{i-1} - \alpha_{i-1}Y_{i-1}2^{-i+1}, \qquad i = 1, 2, \ldots, M, \\
Y_i &= Y_{i-1} + \alpha_{i-1}X_{i-1}2^{-i+1}, \qquad i = 1, 2, \ldots, M,
\end{aligned}
$$

will produce the complex number

$$
(X_M + jY_M) = (X + jY)G_M e^{j\hat{\theta}},
$$

where $\hat{\theta} = \sum_{i=0}^{M-1} \alpha_i\theta_i$ and G_M is real, is positive, and does not depend on θ. That is, the original complex number is rotated in the complex plane by an angle $\hat{\theta}$ and magnified by the constant G_M.

(d) Determine the magnification constant G_M as a function of M.

9.47. In Section 9.3, we developed the decimation-in-frequency FFT algorithm for radix 2, i.e., $N = 2^\nu$. It is possible to formulate a similar algorithm for the general case of $N = m^\nu$, where m is an integer. Such an algorithm is known as a radix-m FFT algorithm. In this problem, we will examine the radix-3 decimation-in-frequency FFT for the case when $N = 9$, i.e., the input sequence $x[n] = 0$ for $n < 0$ and $n > 8$.

(a) Formulate a method of computing the DFT samples $X[3k]$ for $k = 0, 1, 2$. Consider defining $X_1[k] = X(e^{j\omega_k})|_{\omega_k = 2\pi k/3}$. How can you define a time sequence $x_1[n]$ in terms of $x[n]$ such that the 3-point DFT of $x_1[n]$ is $X_1[k] = X[3k]$?

(b) Now define a sequence $x_2[n]$ in terms of $x[n]$ such that the 3-point DFT of $x_2[n]$ is $X_2[k] = X[3k + 1]$ for $k = 0, 1, 2$. Similarly, define $x_3[n]$ such that its 3-point DFT $X_3[k] = X[3k + 2]$ for $k = 0, 1, 2$. Note that we have now defined the 9-point DFT as three 3-point DFTs from appropriately constructed 3-point sequences.

(c) Draw the signal flow graph for the $N = 3$ DFT, i.e., the radix-3 butterfly.

(d) Using the results for parts (a) and (b), sketch the signal flow graph for the system that constructs the sequences $x_1[n]$, $x_2[n]$, and $x_3[n]$, and then use 3-point DFT boxes on these sequences to produce $X[k]$ for $k = 0, \ldots, 8$. Note that in the interest of clarity, you should not draw the signal flow graph for the $N = 3$ DFTs, but simply use boxes labeled "$N = 3$ DFT." The interior of these boxes is the system you drew for part (c).

(e) Appropriate factoring of the powers of W_9 in the system you drew in part (d) allows these systems to be drawn as $N = 3$ DFTs, followed by "twiddle" factors analogous to those in the radix-2 algorithm. Redraw the system in part (d) such that it consists entirely of $N = 3$ DFTs with "twiddle" factors. This is the complete formulation of the radix-3 decimation-in-frequency FFT for $N = 9$.

(f) How many complex multiplications are required to compute a 9-point DFT using a direct implementation of the DFT equation? Contrast this with the number of complex multiplications required by the system you drew in part (e). In general, how many complex multiplications are required for the radix-3 FFT of a sequence of length $N = 3^\nu$?

9.48. Bluestein (1970) showed that if $N = M^2$, then the chirp transform algorithm has a recursive implementation.

(a) Show that the DFT can be expressed as the convolution

$$X[k] = h^*[k] \sum_{n=0}^{N-1} (x[n]h^*[n])h[k-n],$$

where * denotes complex conjugation and

$$h[n] = e^{j(\pi/N)n^2}, \qquad -\infty < n < \infty.$$

(b) Show that the desired values of $X[k]$ (i.e., for $k = 0, 1, \ldots, N-1$) can also be obtained by evaluating the convolution of part (a) for $k = N, N+1, \ldots, 2N-1$.

(c) Use the result of part (b) to show that $X[k]$ is also equal to the output of the system shown in Figure P9.48 for $k = N, N+1, \ldots, 2N-1$, where $\hat{h}[k]$ is the finite-duration sequence

$$\hat{h}[k] = \begin{cases} e^{j(\pi/N)k^2}, & 0 \le k \le 2N-1, \\ 0, & \text{otherwise.} \end{cases}$$

(d) Using the fact that $N = M^2$, show that the system function corresponding to the impulse response $\hat{h}[k]$ is

$$\hat{H}(z) = \sum_{k=0}^{2N-1} e^{j(\pi/N)k^2} z^{-k}$$

$$= \sum_{r=0}^{M-1} e^{j(\pi/N)r^2} z^{-r} \frac{1 - z^{-2M^2}}{1 + e^{j(2\pi/M)r} z^{-M}}.$$

Hint: Express k as $k = r + \ell M$.

(e) The expression for $\hat{H}(z)$ obtained in part (d) suggests a recursive realization of the FIR system. Draw the flow graph of such an implementation.

(f) Use the result of part (e) to determine the total numbers of complex multiplications and additions required to compute all of the N desired values of $X[k]$. Compare those numbers with the numbers required for direct computation of $X[k]$.

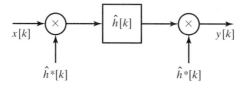

$x[k]$ $\hat{h}[k]$ $y[k]$

$\hat{h}^*[k]$ $\hat{h}^*[k]$ **Figure P9.48**

9.49. In the Goertzel algorithm for computation of the discrete Fourier transform, $X[k]$ is computed as

$$X[k] = y_k[N],$$

where $y_k[n]$ is the output of the network shown in Figure P9.49. Consider the implementation of the Goertzel algorithm using fixed-point arithmetic with rounding. Assume that the register length is B bits plus the sign, and assume that the products are rounded before additions. Also, assume that round-off noise sources are independent.

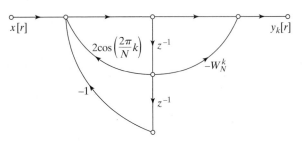

Figure P9.49

(a) Assuming that $x[n]$ is real, draw a flow graph of the linear-noise model for the finite-precision computation of the real and imaginary parts of $X[k]$. Assume that multiplication by ± 1 produces no round-off noise.

(b) Compute the variance of the round-off noise in both the real part and the imaginary part of $X[k]$.

9.50. Consider direct computation of the DFT using fixed-point arithmetic with rounding. Assume that the register length is B bits plus the sign (i.e., a total of $B+1$ bits) and that the round-off noise introduced by any real multiplication is independent of that produced by any other real multiplication. Assuming that $x[n]$ is real, determine the variance of the round-off noise in both the real part and the imaginary part of each DFT value $X[k]$.

9.51. In implementing a decimation-in-time FFT algorithm, the basic butterfly computation is

$$X_m[p] = X_{m-1}[p] + W_N^r X_{m-1}[q],$$
$$X_m[q] = X_{m-1}[p] - W_N^r X_{m-1}[q].$$

In using fixed-point arithmetic to implement the computations, it is commonly assumed that all numbers are scaled to be less than unity. Therefore, to avoid overflow, it is necessary to ensure that the real numbers that result from the butterfly computations do not exceed unity.

(a) Show that if we require

$$|X_{m-1}[p]| < \tfrac{1}{2} \quad \text{and} \quad |X_{m-1}[q]| < \tfrac{1}{2},$$

then overflow cannot occur in the butterfly computation; i.e.,

$$|\mathcal{Re}\{X_m[p]\}| < 1, \qquad |\mathcal{Im}\{X_m[p]\}| < 1,$$

and

$$|\mathcal{Re}\{X_m[q]\}| < 1, \qquad |\mathcal{Im}\{X_m[q]\}| < 1.$$

(b) In practice, it is easier and most convenient to require

$$|\mathcal{Re}\{X_{m-1}[p]\}| < \tfrac{1}{2}, \qquad |\mathcal{Im}\{X_{m-1}[p]\}| < \tfrac{1}{2},$$

and

$$|\mathcal{Re}\{X_{m-1}[q]\}| < \tfrac{1}{2}, \qquad |\mathcal{Im}\{X_{m-1}[q]\}| < \tfrac{1}{2}.$$

Are these conditions sufficient to guarantee that overflow cannot occur in the decimation-in-time butterfly computation? Explain.

9.52. In deriving formulas for the noise-to-signal ratio for the fixed-point radix-2 decimation-in-time FFT algorithm, we assumed that each output node was connected to $(N-1)$ butterfly computations, each of which contributed an amount $\sigma_B^2 = \tfrac{1}{3} \cdot 2^{-2B}$ to the output noise variance. However, when $W_N^r = \pm 1$ or $\pm j$, the multiplications can in fact be done without error. Thus, if the results derived in Section 9.7 are modified to account for this fact, we obtain a less pessimistic estimate of quantization noise effects.

(a) For the decimation-in-time algorithm discussed in Section 9.7, determine, for each stage, the number of butterflies that involve multiplication by either ± 1 or $\pm j$.

(b) Use the result of part (a) to find improved estimates of the output noise variance, Eq. (9.58), and noise-to-signal ratio, Eq. (9.68), for odd values of k. Discuss how these estimates are different for even values of k. Do not attempt to find a closed form expression of these quantities for even values of k.

(c) Repeat parts (a) and (b) for the case where the output of each stage is attenuated by a factor of $\frac{1}{2}$; i.e., derive modified expressions corresponding to Eq. (9.71) for the output noise variance and Eq. (9.72) for the output noise-to-signal ratio, assuming that multiplications by ± 1 and $\pm j$ do not introduce error.

9.53. In Section 9.7 we considered a noise analysis of the decimation-in-time FFT algorithm of Figure 9.11. Carry out a similar analysis for the decimation-in-frequency algorithm of Figure 9.22, obtaining equations for the output noise variance and noise-to-signal ratio for scaling at the input and also for scaling by $\frac{1}{2}$ at each stage of computation.

9.54. In this problem, we consider a procedure for computing the DFT of four real symmetric or antisymmetric N-point sequences using only one N-point DFT computation. Since we are considering only finite-length sequences, by *symmetric* and *antisymmetric,* we explicitly mean *periodic symmetric* and *periodic antisymmetric,* as defined in Section 8.6.4. Let $x_1[n]$, $x_2[n]$, $x_3[n]$, and $x_4[n]$ denote the four real sequences of length N, and let $X_1[k]$, $X_2[k]$, $X_3[k]$, and $X_4[k]$ denote the corresponding DFTs. We assume first that $x_1[n]$ and $x_2[n]$ are symmetric and $x_3[n]$ and $x_4[n]$ are antisymmetric; i.e.,

$$x_1[n] = x_1[N-n], \qquad x_2[n] = x_2[N-n],$$

$$x_3[n] = -x_3[N-n], \qquad x_4[n] = -x_4[N-n],$$

for $n = 1, 2, \ldots, N-1$ and $x_3[0] = x_4[0] = 0$.

(a) Define $y_1[n] = x_1[n] + x_3[n]$ and let $Y_1[k]$ denote the DFT of $y_1[n]$. Determine how $X_1[k]$ and $X_2[k]$ can be recovered from $Y_1[k]$.

(b) $y_1[n]$ as defined in part (a) is a real sequence with symmetric part $x_1[n]$ and antisymmetric part $x_3[n]$. Similarly, we define the real sequence $y_2[n] = x_2[n] + x_4[n]$, and we let $y_3[n]$ be the complex sequence

$$y_3[n] = y_1[n] + jy_2[n].$$

First, determine how $Y_1[k]$ and $Y_2[k]$ can be determined from $Y_3[k]$, and then, using the results of part (a), show how to obtain $X_1[k]$, $X_2[k]$, $X_3[k]$, and $X_4[k]$ from $Y_3[k]$.

The result of part (b) shows that we can compute the DFT of four real sequences simultaneously with only one N-point DFT computation if two sequences are symmetric and the other two are antisymmetric. Now consider the case when all four are symmetric; i.e.,

$$x_i[n] = x_i[N-n], \qquad i = 1, 2, 3, 4,$$

for $n = 0, 1, \ldots, N-1$. For parts (c)–(f), assume $x_3[n]$ and $x_4[n]$ are real and symmetric, not antisymmetric.

(c) Consider a real symmetric sequence $x_3[n]$. Show that the sequence

$$u_3[n] = x_3[((n+1))_N] - x_3[((n-1))_N]$$

is an antisymmetric sequence; i.e., $u_3[n] = -u_3[N-n]$ for $n = 1, 2, \ldots, N-1$ and $u_3[0] = 0$.

(d) Let $U_3[k]$ denote the N-point DFT of $u_3[n]$. Determine an expression for $U_3[k]$ in terms of $X_3[k]$.

(e) By using the procedure of part (c), we can form the real sequence $y_1[n] = x_1[n] + u_3[n]$, where $x_1[n]$ is the symmetric part and $u_3[n]$ is the antisymmetric part of $y_1[n]$. Determine how $X_1[k]$ and $X_3[k]$ can be recovered from $Y_1[k]$.

(f) Now let $y_3[n] = y_1[n] + jy_2[n]$, where

$$y_1[n] = x_1[n] + u_3[n], \qquad y_2[n] = x_2[n] + u_4[n],$$

with

$$u_3[n] = x_3[((n+1))_N] - x_3[((n-1))_N],$$

$$u_4[n] = x_4[((n+1))_N] - x_4[((n-1))_N],$$

for $n = 0, 1, \ldots, N-1$. Determine how to obtain $X_1[k]$, $X_2[k]$, $X_3[k]$, and $X_4[k]$ from $Y_3[k]$. (Note that $X_3[0]$ and $X_4[0]$ cannot be recovered from $Y_3[k]$, and if N is even, $X_3[N/2]$ and $X_4[N/2]$ also cannot be recovered from $Y_3[k]$.)

9.55. The input and output of a linear time-invariant system satisfy a difference equation of the form

$$y[n] = \sum_{k=1}^{N} a_k y[n-k] + \sum_{k=0}^{M} b_k x[n-k].$$

Assume that an FFT program is available for computing the DFT of any finite-length sequence of length $L = 2^\nu$. Describe a procedure that utilizes the available FFT program to compute

$$H(e^{j(2\pi/512)k}) \qquad \text{for } k = 0, 1, \ldots, 511,$$

where $H(z)$ is the system function of the system.

9.56. Suppose that we wish to multiply two very large numbers (possibly thousands of bits long) on a 16-bit computer. In this problem, we will investigate a technique for doing this using FFTs.

(a) Let $p(x)$ and $q(x)$ be the two polynomials

$$p(x) = \sum_{i=0}^{L-1} a_i x^i, \qquad q(x) = \sum_{i=0}^{M-1} b_i x^i.$$

Show that the coefficients of the polynomial $r(x) = p(x)q(x)$ can be computed using circular convolution.

(b) Show how to compute the coefficients of $r(x)$ using a radix-2 FFT program. For what orders of magnitude of $(L + M)$ is this procedure more efficient than direct computation? Assume that $L + M = 2^\nu$ for some integer ν.

(c) Now suppose that we wish to compute the product of two very long positive binary integers u and v. Show that their product can be computed using polynomial multiplication, and describe an algorithm for computing the product using an FFT algorithm. If u is an 8000-bit number and v is a 1000-bit number, approximately how many real multiplications and additions are required to compute the product $u \cdot v$ using this method?

(d) Give a qualitative discussion of the effect of finite-precision arithmetic in implementing the algorithm of part (c).

9.57. The discrete Hartley transform (DHT) of a sequence $x[n]$ of length N is defined as

$$X_H[k] = \sum_{n=0}^{N-1} x[n]H_N[nk], \qquad k = 0, 1, \ldots, N-1,$$

where

$$H_N[a] = C_N[a] + S_N[a],$$

with

$$C_N[a] = \cos(2\pi a/N), \qquad S_N[a] = \sin(2\pi a/N).$$

Problem 8.68 explores the properties of the discrete Hartley transform in detail, particularly its circular convolution property.

(a) Verify that $H_N[a] = H_N[a + N]$, and verify the following useful property of $H_N[a]$:

$$H_N[a + b] = H_N[a]C_N[b] + H_N[-a]S_N[b]$$
$$= H_N[b]C_N[a] + H_N[-b]S_N[a].$$

(b) By decomposing $x[n]$ into its even-numbered points and odd-numbered points, and by using the identity derived in part (a), derive a fast DHT algorithm based on the decimation-in-time principle.

9.58. In this problem, we will write the FFT as a sequence of matrix operations. Consider the 8-point decimation-in-time FFT algorithm shown in Figure P9.58. Let a and f denote the input and output vectors, respectively. Assume that the input is in bit-reversed order and that the output is in normal order (compare with Figure 9.11). Let b, c, d, and e denote the intermediate vectors shown on the flow graph.

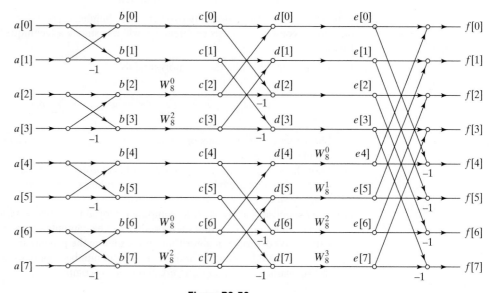

Figure P9.58

(a) Determine the matrices $F_1, T_1, F_2, T_2,$ and F_3 such that

$$b = F_1 a,$$

$$c = T_1 b,$$

$$d = F_2 c,$$

$$e = T_2 d,$$

$$f = F_3 e.$$

(b) The overall FFT, taking input a and yielding output f can be described in matrix notation as $f = Qa$, where

$$Q = F_3 T_2 F_2 T_1 F_1.$$

Let Q^H be the complex (Hermitian) transpose of the matrix Q. Draw the flow graph for the sequence of operations described by Q^H. What does this structure compute?

(c) Determine $(1/N)Q^H Q$.

9.59. In many applications, there is a need to convolve long sequences $x[n]$ and $h[n]$ whose samples are integers. Since the sequences have integer coefficients, the result of the convolution $y[n] = x[n] * h[n]$ will naturally also have integer coefficients as well.

A major drawback of computing the convolution of integer sequences with FFTs is that floating-point arithmetic chips are more expensive than integer arithmetic chips. Also, rounding noise introduced during a floating-point computation may corrupt the result. In this problem, we consider a class of FFT algorithms known as *number-theoretic transforms* (NTTs), which overcome these drawbacks.

(a) Let $x[n]$ and $h[n]$ be N-point sequences and denote their DFTs by $X[k]$ and $H[k]$, respectively. Derive the circular convolution property of the DFT. Specifically, show that $Y[k] = X[k]H[k]$, where $y[n]$ is the N-point circular convolution of $x[n]$ and $h[n]$. Show that the circular convolution property holds as long as W_N in the DFT satisfies

$$\sum_{n=0}^{N-1} W_N^{nk} = \begin{cases} N, & k = 0, \\ 0, & k \neq 0. \end{cases} \tag{P9.59-1}$$

The key to defining NTTs is to find an integer-valued W_N that satisfies Eq. (P9.59-1). This will enforce the orthogonality of the basis vectors required for the DFT to function properly. Unfortunately, no integer-valued W_N exists that has this property for standard integer arithmetic.

In order to overcome this problem, NTTs use integer arithmetic defined modulo some integer P. Throughout the current problem, we will assume that $P = 17$. That is, addition and multiplication are defined as standard integer addition and multiplication, followed by modulo $P = 17$ reduction. For example, $((23+18))_{17} = 7, ((10+7))_{17} = 0,$ $((23 \times 18))_{17} = 6,$ and $((10 \times 7))_{17} = 2.$ (Just compute the sum or product in the normal way, and then take the remainder modulo 17.)

(b) Let $P = 17, N = 4,$ and $W_N = 4.$ Verify that

$$\left(\left(\sum_{n=0}^{N-1} W_N^{nk} \right) \right)_P = \begin{cases} N, & k = 0, \\ 0, & k \neq 0. \end{cases}$$

(c) Let $x[n]$ and $h[n]$ be the sequences

$$x[n] = \delta[n] + 2\delta[n-1] + 3\delta[n-2],$$

$$h[n] = 3\delta[n] + \delta[n-1].$$

Compute the 4-point NTT $X[k]$ of $x[n]$ as follows:

$$X[k] = \left(\left(\sum_{n=0}^{N-1} x[n]W_N^{nk}\right)\right)_P.$$

Compute $H[k]$ in a similar fashion. Also, compute $Y[k] = H[k]X[k]$. Assume the values of P, N, and W_N given in part (a). *Be sure to use modulo 17 arithmetic for each operation throughout the computation, not just for the final result!*

(d) The inverse NTT of $Y[k]$ is defined by the equation

$$y[n] = \left(\left((N)^{-1}\sum_{k=0}^{N-1} Y[k]W_N^{-nk}\right)\right)_P. \qquad (P9.59\text{-}2)$$

In order to compute this quantity properly, we must determine the *integers* $(1/N)^{-1}$ and W_N^{-1} such that

$$\left(\left((N)^{-1}N\right)\right)_P = 1,$$

$$\left(\left(W_N W_N^{-1}\right)\right)_P = 1.$$

Use the values of P, N, and W_N given in part (a), and determine the aforesaid integers.

(e) Compute the inverse NTT shown in Eq. (P9.59-2) using the values of $(N)^{-1}$ and W_N^{-1} determined in part (d). Check your result by manually computing the convolution $y[n] = x[n] * h[n]$.

9.60. Sections 9.2 and 9.3 focus on the fast Fourier transform for sequences where N is a power of 2. However, it is also possible to find efficient algorithms to compute the DFT when the length N has more than one prime factor, i.e., cannot be expressed as $N = m^{\nu}$ for some integer m. In this problem, you will examine the case where $N = 6$. The techniques described extend easily to other composite numbers. Burrus and Parks (1985) discuss such algorithms in more detail.

(a) The key to decomposing the FFT for $N = 6$ is to use the concept of an *index map*, proposed by Cooley and Tukey (1965) in their original paper on the FFT. Specifically, for the case of $N = 6$, we will represent the indices n and k as

$$n = 3n_1 + n_2 \quad \text{for } n_1 = 0, 1; n_2 = 0, 1, 2; \qquad (P9.60\text{-}1)$$

$$k = k_1 + 2k_2 \quad \text{for } k_1 = 0, 1; k_2 = 0, 1, 2; \qquad (P9.60\text{-}2)$$

Verify that using each possible value of n_1 and n_2 produces each value of $n = 0, \ldots, 5$ once and only once. Demonstrate that the same holds for k with each choice of k_1 and k_2.

(b) Substitute Eqs. (P9.60-1) and (P9.60-2) into the definition of the DFT to get a new expression for the DFT in terms of n_1, n_2, k_1, and k_2. The resulting equation should have a double summation over n_1 and n_2 instead of a single summation over n.

(c) Examine the W_6 terms in your equation carefully. You can rewrite some of these as equivalent expressions in W_2 and W_3.

(d) Based on part (c), group the terms in your DFT such that the n_2 summation is outside and the n_1 summation is inside. You should be able to write this expression so that it can be interpreted as three DFTs with $N = 2$, followed by some "twiddle" factors (powers of W_6), followed by two $N = 3$ DFTs.

(e) Draw the signal flow graph implementing your expression from part (d). How many complex multiplications does this require? How does this compare with the number of complex multiplications required by a direct implementation of the DFT equation for $N = 6$?

(f) Find an alternative indexing similar to Eqs. (P9.60-1) and (P9.60-2) that results in a signal flow graph that is two $N = 3$ DFTs followed by three $N = 2$ DFTs.

10
Fourier Analysis
of Signals Using the
Discrete Fourier Transform

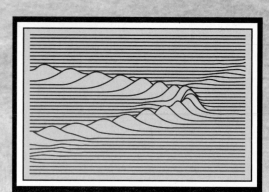

10.0 INTRODUCTION

In Chapter 8, we developed the discrete Fourier transform (DFT) as a Fourier representation of finite-length signals. Because the DFT can be computed efficiently, it plays a central role in a wide variety of signal-processing applications, including filtering and spectrum analysis. In this chapter, we take an introductory look at Fourier analysis of signals using the DFT.

In applications and algorithms based on explicit evaluation of the Fourier transform, it is ideally the discrete-time Fourier transform (DTFT) that is desired, although it is the DFT that can actually be computed. For finite-length signals, the DFT provides frequency-domain samples of the DTFT, and the implications of this sampling must be clearly understood and accounted for. For example, as considered in Section 8.7, in linear filtering or convolution implemented by multiplying DFTs rather than DTFTs, a circular convolution is implemented, and special care must be taken to ensure that the results will be equivalent to a linear convolution. In addition, in many filtering and spectrum analysis applications, the signals do not inherently have finite length. As we will discuss, this inconsistency between the finite-length requirement of the DFT and the reality of indefinitely long signals can be accommodated exactly or approximately through the concepts of *windowing, block processing*, and the *time-dependent Fourier transform*.

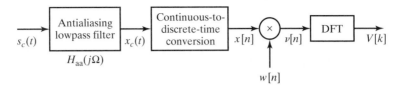

Figure 10.1 Processing steps in the discrete-time Fourier analysis of a continuous-time signal.

10.1 FOURIER ANALYSIS OF SIGNALS USING THE DFT

One of the major applications of the DFT is in analyzing the frequency content of continuous-time signals. For example, as we describe in Section 10.4.1, in speech analysis and processing, frequency analysis is particularly useful in identifying and modeling the resonances of the vocal cavity. Another example, introduced in Section 10.4.2, is Doppler radar, in which the velocity of a target is represented by the frequency shift between the transmitted and received signals.

The basic steps in applying the DFT to continuous-time signals are indicated in Figure 10.1. The antialiasing filter is incorporated to eliminate or minimize the effect of aliasing when the continuous-time signal is converted to a sequence by sampling. The need for multiplication of $x[n]$ by $w[n]$, i.e., windowing, is a consequence of the finite-length requirement of the DFT. In many cases of practical interest, $s_c(t)$ and, consequently, $x[n]$ are very long or even indefinitely long signals (such as with speech or music). Therefore, a finite-duration window $w[n]$ is applied to $x[n]$ prior to computation of the DFT. Figure 10.2 illustrates the Fourier transforms of the signals in Figure 10.1. Figure 10.2(a) shows a continuous-time spectrum that tapers off at high frequencies but is not bandlimited. It also indicates the presence of some narrowband signal energy, represented by the narrow peaks. The frequency response of an antialiasing filter is illustrated in Figure 10.2(b). As indicated in Figure 10.2(c), the resulting continuous-time Fourier transform $X_c(j\Omega)$ contains little useful information about $S_c(j\Omega)$ for frequencies above the cutoff frequency of the filter. Since $H_{aa}(j\Omega)$ cannot be ideal, the Fourier components of the input in the passband and the transition band also will be modified by the frequency response of the filter.

The conversion of $x_c(t)$ to the sequence of samples $x[n]$ is represented in the frequency domain by periodic replication, frequency normalization, and amplitude scaling i.e.,

$$X(e^{j\omega}) = \frac{1}{T} \sum_{r=-\infty}^{\infty} X_c\left(j\frac{\omega}{T} + j\frac{2\pi r}{T}\right). \tag{10.1}$$

This is illustrated in Figure 10.2(d). In a practical implementation, the antialiasing filter cannot have infinite attenuation in the stopband. Therefore, some nonzero overlap of the terms in Eq. (10.1), i.e., aliasing, can be expected; however, this source of error can be made negligibly small either with a high-quality continuous-time filter or through the use of initial oversampling followed by more effective discrete-time lowpass filtering and decimation, as discussed in Section 4.8.1. If $x[n]$ is a digital signal, so that A/D conversion is incorporated in the second system in Figure 10.1, then quantization error is also introduced. As we have seen in Section 4.8.2, this error can be modeled as a

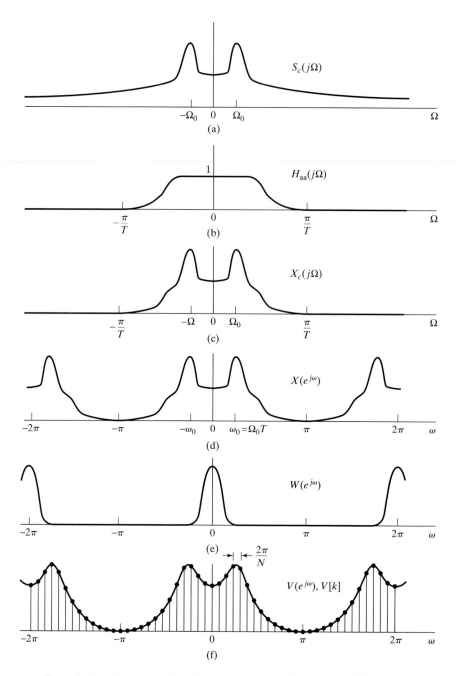

Figure 10.2 Illustration of the Fourier transforms of the system of Figure 10.1. (a) Fourier transform of continuous-time input signal. (b) Frequency response of antialiasing filter. (c) Fourier transform of output of antialiasing filter. (d) Fourier transform of sampled signal. (e) Fourier transform of window sequence. (f) Fourier transform of windowed signal segment and frequency samples obtained using DFT samples.

noise sequence added to $x[n]$. The noise can be made negligible through the use of fine-grained quantization.

The sequence $x[n]$ is typically multiplied by a finite-duration window $w[n]$, since the input to the DFT must be of finite duration. This produces the finite-length sequence $v[n] = w[n]x[n]$. The effect in the frequency domain is a periodic convolution, i.e.,

$$V(e^{j\omega}) = \frac{1}{2\pi} \int_{-\pi}^{\pi} X(e^{j\theta}) W(e^{j(\omega-\theta)}) d\theta. \tag{10.2}$$

Figure 10.2(e) illustrates the Fourier transform of a typical window sequence. Note that the main lobe is assumed to be concentrated around $\omega = 0$, and, in this illustration, the side lobes are very small, suggesting that the window tapers at its edges. The properties of windows such as the Bartlett, Hamming, Hanning, Blackman, and Kaiser windows are discussed in Chapter 7 and in Section 10.2. At this point, it is sufficient to observe that convolution of $W(e^{j\omega})$ with $X(e^{j\omega})$ will tend to smooth sharp peaks and discontinuities in $X(e^{j\omega})$. This is depicted by the continuous curve plotted in Figure 10.2(f).

The final operation in Figure 10.1 is the computation of the DFT. The DFT of the windowed sequence $v[n] = w[n]x[n]$ is

$$V[k] = \sum_{n=0}^{N-1} v[n]e^{-j(2\pi/N)kn}, \qquad k = 0, 1, \ldots, N-1, \tag{10.3}$$

where we assume that the window length L is less than or equal to the DFT length N. $V[k]$, the DFT of the finite-length sequence $v[n]$, corresponds to equally spaced samples of the DTFT of $v[n]$; i.e.,

$$V[k] = V(e^{j\omega})\big|_{\omega=2\pi k/N}. \tag{10.4}$$

Figure 10.2(f) also shows $V[k]$ as the samples of $V(e^{j\omega})$. Since the spacing between DFT frequencies is $2\pi/N$, and the relationship between the normalized discrete-time frequency variable and the continuous-time frequency variable is $\omega = \Omega T$, the DFT frequencies correspond to the continuous-time frequencies

$$\Omega_k = \frac{2\pi k}{NT}. \tag{10.5}$$

The use of this relationship between continuous-time frequencies and DFT frequencies is illustrated by Examples 10.1 and 10.2.

Example 10.1   Fourier Analysis Using the DFT

Consider a bandlimited continuous-time signal $x_c(t)$ such that $X_c(j\Omega)=0$ for $|\Omega| \geq 2\pi(2500)$. We wish to use the system of Figure 10.1 to estimate the continuous-time spectrum $X_c(j\Omega)$. Assume that the antialiasing filter $H_{aa}(j\Omega)$ is ideal, and the sampling rate for the C/D converter is $1/T = 5000$ samples/s. If we want the DFT samples $V[k]$ to be equivalent to samples of $X_c(j\Omega)$ that are at most $2\pi(10)$ rad/s or 10 Hz apart, what is the minimum value that we should use for the DFT size N?

From Eq. (10.5), we see that adjacent samples in the DFT correspond to continuous-time frequencies separated by $2\pi/(NT)$. Therefore, we require that

$$\frac{2\pi}{NT} \leq 20\pi,$$

which implies that

$$N \geq 500$$

satisfies the condition. If we wish to use a radix-2 FFT algorithm to compute the DFT in Figure 10.1, we would choose $N = 512$ for an equivalent continuous-time frequency spacing of $\Delta\Omega = 2\pi(5000/512) = 2\pi(9.77)$ rad/s.

Example 10.2　Relationship Between DFT Values

Consider the problem posed in Example 10.1, in which $1/T = 5000$, $N = 512$, and $x_c(t)$ is real-valued and is sufficiently bandlimited to avoid aliasing with the given sampling rate. If it is determined that $V[11] = 2000(1 + j)$, what can be said about other values of $V[k]$ or about $X_c(j\Omega)$?

Referring to the symmetry properties of the DFT given in Table 8.2, $V[k] = V^*[((-k))_N]$, $k = 0, 1, \ldots, N - 1$, and consequently, $V[N - k] = V^*[k]$, so it follows in this case that

$$V[512 - 11] = V[501] = V^*[11] = 2000(1 - j).$$

We also know that the DFT sample $k = 11$ corresponds to the continuous-time frequency $\Omega_{11} = 2\pi(11)(5000)/512 = 2\pi(107.4)$, and similarly, $k = 501$ corresponds to the frequency $-2\pi(11)(5000)/512 = -2\pi(107.4)$. Although windowing smooths the spectrum, we can say that

$$X_c(j\Omega_{11}) = X_c(j2\pi(107.4)) \approx T \cdot V[11] = 0.4(1 + j).$$

Note that the factor T is required to compensate for the factor $1/T$ introduced by sampling, as in Eq. (10.1). We can again exploit symmetry to conclude that

$$X_c(-j\Omega_{11}) = X_c(-j2\pi(107.4)) \approx T \cdot V^*[11] = 0.4(1 - j).$$

Many commercial real-time spectrum analyzers are based on the principles embodied in Figures 10.1 and 10.2. It should be clear from the preceding discussion, however, that numerous factors affect the interpretation the DFT of a windowed segment of the sampled signal in terms of the continuous-time Fourier transform of the original input $s_c(t)$. To accommodate and mitigate the effects of these factors, care must be taken in filtering and sampling the input signal. Furthermore, to interpret the results correctly, the effects of the time-domain windowing and of the frequency-domain sampling inherent in the DFT must be clearly understood. For the remainder of the discussion, we will assume that the issues of antialiasing filtering and continuous-to-discrete-time conversion have been satisfactorily handled and are negligible. In the next section, we concentrate specifically on the effects of windowing and of the frequency-domain sampling imposed by the DFT. We choose sinusoidal signals as the specific class of examples to discuss, because sinusoids are perfectly bandlimited and they are easily computed. However, most of the issues raised by the examples apply more generally.

10.2 DFT ANALYSIS OF SINUSOIDAL SIGNALS

The DTFT of a sinusoidal signal $A \cos(\omega_0 n + \phi)$ (existing for all n) is a pair of impulses at $+\omega_0$ and $-\omega_0$ (repeating periodically with period 2π). In analyzing sinusoidal signals using the DFT, windowing and spectral (frequency-domain) sampling have important effects. As we will see in Section 10.2.1, windowing smears or broadens the impulses of the Fourier representation, thus, the exact frequency is less sharply defined. Windowing also reduces the ability to resolve sinusoidal signals that are close together in frequency. The spectral sampling inherent in the DFT has the effect of potentially giving a misleading or inaccurate picture of the true spectrum of the sinusoidal signal. This effect is discussed in Section 10.2.3.

10.2.1 The Effect of Windowing

Consider a continuous-time signal consisting of the sum of two sinusoidal components; i.e.,

$$s_c(t) = A_0 \cos(\Omega_0 t + \theta_0) + A_1 \cos(\Omega_1 t + \theta_1), \qquad -\infty < t < \infty. \tag{10.6}$$

Assuming ideal sampling with no aliasing and no quantization error, we obtain the discrete-time signal

$$x[n] = A_0 \cos(\omega_0 n + \theta_0) + A_1 \cos(\omega_1 n + \theta_1), \qquad -\infty < n < \infty, \tag{10.7}$$

where $\omega_0 = \Omega_0 T$ and $\omega_1 = \Omega_1 T$. The windowed sequence $v[n]$ in Figure 10.1 is then

$$v[n] = A_0 w[n] \cos(\omega_0 n + \theta_0) + A_1 w[n] \cos(\omega_1 n + \theta_1). \tag{10.8}$$

To obtain the DTFT of $v[n]$, we can expand Eq. (10.8) in terms of complex exponentials and use the frequency-shifting property of Eq. (2.158) in Section 2.9.2. Specifically, we rewrite $v[n]$ as

$$v[n] = \frac{A_0}{2} w[n] e^{j\theta_0} e^{j\omega_0 n} + \frac{A_0}{2} w[n] e^{-j\theta_0} e^{-j\omega_0 n}$$
$$+ \frac{A_1}{2} w[n] e^{j\theta_1} e^{j\omega_1 n} + \frac{A_1}{2} w[n] e^{-j\theta_1} e^{-j\omega_1 n}, \tag{10.9}$$

from which, with Eq. (2.158), it follows that the Fourier transform of the windowed sequence is

$$V(e^{j\omega}) = \frac{A_0}{2} e^{j\theta_0} W(e^{j(\omega-\omega_0)}) + \frac{A_0}{2} e^{-j\theta_0} W(e^{j(\omega+\omega_0)})$$
$$+ \frac{A_1}{2} e^{j\theta_1} W(e^{j(\omega-\omega_1)}) + \frac{A_1}{2} e^{-j\theta_1} W(e^{j(\omega+\omega_1)}). \tag{10.10}$$

According to Eq. (10.10), the Fourier transform of the windowed signal consists of the Fourier transform of the window, shifted to the frequencies $\pm\omega_0$ and $\pm\omega_1$ and scaled by the complex amplitudes of the individual complex exponentials that make up the signal.

Example 10.3   Effect of Windowing on Fourier Analysis of Sinusoidal Signals

In this example, we consider the system of Figure 10.1 and, in particular, $W(e^{j\omega})$ and $V(e^{j\omega})$ for $s_c(t)$ of the form of Eq. (10.6), a sampling rate $1/T = 10$ kHz and a rectangular window $w[n]$ of length 64. The signal amplitude and phase parameters are $A_0 = 1$, $A_1 = 0.75$, and $\theta_0 = \theta_1 = 0$, respectively. To illustrate the essential features, we specifically display only the magnitudes of the Fourier transforms.

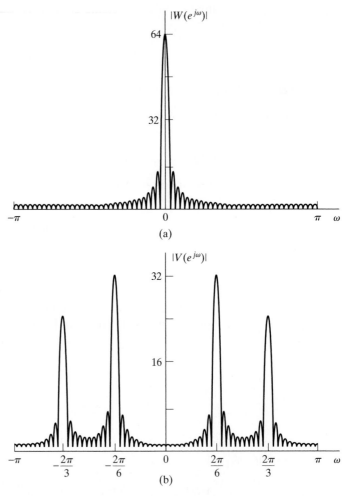

Figure 10.3   Illustration of Fourier analysis of windowed cosines with a rectangular window. (a) Fourier transform of window. (b)–(e) Fourier transform of windowed cosines as $\Omega_1 - \Omega_0$ becomes progressively smaller. (b) $\Omega_0 = (2\pi/6) \times 10^4$, $\Omega_1 = (2\pi/3) \times 10^4$.

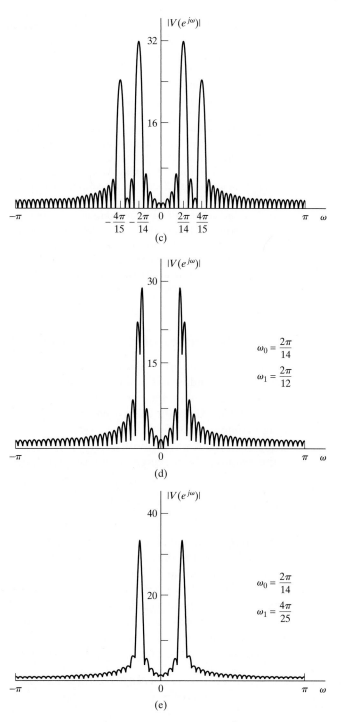

Figure 10.3 (*continued*) (c) $\Omega_0 = (2\pi/14) \times 10^4$, $\Omega_1 = (4\pi/15) \times 10^4$. (d) $\Omega_0 = (2\pi/14) \times 10^4$, $\Omega_1 = (2\pi/12) \times 10^4$. (e) $\Omega_0 = (2\pi/14) \times 10^4$, $\Omega_1 = (4\pi/25) \times 10^4$.

In Figure 10.3(a), we show $|W(e^{j\omega})|$, and in Figures 10.3(b), (c), (d), and (e), we show $|V(e^{j\omega})|$ for several choices of Ω_0 and Ω_1 in Eq. (10.6) or, equivalently, ω_0 and ω_1 in Eq. (10.7). In Figure 10.3(b), $\Omega_0 = (2\pi/6) \times 10^4$ and $\Omega_1 = (2\pi/3) \times 10^4$, or, equivalently, $\omega_0 = 2\pi/6$ and $\omega_1 = 2\pi/3$. In Figure 10.3(c)–(e), the frequencies become progressively closer. For the parameters in Figure 10.3(b), the frequency and amplitude of the individual components are evident. Specifically, Eq. (10.10) suggests that, with no overlap between the replicas of $W(e^{j\omega})$ at ω_0 and ω_1, there will be a peak of height $32A_0$ at ω_0 and $32A_1$ at ω_1, since $W(e^{j\omega})$ has a peak height of 64. In Figure 10.3(b), the two peaks are at approximately $\omega_0 = 2\pi/6$ and $\omega_1 = 2\pi/3$, with peak amplitudes in the correct ratio. In Figure 10.3(c), there is more overlap between the window replicas at ω_0 and ω_1, and while two distinct peaks are present, the amplitude of the spectrum at $\omega = \omega_0$ is affected by the amplitude of the sinusoidal signal at frequency ω_1 and vice versa. This interaction is called *leakage*: The component at one frequency leaks into the vicinity of another component owing to the spectral smearing introduced by the window. Figure 10.3(d) shows the case where the leakage is even greater. Notice how side lobes adding out of phase can *reduce* the heights of the peaks. In Figure 10.3(e), the overlap between the spectrum windows at ω_0 and ω_1 is so significant that the two peaks visible in (b)–(d) have merged into one. In other words, with this window, the two frequencies corresponding to Figure 10.3(e) will not be resolved in the spectrum.

10.2.2 Properties of the Windows

Reduced resolution and leakage are the two primary effects on the spectrum as a result of applying a window to the sinusoidal signal. The resolution is influenced primarily by the width of the main lobe of $W(e^{j\omega})$, whereas the degree of leakage depends on the relative amplitude of the main lobe to the side lobes of $W(e^{j\omega})$. In Chapter 7, in a filter design context, we showed that the width of the main lobe and the relative side-lobe amplitude depend primarily on the window length L and the shape (amount of tapering) of the window. The rectangular window, which has Fourier transform

$$W_r(e^{j\omega}) = \sum_{n=0}^{L-1} e^{-j\omega n} = e^{-j\omega(L-1)/2} \frac{\sin(\omega L/2)}{\sin(\omega/2)}, \tag{10.11}$$

has the narrowest main lobe for a given length ($\Delta_{ml} = 4\pi/L$), but it has the largest side lobes of all the commonly used windows. Other windows discussed in Chapter 7 include the Bartlett, Hann, and Hamming windows. The DTFTs of all these windows have main-lobe width $\Delta_{ml} = 8\pi/(L-1)$, which is approximately twice that of the rectangular window, but they have significantly smaller side-lobe amplitudes. The problem with all these windows is that there is no possibility of trade-off between main-lobe width and side-lobe amplitude, since the window length is the only variable parameter.

As we saw in Chapter 7, the Kaiser window is defined by

$$w_K[n] = \begin{cases} \dfrac{I_0[\beta(1 - [(n-\alpha)/\alpha]^2)^{1/2}]}{I_0(\beta)}, & 0 \le n \le L-1, \\ 0, & \text{otherwise,} \end{cases} \tag{10.12}$$

where $\alpha = (L-1)/2$ and $I_0(\cdot)$ is the zeroth-order modified Bessel function of the first kind. (Note that the notation of Eq. (10.12) differs slightly from that of Eq. (7.72) in that L denotes the length of the window in Eq. (10.12), whereas the length of the filter design window in Eq. (7.72) is denoted $M + 1$.) We have already seen in the context of the filter design problem that this window has two parameters, β and L, which can be used to trade between main-lobe width and relative side-lobe amplitude. (Recall that the Kaiser window reduces to the rectangular window when $\beta = 0$.) The main-lobe width Δ_{ml} is defined as the symmetric distance between the central zero-crossings. The relative side-lobe level A_{sl} is defined as the ratio in dB of the amplitude of the main lobe to the amplitude of the largest side lobe. Figure 10.4, which is a duplicate of Figure 7.32, shows Fourier transforms of Kaiser windows for different lengths and different values of β. In designing a Kaiser window for spectrum analysis, we want to specify a desired value of A_{sl} and determine the required value of β. Figure 10.4(c) shows that the relative side-lobe amplitude is essentially independent of the window length and thus depends only on β. This was confirmed by Kaiser and Schafer (1980), who obtained the following least squares approximation to β as a function of A_{sl}:

$$\beta = \begin{cases} 0, & A_{sl} \leq 13.26, \\ 0.76609(A_{sl} - 13.26)^{0.4} + 0.09834(A_{sl} - 13.26), & 13.26 < A_{sl} \leq 60, \\ 0.12438(A_{sl} + 6.3), & 60 < A_{sl} \leq 120. \end{cases} \quad (10.13)$$

Using values of β from Eq. (10.13) gives windows with actual side-lobe values that differ by less than 0.36 from the value of A_{sl} used in Eq. (10.13) for the entire range of $13.26 < A_{sl} < 120$. (Note that the value 13.26 is the relative side-lobe amplitude of the rectangular window, to which the Kaiser window reduces for $\beta = 0$.)

Figure 10.4(c) also shows that the main-lobe width is inversely proportional to the length of the window. The trade-off between main-lobe width, relative side-lobe amplitude, and window length is displayed by the approximate relationship

$$L \simeq \frac{24\pi(A_{sl} + 12)}{155\Delta_{ml}} + 1, \quad (10.14)$$

which was also given by Kaiser and Schafer (1980).

Equations (10.12), (10.13), and (10.14) are the necessary equations for determining a Kaiser window with desired values of main-lobe width and relative side-lobe amplitude. To design a window for prescribed values of A_{sl} and Δ_{ml} requires simply the computation of β from Eq. (10.13), the computation of L from Eq. (10.14), and the computation of the window using Eq. (10.12). Many of the remaining examples of this chapter use the Kaiser window. Other spectrum analysis windows are considered by Harris (1978).

10.2.3 The Effect of Spectral Sampling

As mentioned previously, the DFT of the windowed sequence $v[n]$ provides samples of $V(e^{j\omega})$ at the N equally spaced discrete-time frequencies $\omega_k = 2\pi k/N, k = 0, 1, \ldots, N - 1$. These are equivalent to the continuous-time frequencies $\Omega_k = (2\pi k)/(NT)$, for $k = 0, 1, \ldots, N/2$ (assuming that N is even). The indices $k = N/2 + 1, \ldots, N - 1$ correspond to the negative continuous-time frequencies $-2\pi(N - k)/(NT)$. Spectral sampling, as imposed by the DFT, can sometimes produce misleading results. This effect is best illustrated by example.

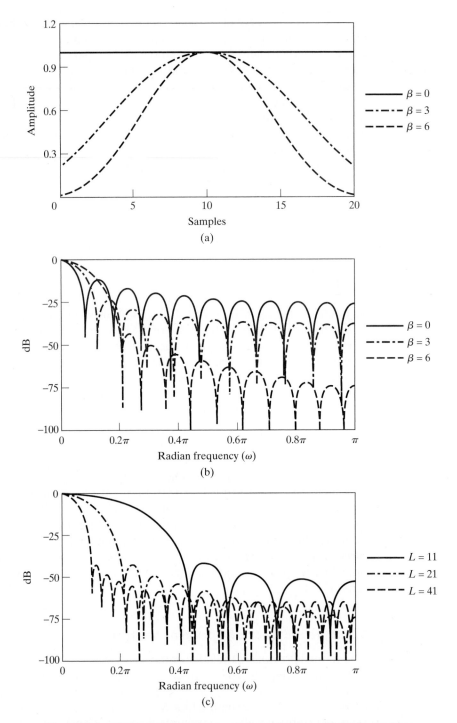

Figure 10.4 (a) Kaiser windows for $\beta = 0, 3$, and 6 and $L = 21$. (b) Fourier transform corresponding to windows in (a). (c) Fourier transforms of Kaiser windows with $\beta = 6$ and $L = 11, 21$, and 41.

Example 10.4   Illustration of the Effect of Spectral Sampling

Consider the same parameters as in Figure 10.3(c) in Example 10.3, i.e., $A_0 = 1$, $A_1 = 0.75$, $\omega_0 = 2\pi/14$, $\omega_1 = 4\pi/15$, and $\theta_1 = \theta_2 = 0$ in Eq. (10.8). $w[n]$ is a rectangular window of length 64. Then

$$v[n] = \begin{cases} \cos\left(\dfrac{2\pi}{14}n\right) + 0.75\cos\left(\dfrac{4\pi}{15}n\right), & 0 \le n \le 63, \\ 0, & \text{otherwise.} \end{cases} \tag{10.15}$$

Figure 10.5(a) shows the windowed sequence $v[n]$. Figures 10.5(b), (c), (d), and (e) show the corresponding real part, imaginary part, magnitude, and phase, respectively, of the DFT of length $N = 64$. Observe that since $x[n]$ is real, $X[N-k] = X^*[k]$ and

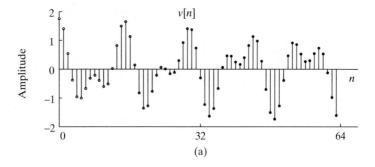

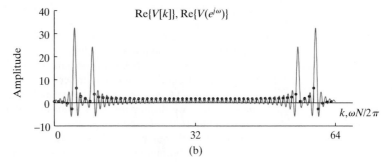

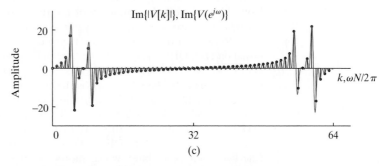

Figure 10.5   Cosine sequence and DFT with a rectangular window for $N = 64$. (a) Windowed signal. (b) Real part of DFT. (c) Imaginary part of DFT. Note that the DTFT is superimposed as the light continuous line.

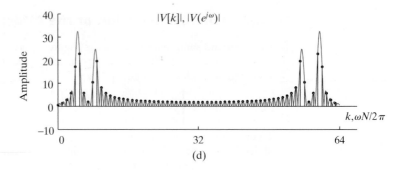

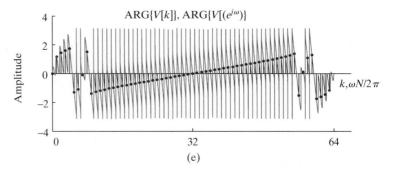

Figure 10.5 (*continued*) (d) Magnitude of DFT. (e) Phase of DFT.

$X(e^{j(2\pi-\omega)}) = X^*(e^{j\omega})$; i.e., the real part and the magnitude are even functions and the imaginary part and phase are odd functions of k and ω.

In Figures 10.5(b)–(e), the horizontal (frequency) axis is labeled in terms of the DFT index or frequency sample number k. The value $k = 32$ corresponds to $\omega = \pi$ or, equivalently, $\Omega = \pi/T$. As is the usual convention in displaying the DFT of a time sequence, we display the DFT values in the range from $k = 0$ to $k = N - 1$, corresponding to displaying samples of the DTFT in the frequency range 0 to 2π. Because of the inherent periodicity of the DTFT, the first half of this range corresponds to the positive continuous-time frequencies, i.e., Ω between zero and π/T, and the second half of the range to the negative frequencies, i.e., Ω between $-\pi/T$ and zero. Note the even periodic symmetry of the real part and the magnitude and the odd periodic symmetry of the imaginary part and the phase.

Recall that the DFT $V[k]$ is a sampled version of the DTFT $V(e^{j\omega})$. Superimposed on each DFT with a light gray line in Figures 10.5(b)–(e) is the corresponding DTFT, i.e., $\mathcal{R}e\{V(e^{j\omega})\}$, $\mathcal{I}m\{V(e^{j\omega})\}$, $|V(e^{j\omega})|$, and $\text{ARG}\{V(e^{j\omega})\}$ respectively. The frequency scale for these functions is the specially defined normalized scale denoted $\omega N/(2\pi)$; i.e., N on the DFT index scale corresponds to $\omega = 2\pi$ on the conventional frequency scale of the DTFT. We also follow this convention of superimposing the DTFT in Figures 10.6, 10.7, 10.8, and 10.9.

The magnitude of the DFT in Figure 10.5(d) corresponds to samples of $|V(e^{j\omega})|$ (the light continuous line), which shows the expected concentration around $\omega_1 = 2\pi/7.5$ and $\omega_0 = 2\pi/14$, the frequencies of the two sinusoidal components of the input. Specifically, the frequency $\omega_1 = 4\pi/15 = 2\pi(8.533\ldots)/64$ lies between the DFT samples corresponding to $k = 8$ and $k = 9$. Likewise, the frequency $\omega_0 = 2\pi/14 = 2\pi(4.5714\ldots)/64$ lies between the DFT samples corresponding to $k = 4$ and $k = 5$. Note that the frequency locations of the peaks of the gray curve in Figure 10.5(d) are between spectrum samples obtained from the DFT. In general, the locations of peaks in the DFT values do not necessarily coincide with the exact frequency locations of the peaks in the DTFT, since the true spectrum peaks can lie between spectrum samples. Correspondingly, as evidenced in Figure 10.5(d), the relative amplitudes of peaks in the DFT will not necessarily reflect the relative amplitudes of the spectrum peaks of $|V(e^{j\omega})|$.

Example 10.5 Signal Frequencies Matching DFT Frequencies Exactly

Consider the sequence

$$v[n] = \begin{cases} \cos\left(\dfrac{2\pi}{16}n\right) + 0.75\cos\left(\dfrac{2\pi}{8}n\right), & 0 \le n \le 63, \\ 0, & \text{otherwise,} \end{cases} \tag{10.16}$$

as shown in Figure 10.6(a). Again, a rectangular window is used with $N = L = 64$. This is very similar to the previous example, except that in this case, the frequencies of the cosines coincide exactly with two of the DFT frequencies. Specifically, the frequency $\omega_1 = 2\pi/8 = 2\pi 8/64$ corresponds exactly to the DFT sample $k = 8$ and the frequency $\omega_0 = 2\pi/16 = 2\pi 4/64$ to the DFT sample $k = 4$.

The magnitude of the 64-point DFT of $v[n]$ for this example is shown in Figure 10.6(b) and corresponds to samples of $|V(e^{j\omega})|$ (which again is superimposed

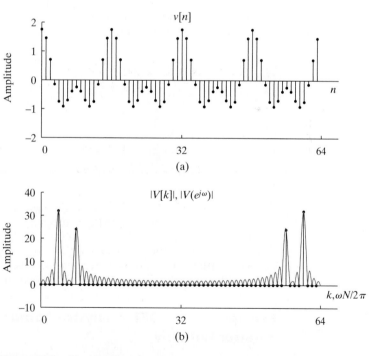

Figure 10.6 Discrete Fourier analysis of the sum of two sinusoids for a case in which the Fourier transform is zero at all DFT frequencies except those corresponding to the frequencies of the two sinusoidal components. (a) Windowed signal. (b) Magnitude of DFT. Note that ($|V(e^{j\omega})|$) is superimposed as the light continuous line.

with a light line) at a frequency spacing of $2\pi/64$. Although the signal parameters in Example 10.4 are very similar, the appearance of the DFT is for this example and strikingly different. In particular, for this example, the DFT has two strong spectral lines at the samples corresponding to the frequencies of the two sinusoidal components in the signal and no frequency content at the other DFT values. In fact, this clean appearance of the DFT in Figure 10.6(b) is largely an illusion resulting from the sampling of the spectrum. Comparing Figures 10.6(b) and (c), we can see that the reason for the clean appearance of Figure 10.6(b) is that for this choice of parameters, the Fourier transform is exactly zero at the frequencies that are sampled by the DFT, except those corresponding to $k = 4, 8, 64 - 8$, and $64 - 4$. Although the signal of Figure 10.6(a) has significant content at almost all frequencies, as evidenced by the gray curve in Figure 10.6(b), we do not see this in the DFT, because of the sampling of the spectrum. Another way of viewing this is to note that the 64-point rectangular window selects exactly an integer number of periods of the two sinusoidal components in Eq. (10.16). The 64-point DFT then corresponds to the DFS of this signal replicated with period 64. This replicated signal will have only four nonzero DFS coefficients corresponding to the two sinusoidal components on Eq. (10.16). This is an example of how the inherent assumption of periodicity gives a correct answer to a different problem. We are interested in the finite-length case and the results are quite misleading from that point of view.

To illustrate this point further, we can extend $v[n]$ in Eq. (10.16) by zero-padding to obtain a 128-point sequence. The corresponding 128-point DFT is shown in Figure 10.7. With this finer sampling of the spectrum, the presence of significant content at other frequencies becomes apparent. In this case, the windowed signal is *not* naturally periodic with period 128.

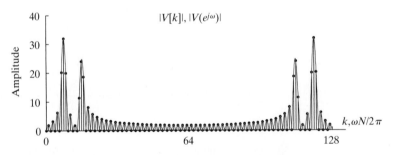

Figure 10.7   DFT of the signal as in Figure 10.6(a), but with twice the number of frequency samples used in Figure 10.6(b).

In Figures 10.5, 10.6, and 10.7, the windows were rectangular. In the next set of examples, we illustrate the effect of different choices for the window.

Example 10.6   DFT Analysis of Sinusoidal Signals Using a Kaiser Window

In this example we return to the frequency, amplitude, and phase parameters of Example 10.4, but now with a Kaiser window applied, so that

$$v[n] = w_K[n] \cos\left(\frac{2\pi}{14}n\right) + 0.75 w_K[n] \cos\left(\frac{4\pi}{15}n\right), \tag{10.17}$$

where $w_K[n]$ is the Kaiser window as given by Eq. (10.12). We will select the Kaiser window parameter β to be equal to 5.48, which, according to Eq. (10.13), results in a window for which the relative side-lobe amplitude is $A_{sl} = 40$ dB. Figure 10.8(a) shows

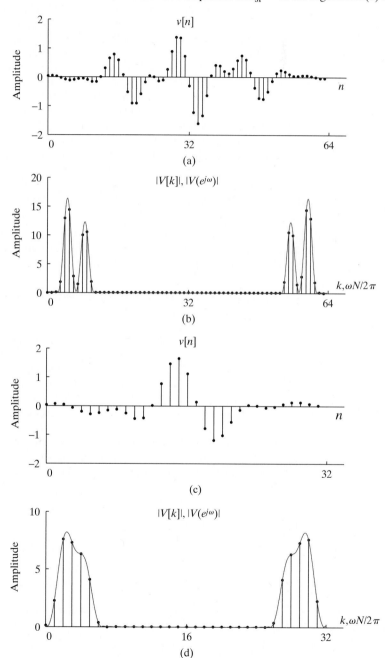

Figure 10.8   Discrete Fourier analysis with Kaiser window. (a) Windowed sequence for $L = 64$. (b) Magnitude of DFT for $L = 64$. (c) Windowed sequence for $L = 32$. (d) Magnitude of DFT for $L = 32$.

the windowed sequence $v[n]$ for a window length of $L = 64$, and Figure 10.8(b) shows the magnitude of the corresponding DFT. From Eq. (10.17), we see that the difference between the two frequencies is $\omega_1 - \omega_0 = 2\pi/7.5 - 2\pi/14 = 0.389$. From Eq. (10.14), it follows that the width of the main lobe of the Fourier transform of the Kaiser window with $L = 64$ and $\beta = 5.48$ is $\Delta_{\mathrm{ml}} = 0.401$. Thus, the main lobes of the two replicas of $W_K(e^{j\omega})$ centered at ω_0 and ω_1 will just slightly overlap in the frequency interval between the two frequencies. This is evident in Figure 10.8(b), where we see that the two frequency components are clearly resolved.

Figure 10.8(c) shows the same signal, multiplied by a Kaiser window with $L = 32$ and $\beta = 5.48$. Since the window is half as long, we expect the width of the main lobe of the Fourier transform of the window to double, and Figure 10.8(d) confirms this. Specifically, Eqs. (10.13) and (10.14) confirm that for $L = 32$ and $\beta = 5.48$, the main-lobe width is $\Delta_{\mathrm{ml}} = 0.815$. Now, the main lobes of the two copies of the Fourier transform of the window overlap throughout the region between the two cosine frequencies, and we do not see two distinct peaks.

In all the previous examples except in Figure 10.7, the DFT length N was equal to the window length L. In Figure 10.7, zero-padding was applied to the windowed sequence before computing the DFT to obtain the Fourier transform on a more finely divided set of frequencies. However, we must realize that this zero-padding will not improve the ability to resolve close frequencies, which depends on the length and shape of the window. This is illustrated by the next example.

Example 10.7   DFT Analysis with 32-point Kaiser Window and Zero-Padding

In this example, we repeat Example 10.6 using the Kaiser window with $L = 32$ and $\beta = 5.48$, and with the DFT length varying. Figure 10.9(a) shows the DFT magnitude for $N = L = 32$ as in Figure 10.8(d), and Figures 10.9(b) and (c) show the DFT magnitude again with window length $L = 32$, but with DFT lengths $N = 64$ and $N = 128$, respectively. As with Example 10.5, this zero-padding of the 32-point sequence results in finer spectral sampling of the DTFT. As shown by the light continuous curve, the underlying envelope of each DFT magnitude in Figure 10.9 is the same. Consequently, increasing the DFT size by zero-padding does not change the ability to resolve the two sinusoidal frequency components, but it does change the spacing of the frequency samples. If N were increased beyond 128, the dots denoting the DFT sample

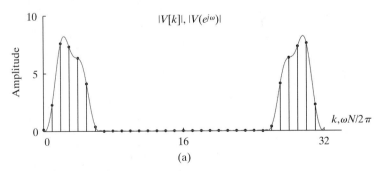

Figure 10.9   Illustration of effect of DFT length for Kaiser window of length $L = 32$. (a) Magnitude of DFT for $N = 32$.

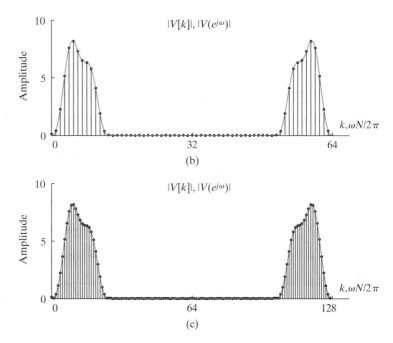

Figure 10.9   (*continued*) (b) Magnitude of DFT for $N = 64$. (c) Magnitude of DFT for $N = 128$.

values would tend to merge together and become indistinct. Consequently, DFT values are often plotted by connecting consecutive points by straight-line segments without indicating each individual point. For example, in Figures 10.5 through 10.8, we have shown a light continuous line as the DTFT $|V(e^{j\omega})|$ of the finite-length sequence $v[n]$. In fact, this curve is a plot of the DFT of the sequence after zero-padding to $N = 2048$. In these examples, this sampling of the DTFT is sufficiently dense so as to be indistinguishable from the function of the continuous variable ω.

For a complete representation of a sequence of length L, the L-point DFT is sufficient, since the original sequence can be recovered exactly from it. However, as we saw in the preceding examples, simple examination of the L-point DFT can result in misleading interpretations. For this reason, it is common to apply zero-padding, so that the spectrum is sufficiently oversampled and important features are therefore readily apparent. With a high degree of time-domain zero-padding or frequency-domain oversampling, simple interpolation (e.g., linear interpolation) between the DFT values provides a reasonably accurate picture of the Fourier spectrum, which can then be used, for example, to estimate the locations and amplitudes of spectrum peaks. This is illustrated in the following example.

Example 10.8   Oversampling and Linear Interpolation for Frequency Estimation

Figure 10.10 shows how a 2048-point DFT can be used to obtain a finely spaced evaluation of the Fourier transform of a windowed signal and how increasing the window width improves the ability to resolve closely spaced sinusoidal components. The signal of Example 10.6 having frequencies $2\pi/14$ and $4\pi/15$ was windowed with Kaiser windows of lengths $L = 32, 42, 54,$ and 64 with $\beta = 5.48$. First, note that in all cases, the 2048-point DFT gives a smooth result when the points are connected by straight lines. In Figure 10.10(a), where $L = 32$, the two sinusoidal components are not resolved, and, of course, increasing the DFT length will only result in a smoother curve. As the window length increases from $L = 32$ to $L = 42$, however, we see improvement in our ability to distinguish the two frequencies and the approximate relative amplitudes of each sinusoidal component. The dashed lines in all the figures indicate the DFT indices $k_0 = 146 \approx 2048/14$ and $k_1 = 273 \approx 4096/15$, which correspond to the nearest DFT frequencies ($N = 2048$) for the cosine components. Note that the 2048-point DFT in Figure 10.10(c) would be much more effective for precisely locating the peak of the windowed Fourier transform than the coarsely sampled DFT in Figure 10.8(b), which is also computed with a 64-point Kaiser window. Note also that the amplitudes of the two peaks in Figure 10.10 are very close to being in the correct ratio of 0.75 to 1.

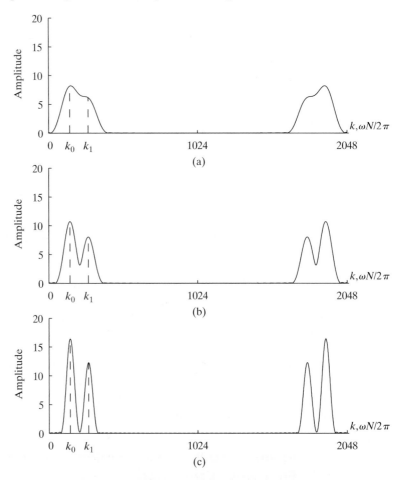

Figure 10.10 Illustration of the computation of the DFT for $N \gg L$ with linear interpolation to create a smooth curve: (a) $N = 1024$, $L = 32$. (b) $N = 1024$, $L = 42$. (c) $N = 1024$, $L = 64$. (The values $k_0 = 146 \approx 2048/14$ and $k_1 = 273 \approx 4096/15$ are the closest DFT frequencies to $\omega_0 = 2\pi/14$ and $\omega_1 = 4\pi/15$ when the DFT length is $N = 2048$.)

10.3 THE TIME-DEPENDENT FOURIER TRANSFORM

In Section 10.2, we illustrated the use of the DFT for obtaining a frequency-domain representation of a signal composed of sinusoidal components. In that discussion, we assumed that the frequencies of the cosines did not change with time, so that no matter how long the window, the signal properties (amplitudes, frequencies, and phases) would be the same from the beginning to the end of the window. Long windows give better frequency resolution, but in practical applications of sinusoidal signal models, the signal properties (e.g., amplitude, frequency) often change with time. For example, nonstationary signal models of this type are required to describe radar, sonar, speech, and data communication signals. This conflicts with the use of long analysis windows. A single DFT estimate is not sufficient to describe such signals, and as a result, we are led to the concept of the *time-dependent Fourier transform,* also referred to as the short-time Fourier transform.[1]

We define the time-dependent Fourier transform of a signal $x[n]$ as

$$X[n, \lambda] = \sum_{m=-\infty}^{\infty} x[n+m]w[m]e^{-j\lambda m}, \tag{10.18}$$

where $w[n]$ is a window sequence. In the time-dependent Fourier representation, the one-dimensional sequence $x[n]$, a function of a single discrete variable, is converted into a two-dimensional function of the time variable n, which is discrete, and the frequency variable λ, which is continuous.[2] Note that the time-dependent Fourier transform is periodic in λ with period 2π; therefore, we need consider only values of λ for $0 \leq \lambda < 2\pi$ or any other interval of length 2π.

Equation (10.18) can be interpreted as the DTFT of the shifted signal $x[n+m]$, as viewed through the window $w[m]$. The window has a stationary origin, and as n changes, the signal slides past the window, so that at each value of n, a different portion of the signal is extracted by the window for Fourier analysis. As an illustration, consider the following example.

Example 10.9   Time-Dependent Fourier Transform of a Linear Chirp Signal

A continuous-time linear chirp signal is defined as

$$x_c(t) = \cos(\theta(t)) = \cos(A_0 t^2), \tag{10.19}$$

---

[1] Further discussion of the time-dependent Fourier transform can be found in a variety of references, including Allen and Rabiner (1977), Rabiner and Schafer (1978), Crochiere and Rabiner (1983) and Quatieri (2002).

[2] We denote the frequency variable of the time-dependent Fourier transform by λ to maintain a distinction from the frequency variable of the conventional DTFT, which we always denote by ω. We use the mixed bracket–parenthesis notation $X[n, \lambda)$ as a reminder that n is a discrete variable, and λ is a continuous variable.

where A_0 has units of radians/s^2. (Such signals are called chirps because, in the auditory frequency range, short pulses sound like bird chirps.) The signal $x_c(t)$ in Eq. (10.19) is a member of the more general class of frequency modulation (FM) signals for which the *instantaneous frequency* is defined as the time derivative of the cosine argument $\theta(t)$. Therefore, in this case, the instantaneous frequency is

$$\Omega_i(t) = \frac{d\theta(t)}{dt} = \frac{d}{dt}\left(A_0 t^2\right) = 2A_0 t, \tag{10.20}$$

which varies in proportion to time; hence, the designation as a *linear* chirp signal. If we sample $x_c(t)$, we obtain the discrete-time linear chirp signal[3]

$$x[n] = x_c(nT) = \cos(A_0 T^2 n^2) = \cos(\alpha_0 n^2), \tag{10.21}$$

where $\alpha_0 = A_0 T^2$ has units of radians. The instantaneous frequency of the sampled chirp signal is a frequency-normalized, sampled version of the instantaneous frequency of the continuous-time signal; i.e.,

$$\omega_i[n] = \Omega_i(nT) \cdot T = 2A_0 T^2 n = 2\alpha_0 n, \tag{10.22}$$

which displays the same proportional increase with sample index n, with α_0 controlling the rate of increase. Figure 10.11 shows two 1201-sample segments of the sampled chirp signal in Eq. (10.21) with $\alpha_0 = 15\pi \times 10^{-6}$. (The samples are connected by straight lines for plotting.) Observe that over a short interval, the signal looks sinusoidal, but the spacing between peaks becomes smaller and smaller as time progresses, indicating increasing frequency with time.

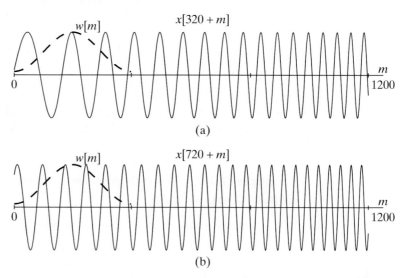

Figure 10.11   Two segments of the linear chirp signal $x[n] = \cos(\alpha_0 n^2)$ for $\alpha_0 = 15\pi \times 10^{-6}$ with a 400-sample Hamming window superimposed. (a) $X[n, \lambda)$ at $n = 320$ would be the DTFT of the top trace multiplied by the window. (b) $X[720, \lambda)$ would be the DTFT of the bottom trace multiplied by the window.

---

[3]We have seen discrete-time linear complex exponential chirp signals in Chapter 9 in the context of the chirp transform algorithm.

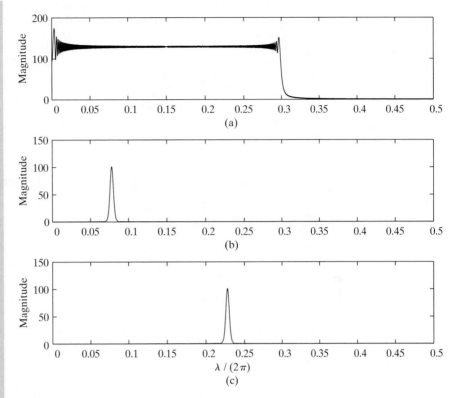

Figure 10.12   DTFTs of segments of a linear chirp signal: (a) DTFT of 20,000 samples of the signal $x[n] = \cos(\alpha_0 n^2)$. (b) DTFT of $x[5000 + m]w[m]$ where $w[m]$ is a Hamming window of length $L = 401$; i.e., $X[5000, \lambda)$. (c) DTFT of $x[15,000 + m]w[m]$ where $w[m]$ is a Hamming window of length $L = 401$; i.e., $X[15,000, \lambda)$.

The relationship of the shifted signal to the window in time-dependent Fourier analysis is also illustrated in Figure 10.11. Typically, $w[m]$ in Eq. (10.18) has finite length around $m = 0$, so that $X[n, \lambda)$ displays the frequency characteristics of the signal around time n. Figure 10.11(a) shows $x[320 + m]$ as a function of m for $0 \leq m \leq 1200$ together with a Hamming window $w[m]$ of length $L = 401$ samples. The time-dependent transform at time $n = 320$ is the DTFT of $w[m]x[320 + m]$. Similarly, Figure 10.11(b) shows the window and a later segment of the chirp signal beginning at sample $n = 720$.

Figure 10.12 illustrates the importance of the window in discrete-time Fourier analysis of time-varying signals. Figure 10.12(a) shows the DTFT of 20,000 samples (with a rectangular window) of the discrete-time chirp. Over this interval, the normalized instantaneous frequency of the chirp,

$$f_i[n] = \omega_i[n]/(2\pi) = 2\alpha_0 n/(2\pi),$$

goes from 0 to $0.00003\pi(20,000)/(2\pi) = 0.3$. This variation of the instantaneous frequency forces the DTFT representation, which involves only fixed frequencies acting over all n, to include all frequencies in that range and beyond as is evident in Figure 10.12(a). Thus, the DTFT of a long segment of the signal shows only that the signal has a wide bandwidth in the conventional DTFT sense. On the other hand, Figures 10.12(b) and (c) show DTFTs using a 401 sample Hamming window for segments of the chirp waveform at $n = 5000$ and 15,000, respectively. Thus, Figures 10.12(b) (c) are plots [as functions of $\lambda/(2\pi)$] of the time-dependent Fourier transform values $|X[5000, \lambda]|$ and $|X[15,000, \lambda]|$, respectively. Since the window length $L = 401$ is such that the signal does not change frequency very much across the window interval, the time-dependent Fourier transform tracks the frequency variation very well. Note that at samples 5000 and 15,000, we would expect a peak in the time-dependent transform at $\lambda/(2\pi) = 0.00003\pi(5000)/(2\pi) = 0.075$ and $\lambda/(2\pi) = 0.00003\pi(15,000)/(2\pi) = 0.225$, respectively. This is confirmed by examination of Figures 10.12(b) and (c).

Example 10.10   Plotting $X[n,\lambda]$: The Spectrogram

In Figure 10.13, we show a display as a function of both time index n and frequency $\lambda/(2\pi)$ of the magnitude of the time-dependent Fourier transform, $|Y[n, \lambda]|$, for the signal

$$y[n] = \begin{cases} 0 & n < 0 \\ \cos(\alpha_0 n^2) & 0 \le n \le 20,000 \\ \cos(0.2\pi n) & 20,000 < n \le 25,000 \\ \cos(0.2\pi n) + \cos(0.23\pi n) & 25,000 < n. \end{cases} \quad (10.23)$$

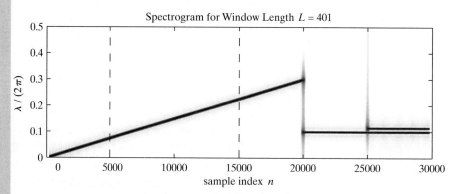

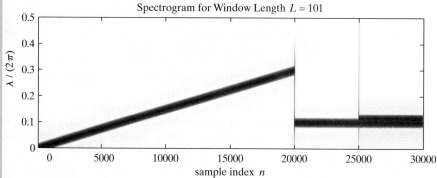

Figure 10.13   The magnitude of the time-dependent Fourier transform of $y[n]$ in Eq. (10.23): (a) Using a Hamming window of length $L = 401$. (b) Using a Hamming window of length $L = 101$.

Note that the signal $y[n]$ is equal to $x[n]$ in Eq. (10.21) in Example 10.9 for $0 \leq n \leq 20{,}000$, and then it abruptly changes to cosine components with fixed frequencies for $n > 20{,}000$. This signal was designed to make several important points about time-dependent Fourier analysis. First, consider Figure 10.13(a), which shows the time-dependent Fourier transform of $y[n]$ over the interval $0 \leq n \leq 30{,}000$ with a Hamming window of length $L = 401$. This display, which shows $20 \log_{10} |Y[n, \lambda]|$ as a function of $\lambda/2\pi$ in the vertical dimension, and the time index n in the horizontal dimension is called a *spectrogram*. The value $20 \log_{10} |Y[n, \lambda]|$ over a restricted range of 50 dB is represented by the darkness of the marking at $[n, \lambda]$. The plots in Figures 10.12(b) and (c) are vertical slices (shown in Figure 10.12 as magnitude) through the image at $n = 5000$ and $n = 15{,}000$ respectively at the locations of the dashed lines in Figure 10.13(a). Note the linear progression during the chirp interval. Also, note that during the constant-frequency intervals, the dark line remains horizontal. The width of the dark features in Figure 10.13(a) is dependent on the width of the main lobe Δ_{ml} of the DTFT of the window. Table 7.2 indicates that for the Hamming window, this width is approximately $\Delta_{\mathrm{ml}} = 8\pi/M$ wherein $M+1$ is the window length. For a 401-point window, $\Delta_{\mathrm{ml}}/(2\pi) = 0.01$. Thus, the two close-in-frequency cosines are clearly resolved in the interval $25{,}000 < n \leq 30{,}000$, because their normalized frequency difference is $(0.23\pi - 0.2\pi)/(2\pi) = 0.015$, which is significantly greater than the main-lobe width 0.01. Note that the vertical width of the dark sloping bar for the chirp interval is wider than the horizontal bars representing the constant-frequency intervals. This extra broadening is caused by the frequency variation across the window and is a small-scale version of the effect seen in Figure 10.12(a), wherein the variation across the 20,000-sample window is much greater.

The image in Figure 10.13(a) illustrates another important aspect of time-dependent Fourier analysis. The 401-sample window provides good frequency resolution at almost all points in time. However, note that at $n = 20{,}000$ and $25{,}000$ the signal properties change abruptly, so that for an interval of about 401 samples around these times, the window contains samples from both sides of the change. This leads to the fuzzy area wherein the signal properties are much less clearly represented by the spectrogram. We can improve the ability to resolve events in the time dimension by shortening the window. This is illustrated in Figure 10.13(b) wherein the window length is $L = 101$. The points of change are much better resolved with this window. However, the normalized main-lobe frequency width of a 101-sample Hamming window is $\Delta_{\mathrm{ml}}/(2\pi) = 0.04$, and the two constant-frequency cosines after $n = 25{,}000$ are only separated by 0.015 in normalized frequency. Thus, as is clear from Figure 10.13(b), the two frequencies are not resolved with the 101-sample window, although the location of the abrupt changes in the signal are much more accurately resolved in time.

Examples 10.9 and 10.10 illustrate how the principles of discrete-time Fourier analysis that were discussed in Sections 10.1 and 10.2 can be applied to signals whose properties vary with time. Time-dependent Fourier analysis is widely used both as an analysis tool for displaying signal properties and as a representation for signals. In the latter use, it is important to develop a deeper understanding of the two-dimensional representation in Eq. (10.18).

10.3.1 Invertibility of $X[n,\lambda]$

Since $X[n, \lambda]$ is the DTFT of $x[n + m]w[m]$, the time-dependent Fourier transform is invertible if the window has at least one nonzero sample. Specifically, from the Fourier transform synthesis equation (2.130),

$$x[n+m]w[m] = \frac{1}{2\pi} \int_0^{2\pi} X[n,\lambda]e^{j\lambda m}d\lambda, \qquad -\infty < m < \infty, \qquad (10.24)$$

or equivalently,

$$x[n+m] = \frac{1}{2\pi\, w[m]} \int_0^{2\pi} X[n,\lambda)d\lambda \qquad (10.25)$$

if $w[m] \neq 0.$[4] Thus with m chosen as any one value for which $w[m] \neq 0$, $x[n]$ for all values of n can be recovered from $X[n,\lambda)$ using Eq. (10.25).

While the above discussion shows that the time-dependent Fourier transform is an invertible transformation, Eq. (10.24) and (10.25) do not provide a computable inverse, since evaluating them requires knowing $X[n,\lambda)$ at all λ and also requires evaluating an integral. However, the inverse transform becomes a DFT when $X[n,\lambda)$ is sampled in both the time and frequency dimensions. We will discuss this matter more fully in Section 10.3.4.

10.3.2 Filter Bank Interpretation of $X[n,\lambda)$

A rearrangement of the sum in Eq. (10.18) leads to another useful interpretation of the time-dependent Fourier transform. If we make the substitution $m' = n + m$ in Eq. (10.18), then $X[n,\lambda)$ can be written as

$$X[n,\lambda) = \sum_{m'=-\infty}^{\infty} x[m']w[-(n-m')]e^{j\lambda(n-m')}. \qquad (10.26)$$

Equation (10.26) can be interpreted as the convolution

$$X[n,\lambda) = x[n] * h_\lambda[n], \qquad (10.27a)$$

where

$$h_\lambda[n] = w[-n]e^{j\lambda n}. \qquad (10.27b)$$

From Eq. (10.27a), we see that the time-dependent Fourier transform as a function of n with λ fixed can be interpreted as the output of an LTI filter with impulse response $h_\lambda[n]$ or, equivalently, with frequency response

$$H_\lambda(e^{j\omega}) = W(e^{j(\lambda-\omega)}). \qquad (10.28)$$

In general, a window that is nonzero for positive time will be called a *noncausal window,* since the computation of $X[n,\lambda)$ using Eq. (10.18) requires samples that follow sample n in the sequence. Equivalently, in the linear-filtering interpretation, the impulse response $h_\lambda[n] = w[-n]e^{j\lambda n}$ is noncausal if $w[n] = 0$ for $n < 0$. That is, a window that is nonzero for $n \geq 0$ gives a noncausal impulse response $h_\lambda[n]$ in Eq. (10.27b), whereas if the window is nonzero for $n \leq 0$, the linear filter is causal.

---

[4]Since $X[n,\lambda)$ is periodic in λ with period 2π, the integration in Eqs. (10.24) and (10.25) can be over any interval of length 2π.

   In the definition of Eq. (10.18), the time origin of the window is held fixed, and the signal is considered to be shifted past the interval of support of the window. This effectively redefines the time origin for Fourier analysis to be at sample n of the signal. Another possibility is to shift the window as n changes, keeping the time origin for Fourier analysis fixed at the original time origin of the signal. This leads to a definition for the time-dependent Fourier transform of the form

$$\check{X}[n, \lambda] = \sum_{m=-\infty}^{\infty} x[m]w[m - n]e^{-j\lambda m}. \tag{10.29}$$

The relationship between the definitions of Eqs. (10.18) and (10.29) is easily shown to be

$$\check{X}[n, \lambda] = e^{-j\lambda n} X[n, \lambda]. \tag{10.30}$$

   The definition of Eq. (10.18) is particularly convenient when we consider using the DFT to obtain samples in λ of the time-dependent Fourier transform, since, if $w[m]$ is of finite length in the range $0 \leq m \leq (L - 1)$, then so is $x[n + m]w[m]$. On the other hand, the definition of Eq. (10.29) has some advantages for the interpretation of Fourier analysis in terms of filter banks. Since our primary interest is in applications of the DFT, we will base most of our discussions on Eq. (10.18).

10.3.3 The Effect of the Window

The primary purpose of the window in the time-dependent Fourier transform is to limit the extent of the sequence to be transformed, so that the spectral characteristics are approximately constant over the duration of the window. The more rapidly the signal characteristics change, the shorter the window should be. We saw in Section 10.2 that as the window becomes shorter, frequency resolution decreases. The same effect is true, of course, for $X[n, \lambda]$. On the other hand, as the window length decreases, the ability to resolve changes with time increases. Consequently, the choice of window length becomes a trade-off between frequency resolution and time resolution. This trade-off was illustrated in Example 10.10.

   The effect of the window on the properties of the time-dependent Fourier transform can be seen by assuming that the signal $x[n]$ has a conventional DTFT $X(e^{j\omega})$. First, let us assume that the window is unity for all m; i.e., assume that there is no window at all. Then, from Eq. (10.18),

$$X[n, \lambda] = X(e^{j\lambda})e^{j\lambda n}. \tag{10.31}$$

Of course, a typical window for time-dependent spectrum analysis tapers to zero, so as to select only a portion of the signal for analysis. On the other hand, as discussed in Section 10.2, the length and shape of the window are chosen so that the Fourier transform of the window is narrow in λ compared with variations in λ of the Fourier transform of the signal. Thus, the need for good time resolution and good frequency resolution often requires compromise. The Fourier transform of a typical window is illustrated in Figure 10.14(a).

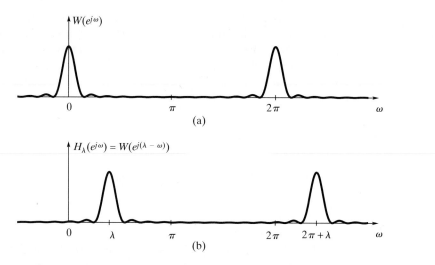

Figure 10.14 (a) Illustration of the Fourier transform of a Bartlett window for time-dependent Fourier analysis. (b) Equivalent bandpass filter for time-dependent Fourier analysis.

If we consider the time-dependent Fourier transform for fixed n, then it follows from the properties of DTFTs that

$$X[n, \lambda] = \frac{1}{2\pi} \int_0^{2\pi} e^{j\theta n} X\,(e^{j\theta}) W(e^{j(\lambda-\theta)}) d\theta; \qquad (10.32)$$

i.e., the Fourier transform of the shifted signal is convolved with the Fourier transform of the window. This is similar to Eq. (10.2), except that in Eq. (10.2), we assumed that the signal was not successively shifted relative to the window. Here, we compute a Fourier transform for each value of n. In Section 10.2, we saw that the ability to resolve two narrowband signal components depends on the width of the main lobe of the Fourier transform of the window, whereas the degree of leakage of one component into the vicinity of the other depends on the relative side-lobe amplitude. The case of no window at all corresponds to $w[n] = 1$ for all n. In this case, $W(e^{j\omega}) = 2\pi \delta(\omega)$ for $-\pi \le \omega \le \pi$, which gives precise frequency resolution but no time resolution.

In the linear-filtering interpretation of Eqs. (10.27a), (10.27b), and (10.28), $W(e^{j\omega})$ typically has the lowpass characteristics depicted in Figure 10.14(a), and consequently, $H_\lambda(e^{j\omega})$ is a bandpass filter whose passband is centered at $\omega = \lambda$, as depicted in Figure 10.14(b). Clearly, the width of the passband of this filter is approximately equal to the width of the main lobe of the Fourier transform of the window. The degree of rejection of adjacent frequency components depends on the relative side-lobe amplitude.

The preceding discussion suggests that if we are using the time-dependent Fourier transform to obtain a time-dependent estimate of the frequency spectrum of a signal, it is desirable to taper the window to lower the side lobes and to use as long a window as feasible to improve the frequency resolution. This has already been illustrated in Examples 10.9 and 10.10, and we will consider other examples in Section 10.4. However, before doing so, we discuss the use of the DFT in explicitly evaluating the time-dependent Fourier transform.

10.3.4 Sampling in Time and Frequency

Explicit computation of $X[n, \lambda)$ can be done only on a finite set of values of λ, corresponding to sampling the time-dependent Fourier transform in the domain of its frequency variable λ. Just as finite-length signals can be exactly represented through samples of the DTFT, signals of indeterminate length can be represented through samples of the time-dependent Fourier transform, if the window in Eq. (10.18) has finite length. As an example, suppose that the window has length L with samples beginning at $m = 0$; i.e.,

$$w[m] = 0 \qquad \text{outside the interval } 0 \le m \le L - 1. \tag{10.33}$$

If we sample $X[n, \lambda)$ at N equally spaced frequencies $\lambda_k = 2\pi k/N$, with $N \ge L$, then we can recover the original windowed sequence from the sampled time-dependent Fourier transform. Specifically, if we define $X[n, k]$ to be

$$X[n, k] = X[n, 2\pi k/N) = \sum_{m=0}^{L-1} x[n + m]w[m]e^{-j(2\pi/N)km}, \qquad 0 \le k \le N - 1, \tag{10.34}$$

then $X[n, k]$ with n fixed is the DFT of the windowed sequence $x[n + m]w[m]$. Using the inverse DFT, we obtain

$$x[n + m]w[m] = \frac{1}{N} \sum_{k=0}^{N-1} X[n, k]e^{j(2\pi/N)km}, \qquad 0 \le m \le L - 1. \tag{10.35}$$

Since we assume that the window $w[m] \ne 0$ for $0 \le m \le L - 1$, the sequence values can be recovered in the interval from n through $(n + L - 1)$ using the equation

$$x[n + m] = \frac{1}{Nw[m]} \sum_{k=0}^{N-1} X[n, k]e^{j(2\pi/N)km}, \qquad 0 \le m \le L - 1. \tag{10.36}$$

The important point is that the window has finite length and that we can take at least as many samples in the λ dimension as there are nonzero samples in the window; i.e., $N \ge L$. While Eq. (10.33) corresponds to a noncausal window, we could have used a causal window with $w[m] \ne 0$ for $-(L - 1) \le m \le 0$ or a symmetric window such that $w[m] = w[-m]$ for $|m| \le (L - 1)/2$, with L an odd integer. The use of a noncausal window in Eq. (10.34) is simply more convenient for our analysis, since it leads very naturally to the interpretation of the sampled time-dependent Fourier transform as the DFT of the windowed block of samples beginning with sample n.

Since Eq. (10.34) corresponds to sampling Eq. (10.18) in λ, it also corresponds to sampling Eqs. (10.26), (10.27a), and (10.27b) in λ. Specifically, Eq. (10.34) can be rewritten as

$$X[n, k] = x[n] * h_k[n], \qquad 0 \le k \le N - 1, \tag{10.37a}$$

where

$$h_k[n] = w[-n]e^{j(2\pi/N)kn}. \tag{10.37b}$$

Equations (10.37a) and (10.37b) can be viewed as a bank of N filters, as depicted in Figure 10.15, with the kth filter having frequency response

$$H_k(e^{j\omega}) = W(e^{j[(2\pi k/N) - \omega]}). \tag{10.38}$$

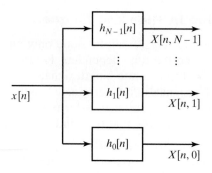

Figure 10.15   Filter bank representation of the time-dependent Fourier transform.

Our discussion suggests that $x[n]$ for $-\infty < n < \infty$ can be reconstructed if $X[n, \lambda)$ or $X[n, k]$ is sampled in the time dimension, as well. Specifically, using Eq. (10.36), we can reconstruct the signal in the interval $n_0 \leq n \leq n_0 + L - 1$ from $X[n_0, k]$, and we can reconstruct the signal in the interval $n_0 + L \leq n \leq n_0 + 2L - 1$ from $X[n_0 + L, k]$, and so on. Thus, $x[n]$ can be reconstructed exactly from the time-dependent Fourier transform sampled in both the frequency and the time dimension. In general, for the region of support of the window as specified in Eq. (10.33), we define this sampled time-dependent Fourier transform as

$$X[rR, k] = X[rR, 2\pi k/N) = \sum_{m=0}^{L-1} x[rR + m]w[m]e^{-j(2\pi/N)km}, \qquad (10.39)$$

where r and k are integers such that $-\infty < r < \infty$ and $0 \leq k \leq N - 1$. To further simplify our notation, we define

$$X_r[k] = X[rR, k] = X[rR, \lambda_k), \qquad -\infty < r < \infty, \quad 0 \leq k \leq N - 1, \quad (10.40)$$

where $\lambda_k = 2\pi k/N$. This notation denotes explicitly that the sampled time-dependent Fourier transform is simply a sequence of N-point DFTs of the windowed signal segments

$$x_r[m] = x[rR + m]w[m], \qquad -\infty < r < \infty, \quad 0 \leq m \leq L - 1, \qquad (10.41)$$

with the window position moving in jumps of R samples in time. Figure 10.16 shows lines in the $[n, \lambda)$-plane corresponding to the region of support of $X[n, \lambda)$ and the grid of sampling points in the $[n, \lambda)$-plane for the case $N = 10$ and $R = 3$. As we have shown, it is possible to uniquely reconstruct the original signal from such a two-dimensional discrete representation for appropriate choice of L.

Equation (10.39) involves the following integer parameters: the window length L; the number of samples in the frequency dimension, or the DFT length N; and the sampling interval in the time dimension, R. Although not all choices of these parameters will permit exact reconstruction of the signal, numerous combinations of N, R, and $w[n]$ and L can be used. The choice $L \leq N$ guarantees that it is possible to reconstruct the windowed segments $x_r[m]$ from the block transforms $X_r[k]$. If $R < L$, the segments overlap, but if $R > L$, some of the samples of the signal are not used and therefore cannot be reconstructed from $X_r[k]$. Thus, as one possibility, if the three sampling parameters satisfy the relation $R \leq L \leq N$, then we can (in principle) recover R samples of $x[n]$ block-by-block for all n from $X_r[k]$. Notice that each block of R samples of the

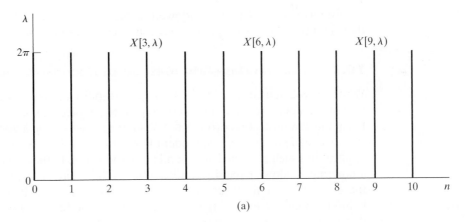

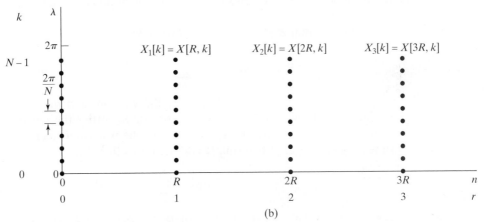

Figure 10.16   (a) Region of support for $X[n, \lambda]$. (b) Grid of sampling points in the $[n, \lambda]$-plane for the sampled time-dependent Fourier transform with $N = 10$ and $R = 3$.

signal is represented by N complex numbers in the sampled time-dependent Fourier representation; or, if the signal is real, only N real numbers are required, due to the symmetry of the DFT.

As a specific example, the signal can be reconstructed exactly from the sampled time-dependent Fourier transform for the special case $R = L = N$. In this case, N samples of a real signal are represented by N real numbers, and this is the minimum that we could expect to achieve for an arbitrarily chosen signal. For $R = L = N$ we can recover $x_r[m] = x[rR + m]w[m]$ for $0 \leq m \leq N - 1$ by computing the inverse DFT of $X_r[k]$. Therefore, we can express $x[n]$ for $rR \leq n \leq [(r + 1)R - 1]$ in terms of the windowed segments $x_r[m]$ as

$$x[n] = \frac{x_r[n - rR]}{w[n - rR]} \qquad rR \leq n \leq [(r + 1)R - 1], \qquad (10.42)$$

i.e., we recover the N-point windowed segments, remove the effect of the window, and then abut the segments together to reconstruct the original sequence.

10.3.5 The Overlap–Add Method of Reconstruction

While the previous discussion verifies the possibility of theoretically exact reconstruction of the signal from its time- and frequency-sampled time-dependent Fourier transform, the demonstration proof is not a viable reconstruction algorithm when modifications are made to the time-dependent Fourier transform as is common, for example, in applications such as audio coding and noise reduction. In these applications, division by a tapering window as required in Eq. (10.42) can greatly enhance errors at the edges; therefore, the signal blocks may not fit together smoothly. In such applications, it is helpful to make R smaller than L and N so that the blocks of samples overlap. Then, if the window is properly chosen, it will not be necessary to undo the windowing as in Eq. (10.42).

Suppose that $R \leq L \leq N$. Then we can write

$$x_r[m] = x[rR + m]w[m] = \frac{1}{N} \sum_{k=0}^{N-1} X_r[k]e^{j(2\pi k/N)m} \quad 0 \leq m \leq L - 1. \quad (10.43)$$

The recovered segments are shaped by the window, and their time origin is at the beginning of the window. A different approach to putting the signal back together that is more robust to changes in $X_r[k]$ is to shift the windowed segments to their original time locations rR and then simply add them together; i.e.,

$$\hat{x}[n] = \sum_{r=-\infty}^{\infty} x_r[n - rR]. \quad (10.44)$$

If we can show that $\hat{x}[n] = x[n]$ for all n, then Eqs. (10.43) and (10.44) together comprise a method for *time-dependent Fourier synthesis* having the capability of perfect reconstruction. Substituting Eq. (10.43) into Eq. (10.44) leads to the following representation of $\hat{x}[n]$:

$$\hat{x}[n] = \sum_{r=-\infty}^{\infty} x[rR + n - rR]w[n - rR]$$

$$= x[n] \sum_{r=-\infty}^{\infty} w[n - rR] \quad (10.45)$$

If we define

$$\tilde{w}[n] = \sum_{r=-\infty}^{\infty} w[n - rR], \quad (10.46a)$$

then the reconstructed signal in Eq. (10.45) can be expressed as

$$\hat{x}[n] = x[n]\tilde{w}[n]. \quad (10.46b)$$

It follows from Eq. (10.46b) that the condition for perfect reconstruction is

$$\tilde{w}[n] = \sum_{r=-\infty}^{\infty} w[n - rR] = C \quad -\infty < n < \infty, \quad (10.47)$$

i.e., the shifted-by-R copies of the window must add to a constant reconstruction gain C for all n.

Note that the sequence $\tilde{w}[n]$ is a periodic sequence (with period R) comprised of time-aliased window sequences. As a simple example, consider a rectangular window $w_{rect}[n]$ of length L samples. If $R = L$, the windowed segments simply fit together block-by-block with no overlap. In this case, the condition of Eq. (10.47) is satisfied with $C = 1$, because the shifted windows fit together with no overlap and no gaps. (A simple sketch will confirm this.) If L for the rectangular window is even, and $R = L/2$ a simple analysis or sketch will again verify that the condition of Eq. (10.47) is satisfied with $C = 2$. In fact, if $L = 2^v$, the signal $x[n]$ can be perfectly reconstructed from $X_r[k]$ by the overlap–add method of Eq. (10.44) when $L \leq N$ and $R = L, L/2, \ldots, 1$. The corresponding reconstruction gains would be $C = 1, 2, \ldots, L$. While this demonstrates that the overlap–add method can perfectly reconstruct the original signal for some rectangular windows, and some window spacings R, the rectangular window is rarely used in time-dependent Fourier analysis/synthesis because of its poor leakage properties. Other tapered windows such as the Bartlett, Hann, Hamming, and Kaiser windows are more commonly used. Fortunately, these windows with their superior spectral isolation properties, can also produce perfect or near-perfect reconstruction from the time-dependent Fourier transform.

Two windows with which perfect reconstruction can be achieved are the Bartlett and Hann windows, which were introduced in Chapter 7 in the context of FIR filter design. They are defined again here in Eqs (10.48) and (10.49), respectively:

Bartlett (triangular)

$$w_{Bart}[n] = \begin{cases} 2n/M, & 0 \leq n \leq M/2, \\ 2 - 2n/M, & M/2 < n \leq M, \\ 0, & \text{otherwise} \end{cases} \tag{10.48}$$

Hann

$$w_{Hann}[n] = \begin{cases} 0.5 - 0.5\cos(2\pi n/M), & 0 \leq n \leq M, \\ 0, & \text{otherwise} \end{cases} \tag{10.49}$$

As these windows are defined, the window length is $L = M + 1$ with the two end samples equal to zero.[5] With M even and $R = M/2$, then it is easily shown for the Bartlett window that the condition of Eq. (10.47) is satisfied with $C = 1$. Figure 10.17(a) shows overlapping Bartlett windows of length $M + 1$ (first and last samples zero) when $R = M/2$. It is clear that these shifted windows add up to the reconstruction gain constant $C = 1$. Figure 10.17(b) shows the same choice of $L = M + 1$ and $R = M/2$ for the Hann window. Although it is less obvious from this plot, it is also true that these shifted windows add up for all n to the constant $C = 1$. A similar statement is also true for the Hamming window and many other windows.

---

[5]With these definitions, the actual number of nonzero samples is $M - 1$ for both the Bartlett and Hann windows, but the inclusion of the zero samples leads to convenient mathematical simplifications.

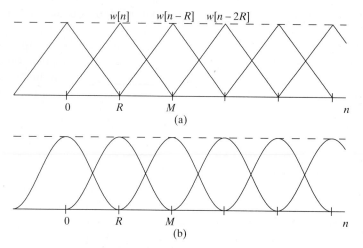

Figure 10.17 (a) Shifted $M + 1$-point Bartlett windows with $R = M/2$. (b) Shifted $M + 1$-point Hann windows with $R = M/2$. The dashed line is the periodic sequence $\tilde{w}[n]$.

Figure 7.30 gives a comparison of the DTFT of the rectangular, Bartlett and Hann windows. Note that the main-lobe width of the Bartlett and Hann windows is twice that of the rectangular window of the same length L, but the side lobes are significantly lower in amplitude for both the Bartlett and Hann windows. Thus, they and the other windows in Figure 7.30 are much preferred over the rectangular window for time-dependent Fourier analysis/synthesis.

While Figure 10.17 is intuitively plausible, it is less obvious that the Bartlett and Hann windows for $M = 2^{\nu}$ can provide perfect reconstruction for values of $R = M/2, M/4, \ldots, 1$ with corresponding reconstruction gains of $M/(2R)$. To see this, it is helpful to recall that the envelope sequence $\tilde{w}[n]$ is inherently periodic with period R, so it can be represented by an inverse DFT as

$$\tilde{w}[n] = \sum_{r=-\infty}^{\infty} w[n - rR] = \frac{1}{R} \sum_{k=0}^{R-1} W(e^{j(2\pi k/R)}) e^{j(2\pi k/R)n}, \qquad (10.50)$$

where $W(e^{j(2\pi k/R)})$ is the DTFT of $w[n]$ sampled at frequencies $(2\pi k/R)$, $k = 0, 1, \ldots, R - 1$. From Eq. (10.50) it is clear that a condition for perfect reconstruction is

$$W(e^{j(2\pi k/R)}) = 0 \qquad k = 1, 2, \ldots, R - 1, \qquad (10.51a)$$

and if Eq. (10.51a) holds, then it follows from Eq. (10.50) that the reconstruction gain is

$$C = \frac{W(e^{j0})}{R}. \qquad (10.51b)$$

Problem 7.43 of Chapter 7 explores the notion that the commonly used Bartlett, Hann, Hamming, and Blackman windows can be represented in terms of rectangular windows for which it is relatively easy to obtain a closed-form expression for the DTFT of the window. In particular, Problem 7.43 gives the result that for M even, the Bartlett window defined as in Eq. (10.48) has DTFT

$$W_{\text{Bart}}(e^{j\omega}) = \left(\frac{2}{M}\right) \left(\frac{\sin(\omega M/4)}{\sin(\omega/2)}\right)^2 e^{-j\omega M/2}. \qquad (10.52)$$

From Eq. (10.52) it follows that the Bartlett window Fourier transform has equally spaced zeros at frequencies $4\pi k/M$, for $k = 1, 2, \ldots, M - 1$. Therefore, if we choose R

so that $2\pi k/R = 4\pi k/M$ or $R = M/2$, the condition Eq. (10.51a) is satisfied. Substituting $\omega = 0$ into Eq. (10.52) gives $W_{\text{Bart}}(e^0) = M/2$, so it follows that perfect reconstruction results with $C = M/(2R) = 1$ if $R = M/2$. Choosing $R = M/2$ aligns the frequencies $2\pi k/R$ with all the zeros of $W_{\text{Bart}}(e^{j\omega})$. If M is divisible by 4, we can use $R = M/4$ and the frequencies $2\pi k/R$ will still align with zeros of $W_{\text{Bart}}(e^{j\omega})$, and the reconstruction gain will be $C = M/(2R) = 2$. If M is a power of two, R can be smaller with concomitant increase in C.

The DTFT $W_{\text{Hann}}(e^{j\omega})$ also has zeros equally spaced at integer multiples of $4\pi/M$, so exact reconstruction is also possible with the Hann window defined as in Eq. (10.49). The equally spaced zeros of $W_{\text{Bart}}(e^{j\omega})$ and $W_{\text{Hann}}(e^{j\omega})$ are evident in the plots in Figure 7.30(b) and (c), respectively. Figure 7.30(d) shows the Hamming window, which is a version of the Hann window that is optimized to minimize the side-lobe levels. As a result of the adjustment of the coefficients from 0.5 and 0.5 to 0.54 and 0.46, the zeros of $W_{\text{Hamm}}(e^{j\omega})$ are slightly displaced, so it is no longer possible to choose R so that the frequencies $2\pi k/R$ fall precisely on zeros of $W_{\text{Hamm}}(e^{j\omega})$. However, as shown in Table 7.2, the maximum side-lobe level for frequencies above $4\pi/M$ is -41 dB. Thus, the condition of Eq. (10.51a) is satisfied approximately at each of the frequencies $2\pi k/R$. Equation 10.50 shows that if Eq. (10.51a) is not satisfied exactly, $\tilde{w}[n]$ will tend to oscillate around C with period R imparting a slight amplitude modulation to the reconstructed signal.

10.3.6 Signal Processing Based on the Time-Dependent Fourier Transform

A general framework for signal processing based on the time-dependent Fourier transform is depicted in Figure 10.18. This system is based on the fact that a signal $x[n]$ can be reconstructed exactly from its time- and frequency-sampled time-dependent Fourier transform $X_r[k]$ if the window and sampling parameters are appropriately chosen, as discussed above. If the processing shown in Figure 10.18 is done so that $Y_r[k]$ maintains its integrity as a time-dependent Fourier transform, then a processed signal $y[n]$ can be reconstructed by a technique of time-dependent Fourier synthesis, such as the overlap–add method or a technique involving a bank of bandpass filters. For example, if $x[n]$ is an audio signal, $X_r[k]$ can be quantized for signal compression. The time-dependent Fourier representation provides a natural and convenient framework, wherein auditory masking phenomena can be exploited to "hide" the quantization noise. (See, for example, Bosi and Goldberg, 2003 and Spanias, Painter and Atti, 2007.) Time-dependent Fourier synthesis is then used to reconstruct a signal $y[n]$ for listening. This is the basis for MP3 audio coding, for example. Another application is audio noise suppression,

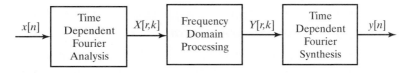

Figure 10.18   Signal processing based on time-dependent Fourier analysis/synthesis.

wherein the acoustic noise spectrum is estimated and then either subtracted from the time-dependent Fourier spectrum of the input signal or used as the basis for Wiener filtering applied to the $X_r[k]$. (See Quatieri, 2002.) These and many other applications are greatly facilitated by the FFT algorithms that are available for efficient computation of the discrete-time-dependent Fourier transform.

A discussion of applications of this type would take us too far afield; however, these kinds of block-processing techniques for discrete-time signals were also introduced in Chapter 8, when we discussed the use of the DFT for implementing the convolution of a finite-length impulse response with an input signal of indefinite length. This method of implementation of LTI systems has a useful interpretation in terms of the definitions and concepts of time-dependent Fourier analysis and synthesis, as discussed so far.

Specifically, assume that $x[n] = 0$ for $n < 0$, and suppose that we compute the time-dependent Fourier transform for $R = L$ and a rectangular window. In other words, the sampled time-dependent Fourier transform $X_r[k]$ consists of a set of N-point DFTs of segments of the input sequence

$$x_r[m] = x[rL + m], \qquad 0 \le m \le L - 1. \tag{10.53}$$

Since each sample of the signal $x[n]$ is included, and the blocks do not overlap, it follows that

$$x[n] = \sum_{r=0}^{\infty} x_r[n - rL]. \tag{10.54}$$

Now, suppose that we define a new time-dependent Fourier transform

$$Y_r[k] = H[k]X_r[k], \qquad 0 \le k \le N - 1, \tag{10.55}$$

where $H[k]$ is the N-point DFT of a finite-length unit sample sequence $h[n]$ such that $h[n] = 0$ for $n < 0$ and for $n > P - 1$. If we compute the inverse DFT of $Y_r[k]$, we obtain

$$y_r[m] = \frac{1}{N} \sum_{k=0}^{N-1} Y_r[k] e^{j(2\pi/N)km} = \sum_{\ell=0}^{N-1} x_r[\ell] h[((m - \ell))_N]. \tag{10.56}$$

That is, $y_r[m]$ is the N-point circular convolution of $h[m]$ and $x_r[m]$. Since $h[m]$ has length P samples and $x_r[m]$ has length L samples, it follows from the discussion of Section 8.7 that if $N \ge L + P - 1$, then $y_r[m]$ will be identical to the linear convolution of $h[m]$ with $x_r[m]$ in the interval $0 \le m \le L + P - 2$, and it will be zero, otherwise. Thus, it follows that if we construct an output signal

$$y[n] = \sum_{r=0}^{\infty} y_r[n - rL], \tag{10.57}$$

then $y[n]$ is the output of an LTI system with impulse response $h[n]$. The procedure just described corresponds exactly to the *overlap–add* method of block convolution. The overlap–save method discussed in Section 8.7 can also be applied within the framework of the time-dependent Fourier transform.

10.3.7 Filter Bank Interpretation of the Time-Dependent Fourier Transform

Another way to see that the time-dependent Fourier transform can be sampled in the time dimension is to recall that for fixed λ (or for fixed k if the analysis frequencies are $\lambda_k = 2\pi k/N$) the time-dependent Fourier transform is a one-dimensional sequence in time that is the output of a bandpass filter with frequency response as in Eq. (10.28).

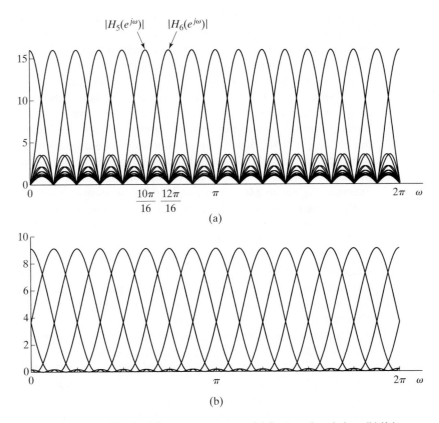

Figure 10.19   Filterbank frequency response. (a) Rectangular window. (b) Kaiser window.

This is illustrated in Figure 10.19. Figure 10.19(a) shows the equivalent set of bandpass filters corresponding to a rectangular window with $L = N = 16$. Figure 10.19 illustrates the filter bank interpretation, even for the case where L and N are much larger. When N increases, the filter bands become narrower, and the side lobes overlap with adjacent channels in the same way. Note that the passbands of the filters corresponding to the rectangular window overlap significantly, and their frequency selectivity is not good by any standard. In fact, the side lobes of any one of the bandpass filters overlap completely with several of the passbands on either side. This suggests that, in general, we might encounter a problem with aliasing in the time dimension, since the Fourier transform of any other finite-length tapering window will not be an ideal filter response either. Our discussion in Section 10.3.5, however, shows that even the rectangular window can provide perfect reconstruction with overlapping windows, in spite of the poor frequency selectivity. Although aliasing occurs in the individual bandpass filter outputs, it can be argued that the aliasing distortion cancels out when all channels are recombined in the overlap–add synthesis. This notion of alias cancellation is one of the important concepts to result from a detailed investigation of the filter bank interpretation.

If a tapering window is used, the side lobes are greatly reduced. Figure 10.19(b) shows the case for a Kaiser window of the same length as the rectangular window used in Figure 10.19(a), i.e., $L = N = 16$. The side lobes are much smaller, but the main lobe is much wider, so the filters overlap even more. Again, the previous argument based on block processing ideas shows conclusively that we can reconstruct the original signal almost exactly from the time- and frequency-sampled time-dependent Fourier transform if R is small enough. Thus, for a Kaiser window such as in Figure 10.19(b), the sampling rate of the sequences representing each of the bandpass analysis channels could be $2\pi/R = \Delta_{ml}$, where Δ_{ml} is the width of the main lobe of the Fourier transform of the window.[6] In the example of Figure 10.19(b), the main lobe width is approximately $\Delta_{ml} = 0.4\pi$, which implies that the time sampling interval could be $R = 5$ for nearly perfect reconstruction of the signal from $X[rR, \lambda_k)$ by the overlap–add method. More generally, in the case of the Hamming window of length $L = M + 1$ samples, for example, $\Delta_{ml} = 8\pi/M$ so nominally, the time sampling interval should be $R = M/4$. With this sampling rate in time, our discussion above shows that the signal $x[n]$ could be reconstructed nearly perfectly from $X[rR, \lambda_k)$ using a Hamming window and the overlap–add method of synthesis with $R = L/4$ and $L \leq N$.

When using the overlap–add method of analysis/synthesis, the parameters generally satisfy the relation $R \leq L \leq N$. This implies that (taking account of symmetries) the effective total number of samples (numbers) per second of the time-dependent Fourier representation $X[rR, \lambda_k)$ is a factor of N/R greater than the sample rate of $x[n]$ itself. This may not be an issue in some applications, but it presents a significant problem in data compression applications, such as audio coding. Fortunately, the filter bank point of view is the basis for showing that it is possible to choose these parameters to satisfy $R = N < L$ and still achieve nearly perfect reconstruction of the signal from its time-dependent Fourier transform. An example of such an analysis/synthesis system was discussed in Section 4.7.6, where $R = N = 2$, and the lowpass and highpass filters have impulse responses of length L, which can be as large as desired to achieve sharp cutoff filters. The two-channel filter bank can be generalized to a higher number of channels with $R = N$, and, as in the example of Section 4.7.6, polyphase techniques can be employed to increase computational efficiency. The advantage of requiring $R = N$ is that the total number of samples/s remains the same as for the input $x[n]$. As an example, Figure 10.20 shows the first few bandpass channels of the basic analysis filter bank specified by the MPEG-II audio coding standard. This filter bank performs time-dependent Fourier analysis with offset center frequencies $\lambda_k = (2k + 1)\pi/64$ using 32 real bandpass filters. Since the real bandpass filters have a pair of passbands centered at frequencies $\pm\lambda_k$, this is equivalent to 64 complex bandpass filters. In this case, the length of the impulse responses (equivalent to the window length) is $L = 513$ with the first and last samples being equal to zero. The downsampling factor is $R = 32$. Observe that the filters overlap significantly at their band edges, and downsampling by $R = 32$ causes significant aliasing distortion. However, a more detailed analysis of the complete analysis/synthesis system shows that the aliasing distortion due to the nonideal frequency responses cancels in the reconstruction process.

---

[6]Since, for our definition, the time-dependent Fourier transform channel signals, $X[n, \lambda_k)$, are bandpass signals centered on frequency λ_k, they can be frequency-downshifted by λ_k, so that the result is a lowpass signal in the band $\pm\Delta_{ml}$. The resulting lowpass signals have highest frequency $\Delta_{ml}/2$, so the lowest sampling rate would be $2\pi/R = \Delta_{ml}$. If $R = N$, the frequency-downshifting occurs automatically as a result of the downsampling operation.

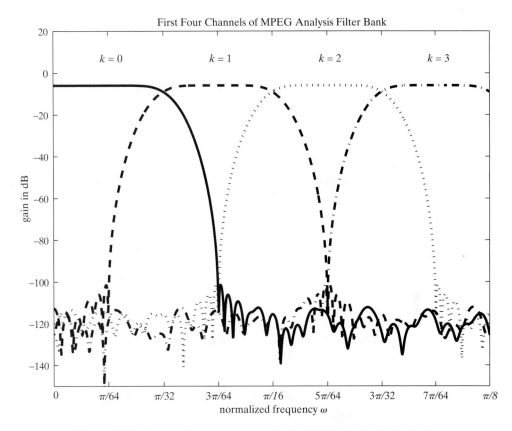

Figure 10.20   Several bandpass channels for the MPEG-II analysis filter bank.

A full-scale discussion of analysis and synthesis filter banks is beyond our scope in this chapter. An outline of such a discussion is given as the basis for Problem 10.46, and detailed discussions can be found in Rabiner and Schafer (1978), Crochiere and Rabiner (1983) and Vaidyanathan (1993).

10.4 EXAMPLES OF FOURIER ANALYSIS OF NONSTATIONARY SIGNALS

In Section 10.3.6, we considered a simple example of how the time-dependent Fourier transform can be used to implement linear filtering. In such applications, we are not so much interested in spectral resolution as in whether it is possible to reconstruct a modified signal from the modified time-dependent Fourier transform. On the other hand, the concept of the time-dependent Fourier transform is often used as a framework for a variety of techniques for obtaining spectrum estimates for nonstationary discrete-time signals, and in these applications spectral resolution, time variation, and other issues are the most important.

A nonstationary signal is a signal whose properties vary with time, for example, a sum of sinusoidal components with time-varying amplitudes, frequencies, or phases. As we will illustrate in Section 10.4.1 for speech signals and in Section 10.4.2 for Doppler radar signals, the time-dependent Fourier transform often provides a useful description of how the signal properties change with time.

When we apply time-dependent Fourier analysis to a sampled signal, the entire discussion of Section 10.1 holds for each DFT that is computed. In other words, for each segment $x_r[n]$ of the signal, the sampled time-dependent Fourier transform $X_r[k]$ would be related to the Fourier transform of the original continuous-time signal by the processes described in Section 10.1. Furthermore, if we were to apply the time-dependent Fourier transform to sinusoidal signals with constant (i.e., nontime-varying) parameters, the discussion of Section 10.2 should also apply to each of the DFTs that we compute. When the signal frequencies do not change with time, it is tempting to assume that the time-dependent Fourier transform would vary only in the frequency dimension in the manner described in Section 10.2, but this would be true only in very special cases. For example, the time-dependent Fourier transform will be constant in the time dimension if the signal is periodic with period N_p and $L = \ell_0 N_p$ and $R = r_0 N_p$, where ℓ_0 and r_0 are integers; i.e., the window includes exactly ℓ_0 periods, and the window is moved by exactly r_0 periods between computations of the DFT. In general, even if the signal is exactly periodic, the varying phase relationships that would result as different segments of the waveform are shifted into the analysis window would cause the time-dependent Fourier transform to vary in the time dimension. However, for stationary signals, if we use a window that tapers to zero at its ends, the magnitude $|X_r[k]|$ will vary only slightly from segment to segment, with most of the variation of the complex time-dependent Fourier transform occurring in the phase.

10.4.1 Time-Dependent Fourier Analysis of Speech Signals

Speech is produced by excitation of an acoustic tube, the *vocal tract,* which is terminated on one end by the lips and on the other end by the glottis. There are three basic classes of speech sounds:

- *Voiced sounds* are produced by exciting the vocal tract with quasi-periodic pulses of airflow caused by the opening and closing of the glottis.
- *Fricative sounds* are produced by forming a constriction somewhere in the vocal tract and forcing air through the constriction so that turbulence is created, thereby producing a noise-like excitation.
- *Plosive sounds* are produced by completely closing off the vocal tract, building up pressure behind the closure, and then abruptly releasing the pressure.

Detailed discussions of models for the speech signal and applications of the time-dependent Fourier transform are found in texts such as Flanagan (1972), Rabiner and Schafer (1978), O'Shaughnessy (1999), Parsons (1986) and Quatieri (2002).

With a constant vocal tract shape, speech can be modeled as the response of an LTI system (the vocal tract) to a quasiperiodic pulse train for voiced sounds or wideband noise for unvoiced sounds. The vocal tract is an acoustic transmission system characterized by its natural frequencies, called *formants,* which correspond to resonances in its frequency response. In normal speech, the vocal tract changes shape relatively slowly with time as the tongue and lips perform the gestures of speech, and thus it can be modeled as a slowly time-varying filter that imposes its frequency-response properties on the spectrum of the excitation. A typical speech waveform is shown in Figure 10.21.

From this brief description of the process of speech production and from Figure 10.21, we see that speech is definitely a nonstationary signal. However, as illustrated

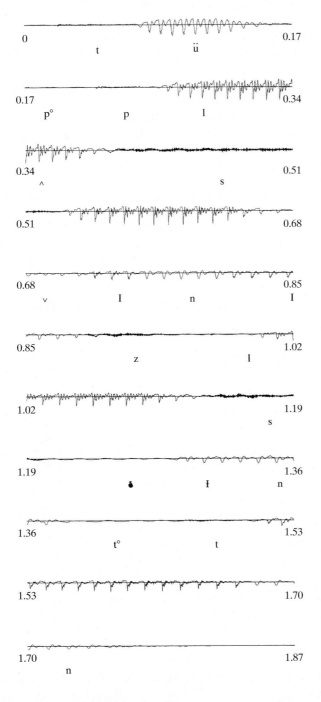

Figure 10.21   Waveform of the speech utterance "Two plus seven is less than ten." Each line is 0.17 s in duration. The time-aligned phonemic transcript is indicated below the waveform. The sampling rate is 16,000 samples/s, so each line represents 2720 samples.

in the figure, the characteristics of the signal can be assumed to remain essentially constant over time intervals on the order of 30 or 40 ms. The frequency content of the speech signal may range up to 15 kHz or higher, but speech is highly intelligible even when bandlimited to frequencies below about 3 kHz. Commercial telephone systems, for example, typically limit the highest transmitted frequency to about 3 kHz. A standard sampling rate for digital telephone communication systems is 8000 samples/s.

Figure 10.21 shows that the waveform consists of a sequence of quasiperiodic *voiced* segments interspersed with noise-like *unvoiced* segments. This figure suggests that if the window length L is not too long, the properties of the signal will not change appreciably from the beginning of the segment to the end. Thus, the DFT of a windowed speech segment should display the frequency-domain properties of the signal at the time corresponding to the window location. For example, if the window length is long enough so that the fundamental frequency and its harmonics are resolved, the DFT of a windowed segment of voiced speech should show a series of peaks at integer multiples of the fundamental frequency of the signal in that interval. This would normally require that the window span several periods of the waveform. If the window is too short, then the harmonics will not be resolved, but the general spectrum shape will still be evident. This is typical of the trade-off between frequency resolution and time resolution that is required in the analysis of nonstationary signals. We saw this before in Example 10.9. If the window is too long, the signal properties may change too much across the window; if the window is too short, resolution of narrowband components will be sacrificed. This trade-off is illustrated in the following example.

Example 10.11   Spectrogram Display of the Time-Dependent Fourier Transform of Speech

Figure 10.22(a) shows a spectrogram display of the time-dependent Fourier transform of the speech signal in Figure 10.21. The time waveform is also shown on the same time scale, below the spectrogram. More specifically, Figure 10.22(a) is a *wideband spectrogram*. A wideband spectrogram representation results from a window that is relatively short in time and is characterized by poor resolution in the frequency dimension and good resolution in the time dimension. The frequency axis is labeled in terms of continuous-time frequency. Since the sampling rate of the signal was 16,000 samples/s, it follows that the frequency $\lambda = \pi$ corresponds to 8 kHz. The specific window used in Figure 10.22(a) was a Hamming window of duration 6.7 ms, corresponding to $L = 108$. The value of R was 16, representing 1-ms time increments.[7] The broad, dark bars that move horizontally across the spectrogram correspond to the resonance frequencies of the vocal tract, which, as we see, change with time. The vertically striated appearance of the spectrogram is due to the quasiperiodic nature of voiced portions of the waveform, as is evident by comparing the variations in the waveform display and the spectrogram. Since the length of the analysis window is on the order of the length of a period of the waveform, as the window slides along in time, it alternately covers high-energy segments of the waveform and then lower energy segments in between, thereby producing the vertical striations in the plot during voiced intervals.

In a *narrowband* time-dependent Fourier analysis, a longer window is used to provide higher frequency resolution, with a corresponding decrease in time resolution.

---

[7]In plotting spectrograms, it is common to use relatively small values of R so that a smoothly varying display is obtained.

Such a narrowband analysis of speech is illustrated by the display in Figure 10.22(b). In this case, the window was a Hamming window of duration 45 ms. This corresponds to $L = 720$. The value of R was again 16.

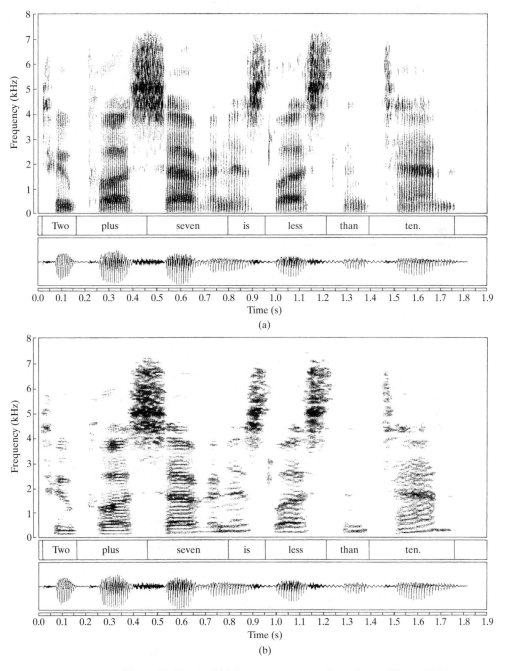

Figure 10.22   (a) Wideband spectrogram of waveform of Figure 10.21. (b) Narrowband spectrogram.

This example only hints at the many reasons that the time-dependent Fourier transform is so important in speech analysis and processing. Indeed, the concept is used directly and indirectly as the basis for acoustic–phonetic analysis and for many fundamental speech-processing applications, such as digital coding, noise and reverberation removal, speech recognition, speaker verification, and speaker identification. For present purposes, our discussion simply serves as an introductory illustration.

10.4.2 Time-Dependent Fourier Analysis of Radar Signals

Another application area in which the time-dependent Fourier transform plays an important role is radar signal analysis. The following are elements of a typical radar system based on the Doppler principle:

- Antennas for transmitting and receiving (often the same).
- A transmitter that generates an appropriate signal at microwave frequencies. In our discussion, we will assume that the signal consists of sinusoidal pulses. While this is often the case, other signals may be used, depending on the specific radar objectives and design.
- A receiver that amplifies and detects echoes of the transmitted pulses that have been reflected from objects illuminated by the antenna.

In such a radar system, the transmitted sinusoidal signal propagates at the speed of light, reflects off the object, and returns at the speed of light to the antenna, thereby undergoing a time delay of the round-trip travel time from the antenna to the object. If we assume that the transmitted signal is a sinusoidal pulse of the form $\cos(\Omega_0 t)$ and the distance from the antenna to the object is $\rho(t)$, then the received signal is a pulse of the form

$$s(t) = \cos[\Omega_0(t - 2\rho(t)/c)], \tag{10.58}$$

where c is the velocity of light. If the object is not moving relative to the antenna, then $\rho(t) = \rho_0$, where ρ_0 is the *range*. Since the time delay between the transmitted and received pulses is $2\rho_0/c$, a measurement of the time delay may be used to estimate the range. If, however, $\rho(t)$ is not constant, the received signal is an angle-modulated sinusoid and the phase difference contains information about both the range and the relative motion of the object with respect to the antenna. Specifically, let us represent the time-varying range in a Taylor's series expansion as

$$\rho(t) = \rho_0 + \dot{\rho}_0 t + \frac{1}{2!}\ddot{\rho}_0 t^2 + \cdots, \tag{10.59}$$

where ρ_0 is the nominal range, $\dot{\rho}_0$ is the velocity, $\ddot{\rho}_0$ is the acceleration, and so on. Assuming that the object moves with constant velocity (i.e., $\ddot{\rho}_0 = 0$), and substituting Eq. (10.59) into Eq. (10.58), we obtain

$$s(t) = \cos[(\Omega_0 - 2\Omega_0\dot{\rho}_0/c)t - 2\Omega_0\rho_0/c]. \tag{10.60}$$

In this case, the frequency of the received signal differs from the frequency of the transmitted signal by the *Doppler frequency*, defined as

$$\Omega_d = -2\Omega_0\dot{\rho}_0/c. \tag{10.61}$$

Thus, the time delay can still be used to estimate the range, and we can determine the speed of the object relative to the antenna if we can determine the Doppler frequency.

In a practical setting, the received signal is generally very weak, and thus a noise term should be added to Eq. (10.60). We will neglect the effects of noise in the simple analysis of this section. Also, in most radar systems, the signal of Eq. (10.60) would be frequency shifted to a lower nominal frequency in the detection process. However, the Doppler shift will still satisfy Eq. (10.61), even if $s(t)$ is demodulated to a lower center frequency.

To apply time-dependent Fourier analysis to such signals, we first bandlimit the signal to a frequency band that includes the expected Doppler frequency shifts and then sample the resulting signal with an appropriate sampling period T, thereby obtaining a discrete-time signal of the form

$$x[n] = \cos[(\omega_0 - 2\omega_0\dot{\rho}_0/c)n - 2\omega_0\rho_0/c], \qquad (10.62)$$

where $\omega_0 = \Omega_0 T$. In many cases, the object's motion would be more complicated than we have assumed, requiring the incorporation of higher order terms from Eq. (10.59) and thereby producing a more complicated angle modulation in the received signal. Another way to represent this more complicated variation of the frequency of the echoes is to use the time-dependent Fourier transform with a window that is short enough, so that the assumption of constant Doppler-shifted frequency is valid across the entire window interval, but not so short as to sacrifice adequate resolution when two or more moving objects create Doppler-shifted return signals that are superimposed at the receiver.

Example 10.12   Time-Dependent Fourier Analysis of Doppler Radar Signals

An example of time-dependent Fourier analysis of Doppler radar signals is shown in Figure 10.23. (See Schaefer, Schafer and Mersereau, 1979.) The radar data had been preprocessed to remove low-velocity Doppler shifts, leaving the variations displayed in the figure. The window for the time-dependent Fourier transform was a Kaiser window with $N = L = 64$ and $\beta = 4$. In the figure, $|X_r[k]|$ is plotted with time as the vertical dimension (increasing upward) and frequency as the horizontal dimension.[8] In this case, the successive DFTs are plotted close together. A hidden-line elimination algorithm is used to create a two-dimensional view of the time-dependent Fourier transform. To the left of the center line is a strong peak that moves in a smooth path through the time-frequency plane. This corresponds to a moving object whose velocity varies in a regular manner. The other broad peaks in the time-dependent Fourier transform are due to noise and spurious returns called *clutter* in radar terminology. An example of motion that might create such a variation of the Doppler frequency is a rocket moving at constant velocity but rotating about its longitudinal axis. A peak moving through the time-dependent Fourier transform might correspond to reflections from a fin on the rocket that is alternately moving toward and then away from the antenna because of the spinning of the rocket. Figure 10.23(b) shows an estimate of the Doppler frequency as a function of time. This estimate was obtained simply by locating the highest peak in each DFT.

[8]The plot shows the negative frequencies on the left of the line through the center of the plot and positive frequencies on the right. This can be achieved by computing the DFT of $(-1)^n x_r[n]$ and noting that the computation effectively shifts the origin of the DFT index to $k = N/2$. Alternatively, the DFT of $x_r[n]$ can be computed and then reindexed.

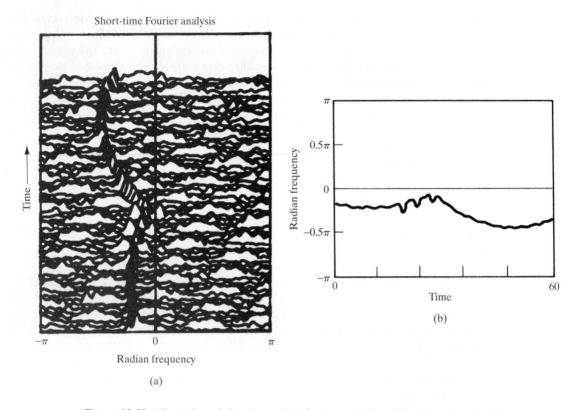

Short-time Fourier analysis

(a)

(b)

Time

Radian frequency

Radian frequency

Figure 10.23   Illustration of time-dependent Fourier analysis of Doppler radar signal. (a) Sequence of Fourier transforms of Doppler radar signal. (b) Doppler frequency estimated by picking the largest peak in the time-dependent Fourier transform.

10.5 FOURIER ANALYSIS OF STATIONARY RANDOM SIGNALS: THE PERIODOGRAM

In previous sections, we discussed and illustrated Fourier analysis for sinusoidal signals with stationary (nontime-varying) parameters and for nonstationary signals such as speech and radar. In cases where the signal can be modeled by a sum of sinusoids or a linear system excited by a periodic pulse train, the Fourier transforms of finite-length segments of the signal have a convenient and natural interpretation in terms of Fourier transforms, windowing, and linear system theory. However, more noise-like signals, such as the example of unvoiced speech in Section 10.4.1, are best modeled as random signals.

As we discussed in Section 2.10 and as shown in Appendix A, random processes are often used to model signals when the process that generates the signal is too complex for a reasonable deterministic model. Typically, when the input to an LTI system is modeled as a stationary random process, many of the essential characteristics of the input and output are adequately represented by averages, such as the mean value (dc level), variance (average power), autocorrelation function, or power density spectrum.

Consequently, it is of particular interest to estimate these for a given signal. As discussed in Appendix A, an estimate of the mean value of a stationary random process from a finite-length segment of data is the *sample mean,* defined as

$$\hat{m}_x = \frac{1}{L} \sum_{n=0}^{L-1} x[n]. \tag{10.63}$$

Similarly, an estimate of the variance is the *sample variance,* defined as

$$\hat{\sigma}_x^2 = \frac{1}{L} \sum_{n=0}^{L-1} (x[n] - \hat{m}_x)^2. \tag{10.64}$$

The sample mean and the sample variance, which are themselves random variables, are *unbiased* and *asymptotically unbiased* estimators, respectively; i.e., the expected value of $\hat{m}_x$ is the true mean m_x and the expected value of $\hat{\sigma}_x^2$ approaches the true variance σ_x^2 as L approaches ∞. Furthermore, they are both *consistent* estimators; i.e., they improve with increasing L, since their variances approach zero as L approaches ∞.

In the remainder of this chapter, we study the estimation of the power spectrum[9] of a random signal using the DFT. We will see that there are two basic approaches to estimating the power spectrum. One approach, which we develop in this section, is referred to as *periodogram analysis* and is based on direct Fourier transformation of finite-length segments of the signal. The second approach, developed in Section 10.6, is to first estimate the autocovariance sequence and then compute the Fourier transform of this estimate. In either case, we are typically interested in obtaining unbiased consistent estimators. Unfortunately, the analysis of such estimators is very difficult, and generally, only approximate analyses can be accomplished. Even approximate analyses are beyond the scope of this text, and we refer to the results of such analyses only in a qualitative way. Detailed discussions are given in Blackman and Tukey (1958), Hannan (1960), Jenkins and Watts (1968), Koopmans (1995), Kay and Marple (1981), Marple (1987), Kay (1988) and Stoica and Moses (2005).

10.5.1 The Periodogram

Let us consider the problem of estimating the power density spectrum $P_{ss}(\Omega)$ of a continuous-time signal $s_c(t)$. An intuitive approach to the estimation of the power spectrum is suggested by Figure 10.1 and the associated discussion in Section 10.1. Based on that approach, we now assume that the input signal $s_c(t)$ is a stationary random signal. The antialiasing lowpass filter creates a new stationary random signal whose power spectrum is bandlimited, so that the signal can be sampled without aliasing. Then, $x[n]$ is a stationary discrete-time random signal whose power density spectrum $P_{xx}(\omega)$ is proportional to $P_{ss}(\Omega)$ over the bandwidth of the antialiasing filter; i.e.,

$$P_{xx}(\omega) = \frac{1}{T} P_{ss}\left(\frac{\omega}{T}\right), \qquad |\omega| < \pi, \tag{10.65}$$

where we have assumed that the cutoff frequency of the antialiasing filter is π/T and that T is the sampling period. (See Problem 10.39 for a further consideration of sampling of

---

[9]The term *power spectrum* is commonly used interchangeably with the more precise term *power density spectrum*.

random signals.) Consequently, a good estimate of $P_{xx}(\omega)$ will provide a useful estimate of $P_{ss}(\Omega)$. The window $w[n]$ in Figure 10.1 selects a finite-length segment (L samples) of $x[n]$, which we denote $v[n]$, the Fourier transform of which is

$$V(e^{j\omega}) = \sum_{n=0}^{L-1} w[n]x[n]e^{-j\omega n}. \tag{10.66}$$

Consider as an estimate of the power spectrum the quantity

$$I(\omega) = \frac{1}{LU}|V(e^{j\omega})|^2, \tag{10.67}$$

where the constant U anticipates a need for normalization to remove bias in the spectrum estimate. When the window $w[n]$ is the rectangular window sequence, this estimator for the power spectrum is called the *periodogram*. If the window is not rectangular, $I(\omega)$ is called the *modified periodogram*. Clearly, the periodogram has some of the basic properties of the power spectrum. It is nonnegative, and for real signals, it is a real and even function of frequency. Furthermore, it can be shown (Problem 10.33) that

$$I(\omega) = \frac{1}{LU} \sum_{m=-(L-1)}^{L-1} c_{vv}[m]e^{-j\omega m}, \tag{10.68}$$

where

$$c_{vv}[m] = \sum_{n=0}^{L-1} x[n]w[n]x[n+m]w[n+m]. \tag{10.69}$$

We note that the sequence $c_{vv}[m]$ is the aperiodic correlation sequence for the finite-length sequence $v[n] = w[n]x[n]$. Consequently, the periodogram is in fact the Fourier transform of the aperiodic correlation of the windowed data sequence.

Explicit computation of the periodogram can be carried out only at discrete frequencies. From Eqs. (10.66) and (10.67), we see that if the DTFT of $w[n]x[n]$ is replaced by its DFT, we will obtain samples at the DFT frequencies $\omega_k = 2\pi k/N$ for $k = 0, 1, \ldots, N-1$. Specifically, samples of the periodogram are given by

$$I[k] = I(\omega_k) = \frac{1}{LU}|V[k]|^2, \tag{10.70}$$

where $V[k]$ is the N-point DFT of $w[n]x[n]$. If we want to choose N to be greater than the window length L, appropriate zero-padding would be applied to the sequence $w[n]x[n]$.

If a random signal has a nonzero mean, its power spectrum has an impulse at zero frequency. If the mean is relatively large, this component will dominate the spectrum estimate, causing low-amplitude, low-frequency components to be obscured by leakage. Therefore, in practice the mean is often estimated using Eq. (10.63), and the resulting estimate is subtracted from the random signal before computing the power spectrum estimate. Although the sample mean is only an approximate estimate of the zero-frequency component, subtracting it from the signal often leads to better estimates at neighboring frequencies.

10.5.2 Properties of the Periodogram

The nature of the periodogram estimate of the power spectrum can be determined by recognizing that, for each value of ω, $I(\omega)$ is a random variable. By computing the mean and variance of $I(\omega)$, we can determine whether the estimate is biased and whether it is consistent.

From Eq. (10.68), the expected value of $I(\omega)$ is

$$\mathcal{E}\{I(\omega)\} = \frac{1}{LU} \sum_{m=-(L-1)}^{L-1} \mathcal{E}\{c_{vv}[m]\}e^{-j\omega m}. \tag{10.71}$$

The expected value of $c_{vv}[m]$ can be expressed as

$$\mathcal{E}\{c_{vv}[m]\} = \sum_{n=0}^{L-1} \mathcal{E}\{x[n]w[n]x[n+m]w[n+m]\}$$
$$= \sum_{n=0}^{L-1} w[n]w[n+m]\mathcal{E}\{x[n]x[n+m]\}. \tag{10.72}$$

Since we are assuming that $x[n]$ is stationary,

$$\mathcal{E}\{x[n]x[n+m]\} = \phi_{xx}[m], \tag{10.73}$$

and Eq. (10.72) can then be rewritten as

$$\mathcal{E}\{c_{vv}[m]\} = c_{ww}[m]\phi_{xx}[m], \tag{10.74}$$

where $c_{ww}[m]$ is the aperiodic autocorrelation of the window, i.e.,

$$c_{ww}[m] = \sum_{n=0}^{L-1} w[n]w[n+m]. \tag{10.75}$$

That is, the mean of the aperiodic autocorrelation of the windowed signal is equal to the aperiodic autocorrelation of the window multiplied by the true autocorrelation function; i.e., *in an average sense*, the autocorrelation function of the data window appears as a window on the true autocorrelation function.

From Eq. (10.71), Eq. (10.74), and the modulation–windowing property of Fourier transforms (Section 2.9.7), it follows that

$$\mathcal{E}\{I(\omega)\} = \frac{1}{2\pi LU} \int_{-\pi}^{\pi} P_{xx}(\theta)C_{ww}(e^{j(\omega-\theta)})d\theta, \tag{10.76}$$

where $C_{ww}(e^{j\omega})$ is the Fourier transform of the aperiodic autocorrelation of the window, i.e.,

$$C_{ww}(e^{j\omega}) = |W(e^{j\omega})|^2. \tag{10.77}$$

According to Eq. (10.76), both the periodogram and the modified periodogram are biased estimates of the power spectrum, since $\mathcal{E}\{I(\omega)\}$ is not equal to $P_{xx}(\omega)$. Indeed, we see that the bias arises as a result of convolution of the true power spectrum with the Fourier transform of the aperiodic autocorrelation of the data window. If we increase the window length, we expect that $W(e^{j\omega})$ should become more concentrated around $\omega = 0$, and thus $C_{ww}(e^{j\omega})$ should look increasingly like a periodic impulse train. If the scale factor $1/(LU)$ is correctly chosen, then $\mathcal{E}\{I(\omega)\}$ should approach $P_{xx}(\omega)$ as

$C_{ww}(e^{j\omega})$ approaches a periodic impulse train. The scale can be adjusted by choosing the normalizing constant U so that

$$\frac{1}{2\pi LU} \int_{-\pi}^{\pi} |W(e^{j\omega})|^2 d\omega = \frac{1}{LU} \sum_{n=0}^{L-1} (w[n])^2 = 1, \qquad (10.78)$$

or

$$U = \frac{1}{L} \sum_{n=0}^{L-1} (w[n])^2. \qquad (10.79)$$

For the rectangular window, we should choose $U = 1$, while other data windows would require a value of $0 < U < 1$ if $w[n]$ is normalized to a maximum value of 1. Alternatively, the normalization can be absorbed into the amplitude of $w[n]$. Therefore, if properly normalized, the periodogram and modified periodogram are both asymptotically unbiased; i.e., the bias approaches zero as the window length increases.

To examine whether the periodogram is a consistent estimate or becomes a consistent estimate as the window length increases, it is necessary to consider the behavior of the variance of the periodogram. An expression for the variance of the periodogram is very difficult to obtain even in the simplest cases. However, it has been shown (see Jenkins and Watts, 1968) that over a wide range of conditions, as the window length increases,

$$\text{var}[I(\omega)] \simeq P_{xx}^2(\omega). \qquad (10.80)$$

That is, the variance of the periodogram estimate is approximately the same size as the square of the power spectrum that we are estimating. Therefore, since the variance does not asymptotically approach zero with increasing window length, the periodogram is not a consistent estimate.

The properties of the periodogram estimate of the power spectrum just discussed are illustrated in Figure 10.24, which shows periodogram estimates of white noise using rectangular windows of lengths $L = 16, 64, 256,$ and 1024. The sequence $x[n]$

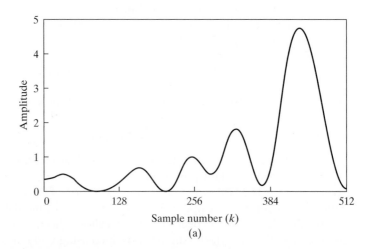

Figure 10.24   Periodograms of pseudorandom white-noise sequence. (a) Window length $L = 16$ and DFT length $N = 1024$.

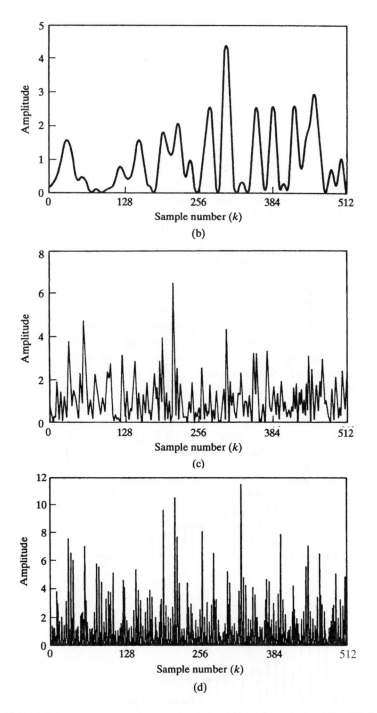

Figure 10.24 (*continued*) (b) $L = 64$ and $N = 1024$. (c) $L = 256$ and $N = 1024$. (d) $L = 1024$ and $N = 1024$.

was obtained from a pseudorandom-number generator whose output was scaled so that $|x[n]| \leq \sqrt{3}$. A good random-number generator produces a uniform distribution of amplitudes, and the measured sample-to-sample correlation is small. Thus, the power spectrum of the output of the random-number generator could be modeled in this case by $P_{xx}(\omega) = \sigma_x^2 = 1$ for all ω. For each of the four rectangular windows, the periodogram was computed with normalizing constant $U = 1$ and at frequencies $\omega_k = 2\pi k/N$ for $N = 1024$ using the DFT. That is,

$$I[k] = I(\omega_k) = \frac{1}{L}|V[k]|^2 = \frac{1}{L}\left|\sum_{n=0}^{L-1} w[n]x[n]e^{-j(2\pi/N)kn}\right|^2. \qquad (10.81)$$

In Figure 10.24, the DFT values are connected by straight lines for purposes of display. Recall that $I(\omega)$ is real and an even function of ω, so we only need to plot $I[k]$ for $0 \leq k \leq N/2$ corresponding to $0 \leq \omega \leq \pi$. Note that the spectrum estimate fluctuates more rapidly as the window length L increases. This behavior can be understood by recalling that, although we view the periodogram method as a direct computation of the spectrum estimate, we have seen that the underlying correlation estimate of Eq. (10.69) is, in effect, Fourier transformed to obtain the periodogram. Figure 10.25 illustrates a windowed sequence, $x[n]w[n]$, and a shifted version, $x[n + m]w[n + m]$, as required in Eq. (10.69). From this figure, we see that $(L - m)$ signal values are involved in computing a particular correlation lag value $c_{vv}[m]$. Thus, when m is close to L, only a few values of $x[n]$ are involved in the computation, and we expect that the estimate of the correlation sequence will be considerably more inaccurate for these values of m and consequently will also show considerable variation between adjacent values of m. On the other hand, when m is small, many more samples are involved, and the variability of $c_{vv}[m]$ with m should not be as great. The variability at large values of m manifests itself in the Fourier transform as fluctuations at all frequencies, and thus, for large L, the periodogram estimate tends to vary rapidly with frequency. Indeed, it can be shown (see Jenkins

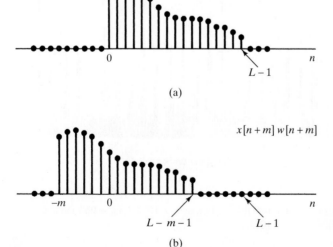

(a)

(b)

Figure 10.25 Illustration of sequences involved in Eq. (10.69). (a) A finite-length sequence. (b) Shifted sequence for $m > 0$.

and Watts, 1968) that if $N = L$, the periodogram estimates at the DFT frequencies $2\pi k/N$ become uncorrelated. Since, as N increases, the DFT frequencies get closer together, this behavior is inconsistent with our goal of obtaining a good estimate of the power spectrum. We would prefer to obtain a smooth spectrum estimate without random variations resulting from the estimation process. This can be accomplished by averaging multiple independent periodogram estimates to reduce the fluctuations.

10.5.3 Periodogram Averaging

The averaging of periodograms in spectrum estimation was first studied extensively by Bartlett (1953); later, after fast algorithms for computing the DFT were developed, Welch (1970) combined these computational algorithms with the use of a data window $w[n]$ to develop the method of averaging modified periodograms. In periodogram averaging, a data sequence $x[n], 0 \le n \le Q-1$, is divided into segments of length-L samples, with a window of length L applied to each; i.e., we form the segments

$$x_r[n] = x[rR + n]w[n], \qquad 0 \le n \le L - 1. \tag{10.82}$$

If $R < L$ the segments overlap, and for $R = L$ the segments are contiguous. Note that Q denotes the length of the available data. The total number of segments depends on the values of, and relationship among, R, L, and Q. Specifically, there will be K full-length segments, where K is the largest integer for which $(K - 1)R + (L - 1) \le Q - 1$. The periodogram of the rth segment is

$$I_r(\omega) = \frac{1}{LU}|X_r(e^{j\omega})|^2, \tag{10.83}$$

where $X_r(e^{j\omega})$ is the DTFT of $x_r[n]$. Each $I_r(\omega)$ has the properties of a periodogram, as described previously. Periodogram averaging consists of averaging the K periodogram estimates $I_r(\omega)$; i.e., we form the time-averaged periodogram defined as

$$\bar{I}(\omega) = \frac{1}{K} \sum_{r=0}^{K-1} I_r(\omega). \tag{10.84}$$

To examine the bias and variance of $\bar{I}(\omega)$, let us take $L = R$, so that the segments do not overlap, and assume that $\phi_{xx}[m]$ is small for $m > L$; i.e., signal samples more than L apart are approximately uncorrelated. If we assume that the periodograms $I_r(\omega)$ are identically distributed independent random variables, then the expected value of $\bar{I}(\omega)$ is

$$\mathcal{E}\{\bar{I}(\omega)\} = \frac{1}{K} \sum_{r=0}^{K-1} \mathcal{E}\{I_r(\omega)\}, \tag{10.85}$$

or, since we assume that the periodograms are independent and identically distributed,

$$\mathcal{E}\{\bar{I}(\omega)\} = \mathcal{E}\{I_r(\omega)\} \qquad \text{for any } r. \tag{10.86}$$

From Eq. (10.76), it follows that

$$\mathcal{E}\{\bar{I}(\omega)\} = \mathcal{E}\{I_r(\omega)\} = \frac{1}{2\pi LU} \int_{-\pi}^{\pi} P_{xx}(\theta)C_{ww}(e^{j(\omega-\theta)})d\theta, \tag{10.87}$$

where L is the window length. When the window $w[n]$ is the rectangular window, the method of averaging periodograms is called *Bartlett's procedure,* and in this case it can be shown that

$$c_{ww}[m] = \begin{cases} L - |m|, & |m| \le (L-1), \\ 0 & \text{otherwise,} \end{cases} \tag{10.88}$$

and, therefore,

$$C_{ww}(e^{j\omega}) = \left(\frac{\sin(\omega L/2)}{\sin(\omega/2)} \right)^2. \tag{10.89}$$

That is, the expected value of the average periodogram spectrum estimate is the convolution of the true power spectrum with the Fourier transform of the triangular sequence $c_{ww}[n]$ that results as the autocorrelation of the rectangular window. Thus, the average periodogram is also a biased estimate of the power spectrum.

To examine the variance, we use the fact that, in general, the variance of the average of K independent identically distributed random variables is $1/K$ times the variance of each individual random variable. (See Bertsekas and Tsitsiklis, 2008.) Therefore, the variance of the average periodogram is

$$\text{var}[\bar{I}(\omega)] = \frac{1}{K} \text{var}[I_r(\omega)], \tag{10.90}$$

or, with Eq. (10.80), it follows that

$$\text{var}[\bar{I}(\omega)] \simeq \frac{1}{K} P_{xx}^2(\omega). \tag{10.91}$$

Consequently, the variance of $\bar{I}(\omega)$ is inversely proportional to the number of periodograms averaged, and as K increases, the variance approaches zero.

From Eq. (10.89), we see that as L, the length of the segment $x_r[n]$, increases, the main lobe of $C_{ww}(e^{j\omega})$ decreases in width, and consequently, from Eq. (10.87), $\mathcal{E}\{\bar{I}(\omega)\}$ more closely approximates $P_{xx}(\omega)$. However, for fixed total data length Q, the total number of segments (assuming that $L = R$) is Q/L; therefore, as L increases, K decreases. Correspondingly, from Eq. (10.91), the variance of $\bar{I}(\omega)$ will increase. Thus, as is typical in statistical estimation problems, for a fixed data length there is a trade off between bias and variance. However, as the data length Q increases, both L and K can be allowed to increase, so that as Q approaches ∞, the bias and variance of $\bar{I}(\omega)$ can approach zero. Consequently, periodogram averaging provides an asymptotically unbiased, consistent estimate of $P_{xx}(\omega)$.

The preceding discussion assumed that nonoverlapping rectangular windows were used in computing the time-dependent periodograms. Welch (1970) showed that if a different window shape is used, the variance of the average periodogram still behaves, as in Eq. (10.91). Welch also considered the case of overlapping windows and showed that if the overlap is one-half the window length, the variance is further reduced by almost a factor of 2, due to the doubling of the number of sections. Greater overlap does not continue to reduce the variance, because the segments become decreasingly independent as the overlap increases.

10.5.4 Computation of Average Periodograms Using the DFT

As with the periodogram, the average periodogram can be explicitly evaluated only at a discrete set of frequencies. Because of the availability of the FFT algorithms for computing the DFT, a particularly convenient and widely used choice is the set of frequencies $\omega_k = 2\pi k/N$ for an appropriate choice of N. From Eq. (10.84), we see that if the DFT of $x_r[n]$ is substituted for the Fourier transform of $x_r[n]$ in Eq. (10.83), we obtain samples of $\bar{I}(\omega)$ at the DFT frequencies $\omega_k = 2\pi k/N$ for $k = 0, 1, \ldots, N-1$. Specifically, with $X_r[k]$ denoting the DFT of $x_r[n]$,

$$I_r[k] = I_r(\omega_k) = \frac{1}{LU}|X_r[k]|^2, \tag{10.92a}$$

$$\bar{I}[k] = \bar{I}(\omega_k) = \frac{1}{K}\sum_{r=0}^{K-1} I_r[k]. \tag{10.92b}$$

It is worthwhile to note the relationship between periodogram averaging and the time-dependent Fourier transform as discussed in detail in Section 10.3. Equation (10.92a) shows that, except for the introduction of the normalizing constant $1/(LU)$, each individual periodogram is simply the magnitude-squared of the time-dependent Fourier transform at time rR and frequency $2\pi k/N$. Thus, for each frequency index k, the average power spectrum estimate at frequency corresponding to k is the time average of the time-sampled time-dependent Fourier transform. This can be visualized by considering the spectrograms in Figure 10.22. The value $\bar{I}[k]$ is simply the average along a horizontal line at frequency $2\pi k/N$ (or $2\pi k/(NT)$ in analog frequency).[10] Averaging the wideband spectrogram implies that the resulting power spectrum estimate will be smooth when considered as a function of frequency, while the narrowband condition corresponds to longer time windows and thus, less smoothness in frequency.

We denote $I_r(2\pi k/N)$ as the sequence $I_r[k]$ and $\bar{I}(2\pi k/N)$ as the sequence $\bar{I}[k]$. According to Eqs. (10.92a) and (10.92b), the average periodogram estimate of the power spectrum is computed at N equally spaced frequencies by averaging the magnitude of the DFTs of the windowed data segments with the normalizing factor LU. This method of power spectrum estimation provides a very convenient framework within which to trade off between resolution and variance of the spectrum estimate. It is particularly simple and efficient to implement using the FFT algorithms discussed in Chapter 9. An important advantage of the method over those to be discussed in Section 10.6 is that the spectrum estimate is always nonnegative.

10.5.5 An Example of Periodogram Analysis

Power spectrum analysis is a valuable tool for modeling signals, and it also can be used to detect signals, particularly when it comes to finding hidden periodicities in sampled

---

[10]Note that the spectrogram is normally computed such that the windowed segments overlap considerably as r varies, while in periodogram averaging R is normally equal to the window length or half the window length.

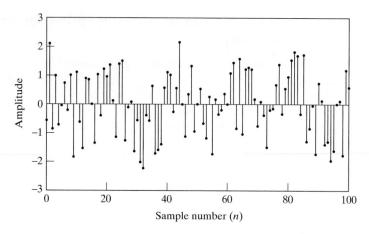

Figure 10.26   Cosine sequence with white noise, as in Eq. (10.93).

signals. As an example of this type of application of the average periodogram method, consider the sequence

$$x[n] = A \cos(\omega_0 n + \theta) + e[n], \qquad (10.93)$$

where θ is a random variable uniformly distributed between 0 and 2π, is independent of $e[n]$, and $e[n]$ is a zero-mean white-noise sequence that has a constant power spectrum; i.e., $P_{ee}(\omega) = \sigma_e^2$ for all ω. In signal models of this form, the cosine is generally the desired component and $e[n]$ is an undesired noise component. Often, in practical signal detection problems, we are interested in the case for which the power in the cosine signal is small compared with the noise power. It can be shown (see Problem 10.40) that over the base period of frequency $|\omega| \leq \pi$, the power spectrum for this signal is

$$P_{xx}(\omega) = \frac{A^2 \pi}{2} [\delta(\omega - \omega_0) + \delta(\omega + \omega_0)] + \sigma_e^2 \qquad \text{for } |\omega| \leq \pi. \qquad (10.94)$$

From Eqs. (10.87) and (10.94), it follows that the expected value of the average periodogram is

$$\mathcal{E}\{\bar{I}(\omega)\} = \frac{A^2}{4LU} [C_{ww}(e^{j(\omega - \omega_0)}) + C_{ww}(e^{j(\omega + \omega_0)})] + \sigma_e^2. \qquad (10.95)$$

Figures 10.26 and 10.27 show the use of the averaging method for a signal of the form of Eq. (10.93), with $A = 0.5$, $\omega_0 = 2\pi/21$, and random phase $0 \leq \theta < 2\pi$. The noise was uniformly distributed in amplitude such that $-\sqrt{3} < e[n] \leq \sqrt{3}$. Therefore, it is easily shown that $\sigma_e^2 = 1$. The mean of the noise component is zero. Figure 10.26 shows 101 samples of the sequence $x[n]$. Since the noise component $e[n]$ has a maximum amplitude $\sqrt{3}$, the cosine component in the sequence $x[n]$ (having period 21) is not visually apparent.

Figure 10.27 shows average periodogram estimates of the power spectrum for rectangular windows with amplitude 1, so that $U = 1$, and of lengths $L = 1024, 256, 64$, and 16, with the total record length $Q = 1024$ in all cases. Except for Figure 10.27(a), the windows overlap by one-half the window length. Figure 10.27(a) is the periodogram of the entire record, and Figures 10.27(b), (c), and (d) show the average periodogram

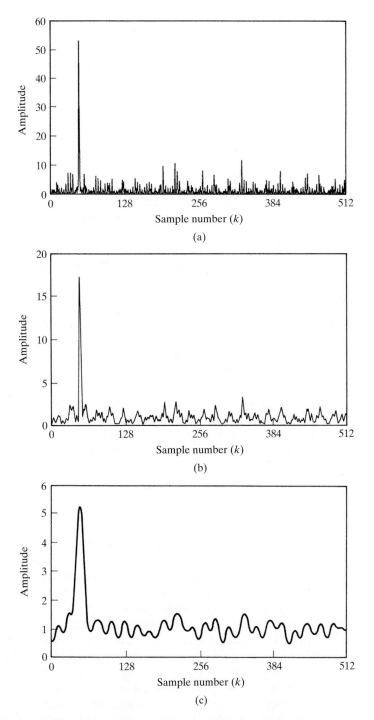

Figure 10.27 Example of average periodogram for signal of length $Q = 1024$. (a) Periodogram for window length $L = Q = 1024$ (only one segment). (b) $K = 7$ and $L = 256$ (overlap by $L/2$). (c) $K = 31$ and $L = 64$.

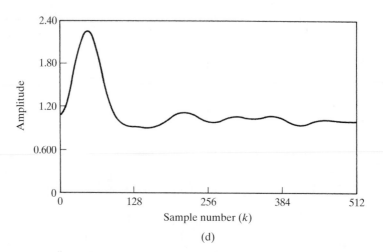

Figure 10.27   (*continued*) (d) $K = 127$ and $L = 16$.

for $K = 7, 31,$ and 127 segments, respectively. In all cases, the average periodogram was evaluated using 1024-point DFTs at frequencies $\omega_k = 2\pi k/1024$. (For window lengths $L < 1024$, the windowed sequence was augmented with zero-samples before computing the DFT.) Therefore, the frequency $\omega_0 = 2\pi/21$ lies between DFT frequencies $\omega_{48} = 2\pi 48/1024$ and $\omega_{49} = 2\pi 49/1024$.

In using such estimates of the power spectrum to detect the presence and/or the frequency of the cosine component, we might search for the highest peaks in the spectrum estimate and compare their size with that of the surrounding spectrum values. From Eqs. (10.89) and (10.95), the expected value of the average periodogram at the frequency ω_0 is

$$\mathcal{E}\{\bar{I}(\omega_0)\} = \frac{A^2 L}{4} + \sigma_e^2. \tag{10.96}$$

Thus, if the peak due to the cosine component is to stand out against the variability of the average periodogram, then in this special case, we must choose L so that $A^2 L/4 \gg \sigma_e^2$. This is illustrated by Figure 10.27(a), where L is as large as it can be for the record length Q. We see that $L = 1024$ gives a very narrow main lobe of the Fourier transform of the autocorrelation of the rectangular window, so it would be possible to resolve very closely spaced sinusoidal signals. Note that for the parameters of this example ($A = 0.5$, $\sigma_e^2 = 1$) and with $L = 1024$, the peak amplitude in the periodogram at frequency $2\pi/21$ is close, but not equal, to the expected value of 65. We also observe additional peaks in the periodogram with amplitudes greater than 10. Clearly, if the cosine amplitude A had been smaller by only a factor of 2, it is possible that its peak would have been confused with the inherent variability of the periodogram.

We have seen that the only sure way to reduce the variance of the spectrum estimate is to increase the record length of the signal. This is not always possible, and even if it is possible, longer records require more processing. We can reduce the variability of the estimate while keeping the record length constant if we use shorter windows and average over more sections. The cost of doing this is illustrated by parts (b), (c), and (d) of Figure 10.27. Note that as more sections are used, the variance of the spectrum estimate decreases, but in accordance with Eq. (10.96), so does the amplitude of the peak as a result of the cosine. Thus, we again face a trade-off. That the shorter windows reduce variability is clear, especially if we compare the high-frequency regions away from the peak in parts (a), (b) and (c) of Figure 10.27. Recall that the idealized power spectrum of the model for the pseudorandom-noise generator is a constant ($\sigma_e^2 = 1$) for all frequencies. In Figure 10.27(a) there are peaks as high as about 10 when the true spectrum is 1. In Figure 10.27(b), the variation away from 1 is less than about 3, and in Figure 10.27(c), the variation around 1 is less than 0.5. However, shorter windows also reduce the peak amplitude of any narrowband component, and they also degrade our ability to resolve closely spaced sinusoids. This reduction in peak amplitude is also clear from Figure 10.27. Again, if we were to reduce A by a factor of 2 in Figure 10.27(b), the peak height would be approximately 4, which is not much different from many of the other peaks in the high-frequency region. In Figure 10.27(c) a reduction of A by a factor of 2 would make the peak approximately 1.25, which would be indistinguishable from the other ripples in the estimate. In Figure 10.27(d), the window is very short, and thus the fluctuations of the spectrum estimate are greatly reduced, but the spectrum peak due to the cosine is very broad and barely above the noise even for $A = 0.5$. If the length were any smaller, spectral leakage from the negative-frequency component would cause there to be no distinct peak in the low-frequency region.

This example confirms that the average periodogram provides a straightforward method of trading off between spectral resolution and reduction of the variance of the spectrum estimate. Although the theme of the example was the detection of a sinusoid in noise, the average periodogram could also be used in signal modeling. The spectrum estimates of Figure 10.27 clearly suggest a signal model of the form of Eq. (10.93), and most of the parameters of the model could be estimated from the average periodogram power spectrum estimate.

10.6 SPECTRUM ANALYSIS OF RANDOM SIGNALS USING ESTIMATES OF THE AUTOCORRELATION SEQUENCE

In the previous section, we considered the periodogram as a direct estimate of the power spectrum of a random signal. The periodogram or the average periodogram is a direct estimate in the sense that it is obtained directly by Fourier transformation of the samples of the random signal. Another approach, based on the fact that the power density spectrum is the Fourier transform of the autocorrelation function, is to first obtain an estimate of the autocorrelation function $\hat{\phi}_{xx}[m]$ for a finite set of lag values $-M \le m \le M$, and then apply a window $w_c[m]$ before computing the DTFT of this estimate. This approach to power spectrum estimation is often referred to as the *Blackman–Tukey method*. (See Blackman and Tukey, 1958.) In this section, we explore some of the important facets of this approach and show how the DFT can be used to implement it.

Let us assume, as before, that we are given a finite record of a random signal $x[n]$. This sequence is denoted

$$v[n] = \begin{cases} x[n] & \text{for } 0 \leq n \leq Q - 1, \\ 0 & \text{otherwise.} \end{cases} \tag{10.97}$$

Consider an estimate of the autocorrelation sequence as

$$\hat{\phi}_{xx}[m] = \frac{1}{Q} c_{vv}[m], \tag{10.98a}$$

where, since $c_{vv}[-m] = c_{vv}[m]$,

$$c_{vv}[m] = \sum_{n=0}^{Q-1} v[n]v[n+m] = \begin{cases} \displaystyle\sum_{n=0}^{Q-|m|-1} x[n]x[n+|m|], & |m| \leq Q - 1, \\ 0 & \text{otherwise,} \end{cases} \tag{10.98b}$$

corresponding to the aperiodic correlation of a rectangularly windowed segment of $x[n]$ of length Q.

To determine the properties of this estimate of the autocorrelation sequence, we consider the mean and variance of the random variable $\hat{\phi}_{xx}[m]$. From Eqs. (10.98a) and (10.98b), it follows that

$$\mathcal{E}\{\hat{\phi}_{xx}[m]\} = \frac{1}{Q} \sum_{n=0}^{Q-|m|-1} \mathcal{E}\{x[n]x[n+|m|]\} = \frac{1}{Q} \sum_{n=0}^{Q-|m|-1} \phi_{xx}[m], \tag{10.99}$$

and since $\phi_{xx}[m]$ does not depend on n for a stationary random process,

$$\mathcal{E}\{\hat{\phi}_{xx}[m]\} = \begin{cases} \left(\dfrac{Q - |m|}{Q}\right) \phi_{xx}[m], & |m| \leq Q - 1, \\ 0 & \text{otherwise.} \end{cases} \tag{10.100}$$

From Eq. (10.100), we see that $\hat{\phi}_{xx}[m]$ is a biased estimate of $\phi_{xx}[m]$, since $\mathcal{E}\{\hat{\phi}_{xx}[m]\}$ is not equal to $\phi_{xx}[m]$, but the bias is small if $|m| \ll Q$. We see also that an unbiased estimator of the autocorrelation sequence for $|m| \leq Q - 1$ is

$$\check{\phi}_{xx}[m] = \left(\frac{1}{Q - |m|}\right) c_{vv}[m]; \tag{10.101}$$

i.e., the estimator is unbiased if we divide by the number of nonzero terms in the sum of lagged products involved in computing each value of $c_{vv}[m]$, rather than by the total number of samples in the data record.

The variance of the autocorrelation function estimates is difficult to compute, even with simplifying assumptions. However, approximate formulas for the variance of both $\hat{\phi}_{xx}[m]$ and $\check{\phi}_{xx}[m]$ can be found in Jenkins and Watts (1968). For our purposes here, it is sufficient to observe from Eq. (10.98b) that as $|m|$ approaches Q, fewer and fewer samples of $x[n]$ are involved in the computation of the autocorrelation estimate; therefore, the variance of the autocorrelation estimate can be expected to increase with increasing $|m|$. In the case of the periodogram, this increased variance affects the spectrum estimate at all frequencies, because all the autocorrelation lag values are implicitly involved in the computation of the periodogram. However, by explicitly computing the autocorrelation estimate, we are free to choose which correlation lag values to include when estimating the power spectrum. Thus, we define the power spectrum estimate

$$S(\omega) = \sum_{m=-(M-1)}^{M-1} \hat{\phi}_{xx}[m]w_c[m]e^{-j\omega m},\tag{10.102}$$

where $w_c[m]$ is a symmetric window of length $(2M - 1)$ applied to the estimated auto-correlation function. We require that the product of the autocorrelation sequence and the window be an even sequence when $x[n]$ is real, so that the power spectrum estimate will be a real, even function of ω. Therefore, the correlation window must be an even sequence. By limiting the length of the correlation window so that $M \ll Q$, we include only autocorrelation estimates for which the variance is low.

The mechanism by which windowing the autocorrelation sequence reduces the variance of the power spectrum estimate is best understood in the frequency domain. From Eqs. (10.68), (10.69), and (10.98b), it follows that, with $w[n] = 1$ for $0 \le n \le (Q - 1)$, i.e., a rectangular window, the periodogram is the Fourier transform of the autocorrelation estimate $\hat{\phi}_{xx}[m]$; i.e.,

$$\hat{\phi}_{xx}[m] = \frac{1}{Q}c_{vv}[m] \overset{\mathcal{F}}{\longleftrightarrow} \frac{1}{Q}|V(e^{j\omega})|^2 = I(\omega).\tag{10.103}$$

Therefore, from Eq. (10.102), the spectrum estimate obtained by windowing of $\hat{\phi}_{xx}[m]$ is the convolution

$$S(\omega) = \frac{1}{2\pi}\int_{-\pi}^{\pi} I(\theta)W_c(e^{j(\omega-\theta)})d\theta.\tag{10.104}$$

From Eq. (10.104), we see that the effect of applying the window $w_c[m]$ to the autocorrelation estimate is to convolve the periodogram with the Fourier transform of the autocorrelation window. This will smooth the rapid fluctuations of the periodogram spectrum estimate. The shorter the correlation window, the smoother the spectrum estimate will be, and vice versa.

The power spectrum $P_{xx}(\omega)$ is a nonnegative function of frequency, and the periodogram and the average periodogram automatically have this property by definition. In contrast, from Eq. (10.104), it is evident that nonnegativity is not guaranteed for $S(\omega)$, unless we impose the further condition that

$$W_c(e^{j\omega}) \ge 0 \qquad \text{for } -\pi < \omega \le \pi.\tag{10.105}$$

This condition is satisfied by the Fourier transform of the triangular (Bartlett) window, but it is not satisfied by the rectangular, Hanning, Hamming, or Kaiser windows. Therefore, although these latter windows have smaller side lobes than the triangular window, spectral leakage may cause negative spectrum estimates in low-level regions of the spectrum.

The expected value of the smoothed periodogram is

$$\mathcal{E}\{S(\omega)\} = \sum_{m=-(M-1)}^{M-1} \mathcal{E}\{\hat{\phi}_{xx}[m]\}w_c[m]e^{-j\omega m}$$

$$= \sum_{m=-(M-1)}^{M-1} \phi_{xx}[m]\left(\frac{Q - |m|}{Q}\right)w_c[m]e^{-j\omega m}.\tag{10.106}$$

If $Q \gg M$, the term $(Q - |m|)/Q$ in Eq. (10.106) can be neglected,[11] so we obtain

$$\mathcal{E}\{S(\omega)\} \cong \sum_{m=-(M-1)}^{M-1} \phi_{xx}[m] w_c[m] e^{-j\omega m} = \frac{1}{2\pi} \int_{-\pi}^{\pi} P_{xx}(\theta) W_c(e^{j(\omega-\theta)}) d\theta. \quad (10.107)$$

Thus, the windowed autocorrelation estimate leads to a biased estimate of the power spectrum. Just as with the average periodogram, it is possible to trade spectral resolution for reduced variance of the spectrum estimate. If the length of the data record is fixed, we can have lower variance if we are willing to accept poorer resolution of closely spaced narrowband spectral components, or we can have better resolution if we can accept higher variance. If we are free to observe the signal for a longer time (i.e., increase the length Q of the data record), then both the resolution and the variance can be improved. The spectrum estimate $S(\omega)$ is asymptotically unbiased if the correlation window is normalized so that

$$\frac{1}{2\pi} \int_{-\pi}^{\pi} W_c(e^{j\omega}) d\omega = 1 = w_c[0]. \quad (10.108)$$

With this normalization, as we increase Q together with the length of the correlation window, the Fourier transform of the correlation window approaches a periodic impulse train and the convolution of Eq. (10.107) duplicates $P_{xx}(\omega)$.

The variance of $S(\omega)$ has been shown (see Jenkins and Watts, 1968) to be of the form

$$\text{var}[S(\omega)] \simeq \left(\frac{1}{Q} \sum_{m=-(M-1)}^{M-1} w_c^2[m] \right) P_{xx}^2(\omega). \quad (10.109)$$

Comparing Eq. (10.109) with the corresponding result in Eq. (10.80) for the periodogram leads to the conclusion that, to reduce the variance of the spectrum estimate, we should choose M and the window shape, possibly subject to the condition of Eq. (10.105), so that the factor

$$\left(\frac{1}{Q} \sum_{m=-(M-1)}^{M-1} w_c^2[m] \right) \quad (10.110)$$

is as small as possible. Problem 10.37 deals with the computation of this variance reduction factor for several commonly used windows.

Estimation of the power spectrum based on the Fourier transform of an estimate of the autocorrelation function is a clear alternative to the method of averaging periodograms. It is not necessarily better in any general sense; it simply has different features, and its implementation would be different. In some situations, it may be desirable to compute estimates of both the autocorrelation sequence and the power spectrum, in which case it would be natural to use the method of this section. Problem 10.43 explores the issue of determining an autocorrelation estimate from the average periodogram.

---

[11] More precisely, we could define an effective correlation window $w_e[m] = w_c[m](Q - |m|)/Q$.

10.6.1 Computing Correlation and Power Spectrum Estimates Using the DFT

The autocorrelation estimate

$$\hat{\phi}_{xx}[m] = \frac{1}{Q} \sum_{n=0}^{Q-|m|-1} x[n]x[n+|m|] \tag{10.111}$$

is required for $|m| \leq M - 1$ in the method of power spectrum estimation that we are considering. Since $\hat{\phi}_{xx}[-m] = \hat{\phi}_{xx}[m]$, it is necessary to compute Eq. (10.111) only for nonnegative values of m, i.e., for $0 \leq m \leq M - 1$. The DFT and its associated fast computational algorithms can be used to advantage in the computation of $\hat{\phi}_{xx}[m]$, if we observe that $\hat{\phi}_{xx}[m]$ is the aperiodic discrete convolution of the finite-length sequence $x[n]$ with $x[-n]$. If we compute $X[k]$, the N-point DFT of $x[n]$, and multiply by $X^*[k]$, we obtain $|X[k]|^2$, which corresponds to the circular convolution of the finite-length sequence $x[n]$ with $x[((-n))_N]$, i.e., a *circular autocorrelation*. As our discussion in Section 8.7 suggests, and as developed in Problem 10.34, it should be possible to augment the sequence $x[n]$ with zero-valued samples and force the circular autocorrelation to be equal to the desired aperiodic autocorrelation over the interval $0 \leq m \leq M - 1$.

To see how to choose N for the DFT, consider Figure 10.28. Figure 10.28(a) shows the two sequences $x[n]$ and $x[n+m]$ as functions of n for a particular positive value of m. Figure 10.28(b) shows the sequences $x[n]$ and $x[((n+m))_N]$ that are involved in the circular autocorrelation corresponding to $|X[k]|^2$. Clearly, the circular autocorrelation will be equal to $Q\hat{\phi}_{xx}[m]$ for $0 \leq m \leq M - 1$ if $x[((n+m))_N]$ does not wrap around and overlap $x[n]$ when $0 \leq m \leq M - 1$. From Figure 10.28(b), it follows that this will be the case whenever $N - (M - 1) \geq Q$ or $N \geq Q + M - 1$.

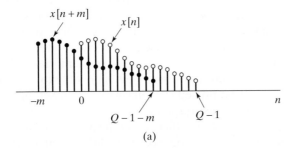

(a)

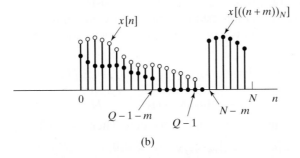

(b)

Figure 10.28   Computation of the circular autocorrelation. (a) $x[n]$ and $x[n+m]$ for a finite-length sequence of length Q. (b) $x[n]$ and $x[((n+m))_N]$ as in circular correlation.

In summary, we can compute $\hat{\phi}_{xx}[m]$ for $0 \leq m \leq M-1$ by the following procedure:

1. Form an N-point sequence by augmenting $x[n]$ with $(M-1)$ zero-samples.
2. Compute the N-point DFT,

$$X[k] = \sum_{n=0}^{N-1} x[n]e^{-j(2\pi/N)kn} \qquad \text{for } k = 0, 1, \ldots, N-1.$$

3. Compute

$$|X[k]|^2 = X[k]X^*[k] \qquad \text{for } k = 0, 1, \ldots, N-1.$$

4. Compute the inverse DFT of $|X[k]|^2$ to obtain

$$\tilde{c}_{vv}[m] = \frac{1}{N}\sum_{k=0}^{N-1} |X[k]|^2 e^{j(2\pi/N)km} \qquad \text{for } m = 0, 1, \ldots, N-1.$$

5. Divide the resulting sequence by Q to obtain the autocorrelation estimate

$$\hat{\phi}_{xx}[m] = \frac{1}{Q}\tilde{c}_{vv}[m] \qquad \text{for } m = 0, 1, \ldots, M-1.$$

This is the desired set of autocorrelation values, which can be extended symmetrically for negative values of m.

If M is small, it may be more efficient simply to evaluate Eq. (10.111) directly. In this case, the amount of computation is proportional to $Q \cdot M$. In contrast, if the DFTs in this procedure are computed using one of the FFT algorithms discussed in Chapter 9 with $N \geq Q+M-1$, the amount of computation will be approximately proportional to $N \log_2 N$ for N a power of 2. Consequently, for sufficiently large values of M, use of the FFT is more efficient than direct evaluation of Eq. (10.111). The exact break-even value of M will depend on the particular implementation of the DFT computations; however, as shown by Stockham (1966), this value would probably be less than $M = 100$.

To reduce the variance of the estimate of the autocorrelation sequence or the power spectrum estimated from it, we must use large values of the record length Q. This is not generally a problem with computers having large memories and fast processors. However, since M is generally much less than Q, it is possible to section the sequence $x[n]$ in a manner similar to the procedures that were discussed in Section 8.7.3 for convolution of a finite-length impulse response with an indefinitely long input sequence. Rader (1970) presented a particularly efficient and flexible procedure that uses many of the properties of the DFT of real sequences to reduce the amount of computation required. The development of this technique is the basis for Problem 10.44.

Once the autocorrelation estimate has been computed, samples of the power spectrum estimate $S(\omega)$ can be computed at frequencies $\omega_k = 2\pi k/N$ by forming the finite-length sequence

$$s[m] = \begin{cases} \hat{\phi}_{xx}[m]w_c[m], & 0 \leq m \leq M-1, \\ 0, & M \leq m \leq N-M, \\ \hat{\phi}_{xx}[N-m]w_c[N-m], & N-M+1 \leq m \leq N-1, \end{cases} \qquad (10.112)$$

where $w_c[m]$ is the symmetric correlation window. Then the DFT of $s[m]$ is

$$S[k] = S(\omega)|_{\omega=2\pi k/N}, \qquad k = 0, 1, \ldots, N-1, \qquad (10.113)$$

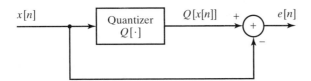

Figure 10.29   Procedure for obtaining quantization noise sequence.

where $S(\omega)$ is the Fourier transform of the windowed autocorrelation sequence as defined by Eq. (10.102). Note that N can be chosen as large as is convenient and practical, thereby providing samples of $S(\omega)$ at closely spaced frequencies. However, as our discussions in this chapter have consistently shown, the frequency resolution is always determined by the length and shape of the window $w_c[m]$.

10.6.2 Estimating the Power Spectrum of Quantization Noise

In Chapter 4, we assumed that the error introduced by quantization has the properties of a white-noise random process. The techniques discussed so far in this chapter were used to compute the power spectrum estimates of Figure 4.60 that were used to suggest the validity of this approximation. In this section, we provide additional examples of the use of estimates of the autocorrelation sequence and power spectrum estimation in studying the properties of quantization noise. The discussion will reinforce our confidence in the white-noise model, and it will also offer an opportunity to point out some practical aspects of power spectrum estimation.

Consider the experiment depicted in Figure 10.29. A lowpass-filtered speech signal $x_c(t)$ was sampled at a 16-KHz rate, yielding the sequence of samples $x[n]$ that were plotted in Figure 10.21.[12] These samples were quantized with a 10-bit linear quantizer $(B = 9)$, and the corresponding error sequence $e[n] = Q[x[n]] - x[n]$ was computed. Figure 10.30 shows 2000 consecutive samples of the speech signal plotted on the first and third lines of the graph. The second and fourth lines show the corresponding quantization error sequence. Visual inspection and comparison of these two plots tends to strengthen our belief in the previously assumed model; i.e., that the noise appears to vary randomly throughout the range $-2^{-(B+1)} < e[n] \leq 2^{-(B+1)}$. However, such qualitative observations can be misleading. The flatness of the quantization noise spectrum can be verified only by estimating the autocorrelation sequence and power spectrum of the quantization noise $e[n]$.

Figure 10.31 shows estimates of the autocorrelation and power spectrum of the quantization noise for a record length of $Q = 3000$ samples. The autocorrelation sequence estimate was calculated over the range of lags $|m| \leq 100$ using Eqs. (10.98a) and (10.98b). The resulting estimate is shown in Figure 10.31(a). Over this range, $-1.45 \times 10^{-8} \leq \hat{\phi}[m] \leq 1.39 \times 10^{-8}$ except for $\hat{\phi}[0] = 3.17 \times 10^{-7}$. The autocorrelation estimate suggests that the sample-to-sample correlation of the noise sequence is quite low. The resulting autocorrelation estimate was multiplied by Bartlett windows

---

[12] Although the samples were originally quantized to 12 bits by the A/D converter, for purposes of this experiment, they were scaled to a maximum value of 1, and a small amount of random noise was added to the samples. We assume that these samples are "unquantized," i.e., we consider the 12-bit samples to effectively be unquantized relative to the subsequent quantization that we are applying in this discussion.

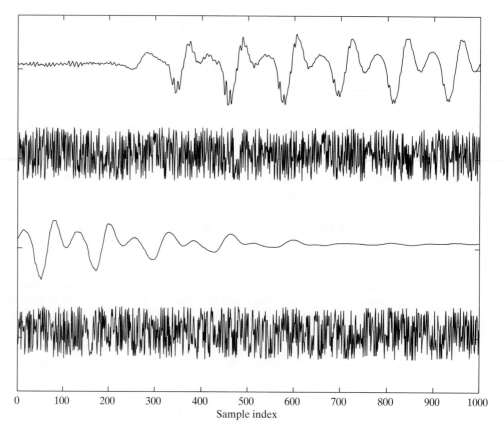

Figure 10.30   Speech waveform (first and third lines) and the corresponding quantization error (second and fourth lines) for 10-bit quantization (magnified 2^9 times). Each line corresponds to 1000 consecutive samples connected by straight lines for convenience in plotting.

with $M = 100$ and $M = 50$. The windows are shown in Figure 10.31 superimposed on $\hat{\phi}[m]$ (with scaling so that they can be plotted on the same axes) and the corresponding spectrum estimates, computed as discussed in Section 10.6.1, are shown in Figure 10.31(b).

As seen in Figure 10.31(b), the Blackman–Tukey spectrum estimate for $M = 100$ (the thin continuous line) shows somewhat erratic fluctuations about the dashed line plotted at the spectrum level $10 \log_{10}(2^{-18}/12) = -64.98$ dB (the value of the white power spectrum with $\sigma_e^2 = 2^{-2B}/12$ for $B = 9$). The heavy line shows the power spectrum estimate for $M = 50$. We see from Figure 10.31(b) that the spectrum estimate is within ± 2 dB of the spectrum of the white-noise approximation for $B + 1 = 10$ for all frequencies. As discussed in Section 10.6, the shorter window gives smaller variance and a smoother spectrum estimate resulting from the lower frequency resolution of the shorter window. In either case, the spectrum estimate seems to support the validity of the white-noise model for quantization noise.

Although we have computed quantitative estimates of the autocorrelation and the power spectrum, our interpretation of these measurements has been only qualitative. It

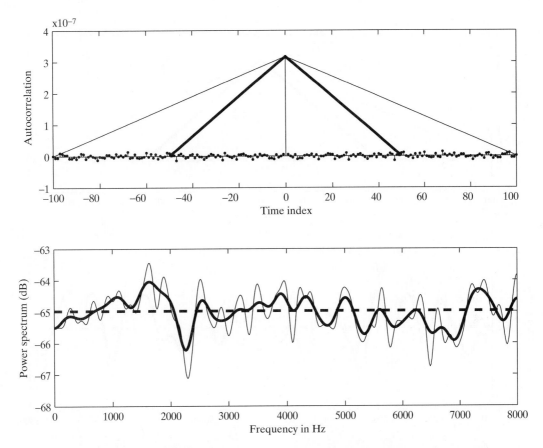

Figure 10.31 (a) Autocorrelation estimate for 10-bit quantization noise for $|m| \leq 100$ with record length $Q = 3,000$. (b) Power spectrum estimates by the Blackman–Tukey method using Bartlett windows with $M = 100$ and $M = 50$. (Dashed line shows level of $10 \log_{10}(2^{-18}/12)$.)

is reasonable now to wonder how small the autocorrelation would be if $e[n]$ were really a white-noise process. To give quantitative answers to such questions, confidence intervals for our estimates could be computed and statistical decision theory applied. (See Jenkins and Watts (1968), for some tests for white noise.) In many cases, however, this additional statistical treatment is not necessary. In a practical setting, we are often comfortable and content simply with the observation that the normalized autocorrelation is very small everywhere, except at $m = 0$.

Among the many important insights of this chapter is that the estimate of the autocorrelation and power spectrum of a stationary random process should improve if the record length is increased. This is illustrated by Figure 10.32, which corresponds to Figure 10.31, except that Q was increased to 30,000 samples. Recall that the variance of the autocorrelation estimate is proportional to $1/Q$. Thus, increasing Q from 3000 to 30,000 should bring about a tenfold reduction in the variance of the estimate. A comparison of Figures 10.31(a) and 10.32(a) seems to verify this result. For $Q = 3000$, the estimate

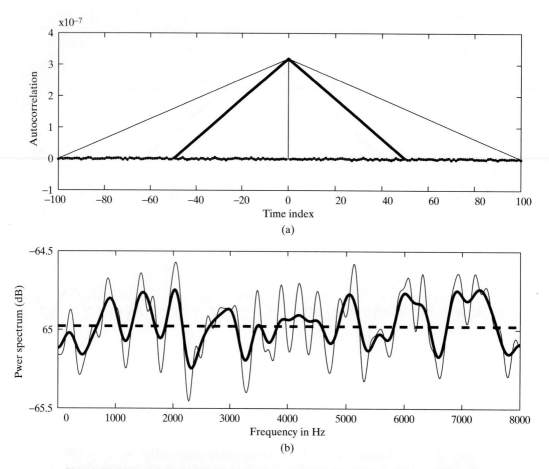

Figure 10.32 (a) Autocorrelation estimate for 10-bit quantization noise; record length $Q = 30,000$. (b) Power spectrum estimates by the Blackman–Tukey method using Bartlett windows with $M = 100$ and $M = 50$.

falls between the limits $-1.45 \times 10^{-8} \leq \hat{\phi}[m] \leq 1.39 \times 10^{-8}$, while for $Q = 30,000$, the limits are $-4.5 \times 10^{-9} \leq \hat{\phi}[m] \leq 4.15 \times 10^{-9}$. Comparing the range of variation for $Q = 3000$ with the range for $Q = 30,000$ indicates that the reduction is consistent with the tenfold reduction in variance that we expected.[13] We note from Eq. (10.110) that a similar reduction in variance of the spectrum estimate is also expected. This is again evident in comparing Figure 10.31(b) with Figure 10.32(b). (Be sure to note that the scales are different between the two sets of plots.) The variation about the white-noise approximate spectrum level is only ±0.5 dB in the case of the longer record length. Note that the spectrum estimates in Figure 10.32(b) display the same trade off between variance and resolution.

In Chapter 4 we argued that the white-noise model was reasonable, as long as the quantization step size was small. When the number of bits is small, this condition does

---

[13]Recall that a reduction in variance by a factor of 10 corresponds to a reduction in amplitude by a factor of $\sqrt{10} \approx 3.16$.

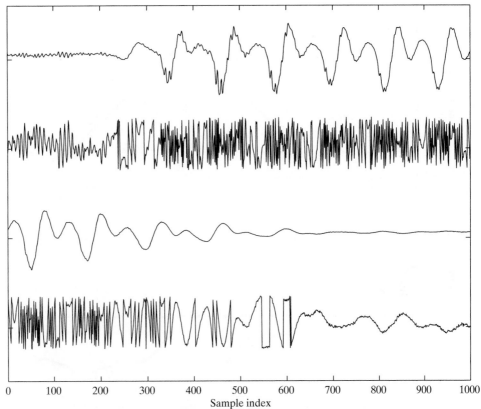

Figure 10.33   Speech waveform (first and third lines) and the corresponding quantization error (second and fourth lines) for 4-bit quantization (magnified 2^3 times). Each line corresponds to 1000 consecutive samples connected by straight lines for convenience in plotting.

not hold. To see the effect on the quantization noise spectrum, the previous experiment was repeated using only 16 quantization levels, or 4 bits. Figure 10.33 shows the speech waveform and quantization error for 4-bit quantization. Note that some portions of the error waveform tend to look very much like the original speech waveform. We would expect this to be reflected in the estimate of the power spectrum.

Figure 10.34 shows the autocorrelation and power spectrum estimates of the error sequence for 4-bit quantization for a record length of 30,000 samples. In this case, the autocorrelation shown in Figures 10.34(a) is much less like the ideal autocorrelation for white noise. This is not surprising in view of the obvious correlation between the signal and noise displayed in Figure 10.33. Figure 10.34(b) shows the power spectrum estimates for Bartlett windows with $M = 100$ and $M = 50$, respectively. Clearly, the spectrum is not flat, although the general level reflects the average noise power. In fact, as we shall see, the quantization noise tends to have the general shape of the speech spectrum. Thus, the white-noise model for quantization noise can be viewed only as a rather crude approximation in this case, and it would be less valid for coarser quantization.

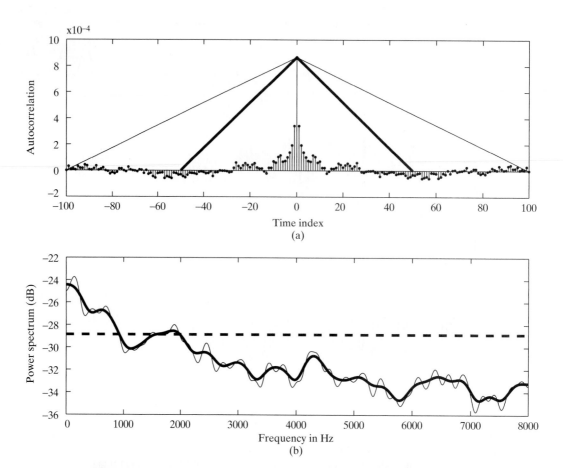

Figure 10.34   (a) Autocorrelation estimate for 4-bit quantization noise; record length $Q = 30,000$. (b) Power spectrum estimates by the Blackman–Tukey method using Bartlett windows with $M = 100$ and $M = 50$. (Dashed line shows level of $10 \log_{10}(2^{-6}/12)$.)

The example of this section illustrates how autocorrelation and power spectrum estimates can be used to support theoretical models. Specifically, we have demonstrated the validity of some of our basic assumptions in Chapter 4, and we have given an indication of how these assumptions break down for very crude quantization. This is only a rather simple, but useful, example that shows how the techniques of the current chapter can be applied in practice.

10.6.3 Estimating the Power Spectrum of Speech

We have seen that the time-dependent Fourier transform is particularly well-suited to the representation of speech signals, since it can track the time-varying nature of the speech signal. However, in some cases, it is useful to take a different point of view. In particular, even though the waveform of speech as in Figure 10.21 shows significant variability in time, as does its time-dependent Fourier transform in Figure 10.22, it is nevertheless possible to *assume* that it is a stationary random signal and apply our

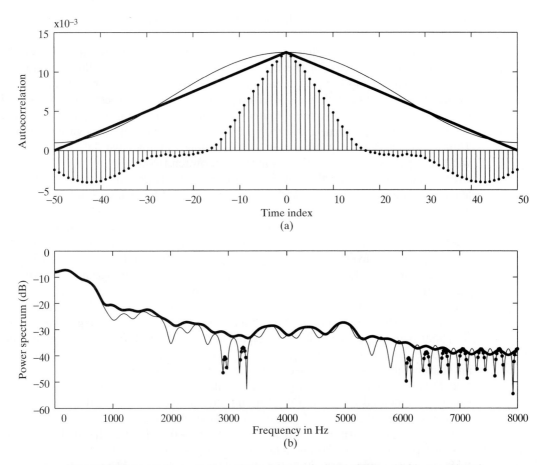

Figure 10.35   (a) Autocorrelation estimate for speech signal of Figure 10.21; record length $Q = 30,000$. (b) Power spectrum estimates by the Blackman–Tukey method using Bartlett window (heavy line) and Hamming window (light line) with $M = 50$.

long-term spectrum analysis techniques. These methods average over a time interval that is much longer than the changing events of speech. This gives a general spectrum shape that can be useful in designing speech coders and in determining the bandwidth requirements for speech transmission.

    Figure 10.35 shows an example of estimating the power spectrum of speech using the Blackman–Tukey method. The autocorrelation sequence estimated from $Q = 30,000$ samples of the speech signal in Figure 10.21 is shown in Figure 10.35(a), together with Bartlett and Hamming windows of length $2M + 1 = 101$. Figure 10.35(b) shows the corresponding power spectrum estimates. The two estimates are grossly similar but dramatically different in detail. This is because of the nature of the DTFTs of the windows. Both have the same main-lobe width $\Delta\omega_m = 8\pi/M$, however their side lobes are very different. The side lobes of the Bartlett window are strictly nonnegative, while those of the symmetric Hamming window (which are smaller than those of the Bartlett window) are negative for some frequencies. When convolved with the periodogram corresponding to the autocorrelation estimate, this yields the dramatically different results shown.

The Bartlett window guarantees a positive spectrum estimate for all frequencies. However, this is not true for the Hamming window. The effect of this is particularly pronounced in regions of rapid variability of the periodogram, where side lobes due to adjacent frequencies can cancel or interfere to produce negative spectrum estimates. The dots in Figure 10.35(b) show the frequencies where the spectrum estimate was negative. When plotting in dB, it is necessary to take the absolute value of the negative estimates. Thus, while the Bartlett window and the Hamming window have the same main-lobe width, the positive side lobes of the Bartlett window tend to fill in gaps between relatively strong frequencies, while the lower side lobes of the Hamming window lead to less leakage between frequencies, but the danger of negative spectrum estimates as positive and negative side lobes interact.

The Hamming window (or other windows such as the Kaiser window) can be used in spectrum estimation without danger of negative estimates if they are used in the method of averaging periodograms that are discussed in Section 10.5.3. This method guarantees positive estimates, because positive periodograms are averaged. Figure 10.36 shows a comparison of the Blackman–Tukey estimates of Figure 10.35(b) with an estimate obtained by the Welch method of averaging modified periodograms. The heavy dashed line is the Welch estimate. Note that it follows the general shape of the other two estimates, but it differs significantly in the high frequency region, where the speech spectrum is naturally small, and where the frequency response of the analog antialiasing filter causes the spectrum to be very small. Because of its superior ability to deliver consistent resolution for spectra with wide dynamic range, and because it is easily implemented using the DFT, the method of averaging periodograms is widely used in many practical applications of spectrum estimation.

All the spectrum estimates in Figure 10.36 show that the speech signal is characterized by a peak below 500 Hz and a fall-off with increasing frequency by 30 to 40 dB at 6 KHz. Several prominent peaks between 3 KHz and 5 KHz could be due to higher vocal tract resonances that do not vary with time. A different speaker or different speech material would certainly produce a different spectrum estimate, but the general nature of the spectrum estimates would be similar to those of Figure 10.36.

10.7 SUMMARY

One of the important applications of signal processing is spectrum analysis of signals. Because of the computational efficiency of the FFT, many of the techniques for spectrum analysis of continuous-time or discrete-time signals use the DFT either directly or indirectly. In this chapter, we explored and illustrated some of these techniques.

Many of the issues associated with spectrum analysis are best understood in the context of the analysis of sinusoidal signals. Since the use of the DFT requires finite-length signals, windowing must be applied in advance of the analysis. For sinusoidal signals, the width of the spectral peak observed in the DFT is dependent on the window length, with an increasing window length resulting in the sharpening of the peak. Consequently, the ability to resolve closely spaced sinusoids in the spectrum estimate

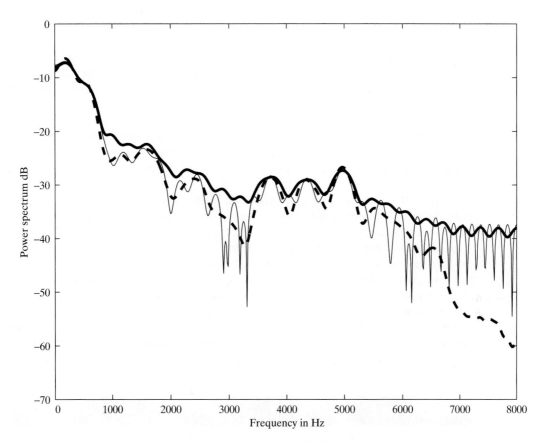

Figure 10.36   Power spectrum estimates by the Blackman–Tukey method using Bartlett window (heavy line) and Hamming window (light line) with $M = 50$. The dashed line shows the power spectrum obtained by averaging overlapping periodograms using a Hamming window with $M = 50$.

decreases as the window becomes shorter. A second, independent effect inherent in spectrum analysis using the DFT is the associated spectral sampling. Specifically, since the spectrum can be computed only at a set of sample frequencies, the observed spectrum can be misleading if we are not careful in its interpretation. For example, important features in the spectrum may not be directly evident in the sampled spectrum. To avoid this, the spectral sample spacing can be reduced by increasing the DFT size in one of two ways. One method is to increase the DFT size while keeping the window length fixed (requiring zero-padding of the windowed sequence). This does not increase resolution. The second method is to increase both the window length and the DFT size. In this case, spectral sample spacing is decreased, and the ability to resolve closely spaced sinusoidal components is increased.

While increased window length and resolution are typically beneficial in the spectrum analysis of stationary data, for time-varying data, it is generally preferable to keep the window length sufficiently short, so that over the window duration, the signal characteristics are approximately stationary. This leads to the concept of the time-dependent

Fourier transform, which, in effect, is a sequence of Fourier transforms obtained as the signal slides past a finite-duration window. A common and useful interpretation of the time-dependent Fourier transform is as a bank of filters, with the frequency response of each filter corresponding to the transform of the window, frequency shifted to one of the DFT frequencies. The time-dependent Fourier transform has important applications both as an intermediate step in filtering signals and for analyzing and interpreting time-varying signals, such as speech and radar signals. Spectral analysis of nonstationary signals typically involves a trade-off between time and frequency resolution. Specifically, our ability to track spectral characteristics in time increases as the length of the analysis window decreases. However, a shorter analysis window results in decreased frequency resolution.

The DFT also plays an important role in the analysis of stationary random signals. An intuitive approach to estimating the power spectrum of random signals is to compute the squared magnitude of the DFT of a segment of the signal. The resulting estimate, called the periodogram, is asymptotically unbiased. The variance of the periodogram estimate, however, does not decrease to zero as the length of the segment increases; consequently, the periodogram is not a good estimate. However, by dividing the available signal sequence into shorter segments and averaging the associated periodograms, we can obtain a well-behaved estimate. An alternative approach is to first estimate the autocorrelation function. This can be done either directly or with the DFT. If a window is then applied to the autocorrelation estimates followed by the DFT, the result, referred to as the smoothed periodogram, is a good spectrum estimate.

Problems

Basic Problems with Answers

10.1. A real continuous-time signal $x_c(t)$ is bandlimited to frequencies below 5 kHz; i.e., $X_c(j\Omega) = 0$ for $|\Omega| \geq 2\pi(5000)$. The signal $x_c(t)$ is sampled with a sampling rate of 10,000 samples per second (10 kHz) to produce a sequence $x[n] = x_c(nT)$ with $T = 10^{-4}$. Let $X[k]$ be the 1000-point DFT of $x[n]$.

   (a) To what continuous-time frequency does the index $k = 150$ in $X[k]$ correspond?
   (b) To what continuous-time frequency does the index $k = 800$ in $X[k]$ correspond?

10.2. A continuous-time signal $x_c(t)$ is bandlimited to 5 kHz; i.e., $X_c(j\Omega) = 0$ for $|\Omega| \geq 2\pi(5000)$. $x_c(t)$ is sampled with period T, producing the sequence $x[n] = x_c(nT)$. To examine the spectral properties of the signal, we compute the N-point DFT of a segment of N samples of $x[n]$ using a computer program that requires $N = 2^v$, where v is an integer.
   Determine the *minimum* value for N and the range of sampling rates

$$F_{\min} < \frac{1}{T} < F_{\max}$$

such that aliasing is avoided, and the effective spacing between DFT values is *less* than 5 Hz; i.e., the equivalent continuous-time frequencies at which the Fourier transform is evaluated are separated by less than 5 Hz.

10.3. A continuous-time signal $x_c(t) = \cos(\Omega_0 t)$ is sampled with period T to produce the sequence $x[n] = x_c(nT)$. An N-point rectangular window is applied to $x[n]$ for $0, 1, \ldots, N-1$, and $X[k]$, for $k = 0, 1, \ldots, N-1$, is the N-point DFT of the resulting sequence.

(a) Assuming that Ω_0, N, and k_0 are fixed, how should T be chosen so that $X[k_0]$ and $X[N-k_0]$ are nonzero, and $X[k] = 0$ for all other values of k?

(b) Is your answer unique? If not, give another value of T that satisfies the conditions of part (a).

10.4. Let $x_c(t)$ be a real-valued, bandlimited signal whose Fourier transform $X_c(j\Omega)$ is zero for $|\Omega| \geq 2\pi(5000)$. The sequence $x[n]$ is obtained by sampling $x_c(t)$ at 10 kHz. Assume that the sequence $x[n]$ is zero for $n < 0$ and $n > 999$.

Let $X[k]$ denote the 1000-point DFT of $x[n]$. It is known that $X[900] = 1$ and $X[420] = 5$. Determine $X_c(j\Omega)$ for as many values of Ω as you can in the region $|\Omega| < 2\pi(5000)$.

10.5. Consider estimating the spectrum of a discrete-time signal $x[n]$ using the DFT with a Hamming window applied to $x[n]$. A conservative rule of thumb for the frequency resolution of windowed DFT analysis is that the frequency resolution is equal to the width of the main lobe of $W(e^{j\omega})$. You wish to be able to resolve sinusoidal signals that are separated by as little as $\pi/100$ in ω. In addition, your window length L is constrained to be a power of 2. What is the minimum length $L = 2^\nu$ that will meet your resolution requirement?

10.6. The following are three different signals $x_i[n]$ that are the sum of two sinusoids:

$$x_1[n] = \cos(\pi n/4) + \cos(17\pi n/64),$$

$$x_2[n] = \cos(\pi n/4) + 0.8\cos(21\pi n/64),$$

$$x_3[n] = \cos(\pi n/4) + 0.001\cos(21\pi n/64).$$

We wish to estimate the spectrum of each of these signals using a 64-point DFT with a 64-point rectangular window $w[n]$. Indicate which of the signals' 64-point DFTs you would expect to have two distinct spectral peaks after windowing.

10.7. Let $x[n]$ be a 5000-point sequence obtained by sampling a continuous-time signal $x_c(t)$ at $T = 50\ \mu s$. Suppose $X[k]$ is the 8192-point DFT of $x[n]$. What is the equivalent frequency spacing in continuous time of adjacent DFT samples?

10.8. Assume that $x[n]$ is a 1000-point sequence obtained by sampling a continuous-time signal $x_c(t)$ at 8 kHz and that $X_c(j\Omega)$ is sufficiently bandlimited to avoid aliasing. What is the minimum DFT length N such that adjacent samples of $X[k]$ correspond to a frequency spacing of 5 Hz or less in the original continuous-time signal?

10.9. $X_r[k]$ denotes the time-dependent Fourier transform (TDFT) defined in Eq. (10.40). For this problem, consider the TDFT when both the DFT length $N = 36$ and the sampling interval $R = 36$. The window $w[n]$ is a rectangular window of length $L = 36$. Compute the TDFT $X_r[k]$ for $-\infty < r < \infty$ and $0 \leq k \leq N-1$ for the signal

$$x[n] = \begin{cases} \cos(\pi n/6), & 0 \leq n \leq 35, \\ \cos(\pi n/2), & 36 \leq n \leq 71, \\ 0, & \text{otherwise.} \end{cases}$$

10.10. Figure P10.10 shows the spectrogram of a chirp signal of the form

$$x[n] = \sin\left(\omega_0 n + \frac{1}{2}\lambda n^2\right).$$

Note that the spectrogram is a representation of the magnitude of $X[n, k]$, as defined in Eq. (10.34), where the dark regions indicate large values of $|X[n, k]|$. Based on the figure, estimate ω_0 and λ.

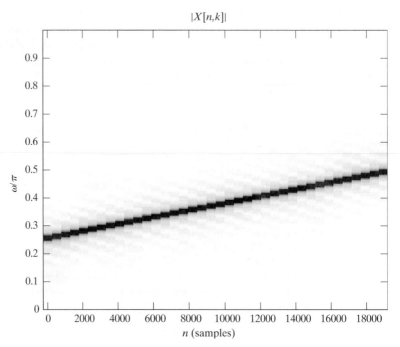

Figure P10.10

10.11. A continuous-time signal is sampled at a sampling rate of 10 kHz, and the DFT of 1024 samples is computed. Determine the continuous-time frequency spacing between spectral samples. Justify your answer.

10.12. Let $x[n]$ be a signal with a single sinusoidal component. The signal $x[n]$ is windowed with an L-point Hamming window $w[n]$ to obtain $v_1[n]$ before computing $V_1(e^{j\omega})$. The signal $x[n]$ is also windowed with an L-point rectangular window to obtain $v_2[n]$, which is used to compute $V_2(e^{j\omega})$. Will the peaks in $|V_2(e^{j\omega})|$ and $|V_1(e^{j\omega})|$ have the same height? If so, justify your answer. If not, which should have a larger peak?

10.13. It is desired to estimate the spectrum of $x[n]$ by applying a 512-point Kaiser window to the signal before computing $X(e^{j\omega})$.

   (a) The requirements for the frequency resolution of the system specify that the largest allowable main lobe for the Kaiser window is $\pi/100$. What is the best side-lobe attenuation expected under these constraints?

   (b) Suppose that you know that $x[n]$ contains two sinusoidal components at least $\pi/50$ apart, and that the amplitude of the stronger component is 1. Based on your answer to part (a), give a threshold on the smallest value of the weaker component you would expect to see over the side lobe of the stronger sinusoid.

10.14. A speech signal is sampled with a sampling rate of 16,000 samples/s (16 kHz). A window of 20-ms duration is used in time-dependent Fourier analysis of the signal, as described in Section 10.3, with the window being advanced by 40 samples between computations of the DFT. Assume that the length of each DFT is $N = 2^v$.

   (a) How many samples are there in each segment of speech selected by the window?

   (b) What is the "frame rate" of the time-dependent Fourier analysis; i.e., how many DFT computations are done per second of input signal?

(c) What is the minimum size N of the DFT such that the original input signal can be reconstructed from the time-dependent Fourier transform?

(d) What is the spacing (in Hz) between the DFT samples for the minimum N from part (c)?

10.15. A real-valued continuous-time segment of a signal $x_c(t)$ is sampled at a rate of 20,000 samples/s, yielding a 1000-point finite-length discrete-time sequence $x[n]$ that is nonzero in the interval $0 \leq n \leq 999$. It is known that $x_c(t)$ is also bandlimited such that $X_c(j\Omega) = 0$ for $|\Omega| \geq 2\pi(10,000)$; i.e., assume that the sampling operation does not introduce any distortion due to aliasing.

$X[k]$ denotes the 1000-point DFT of $x[n]$. $X[800]$ is known to have the value $X[800] = 1 + j$.

(a) From the information given, can you determine $X[k]$ at any other values of k? If so, state which value(s) of k and what the corresponding value of $X[k]$ is. If not, explain why not.

(b) From the information given, state the value(s) of Ω for which $X_c(j\Omega)$ is known and the corresponding value(s) of $X_c(j\Omega)$.

10.16. Let $x[n]$ be a discrete-time signal whose spectrum you wish to estimate using a windowed DFT. You are required to obtain a frequency resolution of at least $\pi/25$ and are also required to use a window length $N = 256$. A safe estimate of the frequency resolution of a spectral estimate is the main-lobe width of the window used. Which of the windows in Table 7.2 will satisfy the criteria given for frequency resolution?

10.17. Let $x[n]$ be a discrete-time signal obtained by sampling a continuous-time signal $x_c(t)$ with some sampling period T so that $x[n] = x_c(nT)$. Assume $x_c(t)$ is bandlimited to 100 Hz, i.e, $X_c(j\Omega) = 0$ for $|\Omega| \geq 2\pi(100)$. We wish to estimate the continuous-time spectrum $X_c(j\Omega)$ by computing a 1024-point DFT of $x[n]$, $X[k]$. What is the smallest value of T such that the equivalent frequency spacing between consecutive DFT samples $X[k]$ corresponds to 1 Hz or less in continuous-time frequency?

10.18. Figure P10.18 shows the magnitude $|V[k]|$ of the 128-point DFT $V[k]$ for a signal $v[n]$. The signal $v[n]$ was obtained by multiplying $x[n]$ by a 128-point rectangular window $w[n]$; i.e., $v[n] = x[n]w[n]$. Note that Figure P10.18 shows $|V[k]|$ only for the interval $0 \leq k \leq 64$. Which of the following signals could be $x[n]$? That is, which are consistent with the information shown in the figure?

$$x_1[n] = \cos(\pi n/4) + \cos(0.26\pi n),$$

$$x_2[n] = \cos(\pi n/4) + (1/3)\sin(\pi n/8),$$

$$x_3[n] = \cos(\pi n/4) + (1/3)\cos(\pi n/8),$$

$$x_4[n] = \cos(\pi n/8) + (1/3)\cos(\pi n/16),$$

$$x_5[n] = (1/3)\cos(\pi n/4) + \cos(\pi n/8),$$

$$x_6[n] = \cos(\pi n/4) + (1/3)\cos(\pi n/8 + \pi/3).$$

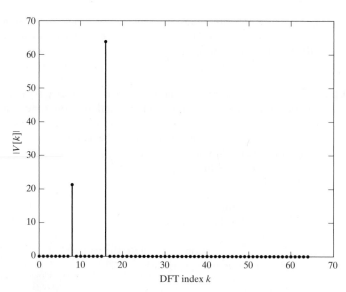

10.19. A signal $x[n]$ is analyzed using the time-dependent Fourier transform $X_r[k]$, as defined in Eq. (10.40). Initially, the analysis is performed with an $N = 128$ DFT using an $L = 128$-point Hamming window $w[n]$. The time-domain sampling of adjacent blocks is $R = 128$; i.e., the windowed segments are offset by 128 samples in time. The frequency resolution obtained with this analysis is not sufficient, and it is desired to improve the resolution. Several methods of modifying the analysis are suggested to accomplish this goal. Which of the following methods will improve the frequency resolution of the time-dependent Fourier transform $X_r[k]$?

> **METHOD 1:**   Increase N to 256 while maintaining L and R at the same values.
> **METHOD 2:**   Increase both N and L to 256, while maintaining R the same.
> **METHOD 3:**   Decrease R to 64 while maintaining the same N and L.
> **METHOD 4:**   Decrease L to 64 while maintaining the same N and R.
> **METHOD 5:**   Maintain N, R and L the same, but change $w[n]$ to be a rectangular window.

10.20. Assume that you wish to estimate the spectrum of $x[n]$ by applying a Kaiser window to the signal before computing the DTFT. You require that the side lobe of the window be 30 dB below the main lobe and that the frequency resolution be $\pi/40$. The width of the main lobe of the window is a safe estimate of the frequency resolution. Estimate the minimum window length L that will meet these requirements.

Basic Problems

10.21. Let $x[n] = \cos(2\pi n/5)$ and $v[n]$ be the sequence obtained by applying a 32-point rectangular window to $x[n]$ before computing $V(e^{j\omega})$. Sketch $|V(e^{j\omega})|$ for $-\pi \leq \omega \leq \pi$, labeling the frequencies of all peaks and the first nulls on either side of the peak. In addition, label the amplitudes of the peaks and the strongest side lobe of each peak.

10.22. In this problem we are interested in estimating the spectra of three very long real-valued data sequences $x_1[n]$, $x_2[n]$, and $x_3[n]$, each consisting of the sum of two sinusoidal components. However, we only have a 256-point segment of each sequence available for analysis.

Let $\bar{x}_1[n]$, $\bar{x}_2[n]$, and $\bar{x}_3[n]$ denote the 256-point segments of $x_1[n]$, $x_2[n]$, and $x_3[n]$, respectively. We have some information about the nature of the spectra of the infinitely long sequences, as indicated in Eqs. (P10.22-1) through (P10.22-3). Two different spectral analysis procedures are being considered for use, one using a 256-point rectangular window and the other a 256-point Hamming window. These procedures are described below. In the descriptions, the signal $\mathcal{R}_N[n]$ denotes the N-point rectangular window and $\mathcal{H}_N[n]$ denotes the N-point Hamming window. The operator $\text{DFT}_{2048}\{\cdot\}$ indicates taking the 2048-point DFT of its argument after zero-padding the end of the input sequence. This will give a good interpolation of the DTFT from the frequency samples of the DFT.

$$X_1(e^{j\omega}) \approx \delta(\omega + \frac{17\pi}{64}) + \delta(\omega + \frac{\pi}{4})$$

$$+\delta(\omega - \frac{\pi}{4}) + \delta(\omega - \frac{17\pi}{64}) \tag{P10.22-1}$$

$$X_2(e^{j\omega}) \approx 0.017\delta(\omega + \frac{11\pi}{32}) + \delta(\omega + \frac{\pi}{4})$$

$$+\delta(\omega - \frac{\pi}{4}) + 0.017\delta(\omega - \frac{11\pi}{32}) \tag{P10.22-2}$$

$$X_3(e^{j\omega}) \approx 0.01\delta(\omega + \frac{257\pi}{1024}) + \delta(\omega + \frac{\pi}{4})$$

$$+\delta(\omega - \frac{\pi}{4}) + 0.01\delta(\omega - \frac{257\pi}{1024}) \tag{P10.22-3}$$

Based on Eqs. (P10.22-1) through (P10.22-3), indicate which of the spectral analysis procedures described below would allow you to conclude responsibly whether the anticipated frequency components were present. A good justification at a minimum will include a quantitative consideration of both resolution and side-lobe behavior of the estimators. Note that it is possible that both or neither of the algorithms will work for any given data sequence. Table 7.2 may be useful in deciding which algorithm(s) to use with which sequence.

Spectral Analysis Algorithms
Algorithm 1: Use the entire data segment with a rectangular window.
$$v[n] = \mathcal{R}_{256}[n]\bar{x}[n]$$

$$\left|V(e^{j\omega})\right|_{\omega=\frac{2\pi k}{2048}} = \left|\text{DFT}_{2048}\{v[n]\}\right|.$$

Algorithm 2: Use the entire data segment with a Hamming window.
$$v[n] = \mathcal{H}_{256}[n]\bar{x}[n]$$

$$\left|V(e^{j\omega})\right|_{\omega=\frac{2\pi k}{2048}} = \left|\text{DFT}_{2048}\{v[n]\}\right|.$$

10.23. Sketch the spectrogram obtained by using a 256-point rectangular window and 256-point DFTs with no overlap ($R = 256$) on the signal
$$x[n] = \cos\left[\frac{\pi n}{4} + 1000\sin\left(\frac{\pi n}{8000}\right)\right]$$
for the interval $0 \le n \le 16{,}000$.

10.24. (a) Consider the system of Figure P10.24-1 with input $x(t) = e^{j(3\pi/8)10^4 t}$, sampling period $T = 10^{-4}$, and
$$w[n] = \begin{cases} 1, & 0 \le n \le N-1, \\ 0, & \text{otherwise.} \end{cases}$$
What is the smallest nonzero value of N such that $X_w[k]$ is nonzero at exactly one value of k?

(b) Suppose now that $N = 32$, the input signal is $x(t) = e^{j\Omega_0 t}$, and the sampling period T is chosen such that no aliasing occurs during the sampling process. Figures P10.24-2 and P10.24-3 show the magnitude of the sequence $X_w[k]$ for $k = 0, \ldots, 31$ for the following two different choices of $w[n]$:

$$w_1[n] = \begin{cases} 1, & 0 \le n \le 31, \\ 0, & \text{otherwise}, \end{cases}$$

$$w_2[n] = \begin{cases} 1, & 0 \le n \le 7, \\ 0, & \text{otherwise}. \end{cases}$$

Indicate which figure corresponds to which choice of $w[n]$. State your reasoning clearly.

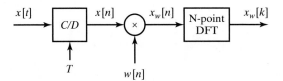

Figure P10.24-1

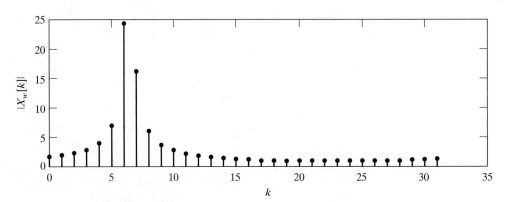

Figure P10.24-2

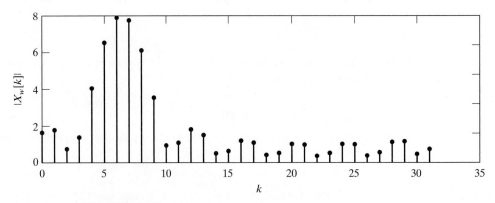

Figure P10.24-3

(c) For the input signal and system parameters of part (b), we would like to estimate the value of Ω_0 from Figure P10.24-3 when the sampling period is $T = 10^{-4}$. Assuming that the sequence

$$w[n] = \begin{cases} 1, & 0 \le n \le 31, \\ 0, & \text{otherwise}, \end{cases}$$

and that the sampling period is sufficient to ensure that no aliasing occurs during sampling, estimate the value of Ω_0. Is your estimate exact? If it is not, what is the maximum possible error of your frequency estimate?

(d) Suppose you were provided with the exact values of the 32-point DFT $X_w[k]$ for the window choices $w_1[n]$ and $w_2[n]$. Briefly describe a procedure to obtain a precise estimate of Ω_0.

Advanced Problems

10.25. In Figure P10.25, a filter bank is shown for which

$$h_0[n] = 3\delta[n+1] + 2\delta[n] + \delta[n-1],$$

and

$$h_q[n] = e^{j\frac{2\pi qn}{M}} h_0[n], \qquad \text{for } q = 1, \ldots, N-1.$$

The filter bank consists of N filters, modulated by a fraction $1/M$ of the total frequency band. Assume M and N are both greater than the length of $h_0[n]$.

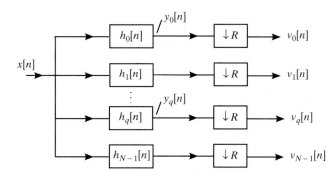

Figure P10.25   Filter bank

(a) Express $y_q[n]$ in terms of the time-dependent Fourier transform $X[n, \lambda)$ of $x[n]$, and sketch and label explicitly the values for the associated window in the time-dependent Fourier transform.

For parts (b) and (c), assume that $M = N$. Since $v_q[n]$ depends on the two integer variables q and n, we alternatively write it as the two-dimensional sequence $v[q, n]$.

(b) For $R = 2$, describe a procedure to recover $x[n]$ for all values of n if $v[q, n]$ is available for all integer values of q and n.

(c) Will your procedure in (b) work if $R = 5$? Clearly explain.

10.26. The system in Figure P10.26-1 uses a modulated filter bank for spectral analysis. (For further illustration, Figure P10.26-2 shows how the frequency responses $H_k(e^{j\omega})$ relate.) The impulse response of the prototype filter $h_0[n]$ is sketched in Figure P10.26-3.

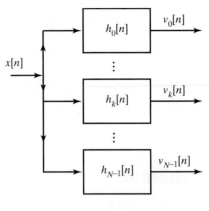

$h_k[n] = e^{j\omega_k n} h_0[n], \qquad \omega_k = \dfrac{2\pi k}{N}, \qquad \text{where } k = 0, 1, \dots, N-1$

$h_0[n] = \text{lowpass prototype filter} \qquad H_k(z) = H_0(e^{-j2\pi k/N} z) \qquad$ **Figure P10.26-1**

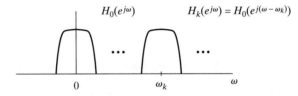

Figure P10.26-2

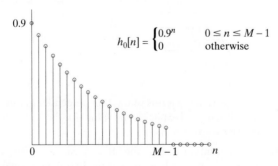

$$h_0[n] = \begin{cases} 0.9^n & 0 \le n \le M-1 \\ 0 & \text{otherwise} \end{cases}$$

Figure P10.26-3

An alternative system for spectral analysis is shown in Figure P10.26-4. Determine $w[n]$ so that $G[k] = v_k[0]$, for $k = 0, 1, \dots, N-1$.

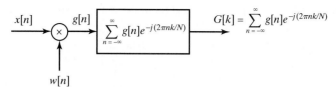

Figure P10.26-4

10.27. We are interested in obtaining 256 equally spaced samples of the z-transform of $x_w[n]$. $x_w[n]$ is a windowed version of an arbitrary sequence $x[n]$ where $x_w[n] = x[n]w[n]$ and $w[n] = 1, 0 \leq n \leq 255$ and $w[n] = 0$ otherwise. The z-transform of $x_w[n]$ is defined as

$$X_w(z) = \sum_{n=0}^{255} x[n]z^{-n}.$$

The samples $X_w[k]$ that we would like to compute are

$$X_w[k] = X_w(z)\big|_{z=0.9e^{j\frac{2\pi}{256}k}} \qquad k = 0, 1, \dots, 255.$$

We would like to process the signal $x[n]$ with a modulated filter bank, as indicated in Figure P10.27.

Each filter in the filter bank has an impulse response that is related to the prototype *causal* lowpass filter $h_0[n]$ as follows:

$$h_k[n] = h_0[n]e^{-j\omega_k n} \qquad k = 1, 2, \dots, 255.$$

Each output of the filter bank is sampled once, at time $n = N_k$, to obtain $X_w[k]$, i.e.,

$$X_w[k] = v_k[N_k].$$

Determine $h_0[n]$, ω_k and N_k so that

$$X_w[k] = v_k[N_k] = X_w(z)\big|_{z=0.9e^{j\frac{2\pi}{256}k}} \qquad k = 0, 1, \dots, 255.$$

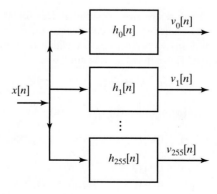

Figure P10.27

10.28. **(a)** In Figure P10.28-1, we show a system for spectral analysis of a signal $x_c(t)$, where

$$G_k[n] = \sum_{l=0}^{N-1} g_l[n]e^{-j\frac{2\pi}{N}lk},$$

$$N = 512, \quad \text{and} \quad LR = 256.$$

For the most general choice of the multiplier coefficient a_l, determine the choice for L and R which will result in the smallest number of multiplies per second.

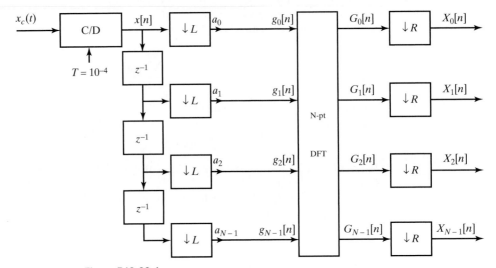

Figure P10.28-1

(b) In Figure P10.28-2, we show another system for spectral analysis of a signal $x_c(t)$, where

$$h[n] = \begin{cases} (0.93)^n & 0 \leq n \leq 255 \\ 0 & \text{otherwise} \end{cases},$$

$$h_k[n] = h[n]e^{-j\omega_k n}, \quad k = 0, 1, \cdots, N-1, \quad \text{and} \quad N = 512.$$

Listed below are **two** possible choices for M, **four** possible choices for ω_k, and **six** possible choices for the coefficients a_l. From this set identify all combinations for which $Y_k[n] = X_k[n]$, i.e., for which both systems will provide the same spectral analysis. There may be more than one.

M:　　(a) 256　　　(b) 512

ω_k :　　(a) $\frac{2\pi k}{256}$　　(b) $\frac{2\pi k}{512}$　　(c) $\frac{-2\pi k}{256}$　　(d) $\frac{-2\pi k}{512}$

a_l:			
(a)	$(0.93)^l$	$l=0, 1, \cdots, 255,$	zero otherwise
(b)	$(0.93)^{-l}$	$l=0, 1, \cdots, 511$	
(c)	$(0.93)^l$	$l=0, 1, \cdots, 511$	
(d)	$(0.93)^{-l}$	$l=0, 1, \cdots, 255,$	zero otherwise
(e)	$(0.93)^l$	$l=256, 257, \cdots, 511,$	zero otherwise
(f)	$(0.93)^{-l}$	$l=256, 257, \cdots, 511,$	zero otherwise

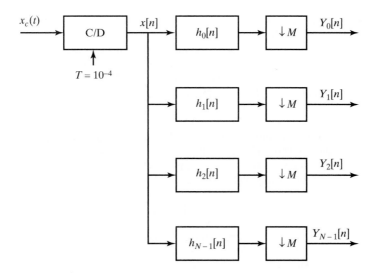

Figure P10.28-2

10.29. The system shown in Figure P10.29 is proposed as a spectrum analyzer. The basic operation is as follows: The spectrum of the sampled input is frequency-shifted; the lowpass filter selects the lowpass band of frequencies; the downsampler "spreads" the selected frequency band back over the entire range $-\pi < \omega < \pi$; and the DFT samples that frequency band uniformly at N frequencies.

Assume that the input is bandlimited so that $X_c(j\Omega) = 0$ for $|\Omega| \geq \pi/T$. The LTI system with frequency response $H(e^{j\omega})$ is an ideal lowpass filter with gain of one and cutoff frequency π/M. Furthermore, assume that $0 < \omega_1 < \pi$ and the data window $w[n]$ is a rectangular window of length N.

(a) Plot the DTFTs, $X(e^{j\omega})$, $Y(e^{j\omega})$, $R(e^{j\omega})$, and $R_d(e^{j\omega})$ for the given $X_c(j\Omega)$ and for $\omega_1 = \pi/2$ and $M = 4$. Give the relationship between the input and output Fourier transforms for each stage of the process; e.g., in the fourth plot, you would indicate $R(e^{j\omega}) = H(e^{j\omega})Y(e^{j\omega})$.

(b) Using your result in part (a), generalize to determine the band of continuous-time frequencies in $X_c(j\Omega)$ that falls within the passband of the lowpass discrete-time filter. Your answer will depend on M, ω_1 and T. For the specific case of $\omega_1 = \pi/2$ and $M = 4$, indicate this band of frequencies on the plot of $X_c(j\Omega)$ given for part (a).

(c) (i) What continuous-time frequencies in $X_c(j\Omega)$ are associated with the DFT values $V[k]$ for $0 \leq k \leq N/2$?

(ii) What continuous-time frequencies in $X_c(j\Omega)$ do the values for $N/2 < k \leq N-1$ correspond to? In each case, give a formula for the frequencies Ω_k.

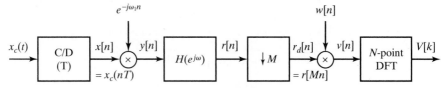

Figure P10.29

10.30. Consider a real time-limited continuous-time signal $x_c(t)$ whose duration is 100 ms. Assume that this signal has a bandlimited Fourier transform such that $X_c(j\Omega) = 0$ for $|\Omega| \geq 2\pi(10,000)$ rad/s; i.e., assume that aliasing is negligible. We want to compute samples of $X_c(j\Omega)$ with 5-Hz spacing over the interval $0 \leq \Omega \leq 2\pi(10,000)$. This can be done with a 4000-point DFT. Specifically, we want to obtain a 4000-point sequence $x[n]$ for which the 4000-point DFT is related to $X_c(j\Omega)$ by

$$X[k] = \alpha X_c(j2\pi \cdot 5 \cdot k), \qquad k = 0, 1, \ldots, 1999,$$

where α is a known scale factor. Three methods are proposed to obtain a 4000-point sequence whose DFT gives the desired samples of $X_c(j\Omega)$.

    METHOD 1: $x_c(t)$ is sampled with a sampling period $T = 25 \ \mu s$; i.e., we compute $X_1[k]$, the DFT of the sequence

$$x_1[n] = \begin{cases} x_c(nT), & n = 0, 1, \ldots, 3999, \\ 0, & \text{otherwise.} \end{cases}$$

Since $x_c(t)$ is time limited to 100 ms, $x_1[n]$ is a finite-length sequence of length 4000 (100 ms/25 μs).

    METHOD 2: $x_c(t)$ is sampled with a sampling period of $T = 50 \ \mu s$. Since $x_c(t)$ is time limited to 100 ms, the resulting sequence will have only 2000 (100 ms/50 μs) nonzero samples; i.e.,

$$x_2[n] = \begin{cases} x_c(nT), & n = 0, 1, \ldots, 1999, \\ 0, & \text{otherwise.} \end{cases}$$

In other words, the sequence is padded with zero-samples to create a 4000-point sequence for which the 4000-point DFT $X_2[k]$ is computed.

    METHOD 3: $x_c(t)$ is sampled with a sampling period of $T = 50 \ \mu s$, as in Method 2. The resulting 2000-point sequence is used to form the sequence $x_3[n]$ as follows:

$$x_3[n] = \begin{cases} x_c(nT), & 0 \leq n \leq 1999, \\ x_c((n-2000)T), & 2000 \leq n \leq 3999, \\ 0, & \text{otherwise.} \end{cases}$$

The 4000-point DFT $X_3[k]$ of this sequence is computed.

For each of the three methods, determine how each 4000-point DFT is related to $X_c(j\Omega)$. Indicate this relationship in a sketch for a "typical" Fourier transform $X_c(j\Omega)$. State explicitly which method(s) provide the desired samples of $X_c(j\Omega)$.

10.31. A continuous-time finite-duration signal $x_c(t)$ is sampled at a rate of 20,000 samples/s, yielding a 1000-point finite-length sequence $x[n]$ that is nonzero in the interval $0 \leq n \leq 999$. Assume for this problem that the continuous-time signal is also bandlimited such that $X_c(j\Omega) = 0$ for $|\Omega| \geq 2\pi(10,000)$; i.e., assume that negligible aliasing distortion occurs in sampling. Assume also that a device or program is available for computing 1000-point DFTs and inverse DFTs.

(a) If $X[k]$ denotes the 1000-point DFT of the sequence $x[n]$, how is $X[k]$ related to $X_c(j\Omega)$? What is the effective continuous-time frequency spacing between DFT samples?

The following procedure is proposed for obtaining an expanded view of the Fourier transform $X_c(j\Omega)$ in the interval $|\Omega| \leq 2\pi(5000)$, starting with the 1000-point DFT $X[k]$.
  Step 1. Form the new 1000-point DFT

$$W[k] = \begin{cases} X[k], & 0 \leq k \leq 250, \\ 0, & 251 \leq k \leq 749, \\ X[k], & 750 \leq k \leq 999. \end{cases}$$

Step 2. Compute the inverse 1000-point DFT of $W[k]$, obtaining $w[n]$ for $n = 0, 1, \ldots, 999$.

Step 3. Decimate the sequence $w[n]$ by a factor of 2 and augment the result with 500 consecutive zero samples, obtaining the sequence

$$y[n] = \begin{cases} w[2n], & 0 \le n \le 499, \\ 0, & 500 \le n \le 999. \end{cases}$$

Step 4. Compute the 1000-point DFT of $y[n]$, obtaining $Y[k]$.

(b) The designer of this procedure asserts that

$$Y[k] = \alpha X_c(j2\pi \cdot 10 \cdot k), \qquad k = 0, 1, \ldots, 500,$$

where α is a constant of proportionality. Is this assertion correct? If not, explain why not.

10.32. An analog signal consisting of a sum of sinusoids was sampled with a sampling rate of $f_s = 10000$ samples/s to obtain $x[n] = x_c(nT)$. Four spectrograms showing the time-dependent Fourier transform $|X[n, \lambda]|$ were computed using either a rectangular or a Hamming window. They are plotted in Figure P10.32. (A log amplitude scale is used, and only the top 35 dB is shown.)

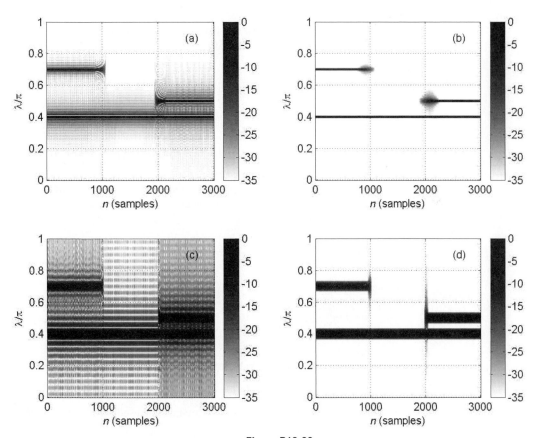

Figure P10.32

(a) Which spectrograms were computed with a rectangular window?
    (a)   (b)   (c)   (d)
(b) Which pair (or pairs) of spectrograms have approximately the same frequency resolution?
    (a&b)   (b&d)   (c&d)   (a&d)   (b&c)
(c) Which spectrogram has the shortest time window? (a)   (b)   (c)   (d)
(d) To the nearest 100 samples, estimate the window length L (in samples) of the window in spectrogram (b).
(e) Use the spectrographic data in Figure P10.32 to assist you in writing an equation (or equations) for an analog sum of sinusoids $x_c(t)$, which when sampled at a sampling rate of $f_s = 10000$, would produce the above spectrograms. Be as complete as you can in your description of the signal. Indicate any parameters that cannot be obtained from the spectrogram.

10.33. The periodogram $I(\omega)$ of a discrete-time random signal $x[n]$ was defined in Eq. (10.67) as

$$I(\omega) = \frac{1}{LU}|V(e^{j\omega})|^2,$$

where $V(e^{j\omega})$ is the DTFT of the finite-length sequence $v[n] = w[n]x[n]$, with $w[n]$ a finite-length window sequence of length L, and U is a normalizing constant. Assume that $x[n]$ and $w[n]$ are real.

Show that the periodogram is also equal to $1/LU$ times the Fourier transform of the aperiodic autocorrelation sequence of $v[n]$; i.e.,

$$I(\omega) = \frac{1}{LU} \sum_{m=-(L-1)}^{L-1} c_{vv}[m]e^{-j\omega m},$$

where

$$c_{vv}[m] = \sum_{n=0}^{L-1} v[n]v[n+m].$$

10.34. Consider a finite-length sequence $x[n]$ such that $x[n] = 0$ for $n < 0$ and $n \geq L$. Let $X[k]$ be the N-point DFT of the sequence $x[n]$, where $N > L$. Define $c_{xx}[m]$ to be the aperiodic autocorrelation function of $x[n]$; i.e.,

$$c_{xx}[m] = \sum_{n=-\infty}^{\infty} x[n]x[n+m].$$

Define

$$\tilde{c}_{xx}[m] = \frac{1}{N}\sum_{m=0}^{N-1} |X[k]|^2 e^{j(2\pi/N)km}, \qquad m = 0, 1, \ldots, N-1.$$

(a) Determine the minimum value of N that can be used for the DFT if we require that

$$c_{xx}[m] = \tilde{c}_{xx}[m], \qquad 0 \leq m \leq L - 1.$$

(b) Determine the minimum value of N that can be used for the DFT if we require that

$$c_{xx}[m] = \tilde{c}_{xx}[m], \qquad 0 \leq m \leq M - 1,$$

where $M < L$.

10.35. The symmetric Bartlett window, which arises in many aspects of power spectrum estimation, is defined as

$$w_B[m] = \begin{cases} 1 - |m|/M, & |m| \le M - 1, \\ 0, & \text{otherwise.} \end{cases} \tag{P10.35-1}$$

The Bartlett window is particularly attractive for obtaining estimates of the power spectrum by windowing an estimated autocorrelation function, as discussed in Section 10.6. This is because its Fourier transform is nonnegative, which guarantees that the smoothed spectrum estimate will be nonnegative at all frequencies.

(a) Show that the Bartlett window as defined in Eq. (P10.35-1) is $(1/M)$ times the aperiodic autocorrelation of the sequence $(u[n] - u[n - M])$.

(b) From the result of part (a), show that the Fourier transform of the Bartlett window is

$$W_B(e^{j\omega}) = \frac{1}{M} \left[\frac{\sin(\omega M/2)}{\sin(\omega/2)} \right]^2, \tag{P10.35-2}$$

which is clearly nonnegative.

(c) Describe a procedure for generating other finite-length window sequences that have nonnegative Fourier transforms.

10.36. Consider a signal

$$x[n] = \left[\sin\left(\frac{\pi n}{2}\right) \right]^2 u[n]$$

whose time-dependent discrete Fourier transform is computed using the analysis window

$$w[n] = \begin{cases} 1, & 0 \le n \le 13, \\ 0, & \text{otherwise.} \end{cases}$$

Let $X[n, k] = X[n, 2\pi k/7)$ for $0 \le k \le 6$, where $X[n, \lambda)$ is defined as in Section 10.3.

(a) Determine $X[0, k]$ for $0 \le k \le 6$.

(b) Evaluate $\sum_{k=0}^{6} X[n, k]$ for $0 \le n < \infty$.

Extension Problems

10.37. In Section 10.6, we showed that a smoothed estimate of the power spectrum can be obtained by windowing an estimate of the autocorrelation sequence. It was stated (see Eq. (10.109)) that the variance of the smoothed spectrum estimate is

$$\text{var}[S(\omega)] \simeq F P_{xx}^2(\omega),$$

where F, the *variance ratio* or *variance reduction factor*, is

$$F = \frac{1}{Q} \sum_{m=-(M-1)}^{M-1} (w_c[m])^2 = \frac{1}{2\pi Q} \int_{-\pi}^{\pi} |W_c(e^{j\omega})|^2 d\omega.$$

As discussed in Section 10.6, Q is the length of the sequence $x[n]$ and $(2M-1)$ is the length of the symmetric window $w_c[m]$ that is applied to the autocorrelation estimate. Thus, if Q is fixed, the variance of the smoothed spectrum estimate can be reduced by adjusting the shape and duration of the window applied to the correlation function.

In this problem we will show that F decreases as the window length decreases, but we also know from the previous discussion of windows in Chapter 7 that the width of the main lobe of $W_c(e^{j\omega})$ increases with decreasing window length, so that the ability to resolve two adjacent frequency components is reduced as the window width decreases. Thus, there is a trade-off between variance reduction and resolution. We will study this trade-off for the following commonly used windows:

Rectangular

$$w_R[m] = \begin{cases} 1, & |m| \le M - 1, \\ 0, & \text{otherwise.} \end{cases}$$

Bartlett (triangular)

$$w_B[m] = \begin{cases} 1 - |m|/M, & |m| \le M - 1, \\ 0, & \text{otherwise.} \end{cases}$$

Hanning/Hamming

$$w_H[m] = \begin{cases} \alpha + \beta \cos[\pi m/(M-1)], & |m| \le M - 1, \\ 0, & \text{otherwise.} \end{cases}$$

($\alpha = \beta = 0.5$ for the Hanning window, and $\alpha = 0.54$ and $\beta = 0.46$ for the Hamming window.)

(a) Find the Fourier transform of each of the foregoing windows; i.e., compute $W_R(e^{j\omega})$, $W_B(e^{j\omega})$, and $W_H(e^{j\omega})$. Sketch each of these Fourier transforms as functions of ω.
(b) For each of the windows, show that the entries in the following table are approximately true when $M \gg 1$:

Window Name	Approximate Main-lobe Width	Approximate Variance Ratio (F)
Rectangular	$2\pi/M$	$2M/Q$
Bartlett	$4\pi/M$	$2M/(3Q)$
Hanning/Hamming	$3\pi/M$	$2M(\alpha^2 + \beta^2/2)/Q$

10.38. Show that the time-dependent Fourier transform, as defined by Eq. (10.18), has the following properties:

(a) *Linearity:*

$$\text{If} \quad x[n] = ax_1[n] + bx_2[n], \quad \text{then} \quad X[n, \lambda) = aX_1[n, \lambda) + bX_2[n, \lambda).$$

(b) *Shifting:* If $y[n] = x[n - n_0]$, then $Y[n, \lambda) = X[n - n_0, \lambda)$.
(c) *Modulation:* If $y[n] = e^{j\omega_0 n}x[n]$, then $Y[n, \lambda) = e^{j\omega_0 n}X[n, \lambda - \omega_0)$.
(d) *Conjugate Symmetry:* If $x[n]$ is real, then $X[n, \lambda) = X^*[n, -\lambda)$.

10.39. Suppose that $x_c(t)$ is a real, continuous-time stationary random signal with autocorrelation function

$$\phi_c(\tau) = \mathcal{E}\{x_c(t)x_c(t + \tau)\}$$

and power density spectrum

$$P_c(\Omega) = \int_{-\infty}^{\infty} \phi_c(\tau)e^{-j\Omega\tau}d\tau.$$

Consider a discrete-time stationary random signal $x[n]$ that is obtained by sampling $x_c(t)$ with sampling period T; i.e., $x[n] = x_c(nT)$.

(a) Show that $\phi[m]$, the autocorrelation sequence for $x[n]$, is

$$\phi[m] = \phi_c(mT).$$

(b) What is the relationship between the power density spectrum $P_c(\Omega)$ for the continuous-time random signal and the power density spectrum $P(\omega)$ for the discrete-time random signal?

(c) What condition is necessary such that

$$P(\omega) = \frac{1}{T} P_c\left(\frac{\omega}{T}\right), \qquad |\omega| < \pi?$$

10.40. In Section 10.5.5, we considered the estimation of the power spectrum of a sinusoid plus white noise. In this problem, we will determine the true power spectrum of such a signal. Suppose that

$$x[n] = A\cos(\omega_0 n + \theta) + e[n],$$

where θ is a random variable that is uniformly distributed on the interval from 0 to 2π and $e[n]$ is a sequence of zero-mean random variables that are independent of each other and also independent of θ. In other words, the cosine component has a randomly selected phase, and $e[n]$ represents white noise.

(a) Show that for the preceding assumptions, the autocorrelation function for $x[n]$ is

$$\phi_{xx}[m] = \mathcal{E}\{x[n]x[m+n]\} = \frac{A^2}{2}\cos(\omega_0 m) + \sigma_e^2\delta[m],$$

where $\sigma_e^2 = \mathcal{E}\{(e[n])^2\}$.

(b) From the result of part (a), show that over one period in frequency, the power spectrum of $x[n]$ is

$$P_{xx}(\omega) = \frac{A^2\pi}{2}[\delta(\omega - \omega_0) + \delta(\omega + \omega_0)] + \sigma_e^2, \qquad |\omega| \le \pi.$$

10.41. Consider a discrete-time signal $x[n]$ of length N samples that was obtained by sampling a stationary, white, zero-mean continuous-time signal. It follows that

$$\mathcal{E}\{x[n]x[m]\} = \sigma_x^2\delta[n - m],$$

$$\mathcal{E}\{x[n]\} = 0.$$

Suppose that we compute the DFT of the finite-length sequence $x[n]$, thereby obtaining $X[k]$ for $k = 0, 1, \ldots, N-1$.

(a) Determine the approximate variance of $|X[k]|^2$ using Eqs. (10.80) and (10.81).

(b) Determine the cross-correlation between values of the DFT; i.e., determine $\mathcal{E}\{X[k]X^*[r]\}$ as a function of k and r.

10.42. A bandlimited continuous-time signal has a bandlimited power spectrum that is zero for $|\Omega| \ge 2\pi(10^4)$ rad/s. The signal is sampled at a rate of 20,000 samples/s over a time interval of 10 s. The power spectrum of the signal is estimated by the method of averaging periodograms as described in Section 10.5.3.

(a) What is the length Q (number of samples) of the data record?

(b) If a radix-2 FFT program is used to compute the periodograms, what is the minimum length N if we wish to obtain estimates of the power spectrum at equally spaced frequencies no more than 10 Hz apart?

(c) If the segment length L is equal to the FFT length N in part (b), how many segments K are available if the segments do not overlap?

(d) Suppose that we wish to reduce the variance of the spectrum estimates by a factor of 10 while maintaining the frequency spacing of part (b). Give two methods of doing this. Do these two methods give the same results? If not, explain how they differ.

10.43. Suppose that an estimate of the power spectrum of a signal is obtained by the method of averaging periodograms, as discussed in Section 10.5.3. That is, the spectrum estimate is

$$\bar{I}(\omega) = \frac{1}{K} \sum_{r=0}^{K-1} I_r(\omega),$$

where the K periodograms $I_r(\omega)$ are computed from L-point segments of the signal using Eqs. (10.82) and (10.83). We define an estimate of the autocorrelation function as the inverse Fourier transform of $\bar{I}(\omega)$; i.e.,

$$\bar{\phi}[m] = \frac{1}{2\pi} \int_{-\pi}^{\pi} \bar{I}(\omega) e^{j\omega m} d\omega.$$

(a) Show that

$$\mathcal{E}\{\bar{\phi}[m]\} = \frac{1}{LU} c_{ww}[m] \phi_{xx}[m],$$

where L is the length of the segments, U is a normalizing factor given by Eq. (10.79), and $c_{ww}[m]$, given by Eq. (10.75), is the aperiodic autocorrelation function of the window that is applied to the signal segments.

(b) In the application of periodogram averaging, we normally use an FFT algorithm to compute $\bar{I}(\omega)$ at N equally spaced frequencies; i.e.,

$$\bar{I}[k] = \bar{I}(2\pi k/N), \qquad k = 0, 1, \ldots, N-1,$$

where $N \geq L$. Suppose that we compute an estimate of the autocorrelation function by computing the inverse DFT of $\bar{I}[k]$, as in

$$\bar{\phi}_p[m] = \frac{1}{N} \sum_{k=0}^{N-1} \bar{I}[k] e^{j(2\pi/N)km}, \qquad m = 0, 1, \ldots, N-1.$$

Obtain an expression for $\mathcal{E}\{\bar{\phi}_p[m]\}$.

(c) How should N be chosen so that

$$\mathcal{E}\{\bar{\phi}_p[m]\} = \mathcal{E}\{\bar{\phi}[m]\}, \qquad m = 0, 1, \ldots, L-1?$$

10.44. Consider the computation of the autocorrelation estimate

$$\hat{\phi}_{xx}[m] = \frac{1}{Q} \sum_{n=0}^{Q-|m|-1} x[n]x[n + |m|], \qquad \text{(P10.44-1)}$$

where $x[n]$ is a real sequence. Since $\hat{\phi}_{xx}[-m] = \hat{\phi}_{xx}[m]$, it is necessary only to evaluate Eq. (P10.44-1) for $0 \leq m \leq M - 1$ to obtain $\hat{\phi}_{xx}[m]$ for $-(M - 1) \leq m \leq M - 1$, as is required to estimate the power density spectrum using Eq. (10.102).

(a) When $Q \gg M$, it may not be feasible to compute $\hat{\phi}_{xx}[m]$ using a single FFT computation. In such cases, it is convenient to express $\hat{\phi}_{xx}[m]$ as a sum of correlation estimates based on shorter sequences. Show that if $Q = KM$,

$$\hat{\phi}_{xx}[m] = \frac{1}{Q} \sum_{i=0}^{K-1} c_i[m],$$

where

$$c_i[m] = \sum_{n=0}^{M-1} x[n + iM]x[n + iM + m],$$

for $0 \leq m \leq M - 1$.

(b) Show that the correlations $c_i[m]$ can be obtained by computing the N-point *circular* correlations

$$\tilde{c}_i[m] = \sum_{n=0}^{N-1} x_i[n]y_i[((n+m))_N],$$

where the sequences

$$x_i[n] = \begin{cases} x[n+iM], & 0 \le n \le M-1, \\ 0, & M \le n \le N-1, \end{cases}$$

and

$$y_i[n] = x[n+iM], \qquad 0 \le n \le N-1. \qquad \text{(P10.44-2)}$$

What is the *minimum* value of N (in terms of M) such that $c_i[m] = \tilde{c}_i[m]$ for $0 \le m \le M-1$?

(c) State a procedure for computing $\hat{\phi}_{xx}[m]$ for $0 \le m \le M-1$ that involves the computation of $2K$ N-point DFTs of real sequences and *one* N-point inverse DFT. How many complex multiplications are required to compute $\hat{\phi}_{xx}[m]$ for $0 \le m \le M-1$ if a radix-2 FFT is used?

(d) What modifications to the procedure developed in part (c) would be necessary to compute the cross-correlation estimate

$$\hat{\phi}_{xy}[m] = \frac{1}{Q} \sum_{n=0}^{Q-|m|-1} x[n]y[n+m], \qquad -(M-1) \le m \le M-1,$$

where $x[n]$ and $y[n]$ are real sequences known for $0 \le n \le Q-1$?

(e) Rader (1970) showed that, for computing the autocorrelation estimate $\hat{\phi}_{xx}[m]$ for $0 \le m \le M-1$, significant savings of computation can be achieved if $N = 2M$. Show that the N-point DFT of a segment $y_i[n]$ as defined in Eq. (P10.44-2) can be expressed as

$$Y_i[k] = X_i[k] + (-1)^k X_{i+1}[k], \qquad k = 0, 1, \ldots, N-1.$$

State a procedure for computing $\hat{\phi}_{xx}[m]$ for $0 \le m \le M-1$ that involves the computation of K N-point DFTs and one N-point inverse DFT. Determine the total number of complex multiplications in this case if a radix-2 FFT is used.

10.45. In Section 10.3 we defined the time-dependent Fourier transform of the signal $x[m]$ so that, for fixed n, it is equivalent to the regular DTFT of the sequence $x[n+m]w[m]$, where $w[m]$ is a window sequence. It is also useful to define a time-dependent autocorrelation function for the sequence $x[n]$ such that, for fixed n, its regular Fourier transform is the magnitude squared of the time-dependent Fourier transform. Specifically, the time-dependent autocorrelation function is defined as

$$c[n, m] = \frac{1}{2\pi} \int_{-\pi}^{\pi} |X[n, \lambda)|^2 e^{j\lambda m} d\lambda,$$

where $X[n, \lambda)$ is defined by Eq. (10.18).

(a) Show that if $x[n]$ is real

$$c[n, m] = \sum_{r=-\infty}^{\infty} x[n+r]w[r]x[m+n+r]w[m+r];$$

i.e., for fixed n, $c[n, m]$ is the aperiodic autocorrelation of the sequence $x[n+r]w[r]$, $-\infty < r < \infty$.

(b) Show that the time-dependent autocorrelation function is an even function of m for n fixed, and use this fact to obtain the equivalent expression

$$c[n, m] = \sum_{r=-\infty}^{\infty} x[r]x[r - m]h_m[n - r],$$

where

$$h_m[r] = w[-r]w[-(m + r)]. \qquad (P10.45\text{-}1)$$

(c) What condition must the window $w[r]$ satisfy so that Eq. (P10.45-1) can be used to compute $c[n, m]$ for fixed m and $-\infty < n < \infty$ by causal operations?

(d) Suppose that

$$w[-r] = \begin{cases} a^r, & r \geq 0, \\ 0, & r < 0. \end{cases} \qquad (P10.45\text{-}2)$$

Find the impulse response $h_m[r]$ for computing the mth autocorrelation lag value, and find the corresponding system function $H_m(z)$. From the system function, draw the block diagram of a causal system for computing the mth autocorrelation lag value $c[n, m]$ for $-\infty < n < \infty$ for the window of Eq. (P10.45-2).

(e) Repeat part (d) for

$$w[-r] = \begin{cases} ra^r, & r \geq 0, \\ 0, & r < 0. \end{cases}$$

10.46. Time-dependent Fourier analysis is sometimes implemented as a bank of filters, and even when FFT methods are used, the filter bank interpretation may provide useful insight. This problem examines that interpretation, the basis of which is the fact that when λ is fixed, the time-dependent Fourier transform $X[n, \lambda)$, defined by Eq. (10.18), is simply a sequence that can be viewed as the result of a combination of filtering and modulation operations.

(a) Show that $X[n, \lambda)$ is the output of the system of Figure P10.46-1 if the impulse response of the LTI system is $h_0[n] = w[-n]$. Show also that if λ is fixed, the overall system in Figure P10.46-1 behaves as an LTI system, and determine the impulse response and frequency response of the equivalent LTI system.

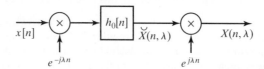

Figure P10.46-1

(b) Assuming λ fixed in Figure P10.46-1, show that, for typical window sequences and for fixed λ, the sequence $s[n] = \check{X}[n, \lambda)$ has a lowpass DTFT. Show also that, for typical window sequences, the frequency response of the overall system in Figure P10.46 is a bandpass filter centered at $\omega = \lambda$.

(c) Figure P10.46-2 shows a bank of N bandpass filter channels, where each channel is implemented as in Figure P10.46-1. The center frequencies of the channels are $\lambda_k = 2\pi k/N$, and $h_0[n] = w[-n]$ is the impulse response of a lowpass filter. Show that the individual outputs $y_k[n]$ are samples (in the λ-dimension) of the time-dependent Fourier transform. Show also that the overall output is $y[n] = Nw[0]x[n]$; i.e., show that the system of Figure P10.46-2 reconstructs the input exactly (within a scale factor) from the sampled time-dependent Fourier transform.

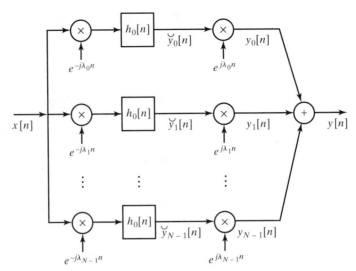

Figure P10.46-2

The system of Figure P10.46-2 converts the single input sequence $x[n]$ into N sequences, thereby increasing the total number of samples per second by the factor N. As shown in part (b), for typical window sequences, the channel signals $\breve{y}_k[n]$ have lowpass Fourier transforms. Thus, it should be possible to reduce the sampling rate of these signals, as shown in Figure P10.46-3. In particular, if the sampling rate is reduced by a factor $R = N$, the total number of samples per second is the same as for $x[n]$. In this case, the filter bank is said to be *critically sampled*. (See Crochiere and Rabiner, 1983.) Reconstruction of the original signal from the decimated channel signals requires interpolation as shown. Clearly, it is of interest to determine how well the original input $x[n]$ can be reconstructed by the system.

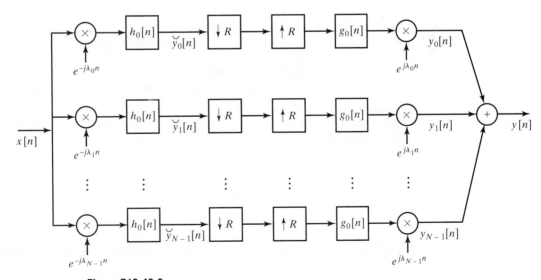

Figure P10.46-3

(d) For the system of Figure P10.46-3, show that the regular DTFT of the output is given by the relation

$$Y(e^{j\omega}) = \frac{1}{R} \sum_{\ell=0}^{R-1} \sum_{k=0}^{N-1} G_0(e^{j(\omega-\lambda_k)}) H_0(e^{j(\omega-\lambda_k-2\pi\ell/R)}) X(e^{j(\omega-2\pi\ell/R)}),$$

where $\lambda_k = 2\pi k/N$. This expression clearly shows the aliasing resulting from the decimation of the channel signals $\breve{y}[n]$. From the expression for $Y(e^{j\omega})$, determine a relation or set of relations that must be satisfied jointly by $H_0(e^{j\omega})$ and $G_0(e^{j\omega})$ such that the aliasing cancels and $y[n] = x[n]$.

(e) Assume that $R = N$ and the frequency response of the lowpass filter is an ideal lowpass filter with frequency response

$$H_0(e^{j\omega}) = \begin{cases} 1, & |\omega| < \pi/N, \\ 0, & \pi/N < |\omega| \leq \pi. \end{cases}$$

For this frequency response $H_0(e^{j\omega})$, determine whether it is possible to find a frequency response of the interpolation filter $G_0(e^{j\omega})$ such that the condition derived in part (d) is satisfied. If so, determine $G_0(e^{j\omega})$.

(f) *Optional:* Explore the possibility of exact reconstruction when the frequency response of the lowpass filter $H_0(e^{j\omega})$ (the Fourier transform of $w[-n]$) is nonideal and nonzero in the interval $|\omega| < 2\pi/N$.

(g) Show that the output of the system of Figure P10.46-3 is

$$y[n] = N \sum_{r=-\infty}^{\infty} x[n-rN] \sum_{\ell=-\infty}^{\infty} g_0[n-\ell R] h_0[\ell R + rN - n].$$

From this expression, determine a relation or set of relations that must be satisfied jointly by $h_0[n]$ and $g_0[n]$ such that $y[n] = x[n]$.

(h) Assume that $R = N$ and the impulse response of the lowpass filter is

$$h_0[n] = \begin{cases} 1, & -(N-1) \leq n \leq 0, \\ 0, & \text{otherwise.} \end{cases}$$

For this impulse response $h_0[n]$, determine whether it is possible to find an impulse response of the interpolation filter $g_0[n]$ such that the condition derived in part (g) is satisfied. If so, determine $g_0[n]$.

(i) *Optional:* Explore the possibility of exact reconstruction when the impulse response of the lowpass filter $h_0[n] = w[-n]$ is a tapered window with length greater than N.

10.47. Consider a stable LTI system with a real input $x[n]$, a real impulse response $h[n]$, and output $y[n]$. Assume that the input $x[n]$ is white noise with zero mean and variance σ_x^2. The system function is

$$H(z) = \frac{\displaystyle\sum_{k=0}^{M} b_k z^{-k}}{1 - \displaystyle\sum_{k=1}^{N} a_k z^{-k}},$$

where we assume the a_ks and b_ks are real for this problem. The input and output satisfy the following difference equation with constant coefficients:

$$y[n] = \sum_{k=1}^{N} a_k y[n-k] + \sum_{k=0}^{M} b_k x[n-k].$$

If all the a_ks are zero, $y[n]$ is called a *moving-average* (MA) linear random process. If all the b_ks are zero, except for b_0, then $y[n]$ is called an *autoregressive* (AR) linear random process. If both N and M are nonzero, then $y[n]$ is an *autoregressive moving-average* (ARMA) linear random process.

(a) Express the autocorrelation of $y[n]$ in terms of the impulse response $h[n]$ of the linear system.

(b) Use the result of part (a) to express the power density spectrum of $y[n]$ in terms of the frequency response of the system.

(c) Show that the autocorrelation sequence $\phi_{yy}[m]$ of an MA process is nonzero only in the interval $|m| \leq M$.

(d) Find a general expression for the autocorrelation sequence for an AR process.

(e) Show that if $b_0 = 1$, the autocorrelation function of an AR process satisfies the difference equation

$$\phi_{yy}[0] = \sum_{k=1}^{N} a_k \phi_{yy}[k] + \sigma_x^2,$$

$$\phi_{yy}[m] = \sum_{k=1}^{N} a_k \phi_{yy}[m-k], \qquad m \geq 1.$$

(f) Use the result of part (e) and the symmetry of $\phi_{yy}[m]$ to show that

$$\sum_{k=1}^{N} a_k \phi_{yy}[|m-k|] = \phi_{yy}[m], \qquad m = 1, 2, \ldots, N.$$

It can be shown that, given $\phi_{yy}[m]$ for $m = 0, 1, \ldots, N$, we can always solve uniquely for the values of the a_ks and σ_x^2 for the random-process model. These values may be used in the result in part (b) to obtain an expression for the power density spectrum of $y[n]$. This approach is the basis for a number of parametric spectrum estimation techniques. (For further discussion of these methods, see Gardner, 1988; Kay, 1988; and Marple, 1987.)

10.48. This problem illustrates the basis for an FFT-based procedure for interpolating the samples (obtained at a rate satisfying the Nyquist theorem) of a periodic continuous-time signal. Let

$$x_c(t) = \frac{1}{16} \sum_{k=-4}^{4} \left(\frac{1}{2}\right)^{|k|} e^{jkt}$$

be a periodic signal that is processed by the system in Figure P10.48.

(a) Sketch the 16-point sequence $G[k]$.

(b) Specify how you would change $G[k]$ into a 32-point sequence $Q[k]$ so that the 32-point inverse DFT of $Q[k]$ is a sequence

$$q[n] = \alpha x_c\left(\frac{n2\pi}{32}\right), \qquad 0 \leq n \leq 31,$$

for some nonzero constant α. You need not specify the value of α.

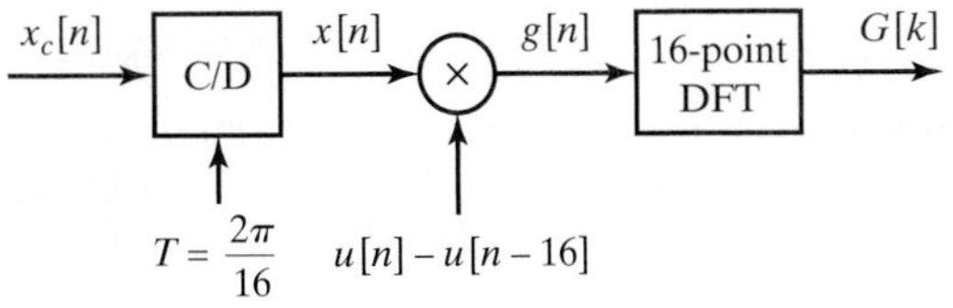

$$T = \frac{2\pi}{16} \qquad u[n] - u[n-16]$$

Figure P10.48

10.49. In many real applications, practical constraints do not allow long time sequences to be processed. However, significant information can be gained from a windowed section of the sequence. In this problem, you will look at computing the Fourier transform of an infinite-duration signal $x[n]$, given only a block of 256 samples in the range $0 \leq n \leq 255$. You decide to use a 256-point DFT to estimate the transform by defining the signal

$$\hat{x}[n] = \begin{cases} x[n], & 0 \leq n \leq 255, \\ 0, & \text{otherwise,} \end{cases}$$

and computing the 256-point DFT of $\hat{x}[n]$.

(a) Suppose the signal $x[n]$ came from sampling a continuous-time signal $x_c(t)$ with sampling frequency $f_s = 20$ kHz; i.e.,

$$x[n] = x_c(nT_s),$$

$$1/T_s = 20 \text{ kHz}.$$

Assume that $x_c(t)$ is bandlimited to 10 kHz. If the DFT of $\hat{x}[n]$ is written $\hat{X}[k]$, $k = 0, 1, \ldots, 255$, what are the continuous-time frequencies corresponding to the DFT indices $k = 32$ and $k = 231$? Be sure to express your answers in Hertz.

(b) Express the DTFT of $\hat{x}[n]$ in terms of the DTFT of $x[n]$ and the DTFT of a 256-point rectangular window $w_R[n]$. Use the notation $X(e^{j\omega})$ and $W_R(e^{j\omega})$ to represent the DTFTs of $x[n]$ and $w_R[n]$, respectively.

(c) Suppose you try an averaging technique to estimate the transform for $k = 32$:

$$X_{\text{avg}}[32] = \alpha \hat{X}[31] + \hat{X}[32] + \alpha \hat{X}[33].$$

Averaging in this manner is equivalent to multiplying the signal $\hat{x}[n]$ by a new window $w_{\text{avg}}[n]$ before computing the DFT. Show that $W_{\text{avg}}(e^{j\omega})$ must satisfy

$$W_{\text{avg}}(e^{j\omega}) = \begin{cases} 1, & \omega = 0, \\ \alpha, & \omega = \pm 2\pi/L, \\ 0, & \omega = 2\pi k/L, \quad \text{for } k = 2, 3, \ldots, L-2, \end{cases}$$

where $L = 256$.

(d) Show that the DTFT of this new window can be written in terms of $W_R(e^{j\omega})$ and two shifted versions of $W_R(e^{j\omega})$.

(e) Derive a simple formula for $w_{\text{avg}}[n]$, and sketch the window for $\alpha = -0.5$ and $0 \leq n \leq 255$.

10.50. It is often of interest to zoom in on a region of a DFT of a signal to examine it in more detail. In this problem, you will explore two algorithms for implementing this process of obtaining additional samples of $X(e^{j\omega})$ in a frequency region of interest.

Suppose $X_N[k]$ is the N-point DFT of a finite-length signal $x[n]$. Recall that $X_N[k]$ consists of samples of $X(e^{j\omega})$ every $2\pi/N$ in ω. Given $X_N[k]$, we would like to compute N samples of $X(e^{j\omega})$ between $\omega = \omega_c - \Delta\omega$ and $\omega = \omega_c + \Delta\omega$ with spacing $2\Delta\omega/N$, where

$$\omega_c = \frac{2\pi k_c}{N}$$

and

$$\Delta\omega = \frac{2\pi k_\Delta}{N}.$$

This is equivalent to zooming in on $X(e^{j\omega})$ in the region $\omega_c - \Delta\omega < \omega < \omega_c + \Delta\omega$. One system used to implement the zoom is shown in Figure P10.50-1. Assume that $x_z[n]$ is zero-padded as necessary before the N-point DFT and $h[n]$ is an ideal lowpass filter with a cutoff frequency $\Delta\omega$.

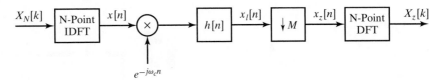

Figure P10.50-1

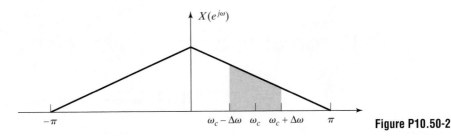

Figure P10.50-2

(a) In terms of k_Δ and the transform length N, what is the largest (possibly noninteger) value of M that can be used if aliasing is to be avoided in the downsampler?

(b) Consider $x[n]$ with the Fourier transform shown in Figure P10.50-2. Using the maximum value of M from part (a), sketch the Fourier transforms of the intermediate signals $x_\ell[n]$ and $x_z[n]$ when $\omega_c = \pi/2$ and $\Delta\omega = \pi/6$. Demonstrate that the system provides the desired frequency samples.

Another procedure for obtaining the desired samples can be developed by viewing the finite-length sequence $X_N[k]$ indexed on k as a discrete-time data sequence to be processed as shown in Figure P10.50-3. The impulse response of the first system is

$$p[n] = \sum_{r=-\infty}^{\infty} \delta[n + rN],$$

and the filter $h[n]$ has the frequency response

$$H(e^{j\omega}) = \begin{cases} 1, & |\omega| \le \pi/M, \\ 0, & \text{otherwise.} \end{cases}$$

The zoomed output signal is defined as

$$X_z[n] = \tilde{X}_{NM}[Mk_c - Mk_\Delta + n], \quad 0 \le n \le N - 1,$$

for appropriate values of k_c and k_Δ. Assume that k_Δ is chosen so that M is an integer in the following parts.

(c) Suppose that the ideal lowpass filter $h[n]$ is approximated by a causal Type I linear-phase filter of length 513 (nonzero for $0 \le n \le 512$). Indicate which samples of $\tilde{X}_{NM}[n]$ provide the desired frequency samples.

(d) Using sketches of a typical spectrum for $X_N[k]$ and $X(e^{j\omega})$, demonstrate that the system in Figure P10.50-3 produces the desired samples.

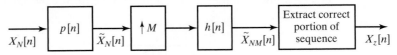

Figure P10.50-3

11

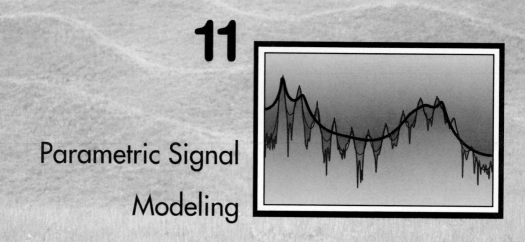

Parametric Signal Modeling

11.0 INTRODUCTION

Throughout this text, we have found it convenient to use several different representations of signals and systems. For example, the representation of a discrete-time signal as a sequence of scaled impulses was used in Eq. (2.5) of Chapter 2 to develop the convolution sum for LTI systems. The representation as a linear combination of sinusoidal and complex exponential signals led to the Fourier series, the Fourier transform, and the frequency domain characterization of signals and LTI systems. Although these representations are particularly useful because of their generality, they are not always the most efficient representation for signals with a known structure.

This chapter introduces another powerful approach to signal representation called *parametric signal modeling*. In this approach. a signal is represented by a mathematical model that has a predefined structure involving a limited number of parameters. A given signal $s[n]$ is represented by choosing the specific set of parameters that results in the model output $\hat{s}[n]$ being as close as possible in some prescribed sense to the given signal. A common example is to model the signal as the output of a discrete-time linear system as shown in Figure 11.1. Such models, which are comprised of the input signal $v[n]$ and the system function $H(z)$ of the linear system, become useful with the addition of constraints that make it possible to solve for the parameters of $H(z)$ given the signal

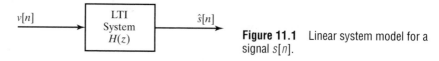

Figure 11.1 Linear system model for a signal $s[n]$.

to be represented. For example, if the input $v[n]$ is specified, and the system function is assumed to be a rational function of the form

$$H(z) = \frac{\displaystyle\sum_{k=0}^{q} b_k z^{-k}}{1 - \displaystyle\sum_{k=1}^{p} a_k z^{-k}}, \tag{11.1}$$

then the signal is modeled by the values of the a_ks and b_ks or equivalently, by the poles and zeros of $H(z)$, along with knowledge of the input. The input signal $v[n]$ is generally assumed to be a unit impulse $\delta[n]$ for deterministic signals, or white noise if the signal $s[n]$ is viewed as a random signal. When the model is appropriately chosen, it is possible to represent a large number of signal samples by a relatively small set of parameters.

Parametric signal modeling has a wide range of applications, including data compression, spectrum analysis, signal prediction, deconvolution, filter design, system identification, signal detection, and signal classification. In data compression, for example, it is the set of model parameters that is transmitted or stored, and the receiver then uses the model with those parameters to regenerate the signal. In filter design, the model parameters are chosen to best approximate, in some sense, the desired frequency response, or equivalently, the desired impulse response, and the model with these parameters then corresponds to the designed filter. The two key elements for success in all of the applications are an appropriate choice of model and an accurate estimate of the parameters for the model.

11.1 ALL-POLE MODELING OF SIGNALS

The model represented by Eq. (11.1) in general has both poles and zeros. While there are a variety of techniques for determining the full set of numerator and denominator coefficients in Eq. (11.1), the most successful and most widely used have concentrated on restricting q to be zero, in which case, $H(z)$ in Figure 11.1 has the form

$$H(z) = \frac{G}{1 - \displaystyle\sum_{k=1}^{p} a_k z^{-k}} = \frac{G}{A(z)}, \tag{11.2}$$

where we have replaced the parameter b_0 by the parameter G to emphasize its role as an overall gain factor. Such models are aptly termed "all-pole" models.[1] By its very nature, it would appear that an all-pole model would be appropriate only for modeling signals of infinite duration. While this may be true in a theoretical sense, this choice for the system function of the model works well for signals found in many applications, and as we will show, the parameters can be computed in a straightforward manner from finite-duration segments of the given signal.

---

[1]Detailed discussion of this case and the general pole/zero case are given in Kay (1988), Thierrien (1992), Hayes (1996) and Stoica and Moses (2005).

The input and output of the all-pole system in Eq. (11.2) satisfy the linear constant–coefficient difference equation

$$\hat{s}[n] = \sum_{k=1}^{p} a_k \hat{s}[n-k] + Gv[n], \tag{11.3}$$

which indicates that the model output at time n is comprised of a linear combination of past samples plus a scaled input sample. As we will see, this structure suggests that the all-pole model is equivalent to the assumption that the signal can be approximated as a linear combination of (or equivalently, is linearly predictable from) its previous values. Consequently, this method for modeling a signal is often also referred to as *linear predictive analysis* or *linear prediction*.[2]

11.1.1 Least-Squares Approximation

The goal in all-pole modeling is to choose the input $v[n]$ and the parameters G, and $a_1, \ldots, a_p$ in Eq. (11.3) such that $\hat{s}[n]$ is a close approximation in some sense to $s[n]$, the signal to be modeled. If, as is usually the case, $v[n]$ is specified in advance (e.g., $v[n] = \delta[n]$), a direct approach to determining the best values for the parameters might be to minimize the total energy in the error signal $e_{se}[n] = (s[n] - \hat{s}[n])$, thereby obtaining a least-squares approximation to $s[n]$. Specifically, for deterministic signals, the model parameters might be chosen to minimize the total squared error

$$\sum_{n=-\infty}^{\infty} (s[n] - \hat{s}[n])^2 = \sum_{n=-\infty}^{\infty} \left(s[n] - \sum_{k=1}^{p} a_k \hat{s}[n-k] - Gv[n] \right)^2. \tag{11.4}$$

In principle, the a_ks minimizing this squared error can be found by differentiating the expression in Eq. (11.4) with respect to each parameter, setting that derivative to zero, and solving the resulting equations. However, this results in a nonlinear system of equations, the solution of which is computationally difficult, in general. Although this least-squares problem is too difficult for most practical applications, the basic least-squares principle can be applied to slightly different formulations with considerable success.

11.1.2 Least-Squares Inverse Model

A formulation based on inverse filtering provides a relatively straightforward and tractable solution for the parameter values in the all-pole model. In any approach to approximation, it is recognized at the outset that the model output will in most cases not be exactly equal to the signal to be modeled. The inverse filtering approach is based on the recognition that if the given signal $s[n]$ is in fact the output of the filter $H(z)$ in the model of Figure 11.1 then with $s[n]$ as the input to the inverse of $H(z)$, the output will be $v[n]$. Consequently, as depicted in Figure 11.2 and with $H(z)$ assumed to be an all-pole system as specified in Eq.(11.2), the inverse filter, whose system function

$$A(z) = 1 - \sum_{k=1}^{p} a_k z^{-k}, \tag{11.5}$$

---

[2]When used in the context of speech processing, linear predictive analysis is often referred to as *linear predictive coding* (LPC). (See Rabner and Schafer, 1978 and Quatieri, 2002.)

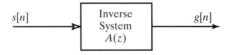

Figure 11.2   Inverse filter formulation for all-pole signal modeling.

is sought so that its output $g[n]$ would be equal to the scaled input $Gv[n]$. In this formulation, then, we choose the parameters of the inverse filter (and therefore implicitly the parameters of the model system) to minimize the mean-squared error between $g[n]$ and $Gv[n]$. As we will see, this leads to a set of well-behaved linear equations.

From Figure 11.2 and Eq. (11.5) it follows that $g[n]$ and $s[n]$ satisfy the difference equation

$$g[n] = s[n] - \sum_{k=1}^{p} a_k s[n-k]. \tag{11.6}$$

The modeling error $\hat{e}[n]$ is now defined as

$$\hat{e}[n] = g[n] - Gv[n] = s[n] - \sum_{k=1}^{p} a_k s[n-k] - Gv[n]. \tag{11.7}$$

If $v[n]$ is an impulse, then, for $n > 0$, the error $\hat{e}[n]$ corresponds to the error between $s[n]$ and the linear prediction of $s[n]$ using the model parameters. Thus, it is convenient to also express Eq. (11.7) as

$$\hat{e}[n] = e[n] - Gv[n], \tag{11.8}$$

where $e[n]$ is the prediction error given by

$$e[n] = s[n] - \sum_{k=1}^{p} a_k s[n-k]. \tag{11.9}$$

For a signal that exactly fits the all-pole model of Eq. (11.3), the modeling error $\hat{e}[n]$ will be zero, and the prediction error $e[n]$ will be the scaled input, i.e.,

$$e[n] = Gv[n]. \tag{11.10}$$

This formulation in terms of inverse filtering leads to considerable simplification, since $v[n]$ is assumed known and $e[n]$ can be computed from $s[n]$ using Eq. (11.9). The parameter values a_k are then chosen to minimize

$$\mathcal{E} = \left\langle |\hat{e}[n]|^2 \right\rangle, \tag{11.11}$$

where the notation $\langle \cdot \rangle$ denotes a summing operation for finite energy deterministic signals and an ensemble averaging operation for random signals. Minimizing $\mathcal{E}$ in Eq. (11.11) results in an inverse filter that minimizes the total energy in the modeling error in the case of deterministic signals or the mean-squared value of the modeling error in the case of random signals. For convenience, we will often refer to $\langle \cdot \rangle$ as the averaging operator where its interpretation as a sum or as an ensemble average should be clear from the context. Again, note that in solving for the parameters a_k specifying the inverse system of Figure 11.2, the all-pole system is implicitly specified, as well.

To find the optimal parameter values, we substitute Eq. (11.8) into Eq. (11.11) to obtain

$$\mathcal{E} = \left\langle (e[n] - Gv[n])^2 \right\rangle, \tag{11.12}$$

or equivalently,

$$\mathcal{E} = \left\langle e^2[n] \right\rangle + G^2 \left\langle v^2[n] \right\rangle - 2G \left\langle v[n]e[n] \right\rangle. \tag{11.13}$$

To find the parameters that minimize $\mathcal{E}$, we differentiate Eq. (11.12) with respect to the ith filter coefficient a_i and set the derivative equal to zero, leading to the set of equations

$$\frac{\partial \mathcal{E}}{\partial a_i} = \frac{\partial}{\partial a_i} \left[\left\langle e^2[n] \right\rangle - 2G \left\langle v[n]s[n-i] \right\rangle \right] = 0, \quad i = 1, 2, \ldots, p, \tag{11.14}$$

where we have assumed that G is independent of a_i and, of course, so is $v[n]$, and consequently that

$$\frac{\partial}{\partial a_i} \left[G^2 \left\langle v^2[n] \right\rangle \right] = 0. \tag{11.15}$$

For models that will be of interest to us, $v[n]$ will be an impulse if $s[n]$ is a causal finite-energy signal and white noise if $s[n]$ is a wide-sense stationary random process. With $v[n]$ an impulse and $s[n]$ zero for $n < 0$, the product $v[n]s[n-i] = 0$ for $i = 1, 2, \ldots p$. With $v[n]$ as white noise,

$$\langle v[n]s[n-i] \rangle = 0, \quad i = 1, 2, \ldots p, \tag{11.16}$$

since for any value of n, the input of a causal system with white-noise input is uncorrelated with the output values prior to time n. Thus, for both cases, Eq. (11.14) reduces to

$$\frac{\partial \mathcal{E}}{\partial a_i} = \frac{\partial}{\partial a_i} \left\langle e^2[n] \right\rangle = 0 \quad i = 1, 2, , \ldots, p \tag{11.17}$$

In other words, choosing the coefficients to minimize the average squared modeling error $\left\langle \hat{e}^2[n] \right\rangle$ is equivalent to minimizing the average squared prediction error $\left\langle e^2[n] \right\rangle$. Expanding Eq. (11.17) and invoking the linearity of the averaging operator, we obtain from Eq. (11.17) the equations

$$\langle s[n]s[n-i] \rangle - \sum_{k=1}^{p} a_k \langle s[n-k]s[n-i] \rangle = 0, \quad i = 1, \ldots, p. \tag{11.18}$$

Defining

$$\phi_{ss}[i, k] = \langle s[n-i]s[n-k] \rangle, \tag{11.19}$$

Eqs. (11.18) can be rewritten more compactly as

$$\sum_{k=1}^{p} a_k \phi_{ss}[i, k] = \phi_{ss}[i, 0], \quad i = 1, 2, \ldots, p. \tag{11.20}$$

Equations (11.20) comprise a system of p linear equations in p unknowns. Computation of the parameters of the model can be achieved by solving the set of linear equations for the parameters a_k for $k = 1, 2, \ldots, p$, using known values for $\phi_{ss}[i, k]$ for $i = 1, 2, \ldots, p$ and $k = 0, 1, \ldots, p$ or first computing them from $s[n]$.

11.1.3 Linear Prediction Formulation of All-Pole Modeling

As suggested earlier, an alternative and useful interpretation of all-pole signal modeling stems from the interpretation of Eq. (11.3) as a linear prediction of the output in terms of past values, with the prediction error $e[n]$ being the scaled input $Gv[n]$, i.e.,

$$e[n] = s[n] - \sum_{k=1}^{p} a_k s[n - k] = Gv[n]. \tag{11.21}$$

As indicated by Eq. (11.17), minimizing the inverse modeling error $\mathcal{E}$ in Eq. (11.11) is equivalent to minimizing the averaged prediction error $\langle e^2[n] \rangle$. If the signal $s[n]$ were produced by the model system, and if $v[n]$ is an impulse, and if $s[n]$ truly fits the all-pole model, then the signal at any $n > 0$ is linearly predictable from past values, i.e., the prediction error is zero. If $v[n]$ is white noise, then the prediction error is white.

The interpretation in terms of prediction is depicted in Figure 11.3, where the transfer function of the prediction filter $P(z)$ is

$$P(z) = \sum_{k=1}^{p} a_k z^{-k}. \tag{11.22}$$

This system is referred to as the pth-order *linear predictor* for the signal $s[n]$. Its output is

$$\tilde{s}[n] = \sum_{k=1}^{p} a_k s[n - k], \tag{11.23}$$

and as Figure 11.3 shows, the prediction error signal is $e[n] = s[n] - \tilde{s}[n]$. The sequence $e[n]$ represents the amount by which the linear predictor fails to exactly predict the signal $s[n]$. For this reason, $e[n]$ is also sometimes called the *prediction error residual* or simply the *residual*. With this point of view, the coefficients a_k are called the *prediction coefficients*. As is also shown in Figure 11.3, the prediction error filter is related to the linear predictor by

$$A(z) = 1 - P(z) = 1 - \sum_{k=1}^{p} a_k z^{-k}. \tag{11.24}$$

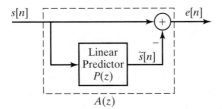

Figure 11.3 Linear prediction formulation for all-pole signal modeling.

11.2 DETERMINISTIC AND RANDOM SIGNAL MODELS

To use the optimum inverse filter or equivalently the optimum linear predictor as a basis for parametric signal modeling, it is necessary to be more specific about the assumed input $v[n]$ and about the method of computing the averaging operator $\langle \cdot \rangle$. To this end, we consider separately the case of deterministic signals and the case of random signals. In both cases, we will use averaging operations that assume knowledge of the signal to be modeled over all time $-\infty < n < \infty$. In Section 11.3, we discuss some of the practical considerations when only a finite-length segment of the signal $s[n]$ is available.

11.2.1 All-Pole Modeling of Finite-Energy Deterministic Signals

In this section, we assume an all-pole model that is causal and stable and also that both the input $v[n]$ and the signal $s[n]$ to be modeled are zero for $n < 0$. We further assume that $s[n]$ has finite energy and is known for all $n \geq 0$. We choose the operator $\langle \cdot \rangle$ in Eq. (11.11) as the total energy in the modeling error sequence $\hat{e}[n]$, i.e.,

$$\mathcal{E} = \left\langle |\hat{e}[n]|^2 \right\rangle = \sum_{n=-\infty}^{\infty} |\hat{e}[n]|^2. \tag{11.25}$$

With this definition of the averaging operator, $\phi_{ss}[i, k]$ in Eq. (11.19) is given by

$$\phi_{ss}[i, k] = \sum_{n=-\infty}^{\infty} s[n - i]s[n - k], \tag{11.26}$$

and equivalently,

$$\phi_{ss}[i, k] = \sum_{n=-\infty}^{\infty} s[n]s[n - (i - k)]. \tag{11.27}$$

The coefficients $\phi_{ss}[i, k]$ in Eq. (11.20) are now

$$\phi_{ss}[i, k] = r_{ss}[i - k], \tag{11.28}$$

where for real signals $s[n]$, $r_{ss}[m]$ is the deterministic autocorrelation function

$$r_{ss}[m] = \sum_{n=-\infty}^{\infty} s[n + m]s[n] = \sum_{n=-\infty}^{\infty} s[n]s[n - m]. \tag{11.29}$$

Therefore, Eq. (11.20) takes the form

$$\sum_{k=1}^{p} a_k r_{ss}[i - k] = r_{ss}[i] \quad i = 1, 2, \ldots, p. \tag{11.30}$$

These equations are called the *autocorrelation normal equations* and also the *Yule–Walker equations*. They provide a basis for computing the parameters $a_1, \ldots, a_p$ from the autocorrelation function of the signal. In Section 11.2.5, we discuss an approach to choosing the gain factor G.

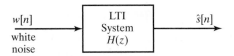

Figure 11.4 Linear system model for a random signal $s[n]$.

11.2.2 Modeling of Random Signals

For all-pole modeling of zero-mean, wide-sense stationary, random signals, we assume that the input to the all-pole model is zero-mean, unit-variance, white noise as indicated in Figure 11.4. The difference equation for this system is

$$\hat{s}[n] = \sum_{k=1}^{p} a_k \hat{s}[n - k] + Gw[n], \qquad (11.31)$$

where the input has autocorrelation function $E\{w[n + m]w[n]\} = \delta[m]$, zero mean $(E\{w[n]\} = 0)$, and unit average power $(E\{(w[n])^2\} = \delta[0] = 1)$, with $E\{\cdot\}$ representing the expectation or probability average operator.[3]

The resulting model for analysis is the same as that depicted in Figure 11.2, but the desired output $g[n]$ changes. In the case of random signals, we want to make $g[n]$ as much like a white-noise signal as possible, rather than the unit sample sequence that was desired in the deterministic case. For this reason, the optimal inverse filter for random signals is often referred to as a *whitening filter*.

We also choose the operator $\langle\cdot\rangle$ in Eq. (11.11) as an appropriate one for random signals, specifically the mean-squared value or equivalently the average power. Then Eq. (11.11) becomes

$$\mathcal{E} = E\{(\hat{e}[n])^2\}. \qquad (11.32)$$

If $s[n]$ is assumed to be a sample function of a stationary random process, then $\phi_{ss}[i, k]$ in Eq. (11.19) would be the autocorrelation function

$$\phi_{ss}[i, k] = E\{s[n - i]s[n - k]\} = r_{ss}[i - k]. \qquad (11.33)$$

The system coefficients can be found as before from Eq. (11.20). Thus, the system coefficients satisfy a set of equations of the same form as Eq. (11.30), i.e.,

$$\sum_{k=1}^{p} a_k r_{ss}[i - k] = r_{ss}[i], \quad i = 1, 2, \ldots, p. \qquad (11.34)$$

Therefore, modeling random signals again results in the Yule–Walker equations, with the autocorrelation function in this case being defined by the probabilistic average

$$r_{ss}[m] = E\{s[n + m]s[n]\} = E\{s[n]s[n - m]\}. \qquad (11.35)$$

---

[3]Computation of $E\{\cdot\}$ requires knowledge of the probability densities. In the case of stationary random signals, only one density is required. In the case of ergodic random processes, a single infinite time average could be used. In practical applications, however, such averages must be approximated by estimates obtained from finite time averages.

11.2.3 Minimum Mean-Squared Error

For modeling of either deterministic signals (Section 11.2.1) or random signals (Section 11.2.2) the minimum value of the prediction error $e[n]$ in Figure 11.3 can be expressed in terms of the corresponding correlation values in Eq. (11.20) to find the optimum predictor coefficients. To see this, we write $\mathcal{E}$ as

$$\mathcal{E} = \left\langle \left(s[n] - \sum_{k=1}^{p} a_k s[n-k] \right)^2 \right\rangle. \tag{11.36}$$

As outlined in more detail in Problem 11.2, if Eq. (11.36) is expanded, and Eq. (11.20) is substituted into the result, it follows that in general,

$$\mathcal{E} = \phi_{ss}[0, 0] - \sum_{k=1}^{p} a_k \phi_{ss}[0, k]. \tag{11.37}$$

Equation (11.37) is true for any appropriate choice of the averaging operator. In particular, for averaging definitions for which $\phi_{ss}[i, k] = r_{ss}[i - k]$, Eq. (11.37) becomes

$$\mathcal{E} = r_{ss}[0] - \sum_{k=1}^{p} a_k r_{ss}[k]. \tag{11.38}$$

11.2.4 Autocorrelation Matching Property

An important and useful property of the all-pole model resulting from the solution of Eq. (11.30) for deterministic signals and Eq. (11.34) for random signals is referred to as the autocorrelation matching property (Makhoul, 1973). Equations (11.30) and (11.34) represent a set of p equations to be solved for the model parameters a_k for $k = 1, \ldots, p$. The coefficients in these equations on both the left- and right-hand sides of the equations are comprised of the $(p+1)$ correlation values $r_{ss}[m], m = 0, 1, \ldots, p$, where the correlation function is appropriately defined, depending on whether the signal to be modeled is deterministic or random.

The basis for verifying the autocorrelation matching property is to observe that the signal $\hat{s}[n]$ obviously fits the model when the model system $H(z)$ in Figure 11.1 is specified as the all-pole system in Eq. (11.2). If we were to consider again applying all-pole modeling to $\hat{s}[n]$, we would of course again obtain Eqs. (11.30) or (11.34), but this time, with $r_{\hat{s}\hat{s}}[m]$ in place of $r_{ss}[m]$. The solution must again be the same parameter values $a_k, k = 1, 2, \ldots, p$, since $\hat{s}[n]$ fits the model, and this solution will result if

$$r_{ss}[m] = c r_{\hat{s}\hat{s}}[m] \quad 0 \le m \le p, \tag{11.39}$$

where c is any constant. The fact that the equality in Eq. (11.39) is required follows from the form of the recursive solution of the Yule–Walker equations as developed in Section 11.6. In words, the autocorrelation normal equations require that for the lags $|m| = 0, 1, \ldots, p$ the autocorrelation functions of the model output and the signal being modeled are proportional.

11.2.5 Determination of the Gain Parameter G

With the approach that we have taken, determination of the optimal choice for the coefficients a_k of the model does not depend on the system gain G. From the perspective of the inverse filtering formulation in Figure 11.2, one possibility is to choose G so that $\langle (\hat{s}[n])^2 \rangle = \langle (s[n])^2 \rangle$. For finite-energy deterministic signals, this corresponds to matching the total energy in the model output to the total energy in the signal that is being modeled. For random signals, it is the average power that is matched. In both cases, this corresponds to choosing G, so that $r_{\hat{s}\hat{s}}[0] = r_{ss}[0]$. With this choice, the proportionality factor c in Eq. (11.39) is unity.

Example 11.1 1st-Order System

Figure 11.5 shows two signals, both of which are outputs of a 1st-order system with system function

$$H(z) = \frac{1}{1 - \alpha z^{-1}}. \tag{11.40}$$

The signal $s_d[n] = h[n] = \alpha^n u[n]$ is the output when the input is a unit impulse $\delta[n]$, while the signal $s_r[n]$ is the output when the input to the system is a zero mean, unit variance white-noise sequence. Both signals extend over the range $-\infty < n < \infty$, as suggested by Figure 11.5.

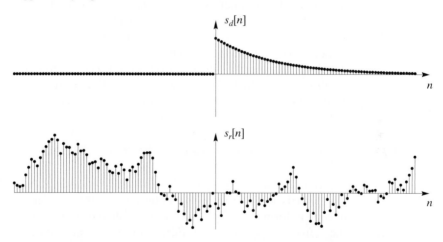

Figure 11.5   Examples of deterministic and random outputs of a 1st-order all-pole system.

The autocorrelation function for the signal $s_d[n]$ is

$$r_{s_d s_d}[m] = r_{hh}[m] = \sum_{n=0}^{\infty} \alpha^{n+m} \alpha^n = \frac{\alpha^{|m|}}{1 - \alpha^2}, \tag{11.41}$$

the autocorrelation function of $s_r[n]$ is also given by Eq. (11.41) since $s_r[n]$ is the response of the system to white noise, for which the autocorrelation function is a unit impulse.

Since both signals were generated with a 1st-order all-pole system, a 1st-order all-pole model will be an exact fit. In the deterministic case, the output of the optimum inverse filter will be a unit impulse, and in the random signal case, the output of the optimum inverse filter will be a zero-mean white-noise sequence with unit average power. To show that the optimum inverse filter will be exact, note that for a 1st-order model, Eqs. (11.30) or (11.34) reduce to

$$r_{s_d s_d}[0]a_1 = r_{s_d s_d}[1], \tag{11.42}$$

so from Eq. (11.41), it follows that the optimum predictor coefficient for both the deterministic and the random signal is

$$a_1 = \frac{r_{s_d s_d}[1]}{r_{s_d s_d}[0]} = \frac{\dfrac{\alpha}{1-\alpha^2}}{\dfrac{1}{1-\alpha^2}} = \alpha. \tag{11.43}$$

From Eq. (11.38), the minimum mean-squared error is

$$\mathcal{E} = \frac{1}{1-\alpha^2} - a_1\frac{\alpha}{1-\alpha^2} = \frac{1-\alpha^2}{1-\alpha^2} = 1, \tag{11.44}$$

which is the size of the unit impulse in the deterministic case and the average power of the white-noise sequence in the random case.

As mentioned earlier, and as is clear in this example, when the signal is generated by an all-pole system excited by either an impulse or white noise, all-pole modeling can determine the parameters of the all-pole system exactly. This requires prior knowledge of the model order p and the autocorrelation function. This was possible to obtain for this example, because a closed-form expression was available for the infinite sum required to compute the autocorrelation function. In a practical setting, it is generally necessary to estimate the autocorrelation function from a finite-length segment of the given signal. Problem 11.14 considers the effect of finite autocorrelation estimates (to be discussed next) for the deterministic signal $s_d[n]$ of this section.

11.3 ESTIMATION OF THE CORRELATION FUNCTIONS

To use the results of Sections 11.1 and 11.2 for modeling of either deterministic or random signals, we require *apriori* knowledge of the correlation functions $\phi_{ss}[i, k]$ that are needed to form the system equations satisfied by the coefficients a_k, or we must estimate these from the given signal. Furthermore, we may want to apply block processing or short-time analysis techniques to represent the time-varying properties of a nonstationary signal, such as speech. In this section, we will discuss two distinct approaches to the computation of the correlation estimates for practical application of the concepts of parametric signal modeling. These two approaches have come to be known as the *autocorrelation method* and the *covariance method*.

11.3.1 The Autocorrelation Method

Suppose that we have available a set of $M+1$ signal samples $s[n]$ for $0 \le n \le M$, and we wish to compute the coefficients for an all-pole model. In the autocorrelation method, it is assumed that the signal ranges over $-\infty < n < \infty$, with the signal samples taken to be zero for all n outside the interval $0 \leq n \leq M$, even if they have been extracted from a longer sequence. This, of course, imposes a limit to the exactness that can be expected of the model, since the IIR impulse response of an all-pole model will be used to model the finite-length segment of $s[n]$.

Although the prediction error sequence need not be computed explicitly to solve for the filter coefficients, it is nevertheless informative to consider its computation in some detail. The impulse response of the prediction error filter is, by the definition of $A(z)$, in Eq. (11.24),

$$h_A[n] = \delta[n] - \sum_{k=1}^{p} a_k \delta[n - k]. \tag{11.45}$$

It can be seen that since the signal $s[n]$ has finite length $M + 1$ and $h_A[n]$, the impulse response of the prediction filter $A[z]$, has length $p + 1$, the prediction error sequence $e[n] = h_A[n] * s[n]$ will always be identically zero outside the interval $0 \leq n \leq M + p$. Figure 11.6 shows an example of the prediction error signal for a linear predictor with $p = 5$. In the upper plot, $h_A[n - m]$ the (time-reversed and shifted) impulse response of the prediction error filter, is shown as a function of m for three different values of n. The dark lines with square dots depict $h_A[n - m]$, and the lighter lines with round dots show the sequence $s[m]$ for $0 \leq m \leq 30$. On the left side is $h_A[0 - m]$, which shows that the first nonzero prediction error sample is $e[0] = s[0]$. This, of course, is consistent with Eq. (11.9). On the extreme right is $h_A[M + p - m]$, which shows that the last nonzero error sample is $e[M + p] = -a_p s[M]$. The second plot in Figure 11.6 shows the error signal $e[n]$ for $0 \leq n \leq M + p$. From the point of view of linear prediction, it follows that the first p samples (dark lines and dots) are predicted from samples that are assumed to be zero. Similarly, the samples of the input for $n \geq M + 1$ are assumed to be zero to obtain a finite-length signal. The linear predictor attempts to predict the zero samples

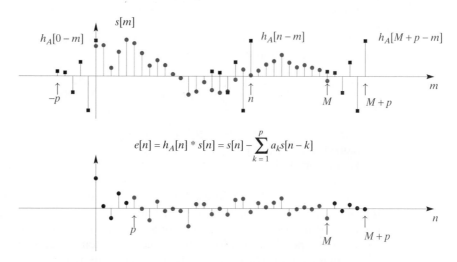

Figure 11.6   Illustration (for $p = 5$) of computation of prediction error for the autocorrelation method. (Square dots denote samples of $h_A[n - m]$ and light round dots denote samples of $s[m]$ for the upper plot and $e[n]$ for the lower plot.)

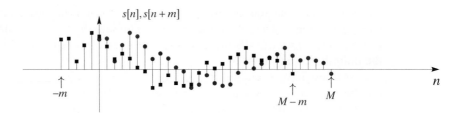

Figure 11.7   Illustration of computation of the autocorrelation function for a finite-length sequence. (Square dots denote samples of $s[n+m]$, and light round dots denote samples of $s[n]$.)

in the interval $M+1 \leq n \leq M+p$ from prior samples that are nonzero and part of the original signal. Indeed, if $s[0] \neq 0$ and $s[M] \neq 0$, then it will be true that both $e[0] = s[0]$ and $e[M+p] = -a_p s[M]$ will be nonzero. That is, the prediction error (total-squared error $\mathcal{E}$) can never be exactly zero if the signal is defined to be zero outside the interval $0 \leq n \leq M$. Furthermore, the total-squared prediction error for a pth-order predictor would be

$$\mathcal{E}^{(p)} = \left\langle e[n]^2 \right\rangle = \sum_{n=-\infty}^{\infty} e[n]^2 = \sum_{n=0}^{M+p} e[n]^2, \tag{11.46}$$

i.e., the limits of summation can be infinite for convenience, but practically speaking, they are finite.

When the signal is assumed to be identically zero outside the interval $0 \leq n \leq M$, the correlation function $\phi_{ss}[i,k]$ reduces to the autocorrelation function $r_{ss}[m]$ where the values needed in Eq. (11.30) are for $m = |i - k|$. Figure 11.7 shows the shifted sequences used in computing $r_{ss}[m]$ with $s[n]$ denoted by round dots and $s[n+m]$ by square dots. Note that for a finite-length signal, the product $s[n]s[n+m]$ is nonzero only over the interval $0 \leq n \leq M - m$ when $m \geq 0$. Since r_{ss} is an even function, i.e., $r_{ss}[-m] = r_{ss}[m] = r_{ss}[|m|]$ it follows that the autocorrelation values needed for the Yule–Walker equations can be computed as,

$$r_{ss}[|m|] = \sum_{n=-\infty}^{\infty} s[n]s[n+|m|] = \sum_{n=0}^{M-|m|} s[n]s[n+|m|]. \tag{11.47}$$

For the finite-length sequence $s[n]$, Eq. (11.47) has all the necessary properties of an autocorrelation function and $r_{ss}[m] = 0$ for $m > M$. But of course $r_{ss}[m]$ is not the same as the autocorrelation function of the infinite length signal from which the segment was extracted.

Equation (11.47) can be used to compute estimates of the autocorrelation function for either deterministic or random signals.[4] Often, the finite-length input signal is extracted from a longer sequence of samples. This is the case, for example, in applications to speech processing, where voiced segments (e.g., vowel sounds) of speech are treated as deterministic and unvoiced segments (fricative sounds) are treated as

---

[4]In the context of random signals, it was shown in Section 10.6 that Eq. (11.47) is a biased estimate of the autocorrelation function. When $p \ll M$ as is often the case, this statistical bias is generally negligible.

random signals.[5] According to the previous discussion, the first p and last p samples of the prediction error can be large due to the attempt to predict nonzero samples from zero samples and to predict zero samples from nonzero samples. Since this can bias the estimation of the predictor coefficients, a signal-tapering window, such as a Hamming window is generally applied to the signal before computation of the autocorrelation function.

11.3.2 The Covariance Method

An alternative choice for the averaging operator for the prediction error for a pth-order predictor is

$$\mathcal{E}_{\text{cov}}^{(p)} = \left\langle (e[n])^2 \right\rangle = \sum_{n=p}^{M} (e[n])^2. \tag{11.48}$$

As in the autocorrelation method, the averaging is over a finite interval ($p \le n \le M$), but the difference is that the signal to be modeled is known over the larger interval $0 \le n \le M$. The total-squared prediction error only includes values of $e[n]$ that can be computed from samples within the interval $0 \le n \le M$. Consequently, the averaging takes place over a shorter interval $p \le n \le M$. This is significant, since it relieves the inconsistency between the all-pole model and the finite-length signal.[6] In this case, we only seek to match the signal over a finite interval rather than over all n as in the autocorrelation method. The upper plot in Figure 11.8 shows the same signal $s[m]$ as

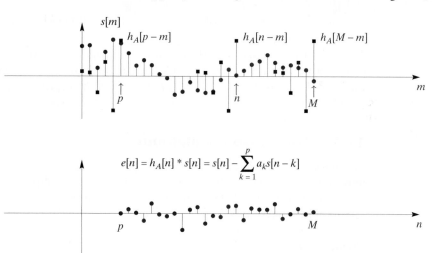

Figure 11.8 Illustration (for $p = 5$) of computation of prediction error for the covariance method. (In upper plot, square dots denote samples of $h_A[n-m]$, and light round dots denote samples of $s[m]$.)

---

[5]In both cases, the deterministic autocorrelation function in Eq. (11.47) is used as an estimate.

[6]The definitions of total-squared prediction error in Eqs. (11.48) and (11.46) are distinctly different, so we use the subscript $_{\text{cov}}$ to distinguish them.

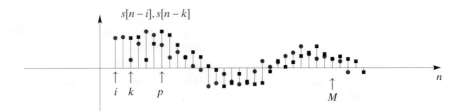

Figure 11.9   Illustration of computation of covariance function for a finite-length sequence. (Square dots denote samples of $s[n - k]$ and light round dots denote samples of $s[n - i]$.)

in the upper part of Figure 11.6, but in this case, the prediction error is only computed over the interval $p \leq n \leq M$ as needed in Eq. (11.48). As shown by the prediction error filter impulse responses $h_A[n - m]$ in the upper plot, there are no end effects when the prediction error is computed in this way, since all the signal samples needed to compute the prediction error are available. Because of this, it is possible for the prediction error to be precisely zero over the entire interval $p \leq n \leq M$, if the signal from which the finite length segment was extracted was generated as the output of an all-pole system. Seen another way, if $s[n]$ is the output of an all-pole system with an input that is zero for $n > 0$, then as seen from Eqs. (11.9) and (11.10) the prediction error will be zero for $n > 0$.

The covariance function inherits the same definition of the averaging operator, i.e.,

$$\phi_{ss}[i, k] = \sum_{n=p}^{M} s[n - i]s[n - k]. \tag{11.49}$$

The shifted sequences $s[n - i]$ (light lines and round dots) and $s[n - k]$ (dark lines and square dots) are shown in Figure 11.9. This figure shows that since we need $\phi_{ss}[i, k]$ only for $i = 0, 1, \ldots, p$ and $k = 1, 2, \ldots, p$, the segment $s[n]$ for $0 \leq n \leq M$ contains all the samples that are needed to compute $\phi_{ss}[i, k]$ in Eq. (11.49).

11.3.3 Comparison of Methods

The autocorrelation and covariance methods have many similarities, but there are many important differences in the methods and the resulting all-pole models. In this section, we summarize some of the differences that we have already demonstrated and call attention to some others.

Prediction Error

Both the averaged prediction error $\langle e^2[n] \rangle$ and averaged modeling error $\langle \hat{e}^2[n] \rangle$ are nonnegative and nonincreasing with increasing model order p. In the autocorrelation method based on estimates obtained from finite-length signals, the averaged modeling or prediction error will never be zero, because the autocorrelation values will not be exact. Furthermore, the minimum value of the prediction error even with an exact model is $Gv[n]$ as indicated by Eq. (11.10). In the covariance method, the prediction error for $n > 0$ can be exactly zero if the original signal was generated by an all-pole model. This will be demonstrated in Example 11.2.

Equations for Predictor Coefficients

In both methods, the predictor coefficients that minimize the averaged prediction error satisfy a general set of linear equations expressed in matrix form as $\boldsymbol{\Phi a} = \boldsymbol{\psi}$. The coefficients of the all-pole model are obtained by inverting the matrix $\boldsymbol{\Phi}$; i.e., $\boldsymbol{a} = \boldsymbol{\Phi}^{-1}\boldsymbol{\psi}$. In the covariance method, the elements $\phi_{ss}[i,k]$ of the matrix $\boldsymbol{\Phi}$ are computed using Eq. (11.49). In the autocorrelation method, the covariance values become autocorrelation values, i.e., $\phi_{ss}[i,k] = r_{ss}[|i-k|]$ and are computed using Eq. (11.47). In both cases, the matrix $\boldsymbol{\Phi}$ is symmetric and positive-definite, but in the autocorrelation method, the matrix $\boldsymbol{\Phi}$ is also a Toeplitz matrix. This implies numerous special properties of the solution, and it implies that the solution of the equations can be done more efficiently than would be true in general. In Section 11.6, we will explore some of these implications for the autocorrelation method.

Stability of the Model System

The prediction error filter has a system function $A(z)$ that is a polynomial in z^{-1}. Therefore, it can be represented in terms of its zeros as

$$A(z) = 1 - \sum_{k=1}^{p} a_k z^{-k} = \prod_{k=1}^{p}(1 - z_k z^{-1}). \tag{11.50}$$

In the autocorrelation method, the zeros of the prediction error filter $A(z)$ are guaranteed to lie strictly within the unit circle of the z plane; i.e., $|z_k| < 1$. This means that the poles of the causal system function $H(z) = G/A(z)$ of the model lie inside the unit circle, which implies that the model system is stable. A simple proof of this assertion is given by Lang and McClellan (1979) and McClellan (1988). Problem 11.10 develops a proof that depends on the lattice filter interpretation of the prediction error system to be discussed in Section 11.7.1. In the covariance method as we have formulated it, no such guarantee can be given.

11.4 MODEL ORDER

An important issue in parametric signal modeling is the model order p, the choice of which has a major impact on the accuracy of the model. A common approach to choosing p is to examine the averaged prediction error (often referred to as the residual) from the optimum p^{th}-order model. Let $a_k^{(p)}$ be the parameters for the optimal p^{th}-order predictor found using Eq. (11.30). The prediction error energy for the pth-order model using the autocorrelation method is[7]

$$\mathcal{E}^{(p)} = \sum_{n=-\infty}^{\infty}\left(s[n] - \sum_{k=1}^{p} a_k^{(p)} s[n-k]\right)^2. \tag{11.51}$$

For the zeroth-order predictor, $(p = 0)$, there are no delay terms in Eq. (11.51), i.e., the "predictor" is just the identity system so $e[n] = s[n]$. Consequently, for $p = 0$,

$$\mathcal{E}^{(0)} = \sum_{n=-\infty}^{\infty} s^2[n] = r_{ss}[0]. \tag{11.52}$$

---

[7]Recall that $\mathcal{E}_{\text{cov}}^{(p)}$ denotes the total-squared prediction error for the covariance method, while we use $\mathcal{E}^{(p)}$ with no subscript to denote the total-squared prediction error for the autocorrelation method.

Plotting the normalized mean-squared prediction error $\mathcal{V}^{(p)} = \mathcal{E}^{(p)}/\mathcal{E}^{(0)}$ as a function of p shows how increasing p changes this error energy. In the autocorrelation method, we showed that the averaged prediction error can never be precisely zero, even if the signal $s[n]$ was generated by an all-pole system, and the model order is the same as the order of the generating system. In the covariance method, however, if the all-pole model is a perfect model for the signal $s[n]$, $\mathcal{E}_{\text{cov}}^{(p)}$ will become identically zero at the correct choice of p, since the averaged prediction error only considers values for $p \leq n \leq M$. Even if $s[n]$ is not perfectly modeled by an all-pole system, there is often a value of p above which increasing p has little or no effect on either $\mathcal{V}^{(p)}$ or $\mathcal{V}_{\text{cov}}^{(p)} = \mathcal{E}_{\text{cov}}^{(p)}/\mathcal{E}_{\text{cov}}^{(0)}$. This threshold is an efficient choice of model order for representing the signal as an all-pole model.

Example 11.2　Model Order Selection

To demonstrate the effect of model order, consider a signal $s[n]$ generated by exciting a 10th-order system

$$H(z) = \frac{0.6}{\begin{array}{l}(1 - 1.03z^{-1} + 0.79z^{-2} - 1.34z^{-3} + 0.78z^{-4} - 0.92z^{-5} \\ + 1.22z^{-6} - 0.43z^{-7} + 0.6z^{-8} - 0.29z^{-9} - 0.23z^{-10})\end{array}} \tag{11.53}$$

with an impulse $v[n] = \delta[n]$. The samples of $s[n]$ for $0 \leq n \leq 30$ are shown as the sequence in the upper plots in Figures 11.6 and 11.8. This signal was used as the signal to be modeled by an all-pole model with both the autocorrelation method and the covariance method. Using the 31 samples of $s[n]$, the appropriate autocorrelation and covariance values were computed and the predictor coefficients computed by solving Eqs. (11.30) and (11.34) respectively. The normalized mean-squared prediction errors are plotted in Figure 11.10. Note that in both the autocorrelation and covariance methods the normalized error decreases abruptly at $p = 1$ in both plots, then decreasing more slowly as p increases. At $p = 10$, the covariance method gives zero error, while the autocorrelation method gives a nonzero averaged error for $p \geq 10$. This is consistent with our discussion of the prediction error in Section 11.3.

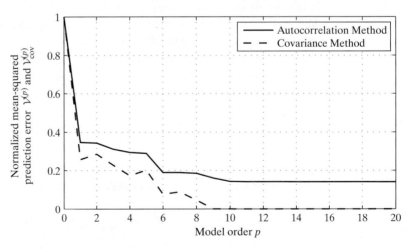

Figure 11.10　Normalized mean-squared prediction error $V^{(p)}$ as a function of model order p in Example 11.2.

While Example 11.2 is an ideal simulation, the general nature of the dependence of averaged prediction error as a function of p is typical of what happens when all-pole modeling is applied to sampled signals. The graph of $\mathcal{V}^{(p)}$ as a function of p tends to flatten out at some point, and that value of p is often selected as the value to be used in the model. In applications such as speech analysis, it is possible to choose the model order based on physical models for the production of the signal to be modeled. (See Rabiner and Schafer, 1978.)

11.5 ALL-POLE SPECTRUM ANALYSIS

All-pole signal modeling provides a method of obtaining high-resolution estimates of a signal's spectrum from truncated or windowed data. The use of parametric signal modeling in spectrum analysis is based on the fact that if the data fits the model, then a finite segment of the data can be used to determine the model parameters and, consequently, also its spectrum. Specifically, in the deterministic case

$$|\hat{S}(e^{j\omega})|^2 = |H(e^{j\omega})|^2|V(e^{j\omega})|^2 = |H(e^{j\omega})|^2 \tag{11.54}$$

since $|V(e^{j\omega})|^2 = 1$ for a unit impulse excitation to the model system. Likewise, for random signals the power spectrum of the output of the model is

$$P_{\hat{s}\hat{s}}(e^{j\omega}) = |H(e^{j\omega})|^2 P_{ww}(e^{j\omega}) = |H(e^{j\omega})|^2, \tag{11.55}$$

since $P_{ww}(e^{j\omega}) = 1$ for the white-noise input. Thus, we can obtain an estimate of the spectrum of a signal $s[n]$ by computing an all-pole model for the signal and then computing the magnitude-squared of the frequency response of the model system. For both the deterministic and random cases, the spectrum estimate takes the form

$$\text{Spectrum estimate} = |H(e^{j\omega})|^2 = \left| \frac{G}{1 - \sum_{k=1}^{p} a_k e^{-j\omega k}} \right|^2. \tag{11.56}$$

To obtain an understanding of the nature of the spectrum estimate in Eq. (11.56) for the deterministic case, it is useful to recall that the DTFT of the finite-length signal $s[n]$ is

$$S(e^{j\omega}) = \sum_{n=0}^{M} s[n]e^{-j\omega n}. \tag{11.57}$$

Furthermore, note that

$$r_{ss}[m] = \sum_{n=0}^{M-|m|} s[n+m]s[n] = \frac{1}{2\pi} \int_{-\pi}^{\pi} |S(e^{j\omega})|^2 e^{j\omega m} d\omega, \tag{11.58}$$

where, due to the finite length of $s[n]$, $r_{ss}[m] = 0$ for $|m| > M$. The values of $r_{ss}[m]$ for $m = 0, 1, 2, \ldots, p$ are used in the computation of the all-pole model using the autocorrelation method. Thus, it is reasonable to suppose that there is a relationship between the Fourier spectrum of the signal, $|S(e^{j\omega})|^2$, and the all-pole model spectrum, $|\hat{S}(e^{j\omega})|^2 = |H(e^{j\omega})|^2$.

One approach to illuminating this relationship is to obtain an expression for the averaged prediction error in terms of the DTFT of the signal $s[n]$. Recall that the prediction error is $e[n] = h_A[n] * s[n]$, where $h_A[n]$ is the impulse response of the prediction error filter. From Parseval's Theorem, the averaged prediction error is

$$\mathcal{E} = \sum_{n=0}^{M+p} (e[n])^2 = \frac{1}{2\pi} \int_{-\pi}^{\pi} |S(e^{j\omega})|^2 |A(e^{j\omega})|^2 d\omega, \qquad (11.59)$$

where $S(e^{j\omega})$ is the DTFT of $s[n]$ as given by Eq. (11.57). Since $H(z) = G/A(z)$, Eq. (11.59) can be expressed in terms of $H(e^{j\omega})$ as

$$\mathcal{E} = \frac{G^2}{2\pi} \int_{-\pi}^{\pi} \frac{|S(e^{j\omega})|^2}{|H(e^{j\omega})|^2} d\omega. \qquad (11.60)$$

Since the integrand in Eq. (11.60) is positive, and $|H(e^{j\omega})|^2 > 0$ for $-\pi < \omega \leq \pi$, it therefore follows from Eq. (11.60) that minimizing $\mathcal{E}$ is equivalent to minimizing the ratio of the energy spectrum of the signal $s[n]$ to the magnitude-squared of the frequency response of the linear system in the all-pole model. The implication of this is that the all-pole model spectrum will attempt to match the energy spectrum of the signal more closely at frequencies where the signal spectrum is large, since frequencies where $|S(e^{j\omega})|^2 > |H(e^{j\omega})|^2$ contribute more to the mean-squared error than frequencies where the opposite is true. Thus, the all-pole model spectrum estimate favors a good fit around the peaks of the signal spectrum. This will be illustrated by the discussion in Section 11.5.1. Similar analysis and reasoning also applies to the case in which $s[n]$ is random.

11.5.1 All-Pole Analysis of Speech Signals

All-pole modeling is widely used in speech processing both for speech coding, where the term linear predictive coding (LPC) is often used, and for spectrum analysis. (See Atal and Hanauer, 1971, Makhoul, 1975, Rabiner and Schafer, 1978, and Quatieri, 2002.) To illustrate many of the ideas discussed in this chapter, we discuss in some detail the use of all-pole modeling for spectrum analysis of speech signals. This method is typically applied in a time-dependent manner by periodically selecting short segments of the speech signal for analysis in much the same way as is done in time-dependent Fourier analysis as discussed in Section 10.3. Since the time-dependent Fourier transform is essentially a sequence of DTFTs of finite-length segments, the above discussion of the relationship between the DTFT and the all-pole spectrum characterizes the relationship between time-dependent Fourier analysis and time-dependent all-pole model spectrum analysis, as well.

Figure 11.11 shows a 201-point Hamming-windowed segment of a speech signal $s[n]$ in the top panel and the corresponding autocorrelation function $r_{ss}[m]$ below. During this time interval, the speech signal is voiced (vocal cords vibrating), as evidenced by the periodic nature of the signal. This periodicity is reflected in the autocorrelation function as the peak at about 27 samples (27/8 = 3.375 ms for 8 kHz sampling rate) and integer multiples thereof.

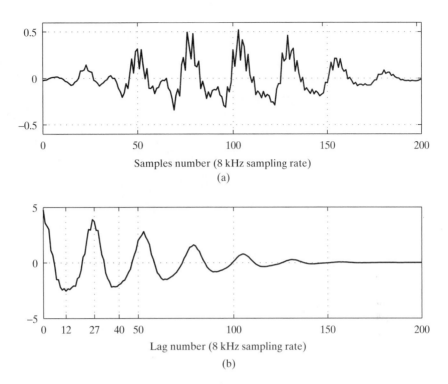

Figure 11.11   (a) Windowed voiced speech waveform. (b) Corresponding auto-correlation function (samples connected by straight lines).

When applying all-pole modeling to voiced speech, it is useful to think of the signal as being deterministic, but with an excitation function that is a periodic train of impulses. This accounts for the periodic nature of the autocorrelation function when several periods of the signal are included in the window as in Figure 11.11(a).

Figure 11.12 shows a comparison of the DTFT of the signal in Figure 11.11(a) with spectra computed from all-pole modeling with two different model orders and using the autocorrelation function in Figure 11.11(b). Note that the DTFT of $s[n]$ shows peaks at multiples of the fundamental frequency $F_0 = 8\,\text{kHz}/27 = 296\,\text{Hz}$, as well as many other less prominent peaks and dips that can be attributed to the windowing effects discussed in Section 10.2.1. If the first 13 samples of $r_{ss}[m]$ in Figure 11.11(b) are used to compute an all-pole model spectrum ($p = 12$), the result is the smooth curve shown with the heavy line in Figure 11.12(a). With the filter order as 12 and the fundamental period of 27 samples, this spectrum estimate in effect ignores the spectral structure owing to the periodicity of the signal and produces a much smoother spectrum estimate. If 41 values of $r_{ss}[m]$ are used, however, we obtain the spectrum plotted with the thin line. Since the period of the signal is 27, a value of $p = 40$ includes the periodicity peak in the autocorrelation function and thus, the all-pole spectrum tends to represent much of the fine detail in the DTFT spectrum. Note that both cases support our assertion above that the all-pole model spectrum estimate tends to favor good representation at the peaks of the DTFT spectrum.

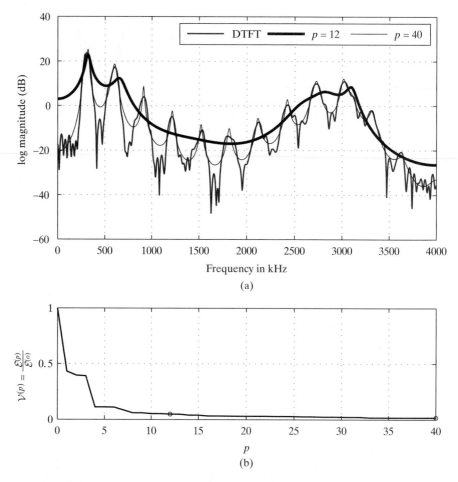

Figure 11.12   (a) Comparison of DTFT and all-pole model spectra for voiced speech segment in Figure 11.11(a). (b) Normalized prediction error as a function of p.

This example illustrates that the choice of the model order p controls the degree of smoothing of the DTFT spectrum. Figure 11.12(b) shows that as p increases, the mean-squared prediction error decreases quickly and then levels off, as in our previous example. Recall that in Sections 11.2.4 and 11.2.5, we argued that the all-pole model with appropriately chosen gain results in a match between the autocorrelation functions of the signal and the all-pole model up to p correlation lags as in Eq. (11.39). This implies that as p increases, the all-pole model spectrum will approach the DTFT spectrum, and when $p \to \infty$, it follows that $r_{hh}[m] = r_{ss}[m]$ for all m, and therefore, $|H(e^{j\omega})|^2 = |S(e^{j\omega})|^2$. However, this does not mean that $H(e^{j\omega}) = S(e^{j\omega})$ because $H(z)$ is an IIR system, and $S(z)$ is the z-transform of a finite-length sequence. Also note that as $p \to \infty$, the averaged prediction error does not approach zero, even though $|H(e^{j\omega})|^2 \to |S(e^{j\omega})|^2$. As we have discussed, this occurs because the total error in Eq. (11.11) is the prediction error $\tilde{e}[n]$ minus $Gv[n]$. Said differently, the linear predictor must always predict the first nonzero sample from the zero-valued samples that precede it.

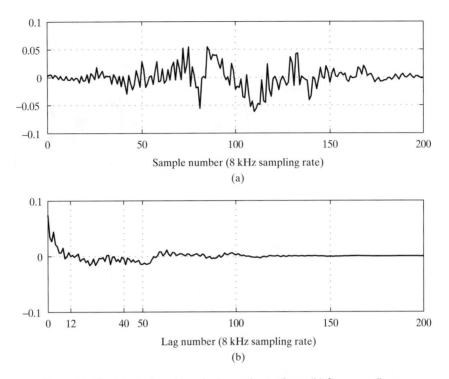

Figure 11.13 (a) Windowed unvoiced speech waveform. (b) Corresponding autocorrelation function (samples connected by straight lines).

The other main class of speech sounds is comprised of the unvoiced sounds such as fricatives. These sounds are produced by creating random turbulent air flow in the vocal tract; therefore, they are best modeled in terms of an all-pole system excited by white noise. Figure 11.13 shows an example of a 201-point Hamming-windowed segment of unvoiced speech and its corresponding autocorrelation function. Note that the autocorrelation function shows no indication of periodicity in either the signal waveform or the autocorrelation function. A comparison of the DTFT of the signal in Figure 11.13(a) with two all-pole model spectra computed from the autocorrelation function in Figure 11.13(b) is shown in Figure 11.14(a). From the point of view of spectrum analysis of random signals, the magnitude-squared of the DTFT is a periodogram. Thus, it contains a component that is randomly varying with frequency. Again, by choice of the model order, the periodogram can be smoothed to any desired degree.

11.5.2 Pole Locations

In speech processing, the poles of the all-pole model have a close relationship to the resonance frequencies of the vocal tract, thus, it is often useful to factor the polynomial $A(z)$ to obtain its zeros for representation as in Eq. (11.50). As discussed in Section 11.3.3, the zeros z_k of the prediction error filter are the poles of the all-pole model system function. It is the poles of the system function that are responsible for the peaks in the spectrum estimates discussed in Section 11.5.1. The closer a pole is to the unit circle, the more peaked is the spectrum for frequencies close to the angle of the pole.

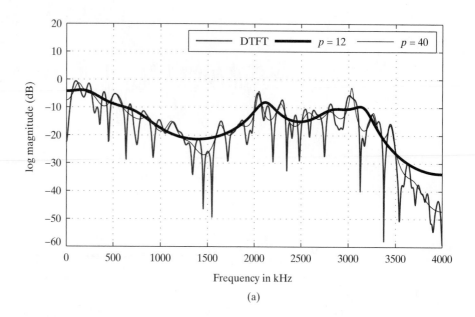

(a)

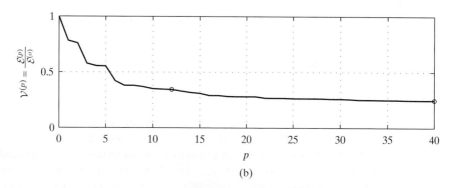

(b)

Figure 11.14   (a) Comparison of DTFT and all-pole model spectra for unvoiced speech segment in Figure 11.13(a). (b) Normalized prediction error as a function of p.

Figure 11.15 shows the zeros of the prediction error system function $A(z)$ (poles of the model system) for the two spectrum estimates in Figure 11.12(a). For $p = 12$, the zeros of $A(z)$ are denoted by the open circles. Five complex conjugate pairs of zeros are close to the unit circle, and their manifestations as poles are clearly evident in heavy line curve of Figure 11.12(a). For the case $p = 40$, the zeros of $A(z)$ are denoted by the large filled dots. Observe that most of the zeros are close to the unit circle, and they are more or less evenly distributed around the unit circle. This produces the peaks in the model spectrum that are spaced approximately at multiples of the normalized radian frequency corresponding to the fundamental frequency of the speech signal; i.e., at angles $2\pi (296 \text{ Hz})/8 \text{ kHz}$.

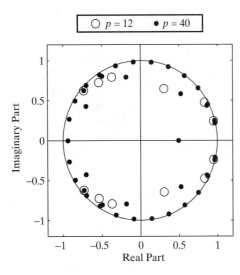

Figure 11.15 Zeros of prediction error filters (poles of model systems) used to obtain the spectrum estimates in Figure 11.12.

11.5.3 All-Pole Modeling of Sinusoidal Signals

As another important example, we consider the use of the poles of an all-pole model to estimate frequencies of sinusoidal signals. To see why this is possible, consider the sum of two sinusoids

$$s[n] = [A_1 \cos(\omega_1 n + \theta_1) + A_2 \cos(\omega_2 n + \theta_2)] u[n]. \tag{11.61}$$

The z-transform of $s[n]$ has the form

$$S(z) = \frac{b_0 + b_1 z^{-1} + b_2 z^{-2} + b_3 z^{-3}}{(1 - e^{j\omega_1} z^{-1})(1 - e^{-j\omega_1} z^{-1})(1 - e^{j\omega_2} z^{-1})(1 - e^{-j\omega_2} z^{-1})}. \tag{11.62}$$

That is, the sum of two sinusoids can be represented as the impulse response of an LTI system whose system function has both poles and zeros. The numerator polynomial would be a somewhat complicated function of the amplitudes, frequencies, and phase shifts. What is important for our discussion is that the numerator is a 3rd-order polynomial and the denominator is a 4th-order polynomial, the roots of which are all on the unit circle at angles equal to $\pm\omega_1$ and $\pm\omega_2$. The difference equation describing this system with impulse excitation has the form

$$s[n] - \sum_{k=1}^{4} a_k s[n-k] = \sum_{k=1}^{3} b_k \delta[n-k] \tag{11.63}$$

where the coefficients a_k would result from multiplying the denominator factors. Note that

$$s[n] - \sum_{k=1}^{4} a_k s[n-k] = 0 \quad \text{for } n \geq 4, \tag{11.64}$$

which suggests that the signal $s[n]$ can be predicted with no error by a 4th-order predictor except at the very beginning ($0 \leq n \leq 3$). The coefficients of the denominator can be

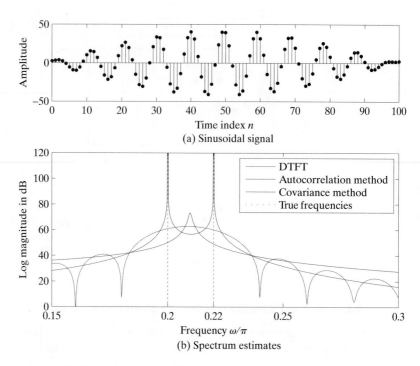

Figure 11.16   Spectrum estimation for a sinusoidal signal.

estimated from the signal by applying the covariance method to a short segment of the signal selected so as not to include the first four samples. In the ideal case for which Eq. (11.61) accurately represents the signal (e.g., high SNR), the roots of the resulting polynomial provide good estimates of the frequencies of the component sinusoids.

Figure 11.16(a) shows a plot of 101 samples of the signal[8]

$$s[n] = 20\cos(0.2\pi n - 0.1\pi) + 22\cos(0.22\pi n + 0.9\pi). \qquad (11.65)$$

Because the two frequencies are close together, it is necessary to use a large number of samples to resolve the two frequencies by Fourier analysis. However, since the signal fits the all-pole model perfectly, the covariance method can be used to obtain very accurate estimates of the frequencies from very short segments of the signal. This is illustrated in Figure 11.16(b).

The DTFT of the 101 samples (with rectangular window) shows no indication that there are two distinct sinusoid frequencies around $\omega = 0.21\pi$. Recall that the main lobe width for an $(M + 1)$-point rectangular window is $\Delta\omega = 4\pi/(M + 1)$. Consequently, a 101-point rectangular window can clearly resolve two frequencies only if they are no closer than about $.04\pi$ rad/s. Correspondingly, the DTFT does not show two spectral peaks.

Similarly, use of the autocorrelation method results in the spectrum estimate shown by the heavy line. This estimate also contains only one spectral peak. The predic-

---

[8]The tapering of the segment of the signal in Figure 11.16(a) is not a result of windowing. It is caused by the "beating" of the two cosines of nearly the same frequency. The period of the beat frequency (difference between 0.22π and 0.2π) is 100 samples.

tion error polynomial (in factored form) obtained with the autocorrelation method is

$$A_a(z) = (1 - 0.998e^{j0.21\pi}z^{-1})(1 - 0.998e^{-j0.21\pi}z^{-1})$$
$$\cdot (1 - 0.426z^{-1})(1 - 0.1165z^{-1}) \tag{11.66}$$

The two real poles contribute no peaks, and the complex poles are close to the unit circle, but at $\pm 0.21\pi$, which is halfway between the two frequencies. Thus, the windowing inherent in the autocorrelation method causes the resulting model to lock onto the average frequency 0.21π.

On the other hand, the factored prediction error polynomial obtained with the covariance method is (with rounding of the magnitudes and angles) given by

$$A_c(z) = (1 - e^{j0.2\pi}z^{-1})(1 - e^{-j0.2\pi}z^{-1})$$
$$\cdot (1 - e^{j0.22\pi}z^{-1})(1 - e^{-j0.22\pi}z^{-1}). \tag{11.67}$$

In this case, the angles of the zeros are almost exactly equal to the frequencies of the two sinusoids. Also shown in Figure 11.16(b) is the frequency response of the model, i.e.,

$$|H_{\text{cov}}(e^{j\omega})|^2 = \frac{1}{|A_{\text{cov}}(e^{j\omega})|^2}. \tag{11.68}$$

plotted in dB. In this case, the prediction error is very close to zero, which, if used to estimate the gain of the all-pole model, would lead to an indeterminant estimate. Therefore, the gain is arbitrarily set to one, which leads to a plot of Eq. (11.68) on a similar scale to the other estimates. Since the poles are almost exactly on the unit circle, the magnitude spectrum becomes exceedingly large at the pole frequencies. Note that the roots of the prediction error polynomial give an accurate estimate of the frequencies. This method, of course, does not provide accurate information about the amplitudes and phases of the sinusoidal components.

11.6 SOLUTION OF THE AUTOCORRELATION NORMAL EQUATIONS

In both the autocorrelation and covariance methods of computing the correlation values, the predictor coefficients that minimize the mean-squared inverse filter error and equivalently the prediction error satisfy a set of linear equations of the general form:

$$\begin{bmatrix} \phi_{ss}[1,1] & \phi_{ss}[1,2] & \phi_{ss}[1,3] & \cdots & \phi_{ss}[1,p] \\ \phi_{ss}[2,1] & \phi_{ss}[2,2] & \phi_{ss}[2,3] & \cdots & \phi_{ss}[2,p] \\ \phi_{ss}[3,1] & \phi_{ss}[3,2] & \phi_{ss}[3,3] & \cdots & \phi_{ss}[3,p] \\ \vdots & \vdots & \vdots & \cdots & \vdots \\ \phi_{ss}[p,1] & \phi_{ss}[p,2] & \phi_{ss}[p,3] & \cdots & \phi_{ss}[p,p] \end{bmatrix} \begin{bmatrix} a_1 \\ a_2 \\ a_3 \\ \vdots \\ a_p \end{bmatrix} = \begin{bmatrix} \phi_{ss}[1,0] \\ \phi_{ss}[2,0] \\ \phi_{ss}[3,0] \\ \vdots \\ \phi_{ss}[p,0] \end{bmatrix}. \tag{11.69}$$

In matrix notation, these linear equations have the representation

$$\Phi a = \psi.$$

(11.70)

Since $\phi[i, k] = \phi[k, i]$, in both the autocorrelation and covariance methods, the matrix Φ is symmetric, and, because it arises in a least-squares problem, it is also positive-definite, which guarantees that it is invertible. In general, this leads to efficient solution methods, such as the Cholesky decomposition (see Press, et al., 2007), that are based on matrix factorization and applicable when Φ is symmetric and positive definite. However, in the specific case of the autocorrelation method or any method for which $\phi_{ss}[i, k] = r_{ss}[|i - k|]$, Eqs. (11.69) become the autocorrelation normal equations (also referred to as the Yule–Walker equations).

$$
\begin{bmatrix}
r_{ss}[0] & r_{ss}[1] & r_{ss}[2] & \cdots & r_{ss}[p-1] \\
r_{ss}[1] & r_{ss}[0] & r_{ss}[1] & \cdots & r_{ss}[p-2] \\
r_{ss}[2] & r_{ss}[1] & r_{ss}[0] & \cdots & r_{ss}[p-3] \\
\vdots & \vdots & \vdots & \cdots & \vdots \\
r_{ss}[p-1] & r_{ss}[p-2] & r_{ss}[p-3] & \cdots & r_{ss}[0]
\end{bmatrix}
\begin{bmatrix}
a_1 \\ a_2 \\ a_3 \\ \vdots \\ a_p
\end{bmatrix}
=
\begin{bmatrix}
r_{ss}[1] \\ r_{ss}[2] \\ r_{ss}[3] \\ \vdots \\ r_{ss}[p]
\end{bmatrix}.
$$

(11.71)

In this case, in addition to the matrix Φ being symmetric and positive-definite, it is also a Toeplitz matrix, i.e., all the elements on each subdiagonal are equal. This property leads to an efficient algorithm, referred to as the Levinson–Durbin recursion, for solving the equations.

11.6.1 The Levinson–Durbin Recursion

The Levinson–Durbin algorithm for computing the predictor coefficients that minimize the total-squared prediction error results from the high degree of symmetry in the matrix Φ and furthermore, as Eq. (11.71) confirms, the elements of the right-hand side vector ψ are primarily the same values that populate the matrix Φ. Equations (L–D.1) to (L–D.6) in Figure 11.17 define the computations. A derivation of these equations is given in Section 11.6.2, but before developing the details of the derivation, it is helpful to simply examine the steps of the algorithm.

(L–D.1) This step initializes the mean-squared prediction error to be the energy of the signal. That is, a zero[th]-order predictor (no predictor) yields no reduction in prediction error energy, since the prediction error $e[n]$ is identical to the signal $s[n]$.

The next line in Figure 11.17 states that steps (L–D.2) through (L–D.5) are repeated p times, with each repetition of those steps increasing the order of the predictor by one. In other words, the algorithm computes a predictor of order i from the predictor of order $i - 1$ starting with $i - 1 = 0$.

(L–D.2) This step computes a quantity k_i. The sequence of parameters $k_i, i = 1, 2, \ldots, p$ which we refer to as the k-parameters, plays a key role in generating the next set of predictor coefficients.[9]

---

[9]For reasons to be discussed in Section 11.7, the k-parameters are also called *PARCOR* (for PARtial CORrelation) *coefficients* or also, *reflection coefficients*.

Levinson–Durbin Algorithm

$$\mathcal{E}^{(0)} = r_{ss}[0] \tag{L–D.1}$$

for $i = 1, 2, \ldots, p$

$$k_i = \left(r_{ss}[i] - \sum_{j=1}^{i-1} a_j^{(i-1)} r_{ss}[i-j] \right) / \mathcal{E}^{(i-1)} \tag{L–D.2}$$

$$a_i^{(i)} = k_i \tag{L–D.3}$$

if $i > 1$ then for $j = 1, 2, \ldots, i-1$

$$a_j^{(i)} = a_j^{(i-1)} - k_i a_{i-j}^{(i-1)} \tag{L–D.4}$$

end

$$\mathcal{E}^{(i)} = (1 - k_i^2)\mathcal{E}^{(i-1)} \tag{L–D.5}$$

end

$$a_j = a_j^{(p)} \quad j = 1, 2, \ldots, p \tag{L–D.6}$$

Figure 11.17   Equations defining the Levinson–Durbin algorithm.

(L–D.3) This equation states that $a_i^{(i)}$, the ith coefficient of the ith-order predictor, is equal to k_i.

(L–D.4) In this equation, k_i is used to compute the remaining coefficients of the ith-order predictor as a combination of the coefficients of the predictor of order $(i - 1)$ with those same coefficients in reverse order.

(L–D.5) This equation updates the prediction error for the ith-order predictor.

(L–D.6) This is the final step where the pth-order predictor is defined to be the result after p iterations of the algorithm.

The Levinson–Durbin algorithm is valuable because it is an efficient method of solution of the autocorrelation normal equations and also for the insight that it provides about the properties of linear prediction and all-pole models. For example, from Eq. (L–D.5), it can be shown that the averaged prediction error for a pth-order predictor is the product of the prediction errors for all lower-order predictors, from which it follows that $0 < \mathcal{E}^{(i)} \leq \mathcal{E}^{(i-1)} < \mathcal{E}^{(p)}$ and

$$\mathcal{E}^{(p)} = \mathcal{E}^{(0)} \prod_{i=1}^{p}(1 - k_i^2) = r_{ss}[0] \prod_{i=1}^{p}(1 - k_i^2). \tag{11.72}$$

Since $\mathcal{E}^{(i)} > 0$, it must be true that $-1 < k_i < 1$ for $i = 1, 2, \ldots, p$. That is, the k-parameters are strictly less than one in magnitude.

11.6.2 Derivation of the Levinson–Durbin Algorithm

From Eq. (11.30), the optimum predictor coefficients satisfy the set of equations

$$r_{ss}[i] - \sum_{k=1}^{p} a_k r_{ss}[i-k] = 0 \quad i = 1, 2, \ldots, p, \tag{11.73a}$$

and the minimum mean-squared prediction error is given by

$$r_{ss}[0] - \sum_{k=1}^{p} a_k r_{ss}[k] = \mathcal{E}^{(p)}. \tag{11.73b}$$

Since Eq. (11.73b) contains the same correlation values as in Eq. (11.73a), it is possible to take them together and write a new set of $p + 1$ equations that are satisfied by the p unknown predictor coefficients and the corresponding unknown mean-squared prediction error $\mathcal{E}^{(p)}$. These equations have the matrix form

$$\begin{bmatrix} r_{ss}[0] & r_{ss}[1] & r_{ss}[2] & \cdots & r_{ss}[p] \\ r_{ss}[1] & r_{ss}[0] & r_{ss}[1] & \cdots & r_{ss}[p-1] \\ r_{ss}[2] & r_{ss}[1] & r_{ss}[0] & \cdots & r_{ss}[p-2] \\ \vdots & \vdots & \vdots & \cdots & \vdots \\ r_{ss}[p] & r_{ss}[p-1] & r_{ss}[p-2] & \cdots & r_{ss}[0] \end{bmatrix} \begin{bmatrix} 1 \\ -a_1^{(p)} \\ -a_2^{(p)} \\ \vdots \\ -a_p^{(p)} \end{bmatrix} = \begin{bmatrix} \mathcal{E}^{(p)} \\ 0 \\ 0 \\ \vdots \\ 0 \end{bmatrix}. \tag{11.74}$$

It is this set of equations that can be solved recursively by the Levinson–Durbin algorithm. This is done by successively incorporating a new correlation value at each iteration and solving for the next higher-order predictor in terms of the new correlation value and the previously found predictor.

For any order i, the set of equations in Eq. (11.74) can be represented in matrix notation as

$$\mathbf{R}^{(i)} \mathbf{a}^{(i)} = \mathbf{e}^{(i)}. \tag{11.75}$$

We wish to show how the ith solution can be derived from the $(i-1)^{\text{st}}$ solution. In other words, given $\mathbf{a}^{(i-1)}$, the solution to $\mathbf{R}^{(i-1)} \mathbf{a}^{(i-1)} = \mathbf{e}^{(i-1)}$, we wish to derive the solution to $\mathbf{R}^{(i)} \mathbf{a}^{(i)} = \mathbf{e}^{(i)}$.

First, write the equations $\mathbf{R}^{(i-1)} \mathbf{a}^{(i-1)} = \mathbf{e}^{(i-1)}$ in expanded form as

$$\begin{bmatrix} r_{ss}[0] & r_{ss}[1] & r_{ss}[2] & \cdots & r_{ss}[i-1] \\ r_{ss}[1] & r_{ss}[0] & r_{ss}[1] & \cdots & r_{ss}[i-2] \\ r_{ss}[2] & r_{ss}[1] & r_{ss}[0] & \cdots & r_{ss}[i-3] \\ \vdots & \vdots & \vdots & \cdots & \vdots \\ r_{ss}[i-1] & r_{ss}[i-2] & r_{ss}[i-3] & \cdots & r_{ss}[0] \end{bmatrix} \begin{bmatrix} 1 \\ -a_1^{(i-1)} \\ -a_2^{(i-1)} \\ \vdots \\ -a_{i-1}^{(i-1)} \end{bmatrix} = \begin{bmatrix} \mathcal{E}^{(i-1)} \\ 0 \\ 0 \\ \vdots \\ 0 \end{bmatrix}. \tag{11.76}$$

Then append a 0 to the vector $\mathbf{a}^{(i-1)}$ and multiply by the matrix $\mathbf{R}^{(i)}$ to obtain

$$\begin{bmatrix} r_{ss}[0] & r_{ss}[1] & r_{ss}[2] & \cdots & r_{ss}[i] \\ r_{ss}[1] & r_{ss}[0] & r_{ss}[1] & \cdots & r_{ss}[i-1] \\ r_{ss}[2] & r_{ss}[1] & r_{ss}[0] & \cdots & r_{ss}[i-2] \\ \vdots & \vdots & \vdots & \cdots & \vdots \\ r_{ss}[i-1] & r_{ss}[i-2] & r_{ss}[i-3] & \cdots & r_{ss}[1] \\ r_{ss}[i] & r_{ss}[i-1] & r_{ss}[i-2] & \cdots & r_{ss}[0] \end{bmatrix} \begin{bmatrix} 1 \\ -a_1^{(i-1)} \\ -a_2^{(i-1)} \\ \vdots \\ -a_{i-1}^{(i-1)} \\ 0 \end{bmatrix} = \begin{bmatrix} \mathcal{E}^{(i-1)} \\ 0 \\ 0 \\ \vdots \\ 0 \\ \gamma^{(i-1)} \end{bmatrix}. \tag{11.77}$$

where, to satisfy Eq. (11.77),

$$\gamma^{(i-1)} = r_{ss}[i] - \sum_{j=1}^{i-1} a_j^{(i-1)} r_{ss}[i-j]. \tag{11.78}$$

It is in Eq. (11.78) that the new autocorrelation value $r_{ss}[i]$ is introduced. However, Eq. (11.77) is not yet in the desired form $\mathbf{R}^{(i)}\mathbf{a}^{(i)} = \mathbf{e}^{(i)}$. The key step in the derivation is to recognize that due to the special symmetry of the Toeplitz matrix $\mathbf{R}^{(i)}$, the equations can be written in reverse order (first equation last and last equation first, and so on) and the matrix for the resulting set of equations is still $\mathbf{R}^{(i)}$; i.e.,

$$
\begin{bmatrix}
r_{ss}[0] & r_{ss}[1] & r_{ss}[2] & \cdots & r_{ss}[i] \\
r_{ss}[1] & r_{ss}[0] & r_{ss}[1] & \cdots & r_{ss}[i-1] \\
r_{ss}[2] & r_{ss}[1] & r_{ss}[0] & \cdots & r_{ss}[i-2] \\
\vdots & \vdots & \vdots & \cdots & \vdots \\
r_{ss}[i-1] & r_{ss}[i-2] & r_{ss}[i-3] & \cdots & r_{ss}[1] \\
r_{ss}[i] & r_{ss}[i-1] & r_{ss}[i-2] & \cdots & r_{ss}[0]
\end{bmatrix}
\begin{bmatrix}
0 \\
-a_{i-1}^{(i-1)} \\
-a_{i-2}^{(i-1)} \\
\vdots \\
-a_1^{(i-1)} \\
1
\end{bmatrix}
=
\begin{bmatrix}
\gamma^{(i-1)} \\
0 \\
0 \\
\vdots \\
0 \\
\mathcal{E}^{(i-1)}
\end{bmatrix}.
\tag{11.79}
$$

Now Eq. (11.77) is combined with Eq. (11.79) according to

$$
\mathbf{R}^{(i)}
\left[
\begin{bmatrix}
1 \\
-a_1^{(i-1)} \\
-a_2^{(i-1)} \\
\vdots \\
-a_{i-1}^{(i-1)} \\
0
\end{bmatrix}
- k_i
\begin{bmatrix}
0 \\
-a_{i-1}^{(i-1)} \\
-a_{i-2}^{(i-1)} \\
\vdots \\
-a_1^{(i-1)} \\
1
\end{bmatrix}
\right]
=
\left[
\begin{bmatrix}
\mathcal{E}^{(i-1)} \\
0 \\
0 \\
\vdots \\
\gamma^{(i-1)}
\end{bmatrix}
- k_i
\begin{bmatrix}
\gamma^{(i-1)} \\
0 \\
0 \\
\vdots \\
\mathcal{E}^{(i-1)}
\end{bmatrix}
\right].
\tag{11.80}
$$

Equation (11.80) is now approaching the desired form $\mathbf{R}^{(i)}\mathbf{a}^{(i)} = \mathbf{e}^{(i)}$. All that remains is to choose $\gamma^{(i-1)}$, so that the right hand vector has only a single nonzero entry. This requires that

$$
k_i = \frac{\gamma^{(i-1)}}{\mathcal{E}^{(i-1)}} = \frac{r_{ss}[i] - \sum_{j=1}^{i-1} a_j^{(i-1)} r_{ss}[i-j]}{\mathcal{E}^{(i-1)}},
\tag{11.81}
$$

which ensures cancelation of the last element of the right hand side vector, and causes the first element to be

$$
\mathcal{E}^{(i)} = \mathcal{E}^{(i-1)} - k_i \gamma^{(i-1)} = \mathcal{E}^{(i-1)}(1 - k_i^2).
\tag{11.82}
$$

With this choice of $\gamma^{(i-1)}$, it follows that the vector of ith-order prediction coefficients is

$$
\begin{bmatrix}
1 \\
-a_1^{(i)} \\
-a_2^{(i)} \\
\vdots \\
-a_{i-1}^{(i)} \\
-a_i^{(i)}
\end{bmatrix}
=
\begin{bmatrix}
1 \\
-a_1^{(i-1)} \\
-a_2^{(i-1)} \\
\vdots \\
-a_{i-1}^{(i-1)} \\
0
\end{bmatrix}
- k_i
\begin{bmatrix}
0 \\
-a_{i-1}^{(i-1)} \\
-a_{i-2}^{(i-1)} \\
\vdots \\
-a_1^{(i-1)} \\
1
\end{bmatrix}
\tag{11.83}
$$

From Eq. (11.83), we can write the set of equations for updating the coefficients as

$$
a_j^{(i)} = a_j^{(i-1)} - k_i a_{i-j}^{(i-1)} \qquad j = 1, 2, \ldots, i-1,
\tag{11.84a}
$$

and

$$a_i^{(i)} = k_i. \tag{11.84b}$$

Equations (11.81), (11.84b), (11.84a), and (11.82) are the key equations of the Levinson–Durbin algorithm. They correspond to Eqs. (L–D.2), (L–D.3), (L–D.4), and (L–D.5) in Figure 11.17, which shows how they are used order-recursively to compute the optimum prediction coefficients as well as the corresponding mean-squared prediction errors and coefficients k_i for all linear predictors up to order p.

11.7 LATTICE FILTERS

Among the many interesting and useful concepts that emerge from the Levinson–Durbin algorithm is its interpretation in terms of the lattice structures introduced in Section 6.6. There, we showed that any FIR filter with system function of the form

$$A(z) = 1 - \sum_{k=1}^{M} \alpha_k z^{-k} \tag{11.85}$$

can be implemented by a lattice structure as depicted in Figure 6.37. Furthermore, we showed that the coefficients of the FIR system function are related to the k-parameters of a corresponding lattice filter by a recursion given in Figure 6.38, which is repeated for convenience in the bottom half of Figure 11.18. By reversing the steps in the k-to-α algorithm, we obtained an algorithm given in Figure 6.39 for computing the k-parameters from the coefficients $\alpha_j, j = 1, 2, \ldots, M$. Thus, there is a unique relationship between the coefficients of the direct form representation and the lattice representation of an FIR filter.

In this chapter, we have shown that a pth-order prediction error filter is an FIR filter with system function

$$A^{(p)}(z) = 1 - \sum_{k=1}^{p} a_k^{(p)} z^{-k},$$

whose coefficients can be computed from the autocorrelation function of a signal through a process that we have called the Levinson–Durbin algorithm. A by-product of the Levinson–Durbin computation is a set of parameters that we have also denoted k_i and called the k-parameters. A comparison of the two algorithms in Figure 11.18 shows that their steps are identical except for one important detail. In the algorithm derived in Chapter 6, we started with the lattice filter with known coefficients k_i and derived the recursion for obtaining the coefficients of the corresponding direct form FIR filter. In the Levinson–Durbin algorithm, we begin with the autocorrelation function of a signal and compute the k-parameters recursively as an intermediate result in computing the coefficients of the FIR prediction error filter. Since both algorithms give a unique result after p iterations, and since there is a unique relationship between the k-parameters and the coefficients of an FIR filter, it follows that if $M = p$ and $a_j = \alpha_j$ for $j = 1, 2, \ldots, p$, the k-parameters produced by the Levinson–Durbin algorithm must be the k-parameters of a lattice filter implementation of the FIR prediction error filter $A^{(p)}(z)$.

Levinson–Durbin Algorithm

$$\mathcal{E}^{(0)} = r_{ss}[0]$$
for $i = 1, 2, \ldots, p$

$$k_i = \left(r_{ss}[i] - \sum_{j=1}^{i-1} a_j^{(i-1)} r_{ss}[i-j] \right) / \mathcal{E}^{(i-1)} \quad \text{Eq. (11.81)}$$

$$a_i^{(i)} = k_i \qquad\qquad\qquad\qquad\qquad\qquad\qquad \text{Eq. (11.84b)}$$
if $i > 1$ then for $j = 1, 2, \ldots, i - 1$

$$a_j^{(i)} = a_j^{(i-1)} - k_i a_{i-j}^{(i-1)} \qquad\qquad\qquad \text{Eq. (11.84a)}$$
end

$$\mathcal{E}^{(i)} = (1 - k_i^2)\mathcal{E}^{(i-1)} \qquad\qquad\qquad \text{Eq. (11.82)}$$
end

$$a_j = a_j^{(p)} \quad j = 1, 2, \ldots, p$$

Lattice k-to-α Algorithm

Given $k_1, k_2, \ldots, k_M$
for $i = 1, 2, \ldots, M$

$$\alpha_i^{(i)} = k_i \qquad\qquad\qquad\qquad\qquad\qquad\qquad \text{Eq. (6.66b)}$$
if $i > 1$ then for $j = 1, 2, \ldots, i - 1$

$$\alpha_j^{(i)} = \alpha_j^{(i-1)} - k_i \alpha_{i-j}^{(i-1)} \qquad\qquad\qquad \text{Eq. (6.66a)}$$
end
end

$$\alpha_j = \alpha_j^{(M)} \quad j = 1, 2, \ldots, M \qquad\qquad \text{Eq. (6.68b)}$$

Figure 11.18 Comparison of the Levinson–Durbin algorithm and the algorithm for converting from k-parameters of a lattice structure to the FIR impulse response coefficients in Eq. (11.85).

11.7.1 Prediction Error Lattice Network

To explore the lattice filter interpretation further, suppose that we have an ith-order prediction error system function

$$A^{(i)}(z) = 1 - \sum_{k=1}^{i} a_k^{(i)} z^{-k}. \tag{11.86}$$

The z-transform representation of the prediction error[10] would be

$$E^{(i)}(z) = A^{(i)}(z) S(z), \tag{11.87}$$

---

[10]The z-transform equations are used assuming that the z-transforms of $e[n]$ and $s[n]$ exist. Although this would not be true for random signals, the relationships between the variables remain in effect for the system. The z-transform notation facilitates the development of these relationships.

and the time-domain difference equation for this FIR filter is

$$e^{(i)}[n] = s[n] - \sum_{k=1}^{i} a_k^{(i)} s[n-k]. \tag{11.88}$$

The sequence $e^{(i)}[n]$ is given the more specific name *forward prediction error* because it is the error in predicting $s[n]$ from i *previous* samples.

The source of the lattice filter interpretation is Eqs. (11.84a) and (11.84b), which, if substituted into Eq. (11.86), yield the following relation between $A^{(i)}(z)$ and $A^{(i-1)}(z)$:

$$A^{(i)}(z) = A^{(i-1)}(z) - k_i z^{-i} A^{(i-1)}(z^{-1}). \tag{11.89}$$

This is not a surprising result if we consider the matrix representation of the polynomial $A^{(i)}(z)$ in Eq. (11.83).[11] Now, if Eq. (11.89) is substituted for $A^{(i)}(z)$ in Eq. (11.87), the result is

$$E^{(i)}(z) = A^{(i-1)}(z)S(z) - k_i z^{-i} A^{(i-1)}(z^{-1})S(z). \tag{11.90}$$

The first term in Eq. (11.90) is $E^{(i-1)}(z)$, i.e., the prediction error for an $(i-1)^{\text{st}}$-order filter. The second term has a similar interpretation, if we define

$$\tilde{E}^{(i)}(z) = z^{-i} A^{(i)}(z^{-1})S(z) = B^{(i)}(z)S(z), \tag{11.91}$$

where we have defined $B^{(i)}(z)$ as

$$B^{(i)}(z) = z^{-i} A^{(i)}(z^{-1}) \tag{11.92}$$

The time-domain interpretation of Eq. (11.91) is

$$\tilde{e}^{(i)}[n] = s[n-i] - \sum_{k=1}^{i} a_k^{(i)} s[n-i+k]. \tag{11.93}$$

The sequence $\tilde{e}^{(i)}[n]$ is called the *backward prediction error*, since Eq. (11.93) suggests that $s[n-i]$ is "predicted" (using coefficients $a_k^{(i)}$) from the i samples that *follow* sample $n-i$.

With these definitions, it follows from Eq. (11.90) that

$$E^{(i)}(z) = E^{(i-1)}(z) - k_i z^{-1} \tilde{E}^{(i-1)}(z). \tag{11.94}$$

and hence,

$$e^{(i)}[n] = e^{(i-1)}[n] - k_i \tilde{e}^{(i-1)}[n-1]. \tag{11.95}$$

By substituting Eq. (11.89) into Eq. (11.91), we obtain

$$\tilde{E}^{(i)}(z) = z^{-1} \tilde{E}^{(i-1)}(z) - k_i E^{(i-1)}(z), \tag{11.96}$$

which, in the time domain, corresponds to

$$\tilde{e}^{(i)}[n] = \tilde{e}^{(i-1)}[n-1] - k_i e^{(i-1)}[n]. \tag{11.97}$$

---

[11]The algebraic manipulations to derive this result are suggested as an exercise in Problem 11.21.

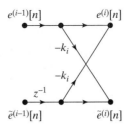

Figure 11.19 Signal flow graph of prediction error computation.

The difference equations in Eq. (11.95) and Eq. (11.97) express the ith-order forward and backward prediction errors in terms of k_i and the $(i-1)^{\text{st}}$-order forward and backward prediction errors. This pair of difference equations is represented by the flow graph of Figure 11.19. Therefore, Figure 11.19 represents a pair of difference equations that embody one iteration of the Levinson–Durbin recursion. As in the Levinson–Durbin recursion, we start with a zeroth-order predictor for which

$$e^{(0)}[n] = \tilde{e}^{(0)}[n] = s[n]. \tag{11.98}$$

With $e^{(0)}[n] = s[n]$ and $\tilde{e}^{(0)}[n] = s[n]$ as inputs to a first stage as depicted in Figure 11.19 with k_1 as coefficient, we obtain $e^{(1)}[n]$ and $\tilde{e}^{(1)}[n]$ as outputs. These are the required inputs for stage 2. We can use p successive stages of the structure in Figure 11.19 to build up a system whose output will be the desired p^{th}-order prediction error signal $e[n] = e^{(p)}[n]$. Such a system, as depicted in Figure 11.20, is identical to the lattice network in Figure 6.37 of Section 6.6.[12] In summary, Figure 11.20 is a signal flow graph representation of the equations

$$e^{(0)}[n] = \tilde{e}^{(0)}[n] = s[n] \tag{11.99a}$$

$$e^{(i)}[n] = e^{(i-1)}[n] - k_i \tilde{e}^{(i-1)}[n-1] \quad i = 1, 2, \ldots, p \tag{11.99b}$$

$$\tilde{e}^{(i)}[n] = \tilde{e}^{(i-1)}[n-1] - k_i e^{(i-1)}[n] \quad i = 1, 2, \ldots, p \tag{11.99c}$$

$$e[n] = e^{(p)}[n], \tag{11.99d}$$

where, if the coefficients k_i are determined by the Levinson–Durbin recursion, the variables $e^{(i)}[n]$ and $\tilde{e}^{(i)}[n]$ are the forward and backward prediction errors for the ith-order optimum predictor.

11.7.2 All-Pole Model Lattice Network

In Section 6.6.2, we showed that the lattice network of Figure 6.42 is an implementation of the all-pole system function $H(z) = 1/A(z)$, where $A(z)$ is the system function of an FIR system; i.e., $H(z)$ is the exact inverse of $A(z)$, and in the present context, it is the system function of the all-pole model with $G = 1$. In this section, we review the all-pole lattice structure in terms of the notation of forward and backward prediction error.

---

[12]Note that in Figure 6.37 the node variables were denoted $a^{(i)}[n]$ and $b^{(i)}[n]$ instead of $e^{(i)}[n]$ and $\tilde{e}^{(i)}[n]$, respectively.

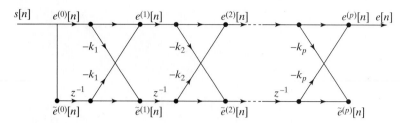

Figure 11.20   Signal flow graph of lattice network implementation of pth-order prediction error computation.

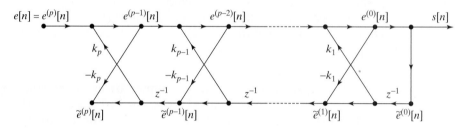

Figure 11.21   All-pole lattice system.

If we replace the node variable labels $a^{(i)}[n]$ and $b^{(i)}[n]$ in Figure 6.42 with the corresponding $e^{(i)}[n]$ and $\tilde{e}^{(i)}[n]$ we obtain the flow graph of Figure 11.21, which represents the set of equations

$$e^{(p)}[n] = e[n] \tag{11.100a}$$

$$e^{(i-1)}[n] = e^{(i)}[n] + k_i \tilde{e}^{(i-1)}[n-1] \quad i = p, p-1, \dots, 1 \tag{11.100b}$$

$$\tilde{e}^{(i)}[n] = \tilde{e}^{(i-1)}[n-1] - k_i e^{(i-1)}[n] \quad i = p, p-1, \dots, 1 \tag{11.100c}$$

$$s[n] = e^{(0)}[n] = e^{(0)}[n]. \tag{11.100d}$$

As we discussed in Section 6.6.2, any stable all-pole system can be implemented by a lattice structure such as Figure 11.21. For such systems, the guarantee of stability inherent in the condition $|k_i| < 1$ is particularly important. Even though the lattice structure requires twice the number of multiplications per output sample as the direct form, it may be the preferred implementation when coefficients must be coarsely quantized. The frequency response of the direct form is exceedingly sensitive to quantization of the coefficients. Furthermore, we have seen that high-order direct form IIR systems can become unstable owing to quantization of their coefficients. This is not the case for the lattice network, as long as the condition $|k_i| < 1$ is maintained for the quantized k-parameters. Furthermore, the frequency response of the lattice network is relatively insensitive to quantization of the k-parameters.

11.7.3 Direct Computation of the *k*-Parameters

The structure of the flow graph in Figure 11.20 is a direct consequence of the Levinson–Durbin recursion, and the parameters k_i, $i = 1, 2, \ldots, p$ can be obtained from the autocorrelation values $r_{ss}[m]$, $m = 0, 1, \ldots, p$ through iterations of the algorithm of Figure 11.17. From our discussion so far, the k_i parameters have been an ancillary consequence of computing the predictor parameters. However, Itakura and Saito (1968, 1970), showed that the k_i parameters can be computed directly from the forward and backward prediction errors in Figure 11.20. And because of the iterative structure as a cascade of the stages in Figure 11.19, the k_i parameters can be computed sequentially from signals available from previous stages of the lattice. The direct computation of the parameter k_i is achieved with the following equation:

$$
k_i^P = \frac{\displaystyle\sum_{n=-\infty}^{\infty} e^{(i-1)}[n]\tilde{e}^{(i-1)}[n-1]}{\left\{ \displaystyle\sum_{n=-\infty}^{\infty} (e^{(i-1)}[n])^2 \sum_{n=-\infty}^{\infty} (\tilde{e}^{(i-1)}[n-1])^2 \right\}^{1/2}}. \tag{11.101}
$$

Observe that Eq. (11.101) is in the form of the energy-normalized cross-correlation between the forward and backward prediction errors at the output of the ith stage. For this reason k_i^P computed using Eq. (11.101) is called a PARCOR coefficient, or more precisely *PARtial CORrelation coefficient*. Figure 11.20 has the interpretation that the correlation in $s[n]$ represented by the autocorrelation function $r_{ss}[m]$ is removed step-by-step by the lattice filter. For a more detailed discussion of the concept of partial correlation, see Stoica and Moses (2005) or Markel and Gray (1976).

Equation (11.101) for computing k_i^P is the geometric mean between a value k_i^f that minimizes the mean-squared forward prediction error and a value k_i^b that minimizes the mean-squared backward prediction error. The derivation of this result is considered in Problem 11.28. Note that we have shown the limits on the sums as infinite simply to emphasize that *all* error samples are involved in the sum. To be more specific, all the sums in Eq. (11.101) could start at $n = 0$ and end at $n = M + i$, since this is the range over which the error signal output of both the forward and backward i^{th}-order predictors would be nonzero. This is the same assumption that was made in setting up the autocorrelation method for finite-length sequences. Indeed, Problem 11.29 outlines a proof that k_i^P computed by Eq. (11.101) gives identically the same result as k_i computed by Eq. (11.81) or Eq. (L–D.2) in Figure 11.17. Therefore, Eq. (11.101) can be substituted for Eq. (L–D.2) in Figure 11.17, and the resulting set of prediction coefficients will be identical to those computed from the autocorrelation function.

To use Eq. (11.101), it is necessary to actually compute the forward and backward prediction errors by employing the computations of Figure 11.19. In summary, the following steps result in computation of the PARCOR coefficients k_i^P for $i = 1, 2, \ldots, p$:

PARCOR.0 Initialize with $e^{(0)}[n] = \tilde{e}^{(0)}[n] = s[n]$ for $0 \leq n \leq M$.

For $i = 1, 2, \ldots, p$ repeat the following steps.

PARCOR.1 Compute $e^{(i)}[n]$ and $\tilde{e}^{(i-1)}[n]$ using Eq. (11.99b) and Eq. (11.99c) respectively for $0 \leq n \leq M + i$. Save these two sequences as input for the next stage.

PARCOR.2 Compute k_i^P using Eq. (11.101).

Another approach to computing the coefficients in Figure 11.20 was introduced by Burg, 1975, who formulated the all-pole modeling problem in terms of the maximum entropy principle. He proposed to use the structure of Figure 11.20, which embodies the Levinson–Durbin algorithm, with coefficients k_i^B that minimize the sum of the mean-squared forward and backward prediction errors at the output of each stage. The result is given by the equation

$$
k_i^B = \frac{2 \sum_{n=i}^{N} e^{(i-1)}[n]\tilde{e}^{(i-1)}[n-1]}{\sum_{n=i}^{N}(e^{(i-1)}[n])^2 + \sum_{n=i}^{N}(\tilde{e}^{(i-1)}[n-1])^2} \tag{11.102}
$$

The procedure for using this equation to obtain the sequence k_i^B, $i = 1, 2, \ldots, p$ is the same as the PARCOR method. In statement PARCOR.2, k_i^P is simply replaced by k_i^B from Eq. (11.102). In this case, the averaging operator is the same as in the covariance method, which means that very short segments of $s[n]$ can be used, while maintaining high spectral resolution.

Even though the Burg method uses a covariance-type analysis, the condition $|k_i^B| < 1$ holds, implying that the all-pole model implemented by the lattice filter will be stable. (See Problem 11.30.) Just as in the case of the PARCOR method, Eq. (11.102) can be substituted for Eq. (L–D.2) in Figure 11.17 to compute the prediction coefficients. While the resulting coefficients will differ from those obtained from the autocorrelation function or from Eq. (11.101), the resulting all-pole model will still be stable. The derivation of Eq. (11.102) is the subject of Problem 11.30.

11.8 SUMMARY

This chapter provides an introduction to parametric signal modeling. We have emphasized all-pole models, but many of the concepts discussed apply to more general techniques involving rational system functions. We have shown that the parameters of an all-pole model can be computed by a two-step process. The first step is the computation of correlation values from a finite-length signal. The second step is solving a set of linear equations, where the correlation values comprise the coefficients. We showed that the solutions obtained depend on how the correlation values are computed, and we showed that if the correlation values are true autocorrelation values, a particularly useful algorithm, called the Levinson–Durbin algorithm, can be derived for the solution of the equations. Furthermore, the structure of the Levinson–Durbin algorithm was shown to illuminate many useful properties of the all-pole model. The subject of parametric signal modeling has a rich history, a voluminous literature, and abundant applications, all of which make it a subject worthy of further advanced study.

Problems

Basic Problems

11.1. $s[n]$ is a finite-energy signal known for all n. $\phi_{ss}[i, k]$ is defined as

$$\phi_{ss}[i, k] = \sum_{n=-\infty}^{\infty} s[n - i]s[n - k].$$

Show that $\phi_{ss}[i, k]$ can be expressed as a function of $|i - k|$.

11.2. In general, the mean-squared prediction error is defined in Eq. (11.36) as

$$\mathcal{E} = \left\langle \left(s[n] - \sum_{k=1}^{p} a_k s[n - k] \right)^2 \right\rangle. \tag{P11.2-1}$$

(a) Expand Eq. (P11.2-1) and use the fact that $\langle s[n - i]s[n - k] \rangle = \phi_{ss}[i, k] = \phi_{ss}[k, i]$ to show that

$$\mathcal{E} = \phi_{ss}[0, 0] - 2 \sum_{k=1}^{p} a_k \phi_{ss}[0, k] + \sum_{i=1}^{p} a_i \sum_{k=1}^{p} a_k \phi_{ss}[i, k] \tag{P11.2-2}$$

(b) Show that for the optimum predictor coefficients, which satisfy Eqs. (11.20), Eq. (P11.2-2) becomes

$$\mathcal{E} = \phi_{ss}[0, 0] - \sum_{k=1}^{p} a_k \phi_{ss}[0, k]. \tag{P11.2-3}$$

11.3. The impulse response of a causal all-pole model of the form of Figure 11.1 and Eq. (11.3) with system parameters G and $\{a_k\}$ satisfies the difference equation

$$h[n] = \sum_{k=1}^{p} a_k h[n - k] + G\delta[n] \tag{P11.3-1}$$

(a) The autocorrelation function of the impulse response of the system is

$$r_{hh}[m] = \sum_{n=-\infty}^{\infty} h[n]h[n + m]$$

By substituting Eq. (P11.3-1) into the equation for $r_{hh}[-m]$, and using the fact that $r_{hh}[-m] = r_{hh}[m]$ show that

$$\sum_{k=1}^{p} a_k r_{hh}[|m - k|] = r_{hh}[m], \quad m = 1, 2, \ldots, p \tag{P11.3-2}$$

(b) Using the same approach as in (a), now show that

$$r_{hh}[0] - \sum_{k=1}^{p} a_k r_{hh}[k] = G^2. \tag{P11.3-3}$$

11.4. Consider a signal $x[n] = s[n] + w[n]$, where $s[n]$ satisfies the difference equation

$$s[n] = 0.8s[n-1] + v[n].$$

$v[n]$ is a zero-mean white-noise sequence with variance $\sigma_v^2 = 0.49$ and $w[n]$ is a zero-mean white-noise sequence with variance $\sigma_w^2 = 1$. The processes $v[n]$ and $w[n]$ are uncorrelated. Determine the autocorrelation sequences $\phi_{ss}[m]$ and $\phi_{xx}[m]$.

11.5. The inverse filter approach to all-pole modeling of a deterministic signal $s[n]$ is discussed in Section 11.1.2 and depicted in Fig. 11.2. The system function of the inverse filter is given in Eq. (11.5).

  (a) Based on this approach, determine the coefficients a_1 and a_2 of the best all-pole model for $s[n] = \delta[n] + \delta[n-2]$ with $p = 2$.
  (b) Again, based on this approach, determine the coefficients a_1, a_2 and a_3 of the best all-pole model for $s[n] = \delta[n] + \delta[n-2]$ with $p = 3$.

11.6. Suppose that you have computed the parameters G and a_k, $k = 1, 2, \ldots, p$ of the all-pole model

$$H(z) = \frac{G}{1 - \displaystyle\sum_{k=1}^{p} a_k z^{-k}}.$$

Explain how you might use the DFT to evaluate the all-pole spectrum estimate $|H(e^{j\omega_k})|$ at N frequencies $\omega_k = 2\pi k/N$ for $k = 0, 1, \ldots, N - 1$.

11.7. Consider a desired causal impulse response $h_d[n]$ that we wish to approximate by a system having impulse response $h[n]$ and system function

$$H(z) = \frac{b}{1 - az^{-1}}.$$

Our optimality criterion is to minimize the error function given by

$$\mathcal{E} = \sum_{n=0}^{\infty} (h_d[n] - h[n])^2.$$

  (a) Suppose a is given, and we wish to determine the unknown parameter b which minimizes $\mathcal{E}$. Assume that $|a| < 1$. Does this result in a nonlinear set of equations? If so, show why. If not, determine b.
  (b) Suppose b is given, and we wish to determine the unknown parameter a which minimizes $\mathcal{E}$. Is this a nonlinear problem? If so, show why. If not, determine a.

11.8. Assume that $s[n]$ is a finite-length (windowed) sequence that is zero *outside* the interval $0 \le n \le M - 1$. The pth-order *backward* linear prediction error sequence for this signal is defined as

$$\tilde{e}[n] = s[n] - \sum_{k=1}^{p} \beta_k s[n+k]$$

That is, $s[n]$ is "predicted" from the p samples that *follow* sample n. The mean-squared backward prediction error is defined as

$$\tilde{\mathcal{E}} = \sum_{m=-\infty}^{\infty} (\tilde{e}[m])^2 = \sum_{m=-\infty}^{\infty} \left(s[m] - \sum_{k=1}^{p} \beta_k s[m+k] \right)^2$$

where the infinite limits indicate that the sum is over all nonzero values of $(\tilde{e}[m])^2$ as in the autocorrelation method used in "forward prediction."

(a) The prediction error sequence $\tilde{e}[n]$ is zero outside a finite interval $N_1 \le n \le N_2$. Determine N_1 and N_2.

(b) Following the approach used in this chapter to derive the forward linear predictor, derive the set of normal equations that are satisfied by the β_ks that minimize the mean-squared prediction error $\tilde{\mathcal{E}}$. Give your final answer in a concise, well-defined form in terms of autocorrelation values.

(c) Based on the result in part (b), describe how the backward predictor coefficients $\{\beta_k\}$ related to the forward predictor coefficients $\{\alpha_k\}$?

Advanced Problems

11.9. Consider a signal $s[n]$ that we model as the impulse response of an all-pole system of order p. Denote the system function of the pth-order all-pole model as $H^{(p)}(z)$ and the corresponding impulse response as $h^{(p)}[n]$. Denote the inverse of $H^{(p)}(z)$ as $H_{\text{inv}}^{(p)}(z) = 1/H^{(p)}(z)$. The corresponding impulse response is $h_{\text{inv}}^{(p)}[n]$. The inverse filter, characterized by $h_{\text{inv}}^{(p)}[n]$, is chosen to minimize the total squared error $\mathcal{E}^{(p)}$ given by

$$\mathcal{E}^{(p)} = \sum_{n=-\infty}^{\infty} \left[\delta[n] - g^{(p)}[n] \right]^2,$$

where $g^{(p)}[n]$ is the output of the filter $H_{\text{inv}}^{(p)}(z)$ when the input is $s[n]$.

(a) Figure P11.9 depicts a signal flow graph of the lattice filter implementation of $H_{\text{inv}}^{(4)}(z)$. Determine $h_{\text{inv}}^{(4)}[1]$, the impulse response at $n = 1$.

(b) Suppose we now wish to model the signal $s[n]$ as the impulse response of a 2nd-order all-pole filter. Draw a signal flow graph of the lattice filter implementation of $H_{\text{inv}}^{(2)}(z)$.

(c) Determine the system function $H^{(2)}(z)$ of the 2nd-order all-pole filter.

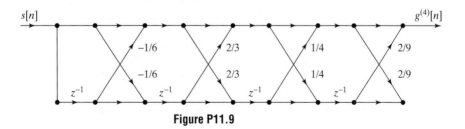

Figure P11.9

11.10. Consider an ith-order predictor with prediction error system function

$$A^{(i)}(z) = 1 - \sum_{j=1}^{i} a_j^{(i)} z^{-j} = \prod_{j=1}^{i} (1 - z_j^{(i)} z^{-1}) \tag{P11.10-1}$$

From the Levinson–Durbin recursion, it follows that $a_i^{(i)} = k_i$. Use this fact with Eq. (P11.10-1) to show that if $|k_i| \ge 1$, it must be true that $|z_j^{(i)}| \ge 1$ for some j. That is, show that the condition $|k_i| < 1$ is a *necessary* condition for $A^{(p)}(z)$ to have all its zeros strictly *inside* the unit circle.

11.11. Consider an LTI system with system function $H(z) = h_0 + h_1 z^{-1}$. The signal $y[n]$ is the output of this system due to an input that is white noise with zero mean and unit variance.

    (a) What is the autocorrelation function $r_{yy}[m]$ of the output signal $y[n]$?

    (b) The 2nd-order forward prediction error is defined as
$$e[n] = y[n] - a_1 y[n-1] - a_2 y[n-2].$$
Without using the Yule–Walker equations directly, find a_1 and a_2, such that the variance of $e[n]$ is minimized.

    (c) The backward prediction filter for $y[n]$ is defined as
$$\tilde{e}[n] = y[n] - b_1 y[n+1] - b_2 y[n+2].$$
Find b_1 and b_2 such that the variance of $\tilde{e}[n]$ is minimized. Compare these coefficients to those determined in part (b).

11.12. **(a)** The autocorrelation function, $r_{yy}[m]$ of a zero-mean wide-sense stationary random process $y[n]$ is given. In terms of $r_{yy}[m]$, write the Yule–Walker equations that result from modeling the random process as the response to a white noise sequence of a 3rd-order all-pole model with system function
$$H(z) = \frac{A}{1 - az^{-1} - bz^{-3}}.$$

    (b) A random process $v[n]$ is the output of the system shown in Figure P11.12-1, where $x[n]$ and $z[n]$ are independent, unit variance, zero mean, white noise signals, and $h[n] = \delta[n-1] + \frac{1}{2}\delta[n-2]$. Find $r_{vv}[m]$, the autocorrelation of $v[n]$.

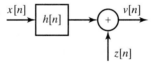

Figure P11.12-1

    (c) Random process $y_1[n]$ is the output of the system shown in Figure P11.12-2, where $x[n]$ and $z[n]$ are independent, unit variance, zero-mean, white noise signals, and
$$H_1(z) = \frac{1}{1 - az^{-1} - bz^{-3}}.$$
The same a and b as found in part (a) are used for all-pole modeling of $y_1[n]$. The inverse modeling error, $w_1[n]$, is the output of the system shown in Figure P11.12-3. Is $w_1[n]$ white? Is $w_1[n]$ zero mean? Explain.

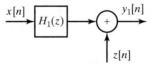

Figure P11.12-2

Figure P11.12-3

    (d) What is the variance of $w_1[n]$?

11.13. We have observed the first six samples of a causal signal $s[n]$ given by $s[0] = 4$, $s[1] = 8$, $s[2] = 4$, $s[3] = 2$, $s[4] = 1$, and $s[5] = 0.5$. For the first parts of this problem, we will model the signal using a stable, causal, minimum-phase, two-pole system having impulse response $\hat{s}[n]$ and system function

$$H(z) = \frac{G}{1 - a_1 z^{-1} - a_2 z^{-2}}.$$

The approach is to minimize the modeling error $\mathcal{E}$ given by

$$\mathcal{E} = \min_{a_1, a_2, A} \sum_{n=0}^{5} (g[n] - G\delta[n])^2,$$

where $g[n]$ is the response of the inverse system to $s[n]$, and the inverse system has system function

$$A(z) = 1 - a_1 z^{-1} - a_2 z^{-2}.$$

(a) Write $g[n] - G\delta[n]$ for $0 \le n \le 5$.
(b) Based on your work in part (a), write the linear equations for the desired parameters a_1, a_2, and G.
(c) What is G?
(d) For this $s[n]$, without solving the linear equations in part (b), discuss whether you expect that the modeling error $\mathcal{E}$ will be zero.

For the rest of this problem, we will model the signal using a different stable, causal, minimum-phase system having impulse response $\hat{s}_2[n]$ and system function

$$H_2(z) = \frac{b_0 + b_1 z^{-1}}{1 - az^{-1}}.$$

The modeling error to be minimized in this case is $\mathcal{E}_2$ given by

$$\mathcal{E}_2 = \min_{a, b_0, b_1} \sum_{n=0}^{5} (g[n] - r[n])^2,$$

where $g[n]$ is the response of the inverse system to $s[n]$, and the inverse system now has system function

$$A(z) = 1 - az^{-1}.$$

Furthermore, $r[n]$ is the impulse response of a system with system function

$$B(z) = b_0 + b_1 z^{-1}.$$

(e) For this model, write $g[n] - r[n]$ for $0 \le n \le 5$.
(f) Calculate the parameter values a, b_0, and b_1 that minimize the modeling error.
(g) Calculate the modeling error $\mathcal{E}_2$ in part (f).

11.14. In Example 11.1, we considered the sequence $s_d[n] = \alpha^n u[n]$, which is the impulse response of a 1st-order all-pole system having system function

$$H(z) = \frac{1}{1 - \alpha z^{-1}}.$$

In this problem we consider the estimation of the parameters of an all-pole model for the signal $s_d[n]$ known only over the interval $0 \le n \le M$.

(a) First, consider the estimation of a 1st-order model by the autocorrelation method. To begin, show that the autocorrelation function of the finite-length sequence $s[n] = s_d[n](u[n] - u[n - M - 1]) = \alpha^n(u[n] - u[n - M - 1])$ is

$$r_{ss}[m] = \alpha^{|m|} \frac{1 - \alpha^{2(M-|m|+1)}}{1 - \alpha^2}. \qquad (P11.14\text{-}1)$$

(b) Use the autocorrelation function determined in (a) in Eq. (11.34), and solve for the coefficient a_1 of the 1st-order predictor.

(c) You should find that the result obtained in (b) is not the exact value (i.e., $a_1 \neq \alpha$) as obtained in Example 11.1, when the autocorrelation function was computed using the infinite sequence. Show, however, that $a_1 \to \alpha$ for $M \to \infty$.

(d) Use the results of (a) and (b) in Eq. (11.38) to determine the minimum mean-squared prediction error for this example. Show that for $M \to \infty$ the error approaches the minimum mean-squared error found in Example 11.1 for the exact autocorrelation function.

(e) Now, consider the covariance method for estimating the correlation function. Show that for $p = 1$, $\phi_{ss}[i, k]$ in Eq. (11.49) is given by

$$\phi_{ss}[i, k] = \alpha^{2-i-k} \frac{1 - \alpha^{2M}}{1 - \alpha^2} \qquad 0 \le (i, k) \le 1. \qquad (P11.14\text{-}2)$$

(f) Use the result of (e) in Eq. (11.20) to solve for the coefficient of the optimum 1st-order predictor. Compare your result to the result in (b) and to the result in Example 11.1.

(g) Use the results of (e) and (f) in Eq. (11.37) to find the minimum mean-squared prediction error. Compare your result to the result in (d) and to the result in Example 11.1.

11.15. Consider the signal

$$s[n] = 3\left(\frac{1}{2}\right)^n u[n] + 4\left(-\frac{2}{3}\right)^n u[n].$$

(a) We want to use a causal, 2nd-order all-pole model, i.e., a model of the form

$$H(z) = \frac{A}{1 - a_1 z^{-1} - a_2 z^{-2}},$$

to optimally represent the signal $s[n]$, in the least-square error sense. Find a_1, a_2, and A.

(b) Now, suppose we want to use a causal, 3rd-order all-pole model, i.e., a model of the form

$$H(z) = \frac{B}{1 - b_1 z^{-1} - b_2 z^{-2} - b_3 z^{-3}},$$

to optimally represent the signal $s[n]$, in the least-square error sense. Find, b_1, b_2, b_3, and B.

11.16. Consider the signal

$$s[n] = 2\left(\frac{1}{3}\right)^n u[n] + 3\left(-\frac{1}{2}\right)^n u[n]. \qquad (P11.16\text{-}1)$$

We wish to model this signal using a 2nd-order ($p = 2$) all-pole model or, equivalently, using 2nd-order linear prediction.

For this problem, since we are given an analytical expression for $s[n]$ and $s[n]$ is the impulse response of an all-pole filter, we can obtain the linear prediction coefficients directly from the z-transform of $s[n]$. (You are asked to do this in part (a).) In practical situations, we are typically given data, i.e., a set of signal values, and not an analytical expression. In this case, even when the signal to be modeled is the impulse response of an all-pole filter, we need to perform some computation on the data, using methods such as those discussed in Section 11.3, to determine the linear prediction coefficients.

There are also situations in which an analytical expression is available for the signal, but the signal is not the impulse response of an all-pole filter, and we would like to model it as such. In this case, we again need to carry out computations such as those discussed in Section 11.3.

(a) For $s[n]$ as given in Eq. (P11.16-1), determine the linear prediction coefficients a_1, a_2 directly from the z-transform of $s[n]$.

(b) Write the normal equations for $p = 2$ to obtain equations for a_1, a_2 in terms of $r_{ss}[m]$.

(c) Determine the values of $r_{ss}[0]$, $r_{ss}[1]$, and $r_{ss}[2]$ for the signal $s[n]$ given in Eq. (P11.16-1).

(d) Solve the system of equations from part (a) using the values you found in part (b) to obtain values for the a_ks.

(e) Are the values of a_k from part (c) what you would expect for this signal? Justify your answer clearly.

(f) Suppose you wish to model the signal now with $p = 3$. Write the normal equations for this case.

(g) Find the value of $r_{ss}[3]$.

(h) Solve for the values of a_k when $p = 3$.

(i) Are the values of a_k found in part (h) what you would expect given $s[n]$? Justify your answer clearly.

(j) Would the values of a_1, a_2 you found in (h) change if we model the signal with $p = 4$?

11.17. $x[n]$ and $y[n]$ are sample sequences of jointly wide-sense stationary, zero-mean random processes. The following information is known about the autocorrelation function $\phi_{xx}[m]$ and cross correlation $\phi_{yx}[m]$:

$$\phi_{xx}[m] = \begin{cases} 0 & m \text{ odd} \\ \frac{1}{2^{|m|}} & m \text{ even} \end{cases}$$

$$\phi_{yx}[-1] = 2 \qquad \phi_{yx}[0] = 3 \qquad \phi_{yx}[1] = 8 \qquad \phi_{yx}[2] = -3$$

$$\phi_{yx}[3] = 2 \qquad \phi_{yx}[4] = -0.75$$

(a) The linear estimate of y given x is denoted $\hat{y}_x$. It is designed to minimize

$$\mathcal{E} = E\left(\,|\,y[n] - \hat{y}_x[n]\,|^2\,\right), \tag{P11.17-1}$$

where the $\hat{y}_x[n]$ is formed by processing $x[n]$ with an FIR filter whose impulse response $h[n]$ is of length 3 and is given by

$$h[n] = h_0\delta[n] + h_1\delta[n-1] + h_2\delta[n-2].$$

Determine h_0, h_1, and h_2 to minimize $\mathcal{E}$.

(b) In this part, $\hat{y}_x$, the linear estimate of y given x, is again designed to minimize $\mathcal{E}$ in Eq. (P11.17-1), but with different assumptions on the structure of the linear filter. Here the estimate is formed by processing $x[n]$ with an FIR filter whose impulse response $g[n]$ is of length 2 and is given by

$$g[n] = g_1\delta[n-1] + g_2\delta[n-2].$$

Determine the g_1 and g_2 to minimize $\mathcal{E}$.

(c) The signal, $x[n]$ can be modeled as the output from a two-pole filter $H(z)$ whose input is $w[n]$, a wide-sense stationary, zero-mean, unit-variance white-noise signal.

$$H(z) = \frac{1}{1 - a_1 z^{-1} - a_2 z^{-2}}$$

Determine a_1 and a_2 based on the least-squares inverse model in Section 11.1.2.

(d) We want to implement the system shown in Figure P11.17 where the coefficients a_i are from all-pole modeling in part (c) and the coefficients h_i are the values of the impulse response of the linear estimator in part (a). Draw an implementation that minimizes the total cost of delays, where the cost of each individual delay is weighted linearly by its clock rate.

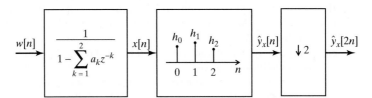

Figure P11.17

(e) Let $\mathcal{E}_a$ be the cost in part (a) and let $\mathcal{E}_b$ be the cost in part (b), where each $\mathcal{E}$ is defined as in Eq. (P11.17-1). Is $\mathcal{E}_a$ larger than, equal to, or smaller than $\mathcal{E}_b$, or is there not enough information to compare them?

(f) Calculate $\mathcal{E}_a$ and $\mathcal{E}_b$ when $\phi_{yy}[0] = 88$. (*Hint*: The optimum FIR filters calculated in parts (a) and (b) are such that $E\left[\hat{y}_x[n](y[n] - \hat{y}_x[n])\right] = 0$.)

11.18. A discrete-time communication channel with impulse response $h[n]$ is to be compensated for with an LTI system with impulse response $h_c[n]$ as indicated in Figure P11.18. The channel $h[n]$ is known to be a one-sample delay, i.e.,

$$h[n] = \delta[n - 1].$$

The compensator $h_c[n]$ is an N-point causal FIR filter, i.e.,

$$H_C(z) = \sum_{k=0}^{N-1} a_k z^{-k}.$$

The compensator $h_c[n]$ is designed to invert (or compensate for) the channel. Specifically, $h_c[n]$ is designed so that with $s[n] = \delta[n]$, $\hat{s}[n]$ is as "close" as possible to an impulse; i.e., $h_c[n]$ is designed so that the error

$$\mathcal{E} = \sum_{n=-\infty}^{\infty} |\hat{s}[n] - \delta[n]|^2$$

is minimized. Find the optimal compensator of length N, i.e., determine $a_0, a_1, \ldots, a_{N-1}$ to minimize $\mathcal{E}$.

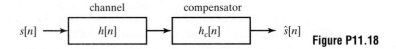

Figure P11.18

11.19. A speech signal was sampled with a sampling rate of 8 kHz. A 300-sample segment was selected from a vowel sound and multiplied by a Hamming window as shown in Figure P11.19.
From this signal a set of linear predictors

$$P^{(i)}(z) = \sum_{k=1}^{i} a_k^{(i)} z^{-k},$$

with orders ranging from $i = 1$ to $i = 11$ was computed using the autocorrelation method.
This set of predictors is shown in Table 11.1 below in a form suggestive of the Levinson–
Durbin recursion.

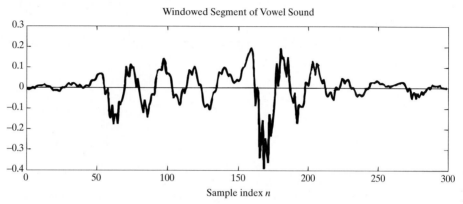

Figure P11.19

(a) Determine the z-transform $A^{(4)}(z)$ of the 4th-order prediction error filter. Draw and
label the flow graph of the direct form implementation of this system.

(b) Determine the set of k-parameters $\{k_1, k_2, k_3, k_4\}$ for the 4th-order prediction error
lattice filter. Draw and label the flow graph of the lattice implementation of this
system.

(c) The minimum mean-squared prediction error for the 2nd-order predictor is $E^{(2)} =$
0.5803. What is the minimum mean-squared prediction error for the 3rd-order predictor? What is the total energy of the signal $s[n]$? What is the value of the autocorrelation
function $r_{ss}[1]$?

TABLE 11.1   PREDICTION COEFFICIENTS FOR A SET OF LINEAR PREDICTORS

i	$a_1^{(i)}$	$a_2^{(i)}$	$a_3^{(i)}$	$a_4^{(i)}$	$a_5^{(i)}$	$a_6^{(i)}$	$a_7^{(i)}$	$a_8^{(i)}$	$a_9^{(i)}$	$a_{10}^{(i)}$	$a_{11}^{(i)}$
1	0.8328										
2	0.7459	0.1044									
3	0.7273	−0.0289	0.1786								
4	0.8047	−0.0414	0.4940	−0.4337							
5	0.7623	0.0069	0.4899	−0.3550	−0.0978						
6	0.6889	−0.2595	0.8576	−0.3498	0.4743	−0.7505					
7	0.6839	−0.2563	0.8553	−0.3440	0.4726	−0.7459	−0.0067				
8	0.6834	−0.3095	0.8890	−0.3685	0.5336	−0.7642	0.0421	−0.0713			
9	0.7234	−0.3331	1.3173	−0.6676	0.7402	−1.2624	0.2155	−0.4544	0.5605		
10	0.6493	−0.2730	1.2888	−0.5007	0.6423	−1.1741	0.0413	−0.4103	0.4648	0.1323	
11	0.6444	−0.2902	1.3040	−0.5022	0.6859	−1.1980	0.0599	−0.4582	0.4749	0.1081	0.0371

(d) The minimum mean-squared prediction errors for these predictors form a sequence $\{E^{(0)}, E^{(1)}, E^{(2)}, \ldots, E^{(11)}\}$. It can be shown that this sequence decreases abruptly in going from $i = 0$ to $i = 1$ and then decreases slowly for several orders and then makes a sharp decrease. At what order i would you expect this to occur?

(e) Sketch carefully the prediction error sequence $e^{(11)}[n]$ for the given input $s[n]$ in Figure P11.19. Show as much detail as possible.

(f) The system function of the 11th-order all-pole model is

$$H(z) = \frac{G}{A^{(11)}(z)} = \frac{G}{1 - \displaystyle\sum_{k=1}^{11} a_k^{(11)} z^{-k}} = \frac{G}{\displaystyle\prod_{i=1}^{11}(1 - z_i z^{-1})}.$$

The following are five of the roots of the 11th-order prediction error filter $A^{(11)}(z)$.

i	$|z_i|$	$\angle z_i$ (rad)
1	0.2567	2.0677
2	0.9681	1.4402
3	0.9850	0.2750
4	0.8647	2.0036
5	0.9590	2.4162

State briefly in words where the other six zeros of $A^{(11)}(z)$ are located. Be as precise as possible.

(g) Use information given in the table and in part (c) of this problem to determine the gain parameter G for the 11th-order all-pole model.

(h) Carefully sketch and label a plot of the frequency response of the 11th-order all-pole model filter for analog frequencies $0 \le F \le 4$ kHz.

11.20. Spectrum analysis is often applied to signals comprised of sinusoids. Sinusoidal signals are particularly interesting, because they share properties with both deterministic and random signals. On the one hand, we can describe them in terms of a simple equation. On the other hand, they have infinite energy, so we often characterize them in terms of their average power, just as with random signals. This problem explores some theoretical issues in modeling sinusoidal signals from the point of view of random signals.

We can consider sinusoidal signals as stationary random signals by assuming that the signal model is $s[n] = A\cos(\omega_0 n + \theta)$ for $-\infty < n < \infty$, where both A and θ can be considered to be random variables. In this model, the signal is considered to be an ensemble of sinusoids described by underlying probability laws for A and θ. For simplicity, assume that A is a constant, and θ is a random variable that is uniformly distributed over $0 \le \theta < 2\pi$.

(a) Show that the autocorrelation function for such a signal is

$$r_{ss}[m] = E\{s[n+m]s[n]\} = \frac{A^2}{2}\cos(\omega_0 m). \qquad \text{(P11.20-1)}$$

(b) Using Eq. (11.34), write the set of equations that is satisfied by the coefficients of a 2nd-order linear predictor for this signal.

(c) Solve the equations in (b) for the optimum predictor coefficients. Your answer should be a function of ω_0.

(d) Factor the polynomial $A(z) = 1 - a_1 z^{-1} - a_2 z^{-2}$ describing the prediction error filter.

(e) Use Eq. (11.37) to determine an expression for the minimum mean-squared prediction error. Your answer should confirm why random sinusoidal signals are called "predictable" and/or "deterministic."

11.21. Using Eqs. (11.84a) and (11.84b) from the Levinson–Durbin recursion, derive the relation between the ith and $i - 1^{\text{st}}$ prediction error filters given in Eq. (11.89).

11.22. In this problem, we consider the construction of lattice filters to implement the inverse filter for the signal

$$s[n] = 2 \left(\frac{1}{3} \right)^n u[n] + 3 \left(-\frac{1}{2} \right)^n u[n].$$

(a) Find the values of the k-parameters k_1 and k_2 for the 2nd-order case (i.e., $p = 2$).

(b) Draw the signal flow graph of a lattice filter implementation of the inverse filter, i.e., the filter that outputs $y[n] = A\delta[n]$ (a scaled impulse) when the input $x[n] = s[n]$.

(c) Verify that the signal flow graph you drew in part (b) has the correct impulse response by showing that the z-transform of this inverse filter is indeed proportional to the inverse of $S(z)$.

(d) Draw the signal flow graph for a lattice filter that implements an all-pole system such that when the input is $x[n] = \delta[n]$, the output is the signal $s[n]$ given above.

(e) Derive the system function of the signal flow graph you drew in part (d) and demonstrate that its impulse response $h[n]$ satisfies $h[n] = s[n]$.

11.23. Consider the signal

$$s[n] = \alpha \left(\frac{2}{3} \right)^n u[n] + \beta \left(\frac{1}{4} \right)^n u[n]$$

where α and β are constants. We wish to linearly predict $s[n]$ from its past p values using the relationship

$$\hat{s}[n] = \sum_{k=1}^{p} a_k s[n - k]$$

where the coefficients a_k are constants. The coefficients a_k are chosen to minimize the prediction error

$$\mathcal{E} = \sum_{n=-\infty}^{\infty} (s[n] - \hat{s}[n])^2.$$

(a) With $r_{ss}[m]$ denoting the autocorrelation function of $s[n]$, write the equations for the case $p = 2$ the solution to which will result in a_1, a_2.

(b) Determine a pair of values for α and β such that when $p = 2$, the solution to the normal equations is $a_1 = \frac{11}{12}$ and $a_2 = -\frac{1}{6}$. Is your answer unique? Explain.

(c) If $\alpha = 8$ and $\beta = -3$, determine k-parameter k_3, resulting from using the Levinson recursion to solve the normal equations for $p = 3$. Is that different from k_3 when solving for $p = 4$?

11.24. Consider the following Yule–Walker equations: $\boldsymbol{\Gamma}_p \, \boldsymbol{a}_p = \boldsymbol{\gamma}_p$, where:

$$\boldsymbol{a}_p = \begin{bmatrix} a_1^p \\ \vdots \\ a_p^p \end{bmatrix} \qquad \boldsymbol{\gamma}_p = \begin{bmatrix} \phi[1] \\ \vdots \\ \phi[p] \end{bmatrix}$$

and

$$\boldsymbol{\Gamma}_p = \begin{bmatrix} \phi[0] & \cdots & \phi[p-1] \\ \vdots & \ddots & \vdots \\ \phi[p-1] & \cdots & \phi[0] \end{bmatrix} \qquad \text{(a Toeplitz matrix)}$$

The Levinson–Durbin algorithm provides the following recursive solution for the normal equation $\mathbf{\Gamma}_{p+1}\,\mathbf{a}_{p+1} = \mathbf{\gamma}_{p+1}$:

$$a_{p+1}^{p+1} = \frac{\phi[p+1] - \left(\mathbf{\gamma}_p^b\right)^T \mathbf{a}_p}{\phi[0] - \left(\mathbf{\gamma}_p\right)^T \mathbf{a}_p} \qquad\qquad a_m^{p+1} = a_m^p - a_{p+1}^{p+1} \cdot a_{p-m+1}^p \quad m = 1, \ldots, p$$

where $\mathbf{\gamma}_p^b$ is the backward version of $\mathbf{\gamma}_p$: $\mathbf{\gamma}_p^b = [\phi[p]\ldots\phi[1]]^T$, and $a_1^1 = \frac{\phi[1]}{\phi[0]}$. Note that for vectors, the model order is shown in the subscript; but for scalars, the model order is shown in the superscript.

Now consider the following normal equation: $\mathbf{\Gamma}_p\,\mathbf{b}_p = \mathbf{c}_p$, where

$$\mathbf{b}_p = \begin{bmatrix} b_1^p \\ \vdots \\ b_p^p \end{bmatrix} \qquad \mathbf{c}_p = \begin{bmatrix} c[1] \\ \vdots \\ c[p] \end{bmatrix}$$

Show that the recursive solution for $\mathbf{\Gamma}_{p+1}\,\mathbf{b}_{p+1} = \mathbf{c}_{p+1}$ is:

$$b_{p+1}^{p+1} = \frac{c[p+1] - \left(\mathbf{\gamma}_p^b\right)^T \mathbf{b}_p}{\phi[0] - \left(\mathbf{\gamma}_p\right)^T \mathbf{a}_p} \qquad\qquad b_m^{p+1} = b_m^p - b_{p+1}^{p+1} \cdot a_{p-m+1}^p \quad m = 1, \ldots, p$$

where $b_1^1 = \frac{c[1]}{\phi[0]}$.

(*Note*: You may find it helpful to note that $\mathbf{a}_p^b = \mathbf{\Gamma}_p^{-1}\mathbf{\gamma}_p^b$.)

11.25. Consider a colored wide-sense stationary random signal $s[n]$ that we desire to whiten using the system in Figure P11.25-1: In designing the optimal whitening filter for a given order p, we pick the coefficient $a_k^{(p)}$, $k = 1, \ldots, p$ that satisfy the autocorrelation normal equations given by Eq. (11.34), where $r_{ss}[m]$ is the autocorrelation of $s[n]$.

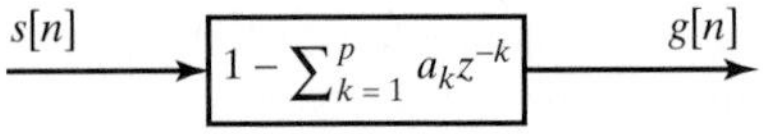

$$s[n] \longrightarrow \boxed{1 - \sum_{k=1}^{P} a_k z^{-k}} \xrightarrow{\;g[n]\;}$$

Figure P11.25-1

It is known that the optimal 2nd-order whitening filter for $s[n]$ is $H_2(z) = 1 + \frac{1}{4}z^{-1} - \frac{1}{8}z^{-2}$, (i.e., $a_1^{(2)} = -\frac{1}{4}$, $a_2^{(2)} = \frac{1}{8}$), which we implement in the 2nd-order lattice structure in Figure P11.25-2. We would also like to use a 4th-order system, with transfer function

$$H_4(z) = 1 - \sum_{k=1}^{4} a_k^{(4)} z^{-k}.$$

We implement this system with the lattice structure in Figure P11.25-3. Determine which, if any of $H_4(z)$, k_1, k_2, k_3, k_4 can be exactly determined from the information given above. Explain why you cannot determine the remaining, if any, parameters.

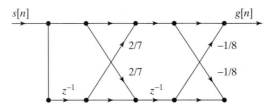

Figure P11.25-2   Lattice structure for 2nd-order system

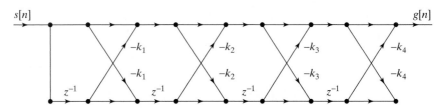

Figure P11.25-3   Lattice structure for 4th-order system

Extension Problems

11.26. Consider a stable all-pole model with system function

$$H(z) = \frac{G}{1 - \displaystyle\sum_{m=1}^{p} a_m z^{-m}} = \frac{G}{A(z)}.$$

Assume that g is positive.

In this problem, we will show that a set of $(p+1)$ samples of the magnitude-squared of $H(z)$ on the unit circle; i.e.,

$$C[k] = |H(e^{j\pi k/p})|^2 \qquad k = 0, 1, \ldots, p,$$

is sufficient to represent the system. Specifically, given $C[k]$, $k = 0, 1, \ldots, p$, show that the parameters G and a_m, $m = 0, 1, \ldots, p$ can be determined.

(a) Consider the z-transform

$$Q(z) = \frac{1}{H(z)H(z^{-1})} = \frac{A(z)A(z^{-1})}{G^2},$$

which corresponds to a sequence $q[n]$. Determine the relationship between $q[n]$ and $h_A[n]$, the impulse response of the prediction error filter whose system function is $A(z)$. Over what range of n will $q[n]$ be nonzero?

(b) Design a procedure based on the DFT for determining $q[n]$ from the given magnitude-squared samples $C[k]$.

(c) Assuming that the sequence $q[n]$ as determined in (b) is known, state a procedure for determining $A(z)$ and G.

11.27. The general IIR lattice system in Figure 11.21 is restricted to all-pole systems. However, both poles and zeros can be implemented by the system of Figure P11.27-1 (Gray and Markel, 1973, 1975). Each of the sections in Figure P11.27-1 is described by the flow graph of Figure P11.27-2. In other words, Figure 11.21 is embedded in Figure P11.27-1 with the output formed as a linear combination of the backward prediction error sequences.

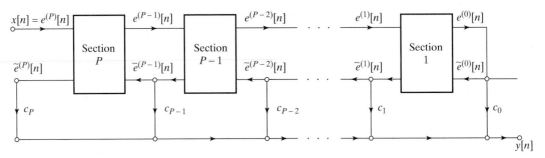

Figure P11.27-1

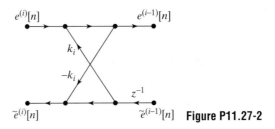

Figure P11.27-2

(a) Show that the system function between the input $X(z) = E^{(P)}(z)$ and $\tilde{E}^{(i)}(z)$ is

$$\tilde{H}^{(i)}(z) = \frac{\tilde{E}^{(i)}(z)}{X(z)} = \frac{z^{-1}A^{(i)}(z^{-1})}{A^{(P)}(z)}. \qquad (P11.27\text{-}1)$$

(b) Show that $\tilde{H}^{(P)}(z)$ is an all-pass system. (This result is not needed for the rest of the problem.)

(c) The overall system function from $X(z)$ to $Y(z)$ is

$$H(z) = \frac{Y(z)}{X(z)} = \sum_{i=0}^{P} \frac{c_i z^{-1} A^{(i)}(z^{-1})}{A^{(P)}(z)} = \frac{Q(z)}{A^{(P)}(z)}. \qquad (P11.27\text{-}2)$$

Show that the numerator $Q(z)$ in Eq. (P11.27-2) is a pth-order polynomial of the form

$$Q(z) = \sum_{m=0}^{p} q_m z^{-m} \qquad (P11.27\text{-}3)$$

where the coefficients c_m in Figure P11.27 are given by the equation

$$c_m = q_m + \sum_{i=m+1}^{p} c_i a_{i-m}^{(i)} \quad m = p, p-1, \ldots, 1, 0. \qquad (P11.27\text{-}4)$$

(d) Give a procedure for computing all the parameters needed to implement a system function such as Eq. (P11.27-2) using the lattice structure of Figure P11.27.

(e) Using the procedure described in (c), draw the complete flow graph of the lattice implementation of the system

$$H(z) = \frac{1 + 3z^{-1} + 3z^{-2} + z^{-3}}{1 - 0.9z^{-1} + 0.64z^{-2} - 0.576z^{-3}}. \qquad \text{(P11.27-5)}$$

11.28. In Section 11.7.3, the k-parameters are computed by Eqs. (11.101). Using the relations $e^{(i)}[n] = e^{(i-1)}[n] - k_i \tilde{e}^{(i-1)}[n-1]$ and $\tilde{e}^{(i)}[n] = \tilde{e}^{(i-1)}[n-1] - k_i e^{(i-1)}[n]$, show that

$$k_i^P = \sqrt{k_i^f k_i^b},$$

where k_i^f is the value of k_i that minimizes the mean-squared forward prediction error

$$\mathcal{E}^{(i)} = \sum_{n=-\infty}^{\infty} (e^{(i)}[n])^2,$$

and k_i^b is the value of k_i that minimizes the mean-squared backward prediction error

$$\tilde{\mathcal{E}}^{(i)} = \sum_{n=-\infty}^{\infty} (\tilde{e}^{(i)}[n])^2.$$

11.29. Substitute Eq. (11.88) and Eq. (11.93) into Eq. (11.101) to show that

$$k_i^P = \frac{\displaystyle\sum_{n=-\infty}^{\infty} e^{(i-1)}[n]\tilde{e}^{(i-1)}[n-1]}{\left\{ \displaystyle\sum_{n=-\infty}^{\infty} (e^{(i-1)}[n])^2 \sum_{n=-\infty}^{\infty} (\tilde{e}^{(i-1)}[n-1])^2 \right\}^{1/2}}$$

$$= \frac{r_{ss}[i] - \displaystyle\sum_{j=1}^{i-1} a_j^{(i-1)} r_{ss}[i-j]}{\mathcal{E}^{(i-1)}} = k_i.$$

11.30. As discussed in Section 11.7.3, Burg (1975) proposed computing the k parameters so as to minimize the sum of the forward and backward prediction errors at the ith stage of the lattice filter; i.e.,

$$\mathcal{B}^{(i)} = \sum_{n=i}^{M} \left[(e^{(i)}[n])^2 + (\tilde{e}^{(i)}[n])^2 \right] \qquad \text{(P11.30-1)}$$

where the sum is over the fixed interval $i \leq n \leq M$.

(a) Substitute the lattice filter signals $e^{(i)}[n] = e^{(i-1)}[n] - k_i \tilde{e}^{(i-1)}[n-1]$ and $\tilde{e}^{(i)}[n] = \tilde{e}^{(i-1)}[n-1] - k_i e^{(i-1)}[n]$ into (P11.30-1) and show that the value of k_i that minimizes $\mathcal{B}^{(i)}$ is

$$k_i^B = \frac{2 \displaystyle\sum_{n=i}^{M} e^{(i-1)}[n]\tilde{e}^{(i-1)}[n-1]}{\left\{ \displaystyle\sum_{n=i}^{M} (e^{(i-1)}[n])^2 + \sum_{n=i}^{M} (\tilde{e}^{(i-1)}[n-1])^2 \right\}}. \qquad \text{(P11.30-2)}$$

(b) Prove that $-1 < k_i^B < 1$.

Hint: Consider the expression $\displaystyle\sum_{n=i}^{M} (x[n] \pm y[n])^2 > 0$ where $x[n]$ and $y[n]$ are two distinct sequences.

(c) Given a set of Burg coefficients k_i^B, $i = 1, 2, \ldots, p$, how would you obtain the coefficients of the corresponding prediction error filter $A^{(p)}(z)$?

12

Discrete Hilbert Transforms

12.0 INTRODUCTION

In general, the specification of the Fourier transform of a sequence requires complete knowledge of both the real and imaginary parts or of the magnitude and phase at all frequencies in the range $-\pi < \omega \leq \pi$. However, we have seen that under certain conditions, there are constraints on the Fourier transform. For example, in Section 2.8, we saw that if $x[n]$ is real, then its Fourier transform is conjugate symmetric, i.e., $X(e^{j\omega}) = X^*(e^{-j\omega})$. From this, it follows that for real sequences, specification of $X(e^{j\omega})$ for $0 \leq \omega \leq \pi$ also specifies it for $-\pi \leq \omega \leq 0$. Similarly, we saw in Section 5.4 that under the constraint of minimum phase, the Fourier transform magnitude and phase are not independent; i.e., specification of magnitude determines the phase and specification of phase determines the magnitude to within a scale factor. In Section 8.5, we saw that for sequences of finite length N, specification of $X(e^{j\omega})$ at N equally spaced frequencies determines $X(e^{j\omega})$ at all frequencies.

In this chapter, we will see that the constraint of causality of a sequence implies unique relationships between the real and imaginary parts of the Fourier transform. Relationships of this type between the real and imaginary parts of complex functions arise in many fields besides signal processing, and they are commonly known as *Hilbert transform relationships*. In addition to developing these relationships for the Fourier transform of causal sequences, we will develop related results for the DFT and for sequences with one-sided Fourier transforms. Also, in Section 12.3 we will indicate how the relationship between magnitude and phase for minimum-phase sequences can be interpreted in terms of the Hilbert transform.

Although we will take an intuitive approach in this chapter (see Gold, Oppenheim and Rader, 1970) it is important to be aware that the Hilbert transform relationships follow formally from the properties of analytic functions. (See Problem 12.21.) Specifically, the complex functions that arise in the mathematical representation of discrete-time signals and systems are generally very well-behaved functions. With few exceptions, the z-transforms that have concerned us have had well-defined regions in which the power series is absolutely convergent. Since a power series represents an analytic function within its ROC, it follows that z-transforms are analytic functions inside their ROCs. By the definition of an analytic function, this means that the z-transform has a well-defined derivative at every point inside the ROC. Furthermore, for analytic functions the z-transform and all its derivatives are continuous functions within the ROC.

The properties of analytic functions imply some rather powerful constraints on the behavior of the z-transform within its ROC. Since the Fourier transform is the z-transform evaluated on the unit circle, these constraints also restrict the behavior of the Fourier transform. One such constraint is that the real and imaginary parts satisfy the Cauchy–Riemann conditions, which relate the partial derivatives of the real and imaginary parts of an analytic function. (See, for example, Churchill and Brown, 1990.) Another constraint is the Cauchy integral theorem, through which the value of a complex function is specified everywhere inside a region of analyticity in terms of the values of the function on the boundary of the region. On the basis of these relations for analytic functions, it is possible, under certain conditions, to derive explicit integral relationships between the real and imaginary parts of a z-transform on a closed contour within the ROC. In the mathematics literature, these relations are often referred to as *Poisson's formulas*. In the context of system theory, they are known as the *Hilbert transform relations*.

Rather than following the mathematical approach just discussed, we will develop the Hilbert transform relations by exploiting the fact that the real and imaginary parts of the Fourier transform of a causal sequence are the transforms of the even and odd components, respectively, of the sequence (properties 5 and 6, Table 2.1). As we will show, a causal sequence is completely specified by its even part, implying that the Fourier transform of the original sequence is completely specified by its real part. In addition to applying this argument to specifying the Fourier transform of a particular causal sequence in terms of its real part, we can also apply it, under certain conditions, to specify the Fourier transform of a sequence in terms of its magnitude.

The notion of an analytic signal is an important concept in continuous-time signal processing. An analytic signal is a complex time function (which is analytic) having a Fourier transform that vanishes for negative frequencies. A complex sequence cannot be considered in any formal sense to be analytic, since it is a function of an integer variable. However, in a style similar to that described in the previous paragraph, it is possible to relate the real and imaginary parts of a complex sequence whose spectrum is zero on the unit circle for $-\pi < \omega < 0$. A similar approach can also be taken in relating the real and imaginary parts of the DFT for a periodic or, equivalently, a finite-length sequence. In this case, the "causality" condition is that the periodic sequence be zero in the second half of each period.

Thus, in this chapter, a notion of causality will be applied to relate the even and odd components of a function or, equivalently, the real and imaginary parts of its transform. We will apply this approach in four situations. First, we relate the real and imaginary parts of the Fourier transform $X(e^{j\omega})$ of a sequence $x[n]$ that is zero for $n < 0$. In the second situation, we obtain a relationship between the real and imaginary parts of the DFT for periodic sequences or, equivalently, for a finite-length sequence considered to be of length N, but with the last $(N/2) - 1$ points restricted to zero. In the third case, we relate the real and imaginary parts of the *logarithm* of the Fourier transform under the condition that the inverse transform of the logarithm of the transform is zero for $n < 0$. Relating the real and imaginary parts of the logarithm of the Fourier transform corresponds to relating the log magnitude and phase of $X(e^{j\omega})$. Finally, we relate the real and imaginary parts of a complex sequence whose Fourier transform, considered as a periodic function of ω, is zero in the second half of each period.

12.1 REAL- AND IMAGINARY-PART SUFFICIENCY OF THE FOURIER TRANSFORM FOR CAUSAL SEQUENCES

Any sequence can be expressed as the sum of an even sequence and an odd sequence. Specifically, with $x_e[n]$ and $x_o[n]$ denoting the even and odd parts, respectively, of $x[n]$,[1] we have

$$x[n] = x_e[n] + x_o[n], \tag{12.1}$$

where

$$x_e[n] = \frac{x[n] + x[-n]}{2} \tag{12.2}$$

and

$$x_o[n] = \frac{x[n] - x[-n]}{2}. \tag{12.3}$$

Equations (12.1) to (12.3) apply to an arbitrary sequence, whether or not it is causal and whether or not it is real. However, if $x[n]$ is causal, i.e., if $x[n] = 0, n < 0$, then it is possible to recover $x[n]$ from $x_e[n]$ or to recover $x[n]$ for $n \neq 0$ from $x_o[n]$. Consider, for example, the causal sequence $x[n]$ and its even and odd components, as shown in Figure 12.1. Because $x[n]$ is causal, $x[n] = 0$ for $n < 0$ and $x[-n] = 0$ for $n > 0$. Therefore, the nonzero portions of $x[n]$ and $x[-n]$ do not overlap except at $n = 0$. For this reason, it follows from Eqs. (12.2) and (12.3) that

$$x[n] = 2x_e[n]u[n] - x_e[0]\delta[n] \tag{12.4}$$

and

$$x[n] = 2x_o[n]u[n] + x[0]\delta[n]. \tag{12.5}$$

The validity of these relationships is easily seen in Figure 12.1. Note that $x[n]$ is completely determined by $x_e[n]$. On the other hand, $x_o[0] = 0$, so we can recover $x[n]$ from $x_o[n]$ only for $n \neq 0$.

[1]If $x[n]$ is real, then $x_e[n]$ and $x_o[n]$ in Eqs. (12.2) and (12.3) are the even and odd parts, respectively, of $x[n]$ as considered in Chapter 2. If $x[n]$ is complex, for the purposes of this discussion we still define $x_e[n]$ and $x_o[n]$ as in Eqs. (12.2) and (12.3), which do not correspond to the conjugate-symmetric and conjugate-antisymmetric parts of a complex sequence as considered in Chapter 2.

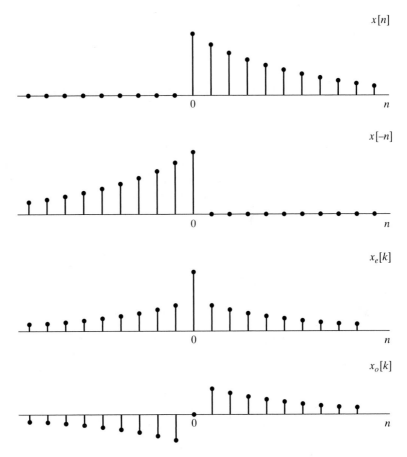

Figure 12.1   Even and odd parts of a real causal sequence.

Now, if $x[n]$ is also stable, i.e., absolutely summable, then its Fourier transform exists. We denote the Fourier transform of $x[n]$ as

$$X(e^{j\omega}) = X_R(e^{j\omega}) + jX_I(e^{j\omega}), \qquad (12.6)$$

where $X_R(e^{j\omega})$ is the real part and $X_I(e^{j\omega})$ is the imaginary part of $X(e^{j\omega})$. Recall that if $x[n]$ is a *real* sequence, then $X_R(e^{j\omega})$ is the Fourier transform of $x_e[n]$ and $jX_I(e^{j\omega})$ is the Fourier transform of $x_o[n]$. Therefore, for a *causal, stable, real* sequence, $X_R(e^{j\omega})$ completely determines $X(e^{j\omega})$, since, if we are given $X_R(e^{j\omega})$, we can find $X(e^{j\omega})$ by the following process:

1. Find $x_e[n]$ as the inverse Fourier transform of $X_R(e^{j\omega})$.
2. Find $x[n]$ using Eq. (12.4).
3. Find $X(e^{j\omega})$ as the Fourier transform of $x[n]$.

This also implies, of course, that $X_I(e^{j\omega})$ can be determined from $X_R(e^{j\omega})$. In Example 12.1, we illustrate how this procedure can be applied to obtain $X(e^{j\omega})$ and $X_I(e^{j\omega})$ from $X_R(e^{j\omega})$.

Example 12.1   Finite-Length Sequence

Consider a real, causal sequence $x[n]$ for which $X_R(e^{j\omega})$, the real part of the DTFT, is

$$X_R(e^{j\omega}) = 1 + \cos 2\omega. \tag{12.7}$$

We would like to determine the original sequence $x[n]$, its Fourier transform $X(e^{j\omega})$, and the imaginary part of the Fourier transform, $X_I(e^{j\omega})$. As a first step, we rewrite Eq. (12.7) expressing the cosine as a sum of complex exponentials:

$$X_R(e^{j\omega}) = 1 + \frac{1}{2}e^{-j2\omega} + \frac{1}{2}e^{j2\omega}. \tag{12.8}$$

We know that $X_R(e^{j\omega})$ is the Fourier transform of $x_e[n]$, the even part of $x[n]$ as defined in Eq. (12.2). Comparing Eq. (12.8) with the definition of the Fourier transform, Eq. (2.131), we can match terms to obtain

$$x_e[n] = \delta[n] + \frac{1}{2}\delta[n-2] + \frac{1}{2}\delta[n+2].$$

Now that we have obtained the even part, we can use the relation in Eq. (12.4) to obtain

$$x[n] = \delta[n] + \delta[n-2]. \tag{12.9}$$

From $x[n]$, we get

$$X(e^{j\omega}) = 1 + e^{-j2\omega}$$

$$= 1 + \cos 2\omega - j \sin 2\omega. \tag{12.10}$$

From Eq. (12.10), we can both confirm that $X_R(e^{j\omega})$ is as specified in Eq. (12.7) and also obtain

$$X_I(e^{j\omega}) = -\sin 2\omega. \tag{12.11}$$

As an alternative path to obtaining $X_I(e^{j\omega})$, we can first use Eq. (12.3) to get $x_o[n]$ from $x[n]$. Substituting Eq. (12.9) into Eq. (12.3) then yields

$$x_o[n] = \frac{1}{2}\delta[n-2] - \frac{1}{2}\delta[n+2].$$

The Fourier transform of $x_o[n]$ is $jX_I(e^{j\omega})$, so we find

$$jX_I(e^{j\omega}) = \frac{1}{2}e^{-j2\omega} - \frac{1}{2}e^{j2\omega}$$

$$= -j \sin 2\omega,$$

so that

$$X_I(e^{j\omega}) = -\sin 2\omega,$$

which is consistent with Eq. (12.11).

Example 12.2   Exponential Sequence

Let

$$X_R(e^{j\omega}) = \frac{1 - \alpha \cos \omega}{1 - 2\alpha \cos \omega + \alpha^2}, \qquad |\alpha| < 1, \tag{12.12}$$

or equivalently,

$$X_R(e^{j\omega}) = \frac{1 - (\alpha/2)(e^{j\omega} + e^{-j\omega})}{1 - \alpha(e^{j\omega} + e^{-j\omega}) + \alpha^2}, \qquad |\alpha| < 1, \tag{12.13}$$

with α real. We first determine $x_e[n]$ and then $x[n]$ using Eq. (12.4).

To obtain $x_e[n]$, the inverse Fourier transform of $X_R(e^{j\omega})$, it is convenient to first obtain $X_R(z)$, the z-transform of $x_e[n]$. This follows directly from Eq. (12.13), given that

$$X_R(e^{j\omega}) = X_R(z)\big|_{z=e^{j\omega}}.$$

Consequently, by replacing $e^{j\omega}$ by z in Eq. (12.13), we obtain

$$X_R(z) = \frac{1 - (\alpha/2)(z + z^{-1})}{1 - \alpha(z + z^{-1}) + \alpha^2} \tag{12.14}$$

$$= \frac{1 - \frac{\alpha}{2}(z + z^{-1})}{(1 - \alpha z^{-1})(1 - \alpha z)}. \tag{12.15}$$

Since we began with the Fourier transform $X_R(e^{j\omega})$ and obtained $X_R(z)$ by extending $X_R(e^{j\omega})$ into the z-plane, the ROC of $X_R(z)$ must, of course, include the unit circle and is then bounded on the inside by the pole at $z = \alpha$ and on the outside by the pole at $z = 1/\alpha$.

From Eq. (12.15), we now want to obtain $x_e[n]$, the inverse z-transform of $X_R(z)$. We do this by expanding Eq. (12.15) in partial fractions, yielding

$$X_R(z) = \frac{1}{2}\left[\frac{1}{1 - \alpha z^{-1}} + \frac{1}{1 - \alpha z}\right], \tag{12.16}$$

with the ROC specified to include the unit circle. The inverse z-transform of Eq. (12.16) can then be applied separately to each term to obtain

$$x_e[n] = \frac{1}{2}\alpha^n u[n] + \frac{1}{2}\alpha^{-n} u[-n]. \tag{12.17}$$

Consequently, from Eq. (12.4),

$$x[n] = \alpha^n u[n] + \alpha^{-n} u[-n]u[n] - \delta[n]$$

$$= \alpha^n u[n].$$

$X(e^{j\omega})$ is then given by

$$X(e^{j\omega}) = \frac{1}{1 - \alpha e^{-j\omega}}, \tag{12.18}$$

and $X(z)$ is given by

$$X(z) = \frac{1}{1 - \alpha z^{-1}} \qquad |z| > |\alpha|. \tag{12.19}$$

The constructive procedure illustrated in Example 12.1 can be interpreted analytically to obtain a general relationship that expresses $X_I(e^{j\omega})$ directly in terms of $X_R(e^{j\omega})$. From Eq. (12.4), the complex convolution theorem, and the fact that $x_e[0] = x[0]$, it follows that

$$X(e^{j\omega}) = \frac{1}{\pi} \int_{-\pi}^{\pi} X_R(e^{j\theta}) U(e^{j(\omega-\theta)}) d\theta - x[0], \tag{12.20}$$

where $U(e^{j\omega})$ is the Fourier transform of the unit step sequence. As stated in Section 2.7, although the unit step is neither absolutely summable nor square summable, it can be represented by the Fourier transform

$$U(e^{j\omega}) = \sum_{k=-\infty}^{\infty} \pi\delta(\omega - 2\pi k) + \frac{1}{1 - e^{-j\omega}}, \tag{12.21}$$

or, since the term $1/(1 - e^{-j\omega})$ can be rewritten as

$$\frac{1}{1 - e^{-j\omega}} = \frac{1}{2} - \frac{j}{2} \cot\left(\frac{\omega}{2}\right), \tag{12.22}$$

Eq. (12.21) becomes

$$U(e^{j\omega}) = \sum_{k=-\infty}^{\infty} \pi\delta(\omega - 2\pi k) + \frac{1}{2} - \frac{j}{2} \cot\left(\frac{\omega}{2}\right). \tag{12.23}$$

Using Eq. (12.23), we can express Eq. (12.20) as

$$X(e^{j\omega}) = X_R(e^{j\omega}) + jX_I(e^{j\omega})$$

$$= X_R(e^{j\omega}) + \frac{1}{2\pi} \int_{-\pi}^{\pi} X_R(e^{j\theta}) d\theta \tag{12.24}$$

$$- \frac{j}{2\pi} \int_{-\pi}^{\pi} X_R(e^{j\theta}) \cot\left(\frac{\omega - \theta}{2}\right) d\theta - x[0].$$

Equating real and imaginary parts in Eq. (12.24) and noting that

$$x[0] = \frac{1}{2\pi} \int_{-\pi}^{\pi} X_R(e^{j\theta}) d\theta, \tag{12.25}$$

we obtain the relationship

$$X_I(e^{j\omega}) = -\frac{1}{2\pi} \int_{-\pi}^{\pi} X_R(e^{j\theta}) \cot\left(\frac{\omega - \theta}{2}\right) d\theta. \tag{12.26}$$

A similar procedure can be followed to obtain $x[n]$ and $X(e^{j\omega})$ from $X_I(e^{j\omega})$ and $x[0]$ using Eq. (12.5). This process results in the following equation for $X_R(e^{j\omega})$ in terms of $X_I(e^{j\omega})$:

$$X_R(e^{j\omega}) = x[0] + \frac{1}{2\pi} \int_{-\pi}^{\pi} X_I(e^{j\theta}) \cot\left(\frac{\omega - \theta}{2}\right) d\theta. \tag{12.27}$$

Equations (12.26) and (12.27), which are called *discrete Hilbert transform relationships,* hold for the real and imaginary parts of the Fourier transform of a causal, stable, real sequence. They are improper integrals, since the integrand is singular at

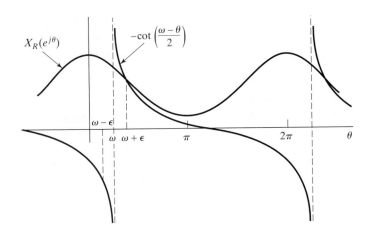

Figure 12.2 Interpretation of the Hilbert transform as a periodic convolution.

$\omega - \theta = 0$. Such integrals must be evaluated carefully to obtain a consistent finite result. This can be done formally by interpreting the integrals as *Cauchy principal values*. That is, Eq. (12.26) becomes

$$X_I(e^{j\omega}) = -\frac{1}{2\pi}\mathcal{P}\int_{-\pi}^{\pi} X_R(e^{j\theta})\cot\left(\frac{\omega-\theta}{2}\right)d\theta, \qquad (12.28\text{a})$$

and Eq. (12.27) becomes

$$X_R(e^{j\omega}) = x[0] + \frac{1}{2\pi}\mathcal{P}\int_{-\pi}^{\pi} X_I(e^{j\theta})\cot\left(\frac{\omega-\theta}{2}\right)d\theta, \qquad (12.28\text{b})$$

where $\mathcal{P}$ denotes the Cauchy principal value of the integral that follows. The meaning of the Cauchy principal value in Eq. (12.28a), for example, is

$$X_I(e^{j\omega}) = -\frac{1}{2\pi}\lim_{\varepsilon\to 0}\left[\int_{\omega+\varepsilon}^{\pi} X_R(e^{j\theta})\cot\left(\frac{\omega-\theta}{2}\right)d\theta\right.$$

$$\left. + \int_{-\pi}^{\omega-\varepsilon} X_R(e^{j\theta})\cot\left(\frac{\omega-\theta}{2}\right)d\theta\right]. \qquad (12.29)$$

Equation (12.29) shows that $X_I(e^{j\omega})$ is obtained by the periodic convolution of $-\cot(\omega/2)$ with $X_R(e^{j\omega})$, with special care being taken in the vicinity of the singularity at $\theta = \omega$. In a similar manner, Eq. (12.28b) involves the periodic convolution of $\cot(\omega/2)$ with $X_I(e^{j\omega})$.

The two functions involved in the convolution integral of Eq. (12.28a) or, equivalently, Eq. (12.29) are illustrated in Figure 12.2. The limit in Eq. (12.29) exists because the function $\cot[(\omega-\theta)/2]$ is antisymmetric at the singular point $\theta = \omega$ and the limit is taken symmetrically about the singularity.

12.2 SUFFICIENCY THEOREMS FOR FINITE-LENGTH SEQUENCES

In Section 12.1, we showed that causality or one-sidedness of a real sequence implies some strong constraints on the Fourier transform of the sequence. The results of the previous section apply, of course, to finite-length causal sequences, but since the finite-length property is more restrictive, it is perhaps reasonable to expect the Fourier transform of a finite-length sequence to be more constrained. We will see that this is indeed the case.

One way to take advantage of the finite-length property is to recall that finite-length sequences can be represented by the DFT. Since the DFT involves sums rather than integrals, the problems associated with improper integrals disappear.

Since the DFT is, in reality, a representation of a periodic sequence, any results we obtain must be based on corresponding results for periodic sequences. Indeed, it is important to keep the inherent periodicity of the DFT firmly in mind in deriving the desired Hilbert transform relation for finite-length sequences. Therefore, we will consider the periodic case first and then discuss the application to the finite-length case.

Consider a periodic sequence $\tilde{x}[n]$ with period N that is related to a finite-length sequence $x[n]$ of length N by

$$\tilde{x}[n] = x[((n))_N]. \tag{12.30}$$

As in Section 12.1, $\tilde{x}[n]$ can be represented as the sum of an even and odd periodic sequence,

$$\tilde{x}[n] = \tilde{x}_e[n] + \tilde{x}_o[n], \qquad n = 0, 1, \ldots, (N-1), \tag{12.31}$$

where

$$\tilde{x}_e[n] = \frac{\tilde{x}[n] + \tilde{x}[-n]}{2}, \qquad n = 0, 1, \ldots, (N-1), \tag{12.32a}$$

and

$$\tilde{x}_o[n] = \frac{\tilde{x}[n] - \tilde{x}[-n]}{2}, \qquad n = 0, 1, \ldots, (N-1). \tag{12.32b}$$

A periodic sequence cannot, of course, be causal in the sense used in Section 12.1. We can, however, define a "periodically causal" sequence to be a periodic sequence for which $\tilde{x}[n] = 0$ for $N/2 < n < N$. That is, $\tilde{x}[n]$ is identically zero over the last half of the period. We assume henceforth that N is even; the case of N odd is considered in Problem 12.25. Note that because of the periodicity of $\tilde{x}[n]$, it is also true that $\tilde{x}[n] = 0$ for $-N/2 < n < 0$. For finite-length sequences, this restriction means that although the sequence is considered to be of length N, the last $(N/2) - 1$ points are in fact zero. In Figure 12.3, we show an example of a periodically causal sequence and its even and odd parts with $N = 8$. Because $\tilde{x}[n]$ is zero in the second half of each period, $\tilde{x}[-n]$ is zero in the first half of each period, and, consequently, except for $n = 0$ and $n = N/2$, there is no overlap between the nonzero portions of $\tilde{x}[n]$ and $\tilde{x}[-n]$. Therefore, for periodically causal periodic sequences,

$$\tilde{x}[n] = \begin{cases} 2\tilde{x}_e[n], & n = 1, 2, \ldots, (N/2) - 1, \\ \tilde{x}_e[n], & n = 0, N/2, \\ 0, & n = (N/2) + 1, \ldots, N - 1, \end{cases} \tag{12.33}$$

and

$$\tilde{x}[n] = \begin{cases} 2\tilde{x}_o[n], & n = 1, 2, \ldots, (N/2) - 1, \\ 0, & n = (N/2) + 1, \ldots, N - 1, \end{cases} \tag{12.34}$$

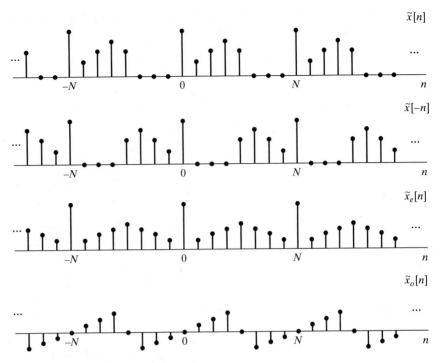

Figure 12.3   Even and odd parts of a periodically causal, real, periodic sequence of period $N = 8$.

where $\tilde{x}[n]$ cannot be recovered from $\tilde{x}_o[n]$ because $\tilde{x}_o[0] = \tilde{x}_o[N/2] = 0$. If we define the periodic sequence

$$\tilde{u}_N[n] = \begin{cases} 1, & n = 0, N/2, \\ 2, & n = 1, 2, \ldots, (N/2) - 1, \\ 0, & n = (N/2) + 1, \ldots, N - 1, \end{cases} \quad (12.35)$$

then it follows that, for N even, we can express $\tilde{x}[n]$ as

$$\tilde{x}[n] = \tilde{x}_e[n]\tilde{u}_N[n] \quad (12.36)$$

and

$$\tilde{x}[n] = \tilde{x}_o[n]\tilde{u}_N[n] + \tilde{x}[0]\tilde{\delta}[n] + \tilde{x}[N/2]\tilde{\delta}[n - (N/2)], \quad (12.37)$$

where $\tilde{\delta}[n]$ is a periodic unit-impulse sequence with period N. Thus, the sequence $\tilde{x}[n]$ can be completely recovered from $\tilde{x}_e[n]$. On the other hand, $\tilde{x}_o[n]$ will always be zero at $n = 0$ and $n = N/2$, and consequently, $\tilde{x}[n]$ can be recovered from $\tilde{x}_o[n]$ only for $n \neq 0$ and $n \neq N/2$.

If $\tilde{x}[n]$ is a real periodic sequence of period N with DFS $\tilde{X}[k]$, then $\tilde{X}_R[k]$, the real part of $\tilde{X}[k]$, is the DFS of $\tilde{x}_e[n]$ and $j\tilde{X}_I[k]$ is the DFS of $\tilde{x}_o[n]$. Hence, Eqs. (12.36) and (12.37) imply that, for a periodic sequence of period N, which is periodically causal in the sense defined earlier, $\tilde{X}[k]$ can be recovered from its real part or (almost) from its imaginary part. Equivalently, $\tilde{X}_I[k]$ can be obtained from $\tilde{X}_R[k]$, and $\tilde{X}_R[k]$ can (almost) be obtained from $\tilde{X}_I[k]$.

Specifically, suppose that we are given $\tilde{X}_R[k]$. Then, we can obtain $\tilde{X}[k]$ and $\tilde{X}_I[k]$ by the following procedure:

1. Compute $\tilde{x}_e[n]$ using the DFS synthesis equation

$$\tilde{x}_e[n] = \frac{1}{N} \sum_{k=0}^{N-1} \tilde{X}_R[k] e^{j(2\pi/N)kn}. \tag{12.38}$$

2. Compute $\tilde{x}[n]$ using Eq. (12.36).
3. Compute $\tilde{X}[k]$ using the DFS analysis equation

$$\tilde{X}[k] = \sum_{n=0}^{N-1} \tilde{x}[n] e^{-j(2\pi/N)kn} = \tilde{X}_R[k] + j\tilde{X}_I[k]. \tag{12.39}$$

In contrast to the general causal case discussed in Section 12.1, the procedure just outlined can be implemented on a computer, since Eqs. (12.38) and (12.39) can be evaluated accurately and efficiently using an FFT algorithm.

To obtain an explicit relation between $\tilde{X}_R[k]$, and $\tilde{X}_I[k]$, we can carry out the procedure analytically. From Eq. (12.36) and Eq. (8.34), it follows that

$$\tilde{X}[k] = \tilde{X}_R[k] + j\tilde{X}_I[k]$$

$$= \frac{1}{N} \sum_{m=0}^{N-1} \tilde{X}_R[m] \tilde{U}_N[k-m]; \tag{12.40}$$

i.e., $\tilde{X}[k]$ is the periodic convolution of $\tilde{X}_R[k]$, the DFS of $\tilde{x}_e[n]$, with $\tilde{U}_N[k]$ the DFS of $\tilde{u}_N[n]$. The DFS of $\tilde{u}_N[n]$ can be shown to be (see Problem 12.24)

$$\tilde{U}_N[k] = \begin{cases} N, & k = 0, \\ -j2\cot(\pi k/N), & k \text{ odd}, \\ 0, & k \text{ even}. \end{cases} \tag{12.41}$$

If we define

$$\tilde{V}_N[k] = \begin{cases} -j2\cot(\pi k/N), & k \text{ odd}, \\ 0, & k \text{ even}, \end{cases} \tag{12.42}$$

then Eq. (12.40) can be expressed as

$$\tilde{X}[k] = \tilde{X}_R[k] + \frac{1}{N} \sum_{m=0}^{N-1} \tilde{X}_R[m] \tilde{V}_N[k-m]. \tag{12.43}$$

Therefore,

$$j\tilde{X}_I[k] = \frac{1}{N} \sum_{m=0}^{N-1} \tilde{X}_R[m] \tilde{V}_N[k-m], \tag{12.44}$$

which is the desired relation between the real and imaginary parts of the DFS of a periodically causal, real, and periodic sequence. Similarly, beginning with Eq. (12.37) we can show that

$$\tilde{X}_R[k] = \frac{1}{N} \sum_{m=0}^{N-1} j\tilde{X}_I[m] \tilde{V}_N[k-m] + \tilde{x}[0] + (-1)^k \tilde{x}[N/2]. \tag{12.45}$$

Equations (12.44) and (12.45) relate the real and imaginary parts of the DFS representation of the periodic sequence $\tilde{x}[n]$. If $\tilde{x}[n]$ is thought of as the periodic repetition of a finite-length sequence $x[n]$ as in Eq. (12.30), then

$$x[n] = \begin{cases} \tilde{x}[n], & 0 \leq n \leq N-1, \\ 0, & \text{otherwise.} \end{cases} \qquad (12.46)$$

If $x[n]$ has the "periodic causality" property with respect to a period N (i.e., $x[n] = 0$ for $n < 0$ and for $n > N/2$), then all of the preceding discussion applies to the DFT of $x[n]$. In other words, we can remove the tildes from Eqs. (12.44) and (12.45), thereby obtaining the DFT relations

$$jX_I[k] = \begin{cases} \dfrac{1}{N} \displaystyle\sum_{m=0}^{N-1} X_R[m]V_N[k-m], & 0 \leq k \leq N-1, \\ 0, & \text{otherwise,} \end{cases} \qquad (12.47)$$

and

$$X_R[k] = \begin{cases} \dfrac{1}{N} \displaystyle\sum_{m=0}^{N-1} jX_I[m]V_N[k-m] + x[0] + (-1)^k x[N/2], & 0 \leq k \leq N-1, \\ 0, & \text{otherwise.} \end{cases} \qquad (12.48)$$

Note that the sequence $V_N[k-m]$ given by Eq. (12.42) is periodic with period N, so we do not need to worry about computing $((k-m))_N$ in Eqs. (12.47) and (12.48), which are the desired relations between the real and imaginary parts of the N-point DFT of a real sequence whose actual length is less than or equal to $(N/2)+1$ (for N even). These equations are circular convolutions, and, for example, Eq. (12.47) can be evaluated efficiently by the following procedure:

1. Compute the inverse DFT of $X_R[k]$ to obtain the sequence

$$x_{\text{ep}}[n] = \frac{x[n] + x[((-n))_N]}{2}, \qquad 0 \leq n \leq N-1. \qquad (12.49)$$

2. Compute the periodic odd part of $x[n]$ by

$$x_{\text{op}}[n] = \begin{cases} x_{\text{ep}}[n], & 0 < n < N/2, \\ -x_{\text{ep}}[n], & N/2 < n \leq N-1, \\ 0, & \text{otherwise.} \end{cases} \qquad (12.50)$$

3. Compute the DFT of $x_{\text{op}}[n]$ to obtain $jX_I[k]$.

Note that if, instead of computing the odd part of $x[n]$ in step 2, we compute

$$x[n] = \begin{cases} x_{\text{ep}}[0], & n = 0, \\ 2x_{\text{ep}}[n], & 0 < n < N/2, \\ x_{\text{ep}}[N/2], & n = N/2, \\ 0, & \text{otherwise,} \end{cases} \qquad (12.51)$$

then the DFT of the resulting sequence would be $X[k]$, the complete DFT of $x[n]$.

Example 12.3   Periodic Sequence

Consider a sequence that is periodically causal with period $N = 4$ and that has

$$X_R[k] = \begin{cases} 2, & k = 0, \\ 3, & k = 1, \\ 4, & k = 2, \\ 3, & k = 3. \end{cases}$$

We can find the imaginary part of the DFT in one of two ways. The first way is to use Eq. (12.47). For $N = 4$,

$$V_4[k] = \begin{cases} 2j, & k = -1 + 4m, \\ -2j, & k = 1 + 4m, \\ 0, & \text{otherwise}, \end{cases}$$

where m is an integer. Implementing the convolution in Eq. (12.47) yields

$$jX_I[k] = \frac{1}{4} \sum_{m=0}^{3} X_R[m]V_4[k-m], \qquad 0 \le k \le 3$$

$$= \begin{cases} j, & k = 1, \\ -j, & k = 3, \\ 0, & \text{otherwise}. \end{cases}$$

Alternatively, we can follow the three-step procedure that includes Eqs. (12.49) and (12.50). Computing the inverse DFT $X_R[k]$ yields

$$x_e[n] = \frac{1}{4} \sum_{k=0}^{3} X_R[k]W_4^{-kn} = \frac{1}{4}[2 + 3(j)^n + 4(-1)^n + 3(-j)^n]$$

$$= \begin{cases} 3, & n = 0, \\ -\frac{1}{2}, & n = 1, 3, \\ 0, & n = 2. \end{cases}$$

Note that although this sequence is not itself even symmetric, a periodic replication of $x_e[n]$ is even symmetric. Thus, the DFT $X_R[k]$ of $x_e[n]$ is purely real. Equation (12.50) allows us to find the periodically odd part $x_{op}[n]$; specifically,

$$x_{op}[n] = \begin{cases} -\frac{1}{2}, & n = 1, \\ \frac{1}{2}, & n = 3, \\ 0, & \text{otherwise}. \end{cases}$$

Finally, we obtain $jX_I[k]$ from the DFT of $x_{op}[n]$:

$$jX_I[k] = \sum_{n=0}^{3} x_{op}[n]W_4^{nk} = -\frac{1}{2}W_4^k + \frac{1}{2}W_4^{3k}$$

$$= \begin{cases} j, & k = 1, \\ -j, & k = 3, \\ 0, & \text{otherwise}, \end{cases}$$

which is, of course, the same as was obtained from Eq. (12.47).

12.3 RELATIONSHIPS BETWEEN MAGNITUDE AND PHASE

So far, we have focused on the relationships between the real and imaginary parts of the Fourier transform of a sequence. Often, we are interested in relationships between the magnitude and phase of the Fourier transform. In this section, we consider the conditions under which these functions might be uniquely related. Although it might appear on the surface that a relationship between real and imaginary parts implies a relationship between magnitude and phase, that is not the case. This is clearly demonstrated by Example 5.9 in Section 5.4. The two system functions $H_1(z)$ and $H_2(z)$ in that example were assumed to correspond to causal, stable systems. Therefore, the real and imaginary parts of $H_1(e^{j\omega})$ are related through the Hilbert transform relations of Eqs. (12.28a) and (12.28b), as are the real and imaginary parts of $H_2(e^{j\omega})$. However, $\angle H_1(e^{j\omega})$ could not be obtained from $|H_1(e^{j\omega})|$, since $H_1(e^{j\omega})$ and $H_2(e^{j\omega})$ have the same magnitude but a different phase.

The Hilbert transform relationship between the real and imaginary parts of the Fourier transform of a sequence $x[n]$ was based on the causality of $x[n]$. We can obtain a Hilbert transform relationship between magnitude and phase by imposing causality on a sequence $\hat{x}[n]$ derived from $x[n]$ for which the Fourier transform $\hat{X}(e^{j\omega})$ is the logarithm of the Fourier transform of $x[n]$. Specifically, we define $\hat{x}[n]$ so that

$$x[n] \xleftrightarrow{\mathcal{F}} X(e^{j\omega}) = |X(e^{j\omega})|e^{j\arg[X(e^{j\omega})]}, \tag{12.52a}$$

$$\hat{x}[n] \xleftrightarrow{\mathcal{F}} \hat{X}(e^{j\omega}), \tag{12.52b}$$

where

$$\hat{X}(e^{j\omega}) = \log[X(e^{j\omega})] = \log|X(e^{j\omega})| + j\arg[X(e^{j\omega})] \tag{12.53}$$

and, as defined in Section 5.1, $\arg[X(e^{j\omega})]$ denotes the continuous phase of $X(e^{j\omega})$. The sequence $\hat{x}[n]$ is commonly referred to as the *complex cepstrum* of $x[n]$, the properties and applications of which are discussed in detail in Chapter 13.[2]

If we now require that $\hat{x}[n]$ be causal, then the real and imaginary parts of $\hat{X}(e^{j\omega})$, corresponding to $\log|X(e^{j\omega})|$ and $\arg[X(e^{j\omega})]$, respectively, will be related through Eqs. (12.28a) and (12.28b); i.e.,

$$\arg[X(e^{j\omega})] = -\frac{1}{2\pi}\mathcal{P}\int_{-\pi}^{\pi}\log|X(e^{j\theta})|\cot\left(\frac{\omega-\theta}{2}\right)d\theta \tag{12.54}$$

and

$$\log|X(e^{j\omega})| = \hat{x}[0] + \frac{1}{2\pi}\mathcal{P}\int_{-\pi}^{\pi}\arg[X(e^{j\theta})]\cot\left(\frac{\omega-\theta}{2}\right)d\theta, \tag{12.55a}$$

where, in Eq. (12.55a), $\hat{x}[0]$ is

$$\hat{x}[0] = \frac{1}{2\pi}\int_{-\pi}^{\pi}\log|X(e^{j\omega})|d\omega. \tag{12.55b}$$

---

[2] Although $\hat{x}[n]$ is referred to as the complex cepstrum it is real valued since $x(e^{j\omega})$ is defined in Eq. (12.53) is conjugate symmetric.

Although it is not at all obvious at this point, in Problem 12.35 and in Chapter 13 we develop the fact that the minimum-phase condition defined in Section 5.6, namely, that $X(z)$ have all its poles and zeros inside the unit circle, guarantees causality of the complex cepstrum. Thus, the minimum-phase condition in Section 5.6 and the condition of causality of the complex cepstrum turn out to be the same constraint developed from different perspectives. Note that when $\hat{x}[n]$ is causal, $\arg[X(e^{j\omega})]$ is completely determined through Eq. (12.54) by $\log|X(e^{j\omega})|$; however, the complete determination of $\log|X(e^{j\omega})|$ by Eq. (12.55a) requires both $\arg[X(e^{j\omega})]$ and the quantity $\hat{x}[0]$. If $\hat{x}[0]$ is not known, then $\log|X(e^{j\omega})|$ is determined only to within an additive constant, or equivalently, $|X(e^{j\omega})|$ is determined only to within a multiplicative (gain) constant.

Minimum phase and causality of the complex cepstrum are not the only constraints that provide a unique relationship between the magnitude and phase of the DTFT. As one example of another type of constraint, it has been shown (Hayes, Lim and Oppenheim, 1980) that if a sequence is of finite length and if its z-transform has no zeros in conjugate reciprocal pairs, then, to within a scale factor, the sequence (and consequently, also the magnitude of the DTFT) is uniquely determined by the phase of the Fourier transform.

12.4 HILBERT TRANSFORM RELATIONS FOR COMPLEX SEQUENCES

Thus far, we have considered Hilbert transform relations for the Fourier transform of causal sequences and the DFT of periodic sequences that are "periodically causal" in the sense that they are zero in the second half of each period. In this section, we consider *complex sequences* for which the real and imaginary components can be related through a discrete convolution similar to the Hilbert transform relations derived in the previous sections. These relations are particularly useful in representing bandpass signals as complex signals in a manner completely analogous to the analytic signals of continuous-time signal theory (Papoulis, 1977).

As mentioned previously, it is possible to base the derivation of the Hilbert transform relations on a notion of causality or one-sidedness. Since we are interested in relating the real and imaginary parts of a complex sequence, one-sidedness will be applied to the DTFT of the sequence. We cannot, of course, require that the DTFT be zero for $\omega < 0$, since it must be periodic. Instead, we consider sequences for which the Fourier transform is zero in the second half of each period; i.e., the z-transform is zero on the bottom half $(-\pi \leq \omega < 0)$ of the unit circle. Thus, with $x[n]$ denoting the sequence and $X(e^{j\omega})$ its Fourier transform, we require that

$$X(e^{j\omega}) = 0, \qquad -\pi \leq \omega < 0. \tag{12.56}$$

(We could just as well assume that $X(e^{j\omega})$ is zero for $0 < \omega \leq \pi$.) The sequence $x[n]$ corresponding to $X(e^{j\omega})$ must be complex, since, if $x[n]$ were real, $X(e^{j\omega})$ would be conjugate symmetric, i.e., $X(e^{j\omega}) = X^*(e^{-j\omega})$. Therefore, we express $x[n]$ as

$$x[n] = x_r[n] + jx_i[n], \tag{12.57}$$

where $x_r[n]$ and $x_i[n]$ are real sequences. In continuous-time signal theory, the comparable signal is an analytic function and thus is called an *analytic signal*. Although analyticity has no formal meaning for sequences, we will nevertheless apply the same terminology to complex sequences whose Fourier transforms are one-sided.

If $X_r(e^{j\omega})$ and $X_i(e^{j\omega})$ denote the Fourier transforms of the real sequences $x_r[n]$ and $x_i[n]$, respectively, then

$$X(e^{j\omega}) = X_r(e^{j\omega}) + jX_i(e^{j\omega}),$$ (12.58a)

and it follows that

$$X_r(e^{j\omega}) = \frac{1}{2}[X(e^{j\omega}) + X^*(e^{-j\omega})],$$ (12.58b)

and

$$jX_i(e^{j\omega}) = \frac{1}{2}[X(e^{j\omega}) - X^*(e^{-j\omega})].$$ (12.58c)

Note that Eq. (12.58c) gives an expression for $jX_i(e^{j\omega})$, which is the Fourier transform of the imaginary signal $jx_i[n]$. Note also that $X_r(e^{j\omega})$ and $X_i(e^{j\omega})$, the Fourier transforms of the real and imaginary parts, respectively, of $x[n]$ are both complex-valued functions. In general, the complex transforms $X_r(e^{j\omega})$ and $jX_i(e^{j\omega})$ play a role similar to that played in the previous sections by the even and odd parts, respectively, of causal sequences. However, $X_r(e^{j\omega})$ is conjugate symmetric, i.e., $X_r(e^{j\omega}) = X_r^*(e^{-j\omega})$. Similarly, $jX_i(e^{j\omega})$ is conjugate antisymmetric, i.e., $jX_i(e^{j\omega}) = -jX_i^*(e^{-j\omega})$.

Figure 12.4 depicts an example of a complex one-sided Fourier transform of a complex sequence $x[n] = x_r[n] + jx_i[n]$, and the corresponding two-sided transforms of the real sequences $x_r[n]$ and $x_i[n]$. This figure shows pictorially the cancellation implied by Eqs. (12.58).

If $X(e^{j\omega})$ is zero for $-\pi \le \omega < 0$, then there is no overlap between the nonzero portions of $X(e^{j\omega})$ and $X^*(e^{-j\omega})$ except at $\omega = 0$. Thus, $X(e^{j\omega})$ can be recovered except at $\omega = 0$ from either $X_r(e^{j\omega})$ or $X_i(e^{j\omega})$. Since $X(e^{j\omega})$ is assumed to be zero at $\omega = \pm\pi$, $X(e^{j\omega})$ is totally recoverable except at $\omega = 0$ from $jX_i(e^{j\omega})$. This is in contrast to the situation in Section 12.2, in which the causal sequence could be recovered from its odd part, except at the endpoints.

In particular,

$$X(e^{j\omega}) = \begin{cases} 2X_r(e^{j\omega}), & 0 < \omega < \pi, \\ 0, & -\pi \le \omega < 0, \end{cases}$$ (12.59)

and

$$X(e^{j\omega}) = \begin{cases} 2jX_i(e^{j\omega}), & 0 < \omega < \pi, \\ 0, & -\pi \le \omega < 0. \end{cases}$$ (12.60)

Alternatively, we can relate $X_r(e^{j\omega})$ and $X_i(e^{j\omega})$ directly by

$$X_i(e^{j\omega}) = \begin{cases} -jX_r(e^{j\omega}), & 0 < \omega < \pi, \\ jX_r(e^{j\omega}), & -\pi \le \omega < 0, \end{cases}$$ (12.61)

or

$$X_i(e^{j\omega}) = H(e^{j\omega})X_r(e^{j\omega}),$$ (12.62a)

where

$$H(e^{j\omega}) = \begin{cases} -j, & 0 < \omega < \pi, \\ j, & -\pi < \omega < 0. \end{cases}$$ (12.62b)

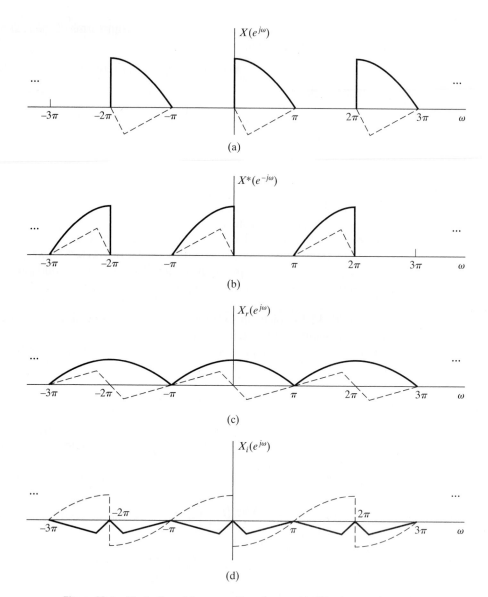

Figure 12.4 Illustration of decomposition of a one-sided Fourier transform. (Solid curves are real parts and dashed curves are imaginary parts.)

Equations (12.62) are illustrated by comparing Figures 12.4(c) and 12.4(d). $X_i(e^{j\omega})$ is the Fourier transform of $x_i[n]$, the imaginary part of $x[n]$, and $X_r(e^{j\omega})$ is the Fourier transform of $x_r[n]$, the real part of $x[n]$. Thus, according to Eqs. (12.62), $x_i[n]$ can be obtained by processing $x_r[n]$ with an LTI discrete-time system with frequency response $H(e^{j\omega})$, as given by Eq. (12.62b). This frequency response has unity magnitude, a phase angle of $-\pi/2$ for $0 < \omega < \pi$, and a phase angle of $+\pi/2$ for $-\pi < \omega < 0$. Such a system is called an ideal 90-degree phase shifter or a *Hilbert transformer*. From Eqs. (12.62), it

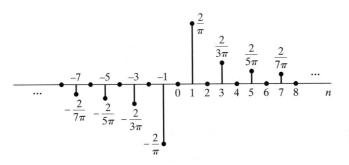

Figure 12.5 Impulse response of an ideal Hilbert transformer or 90-degree phase shifter.

follows that

$$X_r(e^{j\omega}) = \frac{1}{H(e^{j\omega})} X_i(e^{j\omega}) = -H(e^{j\omega}) X_i(e^{j\omega}). \tag{12.63}$$

Thus, $-x_r[n]$ can also be obtained from $x_i[n]$ with a 90-degree phase shifter.

The impulse response $h[n]$ of a 90-degree phase shifter, corresponding to the frequency response $H(e^{j\omega})$ given in Eq. (12.62b), is

$$h[n] = \frac{1}{2\pi} \int_{-\pi}^{0} je^{j\omega n} d\omega - \frac{1}{2\pi} \int_{0}^{\pi} je^{j\omega n} d\omega,$$

or

$$h[n] = \begin{cases} \dfrac{2}{\pi} \dfrac{\sin^2(\pi n/2)}{n}, & n \neq 0, \\ 0, & n = 0. \end{cases} \tag{12.64}$$

The impulse response is plotted in Figure 12.5. Using Eqs. (12.62) and (12.63), we obtain the expressions

$$x_i[n] = \sum_{m=-\infty}^{\infty} h[n-m]x_r[m] \tag{12.65a}$$

and

$$x_r[n] = -\sum_{m=-\infty}^{\infty} h[n-m]x_i[m]. \tag{12.65b}$$

Equations (12.65) are the desired Hilbert transform relations between the real and imaginary parts of a discrete-time analytic signal. Figure 12.6 shows how a discrete-time Hilbert transformer system can be used to form a complex analytic signal, which is simply a pair of real signals.

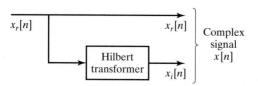

Figure 12.6 Block diagram representation of the creation of a complex sequence whose Fourier transform is one-sided.

12.4.1 Design of Hilbert Transformers

The impulse response of the Hilbert transformer, as given in Eq. (12.64), is not absolutely summable. Consequently,

$$H(e^{j\omega}) = \sum_{n=-\infty}^{\infty} h[n]e^{-j\omega n} \tag{12.66}$$

converges to Eq. (12.62b) only in the mean-square sense. Thus, the ideal Hilbert transformer or 90-degree phase shifter takes its place alongside the ideal lowpass filter and ideal bandlimited differentiator as a valuable theoretical concept that corresponds to a noncausal system and for which the system function exists only in a restricted sense.

Approximations to the ideal Hilbert transformer can, of course, be obtained. FIR approximations with constant group delay can be designed using either the window method or the equiripple approximation method. In such approximations, the 90-degree phase shift is realized exactly, with an additional linear phase component required for a causal FIR system. The properties of these approximations are illustrated by examples of Hilbert transformers designed with Kaiser windows.

Example 12.4   Kaiser Window Design of Hilbert Transformers

The Kaiser window approximation for an FIR discrete Hilbert transformer of order M (length $M + 1$) would be of the form

$$h[n] = \begin{cases} \left(\dfrac{I_o\{\beta(1 - [(n - n_d)/n_d]^2)^{1/2}\}}{I_o(\beta)} \right) \left(\dfrac{2}{\pi} \dfrac{\sin^2[\pi(n - n_d)/2]}{n - n_d} \right), & 0 \le n \le M, \\ 0, & \text{otherwise,} \end{cases} \tag{12.67}$$

where $n_d = M/2$. If M is even, the system is a type III FIR generalized linear-phase system, as discussed in Section 5.7.3.

Figure 12.7(a) shows the impulse response, and Figure 12.7(b) shows the magnitude of the frequency response, for $M = 18$ and $\beta = 2.629$. Because $h[n]$ satisfies the symmetry condition $h[n] = -h[M - n]$ for $0 \le n \le M$, the phase is exactly 90 degrees plus a linear-phase component corresponding to a delay of $n_d = 18/2 = 9$ samples; i.e.,

$$\angle H(e^{j\omega}) = \frac{-\pi}{2} - 9\omega, \qquad 0 < \omega < \pi. \tag{12.68}$$

From Figure 12.7(b), we see that, as required for a type III system, the frequency response is zero at $z = 1$ and $z = -1$ ($\omega = 0$ and $\omega = \pi$). Thus, the magnitude response cannot approximate unity very well, except in some middle band $\omega_L < |\omega| < \omega_H$.

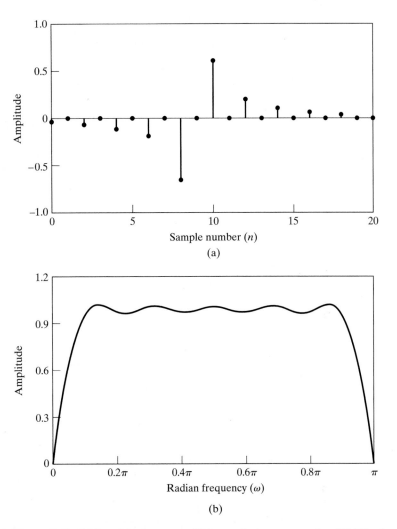

Figure 12.7   (a) Impulse response and (b) magnitude response of an FIR Hilbert transformer designed using the Kaiser window. ($M = 18$ and $\beta = 2.629$.)

If M is an odd integer, we obtain a type IV system, as shown in Figure 12.8, where $M = 17$ and $\beta = 2.44$. For type IV systems, the frequency response is forced to be zero only at $z = 1$ ($\omega = 0$). Therefore, a better approximation to a constant-magnitude response is obtained for frequencies around $\omega = \pi$. The phase response is exactly 90 degrees at all frequencies, plus a linear-phase component corresponding to $n_d = 17/2 = 8.5$ samples delay; i.e.,

$$\angle H(e^{j\omega}) = \frac{-\pi}{2} - 8.5\omega, \qquad 0 < \omega < \pi. \tag{12.69}$$

From a comparison of Figures 12.7(a) and 12.8(a), we see that type III FIR Hilbert transformers have a significant computational advantage over type IV systems when it is not necessary to approximate constant magnitude at $\omega = \pi$. This is because, for type III systems, the even-indexed samples of the impulse response are all exactly zero.

Thus, taking advantage of the antisymmetry in both cases, the system with $M = 17$ would require eight multiplications to compute each output sample, while the system with $M = 18$ would require only five multiplications per output sample.

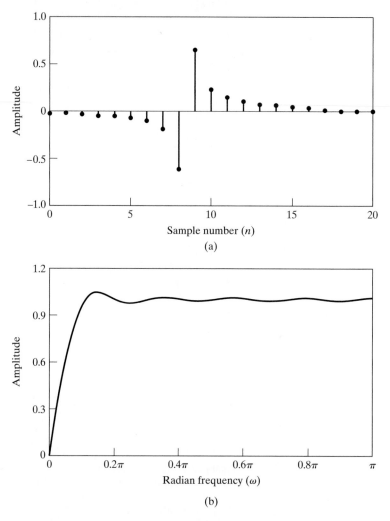

Figure 12.8 (a) Impulse response and (b) magnitude response of an FIR Hilbert transformer designed using the Kaiser window. ($M = 17$ and $\beta = 2.44$.)

Type III and IV FIR linear-phase Hilbert transformer approximations with equirip-ple magnitude approximation and exactly 90-degree phase can be designed using the Parks–McClellan algorithm as described in Sections 7.7 and 7.8, with the expected im-provements in magnitude approximation error over window-designed filters of the same length (see Rabiner and Schafer, 1974).

The exactness of the phase of type III and IV FIR systems is a compelling motiva-tion for their use in approximating Hilbert transformers. IIR systems must have some

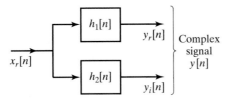

Figure 12.9 Block diagram representation of the allpass phase splitter method for the creation of a complex sequence whose Fourier transform is one-sided.

phase response error as well as magnitude response error in approximating a Hilbert transformer. The most successful approach to designing IIR Hilbert transformers is to design a "phase splitter," which consists of two allpass systems whose phase responses differ by approximately 90 degrees over some portion of the band $0 < |\omega| < \pi$. Such systems can be designed by using the bilinear transformation to transform a continuous-time phase-splitting system to a discrete-time system. (For an example of such a system, see Gold, Oppenheim and Rader, 1970.)

Figure 12.9 depicts a 90-degree phase-splitting system. If $x_r[n]$ denotes a real input signal and $x_i[n]$ its Hilbert transform, then the complex sequence $x[n] = x_r[n] + jx_i[n]$ has a Fourier transform that is identically zero for $-\pi \le \omega < 0$; i.e., $X(z)$ is zero on the bottom half of the unit circle of the z-plane. In the system of Figure 12.6, a Hilbert transformer was used to form the signal $x_i[n]$ from $x_r[n]$. In Figure 12.9, we process $x_r[n]$ through two systems: $H_1(e^{j\omega})$ and $H_2(e^{j\omega})$. Now, if $H_1(e^{j\omega})$ and $H_2(e^{j\omega})$ are allpass systems whose phase responses differ by 90 degrees, then the complex signal $y[n] = y_r[n] + jy_i[n]$ has a Fourier transform that also vanishes for $-\pi \le \omega < 0$. Furthermore, $|Y(e^{j\omega})| = |X(e^{j\omega})|$, since the phase-splitting systems are allpass systems. The phases of $Y(e^{j\omega})$ and $X(e^{j\omega})$ will differ by the phase component common to $H_1(e^{j\omega})$ and $H_2(e^{j\omega})$.

12.4.2 Representation of Bandpass Signals

Many of the applications of analytic signals concern narrowband communication. In such applications, it is sometimes convenient to represent a bandpass signal in terms of a lowpass signal. To see how this may be done, consider the complex lowpass signal

$$x[n] = x_r[n] + jx_i[n],$$

where $x_i[n]$ is the Hilbert transform of $x_r[n]$ and

$$X(e^{j\omega}) = 0, \qquad -\pi \le \omega < 0.$$

The Fourier transforms $X_r(e^{j\omega})$ and $jX_i(e^{j\omega})$ are depicted in Figures 12.10(a) and 12.10(b), respectively, and the resulting transform $X(e^{j\omega}) = X_r(e^{j\omega}) + jX_i(e^{j\omega})$ is shown in Figure 12.10(c). (Solid curves are real parts and dashed curves are imaginary parts.) Now, consider the sequence

$$s[n] = x[n]e^{j\omega_c n} = s_r[n] + js_i[n], \qquad (12.70)$$

where $s_r[n]$ and $s_i[n]$ are real sequences. The corresponding Fourier transform is

$$S(e^{j\omega}) = X(e^{j(\omega - \omega_c)}), \qquad (12.71)$$

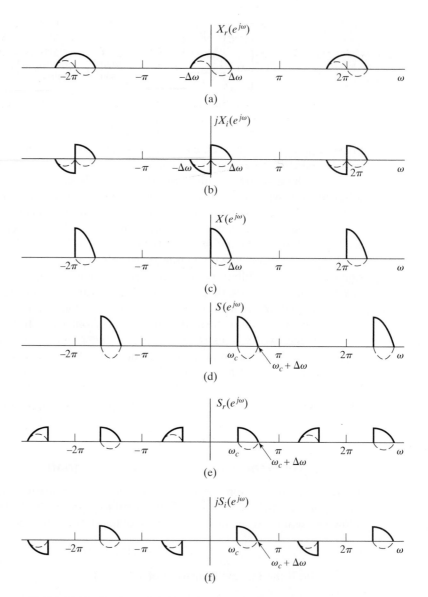

Figure 12.10 Fourier transforms for representation of bandpass signals. (Solid curves are real parts and dashed curves are imaginary parts.) (Note that in parts (b) and (f) the functions $jX_i(e^{j\omega})$ and $jS_i(e^{j\omega})$ are plotted, where $X_i(e^{j\omega})$ and $S_i(e^{j\omega})$ are the Fourier transforms of $x_i[n]$ and $s_i[n]$, respectively.)

which is depicted in Figure 12.10(d). Applying Eqs. (12.58) to $S(e^{j\omega})$ leads to the equations

$$S_r(e^{j\omega}) = \tfrac{1}{2}[S(e^{j\omega}) + S^*(e^{-j\omega})], \tag{12.72a}$$

$$jS_i(e^{j\omega}) = \tfrac{1}{2}[S(e^{j\omega}) - S^*(e^{-j\omega})]. \tag{12.72b}$$

For the example of Figure 12.10, $S_r(e^{j\omega})$ and $jS_i(e^{j\omega})$ are illustrated in Figures 12.10(e) and 12.10(f), respectively. It is straightforward to show that if $X_r(e^{j\omega}) = 0$ for $\Delta\omega < |\omega| \leq \pi$, and if $\omega_c + \Delta\omega < \pi$, then $S(e^{j\omega})$ will be a one-sided bandpass signal such that $S(e^{j\omega}) = 0$ *except* in the interval $\omega_c < \omega \leq \omega_c + \Delta\omega$. As the example of Figure 12.10 illustrates, and as can be shown using Eqs. (12.57) and (12.58), $S_i(e^{j\omega}) = H(e^{j\omega})S_r(e^{j\omega})$, i.e., $s_i[n]$ is the Hilbert transform of $s_r[n]$.

An alternative representation of a complex signal is in terms of magnitude and phase; i.e., $x[n]$ can be expressed as

$$x[n] = A[n]e^{j\phi[n]}, \tag{12.73a}$$

where

$$A[n] = (x_r^2[n] + x_i^2[n])^{1/2} \tag{12.73b}$$

and

$$\phi[n] = \arctan\left(\frac{x_i[n]}{x_r[n]}\right). \tag{12.73c}$$

Therefore, from Eqs. (12.70) and (12.73), we can express $s[n]$ as

$$s[n] = (x_r[n] + jx_i[n])e^{j\omega_c n} \tag{12.74a}$$

$$= A[n]e^{j(\omega_c n + \phi[n])}, \tag{12.74b}$$

from which we obtain the expressions

$$s_r[n] = x_r[n]\cos\omega_c n - x_i[n]\sin\omega_c n, \tag{12.75a}$$

or

$$s_r[n] = A[n]\cos(\omega_c n + \phi[n]), \tag{12.75b}$$

and

$$s_i[n] = x_r[n]\sin\omega_c n + x_i[n]\cos\omega_c n, \tag{12.76a}$$

or

$$s_i[n] = A[n]\sin(\omega_c n + \phi[n]). \tag{12.76b}$$

Equations (12.75a) and (12.76a) are depicted in Figures 12.11(a) and 12.11(b), respectively. These diagrams illustrate how a complex bandpass (single-sideband) signal can be formed from a real lowpass signal.

Taken together, Eqs. (12.75) and (12.76) are the desired time-domain representations of a general complex bandpass signal $s[n]$ in terms of the real and imaginary parts of a complex lowpass signal $x[n]$. Generally, this complex representation is a convenient mechanism for representing a real bandpass signal. For example, Eq. (12.75a) provides a time-domain representation of the real bandpass signal in terms of an "in-phase" component $x_r[n]$ and a "quadrature" (90-degree phase-shifted) component $x_i[n]$. Indeed, as illustrated in Figure 12.10(e), Eq. (12.75a) permits the representation of real bandpass signals (or filter impulse responses) whose Fourier transforms are not conjugate symmetric about the center of the passband (as would be the case for signals of the form $x_r[n]\cos\omega_c n$).

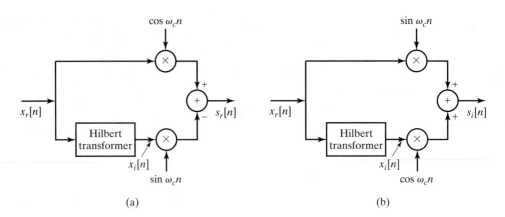

Figure 12.11 Block diagram representation of Eqs. (12.75a) and (12.76a) for obtaining a single-sideband signal.

It is clear from the form of Eqs. (12.75) and (12.76) and from Figure 12.11 that a general bandpass signal has the form of a sinusoid that is both amplitude and phase modulated. The sequence $A[n]$ is called the envelope and $\phi[n]$ the phase. This narrow-band signal representation can be used to represent a variety of amplitude and phase modulation systems. The example of Figure 12.10 is an illustration of single-sideband modulation. If we consider the real signal $s_r[n]$ as resulting from single-sideband modulation with the lowpass real signal $x_r[n]$ as the input, then Figure 12.11(a) represents a scheme for implementing the single-sideband modulation system. Single-sideband modulation systems are useful in frequency-division multiplexing, since they can represent a real bandpass signal with minimum bandwidth.

12.4.3 Bandpass Sampling

Another important use of analytic signals is in the sampling of bandpass signals. In Chapter 4, we saw that, in general, if a continuous-time signal has a bandlimited Fourier transform such that $S_c(j\Omega) = 0$ for $|\Omega| \geq \Omega_N$, then the signal is exactly represented by its samples if the sampling rate satisfies the inequality $2\pi/T \geq 2\Omega_N$. The key to the proof of this result is to avoid overlapping the replicas of $S_c(j\Omega)$ that form the DTFT of the sequence of samples. A bandpass continuous-time signal has a Fourier transform such that $S_c(j\Omega) = 0$ for $0 \leq |\Omega| \leq \Omega_c$ and for $|\Omega| \geq \Omega_c + \Delta\Omega$. Thus, its bandwidth, or region of support, is really only $2\Delta\Omega$ rather than $2(\Omega_c + \Delta\Omega)$, and with a proper sampling strategy, the region $-\Omega_c \leq \Omega \leq \Omega_c$ can be filled with images of the nonzero part of $S_c(j\Omega)$ without overlapping. This is greatly facilitated by using a complex representation of the bandpass signal.

As an illustration, consider the system of Figure 12.12 and the signal shown in Figure 12.13(a). The highest frequency of the input signal is $\Omega_c + \Delta\Omega$. If this signal is sampled at exactly the Nyquist rate, $2\pi/T = 2(\Omega_c + \Delta\Omega)$, then the resulting sequence of samples, $s_r[n] = s_c(nT)$, has the Fourier transform $S_r(e^{j\omega})$ plotted in Figure 12.13(b). Using a discrete-time Hilbert transformer, we can form the complex sequence $s[n] = s_r[n] + js_i[n]$ whose Fourier transform is $S(e^{j\omega})$ in Figure 12.13(c). The

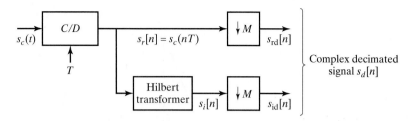

Figure 12.12   System for reduced-rate sampling of a real bandpass signal by decimation of the equivalent complex bandpass signal.

width of the nonzero region of $S(e^{j\omega})$ is $\Delta\omega = (\Delta\Omega)T$. Defining M as the largest integer less than or equal to $2\pi/\Delta\omega$, we see that M copies of $S(e^{j\omega})$ would fit into the interval $-\pi < \omega < \pi$. (In the example of Figure 12.13(c), $2\pi/\Delta\omega = 5$.) Thus, the sampling rate of $s[n]$ can be reduced by decimation as shown in Figure 12.12, yielding the reduced-rate complex sequence $s_d[n] = s_{rd}[n] + js_{id}[n] = s[Mn]$ whose Fourier transform is

$$S_d(e^{j\omega}) = \frac{1}{M} \sum_{k=0}^{M-1} S(e^{j[(\omega - 2\pi k)/M]}). \tag{12.77}$$

Figure 12.13(d) shows $S_d(e^{j\omega})$ with $M = 5$ in Eq. (12.77). $S(e^{j\omega})$ and two of the frequency-scaled and translated copies of $S(e^{j\omega})$ are indicated explicitly in Figure 12.13(d). It is clear that aliasing has been avoided and that all the information necessary to reconstruct the original sampled real bandpass signal now resides in the discrete-time frequency interval $-\pi < \omega \leq \pi$. A complex filter applied to $s_d[n]$ can transform this information in useful ways, such as by further bandlimiting, amplitude or phase compensation, etc., or the complex signal can be coded for transmission or digital storage. This processing takes place at the low sampling rate, and this is, of course, the motivation for reducing the sampling rate.

The original real bandpass signal $s_r[n]$ can be reconstructed ideally by the following procedure:

1. Expand the complex sequence by a factor M; i.e., obtain

$$s_e[n] = \begin{cases} s_{rd}[n/M] + js_{id}[n/M], & n = 0, \pm M, \pm 2M, \ldots, \\ 0, & \text{otherwise.} \end{cases} \tag{12.78}$$

2. Filter the signal $s_e[n]$ using an ideal *complex* bandpass filter with impulse response $h_i[n]$ and frequency response

$$H_i(e^{j\omega}) = \begin{cases} 0, & -\pi < \omega < \omega_c, \\ M, & \omega_c < \omega < \omega_c + \Delta\omega, \\ 0, & \omega_c + \Delta\omega < \omega < \pi. \end{cases} \tag{12.79}$$

(In our example, $\omega_c + \Delta\omega = \pi$.)

3. Obtain $s_r[n] = \mathcal{R}e\{s_e[n] * h_i[n]\}$.

A useful exercise is to plot the Fourier transform $S_e(e^{j\omega})$ for the example of Figure 12.13 and verify that the filter of Eq. (12.79) does indeed recover $s[n]$.

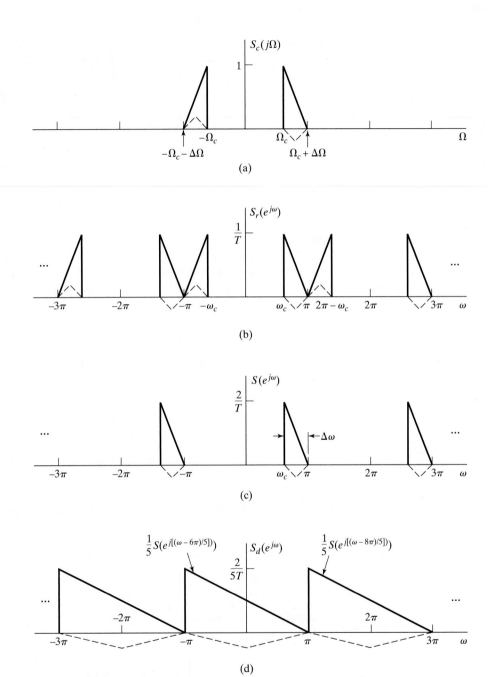

Figure 12.13 Example of reduced-rate sampling of a bandpass signal using the system of Figure 12.12. (a) Fourier transform of continuous-time bandpass signal. (b) Fourier transform of sampled signal. (c) Fourier transform of complex bandpass discrete-time signal derived from the signal of part (a). (d) Fourier transform of decimated complex bandpass of part (c). (Solid curves are real parts and dashed curves are imaginary parts.)

Another useful exercise is to consider a complex continuous-time signal with a one-sided Fourier transform equal to $S_c(j\Omega)$ for $\Omega \geq 0$. It can be shown that such a signal can be sampled with sampling rate $2\pi/T = \Delta\Omega$, directly yielding the complex sequence $s_d[n]$.

12.5 SUMMARY

In this chapter, we have discussed a variety of relations between the real and imaginary parts of Fourier transforms and the real and imaginary parts of complex sequences. These relationships are collectively referred to as *Hilbert transform relationships*. Our approach to deriving all the Hilbert transform relations was to apply a basic causality principle that allows a sequence or function to be recovered from its even part. We showed that, for a causal sequence, the real and imaginary parts of the Fourier transform are related through a convolution-type integral. Also, for the special case when the complex cepstrum of a sequence is causal or, equivalently, both the poles and zeros of its z-transform lie inside the unit circle (the minimum-phase condition), the logarithm of the magnitude and the phase of the Fourier transform are a Hilbert transform pair of each other.

Hilbert transform relations were derived for periodic sequences that satisfy a modified causality constraint and for complex sequences whose Fourier transforms vanish on the bottom half of the unit circle. Applications of complex analytic signals to the representation and efficient sampling of bandpass signals were also discussed.

Problems

Basic Problems

12.1. Consider a sequence $x[n]$ with DTFT $X(e^{j\omega})$. The sequence $x[n]$ is real valued and causal, and

$$\mathcal{R}e\{X(e^{j\omega})\} = 2 - 2a\cos\omega.$$

Determine $\mathcal{I}m\{X(e^{j\omega})\}$.

12.2. Consider a sequence $x[n]$ and its DTFT $X(e^{j\omega})$. The following is known:

$$x[n] \text{ is real and causal,}$$
$$\mathcal{R}e\{X(e^{j\omega})\} = \tfrac{5}{4} - \cos\omega.$$

Determine a sequence $x[n]$ consistent with the given information.

12.3. Consider a sequence $x[n]$ and its DTFT $X(e^{j\omega})$. The following is known:

$$x[n] \text{ is real,}$$
$$x[0] = 0,$$
$$x[1] > 0,$$
$$|X(e^{j\omega})|^2 = \tfrac{5}{4} - \cos\omega.$$

Determine two distinct sequences $x_1[n]$ and $x_2[n]$ consistent with the given information.

12.4. Consider a complex sequence $x[n] = x_r[n] + jx_i[n]$, where $x_r[n]$ and $x_i[n]$ are the real part and imaginary part, respectively. The z-transform $X(z)$ of the sequence $x[n]$ is zero on the bottom half of the unit circle; i.e., $X(e^{j\omega}) = 0$ for $\pi \leq \omega < 2\pi$. The real part of $x[n]$ is

$$x_r[n] = \begin{cases} 1/2, & n = 0, \\ -1/4, & n = \pm 2, \\ 0, & \text{otherwise.} \end{cases}$$

Determine the real and imaginary parts of $X(e^{j\omega})$.

12.5. Find the Hilbert transforms $x_i[n] = \mathcal{H}\{x_r[n]\}$ of the following sequences:

(a) $x_r[n] = \cos \omega_0 n$
(b) $x_r[n] = \sin \omega_0 n$
(c) $x_r[n] = \dfrac{\sin(\omega_0 n)}{\pi n}$

12.6. The imaginary part of $X(e^{j\omega})$ for a causal, real sequence $x[n]$ is

$$X_I(e^{j\omega}) = 2 \sin \omega - 3 \sin 4\omega.$$

Additionally, it is known that $X(e^{j\omega})|_{\omega=0} = 6$. Find $x[n]$.

12.7. **(a)** $x[n]$ is a real, causal sequence with the imaginary part of its DTFT $X(e^{j\omega})$ given by

$$\mathcal{I}m\{X(e^{j\omega})\} = \sin \omega + 2 \sin 2\omega.$$

Determine a choice for $x[n]$.
(b) Is your answer to part (a) unique? If so, explain why. If not, determine a second, distinct choice for $x[n]$ satisfying the relationship given in part (a).

12.8. Consider a real, causal sequence $x[n]$ with DTFT $X(e^{j\omega}) = X_R(e^{j\omega}) + jX_I(e^{j\omega})$. The imaginary part of the DTFT is

$$X_I(e^{j\omega}) = 3 \sin(2\omega).$$

Which of the real parts $X_{Rm}(e^{j\omega})$ listed below are consistent with this information:

$$X_{R1}(e^{j\omega}) = \frac{3}{2} \cos(2\omega),$$

$$X_{R2}(e^{j\omega}) = -3 \cos(2\omega) - 1,$$

$$X_{R3}(e^{j\omega}) = -3 \cos(2\omega),$$

$$X_{R4}(e^{j\omega}) = 2 \cos(3\omega),$$

$$X_{R5}(e^{j\omega}) = \frac{3}{2} \cos(2\omega) + 1.$$

12.9. The following information is known about a real, causal sequence $x[n]$ and its DTFT $X(e^{j\omega})$:

$$\mathcal{I}m\{X(e^{j\omega})\} = 3\sin(\omega) + \sin(3\omega),$$

$$X(e^{j\omega})|_{\omega=\pi} = 3.$$

Determine a sequence $x[n]$ consistent with this information. Is the sequence unique?

12.10. Consider $h[n]$, the real-valued impulse response of a stable, causal LTI system with frequency response $H(e^{j\omega})$. The following is known:

(i) The system has a stable, causal inverse.

(ii) $\left|H(e^{j\omega})\right|^2 = \dfrac{\frac{5}{4} - \cos\omega}{5 + 4\cos\omega}.$

Determine $h[n]$ in as much detail as possible.

12.11. Let $x[n] = x_r[n] + jx_i[n]$ be a complex-valued sequence such that $X(e^{j\omega}) = 0$ for $-\pi \leq \omega < 0$. The imaginary part is

$$x_i[n] = \begin{cases} 4, & n = 3, \\ -4, & n = -3. \end{cases}$$

Specify the real and imaginary parts of $X(e^{j\omega})$.

12.12. $h[n]$ is a causal, real-valued sequence with $h[0]$ nonzero and positive. The magnitude squared of the frequency response of $h[n]$ is given by

$$\left|H(e^{j\omega})\right|^2 = \frac{10}{9} - \frac{2}{3}\cos(\omega).$$

(a) Determine a choice for $h[n]$.

(b) Is your answer to part (b) unique? If so, explain why. If not, determine a second, distinct choice for $h[n]$ satisfying the given conditions.

12.13. Let $x[n]$ denote a causal, complex-valued sequence with Fourier transform

$$X(e^{j\omega}) = X_R(e^{j\omega}) + jX_I(e^{j\omega}).$$

If $X_R(e^{j\omega}) = 1 + \cos(\omega) + \sin(\omega) - \sin(2\omega)$, determine $X_I(e^{j\omega})$.

12.14. Consider a real, anticausal sequence $x[n]$ with DTFT $X(e^{j\omega})$. The real part of $X(e^{j\omega})$ is

$$X_R(e^{j\omega}) = \sum_{k=0}^{\infty}(1/2)^k \cos(k\omega).$$

Find $X_I(e^{j\omega})$, the imaginary part of $X(e^{j\omega})$. (Remember that a sequence is said to be anticausal if $x[n] = 0$ for $n > 0$.)

12.15. $x[n]$ is a real, causal sequence with DTFT $X(e^{j\omega})$. The imaginary part of $X(e^{j\omega})$ is

$$\mathcal{I}m\{X(e^{j\omega})\} = \sin\omega,$$

and it is also known that

$$\sum_{n=-\infty}^{\infty} x[n] = 3.$$

Determine $x[n]$.

12.16. Consider a real, causal sequence $x[n]$ with DTFT $X(e^{j\omega})$, where the following two facts are given about $X(e^{j\omega})$:

$$X_R(e^{j\omega}) = 2 - 4\cos(3\omega),$$

$$X(e^{j\omega})|_{\omega=\pi} = 7.$$

Are these facts consistent? That is, can a sequence $x[n]$ satisfy both? If so, give one choice for $x[n]$. If not, explain why not.

12.17. Consider a real, causal, finite-length signal $x[n]$ with length $N = 2$ and with a 2-point DFT $X[k] = X_R[k] + jX_I[k]$ for $k = 0, 1$. If $X_R[k] = 2\delta[k] - 4\delta[k-1]$, is it possible to determine $x[n]$ uniquely? If so, give $x[n]$. If not, give several choices for $x[n]$ satisfying the stated condition on $X_R[k]$.

12.18. Let $x[n]$ be a real-valued, causal, finite-length sequence with length $N = 3$. Find two choices for $x[n]$ such that the real part of the DFT $X_R[k]$ matches that shown in Figure P12.18. Note that only one of your sequences is "periodically causal" according to the definition in Section 10.2, where $x[n] = 0$ for $N/2 < n \le N - 1$.

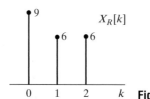

Figure P12.18

12.19. Let $x[n]$ be a real, causal, finite-length sequence with length $N = 4$ that is also periodically causal. The real part of the 4-point DFT $X_R[k]$ for this sequence is shown in Figure P12.19. Determine the imaginary part of the DFT $jX_I[k]$.

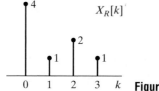

Figure P12.19

12.20. Consider a sequence $x[n]$ that is real, causal, and of finite length with $N = 6$. The imaginary part of the 6-point DFT of this sequence is

$$jX_I[k] = \begin{cases} -j2/\sqrt{3}, & k = 2, \\ j2/\sqrt{3}, & k = 4, \\ 0, & \text{otherwise.} \end{cases}$$

Additionally, it is known that

$$\frac{1}{6}\sum_{k=0}^{5} X[k] = 1.$$

Which of the sequences shown in Figure P12.20 are consistent with the information given?

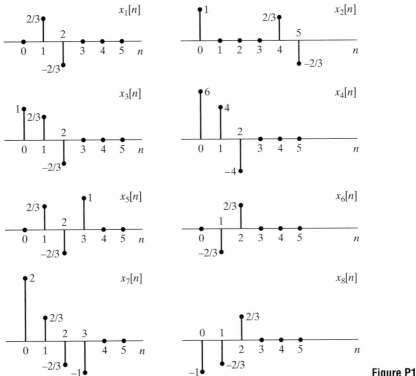

Figure P12.20

12.21. Let $x[n]$ be a real causal sequence for which $|x[n]| < \infty$. The z-transform of $x[n]$ is

$$X(z) = \sum_{n=0}^{\infty} x[n]z^{-n},$$

which is a Taylor's series in the variable z^{-1} and therefore converges to an analytic function everywhere outside some circular disc centered at $z = 0$. (The ROC includes the point $z = \infty$, and, in fact, $X(\infty) = x[0]$.) The statement that $X(z)$ is analytic (in its ROC) implies strong constraints on the function $X(z)$. (See Churchill and Brown, 1990.) Specifically, its real and imaginary parts each satisfy Laplace's equation, and the real and imaginary parts are related by the Cauchy–Riemann equations. We will use these properties to determine $X(z)$ from its real part when $x[n]$ is a real, finite-valued, causal sequence.

Let the z-transform of such a sequence be

$$X(z) = X_R(z) + jX_I(z).$$

where $X_R(z)$ and $X_I(z)$ are real-valued functions of z. Suppose that $X_R(z)$ is

$$X_R(\rho e^{j\omega}) = \frac{\rho + \alpha \cos \omega}{\rho}, \qquad \alpha \text{ real,}$$

for $z = \rho e^{j\omega}$. Then find $X(z)$ (as an explicit function of z), assuming that $X(z)$ is analytic everywhere except at $z = 0$. Do this using both of the following methods.

(a) *Method 1, Frequency Domain.* Use the fact that the real and imaginary parts of $X(z)$ must satisfy the Cauchy–Riemann equations everywhere that $X(z)$ is analytic. The Cauchy–Riemann equations are the following:

1. In Cartesian coordinates,

$$\frac{\partial U}{\partial x} = \frac{\partial V}{\partial y}, \qquad \frac{\partial V}{\partial x} = -\frac{\partial U}{\partial y},$$

where $z = x + jy$ and $X(x + jy) = U(x, y) + jV(x, y)$.

2. In polar coordinates,

$$\frac{\partial U}{\partial \rho} = \frac{1}{\rho}\frac{\partial V}{\partial \omega}, \qquad \frac{\partial V}{\partial \rho} = -\frac{1}{\rho}\frac{\partial U}{\partial \omega},$$

where $z = \rho e^{j\omega}$ and $X(\rho e^{j\omega}) = U(\rho, \omega) + jV(\rho, \omega)$.

Since we know that $U = X_R$, we can integrate these equations to find $V = X_I$ and hence X. (Be careful to treat the constant of integration properly.)

(b) *Method 2, Time Domain.* The sequence $x[n]$ can be represented as $x[n] = x_e[n] + x_o[n]$, where $x_e[n]$ is real and even with Fourier transform $X_R(e^{j\omega})$ and the sequence $x_o[n]$ is real and odd with Fourier transform $jX_I(e^{j\omega})$. Find $x_e[n]$ and, using causality, find $x_o[n]$ and hence $x[n]$ and $X(z)$.

12.22. $x[n]$ is a causal, real-valued sequence with Fourier transform $X(e^{j\omega})$. It is known that

$$\text{Re}\{X(e^{j\omega})\} = 1 + 3\cos\omega + \cos 3\omega.$$

Determine a choice for $x[n]$ consistent with this information, and specify whether or not your choice is unique.

12.23. $x[n]$ is a real-valued, causal sequence with DTFT $X(e^{j\omega})$. Determine a choice for $x[n]$ if the imaginary part of $X(e^{j\omega})$ is given by:

$$\text{Im}\{X(e^{j\omega})\} = 3\sin(2\omega) - 2\sin(3\omega).$$

12.24. Show that the sequence of DFS coefficients for the sequence

$$\tilde{u}_N[n] = \begin{cases} 1, & n = 0, \quad N/2, \\ 2, & n = 1, 2, \ldots, N/2 - 1, \\ 0, & n = N/2 + 1, \ldots, N - 1, \end{cases}$$

is

$$\tilde{U}_N[k] = \begin{cases} N, & k = 0, \\ -j2\cot(\pi k/N), & k \text{ odd}, \\ 0, & k \text{ even}, k \neq 0. \end{cases}$$

Hint: Find the z-transform of the sequence

$$u_N[n] = 2u[n] - 2u[n - N/2] - \delta[n] + \delta[n - N/2],$$

and sample it to obtain $\tilde{U}[k]$.

Advanced Problems

12.25. Consider a real-valued finite-duration sequence $x[n]$ of length M. Specifically, $x[n] = 0$ for $n < 0$ and $n > M - 1$. Let $X[k]$ denote the N-point DFT of $x[n]$ with $N \geq M$ and N odd. The real part of $X[k]$ is denoted $X_R[k]$.

(a) Determine, in terms of M, the smallest value of N that will permit $X[k]$ to be uniquely determined from $X_R[k]$.

(b) With N satisfying the condition determined in part (a), $X[k]$ can be expressed as the circular convolution of $X_R[k]$ with a sequence $U_N[k]$. Determine $U_N[k]$.

12.26. $y_r[n]$ is a real-valued sequence with DTFT $Y_r(e^{j\omega})$. The sequences $y_r[n]$ and $y_i[n]$ in Figure P12.26 are interpreted as the real and imaginary parts of a complex sequence $y[n]$, i.e., $y[n] = y_r[n] + jy_i[n]$. Determine a choice for $H(e^{j\omega})$ in Figure P12.26 so that $Y(e^{j\omega})$ is $Y_r(e^{j\omega})$ for *negative* frequencies and zero for *positive* frequencies between $-\pi$ and π, i.e.,

$$Y(e^{j\omega}) = \begin{cases} Y_r(e^{j\omega}), & -\pi < \omega < 0 \\ 0, & 0 < \omega < \pi \end{cases}$$

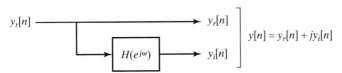

Figure P12.26   System for obtaining $y[n]$ from $y_r[n]$.

12.27. Consider a complex sequence $h[n] = h_r[n] + jh_i[n]$, where $h_r[n]$ and $h_i[n]$ are both real sequences, and let $H(e^{j\omega}) = H_R(e^{j\omega}) + jH_I(e^{j\omega})$ denote the Fourier transform of $h[n]$, where $H_R(e^{j\omega})$ and $H_I(e^{j\omega})$ are the real and imaginary parts, respectively, of $H(e^{j\omega})$.

   Let $H_{ER}(e^{j\omega})$ and $H_{OR}(e^{j\omega})$ denote the even and odd parts, respectively, of $H_R(e^{j\omega})$, and let $H_{EI}(e^{j\omega})$, and $H_{OI}(e^{j\omega})$ denote the even and odd parts, respectively, of $H_I(e^{j\omega})$. Furthermore, let $H_A(e^{j\omega})$ and $H_B(e^{j\omega})$ denote the real and imaginary parts of the Fourier transform of $h_r[n]$, and let $H_C(e^{j\omega})$ and $H_D(e^{j\omega})$ denote the real and imaginary parts of the Fourier transform of $h_i[n]$. Express $H_A(e^{j\omega})$, $H_B(e^{j\omega})$, $H_C(e^{j\omega})$, and $H_D(e^{j\omega})$ in terms of $H_{ER}(e^{j\omega})$, $H_{OR}(e^{j\omega})$, $H_{EI}(e^{j\omega})$, and $H_{OI}(e^{j\omega})$.

12.28. The ideal Hilbert transformer (90-degree phase shifter) has frequency response (over one period)

$$H(e^{j\omega}) = \begin{cases} -j, & \omega > 0, \\ j, & \omega < 0. \end{cases}$$

Figure P12.28-1 shows $H(e^{j\omega})$, and Figure P12.28-2 shows the frequency response of an ideal lowpass filter $H_{lp}(e^{j\omega})$ with cutoff frequency $\omega_c = \pi/2$. These frequency responses are clearly similar, each having discontinuities separated by π.

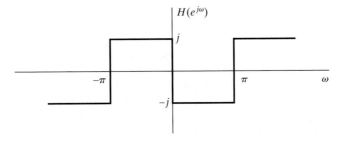

Figure P12.28-1

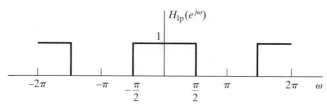

Figure P12.28-2

(a) Obtain a relationship that expresses $H(e^{j\omega})$ in terms of $H_{lp}(e^{j\omega})$. Solve this equation for $H_{lp}(e^{j\omega})$ in terms of $H(e^{j\omega})$.

(b) Use the relationships in part (a) to obtain expressions for $h[n]$ in terms of $h_{lp}[n]$ and for $h_{lp}[n]$ in terms of $h[n]$.

The relationships obtained in parts (a) and (b) were based on definitions of ideal systems with zero phase. However, similar relationships hold for nonideal systems with generalized linear phase.

(c) Use the results of part (b) to obtain a relationship between the impulse response of a causal FIR approximation to the Hilbert transformer and the impulse response of a causal FIR approximation to the lowpass filter, both of which are designed by (1) incorporating an appropriate linear phase, (2) determining the corresponding ideal impulse response, and (3) multiplying by the same window of length $(M + 1)$ samples, i.e., by the window method discussed in Chapter 7. (If necessary, consider the cases of M even and M odd separately.)

(d) For the Hilbert transformer approximations of Example 12.4, sketch the magnitude of the frequency responses of the corresponding lowpass filters.

12.29. In Section 12.4.3, we discussed an efficient scheme for sampling a bandpass continuous-time signal with Fourier transform such that

$$S_c(j\Omega) = 0 \quad \text{for} \quad |\Omega| \le \Omega_c \quad \text{and} \quad |\Omega| \ge \Omega_c + \Delta\Omega.$$

In that discussion, it was assumed that the signal was initially sampled with sampling frequency $2\pi/T = 2(\Omega_c + \Delta\Omega)$. The bandpass sampling scheme is depicted in Figure 12.12. After we form a complex bandpass discrete-time signal $s[n]$ with one-sided Fourier transform $S(e^{j\omega})$, the complex signal is decimated by a factor M, which is assumed to be the largest integer less than or equal to $2\pi/(\Delta\Omega T)$.

(a) By carrying through an example such as the one depicted in Figure 12.13, show that if the quantity $2\pi/(\Delta\Omega T)$ is not an integer for the initial sampling rate chosen, then the resulting decimated signal $s_d[n]$ will have regions of nonzero length where its Fourier transform $S_d(e^{j\omega})$ is identically zero.

(b) How should the initial sampling frequency $2\pi/T$ be chosen so that a decimation factor M can be found such that the decimated sequence $s_d[n]$ in the system of Figure 12.12 will have a Fourier transform $S_d(e^{j\omega})$ that is not aliased yet has no regions where it is zero over an interval of nonzero length?

12.30. Consider an LTI system with frequency response,

$$H(e^{j\omega}) = \begin{cases} 1, & 0 \le \omega \le \pi, \\ 0, & -\pi < \omega < 0. \end{cases}$$

The input $x[n]$ to the system is restricted to be real valued and to have a Fourier transform (i.e., $x[n]$ is absolutely summable). Determine whether or not it is possible to always uniquely recover the system input from the system output. If it is possible, describe how. If it is not possible, explain why not.

Extension Problems

12.31. Derive an integral expression for $H(z)$ *outside* the unit circle in terms of $\mathcal{R}e\{H(e^{j\omega})\}$ when $h[n]$ is a real, stable, and causal sequence, i.e., $h[n] = 0$ for $n > 0$.

12.32. Let $\mathcal{H}\{\cdot\}$ denote the (ideal) operation of Hilbert transformation; that is,

$$\mathcal{H}\{x[n]\} = \sum_{k=-\infty}^{\infty} x[k]h[n-k],$$

where $h[n]$ is

$$h[n] = \begin{cases} \dfrac{2\sin^2(\pi n/2)}{\pi n}, & n \neq 0, \\ 0, & n = 0. \end{cases}$$

Prove the following properties of the ideal Hilbert transform operator.

(a) $\mathcal{H}\{\mathcal{H}\{x[n]\}\} = -x[n]$

(b) $\displaystyle\sum_{n=-\infty}^{\infty} x[n]\mathcal{H}\{x[n]\} = 0$ [*Hint:* Use Parseval's theorem.]

(c) $\mathcal{H}\{x[n]*y[n]\} = \mathcal{H}\{x[n]\}*y[n] = x[n]*\mathcal{H}\{y[n]\}$, where $x[n]$ and $y[n]$ are any sequences.

12.33. An ideal Hilbert transformer with impulse response

$$h[n] = \begin{cases} \dfrac{2\sin^2(\pi n/2)}{\pi n}, & n \neq 0, \\ 0, & n = 0, \end{cases}$$

has input $x_r[n]$ and output $x_i[n] = x_r[n] * h[n]$, where $x_r[n]$ is a discrete-time random signal.

(a) Find an expression for the autocorrelation sequence $\phi_{x_i x_i}[m]$ in terms of $h[n]$ and $\phi_{x_r x_r}[m]$.

(b) Find an expression for the cross-correlation sequence $\phi_{x_r x_i}[m]$. Show that in this case, $\phi_{x_r x_i}[m]$ is an odd function of m.

(c) Find an expression for the autocorrelation function of the complex analytic signal $x[n] = x_r[n] + jx_i[n]$.

(d) Determine the power spectrum $P_{xx}(\omega)$ for the complex signal in part (c).

12.34. In Section 12.4.3, we discussed an efficient scheme for sampling a bandpass continuous-time signal with Fourier transform such that

$$S_c(j\Omega) = 0 \qquad \text{for} \quad |\Omega| \leq \Omega_c \quad \text{and} \quad |\Omega| \geq \Omega_c + \Delta\Omega.$$

The bandpass sampling scheme is depicted in Figure 12.12. At the end of the section, a scheme for reconstructing the original sampled signal $s_r[n]$ was given. The original continuous-time signal $s_c(t)$ in Figure 12.12 can, of course, be reconstructed from $s_r[n]$ by ideal bandlimited interpolation (ideal D/C conversion). Figure P12.34-1 shows a block diagram of the system for reconstructing a real continuous-time bandpass signal from a decimated complex signal. The complex bandpass filter $H_i(e^{j\omega})$ in the figure has a frequency response given by Eq. (12.79).

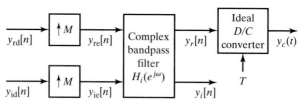

Figure P12.34-1

(a) Using the example depicted in Figure 12.13, show that the system of Figure P12.34-1 will reconstruct the original real bandpass signal (i.e., $y_c(t) = s_c(t)$) if the inputs to the reconstruction system are $y_{rd}[n] = s_{rd}[n]$ and $y_{id}[n] = s_{id}[n]$.

(b) Determine the impulse response $h_i[n] = h_{ri}[n] + jh_{ii}[n]$ of the complex bandpass filter in Figure P12.34-1.

(c) Draw a more detailed block diagram of the system of Figure P12.34-1 in which only real operations are shown. Eliminate any parts of the diagram that are not necessary to compute the final output.

(d) Now consider placing a complex LTI system between the system of Figure 12.12 and the system of Figure P12.34-1. This is depicted in Figure P12.34-2, where the frequency response of the system is denoted $H(e^{j\omega})$. Determine how $H(e^{j\omega})$ should be chosen if it is desired that

$$Y_c(j\Omega) = H_{\text{eff}}(j\Omega)S_c(j\Omega),$$

where

$$H_{\text{eff}}(j\Omega) = \begin{cases} 1, & \Omega_c < |\Omega| < \Omega_c + \Delta\Omega/2, \\ 0, & \text{otherwise.} \end{cases}$$

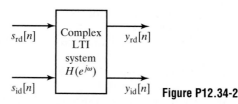

Figure P12.34-2

12.35. In Section 12.3, we defined a sequence $\hat{x}[n]$ referred to as the complex cepstrum of a sequence $x[n]$, and indicated that a causal complex cepstrum $\hat{x}[n]$ is equivalent to the minimum-phase condition of Section 5.4 on $x[n]$. The sequence $\hat{x}[n]$ is the inverse Fourier transform of $\hat{X}(e^{j\omega})$ as defined in Eq. (12.53). Note that because $X(e^{j\omega})$ and $\hat{X}(e^{j\omega})$ are defined, the ROC of both $X(z)$ and $\hat{X}(z)$ must include the unit circle.

(a) Justify the statement that the singularities (poles) of $\hat{X}(z)$ will occur wherever $X(z)$ has either poles or zeros. Use this fact to prove that if $\hat{x}[n]$ is causal, $x[n]$ is minimum phase.

(b) Justify the statement that if $x[n]$ is minimum phase the constraints of the ROC require $\hat{x}[n]$ to be causal.

We can examine this property for the case when $x[n]$ can be written as a superposition of complex exponentials. Specifically, consider a sequence $x[n]$ whose z-transform is

$$X(z) = A \frac{\prod_{k=1}^{M_i}(1 - a_k z^{-1}) \prod_{k=1}^{M_o}(1 - b_k z)}{\prod_{k=1}^{N_i}(1 - c_k z^{-1}) \prod_{k=1}^{N_o}(1 - d_k z)},$$

where $A > 0$ and a_k, b_k, c_k and d_k all have magnitude less than one.

(c) Write an expression for $\hat{X}(z) = \log X(z)$.

(d) Solve for $\hat{x}[n]$ by taking the inverse z-transform of your answer in part (c).

(e) Based on part (d) and the expression for $X(z)$, argue that for sequences $x[n]$ of this form, a causal complex cepstrum is equivalent to having minimum phase.

13

Cepstrum Analysis and Homomorphic Deconvolution

13.0 INTRODUCTION

Throughout this text, we have focused primarily on linear signal processing methods. In this chapter, we introduce a class of nonlinear techniques referred to as *cepstrum analysis* and *homomorphic deconvolution*. These methods have proven to be effective and useful in a variety of applications. In addition, they further illustrate the considerable flexibility and sophistication offered by discrete-time signal processing technologies.

In 1963, Bogert, Healy, and Tukey published a paper with the unusual title "The Quefrency Analysis of Time Series for Echoes: Cepstrum, Pseudoautocovariance, Cross-Cepstrum, and Saphe Cracking."(See Bogert, Healy and Tukey, 1963.) They observed that the logarithm of the power spectrum of a signal containing an echo has an additive periodic component due to the echo, and thus, the power spectrum of the logarithm of the power spectrum should exhibit a peak at the echo delay. They called this function the *cepstrum*, interchanging letters in the word *spectrum* because "in general, we find ourselves operating on the frequency side in ways customary on the time side and vice versa." Bogert et al. went on to define an extensive vocabulary to describe this new signal processing technique; however, only the terms cepstrum and quefrency have been widely used.

At about the same time, Oppenheim (1964, 1967, 1969a) proposed a new class of systems called *homomorphic systems*. Although nonlinear in the classic sense, these systems satisfy a generalization of the principle of superposition; i.e., input signals and their corresponding responses are superimposed (combined) by an operation having the same algebraic properties as addition. The concept of homomorphic systems is very general, but it has been studied most extensively for the combining operations of multiplication and convolution, because many signal models involve these operations. The transformation of a signal into its cepstrum is a homomorphic transformation that maps convolution into addition, and a refined version of the cepstrum is a fundamental part of the theory of homomorphic systems for processing signals that have been combined by convolution.

Since the introduction of the cepstrum, the concepts of the cepstrum and homomorphic systems have proved useful in signal analysis and have been applied with success in processing speech signals (Oppenheim, 1969b, Oppenheim and Schafer, 1968 and Schafer and Rabiner, 1970), seismic signals (Ulrych, 1971 and Tribolet, 1979), biomedical signals (Senmoto and Childers, 1972), old acoustic recordings (Stockham, Cannon and Ingebretsen, 1975), and sonar signals (Reut, Pace and Heator, 1985). The cepstrum has also been proposed as the basis for spectrum analysis (Stoica and Moses, 2005). This chapter provides a detailed treatment of the properties and computational issues associated with the cepstrum and with deconvolution based on homomorphic systems. A number of these concepts are illustrated in Section 13.10 in the context of speech processing.

13.1 DEFINITION OF THE CEPSTRUM

The original motivation for the cepstrum as defined by Bogert et al. is illustrated by the following simple example. Consider a sampled signal $x[n]$ that consists of the sum of a signal $v[n]$ and a shifted and scaled copy (echo) of that signal; i.e.,

$$x[n] = v[n] + \alpha v[n - n_0] = v[n] * (\delta[n] + \alpha\delta[n - n_0]). \tag{13.1}$$

Noting that $x[n]$ can be represented as a *convolution*, it follows that the discrete-time Fourier transform of such a signal has the form of a product

$$X(e^{j\omega}) = V(e^{j\omega})[1 + \alpha e^{-j\omega n_0}]. \tag{13.2}$$

The magnitude of $X(e^{j\omega})$ is

$$|X(e^{j\omega})| = |V(e^{j\omega})|(1 + \alpha^2 + 2\alpha \cos(\omega n_0))^{1/2}, \tag{13.3}$$

a real even function of ω. The basic observation motivating the cepstrum was that the logarithm of the product such as in Eq. (13.3) would be a sum of two corresponding terms, specifically

$$\log|X(e^{j\omega})| = \log|V(e^{j\omega})| + \tfrac{1}{2}\log(1 + \alpha^2 + 2\alpha \cos(\omega n_0)). \tag{13.4}$$

For convenience, we define $C_x(e^{j\omega}) = \log|X(e^{j\omega})|$. Also, in anticipation of a discussion in which we will want to stress the duality between the time- and frequency- domains, we substitute $\omega = 2\pi f$ to obtain

$$C_x(e^{j2\pi f}) = \log|X(e^{j2\pi f})| = \log|V(e^{j2\pi f})| + \tfrac{1}{2}\log(1 + \alpha^2 + 2\alpha \cos(2\pi f n_0)). \tag{13.5}$$

There are two components to this real function of normalized frequency f. The term $\log|V(e^{j2\pi f})|$ is due solely to the signal $v[n]$, and the second term, $\log(1 + \alpha^2 + 2\alpha \cos(2\pi f n_0))$ is due to the combination (echoing) of the signal with itself. We can think of $C_x(e^{j2\pi f})$ as a waveform with continuous independent variable f. The part due to the echo will be periodic in f with period $1/n_0$.[1] We are used to the notion

---

[1]Because $\log(1 + \alpha^2 + 2\alpha \cos(2\pi f n_0))$ is the log-magnitude of a DTFT, it is also periodic in f with period one (2π in ω), as well as $1/n_0$.

that a periodic time waveform has a line spectrum, i.e., its spectrum is concentrated at integer multiples of a common fundamental frequency, which is the reciprocal of the fundamental period. In this case, we have a "waveform" that is a real, even function of f (i.e., frequency). Fourier analysis appropriate for a continuous-variable periodic function such as $C_x(e^{j2\pi f})$ would naturally be the inverse DTFT; i.e.,

$$c_x[n] = \frac{1}{2\pi} \int_{-\pi}^{\pi} C_x(e^{j\omega}) e^{j\omega n} d\omega = \int_{-1/2}^{1/2} C_x(e^{j2\pi f}) e^{j2\pi f n} df. \tag{13.6}$$

In the terminology of Bogert et al., $c_x[n]$ is referred to as the *cepstrum* of $C_x(e^{j2\pi f})$ (or equivalently, of $x[n]$ since $C_x(e^{j2\pi f})$ is derived directly from $x[n]$). Although the cepstrum defined as in Eq. (13.6) is clearly a function of a discrete-time index n, Bogert et al. introduced the term "quefrency" to draw a distinction between the cepstrum time domain and that of the original signal. Because the term $\log(1 + \alpha^2 + 2\alpha \cos(2\pi f n_0))$ in $C_x(e^{j2\pi f})$ is periodic in f with period $1/n_0$, the corresponding component in $c_x[n]$ will be nonzero only at integer multiples of n_0, the fundamental quefrency of the term $\log(1 + \alpha^2 + 2\alpha \cos(2\pi f n_0))$. Later in this chapter, we will show that for this example of a simple echo with $|\alpha| < 1$, the cepstrum has the form

$$c_x[n] = c_v[n] + \sum_{k=1}^{\infty} (-1)^{k+1} \frac{\alpha^k}{2k} (\delta[n + kn_0] + \delta[n - kn_0]), \tag{13.7}$$

where $c_v[n]$ is the inverse DTFT of $\log|V(e^{j\omega})|$, (i.e., the cepstrum of $v[n]$), and the discrete impulses involve only the echo parameters α and n_0. It was this result that led Bogert et al. to observe that the cepstrum of a signal with an echo had a "peak" at the echo delay time n_0 that stands out clearly from $c_v[n]$. Thus the cepstrum could be used as the basis for *detecting* echoes. As mentioned above, the strange-sounding terms "cepstrum" and "quefrency" and other terms were created to call attention to a new way of thinking about Fourier analysis of signals wherein the time and frequency domains were interchanged. In the remainder of this chapter, we will generalize the concept of cepstrum by using the *complex* logarithm, and we will show many interesting properties of the resulting mathematical definition. Furthermore, we will see that the complex cepstrum can also serve as the basis for *separating* signals that are combined by convolution.

13.2 DEFINITION OF THE COMPLEX CEPSTRUM

As the basis for generalizing the concept of the cepstrum, consider a stable sequence $x[n]$ whose z-transform expressed in polar form is

$$X(z) = |X(z)| e^{j\angle X(z)}, \tag{13.8}$$

where $|X(z)|$ and $\angle X(z)$ are the magnitude and angle, respectively, of the complex function $X(z)$. Since $x[n]$ is stable, the ROC for $X(z)$ includes the unit circle, and the DTFT of $x[n]$ exists and is equal to $X(e^{j\omega})$. The *complex cepstrum* associated with $x[n]$

is defined to be the stable sequence $\hat{x}[n]$,[2] whose z-transform is

$$\hat{X}(z) = \log[X(z)]. \tag{13.9}$$

Although any base can be used for the logarithm, the natural logarithm (i.e., base e) is typically used and will be assumed throughout the remainder of the discussion. The logarithm of a complex quantity $X(z)$ expressed as in Eq. (13.8) is defined as

$$\log[X(z)] = \log[|X(z)|e^{j\angle X(z)}] = \log|X(z)| + j\angle X(z). \tag{13.10}$$

Since in the polar representation of a complex number the angle is unique only to within integer multiples of 2π, the imaginary part of Eq. (13.10) is not well defined. We will address that issue shortly; for now we assume that an appropriate definition is possible and has been used.

The complex cepstrum exists if $\log[X(z)]$ has a convergent power series representation of the form

$$\hat{X}(z) = \log[X(z)] = \sum_{n=-\infty}^{\infty} \hat{x}[n]z^{-n}, \qquad |z| = 1, \tag{13.11}$$

i.e., $\hat{X}(z) = \log[X(z)]$ must have all the properties of the z-transform of a stable sequence. Specifically, the ROC for the power series representation of $\log[X(z)]$ must be of the form

$$r_R < |z| < r_L, \tag{13.12}$$

where $0 < r_R < 1 < r_L$. If this is the case, $\hat{x}[n]$, the sequence of coefficients of the power series, is what we call the *complex cepstrum* of $x[n]$.

Since we require $\hat{x}[n]$ to be stable, the ROC of $\hat{X}(z)$ includes the unit circle, and the complex cepstrum can be represented using the inverse DTFT as

$$\begin{aligned} \hat{x}[n] &= \frac{1}{2\pi} \int_{-\pi}^{\pi} \log[X(e^{j\omega})]e^{j\omega n} d\omega \\ &= \frac{1}{2\pi} \int_{-\pi}^{\pi} [\log|X(e^{j\omega})| + j\angle X(e^{j\omega})]e^{j\omega n} d\omega. \end{aligned} \tag{13.13}$$

The term complex cepstrum distinguishes our more general definition from the original definition of the cepstrum by Bogert et al. (1963), which was originally stated in terms of the power spectrum of continuous-time signals. The use of the word *complex* in this context implies that the complex logarithm is used in the definition. It does not imply that the complex cepstrum is necessarily a complex-valued sequence. Indeed, as we will see shortly, the definition we choose for the complex logarithm ensures that the complex cepstrum of a real sequence will also be a real sequence.

The operation of mapping a sequence $x[n]$ into its complex cepstrum $\hat{x}[n]$ is denoted as a discrete-time system operator $D_*[\cdot]$; i.e., $\hat{x} = D_*[x]$. This operation is depicted as the block diagram on the left in Figure 13.1. Similarly, since Eq. (13.9) is invertible with the complex exponential function, we can also define the inverse system $D_*^{-1}[\cdot]$

---

[2]In a somewhat more general definition of the complex cepstrum, $x[n]$ and $\hat{x}[n]$ need not be restricted to be stable. However, with the restriction of stability the important concepts can be illustrated with simpler notation than in the general case.

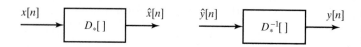

Figure 13.1 System notation for the mapping and inverse mapping between a signal and its complex cepstrum.

which recovers $x[n]$ from $\hat{x}[n]$. The block diagram representation of $D_*^{-1}[\cdot]$ is shown on the right in Figure 13.1. Specifically, $D_*[\cdot]$ and $D_*^{-1}[\cdot]$ in Figure 13.1 are defined so that if $\hat{y}[n] = \hat{x}[n]$ in Figure 13.1, then $y[n] = x[n]$. In the context of homomorphic filtering of convolved signals to be discussed in Section 13.8, $D_*[\cdot]$ is called the *characteristic system* for convolution.

As introduced in Section 13.1, the cepstrum $c_x[n]$ of a signal[3] is defined as the inverse Fourier transform of the logarithm of the magnitude of the Fourier transform; i.e.,

$$c_x[n] = \frac{1}{2\pi} \int_{-\pi}^{\pi} \log |X(e^{j\omega})| e^{j\omega n} d\omega. \tag{13.14}$$

Since the Fourier transform magnitude is real and nonnegative, no special considerations are involved in defining the logarithm in Eq. (13.14). By comparing Eq. (13.14) and Eq. (13.13), we see that $c_x[n]$ is the inverse transform of the real part of $\hat{X}(e^{j\omega})$. Consequently $c_x[n]$ is equal to the conjugate-symmetric part of $\hat{x}[n]$; i.e.,

$$c_x[n] = \frac{\hat{x}[n] + \hat{x}^*[-n]}{2}. \tag{13.15}$$

The cepstrum is useful in many applications, and since it does not depend on the phase of $X(e^{j\omega})$, it is much easier to compute than the complex cepstrum. However, since it is based on only the Fourier transform magnitude, it is not invertible, i.e., $x[n]$ cannot in general be recovered from $c_x[n]$, except in special cases. The complex cepstrum is somewhat more difficult to compute, but it is invertible. Since the complex cepstrum is a more general concept than the cepstrum, and since the properties of the cepstrum can be derived from the properties of the complex cepstrum using Eq. (13.15), we will emphasize the complex cepstrum in this chapter.

The additional difficulties encountered in defining and computing the complex cepstrum are worthwhile for a variety of reasons. First, we see from Eq. (13.10) that the complex logarithm has the effect of creating a new Fourier transform whose real and imaginary parts are $\log |X(e^{j\omega})|$ and $\angle X(e^{j\omega})$, respectively. Thus, we can obtain Hilbert transform relations between these two quantities when the complex cepstrum is causal. We discuss this point further in Section 13.5.2 and see in particular how it relates to minimum-phase sequences. A second more general motivation, developed in Section 13.8, stems from the role that the complex cepstrum plays in defining a class of systems for separating and filtering signals that are combined by convolution.

13.3 PROPERTIES OF THE COMPLEX LOGARITHM

Since the complex logarithm plays a key role in the definition of the complex cepstrum, it is important to understand its definition and properties. Ambiguity in the definition

---

[3]$c_x[n]$ is also referred to as the *real cepstrum* to emphasize that it corresponds to only the real part of the complex logarithm.

of the complex logarithm causes serious computational issues. These will be discussed in detail in Section 13.6. A sequence has a complex cepstrum if the logarithm of its z-transform has a power series expansion, as in Eq. (13.11), where we have specified the ROC to include the unit circle. This means that the Fourier transform

$$\hat{X}(e^{j\omega}) = \log |X(e^{j\omega})| + j\angle X(e^{j\omega}) \tag{13.16}$$

must be a continuous, periodic function of ω, and consequently, both $\log |X(e^{j\omega})|$ and $\angle X(e^{j\omega})$ must be continuous functions of ω. Provided that $X(z)$ does not have zeros on the unit circle, the continuity of $\log |X(e^{j\omega})|$ is guaranteed, since $X(e^{j\omega})$ is assumed to be analytic on the unit circle. However, as previously discussed in Section 5.1.1, $\angle X(e^{j\omega})$ is in general ambiguous, since at each ω, any integer multiple of 2π can be added, and continuity of $\angle X(e^{j\omega})$ is dependent on how the ambiguity is resolved. Since $\mathrm{ARG}[X(e^{j\omega})]$ can be discontinuous, it is generally necessary to specify $\angle X(e^{j\omega})$ explicitly in Eq. (13.16) as the unwrapped (i.e., continuous) phase curve $\arg[X(e^{j\omega})]$.

It is important to note that if $X(z) = X_1(z)X_2(z)$, then

$$\arg[X(e^{j\omega})] = \arg[X_1(e^{j\omega})] + \arg[X_2(e^{j\omega})]. \tag{13.17}$$

A similar additive property will not hold for $\mathrm{ARG}[X(e^{j\omega})]$, i.e., in general,

$$\mathrm{ARG}[X(e^{j\omega})] \neq \mathrm{ARG}[X_1(e^{j\omega})] + \mathrm{ARG}[X_2(e^{j\omega})]. \tag{13.18}$$

Therefore, in order that $\hat{X}(e^{j\omega})$ be analytic (continuous) and have the property that if $X(e^{j\omega}) = X_1(e^{j\omega})X_2(e^{j\omega})$, then

$$\hat{X}(e^{j\omega}) = \hat{X}(e^{j\omega}) + \hat{X}_2(e^{j\omega}), \tag{13.19}$$

we must define $\hat{X}(e^{j\omega})$ as

$$\hat{X}(e^{j\omega}) = \log |X(e^{j\omega})| + j\arg[X(e^{j\omega})]. \tag{13.20}$$

With $x[n]$ real, $\arg[X(e^{j\omega})]$ can always be specified so that it is an odd periodic function of ω. With $\arg[X(e^{j\omega})]$ an odd function of ω and $\log |X(e^{j\omega})|$ an even function of ω, the complex cepstrum $\hat{x}[n]$ is guaranteed to be real.[4]

13.4 ALTERNATIVE EXPRESSIONS FOR THE COMPLEX CEPSTRUM

So far we have defined the complex cepstrum as the sequence of coefficients in the power series representation of $\hat{X}(z) = \log[X(z)]$, and we have also given an integral formula in Eq. (13.13) for determining $\hat{x}[n]$ from $\hat{X}(e^{j\omega}) = \log |X(e^{j\omega})| + \angle X(e^{j\omega})$, where $\angle X(e^{j\omega})$ is the unwrapped phase function $\arg[X(e^{j\omega})]$. The logarithmic derivative can be used to derive other relations for the complex cepstrum that do not explicitly involve the complex logarithm. Assuming that $\log[X(z)]$ is analytic, then

$$\hat{X}'(z) = \frac{X'(z)}{X(z)} \tag{13.21}$$

---

[4]The approach outlined above to the problems presented by the complex logarithm can be developed more formally through the concept of the Riemann surface (Brown and Churchill, 2008).

where $'$ denotes differentiation with respect to z. From property 4 in Table 3.2, $z\hat{X}'(z)$ is the z-transform of $-n\hat{x}[n]$, i.e.,

$$-n\hat{x}[n] \overset{z}{\longleftrightarrow} z\hat{X}'(z) \tag{13.22}$$

Consequently, from Eq. (13.21),

$$-n\hat{x}[n] \overset{z}{\longleftrightarrow} \frac{zX'(z)}{X(z)}. \tag{13.23}$$

Beginning with Eq. (13.21) we can also derive a difference equation that is satisfied by $x[n]$ and $\hat{x}[n]$. Rearranging Eq. (13.21) and multiplying by z, we obtain

$$zX'(z) = z\hat{X}'(z) \cdot X(z). \tag{13.24}$$

Using Eq. (13.22), the inverse z-transform of this equation is

$$-nx[n] = \sum_{k=-\infty}^{\infty} (-k\hat{x}[k])x[n-k]. \tag{13.25}$$

Dividing both sides by $-n$, we obtain

$$x[n] = \sum_{k=-\infty}^{\infty} \left(\frac{k}{n}\right) \hat{x}[k]x[n-k], \qquad n \neq 0. \tag{13.26}$$

The value of $\hat{x}[0]$ can be obtained by noting that

$$\hat{x}[0] = \frac{1}{2\pi} \int_{-\pi}^{\pi} \hat{X}(e^{j\omega}) d\omega. \tag{13.27}$$

Since the imaginary part of $\hat{X}(e^{j\omega})$ is an odd function of ω, Eq. (13.27) becomes

$$\hat{x}[0] = \frac{1}{2\pi} \int_{-\pi}^{\pi} \log |X(e^{j\omega})| d\omega. \tag{13.28}$$

In summary, a signal and its complex cepstrum satisfy a nonlinear difference equation (Eq. (13.26)). Under certain conditions, this implicit relation between $\hat{x}[n]$ and $x[n]$ can be rearranged into a recursion formula that can be used in computation. Formulas of this type are discussed in Section 13.6.4.

13.5 THE COMPLEX CEPSTRUM FOR EXPONENTIAL, MINIMUM-PHASE AND MAXIMUM-PHASE SEQUENCES

13.5.1 Exponential Sequences

If a sequence $x[n]$ consists of a sum of complex exponential sequences, its z-transform $X(z)$ is a rational function of z. Such sequences are both useful and amenable to analysis. In this section, we consider the complex cepstrum for stable sequences $x[n]$ whose z-transforms are of the form

$$X(z) = \frac{Az^r \displaystyle\prod_{k=1}^{M_i}(1 - a_k z^{-1}) \prod_{k=1}^{M_o}(1 - b_k z)}{\displaystyle\prod_{k=1}^{N_i}(1 - c_k z^{-1}) \prod_{k=1}^{N_o}(1 - d_k z)}, \tag{13.29}$$

where $|a_k|$, $|b_k|$, $|c_k|$, and $|d_k|$ are all less than unity, so that factors of the form $(1 - a_k z^{-1})$ and $(1 - c_k z^{-1})$ correspond to the M_i zeros and the N_i poles inside the unit circle, and the factors $(1 - b_k z)$ and $(1 - d_k z)$ correspond to the M_o zeros and the N_o poles outside the unit circle. Such z-transforms are characteristic of sequences composed of a sum of stable exponential sequences. In the special case where there are no poles (i.e., the denominator of Eq. (13.29) is unity), then the corresponding sequence $x[n]$ is a sequence of finite length ($M + 1 = M_o + M_i + 1$).

Through the properties of the complex logarithm, the product of terms in Eq. (13.29) is transformed to the sum of logarithmic terms:

$$\hat{X}(z) = \log(A) + \log(z^r) + \sum_{k=1}^{M_i} \log(1 - a_k z^{-1}) + \sum_{k=1}^{M_o} \log(1 - b_k z)$$

$$- \sum_{k=1}^{N_i} \log(1 - c_k z^{-1}) - \sum_{k=1}^{N_o} \log(1 - d_k z). \tag{13.30}$$

The properties of $\hat{x}[n]$ depend on the composite properties of the inverse transforms of each term.

For real sequences, A is real, and if A is positive, the first term $\log(A)$ contributes only to $\hat{x}[0]$. Specifically, (see Problem 13.15),

$$\hat{x}[0] = \log |A|. \tag{13.31}$$

If A is negative, it is less straightforward to determine the contribution to the complex cepstrum due to the term $\log(A)$. The term z^r corresponds only to a delay or advance of the sequence $x[n]$. If $r = 0$, this term vanishes from Eq. (13.30). However, if $r \neq 0$, then the unwrapped phase function $\arg[X(e^{j\omega})]$ will include a linear term with slope r. Consequently, with $\arg[X(e^{j\omega})]$ defined to be odd and periodic in ω and continuous for $|\omega| < \pi$, this linear-phase term will force a discontinuity in $\arg[X(e^{j\omega})]$ at $\omega = \pm\pi$, and $\hat{X}(z)$ will no longer be analytic on the unit circle. Although the cases of A negative and/or $r \neq 0$ can be formally accommodated, doing so seems to offer no real advantage, because if two transforms of the form of Eq. (13.29) are multiplied together, we would not expect to be able to determine how much of either A or r was contributed by each component. This is analogous to the situation in ordinary linear filtering where two signals, each with dc levels, have been added. Therefore, this question can be avoided in practice by first determining the algebraic sign of A and the value of r and then altering the input, so that its z-transform is of the form

$$X(z) = \frac{|A| \displaystyle\prod_{k=1}^{M_i}(1 - a_k z^{-1}) \prod_{k=1}^{M_o}(1 - b_k z)}{\displaystyle\prod_{k=1}^{N_i}(1 - c_k z^{-1}) \prod_{k=1}^{N_o}(1 - d_k z)}. \tag{13.32}$$

Correspondingly, Eq. (13.30) becomes

$$\hat{X}(z) = \log |A| + \sum_{k=1}^{M_i} \log(1 - a_k z^{-1}) + \sum_{k=1}^{M_o} \log(1 - b_k z)$$

$$- \sum_{k=1}^{N_i} \log(1 - c_k z^{-1}) - \sum_{k=1}^{N_o} \log(1 - d_k z). \tag{13.33}$$

With the exception of the term $\log |A|$, which we have already considered, all the terms in Eq. (13.33) are of the form $\log(1 - \alpha z^{-1})$ and $\log(1 - \beta z)$. Bearing in mind that these factors represent z-transforms with regions of convergence that include the unit circle, we can make the power series expansions

$$\log(1 - \alpha z^{-1}) = -\sum_{n=1}^{\infty} \frac{\alpha^n}{n} z^{-n}, \qquad |z| > |\alpha|, \tag{13.34}$$

$$\log(1 - \beta z) = -\sum_{n=1}^{\infty} \frac{\beta^n}{n} z^n, \qquad |z| < |\beta^{-1}|. \tag{13.35}$$

Using these expressions, we see that for signals with rational z-transforms as in Eq. (13.32), $\hat{x}[n]$ has the general form

$$\hat{x}[n] = \begin{cases} \log |A|, & n = 0, \tag{13.36a} \\[2mm] -\sum_{k=1}^{M_i} \frac{a_k^n}{n} + \sum_{k=1}^{N_i} \frac{c_k^n}{n}, & n > 0, \tag{13.36b} \\[2mm] \sum_{k=1}^{M_o} \frac{b_k^{-n}}{n} - \sum_{k=1}^{N_o} \frac{d_k^{-n}}{n}, & n < 0. \tag{13.36c} \end{cases}$$

Note that for the special case of a finite-length sequence, the second term would be missing in each of Eqs. (13.36b) and (13.36c). Equations (13.36a) to (13.36c) suggest the following general properties of the complex cepstrum:

Property 1: The complex cepstrum decays at least as fast as $1/|n|$: Specifically,

$$|\hat{x}[n]| < C \frac{\alpha^{|n|}}{|n|}, \qquad -\infty < n < \infty,$$

where C is a constant and α equals the maximum of $|a_k|, |b_k|, |c_k|,$ and $|d_k|$.[5]

Property 2: $\hat{x}[n]$ will have infinite duration, even if $x[n]$ has finite duration.

Property 3: If $x[n]$ is real, $\hat{x}[n]$ is also real.

---

[5] In practice, we generally deal with finite-length signals, which are represented by polynomials in z^{-1}; i.e., the numerator in Eq. (13.32). In many cases, the sequence may be hundreds or thousands of samples long. For such sequences, as the sequence length increases, it is increasingly likely that almost all of the zeros of the polynomial will cluster around the unit circle (Hughes and Nikeghbali, 2005). This implies that for long finite-length sequences, the decay of the complex cepstrum is due primarily to the factor $1/n$.

Properties 1 and 2 follow directly from Eqs. (13.36a) to (13.36c). We have suggested property 3 earlier on the basis that for $x[n]$ real, $\log|X(e^{j\omega})|$ is even and $\arg[X(e^{j\omega})]$ is odd, so that the inverse transform of

$$\hat{X}(e^{j\omega}) = \log|X(e^{j\omega})| + j\arg[X(e^{j\omega})]$$

is real. To see property 3 in the context of this section, we note that if $x[n]$ is real, then the poles and zeros of $X(z)$ are in complex conjugate pairs. Therefore, for every complex term of the form α^n/n in Eqs. (13.36a) to (13.36c) there will be a complex conjugate term $(\alpha^*)^n/n$, so that their sum will be real.

13.5.2 Minimum-Phase and Maximum-Phase Sequences

As discussed in Chapters 5 and 12, a minimum-phase sequence is a real, causal, and stable sequence with all the poles and zeros of the z-transform inside the unit circle. Note that $\log[X(z)]$ has singularities at both the poles and the zeros of $X(z)$. Since we require that the ROC of $\log[X(z)]$ include the unit circle so that $\hat{x}[n]$ is stable, and since causal sequences have an ROC of the form $r_R < |z|$, it follows that there can be no singularities of $\log[X(z)]$ on or outside the unit circle if $\hat{x}[n] = 0$ for $n < 0$. Conversely, if all the singularities of $\hat{X}(z) = \log[X(z)]$ are inside the unit circle, then it follows that $\hat{x}[n] = 0$ for $n < 0$. Since the singularities of $\hat{X}(z)$ are the poles and the zeros of $X(z)$, the complex cepstrum of $x[n]$ will be causal ($\hat{x}[n] = 0$ for $n < 0$) if and only if the poles and zeros of $X(z)$ are inside the unit circle. In other words, $x[n]$ is a minimum-phase sequence if and only if its complex cepstrum is causal.

This is easily seen for the case of exponential or finite-length sequences by considering Eqs. (13.36a)–(13.36c). Clearly, all terms in Eq. (13.36c) will be zero if all the coefficients b_k and d_k are zero, i.e., if there are no poles or zeros outside or on the unit circle. Thus, another property of the complex cepstrum is

Property 4: The complex cepstrum $\hat{x}[n] = 0$ for $n < 0$ if and only if $x[n]$ is minimum phase, i.e., $X(z)$ has all its poles and zeros inside the unit circle.

Therefore, causality of the complex cepstrum is equivalent to the minimum phase lag, minimum group delay, and minimum energy delay properties that also characterize minimum-phase sequences.

Example 13.1   Complex Cepstrum of a Minimum-Phase Echo System

The concept of the cepstrum arose initially from a consideration of echoes. As we showed in Section 13.1, a signal with an echo is represented by a convolution $x[n] = v[n] * p[n]$, where

$$p[n] = \delta[n] + \alpha\delta[n - n_0] \xleftrightarrow{\mathcal{Z}} P(z) = 1 + \alpha z^{-n_0}. \tag{13.37}$$

The zeros of $P(z)$ are at locations $z_k = \alpha^{1/n_0}e^{j2\pi(k+1/2)/n_0}$, and if $|\alpha| < 1$, all the zeros will lie inside the unit circle, in which case $p[n]$ is a minimum-phase system. To find the complex cepstrum $\hat{p}[n]$, we can use the power series expansion of $\log[P(z)]$ as in Section 13.5.1 to obtain

$$\hat{P}(z) = \log[1 + \alpha z^{-n_0}] = -\sum_{n=1}^{\infty} \frac{(-\alpha)^n}{n} z^{-nn_0}, \tag{13.38}$$

from which it follows that

$$\hat{p}[n] = \sum_{m=1}^{\infty} (-1)^{m+1} \frac{\alpha^m}{m} \delta[n - mn_0]. \tag{13.39}$$

From Eq. (13.39), we see that $\hat{v}[n] = 0$ for $n < 0$ for $|\alpha| < 1$ as it should be for a minimum-phase system. Furthermore, we see that the nonzero values of the complex cepstrum for the minimum-phase echo system occur at positive integer multiples of n_0.

Maximum-phase sequences are stable sequences whose poles and zeros are all *outside* the unit circle. Thus, maximum-phase sequences are left-sided, and, by analogous arguments, it follows that the complex cepstrum of a maximum-phase sequence is also left-sided. Thus, another property of the complex cepstrum is:

Property 5: The complex cepstrum $\hat{x}[n] = 0$ for $n > 0$ if and only if $x[n]$ is maximum phase; i.e., $X(z)$ has all its poles and zeros outside the unit circle.

This property of the complex cepstrum is easily verified for exponential or finite-length sequences by noting that if all the c_ks and a_ks are zero (i.e., no poles or zeros *inside* the unit circle), then Eq. (13.36b) shows that $\hat{x}[n] = 0$ for $n > 0$.

In Example 13.1, we determined the complex cepstrum of the impulse response of the echo system when $|\alpha| < 1$; i.e., when the echo is smaller than the direct signal. If $|\alpha| > 1$, the echo is larger than the direct signal, and the zeros of the system function $P(z) = 1 + \alpha z^{-n_0}$ lie outside the unit circle. In this case, the echo system is a maximum-phase system.[6] The corresponding complex cepstrum is

$$\hat{p}[n] = \log|\alpha|\delta[n] + \sum_{m=1}^{\infty} (-1)^{m+1} \frac{\alpha^{-m}}{m} \delta[n + mn_0]. \tag{13.40}$$

From Eq. (13.40) we see that $\hat{p}[n] = 0$ for $n > 0$ for $|\alpha| > 1$ as it should be for a maximum-phase system. In this case, we see that the nonzero values of the complex cepstrum for the maximum-phase echo system occur at negative integer multiples of n_0.

13.5.3 Relationship Between the Real Cepstrum and the Complex Cepstrum

As discussed in Sections 13.1 and 13.2, the Fourier transform of the real cepstrum $c_x[n]$ is the real part of the Fourier transform of the complex cepstrum $\hat{x}[n]$, and equivalently, $c_x[n]$ corresponds to the even part of $\hat{x}[n]$. i.e.,

$$c_x[n] = \frac{\hat{x}[n] + \hat{x}[-n]}{2}. \tag{13.41}$$

---

[6] $P(z) = z^{-n_0}(\alpha + z^{n_0})$ has n_0 poles at $z = 0$, which are ignored in computing $\hat{p}[n]$

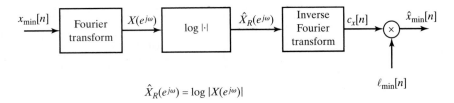

$$\hat{X}_R(e^{j\omega}) = \log |X(e^{j\omega})|$$

Figure 13.2   Determination of the complex cepstrum for minimum-phase signals.

If $\hat{x}[n]$ is causal, as it is if $x[n]$ is minimum phase, then Eq. (13.41) is reversible, i.e., $\hat{x}[n]$ can be recovered from $c_x[n]$ by applying an appropriate window to $c_x[n]$. Specifically,

$$\hat{x}[n] = c_x[n]\ell_{min}[n], \tag{13.42a}$$

where

$$\ell_{min}[n] = 2u[n] - \delta[n] = \begin{cases} 2 & n > 0 \\ 1 & n = 0 \\ 0 & n < 0 \end{cases}. \tag{13.42b}$$

Equations (13.42a) and (13.42b) indicate how the complex cepstrum can be obtained from the cepstrum and consequently also from the log magnitude alone if $x[n]$ is known to be minimum phase. This is also illustrated in block diagram form in Figure 13.2.

In the following example, we illustrate Eqs. (13.41) and (13.42a) for the minimum-phase echo system of Example 13.1.

Example 13.2   Real Cepstrum of a Minimum-Phase Echo System

Consider the complex cepstrum of the minimum-phase echo system as given in Eq. (13.39) in Example 13.1. From Eq. (13.41) it follows that the real cepstrum for the minimum-phase echo system is

$$c_p[n] = \frac{1}{2}\left(\sum_{m=1}^{\infty} (-1)^{m+1} \frac{\alpha^m}{m} \delta[n - mn_0] \right.$$

$$\left. + \sum_{m=1}^{\infty} (-1)^{m+1} \frac{\alpha^m}{m} \delta[-n - mn_0] \right). \tag{13.43}$$

Since $\delta[-n] = \delta[n]$, Eq. (13.43) can be written in the more compact form

$$c_p[n] = \sum_{m=1}^{\infty} (-1)^{m+1} \frac{\alpha^m}{2m} \left(\delta[n - mn_0] + \delta[n + mn_0] \right). \tag{13.44}$$

Also note that if $c_p[n]$ is given by Eq. (13.44) and $\ell_{min}[n]$ is given by Eq. (13.42b), then $\ell_{min}[n]c_p[n]$ is equal to $\hat{p}[n]$ in Eq. (13.39).

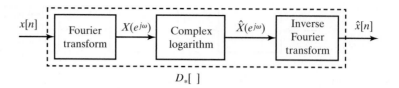

Figure 13.3   Cascade of three systems implementing the computation of the complex cepstrum operation $D_*[\ ]$.

13.6 COMPUTATION OF THE COMPLEX CEPSTRUM

The practical use of the complex cepstrum requires accurate and efficient computational methods to obtain it from a sampled signal. Implicit in all of the previous discussions has been the assumption of uniqueness and continuity of the complex logarithm of the Fourier transform of the input signal. If the mathematical representations obtained above are to serve as the basis for computation of the complex cepstrum, or equivalently, as the basis for realizations of the system $D_*[\cdot]$, then we must deal with the issues associated with computing the Fourier transform and the complex logarithm.

The system $D_*[\cdot]$ is represented in terms of the Fourier transform by the equations

$$X(e^{j\omega}) = \sum_{n=-\infty}^{\infty} x[n]e^{-j\omega n}, \tag{13.45a}$$

$$\hat{X}(e^{j\omega}) = \log[X(e^{j\omega})], \tag{13.45b}$$

$$\hat{x}[n] = \frac{1}{2\pi} \int_{-\pi}^{\pi} \hat{X}(e^{j\omega})e^{j\omega n} d\omega. \tag{13.45c}$$

These equations correspond to the cascade of three systems as depicted in Figure 13.3.

In computing the complex cepstrum numerically, we are limited to finite-length input sequences, and we can compute the Fourier transform at only a finite number of frequencies. That is, instead of using the DTFT, we must use the DFT. Thus, instead of Eqs. (13.45a) to (13.45c), we have the computational realization

$$X[k] = X(e^{j\omega})\Big|_{\omega=(2\pi/N)k} = \sum_{n=0}^{N-1} x[n]e^{-j(2\pi/N)kn}, \tag{13.46a}$$

$$\hat{X}[k] = \log[X(e^{j\omega})]\Big|_{\omega=(2\pi/N)k}, \tag{13.46b}$$

$$\hat{x}_p[n] = \frac{1}{N} \sum_{k=0}^{N-1} \hat{X}[k]e^{j(2\pi/N)kn}. \tag{13.46c}$$

These operations are depicted in Figure 13.4(a), and the corresponding operations for realizing the inverse system are depicted in Figure 13.4(b).

Since in Eq. (13.46b) $\hat{X}[k]$ is a sampled version of $\hat{X}(e^{j\omega})$, it follows from the discussion in Section 8.4 that $\hat{x}_p[n]$ will be a time-aliased version of $\hat{x}[n]$, i.e., that $\hat{x}_p[n]$

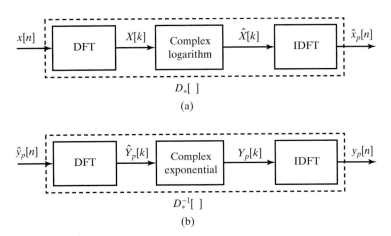

Figure 13.4   Approximate realization using the DFT of (a) $D_*[\,\cdot\,]$ and (b) $D_*^{-1}[\,\cdot\,]$.

is related to the desired $\hat{x}[n]$ by

$$\hat{x}_p[n] = \sum_{r=-\infty}^{\infty} \hat{x}[n+rN]. \tag{13.47}$$

However, we noted in Property 1 in Section 13.5 that $\hat{x}[n]$ decays faster than an exponential sequence, so it is to be expected that the approximation would become increasingly better as N increases. By appending zeros to an input sequence, it is generally possible to increase the sampling rate of the complex logarithm of the Fourier transform so that severe time aliasing does not occur in the computation of the complex cepstrum.

13.6.1 Phase Unwrapping

Samples of $\hat{X}(e^{j\omega})$ as given by Eq. (13.46b) require samples of $\log|X(e^{j\omega})|$ and $\arg[X(e^{j\omega})]$. Samples of $\log|X(e^{j\omega})|$ at a suitable sampling rate can be computed by computing the DFT of $x[n]$ with zero padding. Samples $\mathrm{ARG}[X(e^{j\omega})]$, i.e., the phase modulo 2π are likewise straightforward to compute from samples of $X(e^{j\omega})$ by using standard inverse tangent routines available in most high-level computer languages. However, to obtain the complex cepstrum or its aliased version $\hat{x}_p[n]$, we require samples of the unwrapped phase $\arg[X(e^{j\omega})]$. Consequently, effective procedures for *unwrapping* the phase, that is, obtaining samples of the unwrapped phase from samples of the phase modulo 2π, become an important computational aspect of obtaining the complex cepstrum.

To illustrate the issues, consider a finite-length causal input sequence whose Fourier transform is of the form

$$X(e^{j\omega}) = \sum_{n=0}^{M} x[n]e^{-j\omega n}$$

$$= Ae^{-j\omega M_o} \prod_{k=1}^{M_i}(1 - a_k e^{-j\omega}) \prod_{k=1}^{M_o}(1 - b_k e^{j\omega}), \tag{13.48}$$

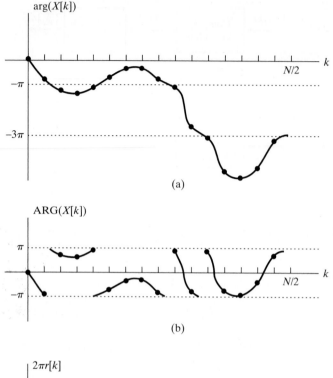

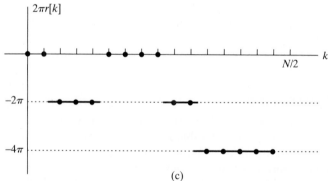

Figure 13.5 (a) Samples of $\arg[X(e^{j\omega})]$. (b) Principal value of part (a). (c) Correction sequence for obtaining arg from ARG.

where $|a_k|$ and $|b_k|$ are less than unity, $M = M_o + M_i$, and A is positive. A continuous-phase curve for a sequence of this form is shown in Figure 13.5(a). The dots indicate samples at frequencies $\omega_k = (2\pi/N)k$. Figure 13.5(b) shows the principal value and its samples as computed from the DFT of the input sequence. One approach to unwrapping the principal-value phase is based on the relation

$$\arg(X[k]) = \text{ARG}(X[k]) + 2\pi r[k], \tag{13.49}$$

where $r[k]$ denotes an integer that determines the appropriate multiple of 2π to add to the principal value at frequency $\omega_k = 2\pi k/N$. Figure 13.5(c) shows $2\pi r[k]$ required to obtain Figure 13.5(a) from 13.5(b). This example suggests the following algorithm for computing $r[k]$ from $\text{ARG}(X[k])$ starting with $r[0] = 0$:

1. If $\text{ARG}(X[k]) - \text{ARG}(X[k-1]) > 2\pi - \varepsilon_1$, then $r[k] = r[k-1] - 1$.
2. If $\text{ARG}(X[k]) - \text{ARG}(X[k-1]) < -(2\pi - \varepsilon_1)$, then $r[k] = r[k-1] + 1$.
3. Otherwise, $r[k] = r[k-1]$.
4. Repeat steps $1 - 3$ for $1 \leq k < N/2$.

After $r[k]$ is determined, Eq. (13.49) can be used to compute $\arg(X[k])$ for $0 \leq k < N/2$. At this stage, $\arg(X[k])$ will contain a large linear-phase component due to the factor $e^{-j\omega M_o}$ in Eq. (13.48). This can be removed by adding $2\pi k M_o/N$ to the unwrapped phase over the interval $0 \leq k < N/2$. The values of $\arg(X[k])$ for $N/2 < k \leq N - 1$ can be obtained by using symmetry. Finally, $\arg(X[N/2]) = 0$.

The above algorithm works well if the samples of $\text{ARG}(X[k])$ are close enough together so that the discontinuities can be detected reliably. The parameter ε_1 is a tolerance recognizing that the magnitude of the difference between adjacent samples of the principal-value phase will always be less than 2π. If ε_1 is too large, a discontinuity will be indicated where there is none. If ε_1 is too small, the algorithm will miss a discontinuity falling between two adjacent samples of a rapidly varying unwrapped phase function $\arg[X(e^{j\omega})]$. Obviously, increasing the sampling rate of the DFT by increasing N improves the chances of correctly detecting discontinuities, and thus, correctly computing $\arg(X[k])$. If $\arg[X(e^{j\omega})]$ varies rapidly, then we expect $\hat{x}[n]$ to decay less rapidly than if $\arg[X(e^{j\omega})]$ varied more slowly. Therefore, aliasing of $\hat{x}[n]$ is more of a problem for rapidly varying phase. Increasing the value of N reduces the aliasing of the complex cepstrum and also improves the chances of being able to correctly unwrap the phase of $X[k]$ by the previously described algorithm.

In some cases, the simple algorithm we just developed may fail because it is impossible or impractical to use a large enough value for N. Often, the aliasing for a given N is acceptable, but principal-value discontinuities cannot be reliably detected. Tribolet (1977, 1979) proposed a modification of the algorithm that uses both the principal value of the phase and the phase derivative to compute the unwrapped phase. As above, Eq. (13.49) gives the set of permissible values at frequency $\omega_k = (2\pi/N)k$, and we seek to determine $r[k]$. It is assumed that we know the phase derivative,

$$\arg'(X[k]) = \frac{d}{d\omega}\arg[X(e^{j\omega})]\Big|_{\omega=2\pi k/N}$$

at all values of k. (A procedure for computing these samples of the phase derivative will be developed in Section 13.6.2.) To compute $\arg(X[k])$ we further assume that $\arg(X[k-1])$ is known. Then, $\widetilde{\arg}(X[k])$, the estimate of $\arg(X[k])$, is defined as

$$\widetilde{\arg}(X[k]) = \arg(X[k-1]) + \frac{\Delta\omega}{2}\{\arg'(X[k]) + \arg'(X[k-1])\}. \tag{13.50}$$

Equation (13.50) is obtained by applying trapezoidal numerical integration to the samples of the phase derivative. This estimate is said to be *consistent* if for some ε_2 an integer $r[k]$ exists such that

$$|\widetilde{\arg}(X[k]) - \text{ARG}(X[k]) - 2\pi r[k]| < \varepsilon_2 < \pi. \tag{13.51}$$

Obviously, the estimate improves with decreasing numerical integration step size $\Delta\omega$. Initially, $\Delta\omega = 2\pi/N$ as provided by the DFT. If Eq. (13.51) cannot be satisfied by an integer $r[k]$, then $\Delta\omega$ is halved, and a new estimate of $\arg(X[k])$ is computed with the new step size. Then, Eq. (13.51) is evaluated with the new estimate. Increasingly accurate estimates of $\arg(X[k])$ are computed by numerical integration until Eq. (13.51) can be satisfied by an integer $r[k]$. That resulting $r[k]$ is used in Eq. (13.49) to finally compute $\arg(X[k])$. This unwrapped phase is then used to compute $\arg(X[k+1])$, and so on.

Another approach to phase unwrapping for a finite-length sequence is based on the fact that the z-transform of a finite-length sequence is a finite-order polynomial, and therefore can be viewed as consisting of a product of 1st-order factors. For each such factor, $\text{ARG}[X(e^{j\omega})]$ and $\arg[X(e^{j\omega})]$ are equal, i.e., the phase for a single factor will never require unwrapping. Furthermore, the unwrapped phase for the product of the individual factors is the sum of the unwrapped phases of the individual factors. Consequently, by treating a finite-length sequence of length N as the coefficients in an Nth-order polynomial, and by first factoring that polynomial into its 1st-order factors, the unwrapped phase can be easily computed. For small values of N, conventional polynomial-rooting algorithms can be applied. For large values, an effective algorithm has been developed by Sitton et al. (2003) and has been successfully demonstrated with polynomials of order in the millions. However, there are cases in which that algorithm also fails, particularly in identifying roots that are not close to the unit circle.

In the discussion above, we have briefly described several algorithms for obtaining the unwrapped phase. Karam and Oppenheim (2007) have also proposed combining these algorithms to exploit their various advantages.

Other issues in computing the complex cepstrum from a sampled input signal $x[n]$ relate to the linear-phase term in $\arg[X(e^{j\omega})]$ and the sign of the overall scale factor A. In our definition of the complex cepstrum, $\arg[X(e^{j\omega})]$ is required to be continuous, odd and periodic in ω. Therefore, the sign of A must be positive, since if negative, a phase discontinuity would occur at $\omega = 0$. Furthermore, $\arg[X(e^{j\omega})]$ cannot contain a linear term, since that would impose a discontinuity at $\omega = \pi$. Consider, for example, a finite-length causal sequence of length $M+1$. The corresponding z-transform will be of the form of Eq. (13.29) with $N_o=N_i=0$, and $M=M_o+M_i$. Also, since $x[n] = 0, n < 0$, it follows that $r = -M_o$. Consequently, the Fourier transform takes the form

$$X(e^{j\omega}) = \sum_{n=0}^{M} x[n]e^{-j\omega n}$$

$$= Ae^{-j\omega M_o} \prod_{k=1}^{M_i}(1 - a_k e^{-j\omega}) \prod_{k=1}^{M_o}(1 - b_k e^{j\omega}),$$

(13.52)

with $|a_k|$ and $|b_k|$ less than unity. The sign of A is easily determined, since it will correspond to the sign of $X(e^{j\omega})$ at $\omega = 0$, which, in turn, is easily computed as the sum of all the terms in the input sequence.

13.6.2 Computation of the Complex Cepstrum Using the Logarithmic Derivative

As an alternative to the explicit computation of the complex logarithm, a mathematical representation based on the logarithmic derivative can be exploited. For real sequences, the derivative of $\hat{X}(e^{j\omega})$ can be represented in the equivalent forms

$$\hat{X}'(e^{j\omega}) = \frac{d\hat{X}(e^{j\omega})}{d\omega} = \frac{d}{d\omega}\log|X(e^{j\omega})| + j\frac{d}{d\omega}\arg[X(e^{j\omega})] \qquad (13.53a)$$

and

$$\hat{X}'(e^{j\omega}) = \frac{X'(e^{j\omega})}{X(e^{j\omega})}, \qquad (13.53b)$$

where $'$ represents differentiation with respect to ω. Since the DTFT of $x[n]$ is

$$X(e^{j\omega}) = \sum_{n=-\infty}^{\infty} x[n]e^{-j\omega n}, \qquad (13.54)$$

its derivative with respect to ω is

$$X'(e^{j\omega}) = \sum_{n=-\infty}^{\infty} (-jnx[n])e^{-j\omega n}; \qquad (13.55)$$

i.e., $X'(e^{j\omega})$ is the DTFT of $-jnx[n]$. Likewise, $\hat{X}'(e^{j\omega})$ is the Fourier transform of $-jn\hat{x}[n]$. Thus, $\hat{x}[n]$ can be determined for $n \neq 0$ from

$$\hat{x}[n] = \frac{-1}{2\pi nj}\int_{-\pi}^{\pi}\frac{X'(e^{j\omega})}{X(e^{j\omega})}e^{j\omega n}d\omega, \qquad n \neq 0. \qquad (13.56)$$

The value of $\hat{x}[0]$ can be determined from the log magnitude as

$$\hat{x}[0] = \frac{1}{2\pi}\int_{-\pi}^{\pi}\log|X(e^{j\omega})|d\omega. \qquad (13.57)$$

Equations (13.54) to (13.57) represent the complex cepstrum in terms of the DTFTs of $x[n]$ and $nx[n]$ and thus do not explicitly involve the unwrapped phase. For finite-length sequences, samples of these transforms can be computed using the DFT, thereby leading to the corresponding equations

$$X[k] = \sum_{n=0}^{N-1} x[n]e^{-j(2\pi/N)kn} = X(e^{j\omega})\bigg|_{\omega=(2\pi/N)k}, \qquad (13.58a)$$

$$X'[k] = -j\sum_{n=0}^{N-1} nx[n]e^{-j(2\pi/N)kn} = X'(e^{j\omega})\bigg|_{\omega=(2\pi/N)k}, \qquad (13.58b)$$

$$\hat{x}_{dp}[n] = -\frac{1}{jnN}\sum_{k=0}^{N-1}\frac{X'[k]}{X[k]}e^{j(2\pi/N)kn}, \qquad 1 \leq n \leq N-1, \qquad (13.58c)$$

$$\hat{x}_{dp}[0] = \frac{1}{N}\sum_{k=0}^{N-1}\log|X[k]|, \qquad (13.58d)$$

where the subscript d refers to the use of the logarithmic derivative and the subscript p is a reminder of the inherent periodicity of the DFT calculations. With the use of

Eqs. (13.58a) to (13.58d), we avoid the problems of computing the complex logarithm at the cost, however, of more severe aliasing, since now

$$\hat{x}_{dp}[n] = \frac{1}{n} \sum_{r=-\infty}^{\infty} (n + rN)\hat{x}[n + rN], \quad n \neq 0. \tag{13.59}$$

Thus, assuming that the sampled continuous phase curve is accurately computed, we would expect that for a given value of N, $\hat{x}_p[n]$ in Eq. (13.46c) would be a better approximation to $\hat{x}[n]$ than would $\hat{x}_{dp}[n]$ in Eq. (13.58c).

13.6.3 Minimum-Phase Realizations for Minimum-Phase Sequences

In the special case of minimum-phase sequences, the mathematical representation is simplified, as indicated in Figure 13.2. A computational realization based on using the DFT in place of the Fourier transform in Figure 13.2 is given by the equations

$$X[k] = \sum_{n=0}^{N-1} x[n]e^{-j(2\pi/N)kn}, \tag{13.60a}$$

$$c_{xp}[n] = \frac{1}{N} \sum_{k=0}^{N-1} \log|X[k]|e^{j(2\pi/N)kn}. \tag{13.60b}$$

In this case, it is the cepstrum that is aliased; i.e.,

$$c_{xp}[n] = \sum_{r=-\infty}^{\infty} c_x[n + rN]. \tag{13.61}$$

To compute the complex cepstrum from $c_{xp}[n]$ based on Figure 13.2, we write:

$$\hat{x}_{cp}[n] = \begin{cases} c_{xp}[n], & n = 0, \quad N/2, \\ 2c_{xp}[n], & 1 \leq n < N/2, \\ 0, & N/2 < n \leq N - 1. \end{cases} \tag{13.62}$$

Clearly, $\hat{x}_{cp}[n] \neq \hat{x}_p[n]$, since it is the even part of $\hat{x}[n]$ that is aliased, rather than $\hat{x}[n]$ itself. Nevertheless, for large N, $\hat{x}_{cp}[n]$ can be expected to be a reasonable approximation to $\hat{x}[n]$ over the finite interval $0 \leq n < N/2$. Similarly, if $x[n]$ is maximum phase, an approximation to the complex cepstrum would be obtained from

$$\hat{x}_{cp}[n] = \begin{cases} c_{xp}[n], & n = 0, \quad N/2, \\ 0, & 1 \leq n < N/2, \\ 2c_{xp}[n], & N/2 < n \leq N - 1. \end{cases} \tag{13.63}$$

13.6.4 Recursive Computation of the Complex Cepstrum for Minimum- and Maximum-Phase Sequences

For minimum-phase sequences, the difference Eq. (13.26) can be rearranged to provide a recursion formula for $\hat{x}[n]$. Since for minimum-phase sequences both $\hat{x}[n] = 0$ and

$x[n] = 0$ for $n < 0$, Eq. (13.26) becomes

$$x[n] = \sum_{k=0}^{n} \left(\frac{k}{n}\right) \hat{x}[k]x[n-k], \qquad n > 0,$$

$$= \hat{x}[n]x[0] + \sum_{k=0}^{n-1} \left(\frac{k}{n}\right) \hat{x}[k]x[n-k],$$

(13.64)

which is a recursion for $D_*[\ ]$ for minimum-phase signals. Solving for $\hat{x}[n]$ yields the recursion formula

$$\hat{x}[n] = \begin{cases} 0, & n < 0, \\ \dfrac{x[n]}{x[0]} - \displaystyle\sum_{k=0}^{n-1} \left(\frac{k}{n}\right) \hat{x}[k]\dfrac{x[n-k]}{x[0]}, & n > 0. \end{cases}$$

(13.65)

Assuming $x[0] > 0$, the value of $\hat{x}[0]$ can be shown to be (see Problem 13.15)

$$\hat{x}[0] = \log(|A|) = \log(|x[0]|).$$

(13.66)

Therefore. Eqs. (13.65) and (13.66) constitute a procedure for computing the complex cepstrum for minimum-phase signals. It also follows from Eq. (13.65) that this computation is causal for minimum-phase inputs; i.e., the output at time n_0 is dependent only on the input for $n \leq n_0$, where n_0 is arbitrary (see Problem 13.20). Similarly, Eqs. (13.64) and (13.66) represent the computation of the minimum-phase sequence from its complex cepstrum.

For maximum-phase signals, $\hat{x}[n] = 0$, and $x[n] = 0$ for $n > 0$. Thus, in this case Eq. (13.26) becomes

$$x[n] = \sum_{k=n}^{0} \left(\frac{k}{n}\right) \hat{x}[k]x[n-k], \qquad n < 0,$$

$$= \hat{x}[n]x[0] + \sum_{k=n+1}^{0} \left(\frac{k}{n}\right) \hat{x}[k]x[n-k].$$

(13.67)

Solving for $\hat{x}[n]$, we have

$$\hat{x}[n] = \begin{cases} \dfrac{x[n]}{x[0]} - \displaystyle\sum_{k=n+1}^{0} \left(\frac{k}{n}\right) \hat{x}[k]\dfrac{x[n-k]}{x[0]}, & n < 0, \\ \log(x[0]), & n = 0, \\ 0, & n > 0. \end{cases}$$

(13.68)

Equation (13.68) serves as a procedure for computing the complex cepstrum for a maximum-phase sequence and Eq. (13.67) is a computational procedure for the inverse characteristic system for convolution.

Thus we see that in the case of minimum-phase or maximum-phase sequences, we also have the recursion formulas of Eqs. (13.64)–(13.68) as possible realizations of the characteristic system and its inverse. These equations can be quite useful when the input sequence is very short or when only a few samples of the complex cepstrum are desired. With these formulas, of course, there is no aliasing error.

13.6.5 The Use of Exponential Weighting

Exponential weighting of a sequence can be used to avoid or mitigate some of the problems encountered in computing the complex cepstrum. Exponential weighting of a sequence $x[n]$ is defined by

$$w[n] = \alpha^n x[n]. \tag{13.69}$$

The corresponding z-transform is

$$W(z) = X(\alpha^{-1} z). \tag{13.70}$$

If the ROC of $X(z)$ is $r_R < |z| < r_L$, then the ROC of $W(z)$ is $|\alpha| r_R < |z| < |\alpha| r_L$, and the poles and zeros of $X(z)$ are shifted radially by the factor $|\alpha|$; i.e., if z_0 is a pole or zero of $X(z)$, then $z_0 \alpha$ is the corresponding pole or zero of $W(z)$.

A convenient property of exponential weighting is that it commutes with convolution. That is, if $x[n] = x_1[n] * x_2[n]$ and $w[n] = a^n x[n]$, then

$$W(z) = X(\alpha^{-1} z) = X_1(\alpha^{-1} z) X_2(\alpha^{-1} z), \tag{13.71}$$

so that

$$w[n] = (a^n x_1[n]) * (a^n x_2[n])$$
$$= w_1[n] * w_2[n]. \tag{13.72}$$

Thus, in computing the complex cepstrum, if $X(z) = X_1(z) X_2(z)$,

$$\hat{W}(z) = \log[W(z)]$$
$$= \log[W_1(z)] + \log[W_2(z)]. \tag{13.73}$$

Exponential weighting can be exploited with cepstrum computation in a variety of ways. For example, poles or zeros of $X(z)$ on the unit circle require special care in computing the complex cepstrum. It can be shown (Carslaw, 1952) that a factor $\log(1 - e^{j\theta} e^{-j\omega})$ has a Fourier series

$$\log(1 - e^{j\theta} e^{-j\omega}) = -\sum_{n=1}^{\infty} \frac{e^{j\theta n}}{n} e^{-j\omega n} \tag{13.74}$$

and thus, the contribution of such a term to the complex cepstrum is $(e^{j\theta n}/n) u[n-1]$. However, the log magnitude is infinite, and the phase is discontinuous with a jump of π radians at $\omega = \theta$. This presents obvious computational difficulties that we would prefer to avoid. By exponential weighting with $0 < \alpha < 1$, all poles and zeros are moved radially inward. Therefore, a pole or zero on the unit circle will move inside the unit circle.

As another example, consider a causal, stable signal $x[n]$ that is nonminimum phase. The exponentially weighted signal, $w[n] = \alpha^n x[n]$, can be converted into a minimum-phase sequence if α is chosen, so that $|z_{max} \alpha| < 1$, where z_{max} is the location of the zero with the greatest magnitude.

13.7 COMPUTATION OF THE COMPLEX CEPSTRUM USING POLYNOMIAL ROOTS

In Section 13.6.1, we discussed the fact that for finite-length sequences, we could exploit the fact that the z-transform is a finite-order polynomial, and that the total unwrapped phase can be obtained by summing the unwrapped phases for each of the factors. If the polynomial is first factored into its 1st-order terms using a polynomial rooting algorithm, then the unwrapped phase for each factor is easily specified analytically. In a similar manner the complex cepstrum for the finite-length sequence can be obtained by first factoring the polynomial, and then summing the complex cepstra for each of the factors.

The basic approach is suggested by Section 13.5.1. If the sequence $x[n]$ has finite length, as is essentially always the case with signals obtained by sampling, then its z-transform is a polynomial in z^{-1} of the form

$$X(z) = \sum_{n=0}^{M} x[n] z^{-n}. \tag{13.75}$$

Such an Mth-order polynomial in z^{-1} can be represented as

$$X(z) = x[0] \prod_{m=1}^{M_i} (1 - a_m z^{-1}) \prod_{m=1}^{M_o} (1 - b_m^{-1} z^{-1}), \tag{13.76}$$

where the quantities a_m are the (complex) zeros that lie inside the unit circle, and the quantities b_m^{-1} are the zeros that are outside the unit circle; i.e., $|a_m| < 1$ and $|b_m| < 1$. We assume that no zeros lie precisely on the unit circle. If we factor a term $-b_m^{-1} z^{-1}$ out of each factor of the product on the right in Eq. (13.76), that equation can be expressed as

$$X(z) = A z^{-M_o} \prod_{m=1}^{M_i} (1 - a_m z^{-1}) \prod_{m=1}^{M_o} (1 - b_m z), \tag{13.77a}$$

where

$$A = x[0](-1)^{M_o} \prod_{m=1}^{M_o} b_m^{-1}. \tag{13.77b}$$

This representation can be computed by using a polynomial rooting algorithm to find the zeros a_m and $1/b_m$ that lie inside and outside the unit circle, respectively, for the polynomial whose coefficients are the sequence $x[n]$.[7]

Given the numeric representation of the z-transform polynomial as in Eqs. (13.77a) and (13.77b), numeric values of the complex cepstrum sequence can be computed from

---

[7]Perhaps not surprisingly, it is rare that a computed root of a polynomial is precisely on the unit circle. In cases where this occurs, such roots can be moved by exponential weighting, as described in Section 13.6.5.

Eqs. (13.36a)–(13.36c) as

$$
\hat{x}[n] = \begin{cases} \log|A|, & n = 0, \\[2mm] -\displaystyle\sum_{m=1}^{M_i} \frac{a_m^n}{n}, & n > 0, \\[4mm] \displaystyle\sum_{m=1}^{M_o} \frac{b_m^{-n}}{n}, & n < 0. \end{cases}
\tag{13.78}
$$

If $A < 0$, this fact can be recorded separately, along with the value of M_o, the number of roots that are outside the unit circle. With this information and $\hat{x}[n]$, we have all that is needed to reconstruct the original signal $x[n]$. Indeed, in Section 13.8.2, it will be shown that, in principle, $x[n]$ can be computed recursively from just $M + 1 = M_o + M_i + 1$ samples of $\hat{x}[n]$.

This method of computation is particularly useful when $M = M_o + M_i$ is small, but it is not limited to small M. Steiglitz and Dickinson (1982) first proposed this method and reported successful rooting of polynomials with degree as high as $M = 256$, which was a practical limit imposed by computational resources readily available at that time. With the polynomial rooting algorithm of Sitton et al. (2003), the complex cepstrum of extremely long finite-length sequences can be accurately computed. Among the advantages of this method are the fact that there is no aliasing and there are none of the uncertainties associated with phase unwrapping.

13.8 DECONVOLUTION USING THE COMPLEX CEPSTRUM

The complex cepstrum operator $D_*[\ ]$, plays a key role in the theory of homomorphic systems, which is based on a generalization of the principle of superposition (Oppenheim, 1964, 1967, 1969a, Schafer, 1969 and Oppenheim, Schafer and Stockham, 1968). In homomorphic filtering of convolved signals, the operator $D_*[\ ]$ is termed the *characteristic system for convolution* since it has the special property of transforming convolution into addition. To see this, suppose

$$
x[n] = x_1[n] * x_2[n]
\tag{13.79}
$$

so that the corresponding z-transform is

$$
X(z) = X_1(z) \cdot X_z(z).
\tag{13.80}
$$

If the complex logarithm is computed as we have prescribed in the definition of the complex cepstrum, then

$$
\begin{aligned}
\hat{X}(z) &= \log[X(z)] = \log[X_1(z)] + \log[X_2(z)] \\
&= \hat{X}_1(z) + \hat{X}_2(z),
\end{aligned}
\tag{13.81}
$$

which implies that the complex cepstrum is

$$
\hat{x}[n] = D_*[x_1[n] * x_2[n]] = \hat{x}_1[n] + \hat{x}_2[n].
\tag{13.82}
$$

A similar analysis shows that if $\hat{y}[n] = y_1[n] + y_2[n]$. then it follows that $D_*^{-1}[\hat{y}_1[n] + \hat{y}_2[n]] = \hat{y}_1[n] * \hat{y}_2[n]$. If the cepstral components $\hat{x}_1[n]$ and $\hat{x}_2[n]$ occupy

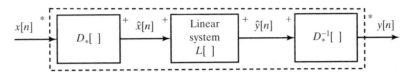

Figure 13.6 Canonic form for homomorphic systems where inputs and corresponding outputs are combined by convolution.

different quefrency ranges, linear filtering can be applied to the complex cepstrum to remove either $x_1[n]$ or $x_2[n]$. If this is followed by transformation through the inverse system $D_*^{-1}[\ ]$, the corresponding component will be removed in the output. This procedure for separating convolved signals (deconvolution) is depicted in Figure 13.6, where the system $L[\ ]$ is a linear (although not necessarily time invariant) system. The symbols $*$ and $+$ at the inputs and outputs of the component systems in Figure 13.6 denote the operations of superposition that hold at each point in the diagram. Figure 13.6 is a general representation of a class of systems that obey a generalized principle of superposition with convolution as the operation for combining signals. All members of this class of systems differ only in the linear part $L[\ ]$.

In the remainder of this section, we illustrate how cepstral analysis can be used for the special deconvolution problems of decomposing a signal into either a convolution of a minimum-phase and allpass component or minimum-phase and maximum-phase component. In Section 13.9, we illustrate how cepstral analysis can be applied to deconvolution of a signal convolved with an impulse train, representing for example, an idealization of a multipath environment. In Section 13.10, we generalize this example to illustrate how cepstral analysis has been successfully applied to speech processing.

13.8.1 Minimum-Phase/Allpass Homomorphic Deconvolution

Any sequence $x[n]$ for which the complex cepstrum exists can always be expressed as the convolution of a minimum-phase and an allpass sequence as in

$$x[n] = x_{min}[n] * x_{ap}[n]. \tag{13.83}$$

In Eq. (13.83) $x_{min}[n]$ and $x_{ap}[n]$ denote minimum-phase and allpass components respectively.

If $x[n]$ is not minimum phase, then the system of Figure 13.2 with input $x[n]$ and $\ell_{min}[n]$ given by Eq. (13.42b) produces the complex cepstrum of the minimum-phase sequence that has the same Fourier transform magnitude as $x[n]$. If $\ell_{max}[n] = \ell_{min}[-n]$ is used, the output will be the complex cepstrum of the maximum-phase sequence having the same Fourier transform magnitude as $x[n]$.

We can obtain the complex cepstrum $\hat{x}_{min}[n]$ of the sequence $x_{min}[n]$ in Eq. (13.83) through the operations of Figure 13.2. The complex cepstrum $\hat{x}_{ap}[n]$ can be obtained from $\hat{x}[n]$ by subtracting $\hat{x}_{min}[n]$ from $\hat{x}[n]$, i.e.,

$$\hat{x}_{ap}[n] = \hat{x}[n] - \hat{x}_{min}[n].$$

To obtain $x_{min}[n]$ and $x_{ap}[n]$, we apply the transformation D_*^{-1} to $\hat{x}_{min}[n]$ and $\hat{x}_{ap}[n]$.

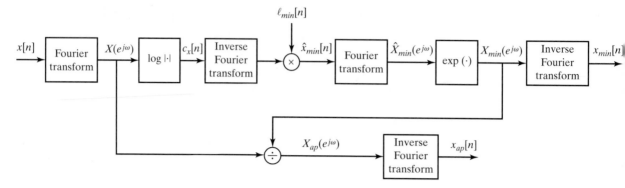

Figure 13.7 Deconvolution of a sequence into minimum-phase and allpass components using the cepstrum.

Although the approach outlined above to obtain $x_{min}[n]$ and $x_{ap}[n]$ is theoretically correct, explicit evaluation of the complex cepstrum $\hat{x}[n]$ is required in its implementation. If we are interested only in obtaining $x_{min}[n]$ and $x_{ap}[n]$, evaluation of the complex cepstrum and the associated need for phase unwrapping can be avoided. The basic strategy is incorporated in the block diagram of Figure 13.7. This system relies on the fact that

$$X_{ap}(e^{j\omega}) = \frac{X(e^{j\omega})}{X_{min}(e^{j\omega})}. \tag{13.84a}$$

The magnitude of $X_{ap}(e^{j\omega})$ is therefore

$$|X_{ap}(e^{j\omega})| = \frac{|X(e^{j\omega})|}{|X_{min}(e^{j\omega})|} = 1 \tag{13.84b}$$

and

$$\angle X_{ap}(e^{j\omega}) = \angle X(e^{j\omega}) - \angle X_{min}(e^{j\omega}). \tag{13.84c}$$

Since $x_{ap}[n]$ is obtained as the inverse Fourier transform of $e^{j\angle X_{ap}(e^{j\omega})}$, (that is, $|X_{ap}(e^{j\omega})| = 1$), each of the phase functions in Eq. (13.84c) need only be known or specified to within integer multiples of 2π. Therefore, even though as a natural consequence of the procedure outlined in Figure 13.7, $\angle X_{min}(e^{j\omega}) = \mathcal{I}m\{\hat{X}_{min}(e^{j\omega})\}$ will be an unwrapped phase function, $\angle X(e^{j\omega})$ in Eq. (13.84c) can be computed modulo 2π.

13.8.2 Minimum-Phase/Maximum-Phase Homomorphic Deconvolution

Another representation of a sequence is as the convolution of a minimum-phase sequence with a maximum-phase sequence as in

$$x[n] = x_{mn}[n] * x_{mx}[n], \tag{13.85}$$

where $x_{mn}[n]$ and $x_{mx}[n]$ denote minimum-phase and maximum-phase components, respectively.[8] In this case, the corresponding complex cepstrum is

$$\hat{x}[n] = \hat{x}_{mn}[n] + \hat{x}_{mx}[n]. \tag{13.86}$$

---

[8]In general the minimum-phase component $x_{mn}[n]$ in Eq. (13.85) will be different from $x_{min}[n]$ in Eq. (13.83).

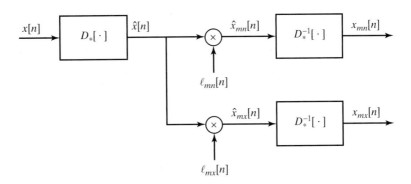

Figure 13.8 The use of homomorphic deconvolution to separate a sequence into minimum-phase and maximum-phase components.

To extract $x_{mn}[n]$ and $x_{mx}[n]$ from $x[n]$, we specify $\hat{x}_{mn}[n]$ as

$$\hat{x}_{mn}[n] = \ell_{mn}[n]\hat{x}[n], \tag{13.87a}$$

where

$$\ell_{mn}[n] = u[n]. \tag{13.87b}$$

Similarly, we specify $\hat{x}_{mx}[n]$ as

$$\hat{x}_{mx}[n] = \ell_{mx}[n]\hat{x}[n] \tag{13.88a}$$

where

$$\ell_{mx}[n] = u[-n-1]. \tag{13.88b}$$

$x_{mn}[n]$ and $x_{mx}[n]$ can be obtained from $\hat{x}_{mn}[n]$ and $\hat{x}_{mx}[n]$, respectively, as the output of the inverse characteristic system $D_*^{-1}[\cdot]$. The operations required for the decomposition of Eq. (13.85) are depicted in Figure 13.8. This method of factoring a sequence into its minimum- and maximum-phase parts has been used by Smith and Barnwell (1986) in the design of filter banks. Note that we have arbitrarily assigned all of $\hat{x}[0]$ to $\hat{x}_{mn}[0]$, and we have set $\hat{x}_{mx}[0] = 0$. Obviously, other combinations are possible, since all that is required is that $\hat{x}_{mn}[0] + \hat{x}_{mx}[0] = \hat{x}[0]$.

The recursion formulas of Section 13.6.4 can be combined with the representation of Eq. (13.85) to yield an interesting result for finite-length sequences. Specifically, in spite of the infinite extent of the complex cepstrum of a finite-length sequence, we can show that for an input sequence of length $M + 1$, we need only $M + 1$ samples of $\hat{x}[n]$ to determine $x[n]$. To see this, consider the z-transform of Eq. (13.85), i.e.,

$$X(z) = X_{mn}(z)X_{mx}(z), \tag{13.89a}$$

where

$$X_{mn}(z) = A \prod_{k=1}^{M_i}(1 - a_k z^{-1}), \tag{13.89b}$$

$$X_{mx}(z) = \prod_{k=1}^{M_o}(1 - b_k z), \tag{13.89c}$$

with $|a_k| < 1$ and $|b_k| < 1$. Note that we have neglected the delay of M_o samples that would be needed for a causal sequence, so that $x_{mn}[n] = 0$ outside the interval $0 \leq n \leq M_i$ and $x_{mx}[n] = 0$ outside the interval $-M_o \leq n \leq 0$. Since the sequence $x[n]$ is the convolution of $x_{mn}[n]$ and $x_{mx}[n]$, it is nonzero in the interval $-M_o \leq n \leq M_i$. Using the previous recursion formulas, we can write

$$
x_{mn}[n] = \begin{cases} 0, & n < 0, \\ e^{\hat{x}[0]}, & n = 0, \\ \hat{x}[n]x_{mn}[0] + \sum_{k=0}^{n-1} \left(\dfrac{k}{n}\right) \hat{x}[k]x_{mn}[n-k], & n > 0, \end{cases} \tag{13.90}
$$

and

$$
x_{mx}[n] = \begin{cases} \hat{x}[n] + \sum_{k=n+1}^{0} \left(\dfrac{k}{n}\right) \hat{x}[k]x_{mx}[n-k], & n < 0, \\ 1, & n = 0, \\ 0, & n > 0. \end{cases} \tag{13.91}
$$

Clearly, we require $M_i + 1$ values of $\hat{x}[n]$ to compute $x_{mn}[n]$ and M_o values of $\hat{x}[n]$ to compute $x_{mx}[n]$. Thus, only $M_i + M_o + 1$ values of the infinite sequence $\hat{x}[n]$ are required to completely recover the minimum-phase and maximum-phase components of the finite-length sequence $x[n]$.

As mentioned in Section 13.7, the result that we have just obtained could be used to implement the inverse characteristic system for convolution when the cepstrum has been computed by polynomial rooting. We simply need to compute $x_{mn}[n]$ and $x_{mx}[n]$ by the recursions of Eqs. (13.90) and (13.91) and then reconstruct the original signal by the convolution $x[n] = x_{mn}[n] * x_{mx}[n]$.

13.9 THE COMPLEX CEPSTRUM FOR A SIMPLE MULTIPATH MODEL

As discussed in Example 13.1, a highly simplified model of multipath or reverberation consists of representing the received signal as the convolution of the transmitted signal with an impulse train. Specifically, with $v[n]$ denoting a transmitted signal and $p[n]$ the impulse response of a multipath channel or other system generating multiple echoes,

$$
x[n] = v[n] * p[n], \tag{13.92a}
$$

or, in the z-transform domain,

$$
X(z) = V(z)P(z). \tag{13.92b}
$$

In our analysis in this section, we choose $p[n]$ to be of the form

$$
p[n] = \delta[n] + \beta\delta[n - N_0] + \beta^2\delta[n - 2N_0], \tag{13.93a}
$$

and its z-transform is then

$$
P(z) = 1 + \beta z^{-N_0} + \beta^2 z^{-2N_0} = \frac{1 - \beta^3 z^{-3N_0}}{1 - \beta z^{-N_0}}. \tag{13.93b}
$$

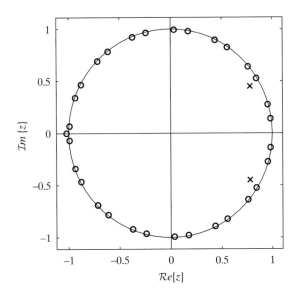

Figure 13.9   Pole–zero plot of the z-transform $X(z) = V(z)P(z)$ for the example signal of Figure 13.10.

For example, $p[n]$ might correspond to the impulse response of a multipath channel or other system that generates multiple echoes at a spacing of N_0 and $2N_0$. The component $v[n]$ will be taken to be the response of a 2nd-order system, such that

$$V(z) = \frac{b_0 + b_1 z^{-1}}{(1 - re^{j\theta}z^{-1})(1 - re^{-j\theta}z^{-1})}, \qquad |z| > |r|. \tag{13.94a}$$

In the time domain, $v[n]$ can be expressed as

$$v[n] = b_0 w[n] + b_1 w[n-1], \tag{13.94b}$$

where

$$w[n] = \frac{r^n}{4\sin^2\theta}\{\cos(\theta n) - \cos[\theta(n+2)]\}u[n], \qquad \theta \neq 0, \pi. \tag{13.94c}$$

Figure 13.9 shows the pole–zero plot of the z-transform $X(z) = V(z)P(z)$ for the specific set of parameters $b_0 = 0.98$, $b_1 = 1$, $\beta = r = 0.9$, $\theta = \pi/6$, and $N_0 = 15$. Figure 13.10 shows the signals $v[n]$, $p[n]$, and $x[n]$ for these parameters. As seen in Figure 13.10, the convolution of the pulse-like signal $v[n]$ with the impulse train $p[n]$ results in a series of superimposed delayed copies (echoes) of $v[n]$.

This signal model is a simplified version of models that are used in the analysis and processing of signals in a variety of contexts, including communications systems, speech processing, sonar, and seismic data analysis. In a communications context, $v[n]$ in Eqs. (13.92a) and (13.92b) might represent a signal transmitted over a multipath channel, $x[n]$ the received signal, and $p[n]$ the channel impulse response. In speech processing, $v[n]$ would represent the combined effects of the glottal pulse shape and the resonance effects of the human vocal tract, while $p[n]$ would represent the periodicity of the vocal excitation during voiced speech such as a vowel sound (Flanagan, 1972; Rabiner and Schafer, 1978; Quatieri, 2002). Equation (13.94a) incorporates only one resonance, while in the general speech model, the denominator would generally include at least ten complex poles. In seismic data analysis, $v[n]$ would represent the waveform

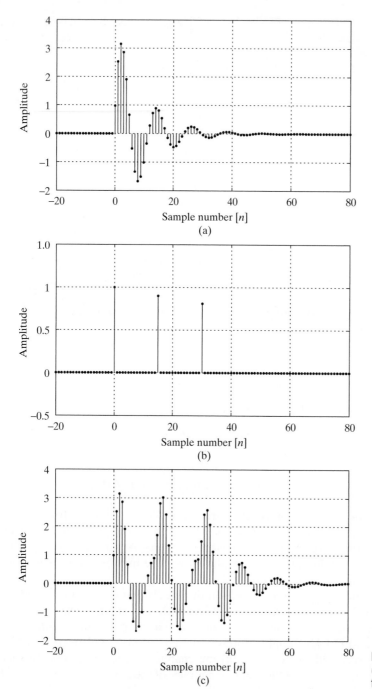

Figure 13.10 The sequences: (a) $v[n]$, (b) $p[n]$, and (c) $x[n]$ corresponding to the pole–zero plot of Figure 13.9.

of an acoustic pulse propagating in the earth due to a dynamite explosion or similar disturbance. The impulsive component $p[n]$ would represent reflections at boundaries between layers having different propagation characteristics. In the practical use of such a model, there would be more impulses in $p[n]$ than we assumed in Eq. (13.93a), and they would be unequally spaced. Also, the component $V(z)$ would generally involve many more zeros, and often no poles are included in the model (Ulrych, 1971; Tribolet, 1979; Robinson and Treitel, 1980).

Although the model discussed above is a highly simplified representation of that encountered in typical applications, it is analytically convenient and useful to obtain exact formulas to compare with computed results obtained for sampled signals. Furthermore, we will see that this simple model illustrates all the important properties of the cepstrum of a signal with a rational z-transform.

In Section 13.9.1, we evaluate analytically the complex cepstrum for the received signal $x[n]$. In Section 13.9.2, we illustrate the computation of the complex cepstrum using the DFT, and in Section 13.9.3 illustrate the technique of homomorphic deconvolution.

13.9.1 Computation of the Complex Cepstrum by z-Transform Analysis

To determine an equation for $\hat{x}[n]$, the complex cepstrum of $x[n]$ for the simple model of Eq. (13.92a), we use the relations

$$\hat{x}[n] = \hat{v}[n] + \hat{p}[n], \tag{13.95a}$$

$$\hat{X}(z) = \hat{V}(z) + \hat{P}(z), \tag{13.95b}$$

$$\hat{X}(z) = \log[X(z)], \tag{13.96a}$$

$$\hat{V}(z) = \log[V(z)], \tag{13.96b}$$

and

$$\hat{P}(z) = \log[P(z)]. \tag{13.96c}$$

To determine $\hat{v}[n]$, we can directly apply the results in Section 13.5. Specifically, to express $V(z)$ in the form of Eq. (13.29), we first note that for the specific signal $X(z)$ in Figure 13.9, the poles of $V(z)$ are inside the unit circle and the zero is outside ($r = 0.9$ and $b_0/b_1 = 0.98$), so that in accordance with Eq. (13.29), we rewrite $V(z)$ as

$$V(z) = \frac{b_1 z^{-1}(1 + (b_0/b_1)z)}{(1 - re^{j\theta}z^{-1})(1 - re^{-j\theta}z^{-1})}, \qquad |z| > |r|. \tag{13.97}$$

As discussed in Section 13.5, the factor z^{-1} contributes a linear component to the unwrapped phase that will force a discontinuity at $\omega = \pm\pi$ in the Fourier transform of $\hat{v}[n]$, so $\hat{V}(z)$ will not be analytic on the unit circle. To avoid this problem, we can alter $v[n]$ (and therefore also $x[n]$) with a one-sample time shift so that we evaluate instead the complex cepstrum of $v[n+1]$ and, consequently, also $x[n+1]$. If $x[n]$ or $v[n]$ is to be resynthesized after some processing of the complex cepstrum, we can remember this time shift and compensate for it at the final output.

With $v[n]$ replaced by $v[n+1]$, and correspondingly $V(z)$ replaced by $zV(z)$, we now consider $V(z)$ to have the form

$$V(z) = \frac{b_1(1 + (b_0/b_1)z)}{(1 - re^{j\theta}z^{-1})(1 - re^{-j\theta}z^{-1})}. \tag{13.98}$$

From Eqs. (13.36a) to (13.36c), we can write $\hat{v}[n]$ exactly as

$$\hat{v}[n] = \begin{cases} \log b_1, & n = 0, & (13.99a) \\[2mm] \dfrac{1}{n}[(re^{j\theta})^n + (re^{-j\theta})^n], & n > 0, & (13.99b) \\[2mm] \dfrac{1}{n}\left(\dfrac{-b_0}{b_1}\right)^{-n}, & n < 0. & (13.99c) \end{cases}$$

To determine $\hat{p}[n]$, we can evaluate the inverse z-transform of $\hat{P}(z)$, which, from Eq. (13.93b), is

$$\hat{P}(z) = \log(1 - \beta^3 z^{-3N_0}) - \log(1 - \beta z^{-N_0}), \qquad (13.100)$$

where for our example $\beta = 0.9$, and consequently, $|\beta| < 1$. One approach to determining the inverse z-transform of Eq. (13.100) is to use the power series expansion of $\hat{P}(z)$. Specifically, since $|\beta| < 1$,

$$\hat{P}(z) = -\sum_{k=1}^{\infty} \frac{\beta^{3k}}{k} z^{-3N_0 k} + \sum_{k=1}^{\infty} \frac{\beta^k}{k} z^{-N_0 k}, \qquad (13.101)$$

from which it follows that $\hat{p}[n]$ is

$$\hat{p}[n] = -\sum_{k=1}^{\infty} \frac{\beta^{3k}}{k} \delta[n - 3N_0 k] + \sum_{k=1}^{\infty} \frac{\beta^k}{k} \delta[n - N_0 k]. \qquad (13.102)$$

An alternative approach to obtaining $\hat{p}[n]$ is to use the property developed in Problem 13.28.

From Eq. (13.95a), the complex cepstrum of $x[n]$ is

$$\hat{x}[n] = \hat{v}[n] + \hat{p}[n], \qquad (13.103)$$

where $\hat{v}[n]$ and $\hat{p}[n]$ are given by Eqs. (13.99a) to (13.99c) and (13.102), respectively. The sequences $\hat{v}[n]$, $\hat{p}[n]$, and $\hat{x}[n]$ are shown in Figure 13.11.

The cepstrum of $x[n]$, $c_x[n]$, is the even part of $\hat{x}[n]$, i.e.,

$$c_x[n] = \tfrac{1}{2}(\hat{x}[n] + \hat{x}[-n]) \qquad (13.104)$$

and furthermore

$$c_x[n] = c_v[n] + c_p[n]. \qquad (13.105)$$

From Eqs. (13.99a) to (13.99c),

$$c_v[n] = \log(b_1)\delta[n] + \sum_{k=1}^{\infty} \frac{(-1)^k (b_0/b_1)^{-k}}{2k} (\delta[n - k] + \delta[n + k])$$

$$\qquad\qquad\qquad\qquad\qquad\qquad (13.106a)$$

$$+ \sum_{k=1}^{\infty} \frac{r^k \cos(\theta k)}{k} (\delta[n - k] + \delta[n + k]).$$

and from Eq. (13.102),

$$c_p[n] = -\frac{1}{2}\sum_{k=1}^{\infty} \frac{\beta^{3k}}{k}\{\delta[n - 3N_0 k] + \delta[n + 3N_0 k]\}$$

$$\qquad\qquad\qquad\qquad\qquad\qquad (13.106b)$$

$$+ \frac{1}{2}\sum_{k=1}^{\infty} \frac{\beta^k}{k}\{\delta[n - N_0 k] + \delta[n + N_0 k]\}.$$

The sequences $c_v[n]$, $c_p[n]$, and $c_x[n]$ for this example are shown in Figure 13.12.

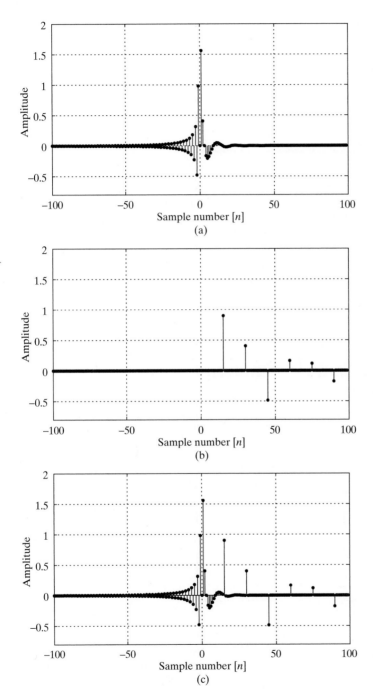

Figure 13.11 The sequences: (a) $\hat{v}[n]$, (b) $\hat{p}[n]$, and (c) $\hat{x}[n]$.

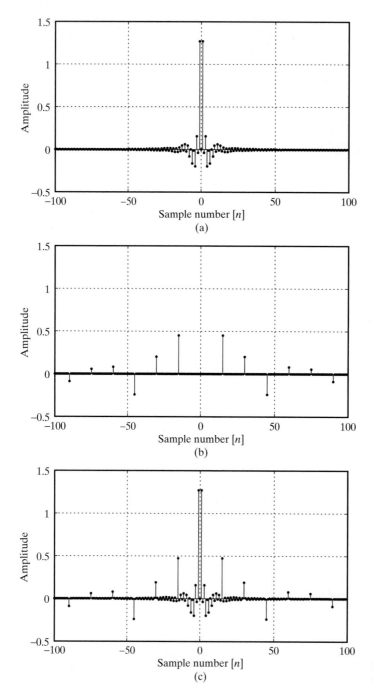

Figure 13.12 The sequences: (a) $c_V[n]$, (b) $c_p[n]$, and (c) $c_X[n]$.

13.9.2 Computation of the Cepstrum Using the DFT

In Figures 13.11 and 13.12, we showed the complex cepstra and the cepstra corresponding to evaluating the analytical expressions obtained in Section 13.9.1. In most applications, we do not have simple mathematical formulas for the signal values, and consequently, we cannot analytically determine $\hat{x}[n]$ or $c_x[n]$. However, for finite-length sequences, we can use either polynomial rooting or the DFT to compute the complex cepstrum. In this section, we illustrate the use of the DFT in the computation of the complex cepstrum and the cepstrum of $x[n]$ for the example of this section.

To compute the complex cepstrum or the cepstrum using the DFT as in Figure 13.4(a), it is necessary that the input be of finite extent. Thus, for the signal model discussed at the beginning of this section, $x[n]$ must be truncated. In the examples discussed in this section, the signal $x[n]$ in Figure 13.10(c) was truncated to $N = 1024$ samples and 1024-point DFTs were used in the system of Figure 13.4(a) to compute the complex cepstrum and the cepstrum of the signal. Figure 13.13 shows the Fourier transforms that are involved in the computation of the complex cepstrum. Figure 13.13(a) shows the logarithm of the magnitude of the DFT of 1024 samples of $x[n]$ in Figure 13.10, with the DFT samples connected in the plot to suggest the appearance of the DTFT of the finite-length input sequence. Figure 13.13(b) shows the principal value of the phase. Note the discontinuities as the phase exceeds $\pm\pi$ and wraps around modulo 2π. Figure 13.13(c) shows the continuous "unwrapped" phase curve obtained as discussed in Section 13.6.1. As discussed above, and as is evident by carefully comparing Figures 13.13(b) and 13.13(c), a linear-phase component corresponding to a delay of one sample has been removed so that the unwrapped phase curve is continuous at 0 and π. Thus, the unwrapped phase of Figure 13.13(c) corresponds to $x[n + 1]$ rather than $x[n]$.

Figures 13.13(a) and 13.13(c) correspond to the computation of samples of the real and imaginary parts, respectively, of the DTFT of the complex cepstrum. Only the frequency range $0 \leq \omega \leq \pi$ is shown, since the function of Figure 13.13(a) is even and periodic with period 2π, and the function of Figure 13.13(c) is odd and periodic with period 2π. In examining the plots in Figures 13.13(a) and 13.13(c), we note that they have the general appearance of a rapidly varying, periodic (in frequency) component added to a more slowly varying component. The periodically varying component in fact corresponds to $\hat{P}(e^{j\omega})$ and the more slowly varying component to $\hat{V}(e^{j\omega})$.

In Figure 13.14(a), we show the inverse Fourier transform of the complex logarithm of the DFT, i.e., the time-aliased complex cepstrum $\hat{x}_p[n]$. Note the impulses at integer multiples of $N_0 = 15$. These are contributed by $\hat{p}[n]$ and correspond to the rapidly varying periodic component observed in the logarithm of the DFT. We see also that since the input signal is not minimum phase, the complex cepstrum is nonzero for $n < 0$.[9]

Since a large number of points were used in computing the DFTs, the time-aliased complex cepstrum differs very little from the exact values that would be obtained by

---

[9]In using the DFT to obtain the inverse Fourier transform of Figures 13.13(a) and 13.13(c), the values associated with $n < 0$ would normally appear in the interval $N/2 < n \leq N - 1$. Traditionally, time sequences are displayed with $n = 0$ in the center, so we have repositioned $\hat{x}_p[n]$ accordingly and have shown only a total of 201 points symmetrically about $n = 0$.

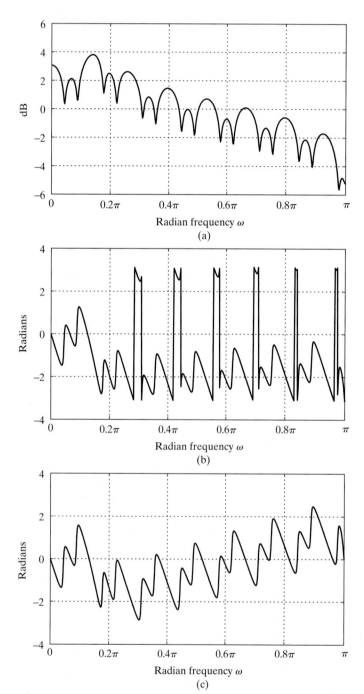

Figure 13.13 Fourier transforms of $x[n]$ in Figure 13.10. (a) Log magnitude. (b) Principal value of the phase. (c) Continuous "unwrapped" phase after removing a linear-phase component from part (b). The DFT samples are connected by straight lines.

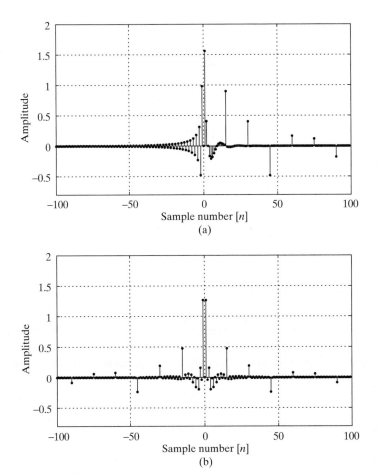

Figure 13.14 (a) Complex cepstrum $\hat{x}_p[n]$ of sequence in Figure 13.10(c). (b) Cepstrum $c_x[n]$ of sequence in Figure 13.10(c).

evaluating Eqs. (13.99a) to (13.99c), (13.102), and (13.103) for the specific values of the parameters used to generate the input signal of Figure 13.10.

The time-aliased cepstrum $c_{xp}[n]$ for this example is shown in Figure 13.14(b). As with the complex cepstrum, impulses at multiples of 15 are evident, corresponding to the periodic component of the logarithm of the magnitude of the Fourier transform.

As mentioned at the beginning of this section, convolution of a signal $v[n]$ with an impulse train such as $p[n]$ is a model for a signal containing multiple echoes. Since $x[n]$ is a convolution of $v[n]$ and $p[n]$, the echo times are often not easily detected by examining $x[n]$. In the cepstral domain, however, the effect of $p[n]$ is present as an additive impulse train, and consequently, the presence and location of the echoes are often more evident. As discussed in Section 13.1, it was this observation that motivated the proposal by Bogert, Healy and Tukey (1963) that the cepstrum be used as a means for detecting echoes. This same idea was later used by Noll (1967) as a basis for detecting vocal pitch in speech signals.

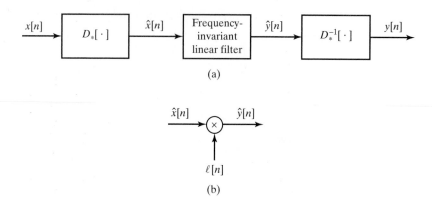

Figure 13.15   (a) System for homomorphic deconvolution. (b) Time-domain representation of frequency-invariant filtering.

13.9.3 Homomorphic Deconvolution for the Multipath Model

For the multipath model that is the basis for Section 13.9, the slowly varying component of the complex logarithm, and equivalently the "low-time" (low-quefrency) portion of the complex cepstrum, were mainly due to $v[n]$. Correspondingly, the more rapidly varying component of the complex logarithm and the "high-time" (high-quefrency) portion of the complex cepstrum were due primarily to $p[n]$. This suggests that the two convolved components of $x[n]$ can be separated by applying linear filtering to the logarithm of the Fourier transform (i.e., frequency invariant filtering), or, equivalently the complex cepstrum components can be separated by windowing or time gating the complex cepstrum.

Figure 13.15(a) depicts the operations involved in separation of the components of a convolution by filtering the complex logarithm of the Fourier transform of a signal. The frequency-invariant linear filter can be implemented by convolution in the frequency domain or, as indicated in Figure 13.15(b), by multiplication in the time domain. Figure 13.16(a) shows the time response of a lowpass frequency-invariant linear system as required for recovering an approximation to $v[n]$, and Figure 13.16(b) shows the time response of a highpass frequency-invariant linear system for recovering an approximation to $p[n]$.[10]

Figure 13.17 shows the result of lowpass frequency-invariant filtering. The more rapidly varying curves in Figures 13.17(a) and 13.17(b) are the complex logarithm of the Fourier transform of the input signal, i.e., the Fourier transform of the complex cepstrum. The slowly varying (dashed) curves in Figures 13.17(a) and 13.17(b) are the real and imaginary parts, respectively, of the Fourier transform of $\hat{y}[n]$, when the frequency-invariant linear system $\ell[n]$ is of the form of Figure 13.16(a) with $N_1 = 14$, $N_2 = 14$, and with the system of Figure 13.15 implemented using DFTs of length $N = 1024$. Figure 13.17(c) shows the corresponding output $y[n]$. This sequence is the approximation to

---

[10]Figure 13.16 assumes that the systems $D_*[\cdot]$ and $D_*^{-1}[\cdot]$ are implemented using the DFT as in Figure 13.4.

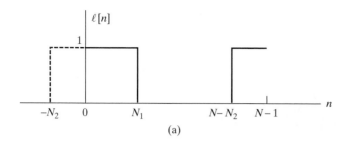

(a)

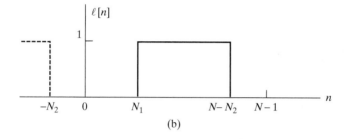

(b)

Figure 13.16 Time response of frequency-invariant linear systems for homomorphic deconvolution. (a) Lowpass system. (b) Highpass system. (Solid line indicates envelope of the sequence $\ell[n]$ as it would be applied in a DFT implementation. The dashed line indicates the periodic extension.)

$v[n]$ obtained by homomorphic deconvolution. To relate this output $y[n]$ to $v[n]$, recall that in computing the unwrapped phase, a linear-phase component was removed, corresponding to a one-sample time shift of $v[n]$. Consequently, $y[n]$ in Figure 13.17(c) corresponds to an approximation to $v[n+1]$ obtained by homomorphic deconvolution.

This type of filtering has been successfully used in speech processing to recover the vocal tract response information (Oppenheim, 1969b; Schafer and Rabiner, 1970) and in seismic signal analysis to recover seismic wavelets (Ulrych, 1971; Tribolet, 1979).

Figure 13.18 shows the result of highpass frequency-invariant filtering. The rapidly varying curves in Figures 13.18(a) and (b) are the real and imaginary parts, respectively, of the Fourier transform of $\hat{y}[n]$ when the frequency-invariant linear system $\ell[n]$ is of the form of Figure 13.16(b) with $N_1 = 14$ and $N_2 = 512$ (i.e., the negative-time parts are completely removed). Again, the system is implemented using a 1024-point DFT. Figure 13.18(c) shows the corresponding output $y[n]$. This sequence is the approximation to $p[n]$ obtained by homomorphic deconvolution. In contrast to the use of the cepstrum to *detect* echoes or periodicity, this approach seeks to obtain the impulse train that specifies the location and size of the repeated copies of $v[n]$.

13.9.4 Minimum-Phase Decomposition

In Section 13.8.1, we discussed ways that homomorphic deconvolution could be used to decompose a sequence into minimum-phase and allpass components or minimum-phase and maximum-phase components. We will apply these techniques to the signal model of Section 13.9. Specifically, for the parameters of the example, the z-transform of the input is

$$X(z) = V(z)P(z) = \frac{(0.98 + z^{-1})(1 + 0.9z^{-15} + 0.81z^{-30})}{(1 - 0.9e^{j\pi/6}z^{-1})(1 - 0.9e^{-j\pi/6}z^{-1})}. \qquad (13.107)$$

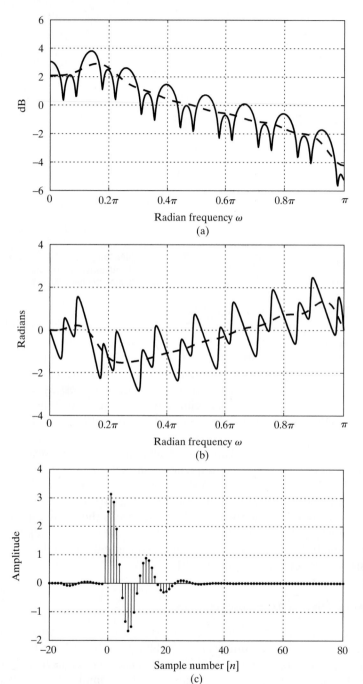

Figure 13.17 Lowpass frequency-invariant linear filtering in the system of Figure 13.15. (a) Real parts of the Fourier transforms of the input (solid line) and output (dashed line) of the lowpass system with $N_1 = 14$ and $N_2 = 14$ in Figure 13.16(a). (b) Imaginary parts of the input (solid line) and output (dashed line). (c) Output sequence $y[n]$ for the input of Figure 13.10(c).

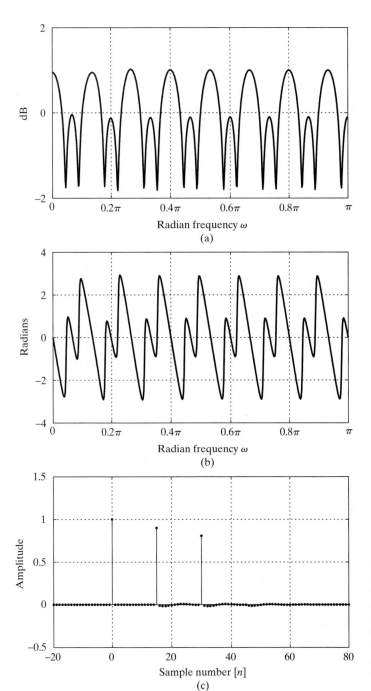

Figure 13.18 Illustration of highpass frequency-invariant linear filtering in the system of Figure 13.15. (a) Real part of the Fourier transform of the output of the highpass frequency-invariant system with $N_1 = 14$ and $N_2 = 512$ in Figure 13.16(b). (b) Imaginary part for conditions of part (a). (c) Output sequence $y[n]$ for the input of Figure 13.10.

First, we can write $X(z)$ as the product of a minimum-phase z-transform and an allpass z-transform; i.e.,

$$X(z) = X_{min}(z)X_{ap}(z), \tag{13.108}$$

where

$$X_{min}(z) = \frac{(1 + 0.98z^{-1})(1 + 0.9z^{-15} + 0.81z^{-30})}{(1 - 0.9e^{j\pi/6}z^{-1})(1 - 0.9e^{-j\pi/6}z^{-1})} \tag{13.109}$$

and

$$X_{ap}(z) = \frac{0.98 + z^{-1}}{1 + 0.98z^{-1}}. \tag{13.110}$$

The sequences $x_{min}[n]$ and $x_{ap}[n]$ can be found using the partial fraction expansion methods of Chapter 3, and the corresponding complex cepstra $\hat{x}_{min}[n]$ and $\hat{x}_{ap}[n]$ can be found using the power series technique of Section 13.5 (see Problem 13.25). Alternatively, $\hat{x}_{min}[n]$ and $\hat{x}_{ap}[n]$ can be obtained exactly from $\hat{x}[n]$ by the operations discussed in Section 13.8.1 and as depicted in Figure 13.7. If the characteristic systems in Figure 13.7 are implemented using the DFT, then the separation is only approximate since $x_{ap}[n]$ is infinitely long, but the approximation error can be small over the interval where $x_{ap}[n]$ is large if the DFT length is large enough. Figure 13.19(a) shows the complex cepstrum for $x[n]$ as computed using a 1024-point DFT, again with a one-sample time delay removed from $v[n]$ so that the phase is continuous at π. Figure 13.19(b) shows the complex cepstrum of the minimum-phase component $\hat{x}_{min}[n]$, and Figure 13.19(c) shows the complex cepstrum of the allpass component $\hat{x}_{ap}[n]$ as obtained by the operations of Figure 13.7 with $D_*[\cdot]$ implemented as in Figure 13.4(a).

Using the DFT as in Figure 13.4(b) to implement the system $D_*^{-1}[\cdot]$ gives the approximations to the minimum-phase and allpass components shown in Figures 13.20(a) and 13.20(b), respectively. Since all the zeros of $P(z)$ are inside the unit circle, all of $P(z)$ is included in the minimum-phase z-transform or, equivalently, $\hat{p}[n]$ is entirely included in $\hat{x}_{min}[n]$. Thus, the minimum-phase component consists of delayed and scaled replicas of the minimum-phase component of $v[n]$. Therefore, the minimum-phase component of Figure 13.20(a) appears very similar to the input shown in Figure 13.10(c). From Eq. (13.110), the allpass component can be shown to be

$$x_{ap}[n] = 0.98\delta[n] + 0.0396(-0.98)^{n-1}u[n-1]. \tag{13.111}$$

The result of Figure 13.20(b) is very close to this ideal result for small values of n where the sequence values are of significant amplitude. This example illustrates a technique of decomposition that has been applied by Bauman, Lipshitz and Vanderkooy (1985) in the analysis and characterization of the response of electroacoustic transducers. A similar decomposition technique can be used to factor magnitude-squared functions as required in digital filter design (see Problem 13.27).

As an alternative to the minimum-phase/allpass decomposition, we can express $X(z)$ as the product of a minimum-phase z-transform and a maximum-phase z-transform; i.e.,

$$X(z) = X_{mn}(z)X_{mx}(z), \tag{13.112}$$

where

$$X_{mn}(z) = \frac{z^{-1}(1 + 0.9z^{-15} + 0.81z^{-30})}{(1 - 0.9e^{j\pi/6}z^{-1})(1 - 0.9e^{-j\pi/6}z^{-1})} \tag{13.113}$$

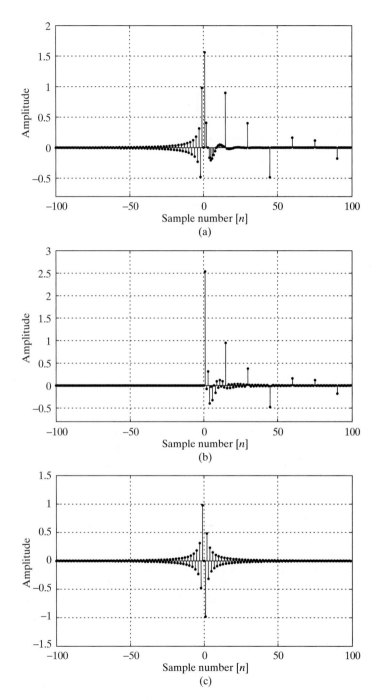

Figure 13.19 (a) Complex cepstrum of $x[n] = x_{min}n * x_{ap}[n]$. (b) Complex cepstrum of $x_{min}[n]$. (c) Complex cepstrum of $x_{ap}[n]$.

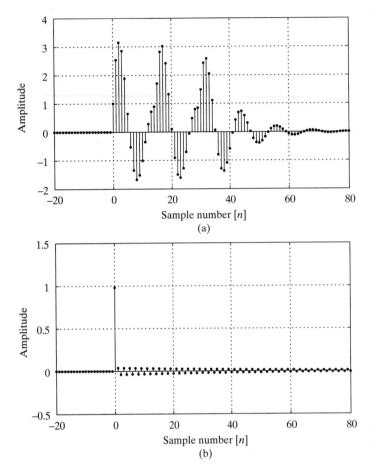

Figure 13.20   (a) Minimum-phase output. (b) Allpass output obtained as depicted in Figure 13.7.

and

$$X_{mx}(z) = 0.98z + 1. \tag{13.114}$$

The sequences $x_{mn}[n]$ and $x_{mx}[n]$ can be found using the partial fraction expansion methods of Chapter 3, and the corresponding complex cepstra $\hat{x}_{mn}[n]$ and $\hat{x}_{mx}[n]$ can be found using the power series technique of Section 13.5 (see Problem 13.25). Alternatively, $\hat{x}_{mn}[n]$ and $\hat{x}_{mx}[n]$ can be obtained exactly from $\hat{x}[n]$ by the operations discussed in Section 13.8.2 and as depicted in Figure 13.8, where

$$\ell_{mn}[n] = u[n] \tag{13.115}$$

and

$$\ell_{mx}[n] = u[-n-1]. \tag{13.116}$$

That is, the minimum-phase sequence is now defined by the positive time part of the complex cepstrum and the maximum-phase part is defined by the negative time part of the complex cepstrum. If the characteristic systems in Figure 13.8 are implemented using the DFT, the negative time part of the complex cepstrum is positioned in the last half of the DFT interval. In this case, the separation of the minimum-phase and maximum-phase components is only approximate because of time aliasing, but the time-aliasing error can be made small by choosing a sufficiently large DFT length. Figure 13.19(a) shows the complex cepstrum of $x[n]$ as computed using a 1024-point DFT. Figure 13.21 shows the two output sequences that are obtained from the complex cepstrum of Figure 13.19(a) using Eqs. (13.87) and (13.88) as in Fig 13.8 with the inverse characteristic system being implemented using the DFT as in Figure 13.4(b). As before, since $\hat{p}[n]$ is entirely included in $\hat{x}_{mn}[n]$, the corresponding output $x_{mn}[n]$ consists of delayed and scaled replicas of a minimum-phase sequence, thus, it also looks very much like the input sequence. However, a careful comparison of Figures 13.20(a) and 13.21(a) shows that $x_{min}[n] \neq x_{mn}[n]$. From Eq. (13.114), the maximum-phase sequence is

$$x_{mx}[n] = 0.98\delta[n+1] + \delta[n]. \tag{13.117}$$

Figure 13.21(b) is very close to this ideal result. (Note the shift due to the linear phase removed in the phase unwrapping.) This technique of minimum-phase/maximum-phase decomposition was used by Smith and Barnwell (1984) in the design and implementation of exact reconstruction filter banks for speech analysis and coding.

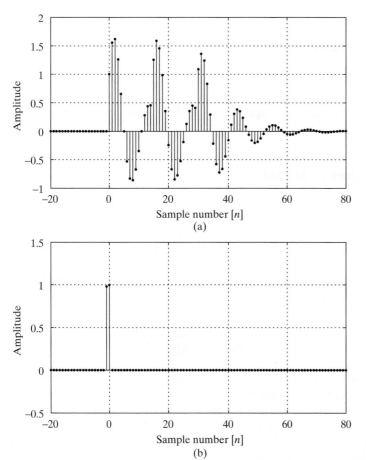

Figure 13.21   (a) Minimum-phase output. (b) Maximum-phase output obtained as depicted in Figure 13.8.

13.9.5 Generalizations

The example in Section 13.9 considered a simple exponential signal that was convolved with an impulse train to produce a series of delayed and scaled replicas of the exponential signal. This model illustrates many of the features of the complex cepstrum and of homomorphic filtering.

In particular, in more general models associated with speech, communication, and seismic applications an appropriate signal model consists of the convolution of two components. One component has the characteristics of $v[n]$, specifically a Fourier transform that is slowly varying in frequency. The second has the characteristics of $p[n]$, i.e., an echo pattern or impulse train for which the Fourier transform is more rapidly varying and quasiperiodic in frequency. Thus, the contributions of the two components would be separated in the complex cepstrum or the cepstrum, and, furthermore, the complex cepstrum or the cepstrum would contain impulses at multiples of the echo delays. Thus, homomorphic filtering can be used to separate the convolutional components of the signal, or the cepstrum can be used to detect echo delays. In the next section, we will illustrate the use of these general properties of the cepstrum in applications to speech analysis.

13.10 APPLICATIONS TO SPEECH PROCESSING

Cepstrum techniques have been applied successfully to speech analysis in a variety of ways. As discussed briefly in this section, the previous theoretical discussion and the extended example of Section 13.9 apply in a relatively straightforward way to speech analysis.

13.10.1 The Speech Model

As we briefly described in Section 10.4.1, there are three basic classes of speech sounds corresponding to different forms of excitation of the vocal tract. Specifically:

- *Voiced sounds* are produced by exciting the vocal tract with quasiperiodic pulses of airflow caused by the opening and closing of the glottis.
- *Fricative sounds* are produced by forming a constriction somewhere in the vocal tract and forcing air through the constriction so that turbulence is created, thereby producing a noise-like excitation.
- *Plosive sounds* are produced by completely closing off the vocal tract, building up pressure behind the closure, and then abruptly releasing the pressure.

In each case, the speech signal is produced by exciting the vocal tract system (an acoustic transmission system) with a wideband excitation. The vocal tract shape changes relatively slowly with time, thus, it can be modeled as a slowly time-varying filter that imposes its frequency-response properties on the spectrum of the excitation. The vocal tract is characterized by its natural frequencies (called *formants*), which correspond to resonances in its frequency response.

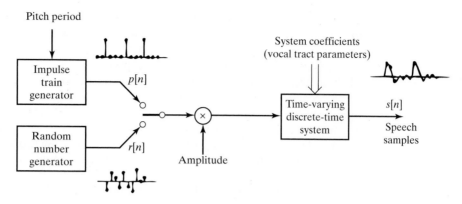

Figure 13.22   Discrete-time model of speech production.

If we assume that the excitation sources and the vocal tract shape are independent, we arrive at the discrete-time model of Figure 13.22 as a representation of the sampled speech waveform. In this model, samples of the speech signal are assumed to be the output of a time-varying discrete-time system that models the resonances of the vocal tract system. The mode of excitation of the system switches between periodic impulses and random noise, depending on the type of sound being produced.

Since the vocal tract shape changes rather slowly in continuous speech, it is reasonable to assume that the discrete-time system in the model has fixed properties over a time interval on the order of 10 ms. Thus, the discrete-time system may be characterized in each such time interval by an impulse response or a frequency response or a set of coefficients for an IIR system. Specifically, a model for the system function of the vocal tract takes the form

$$V(z) = \frac{\displaystyle\sum_{k=0}^{K} b_k z^{-k}}{\displaystyle\sum_{k=0}^{P} a_k z^{-k}} \tag{13.118}$$

or, equivalently,

$$V(z) = \frac{A z^{-K_o} \displaystyle\prod_{k=1}^{K_i}(1 - \alpha_k z^{-1}) \prod_{k=1}^{K_o}(1 - \beta_k z)}{\displaystyle\prod_{k=1}^{[P/2]}(1 - r_k e^{j\theta_k} z^{-1})(1 - r_k e^{-j\theta_k} z^{-1})}, \tag{13.119}$$

where the quantities $r_k e^{j\theta_k}$ (with $|r_h| < 1$ are the complex natural frequencies of the vocal tract, which, of course, are dependent on the vocal tract shape and consequently are time varying. The zeros of $V(z)$ account for the finite-duration glottal pulse waveform and for the zeros of transmission caused by the constrictions of the vocal tract in the creation of nasal voiced sounds and fricatives. Such zeros are often not included, because it is very difficult to estimate their locations from only the speech waveform. Also, it has been shown (Atal and Hanauer, 1971) that the spectral shape of the speech signal can be accurately modeled using no zeros, if we include extra poles beyond the number needed just to account for the vocal tract resonances. The zeros are included in our analysis, because they are necessary for an accurate representation of the complex cepstrum of speech. Note that we include the possibility of zeros outside the unit circle.

The vocal tract system is excited by an excitation sequence $p[n]$, which is a train of impulses when modeling voiced speech sounds and $r[n]$, which is a pseudorandom noise sequence when modeling unvoiced speech sounds, such as fricatives and plosives.

Many of the fundamental problems of speech processing reduce to the estimation of the parameters of the model of Figure 13.22. These parameters are as follows:

- The coefficients of $V(z)$ in Eq. (13.118) or the pole and zero locations in Eq. (13.119)
- The mode of excitation of the vocal tract system; i.e., a *periodic impulse train* or *random noise*
- The amplitude of the excitation signal
- The pitch period of the speech excitation for voiced speech.

Homomorphic deconvolution can be applied to the estimation of the parameters if it is assumed that the model is valid over a short time interval, so that a short segment of length L samples of the sampled speech signal can be thought of as the convolution

$$s[n] = v[n] * p[n] \qquad \text{for } 0 \leq n \leq L - 1, \tag{13.120}$$

where $v[n]$ is the impulse response of the vocal tract and $p[n]$ is either periodic (for voiced speech) or random noise (for unvoiced speech). Obviously, Eq. (13.120) is not valid at the edges of the interval, because of pulses that occur before the beginning of the analysis interval and pulses that end after the end of the interval. To mitigate the effect of the "discontinuities" of the model at the ends of the interval, the speech signal $s[n]$ can be multiplied by a window $w[n]$ that tapers smoothly to zero at both ends. Thus, the input to the homomorphic deconvolution system is

$$x[n] = w[n]s[n]. \tag{13.121}$$

Let us first consider the case of voiced speech. If $w[n]$ varies slowly with respect to the variations of $v[n]$, then the analysis is greatly simplified if we assume that

$$x[n] = v[n] * p_w[n], \tag{13.122}$$

where

$$p_w[n] = w[n]p[n]. \tag{13.123}$$

(See Oppenheim and Schafer, 1968.) A more detailed analysis without this assumption leads to essentially the same conclusions as below (Verhelst and Steenhaut, 1986). For voiced speech, $p[n]$ is a train of impulses of the form

$$p[n] = \sum_{k=0}^{M-1} \delta[n - kN_0] \tag{13.124}$$

so that

$$p_w[n] = \sum_{k=0}^{M-1} w[kN_0]\delta[n - kN_0], \tag{13.125}$$

where we have assumed that the pitch period is N_0 and that M periods are spanned by the window.

The complex cepstra of $x[n]$, $v[n]$, and $p_w[n]$ are related by

$$\hat{x}[n] = \hat{v}[n] + \hat{p}_w[n]. \tag{13.126}$$

To obtain $\hat{p}_w[n]$, we define a sequence

$$w_{N_0}[k] = \begin{cases} w[kN_0], & k = 0, 1, \ldots, M-1, \\ 0, & \text{otherwise}, \end{cases} \tag{13.127}$$

whose Fourier transform is

$$P_w(e^{j\omega}) = \sum_{k=0}^{M-1} w[kN_0]e^{-\omega kN_0} = W_{N_0}(e^{j\omega N_0}). \tag{13.128}$$

Thus, $P_w(e^{j\omega})$ and $\hat{P}_w(e^{j\omega})$ are both periodic with period $2\pi/N_0$, and the complex cepstrum of $p_w[n]$ is

$$\hat{p}_w[n] = \begin{cases} \hat{w}_{N_0}[n/N_0], & n = 0, \pm N_0, \pm 2N_0, \ldots, \\ 0, & \text{otherwise}. \end{cases} \tag{13.129}$$

The periodicity of the complex logarithm resulting from the periodicity of the voiced speech signal is manifest in the complex cepstrum as impulses spaced at integer multiples of N_0 samples (the pitch period). If the sequence $w_{N_0}[n]$ is minimum phase, then $\hat{p}_w[n]$ will be zero for $n < 0$. Otherwise, $\hat{p}_w[n]$ will have impulses spaced at intervals of N_0 samples for both positive and negative values of n. In either case, the contribution of $\hat{p}_w[n]$ to $\hat{x}[n]$ will be found in the interval $|n| \geq N_0$.

From the power series expansion of the complex logarithm of $V(z)$, it can be shown that the contribution to the complex cepstrum due to $v[n]$ is

$$\hat{v}[n] = \begin{cases} \displaystyle\sum_{k=1}^{K_0} \frac{\beta_k^{-n}}{n}, & n < 0, \\ \log|A|, & n = 0, \\ \displaystyle -\sum_{k=1}^{K_i} \frac{\alpha_k^n}{n} + \sum_{k=1}^{[P/2]} \frac{2r_k^n}{n}\cos(\theta_k n), & n > 0. \end{cases} \tag{13.130}$$

As with the simpler example in Section 13.9.1, the term z^{-K_0} in Eq. (13.119) represents a linear-phase factor that would be removed in obtaining the unwrapped phase and the complex cepstrum. Consequently, $\hat{v}[n]$ in Eq. (13.130) more accurately is the complex cepstrum of $v[n + K_0]$.

From Eq. (13.130), we see that the contributions of the vocal tract response to the complex cepstrum occupy the full range $-\infty < n < \infty$, but they are concentrated around $n = 0$. We note also that since the vocal tract resonances are represented by poles inside the unit circle, their contribution to the complex cepstrum is zero for $n < 0$.

13.10.2 **Example of Homomorphic Deconvolution of Speech**

For speech sampled at 10,000 samples/s, the pitch period N_0 will range from about 25 samples for a high-pitched voice up to about 150 samples for a very low-pitched voice. Since the vocal tract component of the complex cepstrum $\hat{v}[n]$ decays rapidly, the peaks of $\hat{p}_w[n]$ stand out from $\hat{v}[n]$. In other words, in the complex logarithm, the vocal tract components are slowly varying, and the excitation components are rapidly varying. This is illustrated by the following example. Figure 13.23(a) shows a segment of a speech wave

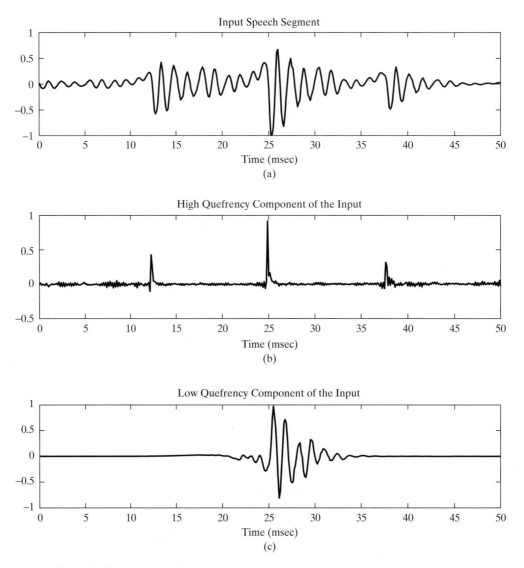

Figure 13.23   Homomorphic deconvolution of speech. (a) Segment of speech weighted by a Hamming window. (b) High quefrency component of the signal in (a). (c) Low quefrency component of the signal in (a).

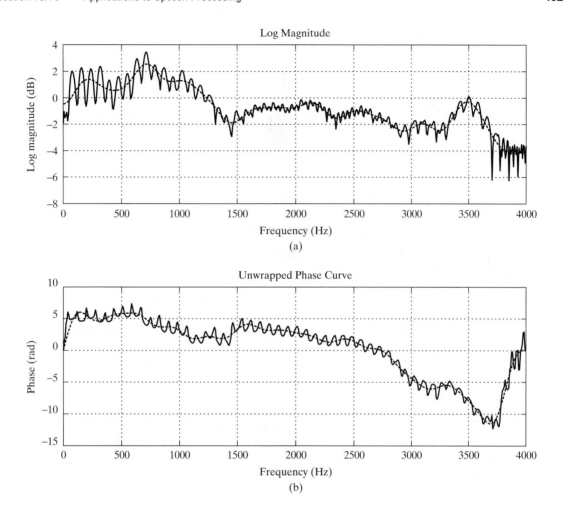

Figure 13.24 Complex logarithm of the signal of Figure 13.23(a): (a) Log magnitude. (b) Unwrapped phase.

multiplied by a Hamming window of length 401 samples (50 ms time duration at a sampling rate of 8000 samples/s). Figure 13.24 shows the complex logarithm (log magnitude and unwrapped phase) of the DFT of the signal in Figure 13.23(a).[11] Note the rapidly varying, almost periodic component due to $p_w[n]$ and the slowly varying component due to $v[n]$. These properties are manifest in the complex cepstrum of Figure 13.25 in the form of impulses at multiples of approximately 13 ms (the period of the input speech segment) due to $\hat{p}_w[n]$ and in the samples in the region $|nT| < 5$ ms, which we attribute to $\hat{v}[n]$. As in the previous section, frequency-invariant filtering can be used

---

[11] In all the figures of this section, the samples of all sequences were connected for ease in plotting.

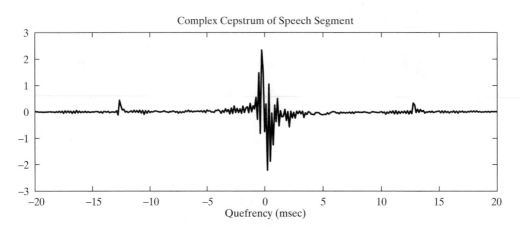

Figure 13.25   Complex cepstrum of the signal in Figure 13.23(a) (inverse DTFT of the complex logarithm in Figure 13.24).

to separate the components of the convolutional model of the speech signal. Lowpass filtering of the complex logarithm can be used to recover an approximation to $v[n]$, and highpass filtering can be used to obtain $p_w[n]$. Figure 13.23(c) shows an approximation to $v[n]$ obtained by using a lowpass frequency-invariant filter as in Figure 13.16(a) with $N_1 = 30$ and $N_2 = 30$. The slowly varying dotted curves in Figure 13.24 show the complex logarithm of the DTFT of the low quefrency component shown in Figure 13.23(c). On the other hand, Figure 13.23(b) is an approximation to $p_w[n]$ obtained by applying to the complex cepstrum a symmetrical highpass frequency-invariant filter as in Figure 13.16(b) with $N_1 = 95$ and $N_2 = 95$. In both cases, the inverse characteristic system was implemented by using 1024-point DFTs, as in Figure 13.4(b).

13.10.3 Estimating the Parameters of the Speech Model

Although homomorphic deconvolution can be successfully applied in *separating* the components of a speech waveform, in many speech processing applications we are interested only in *estimating* the parameters in a parametric representation of the speech signal. Since the properties of the speech signal change rather slowly with time, it is common to estimate the parameters of the model of Figure 13.22 at intervals of about 10 ms (100 times/s). In this case, the time-dependent Fourier transform discussed in Chapter 10 serves as the basis for time-dependent homomorphic analysis. For example, it may be sufficient to examine segments of speech selected about every 10 ms (100 samples at 10,000 Hz sampling rate) to determine the mode of excitation of the model (voiced or unvoiced) and, for voiced speech, the pitch period. Or we may wish to track the variation of the vocal tract resonances (formants). For such problems, the phase computation can be avoided by using the cepstrum, which requires only the logarithm of the magnitude of the Fourier transform. Since the cepstrum is the even part of the

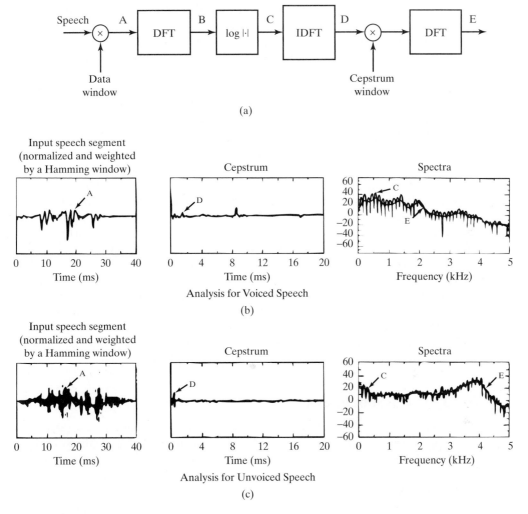

Figure 13.26   (a) System for cepstrum analysis of speech signals. (b) Analysis for voice speech. (c) Analysis for unvoiced speech.

complex cepstrum, our previous discussion suggests that the low-time portion of $c_x[n]$ should correspond to the slowly varying components of the log magnitude of the Fourier transform of the speech segment, and for voiced speech, the cepstrum should contain impulses at multiples of the pitch period. An example is shown in Figure 13.26.

Figure 13.26(a) shows the operations involved in estimating the speech parameters using the cepstrum. Figure 13.26(b) shows a typical result for voiced speech. The windowed speech signal is labeled A, $\log |X[k]|$ is labeled C, and the cepstrum $c_x[n]$ is labeled D. The peak in the cepstrum at about 8 ms indicates that this segment of speech is voiced with that period. The smoothed spectrum, or *spectrum envelope*, obtained by frequency-invariant lowpass filtering with cutoff below 8 ms is labeled E and is superimposed on C. The situation for unvoiced speech, shown in Figure 13.26(c), is similar, except that the random nature of the excitation component of the input speech segment causes a rapidly varying random component in $\log|X[k]|$ instead of a periodic component. Thus, in the cepstrum the low-time components correspond as before to the vocal tract system function; however, since the rapid variations in $\log|X[k]|$ are not periodic, no strong peak appears in the cepstrum. Therefore, the presence or absence of a peak in the cepstrum in the normal pitch period range serves as a very good voiced/unvoiced detector and pitch period estimator. The result of lowpass frequency-invariant filtering in the unvoiced case is similar to that in the voiced case. A smoothed spectrum envelope estimate is obtained as in E.

In speech analysis applications, the operations of Figure 13.26(a) are applied repeatedly to sequential segments of the speech waveform. The length of the segments must be carefully selected. If the segments are too long, the properties of the speech signal will change too much across the segment. If the segments are too short, there will not be enough of the signal to obtain a strong indication of periodicity. Usually the segment length is set at about three to four times the average pitch period of the speech signal. Figure 13.27 shows an example of how the cepstrum can be used for pitch detection and for estimation of the vocal tract resonance frequencies. Figure 13.27(a) shows a sequence of cepstra computed for speech waveform segments selected at 20-ms intervals. The existence of a prominent peak throughout the sequence of speech segments indicates that the speech was voiced throughout. The location of the cepstrum peak indicates the value of the pitch period in each corresponding time interval. Figure 13.27(b) shows the log magnitude with the corresponding smoothed spectra superimposed. The lines connect estimates of the vocal tract resonances obtained by a heuristic peak-picking algorithm. (See Schafer and Rabiner, 1970.)

13.10.4 Applications

As indicated previously, cepstrum analysis methods have found widespread application in speech processing problems. One of the most successful applications is in pitch detection (Noll, 1967). They also have been used successfully in speech analysis/synthesis systems for low bit-rate coding of the speech signal (Oppenheim, 1969b; Schafer and Rabiner, 1970).

Cepstrum representations of speech have also been used with considerable success in pattern recognition problems associated with speech processing such as speaker identification (Atal, 1976), speaker verification (Furui, 1981) and speech recognition (Davis and Mermelstein, 1980). Although the technique of linear predictive analysis of Chapter 11 is the most widely used method of obtaining a representation of the vocal tract component of the speech model, the linear predictive model representation is often transformed to a cepstrum representation for use in pattern recognition problems (Schroeder, 1981; Juang, Rabiner and Wilpon 1987). This transformation is explored in Problem 13.30.

13.11 SUMMARY

In this chapter, we discussed the technique of cepstrum analysis and homomorphic deconvolution. We focused primarily on definitions and properties of the complex cepstrum and on the practical problems in the computation of the complex cepstrum. An

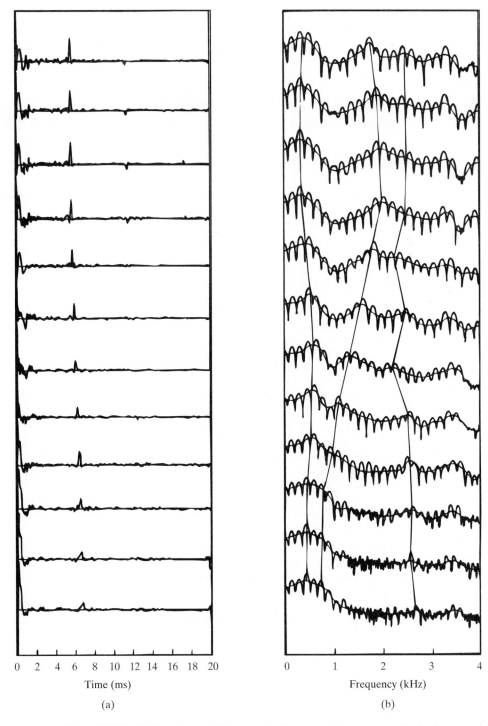

0   2   4   6   8   10 12 14 16 18 20

Time (ms)

(a)

0   1   2   3   4

Frequency (kHz)

(b)

Figure 13.27   (a) Cepstra and (b) log spectra for sequential segments of voiced speech.

idealized example was discussed to illustrate the use of cepstrum analysis and homo-morphic deconvolution for separating components of a convolution. The application of cepstrum analysis techniques to speech processing problems was discussed in some detail as an illustration of their use in a real application.

Problems

Basic Problems

13.1. **(a)** Consider a discrete-time system that is linear in the conventional sense. If $y[n] = T\{x[n]\}$ is the output when the input is $x[n]$, then the *zero signal* $\mathbf{0}[n]$ is the signal that can be added to $x[n]$ such that $T\{x[n] + \mathbf{0}[n]\} = y[n] + T\{\mathbf{0}[n]\} = y[n]$. What is the zero signal for conventional linear systems?

   (b) Consider a discrete-time system $y[n] = T\{x[n]\}$ that is homomorphic, with con-volution as the operation for combining signals at both the input and the output. What is the zero signal for such a system; i.e., what is the signal $\mathbf{0}[n]$ such that $T\{x[n] * \mathbf{0}[n]\} = y[n] * T\{\mathbf{0}[n]\} = y[n]$?

   (c) Consider a discrete-time system $y[n] = T\{x[n]\}$ that is homomorphic, with con-volution as the operation for combining signals at both the input and the output. What is the zero signal for such a system; i.e., what is the signal $\mathbf{0}[n]$ such that $T\{x[n] * \mathbf{0}[n]\} = y[n] * T\{\mathbf{0}[n]\} = y[n]$?

13.2. Let $x_1[n]$ and $x_2[n]$ denote two sequences and $\hat{x}_1[n]$ and $\hat{x}_2[n]$ their corresponding complex cepstra. If $x_1[n] * x_2[n] = \delta[n]$, determine the relationship between $\hat{x}_1[n]$ and $\hat{x}_2[n]$.

13.3. In considering the implementation of homomorphic systems for convolution, we restricted our attention to input signals with rational z-transforms of the form of Eq. (13.32). If an input sequence $x[n]$ has a rational z-transform but has either a negative gain constant or an amount of delay not represented by Eq. (13.32), then we can obtain a z-transform of the form of Eq. (13.32) by shifting $x[n]$ appropriately and multiplying by -1. The complex cepstrum may then be computed using Eq. (13.33).

   Suppose that $x[n] = \delta[n] - 2\delta[n-1]$, and define $y[n] = \alpha x[n-r]$, where $\alpha = \pm 1$ and r is an integer. Find α and r such that $Y(z)$ is in the form of Eq. (13.32), and then find $\hat{y}[n]$.

13.4. In Section 13.5.1, we stated that linear-phase contributions should be removed from the unwrapped phase curve before computation of the complex cepstrum. This problem is concerned with the effect of not removing the linear-phase component due to the factor z^r in Eq. (13.29).

   Specifically, assume that the input to the characteristic system for convolution is $x[n] = \delta[n+r]$. Show that formal application of the Fourier transform definition

$$\hat{x}[n] = \frac{1}{2\pi} \int_{-\pi}^{\pi} \log[X(e^{j\omega})]e^{j\omega n} d\omega \tag{P13.4-1}$$

leads to

$$\hat{x}[n] = \begin{cases} r\dfrac{\cos(\pi n)}{n}, & n \neq 0, \\ 0, & n = 0. \end{cases}$$

The advantage of removing the linear-phase component of the phase is clear from this result, since for large r such a component would dominate the complex cepstrum.

13.5. Suppose that the z-transform of $s[n]$ is

$$S(z) = \frac{(1 - \frac{1}{2}z^{-1})(1 - \frac{1}{4}z)}{(1 - \frac{1}{3}z^{-1})(1 - \frac{1}{5}z)}.$$

Determine the pole locations of the z-transform of $n\hat{s}[n]$, other than poles at $|z| = 0$ or ∞.

13.6. Suppose that the complex cepstrum of $y[n]$ is $\hat{y}[n] = \hat{s}[n] + 2\delta[n]$. Determine $y[n]$ in terms of $s[n]$.

13.7. Determine the complex cepstrum of $x[n] = 2\delta[n] - 2\delta[n-1] + 0.5\delta[n-2]$, shifting $x[n]$ or changing its sign, if necessary.

13.8. Suppose that the z-transform of a stable sequence $x[n]$ is given by

$$X(z) = \frac{1 - \frac{1}{2}z^{-1}}{1 + \frac{1}{2}z},$$

and that a stable sequence $y[n]$ has complex cepstrum $\hat{y}[n] = \hat{x}[-n]$, where $\hat{x}[n]$ is the complex cepstrum of $x[n]$. Determine $y[n]$.

13.9. Equations (13.65) and (13.68) are recursive relationships that can be used to compute the complex cepstrum $\hat{x}[n]$ when the input sequence $x[n]$ is minimum phase and maximum phase, respectively.

    (a) Use Eq. (13.65) to compute recursively the complex cepstrum of the sequence $x[n] = a^n u[n]$, where $|a| < 1$.

    (b) Use Eq. (13.68) to compute recursively the complex cepstrum of the sequence $x[n] = \delta[n] - a\delta[n+1]$, where $|a| < 1$.

13.10. ARG$\{X(e^{j\omega})\}$ represents the principal value of the phase of $X(e^{j\omega})$, and arg$\{X(e^{j\omega})\}$ represents the continuous phase of $X(e^{j\omega})$. Suppose that ARG$\{X(e^{j\omega})\}$ has been sampled at frequencies $\omega_k = 2\pi k/N$ to obtain ARG$\{X[k]\}$ = ARG$\{X(e^{j(2\pi/N)k})\}$ as shown in Figure P13.10. Assuming that $|\text{arg}\{X[k]\} - \text{arg}\{X[k-1]\}| < \pi$ for all k, determine and plot the sequence $r[k]$ as in Eq. (13.49) and arg$\{X[k]\}$ for $0 \le k \le 10$.

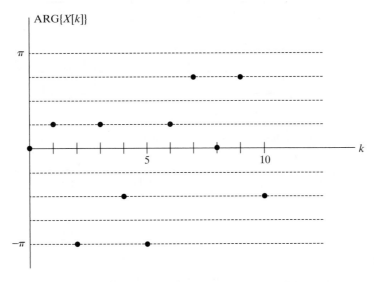

Figure P13.10

13.11. Let $\hat{x}[n]$ be the complex cepstrum of a real-valued sequence $x[n]$. Specify whether each of the following statements is true or false. Give brief justifications for your answers.

Statement 1: If $x_1[n] = x[-n]$ then $\hat{x}_1[n] = \hat{x}[-n]$.

Statement 2: Since $x[n]$ is real-valued, the complex cepstrum $\hat{x}[n]$ must also be real-valued.

Advanced Problems

13.12. Consider the system depicted in Figure P13.12, where S_1 is an LTI system with impulse response $h_1[n]$ and S_2 is a homomorphic system with convolution as the input and output operations; i.e., the transformation $T_2\{\cdot\}$ satisfies

$$T_2\{w_1[n] * w_2[n]\} = T_2\{w_1[n]\} * T_2\{w_2[n]\}.$$

Suppose that the complex cepstrum of the input $x[n]$ is $\hat{x}[n] = \delta[n] + \delta[n-1]$. Find a closed-form expression for $h_1[n]$ such that the output is $y[n] = \delta[n]$.

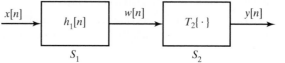

Figure P13.12

13.13. The complex cepstrum of a *finite length* signal $x[n]$ is computed as shown in Figure P13.13-1. Suppose we know that $x[n]$ is minimum phase (all poles and zeros are inside the unit circle). We use the system shown in Figure P13.13-2 to find the real cepstrum of $x[n]$. Explain how to construct $\hat{x}[n]$ from $c_x[n]$.

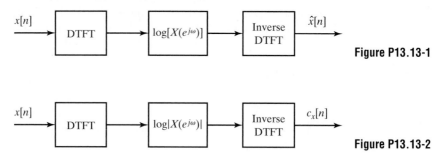

Figure P13.13-1

Figure P13.13-2

13.14. Consider the class of sequences that are real and stable and whose z-transforms are of the form

$$X(z) = |A| \frac{\displaystyle\prod_{k=1}^{M_i}(1 - a_k z^{-1}) \prod_{k=1}^{M_o}(1 - b_k z)}{\displaystyle\prod_{k=1}^{N_i}(1 - c_k z^{-1}) \prod_{k=1}^{N_o}(1 - d_k z)},$$

where $|a_k|, |b_k|, |c_k|, |d_k| < 1$. Let $\hat{x}[n]$ denote the complex cepstrum of $x[n]$.

(a) Let $y[n] = x[-n]$. Determine $\hat{y}[n]$ in terms of $\hat{x}[n]$.

(b) If $x[n]$ is causal, is it also minimum phase? Explain.

(c) Suppose that $x[n]$ is a finite-duration sequence such that

$$X(z) = |A| \prod_{k=1}^{M_i}(1 - a_k z^{-1}) \prod_{k=1}^{M_o}(1 - b_k z),$$

with $|a_k| < 1$ and $|b_k| < 1$. The function $X(z)$ has zeros inside and outside the unit circle. Suppose that we wish to determine $y[n]$ such that $|Y(e^{j\omega})| = |X(e^{j\omega})|$ and $Y(z)$ has no zeros outside the unit circle. One approach that achieves this objective is depicted in Figure P13.14. Determine the required sequence $\ell[n]$. A possible application of the system in Figure P13.14 is to stabilize an unstable system by applying the transformation of Figure P13.14 to the sequence of coefficients of the denominator of the system function.

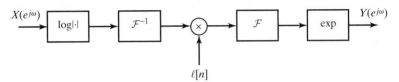

Figure P13.14

13.15. It can be shown (see Problem 3.50) that if $x[n] = 0$ for $n < 0$, then

$$x[0] = \lim_{z \to \infty} X(z).$$

This result was called the *initial value theorem for right-sided sequences*.

(a) Prove a similar result for *left-sided sequences*, i.e., for sequences such that $x[n] = 0$ for $n > 0$.

(b) Use the initial value theorems to prove that $\hat{x}[0] = \log(x[0])$ if $x[n]$ is a minimum-phase sequence.

(c) Use the initial value theorems to prove that $\hat{x}[0] = \log(x[0])$ if $x[n]$ is a maximum-phase sequence.

(d) Use the initial value theorems to prove that $\hat{x}[0] = \log|A|$ when $X(z)$ is given by Eq. (13.32). Is this result consistent with the results of parts (b) and (c)?

13.16. Consider a sequence $x[n]$ with complex cepstrum $\hat{x}[n]$, such that $\hat{x}[n] = -\hat{x}[-n]$. Determine the quantity

$$E = \sum_{n=-\infty}^{\infty} x^2[n].$$

13.17. Consider a real, stable, even, two-sided sequence $h[n]$. The Fourier transform of $h[n]$ is positive for all ω, i.e.,

$$H(e^{j\omega}) > 0, \qquad -\pi < \omega \le \pi.$$

Assume that the z-transform of $h[n]$ exists. Do not assume that $H(z)$ is rational.

(a) Show that there exists a minimum-phase signal $g[n]$, such that

$$H(z) = G(z)G(z^{-1}),$$

where $G(z)$ is the z-transform of a sequence $g[n]$, which has the property that $g[n] = 0$ for $n < 0$. State explicitly the relationship between $\hat{h}[n]$ and $\hat{g}[n]$, the complex cepstra of $h[n]$ and $g[n]$, respectively.

(b) Given a stable signal $s[n]$, with rational z-transform

$$S(z) = \frac{(1 - 2z^{-1})(1 - \frac{1}{2}z^{-1})}{(1 - 4z^{-1})(1 - \frac{1}{3}z^{-1})}.$$

Define $h[n] = s[n] * s[-n]$. Find $G(z)$ (as in part (a)) in terms of $S(z)$.

(c) Consider the system in Figure P13.17, where $\ell[n]$ is defined as

$$\ell[n] = u[n - 1] + (-1)^n u[n - 1].$$

Determine the *most general* conditions on $x[n]$ such that $y[n] = x[n]$ for all n.

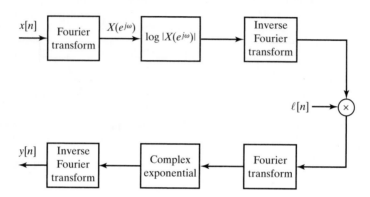

Figure P13.17

13.18. Consider a maximum-phase signal $x[n]$.

(a) Show that the complex cepstrum $\hat{x}[n]$ of a maximum-phase signal is related to its cepstrum $c_x[n]$ by

$$\hat{x}[n] = c_x[n]\ell_{max}[n],$$

where $\ell_{max}[n] = 2u[-n] - \delta[n]$.

(b) Using the relationships in part (a), show that

$$\arg\{X(e^{j\omega})\} = \frac{1}{2\pi}P\int_{-\pi}^{\pi} \log|X(e^{j\theta})| \cot\left(\frac{\omega - \theta}{2}\right) d\theta.$$

(c) Also show that

$$\log|X(e^{j\omega})| = \hat{x}[0] - \frac{1}{2\pi}P\int_{-\pi}^{\pi} \arg\{X(e^{j\theta})\} \cot\left(\frac{\omega - \theta}{2}\right) d\theta.$$

13.19. Consider a sequence $x[n]$ with Fourier transform $X(e^{j\omega})$ and complex cepstrum $\hat{x}[n]$. A new signal $y[n]$ is obtained by homomorphic filtering where

$$\hat{y}[n] = (\hat{x}[n] - \hat{x}[-n])u[n - 1].$$

(a) Show that $y[n]$ is a minimum-phase sequence.

(b) What is the phase of $Y(e^{j\omega})$?

(c) Obtain a relationship between $\arg[Y(e^{j\omega})]$ and $\log|Y(e^{j\omega})|$.

(d) If $x[n]$ is minimum phase, how is $y[n]$ related to $x[n]$?

13.20. Equation (13.65) represents a recursive relationship between a sequence $x[n]$ and its complex cepstrum $\hat{x}[n]$. Show from Eq. (13.65) that the characteristic system $D_*[\cdot]$ behaves as a causal system for minimum-phase inputs; i.e., show that for minimum-phase inputs, $\hat{x}[n]$ is dependent only on $x[k]$ for $k \leq n$.

13.21. Describe a procedure for computing a causal sequence $x[n]$, for which

$$X(z) = -z^3 \frac{(1 - 0.95z^{-1})^{2/5}}{(1 - 0.9z^{-1})^{7/13}}.$$

13.22. The sequence

$$h[n] = \delta[n] + \alpha\delta[n - n_0]$$

is a simplified model for the impulse response of a system that introduces an echo.

(a) Determine the complex cepstrum $\hat{h}[n]$ for this sequence. Sketch the result.

(b) Determine and sketch the cepstrum $c_h[n]$.

(c) Suppose that an approximation to the complex cepstrum is computed using N-point DFTs as in Eqs. (13.46a) to (13.46c). Obtain a closed-form expression for the approximation $\hat{h}_p[n], 0 \leq n \leq N - 1$, for the case $n_0 = N/6$. Assume that phase unwrapping can be accurately done. What happens if N is not divisible by n_0?

(d) Repeat part (c) for the cepstrum approximation $c_{xp}[n], 0 \leq n \leq N - 1$, as computed using Eqs. (13.60a) and (13.60b).

(e) If the largest impulse in the cepstrum approximation $c_{xp}[n]$ is to be used to detect the value of the echo delay n_0, how large must N be to avoid ambiguity? Assume that accurate phase unwrapping can be achieved with this value of N.

13.23. Let $x[n]$ be a *finite-length* minimum-phase sequence with complex cepstrum $\hat{x}[n]$, and define $y[n]$ as

$$y[n] = \alpha^n x[n]$$

with complex cepstrum $\hat{y}[n]$.

(a) If $0 < \alpha < 1$, how is $\hat{y}[n]$ related to $\hat{x}[n]$?

(b) How should α be chosen so that $y[n]$ is no longer minimum phase?

(c) How should α be chosen so that if linear-phase terms are removed before computing the complex cepstrum, then $\hat{y}[n] = 0$ for $n > 0$?

13.24. Consider a minimum-phase sequence $x[n]$ with z-transform $X(z)$ and complex cepstrum $\hat{x}[n]$. A new complex cepstrum is defined by the relation

$$\hat{y}[n] = (\alpha^n - 1)\hat{x}[n].$$

Determine the z-transform $Y(z)$. Is the result also minimum phase?

13.25. Section 13.9.4 contains an example of how the complex cepstrum can be used to obtain two different decompositions involving convolution of a minimum-phase sequence with another sequence. In that example,

$$X(z) = \frac{(0.98 + z^{-1})(1 + 0.9z^{-15} + 0.81z^{-30})}{(1 - 0.9e^{j\pi/6}z^{-1})(1 - 0.9e^{-j\pi/6}z^{-1})}.$$

(a) In one decomposition, $X(z) = X_{min}(z)X_{ap}(z)$ where

$$X_{min}(z) = \frac{(1 + 0.98z^{-1})(1 + 0.9z^{-15} + 0.81z^{-30})}{(1 - 0.9e^{j\pi/6}z^{-1})(1 - 0.9e^{-j\pi/6}z^{-1})}$$

and

$$X_{ap}(z) = \frac{(0.98 + z^{-1})}{(1 + 0.98z^{-1})}.$$

Use the power series expansion of the logarithmic terms to find the complex cepstra $\hat{x}_{min}[n]$, $\hat{x}_{ap}[n]$, and $\hat{x}[n]$. Plot these sequences and compare your plots with those in Figure 13.19.

(b) In the second decomposition, $X(z) = X_{mn}(z)X_{mx}(z)$ where

$$X_{mn}(z) = \frac{z^{-1}(1 + 0.9z^{-15} + 0.81z^{-30})}{(1 - 0.9e^{j\pi/6}z^{-1})(1 - 0.9e^{-j\pi/6}z^{-1})}$$

and

$$X_{mx}(z) = (0.98z + 1).$$

Use the power series expansion of the logarithmic terms to find the complex cepstra and show that $\hat{x}_{mn}[n] \neq \hat{x}_{min}[n]$ but that $\hat{x}[n] = \hat{x}_{mn}[n] + \hat{x}_{mx}[n]$ is the same as in part (a). Note that

$$(1 + 0.9z^{-15} + 0.81z^{-30}) = \frac{(1 - (0.9)^3 z^{-45})}{(1 - 0.9z^{-15})}.$$

13.26. Suppose that $s[n] = h[n] * g[n] * p[n]$, where $h[n]$ is a minimum-phase sequence, $g[n]$ is a maximum-phase sequence, and $p[n]$ is

$$p[n] = \sum_{k=0}^{4} \alpha_k \delta[n - kn_0],$$

where α_k and n_0 are not known. Develop a method to separate $h[n]$ from $s[n]$.

Extension Problems

13.27. Let $x[n]$ be a sequence with z-transform $X(z)$ and complex cepstrum $\hat{x}[n]$. The magnitude-squared function for $X(z)$ is

$$V(z) = X(z)X^*(1/z^*).$$

Since $V(e^{j\omega}) = |X(e^{j\omega})|^2 \geq 0$, the complex cepstrum $\hat{v}[n]$ corresponding to $V(z)$ can be computed without phase unwrapping.

(a) Obtain a relationship between the complex cepstrum $\hat{v}[n]$ and the complex cepstrum $\hat{x}[n]$.

(b) Express the complex cepstrum $\hat{v}[n]$ in terms of the cepstrum $c_x[n]$.

(c) Determine the sequence $\ell[n]$ such that

$$\hat{x}_{min}[n] = \ell[n]\hat{v}[n]$$

is the complex cepstrum of a minimum-phase sequence $x_{min}[n]$ for which

$$|X_{min}(e^{j\omega})|^2 = V(e^{j\omega}).$$

(d) Suppose that $X(z)$ is as given by Eq. (13.32). Use the result of part (c) and Eqs. (13.36a), (13.36b), and (13.36c) to find the complex cepstrum of the minimum-phase sequence, and work backward to find $X_{min}(z)$.

The technique employed in part (d) may be used in general to obtain a minimum-phase factorization of a magnitude-squared function.

13.28. Let $\hat{x}[n]$ be the complex cepstrum of $x[n]$. Define a sequence $x_e[n]$ to be

$$x_e[n] = \begin{cases} x[n/N], & n = 0, \pm N, \pm 2N, \ldots, \\ 0, & \text{otherwise.} \end{cases}$$

Show that the complex cepstrum of $x_e[n]$ is given by

$$\hat{x}_e[n] = \begin{cases} \hat{x}[n/N], & n = 0, \pm N, \pm 2N, \ldots, \\ 0, & \text{otherwise.} \end{cases}$$

13.29. In speech analysis, synthesis, and coding, the speech signal is commonly modeled over a short time interval as the response of an LTI system excited by an excitation that switches between a train of equally spaced pulses for voiced sounds and a wideband random noise source for unvoiced sounds. To use homomorphic deconvolution to separate the components of the speech model, the speech signal $s[n] = v[n] * p[n]$ is multiplied by a window sequence $w[n]$ to obtain $x[n] = s[n]w[n]$. To simplify the analysis, $x[n]$ is approximated by

$$x[n] = (v[n] * p[n]) \cdot w[n] \simeq v[n] * (p[n] \cdot w[n]) = v[n] * p_w[n]$$

where $p_w[n] = p[n]w[n]$ as in Eq. (13.123).

(a) Give an example of $p[n]$, $v[n]$, and $w[n]$ for which the above assumption may be a poor approximation.

(b) One approach to estimating the excitation parameters (voiced/unvoiced decision and pulse spacing for voiced speech) is to compute the real cepstrum $c_x[n]$ of the windowed segment of speech $x[n]$ as depicted in Figure P13.29-1. For the model of Section 13.10.1, express $c_x[n]$ in terms of the complex cepstrum $\hat{x}[n]$. How would you use $c_x[n]$ to estimate the excitation parameters?

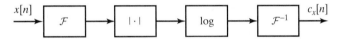

Figure P13.29-1

(c) Suppose that we replace the log operation in Figure P13.29-1 with the "squaring" operation so that the resulting system is as depicted in Figure P13.29-2. Can the new "cepstrum" $q_x[n]$ be used to estimate the excitation parameters? Explain.

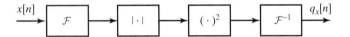

Figure P13.29-2

13.30. Consider a stable LTI system with impulse response $h[n]$ and all-pole system function

$$H(z) = \frac{G}{1 - \displaystyle\sum_{k=1}^{N} a_k z^{-k}}.$$

Such all-pole systems arise in linear-predictive analysis. It is of interest to compute the complex cepstrum directly from the coefficients of $H(z)$.

(a) Determine $\hat{h}[0]$.

(b) Show that

$$\hat{h}[n] = a_n + \sum_{k=1}^{n-1} \left(\frac{k}{n}\right) \hat{h}[k]a_{n-k}, \qquad n \geq 1.$$

With the relations in parts (a) and (b), the complex cepstrum can be computed without phase unwrapping and without solving for the roots of the denominator of $H(z)$.

13.31. A somewhat more general model for echo than the system in Problem 13.22 is the system depicted in Figure P13.31. The impulse response of this system is

$$h[n] = \delta[n] + \alpha g[n - n_0],$$

where $\alpha g[n]$ is the impulse response of the echo path.

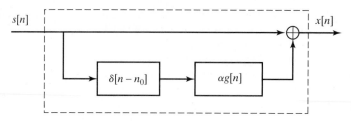

Figure P13.31

(a) Assuming that

$$\max_{-\pi < \omega < \pi} |\alpha G(e^{j\omega})| < 1,$$

show that the complex cepstrum $\hat{h}[n]$ has the form

$$\hat{h}[n] = \sum_{k=1}^{\infty} (-1)^{k+1} \frac{\alpha^k}{k} g_k[n - kn_0],$$

and determine an expression for $g_k[n]$ in terms of $g[n]$.

(b) For the conditions of part (a), determine and sketch the complex cepstrum $\hat{h}[n]$ when $g[n] = \delta[n]$.

(c) For the conditions of part (a), determine and sketch the complex cepstrum $\hat{h}[n]$ when $g[n] = a^n u[n]$. What condition must be satisfied by α and a so that the result of part (a) applies?

(d) For the conditions of part (a), determine and sketch the complex cepstrum $\hat{h}[n]$ when $g[n] = a_0 \delta[n] + a_1 \delta[n - n_1]$. What condition must be satisfied by α, a_0, a_1, and n_1 so that the result of part (a) applies?

13.32. An interesting use of exponential weighting is in computing the complex cepstrum without phase unwrapping. Assume that $X(z)$ has no poles and zeros on the unit circle. Then it is possible to find an exponential weighting factor α in the product $w[n] = \alpha^n x[n]$, such that none of the poles or zeros of $X(z)$ are shifted across the unit circle in forming $W(z) = X(\alpha^{-1}z)$.

(a) Assuming that no poles or zeros of $X(z)$ move across the unit circle, show that

$$\hat{w}[n] = \alpha^n \hat{x}[n]. \tag{P13.32-1}$$

(b) Now suppose that instead of the complex cepstrum, we compute $c_x[n]$ and $c_w[n]$. Use the result of part (a) to obtain expressions for both $c_x[n]$ and $c_w[n]$ in terms of $\hat{x}[n]$.

(c) Now show that

$$\hat{x}[n] = \frac{2(c_x[n] - \alpha^n c_w[n])}{1 - \alpha^{2n}}, \qquad n \neq 0. \tag{P13.32-2}$$

(d) Since $c_x[n]$ and $c_w[n]$ can be computed from $\log |X(e^{j\omega})|$ and $\log |W(e^{j\omega})|$, respectively, Eq. (P13.32-2) is the basis for computing the complex cepstrum without computing the phase of $X(e^{j\omega})$. Discuss some potential problems that might arise with this approach.

A

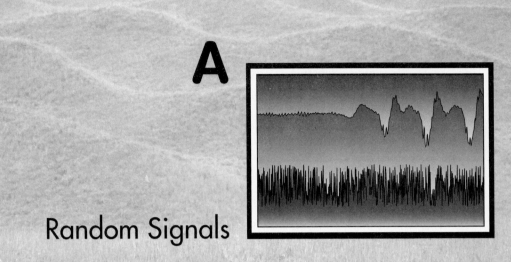

Random Signals

In this appendix, we collect and summarize a number of results and establish the notation relating to the representation of random signals. We make no attempt here to provide a detailed discussion of the difficult and subtle mathematical issues of the underlying theory. Although our approach is not rigorous, we have summarized the important results and the mathematical assumptions implicit in their derivation. Detailed presentation of the theory of random signals are found in texts such as Davenport (1970), Papoulis (1984), Gray and Davidson (2004), Kay (2006), and Bertsekas and Tsitsiklis (2008).

A.1 DISCRETE-TIME RANDOM PROCESSES

The fundamental concept in the mathematical representation of random signals is that of a *random process*. In our discussion of random processes as models for discrete-time signals, we assume that the reader is familiar with the basic concepts of probability, such as random variables, probability distributions, and averages.

In using the random-process model in practical signal-processing applications, we consider a particular sequence to be one of an ensemble of sample sequences. Given a discrete-time signal, the structure, i.e., the underlying probability law, of the corresponding random process is generally not known and must somehow be inferred. It may be possible to make reasonable assumptions about the structure of the process, or it may be possible to estimate the properties of a random-process representation from a finite segment of a typical sample sequence.

Formally, a random process is an indexed family of random variables $\{\mathbf{x}_n\}$ characterized by a set of probability distribution functions that, in general, may be a function of the index n. In using the concept of a random process as a model for discrete-time signals, the index n is associated with the time index. In other words, each sample value $x[n]$ of a random signal is assumed to have resulted from a mechanism that is governed by a probability law. An individual random variable $\mathbf{x}_n$ is described by the probability distribution function

$$P_{\mathbf{x}_n}(x_n, n) = \text{Probability} [\mathbf{x}_n \leq x_n], \tag{A.1}$$

where $\mathbf{x}_n$ denotes the random variable and x_n is a particular value of $\mathbf{x}_n$.[1] If $\mathbf{x}_n$ takes on a continuous range of values, it is equivalently specified by the *probability density function*

$$p_{\mathbf{x}_n}(x_n, n) = \frac{\partial P_{\mathbf{x}_n}(x_n, n)}{\partial x_n}, \tag{A.2}$$

or the *probability distribution function*

$$P_{\mathbf{x}_n}(x_n, n) = \int_{-\infty}^{x_n} p_{\mathbf{x}_n}(x, n)dx. \tag{A.3}$$

The interdependence of two random variables $\mathbf{x}_n$ and $\mathbf{x}_m$ of a random process is described by the joint probability distribution function

$$P_{\mathbf{x}_n,\mathbf{x}_m}(x_n, n, x_m, m) = \text{Probability} [\mathbf{x}_n \leq x_n \text{ and } \mathbf{x}_m \leq x_m] \tag{A.4}$$

and by the joint probability density

$$p_{\mathbf{x}_n,\mathbf{x}_m}(x_n, n, x_m, m) = \frac{\partial^2 P_{\mathbf{x}_n,\mathbf{x}_m}(x_n, n, x_m, m)}{\partial x_n \partial x_m}. \tag{A.5}$$

Two random variables are *statistically independent* if knowledge of the value of one does not affect the probability density of the other. If all the random variables of a collection of random variables, $\{\mathbf{x}_n\}$, are statistically independent, then

$$P_{\mathbf{x}_n,\mathbf{x}_m}(x_n, n, x_m, m) = P_{\mathbf{x}_n}(x_n, n) \cdot P_{\mathbf{x}_m}(x_m, m) \qquad m \neq n. \tag{A.6}$$

A complete characterization of a random process requires the specification of all possible joint probability distributions. As we have indicated, these probability distributions may be a function of the time indices m and n. In the case where all the probability distributions are independent of a shift of time origin, the random process is said to be *stationary*. For example, the 2nd-order distribution of a stationary process satisfies

$$P_{\mathbf{x}_{n+k},\mathbf{x}_{m+k}}(x_{n+k}, n+k, x_{m+k}, m+k) = P_{\mathbf{x}_n\mathbf{x}_m}(x_n, n, x_m, m) \qquad \text{for all } k. \tag{A.7}$$

In many of the applications of discrete-time signal processing, random processes serve as models for signals in the sense that a particular signal can be considered a sample sequence of a random process. Although the details of such signals are unpredictable—making a deterministic approach to signal representation inappropriate—certain average properties of the ensemble can be determined, given the probability law of the process. These average properties often serve as a useful, although incomplete, characterization of such signals.

---

[1] In this appendix, boldface type is used to denote the random variables and regular type denotes dummy variables of probability functions.

A.2 AVERAGES

It is often useful to characterize a random variable by averages such as the mean and variance. Since a random process is an indexed set of random variables, we may likewise characterize the process by statistical averages of the random variables making up the random process. Such averages are called *ensemble averages*. We begin the discussion of averages with some definitions.

A.2.1 Definitions

The average, or mean, of a random process is defined as

$$m_{\mathbf{x}_n} = \mathcal{E}\{\mathbf{x}_n\} = \int_{-\infty}^{\infty} x p_{\mathbf{x}_n}(x, n)dx, \tag{A.8}$$

where $\mathcal{E}$ denotes an operator called *mathematical expectation*. In general, the mean (expected value) may depend on n. In addition, if $g(\cdot)$ is a single-valued function, then $g(\mathbf{x}_n)$ is a random variable, and the set of random variables $\{g(\mathbf{x}_n)\}$ defines a new random process. To compute averages of this new process, we can derive probability distributions of the new random variables. Alternatvely, it can be shown that

$$\mathcal{E}\{g(\mathbf{x}_n)\} = \int_{-\infty}^{\infty} g(x) p_{\mathbf{x}_n}(x, n)dx. \tag{A.9}$$

If the random variables are discrete—i.e., if they have quantized values—the integrals become summations over all possible values of the random variable. In that case $\mathcal{E}\{g(x)\}$ has the form

$$\mathcal{E}\{g(\mathbf{x}_n)\} = \sum_{x} g(x) \hat{p}_{\mathbf{x}_n}(x, n). \tag{A.10}$$

In cases where we are interested in the relationship between multiple random processes, we must be concerned with multiple sets of random variables. For example, for two sets of random variables, $\{\mathbf{x}_n\}$ and $\{\mathbf{y}_m\}$, the expected value of a function of the two random variables is defined as

$$\mathcal{E}\{g(\mathbf{x}_n, \mathbf{y}_m)\} = \int_{-\infty}^{\infty}\int_{-\infty}^{\infty} g(x, y) p_{\mathbf{x}_n, \mathbf{y}_m}(x, n, y, m)dx\, dy, \tag{A.11}$$

where $p_{\mathbf{x}_n, \mathbf{y}_m}(x_m, n, y_m, m)$ is the joint probability density of the random variables $\mathbf{x}_n$ and $\mathbf{y}_m$.

The mathematical expectation operator is a linear operator; that is, it can be shown that

1. $\mathcal{E}\{\mathbf{x}_n + \mathbf{y}_m\} = \mathcal{E}\{\mathbf{x}_n\} + \mathcal{E}\{\mathbf{y}_m\}$; i.e., the average of a sum is the sum of the averages.
2. $\mathcal{E}\{a\mathbf{x}_n\} = a\mathcal{E}\{\mathbf{x}_n\}$; i.e., the average of a constant times $\mathbf{x}_n$ is equal to the constant times the average of $\mathbf{x}_n$.

In general, the average of a product of two random variables is not equal to the product of the averages. When this property holds, however, the two random variables are said to be *linearly independent* or *uncorrelated*. That is, $\mathbf{x}_n$ and $\mathbf{y}_m$ are linearly independent or uncorrelated if

$$\mathcal{E}\{\mathbf{x}_n\mathbf{y}_m\} = \mathcal{E}\{\mathbf{x}_m\} \cdot \mathcal{E}\{\mathbf{y}_m\}. \tag{A.12}$$

It is easy to see from Eqs. (A.11) and (A.12) that a sufficient condition for linear independence is

$$p_{\mathbf{x}_n, \mathbf{y}_m}(x_n, n, y_m, m) = p_{\mathbf{x}_n}(x_n, n) \cdot p_{\mathbf{y}_m}(y_m, m). \tag{A.13}$$

However, Eq. (A.13) is a stronger statement of independence than Eq. (A.12). As previously stated, random variables satisfying Eq. (A.13) are said to be *statistically independent*. If Eq. (A.13) holds for all values of n and m, the random processes $\{\mathbf{x}_n\}$ and $\{\mathbf{y}_m\}$ are said to be statistically independent. Statistically independent random processes are also linearly independent; but the converse is not true: Linear independence does not imply statistical independence.

It can be seen from Eqs. (A.9)–(A.11) that averages generally are functions of the time index. For stationary processes, the mean is the same for all the random variables that constitute the process; i.e., the mean of a stationary process is a constant, which we denote simply m_x.

In addition to the mean of a random process, as defined in Eq. (A.8), a number of other averages are particularly important within the context of signal processing. These are defined next. For notational convenience, we assume that the probability distributions are continuous. Corresponding definitions for discrete random processes can be obtained by applying Eq. (A.10).

The *mean-square* value of $\mathbf{x}_n$ is the average of $|\mathbf{x}_n|^2$; i.e.,

$$\mathcal{E}\{|\mathbf{x}_n|^2\} = \text{mean square} = \int_{-\infty}^{\infty} |x|^2 p_{\mathbf{x}_n}(x, n) dx. \tag{A.14}$$

The mean-square value is sometimes referred to as the *average power*.

The *variance* of $\mathbf{x}_n$ is the mean-square value of $[\mathbf{x}_n - m_{x_n}]$; i.e.,

$$\text{var}[\mathbf{x}_n] = \mathcal{E}\{|(\mathbf{x}_n - m_{x_n})|^2\} = \sigma_{\mathbf{x}_n}^2. \tag{A.15}$$

Since the average of a sum is the sum of the averages, it follows that Eq. (A.15) can be written as

$$\text{var}[\mathbf{x}_n] = \mathcal{E}\{|\mathbf{x}_n|^2\} - |m_{x_n}|^2. \tag{A.16}$$

In general, the mean-square value and the variance are functions of time; however, they are constant for stationary processes.

The mean, mean square, and variance are simple averages that provide only a small amount of information about a process. A more useful average is the *autocorrelation sequence*, which is defined as

$$\phi_{xx}[n, m] = \mathcal{E}\{\mathbf{x}_n \mathbf{x}_m^*\}$$

$$= \int_{-\infty}^{\infty} \int_{-\infty}^{\infty} x_n x_m^* p_{\mathbf{x}_n, \mathbf{x}_m}(x_n, n, x_m, m) dx_n \, dx_m, \tag{A.17}$$

where $*$ denotes complex conjugation. The autocovariance sequence of a random process is defined as

$$\gamma_{xx}[n, m] = \mathcal{E}\{(\mathbf{x}_n - m_{x_n})(\mathbf{x}_m - m_{x_m})^*\}, \tag{A.18}$$

which can be written as

$$\gamma_{xx}[n, m] = \phi_{xx}[n, m] - m_{x_n} m_{x_m}^*. \tag{A.19}$$

Note that, in general, both the autocorrelation and autocovariance are two-dimensional sequences, i.e., functions of two discrete variables.

The autocorrelation sequence is a measure of the dependence between values of the random processes at different times. In this sense, it partially describes the time variation of a random signal. A measure of the dependence between two different random signals is obtained from the cross-correlation sequence. If $\{\mathbf{x}_n\}$ and $\{\mathbf{y}_m\}$ are two random processes, their cross-correlation is

$$\phi_{xy}[n, m] = \mathcal{E}\{\mathbf{x}_n \mathbf{y}_m^*\}$$
$$= \int_{-\infty}^{\infty} \int_{-\infty}^{\infty} xy^* p_{\mathbf{x}_n, \mathbf{y}_m}(x, n, y, m) dx\, dy, \tag{A.20}$$

where $p_{\mathbf{x}_n, \mathbf{y}_m}(x, n, y, m)$ is the joint probability density of $\mathbf{x}_n$ and $\mathbf{y}_m$. The cross-covariance function is defined as

$$\gamma_{xy}[n, m] = \mathcal{E}\{(\mathbf{x}_n - m_{x_n})(\mathbf{y}_m - m_{y_m})^*\}$$
$$= \phi_{xy}[n, m] - m_{x_n} m_{y_m}^*. \tag{A.21}$$

As we have pointed out, the statistical properties of a random process generally vary with time. However, a stationary random process is characterized by an equilibrium condition in which the statistical properties are invariant to a shift of time origin. This means that the 1st-order probability distribution is independent of time. Similarly, all the joint probability functions are also invariant to a shift of time origin; i.e., the 2nd-order joint probability distributions depend only on the time difference $(m - n)$. First-order averages such as the mean and variance are independent of time; 2nd-order averages, such as the autocorrelation $\phi_{xx}[n, m]$, are dependent on the time difference $(m - n)$. Thus, for a stationary process, we can write

$$m_x = \mathcal{E}\{\mathbf{x}_n\}, \tag{A.22}$$
$$\sigma_x^2 = \mathcal{E}\{|(\mathbf{x}_n - m_x)|^2\}, \tag{A.23}$$

both independent of n. If we now denote the time difference by m, we have

$$\phi_{xx}[n + m, n] = \phi_{xx}[m] = \mathcal{E}\{\mathbf{x}_{n+m} \mathbf{x}_n^*\}. \tag{A.24}$$

That is, the autocorrelation sequence of a stationary random process is a one-dimensional sequence, a function of the time difference m.

In many instances, we encounter random processes that are not stationary in the *strict sense*—i.e., their probability distributions are not time invariant—but Eqs. (A.22)–(A.24) still hold. Such random processes are said to be *wide-sense stationary*.

A.2.2 Time Averages

In a signal-processing context, the notion of an ensemble of signals is a convenient mathematical concept that allows us to use the theory of probability to represent the signals. However, in a practical situation, we always have available at most a finite number of finite-length sequences rather than an infinite ensemble of sequences. For example, we might wish to infer the probability law or certain averages of the random-process representation from measurements on a single member of the ensemble. When the probability distributions are independent of time, intuition suggests that the amplitude distribution (histogram) of a long segment of an individual sequence of samples should be approximately equal to the single probability density that describes each of the random variables of the random-process model. Similarly, the arithmetic average of a large number of samples of a single sequence should be very close to the mean of the process. To formalize these intuitive notions, we define the time average of a random process as

$$\langle \mathbf{x}_n \rangle = \lim_{L \to \infty} \frac{1}{2L+1} \sum_{n=-L}^{L} \mathbf{x}_n. \tag{A.25}$$

Similarly, the time autocorrelation sequence is defined as

$$\langle \mathbf{x}_{n+m} \mathbf{x}_n^* \rangle = \lim_{L \to \infty} \frac{1}{2L+1} \sum_{n=-L}^{L} \mathbf{x}_{n+m} \mathbf{x}_n^*. \tag{A.26}$$

It can be shown that the preceding limits exist if $\{\mathbf{x}_n\}$ is a stationary process with finite mean. As defined in Eqs. (A.25) and (A.26), these time averages are functions of an infinite set of random variables and thus are properly viewed as random variables themselves. However, under the condition known as *ergodicity*, the time averages in Eqs. (A.25) and (A.26) are equal to constants in the sense that the time averages of almost all possible sample sequences are equal to the same constant. Furthermore, they are equal to the corresponding ensemble average.[2] That is, for any single sample sequence $\{x[n]\}$ for $-\infty < n < \infty$,

$$\langle x[n] \rangle = \lim_{L \to \infty} \frac{1}{2L+1} \sum_{n=-L}^{L} x[n] = \mathcal{E}\{\mathbf{x}_n\} = m_x \tag{A.27}$$

and

$$\langle x[n+m]x^*[n] \rangle = \lim_{L \to \infty} \frac{1}{2L+1} \sum_{n=-L}^{L} x[n+m]x^*[n] = \mathcal{E}\{\mathbf{x}_{n+m}\mathbf{x}_n^*\} = \phi_{xx}[m]. \tag{A.28}$$

The time-average operator $\langle \cdot \rangle$ has the same properties as the ensemble-average operator $\mathcal{E}\{\cdot\}$. Thus, we generally do not distinguish between the random variable $\mathbf{x}_n$ and its value in a sample sequence, $x[n]$. For example, the expression $\mathcal{E}\{x[n]\}$ should be interpreted as $\mathcal{E}\{\mathbf{x}_n\} = \langle x[n] \rangle$. In general, for *ergodic processes*, time averages equal ensemble averages.

In practice, it is common to assume that a given sequence is a sample sequence of an ergodic random process so that averages can be computed from a single sequence. Of course, we generally cannot compute with the limits in Eqs. (A.27) and (A.28), but instead the quantities

$$\hat{m}_x = \frac{1}{L} \sum_{n=0}^{L-1} x[n], \tag{A.29}$$

$$\hat{\sigma}_x^2 = \frac{1}{L} \sum_{n=0}^{L-1} |x[n] - \hat{m}_x|^2, \tag{A.30}$$

---

[2] A more precise statement is that the random variables $\langle \mathbf{x}_n \rangle$ and $\langle \mathbf{x}_{n+m} \mathbf{x}_n^* \rangle$ have means equal to m_x and $\phi_{xx}[m]$, respectively, and their variances are zero.

and

$$\langle x[n+m]x^*[n]\rangle_L = \frac{1}{L}\sum_{n=0}^{L-1}x[n+m]x^*[n] \tag{A.31}$$

or similar quantities are often computed as *estimates* of the mean, variance, and autocorrelation. $\hat{m}_x$ and $\hat{\sigma}_x^2$ are referred to as the sample mean and sample variance, respectively. The estimation of averages of a random process from a finite segment of data is a problem of statistics, which we touch on briefly in Chapter 10.

A.3 PROPERTIES OF CORRELATION AND COVARIANCE SEQUENCES OF STATIONARY PROCESSES

Several useful properties of correlation and covariance functions follow in a straightforward way from the definitions. These properties are given in this section.

Consider two real stationary random processes $\{x_n\}$ and $\{y_n\}$ with autocorrelation, autocovariance, cross-correlation, and cross-covariance being given, respectively, by

$$\phi_{xx}[m] = \mathcal{E}\{x_{n+m}x_n^*\}, \tag{A.32}$$

$$\gamma_{xx}[m] = \mathcal{E}\{(x_{n+m}-m_x)(x_n-m_x)^*\}, \tag{A.33}$$

$$\phi_{xy}[m] = \mathcal{E}\{x_{n+m}y_n^*\}, \tag{A.34}$$

$$\gamma_{xy}[m] = \mathcal{E}\{(x_{n+m}-m_x)(y_n-m_y)^*\}, \tag{A.35}$$

where m_x and m_y are the means of the two processes. The following properties are easily derived by simple manipulations of the definitions:

Property 1

$$\gamma_{xx}[m] = \phi_{xx}[m] - |m_x|^2, \tag{A.36a}$$

$$\gamma_{xy}[m] = \phi_{xy}[m] - m_x m_y^*. \tag{A.36b}$$

These results follow directly from Eqs. (A.19) and (A.21), and they indicate that the correlation and covariance sequences are identical for zero-mean processes.

Property 2

$$\phi_{xx}[0] = \mathcal{E}[|x_n|^2] = \text{Mean-square value}, \tag{A.37a}$$

$$\gamma_{xx}[0] = \sigma_x^2 = \text{Variance}. \tag{A.37b}$$

Property 3

$$\phi_{xx}[-m] = \phi_{xx}^*[m], \tag{A.38a}$$

$$\gamma_{xx}[-m] = \gamma_{xx}^*[m], \tag{A.38b}$$

$$\phi_{xy}[-m] = \phi_{yx}^*[m], \tag{A.38c}$$

$$\gamma_{xy}[-m] = \gamma_{yx}^*[m]. \tag{A.38d}$$

Property 4

$$|\phi_{xy}[m]|^2 \le \phi_{xx}[0]\phi_{yy}[0], \qquad (A.39a)$$

$$|\gamma_{xy}[m]|^2 \le \gamma_{xx}[0]\gamma_{yy}[0]. \qquad (A.39b)$$

In particular,

$$|\phi_{xx}[m]| \le \phi_{xx}[0], \qquad (A.40a)$$

$$|\gamma_{xx}[m]| \le \gamma_{xx}[0]. \qquad (A.40b)$$

Property 5. If $\mathbf{y}_n = \mathbf{x}_{n-n_0}$, then

$$\phi_{yy}[m] = \phi_{xx}[m], \qquad (A.41a)$$

$$\gamma_{yy}[m] = \gamma_{xx}[m]. \qquad (A.41b)$$

Property 6. For many random processes, the random variables become uncorrelated as they become more separated in time. If this is true,

$$\lim_{m \to \infty} \gamma_{xx}[m] = 0, \qquad (A.42a)$$

$$\lim_{m \to \infty} \phi_{xx}[m] = |m_x|^2, \qquad (A.42b)$$

$$\lim_{m \to \infty} \gamma_{xy}[m] = 0, \qquad (A.42c)$$

$$\lim_{m \to \infty} \phi_{xy}[m] = m_x m_y^*. \qquad (A.42d)$$

The essence of these results is that the correlation and covariance are finite-energy sequences that tend to die out for large values of m. Thus, it is often possible to represent these sequences in terms of their Fourier transforms or z-transforms.

A.4 FOURIER TRANSFORM REPRESENTATION OF RANDOM SIGNALS

Although the Fourier transform of a random signal does not exist except in a generalized sense, the autocovariance and autocorrelation sequences of such a signal are aperiodic sequences for which the transform does exist. The spectral representation of the correlation functions plays an important role in describing the input–output relations for a linear time-invariant system when the input is a random signal. Therefore, it is of interest to consider the properties of correlation and covariance sequences and their corresponding Fourier and z-transforms.

We define $\Phi_{xx}(e^{j\omega})$, $\Gamma_{xx}(e^{j\omega})$, $\Phi_{xy}(e^{j\omega})$, and $\Gamma_{xy}(e^{j\omega})$ as the DTFTs of $\phi_{xx}[m]$, $\gamma_{xx}[m]$, $\phi_{xy}[m]$, and $\gamma_{xy}[m]$, respectively. Since these functions are all DTFTs of sequences, they must be periodic with period 2π. From Eqs. (A.36a) and (A.36b), it follows that, over one period $|\omega| \le \pi$,

$$\Phi_{xx}(e^{j\omega}) = \Gamma_{xx}(e^{j\omega}) + 2\pi |m_x|^2 \delta(\omega), \qquad |\omega| \le \pi, \qquad (A.43a)$$

and

$$\Phi_{xy}(e^{j\omega}) = \Gamma_{xy}(e^{j\omega}) + 2\pi m_x m_y^* \delta(\omega), \qquad |\omega| \le \pi. \qquad (A.43b)$$

In the case of zero-mean processes ($m_x = 0$ and $m_y = 0$), the correlation and covariance functions are identical so that $\Phi_{xx}(e^{j\omega}) = \Gamma_{xx}(e^{j\omega})$ and $\Phi_{xy}(e^{j\omega}) = \Gamma_{xy}(e^{j\omega})$.

From the inverse Fourier transform equation, it follows that

$$\gamma_{xx}[m] = \frac{1}{2\pi} \int_{-\pi}^{\pi} \Gamma_{xx}(e^{j\omega}) e^{j\omega m} d\omega, \tag{A.44a}$$

$$\phi_{xx}[m] = \frac{1}{2\pi} \int_{-\pi}^{\pi} \Phi_{xx}(e^{j\omega}) e^{j\omega m} d\omega, \tag{A.44b}$$

and, consequently,

$$\mathcal{E}\{|x[n]|^2\} = \phi_{xx}[0] = \sigma_x^2 = \frac{1}{2\pi} \int_{-\pi}^{\pi} \Phi_{xx}(e^{j\omega}) d\omega, \tag{A.45a}$$

$$\sigma_x^2 = \gamma_{xx}[0] = \frac{1}{2\pi} \int_{-\pi}^{\pi} \Gamma_{xx}(e^{j\omega}) d\omega. \tag{A.45b}$$

Sometimes it is notationally convenient to define the quantity

$$P_{xx}(\omega) = \Phi_{xx}(e^{j\omega}), \tag{A.46}$$

in which case Eqs. (A.45a) and (A.45b) are expressed as

$$\mathcal{E}\{|x[n]|^2\} = \frac{1}{2\pi} \int_{-\pi}^{\pi} P_{xx}(\omega) d\omega, \tag{A.47a}$$

$$\sigma_x^2 = \frac{1}{2\pi} \int_{-\pi}^{\pi} P_{xx}(\omega) d\omega. \tag{A.47b}$$

Thus, the area under $P_{xx}(\omega)$ for $-\pi \leq \omega \leq \pi$ is proportional to the average power in the signal. In fact, as we discussed in Section 2.10, the integral of $P_{xx}(\omega)$ over a band of frequencies is proportional to the power in the signal in that band. For this reason, the function $P_{xx}(\omega)$ is called the *power density spectrum,* or simply, the *power spectrum.* When $P_{xx}(\omega)$ is a constant independent of ω, the random process is referred to as a white-noise process, or simply, white noise. When $P_{xx}(\omega)$ is constant over a band and zero otherwise, we refer to it as bandlimited white noise.

From Eq. (A.38a), it can be shown that $P_{xx}(\omega) = P_{xx}^*(\omega)$; i.e., $P_{xx}(\omega)$ is always real valued. Furthermore, for real random processes, $\phi_{xx}[m] = \phi_{xx}[-m]$, so in the real case, $P_{xx}(\omega)$ is both real and even; i.e.,

$$P_{xx}(\omega) = P_{xx}(-\omega). \tag{A.48}$$

An additional important property is that the power density spectrum is nonnegative; i.e., $P_{xx}(\omega) \geq 0$ for all ω. This point is discussed in Section 2.10.

The *cross power density spectrum* is defined as

$$P_{xy}(\omega) = \Phi_{xy}(e^{j\omega}). \tag{A.49}$$

This function is generally complex, and from Eq. (A.38c), it follows that

$$P_{xy}(\omega) = P_{yx}^*(\omega). \tag{A.50}$$

Finally, as shown in Section 2.10, if $x[n]$ is a random signal input to a linear time-invariant discrete-time system with frequency response $H(e^{j\omega})$, and if $y[n]$ is the corresponding output, then

$$\Phi_{yy}(e^{j\omega}) = |H(e^{j\omega})|^2 \Phi_{xx}(e^{j\omega}) \tag{A.51}$$

and

$$\Phi_{xy}(e^{j\omega}) = H(e^{j\omega})\Phi_{xx}(e^{j\omega}). \tag{A.52}$$

Example A.1   Noise Power Output of Ideal Lowpass Filter

Suppose that $x[n]$ is a zero-mean white-noise sequence with $\phi_{xx}[m] = \sigma_x^2 \delta[m]$ and power spectrum $\Phi_{xx}(e^{j\omega}) = \sigma_x^2$ for $|\omega| \leq \pi$, and furthermore, assume that $x[n]$ is the input to an ideal lowpass filter with cutoff frequency ω_c. Then from Eq. (A.51), it follows that the output $y[n]$ would be a bandlimited white noise process whose power spectrum would be

$$\Phi_{yy}(e^{j\omega}) = \begin{cases} \sigma_x^2, & |\omega| < \omega_c, \\ 0, & \omega_c < |\omega| \leq \pi. \end{cases} \tag{A.53}$$

Using the inverse Fourier transform, we obtain the autocorrelation sequence

$$\phi_{yy}[m] = \frac{\sin(\omega_c m)}{\pi m}\sigma_x^2. \tag{A.54}$$

Now, using Eq. (A.45a), we get for the average power of the output,

$$\mathcal{E}\{y^2[n]\} = \phi_{yy}[0] = \frac{1}{2\pi}\int_{-\omega_c}^{\omega_c} \sigma_x^2 d\omega = \sigma_x^2 \frac{\omega_c}{\pi}. \tag{A.55}$$

A.5 USE OF THE z-TRANSFORM IN AVERAGE POWER COMPUTATIONS

To carry out average power calculations using Eq. (A.45a), we must evaluate an integral of the power spectrum as was done in Example A.1. While the integral in that example was easy to evaluate, such integrals in general are difficult to evaluate as real integrals. However, a result based on the z-transform makes the calculation of average output power straightforward in the important case of systems that have rational system functions.

In general, the z-transform can be used to represent the covariance function but not a correlation function. This is because when a signal has nonzero average value, its correlation function will contain an additive constant component that does not have a z-transform representation. When the average value is zero, however, the covariance and correlation functions are, of course, equal. If the z-transform of $\gamma_{xx}[m]$ exists, then since $\gamma_{xx}[-m] = \gamma_{xx}^*[m]$ it follows that in general

$$\Gamma_{xx}(z) = \Gamma_{xx}^*(1/z^*). \tag{A.56}$$

Furthermore, since $\gamma_{xx}[m]$ is two sided and conjugate-symmetric, it follows that the region of convergence of $\Gamma_{xx}(z)$ must be of the form

$$r_a < |z| < \frac{1}{r_a}$$

where necessarily $0 < r_a < 1$. In the important case when $\Gamma_{xx}(z)$ is a rational function of z, Eq. (A.56) implies that the poles and zeros of $\Gamma_{xx}(z)$ must occur in complex-conjugate reciprocal pairs.

The major advantage of the *z*-transform representation is that when $\Gamma_{xx}(z)$ is a rational function, the average power of the random signal can be computed easily using the relation

$$\mathcal{E}\{|x[n] - m_x|^2\} = \sigma_x^2 = \gamma_{xx}[0] = \left\{ \begin{array}{c} \text{Inverse } z\text{-transform} \\ \text{of } \Gamma_{xx}(z), \\ \text{evaluated for } m = 0 \end{array} \right\}. \tag{A.57}$$

It is straightforward to evaluate the right-hand side of this equation using a method based on the observation that when $\Gamma_{xx}(z)$ is a rational function of z, $\gamma_{xx}[m]$ can be computed for all m by employing a partial fraction expansion. Then to obtain the average power, we can simply evaluate $\gamma_{xx}[m]$ for $m = 0$.

The *z*-transform is also useful in determining the autocovariance and average power of the output of an LTI system when the input is a random signal. Generalizing Eq. (A.51) leads to

$$\Gamma_{yy}(z) = H(z)H^*(1/z^*)\Gamma_{xx}(z), \tag{A.58}$$

and from the properties of the *z*-transform and Eq. (A.58), it follows that the autocovariance of the output is the convolution

$$\gamma_{yy}[m] = h[m] * h^*[-m] * \gamma_{xx}[m]. \tag{A.59}$$

This result is particularly useful in quantization noise analysis where we need to compute the average output power when the input to a linear difference equation is a zero-mean white noise signal with average power σ_x^2. Since the autocovariance of such an input is $\gamma_{xx}[m] = \sigma_x^2 \delta[m]$, it follows that the autocovariance of the output is $\gamma_{yy}[m] = \sigma_x^2(h[m] * h^*[-m])$, i.e., the covariance of the output is proportional to the deterministic autocorrelation of the impulse response of the LTI system. From this result it follows that

$$\mathcal{E}\{y^2[n]\} = \gamma_{yy}[0] = \sigma_x^2 \sum_{n=-\infty}^{\infty} |h[n]|^2. \tag{A.60}$$

As an alternative to computing the sum of squares of the impulse response sequence, which can be rather difficult for IIR systems, we can apply the method suggested in Eq. (A.57) to obtain $\mathcal{E}\{y^2[n]\}$ from a partial fraction expansion of $\Gamma_{yy}(z)$. Recall that for a white noise input with $\gamma_{xx}[m] = \sigma_x^2 \delta[m]$, the *z*-transform is $\Gamma_{xx}(z) = \sigma_x^2$ so $\Gamma_{yy}(z) = \sigma_x^2 H(z)H^*(1/z^*)$. Therefore, Eq. (A.57) applied to the output of the system gives

$$\mathcal{E}\{y^2[n]\} = \gamma_{yy}[0] = \left\{ \begin{array}{c} \text{Inverse } z\text{-transform of} \\ \Gamma_{yy}(z) = H(z)H^*(1/z^*)\sigma_x^2, \\ \text{evaluated for } m = 0 \end{array} \right\}. \tag{A.61}$$

Now consider the special case of a stable and causal system having a rational system function of the form

$$H(z) = A \frac{\displaystyle\prod_{m=1}^{M}(1 - c_m z^{-1})}{\displaystyle\prod_{k=1}^{N}(1 - d_k z^{-1})} \qquad |z| > \max_{k}\{|d_k|\}, \tag{A.62}$$

where $\max_k\{|d_k|\} < 1$ and $M < N$. Such a system function might describe the relationship between an internal round-off noise source and the output of a system implemented with fixed-point arithmetic. Substituting Eq. (A.62) for $H(z)$ in Eq. (A.58) gives

$$\Gamma_{yy}(z) = \sigma_x^2 H(z) H^*(1/z^*) = \sigma_x^2 |A|^2 \frac{\displaystyle\prod_{m=1}^{M}(1 - c_m z^{-1})(1 - c_m^* z)}{\displaystyle\prod_{k=1}^{N}(1 - d_k z^{-1})(1 - d_k^* z)}. \tag{A.63}$$

Since we have assumed that $|d_k| < 1$ for all k, all of the original poles are inside the unit circle and therefore the other poles at $(d_k^*)^{-1}$ are at conjugate reciprocal locations outside the unit circle. The region of convergence for $\Gamma_{yy}(z)$ is therefore $\max_k |d_k| < |z| < \min_k |(d_k^*)^{-1}|$. For such rational functions, it can be shown that since $M < N$, the partial fraction expansion has the form

$$\Gamma_{yy}(z) = \sigma_x^2 \left(\sum_{k=1}^{N} \left(\frac{A_k}{1 - d_k z^{-1}} - \frac{A_k^*}{1 - (d_k^*)^{-1} z^{-1}} \right) \right), \tag{A.64}$$

where the coefficients are found from

$$A_k = H(z) H^*(1/z^*)(1 - d_k z^{-1}) \Big|_{z=d_k}. \tag{A.65}$$

Since the poles at $z = d_k$ are inside the inner boundary of the region of convergence, each of them corresponds to a right-sided sequence, while the poles at $z = (d_k^*)^{-1}$ each correspond to a left-sided sequence. Thus, the autocovariance function corresponding to Eq. (A.64) is

$$\gamma_{yy}[n] = \sigma_x^2 \sum_{k=1}^{N} (A_k (d_k)^n u[n] + A_k^* (d_k^*)^{-n} u[-n-1]),$$

from which it follows that we can obtain the average power from

$$\sigma_y^2 = \gamma_{yy}[0] = \sigma_x^2 \left(\sum_{k=1}^{N} A_k \right), \tag{A.66}$$

where the quantities A_k are given by Eq. (A.65).

Thus, the computation of the total average power of the output of a system with rational system function and white noise input reduces to the straightforward problem of finding partial fraction expansion coefficients for the z-transform of the output autocorrelation function. The utility of this approach is illustrated by the following example.

Example A.2   Noise Power Output of a 2nd-Order IIR Filter

Consider a system with impulse response

$$h[n] = \frac{r^n \sin\theta(n+1)}{\sin\theta} u[n] \tag{A.67}$$

and system function

$$H(z) = \frac{1}{(1 - re^{j\theta}z^{-1})(1 - re^{-j\theta}z^{-1})}. \tag{A.68}$$

When the input is white noise with total average power σ_x^2, the z-transform of the autocovariance function of the output is

$$\Gamma_{yy}(z) = \sigma_x^2 \left(\frac{1}{(1 - re^{j\theta}z^{-1})(1 - re^{-j\theta}z^{-1})} \right) \left(\frac{1}{(1 - re^{-j\theta}z)(1 - re^{j\theta}z)} \right) \tag{A.69}$$

from which we obtain, using Eq. (A.65),

$$\mathcal{E}\{y^2[n]\} = \sigma_x^2 \left[\left(\frac{1}{(1 - re^{-j\theta}z^{-1})} \right) \left(\frac{1}{(1 - re^{-j\theta}z)(1 - re^{j\theta}z)} \right) \bigg|_{z=re^{j\theta}} \right.$$

$$\left. + \left(\frac{1}{(1 - re^{j\theta}z^{-1})} \right) \left(\frac{1}{(1 - re^{-j\theta}z)(1 - re^{j\theta}z)} \right) \bigg|_{z=re^{-j\theta}} \right]. \tag{A.70}$$

Making the indicated substitutions, placing both terms over a common denominator, and doing some algebra leads to

$$\mathcal{E}\{y^2[n]\} = \sigma_x^2 \left(\frac{1 + r^2}{1 - r^2} \right) \left(\frac{1}{1 - 2r^2 \cos(2\theta) + r^4} \right). \tag{A.71}$$

Thus, using the partial fraction expansion of $\Gamma_{yy}(z)$, we have effectively evaluated the expression

$$\mathcal{E}\{y^2[n]\} = \sigma_x^2 \sum_{n=-\infty}^{\infty} |h[n]|^2 = \sigma_x^2 \sum_{n=0}^{\infty} \left| \frac{r^n \sin\theta(n+1)}{\sin\theta} \right|^2,$$

which would be difficult to sum in closed form, and the expression

$$\mathcal{E}\{y^2[n]\} = \frac{1}{2\pi} \int_{-\pi}^{\pi} \sigma_x^2 |H(e^{j\omega})|^2 d\omega = \frac{\sigma_x^2}{2\pi} \int_{-\pi}^{\pi} \frac{d\omega}{|(1 - re^{j\theta}e^{-j\omega})(1 - re^{-j\theta}e^{-j\omega})|^2},$$

which would be difficult to evaluate as an integral over the real variable ω.

The result of Example A.2 is an illustration of the power of the partial fraction method in evaluating average power formulas. In Chapter 6, we make use of this technique in the analysis of quantization effects in the implementation of digital filters.

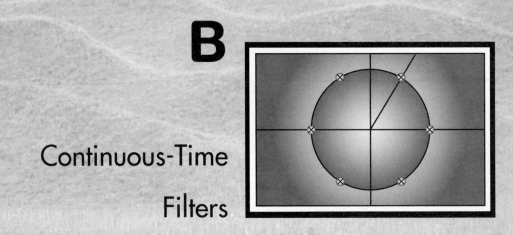

B

Continuous-Time

Filters

The techniques discussed in Chapter 7 for designing IIR digital filters rely on the availability of appropriate continuous-time filter designs. In this appendix, we briefly summarize the characteristics of several classes of lowpass filter approximations that we referred to in Chapter 7. More detailed discussions of these classes of filters appear in Guillemin (1957), Weinberg (1975) and Parks and Burrus (1987), and extensive design tables and formulas are found in Zverev (1967). Design programs for all the common continuous-time approximations and transformations to digital filters are available in MATLAB, Simulink, and LabVIEW.

B.1 BUTTERWORTH LOWPASS FILTERS

Butterworth lowpass filters are defined by the property that the magnitude response is maximally flat in the passband. For an Nth-order lowpass filter, this means that the first $(2N - 1)$ derivatives of the magnitude-squared function are zero at $\Omega = 0$. Another property is that the magnitude response is monotonic in the passband and the stopband. The magnitude-squared function for a continuous-time Butterworth lowpass filter has the form

$$|H_c(j\Omega)|^2 = \frac{1}{1 + (j\Omega/j\Omega_c)^{2N}}. \tag{B.1}$$

This function is plotted in Figure B.1.

As the parameter N in Eq. (B.1) increases, the filter characteristics become sharper: that is, they remain close to unity over more of the passband and become close to zero more rapidly in the stopband, although the magnitude-squared function at the cutoff frequency Ω_c will always be equal to one-half because of the nature of Eq. (B.1). The

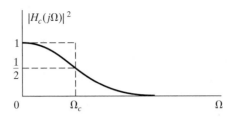

Figure B.1 Magnitude-squared function for continuous-time Butterworth filter.

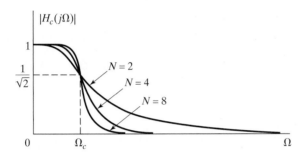

Figure B.2 Dependence of Butterworth magnitude characteristics on the order N.

dependence of the Butterworth filter characteristic on the parameter N is indicated in Figure B.2, which shows $|H_c(j\Omega)|$ for several values of N.

From the magnitude-squared function in Eq. (B.1), we observe by substituting $j\Omega = s$ that $H_c(s)H_c(-s)$ must be of the form

$$H_c(s)H_c(-s) = \frac{1}{1 + (s/j\Omega_c)^{2N}}. \tag{B.2}$$

The roots of the denominator polynomial (the poles of the magnitude-squared function) are therefore located at values of s satisfying $1 + (s/j\Omega_c)^{2N} = 0$; i.e.,

$$s_k = (-1)^{1/2N}(j\Omega_c) = \Omega_c e^{(j\pi/2N)(2k+N-1)}, \qquad k = 0, 1, \ldots, 2N - 1. \tag{B.3}$$

Thus, there are $2N$ poles equally spaced in angle on a circle of radius Ω_c in the s-plane. The poles are symmetrically located with respect to the imaginary axis. A pole never falls on the imaginary axis, and one occurs on the real axis for N odd, but not for N even. The angular spacing between the poles on the circle is π/N radians. For example, for $N = 3$, the poles are spaced by $\pi/3$ radians, or 60 degrees, as indicated in Figure B.3. To determine the system function of the analog filter to associate with the Butterworth magnitude-squared function, we perform the factorization $H_c(s)H_c(-s)$. The poles of the magnitude-squared function always occur in pairs; i.e., if there is a pole at $s = s_k$, then a pole also occurs at $s = -s_k$. Consequently, to construct $H_c(s)$ from the magnitude-squared function, we would choose the one pole from each such pair. To obtain a stable and causal filter, we should choose all the poles on the left-half-plane part of the s-plane. With this approach, $H_c(s)$ would be

$$H_c(s) = \frac{\Omega_c^3}{(s + \Omega_c)(s - \Omega_c e^{j2\pi/3})(s - \Omega_c e^{-j2\pi/3})},$$

which can be written as

$$H_c(s) = \frac{\Omega_c^3}{s^3 + 2\Omega_c s^2 + 2\Omega_c^2 s + \Omega_c^3}.$$

In general the numerator of $H_c(s)$ would be Ω_c^N to ensure that $|H_c(0)| = 1$.

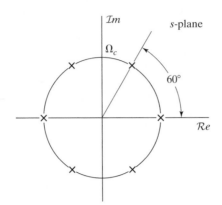

Figure B.3 s-plane pole locations for the magnitude-squared function of 3rd-order Butterworth filter.

B.2 CHEBYSHEV FILTERS

In a Butterworth filter, the magnitude response is monotonic in both the passband and the stopband. Consequently, if the filter specifications are in terms of maximum passband and stopband approximation error, the specifications are exceeded toward the low-frequency end of the passband and above the stopband cutoff frequency. A more efficient approach, which usually leads to a lower order filter, is to distribute the accuracy of the approximation uniformly over the passband or the stopband (or both). This is accomplished by choosing an approximation that has an equiripple behavior rather than a monotonic behavior. The class of Chebyshev filters has the property that the magnitude of the frequency response is either equiripple in the passband and monotonic in the stopband (referred to as a type I Chebyshev filter) or monotonic in the passband and equiripple in the stopband (a type II Chebyshev filter). The frequency response of a type I Chebyshev filter is shown in Figure B.4. The magnitude-squared function for this filter is of the form

$$|H_c(j\Omega)|^2 = \frac{1}{1 + \varepsilon^2 V_N^2(\Omega/\Omega_c)}, \tag{B.4}$$

where $V_N(x)$ is the Nth-order Chebyshev polynomial defined as

$$V_N(x) = \cos(N \cos^{-1} x). \tag{B.5}$$

For example, for $N = 0$, $V_0(x) = 1$; for $N = 1$, $V_1(x) = \cos(\cos^{-1} x) = x$; for $N = 2$, $V_2(x) = \cos(2 \cos^{-1} x) = 2x^2 - 1$; and so on.

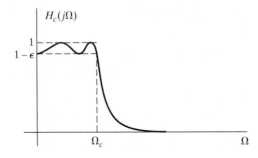

Figure B.4 Type I Chebyshev lowpass filter approximation.

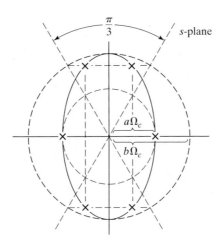

Figure B.5 Location of poles for the magnitude-squared function of 3rd-order type I lowpass Chebyshev filter.

From Eq. (B.5), which defines the Chebyshev polynomials, it is straightforward to obtain a recurrence formula from which $V_{N+1}(x)$ can be obtained from $V_N(x)$ and $V_{N-1}(x)$. By applying trigonometric identities to Eq. (B.5), it follows that

$$V_{N+1}(x) = 2x V_N(x) - V_{N-1}(x). \tag{B.6}$$

From Eq. (B.5), we note that $V_N^2(x)$ varies between zero and unity for $0 < x < 1$. For $x > 1$, $\cos^{-1} x$ is imaginary, so $V_N(x)$ behaves as a hyperbolic cosine and consequently increases monotonically. Referring to Eq. (B.4), we see that $|H_c(j\Omega)|^2$ ripples between 1 and $1/(1 + \varepsilon^2)$ for $0 \le \Omega/\Omega_c \le 1$ and decreases monotonically for $\Omega/\Omega_c > 1$. Three parameters are required to specify the filter: ε, Ω_c, and N. In a typical design, ε is specified by the allowable passband ripple and Ω_c is specified by the desired passband cutoff frequency. The order N is then chosen so that the stopband specifications are met.

The poles of the Chebyshev filter lie on an ellipse in the s-plane. As shown in Figure B.5, the ellipse is defined by two circles whose diameters are equal to the minor and major axes of the ellipse. The length of the minor axis is $2a\Omega_c$, where

$$a = \tfrac{1}{2}(\alpha^{1/N} - \alpha^{-1/N}) \tag{B.7}$$

with

$$\alpha = \varepsilon^{-1} + \sqrt{1 + \varepsilon^{-2}}. \tag{B.8}$$

The length of the major axis is $2b\Omega_c$, where

$$b = \tfrac{1}{2}(\alpha^{1/N} + \alpha^{-1/N}). \tag{B.9}$$

To locate the poles of the Chebyshev filter on the ellipse, we first identify the points on the major and minor circles equally spaced in angle with a spacing of π/N in such a way that the points are symmetrically located with respect to the imaginary axis and such that a point never falls on the imaginary axis and a point occurs on the real axis for N odd but not for N even. This division of the major and minor circles corresponds exactly to the manner in which the circle is divided in locating the poles of a Butterworth filter as in Eq. (B.3). The poles of a Chebyshev filter fall on the ellipse, with the ordinate specified by the points identified on the major circle and the abscissa specified by the points identified on the minor circle. In Figure B.5, the poles are shown for $N = 3$.

A type II Chebyshev lowpass filter can be related to a type I filter through a transformation. Specifically, if in Eq. (B.4) we replace the term $\varepsilon^2 V_N^2(\Omega/\Omega_c)$ by its reciprocal and also replace the argument of V_N^2 by its reciprocal, we obtain

$$|H_c(j\Omega)|^2 = \frac{1}{1 + [\varepsilon^2 V_N^2(\Omega_c/\Omega)]^{-1}}. \tag{B.10}$$

This is the analytic form for the type II Chebyshev lowpass filter. One approach to designing a type II Chebyshev filter is to first design a type I filter and then apply the transformation of Eq. (B.10).

B.3 ELLIPTIC FILTERS

If we distribute the error uniformly across the entire passband or across the entire stopband, as in the Chebyshev cases, we are able to meet the design specifications with a lower order filter than if we permit a monotonically varying error in the passband and stopband, as in the Butterworth case. We note that in the type I Chebyshev approximation, the stopband error decreases monotonically with frequency, raising the possibility of further improvements if we distribute the stopband error uniformly across the stopband. This suggests the lowpass filter approximation in Figure B.6. Indeed, it can be shown (Papoulis, 1957) that this type of approximation (i.e., equiripple error in the passband and the stopband) is the best that can be achieved for a given filter order N, in the sense that for given values of Ω_p, δ_1, and δ_2, the transition band $(\Omega_s - \Omega_p)$ is as small as possible.

This class of approximations, referred to as elliptic filters, has the form

$$|H_c(j\Omega)|^2 = \frac{1}{1 + \varepsilon^2 U_N^2(\Omega)}, \tag{B.11}$$

where $U_N(\Omega)$ is a Jacobian elliptic function. To obtain equiripple error in both the passband and the stopband, elliptic filters must have both poles and zeros. As can be seen from Figure B.6, such a filter will have zeros on the $j\Omega$-axis of the s-plane. A discussion of elliptic filter design, even on a superficial level, is beyond the scope of this appendix. The reader is referred to the texts by Guillemin (1957), Storer (1957), Gold and Rader (1969) and Parks and Burrus (1987) for more detailed discussions.

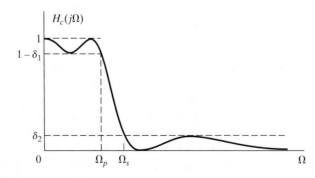

Figure B.6 Equiripple approximation in both passband and stopband.

C

Answers to Selected

Basic Problems

This appendix contains the answers to the first 20 basic problems in Chapter 2 through 10.

Answers to Basic Problems in Chapter 2

2.1. (a) Always (2), (3), (5). If $g[n]$ is bounded, (1).
   (b) (3).
   (c) Always (1), (3), (4). If $n_0 = 0$, (2) and (5).
   (d) Always (1), (3), (4). If $n_0 = 0$, (5). If $n_0 \geq 0$, (2).
   (e) (1), (2), (4), (5).
   (f) Always (1), (2), (4), (5). If $b = 0$, (3).
   (g) (1), (3).
   (h) (1), (5).

2.2. (a) $N_4 = N_0 + N_2$, $N_5 = N_1 + N_3$.
   (b) At most $N + M - 1$ nonzero points.

2.3.

$$y[n] = \begin{cases} \dfrac{a^{-n}}{1-a}, & n < 0, \\ \dfrac{1}{1-a}, & n \geq 0. \end{cases}$$

2.4. $y[n] = 8[(1/2)^n - (1/4)^n]u[n]$.

2.5. (a) $y_h[n] = A_1(2)^n + A_2(3)^n$.

(b) $h[n] = 2(3^n - 2^n)u[n]$.

(c) $s[n] = [-8(2)^{(n-1)} + 9(3)^{(n-1)} + 1]u[n]$.

2.6. (a)

$$H(e^{j\omega}) = \frac{1 + 2e^{-j\omega} + e^{-j2\omega}}{1 - \frac{1}{2}e^{-j\omega}}.$$

(b) $y[n] + \frac{1}{2}y[n-1] + \frac{3}{4}y[n-2] = x[n] - \frac{1}{2}x[n-1] + x[n-3]$.

2.7. (a) Periodic, $N = 12$.

(b) Periodic, $N = 8$.

(c) Not periodic.

(d) Not periodic.

2.8. $y[n] = 3(-1/2)^n u[n] + 2(1/3)^n u[n]$.

2.9. (a)

$$h[n] = 2\left[\left(\frac{1}{2}\right)^n - \left(\frac{1}{3}\right)^n\right]u[n],$$

$$H(e^{j\omega}) = \frac{\frac{1}{3}e^{-j\omega}}{1 - \frac{5}{6}e^{-j\omega} + \frac{1}{6}e^{-j2\omega}},$$

$$s[n] = \left[-2\left(\frac{1}{2}\right)^n + \left(\frac{1}{3}\right)^n + 1\right]u[n].$$

(b) $y_h[n] = A_1(1/2)^n + A_2(1/3)^n$.

(c) $y[n] = 4(1/2)^n - 3(1/3)^n - 2(1/2)^n u[-n-1] + 2(1/3)^n u[-n-1]$. Other answers are possible.

2.10. (a)

$$y[n] = \begin{cases} a^{-1}/(1-a^{-1}), & n \geq -1, \\ a^n/(1-a^{-1}), & n \leq -2. \end{cases}$$

(b)

$$y[n] = \begin{cases} 1, & n \geq 3, \\ 2^{(n-3)}, & n \leq 2. \end{cases}$$

(c)

$$y[n] = \begin{cases} 1, & n \geq 0, \\ 2^n, & n \leq -1. \end{cases}$$

(d)

$$y[n] = \begin{cases} 0, & n \geq 9, \\ 1 - 2^{(n-9)}, & 8 \geq n \geq -1, \\ 2^{(n+1)} - 2^{(n-9)}, & -2 \geq n. \end{cases}$$

2.11. $y[n] = 2\sqrt{2}\sin(\pi(n+1)/4)$.

2.12. (a) $y[n] = n!u[n]$.

(b) The system is linear.

(c) The system is not time invariant.

2.13. **(a)**, **(b)**, and **(e)** are eigenfunctions of stable LTI systems.

2.14. **(a)** (iv).

(b) (i).

(c) (iii), $h[n] = (1/2)^n u[n]$.

2.15. **(a)** Not LTI. Inputs $\delta[n]$ and $\delta[n-1]$ violate TI.

(b) Not causal. Consider $x[n] = \delta[n-1]$.

(c) Stable.

2.16. **(a)** $y_h[n] = A_1(1/2)^n + A_2(-1/4)^n$.

(b) Causal: $h_c[n] = 2(1/2)^n u[n] + (-1/4)^n u[n]$.
Anticausal: $h_{ac}[n] = -2(1/2)^n u[-n-1] - (-1/4)^n u[-n-1]$.

(c) $h_c[n]$ is absolutely summable, $h_{ac}[n]$ is not.

(d) $y_p[n] = (1/3)(-1/4)^n u[n] + (2/3)(1/2)^n u[n] + 4(n+1)(1/2)^{(n+1)} u[n+1]$.

2.17. **(a)**

$$R(e^{j\omega}) = e^{-j\omega M/2} \frac{\sin\left(\omega\left(\frac{M+1}{2}\right)\right)}{\sin\left(\frac{\omega}{2}\right)}.$$

(b) $W(e^{j\omega}) = (1/2)R(e^{j\omega}) - (1/4)R(e^{j(\omega-2\pi/M)}) - (1/4)R(e^{j(\omega+2\pi/M)})$.

2.18. Systems (a) and (b) are causal.

2.19. Systems (b), (c), (e), and (f) are stable.

2.20. **(a)** $h[n] = (-1/a)^{n-1} u[n-1]$.

(b) The system will be stable for $|a| > 1$.

Answers to Basic Problems in Chapter 3

3.1. **(a)** $\dfrac{1}{1 - \frac{1}{2}z^{-1}}, \quad |z| > \frac{1}{2}$.

(b) $\dfrac{1}{1 - \frac{1}{2}z^{-1}}, \quad |z| < \frac{1}{2}$.

(c) $\dfrac{-\frac{1}{2}z^{-1}}{1 - \frac{1}{2}z^{-1}}, \quad |z| < \frac{1}{2}$.

(d) $1, \quad$ all z.

(e) $z^{-1}, \quad z \neq 0$.

(f) $z, \quad |z| < \infty$.

(g) $\dfrac{1 - \left(\frac{1}{2}\right)^{10} z^{-10}}{1 - \frac{1}{2}z^{-1}}, \quad |z| \neq 0$.

3.2. $X(z) = \dfrac{(1 - z^{-N})^2}{(1 - z^{-1})^2}$.

3.3. (a) $X_a(z) = \dfrac{z^{-1}(\alpha - \alpha^{-1})}{(1 - \alpha z^{-1})(1 - \alpha^{-1} z^{-1})}$,   ROC: $|\alpha| < |z| < |\alpha^{-1}|$.

(b) $X_b(z) = \dfrac{1 - z^{-N}}{1 - z^{-1}}$,   ROC: $z \neq 0$.

(c) $X_c(z) = \dfrac{(1 - z^{-N})^2}{(1 - z^{-1})^2}$,   ROC: $z \neq 0$.

3.4. (a) $(1/3) < |z| < 2$, two sided.

(b) Two sequences. $(1/3) < |z| < 2$ and $2 < |z| < 3$.

(c) No. Causal sequence has $|z| > 3$, which does not include the unit circle.

3.5. $x[n] = 2\delta[n + 1] + 5\delta[n] - 4\delta[n - 1] - 3\delta[n - 2]$.

3.6. (a) $x[n] = \left(-\dfrac{1}{2}\right)^n u[n]$, Fourier transform exists.

(b) $x[n] = -(-\dfrac{1}{2})^n u[-n - 1]$, Fourier transform does not exist.

(c) $x[n] = 4\left(-\dfrac{1}{2}\right)^n u[n] - 3\left(-\dfrac{1}{4}\right)^n u[n]$, Fourier transform exists.

(d) $x[n] = \left(-\dfrac{1}{2}\right)^n u[n]$, Fourier transform exists.

(e) $x[n] = -(a^{-(n+1)})u[n] + a^{-(n-1)}u[n - 1]$, Fourier transform exists if $|a| > 1$.

3.7. (a) $H(z) = \dfrac{1 - z^{-1}}{1 + z^{-1}}$, $|z| > 1$.

(b) $\text{ROC}\{Y(z)\} = |z| > 1$.

(c) $y[n] = \left[-\dfrac{1}{3}\left(\dfrac{1}{2}\right)^n + \dfrac{1}{3}(-1)^n\right]u[n]$.

3.8. (a) $h[n] = \left(-\dfrac{3}{4}\right)^n u[n] - \left(-\dfrac{3}{4}\right)^{n-1} u[n - 1]$.

(b) $y[n] = \dfrac{8}{13}\left(-\dfrac{3}{4}\right)^n u[n] - \dfrac{8}{13}\left(\dfrac{1}{3}\right)^n u[n]$.

(c) The system is stable.

3.9. (a) $|z| > (1/2)$.

(b) Yes. The ROC includes the unit circle.

(c) $X(z) = \dfrac{1 - \frac{1}{2}z^{-1}}{1 - 2z^{-1}}$, ROC: $|z| < 2$.

(d) $h[n] = 2\left(\dfrac{1}{2}\right)^n u[n] - \left(-\dfrac{1}{4}\right)^n u[n]$.

3.10. (a) $|z| > \dfrac{3}{4}$.

(b) $0 < |z| < \infty$.

(c) $|z| < 2$.

(d) $|z| > 1$.

(e) $|z| < \infty$.

(f) $\dfrac{1}{2} < |z| < \sqrt{13}$.

3.11. (a) Causal.

   (b) Not causal.

   (c) Causal.

   (d) Not causal.

3.12. (a)

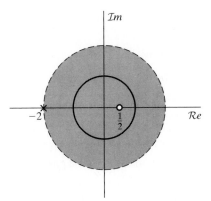

Figure P3.12

(b)

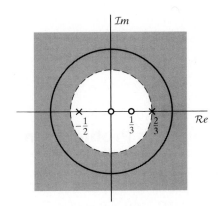

Figure P3.12

(c)

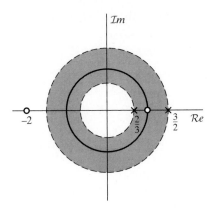

Figure P3.12

3.13. $g[11] = -\dfrac{1}{11!} + \dfrac{3}{9!} - \dfrac{2}{7!}$.

3.14. $A_1 = A_2 = 1/2, \quad \alpha_1 = -1/2, \quad \alpha_2 = 1/2$.

3.15. $h[n] = \left(\dfrac{1}{2}\right)^n (u[n] - u[n-10])$. The system is causal.

3.16. (a) $H(z) = \dfrac{1 - 2z^{-1}}{1 - \frac{2}{3}z^{-1}}, \quad |z| > \frac{2}{3}$.

    (b) $h[n] = \left(\dfrac{2}{3}\right)^n u[n] - 2\left(\dfrac{2}{3}\right)^{(n-1)} u[n-1]$.

    (c) $y[n] - \frac{2}{3}y[n-1] = x[n] - 2x[n-1]$.

    (d) The system is stable and causal.

3.17. $h[0]$ can be 0, 1/3, or 1. To be painstakingly literal, $h[0]$ can also be 2/3, due to the impulse response $h[n] = (2/3)(2)^n u[n] - (1/3)(1/2)^n u[-n-1]$, which satisfies the difference equation but has no ROC. This noncausal system with no ROC can be implemented as the parallel combination of its causal and anticausal components.

3.18. (a) $h[n] = -2\delta[n] + \frac{1}{3}\left(-\frac{1}{2}\right)^n u[n] + \frac{8}{3}u[n]$.

    (b) $y[n] = \dfrac{18}{5}2^n$.

3.19. (a) $|z| > 1/2$.

    (b) $1/3 < |z| < 2$.

    (c) $|z| > 1/3$.

3.20. (a) $|z| > 2/3$.

    (b) $|z| > 1/6$.

Answers to Basic Problems in Chapter 4

4.1. $x[n] = \sin(\pi n/2)$.

4.2. $\Omega_0 = 250\pi, 1750\pi$.

4.3. (a) $T = 1/12{,}000$. **(b)** Not unique. $T = 5/12{,}000$.

4.4. (a) $T = 1/100$. **(b)** Not unique. $T = 11/100$.

4.5. (a) $T \le 1/10{,}000$. **(b)** 625 Hz. **(c)** 1250 Hz.

4.6. (a) $H_c(j\Omega) = 1/(a + j\Omega)$.

    (b) $H_d(e^{j\omega}) = T/(1 - e^{-aT}e^{-j\omega})$.

    (c) $|H_d(e^{j\omega})| = T/(1 + e^{-\alpha T})$.

4.7. (a)

$$X_c(j\Omega) = S_c(j\Omega)(1 + \alpha e^{-j\Omega\tau_d}),$$

$$X(e^{j\omega}) = \left(\frac{1}{T}\right) S_c\left(\frac{j\omega}{T}\right)\left(1 + \alpha e^{-j\omega\tau_d/T}\right) \qquad \text{for } |\omega| \le \pi.$$

    (b) $H(e^{j\omega}) = 1 + \alpha e^{-j\omega\tau_d/T}$.

    (c) (i) $h[n] = \delta[n] + \alpha\delta[n-1]$.

        (ii) $h[n] = \delta[n] + \alpha\dfrac{\sin(\pi(n-1/2))}{\pi(n-1/2)}$.

4.8. (a) $T \leq 1/20,000$.

    (b) $h[n] = Tu[n]$.

    (c) $TX\left(e^{j\omega}\right)|_{\omega=0}$.

    (d) $T \leq 1/10,000$.

4.9. (a) $X\left(e^{j(\omega+\pi)}\right) = X\left(e^{j(\omega+\pi-\pi)}\right) = X\left(e^{j\omega}\right)$.

    (b) $x[3] = 0$.

    (c) $x[n] = \begin{cases} y[n/2], & n \text{ even}, \\ 0, & n \text{ odd}. \end{cases}$

4.10. (a) $x[n] = \cos(2\pi n/3)$.

    (b) $x[n] = -\sin(2\pi n/3)$.

    (c) $x[n] = \sin(2\pi n/5)/(\pi n/5000)$.

4.11. (a) $T = 1/40, T = 9/40$.

    (b) $T = 1/20$, unique.

4.12. (a) (i) $y_c(t) = -6\pi \sin(6\pi t)$.

        (ii) $y_c(t) = -6\pi \sin(6\pi t)$.

    (b) (i) Yes.

        (ii) No.

4.13. (a) $y[n] = \sin\left(\frac{\pi n}{2} - \frac{\pi}{4}\right)$.

    (b) Same $y[n]$.

    (c) $h_c(t)$ has no effect on T.

4.14. (a) No.

    (b) Yes.

    (c) No.

    (d) Yes.

    (e) Yes. (No information is lost; however, the signal cannot be recovered by the system in Figure P3.21.)

4.15. (a) Yes.

    (b) No.

    (c) Yes.

4.16. (a) $M/L = 5/2$, unique.

    (b) $M/L = 2/3$; unique.

4.17. (a) $\tilde{x}_d[n] = (4/3)\sin(\pi n/2)/(\pi n)$.

    (b) $\tilde{x}_d[n] = 0$.

4.18. (a) $\omega_0 = 2\pi/3$.

    (b) $\omega_0 = 3\pi/5$.

    (c) $\omega_0 = \pi$.

4.19. $T \leq \pi/\Omega_0$.

4.20. (a) $F_s \geq 2000$ Hz.

    (b) $F_s \geq 4000$ Hz.

Answers to Basic Problems in Chapter 5

5.1. $x[n] = y[n]$, $\omega_c = \pi$.

5.2. (a) Poles: $z = 3, 1/3$, Zeros: $z = 0, \infty$.

   (b) $h[n] = -(3/8)(1/3)^n u[n] - (3/8)3^n u[-n-1]$.

5.3. (a), **(d)** are the impulse responses.

5.4. (a) $H(z) = \dfrac{1 - 2z^{-1}}{1 - \frac{3}{4}z^{-1}}$, $|z| > 3/4$.

   (b) $h[n] = (3/4)^n u[n] - 2(3/4)^{n-1} u[n-1]$.

   (c) $y[n] - (3/4)y[n-1] = x[n] - 2x[n-1]$.

   (d) Stable and causal.

5.5. (a) $y[n] - (7/12)y[n-1] + (1/12)y[n-2] = 3x[n] - (19/6)x[n-1] + (2/3)x[n-2]$.

   (b) $h[n] = 3\delta[n] - (2/3)(1/3)^{n-1} u[n-1] - (3/4)(1/4)^{n-1} u[n-1]$.

   (c) Stable.

5.6. (a) $X(z) = \dfrac{1}{(1 - \frac{1}{2}z^{-1})(1 - 2z^{-1})}$, $\dfrac{1}{2} < |z| < 2$.

   (b) $\frac{1}{2} < |z| < 2$.

   (c) $h[n] = \delta[n] - \delta[n-2]$.

5.7. (a) $H(z) = \dfrac{1 - z^{-1}}{(1 - \frac{1}{2}z^{-1})(1 + \frac{3}{4}z^{-1})}$, $|z| > \dfrac{3}{4}$.

   (b) $h[n] = -(2/5)(1/2)^n u[n] + (7/5)(-3/4)^n u[n]$.

   (c) $y[n] + (1/4)y[n-1] - (3/8)y[n-2] = x[n] - x[n-1]$.

5.8. (a) $H(z) = \dfrac{z^{-1}}{1 - \frac{3}{2}z^{-1} - z^{-2}}$, $|z| > 2$.

   (b) $h[n] = -(2/5)(-1/2)^n u[n] + (2/5)(2)^n u[n]$.

   (c) $h[n] = -(2/5)(-1/2)^n u[n] - (2/5)(2)^n u[-n-1]$.

5.9.

$$h[n] = \left[-\frac{4}{3}(2)^{n-1} + \frac{1}{3}\left(\frac{1}{2}\right)^{n-1} \right] u[-n], \qquad |z| < \frac{1}{2},$$

$$h[n] = -\frac{4}{3}(2)^{n-1} u[-n] - \frac{1}{3}\left(\frac{1}{2}\right)^{n-1} u[n-1], \qquad \frac{1}{2} < |z| < 2,$$

$$h[n] = \frac{4}{3}(2)^{n-1} u[n-1] - \frac{1}{3}\left(\frac{1}{2}\right)^{n-1} u[n-1], \qquad |z| > 2.$$

5.10. $H_i(z)$ cannot be causal and stable. The zero of a $H(z)$ at $z = \infty$ is a pole of $H_i(z)$. The existence of a pole at $z = \infty$ implies that the system is not causal.

5.11. (a) Cannot be determined.

   (b) Cannot be determined.

   (c) False.

   (d) True.

5.12. (a) Stable.

(b)

$$H_1(z) = -9 \frac{(1 + 0.2z^{-1})\left(1 - \frac{1}{3}z^{-1}\right)\left(1 + \frac{1}{3}z^{-1}\right)}{(1 - j0.9z^{-1})(1 + j0.9z^{-1})},$$

$$H_{\mathrm{ap}}(z) = \frac{\left(z^{-1} - \frac{1}{3}\right)\left(z^{-1} + \frac{1}{3}\right)}{\left(1 - \frac{1}{3}z^{-1}\right)\left(1 + \frac{1}{3}z^{-1}\right)}.$$

5.13. $H_1(z)$, $H_3(z)$, and $H_4(z)$ are allpass systems.

5.14. (a) 5.

(b) $\frac{1}{2}$.

5.15. (a) $\alpha = 1$, $\beta = 0$, $A(e^{j\omega}) = 1 + 4\cos(\omega)$. The system is a generalized linear-phase system but not a linear-phase system, because $A(e^{j\omega})$ is not nonnegative for all ω.

(b) Not a generalized linear-phase or a linear-phase system.

(c) $\alpha = 1$, $\beta = 0$, $A(e^{j\omega}) = 3 + 2\cos(\omega)$. Linear phase, since $|H(e^{j\omega})| = A(e^{j\omega}) \geq 0$ for all ω.

(d) $\alpha = 1/2$, $\beta = 0$, $A(e^{j\omega}) = 2\cos(\omega/2)$. Generalized linear phase, because $A(e^{j\omega})$ is not nonnegative at all ω.

(e) $\alpha = 1$, $\beta = \pi/2$, $A(e^{j\omega}) = 2\sin(\omega)$. Generalized linear phase, because $\beta \neq 0$.

5.16. $h[n]$ is not necessarily causal. Both $h[n] = \delta[n - \alpha]$ and $h[n] = \delta[n + 1] + \delta[n - (2\alpha + 1)]$ will have this phase.

5.17. $H_2(z)$ and $H_3(z)$ are minimum-phase systems.

5.18. (a) $H_{\min}(z) = \dfrac{2\left(1 - \frac{1}{2}z^{-1}\right)}{1 + \frac{1}{3}z^{-1}}.$

(b) $H_{\min}(z) = 3\left(1 - \dfrac{1}{2}z^{-1}\right).$

(c) $H_{\min}(z) = \dfrac{9}{4}\dfrac{\left(1 - \frac{1}{3}z^{-1}\right)\left(1 - \frac{1}{4}z^{-1}\right)}{\left(1 - \frac{3}{4}z^{-1}\right)^2}.$

5.19. $h_1[n] : 2$, $h_2[n] : 3/2$, $h_3[n] : 2$, $h_4[n] : 3$, $h_5[n] : 3$, $h_6[n] : 7/2$.

5.20. Systems $H_1(z)$ and $H_3(z)$ have a linear phase and can be implemented by a real-valued difference equation.

Answers to Basic Problems in Chapter 6

6.1. Network 1:

$$H(z) = \frac{1}{1 - 2r\cos\theta z^{-1} + r^2 z^{-2}}.$$

Network 2:

$$H(z) = \frac{r \sin \theta z^{-1}}{1 - 2r \cos \theta z^{-1} + r^2 z^{-2}}.$$

Both systems have the same denominators and thus the same poles.

6.2. $y[n] - 3y[n-1] - y[n-2] - y[n-3] = x[n] - 2x[n-1] + x[n-2]$.

6.3. The system in Part (d) is the same as that in Part (a).

6.4. (a)

$$H(z) = \frac{2 + \frac{1}{4}z^{-1}}{1 + \frac{1}{4}z^{-1} - \frac{3}{8}z^{-2}}.$$

(b)

$$y[n] + \frac{1}{4}y[n-1] - \frac{3}{8}y[n-2] = 2x[n] + \frac{1}{4}x[n-1].$$

6.5. (a)

$$y[n] - 4y[n-1] + 7y[n-3] + 2y[n-4] = x[n].$$

(b)

$$H(z) = \frac{1}{1 - 4z^{-1} + 7z^{-3} + 2z^{-4}}.$$

(c) Two multiplications and four additions.

(d) No. It requires at least four delays to implement a 4th-order system.

6.6.

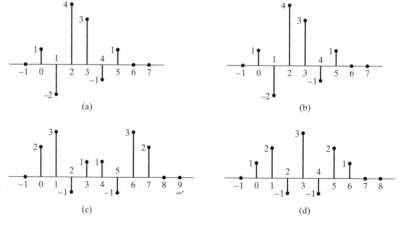

Figure P6.6

6.7.

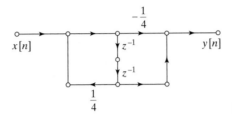

Figure P6.7

6.8. $y[n] - 2y[n-2] = 3x[n-1] + x[n-2]$.

6.9. (a) $h[1] = 2$.

   (b) $y[n] + y[n-1] - 8y[n-2] = x[n] + 3x[n-1] + x[n-2] - 8x[n-3]$.

6.10. (a)

$$y[n] = x[n] + v[n-1].$$

$$v[n] = 2x[n] + \frac{1}{2}y[n] + w[n-1].$$

$$w[n] = x[n] + \frac{1}{2}y[n].$$

(b)

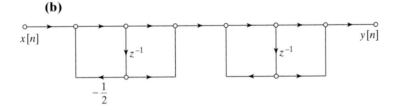

Figure P6.10

   (c) The poles are at $z = -1/2$ and $z = 1$. Since the second pole is on the unit circle, the system is not stable.

6.11. (a)

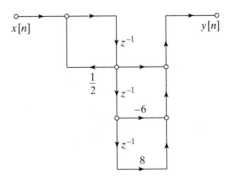

Figure P6.11

(b)

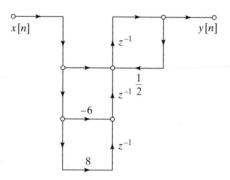

Figure P6.11

6.12. $y[n] - 8y[n-1] = -2x[n] + 6x[n-1] + 2x[n-2]$.

6.13.

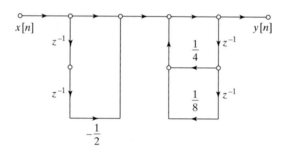

Figure P6.13

6.14.

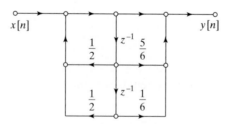

Figure P6.14

6.15.

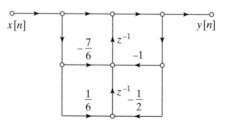

Figure P6.15

6.16. (a)

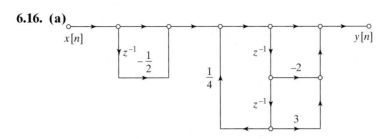

Figure P6.16

(b) Both systems have the system function

$$H(z) = \frac{\left(1 - \frac{1}{2}z^{-1}\right)(1 - 2z^{-1} + 3z^{-2})}{1 - \frac{1}{4}z^{-2}}.$$

6.17. (a)

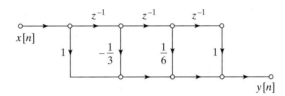

Figure P6.17-1

(b)

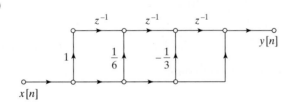

Figure P6.17-2

6.18. If $a = 2/3$, the overall system function is

$$H(z) = \frac{1 + 2z^{-1}}{1 + \frac{1}{4}z^{-1} - \frac{3}{8}z^{-2}}.$$

If $a = -2$, the overall system function is

$$H(z) = \frac{1 - \frac{2}{3}z^{-1}}{1 + \frac{1}{4}z^{-1} - \frac{3}{8}z^{-2}}.$$

6.19.

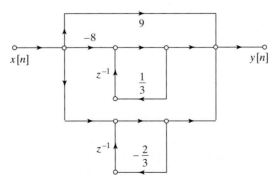

Figure P6.19

6.20.

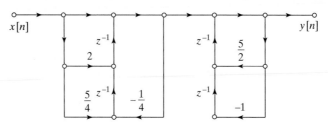

Figure P6.20

Answers to Basic Problems in Chapter 7

7.1. (a)

$$H_1(z) = \frac{1 - e^{-aT}\cos(bT)z^{-1}}{1 - 2e^{-aT}\cos(bT)z^{-1} + e^{-2aT}z^{-2}}, \quad \text{ROC: } |z| > e^{-aT}.$$

(b)

$H_2(z) = (1 - z^{-1})S_2(z)$, ROC: $|z| > e^{-aT}$, where

$$S_2(z) = \frac{a}{a^2 + b^2}\frac{1}{1 - z^{-1}} - \frac{1}{2(a + jb)}\frac{1}{1 - e^{-(a+jb)T}z^{-1}} - \frac{1}{2(a - jb)}\frac{1}{1 - e^{-(a-jb)T}z^{-1}}.$$

(c) They are not equal.

7.2. (a)

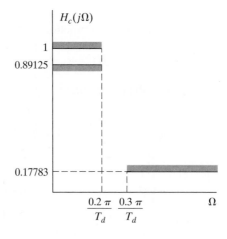

Figure P7.2

(b) $N = 6, \Omega_c T_d = 0.7032$.

(c) The poles in the s-plane are on a circle of radius $R = 0.7032/T_d$. They map to poles in the z-plane at $z = e^{s_k T_d}$. The factors of T_d cancel out, leaving the pole locations in the z-plane for $H(z)$ independent of T_d.

7.3. (a) $\hat{\delta}_2 = \delta_2/(1 + \delta_1)$, $\hat{\delta}_1 = 2\delta_1/(1 + \delta_1)$.

(b)

$$\delta_2 = 0.18806, \delta_1 = 0.05750$$

$$H(z) = \frac{0.3036 - 0.4723z^{-1}}{1 - 1.2971z^{-1} + 0.6949z^{-2}} + \frac{-2.2660 + 1.2114z^{-1}}{1 - 1.0691z^{-1} + 0.3699z^{-2}}$$

$$+ \frac{1.9624 - 0.6665z^{-1}}{1 - 0.9972z^{-1} + 0.2570z^{-2}}$$

(c) Use the same δ_1 and δ_2.

$$H(z) = \frac{0.0007802(1 + z^{-1})^6}{(1 - 1.2686z^{-1} + 0.7051z^{-2})(1 - 1.0106z^{-1} + 0.3583z^{-2})(1 - 0.9044z^{-1} + 0.2155z^{-2})}.$$

7.4. (a)

$$H_c(s) = \frac{1}{s + 0.1} - \frac{0.5}{s + 0.2}.$$

The answer is not unique. Another possibility is

$$H_c(s) = \frac{1}{s + 0.1 + j2\pi} - \frac{0.5}{s + 0.2 + j2\pi}.$$

(b)

$$H_c(s) = \frac{2(1 + s)}{0.1813 + 1.8187s} - \frac{1 + s}{0.3297 + 1.6703s}.$$

This answer is unique.

7.5. (a) $M + 1 = 91$, $\beta = 3.3953$.

(b) $M/2 = 45$.

(c) $h_d[n] = \dfrac{\sin[0.625\pi(n - 45)]}{\pi(n - 45)} - \dfrac{\sin[0.3\pi(n - 45)]}{\pi(n - 45)}$.

7.6. (a) $\delta = 0.03$, $\beta = 2.181$.

(b) $\Delta\omega = 0.05\pi$, $M = 63$.

7.7.

$$0.99 \leq |H(e^{j\omega})| \leq 1.01, \qquad |\omega| \leq 0.2\pi,$$

$$|H(e^{j\omega})| \leq 0.01, \qquad 0.22\pi \leq |\omega| \leq \pi$$

7.8. (a) Six alternations. $L = 5$, so this does not satisfy the alternation theorem and is not optimal.

(b) Seven alternations, which satisfies the alternation theorem for $L = 5$.

7.9. $\omega_c = 0.4\pi$.

7.10. $\omega_c = 2.3842$ rad.

7.11. $\Omega_c = 2\pi(1250)$ rad/sec.

7.12. $\Omega_c = 2000$ rad/sec.

7.13. $T = 50\,\mu$s. This T is unique.

7.14. $T = 1.46$ ms. This T is unique.

7.15. Hamming and Hanning: $M + 1 = 81$, Blackman: $M + 1 = 121$.

7.16. $\beta = 2.6524$, $M = 181$.

7.17.

$$
\begin{aligned}
|H_c(j\Omega)| &< 0.02, & |\Omega| &\leq 2\pi(20) \text{ rad/sec}, \\
0.95 < |H_c(j\Omega)| &< 1.05, & 2\pi(30) \leq |\Omega| &\leq 2\pi(70) \text{ rad/sec}, \\
|H_c(j\Omega)| &< 0.001, & 2\pi(75) \text{ rad/sec} &\leq |\Omega|.
\end{aligned}
$$

7.18.

$$
\begin{aligned}
|H_c(j\Omega)| &< 0.04, & |\Omega| &\leq 324.91 \text{ rad/sec}, \\
0.995 < |H_c(j\Omega)| &< 1.005, & |\Omega| &\geq 509.52 \text{ rad/sec}.
\end{aligned}
$$

7.19. $T = 0.41667$ ms. This T is unique.

7.20. True.

Answers to Basic Problems in Chapter 8

8.1. **(a)** $x[n]$ is periodic with period $N = 6$.

(b) T will not avoid aliasing.

(c)

$$
\tilde{X}[k] = 2\pi \begin{cases}
a_0 + a_6 + a_{-6}, & k = 0, \\
a_1 + a_7 + a_{-5}, & k = 1, \\
a_2 + a_8 + a_{-4}, & k = 2, \\
a_3 + a_9 + a_{-3} + a_{-9}, & k = 3, \\
a_4 + a_{-2} + a_{-8}, & k = 4, \\
a_5 + a_{-1} + a_{-7}, & k = 5.
\end{cases}
$$

8.2. **(a)**

$$
\tilde{X}_3[k] = \begin{cases}
3\tilde{X}[k/3], & \text{for } k = 3\ell, \\
0, & \text{otherwise}.
\end{cases}
$$

(b)

$$
\tilde{X}[k] = \begin{cases}
3, & k = 0, \\
-1, & k = 1.
\end{cases}
$$

$$
\tilde{X}_3[k] = \begin{cases}
9, & k = 0, \\
0, & k = 1, 2, 4, 5, \\
-3, & k = 3.
\end{cases}
$$

8.3. **(a)** $\tilde{x}_2[n]$.

(b) None of the sequences.

(c) $\tilde{x}_1[n]$ and $\tilde{x}_3[n]$.

8.4. (a)

$$X(e^{j\omega}) = \frac{1}{1 - \alpha e^{-j\omega}}.$$

(b)

$$\tilde{X}[k] = \frac{1}{1 - \alpha e^{-j(2\pi/N)k}}.$$

(c)

$$\tilde{X}[k] = X(e^{j\omega})|_{\omega = (2\pi k/N)}.$$

8.5. (a) $X[k] = 1.$

(b) $X[k] = W_N^{kn_0}.$

(c)

$$X[k] = \begin{cases} N/2, & k = 0, N/2, \\ 0, & \text{otherwise.} \end{cases}$$

(d)

$$X[k] = \begin{cases} N/2, & k = 0, \\ e^{-j(\pi k/N)(N/2-1)}(-1)^{(k-1)/2}\frac{1}{\sin(k\pi/N)}, & k \text{ odd}, \\ 0, & \text{otherwise.} \end{cases}$$

(e)

$$X[k] = \frac{1 - a^N}{1 - aW_N^k}.$$

8.6. (a)

$$X(e^{j\omega}) = \frac{1 - e^{j(\omega_0 - \omega)N}}{1 - e^{j(\omega_0 - \omega)}}.$$

(b)

$$X[k] = \frac{1 - e^{j\omega_0 N}}{1 - e^{j\omega_0}W_N^k}.$$

(c)

$$X[k] = \begin{cases} N, & k = k_0 \\ 0, & \text{otherwise.} \end{cases}$$

8.7.

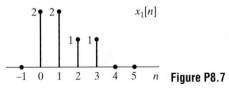

Figure P8.7

8.8.

$$y[n] = \begin{cases} \frac{1024}{1023}\left(\frac{1}{2}\right)^n, & 0 \le n \le 9, \\ 0, & \text{otherwise.} \end{cases}$$

8.9. (a) **1.** Let $x_1[n] = \sum_m x[n + 5m]$ for $n = 0, 1, \ldots 4$.

      2. Let $X_1[k]$ be the five-point FFT of $x_1[n]$. $M = 5$.

      3. $X_1[2]$ is $X(e^{j\omega})$ at $\omega = 4\pi/5$.

(b) Define $x_2[n] = \sum_m W_{27}^{-(n+9m)} x[n + 9m]$ for $n = 0, \ldots, 8$.
Compute $X_2[k]$, the 9-point DFT of $x_2[n]$.
$X_2[2] = X(e^{j\omega})\big|_{\omega=10\pi/27}$.

8.10. $X_2[k] = (-1)^k X_1[k]$.

8.11.

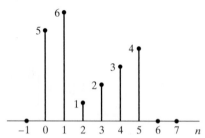

Figure P8.11

8.12. (a)

$$X[k] = \begin{cases} 2, & k = 1, 3, \\ 0, & k = 0, 2. \end{cases}$$

(b)

$$H[k] = \begin{cases} 15, & k = 0, \\ -3 + j6, & k = 1, \\ -5, & k = 2, \\ -3 - j6, & k = 3. \end{cases}$$

(c) $y[n] = -3\delta[n] - 6\delta[n - 1] + 3\delta[n - 2] + 6\delta[n - 3]$.

(d) $y[n] = -3\delta[n] - 6\delta[n - 1] + 3\delta[n - 2] + 6\delta[n - 3]$.

8.13.

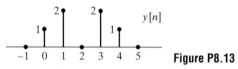

Figure P8.13

8.14. $x_3[2] = 9$.

8.15. $a = -1$. This is unique.

8.16. $b = 3$. This is unique.

8.17. $N = 9$.

8.18. $c = 2$.

8.19. $m = 2$. This is not unique. Any $m = 2 + 6\ell$ for integer ℓ works.

8.20. $N = 5$. This is unique.

Answers to Basic Problems in Chapter 9

9.1. If the input is $(1/N)X[((-n))_N]$, the output of the DFT program will be $x[n]$, the IDFT of $X[k]$.

9.2.

$$X = AD - BD + CA - DA = AC - BD$$

$$Y = AD - BD + BC + BD = BC + AD.$$

9.3.

$$y[32] = X(e^{-j2\pi(7/32)}) = X(e^{j2\pi(25/32)}).$$

9.4. $\omega_k = 7\pi/16.$

9.5.

$$a = -\sqrt{2}$$

$$b = -e^{-j(6\pi/8)}.$$

9.6. (a) The gain is $-W_N^2$.

(b) There is one path. In general, there is only one path from any input sample to any output sample.

(c) By tracing paths, we see

$$X[2] = x[0] \cdot 1 + x[1]W_8^2 - x[2] - x[3]W_8^2 + \ldots$$

$$x[4] + x[5]W_8^2 - x[6] - x[7]W_8^2.$$

9.7. (a) Store $x[n]$ in $A[\cdot]$ in bit-reversed order, and $D[\cdot]$ will contain $X[k]$ in sequential (normal) order.

(b)

$$D[r] = \begin{cases} 8, & r = 3, \\ 0, & \text{otherwise.} \end{cases}$$

(c)

$$C[r] = \begin{cases} 1, & r = 0, 1, 2, 3, \\ 0, & \text{otherwise.} \end{cases}$$

9.8. (a) $N/2$ butterflies with $2^{(m-1)}$ different coefficients.

(b) $y[n] = W_N^{2^{\nu-m}} y[n-1] + x[n].$

(c) Period: 2^m, Frequency: $2\pi 2^{-m}$.

9.9. Statement 1.

9.10.

$$y[n] = X(e^{j\omega})|_{\omega=(2\pi/7)+(2\pi/21)(n-19)}.$$

9.11. (a) 2^{m-1}.

(b) 2^m.

9.12. $r[n] = e^{-j(2\pi/19)n} W^{n^2/2}$ where $W = e^{-j(2\pi/10)}$.

9.13. $x[0], x[8], x[4], x[12], x[2], x[10], x[6], x[14], x[1], x[9], x[5], x[13], x[3], x[11],$ $x[7], x[15].$

9.14. False.

9.15. $m = 1.$

9.16.

$$r = \begin{cases} 0, & m = 1, \\ 0, 4, & m = 2, \\ 0, 2, 4, 6, & m = 3, \\ 0, 1, 2, 3, 4, 5, 6, 7, & m = 4. \end{cases}$$

9.17. $N = 64.$

9.18. $m = 3$ or 4.

9.19. Decimation-in-time.

9.20. 1021 is prime, so the program must implement the full DFT equations and cannot exploit any FFT algorithm. The computation time goes as N^2. Contrastingly, 1024 is a power of 2 and can exploit the $N \log N$ computation time of the FFT.

Answers to Basic Problems in Chapter 10

10.1. **(a)** $f = 1500$ Hz.

   (b) $f = -2000$ Hz.

10.2. $N = 2048$ and 10000 Hz $< f <$ 10240 Hz.

10.3. **(a)** $T = 2\pi k_0 / (N \Omega_0).$

   (b) Not unique. $T = (2\pi / \Omega_0)(1 - k_0/N).$

10.4.

$$X_c(j2\pi(4200)) = 5 \times 10^{-4}$$

$$X_c(-j2\pi(4200)) = 5 \times 10^{-4}$$

$$X_c(j2\pi(1000)) = 10^{-4}$$

$$X_c(-j2\pi(1000)) = 10^{-4}$$

10.5. $L = 1024.$

10.6. $x_2[n]$ will have two distinct peaks.

10.7. $\Delta\Omega = 2\pi(2.44)$ rad/sec.

10.8. $N \geq 1600.$

10.9.

$$X_0[k] = \begin{cases} 18, & k = 3, 33, \\ 0, & \text{otherwise.} \end{cases}$$

$$X_1[k] = \begin{cases} 18, & k = 9, 27, \\ 0, & \text{otherwise.} \end{cases}$$

$$X_r[k] = 0 \text{ for } r \neq 0, 1.$$

10.10. $\omega_0 = 0.25\pi$ rad/sample, $\lambda = \pi/76000$ rad/sample^2.

10.11. $\Delta f = 9.77$ Hz.

10.12. The peaks will not have the same height. The peak from the rectangular window will be bigger.

10.13. **(a)** $A = 21$ dB.

(b) Weak components will be visible if their amplitude exceeds 0.0891.

10.14. **(a)** 320 samples.

(b) 400 DFT/second.

(c) $N = 256$.

(d) 62.5 Hz.

10.15. **(a)** $X[200] = 1 - j$.

(b)

$$X(j2\pi(4000)) = 5 \times 10^{-5}(1 - j)$$

$$X(-j2\pi(4000)) = 5 \times 10^{-5}(1 + j).$$

10.16. Rectangular, Hanning, Hamming, and Bartlett windows work.

10.17. $T > 1/1024$ sec.

10.18. $x_2[n]$, $x_3[n]$, $x_6[n]$.

10.19. Methods 2 and 5 will improve the resolution.

10.20. $L = M + 1 = 262$.

BIBLIOGRAPHY

Adams, J. W., and Wilson, J. A. N., "A New Approach to FIR Digital Filters with Fewer Multiplies and Reduced Sensitivity," *IEEE Trans. of Circuits and Systems*, Vol. 30, pp. 277–283, May 1983.

Ahmed, N., Natarajan, T., and Rao, K. R., "Discrete Cosine Transform," *IEEE Trans. on Computers*, Vol. C-23, pp. 90–93, Jan. 1974.

Allen, J., and Rabiner, L., "A Unified Approach to Short-time Fourier Analysis and Synthesis," *Proc. IEEE Trans. on Computers*, Vol. 65, pp. 1558–1564, Nov. 1977.

Atal, B. S., and Hanauer, S. L., "Speech Analysis and Synthesis by Linear Prediction of the Speech Wave," *J. Acoustical Society of America*, Vol. 50, pp. 637–655, 1971.

Atal, B. S., "Automatic Recognition of Speakers from their Voices," *IEEE Proceedings*, Vol. 64, No. 4, pp. 460–475, Apr. 1976.

Andrews, H. C., and Hunt, B. R., *Digital Image Restoration*, Prentice Hall, Englewood Cliffs, NJ, 1977.

Bagchi, S., and Mitra, S., *The Nonuniform Discrete Fourier Transform and Its Applications in Signal Processing*, Springer, New York, NY, 1999.

Baran, T. A., and Oppenheim, A. V., "Design and Implementation of Discrete-time Filters for Efficient Rate-conversion Systems," *Proceedings of the 41st Annual Asilomar Conference on Signals, Systems, and Computers*, Asilomar, CA, Nov. 4–7, 2007.

Baraniuk, R., "Compressive Sensing," *IEEE Signal Processing Magazine*, Vol. 24, No. 4, pp. 118–121, July 2007.

Barnes, C. W., and Fam, A. T., "Minimum Norm Recursive Digital Filters that are Free of Over-flow Limit Cycles," *IEEE Trans. Circuits and Systems*, Vol. CAS-24, pp. 569–574, Oct. 1977.

Bartels R. H., Beatty, J. C., and Barsky, B. A., *An Introduction to Splines for Use in Computer Graphics and Geometric Modelling*, Morgan Kauffman, San Francisco, CA, 1998.

Bartle, R. G., *The Elements of Real Analysis*, 3rd ed, John Wiley and Sons, New York, NY, 2000.

Bartlett, M. S., *An Introduction to Stochastic Processes with Special Reference to Methods and Applications*, Cambridge University Press, Cambridge, UK, 1953.

Bauman, P., Lipshitz, S., and Vanderkooy, J., "Cepstral Analysis of Electroacoustic Transducers," *Proc. Int. Conf. Acoustics, Speech, and Signal Processing* (ICASSP '85), Vol. 10, pp. 1832–1835, Apr. 1985.

Bellanger, M., *Digital Processing of Signals*, 3rd ed., Wiley, New York, NY, 2000.

Bennett, W. R., "Spectra of Quantized Signals," *Bell System Technical J.*, Vol. 27, pp. 446–472, 1948.

Bertsekas, D. and Tsitsiklis, J., *Introduction to Probability*, 2nd ed., Athena Scientific, Belmont, MA, 2008.

Blackman, R. B., and Tukey, J. W., *The Measurement of Power Spectra*, Dover Publications, New York, NY, 1958.

Blackman, R., *Linear Data-Smoothing and Prediction in Theory and Practice*, Addison-Wesley, Reading, MA, 1965.

Blahut, R. E., *Fast Algorithms for Digital Signal Processing*, Addison-Wesley, Reading, MA, 1985.

Bluestein, L. I., "A Linear Filtering Approach to the Computation of Discrete Fourier Transform," *IEEE Trans. Audio Electroacoustics*, Vol. AU-18, pp. 451–455, 1970.

Bogert, B. P., Healy, M. J. R., and Tukey, J. W., "The Quefrency Alanysis of Times Series for Echos: Cepstrum, Pseudo-autocovariance, Cross-cepstrum, and Saphe Cracking," Chapter 15, *Proc. Symposium on Time Series Analysis*, M. Rosenblatt, ed., John Wiley and Sons, New York, NY, 1963.

Bosi, M., and Goldberg, R. E., *Introduction to Digital Audio Coding and Standards*, Springer Science+Business Media, New York, NY, 2003.

Bovic, A., ed., *Handbook of Image and Video Processing*, 2nd ed., Academic Press, Burlington, MA, 2005.

Bracewell, R. N., "The Discrete Hartley Transform," *J. Optical Society of America*, Vol. 73, pp. 1832–1835, 1983.

Bracewell, R. N., "The Fast Hartley Transform," *IEEE Proceedings*, Vol. 72, No. 8, pp. 1010–1018, 1984.

Bracewell, R. N., *Two-Dimensional Imaging*, Prentice Hall, New York, NY, 1994.

Bracewell, R. N., *The Fourier Transform and Its Applications*, 3rd ed., McGraw-Hill, New York, NY, 1999.

Brigham, E., *Fast Fourier Transform and Its Applications*, Prentice Hall, Upper Saddle River, NJ, 1988.

Brigham, E. O., and Morrow, R. E., "The Fast Fourier Transform," *IEEE Spectrum*, Vol. 4, pp. 63–70, Dec. 1967.

Brown, J. W., and Churchill, R. V., *Introduction to Complex Variables and Applications*, 8th ed., McGraw-Hill, New York, NY, 2008.

Brown, R. C., *Introduction to Random Signal Analysis and Kalman Filtering*, Wiley, New York, NY, 1983.

Burden, R. L., and Faires, J. D., *Numerical Analysis*, 8th ed., Brooks Cole, 2004.

Burg, J. P., "A New Analysis Technique for Time Series Data," *Proc. NATO Advanced Study Institute on Signal Processing*, Enschede, Netherlands, 1968.

Burrus, C. S., "Efficient Fourier Transform and Convolution Algorithms," in *Advanced Topics in Signal Processing*, J. S. Lim and A. V. Oppenheim, eds., Prentice Hall, Englewood Cliffs, NJ, 1988.

Burrus, C. S., and Parks, T. W., *DFT/FFT and Convolution Algorithms Theory and Implementation*, Wiley, New York, NY, 1985.

Burrus, C. S., Gopinath, R. A., and Guo, H., *Introduction to Wavelets and Wavelet Transforms: A Primer*, Prentice Hall, 1997.

Candy, J. C., and Temes, G. C., *Oversampling Delta-Sigma Data Converters: Theory, Design, and Simulation*, IEEE Press, New York, NY, 1992.

Candes, E., "Compressive Sampling," *Int. Congress of Mathematics*, 2006, pp. 1433–1452.

Candes, E., and Wakin, M., "An Introduction to Compressive Sampling," *IEEE Signal Processing Magazine*, Vol. 25, No. 2, pp. 21–30, Mar. 2008.

Capon, J., "Maximum-likelihood Spectral Estimation," in *Nonlinear Methods of Spectral Analysis*, 2nd ed., S. Haykin, ed., Springer-Verlag, New York, NY, 1983.

Carslaw, H. S., *Introduction to the Theory of Fourier's Series and Integrals*, 3rd ed., Dover Publications, New York, NY, 1952.

Castleman, K. R., *Digital Image Processing*, 2nd ed., Prentice Hall, Upper Saddle River, NJ, 1996.

Chan, D. S. K., and Rabiner, L. R., "An Algorithm for Minimizing Roundoff Noise in Cascade Realizations of Finite Impulse Response Digital Filters," *Bell System Technical J.*, Vol. 52, No. 3, pp. 347–385, Mar. 1973.

Chan, D. S. K., and Rabiner, L. R., "Analysis of Quantization Errors in the Direct Form for Finite Impulse Response Digital Filters," *IEEE Trans. Audio Electroacoustics*, Vol. 21, pp. 354–366, Aug. 1973.

Chellappa, R., Girod, B., Munson, D. C., Tekalp, A. M., and Vetterli, M., "The Past, Present, and Future of Image and Multidimensional Signal Processing," *IEEE Signal Processing Magazine*, Vol. 15, No. 2, pp. 21–58, Mar. 1998.

Chen, W. H., Smith, C. H., and Fralick, S. C., "A Fast Computational Algorithm for the Discrete Cosine Transform," *IEEE Trans. Commun.*, Vol. 25, pp. 1004–1009, September 1977.

Chen, X., and Parks, T. W., "Design of FIR Filters in the Complex Domain," *IEEE Trans. Acoustics, Speech, and Signal Processing*, Vol. 35, pp. 144–153, 1987.

Cheney, E. W., *Introduction to Approximation Theory*, 2nd ed., Amer. Math. Society, New York, NY, 2000.

Chow, Y., and Cassignol, E., *Linear Signal Flow Graphs and Applications*, Wiley, New York, NY, 1962.

Cioffi, J. M., and Kailath, T., "Fast Recursive Least-squares Transversal Filters for Adaptive Filtering," *IEEE Trans. Acoustics, Speech, and Signal Processing*, Vol. 32, pp. 607–624, June 1984.

Claasen, T. A., and Mecklenbräuker, W. F., "On the Transposition of Linear Time-varying Discrete-time Networks and its Application to Multirate Digital Systems," *Philips J. Res.*, Vol. 23, pp. 78–102, 1978.

Claasen, T. A. C. M., Mecklenbrauker, W. F. G., and Peek, J. B. H., "Second-order Digital Filter with only One Magnitude-truncation Quantizer and Having Practically no Limit Cycles," *Electronics Letters*, Vol. 9, No. 2, pp. 531–532, Nov. 1973.

Clements, M. A., and Pease, J., "On Causal Linear Phase IIR Digital Filters," *IEEE Trans. Acoustics, Speech, and Signal Processing*, Vol. 3, pp. 479–484, Apr. 1989.

Committee, DSP, ed., *Programs for Digital Signal Processing*, IEEE Press, New York, NY, 1979.

Constantinides, A. G., "Spectral Transformations for Digital Filters," *IEEE Proceedings*, Vol. 117, No. 8, pp. 1585–1590, Aug. 1970.

Cooley, J. W., Lewis, P. A. W., and Welch, P. D., "Historical Notes on the Fast Fourier Transform," *IEEE Trans. Audio Electroacoustics*, Vol. 15, pp. 76–79, June 1967.

Cooley, J. W., and Tukey, J. W., "An Algorithm for the Machine Computation of Complex Fourier Series," *Mathematics of Computation*, Vol. 19, pp. 297–301, Apr. 1965.

Crochiere, R. E., and Oppenheim, A. V., "Analysis of Linear Digital Networks," *IEEE Proceedings*, Vol. 63, pp. 581–595, Apr. 1975.

Crochiere, R. E., and Rabiner, L. R., *Multirate Digital Signal Processing*, Prentice Hall, Englewood Cliffs, NJ, 1983.

Daniels, R. W., *Approximation Methods for Electronic Filter Design*, McGraw-Hill, New York, NY, 1974.

Danielson, G. C., and Lanczos, C., "Some Improvements in Practical Fourier Analysis and their Application to X-ray Scattering from Liquids," *J. Franklin Inst.*, Vol. 233, pp. 365–380 and 435–452, Apr. and May 1942.

Davenport, W. B., *Probability and Random Processes: An Introduction for Applied Scientists and Engineers*, McGraw-Hill, New York, NY, 1970.

Davis, S. B., and Mermelstein, P., "Comparison of Parametric Representations for Monosyllabic Word Recognition," *IEEE Trans. Acoustics, Speech and Signal Processing*, Vol. ASSP-28, No. 4, pp. 357–366, Aug. 1980.

Deller, J. R., Hansen, J. H. L., and Proakis, J. G., *Discrete-Time Processing of Speech Signals*, Wiley-IEEE Press, New York, NY, 2000.

Donoho, D. L., "Compressed Sensing," *IEEE Trans. on Information Theory*, Vol. 52, No. 4, pp. 1289–1306, Apr. 2006.

Dudgeon, D. E., and Mersereau, R. M., *Two-Dimensional Digital Signal Processing*, Prentice Hall, Englewood Cliffs, NJ, 1984.

Duhamel, P., "Implementation of 'Split-radix' FFT Algorithms for Complex, Real, and Real-symmetric Data," *IEEE Trans. Acoustics, Speech, and Signal Processing*, Vol. 34, pp. 285–295, Apr. 1986.

Duhamel, P., and Hollmann, H., "Split Radix FFT Algorithm," *Electronic Letters*, Vol. 20, pp. 14–16, Jan. 1984.

Ebert, P. M., Mazo, J. E., and Taylor, M. C., "Overflow Oscillations in Digital Filters," *Bell System Technical J.*, Vol. 48, pp. 2999–3020, 1969.

Eldar, Y. C., and Oppenheim, A. V., "Filterbank Reconstruction of Bandlimited Signals from Nonuniform and Generalized Samples," *IEEE Trans. on Signal Processing*, Vol. 48, No. 10, pp. 2864–2875, October, 2000.

Elliott, D. F., and Rao, K. R., *Fast Transforms: Algorithms, Analysis, Applications*, Academic Press, New York, NY, 1982.

Feller, W., *An Introduction to Probability Theory and Its Applications*, Wiley, New York, NY, 1950, Vols. 1 and 2.

Fettweis, A., "Wave Digital Filters: Theory and Practice," *IEEE Proceedings*, Vol. 74, No. 2, pp. 270–327, Feb. 1986.

Flanagan, J. L., *Speech Analysis, Synthesis and Perception*, 2nd ed., Springer-Verlag, New York, NY, 1972.

Frerking, M. E., *Digital Signal Processing in Communication Systems*, Kluwer Academic, Boston, MA, 1994.

Friedlander, B., "Lattice Filters for Adaptive Processing," *IEEE Proceedings*, Vol. 70, pp. 829–867, Aug. 1982.

Friedlander, B., "Lattice Methods for Spectral Estimation," *IEEE Proceedings*, Vol. 70, pp. 990–1017, September 1982.

Frigo, M., and Johnson, S. G., "FFTW: An Adaptive Software Architecture for the FFT," *Proc. Int. Conf. Acoustics, Speech, and Signal Processing* (ICASSP '98), Vol. 3, pp. 1381–1384, May 1998.

Frigo, M., and Johnson, S. G., "The Design and Implementation of FFTW3," *Proc. of the IEEE*, Vol. 93, No. 2, pp. 216–231, Feb. 2005.

Furui, S., "Cepstral Analysis Technique for Automatic Speaker Verification," *IEEE Trans. Acoustics, Speech, and Signal Processing*, Vol. ASSP-29, No. 2, pp. 254–272, Apr. 1981.

Gallager, R., *Principles of Digital Communication*, Cambridge University Press, Cambridge, UK, 2008.

Gardner, W., *Statistical Spectral Analysis: A Non-Probabilistic Theory*, Prentice Hall, Englewood Cliffs, NJ, 1988.

Gentleman, W. M., and Sande, G., "Fast Fourier Transforms for Fun and Profit," *1966 Fall Joint Computer Conf., AFIPS Conf. Proc*, Vol. 29., Spartan Books, Washington, D.C., pp. 563–578, 1966.

Goertzel, G., "An Algorithm for the Evaluation of Finite Trigonometric Series," *American Math. Monthly*, Vol. 65, pp. 34–35, Jan. 1958.

Gold, B., Oppenheim, A. V., and Rader, C. M., "Theory and Implementation of the Discrete Hilbert Transform," in *Proc. Symp. Computer Processing in Communications*, Vol. 19, Polytechnic Press, New York, NY, 1970.

Gold, B., and Rader, C. M., *Digital Processing of Signals*, McGraw-Hill, New York, NY, 1969.

Gonzalez, R. C., and Woods, R. E., *Digital Image Processing*, Wiley, 2007.

Goyal, V., "Theoretical Foundations of Transform Coding," *IEEE Signal Processing Magazine*, Vol. 18, No. 5, pp. 9–21, Sept. 2001.

Gray, A. H., and Markel, J. D., "A Computer Program for Designing Digital Elliptic Filters," *IEEE Trans. Acoustics, Speech, and Signal Processing*, Vol. 24, pp. 529–538, Dec. 1976.

Gray, R. M., and Davidson, L. D., *Introduction to Statistical Signal Processing*, Cambridge University Press, 2004.

Griffiths, L. J., "An Adaptive Lattice Structure for Noise Canceling Applications," *Proc. Int. Conf. Acoustics, Speech, and Signal Processing* (ICASSP '78), Tulsa, OK, Apr. 1978, pp. 87–90.

Grossman, S., *Calculus Part 2*, 5th ed., Saunders College Publications, Fort Worth, TX, 1992.

Guillemin, E. A., *Synthesis of Passive Networks*, Wiley, New York, NY, 1957.

Hannan, E. J., *Time Series Analysis*, Methuen, London, UK, 1960.

Harris, F. J., "On the Use of Windows for Harmonic Analysis with the Discrete Fourier Transform," *IEEE Proceedings*, Vol. 66, pp. 51–83, Jan. 1978.

Hayes, M. H., Lim, J. S., and Oppenheim, A. V., "Signal Reconstruction from Phase and Magnitude," *IEEE Trans. Acoustics, Speech, and Signal Processing*, Vol. 28, No. 6, pp. 672–680, Dec. 1980.

Hayes, M., *Statistical Digital Signal Processing and Modeling*, Wiley, New York, NY, 1996.

Haykin, S., *Adaptive Filter Theory*, 4th ed., Prentice Hall, 2002.

Haykin, S., and Widrow, B., *Least-Mean-Square Adaptive Filters*, Wiley-Interscience, Hoboken, NJ, 2003.

Heideman, M. T., Johnson, D. H., and Burrus, C. S., "Gauss and the History of the Fast Fourier Transform," *IEEE ASSP Magazine*, Vol. 1, No. 4, pp. 14–21, Oct. 1984.

Helms, H. D., "Fast Fourier Transform Method of Computing Difference Equations and Simulating Filters," *IEEE Trans. Audio Electroacoustics*, Vol. 15, No. 2, pp. 85–90, 1967.

Herrmann, O., "On the Design of Nonrecursive Digital Filters with Linear Phase," *Elec. Lett.*, Vol. 6, No. 11, pp. 328–329, 1970.

Herrmann, O., Rabiner, L. R., and Chan, D. S. K., "Practical Design Rules for Optimum Finite Impulse Response Lowpass Digital Filters," *Bell System Technical J.*, Vol. 52, No. 6, pp. 769–799, July–Aug. 1973.

Herrmann, O., and Schüssler, W., "Design of Nonrecursive Digital Filters with Minimum Phase," *Elec. Lett.*, Vol. 6, No. 6, pp. 329–330, 1970.

Herrmann, O., and W. Schüssler, "On the Accuracy Problem in the Design of Nonrecursive Digital Filters," *Arch. Electronic Ubertragungstechnik*, Vol. 24, pp. 525–526, 1970.

Hewes, C. R., Broderson, R. W., and Buss, D. D., "Applications of CCD and Switched Capacitor Filter Technology," *IEEE Proceedings*, Vol. 67, No. 10, pp. 1403–1415, Oct. 1979.

Hnatek, E. R., *A User's Handbook of D/A and A/D Converters*, R. E. Krieger Publishing Co., Malabar, 1988.

Hofstetter, E., Oppenheim, A. V., and Siegel, J., "On Optimum Nonrecursive Digital Filters," *Proc. 9th Allerton Conf. Circuit System Theory*, Oct. 1971.

Hughes, C. P., and Nikeghbali, A., "The Zeros of Random Polynomials Cluster Near the Unit Circle," arXiv:math/0406376v3 [math.CV], http://arxiv.org/ PS_cache/math/pdf/0406/0406376v3.pdf.

Hwang, S. Y., "On Optimization of Cascade Fixed Point Digital Filters," *IEEE Trans. Circuits and Systems*, Vol. 21, No. 1, pp. 163–166, Jan. 1974.

Itakura, F. I., and Saito, S., "Analysis-synthesis Telephony Based upon the Maximum Likelihood Method," *Proc. 6th Int. Congress on Acoustics*, pp. C17–20, Tokyo, 1968.

Itakura, F. I., and Saito, S., "A Statistical Method for Estimation of Speech Spectral Density and Formant Frequencies," *Elec. and Comm. in Japan*, Vol. 53-A, No. 1, pp. 36–43, 1970.

Jackson, L. B., "On the Interaction of Roundoff Noise and Dynamic Range in Digital Filters," *Bell System Technical J.*, Vol. 49, pp. 159–184, Feb. 1970.

Jackson, L. B., "Roundoff-noise Analysis for Fixed-point Digital Filters Realized in Cascade or Parallel Form," *IEEE Trans. Audio Electroacoustics*, Vol. 18, pp. 107–122, June 1970.

Jackson, L. B., *Digital Filters and Signal Processing: With MATLAB Exercises*, 3rd ed., Kluwer Academic Publishers, Hingham, MA, 1996.

Jacobsen, E., and Lyons, R., "The Sliding DFT," *IEEE Signal Processing Magazine*, Vol. 20, pp. 74–80, Mar. 2003.

Jain, A. K., *Fundamentals of Digital Image Processing*, Prentice Hall, Englewood Cliffs, NJ, 1989.

Jayant, N. S., and Noll, P., *Digital Coding of Waveforms*, Prentice Hall, Englewood Cliffs, NJ, 1984.

Jenkins, G. M., and Watts, D. G., *Spectral Analysis and Its Applications*, Holden-Day, San Francisco, CA, 1968.

Jolley, L. B. W., *Summation of Series*, Dover Publications, New York, NY, 1961.

Johnston, J., "A Filter Family Designed for Use in Quadrature Mirror Filter Banks," *Proc. Int. Conf. Acoustics, Speech, and Signal Processing* (ICASSP '80), Vol. 5, pp. 291–294, Apr. 1980.

Juang, B.-H., Rabiner, L. R., and Wilpon, J. G., "On the Use of Bandpass Liftering in Speech Recognition," *IEEE Trans. Acoustics, Speech, and Signal Processing*, Vol. ASSP-35, No. 7, pp. 947–954, July 1987.

Kaiser, J. F., "Digital Filters," in *System Analysis by Digital Computer*, Chapter 7, F. F. Kuo and J. F. Kaiser, eds., Wiley, New York, NY, 1966.

Kaiser, J. F., "Nonrecursive Digital Filter Design Using the I_0-sinh Window Function," *Proc. 1974 IEEE International Symp. on Circuits and Systems*, San Francisco, CA, 1974.

Kaiser, J. F., and Hamming, R. W., "Sharpening the Response of a Symmetric Nonrecursive Filter by Multiple Use of the Same Filter," *IEEE Trans. Acoustics, Speech, and Signal Processing*, Vol. 25, No. 5, pp. 415–422, Oct. 1977.

Kaiser, J. F., and Schafer, R. W., "On the Use of the I_0-sinh Window for Spectrum Analysis," *IEEE Trans. Acoustics, Speech, and Signal Processing*, Vol. 28, No. 1, pp. 105–107, Feb. 1980.

Kan, E. P. F., and Aggarwal, J. K., "Error Analysis of Digital Filters Employing Floating Point Arithmetic," *IEEE Trans. Circuit Theory*, Vol. 18, pp. 678–686, Nov. 1971.

Kaneko, T., and Liu, B., "Accumulation of Roundoff Error in Fast Fourier Transforms," *J. Assoc. Comput. Mach.*, Vol. 17, pp. 637–654, Oct. 1970.

Kanwal, R., *Linear Integral Equations*, 2nd ed., Springer, 1997.

Karam, L. J., and McClellan, J. H., "Complex Chebychev Approximation for FIR Filter Design," *IEEE Trans. Circuits and Systems*, Vol. 42, pp. 207–216, Mar. 1995.

Karam, Z. N., and Oppenheim, A. V., "Computation of the One-dimensional Unwrapped Phase," *15th International Conference on Digital Signal Processing*, pp. 304–307, July 2007.

Kay, S. M., *Modern Spectral Estimation Theory and Application*, Prentice Hall, Englewood Cliffs, NJ, 1988.

Kay, S. M., *Intuitive Probability and Random Processes Using MATLAB*, Springer, New York, NY, 2006.

Kay, S. M., and Marple, S. L., "Spectrum Analysis: A Modern Perspective," *IEEE Proceedings*, Vol. 69, pp. 1380–1419, Nov. 1981.

Keys, R., "Cubic Convolution Interpolation for Digital Image Processing," *IEEE Trans. Acoustics, Speech and Signal Processing*, Vol. 29, No. 6, pp. 1153–1160, Dec. 1981.

Kleijn, W., "Principles of Speech Coding," in *Springer Handbook of Speech Processing*, J. Benesty, M. Sondhi, and Y. Huang, eds., Springer, 2008, pp. 283–306.

Knuth, D. E., *The Art of Computer Programming; Seminumerical Algorithms*, 3rd ed., Addison-Wesley, Reading, MA, 1997, Vol. 2.

Koopmanns, L. H., *Spectral Analysis of Time Series*, 2nd ed., Academic Press, New York, NY, 1995.

Korner, T. W., *Fourier Analysis*, Cambridge University Press, Cambridge, UK, 1989.

Lam, H. Y. F., *Analog and Digital Filters: Design and Realization*, Prentice Hall, Englewood Cliffs, NJ, 1979.

Lang, S. W., and McClellan, J. H., "A Simple Proof of Stability for All-pole Linear Prediction Models," *IEEE Proceedings*, Vol. 67, No. 5, pp. 860–861, May 1979.

Leon-Garcia, A., *Probability and Random Processes for Electrical Engineering*, 2nd ed., Addison-Wesley, Reading, MA, 1994.

Lighthill, M. J., *Introduction to Fourier Analysis and Generalized Functions*, Cambridge University Press, Cambridge, UK, 1958.

Lim, J. S., *Two-Dimensional Digital Signal Processing*, Prentice Hall, Englewood Cliffs, NJ, 1989.

Liu, B., and Kaneko, T., "Error Analysis of Digital Filters Realized in Floating-point Arithmetic," *IEEE Proceedings*, Vol. 57, pp. 1735–1747, Oct. 1969.

Liu, B., and Peled, A., "Heuristic Optimization of the Cascade Realization of Fixed Point Digital Filters," *IEEE Trans. Acoustics, Speech, and Signal Processing*, Vol. 23, pp. 464–473, 1975.

Macovski, A., *Medical Image Processing*, Prentice Hall, Englewood Cliffs, NJ, 1983.

Makhoul, J., "Spectral Analysis of Speech by Linear Prediction," *IEEE Trans. Audio and Electroacoustics*, Vol. AU-21, No. 3, pp. 140–148, June 1973.

Makhoul, J., "Linear Prediction: A Tutorial Review," *IEEE Proceedings*, Vol. 62, pp. 561–580, Apr. 1975.

Makhoul, J., "A Fast Cosine Transform in One and Two Dimensions," *IEEE Trans. Acoustics, Speech, and Signal Processing*, Vol. 28, No. 1, pp. 27–34, Feb. 1980.

Maloberti, F., *Data Converters*, Springer, New York, NY, 2007.

Markel, J. D., "FFT Pruning," *IEEE Trans. Audio and Electroacoustics*, Vol. 19, pp. 305–311, L

Markel, J. D., and Gray, A. H., Jr., *Linear Prediction of Speech*, Springer-Verlag, New York, NY,

Marple, S. L., *Digital Spectral Analysis with Applications*, Prentice Hall, Englewood Cliffs, NJ, 1987.

Martucci, S. A., "Symmetrical Convolution and the Discrete Sine and Cosine Transforms," *IEEE Signal Processing*, Vol. 42, No. 5, pp. 1038–1051, May 1994.

Mason, S., and Zimmermann, H. J., *Electronic Circuits, Signals and Systems*, Wiley, New York, NY, 1960.

Mathworks, *Signal Processing Toolbox Users Guide*, The Mathworks, Inc., Natick, MA, 1998.

McClellan, J. H., and Parks, T. W., "A Unified Approach to the Design of Optimum FIR Linear Phase Digital Filters," *IEEE Trans. Circuit Theory*, Vol. 20, pp. 697–701, Nov. 1973.

McClellan, J. H., and Rader, C. M., *Number Theory in Digital Signal Processing*, Prentice Hall, Englewood Cliffs, NJ, 1979.

McClellan, J. H., "Parametric Signal Modeling," Chapter 1, *Advanced Topics in Signal Processing*, J. S. Lim and A. V. Oppenheim, eds., Prentice Hall, Englewood Cliffs, 1988.

Mersereau, R. M., Schafer, R. W., Barnwell, T. P., and Smith, D. L., "A Digital Filter Design Package for PCs and TMS320s," *Proc. MIDCON*, Dallas, TX, 1984.

Mills, W. L., Mullis, C. T., and Roberts, R. A., "Digital Filter Realizations Without Overflow Oscillations," *IEEE Trans. Acoustics, Speech, and Signal Processing*, Vol. 26, pp. 334–338, Aug. 1978.

Mintzer, F., "Filters for Distortion-free Two-band Multirate Filter Banks," *IEEE Trans. Acoustics, Speech and Signal Processing*, Vol. 33, No. 3, pp. 626–630, June 1985.

Mitra, S. K., *Digital Signal Processing*, 3rd ed., McGraw-Hill, New York, NY, 2005.

Moon, T., and Stirling, W., *Mathematical Methods and Algorithms for Signal Processing*, Prentice Hall, 1999.

Nawab, S. H., and Quatieri, T. F., "Short-time Fourier transforms," in *Advanced Topics in Signal Processing*, J. S. Lim and A. V. Oppenheim, eds., Prentice Hall, Englewood Cliffs, NJ, 1988.

Neuvo, Y., Dong, C.-Y., and Mitra, S., "Interpolated Finite Impulse Response Filters," *IEEE Trans. Acoustics, Speech and Signal Processing*, Vol. 32, No. 3, pp. 563–570, June 1984.

Noll, A. M., "Cepstrum Pitch Determination," *J. Acoustical Society of America*, Vol. 41, pp. 293–309, Feb. 1967.

Nyquist, H., "Certain Topics in Telegraph Transmission Theory," *AIEE Trans.*, Vol. 90, No. 2, pp. 280–305, 1928.

Oetken, G., Parks, T. W., and Schüssler, H. W., "New Results in the Design of Digital Interpolators," *IEEE Trans. Acoustics, Speech, and Signal Processing*, Vol. 23, pp. 301–309, June 1975.

Oppenheim, A. V., "Superposition in a Class of Nonlinear Systems," *RLE Technical Report No. 432*, MIT, 1964.

Oppenheim, A. V., "Generalized Superposition," *Information and Control*, Vol. 11, Nos. 5–6, pp. 528–536, Nov.–Dec., 1967.

Oppenheim, A. V., "Generalized Linear Filtering," Chapter 8, *Digital Processing of Signals*, B. Gold and C. M. Rader, eds., McGraw-Hill, New York, 1969a.

Oppenheim, A. V., "A Speech Analysis-synthesis System Based on Homomorphic Filtering," *J. Acoustical Society of America*, Vol. 45, pp. 458–465, Feb. 1969b.

Oppenheim, A. V., and Johnson, D. H., "Discrete Representation of Signals," *IEEE Proceedings*, Vol. 60, No. 6, pp. 681–691, June 1972.

Oppenheim, A. V., and Schafer, R. W., "Homomorphic Analysis of Speech," *IEEE Trans. Audio Electroacoustics*, Vol. AU-16, No. 2, pp. 221–226, June 1968.

Oppenheim, A. V., and Schafer, R. W., *Digital Signal Processing*, Prentice Hall, Englewood Cliffs, NJ, 1975.

Oppenheim, A. V., Schafer, R. W., and Stockam, T. G., Jr., "Nonlinear Filtering of Multiplied and Convolved Signals," *IEEE Proceedings*, Vol. 56, No. 8, pp. 1264–1291, Aug. 1968.

Oppenheim, A. V., and Willsky, A. S., *Signals and Systems*, 2nd ed., Prentice Hall, Upper Saddle River, NJ, 1997.

Oraintara, S., Chen, Y. J., and Nguyen, T., "Integer Fast Fourier Transform," *IEEE Trans. on Signal Processing*, Vol. 50, No. 3, pp. 607–618, Mar. 2001.

O'Shaughnessy, D., *Speech Communication, Human and Machine*, 2nd ed., Addison-Wesley, Reading, MA, 1999.

Pan, D., "A Tutorial on MPEG/audio Compression," *IEEE Multimedia*, pp. 60–74, Summer 1995.

Papoulis, A., "On the Approximation Problem in Filter Design," in *IRE Nat. Convention Record, Part 2*, 1957, pp. 175–185.

Papoulis, A., *The Fourier Integral and Its Applications*, McGraw-Hill, New York, NY, 1962.

ulis, A., *Signal Analysis*, McGraw-Hill Book Company, New York, NY, 1977.

apoulis, A., *Probability, Random Variables and Stochastic Processes*, 4th ed., McGraw-Hill, New York, NY, 2002.

Parks, T. W., and Burrus, C. S., *Digital Filter Design*, Wiley, New York, NY, 1987.

Parks, T. W., and McClellan, J. H., "Chebyshev Approximation for Nonrecursive Digital Filters with Linear Phase," *IEEE Trans. Circuit Theory*, Vol. 19, pp. 189–194, Mar. 1972.

Parks, T. W., and McClellan, J. H., "A Program for the Design of Linear Phase Finite Impulse Response Filters," *IEEE Trans. Audio Electroacoustics*, Vol. 20, No. 3, pp. 195–199, Aug. 1972.

Parsons, T. J., *Voice and Speech Processing*, Prentice Hall, New York, NY, 1986.

Parzen, E., *Modern Probability Theory and Its Applications*, Wiley, New York, NY, 1960.

Pennebaker, W. B., and Mitchell, J. L., *JPEG: Still Image Data Compression Standard*, Springer, New York, NY, 1992.

Phillips, C. L., and Nagle, H. T., Jr., *Digital Control System Analysis and Design*, 3rd ed., Prentice Hall, Upper Saddle River, NJ, 1995.

Pratt, W., *Digital Image Processing*, 4th ed., Wiley, New York, NY, 2007.

Press, W. H. F., Teukolsky, S. A. B. P., Vetterling, W. T., and Flannery, B. P., *Numerical Recipes: The Art of Scientific Computing*, 3rd ed., Cambridge University Press, Cambridge, UK, 2007.

Proakis, J. G., and Manolakis, D. G., *Digital Signal Processing*, Prentice Hall, Upper Saddle River, NJ, 2006.

Quatieri, T. F., *Discrete-Time Speech Signal Processing: Principles and Practice*, Prentice Hall, Englewood Cliffs, NJ, 2002.

Rabiner, L. R., "The Design of Finite Impulse Response Digital Filters Using Linear Programming Techniques," *Bell System Technical J.*, Vol. 51, pp. 1117–1198, Aug. 1972.

Rabiner, L. R., "Linear Program Design of Finite Impulse Response (FIR) Digital Filters," *IEEE Trans. Audio and Electroacoustics*, Vol. 20, No. 4, pp. 280–288, Oct. 1972.

Rabiner, L. R., and Gold, B., *Theory and Application of Digital Signal Processing*, Prentice Hall, Englewood Cliffs, NJ, 1975.

Rabiner, L. R., Kaiser, J. F., Herrmann, O., and Dolan, M. T., "Some Comparisons Between FIR and IIR Digital Filters," *Bell System Technical J.*, Vol. 53, No. 2, pp. 305–331, Feb. 1974.

Rabiner, L. R., and Schafer, R. W., "On the Behavior of Minimax FIR Digital Hilbert Transformers," *Bell System Technical J.*, Vol. 53, No. 2, pp. 361–388, Feb. 1974.

Rabiner, L. R., and Schafer, R. W., *Digital Processing of Speech Signals*, Prentice Hall, Englewood Cliffs, NJ, 1978.

Rabiner, L. R., Schafer, R. W., and Rader, C. M., "The Chirp z-transform Algorithm," *IEEE Trans. Audio Electroacoustics*, Vol. 17, pp. 86–92, June 1969.

Rader, C. M., "Discrete Fourier Transforms when the Number of Data Samples is Prime," *IEEE Proceedings*, Vol. 56, pp. 1107–1108, June 1968.

Rader, C. M., "An Improved Algorithm for High-speed Autocorrelation with Applications to Spectral Estimation," *IEEE Trans. Audio Electroacoustics*, Vol. 18, pp. 439–441, Dec. 1970.

Rader, C. M., and Brenner, N. M., "A New Principle for Fast Fourier Transformation," *IEEE Trans. Acoustics, Speech, and Signal Processing*, Vol. 25, pp. 264–265, June 1976.

Rader, C. M., and Gold, B., "Digital Filter Design Techniques in the Frequency Domain," *IEEE Proceedings*, Vol. 55, pp. 149–171, Feb. 1967.

Ragazzini, J. R., and Franklin, G. F., *Sampled Data Control Systems*, McGraw-Hill, New York, NY, 1958.

Rao, K. R., and Hwang, J. J., *Techniques and Standards for Image, Video, and Audio Coding*, Prentice Hall, Upper Saddle River, NJ, 1996.

Rao, K. R., and Yip, P., *Discrete Cosine Transform: Algorithms, Advantages, Applications*, Academic Press, Boston, MA, 1990.

Rao, S. K., and Kailath, T., "Orthogonal Digital Filters for VLSI Implementation," *IEEE Trans. Circuits and System*, Vol. 31, No. 11, pp. 933–945, Nov. 1984.

Reut, Z., Pace, N. G., and Heaton, M. J. P., "Computer Classification of Sea Beds by Sonar," *Nature*, Vol. 314, pp. 426–428, Apr. 4, 1985.

Robinson, E. A., and Durrani, T. S., *Geophysical Signal Processing*, Prentice Hall, Englewood Cliffs, NJ, 1985.

Robinson, E. A., and Treitel, S., *Geophysical Signal Analysis*, Prentice Hall, Englewood Cliffs, NJ, 1980.

Romberg, J., "Imaging Via Compressive Sampling," *IEEE Signal Processing Magazine*, Vol. 25, No. 2, pp. 14–20, Mar. 2008.

Ross, S., *A First Course in Probability*, 8th ed., Prentice Hall, Upper Saddle River, NJ, 2009.

Runge, C., "Uber die Zerlegung Empirisch Gegebener Periodischer Functionen in Sinuswellen," *Z. Math. Physik*, Vol. 53, pp. 117–123, 1905.

Sandberg, I. W., "Floating-point-roundoff Accumulation in Digital Filter Realizations," *Bell System Technical J.*, Vol. 46, pp. 1775–1791, Oct. 1967.

Sayed, A., *Adaptive Filters*, Wiley, Hoboken, NJ, 2008.

Sayed, A. H., *Fundamentals of Adaptive Filtering*, Wiley-IEEE Press, 2003.

Sayood, K., *Introduction to Data Compression*, 3rd ed., Morgan Kaufmann, 2005.

Schaefer, R. T., Schafer, R. W., and Mersereau, R. M., "Digital Signal Processing for Doppler Radar Signals," *Proc. 1979 IEEE Int. Conf. on Acoustics, Speech, and Signal Processing*, pp. 170–173, 1979.

Schafer, R. W., "Echo Removal by Generalized Linear Filtering," *RLE Tech. Report No. 466*, MIT, Cambridge, MA, 1969.

Schafer, R. W., "Homomorphic Systems and Cepstrum Analysis of Speech," Chapter 9, *Springer Handbook of Speech Processing and Communication*, J. Benesty, M. M. Sondhi, and Y. Huang, eds., Springer-Verlag, Heidelberg, 2007.

Schafer, R. W., and Rabiner, L. R., "System for Automatic Formant Analysis of Voiced Speech," *J. Acoustical Society of America*, Vol. 47, No. 2, pt. 2, pp. 634–648, Feb. 1970.

Schafer, R. W., and Rabiner, L. R., "A Digital Signal Processing Approach to Interpolation," *IEEE Proceedings*, Vol. 61, pp. 692–702, June 1973.

Schmid, H., *Electronic Analog/Digital Conversions*, Wiley, New York, NY, 1976.

Schreier, R., and Temes, G. C., *Understanding Delta-Sigma Data Converters*, IEEE Press and John Wiley and Sons, Hoboken, NJ, 2005.

Schroeder, M. R., "Direct (Nonrecursive) Relations Between Cepstrum and Predictor Coefficients," *IEEE Trans. Acoustics, Speech and Signal Processing*, Vol. 29, No. 2, pp. 297–301, Apr. 1981.

Schüssler, H. W., and Steffen, P., "Some Advanced Topics in Filter Design," in *Advanced Topics in Signal Processing*, S. Lim and A. V. Oppenheim, eds., Prentice Hall, Englewood Cliffs, NJ, 1988.

Senmoto, S., and Childers, D. G., "Adaptive Decomposition of a Composite Signal of Identical Unknown Wavelets in Noise," *IEEE Trans. on Systems, Man, and Cybernetics*, Vol. SMC-2, No. 1, pp. 59, Jan. 1972.

Shannon, C. E., "Communication in the Presence of Noise," *Proceedings of the Institute of Radio Engineers (IRE)*, Vol. 37, No. 1, pp. 10–21, Jan. 1949.

Singleton, R. C., "An Algorithm for Computing the Mixed Radix Fast Fourier Transforms," *IEEE Trans. Audio Electroacoustics*, Vol. 17, pp. 93–103, June 1969.

Sitton, G. A., Burrus, C. S., Fox, J. W., and Treitel, S., "Factoring Very-high-degree Polynomials," *IEEE Signal Processing Magazine*, Vol. 20, No. 6, pp. 27–42, Nov. 2003.

Skolnik, M. I., *Introduction to Radar Systems*, 3rd ed., McGraw-Hill, New York, NY, 2002.

Slepian, D., Landau, H. T., and Pollack, H. O., "Prolate Spheroidal Wave Functions, Fourier Analysis, and Uncertainty Principle (I and II)," *Bell System Technical J.*, Vol. 40, No. 1, pp. 43–80, 1961.

Smith, M., and Barnwell, T., " A Procedure for Designing Exact Reconstruction Filter Banks for Tree-structured Subband Coders," *Proc. Int. Conf. Acoustics, Speech, and Signal Processing* (ICASSP '84), Vol. 9, Pt. 1, pp. 421–424, Mar. 1984.

Spanias, A., Painter, T., and Atti, V., *Audio Signal Processing and Coding*, Wiley, Hoboken, NJ, 2007.

Sripad, A., and Snyder, D., "A Necessary and Sufficient Condition for Quantization Errors to be Uniform and White," *IEEE Trans. Acoustics, Speech and Signal Processing*, Vol. 25, No. 5, pp. 442–448, Oct. 1977.

Stark, H., and Woods, J., *Probability and Random Processes with Applications to Signal Processing*, 3rd ed., Prentice Hall, Englewood Cliffs, NJ, 2001.

Starr, T., Cioffi, J. M., and Silverman, P. J., *Understanding Digital Subscriber Line Technology*, Prentice Hall, Upper Saddle River, NJ, 1999.

Steiglitz, K., "The Equivalence of Analog and Digital Signal Processing," *Information and Control*, Vol. 8, No. 5, pp. 455–467, Oct. 1965.

Steiglitz, K., and Dickinson, B., "Phase Unwrapping by Factorization," *IEEE Trans. Acoustics, Speech and Signal Processing*, Vol. 30, No. 6, pp. 984–991, Dec. 1982.

Stockham, T. G., "High Speed Convolution and Correlation," in *1966 Spring Joint Computer Conference*, *AFIPS Proceedings*, Vol. 28, pp. 229–233, 1966.

Stockham, T. G., Cannon, T. M., and Ingebretsen, R. B., "Blind Deconvolution Through Digital Signal Processing," *IEEE Proceedings*, Vol. 63, pp. 678–692, Apr. 1975.

Stoica, P., and Moses, R., *Spectral Analysis of Signals*, Pearson Prentice Hall, Upper Saddle River, NJ, 2005.

Storer, J. E., *Passive Network Synthesis*, McGraw-Hill, New York, NY, 1957.

Strang, G., "The Discrete Cosine Transforms," *SIAM Review*, Vol. 41, No. 1, pp. 135–137, 1999.

Strang, G., and Nguyen, T., *Wavelets and Filter Banks*, Wellesley–Cambridge Press, Cambridge, MA, 1996.

Taubman D. S., and Marcellin, M. W., *JPEG 2000: Image Compression Fundamentals, Standards, and Practice*, Kluwer Academic Publishers, Norwell, MA, 2002.

Therrien, C. W., *Discrete Random Signals and Statistical Signal Processing*, Prentice Hall, Englewood Cliffs, NJ, 1992.

Tribolet, J. M., "A New Phase Unwrapping Algorithms," *IEEE Trans. Acoustics, Speech, and Signal Processing*, Vol. 25, No. 2, pp. 170–177, Apr. 1977.

Tribolet, J. M., *Seismic Applications of Homomorphic Signal Processing*, Prentice Hall, Englewood Cliffs, NJ, 1979.

Tukey, J. W., *Exploratory Data Analysis*, Addison-Wesley, Reading, MA, 1977.

Ulrych, T. J., "Application of Homomorphic Deconvolution to Seismology," *Geophysics*, Vol. 36, No. 4, pp. 650–660, Aug. 1971.

Unser, M., "Sampling—50 Years after Shannon," *IEEE Proceedings*, Vol. 88, No. 4, pp. 569–587, Apr. 2000.

Vaidyanathan, P. P., *Multirate Systems and Filter Banks*, Prentice Hall, Englewood Cliffs, NJ, 1993.

Van Etten, W. C., *Introduction to Random Signals and Noise*, John Wiley and Sons, Hoboken, NJ, 2005.

Verhelst, W., and Steenhaut, O., "A New Model for the Short-time Complex Cepstrum of Voiced Speech," *IEEE Trans. on Acoustics, Speech, and Signal Processing*, Vol. ASSP-34, No. 1, pp. 43–51, February 1986.

Vernet, J. L., "Real Signals Fast Fourier Transform: Storage Capacity and Step Number Reduction by Means of an Odd Discrete Fourier Transform," *IEEE Proceedings*, Vol. 59, No. 10, pp. 1531–1532, Oct. 1971.

Vetterli, M., "A Theory of Multirate Filter Banks," *IEEE Trans. Acoustics, Speech, and Signal Processing*, Vol. 35, pp. 356–372, Mar. 1987.

Vetterli, M., and Kovačević, J., *Wavelets and Subband Coding*, Prentice Hall, Englewood Cliffs, NJ, 1995.

Volder, J. E., "The Cordic Trigonometric Computing Techniques," *IRE Trans. Electronic Computers*, Vol. 8, pp. 330–334, Sept. 1959.

Walden, R., "Analog-to-digital Converter Survey and Analysis," *IEEE Journal on Selected Areas in Communications*, Vol. 17, No. 4, pp. 539–550, Apr. 1999.

Watkinson, J., *MPEG Handbook*, Focal Press, Boston, MA, 2001.

Weinberg, L., *Network Analysis and Synthesis*, R. E. Kreiger, Huntington, NY, 1975.

Weinstein, C. J., "Roundoff Noise in Floating Point Fast Fourier Transform Computation," *IEEE Trans. Audio Electroacoustics*, Vol. 17, pp. 209–215, Sept. 1969.

Weinstein, C. J., and Oppenheim, A. V., "A Comparison of Roundoff Noise in Floating Point and Fixed Point Digital Filter Realizations," *IEEE Proceedings*, Vol. 57, pp. 1181–1183, June 1969.

Welch, P. D., "A Fixed-point Fast Fourier Transform Error Analysis," *IEEE Trans. Audio Electroacoustics*, Vol. 17, pp. 153–157, June 1969.

Welch, P. D., "The Use of the Fast Fourier Transform for the Estimation of Power Spectra," *IEEE Trans. Audio Electroacoustics*, Vol. 15, pp. 70–73, June 1970.

Widrow, B., "A Study of Rough Amplitude Quantization by Means of Nyquist Sampling Theory," *IRE Trans. Circuit Theory*, Vol. 3, pp. 266–276, Dec. 1956.

Widrow, B., "Statistical Analysis of Amplitude-quantized Sampled-data Systems," *AIEE Trans. (Applications and Industry)*, Vol. 81, pp. 555–568, Jan. 1961.

Widrow, B., and Kollár, I., *Quantization Noise: Roundoff Error in Digital Computation, Signal Processing, Control, and Communications*, Cambridge University Press, Cambridge, UK, 2008.

Widrow, B., and Stearns, S. D., *Adaptive Signal Processing*, Prentice Hall, Englewood Cliffs, NJ, 1985.

Winograd, S., "On Computing the Discrete Fourier Transform," *Mathematics of Computation*, Vol. 32, No. 141, pp. 175–199, Jan. 1978.

Woods, J. W., *Multidimensional Signal, Image, and Video Processing and Coding*, Academic Press, 2006.

Yao, K., and Thomas, J. B., "On Some Stability and Interpolatory Properties of Nonuniform Sampling Expansions," *IEEE Trans. Circuit Theory*, Vol. CT-14, pp. 404–408, Dec. 1967.

Yen, J. L., On Nonuniform Sampling of Bandwidth-limited Signals," *IEEE Trans. Circuit Theory*, Vol. CT-3, pp. 251–257, Dec. 1956.

Zverev, A. I., *Handbook of Filter Synthesis*, Wiley, New York, NY, 1967.

INDEX